国家教学名师学术文库

U0277945

LAISHAOCONG
LUNZHUJI

赖绍聪论著集

赖绍聪 等著

中

西北大学
·西安·
出版社

目 录

中 册

Petrogenesis of late Paleozoic-to-early Mesozoic granitoids and metagabbroic rocks of the Teng-chong Block, SW China: Implications for the evolution of the eastern Paleo-Tethys
·················· Zhu Renzhi, Lai Shaocong, Qin Jiangfeng, Zhao Shaowei / 845

Late Early-Cretaceous quartz diorite-granodiorite-monzogranite association from the Gaoligong belt, southeastern Tibet Plateau: Chemical variations and geodynamic implications
·············· Zhu Renzhi, Lai Shaocong, Qin Jiangfeng, Zhao Shaowei, Wang Jiangbo / 886

Strongly peraluminous fractionated S-type granites in the Baoshan Block, SW China: Implications for two-stage melting of fertile continental materials following the closure of Bangong-Nujiang Tethys
·············· Zhu Renzhi, Lai Shaocong, Qin Jiangfeng, Zhao Shaowei, M Santosh / 918

Early-Cretaceous syenites and granites in the northeastern Tengchong Block, SW China: Petro-genesis and tectonic implications
·················· Zhu Renzhi, Lai Shaocong, Qin Jiangfeng, Zhao Shaowei / 962

Geochemistry and zircon U-Pb-Hf isotopes of the 780 Ma I-type granites in the western Yangtze Block: Petrogenesis and crustal evolution
··················· Zhu Yu, Lai Shaocong, Qin Jiangfeng, Zhu Renzhi,
Zhang Fangyi, Zhang Zezhong / 990

Compositional variations of granitic rocks in continental margin arc: Constraints from the petro-genesis of Eocene granitic rocks in the Tengchong Block, SW China
··················· Zhao Shaowei, Lai Shaocong, Pei Xianzhi, Qin Jiangfeng,
Zhu Renzhi, Tao Ni, Gao Liang / 1024

Middle Permian high Sr/Y monzogranites in central Inner Mongolia: Reworking of the juvenile lower crust of Bainaimiao arc belt during slab break-off of the Palaeo-Asian oceanic lithosphere

.... Liu Min, Lai Shaocong, Zhang Da, Zhu Renzhi,

Qin Jiangfeng, Xiong Guangqiang / 1058

Petrogenesis of high-K calc-alkaline granodiorite and its enclaves from the SE Lhasa Block, Tibet (SW China): Implications for recycled subducted sediments

.... Zhu Renzhi, Lai Shaocong, Qin Jiangfeng, Zhao Shaowei, M Santosh / 1085

Late Triassic biotite monzogranite from the western Litang area, Yidun terrane, SW China: Petrogenesis and tectonic implications

.... Zhu Yu, Lai Shaocong, Qin Jiangfeng, Zhang Zezhong, Zhang Fangyi / 1117

Early-Middle Triassic intrusions in western Inner Mongolia, China: Implications for the final orogenic evolution in southwestern Xing-Meng orogenic belt

.... Liu Min, Lai Shaocong, Zhang Da, Zhu Renzhi,

Qin Jiangfeng, Di Yongjun / 1144

Neoproterozoic peraluminous granites in the western margin of the Yangtze Block, South China: Implications for the reworking of mature continental crust

.... Zhu Yu, Lai Shaocong, Qin Jiangfeng, Zhu Renzhi,

Zhang Fangyi, Zhang Zezhong, Zhao Shaowei / 1176

Petrogenesis and geodynamic implications of Neoproterozoic gabbro-diorites, adakitic granites, and A-type granites in the southwestern margin of the Yangtze Block, South China

.... Zhu Yu, Lai Shaocong, Qin Jiangfeng, Zhu Renzhi,

Zhang Fangyi, Zhang Zezhong, Gan Baoping / 1221

Genesis of high-potassium calc-alkaline peraluminous I-type granite: New insights from the Gaoligong belt granites in southeastern Tibet Plateau

.... Zhu Renzhi, Lai Shaocong, Qin Jiangfeng, M Santosh, Zhao Shaowei,

Zhang Encai, Zong Chunlei, Zhang Xiaoli, Xue Yuze / 1273

Petrogenesis and geochemical diversity of Late Mesoproterozoic S-type granites in the western Yangtze Block, South China: Co-entrainment of peritectic selective phases and accessory minerals

.... Zhu Yu, Lai Shaocong, Qin Jiangfeng, Zhu Renzhi,

Zhang Fangyi, Zhang Zezhong / 1297

Constructing the latest Neoproterozoic to Early Paleozoic multiple crust-mantle interactions in western Bainaimiao arc terrane, southeastern Central Asian Orogenic Belt

.... Liu Min, Lai Shaocong, Zhang Da, Zhu Renzhi, Qin Jiangfeng,

Xiong Guangqiang, Wang Haoran / 1342

Genesis of ca. 850-835 Ma high-Mg$^{\#}$ diorites in the western Yangtze Block, South China: Implications for mantle metasomatism under the subduction process

.... Zhu Yu, Lai Shaocong, Qin Jiangfeng, Zhu Renzhi, Liu Min,

Zhang Fangyi, Zhang Zezhong, Yang Hang / 1375

Vein-plus-wall rock melting model for the origin of Early Paleozoic alkali diabases in the South Qinling Belt, central China

·················· Zhang Fangyi, Lai Shaocong, Qin Jiangfeng, Zhu Renzhi,

Zhao Shaowei, Zhu Yu, Yang Hang / 1416

Early Paleozoic alkaline trachytes in the North Daba Mountains, South Qinling Belt: Petrogenesis and geological implications

····· Yang Hang, Lai Shaocong, Qin Jiangfeng, Zhu Renzhi, Zhao Shaowei, Zhu Yu,

Zhang Fangyi, Zhang Zezhong, Wang Xingying / 1454

Neoproterozoic metasomatized mantle beneath the western Yangtze Block, South China: Evidence from whole-rock geochemistry and zircon U-Pb-Hf isotopes of mafic rocks

················· Zhu Yu, Lai Shaocong, Qin Jiangfeng, Zhu Renzhi, Liu Min,

Zhang Fangyi, Zhang Zezhong, Yang Hang / 1486

Peritectic assemblage entrainment (PAE) model for the petrogenesis of Neoproterozoic high-maficity I-type granitoids in the western Yangtze Block, South China

················· Zhu Yu, Lai Shaocong, Qin Jiangfeng, Zhu Renzhi, Zhao Shaowei,

Liu Min, Zhang Fangyi, Zhang Zezhong, Yang Hang / 1523

High-K calc-alkaline to shoshonitic intrusions in SE Tibet: Implications for metasomatized lithospheric mantle beneath an active continental margin

····· Zhu Renzhi, Ewa Słaby, Lai Shaocong, Chen Lihui, Qin Jiangfeng, Zhang Chao,

Zhao Shaowei, Zhang Fangyi, Liu Wenhang, Mike Fowler / 1570

Petrogenetic evolution of Early Paleozoic trachytic rocks in the South Qinling Belt, central China: Insights from mineralogy, geochemistry, and thermodynamic modeling

················· Yang Hang, Lai Shaocong, Qin Jiangfeng, Zhang Fangyi, Zhu Renzhi,

Zhu Yu, Liu Min, Zhao Shaowei, Zhang Zezhong / 1601

Westward migration of high-magma addition rate events in SE Tibet

···························· Zhu Renzhi, Lai Shaocong, Scott R Paterson,

Peter Luffi, Zhang Bo, Lance R Pompe / 1632

Magma mixing for the genesis of Neoproterozoic Mopanshan granitoids in the western Yangtze Block, South China

················· Zhu Yu, Lai Shaocong, Qin Jiangfeng, Zhu Renzhi, Zhao Shaowei,

Liu Min, Zhang Fangyi, Zhang Zezhong, Yang Hang / 1654

Episodic provenance changes in Middle Permian to Middle Jurassic foreland sediments in southeastern Central Asian Orogenic Belt: Implications for collisional orogenesis in accretionary orogens

···················· Liu Min, Lai Shaocong, Zhang Da, Di Yongjun, Zhou Zhiguang,

Qin Jiangfeng, Zhu Renzhi, Zhu Yu, Zhang Fangyi / 1689

Petrogenesis of Early Cretaceous alkaline basalts in the West Qinling: Constraints from olivine chemistry

.... Yang Zhen, Lai Shaocong, Qin Jiangfeng, Zhu Renzhi,

Liu Min, Zhang Fangyi, Yang Hang, Zhu Yu / 1721

U-Pb zircon geochronology, geochemistry, and Sr-Nd-Pb-Hf isotopic composition of the Late Cretaceous monzogranite from the north of the Yidun Arc, Tibetan Plateau Eastern, SW China: Petrogenesis and tectonic implication

.... Gan Baoping, Lai Shaocong, Qin Jiangfeng, Zhu Renzhi, Zhu Yu / 1747

Petrogenesis of late Paleozoic-to-early Mesozoic granitoids and metagabbroic rocks of the Tengchong Block, SW China: Implications for the evolution of the eastern Paleo-Tethys[①]

Zhu Renzhi　Lai Shaocong[②]　Qin Jiangfeng　Zhao Shaowei

Abstract: This paper presents precise zircon U-Pb, bulk-rock geochemical, and Sr-Nd-Pb isotopic data for metagabbro, quartz diorite, and granite units within the Tengchong Block of SW China, which forms the southeastern extension of the Himalayan orogeny and the southwestern section of the Sanjiang orogenic belt, a key region for furthering our understanding of the evolution of the eastern Paleo-Tethys. These data reveal four groups of zircon U-Pb ages that range from the late Paleozoic to the early Mesozoic, including a 263.6 ± 3.6 Ma quartz diorite, a 218.5 ± 5.4 Ma two-mica granite, a 205.7 ± 3.1 Ma metagabbroic unit, and a 195.5 ± 2.2 Ma biotite granite. The quartz diorite in this area contains low concentrations of SiO_2 ($60.71-64.32$ wt%), is sodium-rich, and is metaluminous, indicating formation from magmas generated by a mixed source of metamafic rocks with a significant meta-pelitic sedimentary material within lower arc crust. The two-mica granites contain high concentrations of SiO_2 ($73.2-74.3$ wt%), are strongly peraluminous, and have evolved Sr-Nd-Pb isotopic compositions, all of which are indicative of a crustal source, most probably from the partial melting of felsic pelite and metagreywacke/psammite material. The metagabbros contain low concentrations of SiO_2 ($50.17-50.96$ wt%), are sodium-rich, contain high concentrations of $Fe_2O_3^T$ ($9.79-10.06$ wt%) and CaO ($6.88-7.12$ wt%), and are significantly enriched in the Sr ($869-894$ ppm) and LREE ($198.14-464.60$ ppm), indicative of derivation from magmas generated by a metasomatized mantle wedge modified by the sedimentary-derived component. The biotite granites are weakly peraluminous and formed from magmas generated by melting of metasedimentary sources dominanted by metagreywacke/psammite material. Combining the petrology and geochemistry of these units with the regional geology of the Indosinian orogenic belt provides evidence for two stages of magmatism: an initial stage that generated magmas during partial melting of mantle-derived material associated with late Permian-to-Early Triassic subduction of the Paleo-Tethys, and a second stage that generated granitoid magmas by the partial melting of crustal-derived sources during the Late Triassic collision between the Lhasa and Tengchong blocks and the northern margin of the Australian continent. These rocks, therefore,

---

①　Published in *International Journal of Earth Sciences*, 2018, 107.
②　Corresponding author.

provide evidence of a systematic late Permian-to-Late Triassic transition from a pre-collision/ volcanic arc setting through a collisional setting to a final within-plate phase of magmatism. The previous research involving bulk-rock Sr-Nd analyses of units from the southern Sanjiang orogenic belt and zircon Hf isotopic analyses of units from the Tengchong Block suggests that these areas may record similar magmatic evolutionary trends from mantle- to crustal-derived sources during the evolution of the eastern Paleo-Tethys.

Introduction

Granitoid magmas are commonly generated throughout the evolution of an orogen from subduction to collision-related orogeny and post-orogenic extension (Chappell and White, 1974, 1992; Pitcher, 1983; Pearce et al., 1984; Frost and Mahood, 1987; Sylvester, 1989; Brown, 1994). Understanding the petrogenesis of such magmas provides important insights into the geodynamic processes that operate in the deeper crust, including mantle-derived magmatism and the anatexis of pre-existing crustal material (Barbarin, 1999, 2005; Bonin, 2007; Koteas et al., 2010).

The Sanjiang orogenic belt incorporates the Jinshajiang-Ailaoshan, Lancangjiang, and Nujiang tectonic belts, and is located in the southeastern segment of the eastern Paleo-Tethys tectonic domain (Zhong, 1998). The orogenic belt varies in orientation from the WNW-ENE trending Himalayan-Tethyan segment of the belt to the N-S trending southeastern Asian segment (Hutchison, 1989; Metcalfe, 1996a, b, 2002, 2013; Zhang et al., 2008, 2012a; Wang et al., 2013, 2014; Deng et al., 2014). The Lancangjiang tectonic zone (also known as the Changning-Menglian suture belt) within this belt represents the eastern Paleo-Tethys main ocean that separated the Baoshan-Tengchong Block to the west from the Simao Block to the east (Mahawat et al., 1990; Cong et al., 1993; Metcalfe, 1996a, b, 2002; Mo et al., 1998; Zhong, 1998; Feng et al., 2005; Peng et al., 2008, 2013; Henning et al., 2009; Wang et al., 2010; Yang et al., 2014). The magmatism in the east of this area generally occurred during orogenesis in the Paleo-Tethyan region between the late Paleozoic and the early Mesozoic. In addition, the westernmost Tengchong Block forms the southwestern part of the Sanjiang orogenic belt, which is thought to represent the southeastern extension of Lhasa Block (Xu et al., 2008, 2012; Li et al., 2011; Huang et al., 2013; Wang et al., 2014). Recent research has also identified Triassic granitoids within the Tengchong Block (Cong et al., 2010; Li et al., 2010, 2011; Zou et al., 2011; Huang et al., 2013), also suggesting that records tectonomagmatism associated with the evolution of the eastern Paleo-Tethys. However, it remains unclear whether these granitoids were derived from partial melting of crustal sources during collisional orogenesis as part of the evolution of the Sanjiang orogenic belt (Cong et al.,

2010; Li et al., 2010; Zou et al., 2011) or whether they were derived from the partial melting of mantle-derived sources during the late Permian-to-Early Triassic subduction- and collision-related orogeny between the Lhasa and Northern Australia blocks (Huang et al., 2013), an event that also caused crustal melting during the late Triassic (Li et al., 2011). Consequently, the evolution of both the petrogenesis and the geodynamic setting of the late Permian-to-Late Triassic tectonic-associated magmatism in the Tengchong Block remains poorly constrained.

Here, we present new precise geochronological (zircon U-Pb ages), bulk-rock geochemical, and Sr-Nd-Pb isotopic data for late Paleozoic-to-early Mesozoic quartz diorite, granitoid, and metagabbroic units within the Tengchong Block. These data provide new insights that further constrain the Paleo-Tethyan evolution of this block and the southeastern Tibetan Paleo-Tethys belt (Yang et al., 2014).

Geological setting and petrography

The Sanjiang orogenic belt is part of the eastern Tethyan tectonic domain and represents the southeastern extension of the Himalayan Orogen (Kou et al., 2012; Fig. 1a,b). From east to west, this region is divided into the Yangtze, Lanping-Simao-Indochina, Baoshan-Shan-Thai, and Tengchong Blocks (Wang et al., 2006, 2010, 2013, 2014; Chen et al., 2007; Metcalfe 2013; Deng et al., 2014), which are separated by the Jinshajiang-Ailaoshan, Changning-Menglian, and Bangong-Nujiang sutures, respectively.

The Tengchong Block forms the northern part of Gondwanan Sibumasu and is located in the southwestern segment of the Sanjiang orogenic belt (Fig. 1b). The presence of Permian-Carboniferous glacio-marine deposits and overlying post-glacial black mudstones and Gondwana-like fossil assemblages suggests that this block was derived from the margin of western Australia within the eastern Gondwana supercontinent and remained in a platform-type setting at a passive continental margin (Jin, 1996; Xiao et al., 2003; Chen et al., 2006; Jin et al., 2002). The block contains Mesoproterozoic metamorphic basement material of the Gaoligong Mountain Group, which is overlain by late Paleozoic clastic sedimentary rocks and carbonates, and Tertiary-Quaternary volcano-sedimentary sequences, intruded by Mesozoic-Tertiary granitoids (YNBGMR, 1990; Zhong, 1998; Zhao et al., 2016). The Gaoligong Mountain Group contains quartzite, two-mica-quartz schist, feldspathic gneiss, migmatite, amphibolite, and marble units, and zircons from paragneiss and orthogneiss of this group yield zircon U-Pb ages of 1 053−635 and 490−470 Ma, respectively (Song et al., 2010). The Paleozoic sedimentary units in this area are dominated by Carboniferous clastic rocks, upper Triassic to Jurassic turbidites, Cretaceous red beds, and Cenozoic sandstones (YNBGMR, 1990; Zhong, 2000).

The Tengchong Block contains voluminous granitic gneiss, migmatite, and leucogranite

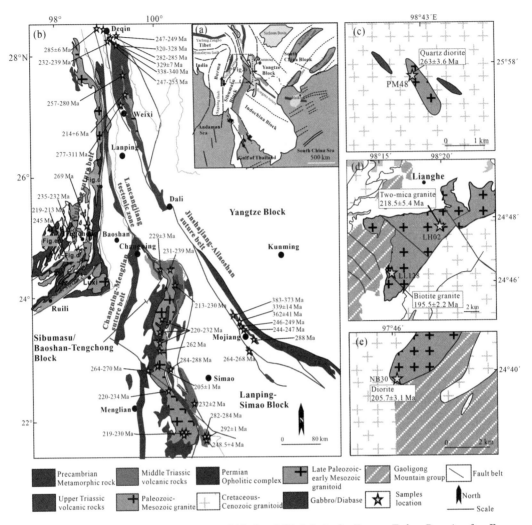

Fig. 1　(a) Distribution of main continental blocks of SE Asia in the Eastern Tethys Domain after Fan
　　et al. (2010). (b) Distribution of principal magma and strata of Sanjiang Tethys tectonic Domain, SW
　　China after YNBGMR (1990). All the age data in the diagram are from above reference therein.
　　Geological map of Pianma area (c), Lianghe area (d), and western Yingjiang area (e) in the
　　Tengchong Block, SW China after YNBGMR (1990).

units that were previously thought to have formed during the Proterozoic (YNBGMR, 1990).
However, recent geochronological data indicate that this block also contains numerous late
Paleozoic to early Mesozoic granitoids with zircon U-Pb ages of 245 - 206 Ma (Cong et al.,
2010; Li et al., 2010, 2011; Zou et al., 2011; Huang et al., 2013). All these granitoids
were emplaced into Paleozoic and Mesozoic units and some emplaced into the Gaoligong
Mountain group. The long axis of these granitic plutons are parallel to the suture belt
(Fig. 1b), but some were controlled by the regional fault (Fig. 1d). The granites at Gaoligong
lie to the west of the Lushui-Luxi-Ruili Fault, have undergone strong shearing, and possess a

schistosity dipping at a shallowly angle to the north and a subhorizontal mineral lineation (Wang et al., 2006; Zhang et al., 2011). Based on similar stratigraphy, paleobiogeography, and magmatism between the Tengchong and Lhasa Terranes, both have experienced similar tectonomagmatic history since the Early Paleozoic (Xie et al., 2016): as part of the northern margin of the Australian Gondwana in the Early Paleozoic, with moveward to the Eura-Asian continent from the Triassic and collision with Qiangtang-Baoshan during the Cretaceous.

In comparison, the Pianma quartz diorites (Fig. 1c) are fine-grained (Fig. 2a), are undeformed, and contain quartz veins. These intrusions contain 0.1-0.4 mm grains of quartz (15%-20%), plagioclase (45%-55%), K-feldspar (15%-20%), biotite (10%-15%), and hornblende (8%-12%), as well as accessory apatite, zircon, ilmenite, titanite, and magnetite (total 3%-5%; Fig. 3a). The Lianghe two-mica granite was emplaced into the Gaoligong Mountain group (Fig. 1d), is underformed coarse-grained and porphyritic (Fig. 2b), and contains quartz (35%-40%), K-feldspar (28%-33%), plagioclase (21%-28%), biotite (5%-10%), muscovite (~5%), and accessory apatite, zircon, and ilmenite (total ~1%; Fig. 3b). The Yingjiang metagabbros were emplaced into the Gaoligong Mountain Group (Fig. 1e), are fine-grained (Fig. 2c), are undeformed, and contain plagioclase (50%-60%), K-feldspar (10%-15%), biotite (10%-15%), hornblende (8%-12%), clinopyroxene (3%-5%), quartz (<1%), calcite (<3%), and accessory apatite, zircon, ilmenite, titanite, and magnetite (total 3%-5%; Fig. 3c). The Lianghe biotite granite was also emplaced into the Gaoligong Mountain Group (Fig. 1d), is coarse-grained and porphyritic (Fig. 2d), and records a little deformation. The granite contains quartz veins and consists of 0.2-0.8 mm grains of quartz (30%-38%), K-feldspar (30%-35%), plagioclase (25%-30%), biotite (5%-8%), and accessory apatite, zircon, ilmenite, and magnetite (total 1%-3%; Fig. 3d).

Analytical methods

Major and trace elements

Prior to analysis, bulk-rock samples were trimmed to remove weathered surfaces, cleaned with deionized water, crushed, and powdered using a tungsten carbide ball mill to pass through a 200 mesh screen. Major element concentrations were determined using a Rikagu RIX 2100 X-ray fluorescence (XRF) spectrometer at the State Key Laboratory of Continental Dynamics, Northwest University, Xi'an, China and the Guizhou Tuopu Resource and Environmental Analysis Center, China. Repeat analyses of USGS and Chinese national rock standards (BCR-2, GSR-1, and GSR-3) indicate that major element analytical precision and accuracy are generally better than 5%.

Fig. 2　Field petrography of quartz diorite in the Pianma area (a), two-mica granite (b),
metagabbroic rock in the Yingjiang area (c), and biotite granite (d) in the
Lianghe area of the Tengchong Block, SW China.

Trace element concentrations were determined using inductively coupled plasma-mass spectrometry (ICP-MS; Bruker Aurora M90) at the State Key Laboratory of Continental Dynamics, Northwest University, and the Guizhou Tuopu Resource and Environmental Analysis Center using the methods of Qi et al. (2000). Prior to analysis, sample powders were dissolved using an $HF + HNO_3$ mixture in a high-pressure PTFE bomb at 185℃ for 36 h. The ICP-MS analyses are estimated to have accuracies better than ±5% to ±10% (relative) for most elements.

Sr-Nd-Pb isotopic analyses

Bulk-rock Sr-Nd-Pb isotopic data were obtained using a Nu Plasma HR multi-collector (MC) mass spectrometer at the State Key Laboratory of Continental Dynamics, Northwest University, and the Guizhou Tuopu Resource and Environmental Analysis Center, using an approach similar to that of Chu et al. (2009). Sr and Nd isotopic fractionation was corrected to $^{87}Sr/^{86}Sr = 0.119\,4$ and $^{146}Nd/^{144}Nd = 0.721\,9$, respectively, and a Neptune MC-ICP-MS instrument was used to measure $^{87}Sr/^{86}Sr$ and $^{143}Nd/^{144}Nd$ isotope ratios. NIST SRM-987 and JMC-Nd certified reference standard solutions were used for $^{87}Sr/^{86}Sr$ and $^{143}Nd/^{144}Nd$ isotopic

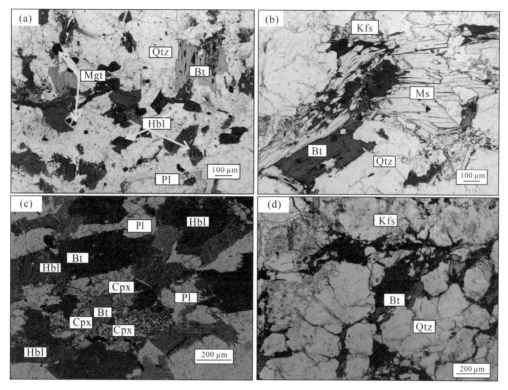

Fig. 3　Microscope petrography of quartz diorite in the Pianma area (a), two-mica granite (b),
metagabbroic rock in the Yingjiang area (c), and biotite granite (d)
in the Lianghe area of the Tengchong Block, SW China.

ratios, respectively, and BCR-1 and BHVO-1 were used as reference materials.

Bulk-rock Pb was separated using an anion exchange and HCl-Br columns, with Pb isotopic fractionation corrected to ^{205}Tl/^{203}Tl = 2.387 5. Thirty measurements of the NBS981 standard during the analytical period yielded average values of ^{206}Pb/^{204}Pb = 16.937 ± 1 (2σ), ^{207}Pb/^{204}Pb = 15.491 ± 1 (2σ), and ^{208}Pb/^{204}Pb = 36.696 ± 1 (2σ), and repeat analyses of the BCR-2 standard yielded values of ^{206}Pb/^{204}Pb = 18.742 ± 1 (2σ), ^{207}Pb/^{204}Pb = 15.620 ± 1 (2σ), and ^{208}Pb/^{204}Pb = 38.705 ± 1 (2σ). Total procedural Pb blanks were 0.1–0.3 ng.

Zircon U-Pb analyses

Prior to analysis, zircons were separated from four samples (each ~4 kg) taken from various sampling locations within the Tengchong Block, using conventional heavy liquid and magnetic techniques. Representative zircons were then handpicked and mounted in epoxy resin disks before polishing and carbon coating. The internal morphology of these zircons was then imaged using cathodoluminescence (CL) prior to U-Pb analysis.

Laser ablation-ICP-MS (LA-ICP-MS) zircon U-Pb analyses were undertaken using an Agilent 7500a ICP-MS instrument equipped with a 193 nm laser at the State Key Laboratory of

Continental Dynamics, Northwest University, following Yuan et al. (2004). The resulting $^{207}Pb/^{206}Pb$ and $^{206}Pb/^{238}U$ ratios were calculated using the GLITTER program and corrected using external calibration by analysis of a standard 91500 Harvard zircon. The resulting factors were then applied to each sample to correct for both instrumental mass bias and depth-dependent elemental and isotopic fractionation, following Yuan et al. (2004). Common Pb contents were evaluated using the approach of Andersen (2002), and age calculations and concordia diagrams were made using ISOPLOT version 3.0 (Ludwig, 2003). Uncertainties are quoted at the 2σ level.

Results

The metagabbroic rocks and granitoids of the Tengchong Block include late Permian quartz diorites in the Pianma area, Late Triassic and earliest Jurassic granites in the Lianghe area, and latest Triassic metagabbroic rocks in the Yingjiang area. The data obtained in this study are given in Tables 1 (major and trace elements), 2 (Sr-Nd isotopic data), and 3 (Pb isotopic data).

Zircon U-Pb ages

The complete data set for the zircon U-Pb analysis is given in the Supplementary Data set Table, with sampling locations, lithologies, and results summarized in the Supplementary Data set Table and in Figs 1 and 2. The zircons are generally euhedral, up to $100-400$ μm long, and have length:width ratios of $2:1-4:1$ (Fig. 4). The majority of these zircons are colorless or light brown, prismatic, transparent to translucent, have clear oscillatory zoning visible during CL imaging, and Th/U ratios >0.4, all of which are indicative of a magmatic origin (Hoskin and Schaltegger, 2003).

The 36 spot analyses of zircons from quartz diorite (sample PM48) included seven inherited zircons with ages of $1\,581-478$ Ma and five discordant analyses that are not discussed here. The youngest 10 analyses are scattered and have a wide range of $^{206}Pb/^{238}U$ ages (240 ± 5 Ma to 197 ± 4 Ma) that may reflect Pb loss. Fourteen coherent analyses of zircons from this sample yield $^{206}Pb/^{238}U$ ages of 251 ± 4.87 Ma to 271 ± 5.31 Ma, with a weighted mean age of 263.3 ± 3.6 Ma (MSWD $=1.4$, $n=14$, 2σ) (Fig. 5a).

Some 42 spot analyses were undertaken on zircons from two-mica granite (sample LH02). Four of these yielded unreliable and discordant results with a further 10 yielding anomalous Th/U values (<0.1), none of which are discussed further. Six inherited zircons within this sample yielded ages between $2\,393\pm41$ Ma and 507 ± 10 Ma, all of which are thought to represent inherited zircons derived from ancient crustal material. The remaining 32 coherent analyses yield $^{206}Pb/^{238}U$ ages from 201 ± 4 Ma to 242 ± 5 Ma and a weighted mean age of 218.5 ± 5.4 Ma (MSWD $=1.7$, $n=22$, 2σ) (Fig. 5b).

Table 1　Major(wt%) and trace(ppm) element analysis result of granitoids and metagabbroic rocks from the Tengchong Block.

Sample	Quartz diorite				Two-mica granite					Metagabbroic rocks								Biotite granite				
	PM37	PM39	PM53	PM55	LH04	LH05	LH06	LH07	LH08	NB29	NB31	NB32	NB33	NB33R	NB34	NB35	NB36	LL127	LL153	LL153-1	LL153-2	LL153-3
SiO_2	60.71	61.69	64.32	61.29	73.45	73.55	73.23	74.35	73.85	50.52	50.35	50.34	50.89	50.96	50.65	50.17	50.66	69.78	75.92	76.32	75.92	75.41
TiO_2	0.90	0.85	0.73	0.80	0.18	0.15	0.19	0.19	0.17	1.65	1.68	1.69	1.62	1.62	1.67	1.68	1.67	0.30	0.13	0.15	0.14	0.13
Al_2O_3	17.26	16.95	15.92	17.50	14.31	14.64	14.20	14.05	14.04	19.4	19.37	19.58	19.31	19.37	19.53	19.46	19.64	15.12	12.89	12.41	12.53	12.85
$Fe_2O_3^T$	6.35	6.05	5.41	5.97	1.31	1.18	1.33	1.26	1.19	9.85	9.84	10.06	9.81	9.79	9.78	10.13	10.0	2.57	1.20	1.36	1.23	1.14
MnO	0.10	0.09	0.09	0.09	0.01	0.02	0.01	0.01	0.02	0.16	0.16	0.17	0.16	0.16	0.16	0.19	0.16	0.04	0.02	0.02	0.02	0.02
MgO	2.56	2.39	2.10	2.33	0.36	0.27	0.34	0.36	0.32	3.02	3.08	3.12	3.03	3.05	3.08	3.09	3.08	0.94	0.34	0.4	0.36	0.35
CaO	6.05	5.68	3.99	4.57	1.11	0.90	0.91	1.00	0.82	7.12	6.91	7.05	6.91	6.9	6.93	6.88	7.02	1.85	1.55	1.51	1.46	1.44
Na_2O	3.49	3.42	3.03	3.37	2.72	2.80	2.53	2.69	2.6	3.97	3.93	4.00	3.91	3.89	4.13	4.01	4.00	2.35	2.26	2.22	2.18	2.16
K_2O	1.86	1.98	2.79	2.71	5.10	5.22	5.38	5.21	5.7	2.41	2.61	2.35	2.56	2.56	2.75	2.43	2.48	5.99	5.04	4.96	5.2	5.48
P_2O_5	0.23	0.22	0.21	0.24	0.15	0.13	0.14	0.14	0.19	0.63	0.65	0.65	0.64	0.64	0.65	0.64	0.64	0.13	0.03	0.02	0.02	0.02
LOI	0.70	0.82	1.35	1.40	0.80	0.66	1.33	0.79	0.89	0.94	0.92	0.92	0.91	0.91	0.92	0.92	0.55	0.57	0.41	0.53	0.5	0.65
Total	100.21	100.14	99.94	100.27	99.50	99.52	99.59	100.05	99.79	99.67	99.5	99.93	99.75	99.85	100.25	99.6	99.9	99.64	99.79	99.9	99.56	99.65
Li	37.7	42.2	23.3	39.9	21.1	22.2	24.3	23.4	14.6	48.7	49.9	52.0	48.5	48.6	47.0	55.5	51.7	22.5	11.7	12.3	12.2	11.7
Be	1.79	1.72	2.38	2.70	3.77	9.50	3.35	4.14	6.61	2.69	2.64	2.80	2.62	2.65	2.65	3.19	2.71	3.59	3.23	3.13	3.09	3.08
Sc	19.3	16.4	14.4	13.8	2.33	2.28	1.97	2.05	1.90	22.16	21.5	22.2	20.9	20.8	21.9	22.5	22.4	6.46	2.14	2.04	2.05	1.89
V	146	132	106	106	6.44	4.92	6.09	5.71	5.15	119	124	129	124	124	127	124	124	37.6	15.2	15.2	14.5	14.2
Cr	13.4	12.3	20.3	19	5.20	4.58	9.71	5.67	4.11	2.24	2.80	3.15	3.57	2.24	6.94	2.85	4.53	13.9	5.75	6.14	11.2	6.23
Co	67.2	42.2	36.2	61.1	139	109	118	122	110	22.7	22.1	23.6	21.5	20.8	21.6	23.8	20.7	110	156	218	189	186
Ni	6.40	6.97	11.8	12.6	2.04	1.88	4.81	2.54	2.02	1.32	1.65	1.76	1.82	1.08	3.72	1.41	3.02	7.23	2.99	3.12	5.74	3.13
Cu	26.31	16.81	13.70	18.91	1.74	1.07	1.19	3.52	1.45	9.31	9.43	10.33	9.65	9.43	9.42	9.88	10.41	2.39	0.91	1.74	4.77	1.77
Zn	113	111	101	102	28.2	33.0	35.0	27.9	32.7	104	106	111	104	104	105	123	108	44.5	18.5	20.3	24.3	20.1
Ga	20.5	19.8	20.3	21.3	19.3	18.8	18.4	18.1	17.9	23.7	24.4	24.7	24.2	23.9	24.2	25.7	23.6	17.1	13.0	12.2	12.4	12.5
Ge	1.21	1.1	1.2	1.17	1.29	1.49	1.22	1.30	1.33	1.38	1.46	1.46	1.42	1.42	1.45	1.49	1.39	1.41	1.19	1.17	1.23	1.21
Rb	138	127	147	153	260	309	299	259	303	110	117	135	108	108	105	173	120	318	182	179	184	191
Sr	372	349	274	312	102	75.8	80.9	101	87.2	894	888	890	891	881	885	869	892	188	172	159	158	160
Y	30.8	21.4	26.5	26.6	14.6	19.7	23.0	13.0	12.0	41.1	43.5	43.2	42.6	42.6	42.5	43.1	41.7	17.0	8.92	9.60	11.7	9.48

Continued

Sample	Quartz diorite				Two-mica granite					Metagabbroic rocks									Biotite granite			
	PM37	PM39	PM53	PM55	LH04	LH05	LH06	LH07	LH08	NB29	NB31	NB32	NB33	NB33R	NB34	NB35	NB36	LL127	LL153	LL153-1	LL153-2	LL153-3
Zr	107	112	143	134	99.5	83.5	86.7	86.1	67.9	233	238	248	220	219	224	228	204	61.2	82.3	79.0	84.7	79.3
Nb	16.8	11.9	14.7	15.1	12.6	13.7	11.3	11.9	10.7	24.6	25.2	25.2	24.2	24.1	24.8	27.9	25.0	16.1	4.24	4.22	4.11	3.73
Cs	5.03	5.39	5.41	4.16	7.99	31.6	11.4	9.21	11.9	3.31	3.73	5.40	2.77	2.80	2.24	10.58	4.16	10.1	5.18	5.20	5.13	4.97
Ba	429	449	610	477	301	240	237	297	301	880	1051	837	962	965	1100	961	1017	850	359	332	345	365
La	36.8	31	46.6	52.7	27.2	20.2	23.4	23.7	17.4	52.9	121	100	98.7	94.7	90.8	66.6	37.5	14.3	15.8	14.3	25.4	16.8
Ce	60.9	45.3	90.2	102	55.1	42.3	47.7	47.5	36.2	108	215	181	180	175	167	130	84	27.2	29.4	27.8	48.1	32.5
Pr	7.45	5.43	9.59	10.5	6.53	4.81	5.62	5.49	4.20	13.9	23.9	20.6	20.6	20.1	19.5	16.3	11.6	3.29	3.40	3.03	5.27	3.54
Nd	29	20.9	33.8	36.4	23.7	17.6	20.3	19.7	15.3	57.9	86.6	76.9	76.3	75.2	73.6	65.2	51.2	12.8	12.1	10.9	18.7	12.6
Sm	5.59	4.13	6.03	6.3	5.32	4.04	4.67	4.51	3.60	12.0	14.5	13.7	13.5	13.4	13.3	12.7	11.5	3.01	2.37	2.23	3.66	2.50
Eu	1.26	0.98	1.25	1.38	0.77	0.58	0.64	0.68	0.65	2.91	3.20	3.00	3.05	3.03	2.96	2.94	2.76	1.10	0.81	0.75	0.83	0.76
Gd	5.34	3.80	4.92	5.27	4.64	3.91	4.46	3.81	3.16	10.1	11.8	11.3	11.1	10.9	11.0	10.7	9.7	3.11	2.03	1.94	3.06	2.13
Tb	0.81	0.58	0.78	0.81	0.65	0.62	0.71	0.56	0.49	1.41	1.54	1.50	1.47	1.46	1.47	1.48	1.39	0.50	0.29	0.30	0.43	0.32
Dy	4.74	3.21	4.53	4.53	3.08	3.42	3.98	2.60	2.34	7.80	8.31	8.17	8.03	7.99	7.99	8.04	7.70	2.96	1.58	1.63	2.19	1.67
Ho	0.91	0.63	0.84	0.85	0.47	0.61	0.72	0.39	0.37	1.46	1.53	1.51	1.49	1.48	1.49	1.51	1.46	0.57	0.29	0.31	0.38	0.31
Er	2.72	1.78	2.43	2.53	1.12	1.59	1.92	0.91	0.87	3.96	4.14	4.10	4.01	3.99	4.00	4.05	3.88	1.50	0.84	0.87	1.00	0.86
Tm	0.36	0.24	0.34	0.35	0.14	0.22	0.25	0.12	0.12	0.54	0.55	0.55	0.53	0.53	0.54	0.54	0.52	0.18	0.12	0.13	0.15	0.13
Yb	2.39	1.6	2.16	2.38	0.81	1.31	1.51	0.70	0.68	3.19	3.35	3.30	3.21	3.19	3.19	3.29	3.09	1.08	0.84	0.87	0.95	0.84
Lu	0.33	0.23	0.3	0.33	0.12	0.18	0.21	0.10	0.098	0.46	0.48	0.48	0.46	0.46	0.47	0.48	0.45	0.15	0.13	0.13	0.14	0.13
Hf	2.42	2.42	3.26	3.12	2.97	2.67	2.64	2.48	2.00	5.04	5.17	5.31	4.73	4.77	4.82	4.96	4.38	1.68	2.70	2.53	2.71	2.54
Ta	1.26	0.75	1.17	1.14	1.97	3.21	1.39	1.85	2.26	1.11	1.09	1.08	1.07	1.06	1.06	1.63	1.06	1.45	0.65	0.74	0.69	0.60
Pb	16	16	20.7	20.6	52.1	51.6	55.5	57.8	60.9	9.76	9.84	9.20	9.81	9.75	10.24	9.14	9.48	51.1	41.0	41.5	43.2	45.0
Th	11.6	6.83	18	17.6	24.0	18.2	20.8	20.6	15.0	7.91	19.06	15.44	15.52	15.04	14.22	10.76	5.16	7.52	14.0	13.7	20.6	15.9
U	3.32	1.99	2.48	2.16	14.4	9.81	6.89	9.05	11.9	1.42	1.52	1.45	1.46	1.46	1.32	1.43	1.13	2.57	2.50	3.43	3.17	8.49
Mg#	48.4	47.9	47.5	47.6	39.0	34.8	37.3	40.0	38.5	41.7	42.2	42.0	41.9	42.1	42.3	41.6	41.8	46.0	39.8	40.7	40.6	41.7
A/CNK	0.92	0.94	1.04	1.04	1.19	1.23	1.22	1.18	1.18	0.88	0.89	0.89	0.89	0.89	0.87	0.90	0.89	1.10	1.07	1.05	1.06	1.06
REE	189	141	230	253	144	121	139	124	97.47	318	540	469	465	454	440	367	268	88.76	78.95	74.79	122	84.6
LREE	141	108	187	209	119	89.5	102	102	77.3	248	465	395	392	382	368	294	198	61.7	63.9	59.0	102	68.7

Continued

Sample	Quartz diorite				Two-mica granite					Metagabbroic rocks								Biotite granite				
	PM37	PM39	PM53	PM55	LH04	LH05	LH06	LH07	LH08	NB29	NB31	NB32	NB33	NB33R	NB34	NB35	NB36	LL127	LL153	LL153-1	LL153-2	LL153-3
HREE	48.4	33.5	42.8	43.6	25.6	31.6	36.8	22.2	20.1	70.0	75.1	74.2	72.9	72.6	72.7	73.1	69.9	27.1	15.0	15.8	20.0	15.9
LREE /HREE	2.91	3.22	4.38	4.79	4.63	2.84	2.78	4.57	3.84	3.54	6.18	5.32	5.38	5.26	5.06	4.02	2.84	2.28	4.25	3.74	5.09	4.33

Table 2　Whole-rock Rb-Sr and Sm-Nd isotopic data for granitoids and metagabbroic rocks from the Tengchong Block.

Sample	$^{87}Sr/^{86}Sr$	2SE	Sr/ppm	Rb/ppm	$^{143}Nd/^{144}Nd$	2SE	Nd/ppm	Sm/ppm	T_{DM2}/Ga	$\varepsilon_{Nd}(t)$	I_{Sr}
PM39	0.710 407	13	349	127	0.512 211	11	21	4	1.31	-5.7	0.706 467
LH-06	0.762 908	77	81	299	0.511 970	15	20	5	1.67	-11.4	0.729 082
LL-153	0.728 653	2	172	182	0.512 191	40	12	2	1.33	-6.8	0.720 066
NB-33	0.708 814	41	891	108	0.512 344	30	76	14	1.10	-3.4	0.707 805
NB-36	0.708 851	44	892	120	0.512 350	37	51	11	1.15	-4.0	0.707 724

$^{87}Rb/^{86}Sr$ and $^{147}Sm/^{144}Nd$ ratios were calculated using Rb, Sr, Sm, and Nd contents analyzed by ICP-MS.

T_{DM2} represent the two-stage model age and were calculated using present-day $(^{147}Sm/^{144}Nd)_{DM} = 0.213\ 7$, $(^{147}Sm/^{144}Nd)_{DM} = 0.513\ 15$, and $(^{147}Sm/^{144}Nd)_{crust} = 0.101\ 2$. $\varepsilon_{Nd}(t)$ values were calculated using present-day $(^{147}Sm/^{144}Nd)_{CHUR} = 0.196\ 7$ and $(^{147}Sm/^{144}Nd)_{CHUR} = 0.512\ 638$. $\varepsilon_{Nd}(t) = \left[(^{143}Nd/^{144}Nd)_S(t) / (^{143}Nd/^{144}Nd)_{CHUR}(t) - 1 \right] \times 10^4$, $T_{DM2} = \frac{1}{\lambda} \left\{ 1 + \left[(^{143}Nd/^{144}Nd)_S - ((^{147}Sm/^{144}Nd)_S - ((^{147}Sm/^{144}Nd)_S - (^{147}Sm/^{144}Nd)_{crust})(e^{\lambda t} - 1) - (^{143}Nd/^{144}Nd)_{DM} \right] / \left[(^{147}Sm/^{144}Nd)_S(t) / (^{147}Sm/^{144}Nd)_{crust} - (^{147}Sm/^{144}Nd)_{DM} \right] \right\}$.

Table 3　Whole-rock Pb isotopic data for granitoids and metagabbroic rocks from the Tengchong Block.

Sample	U	Th	Pb	$^{206}Pb/^{204}Pb$	$^{207}Pb/^{204}Pb$	2SE	$^{208}Pb/^{204}Pb$	2SE	$(^{206}Pb/^{204}Pb)_i$	$(^{207}Pb/^{204}Pb)_i$	$(^{208}Pb/^{204}Pb)_i$
PM39	2.0	6.8	16.0	18.833	15.744	12	39.349	38	18.501	15.727	38.979
LH-06	2.5	14.0	41.0	18.879	15.777	16	39.339	40	18.742	15.770	39.089
LL-153	6.9	20.8	55.5	18.888	15.813	47	39.524	101	18.641	15.801	39.281
NB-33	1.5	15.5	9.8	18.981	15.723	36	39.439	76	18.674	15.707	38.378
NB-36	1.1	5.2	9.5	18.951	15.708	43	39.244	92	18.706	15.696	38.880

U, Th, and Pb concentrations were analyzed by ICP-MS. The initial Pb isotopic ratios were calculated for 263 Ma (PM39), 218 Ma (LH06), 205 Ma (NB33 and NB36), and 195 Ma (LL153) using single-stage model.

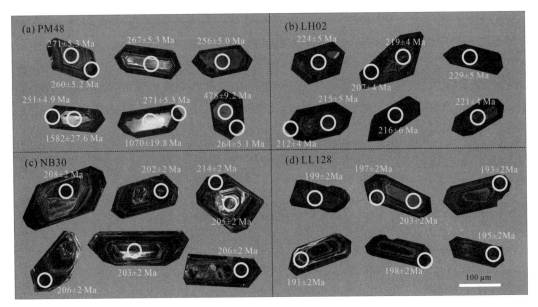

Fig. 4　Zircon CL images for the quartz diorite (a,PM48), two-mica granite (b,LH02), metagabbroic
rock (c,NB30), and biotite granite (d,LL128) in the Tengchong Block, SW China.

Of 36 analyses on zircons from metagabbros (sample NB30), 14 are discordant and are
not discussed further. One analysis yielded a slightly old ^{206}Pb/^{238}U age of 232 ± 2 Ma, with a
further three spots yielding slightly young ^{206}Pb/^{238}U ages (181 ± 2 Ma to 189 ± 2 Ma) that may
reflect radiogenic Pb loss. Eighteen coherent analyses of zircons from this sample yielded
^{206}Pb/^{238}U ages from 195 ± 2 Ma to 216 ± 2 Ma and a weighted mean age of $205. 7 \pm 3. 1$ Ma
(MSWD = 2. 5, $n = 20$, 2σ) (Fig. 5c).

Thirty-six analyses were undertaken on zircons from biotite granites (sample LL128),
with a single analysis (spot 25) yielding a concordant ^{206}Pb/^{238}U age of 982 ± 10 Ma that most
likely reflects an inherited zircon, further four spots yielding scattered ^{206}Pb/^{238}U ages that
range from 208 ± 2 to 252 ± 2 Ma, and two other spots yielding slightly young and scattered
^{206}Pb/^{238}U ages of 174 ± 2 Ma and 179 ± 2 Ma, both of which likely reflect radiogenic lead loss.
A total of 26 coherent analyses yielded ^{206}Pb/^{238}U ages from 187 ± 2 Ma to 205 ± 2 Ma, with a
weighted mean age of $195. 5 \pm 2. 2$ Ma (MSWD = 0. 68, $n = 26$, 2σ) (Fig. 5d).

Major and trace elemental geochemistry

On an FeO^T/MgO vs. SiO$_2$ diagram (Fig. 6a) the metagabbros are typical high-Fe
tholeiites, whereas the quartz diorite, two-mica granite, and biotite granite are medium- to low-
Fe calc-alkaline rocks. On an A/CNK vs. A/NK diagram (Fig. 6b), the quartz diorites are
metaluminous to slightly peraluminous, whereas the two-mica granite and biotite granite are
peraluminous and the metagabbros are metaluminous. These samples are also classified as

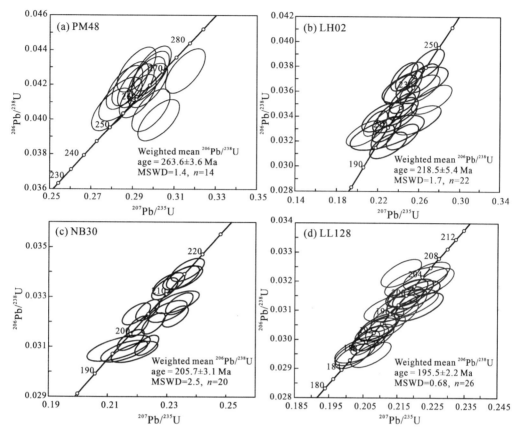

Fig. 5　LA-ICP-MS U-Pb zircon concordia diagram and CL images of representative zircon grains for the quartz diorite (a, PM48), two-mica granite (b, LH02), metagabbroic rock (c, NB30), and biotite granite (d, LL128) in the Tengchong Block, SW China.

gabbros, diorites/quartz diorites, and granites in both TAS and R_1-R_2 diagrams (Fig. 6c, d).

Pianma quartz diorites (263 Ma)

The Pianma quartz diorites have SiO_2 concentrations of 60.71−64.32 wt% and relatively high concentrations of Na_2O (3.03−3.49 wt%) with Na_2O/K_2O ratios of 1.09−1.88, but contain relatively low concentrations of TiO_2 (0.73−0.90 wt%), CaO (3.99−6.05 wt%), $Fe_2O_3^T$ (5.41−6.35 wt%), and MgO (2.10−2.56 wt%), with $Mg^{\#}$ values of 47.5−48.4. These samples also contain 15.92−17.50 wt% Al_2O_3 and have A/CNK (molar Al_2O_3/CaO + $Na_2O + K_2O$) ratios of 0.92−1.04. They contain 107.7−209.3 ppm light rare earth elements (LREE), 33.47−48.40 ppm heavy REE (HREE), have LREE/HREE ratios of 2.91−4.79, and when plotted on a chondrite-normalized rare earth element (REE) variation diagram (Fig. 7a) have insignificant negative Eu anomalies (δEu = 0.70−0.75). These samples also have high $(La/Yb)_N$ (11.04−15.88) and $(Gd/Yb)_N$ (1.83−1.97) values. On a primitive-mantle-normalized trace element variation diagram (Fig. 7b) the samples are enriched in large

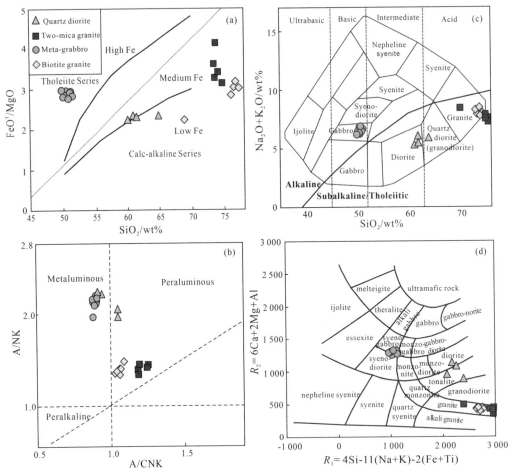

Fig. 6 FeO^T/MgO vs. SiO$_2$ diagram after Miyashiro (1974) (a) ; A/NK vs. A/CNK diagram after
Frost et al. (2001) (b) ; TAS diagram for classification of the rocks Middlemost (1994) (c) ;
R_1 vs. R_2 diagram for classification of the rocks De La Roche et al. (1980) (d).

ion lithophile elements (LILE; Rb, Th, U, and Pb), have positive Nd and Sm anomalies,
and are relatively depleted in Ba (429–610 ppm), Nb, P, and Ti, with Sr = 274–372 ppm,
and Y = 21. 4–30. 8 ppm.

Lianghe two-micagranite (218 Ma)

The samples from the Lianghe area contain high concentrations of SiO$_2$ (73. 23 – 74. 35
wt%), have K$_2$O/Na$_2$O ratios of 1. 86 – 2. 19, and contain low concentrations of Fe$_2$O$_3$^{T}
(1. 18–1. 33 wt%) and MgO (0. 27–0. 36 wt%). They have Al$_2$O$_3$ concentrations of 14. 04–
14. 64 wt% and have high A/CNK ratios (1. 18–1. 23). These samples contain 77. 3–118. 7
ppm LREE, have LREE/HREE values of 2. 78–4. 63, (La/Yb)$_N$ values of 11. 07–24. 44,
(Gd/Yb)$_N$ values of 2. 45–4. 77, and have chondrite-normalized REE patterns (Fig. 7c) that
show sharply negative Eu anomalies (δEu = 0. 43 – 0. 59). They have primitive-mantle-

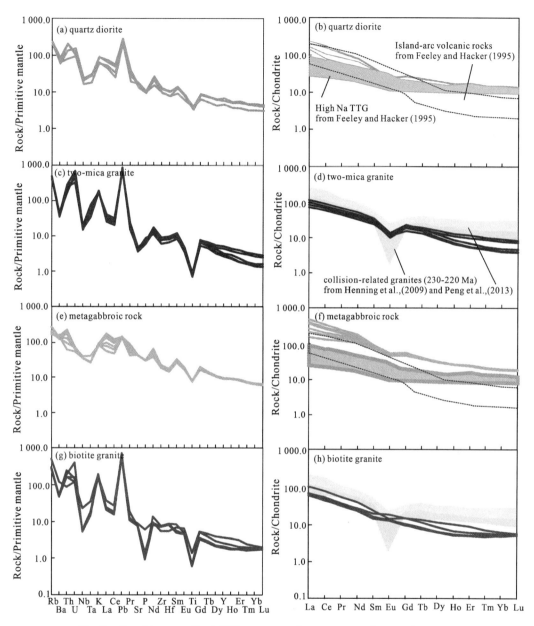

Fig. 7　Chondrite-normalized REE patterns and primitive-mantle-normalized trace
element spider diagram for the rocks in the Tengchong Block.

The primitive mantle and chondrite values are from Sun and McDonough (1989). The high Al TTG and island arc
volcanic rocks field from Feeley and Hacker (1995) and collision-related granites from Henning et al. (2009) and
Peng et al. (2013).

normalized trace element variation diagrams (Fig. 7d) that show positive Rb, Th, U, and Pb
anomalies along with enrichments in Ta and/or the other high field strength elements (HFSE),
and clear depletions in Ba, Nb, Sr, and Eu.

Yingjiang metagabbros (205 Ma)

Samples from the Yingjiang area contain low concentrations of SiO_2 (50. 17–50. 96 wt%) and high concentrations of Na_2O (3. 89–4. 13 wt%), have Na_2O/K_2O values of 1. 50–1. 70, and contain 9. 78–10. 13 wt% $Fe_2O_3^T$, 6. 88–7. 12 wt% CaO, and 19. 31–19. 64 wt% Al_2O_3, with A/CNK ratios of 0. 87–0. 90. These samples contain elevated total REE concentrations (268. 0–539. 8 ppm) with slightly elevated $(La/Yb)_N$ values (8. 70–25. 88) and have chondrite-normalized rare-earth element (REE) patterns (Fig. 7e) with insignificant negative Eu anomalies ($\delta Eu = 0. 74–0. 81$). They also have primitive-mantle-normalized trace element patterns (Fig. 7f) that are enriched in the LILE and depleted in Nb, P, Hf, and Ti, and contain elevated concentrations of Sr (869–894 ppm), Ba (837–1 100 ppm), and Y (41. 1–43. 5 ppm).

Lianghe biotitegranite (195 Ma)

The samples from the Lianghe area have variable concentrations of SiO_2 (69. 78–76. 32 wt%), elevated concentrations of K_2O = (4. 96–5. 99 wt%), K_2O/Na_2O values of 2. 23–2. 55, low concentrations of $Fe_2O_3^T$ (1. 14–1. 36 wt% barring one sample with 2. 57 wt%), moderate concentrations of Al_2O_3 (12. 41–12. 89 wt% barring one sample with 15. 12 wt%), and A/CNK ratios of 1. 05–1. 10. They contain 74. 8–121. 9 ppm total REE, have $(La/Yb)_N$ values of 9. 55–19. 12, and have chondrite-normalized REE patterns (Fig. 7g) characterized by variable Eu anomalies ($\delta Eu = 0. 76–1. 13$). These samples have primitive-mantle-normalized trace element distribution patterns (Fig. 7h) that are enriched in Rb, Th, U, Pb, and the HFSE, and depleted in Ba and Nb.

Bulk-rock Sr-Nd-Pb isotopes

The bulk-rock Sr-Nd-Pb isotopic data for the granitoids and metagabbros in the study area are given in Tables 2 and 3, with all initial $^{87}Sr/^{86}Sr$ isotopic ratios (I_{Sr}) and $\varepsilon_{Nd}(t)$ values calculated for the time of magma crystallization.

The Pianmaquartz diorite has a low I_{Sr} ratio (0. 706 467) but a stable $\varepsilon_{Nd}(t)$ value (−5. 7) and a T_{DM2} age of 1. 31 Ga. This unit has initial $^{206}Pb/^{204}Pb$, $^{207}Pb/^{204}Pb$, and $^{208}Pb/^{204}Pb$ ratios of 18. 501, 15. 727, and 38. 979, respectively.

The Lianghe two-mica granite has an elevated I_{Sr} value (0. 729 082), an $\varepsilon_{Nd}(t)$ value of −11. 4, a T_{DM2} age of 1. 67 Ga, and initial $^{206}Pb/^{204}Pb$, $^{207}Pb/^{204}Pb$, and $^{208}Pb/^{204}Pb$ ratios of 18. 742, 15. 770, and 39. 089, respectively.

The Yingjiang metagabbro has relatively low I_{Sr} values (0. 707 724–0. 707 805), $\varepsilon_{Nd}(t)$ values of −3. 4 to −4. 0, T_{DM2} ages of 1. 15–1. 10 Ga, and initial $^{206}Pb/^{204}Pb$, $^{207}Pb/^{204}Pb$, and $^{208}Pb/^{204}Pb$ ratios of 18. 674–18. 706, 15. 696–15. 707, and 38. 378–38. 880, respectively.

The Lianghe biotite granite has an elevated I_{Sr} value (0. 720 066), an $\varepsilon_{Nd}(t)$ value of -6. 8, a T_{DM2} age of 1. 33 Ga, and initial $^{206}Pb/^{204}Pb$, $^{207}Pb/^{204}Pb$, and $^{208}Pb/^{204}Pb$ ratios of 18. 641, 15. 801, and 39. 281, respectively.

Discussion

Petrogenesis of the multi-stage late Permian to Late Triassic magmatism within the Tengchong Block

Late Permian quartz diorites

The quartz diorites in the study area are characterized by low concentrations of SiO_2 (60. 71 – 64. 32 wt%), are metaluminous to slightly peraluminous, and are Na-rich with Na_2O/K_2O ratios of 1. 09–1. 88, similar to those intermediate to granitic rocks generated by the partial melting of basaltic rocks (Rapp and Watson, 1995). The low $(Na_2O + K_2O)/(FeO^T + MgO + TiO_2)$ and high $Na_2O + K_2O + FeO + MgO + TiO_2$ values of these samples, combined with their slightly elevated $Mg^\#$ values (47. 5–48. 4), are typical of magmas from the basaltic sources (metamafic rocks) (Patiño Douce, 1999; Fig. 8a, b). However, the involvement of pre-existing ancient continental crustal material in the formation of these quartz diorites is indicated by the presence of inherited zircons with U-Pb ages of 1 582–552 Ma that are almost identical to the age of the metamorphic basement in this area (YNBGMR, 1990; Zhong, 1998; Song et al., 2010), which infer the significant involvement of metasedimantery materials. Furthermore, the characters of both low Al_2O_3/TiO_2 and high CaO/Na_2O ratios and low Rb/Ba and moderate Rb/Sr of these samples indicate that a mixture sources between the predominantly basalt- and partly pelite-derived melts (Sylvester, 1998; Janoušek et al., 2004; Fig. 7c, d). The low Rb/Sr and Rb/Ba ratios (Fig. 8d) and insignificant δEu anomalies (0. 70–0. 75) imply the insignificant fractional crystallization of plagioclase (Patiño Douce, 1999). The decreasing Sr/Y ratios (10. 3 – 16. 3) with increasing Y concentrations (21. 4 – 30. 8 ppm) and fractionated REE pattern (LREE/HREE = 2. 91 – 4. 79) may indicate amphibole and/or garnet in the residue (Defant and Drummond, 1990; Petford and Atherton, 1996). These samples have higher total REE concentrations (141. 2–252. 9 ppm) than typical lower crust and have REE patterns that are similar to island arc volcanic rocks, suggesting that the magmas were formed by involvement of sub-arc crustal material (Feeley and Hacker, 1995). Furthermore, the lithospheric mantle-like isotopic composition of these intrusions [I_{Sr} = 0. 706 467, $\varepsilon_{Nd}(t)$ composition = −5. 7] suggests that the magmas were generated as a result of mixing between crust- and mantle-derived components (Fig. 9a-c), before the mixed magmas assimilated enriched mantle (EM Ⅱ) material, as evidenced by the initial Pb isotopic

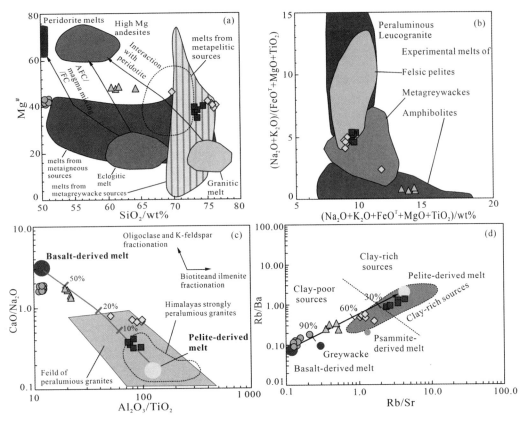

Fig. 8　(a) Plots of SiO_2 vs. Mg number. Marked fields outline experimentally obtained compositions of partial melts by dehydration melting of different source rocks under crustal P-T conditions (0.5–1.5 GPa, 800–1 000 ℃), based on Patiño Douce (1999), and Wolf and Wyllie (1994); (b) Compositional field of experimental melts derived from melting of felsic pelites (muscovite schists), metagreywackes and amphibolites (Patiño Douce, 1999); (c) CaO/Na_2O vs. Al_2O_3/TiO_2 diagram referenced by Sylvester (1998); (d) Rb/Ba vs. Rb/Sr diagram reference by Janoušek et al. (2004). Symbols as Fig 6.

composition of these samples (Fig. 10a, b). In addition, the quartz diorites contain late-stage, interstitial hydrous minerals such as hornblende and biotite, euhedral-subhedral zoned plagioclase, and polycrystalline clots of amphibole, all of which provide evidence for low concentrations of water within the magmatic system. The previous experimental and petrological analyses suggest that the generation of water-poor melts (~2 wt% H_2O) from basaltic material within sub-arc lower crust that incorporated the assimilation of pelitic sediments requires temperatures of >1 000℃ (Castro et al., 2010). Also, Annel et al. (2006) suggested that the low water contents (~2 wt%) of tonalite magmas could be produced by 50% melting of an amphibole-bearing protolith within lower arc crust. These suggest that the quartz diorite were generated by a mixed source of metamafic rocks with significant metapelitic sedimentary material at relatively high temperatures but under water-undersaturated conditions within lower arc crust.

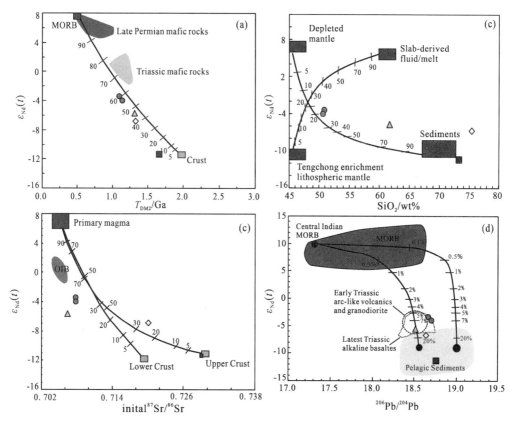

Fig. 9 $\varepsilon_{Nd}(t)$ vs. $T_{DM2}(a)$, initial $^{87}Sr/^{86}Sr$ (b) and $SiO_2(c)$.

Permian mafic rocks are from Zhai et al. (2013); Late Triassic mafics are from Zhang et al. (2011); The curve representing the mixing proportion between two components, mantle-derived magma, and middle/upper crustal melts corresponding to the mafic rocks are from Zhai et al. (2013). Plots $\varepsilon_{Nd}(t)$ vs. $^{206}Pb/^{204}Pb$, illustrating the input of subducted-related sediments into an India MORB source to form the Early Triassic volcanic rocks in the Lancangjiang zone. The compositions of the Indian MORB component is represented by $^{206}Pb/^{204}Pb = 17.31$, $Nd = 10$ ppm, $Pb = 0.03$ ppm and $Nd = 0.8$ ppm (Saunders et al., 1988) and two distinct end-member of the sediments are selected: one is $^{206}Pb/^{204}Pb = 18.56$, $\varepsilon_{Nd}(t) = -9$, $Pb = 9$ ppm and $Nd = 30$ ppm, and the other is $^{206}Pb/^{204}Pb = 19.00$, $\varepsilon_{Nd}(t) = -9$, $Pb = 60$ ppm and $Nd = 30$ ppm (Ben Othman et al., 1989). Symbols as Fig 6.

Late Triassic Lianghe two-mica granite

The two-mica granites in the study area are K-rich (5.10–5.7 wt%) with high K_2O/Na_2O ratios (1.86–2.19) that contrast with the Na-rich, low-K_2O, and low K_2O/Na_2O ratio of the other intrusions in this area, indicating that a model for the formation of these granites via the fractionation of an aluminous-poor magma and the dehydration melting of tholeiitic amphibolite material is not possible. These granites also contain inherited zircons with U-Pb ages of 2 393–507 Ma, suggesting the magmas that formed these granites involved ancient crustal material, most likely during the partial melting of crustal source rocks.

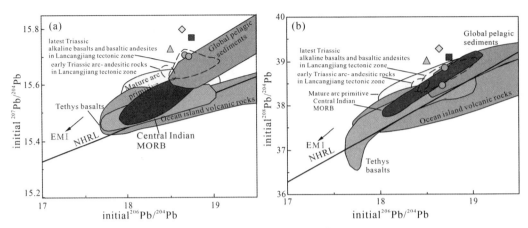

Fig. 10　Initial $^{207}Pb/^{204}Pb$ (a), $^{208}Pb/^{204}Pb$ (b) vs. $^{206}Pb/^{204}Pb$ diagrams for the Early Triassic volcanic rocks in the Lancangjiang zone.

The field of EM1 and EM2 is from Zindle and Hart (1986). The Northern Hemisphere Reference Line (NHRL) is from Hart (1984). The fields of central Indian Ocean MORB and Pacific MORB are from Price et al. (1986) and Ferguson and Klein (1993). The data of the Tethyan Basalts are from Mahoney et al. (1998) and the fields of global pelagic sediments are from Stolz et al. (1988, 1990). The early Triassic arc-like andesites are from Peng et al. (2008) and latest Triassic alkaline basalts and basaltic andesites are from Wang et al. (2010). Symbols as Fig 6.

The two-mica granites have low CaO/Na$_2$O ratios (0.32 – 0.41; Fig. 8c), and high Al$_2$O$_3$/TiO$_2$(73.95–97.60) and A/CNK values (1.18–1.23), Rb/Ba (0.87–1.29), and Rb/Sr (2.55 – 4.08) ratios (Fig. 8d), suggesting that they formed from magmas generally derived from a metapelite source (Sylvester, 1998; Janoušek et al., 2004). These granites have high (Na$_2$O + K$_2$O)/(FeO + MgO + TiO$_2$) values and low Na$_2$O + K$_2$O + FeO + MgO + TiO$_2$ concentrations that imply the magma sources including both felsic pelites and metagreywackes (Patiño Douce, 1999; Fig. 8b), a model that is consistent with the distribution of these samples on a Mg$^{\#}$ vs. SiO$_2$ diagram (Fig. 8a). In addition, these late Triassic granites have negative $\varepsilon_{Nd}(t)$ (−11.4) and high I_{Sr}(0.729 082) values that are typical of magmas derived from crustal sources (Fig. 9), also supported by Pb isotopic compositions that are indicative of derivation from the middle-upper crust. The low Fe$_2$O$_3$ + MgO + TiO$_2$ (1.48–1.73 wt%) and CaO (0.82–1.00 wt%) concentrations within these samples are indicative of anatectic melts that incorporated negligible amounts of peritectic or restitic minerals from the source region (Patiño Douce, 1999). In summary, we suggest that these strongly peraluminous two-mica granites formed from magmas generated by the partial melting of felsic pelite material that also contained some meta-greywacke or psammitic material.

Latest Triassic Yingjiang metagabbros

The Yingjiang metagabbros contain low concentrations of SiO$_2$(50.17 – 50.96 wt%) but high concentrations of TiO$_2$(1.62 – 1.69 wt%) and are sodium-rich (Na$_2$O = 3.89 – 4.13 wt%), suggesting these units are high-Fe series tholeiites (Fig. 6a). The elevated Nb/Ta ratios (22.2-23.5), and low Zr/Sm ratios (16.3-19.4) of these rocks are indicative of formation from melts generated by the partial melting of rutile-bearing eclogitic material (Foley et al., 2002), however, the low Sr/Y ratios (20.2-21.7) preclude melting of an eclogitic source which should generate the high-Mg and adakitic magma rather than mafic magma (Martin et al., 2005). The rocks have low $Mg^{\#}$ values (41.6-42.3) and contain low concentrations of Ni (1.08-3.72 ppm), Cr (2.24-6.94 ppm), and Co (20.7-23.8 ppm), suggesting that they are derived from parental magmas that underwent shallow-level fractional crystallization before emplacement (Wang et al., 2015). These metagabbros have SiO_2 concentrations that negatively correlate with FeO^T, MgO (Fig. 11), and CaO/Al_2O_3 ratios as well as having negative correlations between MgO and compatible elements, such as Ni, Co, and Cr, suggesting that they formed from magmas that fractionated significant amounts of olivine and clinopyroxene, although the Eu anomalies ($\delta Eu = 0.74-0.81$), elevated CaO concentrations, and relatively low $Na_2O + K_2O$ contents of these metagabbros provide a little evidence for plagioclase fractionation. In addition, the occurrence of negative P and Ti, and positive Sr anomalies in primitive-mantle-normalized trace element diagrams (Fig. 7e, f) probably reflects the nature of the source of the parental magmas for these metagabbros, as suggested by the lack of a correlation between SiO_2 concentrations, TiO_2 and P_2O_5 contents, and Rb/Sr ratios. The distribution of the metagabbro samples on La vs. La/Sm and Yb vs. La/Yb diagrams (Fig. 11c, d) is consistent with the model of Wang et al. (2015), who suggested that the composition of the gabbroic magmas in this region was controlled by partial melting and source heterogeneity rather than fractional crystallization. The high Al_2O_3 concentrations (19.31-19.64 wt%), negative $\varepsilon_{Nd}(t)$ values (-4.04 to -3.41), along with Pb isotopic compositions that are similar to those of the global sediment average (Fig. 10) and the presence of negative Nb-Ta and slightly negative Zr-Hf anomalies (Fig. 7e), suggest the involvement of crustal materials in magma-genesis (Wang et al., 2015). However, the mantle-like initial $^{87}Sr/^{86}Sr$ values (0.705-0.708; Ratajeski et al., 2005; Fig. 9b, c) may also indicate the admixture of mantle-like melts (Fig. 9a-d). These feature are very similar with the metagabbroic rocks in the Nabang (Wang et al., 2015), for which three models have been proposed: mantle wedge newly modified by slab-derived fluid/melt, addition of subducted sediments into the depleted mantle, and input of slab-derived fluid/melt into the enriched lithospheric mantle. Our negative $\varepsilon_{Nd}(t)$ values are contrary to the characters of positive $\varepsilon_{Nd}(t)$ values of first model. The low Ba/Th, U/Th, and Sr/La, and high Th/La, Th/Nb, La/Sm do not agree with the input of the slab-derived fluid/melts but the sedimentary-derived materials (Plank and Langmuir, 1998; Rudnick et al., 2000; Elliott, 2003; Plank, 2005; Stern, 2006). Therefore, the model of a depleted mantle modified by the sedimentary-derived component is more suited (Fig. 9a-d). In addition, the presence of plagioclase within the crystallization sequence recorded by these metagabbros, the irregular contacts between hornblende and plagioclase, and the presence of hornblende clots are all indicative of high-temperature conditions (~ 1 000 − 1 100℃) and water-undersaturated melts (Castro et al., 2010; Castro, 2013). These metagabbros, therefore have similar petrogenetic histories to the coeval basaltic andesites and porphyritic intrusions in the Sanjiang belt (Wang et al., 2010; Kou et al., 2012).

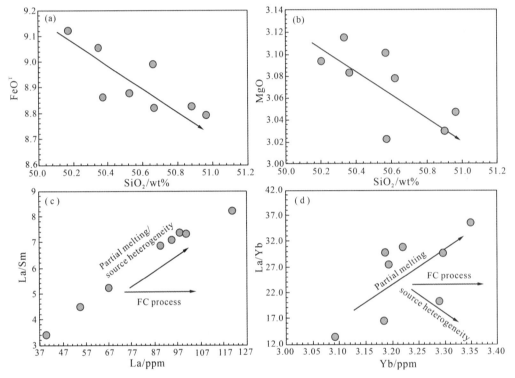

Fig. 11   SiO$_2$ vs. FeO^T(a) and MgO (b) diagram, La vs. La/Sm (c), and Yb vs. La/Yb (d)
for the Yingjiang gabbroic rocks after Wang et al. (2015).

The evolution trends in the c and d suggest that the formation of gabbroic magma is controlled by the partial melting and source heterogeneity rather than crystallization fractionation. Symbols as Fig 6.

Early Jurassic biotite granites

The biotite granites in the study area have high K$_2$O/Na$_2$O ratios (2. 23 − 2. 55) and K$_2$O + Na$_2$O values (7. 18 − 8. 34 wt%), and are compositionally similar to the late Triassic two-mica granites, suggesting that both were derived from fractionated alumina-poor magmas that usually generate metaluminous Na-rich and low-K$_2$O/Na$_2$O acid rocks during closed-system fractionation (Zen, 1986; Gaudemer et al., 1988; Springer and Seck, 1997; Sylvester, 1998; Clemens, 2003). This is in contrast to a previous model for these granites that describes their derivation from magmas generated by the dehydration melting of tholeiitic amphibolite, generating a granulite residue at pressures of 0. 8 − 1. 2 GPa and a garnet-bearing granulite to eclogite residue at pressures of 1. 2 - 3. 2 GPa (Rushmer, 1991; Rapp and Watson, 1995). The relatively high Al_2O_3/TiO_2 ratios (82. 7 - 99. 2), A/CNK values (1. 05 - 1. 10), and slightly low CaO/Na_2O ratios (0. 67 - 0. 79) of these granites are indicative of derivation from a parental magma that was probably generated by the partial melting of a metasedimentary source (Sylvester, 1998; Fig. 8c), as supported by the presence of inherited zircons. In addition, these granites have lower Rb/Ba (0. 37 - 0. 54) and Rb/Sr (1. 06 - 1. 69) ratios than the Late Triassic two-mica granites, suggesting that the former were derived from a sedimentary source that contained more psammite than the latter (Janoušek et al., 2004; Fig. 8d). The high $(Na_2O + K_2O)/(FeO + MgO + TiO_2)$ and low $Na_2O + K_2O + FeO + MgO + TiO_2$ values indicate they were derived from a source containing both felsic pelite and metagreywackes material (Patiño Douce 1999; Fig. 7b), as also indicated by the distribution of data in a $Mg^{\#}$ vs. SiO_2 diagram (Fig. 8a). The fact that these samples have higher (but negative) $\varepsilon_{Nd}(t)$ (-6. 8) and lower initial $^{87}Sr/^{86}Sr$ (0. 720 066) values than the late Triassic granites in this region also supports a crustal-dominated source for the former (Fig. 9), and Pb isotopic compositions that are indicative of middle to lower crustal affinities (Fig. 10). As discussed above, these evidence suggest that the biotite granites were derived from melting of a metasedimentary source which mainly dominated by metagreywacke and/or psammite.

Late Paleozoic-to-early Mesozoic evolution of the deep crust in the southern Sanjiang orogenic belt

Here, we outline the evolution of the deep crustal sources for the magmatism within the southern Sanjiang orogenic belt during subduction- and collision-related orogenesis by plotting the I_{Sr} and $\varepsilon_{Nd}(t)$ data obtained for the late Paleozoic-to-early Mesozoic (ca. 300 - 190 Ma) intrusions with data for the surrounding tectonic zones on I_{Sr} and $\varepsilon_{Nd}(t)$ vs. age diagrams (Fig. 12). These diagrams indicate the following points: ① The I_{Sr} and $\varepsilon_{Nd}(t)$ values of samples from the neighboring Lancanjiang and Jinshajiang-Ailaoshan tectonic zones record three different phases of magmatism that reflect changes in source regions and tectonism, from subduction (> ca. 250 Ma) through collision-related orogenesis (ca. 250 - 218 Ma) to post-orogeny extension (after ca. 218 Ma). ② The first phase of magmatism generated intrusions with positive $\varepsilon_{Nd}(t)$ (generally > 0 barring one sample with a value of around -3) and low I_{Sr} (< 0. 707 5) values, whereas the second phase generated units with generally negative $\varepsilon_{Nd}(t)$ values (generally < 0) and high but variable I_{Sr} values (> 0. 707 5), and the third phase produced magmas with similar (but less variable) isotopic compositions to the first phase. And ③ our samples have similar compositions to the trends defined above, although there are some differences in the composition of magmas generated at around 195 Ma.

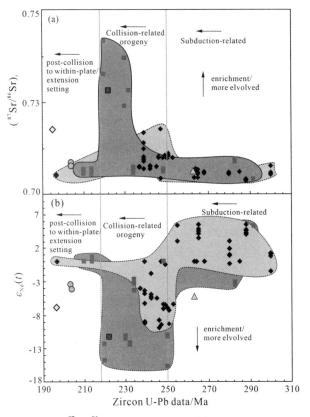

Fig. 12 $(^{87}Sr/^{86}Sr)_i$ vs. age and $\varepsilon_{Nd}(t)$ vs. age diagram.

The blue plots and light grey area represent the data from Jinshajiang-Ailaoshan tectonic zone, and these data from Xiao et al. (2004), Jian et al. (2004, 2008), Fan et al. (2010), Kou et al. (2012), and Zi et al. (2012a,b). The purplish red plots and dark grey area represent the data from Lancangjiang tectonic zone, these data from Peng et al. (2008), Henning et al. (2009), Wang et al. (2010), and Peng et al. (2013). In addition, the magenta designs are plots of our samples.

These comparisons suggest that the study area records three different types of interaction between mantle-derived and pre-existing crustal sources that reflect the changing evolution of the southern Sangjiang orogenic belt from the late Paleozoic to the early Mesozoic. This indicates that mantle-derived magmas dominate the subduction-related magmatism in this area before 250 Ma, whereas magmas derived from ancient crustal sources were involved in the widespread magmatism associated with collision-related orogenesis in the Jinshajiang-Ailaoshan zone between ca. 254 Ma and 234 Ma. The magmatism in this area after ca. 218 Ma involved magmas derived from deeper crustal sources that also mixed with mantle-derived magmas. Given that the metagabbros and granitoids in the study area have enriched I_{Sr} and $\varepsilon_{Nd}(t)$ compositions that are not present in the two neighboring tectonic zones, we infer that the former involved more crustal-derived material than the magmatism in surrounding areas. In contrast, the Tengchong Block also records a similar evolutionary trend in terms of changes in magma sourcing from mantle- to crustal-derived sources during the Early to Late Triassic as evidenced by variations in a $\varepsilon_{Hf}(t)$ vs. zircon U-Pb age diagram (Fig. 13).

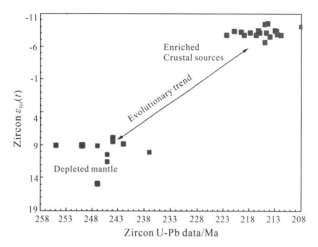

Fig. 13   Zircon $\varepsilon_{Hf}(t)$ vs. zircon U-Pb age diagram for the Triassic magmatisms in the Tengchong Block.
Data from Cong et al. (2010) and Huang et al. (2013).

Geodynamic implications

The Tengchong Block was located within the southern Sanjiang Tethys belt (Yang et al., 2014), although the precise geodynamic setting and the relationship between this and other blocks (e.g., the Lhasa Block and others) during the evolution of the eastern Paleo-Tethys remains unclear. Clarifying these relationships and the setting of the block requires the identification of the relationship between the Permian-Triassic magmatism in the Tengchong-Lhasa Blocks to the west and the Changning-Menglian and Jinshajiang-Ailaoshan and Jinghong belts to the east (Table 4; Fig. 1a,b). Many workers have confirmed the Changning-Menglian suture belt as the Paleo-Tethyan main ocean (Peng et al., 2008, 2013; Henning et al., 2009; Jian et al., 2009a, b; Wang et al., 2010), to the east, the Jinshajiang-Ailaoshan and Jinghong suture belt was regarded as the remnants of a back-arc basin rather than the Paleo-Tethyan main ocean (Wang et al., 2000; Carter et al., 2001; Nam et al., 2001; Xiao et al., 2004; Jian et al., 2004, 2008, 2009a,b; Fan et al., 2010; Kou et al., 2012; Zi et al., 2012a,b, 2013; Wang et al., 2016). Interestingly, in the west, the Permain-Triassic magmatism in the Indosinian orogenic belt from Lhasa to Tengchong Blocks share synchronous igneous rocks on both sides of the Paleo-Tethyan main ocean (Fig. 1a,b and Fig. 14; Table 4). To balance and interpret the interesting distribution, some typical features of both of them should be considered. In the east, there are the following characteristics (Fig. 1b; Table 4): ① Voluminous Na-rich and subduction-arc affinity magmatism, including the formation of mafic-ultramafic complexes, blueschists, ophiolites, andesites, and granodiorites between

Table 4 Concise comparison of magmatic evolution from Permain to early Jurassic in the southern of Sanjiang Tethys.

Nujiang belt (Correlation with Paleo-Tethys? A mature island arc zone during this evolution)			Lancangjiang belt (Main Paleo-Tethys ocean)			Jinshajiang-Ailaoshan belt (a branch of main Paleo-Tethys ocean)		
Petrology	Age /Ma	Geochemical affinity	Petrology	Age /Ma	Geochemical affinity/Setting	Petrology	Age /Ma	Geochemical affinity/Setting
Hornblend-biotite granite	269	Island arc	Mafic-ultramafic complex	292	Sodium-rich and metaluminous, Subduction-related arc affinity/subducted zone magmatisms	Basalt	288	I-type affinity/back-arc basin/subducution zone magmatisms
Quartz diorite	263			288		Trondhjemite	285	
Diorite	245		Granodiorite	284		grabbro	280	
Granite	235	Peraluminous and crustal sources	Granodiorite	284		granodiorite	280	
Granite	234		Bluechist	282		grabbro	277	
Granite	232			274		granodiorite	269	
Two-mica granite	218		Ophiolite	270		Basalt	268	
Granites	219		Ophiolite	264		Basalt	264	
Granites	213		Granodiorite	262		granodiorite	254	
Diorite	204	Within-plate basaltic affinity	Andesites	248.5		granodiorite	249	
Biotite granite	196		Granite	239	S-type affinity, syncollision-related	Pyroxenites	246	Post-orogenic extension/collision-related high Si, S-type felsic rocks
			Granite	234		Rhyolites	249	
			Granite	232		Rhyolites	247	
			Rhyolite	231		Rhyolites	242	
			Granite	230		Rhyolites	239	
			Granite	229		Granitoid	239	
			Granite	220	A-type affinity, postcollision	Granitoid	238	
			Granite	219		Granitoid	231	
			Granite	216		Porphyrites	197	Within-plate Related to upwelling asthenosphere
			Alkaline basalts	214	Post-collision related to extension upwelling asthenosphere	Porphyrites	197	
			Basaltic andesites	210				

The data of magmatic evolution from Permain to early Jurassic in the Nujiang tectonic zone related to the Paleo-Tethys referenced from Zhao et al. (1999), Cong et al. (2010), Li et al. (2010, 2011) and Zou et al. (2011), among them, the data of the quartz diorite, two-mica granite, diorite, biotite granite, and their geochemic affinities from this study. The data of magmatic evolution from Permain to early Jurassic in the Lancangjiang tectonic zone (as the main Paleo-Tethys ocean) from Jian et al. (2009a,b), Henning et al. (2009), Peng et al. (2008,2013), and Wang et al. (2010). The data of magmatic evolution from Permian to early Jurassic in the Jinshjiang-Ailaoshan tectonic zone (as a branch of Paleo-Tethys ocean) from Xiao et al. (2004), Jian et al. (2004,2008,2009a,b), Wang et al. (2000), Carter et al. (2001), Nam et al. (2001), Fan et al. (2010), Kou et al. (2012) and Zi et al. (2012a,b). And later, two geochemic affinity and settings are also from those referenced views.

292 Ma and 248.5 Ma within the Lancangjiang and Jinshajiang tectonic zone (Jian et al., 2009a,b; Wang et al., 2010), whereas the Jinshajiang-Ailaoshan tectonic zone was in a back-arc basin setting between 288 Ma and 249 Ma (Fan et al., 2010; Zi et al., 2012a,b, 2013). ② Typical crustal-source-derived rhyolitic and granitic magmatism in both zones between 247 Ma and 214 Ma (Wang et al., 2010; Fan et al., 2010; Zi et al., 2012a,b, 2013; Peng et al., 2013). And ③ post-orogenic alkaline basaltic to basaltic and andesitic magmatism in an extensional setting at 214−197 Ma (Wang et al., 2010; Kou et al., 2012). The late Paleozoic-to-early Mesozoic magmatism in the Tengchong Block even to the Indosinian orogenic belt from Tengchong to Lhasa Blocks occurred at a similar time to the magmatism in both the Lancangjiang and Jinshajiang-Ailaoshan tectonic zones, but the Indosinian orogenic belt to the west has a different lithological, sedimentological, and paleobiological characters from the other two tectonic zones to the east at this time.

In the west, the Indosinian orogenic belt lies to the northwest of the Sanjiang belt and extend from the Lhasa to Tengchong blocks (Fig. 14). Triassic granitoids (SHRIMP zircon U-Pb ages of 245−206 Ma) in the Tengchong Block (Cong et al., 2010; Li et al., 2011; Zou et al., 2011; Huang et al., 2013) are thought to represent either the southeastern extension of the Gangdese Indosinian magmatism in the Lhasa Block (Li et al., 2011) or magmatism related to collision/within-plate setting in an extensional enviroment as a result of the evolution of the Paleo-Tethyan Ocean in the Sanjiang orogenic belt (Cong et al., 2010; Zou et al., 2011). This can be further clarified by considering the following points: ① Li et al. (2011) suggested that an Indosinian orogenic belt extends from the western Lhasa Block to the southeastern Tengchong Block, as evidenced by magmatism in this area between ca. 262 Ma and 190 Ma (Fig. 14). This magmatism was associated with the formation of subduction-related eclogites, diorites, and granites between 262 Ma and 245 Ma, the majority of which formed from magmas generated by the partial melting of mantle-derived rocks (Xu et al., 2007; Yang et al., 2007, 2009; Wang et al., 2008; Zhu et al., 2009). The later magmatism in this area is dominated by granitoids formed by the partial melting of crustal material between ca. 235 Ma and 190 Ma (He et al., 2006; Liu et al., 2006; Li et al., 2003, 2011; Li,2009; Zheng et al., 2003; Zhang et al., 2007; Han, 2007), all of which occurred during collisional orogenesis. ② The lack of upper Permian to Lower Triassic sediments and the presence of an angular unconformity between the Middle-Upper Triassic and the upper Carboniferous to lower Permian sediments in the Tengchong Block (YNBGMR, 1990; Yin and Harrison, 2000) are consistent with evolving sedimentation within the Lhasa Block but not in the Lancangjiang and Jinshajiang-Ailaoshan tectonic zones. This suggests that both the Tengchong and Lhasa blocks were in a similar tectonic setting at this time. ③ Paleontological and paleomagnetic data for this area, combined with the presence of glacial-marine diamictites and cool/cold-water faunas in

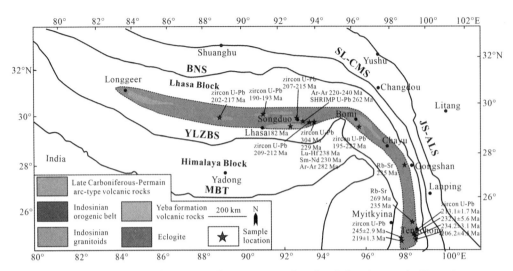

Fig. 14 Distribution characters of the magmatisms from late Paleozoic to early Mesozoic
in the Indosinian orogenic belt.

These age data from Xu et al. (2007), Wang et al. (2008), Zhu et al. (2009), He et al. (2006), Liu et al. (2006),
Li (2009), Li et al. (2003,2010,2011), Zhang et al. (2007), Zheng et al. (2003), Han et al. (2007), Cong et al.
(2010), Zou et al. (2011), Huang et al. (2013), Yang et al. (2007, 2009), Cheng et al. (2015) and Meng et al.
(2016). MBT: Main Boundary Thrust; YLZBS: Yarlung Zangbo suture; BNS: Bangong-Nujiang suture; SL-CMS:
Shuanghu-Longmucuo Changning-Menglian suture; JS-ALS: Jinshajiang-Ailaoshan suture.

the lower Permian within both the Lhasa and Tengchong blocks, are indicative of their Gondwanan affinity at this time (Sengör, 1984; Metcalfe, 2013). In addition, paleomagnetic data for these areas suggest they were located at similar paleolatitudes during the Permian (Van Der Voo, 1993; Li et al., 2004; Chen et al., 2012). And ④ the Tengchong Block was located at an equivalent geotectonic location as the Lhasa Block, with both blocks located between the Bangong-Nujiang and Yarlung-Zangbo-Myitkyina suture belts (Dewey et al., 1988; Yin and Harrison, 2000; Kapp et al., 2005; Li et al., 2011; Xu et al., 2012; Huang et al., 2013; Wang et al., 2014, 2015). These features indicate that it is likely that the Indosinian orogenic belt extended from the Lhasa Block to the Tengchong Block during the late Paleozoic to the early Mesozoic.

As mentioned above, we may infer that it is not easy to conclude that the Indosinian belt as a branch of the Changning-Menglian Paleo-Tethyan main ocean looks the same as the Jinshajiang-Ailaoshan and Jinghong suture belt due to their definitely different data of lithological, sedimentological, paleomagnetic and paleobiological characters. However, the synchronous magmatism on the both sides could be considered a the similar response to the evolution of the Eastern Paleo-Tethys.

The synchronous igneous rocks on the both sides of the Paleo-tethyan main ocean and some differences have been considered during above discussion. However, the view of tectonic evolution should be also proposed: to the east, there was devepoled a Late Paleozoic-to-Early Mesozoic back-arc basin along the Ailaoshan and Jinghong-Nan tectonic zone in response to the northward subduction of the Paleo-tethys main ocean and the final closure of the back-arc basin took place in the uppermost Triassic due to the diachronous amalgamation between the Yangtze and Simao-Indochina blocks (Fan et al., 2010; Wang et al., 2016). At the same time, the large scale Lincang-Sukhothai magmatic arc was located between the Changning-Menglian suture zone and the Jinghong-Nan back-arc basin (Wang et al., 2016). As a result, the Paleotethyan pattern is spatially characterized by the Changning-Menglian-Inthanon suture zone, Lincang-Sukhokai arc and Jinghong-Nan back-arc basin from west to east (Henning et al., 2009; Fan et al., 2010; Peng et al., 2008, 2013; Wang et al., 2010, 2016 reference therein); To the west, the Lhasa-Tengchong Blocks experienced similar tectonomagmatic history since the Early Paleozoic (Xie et al., 2016): As part of the northern margin of the Australian Gondwana in the Early Paleozoic, with moveward to the Eura-Asian continent from Triassic and collision with Qiangtang-Baoshan during the Cretaceous. Then, this magmatism in the Indosinian orogenic belt from Lhasa to Tengchong blocks in response to the evolution of the Paleotethys has been divided into two stages: An early stage involving mantle-derived magmas generated by the late Permian-to-Early Triassic subduction of Paleo-Tethyan oceanic crust (ca. 263-245 Ma; Yang et al., 2009; Huang et al., 2013), and a later stage of crustal-derived granitoid magmatism generated by the Late Triassic (after ca. 235 Ma) collision between the Lhasa-Tengchong blocks and the northern margin of the Australian continent (Li et al., 2011; Zhu et al., 2009; Zou et al., 2011). Therefore, in consideration of mentioned above, both of them share the synchronous igneous rocks, but there are quite different lithological, sedimentological, paleomagnetic and paleobiological characters, especially, two distinct collision and final amalgamation between the Yangtze and Simao-Indochina Blocks and another between the Lhasa-Tengchong blocks and the northern margin of the Australian continent. The Jinghong and Ailaoshan suture zone was regarded as the back-arc basin of the Paleo-tethys main ocean during the Permain-Triassic in the east, but the Lhasa-Tengchong blocks were final amalgamated with Qiangtang-Baoshan blocks during the Cretaceous in the west (Fig. 1a, b). That is why you could see the synchronous igneous rocks appear to the both sides of the Paleo-tethyan main ocean.

According to the recent 1:250 000 regional geological map, their strata are characterized by transition from marine deposits, terrestrial deposits, to marine deposits during early Triassic to latest Triassic. These granitoid magmatism also records the evolution of this tectonism in this area from pre-collision to collision-related stages of orogenesis, as evidenced by the trends in an R_1-R_2 multi-cationic diagram (Batchelor and Bowden, 1985; Fig. 15a) and by changes in the composition of the granitoids that formed at this time from island-arc-related, to syn- and post-collisional settings, as inferred from a Rb vs. Y + Nb diagram (Pearce et al., 1984; Fig. 15b). The metagabbros that formed during this magmatism also provide evidence of a within-plate and/ or spreading center type tectonic setting, as evidenced by their distribution in Ti vs. Zr and FeO^T-MgO-Al_2O_3 diagrams (Fig. 15c,d). All of these data indicate that this area underwent a systematic transition from subduction-related to collisional and finally within-plate tectonic settings (Fig. 16), which is consistent with the evolution of magmatism, sedimentology, and tectonic location in the Indosinian orogenic belt over time Zhu et al. (2011, 2013).

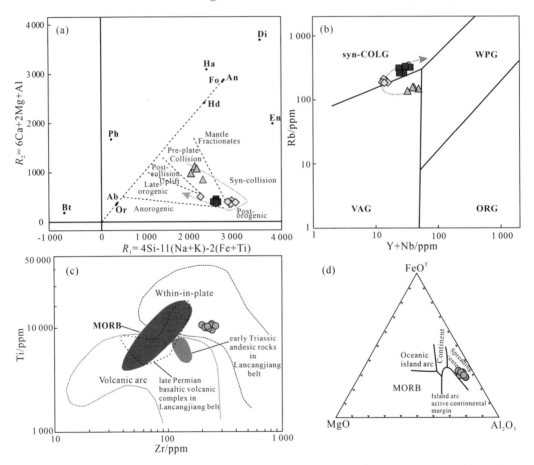

Fig. 15 (a) R_2(6Ca + 2Mg + Al) vs. R_1[4Si − 11(Na + K) − 2(Fe + Ti)] diagram after Batchelor and Bowden (1985); (b) Rb vs. Nb + Y diagram after Pearce et al. (1984); (c)plots of Ti vs. Zr diagram after Pearce et al. (1982) and (d) FeO^T-MgO-Al_2O_3 diagram. Symbols as Fig 6.

Conclusions

(1)New zircon U-Pb geochronological data provide evidence for four stages of magmatism in the Tengchong Block that span the time period between the late Permian and the Late

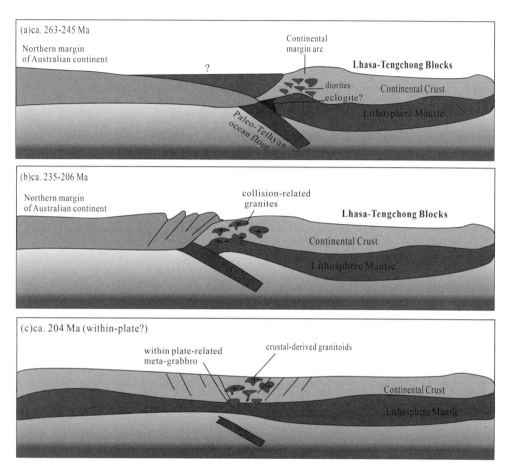

Fig. 16　Tectonic model of the late Paleozoic-to-early Mesozoic magmatisms
in the Tengchong Block, SW China.

Triassic. Combining the new petrological and geochemical data obtained during this study with the geology of this region suggests that these metagabbros and granitoids record the transition from the pre-collision/island arc stage of tectonism through collision-related orogenesis until finally reaching a within-plate setting, essentially reflecting the evolution of the eastern Paleo-Tethys within the southwestern segment of the Sanjiang orogenic belt.

(2)The widespread magmatism within the Indosinian orogenic belt from the western Lhasa Block to southeastern Tengchong Block occurred in two distinct phases: An initial period of mantle-derived magmatism associated with Paleo-Tethyan subduction until the Early Triassic, and a later stage of crustal-derived granitoid magmatism associated with the collision between the Lhasa-Tengchong blocks and the northern margin of the Australian continent, which lasted until the Late Triassic.

(3) Widespread late Permian-to-Late Triassic magmatism occurred in both the Sanjiang and the Indosinian orogenic belts, reflecting the subduction and collisional processes of the

eastern East Paleo-Tethys. The different stages of magmatism in this area are all related to subduction (generating mantle-derived magmas) and later collisional (generating crustal-derived magmas) tectonism that affected the entire region, even though the study area has a somewhat different lithological, sedimentological, and paleobiological history to the other two tectonic zones in this area.

Acknowledgements We thank the constructive and helpful comments from Prof. Wolf-Christan Dullo, Editor-in-Chief, Prof. Wenjiao Xiao and two anonymous reviewers, sincerely. We also thank the English improvement from Dr. Mike Fowler, University of Portsmouth. This study was jointly supported by the National Natural Science Foundation of China (Grants. 41421002, 41372067, and 41190072) and support was also provided by the MOST Special Fund from the State Key Laboratory of Continental Dynamics, Northwest University, and Province Key Laboratory Construction Item (08JZ62).

Electronic supplementary material The online version of this article (doi: 10.1007/s00531-017-1501-x) contains supplementary material, which is available to authorized users.

References

Andersen, T., 2002. Correction of common lead in U-Pb analyses that do not report [204]Pb. Chemical Geology, 192:595-79.

Annen, C., Blundy, J.D., Sparks, R.S.J., 2006. The genesis of intermediate and silicic magmas in deep crustal hot zones. Journal of Petrology, 47(3):505-539.

Barbarin, B., 1999. A review of the relationships between granitoid types, their origins and their geodynamic environments. Lithos, 46:605-626.

Barbarin, B., 2005. Mafic magmatic enclaves and mafic rocks associated with some granitoids of the central Sierra Nevada batholith, California: Nature, origin, and relations with the hosts. Lithos, 80:155-177.

Batchelor, R. A., Bowden, P., 1985. Petrogenetic interpretation of granitoid rock series using multicationic parameters. Chemical Geology, 48(85):43-55.

Ben Othman, D., White, W.M., Patchett, J., 1989. The geochemistry of marine sediments, island arc magma genesis, and crust-mantle recycling. Earth & Planetary Science Letters, 94:1-21.

Bonin, B., 2007. A-type granites and related rocks: Evolution of a concept, Problems and Prospects. Lithos, 97:1-29.

Brown, M., 1994. The generation, segregation, ascent and emplacement of granite magma: The migmatite-to-crustally-derived granite connection in thickened orogens. Earth-Science Reviews, 36:83-130.

Castro, A., 2013. Tonalite-granodiorite suites as cotectic systems: A review of experimental studies with applications to granitoid petrogenesis. Earth-Science Reviews, 124(9):68-95.

Castro, A., Gerya, T., Garcíacasco, A., Fernández, C., Díazalvarado, J., Morenoventas, I., Löw, I., 2010. Melting Relations of MORB-sediment Mélanges in underplated Mantle Wedge Plumes: Implications for the Origin of Cordilleran-type Batholiths. Journal of Petrology, 51(6):1267-1295.

Chappell, B.W., White, A.J.R., 1974. Two contrasting granite types. Pacific Geology, 8:173-174.

Chappell, B.W., White, A.J.R., 1992. I- and S-type granites in Lachlan Fold Belt. Transactions Royal Society Edinburgh, 83:1-26.

Carter, A., Roques ,D., Bristow, C., 2001. Understanding Mesozoic accretion in southeast Asia: Significance of Triassic thermotectonism (Indosinian orogen) in Vietnam. Geology, 29(3):211-214.

Chen, F., Li, X., H., Wang, X.L., Li, Q., L., Siebel, W., 2007. Zircon age and Nd-Hf isotopic composition of the Yunnan Tethyan belt, southwestern China. International Journal of Earth Sciences, 96 (6):1179-1194.

Chen, W., Yang, T., Zhang, S., Yang, Z., Li, H., Wu, H., et al., 2012. Paleo-magnetic results from the early cretaceous Zenong group volcanic rocks, Cuoqin, Tibet, and their paleogeographic implications. Gondwana Research, 22(2):461-469.

Chen, Y.L., Zhang, K.H, Yang, Z.M., Luo, T., 2006. Discovery of a complete ophiolite section in the Jueweng area, Nagqu County, in the central segment of the Bangong Co-Nujiang junction zone, Qinghai-Tibet Plateau. Geological Bulletin of China, 25:694-699.

Cheng, H., Liu, Y., Vervoort, J.D., Lu, H., 2015. Combined U-Pb, Lu-Hf, Sm-Nd and Ar-Ar multichronometric dating on the Bailang eclogite constrains the closure timing of the Paleotethys ocean in the Lhasa Terrane, Tibet. Gondwana Research, 28(4):1482-1499.

Chu, Z.Y., Chen, F.K., Yang, Y.H, Guo, J.H., 2009. Precise determination of Sm, Nd concentrations and Nd isotopic compositions at the nanogram level in geological samples by thermal ionization mass spectrometry. Journal of Analytical Atomic Spectrometry, 24:1534-1544.

Clemens, J.D., 2003. S-type granitic magmas-petrogenetic issues, models and evidence. Earth-Science Reviews, 61:1-18.

Cong, B.L., Wu, G.Y., Zhang, Q., Zhang, R.Y., Zhai, M.G., Zhao, D.S., Zhang, W.H., 1993. Petrotectonic evolution of the Tethys zone in western Yunnan, China. Chinese Bulletin (B), 23(11): 1201-1207 (in Chinese with English abstract).

Cong, F., Lin, S.L., Zou, G.F., Li, Z.H., Xie, T., Peng, Z.M., Liang, T., 2010. Trace elements and Hf isotope compositions and U-Pb age of igneous zircons from the Triassic granite in Lianghe, western Yunnan. Acta Geologica Sinica, 84:1155-1164.

Defant, M.J., Drummond, M.S., 1990. Derivation of some modern arc magmas by melting of young subducted lithosphere. Nature, 347:662-665.

De la Roche, H., Leterrier, J., Grandclaude, P., Marchal, M., 1980. A classification of volcanic and plutonic rocks using R_1-R_2 diagram and major-element analyses: Its relationships with current nomenclature. Chemical Geology, 29:183-210.

Deng, J., Wang, Q.F., Li, G.J., Li, C.S., Wang, C.M., 2014. Tethys tectonic evolution and its bearing on the distribution of important mineral deposits in the Sanjiang region, SW China. Gondwana Research, 26 (2):419-437.

Dewey, J.F., Shackleton, R.M., Chang, C.F., 1988. The tectonic evolution of the Tibetan plateau. Philosophical Transactions of the Royal Society of London (Series A): Mathematical and Physical Sciences, 327:379-413.

Elliott, T., 2003. Tracers of the slab. In: Eiler J. Inside the subduction factory, vol 138. American

Geophysical Union, Geophysical Monograph, Washington DC: 23-45.

Fan, W., Wang, Y., Zhang, A., Zhang, F., Zhang, Y., 2010. Permian arc-back-arc basin development along the Ailaoshan tectonic zone: Geochemical, isotopic and geochronological evidence from the Mojiang volcanic rocks, southwest China. Lithos, 119(3):553-568.

Feeley, T.C., Hacker, M.D., 1995. Intracrustal derivation of Na-rich andesitic and dacitic magmas: An example from volcán ollagüe, Andean central volcanic zone. Journal of Geology, 103(2):213-225.

Feng, Q.L., Chonglakmanib, C.P., Helmckec, D., Ingavat-Helmckec, R., Liua, B.P., 2005. Correlation of Triassic stratigraphy between the Simao and Lampang-Phrae Basins: Implications for the tectonopaleogeography of Southeast Asia. Journal of Asian Earth Sciences, 24:777-785.

Ferguson, E.M., Klein, E.M., 1993. Fresh basalts from the Pacific-Antarctic Ridge extend the Pacific geochemical province. Nature, 366:330-333.

Foley, S., Tiepolo, M., Vannucci, R., 2002. Growth of early continental crust controlled by melting of amphibolite in subduction zones. Nature, 417:837-840.

Frost, T.P., Mahood, G.A., 1987. Field, chemical, and physical constraints on mafic-felsic magma interaction in the Lamarck Granodiorite, Sierra Nevada, California. Geological Society of America Bulletin, 99:272-291.

Frost, B.R., Barnes, C.G., Collins, W.J., Arculus, R.J., Ellis, D.J., Frost, C.D., 2001. A geochemical classification for granitic rocks. Journal of Petrology, 42(11):2033-2048.

Gaudemer, Y., Jaupart, C., Tapponnier, P., 1988. Thermal control on post-orogenic extension in collision belts. Earth & Planetary Science Letters, 89 (1):48-62.

Han, B.F., 2007. Diverse post-collisional granitoids and their tectonic setting discrimination. Earth Science Frontiers, 14(3):64-72 (in Chinese with English abstract).

Hart, S.R., 1984. A large-scale isotope anomaly in the Southern Hemisphere mantle. Nature, 309:753-757.

Hennig, D., Lehmann, B., Frei, D., Belyatsky, B., Zhao, X.F., Cabral, A.R., Zeng, P.S., Zhou, M.F., Schmidt, K., 2009. Early permian seafloor to continental arc magmatism in the eastern Paleo-tethys: U-Pb age and Nd-Sr isotope data from the southern Lancangjiang zone, Yunnan, China. Lithos, 113 (3): 408-422.

Hoskin, P.W.O., Schaltegger, U., 2003. The composition of zircon and igneous and Metamorphic Petrogenesis. Reviews in Mineralogy and Geochemistry, 53:27-62.

He, Z.H., Yang, D.M., Zheng, C.Q., Wang, T.W., 2006. Isotopic dating of the Mamba granitoid in the Gangdese tectonic belt and its constraint on the subduction time of the Neotethys. Geological Review, 52 (1):100-106 (in Chinese with English abstract).

Huang, Z.Y., Qi, X.X., Tang, G.Z., Liu, J.K., Zhu, L.H., Hu, Z.C., Zhao, Y.H., Zhang, C., 2013. The identification of early Indosinian tectonic movement in Tengchong block, western Yunnan: Evidence of zircon U-Pb dating and Lu-Hf isotope for Nabang diorite. Geology in China, 40 (3):730-741 (in Chinese with English abstract).

Hutchison, C.S., 1989. Geological evolution of South-east Asia. Oxford and New York: Clarendon Press, 1-368.

Janoušek, V., Finger, F., Roberts, M., Frýda, J., Pin, C., Dolejš, D., 2004. Deciphering the petrogenesis of deeply buried granites: Whole-rock geochemical constraints on the origin of largely undepleted felsic granulites from the Moldanubian Zone of the Bohemian Massif. Geological Society of America Special Papers, 389:141-159.

Jian, P., Liu, D.Y., Sun, X.M., 2004. SHRIMP dating of Jicha Alaskan-type gabbro in western Yunnan Province: Evidence for the early Permian subduction. Acta Geologica Sinica, 78:165-170.

Jian, P., Liu, D.Y., Sun, X.M., 2008. SHRIMP dating of the Permo-Carboniferous Jinshajiang ophiolite, southwestern China: Geochronological constraints for the evolution of Paleo-Tethys. Journal of Asian Earth Sciences, 32:371-384.

Jian, P., Liu, D.Y., Kröner, A., Zhang, Q., Wang, Y.Z., Sun, X.M., Zhang, W., 2009a. Devonian to Permian plate tectonic cycle of the Paleo-Tethys Orogen in southwest China (Ⅱ): Insights from zircon ages of ophiolites, arc/back-arc assemblages and within-plate igneous rocks and generation of the Emeishan CFB province. Lithos, 113:767-784.

Jian, P., Liu, D.Y., Kröner, A., Zhang, Q., Wang, Y.Z., Sun, X.M., Zhang, W., 2009b. Devonian to Permian plate tectonic cycle of the Paleo-Tethys Orogen in southwest China (Ⅰ): Geochemistry of ophiolites, arc/back-arc assemblages and within-plate igneous rocks. Lithos, 113:748-766.

Jin, X.C., 1996. Tectono-stratigraphic units in western Yunnan and their counterparts in southeast Asia. Discrete And Continuous Dynamical Systems, 1:123-133.

Jin, X., et al., 2002. Permo-Carboniferous sequences of Gondwana affinity in southwest China and their paleogeographic implications. Journal of Asian Earth Sciences, 6:633-646.

Kapp, P., Yin, A., Harrison, T.M., 2005. Cretaceous-Tertiary shortening, basin development, and volcanism in central Tibet. Geological Society of America Bulletin, 117:865-878.

Koteas, G.C., Williams, M.L., Seaman, S.J., Dumond, G., 2010. Granite genesis and mafic-felsic magma interaction in the lower crust. Geology, 38:1067-1070.

Kou, K.H., Zhang, Z.C., Santosh, M., Huang, H., Hou, T., Liao, B.L., Li, H.B., 2012. Picritic porphyrites generated in a slab-window setting: Implications for the transition from Paleo-Tethyan to Neo-Tethyan tectonics. Lithos, 155:375-391.

Li, H.Q., 2009.The geological significance of Indosinian orogenesis occurred in the Lhasa Terrane, Tibet. Ph. D. Dissertation. Beijing: Institute of Geology, Chinese Academy of Geological Sciences (in Chinese with English summary).

Li, C., Wang, T.W., Li, H.M., Zeng, Q.G., 2003. Discovery of Indosinian megaporphyritic granodiorite in the Gangdese area: Evidence for the existence of Paleo-Gangdise. Geological Bulletin of China, 22 (5): 364-366(in Chinese with English abstract).

Li, X.Z., Liu, C.J., Ding, J., 2004. Correlation and connection of the main suture zones in the Greater Mekong subregion. Sedimentary Geologyand Tethyan Geology, 4:001.

Li, Z.H., Lin, S.L., Cong, F., Zou, G.F., Xie, T., 2010. Indosinian orogenesis of the Tengchong-Lianghe block,Western Yunnan: Evidence from zircon U-Pb dating and petrogenesis of granitoids. Acta Petrological et Mineralogica, 29(3):298-312.

Li, H.Q., Xu, Z.Q., Cai, Z.H., Tang, Z.M., Yang, M., 2011. Indosinian epoch magmatic event and geological significance in the Tengchong block, western Yunnan Province. Acta Petrologica Sinica, 27(7):

2165-2172.

Liu, Q.S., Jiang, W., Jian, P., Ye, P.S., Wu, Z.H., Hu, D.G., 2006. Zircon SHRIMP U-Pb age and petrochemical and geochemical features of Mesozoic muscovite monzonitic granite at Ningzhong, Tibet. Acta Petrologica Sinica, 22(3):643-652 (in Chinese with English abstract).

Ludwig, K.R., 2003. ISOPLOT 3.0: A geochronological toolkit for Microsoft Excel. Special Publication, 4, Berkeley Geochronology Center.

Mahawat, C., Atherton, M.P., Brotherton, M.S., 1990. The Tak batholith, Thailand: The evolution of contrasting granite types and implications for tectonic setting. Journal of Southeastern Asian Earth Sciences, 4:11-27.

Mahoney, J.J., Frel, R., Tejada, M.L.G., Mo, X.X., Leat, P.T., Nögler, T.F., 1998. Tracing the Indian Ocean mantle domain through time: Isotopic results from old west Indian, east Tethyan, and south Pacific seafloor. Journal of Petrology, 39:1285-1306

Martin, H., Smithies, R.H., Rapp, R., Moyen, J.F., Champion, D., 2005. An overview of adakite, tonalite-trondhjemite-granodiorite TTG, and sanukitoid: Relationships and some implications for crustal evolution. Lithos, 79:1-24.

Meng, Y., Xu, Z., Santosh, M., Ma, X., Chen, X., Guo Get., al., 2016. Late Triassic crustal growth in southern Tibet: Evidence from the Gangdese magmatic belt. Gondwana Research, 37:449-464.

Metcalfe, I., 1996a. Pre-cretaceous evolution of SE Asian terranes. Geological Society of London, Special Publication, 106(1):97-122.

Metcalfe, I., 1996b. Gondwanaland dispersion, Asian accretion and evolution of eastern Tethys. Aust Journal of Asian Earth Science, (6):605-623.

Metcalfe, I., 2002. Permian tectonic framework and palaeogeography of SE Asia. Journal of Asian Earth Sciences, 20(6):551-566.

Metcalfe, I., 2013. Gondwana dispersion and Asian accretion: Tectonic and paleogeography evolution of eastern Tethys. Journal of Asian Earth Sciences, 66:1-33.

Middlemost, E.A.K., 1994. Naming materials in the magma/igneous rock system. Earth-Science Reviews, 37: 215-224.

Miyashiro, A., 1974. Volcanic rock series in island arcs and active continental margins. American Journal of Science, 274(4):321-355.

Mo, X.X., Shen, S.Y., Zhu, Q.W., 1998. Volcanics-ophiolite and mineralization of middle and southern part in Sanjiang, southern China. Beijing: Geological Publishing House, 1-128 (in Chinese).

Nam, T.N., Sano, Y., Terada, K., Toriumi, M., Quynh, P.V., Le, T.D., 2001. First shrimp U-Pb zircon dating of granulites from the kontum massif (vietnam) and tectonothermal implications. Journal of Asian Earth Sciences, 19(1-2):77-84.

Patiño Douce, A.E., 1999. What do experiments tell us about the relative contributions of crust and mantle to the origin of the granitic magmas. Geology Society of London, 168:55-75.

Pearce, J.A., 1982. Trace element characteristics of lavas from destructive plate boundaries. In: Thorpe, R.S. Andesites. Chichester: Wiley:525-548.

Pearce, J.A., Harris, N.B.W., Tindle, A.G., 1984. Trace element discrimination diagrams for the tectonic interpretation of granitic rocks. Journal of Petrology, 25:956-983.

Peng, T.P., Wang, Y.P., Zhao, G.C., Fan, W.M., Peng, B.X., 2008. Arc-like volcanic rocks from the southern Lancangjiang zone, SW China: Geochronological and geochemical constraints on their petrogenesis and tectonic implications. Lithos, 102:358-73.

Peng, T.P., Wilde, S.A., Wang, Y.J., Fan, W.M., Peng, B.X., 2013. Mid-Triassic felsic igneous rocks from the southern Lancangjiang zone, SW China: Petrogenesis and implications for the evolution of Paleo-tethys. Lithos, 168(2):15-32.

Petrford, N., Atherton, M., 1996. Na-rich partial melts from newly underplated basaltic crust: The Cordillera Blanca Batholith, Peru. Journal of Petrology, 37:1491-1521.

Pitcher, W., 1983. Granite types and tectonic environment. In: Hsu, K. Mountain Building Processes. London: Academic Press:19-40.

Plank, T., 2005. Constraints from thorium/lanthanumon sediment recycling at subduction zones and the evolution of the continents. Journal of Petrology, 46: 921--944.

Plank, T., Langmuir, C.H., 1998. The chemical composition of subducting sediment and its consequences for the crust and mantle. Chemical Geology, 145(3-4): 325-394.

Price, R.C., Kennedy, A.K., Riggs-Sneeringer, M., Frey, F.A., 1986. Geochemistry of basalts from the Indian Ocean Triple Junction: Implication for the generation and evolution of Indian Ocean Ridge basalts. Earth & Planetary Science Letters, 78:379-396.

Qi, L., Hu, J., Gregoire, D.C., 2000. Determination of trace elements in granites by inductively coupled plasma mass spectrometry. Talanta, 51:507-513.

Rapp, R.P., Watson, E.B., 1995. Dehydration melting of metabasalt at 8 – 32 kbar: Implications for continental growth and crust-mantle recycling. Journal of Petrology, 36:891-931.

Ratajeski, K., Sisson, T.W., Glazner, A.F., 2005. Experimental and geochemical evidence for derivation of the el capitan granite, california, by partial melting of hydrous gabbroic lower crust. Contributions to Mineralogy and Petrology, 149(6):713-734.

Rudnick, R.L., Barth, M.G., Horn, I., McDonough, W.F., 2000. Rutile-bearing refractory eclogites: Missing link between continents and depleted mantle. Science, 287:278-281.

Rushmer, T., 1991. Partial melting of two amphibolites: Contrasting experimental results under fluid-absent conditions. Contributions to Mineralogy and Petrology, 107:41-59.

Saunders, A.D., Norry, M.J., Tarney, J., 1988. Origin of MORB and chemically depleted mantle reservoirs: Trace elements constraints. Journal of Petrology Special Lithosphere Issue, (1) :415-445.

Sengör, A.M.C., 1984. The Cimmeride orogenic system and the tectonics of Eurasia. Geological Society of America Special Paper, 195:1-82.

Song, S.G., Niu, Y.L., Wei, C.J., Ji, J.Q., Su, L., 2010. Metamorphism, anatexis, zircon ages and tectonic evolution of the Gongshan block in the northern Indochina continent: An eastern extension of the Lhasa Block. Lithos, 120:327-346.

Springer, W., Seck, H.A., 1997. Partial fusion of basic granulites at 5−15 kbar: Implications for the origin of TTG magmas. Contributions to Mineralogy and Petrology, 127:30-45.

Stern, R.J., Kohut, E., Bloomer, S.H., Leybourne, M., Fouch, M., Vervoort, J., 2006. Subduction factory processes beneath the guguan cross-chain, mariana arc: No role for sediments, are serpentinites important? Contributions to Mineralogy and Petrology, 151(2): 202-221.

Stolz, A.J., Vame, R., Wheller, G.E., Foden, J.D., Abbott, M.J., 1988. The geochemistry and petrogenesis of K-rich alkaline volcanics from the Batu Tara volcano, eastern Sunda Arc. Contributions to Mineralogy and Petrology, 98:374-389.

Stolz, A.J., Vame, R., Davies, G.R., Wheller, G.E., Foden, J.D., 1990. Magma source components in an arc-continent collision zone: The Flores-Lembata sector, Sunda Arc, Indonesia. Contributions to Mineralogy and Petrology, 105:585-601.

Sun, S.S., McDonough, W.F., 1989. Chemical and isotopic systematics of oceanic basalts: Implications or mantle composition and processes. Geological Society, London, Special Publications, 42(1):313-345.

Sylvester, P.J., 1989. Post-collisional alkaline granites. Journal of Geology, 97:261-280.

Sylvester, P.J., 1998. Postcollisional strongly peraluminous granites. Lithos, 45:29-44.

Van Der Voo, R., 1993. Paleomagnetism of the Atlantic, Tethys and Iapetus oceans. Cambridge: Cambridge University Press.

Wang, X., Metalfe, I., Jian, P., He, L., Wang, C., 2000. The Jinshajiang-Ailaoshan Suture Zone, China: Tectonostratigraphy, age and evolution. Journal of Asian Earth Sciences, 18:675-690.

Wang, Y.J., Fan, W.M., Zhang, Y.H., Peng, T.P., Chen, X.Y., Xu, Y.G., 2006. Kinematics and $^{40}Ar/^{39}Ar$ geochronology of the Gaoligong and Chongshan shear systems, western Yunnan, China: Implications for early Oligocene tectonic extrusion of SE Asia.Tectonophysics, 418:235-254

Wang, L., Pan, G., Zhu, D., 2008. Carboniferous-Permian island arc Gangdise belt, Tibet, China: Evidence from volcanic rocks and geochemistry. Geological Bulletin of China, 27(9):1509-1534(in Chinese with English abstract).

Wang, Y.J., Zhang, A.M., Fan, W.M., Peng, T.P., Zhang, F.F., Zhang, Y.H., Bi, X.W., 2010. Petrogenesis of late Triassic post-collisional basaltic rocks of the Lancangjiang tectonic zone, southwest China, and tectonic implications for the evolution of the eastern Paleotethys: Geochronological and geochemical constraints. Lithos, 120(3-4):529-546.

Wang, Y.J., Xing, X.W., Cawood, P.A., Lai, S.C., Xia, X.P., Fan, W.M., Liu, H.C., Zhang, F.F., 2013. Petrogenesis of early Paleozoic peraluminous granite in the Sibumasu Block of SW Yunnan and diachronous accretionary orogenesis along the northern margin of Gondwana. Lithos, 182:67-85.

Wang, C.M., Deng, J., Emmanuel, J.M., Carranza Santosh, M., 2014. Tin metallogenesis associated with granitoids in the southwestern Sanjiang Tethyan Domain: Nature, deposit types, and tectonic setting. Gondwana Research, 26(2):576-593.

Wang, Y., Li, S., Ma, L., Fan, W., Cai, Y., Zhang, Y., et al., 2015. Geochronological and geochemical constraints on the petrogenesis of early Eocene metagabbroic rocks in Nabang (SW Yunnan) and its implications on the NeoTethyan slab subduction. Gondwana Research, 27(4):1474-1486.

Wang, Y., He, H., Cawood, P.A., Fan, W., Srithai, B., Feng, Q., et al., 2016. Geochronological, elemental and Sr-Nd-Hf-O isotopic constraints on the petrogenesis of the Triassic post-collisional granitic rocks in NW Thailand and its Paleotethyan implications. Lithos, 266-267:264-286.

Wolf, M.B., Wyllie, P.J., 1994. Dehydration-melting of amphibolite at 10 kbar: The effects of temperature and time. Contributions to Mineralogy and Petrology, 115(4):369-383.

Xiao, L., Xu, Y.G., Chung, S.L., He, B., Mei, H., 2003. Chemostrati-graphic correlation of upper Permian lavas from Yunnan prov-ince, China: Extent of the Emeishan large igneous province. International Geology Review, 45(8):753-766.

Xiao, L., Xu, Y.G., Chung, S.L., He, B., Mei, H., 2003. Chemostratigraphic correlation of upper Permian lavas from Yunnan province, China: Extent of the Emeishan large igneous province.International Geology Review, 45(8):753-766.

Xiao, L., He, Q., Pirajno, F., Ni, P., Du, J., Wei, Q., 2004. Possible correlation between a mantle plume and the evolution of Paleo-Tethys Jinshajiang ocean: Evidence from a volcanic rifted margin in the Xiaru-Tuoding area, Yunnan, SW China. Lithos, 100(1):112-126.

Xie, J.C., Zhu, D.C., Dong, G., Zhao, Z.D., Wang, Q., Mo, X., 2016. Linking the Tengchong terrane in SW Yunnan with the Lhasa terrane in southern Tibet through magmatic correlation. Gondwana Research, 39:217-229.

Xu, X., Yang, J., Li, T., 2007. SHRIMP U-Pb ages and inclusions of zircons from the Sumdo eclogite in the Lhasa block, Tibet, China. Geological Bulletin of China, 26(10):1340-1355 (in Chinese with English abstract).

Xu, Y.G., Lan, J.B., Yang, Q.J., Huang, X.L., Qiu, H.N., 2008. Eocene breakoff of the Neo-Tethyan slab as inferred from intraplate-type mafic dykes in the Gaoligong orogenic belt, eastern Tibet. Chemical Geology, 255:439-453

Xu, Y.G., Yang, Q.J., Lan, J.B., Luo, Z.Y., Huang, X.L., Shi, Y.R., Xie, L.W., 2012. Temporal-spatial distribution and tectonic implications of the batholiths in the Gaoligong-Tengliang-Yingjiang area, western Yunnan: Constraints from zircon U-Pb ages and Hf isotopes.Journal of Asian Earth Sciences, 53: 151-175

Yang, J.S., Xu, Z.Q., Li, T.F., Li, H.Q., Li, Z.L., Ren, Y.F., Xu, X.Z., Chen, S.Y., 2007. Oceanic subduction-type eclogite in the Lhasa block, Tibet, China: Remains of the Paleo-Tethys ocean basin? Geological Bulletin of China, 26(10):1277-1287.

Yang, J., Xu, Z., Li, Z., Xu, X., Li, T., Ren, Y., Li, H., Chen, S., Robinson, P. T., 2009. Discovery of an eclogite belt in the Lhasa block, tibet: A new border for Paleo-Tethys? Journal of Asian Earth Sciences, 34(1):76-89.

Yang, T.N., Ding, Y., Zhang, H.R., Fan, J.W., Liang, M.J., Wang, X.H., 2014. Two-phase subduction and subsequent collision defines the Paleotethyan tectonics of the southeastern Tibetan Plateau: Evidence from zircon U-Pb dating, geochemistry, and structural geology of the Sanjiang orogenic belt, southwest China. Geological Society of America Bulletin, 126(11/12):1654-1682.

YBGMR(Yunnan Bureau Geological Mineral Resource), 1990. Regional geology of Yunnan Province.Beijing: Geological Publishing House:1-729 (in Chinese with English abstract).

Yin, A., Harrison, T.M., 2000. Geologic evolution of the Himalayan-Tibetan orogeny. Earth & Planetary Sciences Letters, 28:211-280.

Yuan, H.L., Gao, S., Liu, X.M., Li, H.M., Gunther, D., Wu F.Y., 2004. Accurate U-Pb age and trace element determinations of zircon by laser ablation-inductively coupled plasma mass spectrometry. Geostandard Newsletters, 28:353-370.

Zen, E., 1986. Aluminum enrichment in silicate melts by fractional crystallization: Some mineralogical and petrographical constrains. Journal of Petrology, 27:1095-1117.

Zhai, Q.G., Jahn, B.M., Su, L., Wang, J., Mo, X.X., Lee, H.Y., Wang, K.L., Tang, S., 2013. Triassic arc magmatism in the Qiangtang area, northern Tibet: Zircon U-Pb ages, geochemical and Sr-Nd-Hf isotopic characteristics, and tectonic implications. Journal of Asian Earth Sciences, 63:162-178.

Zhang, H.F., Xu, W.C., Guo, J.Q., Zone, K.Q., Cai, H.M., Yuan, H.L., 2007. Indosinian orogenesis of the gangdise terrane: Evidences of zircon U-Pb dating and petrogenesis of granitoids. Earth Science, 32 (2):155-166 (in Chinese with English abstract).

Zhang, Z.M., Wang, J.L., Shen, K., Shi, C., 2008. Paleozoic circus-Gondwana orogens: Petrology and geochronology of the Namche Barwa Complex in the eastern Himalayan syntaxis, Tibet. Acta Petrologica Sinica, 24:1627-1637 (in Chinese with English abstract).

Zhang, K.J., Tang, X.C., Wang, Y., Zhang, Y.X., 2011. Geochronology, geochemistry, and Nd isotopes of early Mesozoic bimodal volcanismin northern Tibet, western China: Constraints on the exhumation of the central Qiangtang metamorphic belt. Lithos, 121:167-175.

Zhang, Z.M., Dong, X., Santosh, M., Liu, F., Wang, W., Yiu, F., He, Z.Y., Shen, K., 2012a. Petrology and geochronology of the Namche Barwa Complex in the eastern Himalayan syntaxis, Tibet: Constrains on the origin and evolution of the north-eastern margin of the Indian craton. Gondwana Research, 21:123-137.

Zhao, S.W., Lai, S.C., Qin, J.F., Zhu, R.Z., 2016. Tectono-magmatic evolution of the gaoligong belt, southeastern margin of the tibetan plateau: Constraints from granitic gneisses and granitoid intrusions. Gondwana Research, 1(1):56-66.

Zhao, C.F., et al., 1999. Variscan and Idosinian granites in northern Tengchong, Yunnan. Regional Geology of China, 18(3):260-263.

Zheng, L.L., Geng, Q.R., Dong, H., Ou, C.S., Wang, X.W., 2003. The discovery and significance of the relicts of ophiolitic mélanges alongthe Parlung Zangbo in the Bomi region, eastern Xizang. Sedimentary Geology and Tethyan Geology, 23(1):27-30 (in Chinese with English abstract).

Zhong, D.L., 1998. The Paleotethys Orogenic Belt in West of Sichuan and Yunnan. Beijing: Science Press: 1-230 (in Chinese).

Zhong, D.L., 2000. Paleotethys sides in West Yunnan and Sichuan, China. Beijing: Science Press: 1-248 (in Chinese with English abstract).

Zhu, D., Mo, X., Niu, Y., 2009. Zircon U-Pb dating and in situ Hf isotopic analysis of Permian peraluminous granite in the Lhasa terrane, southern Tibet: Implications for Permian collisional orogeny and paleogeography. Tectonophysics, 469:48-60.

Zhu, D.C., Zhao, Z.D., Niu, Y.L., Dilek, Y., Hou, Z.Q., Mo, X.X., 2013. The origin and pre-Cenozoic evolution of the Tibetan Plateau. Gondwana Research, 23:1429-1454.

Zhu, D.C., Zhao, Z.D., Niu, Y.L., Mo, X.X., Chung, S.L., Hou, Z.Q., Wang, L.Q., Wu, F.Y., 2011. The Lhasa Terrane: Record of a microcontinent and its histories of drift and growth. Earth & Planetary Science Letters, 301:241-255.

Zi, J.W., Cawood, P.A., Fan, W.M., Tohver, E., Wang, Y.J., Mccuaig, T.C., 2012a. Generation of early Indosinian enriched mantle-derived granitoid pluton in the Sanjiang orogen (SW China) in response to closure of the Paleo-Tethys. Lithos, 140(5):166-182.

Zi, J.W., Cawood, P.A., Fan, W.M., Wang, Y.J., Tohver, E., Mccuaig, T.C., Peng, T.P., 2012b. Triassic collision in the Paleo-Tethys ocean constrained by volcanic activity in SW China. Lithos, 144(7): 145-160.

Zi, J.W., Cawood, P.A., Fan, W.M., Tohver, E., Wang, Y.J., Mccuaig, T.C., et al., 2013. Late permian-triassic magmatic evolution in the Jin-shajiang orogenic belt, SW China and implications for orogenic processes following closure of the paleo-tethys. American Journal of Science, 313(2):81-112.

Zindle, A., Hart, S.R., 1986. Chemical geodynamics. Annual Review of Earth & Planetary Sciences, 14: 493-571.

Zou, G.F., Lin, S.L., Li, Z.H., Cong, F., Xie, T., 2011. Geochronology and geochemistry of the Longtang granite in the Lianghe area, Western Yunnan and its tectonic implications. Geotectonica et Metallogenia, 35:439-451.

Late Early-Cretaceous quartz diorite-granodiorite-monzogranite association from the Gaoligong belt, southeastern Tibet Plateau: Chemical variations and geodynamic implications[①]

Zhu Renzhi Lai Shaocong[②] Qin Jiangfeng Zhao Shaowei Wang Jiangbo

Abstract: Geochemical variations in granitic rocks may be controlled by their source rocks, melting reactions and subsequent magmatic processes, which resulted from various geodynamic processes related to subduction, collision, or slab break-off. Here we report new LA-ICP-MS zircon U-Pb ages and Hf isotopes, whole-rock chemistry and Sr-Nd isotopes for the late Early Cretaceous quartz diorite, granodiorite and monzogranite in the Gaoligong belt, southeastern Tibet Plateau. The zircon U-Pb dating yield ages of 113.9 ± 1.6 Ma, 111.7 ± 0.8 Ma, and 112.8 ± 1.7 Ma for the quartz diorite, granodiorite, and monzogranite, respectively, which are coeval with bimodal magmatism in the central and northern Lhasa sub-terrane. There are the distinct sources regions for the quartz diorite and granodiorite-monzogranite association. The quartz diorites are sodic, calc-alkaline and have high $Mg^{\#}(52-54)$ values. They also have elevated initial $^{87}Sr/^{86}Sr$ ($0.707\,019-0.709\,176$) and low $\varepsilon_{Nd}(t)$ (-7.63 to -5.16), with variable zircon $\varepsilon_{Hf}(t)$ values (-9.02 to $+5.65$). Zircon chemical data indicate a typical crustal-derived character with high Th ($142-1\,260$ ppm) and U ($106-1\,082$ ppm) and moderate U/Yb ratios ($0.30-2.32$) and Y content ($705-1\,888$ ppm). Those data suggest that the quartz diorites were derived from partial melting of ancient basaltic lower crust by a mantle-derived magma in source region. The granodiorite-monzogranite association has high-K calc-alkaline, weakly peraluminous characters. They show lower Nb/Ta ($5.57-13.8$), CaO/Na_2O ($0.62-1.21$), higher Al_2O_3/TiO_2($24.4-44.4$) ratios, more evolved whole-rock Sr-Nd and zircon Hf isotopic signatures, all of which suggest derivation from mixed basaltic and metasedimentary source rocks in a deep crustal zone. We propose that the granitic magmatisms at ca. $113-110$ Ma in the Gaoligong belt was triggered by the slab break-off of Bangong-Nujiang Tethyan oceanic lithosphere.

1 Introduction

The chemical variations in granitic rocks may be produced by partial melting of a wide

---

① Published in *Lithos*, 2017, 288-289.

② Corresponding author.

variety of protolith materials, may be the end-products of extreme fraction of mafic magmas, or the result of peritectic assemblage entrainment (Clemens and Stevens, 2012). Many processes could also contribute (Lai et al., 2015; Qin et al., 2016), for example, different partial melting conditions from a homogeneous source region, different source regions, progressive assimilation of wall rocks (Pitcher, 1997) and few are mutually exclusive. Granitic rocks in orogenic settings are usually geochemically heterogeneous and can supply robust constraints from petrography, whole-rock chemical and isotopes, and zircon Hf isotopic composition (Chappell et al., 2012; Clemens et al., 2010, 2011; Clemens and Stevens, 2012; Lai et al., 2015; Qin et al., 2016). A magmatic flare-up are occurred at ca. 113−110 Ma in the central and northern Lhasa subterrane (Chen et al., 2014; Sui et al., 2013; Zhu et al., 2009, 2011), known to be characterized by contributions from mantle- or juvenile crust-derived melts both in the Tengchong and Lhasa Terranes (Xie et al., 2016). However, its detailed petrogenesis and deep processes are still uncertain. This paper describes geochemical variations between quartz diorites and granodiorite-monzogranite associations and donates the new insights in the related interaction between mantle- and crust-derived materials in the deep crustal zone from Lhasa to Tengchong terranes.

2　Geological setting, field relationships and petrography

The Tibet Plateau contains the Songpan-Garze flysch complex, and several tectonic blocks including Northern Qiangtang, Southern Qiangtang, Lhasa, and the Himalaya blocks, separated by the Jinshajiang, Longmucuo-Shuanghu-Lancangjiang, Bangong-Nujiang, and Yarlung-Zangbo suture zones from north to south, respectively (Zhu et al., 2011). The Lhasa terrane is bordered by the Bangong-Nujiang suture (BNS) in the north and the Yarlung-Zangbo suture (YZS) in the south (Chiu et al., 2009) (Fig. 1a). It dispersed from Gondwana during the Permian to Triassic and then drifted northward to finally collide with the Qiangtang terrane during the late Jurassic to Early Cretaceous (Kapp et al., 2005; Metcalfe, 2013; Zhu et al., 2011). The central and northern Lhasa subterranes are covered by widespread Carboniferous metasedimentary rocks and Jurassic-Early Cretaceous volcano-sedimentary rocks, with minor Ordovician, Silurian, and Triassic limestone (Pan et al., 2004; Zhu et al., 2011). Voluminous Mesozoic volcanic rocks (ca. 143−99 Ma) that consist of andesite, dacite, rhyolite and associated volcaniclastic rocks are exposed within the subterranes (Chen et al., 2014; Kapp et al., 2005, 2007; Sui et al., 2013; Zhu et al., 2009). Mesozoic plutonic rocks (ca. 215−80 Ma) intruded the pre-Ordovician, Carboniferous-Permian metasedimentary successions, mainly in the Early Cretaceous (Guynn et al., 2006; Harris et al., 1990; Qu et al., 2012; Sui et al., 2013; Xu et al., 1985; Zhu et al., 2009; R.Z. Zhu et al., 2015). Abundant dioritic enclaves occur within these Early Cretaceous granitoids (Sui et al., 2013;

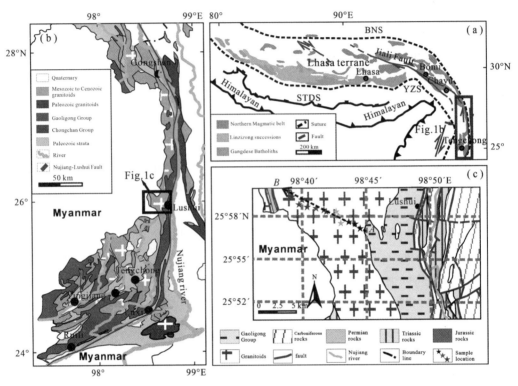

Fig. 1 (a) Distribution of principal magma and strata of central and southern Tibetan Plateau (modified after Chiu et al., 2009). (b) Distribution of principal magma and strata of western Yunnan (modified after Xu et al., 2012; R.Z. Zhu et al., 2015; D.C. Zhu et al., 2015). (c) Study area of the granitic rocks in the middle of Gaoligong belt as the southeastern extension of Bomi-Chayu batholith in the SE Lhasa Block (modified after YNBGMR, 1991).

BNS: Bangong-Nujiang suture; YZS: Yarlung-Zangbo suture; LCJ: Longchuanjiang.

The black star is quartz diorite, lightly blue is granodiorite and gray is monzogranite.

Zhu et al., 2009, 2011; R.Z. Zhu et al., 2015; D.C. Zhu et al., 2015).

In the east, extensive Cretaceous and minor Ordovician, Jurassic, and Cenozoic granitoids are exposed in a NW-SE belt to the southwest of the Bangong-Nujiang suture zone (Chiu et al., 2009; Liang et al., 2008; Pan et al., 2004; Zhu et al., 2009), and crop out in the Carboniferous-Permian, Devonian and Proterozoic metamorphic rocks. The Cretaceous granitoids occur mostly as batholiths in the Bomi, Basu, Ranwu, and Chayu area (so-called Bomi-Chayu batholiths) and consist of monzogranites and granodiorites with minor dioritic veins and dioritic enclaves (Chiu et al., 2009; Liang et al., 2008; Lin et al., 2013; Pan et al., 2004; Zhu et al., 2009).

The area covered by the north-south trending Gaoligong belt is situated to southeastern extension of the eastern Himalayan Syntaxis (Xu et al., 2012) (Fig. 1a). This mountain range marks the divide between the Longchuanjiang (LCJ) in the west and the Nujiang (Salween) in the east (Fig. 1b). The core of the Gaoligong range is composed of Precambrian high-grade metamorphic rocks (Gaoligong Group), late Paleozoic clastic rocks and carbonates, and granitic intrusives. These are intensely deformed and affected by a sub-vertical foliation, dipping toward both east and west directions. Ductile shear sense criteria show a right-lateral motion. Mylonitization is dated between 32 Ma and 12 Ma (Eroğlu et al., 2013; Wang et al., 2006), in agreement with the dextral shearing along the Jiali fault and Karakorum fault (Lee et al., 2002; Searle, 1997). Mesozoic to Cenozoic basic intrusive and volcanic rocks are distributed along the Nujiang fault (R.Z. Zhu et al., 2015; D.C. Zhu et al., 2015; Zhu et al., in press-a,b). To the east, it was separated by Nujiang fault with the Lanping-Simao block; to the west, it was separated by Longchuanjiang fault with the Tengchong-Lianghe terrane.

The N-S trending granites from Gaoligong belt along the western Nujiang river are less deformed, extend northward into the NWW trending Bomi-Chayu magmatic belt, which belongs to the components of the Himalayan Syntaxis (Fig. 1b). It also extends south-westward into the Tengchong-Lianghe-Yingjiang area, and likely extends further into the Shan Scarp in Myanmar (Fig. 1b). The basement in the Gaoligong area is also composed of flat foliated granites and metamorphic rocks. The Gaoligong granites are intruded into the Precambrian Chongshan Group and Gaoligong metamorphic rocks (Gaoligong Group), with massive structure in the middle and partly mylonitized at the margin. It is exposed as a narrow lens between the Nujiang and Longchuanjiang strike-slip faults and characterized by obviously syntectonic emplacement (Wang et al., 2006; Xu et al., 2012).

According to field observations (across *A-B* section, Fig. 1c), quartz diorite, granodiorite, and monzogranite were distributed from east to west-part in the study area. The contact between the fine-grained quartz diorite and the coarse granodiorite-monzogranite associations are sharp but the relationship between granodiorite and monzogranite is gradational. The quartz diorites mainly consist of plagioclase (40 – 50 vol%), alkali feldspar (10 – 15 vol%), amphibole (15 – 18 vol%), biotite (10 – 12 vol%), quartz (8 – 12 vol%), with accessory minerals including zircon, apatite, titanite, and magnetite (Fig. 2a). Amphiboles and most biotites have subhedral shapes, with some felsic inclusions, and apatites are needle-like (Fig. 2b,c). Plagioclase has insignificant resorption and regrowth but clearly two-stage growth with narrow marginal zones separated from the main crystal by a chain of mineral inclusions. Fewer magma mixing processes such as resorption and regrowth are visible than typical samples elsewhere (Słaby and Gotze, 2004; Słaby et al., 2011, 2016) (Fig. 2c). The granodiorites are coarse grained, and mainly contained plagioclase (35 – 40 vol%), alkali feldspar (15 – 20 vol%), quartz (20 – 24 vol%), amphibole (10 – 15 vol%), biotite (10 – 12 vol%), and accessory minerals including titanite, zircon, apatite, and magnetite (Fig. 2d). Amphibole is subhedral to euhedral with some biotite clusters, and sometimes included in quartz phenocrysts. Titanite is euhedral with typical rhombic shape (Fig. 2e,f). The plagioclase exhibits concentric zonation (Fig. 2f), but also has many resorption planes pointing to dissolution and regrowth (Fig. 2f). In addition, the core has sieve textures (Fig. 2f), indicating mixing processes (Słaby et al., 2002; Słaby and Gotze, 2004). The monzogranites are also coarse grained, and mainly consist of alkali feldspar (25−35 vol%), plagioclase (15−28 vol%), quartz (25−35 vol%), biotite (10−15 vol%), amphibole (1−3 vol%), and accessory minerals including titanite, zircon and apatite (Fig. 2g). Biotite is subhedral, and is the predominant mafic mineral in the monzogranites. Amphiboles are sporadic between the felsic minerals (Fig. 2h,i). Similar to granodiorites, the feldspars show resorption and regrowth textures (Fig. 2i), also implying mixing processes (Słaby and Gotze, 2004; Słaby et al., 2011, 2016).

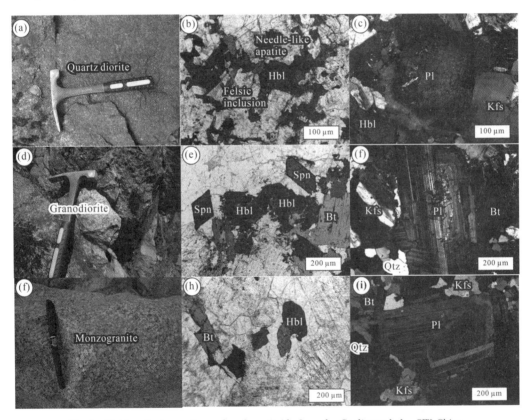

Fig. 2　Field and micro petrography of granitoids from the Gaoligong belt, SW China.

3　Analytical methods

3. 1　Major and trace elements

Whole-rock samples were trimmed to remove weathered surfaces, cleaned with deionized water, crushed, and then powdered through a 200 mesh screen using a tungsten carbide ball mill. Major elements were analyzed using an X-ray fluorescence (XRF) spectrometer (Rikagu RIX 2100) at the Guizhou Tuopu Resource and Environmental Analysis Center, Institute of Geochemistry, Chinese Academy of Sciences, Guiyang, China. Analyses of USGS and Chinese national rock standards (BCR-2, GSR-1, and GSR-3) indicate that both analytical precision and accuracy for major elements are generally better than 5%. Trace elements were determined by using a Bruker Aurora M90 inductively coupled plasma mass spectrometry (ICP-MS) at the Guizhou Tuopu Resource and Environmental Analysis Center, Institute of Geochemistry, Chinese Academy of Sciences, Guiyang, China, following the method of Qi et al. (2000). Sample powders were dissolved using an $HF + HNO_3$ mixture in a high-pressure PTFE bombs at 185℃ for 36 h. The accuracies are > 95% of the ICP-MS analyses and are estimated to be better than ±5%–10% (relative) for most elements (Qi et al., 2000).

3.2 Sr-Nd isotopic analyses

Whole-rock Sr-Nd isotopic data were obtained by using a Neptune Plasma HR multi-collector mass spectrometer at the State Key Laboratory of Continental Dynamics, Northwest University, China. The Sr and Nd isotopes were determined by using a method similar to that of Chiu et al. (2009). Sr and Nd isotopic fractionation was corrected to $^{87}Sr/^{86}Sr = 0.119\ 4$ and $^{146}Nd/^{144}Nd = 0.721\ 9$, respectively. The average $^{143}Nd/^{144}Nd$ ratio of the La Jolla standard measured during the sample runs is $0.511\ 862 \pm 5\ (2\sigma)$, and the average $^{87}Sr/^{86}Sr$ ratio of the NBS987 standard is $0.710\ 236 \pm 16\ (2\sigma)$. The total procedural Sr and Nd blanks are b1 ng and b50 pg, respectively. NIST SRM-987 and JMC-Nd were used as certified reference standard solutions for $^{87}Sr/^{86}Sr$ and $^{143}Nd/^{144}Nd$ isotopic ratios, respectively. BCR-1 and BHVO-1 were used as reference materials.

3.3 Zircon U-Pb and Hf isotopic analyses

Zircon was separated from three ~ 5 kg samples taken from various sampling locations within the Gaoligong belt so that we can take out enough and representative zircons from the samples. The zircon grains were separated by using conventional heavy liquid and magnetic techniques. Representative zircon grains were handpicked and mounted in epoxy resin disks and then polished and coated with carbon. Internal morphology was examined using cathodoluminescent (CL) prior to U-Pb analyses.

Laser ablation ICP-MS zircon U-Pb analyses were conducted on an Agilent 7500a ICP-MS equipped with a 193-nm laser, which is housed at the State Key Laboratory of Continental Dynamics, Northwest University, Xi'an, China, following the method of Yuan et al. (2004). The $^{207}Pb/^{206}Pb$ and $^{206}Pb/^{238}U$ ratios were calculated by using the GLITTER program and corrected using the Harvard zircon 91500 as external calibration. These correction factors were then applied to each sample to correct for both instrumental mass bias and depth-dependent elemental and isotopic fractionation. The detailed analytical technique is described in Yuan et al. (2004). Common Pb contents were therefore evaluated by using the method described in Andersen (2002). The age calculations and plotting of concordia diagrams were made using ISOPLOT (version 3.0; Ludwig, 2003). The errors quoted in tables and figures are at the 2σ level.

In situ zircon Hf isotopic analyses were conducted using a Neptune MC-ICPMS, equipped with a 193-nm laser. During analyses, a laser repetition rate of 10 Hz at 100 mJ was used and spot sizes were 32 μm. The detailed analytical technique is described by Yuan et al. (2008). During the analysis in Xi'an, the measured values of well-characterized zircon standards (91500, GJ-1, and Monastery) were consistent with the recommended values (cf. Yuan et al., 2008) within 2σ. The obtained Hf isotopic compositions were $0.282\,016 \pm 20$ ($2\sigma_n$, $n = 84$) for the GJ-1 standard and $0.282\,735 \pm 24$ ($2\sigma_n$, $n = 84$) for the Monastery standard, respectively, consistent with the recommended values (cf. Yuan et al., 2008) to within 2σ. The initial ^{176}Hf/^{177}Hf ratios and $\varepsilon_{Hf}(t)$ values were calculated with the reference to the chondritic reservoir (CHUR) at the time of zircon growth from the magmas. The decay constant for ^{176}Lu of 1.867×10^{-11} year^{-1} (Soderlund et al., 2004), the chondritic ^{176}Hf/^{177}Hf ratio of $0.282\,785$ and ^{176}Lu/^{177}Hf ratio of $0.033\,6$ were adopted. The depleted mantle model ages that were (T_{DM}) used for basic rocks were calculated with reference to the depleted mantle at the present-day ^{176}Hf/^{177}Hf ratio of $0.283\,25$, similar to that of the average MORB (Nowell et al., 1998) and ^{176}Lu/^{177}Hf = $0.038\,4$ (Griffin et al., 2000). For the zircons from felsic rocks, we also calculated the Hf isotope "crustal" model age (T_{DMC}) by assuming that its parental magma was derived from an average continental crust, with ^{176}Lu/^{177}Hf = 0.015, that originated from the depleted mantle source (Griffin et al., 2000). Our conclusions would not be affected even if other decay constants were used.

4　Results

We chose representative samples from the Gaoligong belt, SE Lhasa block. All the analytical data are listed in Table 1 (major and trace elements), Table 2 (bulk-rock Sr-Nd isotopic compositions), and Supplementary Dataset Table (zircon U-Pb isotopic data).

4.1　Zircon LA-ICP-MS U-Pb Dating

The sampling lithology and zircon U-Pb data are summarized in the Supplementary Dataset Table 1 and in Figs. 1, 2 and 3.

The separated zircons from monzogranites (PM2-57) are subhedral to enhedral, measuring up to 100–300 μm with length to width ratios 2:1 to 3:1. The prismatic crystals

Table 1　Major(wt%) and trace(ppm) elements of granitic rocks from Gaoligong belt, SW China.

Sample	Quartz diorite							Granodiorite					
	N: 25°57.610' E: 98°43.431'			N: 25°57.233' E: 98°44.068'				N: 25°58.312' E: 98°42.719'			N: 25°57.555' E: 98°43.201'		
	PM2-120	PM2-121	PM2-123	PM2-127-1	PM2-127-2	PM2-127-3	PM2-127-4	PM2-97	PM2-98	PM2-99	PM2-115	PM2-116	PM2-117
SiO_2	62.05	62.31	62.47	61.53	60.34	60.66	60.02	67.47	65.81	67.53	64.79	64.68	64.36
TiO_2	0.70	0.68	0.71	0.77	0.81	0.80	0.80	0.53	0.54	0.52	0.64	0.59	0.68
Al_2O_3	17.17	16.90	16.88	16.25	16.35	16.27	16.37	15.34	16.28	15.83	16.52	16.31	16.49
MgO	2.62	2.63	2.64	3.10	3.27	3.10	3.30	1.48	1.54	1.43	1.85	1.84	1.90
$Fe_2O_3^T$	5.75	5.64	5.70	6.50	6.62	6.36	6.81	4.19	4.18	3.85	4.86	4.86	5.04
CaO	4.69	4.87	4.92	4.42	4.42	4.89	4.70	3.20	3.57	3.17	3.80	4.37	4.06
Na_2O	3.34	3.35	3.32	3.59	3.38	3.32	3.20	3.36	3.45	3.40	3.28	3.60	3.43
K_2O	2.15	1.97	1.93	2.23	2.52	2.52	2.67	2.96	3.01	3.30	3.23	2.21	2.71
MnO	0.09	0.10	0.10	0.12	0.12	0.12	0.13	0.10	0.10	0.09	0.09	0.09	0.09
P_2O_5	0.16	0.17	0.17	0.19	0.19	0.19	0.20	0.16	0.15	0.15	0.18	0.18	0.18
LOI	1.19	1.18	1.11	1.31	1.65	1.61	1.75	0.84	1.19	1.09	1.03	1.13	0.93
Total	99.90	99.80	99.92	100.02	99.68	99.85	99.95	99.62	99.82	100.35	100.26	99.87	99.86
Li	33.20	32.50	29.80	32.50	37.10	34.60	38.20	19.30	23.00	20.10	50.10	37.80	47.50
Be	1.70	1.73	1.70	2.14	2.08	2.16	2.11	2.90	2.94	2.80	2.51	2.84	2.69
Sc	19.27	18.39	18.04	19.01	19.18	19.27	19.62	14.52	14.43	14.78	13.46	19.89	17.07
V	115.00	115.00	115.00	118.00	126.00	125.00	129.00	74.20	73.00	68.10	90.10	92.50	93.30
Cr	19.90	20.00	19.30	64.50	50.90	50.20	60.70	12.10	12.90	21.60	15.50	16.10	19.10
Co	33.70	29.50	32.00	39.00	40.50	36.30	36.90	29.50	37.80	29.80	38.70	36.30	35.60
Ni	9.51	9.27	9.15	22.70	20.00	19.20	22.30	5.83	6.30	7.71	8.20	7.13	8.33
Cu	8.55	9.52	10.60	7.57	7.65	8.02	8.22	7.47	8.11	8.91	9.55	7.44	9.00
Zn	116.00	102.00	90.70	127.00	128.00	134.00	129.00	90.10	111.00	95.80	106.00	87.90	100.00
Ga	18.30	17.90	18.40	18.90	18.90	19.00	19.20	18.40	18.80	18.20	18.70	19.10	19.40
Ge	1.11	1.05	1.12	1.28	1.23	1.22	2.43	1.30	1.29	1.24	1.20	1.27	1.20
As	3.15	1.74	1.39	1.28	0.85	2.12	1.23	1.80	3.38	1.36	2.26	1.22	1.21

Continued

| Sample | Quartz diorite | | | | | | | Granodiorite | | | | | |
| | N: 25°57.610′ E: 98°43.431′ | | | N: 25°57.233′ E: 98°44.068′ | | | | N: 25°58.312′ E: 98°42.719′ | | | N: 25°57.555′ E: 98°43.201′ | | |
	PM2-120	PM2-121	PM2-123	PM2-127-1	PM2-127-2	PM2-127-3	PM2-127-4	PM2-97	PM2-98	PM2-99	PM2-115	PM2-116	PM2-117
Rb	93.40	90.90	86.10	155.00	176.00	171.00	185.00	116.00	124.00	116.00	138.00	99.70	125.00
Sr	360.00	356.00	373.00	302.00	297.00	305.00	298.00	278.00	285.00	284.00	326.00	334.00	337.00
Y	17.40	18.70	19.10	23.90	23.60	25.10	26.50	25.20	26.30	24.60	19.40	29.50	26.90
Zr	84.20	104.00	102.00	97.00	123.00	108.00	110.00	131.00	137.00	149.00	153.00	146.00	149.00
Nb	8.71	9.31	9.58	13.30	13.00	14.10	13.60	14.10	15.00	14.00	12.30	13.50	14.60
Mo	0.44	0.55	0.32	0.98	0.38	1.27	1.11	0.47	0.77	0.61	0.62	0.37	0.68
Cs	5.00	4.91	4.69	13.60	15.00	13.80	15.90	3.35	2.86	2.67	5.13	3.82	4.89
Ba	319.00	330.00	303.00	230.00	225.00	244.00	235.00	496.00	482.00	580.00	711.00	402.00	502.00
La	12.30	21.60	13.40	24.10	22.30	23.70	18.10	38.20	36.40	34.90	49.30	43.10	40.40
Ce	27.60	41.40	29.50	50.60	48.80	51.50	44.20	73.40	69.10	67.90	85.70	82.90	76.20
Pr	3.24	4.75	3.72	5.94	6.09	6.23	5.71	7.55	7.46	7.01	9.03	8.95	8.34
Nd	12.70	18.00	15.40	22.50	23.00	24.10	23.80	26.30	25.00	24.60	29.40	31.50	28.80
Sm	3.08	3.70	3.61	4.62	4.95	5.03	5.36	4.85	4.75	4.72	4.69	6.32	5.52
Eu	0.85	0.89	0.91	1.19	1.24	1.28	1.30	1.11	1.07	1.05	1.11	1.37	1.28
Gd	3.25	3.51	3.46	4.42	4.64	4.59	5.03	4.73	4.60	4.26	4.32	6.15	5.53
Tb	0.46	0.55	0.55	0.66	0.67	0.71	0.80	0.68	0.68	0.63	0.56	0.83	0.73
Dy	2.69	3.11	3.23	3.79	4.05	4.12	4.56	4.05	4.20	3.92	3.09	4.67	4.18
Ho	0.54	0.62	0.64	0.74	0.76	0.81	0.87	0.80	0.81	0.74	0.63	0.94	0.82
Er	1.47	1.83	1.82	2.25	2.26	2.37	2.46	2.35	2.44	2.19	1.77	2.67	2.37
Tm	0.21	0.24	0.25	0.32	0.31	0.34	0.35	0.32	0.33	0.31	0.24	0.37	0.33
Yb	1.30	1.57	1.58	2.09	1.96	2.15	2.21	2.15	2.14	1.97	1.60	2.41	2.10
Lu	0.20	0.23	0.23	0.31	0.29	0.31	0.32	0.31	0.32	0.30	0.23	0.33	0.30
Hf	1.77	2.38	2.42	2.46	2.93	2.61	2.63	3.11	3.26	3.44	3.38	3.39	3.30
Ta	0.60	0.87	0.83	1.12	1.07	1.23	1.17	1.40	1.79	1.40	1.19	1.21	1.18

Continued

Sample	Quartz diorite							Granodiorite					
	N: 25°57.610' E: 98°43.431'			N: 25°57.233' E: 98°44.068'				N: 25°58.312' E: 98°42.719'			N: 25°57.555' E: 98°43.201'		
	PM2-120	PM2-121	PM2-123	PM2-127-1	PM2-127-2	PM2-127-3	PM2-127-4	PM2-97	PM2-98	PM2-99	PM2-115	PM2-116	PM2-117
W	136.00	104.00	136.00	170.00	185.00	106.00	145.00	151.00	174.00	154.00	205.00	189.00	164.00
Tl	0.40	0.40	0.38	0.74	0.82	0.79	0.88	0.53	0.54	0.53	0.59	0.40	0.48
Pb	14.40	16.60	16.20	16.20	15.30	16.00	15.10	26.30	25.60	27.10	24.80	16.90	17.00
Bi	0.14	0.12	0.16	0.20	0.21	0.14	0.23	0.17	0.13	0.19	0.09	0.08	0.07
Th	3.93	6.82	5.39	11.43	10.13	7.83	8.10	14.87	14.49	14.32	19.42	14.24	12.81
U	1.43	1.75	1.64	2.29	1.99	1.59	2.12	3.23	3.30	3.64	3.18	2.45	2.32
$Mg^{\#}$	51.51	52.07	51.91	52.64	53.52	53.23	53.06	45.07	46.18	46.32	47.04	46.84	46.71
A/CNK	1.05	1.02	1.02	0.99	1.00	0.95	0.98	1.05	1.06	1.06	1.05	1.00	1.03
Nb/Ta	14.52	10.70	11.54	11.88	12.15	11.46	11.62	10.07	8.38	10.00	10.34	11.16	12.37
Na_2O/K_2O	1.55	1.70	1.72	1.61	1.34	1.32	1.20	1.14	1.14	1.03	1.01	1.63	1.27
REE	69.90	102.00	78.30	123.54	121.32	127.24	115.07	166.80	159.30	154.50	191.67	192.51	176.89
FeO^{T}/MgO	1.97	1.93	1.94	1.89	1.82	1.84	1.86	2.56	2.44	2.43	2.36	2.38	2.39

Sample	Monzogranite														
	N: 25°59.721' E: 98°39.336'		N: 25°59.159' E: 98°40.084'			N: 25°58.646' E: 98°40.415'			N: 25°58.899' E: 98°42.055'				N: 25°58.340' E: 98°42.159'		
	PM2-2	PM2-10	PM2-14	PM2-18	PM2-20	PM2-32	PM2-38	PM2-45	PM2-53	PM2-56	PM2-60	PM2-61	PM2-86	PM2-88	PM2-89
SiO_2	68.66	69.12	66.61	71.07	71.44	68.32	68.04	71.87	70.56	70.67	73.55	74.25	69.03	71.15	69.52
TiO_2	0.41	0.46	0.49	0.35	0.33	0.43	0.39	0.34	0.37	0.37	0.29	0.29	0.37	0.34	0.39
Al_2O_3	15.23	14.72	16.04	14.45	14.31	15.08	15.54	13.50	14.71	14.49	12.85	12.16	14.82	14.58	15.24
MgO	1.05	1.19	1.26	0.88	0.83	1.11	1.15	0.77	0.89	0.92	0.64	0.64	0.93	0.86	1.05
$Fe_2O_3^{T}$	3.11	3.38	3.57	2.67	2.49	3.31	3.16	2.55	2.84	2.88	2.18	2.22	2.77	2.61	3.12
CaO	2.88	2.81	2.98	2.43	2.16	2.91	2.85	2.41	2.39	2.56	2.12	2.09	2.56	2.31	2.64
Na_2O	3.32	3.14	3.42	3.40	3.15	3.41	3.24	3.33	3.29	3.44	3.38	3.38	3.27	3.20	3.45
K_2O	4.16	3.92	3.81	4.12	4.49	3.84	4.26	4.04	3.88	3.17	3.84	3.66	4.17	3.81	3.65

Continued

Monzogranite

Sample	N: 25°59.721' E: 98°39.336'		N: 25°59.159' E: 98°40.084'			N: 25°58.646' E: 98°40.415'			N: 25°58.899' E: 98°42.055'			N: 25°58.340' E: 98°42.159'			
	PM2-2	PM2-10	PM2-14	PM2-18	PM2-20	PM2-32	PM2-38	PM2-45	PM2-53	PM2-56	PM2-60	PM2-61	PM2-86	PM2-88	PM2-89
MnO	0.07	0.07	0.07	0.06	0.06	0.08	0.07	0.06	0.07	0.07	0.06	0.06	0.06	0.06	0.08
P_2O_5	0.13	0.13	0.15	0.11	0.11	0.13	0.12	0.11	0.11	0.12	0.10	0.10	0.11	0.11	0.13
LOI	0.69	0.87	1.25	0.87	0.81	1.09	0.93	1.25	0.95	0.98	0.83	0.88	0.76	0.69	0.48
Total	99.71	99.81	99.66	100.42	100.17	99.72	99.77	100.22	100.06	99.68	99.84	99.73	98.86	99.72	99.76
Li	49.40	45.60	48.30	57.90	55.10	47.80	42.90	47.00	29.00	27.70	38.20	38.80	47.30	40.90	46.60
Be	2.81	2.60	2.83	3.31	2.96	3.20	2.58	2.90	2.85	3.05	2.46	2.55	3.00	2.82	3.15
Sc	14.26	14.78	14.78	13.29	11.88	13.73	14.26	11.62	13.02	12.23	10.91	11.09	11.44	10.91	11.88
V	56.60	64.40	64.00	45.30	41.40	56.80	55.80	42.80	46.10	47.20	35.00	37.40	48.20	45.20	53.10
Cr	13.00	17.10	11.50	11.30	8.48	20.10	11.40	9.11	9.74	9.40	6.99	10.80	12.00	7.99	9.42
Co	43.80	34.10	36.70	39.10	42.10	36.90	27.60	37.70	41.90	38.30	36.30	33.00	39.50	40.00	23.10
Ni	6.72	6.61	5.64	4.95	4.36	11.90	4.85	4.56	4.76	4.98	4.81	5.16	5.88	4.23	4.68
Cu	4.47	5.20	7.39	3.66	3.34	4.61	4.25	3.54	3.91	5.02	3.72	3.43	4.31	3.66	4.59
Zn	71.70	75.10	86.90	68.40	64.60	62.70	76.00	58.20	75.70	81.50	66.90	74.00	77.40	67.60	83.60
Ga	16.80	17.00	18.00	16.00	15.40	16.70	16.60	15.70	16.40	16.50	15.30	15.20	17.10	15.90	17.30
Ge	1.35	1.30	1.31	1.45	1.51	1.47	1.42	1.29	1.23	1.24	1.31	1.32	1.31	1.35	1.40
As	3.24	2.88	2.18	3.24	1.56	4.21	1.55	2.22	1.35	2.62	0.81	1.89	1.66	1.53	3.20
Rb	184.00	176.00	167.00	204.00	216.00	194.00	189.00	165.00	149.00	133.00	174.00	166.00	189.00	184.00	190.00
Sr	231.00	229.00	243.00	185.00	168.00	222.00	222.00	200.00	251.00	228.00	204.00	195.00	238.00	209.00	220.00
Y	23.00	22.10	22.50	23.50	22.50	25.20	22.40	21.40	21.30	19.60	15.20	16.70	21.20	20.20	20.60
Zr	142.00	143.00	160.00	133.00	125.00	152.00	141.00	134.00	136.00	126.00	116.00	115.00	125.00	124.00	110.00
Nb	14.20	13.80	15.00	15.40	15.10	15.50	13.90	13.80	13.50	12.60	13.80	13.40	13.50	13.40	14.60
Mo	1.54	0.64	0.58	0.36	0.24	0.42	0.25	0.36	0.17	0.45	0.31	0.68	0.73	0.37	0.84
Cs	6.30	7.46	5.91	10.70	10.90	7.55	7.19	6.96	3.16	3.22	8.43	8.54	8.40	8.80	9.39

Continued

Monzogranite

Sample	N: 25°59.721' E: 98°39.336'		N: 25°59.159' E: 98°40.084'			N: 25°58.646' E: 98°40.415'			N: 25°58.899' E: 98°42.055'				N: 25°58.340' E: 98°42.159'		
	PM2-2	PM2-10	PM2-14	PM2-18	PM2-20	PM2-32	PM2-38	PM2-45	PM2-53	PM2-56	PM2-60	PM2-61	PM2-86	PM2-88	PM2-89
Ba	425.00	407.00	389.00	355.00	344.00	342.00	404.00	376.00	539.00	394.00	362.00	345.00	539.00	362.00	396.00
La	22.30	26.10	25.90	26.90	20.80	24.10	26.60	28.70	30.00	45.30	22.20	22.30	36.70	26.30	27.50
Ce	45.80	52.20	49.70	53.80	43.00	50.40	53.90	56.80	59.10	83.40	41.70	40.60	71.30	51.40	55.20
Pr	5.00	5.54	5.43	5.58	4.57	5.26	5.68	5.77	6.10	8.91	4.24	4.24	7.42	5.46	5.65
Nd	18.90	19.90	20.50	19.40	15.80	19.50	19.60	20.00	21.10	29.80	14.10	14.60	24.90	18.40	19.60
Sm	4.12	4.17	4.26	3.58	3.31	4.09	3.96	3.73	3.90	4.62	2.64	2.77	4.31	3.45	3.55
Eu	0.86	0.87	0.91	0.70	0.66	0.83	0.80	0.75	0.87	0.89	0.60	0.60	0.91	0.75	0.79
Gd	3.72	3.60	3.85	3.43	3.19	3.82	3.61	3.49	3.68	4.09	2.51	2.65	3.74	3.05	3.16
Tb	0.60	0.60	0.62	0.55	0.51	0.61	0.56	0.54	0.56	0.55	0.36	0.42	0.57	0.50	0.49
Dy	3.52	3.55	3.72	3.33	3.19	3.68	3.33	3.25	3.30	2.98	2.16	2.45	3.27	2.96	2.99
Ho	0.75	0.72	0.76	0.70	0.68	0.77	0.68	0.68	0.69	0.59	0.46	0.51	0.64	0.59	0.61
Er	2.22	2.15	2.23	2.24	2.17	2.38	2.11	2.02	2.07	1.86	1.43	1.58	1.95	1.81	1.88
Tm	0.34	0.32	0.33	0.36	0.35	0.38	0.31	0.31	0.31	0.26	0.22	0.25	0.29	0.28	0.30
Yb	2.31	2.23	2.31	2.66	2.54	2.63	2.14	2.02	2.10	1.79	1.57	1.69	1.93	1.93	1.96
Lu	0.35	0.34	0.36	0.42	0.38	0.40	0.32	0.31	0.30	0.27	0.25	0.26	0.30	0.28	0.30
Hf	3.96	3.77	4.37	4.00	3.70	4.33	3.71	3.55	3.65	3.21	3.27	3.14	3.43	3.35	2.96
Ta	1.67	1.59	1.81	2.71	2.71	2.42	1.66	1.85	1.62	1.34	1.50	1.41	1.51	1.65	1.06
W	168.00	213.00	210.00	248.00	273.00	227.00	170.00	249.00	268.00	234.00	232.00	174.00	215.00	243.00	57.50
Tl	0.73	0.72	0.68	0.96	0.87	0.87	0.84	0.87	0.69	0.61	0.78	0.75	0.87	0.85	0.89
Pb	28.60	25.60	21.80	36.00	33.10	28.50	29.20	29.50	33.00	28.50	30.40	29.50	32.70	31.00	31.10
Bi	0.07	0.09	0.03	0.17	0.18	0.10	0.10	0.13	0.14	0.09	0.16	0.09	0.10	0.12	0.06
Th	15.46	17.89	22.35	32.35	20.96	28.68	23.97	19.67	16.89	27.38	24.33	23.09	23.84	19.50	19.17
U	3.95	4.21	4.92	8.09	5.62	4.76	5.56	4.38	3.90	5.21	4.92	4.08	4.26	4.80	3.89

Continued

Monzogranite

Sample	N: 25°59.721' E: 98°39.336'		N: 25°59.159' E: 98°40.084'			N: 25°58.646' E: 98°40.415'			N: 25°58.899' E: 98°42.055'				N: 25°58.340' E: 98°42.159'		
	PM2-2	PM2-10	PM2-14	PM2-18	PM2-20	PM2-32	PM2-38	PM2-45	PM2-53	PM2-56	PM2-60	PM2-61	PM2-86	PM2-88	PM2-89
$Mg^\#$	44.13	45.12	45.12	43.29	43.75	43.96	45.95	41.20	42.07	42.73	40.75	40.10	43.77	43.44	43.89
A/CNK	1.00	1.01	1.06	1.00	1.02	1.00	1.03	0.95	1.05	1.05	0.95	0.91	1.02	1.07	1.06
Nb/Ta	8.50	8.68	8.29	5.68	5.57	6.40	8.37	7.46	8.33	9.40	9.20	9.50	8.94	8.12	13.77
Na_2O/K_2O	0.80	0.80	0.90	0.82	0.70	0.89	0.76	0.82	0.85	1.08	0.88	0.92	0.78	0.84	0.95
REE	110.79	122.29	120.88	123.65	101.16	118.86	123.60	128.37	134.08	185.31	94.44	94.93	158.23	117.16	123.97
FeO^T/MgO	2.66	2.55	2.55	2.75	2.70	2.67	2.47	2.99	2.89	2.81	3.05	3.13	2.69	2.73	2.68

Table 2 The whole-rock Sr-Nd isotopic compositions of granitic rocks from Gaoligong belt, SW China.

Sample	$^{87}Sr/^{86}Sr$	2SE	Rb /ppm	Sr /ppm	$^{143}Nd/^{144}Nd$	2SE	Nd /ppm	Sm /ppm	T_{DM2} /Ga	$\varepsilon_{Hf}(t)$	$(^{87}Sr/^{86}Sr)_i$	$(^{143}Nd/^{144}Nd)_i$
PM2-120	0.710 403	0.000 015	93.4	360	0.512 209	0.000 009	12.7	3.08	1.32	-7.63	0.709 176	0.512 099
PM2-127-1	0.709 622	0.000 008	155	302	0.512 313	0.000 007	22.5	4.62	1.16	-5.27	0.707 194	0.512 220
PM2-127-3	0.709 670	0.000 010	171	305	0.512 320	0.000 007	24.1	5.03	1.15	-5.16	0.707 019	0.512 225
PM2-116	0.714 391	0.000 017	99.7	334	0.512 033	0.000 008	31.5	6.32	1.53	-10.72	0.713 004	0.511 943
PM2-97	0.715 005	0.000 010	116	278	0.512 059	0.000 007	26.3	4.85	1.49	-10.07	0.713 065	0.511 977
PM2-2	0.716 285	0.000 019	184	231	0.512 087	0.000 007	18.9	4.12	1.47	-9.76	0.712 286	0.511 982
PM2-20	0.718 697	0.000 015	216	168	0.511 893	0.000 006	15.8	3.31	1.73	-13.49	0.712 452	0.511 795
PM2-53	0.715 860	0.000 011	149	251	0.511 845	0.000 007	21.1	3.9	1.77	-14.20	0.712 977	0.511 758
PM2-86	0.716 561	0.000 014	189	238	0.511 821	0.000 008	24.9	4.31	1.80	-14.56	0.712 704	0.511 740

$^{87}Rb/^{86}Sr$ and $^{147}Sm/^{144}Nd$ ratios were calculated using Rb, Sr, Sm and Nd contents analyzed by ICP-MS. T_{DM2} represent the two-stage model age and were calculated using present-day $(^{147}Sm/^{144}Nd)_{DM} = 0.213\ 7$, $(^{147}Sm/^{144}Nd)_{DM} = 0.513\ 15$ and $(^{147}Sm/^{144}Nd)_{crust} = 0.101\ 2$. $\varepsilon_{Nd}(t)$ values were calculated using present-day $(^{147}Sm/^{144}Nd)_{CHUR} = 0.196\ 7$ and $(^{147}Sm/^{144}Nd)_{CHUR} = 0.512\ 638$. $\varepsilon_{Nd}(t) = [(^{143}Nd/^{144}Nd)_S(t)/(^{143}Nd/^{144}Nd)_{CHUR}(t)-1] \times 10^4$, $T_{DM2} = \dfrac{1}{\lambda}\{1+[(^{143}Nd/^{144}Nd)_S - ((^{147}Sm/^{144}Nd)_S - (^{147}Sm/^{144}Nd)_{crust})(e^{\lambda t}-1) - (^{143}Nd/^{144}Nd)_{DM}]/[(^{147}Sm/^{144}Nd)_{crust} - (^{147}Sm/^{144}Nd)_{DM}]\}$.

are colorless to light brown, and transparent to subtransparent with clear oscillatory zoning in CL images (Fig. 3a), which is typical of magmatic origin (Hoskin and Black, 2000). The 13 reliable analytical data from PM2-57 have high Th (192−1 085 ppm) and U (226−1 517 ppm) with Th/U ratios of 0. 50−1. 39. ^{206}Pb/^{238}U ages range from 109 ± 1.0 Ma to 117 ± 1.0 Ma, with a weighted mean age of 112.8 ± 1.7 Ma (MSWD=1. 3, n=13, 2σ) (Fig. 3d).

CL images show the zircons from granodiorites (PM2−110) with euhedral, prismatic and relative clear oscillatory zonation (Fig. 3b). They share similar features with monzogranites.

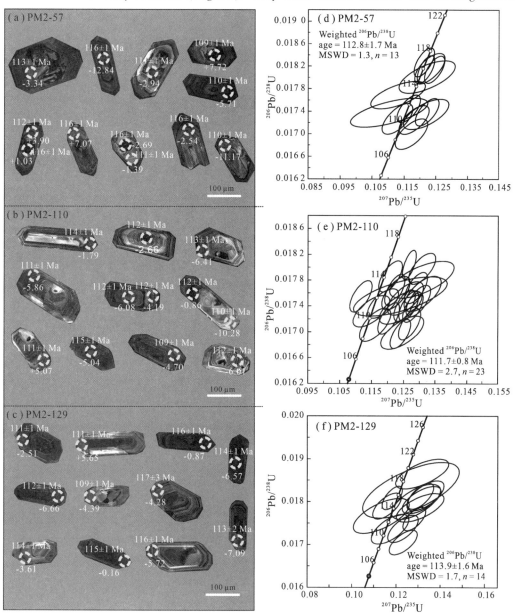

Fig. 3　Zircon CL images and LA-ICP-MS U-Pb data from the Gaoligong belt, SW China.

The twenty-three reliable analytical data from PM2-110 has Th (163−1 639 ppm, except one is extremely high to 10 363 ppm) and U (187−2 621 ppm) contents with Th/U ratios of 0.48− 1.58 (except one is 3.95). The twenty-three reliable analyses yield $^{206}Pb/^{238}U$ ages from 109 ±1.0 Ma to 115±1.0 Ma, with a weighted mean age of 111.7±0.8 Ma (MSWD=2.7, $n=$ 23, 2σ) (Fig. 3e).

Zircons from quartz diorites (PM2-129) are subhedral to euhedral, light yellowish-brown to colorless, with crystal lengths of 150−300 μm and aspect ratios of 2∶1 to 3∶1. Most of the grains are gray and display well-developed oscillatory zoning in CL images (Fig. 3c). They have smaller grains than the granodiorites and monzogranites. The fourteen reliable analytical data spots have Th (142−1 260 ppm) and U (106−1 082 ppm) contents with Th/U ratios of 0.57−1.34. And yield $^{206}Pb/^{238}U$ ages from 109±1.0 Ma to 119±1.0 Ma, with a weighted mean age of 113.9±1.6 Ma (MSWD=1.7, $n=14$, 2σ) (Fig. 3f).

4.2 Major and trace element geochemistry

In the SiO_2 vs. $K_2O + Na_2O$ (Fig. 4a), SiO_2 vs. K_2O (Fig. 4b), and A/CNK vs. A/NK (Fig. 4c), these rocks are plotted as quartz diorite, granodiorite, and monzogranite, which belong to the calc-alkaline to high-K calc-alkaline suit and are weakly peraluminous.

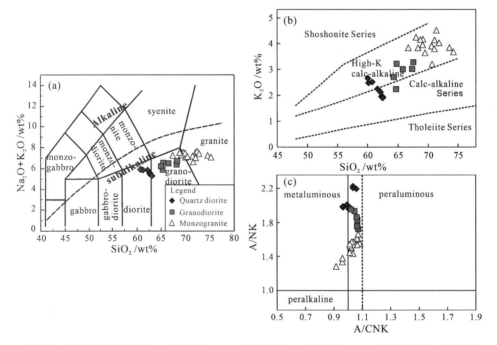

Fig. 4 (a)($Na_2O + K_2O$) vs. SiO_2 diagram; (b)K_2O vs. SiO_2 diagram; (c)A/NK vs.A/CNK diagram (Frost et al., 2001) for the granitic rocks in the Gaoligong belt, SW China.

4. 2. 1　Quartz diorite

The quartz diorites display relatively low $SiO_2 = 60.02 - 62.47$ wt%, high TiO_2 ($0.68 - 0.81$ wt%), $Fe_2O_3^T$ ($5.64 - 6.81$ wt%), and MgO ($2.62 - 3.30$ wt%, mean = ~3.0 wt%) with high $Mg^\#$ values range from 51.5 to 53.5. The relative high Al_2O_3 content ($16.25 - 17.17$ wt%) with A/CNK = $0.95 - 1.05$, show the affinity of meta- to peraluminous. These sample are sodic and have $Na_2O = 3.20 - 3.59$ wt% and low $K_2O = 1.93 - 2.67$ wt%. In the primitive mantle-normalized trace element diagram (Fig. 5a), these samples show toughs in Ba, Nb-Ta, P and Ti, and spikes at Rb, Th- U and Pb. In the chondrite-normalized rare-earth element (REE) diagram (Fig. 5b), the quartz diorites have relatively low $\sum REE = 69.9 - 127.2$ ppm and show enrichment of LREE with $(La/Yb)_N = 5.87 - 9.87$ and flat HREE pattern with low $(Dy/Yb)_N = 1.21 - 1.38$. In addition, they also have small negative Eu anomalies of $0.76 - 0.82$.

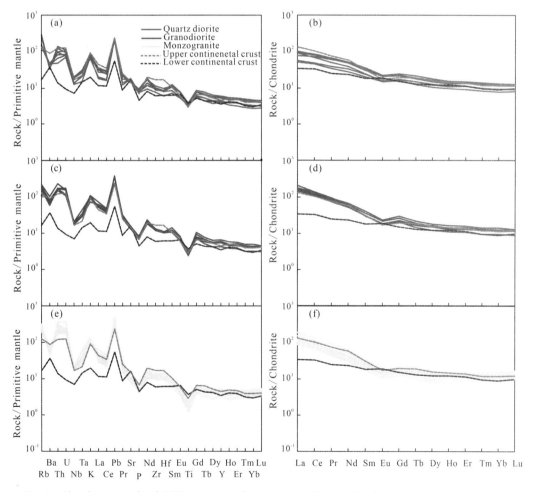

Fig. 5　Chondrite-normalized REE patterns and primitive-mantle-normalized trace element spider diagram (a and b) for the granitic rocks in the Gaoligong belt, SW China.

The primitive mantle, chondrite, lower and upper crust values are from Sun and McDonough (1989).

4. 2. 2　Granodiorite

The granodiorite samples in this study area have relative low $SiO_2 = 64.36 - 67.53$ wt%, lower $TiO_2 = 0.52 - 0.68$ wt%, $MgO = 1.43 - 1.90$ wt% with $Mg^\# = 45.1 - 47.0$, and $Fe_2O_3^T = 3.85 - 5.04$ wt%, and higher $A/CNK = 1.00 - 1.06$ than the quartz diorites. They are also sodium-rich with Na_2O/K_2O ratio $= 1.01 - 1.63$ and higher K_2O content ($2.21 - 3.30$ wt%) than quartz diorite. In the primitive mantle-normalized trace element diagram (Fig. 5c), the rocks are enriched in Rb, Th-U and Pb, and depleted in Ba, Nb-Ta, P and Ti. In the chondrite-normalized rare-earth element (REE) diagram (Fig. 5d), these granodiorites display enrichment in LREE with $(La/Yb)_N = 12.2 - 22.1$ and pronounced negative Eu anomalies ($Eu/Eu^* = 0.67 - 0.75$). They have higher $\sum REE = 154.5 - 192.5$ ppm than the quartz diorites, but they also share similar K/Rb, Ba/Th, Rb/Sr, and Nb/Ta ratios.

4. 2. 3　Monzogranite

The rocks have high SiO_2 content $= 66.61 - 74.25$ wt%, and higher K_2O content $= 3.17 - 4.49$ wt% with Na_2O/K_2O ratio $= 0.70 - 0.95$, except one at 1.08, than granodiorites. The samples show lower TiO_2($0.29 - 0.49$ wt%), MgO ($0.64 - 1.26$ wt%), and $Fe_2O_3^T$($2.18 - 2.67$ wt%) than both quartz diorite and granodiorite. The high Al_2O_3 content $= 12.16 - 16.04$ wt% and $A/CNK = 0.91 - 1.07$, which share similar characters with both quartz diorite and granodiorite. In the primitive mantle-normalized trace element diagram (Fig. 5e), these samples display similar patterns to average upper continental crust (Rudnick and Gao, 2003) with enriched Rb, Th, U, K, and Pb, and depleted Nb, Ta, P, and Ti. In the chondrite-normalized rare-earth element (REE) diagram (Fig. 5f), the rocks share similar patterns with the quartz diorites, with slight enrichment of LREE $[(La/Yb)_N = 5.87 - 18.15]$ and flat HREE patterns with $(Dy/Yb)_N = 0.84 - 1.13$. The latter is lower than quartz diorites and granodiorites, which may indicate fractionation of amphibole. In addition, the obvious negative Eu anomalies ($Eu/Eu^* = 0.61 - 0.72$) possibly imply some fractionation of plagioclase.

4. 3　Whole-rock Sr-Nd isotopes

Whole-rock Sr-Nd isotopic data for the granitoids from the Gaoligong belt are listed in Table 2. All the initial $^{87}Sr/^{86}Sr$ isotopic ratios (I_{Sr}) and $\varepsilon_{Nd}(t)$ values are calculated according to the LA-ICP-MS zircon U-Pb ages for the quartz diorite, granodiorite, and monzogranite. Whole-rock Nd model ages were calculated using the model of DePaolo and Series (1981).

The quartz diorites (samples PM2-120, PM2-127-1, and PM2-127-2) have moderate Sr content ($302 - 360$ ppm) and high Rb content ($93.2 - 171$ ppm), and which have initial

$^{87}Sr/^{86}Sr = 0.707\ 019 - 0.709\ 176$, with high $^{143}Nd/^{144}Nd$ ratios $= 0.510\ 99 - 0.512\ 225$, $\varepsilon_{Nd}(t)$ values are -7.63 to -5.16, and two-stage Nd model ages of $1.15 - 1.32$ Ga. The granodiorites (sample PM2-97 and PM2-116) share similar Sr and Rb contents with the quartz diorite (Sr $= 278 - 334$ ppm and Rb $= 99.7 - 116$ ppm). They have higher intial $^{87}Sr/^{86}Sr$ (0.713 004 and 0.713 065) and slightly lower $^{143}Nd/^{144}Nd$ ratio (0.511 943 and 0.511 977) with $\varepsilon_{Nd}(t)$ values of -10.72 and -10.07, and two-stage Nd model ages of $1.49 - 1.53$ Ga. The monzogrante (samples PM2-2, PM2-20, PM2-53, and PM2-86) have lower Sr content (168 - 251 ppm) and higher Rb content (149 - 216 ppm) than both quartz diorite and granodiorite. They also show relative high initial $^{87}Sr/^{86}Sr$ (0.712 286 - 0.712 977) and lowest $^{143}Nd/^{144}Nd$ (0.511 740 - 0.511 982) with $\varepsilon_{Nd}(t)$ values of -14.56 tp -9.76. The samples have two-stage Nd model ages of $1.47 - 1.80$ Ga.

As shown in the $\varepsilon_{Nd}(t)$ vs. initial $^{87}Sr/^{86}Sr$ diagram (Fig. 6), the quartz diorite and granodiorite samples share the similar characters with the granitoids and volcanic rocks from northern and central Lhasa subterranes (Fig. 6a) (Chen et al., 2014; Qu et al., 2012; Zhu et al., 2009).

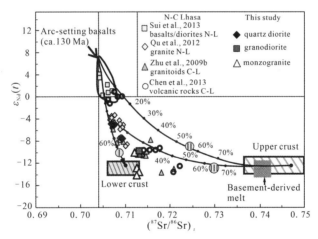

Fig. 6　The $\varepsilon_{Nd}(t)$ values vs. initial $^{87}Sr/^{86}Sr$.

Data of Juvenile lower crust and Back-arc basalts from Chen et al. (2014, referenced Zhu et al. and Wang et al. unpublished data). Basement-derived melts (Zhu et al., 2011). Lower continental crust (Miller et al., 1999) and upper continental crust (Harris et al., 1988). Symbols as in Fig. 4.

4.4　Zircon chemical and Lu-Hf isotopic compositions

We selected typical zircon grains from the quartz diorites, granodiorites and monzogranites for Lu-Hf isotopic analysis. All of these spots display concordant zircon U-Pb dating. The zircons from three dated samples were also analyzed for Lu-Hf isotopes on the same domains, and the results are listed in Supplementary Dataset Table 2. Initial $^{176}Hf/^{177}Hf$ ratios and $\varepsilon_{Hf}(t)$ values of the zircons were calculated according to their LA-ICP-MS zircon U-Pb ages for the quartz diorite, granodiorite, and monzogranite. Fig. 7 shows $\varepsilon_{Hf}(t)$ values vs. number of the quartz diorite, granodiorite and monzogranite from Gaoligong belt.

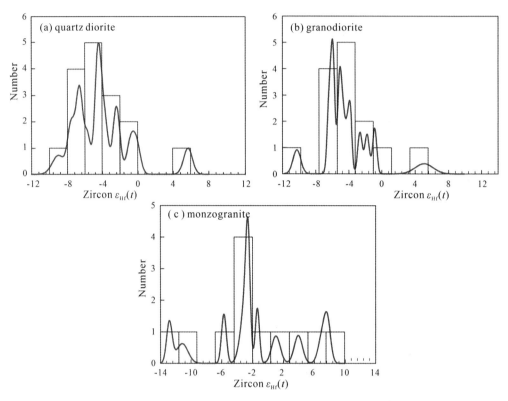

Fig. 7   Histograms of initial Hf isotope ratios for granitic rocks from Gaoligong belt, SW China.

Zircon from quartz diorites (PM2-129) show various Hf isotopic compositions with $\varepsilon_{Hf}(t)$ values ranging from -9.02 to $+5.65$, with corresponding two-stage Hf model ages of $0.81-1.75$ Ga. Zircons from the granodiorite (PM2-110) also show various depleted and enriched Hf isotopic compositions with $\varepsilon_{Hf}(t)$ values of ranging from -10.28 to $+5.07$ and their two-stage Hf model ages of $0.85-1.82$ Ga. The 12 zircon grains from monzogranites share the similar Hf isotopic compositions with the quartz diorite and granodiorite, nine are negative (-12.84 to -1.39) with the two-stage Hf model age of $1.11-1.53$ Ga, and three are positive from $+1.03$ to $+7.72$ with single-stage Hf model age of $0.51-0.79$ Ga. These variable Hf compositions possibly indicate the mixing of between the depleted mantle and more ancient continental crust (Kemp et al., 2007; Yang et al., 2004, 2007).

Zircons from quartz diorite, granodiorite and monzogranite have moderate Th and U contents with Th/U ratios $= 0.48-3.95$, and most of them are distributed in the near the line of Th/U ratio $= 1.0$, which plots in the field of the zircon from the Lhasa Terrane records of continental crustal reworking during Mesozoic-Cenozoic magmatisms (Liu et al., 2014) (Fig. 8). They also display the high U/Yb ratios $= 0.30-3.07$ and moderate Y content $= 366-$

5 379 ppm, suggesting the typical continental crust character rather than the recycled oceanic crust (Grimes et al., 2007) (Fig. 8).

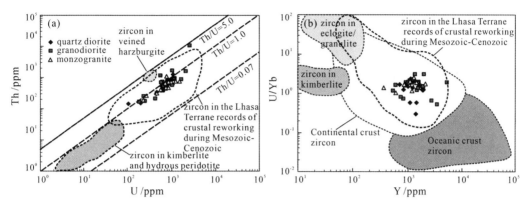

Fig. 8　Discriminant diagrams for zircon origin for granitic rocks in the Gaoligong belt, SW China
(modified after Liu et al. (2014)).
(a)Th vs. U; (b)U/Yb vs. Y (Grimes et al., 2007).

5　Discussion

5.1　Geochemical variations from quartz diorite, granodiorite to monzogranite

As noted above, many models could explain chemical variations in the granitic magmas (Lai et al., 2015; Qin et al., 2016), for example, ① fractional crystallization, magma mixing and assimilation of wall rocks (Pitcher, 1997); ② entrainment peritectic minerals during segregation from crustal source region (Clemens and Stevens, 2012); ③ different partial melting; and ④ distinct sources regions.

The absence of peritectic minerals (eg., garnet and so on) in the quartz diorite and granodiorite-monzogranite associates in the field or this section (Fig. 2a,d and e) argues against the model of peritectic assemblage entrainment. In the primitive mantle spider and chondrite-normalized REE patterns (Fig. 5), the consistent depletion of Ba, Nb, Ta, P and Ti and enrichment of Rb, Th, U, and Pb, and negative Eu anomalies, are all consistent with fractionation of plagioclase, titanite, apatite, and other Ti-bearing accessory minerals during magma evolution. In view of REE pattern, if all these rocks are products of mafic + felsic two-endmember mixing, the granodiorites should lie between quartz diorite and monzogranite, which is not the case (Fig. 5). So there must be some other processes, not single one, for example, melts from different sources evolving individually. The composition gap (Fig. 9), almost different major and trace element evolutionary trend indicate on diagram (Fig. 9) between quartz diorite and granodiorite-monzogranite associations, different grain (Fig. 2) and lower total REE contents of quartz diorites than granodiorite-monzogranite indicate that they have different path of magma evolution and/or source region. The variation zircon ε_{Hf} isotopes may also imply an open system and different sources (Kemp et al., 2007) (Fig. 7). Therefore, we propose that the distinct sources region between the quartz diorites and granodiorite-monzogranite associations. On the other hand, the linear chemical trends between granodiorite and monzogranite on Harker diagrams (Fig. 9) may support a model of fractional crystallization or magma mixing. Increasing $FeO^T + MgO$ with decreasing initial $^{87}Sr/^{86}Sr$ and K_2O (no shown), increasing $^{143}Nd/^{144}Nd$ (Fig. 10a) and TiO_2 content (Fig. 10b), plus the decreasing total REE from granodiorite to monzogranite will in favor of magma mixing model for the chemical variations between granodiorite and monzogranite (Clemens and Stevens, 2012). Petrographic evidence supporting plagioclase in both granodiorite and monzogranite also has many resorption planes pointing to dissolution and regrowth (Fig. 2f, i), and central sieve textures (Fig. 2f), indicating the mixing process (Słaby and Goetze, 2004; Słaby et al., 2011, 2016).

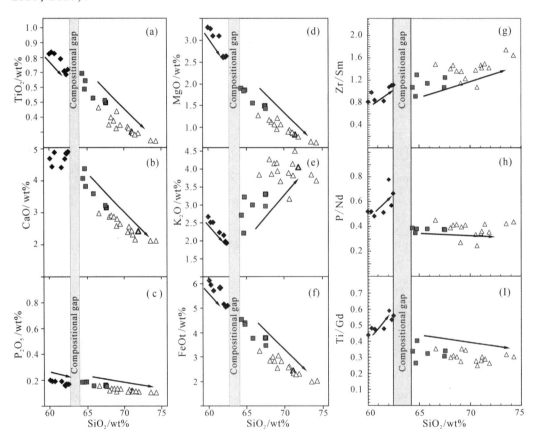

Fig. 9   The Harker and Selected trace element ratios vs. SiO_2 diagram (after Słaby and Martin, 2008)
for granitic rocks in the Gaoligong belt, SW China.
Symbols as in Fig. 4.

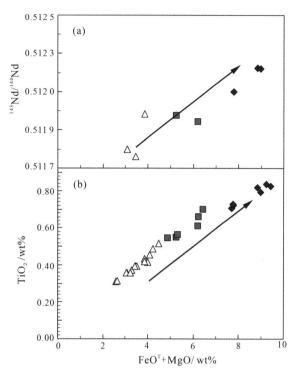

Fig. 10　Isotopic variations and TiO_2 with maficity for granitic rocks in the Gaoligong belt.
Symbols as in Fig. 4.

5. 1. 1　Quartz diorite: Melts from ancient basaltic lower crust

Previous workers have suggested that differentiation of mantle-derived mafic rocks and partial melting of basaltic lower crust could produce sodium-rich intermediate to felsic rocks (Rapp and Watson, 1995; Stern et al., 1989). The significant negative and variable zircon $\varepsilon_{Hf}(t)$ values plus evolved whole-rock Sr-Nd isotopic compositions of the quartz diorites argue against simple differentiation of mantle-derived mafic rocks. The following evidence is consistent with partial melting of ancient basaltic lower crust: ① as mentioned above, the enriched isotopic compositions from not only bulk-rock but also zircon with the Meso- to Neo-Proterozoic model ages (Supplementary Dataset Table 2) indicate the sources in ancient continental crust; ② the samples are weakly peraluminous A/CNK values from 0. 95 to 1. 05 may imply crustal Al-rich sources (Fig. 4c); and ③ the high CaO/Na_2O and low Al_2O_3/TiO_2 ratios infer the significant basalt-derived sources (Sylvester, 1998) (Fig. 11), both of their REE patterns are similar to rocks derived from continental crust (Pitcher, 1997), which also supported by low Nb/Ta and high Zr/Sm (not shown) (Foley et al., 2002; Martin, 1999). Finally, the zircon chemical compositions, including high Th, U content and moderate Th/Yb ratios and Y (Fig. 8) are typical of zircon from continental crust and similar to zircon in the Lhasa Terrane records of crustal reworking during Mesozoic-Cenozoic (Liu et al., 2014).

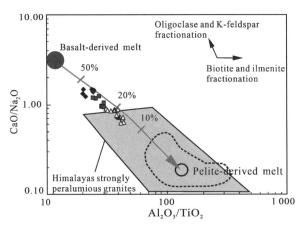

Fig. 11 The diagram of Al_2O_3/TiO_2 vs. CaO/Na_2O referenced by Sylvester (1998).

Symbols as in Fig. 4.

The quartz diorites have higher MgO (mean ca. 3. 0 wt%) and $Mg^{\#}$ (51. 5−53. 5) than melts derived from metabasalts (e.g., MgO < 3 wt% and $Mg^{\#} < 45$, Rapp and Watson, 1995) and so these parental melts are most likely of mantle origin (Guo et al., 2014). Unlike typical high-Mg andesites, however, they have $Mg^{\#}$ that cannot reflect equilibration with the mantle peridotites (c.f. > 60), and this is supported by correspondingly low concentrations of Cr and Ni. Zircon $\varepsilon_{Hf}(t)$ values range up to + 5. 65, which support involvement of depleted mantle-derived magma (Kemp et al., 2007; R.X. Wu et al., 2006; Y.B. Wu et al., 2006; Yang et al., 2007; Zheng et al., 2006). High Nb/Ta ratios (10. 7−14. 5) are consistent with values between the lower crust and mantle (Rudnick and Gao, 2003; McDonough and Sun, 1995) and similar with the western part of the northern Lhasa subterrane and the whole southern Lhasa subterrane (Hou et al., 2015).

We therefore could conclude that the quartz diorite was derived from the melts from partial melting of ancient basaltic lower crust by mantle-like magmas in the source region.

5. 1. 2 Granodiorite and monzogranite: Deep seated magma mixing in the hot zone

The granodiorite and monzogranite associates are high-K, calc-alkaline, and most samples are weakly peraluminous (Fig. 4), which is consistent with the typical granitic magmas from reworking the continental crust. Not only the moderate CaO/Na_2O and Al_2O_3/TiO_2 ratios (Fig. 11) but also the moderate Rb/Ba and Rb/Sr ratios indicate the mixing sources both from basaltic and meta-sedimentary sources (Sylvester, 1998). The rocks are also characterized by typical zircons from continental crust which records of crustal reworking with significant negative zircon $\varepsilon_{Hf}(t)$ values (Figs. 7 and 8), evolved whole-rock Sr-Nd isotopic compositions are plotted into the fields of lower crust and also similar with the granitoids and volcanic rocks from Central Lhasa subterrane (Fig. 6a) (Chen et al., 2014; Zhu et al., 2009). The moderate to low Nb/Ta ratios are so close to the global lower crust (8. 3) and in part similar with boundary of lower crust and mantle imply a deep crustal sources, which suggesting the mixing sources between the basaltic rocks and sediment-derived rocks in the deep crustal zone. Furthermore, several zircon $\varepsilon_{Hf}(t)$ values can up to $+7.72$ (Fig. 7b,c) and some part similarity of moderate Nb/Ta ratios may suggest us the in part significant input from mantle. In addition, in the case of mixing, all of the points on a ($C_{average\ hybrid\ rocks} - C_{parental\ magma}$) vs. ($C_{granodiorite-monzogranite} - C_{parental\ magma}$) diagram (Fig. 12; Fourcade and Allegre, 1981; Słaby and Martin, 2008) could plot on a straight line passing through the origin and with a slope that represents the degree of mixing. The line ($y = 0.467\ 7x - 0.003\ 4$, $R^2 = 0.974\ 5$) of the granodiorite-monzogranite supports the mixing hypothesis and shows an average hybrid (Fig. 12).

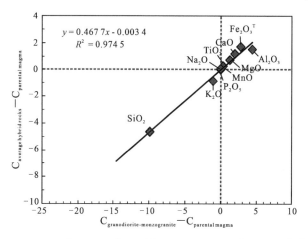

Fig. 12　The ($C_{average\ hybrid\ rocks} - C_{parental\ magma}$) vs. ($C_{granodiorite-monzogranite} - C_{parental\ magma}$) diagram (Fourcade and Allegre, 1981; Słaby and Martin, 2008) for granodiorite-monzogranite associate.

The major element compositions plot on a straight line that passes through the origin, indicating a mixing model.

In summary, the granodiorite and monzogranite associates were derived from the melting of mixed sources between the basaltic and meta-sedimentary rocks in the deep crustal zone.

5.2　The widespread magmatism at ca. 113−110 Ma from Lhasa to Tengchong terranes

Zhu et al. (2009) proposed a magmatic flare-up at ca. 110 Ma with an increased contribution in the generation of the igneous rocks in the central Lhasa subterrane genetically associated with the slab break-off of the subducting Bangong-Nujiang seafloor. Subsequently, more workers agreed that the ca. 113 Ma magmatic event is bimodal in composition, and widely developed in the central and northern Lhasa subterrane (Chen et al., 2014; Hou et al., 2015; Sui et al., 2013; Zhang et al., 2012; Zhu et al., 2011). The felsic rocks are mainly composed of calc-alkaline volcanic rocks with minor granitoid intrusions, with positive zircon $\varepsilon_{Hf}(t)$ values ($+3.6$ to $+7.3$; Li et al, 2013). In contrast, mafic rocks are uncommon, but the presence of mafic enclaves in the host granitoids indicate of the presence of mafic magma (Zhu et al., 2009). Both of these mafic enclaves have elevated bulk-rock $\varepsilon_{Nd}(t)$ values and high MgO and Cr contents (Zhang et al., 2004). Such bimodal magmatism was taken to indicate an extensional setting and the positive zircon $\varepsilon_{Hf}(t)$ values and young model ages to imply significant crustal growth in the northern Lhasa subterrane at that time (Sui et al., 2013). Very recently, the occurrence of Xainza basalts at ca. 113 Ma with within-plate geochemical affinity and high Mg, Cr, and Ni also indicate an extensional setting, while the coeval Raguo rhyolites are A_2-type silicic rocks that generally emplaced in a non-compressional setting during the late phase of collision (Chen et al., 2014).

In the Gaoligong case, Xu et al. (2012) ruled out an intra-continental rifting setting for the Cretaceous and Early Tertiary batholiths in favor of subduction-related or intra-crustal thickening should be considered, the major difference being the extent of mantle involvement in crustal melting. Crustal melting and mixing induced by influx of mantle-derived magma are common in the subduction-related setting but rare in intra-crustal setting. The quartz diorite-granodiorite-monzogranite samples in this study are characterized by a wide range in SiO_2 content (61 – 71 wt%; Fig. 4). All are hornblende-bearing (Fig. 2) and have variation of zircon Hf compositions with $\varepsilon_{Hf}(t)$ values to + 7.72 (Supplementary Dataset and Fig. 7). These features indicate significant mantle involvement and thus further suggest a subduction-related rather than intra-crustal setting (Driver et al., 2000). Of course, delaminated thickened crust can also result in crustal melting and mixing caused by influx of mantle-like magma, but evidence related to delamination model has not been found from Lhasa to Tengchong during the late Early Cretaceous (Chen et al., 2014; Sui et al., 2013; Xie et al., 2016; Xu et al., 2012; Zhu et al., 2011). In comparison, the Tengchong Terrane is linked with the Lhasa Terrane, both of which experienced similar tectonomagmatic histories and extensive magmatism at ca. 120 – 110 Ma with enhanced contributions from mantle- or juvenile crust-derived melts (Xie et al., 2016; Zhu et al., in press-a) (Fig. 13), triggered by the slab break-off and/or detachment of Bangong-Nujiang Tethyan ocean lithosphere (Chen et al., 2014; Sui et al., 2013; Zhu et al., 2009, 2011; R.Z. Zhu et al., 2015; D.C. Zhu et al., 2015). Later, a slab window may result in the partial metling of various sources region including upwelling asthenospheric mantle, enriched lithospheric mantle, and overlying crust (Coulon et al., 2002; Davies and von Blanckenburg, 1995; Ferrari, 2004; Hildebrand and Bowring, 1999; Wortel and Spakman, 1992). In this context, the quartz diorite and granodiorite-monzogranite associations at ca. 113 – 110 Ma may represent the products of partial melting of overlying lower crust by mantle-like magmas in their source region. In the slab break-off setting (Fig. 14), the sub-arc mantle wedge could possibly contain melts from both sub-slab asthenosphere that enriched in HFSE and normal sub-arc mantle that depleted in

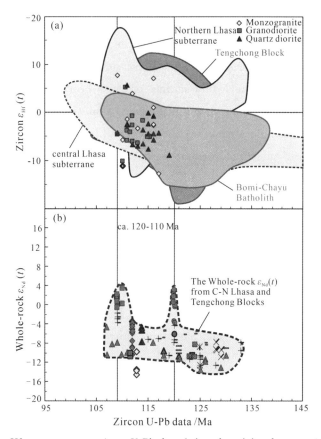

Fig. 13　The zircon Hf components vs. zircon U-Pb data (a) and $\varepsilon_{Nd}(t)$ values vs. zircon U-Pb data (b)
modified after Zhu et al. (submitted for publication).

The fields of central and northern Lhasa and Tengchong terranes referenced to Zhu et al. (submitted for publication).

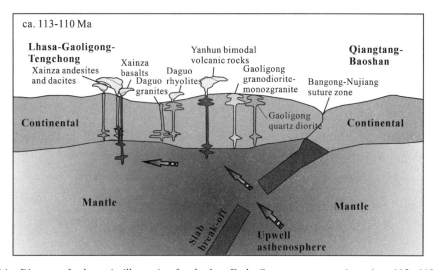

Fig. 14　Diagram of schematic illustration for the late Early-Cretaceous magmatisms (ca. 113−110 Ma)
from Lhasa to Gaoligong (modified after Chen et al., 2014).

HFSE (Ferrari, 2004). A hybrid basaltic magma that had affinity with arc-type and within-plate suites would be produced, for example, the late Early Cretaceous Xainza basalts and bimodal volcanic rocks in central and northern Lhasa subterrane (Chen et al., 2014; Sui et al., 2013) (Fig. 14). Then the hybrid magma provides heat to induce partial melting of ancient lower crust, mature continental basement, and juvenile crust, which resulted in various melts (Fig. 14) including Xainza andesites and dacites (Chen et al., 2014), Daguo rhyolites (Sui et al., 2013), and quartz diorite-granodiorite-monzogranite (this study).

6 Conclusion

The quartz diorite and granodiorite-monzogranite associations from Gaoligong belt were emplaced at ca. 113.9 ± 1.6 Ma, 111.7 ± 0.8 Ma, and 112.8 ± 1.7 Ma, approximately coeval with bimodal magmatisms in the central and northern Lhasa subterrane. There are distinct sources region for the quartz diorites and granodiorite-monzogranite associates. The quartz diorites were derived from partial melting of ancient basaltic source in the lower crust by a mantle-like magma in the source region. The granodiorite-monzogranite associates were derived from the melting of mixed basaltic and meta-sedimentary sources in the deep zone. Both were induced by the upwelling of hybrid basaltic magma due to the slab break-off. Together with data from the literatures, we infer that the widespread magmatism at ca. 113−110 Ma from Lhasa, Gaoligong to Tengchong were triggered by slab break-off of the subducting Bangong-Nujiang Tethyan oceanic lithosphere.

Acknowledgements Thanks for the constructive comments and kindly help from Prof. Nelson Eby, Prof. Ewa Slaby and Dr. Mike Fowler. This study was jointly supported by the National Natural Science Foundation of China (Grant Nos. 41421002, 41372067, and 41190072). Support was also provided by the Foundation for the Author of National Excellent Doctoral Dissertation of China (201324) and the MOST Special Fund from the State Key Laboratory of Continental Dynamics, Northwest University and Province Key Laboratory Construction Item (08JZ62).

Appendix A. Supplementary data Supplementary data to this article can be found online at https://dx.doi.org/10.1016/j.lithos.2017.07.021.

References

Andersen, T., 2002. Correction of common lead in U-Pb analyses that do not report [204]Pb. Chemical Geology, 192, 59-79.

Chappell, B.W., Bryant, C.J., Wyborn, D., 2012. Peraluminous I-type granites. Lithos, 153 (8), 142-153.

Chen, Y., Zhu, D.C., Zhao, Z.D., Meng, F.Y., Wang, Q., Santosh, M., et al., 2014. Slab break-off triggered ca. 113 Ma magmatism around Xainza area of the Lhasa terrane, Tibet. Gondwana Research, 26

(2), 449-463.

Chiu, H.Y., Chung, S.L., Wu, F.Y., Liu, D.Y., Liang, Y.H., Lin, I.J., Lizuka, Y., Xie, L.W., Wang, Y.B., Chu, M.F., 2009. Zircon U-Pb and Hf isotopic constraints from eastern Transhimalayan batholiths on the precollisional magmatic and tectonic evolution in southern Tibet. Tectonophysics, 477, 3-19.

Clemens, J.D., Stevens, G., 2012. What controls chemical variation in granitic magmas? Lithos, 134-135, 317-329.

Clemens, J.D., Helps, P.A., Stevens, G., 2010. Chemical structure in granitic magmas: A signal from the source? Earth and Environmental Science Transactions of the Royal Society of Edinburgh, 100 (1-2), 159-172.

Clemens, J. D., Stevens, G., Farina, F., 2011. The enigmatic sources of I-type granites and the clinopyroxene-ilmenite connexion. Lithos, 126, 174-181.

Coulon, C., Megartsi, M., Fourcade, S., Maury, R., Bellon, H., Louni-Hacini, A., Cotton, J., Coutelle, A., Hermitte, D., 2002. Post-collisional transition from calc-alkaline to alkaline volcanism during the Neogene in Oranie (Algeria): Magmatic expression of a slab break-off. Lithos, 62, 87-110.

Davies, J.H., von Blanckenburg, F., 1995. Slab break-off: A model of lithosphere detachment and its test in the magmatism and deformation of collisional orogens. Earth & Planetary Science Letters, 129, 85-102.

DePaolo, D.J., Series, C.R., 1981. A Neodymium and Strontium Isotopic Study of the Mesozoic Calc-Alkaline Granitic Batholiths of the Sierra Nevada and Peninsular Ranges, California. Granites and Rhyolites. American Geophysical Union.

Driver, L. A., Creaser, R. A., Chacko, T., Erdmer, P., 2000. Petrogenesis of the Cretaceous Cassiar batholith, Yukon-British Columbia, Canada: Implications for magmatism in the North American Cordilleran Interior. Geological Society of America Bulletin, 112, 1119-1133.

Eroğlu, S., Siebel, W., Danišík, M., Pfänder, J.A., Chen, F., 2013. Multi-system geochronological and isotopic constraints on age and evolution of the Gaoligongshan metamorphic belt and shear zone system in western Yunnan, China. Journal of Asian Earth Sciences, 73, 218-239.

Ferrari, L., 2004. Slab detachment control on mafic volcanic pulse and mantle heterogeneity in central Mexico. Geology, 32, 77-80.

Foley, S.F., Tiepolo, M., Vannucci, R., 2002. Growth of early continental crust controlled by melting of amphibolite in subduction zones. Nature, 417, 637-640.

Fourcade, S., Allegre, C.J., 1981. Trace elements behavior in granite genesis: A case study the calc-alkaline plutonic association from the Querigut complex (Pyrénées, France). Contributions to Mineralogy and Petrology, 76 (2), 177-195.

Frost, B.R., Barnes, C.G., Collins, W.J., Arculus, R.J., Ellis, D.J., Frost, C.D., 2001. A geochemical classification for granitic rocks. Journal of Petrology, 42 (11), 2033-2048.

Griffin, W.L., Pearson, N.J., Belousova, E., Jackson, S.E., van Achterbergh, E., O'Reilly, S.Y., Shee, S. R., 2000. The Hf isotope composition of cratonic mantle: LAM-MC-ICPMS analysis of zircon megacrysts in kimberlites. Geochimica et Cosmochimica Acta, 64, 133-147.

Grimes, C.B., John, B.E., Kelemen, P.B., Mazdab, F.K., Wooden, J.L., Cheadle, M.J., Hanghoj, K., Schwartz, J.J., 2007. Trace element chemistry of zircons from oceanic crust: A method for distinguishing detrital zircon provenance. Geology, 35, 643-646.

Guo, F., Fan, W., Li, C., Wang, C.Y., Li, H., Zhao, L., et al., 2014. Hf-Nd-O isotopic evidence for melting of recycled sediments beneath the Sulu orogen, North China. Chemical Geology, 381, 243-258.

Guynn, J.H., Kapp, P., Pullen, A., et al., 2006. Tibetan basement rocks near Amdo reveal "missing" Mesozoic tectonism along the Bangong suture, central Tibet. Geology, 34, 505-508.

Harris, N.B.W., Xu, R.H., Lewis, C.L., Jin, C., 1988. Plutonic rocks of the 1985 Tibet Geotraverse: Lhasa to Golmud. Philosophical Transactions of the Royal Society of London, A327, 145-168.

Harris, N.B.W., Inger, S., Xu, R., 1990. Cretaceous plutonism in Central Tibet: An example of post-collision magmatism? Journal of Volcanology and Geothermal Research, 44, 21-32.

Hildebrand, R.S., Bowring, S.A., 1999. Crustal recycling by slab failure. Geology, 27, 11-14.

Hoskin, P.W.O., Black, L.P., 2000. Metamorphic zircon formation by solid-state recrystallization of protolith igneous zircon. Journal of Metamorphic Geology, 18, 423-439.

Hou, Z., Duan, L., Lu, Y., Zheng, Y., Zhu, D., Yang, Z., et al., 2015. Lithospheric architecture of the Lhasa Terrane and its control on ore deposits in the Himalayan-Tibetan orogen. Economic Geology, 110, 1541-1575.

Kapp, P., Yin, A., Harrison, T.M., Ding, L., 2005. Cretaceous-Tertiary shortening, basin development and volcanism in central Tibet. Geological Society of America Bulletin, 117, 865-878.

Kapp, P., Peter, G.D., George, E.G., Matthew, H., Lin, D., 2007. Geological records of the Lhasa-Qiangtang and Indo-Asian collisions in the Nima area of central Tibet. Geological Society of America Bulletin, 9, 917-993.

Kemp, A.I.S., Hawkesworth, C.J., Foster, G.L., Paterson, B.A., Woodhead, J.D., Hergt, J.M., Whitehouse, M.J., 2007. Magmatic and crustal differentiation history of granitic rocks from Hf-O isotopes in zircon. Science, 315 (5814), 980-983.

Lai, S.C., Qin, J.F., Zhu, R.Z., Zhao, S.W., 2015. Neoproterozoic quartz monzodiorite-granodiorite association from the Luding-Kangding area: Implications for the interpretation of an active continental margin along the Yangtze block (south China block). Precambrian Research, 267, 196-208.

Lee, H.Y., Chung, S.L., Wang, J.R., et al., 2002. Miocene Jiali faulting and implications for Tibet tectonic evolution. Earth & Planetary Science Letters, 205, 185-194.

Li, J.X., Qin, K.Z., Li, G.M., Xiao, B., Zhao, J.X., Cao, M.J., Chen, L., 2013. Petrogenesis of orebearing porphyries from the Duolong porphyry Cu-Au deposit, central Tibet: Evidence from U-Pb geochronology, petrochemistry and Sr-Nd-Hf-O isotope characteristics. Lithos, 161, 216-227.

Liang, Y.H., Chung, S.L., Liu, D., Xu, Y., Wu, F.Y., Yang, J.H., Wang, Y., Lo, C.H., 2008. Detrital zircon evidence from Burma for reorganization of the eastern Himalayan river system. American Journal of Science, 308, 618-638.

Lin, I.J., Chung, S.L., Chu, C.H., Lee, H.Y., Gallet, S., Wu, G., Ji, J., Zhang, Y., 2013. Geochemical and Sr-Nd isotopic characteristics of Cretaceous to Paleocene granitoids and volcanic rocks, SE Tibet: Petrogenesis and tectonic implications. Journal of Asian Earth Sciences, 53, 131-150.

Liu, D., Zhao, Z., Zhu, D.C., Niu, Y., Harrison, T.M., 2014. Zircon xenocrysts in Tibetan ultrapotassic magmas: Imaging the deep crust through time. Geology, 42 (1), 43-46.

Ludwig, K.R., 2003. ISOPLOT 3.0: A geochronological toolkit for Microsoft Excel. Special Publication, 4, Berkeley Geochronology Center.

Martin, H., 1999. Adakitic magmas: Modern analogues of Archaean granitoids. Lithos, 46 (3), 411-429.

McDonough, W., Sun, S., 1995. Chemical evolution of the mantle. Chemical Geology, 123 (3), 223-253.

Metcalfe, I., 2013. Gondwana dispersion and Asian accretion: Tectonic and paleogeography evolution of eastern Tethys. Journal of Asian Earth Sciences, 66, 1-33.

Miller, C., Schuster, R., Kotzli, U., et al., 1999. Post-collisional potassic and ultrapotassic magmatism in SW Tibet: Geochemical and Sr-Nd-Pb-O isotopic constraints for mantle source characteristics and petrogenesis. Journal of Petrology, 40, 1399-1424.

Nowell, G.M., Kempton, P.D., Noble, S.R., Fitton, J.G., Saunders, A.D., Mahoney, J.J., Taylor, R.N., 1998. High precision Hf isotope measurements of MORB and OIB by thermal ionisation mass spectrometry: Insights into the depleted mantle. Chemical Geology, 149, 211-233.

Pan, G., Ding, J., Yao, D., Wang, L., 2004. Guidebook of 1:1 500 000 geologic map of the Qinghai-Xizang (Tibet) plateau and adjacent areas. Chengdu: Chengdu Cartographic Publishing House, 48.

Pitcher, W.S., 1997. The Nature and Origin of Granite. second edition. Netherlands: Springer Science and Business Media, 387.

Qi, L., Hu, J., Gregoire, D.C., 2000. Determination of trace elements in granites by inductively coupled plasma mass spectrometry. Talanta, 51, 507-513.

Qin, J.F., Lai, S.C., Li, Y.F., Ju, Y.J., Zhu, R.Z., Zhao, S.W., 2016. Early Jurassic monzogranite-tonalite association from the southern Zhangguangcai Range: Implications for Paleo-Pacific plate subduction along northeastern China. Lithosphere, 8 (4), 396-411.

Qu, X.M., Wang, R.J., Xin, H.B., Jiang, J.H., Chen, H., 2012. Age and petrogenesis of A-type granites in the middle segment of the Bangonghu-Nujiang suture, Tibetan plateau. Lithos, 146, 264-275.

Rapp, R.P., Watson, E.B., 1995. Dehydration melting of metabasalt at 8 – 32 kbar: Implications for continental growth and crust-mantle recycling. Journal of Petrology, 36, 891-931.

Rudnick, R.L., Gao, S., 2003. Composition of the continental crust. In: Holland, H.D., Turekian, K. K. Treatise on Geochemistry, 3. Oxford: Elsevier-Pergamon, 1-64.

Searle, M., Parrish, R., Hodges, K., 1997. Shisha Pangma leucogranite, south Tibet Himalaya: Field relations, geochemistry, age, origin, and emplacement. Journal of Geology, 105, 295-317.

Słaby, E., Gotze, J., 2004. Feldspar crystallization under magma mixing conditions shown by cathodoluminescence and geochemical modeling: A case study from the Karkonosze pluton (SW Poland). Mineralogical Magazine, 68, 541-557.

Słaby, E., Martin, H., 2008. Mafic and felsic magma interaction in granites: The Hercynian Karkonosze Pluton (Sudetes, Bohemian Massif). Journal of Petrology, 49, 353-391.

Słaby, E., Galbarczyk-Gasiorowska, L., Baszkiewicz, A., 2002. Mantled alkali-feldspar megacrysts from the marginal part of the Karkonosze granitoid massif (SW Poland). Acta Geol Polonica, 5 (2), 501-519.

Słaby, E., Smigielski, M., Smigielski, T., Domonik, A., Simon, K., Kronz, A., 2011. Chaotic three-dimensional distribution of Ba, Rb, and Sr in feldspar megacrysts grown in an open magmatic system. Contributions to Mineralogy and Petrology, 162 (5), 909-927.

Słaby, E., De Campos, C.P., Majzner, K., Simon, K., Gros, K., Moszumańska, I., et al., 2017. Feldspar megacrysts from the Santa Angélica composite pluton-formation/transformation path revealed by combined Cl, Laman and LA-ICP-MS data. Lithos, 277, 269-283.

Soderlund, U., Patchett, P. J., Vervoort, J.D., Isachsen, C.E., 2004. The ^{176}Lu decay constant determined by Lu-Hf and U-Pb isotope systematics of Precambrian mafic intrusions. Earth & Planetary Science Letters, 219, 311-324.

Stern, R.A., Hanson, G.N., Shirey, S.B., 1989. Petrogenesis of mantle-derived, LILE-enriched Archean monzodiorites and trachyandesites (sanukitoids) insouthwestern Superior Province. Canadian Journal of Earth Sciences, 26, 1688-1712.

Sui, Q.L., Wang, Q., Zhu, D.C., Zhao, Z.D., Chen, Y., Santosh, M., Hu, Z.C., Yuan, H.L., Mo, X. X., 2013. Compositional diversity of ca. 110 Ma magmatism in the northern Lhasa Terrane, Tibet: Implications for the magmatic origin and crustal growth in a continent-continent collision zone. Lithos, 168-169 (3), 144-159.

Sun, S.S., McDonough, W.F., 1989. Chemical and isotopic systematics of oceanic basalts: Implications or mantle composition and processes. Geological Society, London, Special Publications, 42 (1), 313-345.

Sylvester, P.J., 1998. Post collisional strongly peraluminous granites. Lithos, 45, 29-44.

Wang, Y., Fan, W., Zhang, Y., Peng, T., Chen, X., Xu, Y., 2006. Kinematics and ^{40}Ar/^{39}Ar geochronology of the Gaoligong and Chongshan shear systems, western Yunnan, China: Implications for early Oligocene tectonic extrusion of SE Asia. Tectonophysics, 418 (3-4), 235-254.

Wortel, M.J.R., Spakman, W., 1992. Structure and dynamics of subducted lithosphere in the Mediterranean region. Proceedings of the Koninklijke Nederlandse Akademie van Wetenschanppen-Biological Chemical Geological Physical and Medical Sciences, 95, 325-347.

Wu, R.X., Zheng, Y.F., Wu, Y.B., Zhao, Z.F., Zhang, S.B., Liu, X., et al., 2006a. Reworking of juvenile crust: Element and isotope evidence from Neoproterozoic granodiorite in South China. Precambrian Research, 146 (3-4), 179-212.

Wu, Y.B., Zheng, Y.F., Zhao, Z.F., Gong, B., Liu, X., Wu, F.Y., 2006b. U-Pb, Hf and O isotope evidence for two episodes of fluid-assisted zircon growth in marble-hosted eclogites from the Dabie Orogen. Geochimica et Cosmochimica Acta, 70 (3), 743-761.

Xie, J.C., Zhu, D.C., Dong, G., Zhao, Z.D., Wang, Q., Mo, X., 2016. Linking the Tengchong terrane in SW Yunnan with the Lhasa terrane in southern Tibet through magmatic correlation. Gondwana Research, 18.

Xu, R., Scharer, U., Allègre, C.J., 1985. Magmatism and metamorphism in the Lhasa Block (Tibet): A geochronological study. Journal of Geology, 93, 41-57.

Xu, Y.G., Yang, Q.J., Lan, J.B., Luo, Z.Y., Huang, X.L., Shi, Y.R., Xie, L.W., 2012. Temporal-spatial distribution and tectonic implications of the batholiths in the Gaoligong-Tengliang-Yingjiang area, western Yunnan: Constraints from zircon U-Pb ages and Hf isotopes. Journal of Asian Earth Sciences, 53, 151-175.

YNBGMR (Yunnan Bureau Geological Mineral Resource), 1991. Regional Geology of Yunnan Province. Beijing: Geological Publishing House, 1-729 (in Chinese with English abstract).

Yang, J.H., Wu, F.Y., Chung, S.L., Wilde, S.A., Chu, M.F., 2004. Multiple sources for the origin of granites: Geochemical and Nd/Sr isotopic evidence from the Gudaoling granite and its mafic enclaves, NE China. Geochimica et Cosmochimica Acta, 68, 4469-4483.

Yang, J.H., Wu, F.Y., Wilde, S.A., Xie, L.W., Yang, Y.H., Liu, X.M., 2007. Tracing magma mixing in granite genesis: In-situ U-Pb dating and Hf-isotope analysis of zircons. Contributions to Mineralogy and Petrology, 135, 177-190.

Yuan, H.L., Gao, S., Liu, X.M., Li, H.M., Gunther, D., Wu, F.Y., 2004. Accurate U-Pb age and trace element determinations of zircon by laser ablation-inductively coupled plasma mass spectrometry. Geo-Standard Newsletters, 28, 353-370.

Yuan, H.L., Gao, S., Dai, M.N., Zong, C.L., Gunther, D., Fontaine, G.H., Liu, X.M., Diwu, C.R., 2008. Simultaneous determinations of U-Pb age, Hf isotopes and trace element compositions of zircon by excimer laser-ablation quadrupole and multiple-collector ICPMS. Chemical Geology, 247, 100-118.

Zhang, K.J., Xia, B.D., Wang, G.M., Li, Y.T., Ye, H.F., 2004. Early Cretaceous stratigraphy, depositional environments, sandstone provenance, and tectonic setting of central Tibet, western China. Geological Society of America Bulletin, 116, 1202-1222.

Zhang, K.J., Zhang, Y.X., Tang, X.C., Xia, B., 2012. Late Mesozoic tectonic evolution and growth of the Tibetan Plateau prior to the Indo-Asian collision. Earth-Science Reviews, 114, 236-249.

Zheng, Y.F., Zhao, Z.F., Wu, Y.B., Zhang, S.B., Liu, X.M., Wu, F.Y., 2006. Zircon U-Pb age, Hf and O isotope constraints on protolith origin of ultrahigh-pressure eclogite and gneiss in the Dabie orogen. Chemical Geology, 231, 135-158.

Zhu, D.C., Mo, X.X., Niu, Y.L., Zhao, Z.D., Wang, L.Q., Liu, Y.S., Wu, F.Y., 2009. Geochemical investigation of Early Cretaceous igneous rocks along an east-west traverse throughout the central Lhasa Terrane, Tibet. Chemical Geology, 268, 298-312.

Zhu, D.C., Zhao, Z.D., Niu, Y.L., Mo, X.X., Chung, S.L., Hou, Z.Q., Wang, L.Q., Wu, F.Y., 2011. The Lhasa Terrane: Record of a microcontinent and its histories of drift and growth. Earth & Planetary Science Letters, 301, 241-255.

Zhu, R.Z., Lai, S.C., Qin, J.F., Zhao, S.W., 2015a. Early-cretaceous highly fractionated I-type granites from the northern Tengchong Block, western Yunnan, SW China: Petrogenesis and tectonic implications. Journal of Asian Earth Sciences, 100, 145-163.

Zhu, D.C., Li, S.M., Cawood, P.A., Wang, Q., Zhao, Z.D., Liu, S.A., et al., 2015b. Assembly of the Lhasa and Qiangtang terranes in central Tibet by divergent double subduction. Lithos, 245, 7-17.

Zhu, R.Z., Lai, S.C., Santosh, M., Qin, J.F., Zhao, S.W., 2017b. Early Cretaceous Na-rich granitoids and their enclaves in the Tengchong Block, SW China: Magmatism in relation to subduction of the Bangong-Nujiang Tethys ocean. Lithos, 286-287:175-190.

Zhu, R.Z., Lai, S.C., Qin, J.F., Zhao, S.W., 2017c. Petrogenesis of late Paleozoic-to-early Mesozoic granitoids and metagabbroic rocks of the Tengchong Block, SW China: Implications for the evolution of the eastern Paleo-Tethys. International Journal of Earth Sciences:1-17.

Zhu, R.Z., Lai, S.C., Qin, J.F., Zhao, S.W., 2017a. Petrogenesis of calc-alkaline granodiorite and associated enclaves from SE Lhasa Block, Tibet (SW China): Implication for recycled subducting sediments. Geological Society of America Bulletin (submitted for publication).

Strongly peraluminous fractionated S-type granites in the Baoshan Block, SW China: Implications for two-stage melting of fertile continental materials following the closure of Bangong-Nujiang Tethys[①]

Zhu Renzhi　Lai Shaocong[②]　Qin Jiangfeng　Zhao Shaowei　M Santosh

Abstract: Strongly peraluminous fractionated S-type granites are important indicators of compositional maturity of the continental crust. Here we report the finding of strongly peraluminous fractionated S-type granites from the western Baoshan Block in the southeastern Tibetan-Himalayan orogenic belt. We present LA ICP-MS zircon U-Pb ages from biotite- and two-mica- granites which reveal two distinct periods of magmatism at 121.8 ± 1.1 Ma and $81.9-79.7$ Ma. These Cretaceous granites are strongly peraluminous (A/CNK = $1.18-1.31$), high-K calc-alkaline, enriched in Rb, Th, K, and Pb, and depleted in Ba, Sr and Ti. They also show high normative-corundum contents of $2.84-4.42$ wt%, moderately high P_2O_5 content ($0.22-0.28$ wt%) and high differentiation index (DI) of $93.1-96.8$, similar to the features of fractionated S-type granites. The low total $Fe_2O_3^T +$ $MgO + TiO_2$ contents (< 3.0 wt%) are also consistent with the results from experimental studies on S-type granite melts. The S-type granites from Baoshan show high Al_2O_3/TiO_2 ($93.8-182.9$), Rb/Ba ($6.19-23.5$), and Rb/Sr ($17.0-30.5$) values, and low CaO/Na_2O ($0.09-1.19$). They also display negative $\varepsilon_{Nd}(t)$ values (-13.6 to -11.7) and $\varepsilon_{Hf}(t)$ ranges (-5.47 to -0.77 for biotite granites and -7.67 to -0.81 for two-mica granites) suggesting evolved and heterogeneous continental crustal sources for the magma. The initial $^{206}Pb/^{204}Pb$ ($17.527-19.277$), $^{207}Pb/^{204}Pb$ ($15.772-15.888$), and $^{208}Pb/^{204}Pb$ ($39.237-39.498$) isotopic ratios are also consistent with lower to upper crustal affinity. Our data, in conjunction with those from experimental studies suggest that the strongly peraluminous S-type granites were derived from partial melting of metasedimentary, pelite-dominated source within continental crustal basement, similar to typical fractionated S-type granites from the Jiangnan Orogen (SE China), Erzgebirge (Central Europe), Eastern Australia and the low-temperature SP granites in the Himalayas. We correlate the Early- to Late- Cretaceous magmatism with two-stage melting of fertile continental materials following the final closure and termination of Bangong-Nujiang Tethys ocean.

---

①　Published in *Lithos*, 2018, 316-317.

②　Corresponding author.

1 Introduction

Granitoids are widespread in the Earth's upper crust, and understanding their genesis provides insights into the magma sources and tectonic setting, such as subduction, collision and post-orogenic extension (Barbarin, 1999; Champion and Bultitude, 2013; Chappell and White, 1992; Sylvester, 1998). Peraluminous granitoids are widely distributed within orogenic belts of various ages (Barbarin, 1999 and reference therein), and fractionated peraluminous granitoids are considered as key indicators of compositional maturity of the continental crust (Wu et al., 2017).

The Tibetan-Himalayan orogenic system preserves long records of the evolution of various terranes, including the Tethyan Himalaya, Lhasa-Tengchong, and Qiangtang-Baoshan (Xu et al., 2012; Zhu et al., 2011; Zhu et al., 2015). Peraluminous granitoids are widely distributed in this region, particularly in the southeastern Himalayan-Tibetan orogenic belt, which also include the Cretaceous peraluminous to strongly peraluminous granitoids (Table 1). Previous studies have suggested contrasting models for the origin of these rocks including crustal anatexis associated with subduction (Cao et al., 2014; Zhu et al., 2009, 2011, 2016; Zhu et al., 2015, 2017a,b), collision (Qu et al., 2012; Xu et al., 2012) and crustal thickening (Chen et al., 2015; Dong et al., 2013a; Xu et al., 2012; Yu et al., 2014). However, few studies have considered two-stage melting of fertile continental materials following the final closure and termination of Bangong-Nujiang Tethyan ocean.

Our recent field investigations identified two Early- to Late-Cretaceous strongly peraluminous fractionated S-type granitoids in the western Baoshan Block. In this study, we characterize these granitoids using LA-ICP-MS zircon U-Pb geochronology and in situ Lu-Hf isotopic data, bulk-rock major and trace element composition, and whole rock Sr-Nd-Pb isotopic data with a view to understand the genesis of these rocks and the two-stage anatectic melting of ancient continental basement following the final closure and termination of the Bangong-Nujiang Tethys ocean.

2 Geological setting and petrography

The western Yunnan region, as the western part of Sanjiang Tethyan orogenic belt, is an important segment of the East Tethyan tectonic domain (Wang et al., 2014; Fig. 1a). From east to west, the region comprises the Simao, Baoshan, and Tengchong blocks, separated by the Changning-Menglian suture and the Lushui-Luxi-Ruili Fault (LLRF), respectively (Cong et al., 2011; Xu et al., 2012; Wang et al., 2014; Deng et al., 2014; Metcalfe, 2013).

The Simao-Indochina block, has a Cathaysian affinity (Wopfner, 1996; Zhong, 1998), and is composed of Proterozoic metamorphic succession comprising volcaniclastic and carbonate rocks (Zhong, 1998). These rocks are unconformably overlain by a suite of Paleozoic carbonates

Table 1　Summary of the magmatism in the Tengchong-Banshan Terranes and Gaoligong belt during the Cretaceous.

Sample	GPS Position	Lithology	SiO_2 /wt%	A/CNK	Age /Ma	Zircon $\varepsilon_{Hf}(t)$	Bulk-rock $(\frac{^{87}Sr}{^{86}Sr})_i$	$\varepsilon_{Nd}(t)$	Mineral assemblage	Reference
Tengchong Terrane										
DT12	25°46'19"N, 98°29'31"E	Syenogranite	76.0–77.9	1.02–1.15	130.6 ±2.5			−9.2	Qtz(35%–40%) + Kf(45%–60%) + Pl(3%–5%) + Bt(5%–8%) + Chl(<3%)	Zhu et al.,2015
ZD1	24°38'28"N, 98°36'38"E	Rhyolite	74.73	1.34	130.0 ±1.7				Phenocryst 11%（Qtz+Pl）+matrix 90%（Se+felsic assemblage）+Lim1%	Bai et al.,2012
	24°38'28"N, 98°36'39"E	Dacite	62.23–65.70	1.12					Phenocryst 6%（Qtz+Pl）+matrix 85%（Se+felsic assemblage）+Mus1%+Lim1%	Bai et al.,2012
D9021	24°49'51"N, 98°23'39"E	Monzogranite	68.06–74.75	1.06–1.44	128.9 ±2.4				Pl(18%–44%) + Kf(25%–56%) + Qtz(20%–25%) + Bt(3%–5%)	Luo et al.,2012
	24°49'51"N, 98°23'40"E	Granodiorite	65.44–66.25	1.1–1.25					Pl(55%–64%) + Kf(6%–12%) + Qtz(20%–25%) + Bt(6%–12%)	Luo et al.,2012
	24°49'51"N, 98°23'41"E	Diorite	64.5	0.91					Pl(53%) + Hbl(42%) + Bt(5%) + apatite(<1%)	Luo et al.,2012
	24°35'50"N, 98°15'00"E	Granite	74.75	1.13	127.4 ±1.0	−9.1 to −5.4			Kf(75%) + Pl(<5%) + Qtz(20%) + Bt(3%)	Cong et al.,2011
	24°30'00"N, 98°17'10"E	Granodiorite	62.6	0.99	115.2 ±1.1	−4.5 to 0			Pl(60%) + Kf(10%) + Qtz(20%) + Hbl(5%) + Bt(5%)	Cong et al.,2011
	24°34'20"N, 98°17'50"E	Dioritic enclaves	51.4	0.67	122.6± 0.8	+3.6 to +6.2			Pl(50%) + Qtz(1%–2%) + Hbl(45%) + Bt(1%)	Cong et al.,2011
D1166	24°37'42"N, 98°17'00"E	Monzogranite	68.67–74.17	0.94–1.17	123.8 ±2.5				Kf(40%) + Pl(30%) + Qtz(25%) + Bt(5%)	Li et al.,2012b

Continued

Sample	GPS Position	Lithology	SiO$_2$ /wt%	A/CNK	Age /Ma	Zircon $\varepsilon_{Hf}(t)$	Bulk-rock (^{87}Sr/^{86}Sr)$_i$	$\varepsilon_{Nd}(t)$	Mineral assemblage	Reference
D2346	24°37'40"N, 98°16'00"E	Quartz diorite	62.39–64.70	0.94–1.00	127.1 ±1.0	−7.6 to −3.9			Pl(60%)+Qtz(20%)+Hbl(15%)+Bt(5%)	Li et al.,2012b
		Dioritic enclaves	51.40–59.8	0.67–0.95						Li et al.,2012b
09QT20	25°05'41"N, 98°30'39"E	Quartz monzonite	61.02–65.28	0.96–1.08	125.0 ±1.3	−7.8 to −4.9			Pl(43%−48%)+Kf(30%−35%)+Qtz(8%−10%)+Hbl(8%−10%)+Spe+Me+Zr(1%−2%)	Qi et al.,2011
09QT33	24°57'31"N, 98°42'44"E	Granodiorite	65.75–69.10	0.95–1.25	122.3 ±1.2	−9.6 to −4.8			Pl(30%−35%)+Kf(35%−40%)+Qtz(15%−20%)+Bt(5%−10%)+Hbl(3%−5%)+(Ap+Me+Zr+Fe-Ti)(1%−2%)	Qi et al.,2011
GD14	25°49'08"N, 98°32'04"E	Monzogranite	74.7–76.8	1.03–1.09	124.1 ±1.6		0.707 9	−9.6 to −10.1	Qtz(30%−40%)+Kf(40%−55%)+Pl(5%−10%)+Bt(3%−7%)+Hbl(1%−3%)	Zhu et al.,2015
GLS8	25°49'52"N, 98°47'19"E	Granite			122.0 ±1.0	−18.9 to −4.3			Kf(40%)+Pl(30%)+Qtz(20%−30%)+Bt(2%−3%)+Mus+Ap+Hbl	Xu et al.,2012
ITCXC4	25°13'20"N, 98°41'26"E	Granite	73.87–75.70	0.98–1.10	121.6 ±0.7		0.710 5–0.712 4	−7.13 to −8.16	Qtz(30%)+Kf(30%)+Pl(25%)+Bt(7%)+accessory mineral(3%)	Chen et al.,2013
ITCXC3	25°13'20"N, 98°41'27"E	Granite	73.87–75.71	0.98–1.11	121.2 ±0.9		0.710 5–0.712 4	−7.13 to −8.16	Qtz(30%)+Kf(30%)+Pl(25%)+Bt(7%)+accessory mineral(4%)	Chen et al.,2013
PM25-3	24°45'51"N, 98°23'39"E	Dacite	70.21	1.09	121.4 ±1.4				Qtz(30%)+(Pl+Kf)(10%)+(Bt+Mc)(20%)+Glassy(10%)	Gao et al.,2012
JJGLZ	25°51'00"N, 98°33'00"E	Monzogranite	72.96	1.06	120.0 ±0.6	−4.9 to −1.4				Gao et al.,2014
D1332	24°36'34"N, 98°20'26"E	Diorite	58.79	0.93	119.8 ±1.0	−6.7 to −3.1		−7.69	Pl(55%)+Qtz(15%)+Hbl(25%)+Mgt(2%)+Spe(2%)+Ap(1%)	Gao et al.,2014

Continued

Sample	GPS Position	Lithology	SiO$_2$ /wt%	A/CNK	Age /Ma	Zircon $\varepsilon_{Hf}(t)$	Bulk-rock (^{87}Sr/ ^{86}Sr)$_i$	$\varepsilon_{Nd}(t)$	Mineral assemblage	Reference
D0032	24°36′53″N, 98°20′48″E	Granodiorite	64.7	1	118.7 ±1.5	-5.6 to -3.3		-8.02	Pl(50%)+Qtz(20%)+Kf(10%)+Bt(18%)+Mgt (2%)	Gao et al.,2014
GD02	25°39′50″N, 98°31′27″E	Monzogranite	71.7- 73.4	1.34- 1.40	119.6 ±0.9		0.706 7	-8.6		Zhu et al.,2015
13TC09	24°46′09″N, 98°32′35″E	Monzogranite	74.26	1.14	117.6 ±0.8	-3.5 to +3.2			Qtz(20%-30%)+Pl(25%-40%)+Kf(25%-30%)+Bt(2%-3%)+Hbl(1%)	Xie et al.,2016
13TC04	24°59′26″N, 98°36′01″E	Monzogranite	67.58	1.06	114.3 ±0.6	-0.1 to +7.1			Qtz(20%-30%)+Pl(25%-35%)+Kf(30%-35%)+Bt(2%-3%)+Ms 1%+Spe+Ap+Fe-Ti oxides	Xie et al.,2016
JJGLZ	25°51′00″N, 98°33′00″E	Monzogranite	71.94- 74.25	1.03- 1.29	120.0 ±0.6	-4.9 to -1.4			Kf(15%-20%)+Pl(10%-15%)+Qtz(10%-15%)+Bt(3%-5%)+Ms(3%)	Cao et al.,2014
XH23		Granodiorite	64.17- 68.22	0.91- 1.04	116.1 ±0.8	-2.02 to +8.90			Kf(5-10%)+Pl(45-55%)+Qtz(20%)+Bt(10%)+Hbl(15%)	Zhu et al.,2017
MD55		Monzodiorite	59.58- 66.39	0.94- 1.06	122.6 ±0.9	-5.55 to +0.58			Kf(12%)+Pl(45-50%)+Qtz(16%)+Bt(15%)+Hbl(8%)	Zhu et al.,2017a,b
TCXL-2	25° 25′22″N, 98°23′40″E	Monzogranite			76±1	-5.76 to -10.08			Kf(35%)+Pl(35%)+Qtz(20%-30%)+Bt(3%)±Hbl±Ap	Xu et al.,2012
TXBH-6	25 °05′52″N, 98°15′21″E	Monzogranite			72	-8.09 to -11.85			Kf(35%)+Pl(35%)+Qtz(20%-30%)+Bt(4%)±Hbl±Ap	Xu et al.,2012
TCGY-11	25°20′56″N, 98°17′43″E	Monzogranite			74.9 ±1.8	-4.61 to -9.15			Kf 35%+Pl 35%+Qtz(20%-30%)+Bt 5%±Hbl±Ap	Xu et al.,2012
TCGY-3	25°18′34″N, 98°15′24″E	Monzogranite			67.8 ±1.4	-8.73 to -13.07			Kf 35%+Pl 35%+Qtz(20%-30%)+Bt 6%±Hbl ±Ap	Xu et al.,2012
GY06	25°22.695′N, 98°12.251′E	Granite	70.64- 72.33	1.01- 1.02	64±1		0.709 9	-9.3	Phenocryst: Kf(5%)+Pl(20%),Matrix: Qz(25%)+Pl(20%)+ Ms(15%)+Kf(20%)+Bi(12%)+Hb(3%)	Zhao et al.,2017a,b

Continued

Sample	GPS Position	Lithology	SiO_2 /wt%	A/CNK	Age /Ma	Zircon $\varepsilon_{Hf}(t)$	Bulk-rock $(^{87}Sr/^{86}Sr)_i$	$\varepsilon_{Nd}(t)$	Mineral assemblage	Reference
GY17	25°21.476′N, 98°13.647′E	Granite	75.04– 75.58	1.03– 1.04	65±1		0.706 5	−9.2	Qz(30%)+Pl(25%)+Ms(20%)+Kf(22%)+Bi(3%)	Zhao et al.,2017a,b
GY46	25°20.715′N, 98°15.840′E	Granite	70.44– 75.77	1.05– 1.12	64±1		0.711 8	−11.6	Qz(22%)+Pl(40%)+Kf(30%)+Bi(8%)	Zhao et al.,2017a,b
LC28	24°27.855′N, 97°45.040′E	Granodiorite	65.37– 66.99	0.96– 0.97	64±1	−4.4 to −18.1	0.716 5	−16.5	Phenocryst: Ms(10%)+Matrix: Qz(20%)+Pl(30%)+Ms(10%)+Kf(15%)+Bi(7%)+Hb(8%)	Zhao et al.,2017a,b
DSP-1	25°25′36″N, 98°22′50″E	Granite	73.79– 76.22	0.92– 1.04	73.3 ±0.3		0.718 4– 0.745 7	−11.4 to −12.1	Kf(30%)+Pl(35%)+Qtz(28%)+Bt(10%)	Chen et al.,2015
XLH-6	25°26′49″N, 98°25′46″E	Granite	74.74– 77.37	1.00– 1.05	73.3 ±0.2	−11.4 to −8.8	0.718 2– 0.732 3	−12.0 to −12.4	Kf(38%)+Pl(32%)+Qtz(20%–30%)+Bt(8%)	Chen et al.,2015
09QG36	25°58′21″N, 98°28′19″E	Two-mica granite	71.93	1.13	73.2 ±0.7	−11.2 to −8.1			Kf(25%–35%)+Pl(20%–30%)+Qtz(35%)+Bt(10%)+MS(0%–3%)	Qi et al.,2015
09QG45	24°57′41″N, 98°19′13″E	Two-mica granite	72.28	1.02	76.6 ±2.0	−8.8 to −5.7			Kf(25%–35%)+Pl(20%–30%)+Qtz(35%)+Bt(10%)+Ms(0%–1%)	Qi et al.,2015
13 MB06	25°00′05″N, 98°19′56″E	Two-mica granite	74.12	1.13	73.8 ±0.5	−8.2 to −4.6			Kf(30%–35%)+Pl(20%–35%)+Qtz(30%)+Bt(2%–3%)+Ms(1%)	Xie et al.,2016
13TC19	25°21′58″N, 98°18′53″E	Two-mica granite	73.53	1.05	74.3 ±0.5	−8.8 to −3.4			Kf(30%–35%)+Pl(20%–35%)+Qtz(30%)+Bt(2%–3%)+Ms(1%)	Xie et al.,2016
13ZX06	25°05′40″N, 98°14′53″E	Two-mica granite	77.4	1.14	65.6 ±0.6	−6.9 to +6.5			Kf(30%–35%)+Pl(20%–30%)+Qtz(35%)+Bt(2%–3%)+Ms(1%)	Xie et al.,2016
09QG60	25°18′30″N, 98°12′45″E	Granite	75.26	0.98	64.8 ±0.7	−10.5 to −4.6			Kf(15%)+Pl(45%)+Qtz(35%)+Bt(5%)	Qi et al.,2015
09QG55	25°13′50″N, 98°12′02″E	Granite	72.73	0.99	64.9 ±0.9	−13.4 to 0.0			Kf(15%)+Pl(45%)+Qtz(35%)+Bt(5%)	Qi et al.,2015

Continued

Sample	GPS Position	Lithology	SiO₂ /wt%	A/CNK	Age /Ma	Zircon $\varepsilon_{Hf}(t)$	Bulk-rock $(^{87}Sr/^{86}Sr)_i$	$\varepsilon_{Nd}(t)$	Mineral assemblage	Reference
09QG50	25°13′50″N, 98°11′42″E	Granite	70.84	0.96	65.3 ±1.0	−13.9 to −5.3			Kf(15%)+Pl(45%)+Qtz(35%)+Bt(5%)	Qi et al.,2015
Gaoligong Belt										
GLS36	27°45′32″N, 98°35′07″E	Monzogranite	64.6– 73.7	0.93– 1.27	124.1 ±2.7	−10.8 to −5.1	0.706 2– 0.715 1	−6.1 to +3.0	Kf(40%)+Pl(30%)+Qtz(20%–30%)+Bt(2%–3%) ±Mc+Ap+Hbl	Xu et al.,2012; Yang et al.,2006
GLS38	27°45′32″N, 98°35′07″E	Monzogranite	67.2– 74.1	1.08– 1.30	126.0 ±2.0	−17.5 to −2.6	0.710 2– 0.711 9	−8.3 to −6.2	Kf(40%)+Pl(30%)+Qtz(20%–30%)+Bt(2%–4%) ±Mc+Ap+Hbl	Xu et al.,2012; Yang et al.,2006
GLS62	27°08′01″N, 98°49′37″E	Monzogranite	65.2– 73.5	1.01– 1.13	121.0 ±4.0	−18.9 to −4.3	0.708 5– 0.718 6	−11.7 to −5.2	Kf(40%)+Pl(30%)+Qtz(20%–30%)+Bt(2%–5%) ±Mc+Ap+Hbl	Xu et al.,2012; Yang et al.,2006
GLS58	27°09′52″N, 98°47′19″E	Monzogranite	65.2– 73.6	1.01– 1.14	122.0 ±1.0	−5.4 to +1.9	0.708 5– 0.718 7	−11.7 to −5.2	Kf(40%)+Pl(30%)+Qtz(20%–30%)+Bt(2%–6%) ±Mc+Ap+Hbl	Xu et al.,2012; Yang et al.,2006
GLS53	27°12′11″N, 98°42′48″E	Monzogranite	68.2– 75.2	1.06– 1.25	76.3 ±2.2	−8.93 to −5.34	0.708 5– 0.712 5	−6.5 to −5.8	Kf(40%)+Pl(30%)+Qtz(20%–30%)+Bt(2%–7%) ±Mc+Ap+Hbl	Xu et al.,2012; Yang et al.,2006
LL234	25°00.280′N, 98°39.064′E	Granodiorite	63.15– 66.0	0.98– 1.18	120.9 ±1.0		0.712	−8.92	Pl(25%)+Kf(40%)+Qz(20%)+Bi(12%)+Hb (3%)	Zhao et al.,2016

Continued

Sample	GPS Position	Lithology	SiO$_2$ /wt%	A/CNK	Age /Ma	Zircon $\varepsilon_{Hf}(t)$	Bulk-rock (^{87}Sr/^{86}Sr)$_i$	$\varepsilon_{Nd}(t)$	Mineral assemblage	Reference
LL265	24°55.606'N, 98°46.133'E	Two-mica granite	72.83–75.24	1.21–1.29	89.9		0.713	−9.58	Qz(75%)+Bi(15%)+Pl(35%)+Kf(30%)+Qz(20%)+Ms(15%)	Zhao et al.,2016
LL64	24°13.758'N, 98°08.628'E	Granitic gneiss	71.16–71.69	1.03–1.09	69.6 ±0.5		0.713 2	−4.41	Phenocryst: Pl(10%)+Kf(5%)+Qz(3%)+Hb(2%) +Matrix: Pl(25%)+Kf(20%)+Qz(20%)+Bi(15%)	Zhao et al.,2016
LL85	24°14.131'N, 97°59.360'E	Granitic gneiss	75.21–75.96	1.08	63.2 ±1.4		0.7147	−10	Phenocryst: Pl(12%)+Kf(8%)+Matrix: Pl (20%)+Kf(25%)+Qz(25%)+Bi(10%)	Zhao et al.,2016
LL259	24°59.803'N, 98°43.600'E	Granitic gneiss	72.10–75.14	1.06–1.15	69.3 ±0.9		0.716 6	−7.41	Phenocryst: Pl(10%)+Kf(10%)+Matrix: Pl (20%)+Kf(25%)+Qz(25%)+Bi(10%)	Zhao et al.,2016
Baoshan Terrane										
CJ12-01	25°39'20"N, 99°05'52"E	Two-mica granite	73.76–74.47	1.15–1.23	73.3 ±0.19	−3.8 to −7.3			Qz+Pl+Ms+Kf+Bi	Yu et al.,2014
CJ12-02					73.4 ±0.20					
BM1148		Two-mica granite	73.35–74.59	1.20–1.22	73.4 ±1.10	−1.2 to −3.5			Qz+Pl+Ms+Kf+Bi	Dong et al.,2013
Z02		Granite	75.1–75.48	1.08–1.52	126.7 ±1.6	−2.6 to −6.6			Qz+Pl+Kf+Bi	Tao et al.,2010
D1684		Two-mica granite	74.62	1.18	73.2				Qz+Pl+Ms+Kf+Bi	Liao et al.,2013
ZA26	24°32.949'N, 98°50.535'E	Biotite granite	70.70–74.70	1.18–1.30	121.8 ±1.1	−5.47 to +6.13		−11.8 to −12.4	Kf(30%)+Pl(25%)+Qtz(38%)+Bt(8%)	This study
XD08	24°31.022'N, 98°47.861E'	Two-mica granite	73.98–74.63	1.29–1.31	81.9 ±0.8			−11.7 to −13.6	Kf(30%)+Pl(25%)+Qtz(35%)+Bt(5%)+Ms (5%)+Grt(1%)	This study
LL39	24°35.551'N, 98°48.629'E;	Two-mica granite	74.17–74.77	1.27–1.31	79.7 ±0.9	−17.23 to −0.81		−13.3	Kf(30%)+Pl(25%)+Qtz(35%)+Bt(5%)+Ms (5%)+Grt(2%)	This study

Pl: Plagioclase; Kf: feldspar; Qtz: Quartz; Bt: Biotite; Hbl: Hornblende; Ap: Apatite.

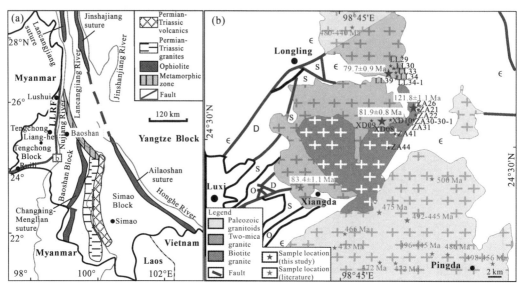

Fig. 1　(a) Distribution of the main continental blocks of Sanjiang Tethyan orogenic belt (after Zhao et al., 2017a); (b) Study area in the western Baoshan Block (after YNBGMR, 1991).

LLFR: Lushui-Longling-Ruili fault. Pentagram represents the location of granite ages. S means Silurian stratum, D means Devonian stratum, O means Ordovician stratum, and ε means Cambrian stratum. Late Cretaceous data in literature (Dong et al., 2013a); Paleozoic granites data in literatures from Liu et al. (2009), Wang et al. (2013), Dong et al. (2013b), Zhao et al. (2014, 2017a), C.M. Wang et al. (2015a, b) and Li et al. (2016).

and siliciclastic rocks with typical Cathaysian fossils [Feng, 2002; Yunnan Bureau of Geology and Mineral Resources (YNBGMR), 1991; Zhong, 1998]. The Baoshan, Tengchong, and Shan-Thai blocks have stratigraphic and paleontological similarities to Gondwana (Feng, 2002; Metcalfe, 2013; Zhong, 1998). The dominant stratigraphic sequences of these blocks include pre-Mesozoic high-grade metamorphic rocks and Mesozoic-Cenozoic sedimentary and igneous rocks (YNBGMR, 1991; Zhong, 1998).

Most of the granitic gneisses, migmatites, and granites that are exposed in the Tengchong Block and the western Baoshan Block occur near the southeastern segment of the Bangong-Nujiang suture (Table 1). The earliest granitoids (ca. 509−445 Ma) are distributed in the Gaoligong belt and western Baoshan Block (Liu et al., 2009; Wang et al., 2013; Dong et al., 2013b; Zhao et al., 2014, 2017a; Wang et al., 2015a, b; Li et al., 2016), including the Ximeng, Xiangda and Pinghe area. Mesozoic granitoids with U-Pb ages of 232−206 Ma occur in the southern part of Tengchong County and the northern part of Lianghe County (Li et al., 2012a; Zhu et al., 2018). Younger granitoids that yield zircon U-Pb ages of 139−115 Ma are also common in the eastern part of the Tengchong Block and the western Baoshan Block near Tengchong-Lianghe-Luxi, and in the Gaoligong (Cao et al., 2014; Cong et al., 2011; Li et al., 2012a, b; Luo et al., 2012; Qi et al., 2011; Xu et al., 2012; Zhao et al., 2016; Zhu et al., 2015, 2017a). The S- and A-type granites with U-Pb ages of 85−60 Ma (mainly Late

Cretaceous to early Eocene) are distributed in the mid-western part of the Tengchong Block along the Guyong-Lailishan belt and on the western Baoshan Block in the southern segment of the Bangong-Nujiang suture (Dong et al., 2013a; Xu et al., 2012; Yu et al., 2014). In addition, I-type granitoids with zircon ages as young as 66−52 Ma were reported from the Tengchong Block in Yingjiang, near the China-Myanmar border (Wang et al., 2015a, b; Xu et al., 2012). These granitoids intruded Proterozoic-Paleozoic strata, including the Gaoligong and Gongyanghe groups (Zhong, 1998).

The Xiangda pluton covering an area of ca. 180 km² occur within the western Baoshan Block (Fig. 1b), and intrude the lower Paleozoic-Middle Cambrian flysch and sandy shale of the Gongyanghe Group, Devonian and Silurian limestone, siltstone and shale. The biotite granites occur in the central part of the pluton, whereas the two-mica granites occur in the periphery (Fig. 2d). The granitic body is undeformed, medium-grained and traversed by quartz veins (Fig. 2d). The granitoid is composed of 30%−38% quartz, 30%−35% K-feldspar, 25%−30% plagioclase, 5%−8% biotite, and 1%−3% accessory minerals including apatite, zircon, ilmenite, and titanite. Quartz crystals are xenomorphic to subhedral, display undulatory

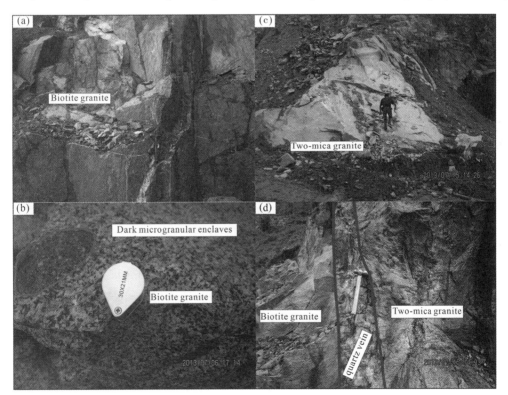

Fig. 2   Field photographs of Cretaceous granites from the Xiangda pluton
in the western Baoshan Block, SW China.

(a-c) Early Cretaceous biotite granites; (d) Late Cretaceous two-mica granites.

extinction, and are intergrown with feldspar. The K-feldspar is subhedral and partly altered to kaolinite; Plagioclase is also subhedral and partly altered to sericite. Brownish biotite occurs between quartz and feldspar crystals (Fig. 2a-c).

The undeformed coarse-grained two-mica granite (Fig. 2c,d) is composed of 33% −35% quartz, 28% −30% K-feldspar, 20% −25% plagioclase, 3% −5% biotite, ~ 5% muscovite, ~1% garnet, and accessory minerals including ilmenite, apatite, zircon, and euhedral to subhedral tourmaline. Quartz occurs as xenomorphic to subhedral crystals. K-feldspar and plagioclase occur as subhedral laths. Reddish brown biotite occurs interstitially between quartz and feldspar grains.

3　Analytical methods

3. 1　Whole-rock geochemistry and Sr-Nd isotopes

The whole rock analyses were performed at the State Key Laboratory of Continental Dynamics, Northwest University, Xi'an, China. Fresh chips of whole-rock samples (ca. 1 kg) were powdered to 200 mesh using a tungsten carbide ball mill. Major elements were analyzed using an X-ray fluorescence (Rikagu RIX 2100). Analyses of U.S. Geological Survey and Chinese national rock standards BCR-2(SiO_2 = 51. 3 wt%, Na_2O = 0. 4 wt%, K_2O = 0. 31 wt%, MgO = 3. 22 wt%, Al_2O_3 = 13. 2 wt%, P_2O_5 = 0. 29 wt%, CaO = 6. 81 wt%, TiO_2 = 1. 95 wt%, MnO = 0. 122 wt% and $Fe_2O_3^T$ = 12. 43 wt%), GSR-1(SiO_2 = 72. 06 wt%, Na_2O = 2. 83 wt%, K_2O = 4. 85 wt%, MgO = 0. 36 wt%, Al_2O_3 = 13. 3 wt%, P_2O_5 = 0. 08 wt%, CaO = 1. 46 wt%, TiO_2 = 0. 172 wt%, MnO = 0. 05 wt% and $Fe_2O_3^T$ = 1. 95 wt%), and GSR-3(SiO_2 = 43. 8 wt%, Na_2O = 2. 866 wt%, K_2O = 2. 04 wt%, MgO = 6. 68 wt%, Al_2O_3 = 0. 14 wt%, P_2O_5 = 0. 09 wt%, CaO = 8. 25 wt%, TiO_2 = 1. 42 wt%, MnO = 0. 14 wt% and $Fe_2O_3^T$ = 13. 06 wt%), which indicate that both analytical precision and accuracy for major elements are generally better than 5%. For trace elements analyses, which were determined by using a Bruker Aurora M90 inductively coupled plasma mass spectrometry (ICP-MS) following the method of Qi et al. (2000). Sample powders were dissolved using an $HF + HNO_3$ mixture in a high-pressure PTFE bomb at 190℃ for 48 h. The analytical precision was better than ±5% −10% (relative) for most of the trace elements, the results of standard samples could be seen in the Supplemental Table 1.

Whole-rock Sr-Nd isotopic data were obtained using a Nu Plasma HR multi-collector (MC) mass spectrometer. The Sr and Nd isotopic fractionation was corrected to $^{87}Sr/^{86}Sr$ = 0. 119 4 and $^{146}Nd/^{144}Nd$ = 0. 721 9, respectively (Chu et al., 2009). During the sample runs, the La Jolla standard yielded an average of $^{143}Nd/^{144}Nd$ = 0. 511 862 ± 5 (2σ), and the NBS987 standard yielded an average of $^{87}Sr/^{86}Sr$ ratio = 0. 710 236 ± 16 (2σ). The total procedural Sr and Nd blanks are b1 ng and b50 pg, and NIST SRM-987 and JMC-Nd were used as certified reference standard solutions for $^{87}Sr/^{86}Sr$ and $^{143}Nd/^{144}Nd$ isotopic ratios, respectively. The BCR-1 and BHVO-1 standards yielded an average of $^{87}Sr/^{86}Sr$ ratio are 0. 705 014 ± 3 (2σ) and 0. 703 477 ± 20 (2σ), the BCR-1 and BHVO-1 standards yielded an average of $^{146}Nd/^{144}Nd$ ratio are 0. 512 615 ± 12 (2σ) and 0. 512 987 ± 23 (2σ).

3. 2   Zircon U-Pb and Hf isotopic analyses

Zircon grains were separated using conventional heavy liquid and magnetic techniques. Representative grains (~5 kg) from Xiangda pluton were hand-picked and mounted on epoxy resin discs, polished and carbon coated. Internal morphology was examined using cathodoluminescent (CL) prior to U-Pb and Lu-Hf isotopic analyses. Laser ablation (LA) ICP-MS zircon U-Pb analyses were conducted on an Agilent 7500a ICP-MS equipped with a 193-nm laser, following method of Bao et al. (2017) at the State Key Laboratory of Continental Dynamics, Northwest University, Xi'an, China. The $^{207}Pb/^{206}Pb$ and $^{206}Pb/^{238}U$ ratios were calculated using the GLITTER data reduction software program (http://www. glitter-gemoc. com/), and calibrated using the Harvard zircon 91500 as external standard. These correction factors were then applied to each sample to correct for both instrumental mass bias and depth-dependent elemental and isotopic fractionation. Common Pb contents were evaluated using the method described by Andersen (2002). The age calculations and plotting of concordia diagrams were performed using ISOPLOT (version 3. 0; Ludwig, 2003). The errors quoted in tables and figures are at the 2σ level. In situ zircon Hf isotopic analyses were conducted using a Neptune MC-ICPMS equipped with a 193 nm laser. During analyses, a laser repetition rate of 10 Hz at 100 mJ was used and spot sizes were 44 μm. The $^{176}Yb/^{172}Yb$ value of 0. 588 7 and mean Yb value obtained during Hf analysis on the same spot were applied for the interference correction of ^{176}Yb on ^{176}Hf (Iizuka and Hirata, 2005). The $^{176}Hf/^{177}Hf$ and $^{176}Lu/^{177}Hf$ ratios of the standard zircon (91500) were 0. 282 294 ± 15 (2σ, $n=20$) and 0. 000 31, similar to the commonly accepted $^{176}Hf/^{177}Hf$ ratio of 0. 282 302 ± 8 and 0. 282 306 ± 8 (2s) measured using the solution method. The obtained Hf isotopic compositions were 0. 282 016 ± 20 (2σ, $n=84$) for the GJ-1 standard and 0. 282 735 ± 24 (2σ, $n=84$) for the Monastery standard, both of which are consistent with the recommended values (cf. Yuan et al., 2008). The initial $^{176}Hf/^{177}Hf$ ratios and $\varepsilon_{Hf}(t)$ values were calculated with reference to the chondritic reservoir (CHUR) at the time of zircon growth from the magmas. The decay constant for ^{176}Lu of 1.867×10^{-11} year^{-1} (Soderlund et al., 2004), the chondritic $^{176}Hf/^{177}Hf$ ratio of 0. 282 785 ± 7 and $^{176}Lu/^{177}Hf$ ratio of 0. 033 6 were adopted. The depleted mantle model ages (T_{DM}) were calculated using the Hf isotope "crustal" model age (T_{DMC}) by assuming that the parental magma was derived from an average continental crust, with $^{176}Lu/^{177}Hf=0.015$, that originated from the depleted mantle source (Griffinet al., 2000, $^{176}Hf/^{177}Hf$ ratio of 0. 283 25 ± 20), similar to that of the average MORB (Nowell et al., 1998). In addition, the notations of $\varepsilon_{Hf}(t)$ value, $f_{Lu/Hf}$ (f: different constant of Lu and Hf), single-stage model age (T_{DM1}: depleted mantle), and two-stage model age (T_{DM2}) are as defined in Yuan et al. (2008).

4 Results

4.1 Zircon U-Pb age data

Zircon grains from three representative samples (ZA26, XD08, and LL39; Fig. 3a-c) from the Xiangda granitic pluton were analyzed by LA-ICP-MS. The CL images, U-Pb concordia diagrams, and weighted mean ages are shown in Figs. 3 and 4, and the U-Pb data are provided in the Supplementary Table. The zircon grains are generally euhedral and up to 400 μm in size, with length/width ratios of 2−4. Most of the crystals are prismatic, colorless to light brown, and transparent to translucent. They exhibit clear oscillatory zoning in CL images, and show Th/U ratios > 0.4, indicating magmatic origin.

4.1.1 Biotite granite

Twenty-six analyses were obtained from zircon grains in the sample ZA26. All the analyzed grains have high in U (174−1 768 ppm) and Th (152−1 565 ppm) with Th/U values in the range of 0.35−1.61. The 26 analyses yield $^{206}Pb/^{238}U$ ages of 116 ± 3.0 Ma to 125 ± 2.0 Ma, with a weighted mean age of 121.8 ± 1.1 Ma (MSWD = 1.6, n = 26, 1σ).

4.1.2 Two-mica granite

Thirty-one analyses were obtained from the sample XD08. All the analyzed grains have very high U (350−9 893 ppm, except for two values that exceed 10 000 ppm) and variable Th (43 − 1 796 ppm) contents, with Th/U ratios of 0.01 − 0.71. The 29 analyses yielded $^{206}Pb/^{238}U$ ages of 79 ± 2.0 Ma to 88 ± 2.0 Ma with a weighted mean age of 81.9 ± 0.8 Ma (MSWD = 1.08, n = 29, 1σ).

Twenty-one analyses were obtained from the sample LL39 and the data show very high and variable U (4 567−7 314 ppm) and Th (116−2 411 ppm, except one value of 51 ppm), with Th/U ratios of 0.31−2.16, except three low values below 0.03. Two grains show $^{206}Pb/^{238}U$ ages of 486 ± 5.0 Ma and 462 ± 5.0 Ma, which were interpreted as inherited grains. The remaining analyses yielded $^{206}Pb/^{238}U$ ages of 77 ± 1.0 Ma to 84 ± 1.0 Ma with a weighted mean age of 79.7 ± 0.9 Ma (MSWD = 3.6, n = 18, 1σ).

4.2 Major and trace element compositions

The results of major and trace element analyses are listed in Table 2. Both granitoids show strongly peraluminous and high K calc-alkaline signature (Fig. 5).

The Early Cretaceous biotite granite is strongly peraluminous (A/CNK = 1.18−1.30) and

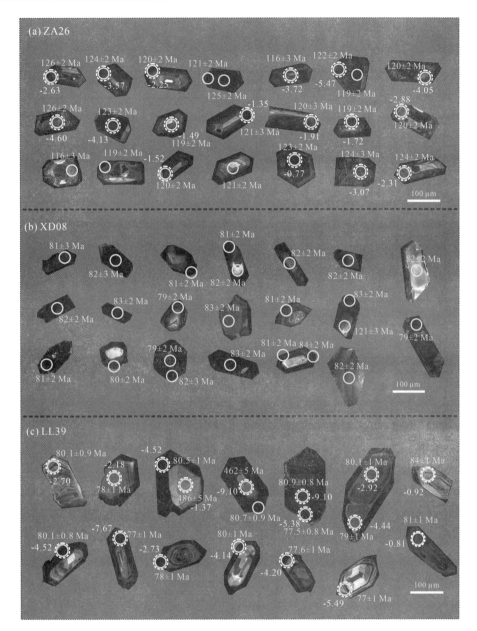

Fig. 3　CL images of representative zircon grains for the Xiangda pluton,
western Baoshan Block, SW China.

(a) the twenty-two typical zircon grains from Early-Cretaceous granites was presented; (b) and (c) there are thirty-four typical zircon grains from Late-Cretaceous granites were presented. The yellow circle presents their in situ U-Pb isotopic analyses spots and the white dotted circle represents the in situ Lu-Hf isotopic analyses spots.

contains 70. 70−74. 70 wt%SiO_2, 4. 77−5. 12 wt% K_2O, and 3. 00−3. 58 wt% Na_2O, with a K_2O/Na_2O ratio of 1. 33 − 1. 62. The total alkali content is 7. 85 − 8. 54 wt% with low CaO (0. 49−0. 63), MgO (0. 19 − 0. 32), $Fe_2O_3^T$(0. 96 − 1. 86), TiO_2(0. 10 − 0. 17), P_2O_5 (0. 23−0. 28), and $Mg^\#$(27. 5−31. 6).

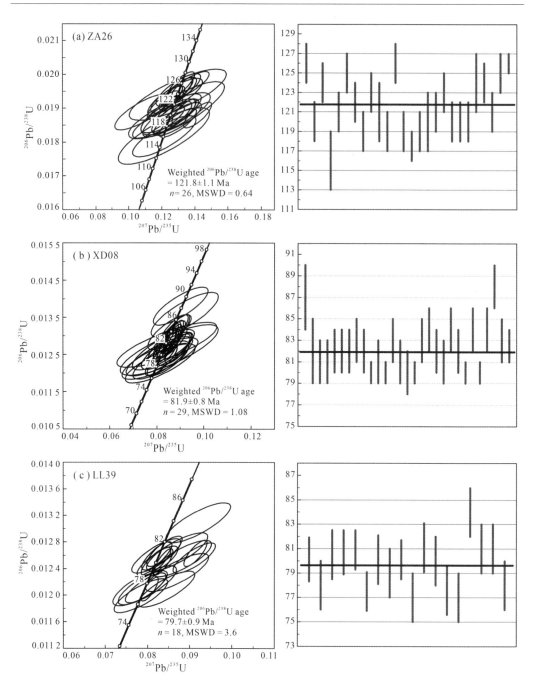

Fig. 4   LA-ICP-MS U-Pb zircon concordia diagram for the Xiangda pluton,
western Baoshan Block, SW China.

The Early Cretaceous biotite granite is enriched in Rb, Th, U, K, and Pb, and depleted in Ba and Sr, as shown by primitive-mantle-normalized trace element spider diagram (Fig. 6a). The rocks show relative depletions in Nb and Ti, and enrichments in Ta, P, Zr, and Hf. Low total REE ($\sum$REE) content of 54.6–94.5, and enrichment in light rare earth

Table 2　Major(wt%) and trace(ppm) element analysis results of granites from Xiangda pluton.

Sample	Late Cretaceous granites									Early Cretaceous granites				
	LL29	LL30	LL33	LL34	LL34-1	XD09	XD10	ZA21	ZA22	ZA30	ZA30-1	ZA31	ZA41	ZA44
SiO_2	74.77	74.52	74.55	74.31	74.17	73.98	74.63	74.73	74.71	73.58	73.50	70.66	73.65	74.01
TiO_2	0.09	0.09	0.10	0.09	0.10	0.08	0.09	0.11	0.10	0.11	0.10	0.17	0.10	0.10
Al_2O_3	14.7	14.9	14.3	14.9	14.8	14.6	14.3	14.1	14.1	14.6	14.5	15.9	14.6	14.51
$Fe_2O_3^T$	0.84	0.79	0.87	0.87	0.85	0.92	0.98	1.05	0.96	1.18	1.17	1.86	1.11	1.15
MnO	0.02	0.02	0.04	0.03	0.03	0.03	0.03	0.03	0.03	0.04	0.04	0.07	0.04	0.04
MgO	0.19	0.19	0.19	0.19	0.19	0.19	0.18	0.20	0.19	0.20	0.19	0.32	0.20	0.20
CaO	0.46	0.43	0.43	0.47	0.47	0.33	0.27	0.49	0.49	0.59	0.58	0.63	0.54	0.59
Na_2O	3.22	2.99	3.02	3.17	3.14	3.13	3.15	3.05	3.00	3.48	3.48	3.40	3.28	3.58
K_2O	4.96	5.25	4.76	4.96	4.98	4.99	4.99	4.83	4.85	5.06	5.04	5.12	4.91	4.77
P_2O_5	0.26	0.24	0.24	0.26	0.26	0.23	0.22	0.26	0.25	0.24	0.23	0.28	0.25	0.25
LOI	0.92	0.95	1.03	0.96	0.96	1.04	1.11	1.02	0.88	0.68	0.69	1.06	0.84	0.74
Total	100.38	100.35	99.51	100.16	99.90	99.55	99.98	99.87	99.54	99.71	99.51	99.52	99.54	99.94
A/CNK	1.27	1.31	1.31	1.30	1.29	1.31	1.29	1.27	1.27	1.19	1.18	1.30	1.25	1.20
A/NK	1.79	1.81	1.84	1.83	1.82	1.80	1.76	1.79	1.79	1.70	1.70	1.87	1.79	1.74
Li	250	254	256	265	262	218	228	171	156	232	221	395	283	242
Be	7.17	7.14	7.22	29.7	27.8	12.5	8.48	17.5	15.2	32.4	33.0	8.29	32.2	23.9
Sc	2.64	2.92	2.94	2.93	2.77	2.35	2.09	3.58	3.31	2.09	2.12	4.12	2.68	2.42
V	3.14	3.58	3.72	3.19	3.60	2.10	1.97	2.93	2.84	3.06	2.92	4.31	2.97	2.79
Cr	2.23	3.18	5.71	2.88	9.14	3.99	2.71	3.77	3.21	3.66	3.16	3.49	4.38	3.96
Co	115	89.1	107	98.8	98.3	150	149	121	162	138	137	93.4	156	144
Ni	2.17	3.45	3.97	3.06	6.66	1.97	1.33	1.76	1.22	1.60	1.28	1.63	1.74	1.66
Cu	8.51	7.80	8.71	6.91	6.90	0.72	0.76	0.71	0.55	0.54	0.57	0.93	8.48	0.72
Zn	22.8	21.4	18.2	21.0	20.2	40.4	38.5	41.3	40.0	55.0	53.7	89.0	55.9	62.4
Ga	26.5	27.7	28.4	28.5	28.3	24.9	24.2	28.4	26.5	24.1	24.0	34.0	26.2	25.1
Ge	2.14	2.05	2.18	2.15	2.24	2.14	2.11	2.01	2.00	2.08	2.04	2.30	2.15	2.06
Rb	531	564	527	544	540	506	497	478	479	485	487	578	516	462
Sr	21.8	22.9	19.0	21.0	20.8	16.6	16.6	23.4	25.1	28.1	28.7	20.7	23.8	25.7
Y	11.5	10.9	11.8	12.6	12.2	13.4	11.8	13.2	11.6	11.3	11.9	18.2	11.3	13.7
Zr	51.1	52.1	52.7	54.5	54.2	41.2	43.3	55.7	51.7	51.2	60.3	69.2	50.9	59.8
Nb	28.0	30.4	31.9	31.8	32.0	26.2	24.8	31.9	28.5	25.3	24.9	48.6	28.0	26.4

Continued

Sample	Late Cretaceous granites							Early Cretaceous granites						
	LL29	LL30	LL33	LL34	LL34-1	XD09	XD10	ZA21	ZA22	ZA30	ZA30-1	ZA31	ZA41	ZA44
Cs	55.6	56.1	52.2	55.7	55.8	52.4	53.3	43.1	44.3	52.4	50.5	73.1	53.8	42.9
Ba	54.6	59.7	49.5	52.4	52.1	21.5	24.7	48.9	53.9	78.0	78.7	52.9	69.1	65.2
La	9.68	8.18	8.57	9.10	9.16	7.21	7.84	9.37	9.35	10.6	10.2	14.6	8.18	12.5
Ce	20.3	17.3	18.2	19.7	19.5	14.6	15.5	19.4	19.0	22.2	21.1	30.5	16.8	25.1
Pr	2.48	2.12	2.24	2.33	2.32	1.79	1.95	2.38	2.32	2.66	2.52	3.69	2.09	3.01
Nd	8.71	7.51	7.90	8.35	8.30	6.39	6.92	8.59	8.26	9.34	9.03	13.1	7.45	10.8
Sm	2.21	1.97	2.04	2.18	2.16	1.76	1.82	2.24	2.09	2.32	2.23	3.41	1.97	2.71
Eu	0.15	0.14	0.13	0.14	0.14	0.11	0.11	0.14	0.15	0.18	0.18	0.15	0.14	0.17
Gd	1.95	1.80	1.89	2.02	1.95	1.78	1.72	2.08	1.94	2.07	2.00	3.12	1.83	2.48
Tb	0.36	0.33	0.36	0.38	0.38	0.36	0.34	0.40	0.36	0.36	0.36	0.58	0.35	0.44
Dy	2.06	1.96	2.08	2.16	2.14	2.27	2.07	2.36	2.10	2.08	2.10	3.34	2.07	2.51
Ho	0.35	0.34	0.36	0.38	0.37	0.42	0.38	0.41	0.35	0.34	0.36	0.56	0.35	0.42
Er	0.94	0.89	0.96	1.02	0.99	1.13	1.02	1.06	0.93	0.93	0.97	1.46	0.92	1.08
Tm	0.14	0.13	0.14	0.15	0.14	0.16	0.14	0.15	0.14	0.13	0.14	0.21	0.13	0.15
Yb	0.88	0.85	0.95	0.95	0.92	1.03	0.92	0.99	0.88	0.86	0.93	1.35	0.84	0.96
Lu	0.12	0.11	0.12	0.13	0.12	0.13	0.12	0.13	0.11	0.11	0.12	0.18	0.11	0.13
Hf	1.88	1.91	1.90	1.95	1.92	1.64	1.66	2.01	1.80	1.77	2.05	2.54	1.83	2.16
Ta	6.61	6.77	7.17	7.03	7.04	8.32	8.02	7.02	6.08	6.69	6.60	9.96	6.39	6.47
Pb	23.8	25.8	20.3	22.4	22.0	29.9	28.5	26.4	27.8	32.5	32.9	28.6	30.9	29.3
Th	9.78	8.78	9.27	9.06	9.25	7.48	8.12	9.85	8.94	11.1	10.2	15.0	9.00	11.3
U	24.6	30.1	24.3	24.1	25.8	6.35	5.70	42.6	28.1	36.6	18.9	49.9	20.7	34.5
$\sum REE$	504	43.6	46.0	49.0	48.6	39.2	40.9	49.7	47.9	54.2	52.2	76.3	43.2	62.46
LREE/HREE	6.40	5.80	5.70	5.81	5.92	4.38	5.09	5.56	6.02	6.86	6.48	6.08	5.56	6.65
K_2O+Na_2O	8.2	8.2	7.8	8.1	8.1	8.1	8.1	7.9	7.9	8.5	8.5	8.5	8.2	8.4
Al_2O_3/TiO_2	163	165	143	165	148	183	159	128	141	132	145	93.8	146	145.10
CaO/Na_2O	0.1	0.1	0.1	0.1	0.1	0.1	0.1	0.2	0.2	0.2	0.2	0.2	0.2	0.2
Rb/Ba	9.72	9.46	10.7	10.4	10.4	23.5	20.2	9.79	8.89	6.21	6.19	10.9	7.47	7.08
Rb/Sr	24.3	24.7	27.8	25.8	25.9	30.5	29.9	20.4	19.1	17.2	17.0	27.9	21.7	18.0
T_{Zr}	706	709	712	712	712	693	696	713	708	700	712	727	704	712.53

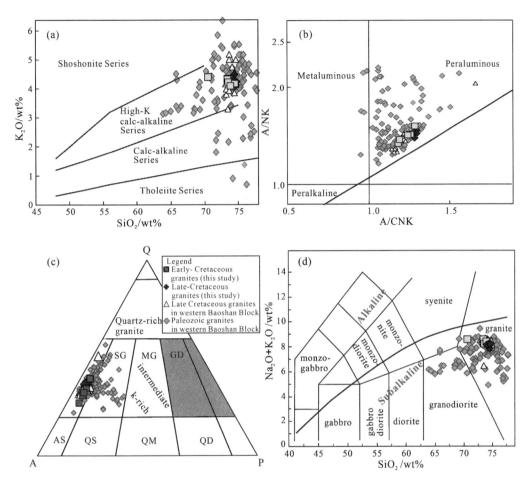

Fig. 5 (a) K_2O vs. SiO_2 referenced from Roberts and Clemens (1993) and (b) A/NK vs.
A/CNK diagram (Frost et al., 2001), (c) Q-A-P referenced from Streckeisen (1979)
and (d) (Na_2O+K_2O) vs. SiO_2 diagram referenced from Middlemost (1994)
for the Cretaceous granites, western Baoshan Block, SW China.
The data source of Cretaceous and Paleozoic granites are the same as in Fig. 1.

elements (LREE; Fig. 6b) with LREE/HREE ratios of 2.03−2.60 and low Gd_N/Yb_N ratio (1.74−2.13) are also seen, similar to the features of the granitoids from the central segment of the Bangong-Nujiang suture (Qu et al., 2012). The biotite granite shows negative Eu anomaly ($\delta Eu = 0.14-0.26$).

Similar to the Early Cretaceous biotite granite, the Late Cretaceous two-mica granite is also strongly peraluminous (A/CNK = 1.27−1.31) with SiO_2 in the range of 74.0−74.8 wt% and K_2O of 4.76−5.25 wt%. The K_2O/Na_2O ratio range from 1.54 to 2.15, and the rock is classified as high K calc-alkaline (Fig. 5). The granite shows low CaO (0.27−0.47 wt%), MgO (0.18−0.19 wt%), $Fe_2O_3^T$(0.79−0.98 wt%), and TiO_2(0.08−0.10 wt%).

The primitive-mantle-normalized trace element spider diagram of Late-Cretaceous two-mica granite is similar to that of the Early Cretaceous biotite granite (Fig. 6c). The two-mica granite

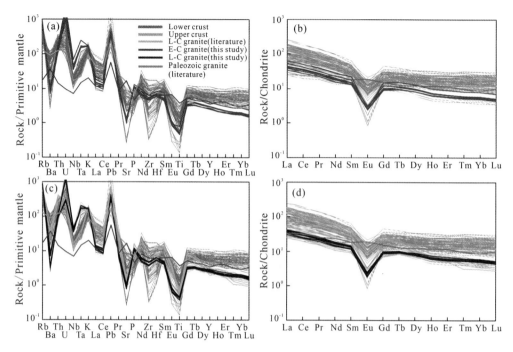

Fig. 6　Chondrite-normalized REE patterns and primitive-mantle-normalized trace element spider diagram.
(a,b) Early Cretaceous granites, and (c,d) Late Cretaceous granites; Yellow line represents Cretaceous granites
(Dong et al., 2013a; Yu et al., 2014). Paleozoic granites data from the primitive mantle and chondrite values are
from Sun and McDonough (1989). The upper and lower averages reference from Rudnick and Gao (2003). The data
of Cretaceous and Paleozoic granites in literature was same as Fig. 1.

shows enrichment in Rb, Th, U, K, Pb, Ta, P, Zr, and Hf, and depletion in Ba, Sr, Nb, and Ti. The REE content is low with ΣREE of 52. 6-61. 9, with depletion in HREE and LREE/HREE ratio of 1. 54-3. 89 (Fig. 6d). A markedly negative Eu anomaly ($\delta Eu = 0.$ 18-0. 23) is also seen.

4. 3　Whole-rock Sr-Nd-Pb isotopes

The whole-rock Sr-Nd-Pb isotopic compositions of the granitoids from the Xiangda plutons in the western Baoshan Block are listed in Tables 3 and 4.

The Early Cretaceous granite shows anomalous initial $^{87}Sr/^{86}Sr$ ratio due to its high Rb/Sr ratio (17. 2-20. 4; Jahn et al., 2001). The rocks are characterized by negative $\varepsilon_{Nd}(t)$ values (-12. 4 to -11. 8) (Fig. 7) and initial $^{206}Pb/^{204}Pb$, $^{207}Pb/^{204}Pb$, and $^{208}Pb/^{204}Pb$ ratios of 17. 521-18. 111, 15. 785-15. 797, and 39. 323-39. 423 (Fig. 8), respectively.

The Late Cretaceous granite also shows anomalous initial $^{87}Sr/^{86}Sr$ ratio due to its high Rb/Sr ratio (24. 3-30. 5; Jahn et al., 2001). The $\varepsilon_{Nd}(t)$ values are negative and show a range of -13. 6 to -11. 7 (Fig. 7), with initial $^{206}Pb/^{204}Pb$, $^{207}Pb/^{204}Pb$, and $^{208}Pb/^{204}Pb$ ratios in the range of 18. 711-19. 277, 15. 772-15. 888, and 39. 237-39. 498 (Fig. 8), respectively.

Table 3 Whole-rock Rb-Sr and Sm-Nd isotopic data for granites from Xiangda pluton.

Sample	$^{87}Sr/^{86}Sr$	2SE	Rb/ppm	Sr/ppm	2SE	$^{143}Nd/^{144}Nd$	2SE	Nd/ppm	Sm/ppm	T_{DM2}/Ga	$\varepsilon_{Nd}(t)$	I_{Sr}
XD-09	0.803 338	71	506	17	80	0.511 923		6	2	1.70	−13.6	0.701 981
XD-10	0.799 363	84	497	17	58	0.512 017		7	2	1.57	−11.7	0.700 179
LL-29	0.750 143	7	531	22	4	0.511 935		9	2	1.68	−13.3	0.669 916
ZA-21	0.782 833	59	478	23	46	0.511 971		9	2	1.65	−12.4	0.681 293
ZA-30	0.770 204	53	485	28	52	0.511 999		9	2	1.61	−11.8	0.684 600
ZA-44	0.773 318	71	462	26	42	0.511 995		11	3	1.61	−11.9	0.684 023

^{87}Rb/^{86}Sr and ^{147}Sm/^{144}Nd ratios were calculated using Rb, Sr, Sm and Nd contents analyzed by ICP-MS, referenced from DePaolo and Series (1981).

T_{DM2} represent the two-stage model age and were calculated using present-day $(^{147}Sm/^{144}Nd)_{DM} = 0.213\ 7$, $(^{147}Sm/^{144}Nd)_{DM} = 0.513\ 15$ and $(^{147}Sm/^{144}Nd)_{crust} = 0.1012$. $\varepsilon_{Nd}(t)$ values were calculated using present-day $(^{147}Sm/^{144}Nd)_{CHUR} = 0.196\ 7$ and $(^{147}Sm/^{144}Nd)_{CHUR} = 0.512\ 638$.

$$\varepsilon_{Nd}(t) = \left[(^{143}Nd/^{144}Nd)_S(t)/(^{143}Nd/^{144}Nd)_{CHUR}(t) - 1 \right] \times 10^4.$$

$$T_{DM2} = \frac{1}{\lambda}\left\{ 1 + \left[(^{143}Nd/^{144}Nd)_S - ((^{147}Sm/^{144}Nd)_S - (^{147}Sm/^{144}Nd)_{crust})(e^{\lambda t}-1) - (^{143}Nd/^{144}Nd)_{DM} \right] / \left[(^{147}Sm/^{144}Nd)_{crust} - (^{147}Sm/^{144}Nd)_{DM} \right] \right\}.$$

Table 4 Whole-rock Pb isotopic data for granites from Xiangda pluton.

Sample	U	Th	Pb	$^{206}Pb/^{204}Pb$	2SE	$^{207}Pb/^{204}Pb$	2SE	$^{208}Pb/^{204}Pb$	2SE	$^{238}U/^{204}Pb$	$^{232}Th/^{204}Pb$	$(^{206}Pb/^{204}Pb)_i$	$(^{207}Pb/^{204}Pb)_i$	$(^{208}Pb/^{204}Pb)_i$
XD-09	6.3	7.5	29.9	19.324	24	15.790	20	39.303	54	13.721	16.593	19.153	15.782	39.237
XD-10	5.7	8.1	28.5	19.439	30	15.896	24	39.574	64	12.999	19.007	19.277	15.888	39.498
LL-29	24.6	9.8	23.8	19.549	5	15.812	4	39.385	11	67.148	27.348	18.711	15.772	39.276
ZA-21	42.6	9.9	26.4	19.496	36	15.893	29	39.572	75	105.122	24.974	17.521	15.797	39.423
ZA-30	36.6	11.1	32.5	19.275	23	15.851	19	39.492	47	72.919	22.708	17.905	15.785	39.356
ZA-44	34.5	11.3	29.3	19.550	21	15.856	15	39.477	40	76.582	25.743	18.111	15.786	39.323

U, Th, Pb concentrations were analyzed by ICP-MS. Initial Pb isotopic ratios were calculated for 80 Ma(XD09-10 and LL29) and 120 Ma(ZA21-44) using single-stage model.

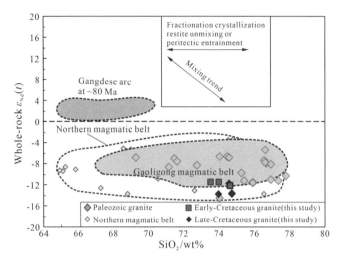

Fig. 7 (a) SiO_2 vs. $\varepsilon_{Nd}(t)$ diagram, northern magmatic belt data from Chen et al. (2015),
Qu et al. (2012), and Zhu et al. (2009a,b) and the Gaoligong belt data are from
Zhu et al. (2015), the Gangdese arc data from Wen et al. (2008).

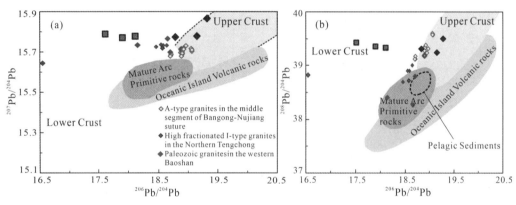

Fig. 8 Pb plumbotectonic framework diagrams (Zartman and Doe, 1981)
of the Cretaceous granites, western Yunnan, SW China.

A-type granites data are from Qu et al. (2012) and I-type granites data are from Zhu et al. (2015). Symbols as Fig. 5.

4. 4 Zircon Lu-Hf isotopes

Zircon grains from the three dated samples were also analyzed for Lu-Hf isotopes on the same domains, and the results are listed in Table 5. Initial $^{176}Hf/^{177}Hf$ ratios and $\varepsilon_{Hf}(t)$ values of the early Cretaceous zircon grains were calculated according their crystallization age. Zircon grains from the Cretaceous granites show evolved Hf isotopic composition (Fig. 9), with $\varepsilon_{Hf}(t)$ values of -5.47 to -0.77 for the Early Cretaceous biotite granites (17 zircons), and -7.67 to -0.81 for the Late-Cretaceous two-mica granites (14 zircons). The two inherited grains display $\varepsilon_{Hf}(t)$ values of -1.37 (486 Ma) and -9.10 (486 Ma). The crustal Hf model ages are in the range of 1 225 - 1 523 Ma for Early Cretaceous biotite granites, and 1 196 - 1 631 Ma for the Late-Cretaceous two-mica granites.

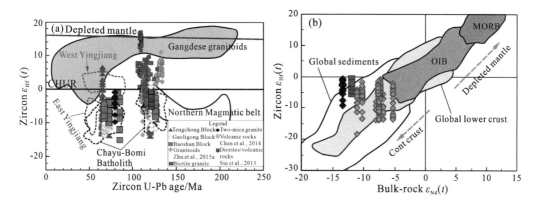

Fig. 9 （a）zircon $\varepsilon_{Hf}(t)$ vs. zircon U-Pb data for the magmatism from Lhasa-Tengchong Block, Baoshan Block and Gaoligong belt（Xu et al., 2012; Zhu et al., 2017a）. （b）zircon $\varepsilon_{Hf}(t)$ vs. zircon U-Pb data for the Cretaceous granites from western Baoshan.

Global sediments referenced from Vervoort and Blichert-Toft（1999）, Global lower crust referenced from Dobosi et al. （2003）, and OIB, and MORB referenced from Kempton and McGill（2002）. The data for Baoshan-Tengchong Block and Gaoligong belt was referenced in Table 1. Symbols as the Fig. 5.

5　Discussion

5. 1.　Petrogenesis of the Cretaceous biotite- and two-mica- granites

5. 1. 1　Fractionated S-type granites

The Early- and Late-Cretaceous granites from Xiangda pluton show similar geochemical characteristics with high silica, high-K calc-alkaline and strongly peraluminous nature （Fig. 5）. The SiO_2 contents show a range of 74. 0 - 74. 8 wt% and A/CNK ratios are in the range of 1. 18 - 1. 31, with normative-CIPW corundum contents of 2. 84 - 4. 42 wt%. These features are similar to those of S-type granites rather than the I- and A-type ones（Chappell and White, 1992; Chappell, 1999; Clemens, 2003）. The petrographic observations（Fig. 2） showing the presence of muscovite, biotite and garnet and absence of hornblende and riebeckite or other alkaline-rich ferro-magnesian minerals also indicate S-type and/or highly fractionated I-type affinity（Chappell and White, 1992）. On the $[Al_2O_3 - (Na_2O + K_2O)]$-CaO-$(FeO^T +$ MgO） diagram（Fig. 10a）, these samples are plotted in the field of S-type. They also plot in the field S-type granites of LFB（Lachlan Fold Belt）on the diagram of SiO_2 vs. P_2O_5 （Fig. 10b）. Since apatite is soluble in peraluminous melts, the concentration of P increases with increasing fractionation in S-type melts（Chappell, 1999）. The Cretaceous granitoids in our study display similar trend with S-type granites of the LFB（Chappell, 1999）, as also

Table 5　Single-grain zircon Hf isotopic data for granites from Xiangda pluton.

Grain spot	Age /Ma	$^{176}Yb/^{177}Hf$	$2SE$	$^{176}Lu/^{177}Hf$	$2SE$	$^{176}Hf/^{177}Hf$	$2SE$	$(^{176}Hf/^{177}Hf)_i$	$f_{Lu/Hf}$	$\varepsilon_{Hf}(t)$	$2SE$	T_{DM1} /Ma	T_{DMC} /Ma
Early-Cretaceous biotite granite													
ZA26-01	121.8	0.044 679	0.000 197	0.001 297	0.000 004	0.282 625	0.000 020	0.282 622	-0.96	-2.63	0.53	895	1 343
ZA26-02	121.8	0.041 518	0.000 195	0.001 183	0.000 004	0.282 598	0.000 020	0.282 596	-0.96	-3.57	0.53	931	1 403
ZA26-03	121.8	0.087 925	0.000 548	0.002 364	0.000 011	0.282 638	0.000 019	0.282 633	-0.93	-2.25	0.53	903	1 319
ZA26-04	121.8	0.106 028	0.001 050	0.002 875	0.000 026	0.282 598	0.000 024	0.282 591	-0.91	-3.72	0.53	975	1 412
ZA26-05	121.8	0.145 928	0.000 983	0.003 899	0.000 023	0.282 551	0.000 025	0.282 542	-0.88	-5.47	0.53	1 076	1 523
ZA26-06	121.8	0.054 008	0.000 261	0.001 498	0.000 006	0.282 585	0.000 020	0.282 582	-0.95	-4.05	0.53	957	1 433
ZA26-07	121.8	0.068 038	0.000 542	0.001 897	0.000 013	0.282 619	0.000 023	0.282 615	-0.94	-2.88	0.53	919	1 359
ZA26-08	121.8	0.074 039	0.000 628	0.002 031	0.000 015	0.282 653	0.000 021	0.282 648	-0.94	-1.72	0.53	874	1 285
ZA26-09	121.8	0.047 872	0.000 385	0.001 344	0.000 010	0.282 646	0.000 021	0.282 643	-0.96	-1.91	0.53	867	1 297
ZA26-10	121.8	0.072 205	0.001 030	0.002 055	0.000 029	0.282 663	0.000 022	0.282 658	-0.94	-1.35	0.53	859	1 262
ZA26-11	121.8	0.043 241	0.000 228	0.001 213	0.000 006	0.282 657	0.000 020	0.282 654	-0.96	-1.49	0.53	848	1 271
ZA26-12	121.8	0.039 473	0.000 151	0.001 127	0.000 004	0.282 582	0.000 023	0.282 580	-0.97	-4.13	0.53	952	1 439
ZA26-13	121.8	0.075 214	0.000 236	0.002 067	0.000 006	0.282 571	0.000 023	0.282 566	-0.94	-4.60	0.53	993	1 468
ZA26-14	121.8	0.068 792	0.000 957	0.001 871	0.000 020	0.282 658	0.000 019	0.282 653	-0.94	-1.52	0.53	862	1 273
ZA26-15	121.8	0.087 050	0.000 417	0.002 413	0.000 011	0.282 680	0.000 023	0.282 675	-0.93	-0.77	0.53	842	1 225
ZA26-16	121.8	0.087 178	0.000 451	0.002 443	0.000 009	0.282 615	0.000 020	0.282 610	-0.93	-3.07	0.53	938	1 371
ZA26-17	121.8	0.051 410	0.000 728	0.001 568	0.000 020	0.282 635	0.000 017	0.282 631	-0.95	-2.31	0.53	888	1 323
Late-Cretaceous two-mica granite													
LL39-01	79.7	0.019 804	0.000 309	0.000 618	0.000 009	0.282 647	0.000 020	0.282 646	-0.98	-2.70	0.53	849	1 316

Continued

Grain spot	Age /Ma	^{176}Yb/^{177}Hf	2SE	^{176}Lu/^{177}Hf	2SE	^{176}Hf/^{177}Hf	2SE	(^{176}Hf/^{177}Hf)$_i$	$f_{\mathrm{Lu/Hf}}$	$\varepsilon_{\mathrm{Hf}}(t)$	2SE	T_{DM1} /Ma	T_{DMC} /Ma
LL39-02	79.7	0.016 252	0.000 263	0.000 502	0.000 007	0.282 662	0.000 020	0.282 661	−0.98	−2.18	0.53	826	1 283
LL39-03	79.7	0.016 013	0.000 091	0.000 471	0.000 002	0.282 596	0.000 021	0.282 595	−0.99	−4.52	0.53	917	1 432
LL39-04	79.7	0.033 180	0.000 617	0.000 901	0.000 011	0.282 596	0.000 019	0.282 595	−0.97	−4.52	0.53	927	1 432
LL39-05	486	0.076 679	0.000 344	0.002 124	0.000 005	0.282 450	0.000 017	0.282 431	−0.94	−1.37	0.53	1 169	1 540
LL39-06	79.7	0.026 495	0.000 479	0.000 756	0.000 013	0.282 572	0.000 020	0.282 570	−0.98	−5.38	0.53	958	1 486
LL39-07	79.7	0.013 936	0.000 180	0.000 425	0.000 006	0.282 641	0.000 018	0.282 640	−0.99	−2.92	0.53	853	1 330
LL39-08	462	0.038 297	0.000 358	0.001 131	0.000 012	0.282 237	0.000 027	0.282 228	−0.97	−9.10	0.53	1 438	2 009
LL39-09	79.7	0.040 461	0.000 156	0.001 225	0.000 006	0.282 508	0.000 026	0.282 506	−0.96	−7.67	0.53	1 060	1 631
LL39-10	79.7	0.026 957	0.000 282	0.000 802	0.000 009	0.282 647	0.000 020	0.282 646	−0.98	−2.73	0.53	854	1 318
LL39-11	79.7	0.021 455	0.000 175	0.000 610	0.000 005	0.282 701	0.000 026	0.282 700	−0.98	−0.81	0.53	774	1 196
LL39-12	79.7	0.036 477	0.000 569	0.001 030	0.000 014	0.282 607	0.000 019	0.282 606	−0.97	−4.14	0.53	915	1 407
LL39-13	79.7	0.045 999	0.001 020	0.001 273	0.000 028	0.282 606	0.000 021	0.282 604	−0.96	−4.20	0.53	922	1 411
LL39-14	79.7	0.028 363	0.000 095	0.000 820	0.000 003	0.282 569	0.000 020	0.282 567	−0.98	−5.49	0.53	964	1 493
LL39-15	79.7	0.031 765	0.000 948	0.000 825	0.000 020	0.282 598	0.000 022	0.282 597	−0.98	−4.44	0.53	922	1 426
LL39-16	79.7	0.034 421	0.000 246	0.000 988	0.000 007	0.282 698	0.000 021	0.282 697	−0.97	−0.92	0.53	785	1 203

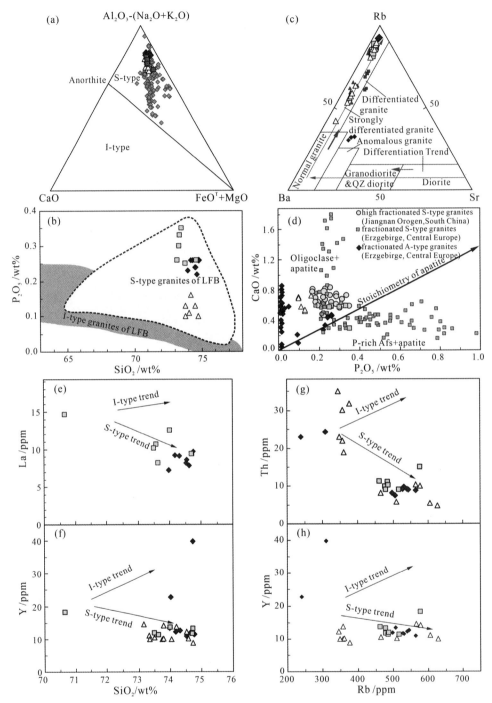

Fig. 10 (a) [Al₂O₃- (Na₂O+K₂O)]-CaO-(FeO^T+MgO) and (b) Rb-Ba-Sr diagram (Wang et al., 2013), (c) SiO₂ vs. P₂O₅(Chappell, 1999), (d) CaO vs. P₂O₅(Breiter, 2012; Huang and Jiang, 2014), and (e,f) SiO₂ vs. La and Y diagrams and (g,h) Rb vs. Th and Y diagrams (Chappell, 1999) for the Cretaceous granites in the western Baoshan Block, SW China.

Symbols as the Fig. 5.

indicated by the variations in Harker diagram were decreasing La and Y against increasing SiO_2 is seen (Fig. 10e,f), together with a decrease in Th and Y with increasing Rb (Fig. 10g,h). The fractionated S-type granites in this study have high differentiation index (DI) of 93.1−96.8. All these features also correlate with those of highly fractionated S-type granites from the Jiangnan Orogen (South China) and Erzgebirge (Central Europe) (Breiter, 2012; Huang and Jiang, 2014) (Fig. 10c,d). Based on the whole-rock CaO/P_2O_5 plot (Fig. 10d), it is inferred that oligoclase crystallized at the same time as apatite (Breiter, 2012). Also, the low total $Fe_2O_3^T + MgO + TiO_2$ contents (1.07−1.45 wt%) are consistent with that from experimental studies which show that S-type granite melts are characterized by felsic compositions with low total Fe+Mg or Fe+Mg+Ti contents (less than 3%−4%) at crustal pressures and temperatures (<1 000 ℃; Stevens et al., 2007; Champion and Bultitude, 2013 reference therein), and is similar to the more felsic S-types that were derived from these melts by crystal fractionation processes (Stevens et al., 2007). In addition, similar trend is also seen in the geochemical variation diagrams for Fe+Mg vs. various elements (Fig. 11) (Champion and Bultitude, 2013). The features include, increasing Rb and ASI with decreasing Fe+Mg, increasing TiO_2, K/Rb, Na_2O and variable Sr with increasing Fe+Mg. Based on the above features, as well as the low Zr, Y, $\sum REE$ and Th contents, elevated U/Th ratios, P_2O_5 content, A/CNK values, and the elevated Rb/Sr and Rb/Ba ratios, we conclude that the Cretaceous granites in this study are similar to typical fractionated S-type granites elsewhere in the word (Breiter, 2012; Champion and Bultitude, 2013; Chappell, 1999; Huang and Jiang, 2014).

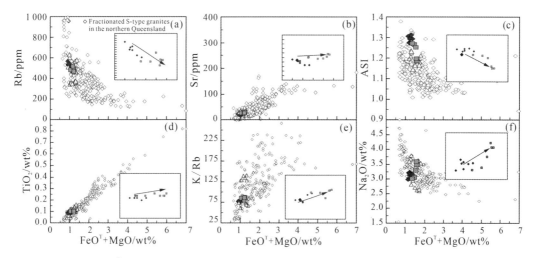

Fig. 11　$FeO^T + MgO$ vs. various elements diagram for the Cretaceous granites in the western
Baoshan Block, SW China, referenced from Champion and Bultitude (2013).

All the data from S-type granites in the northern Queensland was come from
Champion and Bultitude (2013). Symbols as the Figs. 5 and 10.

5. 1. 2 Magma source of the fractionated S-type granites

Experimental data show that dehydration melting of amphibolite and partial melting of metaluminous protoliths such as amphibolites and felsic orthogneisses at low melting fractions and water-poor (anhydrous) conditions can produce peraluminous intermediate to felsic melts with low K_2O/Na_2O ratio leaving a garnet-bearing granulite to eclogitic residue at a pressure of 0. 8 – 3. 2 GPa (Rapp and Watson, 1995; Sylvester, 1998). However, both the Early Cretaceous biotite- and Late Cretaceous two-mica granites in this study have high SiO_2 content (70. 7 – 74. 8 wt%), with strongly peraluminous nature (A/CNK = 1. 18 – 1. 31) and high K_2O/Na_2O ratios (1. 33 – 1. 76), and where biotite is the only ferromagnesian mineral (Fig. 4), which preclude the above possibility. Previous studies suggested that partial melting of metasedimentary protoliths including clay-rich metapelites and clay-poor metagreywackes (psammites) can produce strongly peraluminous granitoids (Clemens, 2003; Sylvester, 1998). The fractionated S-type granites are considered to form from the melting of quartzo-feldspathic and pelitic lithologies (Breiter, 2012), or from unexposed metasedimentary source rocks at depth (Champion and Bultitude, 2013). The partial melting of metamorphosed pelitic rocks forming part of the continental basement can also generate such rocks (Huang and Jiang, 2014).

Some of the geochemical features of the granitoids in our study provide some clues on the nature of magma source. ① The granitoids show high SiO_2 content (70. 7 – 74. 8 wt%), strongly peraluminous (A/CNK = 1. 18 – 1. 31) nature, and high K_2O/Na_2O ratios (1. 33 – 1. 76) (Fig. 5), suggesting typical metasedimentary source rather than metaigneous protoliths. ② Both biotite- and two-mica-granites have low molar $CaO/(MgO + FeO^T)$ (0. 25 – 0. 49) and high molar $Al_2O_3/(MgO + FeO^T)$ (8. 00 – 16. 5), and plot within the metapelitic field (Fig. 12a), similar to the Cretaceous granites (Dong et al., 2013a; Yu et al., 2014) and Paleozoic granites in the western Baoshan Block (Wang et al., 2013; Zhao et al., 2014, 2017a; Wang et al., 2015a,b; Li et al., 2016). On the $(Al_2O_3 + Fe_2O_3^T + MgO + Ti_2O)$ vs. $Al_2O_3/(Fe_2O_3^T + MgO + Ti_2O)$ diagram (Fig. 12b), the composition of the Cretaceous granites are comparable with felsic melts from pelites (Patiño Douce, 1999). ③ Experimental data suggest that the CaO/Na_2O ratio (< 0. 3) of strongly peraluminous granites derived from plagioclase-poor, clay-rich pelitic rocks is significantly lower than that sourced from plagioclase-rich, clay-poor psammitic rocks and meta-igneous rocks (> 0. 3) (Sylvester, 1998). These Cretaceous granites show similar geochemical characteristics with high Al_2O_3/TiO_2 (93. 8 – 182. 9) and low CaO/Na_2O (0. 09 – 1. 19) (Fig. 12c), suggesting that they originated from the partial melting of a metapelite (Sylvester, 1998). On the Rb/Ba vs. Rb/Sr (Fig. 12d) diagram, their high Rb/Ba (6. 19 – 23. 5) and Rb/Sr (17. 0 – 30. 5) ratios also display obvious affinity to melts derived from meta-pelitic source rocks (Janoušek et al.,

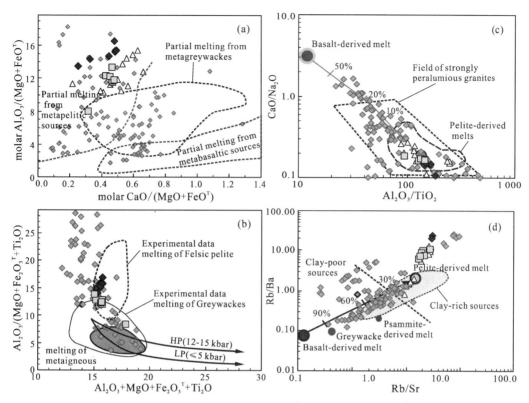

Fig. 12　(a) molar $Al_2O_3/(MgO+FeO^T)$ vs. molar $CaO/(MgO+FeO^T)$ diagram (Altherr et al., 2000), (b) $(Al_2O_3+Fe_2O_3^T+MgO+Ti_2O)$ vs. $Al_2O_3/(Fe_2O_3^T+MgO+Ti_2O)$ diagram (Patiño Douce, 1999), (c) CaO/Na_2O vs. Al_2O_3/TiO_2 diagram (Sylvester, 1998) and (d) Rb/Ba vs. Rb/Sr diagram (Janoušek et al., 2004) for the Cretaceous granites in the western Baoshan Block, SW China. Symbols as the Fig. 5.

2004). ④ The Cretaceous granites display flat HREE and Y contents and low $(Gd/Yb)_N$ ratios (1.43-2.13), with nearly flat HREE patterns, corresponding to fractionation of phosphate minerals such as monazite and xenotime. These features are similar to those of fractionated S-types described by Champion and Bultitude (2013). ⑤ The significant negative $\varepsilon_{Nd}(t)$ values (-13.6 to -11.7) together with the old two-stage model ages range from 1.57 Ga to 1.70 Ga suggesting enriched and evolved ancient crustal source, similar to that of the coeval peraluminous granites in the northern magmatic belt along the Bangong-Nujiang suture zone (Fig. 7). On the initial $^{206}Pb/^{204}Pb$ vs. $^{207}Pb/^{204}Pb$ and $^{206}Pb/^{204}Pb^{204}$ vs. $^{208}Pb/^{204}Pb$ diagrams (Fig. 8), the rocks plot in the field of lower- to upper- crust, similar to the strongly peraluminous granites in the central segment of Bangong-Nujiang suture zone and fractionated peraluminous granites in the eastern Tengchong Block. ⑥ The zircons in the biotite- and two-mica- granite display negative $\varepsilon_{Hf}(t)$ in the range of -5.47 to -0.77 and -7.67 to -0.81 (Fig. 9), indicating the involvement of significant amount of material derived from an enriched and evolved continental source. Champion and Bultitude (2013) suggested that the more than $> 4\varepsilon$ units of isotopic variation could reflect isotopic heterogeneity of a metasedimentary protolith. Villaros et al. (2010) suggested that isotopic variations of zircon $\varepsilon_{Hf}(t)$ values in S-type granites reflect inheritance from a heterogeneous source, with mixing of a range of crustal materials. Therefore, the variation of the zircon Hf isotopic data in this study may also imply the heterogeneity of the source materials, including metapelitic and metagraywacke components. The corresponding model ages of 1. 23 – 1. 52 Ga for Early Cretaceous biotite granites and 1. 20–1. 63 Ga for the Late-Cretaceous two-mica granites suggest magma derivation from Mesoproterozoic evolved crust materials, such as the Gongyanghe Group as the basement of Baoshan Block (YNBGMR, 1991). The model ages of 1. 54–2. 01 Ga from the two inherited grains in the two-mica granites which display negative zircon $\varepsilon_{Hf}(t)$ values (– 9. 10 and – 1. 37) suggest contributions from evolved Paleo- to Mesoproterozoic sources (Table 5). The inherited ages are consistent with Paleozoic granites in the region including Mengmao and Pingda, and they share similar zircon Hf compositions (Fig. 9). In addition, most of these Paleozoic granites in the region were derived from ancient and evolved metasedimentary sources, possibly involving metapelitic components. These features are consistent with those of the other Cretaceous granites in the Baoshan and Tengchong Block (Fig. 9) (Dong et al., 2013a; Xu et al., 2012; Yu et al., 2014; Zhu et al., 2015) and typical fractionated S-type around the world (Breiter, 2012; Champion and Bultitude, 2013; Huang and Jiang, 2014). They also compare with the SP granites in the Himalayas (Sylvester, 1998).

From their study of fractionated S-type granites in the Lachlan orogen, Champion and Bultitude (2013) noted that it is difficult to ascertain restite-unmixing, fractionation or mixing of various components from the geochemical trends (see also Clemens, 2003). In this study, however, several lines of evidence provide some clues as follows: ① Absence of both cumulate and restitic materials, such as metasedimentary enclaves, the rare occurrence of mafic micro-granular enclaves in the granites. ② Nearly consistent negative $\varepsilon_{Nd}(t)$ values (– 13. 6 to – 11. 7) (variation < 4 units) which exclude the possibility of involvement of multiple components, and their notable similarity with the evolved host metasediments of the Hodgkinson province (Champion and Bultitude, 2013), Northern magmatic and Gaoligong magmatic belts (Chen et al., 2015; Qu et al., 2012; Zhu et al., 2015; Zhu et al., 2009) where the magmas were derived from an ancient and evolved continental crustal sources. ③ The low total $Fe_2O_3^T + MgO + TiO_2$ contents (1. 07–1. 45 wt%) which are consistent with those from experimental studies of S-type granite melts where felsic compositions show low total Fe + Mg or Fe + Mg + Ti contents (less than 3%–4%) at crustal pressures and temperatures (Champion and Bultitude, 2013 reference therein). And ④ unlike the case where Champion and Bultitude (2013) interpreted that the Fe + Mg contents < 2. 5 of NQ S-type granites were controlled by crystal fractionation processes, the differentiation (or Fe + Mg) is generally higher, including both negative (markedly increasing Rb/Sr, Na_2O, ASI) and positive (decreasing Ba, Sr) trends with decreasing Fe + Mg (Fig. 11). Also, the Cretaceous granites share a similar geochemical trend with NQ S-type granites, with only Sr in the Late-Cretaceous granites showing various geochemical trends with Fe + Mg, although within a narrow range (Fig. 11). In addition, oligoclase crystallized at the same time as apatite in the S-type magmas of this study (Fig. 10d), similar with high fractionated S-type granites from the Jiangnan Orogen (south China) and Erzgebirge (Central Europe) (Breiter, 2012; Huang and Jiang, 2014). Therefore, we do not exclude the role of fractionation crystallization (Champion and Bultitude, 2013).

Based on the above discussion, we propose that both the strongly peraluminous Early-Cretaceous biotite granites and Late-Cretaceous two-mica granites are fractionated S-type granites, which were formed by the anatectic melting of heterogeneous crustal materials with pelite-dominated sources. They are similar to other typical fractionated S-type granites from the Jiangnan Orogen (SE China), Erzgebirge (Central Europe), Eastern Australia and the low-temperature SP granites in the Himalayas.

5. 2 Geodynamic implications

5. 2. 1 Two-stage melting processes of fertile continental materials

Strongly peraluminous fractionated granites are generally regarded as important indicators of the compositional maturity of the continental crust (Wu et al., 2017). Examples include the fractionated S-type granites from South China (Huang and Jiang, 2014), eastern Australia (Champion and Bultitude, 2013; Chappell, 1999; Hensel et al., 1985; Jeon et al., 2014), the Variscan orogenic belt and Erzgebirge in Europe (Breiter, 2012) and SW China (this study). In this study, the Cretaceous (ca. 120 Ma and 80 Ma) strongly peraluminous fractionated S-type granites in the Baoshan Block were produced by the anatectic melting of metasedimentary, a pelite-dominated source in an evolved continental crust. A two-stage melting process of mature continental crust is envisaged (Table 1 and Fig. 13): In the Early-Cretaceous, the peraluminous S-type granites with negative zircon $\varepsilon_{Hf}(t)$ values (-12 to -2) in the Gaoligong belt represent the products of melting of ancient metapelitic source with very little mantle contribution (Xu et al., 2012; Zhao et al., 2016). The peraluminous granites in the northern Tengchong Block also display evolved isotopic compositions and represent melts derived from dominantly pelitic sedimentary protoliths and/or other ancient continent crustal materials (Cao et al., 2014; Zhu et al., 2015). The strongly peraluminous fractionated S-type granites in the western Baoshan Block in this study indicate similar fertile continental source. In the case of Late-Cretaceous granites, Dong et al. (2013a) and Yu et al. (2014) who

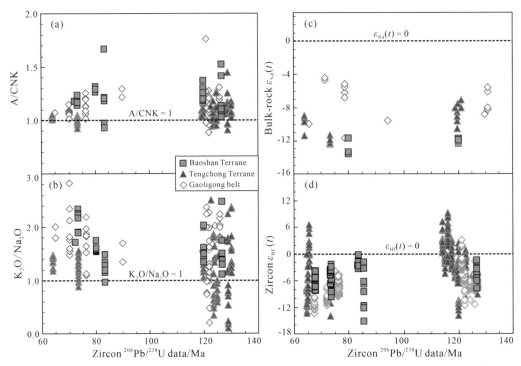

Fig. 13　(a) A/CNK, K_2O/Na_2O, Bulk-rock $\varepsilon_{Nd}(t)$, Zircon $\varepsilon_{Hf}(t)$ vs. Zircon $^{206}Pb/^{238}U$ data
for the Cretaceous granites, western Yunnan, SW China.
The data for Baoshan-Tengchong Terranes and Gaoligong belt was referenced in Table 1.

investigated the 83–73 Ma granites in the southern part of Xiangda pluton (Fig. 1b) and in the northern Baoshan Block noted strongly peraluminous S-type nature, with magma derivation through anatexis of Mesoproterozoic pelitic crustal components. The ca. 89–68 Ma granites in the southwestern part of the Gaoligong belt display a strongly peraluminous affinity, indicating a dominant Mesoproterozoic metapelitic source (Xu et al., 2012; Zhao et al., 2016). The ca. 77–65 Ma granites in the central Tengchong Block are also peraluminous to strongly peraluminous with enriched isotopic compositions, which were derived from melting of ancient mafic and metasedimentary basement rocks of the Tengchong Block (Xu et al., 2012; Chen et al., 2015; Qi et al., 2015; Xie et al., 2016; Zhao et al., 2017a), with limited input of mantle-derived materials and/or juvenile crustal source after ca. 65 Ma (Xie et al., 2016). In all these cases, fertile continental materials rather than depleted rocks from mantle depth are indicated. In addition, as the basement of Baoshan and Tengchong Block, both the Gaoligong and Gongyanghe Group carry abundant metasedimentary rocks including metapelitic to greywacke (YNBGMR, 1991). Crustal melting is a viable scenario to produce peraluminous melts (Bea, 2012). Thus, the strongly peraluminous fractionated S-type granites from ca. 120 Ma to 80 Ma in the western Baoshan Block of this study represent melting of fertile continental materials. Together with data from literature, the former stage ranged from ca. 130 Ma to 115 Ma, and the latter one occurred at ca. 89−65 Ma. Examples of such multi-stage S-type granitoid magmatism are also seen in the southern New England orogen (Hensel et al., 1985; Jeon et al., 2014). In northern Queensland (Champion and Bultitude, 2013), the Carboniferous felsic I-type granites (ca. 300 Ma) were derived from varying degrees of partial melting of andesitic to dacitic protoliths. The Permian S-types to the east of Hodgkinson province (ca. 280−260 Ma) were derived largely from either immature sediments at depth with minor some juvenile input.

The reason for episodic granitic magmatism, particularly the magmatic quiescence at ca. 110−90 Ma is an important aspect to be considered. The magmatic quiescence (ca. 80−70 Ma) in the Lhasa Block has been correlate to low rate of plate convergence (Shellnutt et al., 2014) and a flat subduction angle, which hampered conductive heating from the mantle to the overriding crust (Wen et al., 2008; Chen et al., 2015 reference therein). However, significant ca. 80−70 Ma magmatism is recorded in the Tengchong-Baoshan Block and the magmatic quiescence in the western Yunnan ranges from ca. 110−90 Ma, suggesting that the convergence rate and geometry of the Neo-Tethyan subduction after termination and closure of the Bangong-Nujiang Tethys were different in Tibet and in western Yunnan (Chen et al., 2015; Xu et al., 2012). In addition, Zhao et al. (2017a) suggested that the Tengchong Block is characterized by thin crust, whereas the eastern Himalayan syntaxis and northernmost part of southern Lhasa Block display normal crustal thickness. The southernmost part of the southern Lhasa Block was characterized by thickened crust during Late-Cretaceous, which also confirms the different geometry between the Tibet and its southeastern extension (western Yunnan). In contrast, ca. 110−90 Ma magmatic rocks are abundant in Tibet, with adakitic and/high Sr/Y rocks distributed in the central to northern Lhasa terrane and the southern margin of Qiangtang Blocks, the genesis of which are correlated to delamination of thickened lithosphere (Wang et al., 2014). The coeval Gangdese magmatic rocks in the southern Lhasa terrane are dominantly characterized by high and positive whole-rock $\varepsilon_{Nd}(t)$ (up to +5.5) and zircon $\varepsilon_{Hf}(t)$ (up to +16.5) values, and are regarded as the products of northward subduction of Neo-Tethyan lithosphere beneath the southern Tibet (Wen et al., 2008; Zhu et al., 2011). In the Lhasa Block of Tibet and its southeastern extension in western Yunnan, the Cretaceous magmatism was controlled by the subduction of Bangong-Nujiang Tethys and Neo-Tethys, and/or delamination of thickened lithosphere. However, the Tengchong-Baoshan Blocks show marked difference in their geometry and crustal thickness as compared to the Tibetan region, which might explain the magmatic quiescence at ca. 110−90 Ma in western Yunnan as against the ca. 80−70 Ma quiescence in Tibet.

5. 2. 2　Tectonic implications

Ever since S-type granites were first recognized as strongly peraluminous magmas derived from metasedimentary rocks, their tectonic setting has been debated (Collins and Richards, 2008, reference therein). Barbarin (1999) suggested that a metasedimentary source in the deep crust as a key factor in the genesis of S-type granites, and the formation was correlated to continent-continent collision and post-collision, such as in the case of the Himalayan-type strongly peraluminous granites formed during the syn- and post-collisional stage in a thickened crust (> 50 km) (Sylvester, 1998). Collins and Richards (2008) proposed a different model for the origin of S-type granites in the circum-Pacific orogens, where arc magmatism following thickening of a preexisting, sediment-dominated back-arc basin is envisaged. Hot basaltic magmas intrude into the thickened backarc crust following slab retreat causing extensive melting. The collision model involves heating through crustal thickening, whereas the circum-Pacific model invokes mantle heat input. Crustal melting to produce granite magmas requires three main mechanisms of heating that can extensively melt a fertile crust (Bea, 2012, reference therein): accumulation of radiogenic heat after crustal thickening; increased heat flux from the mantle, mostly due to asthenospheric upwelling or mantle wedge convection; and advection by hot mantle magmas emplaced into crust. In our case (Fig. 14), the coeval rocks from the first-stage contain more hornblende, biotite and plagioclase and less muscovite with variable A/CNK values (0. 88 - 1. 51), K_2O/Na_2O ratios (0. 11 - 2. 51) and display highly variable zircon $\varepsilon_{Hf}(t)$ values (- 17. 5 to + 8. 90), with negative whole-rock $\varepsilon_{Nd}(t)$ values (- 13. 6 to + 3. 00). However, rocks from the later stage contain more K-feldspar and muscovite and rare hornblende. Most of these rocks are peraluminous (A/CNK > 1. 0), K-rich (K_2O/Na_2O ratios > 1), and show negative whole-rock $\varepsilon_{Nd}(t)$ values (< -4. 0) and zircon $\varepsilon_{Hf}(t)$ values (-13. 9 to + 6. 50; only six spots are positive). Several Early-Cretaceous diorites, tonalites, and granodiorites with associated mafic enclaves that show Na-rich, metaluminous and positive zircon $\varepsilon_{Hf}(t)$ values (Fig. 14) were reported in recent studies from this region (Cong et al., 2011; Xie et al., 2016; Zhu et al., 2017a) suggest that mantle contribution might have also played an important role in the genesis of the Early-Cretaceous intrusions (Cong et al., 2011; Xie et al., 2016; Zhu et al., 2017a). However, the geochemical features and negative isotopic components in the Late-Cretaceous magmatic suites (Fig. 14), suggesting that evolved crustal sources were dominant (Chen et al., 2015; Xu et al., 2012; Yu et al., 2014; Zhao et al., 2016, 2017a). Magma mixing and interaction between mantle and crust is considered to have played a significant role in the genesis of widespread magmatism during the Early-Cretaceous (ca. 130-110 Ma) (Cong et al., 2011; Gao et al., 2014; Xie et al., 2016; Zhu et al., 2017a,b). However, the felsic compositions and enriched isotopic components of the Late-Cretaceous suite in this region suggest that the melting of evolved continental crustal materials

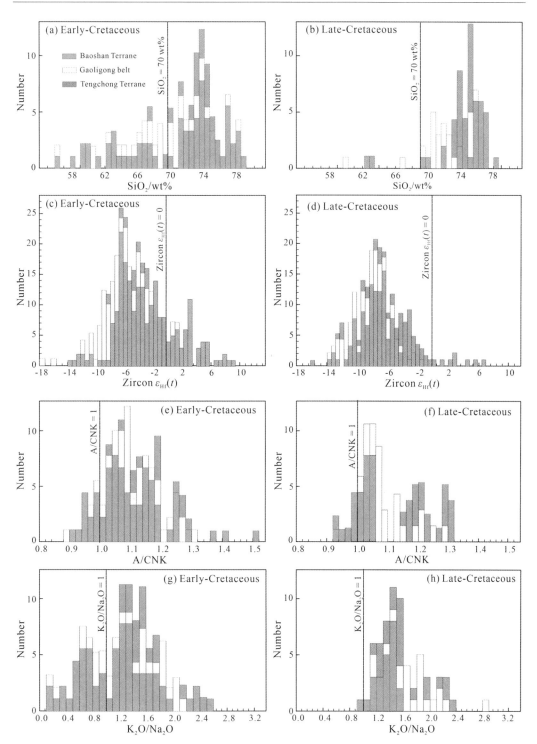

Fig. 14　Number vs. SiO_2(a,b), Zircon $\varepsilon_{Hf}(t)$ (c,d), A/CNK (e,f), and K_2O/Na_2O (g,h)

for the Cretaceous magmatisms in the Tengchong-Baoshan Terranes and Gaoligong belt.

These geochemical data for Baoshan-Tengchong Terranes and Gaoligong belt was referenced in Table 1.

mainly included metapelitic metasedimentary rocks.

For the Early Cretaceous events, the model of Neo-Tethyan slab subduction has been excluded based on several lines of evidence in previous studies (YNBGMR, 1991; Zhu et al., 2015, 2017a,b; Xie et al., 2016). The models relating the granitoids with Bangong-Nujiang Meso-Tethys subduction and/or the collision between the Lhasa-Tengchong and Qiangtang-Baoshan Blocks remains debated (Cao et al., 2014; Qi et al., 2011; Xu et al., 2012; Zhu et al., 2015, 2017a,b). A recent study proposed divergent double subduction (Zhu et al., 2016). Absence of Early Cretaceous high grade metamorphic rocks in the Lhasa-Qiangtang collision zone indicates that the Bangong-Nujiang ocean closed via divergent double-sided subduction resulting in soft collision of opposing arcs rather than continent-continent "hard" collision (Zhu et al., 2016). The abundant late Jurassic to Early Cretaceous (ca. 140 - 110 Ma) magmatic rocks in the Lhasa-Tengchong Block have been interpreted to be the products of the southward subduction of the Bangong-Nujiang Tethyan ocean (D.C. Zhu et al., 2009, 2011, 2016; Zhu et al., 2015, 2017a,b). The Middle Triassic to Early Cretaceous (ca. 240-105 Ma) magmatic rocks in the southern margin of the western Qiangtang and Baoshan Block were considered as products of the northward subduction of the Bangong-Nujiang Tethyan Ocean lithosphere (Chen et al., 2016; Li et al., 2014; Shi et al., 2008). According to available stratigraphic, structural, metamorphic and magmatic features of the region, Zhu et al. (2016) proposed subduction (> 140 Ma), which continued into divergent double subduction ultimately leading to the closure of the Bangong-Nujiang ocean (140 - 130 Ma). They also proposed the rupturing of "soft" arc-arc collision zone, with roll-back and detachment through gravitational instability (130 - 120 Ma), and the rupture propagated from south to north and/or resulted in slab break-off (120 - 110 Ma). This process has been confirmed by evidence from ca. 130 - 110 Ma magmatism along the southeastern extension zone (Tengchong-Gaoligong) (Zhu et al., 2015, 2017a,b). The strongly peraluminous S-type granites in the western Baoshan Block may represent the anatectic melting products of the northward subduction of Bangong-Nujiang Tethys, and can be compared with the peraluminous to strongly peraluminous felsic intrusions in the central Tibet (Eby, 1992; Li et al., 2014; Qu et al., 2012).

The Late Cretaceous events were interpreted as the products of melting of thickened crust in the hinterland, which might have been influenced by the subduction of Neo-Tethys after the final termination of the Bangong-Nujiang Tethys (Xu et al., 2012; Yu et al., 2014). However, Zhao et al. (2017b) suggested that there was no thickened crust during the Late Cretaceous based on crustal thickness calculations using Sr/Y ratios. Also, as mentioned above, different geometry between has been proposed for the Tibet and its southeastern extension (western Yunnan) during Late-Cretaceous (ca. 76-73 Ma) (Chen et al., 2015; Xu et al., 2012). Xu et al. (2012) compared the Late-Cretaceous granites (ca. 76-66 Ma) in the western Yunnan with the inland peraluminous granites in the North American Cordillera (Driver et al., 2000), invoking a similar model of subduction of Neo-Tethys to explain the origin of the extensive S-type granites in the region. Alternately, the subduction of Neo-Tethyan Oceanic plate beneath the Asian continent offers a long-term regional mechanism for late Mesozoic-early Cenozoic crustal deformation and magmatism in a Cordilleran-type setting (Xu et al., 2012). For example, the collision between Lhasa-Tengchong and Qiangtang-Baoshan, Burma and Sundaland Blocks (Xu et al., 2012) is supported by the high-temperature, low-pressure metamorphic event at ca. 84 Ma (Dunning et al., 1995). However, influence from the far-field subduction of Neo-Tethys cannot be ignored, which can explain the widely distributed S-type granitoids in the hinterland from Tengchong, Lianghe, Gaoligong to Baoshan (Xu et al., 2012; Dong et al., 2013; Yu et al., 2014; Chen et al., 2015; Zhao et al., 2016; and this study).

In summary, the model of divergent double subduction of Bangong-Nujiang Tethys would explain the various magmatic pulses in the Early-Cretaceous (Fig. 15a), and the subduction of Neo-Tethyan oceanic plate would account for the production of S-type magmatic rocks in the hinterland from Tengchong to Baoshan during Late-Cretaceous (Fig. 15b).

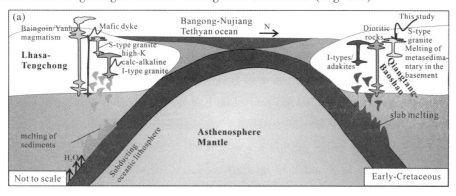

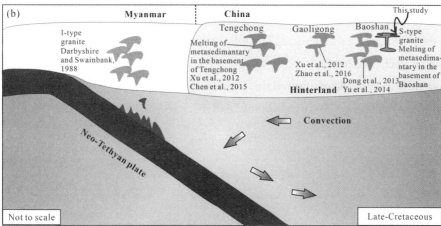

Fig. 15   Tectonic model for the Early- to Late-Cretaceous magmatism was distributed in the southeastern Tibet-Himalayan orogens.

6 Conclusions

The strongly peraluminous fractionated S-type granites in the western Baoshan Block comprises an Early Cretaceous biotite granite (121. 8 ± 1. 1 Ma) and a Late Cretaceous two-mica granite (81. 9−79. 7 Ma).

Both the Early-Cretaceous biotite- and Late-Cretaceous two-mica- granites show high silica, high-K calc-alkaline and strongly peraluminous nature, and normative-CIPW corundum contents of 2. 84−4. 42 wt%. They also display low total $Fe_2O_3^T + MgO + TiO_2$ contents (< 3. 0 wt%) and relative high P_2O_5 content (0. 22−0. 28 wt%) consistent with the features of S-type granite melts. In addition, the rocks display high differentiation index (DI) of 93. 1−96. 8. Thus, we suggest that these Cretaceous granites are fractionated S-type granites. The high ratios of Al_2O_3/TiO_2(93. 8−182. 9), Rb/Ba (6. 19−23. 5), and Rb/Sr (17. 0−30. 5), low ratios of CaO/Na_2O (0. 09 − 1. 19), and evolved whole-rock Sr-Nd-Pb and zircon Hf isotopic compositions indicate that the granites were derived from anatectic melting of a pelite-dominated metasedimentary source, similar to typical fractionated S-type granites,

In conjunction with regional geology and tectono-magmatic history of the Tengchong-Banshan Terranes and Gaoligong belt, we conclude that the strongly peraluminous magmatic suite of Early to Late Cretaceous represent two-stage melting of fertile continental crust following the final closure and termination of Bangong-Nujiang Tethys.

Acknowledgements We thank Prof.Nelson Eby for his valuable editorial comments and Prof. David Champion and Dr. Mike Fowler for constructive reviews which significantly improved the focus and quality of this manuscript. This work was jointly supported by the National Natural Science Foundation of China (Grant Nos. 41421002, 41372067, and 41190072) and program for Changjiang Scholars and Innovative Research Team in University (Grant IRT1281). Support was also provided by the MOST Special Fund from the State Key Laboratory of Continental Dynamics, Northwest University and Province Key Laboratory Construction Item (08JZ62).

Appendix A. Supplementary data Supplementary data to this article can be found online at https://doi. org/10. 1016/j.lithos.2018. 07. 016.

References

Altherr, R., Holl, A., Hegner, E., Langer, C., Kreuzer, H., 2000. High-potassium, calc-alkaline I-type plutonism in the European Variscides: Northern Vosges (France) and northern Schwarzwald (Germany). Lithos, 50, 51-73.

Andersen, T., 2002. Correction of common lead in U-Pb analyses that do not report [204]Pb. Chemical Geology, 192(1-2), 59-79.

Bai, X.Z., Jia, X.C., Yang, X.J., Xiong, C.L., Liang, B., Huang, B.X., Luo, G., 2002. LA-ICP-MS zircon U-Pb dating and geochemical characteristics of Early Cretaceous volcanic rocks in Longli-Ruili fault belt, western Yunnan province. Geological Bulletin of China, 31 (2/3), 297-305.

Bao, Z., Chen, L., Zong, C., Yuan, H., Chen, K., Dai, M., 2017. Development of pressed sulfide powder tablets for in situ, Sulfur and Lead isotope measurement using LA-MC-ICP-MS. International Journal of Mass Spectrometry, 421.

Barbarin, B., 1999. A review of the relationship between granitoid types, their origins and their geodynamic environments. Lithos, 46, 605-626.

Bea, F., 2012. The sources of energy for crustal melting and the geochemistry of heat-producing elements. Lithos, 153 (8), 278-291.

Breiter, K., 2012. Nearly contemporaneous evolution of the A- and S-type fractionated granites in the Krušné Hory/Erzgebirge mts. central Europe. Lithos, 151, 105-121.

Cao, H.W., Zhang, S.T., Lin, J.Z., Zheng, L., Wu, J.D., Li, D., 2014. Geology, geochemistry and geochronology of the Jiaojiguanliangzi Fe-polymetallic Deposit, Tengchong County, western Yunnan (China): Regional tectonic implications. Journal of Asian Earth Sciences, 81, 142-152.

Champion, D.C., Bultitude, R.J., 2013. The geochemical and Sr, Nd isotopic characteristics of Paleozoic fractionated S-types granites of north Queensland: Implications for S-type granite petrogenesis. Lithos, 162-163 (2), 37-56.

Chappell, B.W., 1999. Aluminium saturation in I- and S-type granites and the characterization of fractionated haplogranites. Lithos, 46 (3), 535-551.

Chappell, B.W., White, A.J.R., 1992. I- and S-type granites in the Lachlan Fold Belt. Geological Society of America Special Paper, 272, 1-26.

Chen, Y., Lu., Y., Zhao, H., Cheng, Z., Jiang, C., Liu, Z., 2013. Zircon Shrimp U-Pb geochronology, geochemistry of the Xiaochang monzonitic granite with Mo mineralization and implications for tectonic setting in Tengchong Block, western Yunnan Terrain, southwestern China. Earth Science Frontiers, 20 (5), 1-14.

Chen, X.C., Hu, R.Z., Bi, X.W., Zhong, H., Lan, J.B., Zhao, C.H., et al., 2015. Petrogenesis of metaluminous A-type granitoids in the Tengchong-Lianghe tin belt of southwestern China: Evidences from zircon U-Pb ages and Hf-O isotopes, and whole-rock Sr-Nd isotopes. Lithos, 212 (215), 93-110.

Chen, S.S., Shi, R.D., Fan, W.M., Zou, H.B., Liu, D.L., Huang, Q.S., et al., 2016. Middle Triassic ultrapotassic rhyolites from the Tanggula pass, southern Qiangtang, China: A previously unrecognized stage of silicic magmatism. Lithos, 264, 258-276.

Chu, Z.Y., Chen, F.K., Yang, Y.H., Guo, J.H., 2009. Precise determination of Sm, Nd concentrations and Nd isotopic compositions at the nanogram level in geological samples by thermal ionization mass spectrometry. Journal Atom of Analytical Atomic Spectrometry, 24, 1534-1544.

Clemens, J. D., 2003. S-type granitic magmas-petrogenetic issues, models and evidence. Earth-Science Reviews, 61, 1-18.

Collins, W.J., Richards, S. W., 2008. Geodynamic significance of S-type granites in circumpacific orogens. Geology, 36 (7), 559-562.

Cong, F., Lin, S.L., Zou, G.F., Li, Z.H., Xie, T., Peng, Z.M., Liang, T., 2011. Magma mixing of granites at Lianghe: In-situ zircon analysis for trace elements, U-Pb ages and Hf isotopes. Science China: Earth Science, 54, 1346-1359.

Deng, J., Wang, Q., Li, G., Li, C., Wang, C., 2014. Tethys tectonic evolution and its bearing on the distribution of important mineral deposits in the Sanjiang region, SW China. Gondwana Ressearch, 26 (2), 419-437.

Dobosi, G., Kempton, P.D., Downes, H., Embey-Isztin, A., Thirlwall, M., Greenwood, P., 2003. Lower crustal granulite xenoliths from the Pannonian Basin, Hungary, part 2: Sr-Nd-Pb-Hf and O isotope evidence for formation of continental lower crust by tectonic emplacement of oceanic crust. Contributions to Mineralogy and Petrology, 144 (6), 671-683.

Dong, M.L., Dong, G.C., Mo, X.X., Zhu, D.C., Nie, F., Yu, J.C., Wang, P., Luo, W., 2013a. The Mesozoic-Cenozoic magmatism in Baoshan block, western Yunnan and its tectonic significance. Acta Petrologica Sinca, 29 (11), 3901-3913.

Dong, M., Dong, G., Mo, X., Santosh, M., Zhu, D., Yu, J., et al., 2013b. Geochemistry, zircon U-Pb geochronology and Hf isotopes of granites in the Baoshan block, western Yunnan: Implications for early Paleozoic evolution along the Gondwana margin. Lithos, 179 (5), 36-47.

Driver, L.A., Creaser, R. A., Chacko, T., Erdmer, P., 2000. Petrogenesis of the Cretaceous Cassiar batholith, Yukon-British Columbia, Canada: Implications for magmatism in the North American Cordilleran Interior. Geological Society of America Bulletin, 112, 1119-1133.

Dunning, G.R., MacDonald, A.S., Barr, S.M., 1995. Zircon and monazite U-Pb dating of the Doi Inthanon core complex, northern Thailand: Implications for extension within the Indosinian Orogen. Tectonophys, 251, 197-213.

Eby, G.N., 1992. Chemical subdivision of the A-type granitoids: Petrogenetic and tectonic implications. Geology, 20, 641-644.

Feng, Q.L., 2002. Stratigraphy of volcanic rocks in the Changning-Menglian belt in southwestern Yunnan, China. Journal of Asian Earth Science, 20, 657-664.

Frost, B.R., Barnes, C.G., Collins, W.J., Arculus, R.J., Ellis, D.J., Frost, C.D., 2001. A geochemical classification for granitic rocks. Journal of Petrology, 42 (11), 2033-2048.

Gao, Y.J., Lin, S.J., Cong, F., Zou, G.F., Xie, T., Tang, F.W., Li, Z.H., Liang, T., 2012. LA-ICP-MS zircon U-Pb dating and geological implications for the Early Cretaceous volcanic rocks on the southeastern margin of the Tengchong Block, western Yunnan. Sediment. Geology and Tethyan Geology, 32 (4), 59-64.

Gao, Y.J., Lin, S.L., Cong, F., Zou, G.F., Chen, L., Xie, T., 2014. LA-ICP-MS zircon U-Pb ages and Hf isotope compositions of zircons from lower Cretaceous diorite-dykes in Lianghe area, western Yunnan, and their geological implications. Geological Bulletin of China, 33 (10), 1482-1491.

Griffin, W.L., Pearson, N.J., Belousova, E., Jackson, S.E., van Achterbergh, E., O'Reilly, S.Y., Shee, S. R., 2000. The Hf isotope composition of cratonic mantle: LAM-MC-ICPMS analysis of zircon megacrysts in kimberlites. Geochimica Cosmochimic Acta, 64, 133-147.

Hensel, H.D., McCulloch, M.T., Chappell, B.W., 1985. The New England Batholith: Constraints on its derivation from Nd and Sr isotopic studies of granitoids and country rocks. Geochimica Cosmochimica Acta, 49, 369-384.

Huang, L.C., Jiang, S.Y., 2014. Highly fractionated S-type granites from the giant Dahutang Tungsten deposit in Jiangnan orogen, southeast China: Geochronology, petrogenesis and their relationship with W-mineralization. Lithos, 202-203 (4), 207-226.

Iizuka, T., Hirata, T., 2005. Improvements of precision and accuracy in situ Hf isotope microanalysis of zircon using the Laser ablation-MV-ICPMS technique. Chemical Geology, 220 (1), 121-137.

Jahn, B.M., Wu, F.Y., Capdevila, R., Martineau, F., Wang, Y.X., Zhao, Z.H., 2001. Highly evolved juvenile granites with tetrad REE patterns: The Woduhe and Baerzhe granites from the Great Xing'an (Khingan) Mountains in NE China. Lithos, 59, 171-198.

Janoušek, V., Finger, F., Roberts, M., Frýda, J., Pin, C., Dolejš, D., 2004. Deciphering the petrogenesis of deeply buried granites: Whole-rock geochemical constraints on the origin of largely undepleted felsic granulites from the Moldanubian Zone of the Bohemian Massif. Geological Society of America Special Paper, 389, 141-159.

Jeon, H., Williams, I.S., Bennett, V.C., 2014. Uncoupled O and Hf isotopic systems in zircon from the contrasting granite suites of the New England orogen, eastern Australia: Implications for studies of Phanerozoic magma genesis. Geochimica Cosmochimica Acta, 146, 132-149.

Kempton, P.D., McGill, R., 2002. Procedures for the analysis of common lead at the NERC Isotope Geosciences Laboratory and an assessment of data quality. NIGL Rep. Ser., 178.

Li, Z.H., Lin, S.L., Cong, F., Zou, G.F., Xie, T., 2012a. U-Pb dating and Hf isotopic compositions of quartz diorite and monzonitic granite from the Tengchong-Lianghe block, western Yunnan, and its geological implications. Acta Geological Sinca, 86 (7), 1047-1062 (in Chinese with English abstract).

Li, Z.H., Lin, S.L., Cong, F., Zou, G.F., Xie, T., 2012b. Early Cretaceous magmatism in Tengchong-Lianghe block, western Yunnan. Bulletin of Mineralogy Petrology and Geochemistry, 31, 590-598 (in Chinese with English abstract).

Li, J.X., Qin, K.Z., Li, G.M., Richards, J.P., Zhao, J.X., Cao, M.J., 2014. Geochronology, geochemistry, and zircon Hf isotopic compositions of Mesozoic intermediate-felsic intrusions in central Tibet: Petrogenetic and tectonic implications. Lithos, 198-199, 77-91.

Li, G.J., Wang, Q.F., Huang, Y.H., Gao, L., Yu, L., 2016. Petrogenesis of middle Ordovician peraluminous granites in the Baoshan Bock: Implications for the early Paleozoic tectonic evolution along east Gondwana. Lithos, 245, 76-92.

Liao, S.Y., Yin, H.F., Sun, Z.M., Wang, B.D., Tang, Y., Sun, J., 2013. The discovery of Late Triassic subvolcanic dacite porphyry in the eastern margin of the Baoshan terrane, western Yunnan Province, and its geodynamic implications. Geological Bulletin of China, 32 (7), 1006-1013 (in Chinese with English abstract).

Liu, S., Hu, R.Z., Gao, S., Feng, C.X., Huang, Z.L., Lai, S.C., Yuan, H.L., Liu, X.M., Coulson, I.M., Feng, G.Y., Wang, T., Qi, Y.Q., 2009. U-Pb zircon, geochemical and Sr-Nd-Hf isotopic constraints on the age and origin of Early Palaeozoic I-type granite from the Tengchong-Baoshan Block, Western Yunnan Province, SW China. Journal of Asian Earth Sciences, 36, 168-182.

Ludwig, K.R., 2003. ISOPLOT 3.0: A Geochronological Toolkit for Microsoft Excel. Special Publication, 4. Berkeley Geochronology Center.

Luo, G., Jia, X.C., Yang, X.J., Xiong, C.L., Bai, X.Z., Huang, B.X., Tan, X.L., 2012. LA-ICP-MS zircon U-Pb dating of southern Menglian granite in Tengchong area of western Yunnan Province and its tectonic implications. Geological Bulletin of China, 31, 287-296 (in Chinese with English abstract).

Metcalfe, I., 2013. Gondwana dispersion and Asian accretion: Tectonic and paleogeography evolution of eastern Tethys. Journal of Asian Earth Sciences, 66, 1-33.

Middlemost, E.A.K., 1994. Naming materials in the magma/igneous rock system. Earth-Science Reviews, 37, 215-224.

Nowell, G.M., Kempton, P.D., Noble, S.R., Fitton, J.G., Saunders, A.D., Mahoney, J.J., Taylor, R.N., 1998. High precision Hf isotope measurements of MORB and OIB by thermal ionisation mass spectrometry: Insights into the depleted mantle. Chemical Geology, 149,211-233.

Patiño Douce, A.E., 1999. What do experiments tell us about the relative contributions of crust and mantle to the origin of the granitic magmas. Geological Society of London, 168, 55-75.

Qi, L., Hu, J., Gregoire, D.C., 2000. Determination of trace elements in granites by inductively coupled plasma mass spectrometry. Talanta, 51, 507-513.

Qi, X.X., Zhu, L.H., Hu, Z.C., Li, Z.Q., 2011. Zircon SHRIMP U-Pb dating and Lu-Hf isotopic composition for Early Cretaceous plutonic rocks in Tengchong block, southeastern Tibet, and its tectonic implications. Acta Petrologica Sinca, 27, 3409-3421 (in Chinese with English abstract).

Qu, X.M., Wang, R.J., Xin, H.B., Jiang, J.H., Chen, H., 2012. Age and petrogenesis of A-type granites in the middle segment of the Bangonghu-Nujiang suture, Tibetan plateau. Lithos, 146, 264-275.

Qi, X., Zhu, L., Grimmer, J.C., et al., 2015. Tracing the Transhimalayan magmatic belt and the Lhasa block southward using zircon U-Pb, Lu-Hf isotopic and geochemical data: Cretaceous-Cenozoic granitoids in the Tengchong Block, Yunnan, China. Journal of Asian Earth Sciences, 110, 170-188.

Rapp, R.P., Watson, E.B., 1995. Dehydration melting of metabasalt at 8 − 32 kbar: Implications for continental growth and crust-mantle recycling. Journal of Petrology, 36, 891-931.

Roberts, M.P., Clemens, J.D., 1993. Origin of high-potassium, calc-alkaline, I-type granitoids. Geology, 21 (9), 825.

Rudnick, R.L., Gao, S., 2003. Composition of the continental crust. Treatise Geochem., 3, 1-64.

Shellnutt, J.G., Lee, T.Y., Brookfield, M.E., Chung, S.L., 2014. Correlation between magmatism of the Ladakh batholith and plate convergence rates during the India-Eurasia collision. Gondwana Research, 26 (3-4), 1051-1059.

Shi, R.D., Yang, J.S., Xu, Z.Q., Qi, X.X., 2008. The Bangong Lake ophiolite (NW Tibet) and its bearing on the tectonic evolution of the Bangong-Nujiang suture zone. Journal of Asian Earth Sciences, 32, 438-457.

Soderlund, U., Patchett, P.J., Vervoort, J.D., Isachsen, C.E., 2004. The 176Lu decay constantdetermined by Lu-Hf and U-Pb isotope systematics of Precambrian mafic intrusions. Earth & Planetary Science Letters, 219, 311-324.

Stevens, G., Villaros, A., Moyen, J.F., 2007. Selective peritectic garnet entrainment as the origin of geochemical diversity in S-type granites. Geology, 35, 9-12.

Streckeisen, A., Le Maitre, R.W., 1979. A chemical approximation to the modal QAPF classification of the igneous rocks. Neues Jahrbuch für Mineralogie, 136, 169-206.

Sun, S.S., McDonough, W.F., 1989. Chemical and isotopic systematics of oceanic basalts: Implications or mantle composition and processes. Geological Society of London, Special Publication, 42(1), 313-345.

Sylvester, P.J., 1998. Postcollisional strongly peraluminous granites. Lithos, 45, 29-44.

Tao, Y., Hu, R.Z., Zhu, F.L., Ma, Y.S., Ye, L., Cheng, Z.T., et al., 2010. Ore-forming age and the geodynamic background of the Hetaoping lead-zinc deposit in Baoshan, Yunnan. Acta Petrologica Sinica, 26 (6), 1760-1772.

Vervoort, J.D., Blichert-Toft, J., 1999. Evolution of the depletedmantle: Hf isotope evidence from juvenile rocks through time. Geochimica Cosmochimica Acta, 63 (3-4), 533-556.

Villaros, A., Stevens, G., Moyen, J.F., Buick, I.S., 2010. The trace element compositions of S-type granites: Evidence for disequilbrium melting and accessory phase entrainment in the source. Contributions to Mineralogy and Petrology, 158, 543-561.

Wang, Y.J., Xing, X.X., Cawood, P.A., Lai, S.C., Xia, X.P., Fan, W.M., Liu, H.C., Zhang, F.F., 2013. Petrogenesis of early Paleozoic peraluminous granite in the Sibumasu Block of SW Yunnan and diachronous accretionary orogenesis along the northern margin of Gondwana. Lithos, 182-183:67-85.

Wang, Q., Zhu, D.C., Zhao, Z.D., Liu, S.A., Chung, S.L., Li, S.M., 2014. Origin of the ca. 90 Ma magnesia-rich volcanic rocks in Se Nyima, central Tibet: Products of lithospheric delamination beneath the Lhasa-Qiangtang collision zone. Lithos, 198-199(3), 24-37.

Wang, C., Deng, J., Lu, Y., Bagas, L., Kemp, A.I.S., McCuaig, T.C., 2015a. Age, nature, and origin of Ordovician Zhibenshan granite from the Baoshan Terrane in the Sanjiang region and its significance for understanding Proto-Tethys evolution. International Geology Review, 57 (15), 1922-1939.

Wang, Y., Li, S., Ma, L., Fan, W., Cai, Y., Zhang, Y., Zhang, F., 2015b. Geochronological and geochemical constraints on the petrogenesis of early Eocene metagabbroic rocks in Nabang (SW Yunnan) and its implications on the Neotethyan slab subduction. Gondwana Research, 27, 1474-1486.

Wen, D.R., Chung, S.L., Song, B., Iizuka, Y., Yang, H.J., Ji, J.Q., Liu, D.Y., Gallet, S., 2008. Late Cretaceous Gangdese intrusions of adakitic geochemical characteristics, SE Tibet: Petrogenesis and tectonic implications. Lithos, 105 (1-2), 1-11.

Wopfner, H., 1996. Gondwana origin of the Baoshan and Tengchong terrenes of west Yunnan. Geological Society of London, Special Publication, 106 (1), 539-547.

Wu, F.Y., Liu, X.C., Ji, W.Q., Wang, J.M., Yang, L., 2017. Highly fractionated granites: Recognition and research. Science in China: Series D, Earth Sciences, (7), 1-19.

Xie, J.C., Zhu, D.C., Dong, G., Zhao, Z.D., Wang, Q., Mo, X., 2016. Linking the Tengchong terrane in SW Yunnan with the Lhasa terrane in southern Tibet through magmatic correlation. Gondwana Research, 39, 217-219.

Xu, Y.G., Yang, Q.J., Lan, J.B., Luo, Z.Y., Huang, X.L., Shi, Y.R., Xie, L.W., 2012. Temporal-spatial distribution and tectonic implications of the batholiths in the Gaoligong-Tengliang-Yingjiang area, western Yunnan: Constraints from zircon U-Pb ages and Hf isotopes. Journal of Asian Earth Sciences, 53, 151-175.

Yang, Q.J., Xu, Y.G., Huang, X.L., Luo, Z.Y., 2006. Geochronology and geochemistry granites in the Gaoligong tectonic belt, western Yunnan, tectonic implications. Acta Petrologica Sinica, 22, 817-834 (in

Chinese with English abstract).

YBGMR (Yunnan Bureau Geological Mineral Resource), 1991. Regional Geology of Yunnan Province. Beijing: Geological Publishing House, 1-729 (in Chinese with English abstract).

Yu, L., Li, G.J., Wang, Q.F., Liu, X.F., 2014. Petrogenesis and tectonic significance of the Late Cretaceous magmatism in the northern part of Baoshan Block: Constraints from bulk geochemistry, zircon U-Pb geochronology and Hf isotopic compositions. Acta Petrological Sinica, 30 (9), 2709-2724.

Yuan, H.L., Gao, S., Dai, M.N., Zong, C.L., Gunther, D., Fontaine, G.H., Liu, X.M., Diwu, C.R., 2008. Simultaneous determinations of U-Pb age, Hf isotopes and trace element compositions of zircon by excimer laser-ablation quadrupole and multiple-collector ICP-MS. Chemical Geology, 247, 100-118.

Zartman, R.E., Doe, B.R., 1981. Plumbotectonics-the model. Tectonophysics, 75, 135-162.

Zhao, S.W., Lai, S.C., Qin, J.F., Zhu, R.Z., 2014. Zircon U-Pb ages, geochemistry and Sr-Nd-Pb-Hf isotopic compositions of the Pinghe pluton, southwest China: Implications for the evolution of the early Paleozoic Proto-Tethys in southeast Asia. International Geology Review, 56 (7), 885-904.

Zhao, S.W., Lai, S.C., Qin, J.F., Zhu, R.Z., 2016. Tectono-magmatic evolution of the gaoligong belt, southeastern margin of the Tibetan plateau: Constraints from granitic gneisses and granitoid intrusions. Gondwana Research, 35 (1), 238-256.

Zhao, S., Lai, S., Gao, L., Qin, J., Zhu, R.Z., 2017a. Evolution of the Proto-Tethys in the Baoshan Block along the east Gondwana margin: Constraints from early Palaeozoic magmatism. International Geology Review, 1-15.

Zhao, S.W., Lai, S.C., Qin, J.F., Zhu, R.Z., Wang, J.B., 2017b. Geochemical and geochronological characteristics of late Cretaceous to early Paleocene granitoids in the Tengchong Block, southwestern China: Implications for crustal anatexis and thickness variations along the eastern Neo-Tethys subduction zone. Tectonophysics, 694, 87-100.

Zhong, D.L., 1998. Paleo-Tethyan Orogenic Belt in the Western Parts of the Sichuan and Yunnan provinces. Beijing: Science Press, 231.

Zhu, D., Mo, X., Wang, L., Zhao, Z., Niu, Y., Zhou, C., Yang, Y., 2009. Petrogenesis of highly fractionated I-type granites in the Zayu area of eastern Gangdese, Tibet: Constraints from zircon U-Pb geochronology, geochemistry and Sr-Nd-Hf isotopes. Science in China: Series D, Earth Sciences, 52 (9), 1223-1239.

Zhu, D.C., Zhao, Z.D., Niu, Y., Mo, X.X., Chung, S.L., Hou, Z.Q., 2011. The Lhasa Terrane: Record of a microcontinent and its histories of drift and growth. Earth & Planetary Science Letters, 301, 241-255.

Zhu, R.Z., Lai, S.C., Qin, J.F., Zhao, S.W., 2015. Early-cretaceous highly fractionated I-type granites from the northern Tengchong block, western Yunnan, SW China: Petrogenesis and tectonic implications. Journal of Asian Earth Sciences, 145-163.

Zhu, D.C., Li, S.M., Cawood, P.A., Wang, Q., Zhao, Z.D., Liu, S.A., et al., 2016. Assembly of the Lhasa and Qiangtang terranes in central Tibet by divergent double subduction. Lithos, 245, 7-17.

Zhu, R.Z., Lai, S.C., Santosh, M., Qin, J.F., Zhao, S.W., 2017a. Early Cretaceous Na-rich granitoids and their enclaves in the Tengchong Block, SW China: Magmatism in relation to subduction of the Bangong-Nujiang Tethys ocean. Lithos, 286 (287), 175-190.

Zhu, R.Z., Lai, S.C., Qin, J.F., Zhao, S.W., Wang, J.B., 2017b. Late Early-Cretaceous quartz diorite-granodiorite-monzogranite association from the Gaoligong belt, southeastern Tibet Plateau: Chemical variations and geodynamic implications. Lithos, 288-289, 311-325.

Zhu, R.Z., Lai, S.C., Qin, J.F., Zhao, S.W., 2018. Petrogenesis of late Paleozoic-to-early Mesozoic granitoids and metagabbroic rocks of the Tengchong Block, SW China: Implications for the evolution of the eastern Paleo-Tethys. International Journal of Earth Science, 107 (2), 431-457.

Early-Cretaceous syenites and granites in the northeastern Tengchong Block, SW China: Petrogenesis and tectonic implications[①]

Zhu Renzhi Lai Shaocong[②] Qin Jiangfeng Zhao Shaowei

Abstract: Whole-rock major and trace element and Sr-Nd isotopic data, together with zircon LA-ICP-MS in-situ U-Pb and Hf isotopic data of the syenites and granites in the Tengchong Block are reported in order to understand their petrogenesis and tectonic implications. Zircon U-Pb data gives the emplacement ages of ca. 115.3 ± 0.9 Ma for syenites and 115.7 ± 0.8 Ma for granites, respectively. The syenites are characterized by low SiO_2 content ($62.01-63.03$ wt%) and notably high Na_2O content ($7.04-7.24$ wt%) and Na_2O/K_2O ratios ($2.02-2.10$), low MgO, $Fe_2O_3^T$ and TiO_2, enrichment of LILEs(large-ion lithophile element) such as Rb, Th, U, K, and Pb and obvious depletion HFSE(high field strength element; e.g. Nb, Ta, P, and Ti) with clearly negative Eu anomalies ($\delta Eu = 0.53 - 0.56$). They also display significant negative whole-rock $\varepsilon_{Nd}(t)$ values of -6.8 and zircon $\varepsilon_{Hf}(t)$ values (-9.11 to -0.27, but one is $+5.30$) and high initial $^{87}Sr/^{86}Sr = 0.713\,013$. Based on the data obtained in this study, we suggest that the ca. 115.3Ma syenites were possibly derived from a sodium-rich continental crustal sources, and the fractionation of some ferro-magnesian mineral and plagioclase might occur during the evolution of magma. The granites have high SiO_2 content ($71.35-74.47$ wt%), metaluminous to peraluminous, low Rb/Ba, Rb/Sr, and $Al_2O_3/(MgO+FeO^T+TiO_2)$ ratios and moderate ($Al_2O_3+MgO+FeO^T+TiO_2$) content. They show low initial $^{87}Sr/^{86}Sr$ ($0.703\,408-0.704\,241$) and $\varepsilon_{Nd}(t)$ values (-3.8 to -3.5), plotted into the evolutionary trend between basalts and lower crust. Hence, we suggest that the granites were derived from the melting of mixing sources in the ancient continental crust involving some metabasaltic materials and predominated metasedimantary greywackes. Together with data in the literatures, we infer that the Early Cretaceous magmatism in the Tengchong Block was dominated by magmas generated by the partial melting of ancient crustal material, which represent the products that associated to the closure of Bangong-Nujiang Meso-Tethys.

1　Introduction

Granites are widespread in the upper crust, especially in orogenic belts, and an

---

①　Published in *Acta Geologica Sinica* (*English Edition*),2018,92(4).

②　Corresponding author.

understanding of their genesis provides insight into the evolution of their deep continental source and tectonic setting (Chappell and White, 1992; Sylvester, 1998; Brown, 2013; Hao et al., 2015; Zhang et al., 2015; Wu et al., 2015), such as subduction, collision, or post-orogenic extension. Meanwhile, syenites are also an important objects to decipher the magmatic processes within the continental lithosphere such as the sources in the lower and upper crust (Chen et al., 2017; Yang et al., 2012), which were usually been proposed as the products of: ① partial melting of crustal rocks (Huang and Wyllie, 1975; Lubala et al., 1994); ② differentiation of mantle-derived mafic rocks (Litvinovsky et al., 2002; Yang et al., 2005); and ③ differentiation of the hybrid liquids (mixing between basic and silicic melts or mantle-derived silica-undersaturated alkaline magmas with lower crustal-derived granitic magmas) (Sheppard, 1995; Dorais, 1990; Riishuus et al., 2005; Yang et al., 2008). In contrast to granites, it is commonly developed in the extensive setting including post-orogenic, rift, or intraplate tectonic setting (Sylvester, 1989; Bonin et al., 1998; Yang et al., 2005; Yang et al., 2012 reference therein). Therefore, the petrogenesis of both syenites and granites are crucial to understand the magmatic process and deep sources in the continental crust.

In our recent field investigations, some Early-Cretaceous syenites and granites are found out in the Qushijie and Lushui area in the northeastern Tengchong Block, we all know that the abundant Early-Cretaceous magmatism in the Tengchong Block was related to the subducted and closure process of the Bangong-Nujiang Meso-Tethyan ocean (Xu et al., 2012; Zhu et al., 2015b; Zhu et al., 2017a,b). Hence, we present the zircon LA ICP-MS U-Pb data, whole-rock chemical and Sr-Nd isotopic data, and zircon in-situ Hf isotopic data to constraint the petrogenesis of the syenites and granites, which aim to understand the continent crustal magmatic processes and deep sources during the subducted and closure process of Bangong-Nujiang Tethyan ocean.

2 Geological setting and petrography

The Tengchong Block represents the southeastern extension of the Lhasa Block (Xie et al., 2016; Fig. 1a) and is separated from the eastern Baoshan Block by the Nujiang-Luxi-Ruili fault (NLRF) and from the western Burma Block by the Putao-Myitkyina suture zone (Li et al., 2004; Cong et al., 2011a,b; Xu et al., 2012; Wang et al., 2013; Metcalfe, 2013). Based on the presence of Permo-Carboniferous glacio-marine deposits and overlying post-glacial black mudstones and Gondwana-like fossil assemblages, it has been suggested that the Tengchong Block was derived from the margins of western Australia, in the eastern part of the Gondwana supercontinent (Jin, 1996). The block contains Mesoproterozoic metamorphic basement belonging to the Gaoligong Mountain Group and upper Paleozoic clastic sedimentary rocks and carbonates. The Mesozoic to Tertiary granitoids were emplaced into these strata and

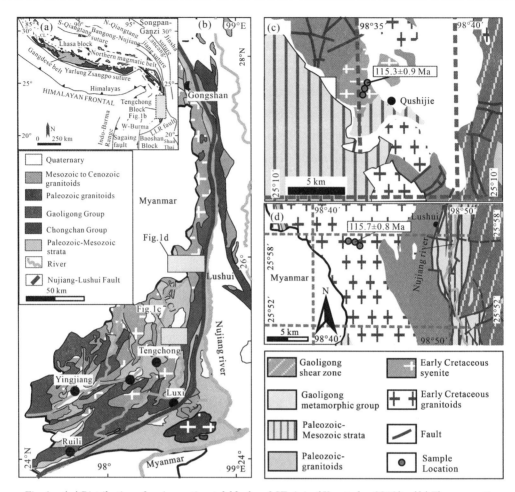

Fig. 1　(a) Distribution of main continental blocks of SE Asia (Xu et al., 2012). (b) The magmatic
distribution in the Tengchong Block (Zhu et al., 2017b). (c) Distribution of the syenites near
Qushijie in the northern part of Menglian batholiths (YNBGMR, 1991). (d) Study area of
the granitic rocks in the northeastern of Tengchong Block (YNBGMR, 1991).

were then covered by Tertiary-Quaternary volcano-sedimentary sequences (YNBGMR, 1991).
The Gaoligong Mountain Group contains quartzites, two-mica-quartz schists, feldspathic
gneisses, migmatites, amphibolites, and marble. Zircons from the paragneiss and orthogneiss
samples in this group yield ages in the range of 1 053−635 Ma and 490−470 Ma, respectively
(Song et al., 2010). The Paleozoic sediments in this area are dominated by Carboniferous
clastic rocks, and others Mesozoic strata including Upper Triassic to Jurassic turbidites,
Cretaceous red beds, and Cenozoic sandstones (YNBGMR, 1991; Zhong, 2000).

The Tengchong Block contains abundant granitic gneiss, migmatite, and leucogranite
units, the vast majority of which were previously thought to have formed during the Proterozoic
(YNBGMR, 1991). However, recent studies have identified several massive granitoids with
zircon U-Pb ages of 114−139 Ma in the eastern part of this block (Yang et al., 2006; Cong et

al., 2011a,b; Xie et al., 2010; Qi et al., 2011; Li et al., 2012; Luo et al., 2012; Xu et al., 2012; Cao et al., 2014; Zhu et al., 2015b; Xie et al., 2016), all of which were emplaced into the Paleozoic and Mesozoic units. The granites in the Gaoligong area are located to the west of the Nujiang-Luxi-Ruili fault and have undergone strong shearing that developed a north-trending foliation and a subhorizontally plunging mineral lineation (Wang and Burchfiel, 1997; Zhang et al., 2012).

This study focuses on the syenite from Menglian batholiths and granite from the Gaoligong belt of the northeastern part of Tengchong Block (Fig. 1b,c; Table 1). The Menglian batholith are located in the eastern Tengchong Terrane and intruded into the Gaoligong Group metamorphic rocks as well as the Paleozoic-Mesozoic strata (Fig. 1b). In most cases, they show faulted contact with the wall-rock although the intrusive contact can be seen in some places as well. Cenozoic volcanic rocks unconformably overly these intrusions and/or are covered by Quaternary sediments (Luo et al., 2012). The intrusions show round or elongate shape, and occur parallel to the ~NNW-SSW orientation with the Gaoligong shear zone and Nujiang-Longling-Ruili fault (NLRF) which are regarded to mark the southeastern extension of the Nujiang suture belt (Cong et al., 2011a). The studied pluton was located on the northern of the Menglian batholith near the Qushijie town, which mainly consists of the syenites that as the intrusions in the Paleozoic-Mesozoic strata and Gaoligong Group metamorphic rocks with sharply and unconformably contact (Fig. 1b). The syenite are medium to coarse grain and some part porphyroid texture and massive structure (Fig. 2a), mainly contained of plagioclase (most of them are albite, 50%−55%), alkaline feldspar (15−20 vol%), quartz(5−10 vol%), hornblend (possible arfvedsonite, 5−10 vol%), biotite(5 vol%), and accessory mineral including titanite, magnetite, zircon, and apatite (Fig. 2b).

Table 1 Summary of petrological characters of the Early-Cretaceous syenite-granites in the Tengchong Block.

Sample	Location	Petrology	Age/Ma	Structure	Minerals
DMJ13	N: 25°16.056′ E: 98°35.525′	Syenite	115.3±0.87	medium to coarse grain; some part porphyroid texture; massive structure	Pl(50%−55%) + Kf(15%−20%) + Qtz(5%−10%) + Hbl (5%−10%) + Bt(5%)
PM2-100	N:25°57.738′ E:98°42.948′	Granite	115.7±0.77	medium grained; massive structure	Pl(18%−25%) + Kf(25%−40%) + Qtz(25%−33%) + Hbl (1%) + Bt(10%−12%)

Then, the N-S trending granites in the Gaoligong belt are less deformed, extends northward into the NWW-trending Bomi-Chayu (TransHimalaya) magmatic belt. It was also extended south-westward into the Tengchong-Lianghe-Yingjiang area, and likely extends further into the Shan Scarp in Myanmar. The basement in the Gaoligong area is also composed of flat

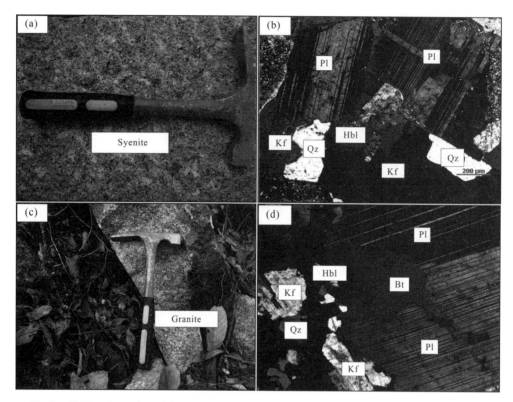

Fig. 2   Field and petrological features of syenites and granites in the Tengchong Block, SW China.

foliated granites and metamorphic rocks. The Gaoligong granites are intruded into the Precambrian Gaoligong Group, massive structure in the middle but slightly mylonitization occurred in the eastern and western margin. It was exposed as the long and narrow lens between the Nujiang and Longchuanjiang strike-slip faults and characterized by the obviously syntectonic emplacement (Fig. 1c). The granites are medium grained and also massive structure (Fig. 2c), and mainly consist of alkaline feldspar (25 – 40 vol%), plagioclase (18 – 25 vol%), quartz (25 – 33 vol%), biotite (10 – 12 vol%), amphibole (1 vol%), and accessory minerals including sphene, zircon, and apatite (Fig. 2d).

3   Samples and analytical methods

3. 1   Major and trace elements

Whole-rock samples were trimmed to remove weathered surfaces, cleaned with deionized water, crushed, and then powdered through a 200 mesh screen using a tungsten carbide ball mill. Major elements were analyzed using an X-ray fluorescence (XRF) spectrometer (Rikagu RIX 2100) at the Guizhou Tuopu Resource and Environmental Analysis Center, Institute of Geochemistry, Chinese Academy of Sciences, Guiyang, China. Analyses of USGS and Chinese national rock standards (BCR-2, GSR-1, and GSR-3) indicate that both analytical precision and accuracy for major elements are generally better than 5%. Trace elements were determined by using a Bruker Aurora M90 inductively coupled plasma mass spectrometry (ICP-MS) at the Guizhou Tuopu Resource and Environmental Analysis Center, Institute of Geochemistry, Chinese Academy of Sciences, Guiyang, China. Following the method of Qi et al. (2000). Sample powders were dissolved using an $HF + HNO_3$ mixture in a high-pressure PTFE bomb at 185 ℃ for 36 h. The accuracies of the ICP-MS analyses are estimated to be better than ±5% - 10% (relative) for most elements.

3.2 Sr-Nd isotopic analyses

Whole-rock Sr-Nd isotopic data were obtained by using a Nu Plasma HR multi-collector mass spectrometer at the State Key Laboratory of Continental Dynamics, Northwest University, China. The Sr and Nd isotopes were determined by using a method similar to that of Chu et al. (2009). Sr and Nd isotopic fractionation was corrected to $^{87}Sr/^{86}Sr = 0.119\ 4$ and $^{146}Nd/^{144}Nd = 0.721\ 9$, respectively. During the period of analysis, a Neptune multi-collector ICP-MS was used to measure the $^{87}Sr/^{86}Sr$ and $^{143}Nd/^{144}Nd$ isotope ratios. NIST SRM-987 and JMC-Nd were used as certified reference standard solutions for $^{87}Sr/^{86}Sr$ and $^{143}Nd/^{144}Nd$ isotopic ratios, respectively. BCR-1 and BHVO-1 were used as the reference materials.

3.3 Zircon U-Pb and Hf isotopic analyses

Zircon was separated from three ~5 kg samples taken from various sampling locations within the Gaoligong belt. The zircon grains were separated by using conventional heavy liquid and magnetic techniques. Representative zircon grains were handpicked and mounted in epoxy resin disks and then polished and coated with carbon. Internal morphology was examined using cathodoluminescent (CL) prior to U-Pb analyses.

Laser ablation ICP-MS zircon U-Pb analyses were conducted on an Agilent 7500a ICP-MS equipped with a 193-nm laser, which is housed at the State Key Laboratory of Continental Dynamics, Northwest University, Xi'an, China, following the method of Yuan et al. (2004). The $^{207}Pb/^{206}Pb$ and $^{206}Pb/^{238}U$ ratios were calculated by using the GLITTER program and corrected using the Harvard zircon 91500 as external calibration. These correction factors were then applied to each sample to correct for both instrumental mass bias and depth-dependent elemental and isotopic fractionation. The detailed analytical technique is described in Yuan et al. (2004). Common Pb contents were therefore evaluated by using the method described in Andersen (2002). The age calculations and plotting of concordia diagrams were made using ISOPLOT (version 3.0) (Ludwig, 2003). The errors quoted in tables and figures are at the 2σ level.

In situ zircon Hf isotopic analyses were conducted using a Nepyune MC-ICP-MS, equipped with a 193-nm laser. During analyses, a laser repetition rate of 10 Hz at 100 mJ was used and spot sizes were 32 μm. During analyses, $^{176}Hf/^{177}Hf$ and $^{176}Lu/^{177}Hf$ ratios of the standard zircon (91500) were $0.282\,294 \pm 15$ (2σ, $n = 20$) and $0.000\,31$, similar to the commonly accepted $^{176}Hf/^{177}Hf$ ratio of $0.282\,302 \pm 8$ and $0.282\,306 \pm 8(2\sigma)$ measured using the solution method (Goolaerts et al., 2004). The notations of $\varepsilon_{Hf}(t)$ value, $f_{Lu/Hf}$, single-stage model age (T_{DM1}) and two-stage model age (T_{DM2}) are defined as in Zheng et al (2007).

4　Results

The zircon LA-ICP-MS U-Pb data of these samples are given in Table 2, major and trace element compositions in Table 3, whole-rock Sr-Nd isotopic data in Table 4, and zircon Hf isotopic data in Table 5.

4.1　Zircon U-Pb data

Zircon grains from syenites (samples DMJ13) are generally euhedral to subhedral, have lengths of 120–300 mm, and have length-to-width ratios of 2∶1 to 3∶1 (Fig. 3a). The majority of these zircons are colorless or light brown, prismatic, and transparent to translucent, and have weak oscillatory zoning visible during cathodoluminescence (CL) imaging. The 12 spot analyses of zircons from sample DMJ13 yielded high Th (199–1 306 ppm) and U (206–1 893 ppm) concentrations with Th/U ratios of 0.58–1.34 that are indicative of a magmatic origin (Hoskin and Schaltegger, 2003). These spots yield $^{206}Pb/^{238}U$ ages ranging from 113.0 ± 1.0 Ma to 117 ± 1.0 Ma, with a weighted mean age of 115.3 ± 0.9 Ma (MSWD = 0.74, $n = 12$, 1σ; Fig. 4a,b). Zircons from the granites (sample PM2-100) (Fig. 3b) are subhedral and partly prismatic, have weak oscillatory zoning visible under CL images. A total of 16 analyses yielded high Th (210–1 318 ppm) and U (314–2 388 ppm) concentrations and Th/U values of 0.38–1.44. Their analytical data yield $^{206}Pb/^{238}U$ ages ranging from 113.0 ± 1.0 Ma to 117 ± 1.0 Ma, with a weighted mean age of 115.7 ± 0.8 Ma (MSWD = 1.5, $n = 16$, 1σ; Fig. 4c,d).

4.2　Whole-rock geochemistry

The samples from northern part of Menglian batholith are peralkaline and ferroan, belongs to the syenite (Fig. 5a-d). The samples from Gaoligong belt in the northeastern Tengchong Block are high-K calc-alkali (no shown in the SiO_2 vs. K_2O diagram) and metaluminous to peraluminous, belong to the granite (Fig. 5a-d). Both of them are similar to the early Cretaceous rocks in the Tengchong Block and early Cretaceous granitoids from Gaoligong belt (Fig. 5a-d).

Table 2 Results of zircon LA-ICP-MS U-Pb data for the syenite and granite in the Tengchong Block.

Analysis	Content/ppm			Ratios								Age/Ma							
	Th	U	Th/U	$\frac{^{207}Pb}{^{206}Pb}$	1σ	$\frac{^{207}Pb}{^{235}U}$	1σ	$\frac{^{206}Pb}{^{238}U}$	1σ	$\frac{^{208}Pb}{^{232}Th}$	1σ	$\frac{^{207}Pb}{^{206}Pb}$	1σ	$\frac{^{207}Pb}{^{235}U}$	1σ	$\frac{^{206}Pb}{^{238}U}$	1σ	$\frac{^{208}Pb}{^{232}Th}$	1σ
Location1 DML13 syenite (N: 25°16.056′ E: 98°35.525′)																			
spot-1	581	843	0.69	0.049 24	0.002 35	0.121 15	0.005 64	0.017 84	0.000 18	0.005 64	0.000 05	159	110	116	5	114	1	114	1
spot-2	1 306	976	1.34	0.049 95	0.002 10	0.126 00	0.005 16	0.018 30	0.000 18	0.005 78	0.000 04	193	99	120	5	117	1	116.4	0.8
spot-3	492	718	0.69	0.050 78	0.001 21	0.125 82	0.001 80	0.017 97	0.000 16	0.005 46	0.000 05	231	17	120	2	115	1	110	1
spot-4	467	468	1.00	0.048 91	0.002 03	0.123 99	0.005 00	0.018 39	0.000 18	0.005 82	0.000 04	144	96	119	5	117	1	117.3	0.8
spot-5	199	211	0.94	0.047 22	0.004 09	0.116 61	0.010 03	0.017 91	0.000 20	0.005 69	0.000 12	60	192	112	9	114	1	115	2
spot-6	436	748	0.58	0.052 15	0.001 50	0.131 82	0.003 60	0.018 33	0.000 17	0.005 76	0.000 04	292	67	126	3	117	1	116	0.9
spot-7	606	811	0.75	0.051 39	0.001 69	0.126 96	0.004 00	0.017 92	0.000 17	0.005 64	0.000 04	258	77	121	4	114	1	113.6	0.9
spot-8	781	1 125	0.69	0.052 03	0.001 72	0.129 97	0.004 12	0.018 12	0.000 17	0.005 69	0.000 04	287	77	124	4	116	1	114.7	0.9
spot-9	718	1 117	0.64	0.049 95	0.001 44	0.122 01	0.003 31	0.017 72	0.000 17	0.005 59	0.000 04	193	68	117	3	113	1	112.7	0.9
spot-10	237	206	1.15	0.050 03	0.001 66	0.125 33	0.003 38	0.018 17	0.000 18	0.005 48	0.000 06	196	44	120	3	116	1	110	1
spot-11	1 265	1 893	0.67	0.050 45	0.001 11	0.124 77	0.001 45	0.017 94	0.000 16	0.005 44	0.000 04	216	13	119	1	115	1	109.7	0.8
spot-12	341	460	0.74	0.046 05	0.004 91	0.115 22	0.012 22	0.018 15	0.000 20	0.005 96	0.000 30		215	111	11	116	1	120	6
Location 2 PM2-100 granite (N:25°57.738′ E:98°42.948′)																			
spot-1	475	584	0.81	0.049 66	0.001 21	0.125 34	0.001 96	0.018 31	0.000 16	0.005 58	0.000 05	179	21	120	2	117	1	112	1
spot-2	377	673	0.56	0.050 54	0.001 21	0.127 15	0.001 90	0.018 24	0.000 16	0.005 69	0.000 06	220	19	122	2	117	1	115	1
spot-3	210	405	0.52	0.050 01	0.001 69	0.125 24	0.004 05	0.018 17	0.000 17	0.005 73	0.000 04	195	80	120	4	116	1	115.6	0.9
spot-4	578	844	0.69	0.049 79	0.001 18	0.121 95	0.001 77	0.017 76	0.000 16	0.005 42	0.000 05	185	18	117	2	113	1	109	1
spot-5	734	630	1.17	0.050 67	0.001 24	0.127 49	0.001 99	0.018 25	0.000 16	0.005 50	0.000 05	226	20	122	2	117	1	111	1

Continued

Analysis	Content/ppm		Th/U	Ratios								$^{207}Pb/^{206}Pb$	1σ	Age/Ma							
	Th	U		$^{207}Pb/^{206}Pb$	1σ	$^{207}Pb/^{235}U$	1σ	$^{206}Pb/^{238}U$	1σ	$^{208}Pb/^{232}Th$	1σ			$^{207}Pb/^{235}U$	1σ	$^{206}Pb/^{238}U$	1σ	$^{208}Pb/^{232}Th$	1σ
spot-6	1 318	2 388	0.55	0.048 46	0.001 28	0.119 57	0.002 95	0.017 90	0.000 17	0.005 67	0.000 04	122	63	115	3	114	1	114.3	0.9
spot-7	666	821	0.81	0.051 02	0.002 79	0.126 00	0.006 76	0.017 91	0.000 20	0.005 64	0.000 04	242	128	120	6	114	1	113.7	0.8
spot-8	758	690	1.10	0.051 97	0.001 24	0.128 91	0.001 9	0.017 99	0.000 16	0.005 57	0.000 05	284	18	123	2	115	1	112	1
spot-9	212	361	0.59	0.052 85	0.001 44	0.132 40	0.002 58	0.018 17	0.000 17	0.005 36	0.000 05	322	27	126	2	116	1	108	1
spot-10	367	668	0.55	0.051 09	0.001 25	0.128 30	0.001 99	0.018 21	0.000 16	0.005 64	0.000 05	245	20	123	2	116	1	114	1
spot-11	498	977	0.51	0.048 04	0.001 35	0.121 13	0.002 48	0.018 29	0.000 17	0.005 64	0.000 07	101	31	116	2	117	1	114	1
spot-12	275	715	0.38	0.050 26	0.001 28	0.126 80	0.002 16	0.018 30	0.000 17	0.005 65	0.000 06	207	23	121	2	117	1	114	1
spot-13	729	861	0.85	0.052 47	0.001 25	0.132 37	0.001 88	0.018 30	0.000 16	0.005 74	0.000 05	306	17	126	2	117	1	116	1
spot-14	403	314	1.28	0.048 30	0.001 15	0.121 36	0.001 73	0.018 22	0.000 16	0.005 67	0.000 05	114	18	116	2	116	1	114	1
spot-15	545	378	1.44	0.049 77	0.001 30	0.121 69	0.002 14	0.017 73	0.000 16	0.005 85	0.000 08	184	24	117	2	113	1	118	2
spot-16	755	970	0.78	0.047 51	0.001 18	0.118 43	0.001 85	0.018 08	0.000 16	0.005 68	0.000 05	75	21	114	2	116	1	114	1

Table 3 Major(wt%) and trace(ppm) element data for the syenite-granite in the Tengchong Block.

Sample	Syenite			Granite		
	DMJ-08	DMJ-12	DMJ-13	PM2-103	PM2-104	PM2-105
SiO_2	63.03	62.01	62.92	74.41	71.35	72.40
TiO_2	0.47	0.51	0.48	0.32	0.30	0.30
Al_2O_3	15.20	15.36	15.17	12.34	14.93	14.20
MgO	0.74	0.85	0.78	0.54	0.64	0.60
$Fe_2O_3^T$	3.64	4.31	4.09	2.32	2.24	2.29
CaO	2.82	2.83	2.88	2.74	2.70	2.56
Na_2O	7.11	7.24	7.04	3.42	3.36	3.15
K_2O	3.47	3.44	3.49	2.79	3.27	3.68
MnO	0.06	0.06	0.06	0.04	0.04	0.04
P_2O_5	0.14	0.16	0.15	0.11	0.11	0.11
LOI	3.12	3.12	3.28	0.94	0.71	0.91
Total	99.80	99.89	100.33	99.98	99.65	100.24
Li	9.01	11.3	10.3	27.1	22.8	25.1
Be	2.74	2.93	2.86	1.72	1.71	1.49
Sc	18.5	19.4	18.0	9.6	10.6	9.4
V	28.7	33.8	30.9	33.5	29.5	30.7
Cr	5.71	6.51	17.4	9.31	7.52	5.68
Co	14.5	9.99	10.1	50.4	47.1	44.5
Ni	2.58	3.08	8.33	8.09	4.82	4.43
Cu	3.9	3.9	4.29	3.72	3.75	3.36
Zn	83.8	84	85	98	49.3	62.9
Ga	17.7	18.1	17.8	15.5	15.7	14.4
Ge	1.19	1.22	1.2	0.96	0.97	0.93
As	1.43	1.34	1.4	1.66	0.89	1.31
Rb	99.3	93.4	96.6	103	122	129
Sr	440	439	446	269	287	287
Y	23.6	23.6	23.5	9.3	9.93	8.13
Zr	222	238	216	194	181	176
Nb	16.5	15.8	15.6	9.05	8.75	8
Mo	0.09	0.16	0.35	0.64	0.26	1.45
Cs	1.51	1.64	1.55	5.74	5.69	5.61
Ba	539	510	531	614	636	796
La	59.5	63.4	59.7	29	35.1	32.1
Ce	112	118	108	50.5	61.5	56.4
Pr	12	13	12	4.71	5.54	5.05
Nd	38.1	41.9	39.1	14.4	16.8	15
Sm	6.08	6.7	6.16	2.14	2.4	2.28
Eu	1.05	1.07	1.02	0.73	0.70	0.71
Gd	5.46	5.62	5.31	1.90	2.02	1.67

Continued

Sample	Syenite			Granite		
	DMJ-08	DMJ-12	DMJ-13	PM2-103	PM2-104	PM2-105
Tb	0.72	0.73	0.72	0.25	0.28	0.24
Dy	3.88	3.81	3.75	1.41	1.57	1.23
Ho	0.75	0.71	0.75	0.29	0.31	0.25
Er	2.29	2.18	2.26	0.84	0.94	0.75
Tm	0.32	0.32	0.33	0.12	0.14	0.11
Yb	2.04	2.14	2.07	0.9	1.02	0.79
Lu	0.31	0.33	0.32	0.14	0.16	0.13
Hf	4.89	5.76	5.25	4.15	4.17	3.82
Ta	1.49	1.36	1.4	0.84	1.08	0.78
W	60.3	34.5	36.8	270	303	269
Tl	0.46	0.48	0.48	0.55	0.5	0.48
Pb	28.6	27.9	25.4	24.9	25.2	24.9
Bi	0.057	0.052	0.046	0.04	0.049	0.007
Th	35.64	36.72	31.93	16.91	18.58	15.98
U	4.51	4.29	3.92	3.10	3.63	2.23
$Mg^{\#}$	32.11	31.43	30.78	35.37	39.83	37.78
A/CNK	0.74	0.74	0.74	0.91	1.07	1.03
$\sum$REE	244.50	259.90	241.49	107.33	128.48	116.71

Table 4 **Whole-rock Sr-Nd isotopic data for the syenite and granite in the Tengchong Block.**

Sample	$^{87}Sr/^{86}Sr$	$2SE$	Sr /ppm	Rb /ppm	$^{143}Nd/^{144}Nd$	$2SE$
DMJ-12	0.714 011	0.000 011	439	93.4	0.512 213	0.000 007
PM2-105	0.712 916	0.000 015	287	129	0.512 122	0.000 012
PM2-105-1	0.712 082	0.000 008	287	129	0.512 138	0.000 004

Sample	Nd /ppm	Sm /ppm	T_{DM2} /Ga	$\varepsilon_{Nd}(t)$	$(^{87}Sr/^{86}Sr)_i$	$(^{143}Nd/^{144}Nd)_i$
DMJ-12	41.9	6.7	1.26	−6.83	0.713 013	0.512 141
PM2-105	15	2.28	1.35	−3.80	0.704 241	0.512 051
PM2-105-1	15	2.28	1.33	−3.48	0.703 408	0.512 067

$^{87}Rb/^{86}Sr$ and $^{147}Sm/^{144}Nd$ ratios were calculated using Rb, Sr, Sm and Nd contents analyzed by ICP-MS. T_{DM2} represent the two-stage model age and were calculated using present-day $(^{147}Sm/^{144}Nd)_{DM} = 0.213\ 7$, $(^{147}Sm/^{144}Nd)_{DM} = 0.513\ 15$ and $(^{147}Sm/^{144}Nd)_{crust} = 0.101\ 2$. $\varepsilon_{Nd}(t)$ values were calculated using present-day $(^{147}Sm/^{144}Nd)_{CHUR} = 0.196\ 7$ and $(^{147}Sm/^{144}Nd)_{CHUR} = 0.512\ 638$.

$$\varepsilon_{Nd}(t) = [(^{143}Nd/^{144}Nd)_S(t)/(^{143}Nd/^{144}Nd)_{CHUR}(t) - 1] \times 10^4.$$

$$T_{DM2} = \frac{1}{\lambda} \{1 + [(^{143}Nd/^{144}Nd)_S - ((^{147}Sm/^{144}Nd)_S - (^{147}Sm/^{144}Nd)_{crust})(e^{\lambda t} - 1) - (^{143}Nd/^{144}Nd)_{DM}]/$$
$$[(^{147}Sm/^{144}Nd)_{crust} - (^{147}Sm/^{144}Nd)_{DM}]\}.$$

Table 5　Single-grain zircon Hf isotopic data for the syenite and granite in the Tengchong Block.

Grain spot	Age/Ma	$^{176}\mathrm{Yb}/^{177}\mathrm{Hf}$	2SE	$^{176}\mathrm{Lu}/^{177}\mathrm{Hf}$	2SE	$^{176}\mathrm{Hf}/^{177}\mathrm{Hf}$	2SE	$(^{176}\mathrm{Hf}/^{177}\mathrm{Hf})_i$	$f_{\mathrm{Lu/Hf}}$	$\varepsilon_{\mathrm{Hf}}(t)$	2SE	$T_{\mathrm{DM1}}/\mathrm{Ma}$	$T_{\mathrm{DMC}}/\mathrm{Ma}$
Syenite													
DMJ13-01	111	0.029 875	0.000 215	0.000 887	0.000 006	0.282 502	0.000 032	0.282 501	−0.97	−7.17	1.15	1 058	1 623
DMJ13-02	114	0.092 500	0.001 004	0.002 974	0.000 032	0.282 857	0.000 021	0.282 851	−0.91	5.30	0.76	591	833
DMJ13-03	117	0.064 597	0.000 324	0.002 154	0.000 012	0.282 625	0.000 050	0.282 620	−0.94	−2.81	1.76	917	1 351
DMJ13-04	119	0.072 887	0.000 643	0.002 387	0.000 023	0.282 657	0.000 031	0.282 652	−0.93	−1.65	1.11	876	1 279
DMJ13-05	115	0.057 979	0.000 287	0.001 607	0.000 009	0.282 653	0.000 013	0.282 649	−0.95	−1.83	0.45	864	1 287
DMJ13-06	111	0.048 472	0.000 208	0.001 799	0.000 017	0.282 699	0.000 025	0.282 695	−0.95	−0.27	0.88	801	1 185
DMJ13-07	110	0.057 137	0.000 235	0.001 723	0.000 019	0.282 642	0.000 016	0.282 638	−0.95	−2.32	0.56	882	1 315
DMJ13-08	117	0.047 868	0.000 160	0.001 316	0.000 005	0.282 636	0.000 017	0.282 633	−0.96	−2.35	0.60	881	1 322
DMJ13-09	114	0.048 990	0.000 289	0.001 611	0.000 007	0.282 625	0.000 028	0.282 621	−0.95	−2.82	1.00	903	1 350
DMJ13-10	107	0.039 665	0.000 382	0.001 372	0.000 009	0.282 655	0.000 053	0.282 652	−0.96	−1.90	1.88	855	1 286
DMJ13-11	104	0.049 328	0.000 362	0.001 746	0.000 013	0.282 584	0.000 040	0.282 581	−0.95	−4.47	1.43	965	1 447
DMJ13-12	117	0.040 564	0.000 242	0.001 165	0.000 005	0.282 657	0.000 017	0.282 654	−0.96	−1.61	0.59	848	1 275
DMJ13-13	114	0.047 802	0.000 174	0.001 352	0.000 005	0.282 636	0.000 014	0.282 633	−0.96	−2.43	0.48	882	1 325
DMJ13-14	116	0.072 374	0.000 232	0.002 424	0.000 010	0.282 602	0.000 027	0.282 596	−0.93	−3.67	0.97	958	1 405
DMJ13-15	113	0.056 633	0.000 182	0.001 582	0.000 004	0.282 689	0.000 016	0.282 686	−0.95	−0.58	0.58	811	1 206
DMJ13-16	110	0.048 512	0.000 295	0.001 349	0.000 009	0.282 680	0.000 013	0.282 678	−0.96	−0.93	0.46	818	1 226
DMJ13-17	116	0.062 370	0.000 072	0.002 177	0.000 004	0.282 680	0.000 025	0.282 676	−0.93	−0.86	0.87	836	1 227
DMJ13-18	115	0.062 080	0.000 740	0.001 864	0.000 039	0.282 599	0.000 016	0.282 595	−0.94	−3.74	0.57	947	1 409
DMJ13-19	116	0.048 355	0.000 044	0.001 691	0.000 002	0.282 446	0.000 039	0.282 443	−0.95	−9.11	1.38	1 161	1 749
DMJ13-20	120	0.079 978	0.000 341	0.002 598	0.000 011	0.282 686	0.000 023	0.282 680	−0.92	−0.62	0.82	838	1 214

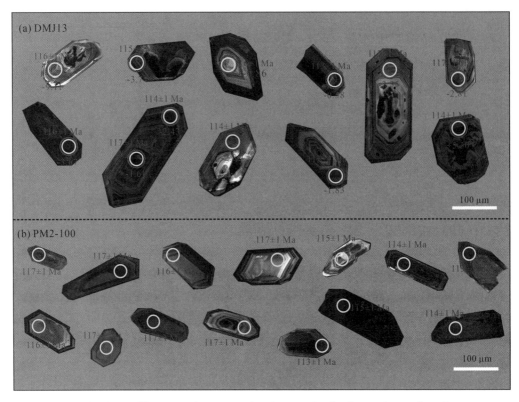

Fig. 3   Zircon CL images of representative zircon grains for the syenites and granites
in the Tengchong Block, SW China.

The syenite have low SiO_2 = 62. 01 − 63. 02 wt%, notably high Na_2O (7. 04 − 7. 24 wt%
with Na_2O/K_2O = 2. 02 − 2. 10) and $Na_2O + K_2O$ = 10. 5 − 10. 7 (Fig. 5a). They display
moderate Al_2O_3 concentrations of 15. 17−15. 36 wt% with A/CNK [molar $Al_2O_3/(CaO + Na_2O$
+ K_2O)] values of 0. 74 (Fig. 5b). The primitive-mantle-normalized multi-element variation
diagrams (Fig. 6a) are characterized by positive large-ion lithophile element (LILE; e.g.,
Rb, Th, U, K, and Pb) anomalies and obvious negative high field strength element (HFSE;
e.g. Nb, Ta, P, and Ti) anomalies. These samples show high total REE concentrations of
241. 5− 259. 9 ppm and enrichment in light REEs (LREEs) (Fig. 6b) with relative high
$(La/Yb)_N$ values of 20. 7−21. 3, and clearly negative Eu anomalies (δEu = 0. 53 − 0. 56),
suggesting insignificant fractionation of plagioclase.

The granites have high SiO_2 content (71. 35 − 74. 47 wt%), relative moderate K_2O =
2. 79−3. 68 wt% with various K_2O/Na_2O ratios = 0. 82−1. 17, and moderate Al_2O_3 = 12. 34−
14. 93 wt% with A/CNK = 0. 91 − 1. 07, which belong to the metaluminous to peraluminous
(Fig. 5b). They also show lower $Fe_2O_3^T$, TiO_2, and MgO content than the syenites. The
primitive-mantle-normalized multi-element variation diagrams (Fig. 6a) are also characterized
by positive large-ion lithophile element (LILE; e.g., Rb, Th, U, K, and Pb) anomalies and

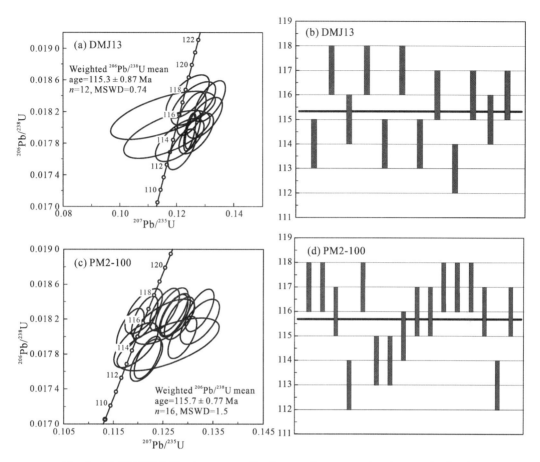

Fig. 4   LA-ICP-MS U-Pb zircon concordia diagram of representative zircon grains for the
syenites and granites in the Tengchong Block, SW China.

more obvious negative high field strength element (HFSE; e.g. Nb, Ta, P, and Ti) anomalies
than syenite. These samples show low total REE concentrations of 107. 3–148. 5 ppm and more
enrichment in light REEs (LREEs) (Fig. 6b) with notably high (La/Yb)$_N$ values of 23. 1–
29. 1, but clearly unsignificant Eu anomalies ($\delta Eu = 0.98 - 1.10$), implying negligible
fractional of plagioclase.

4. 3   Whole-rock Sr-Nd isotopes

Whole-rock Sr-Nd isotopic data for the syenite and granite from the Tengchong Block are
listed in Table 4. All the initial $^{87}Sr/^{86}Sr$ isotopic ratios (I_{Sr}) and $\varepsilon_{Nd}(t)$ values are calculated
according to the LA-ICP-MS zircon U-Pb dates for the syenite and granite.

The syenite (samples DMJ13) are relative high Sr content = 439 ppm and moderate Rb
content = 93. 4 ppm, and which have initial $^{87}Sr/^{86}Sr = 0.713\ 013$, high $^{143}Nd/^{144}Nd$ ratios =
0. 512 121 with $\varepsilon_{Nd}(t)$ values of −6. 8, and two-stage Nd model ages of 1. 26 Ga. The granites

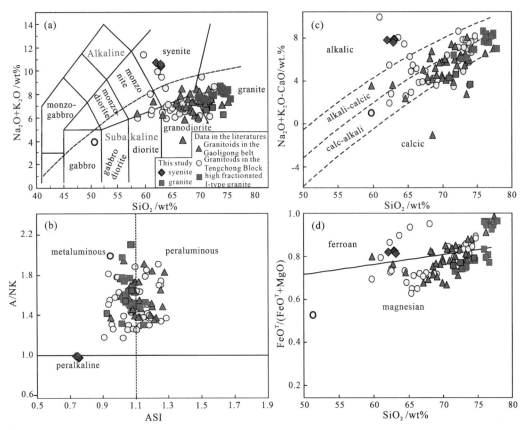

Fig. 5 (a) ($Na_2O + K_2O$) vs. SiO_2 diagram; (b)A/NK vs. A/CNK diagram; (c)($Na_2O +$ K_2O-CaO) vs. SiO_2 diagram; (d)$FeO^T/(FeO^T/MgO)$ vs. SiO_2 diagram.

All of them are referenced by Frost et al. (2001). The data of early Cretaceous granitoids in the Gaoligong belt from Yang et al. (2006). The data of early Cretaceous granitoids in the Tengchong Block from Cong et al. (2011a,b), Qi et al. (2011), Li et al. (2012), Luo et al. (2012), and (Cao et al. 2014). The data of high fractionated I-type granites in the Tengchong Block from Zhu et al. (2015).

(sample PM2-105 and PM2-105-1) share similar Sr and Rb contents including relative low Sr = 287 ppm and Rb = 129 ppm. They have lower inital $^{87}Sr/^{86}Sr$ (0. 703 408 and 0. 704 241) and $^{143}Nd/^{144}Nd$ ratio (0. 512 051 and 0. 512 067) with $\varepsilon_{Nd}(t)$ values of -3.8 and -3.5, and two-stage Nd model ages of 1. 33–1. 35 Ga.

As shown in the $\varepsilon_{Nd}(t)$ vs. initial $^{87}Sr/^{86}Sr$ diagram, the granite samples share the similar characters with the granitoids and volcanic rocks from TransHimalaya batholith and the syenite are similar to the granitoids and volcanic rocks among the Tengchong and central and northern Lhasa Blocks and TransHimalaya batholith (Fig. 6a) (Yang et al., 2006; Zhu et al., 2009; Qu et al., 2012; Lin et al., 2012; Chen et al., 2014; Zhu et al., 2015b). Both of them are plotted into the evolutionary line between the arc-setting basalts and lower- and upper-crust (Fig. 7).

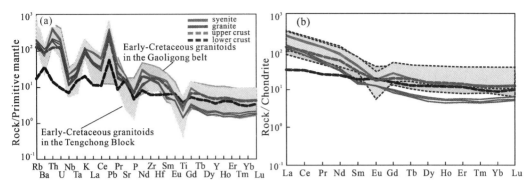

Fig. 6 Chondrite-normalized REE patterns and primitive-mantle-normalized trace element
spider diagram for the syenites and granites in the Tengchong Block, SW China.

The primitive mantle and chondrite values are from Sun and McDonough (1989).

The data in the Tengchong Block and Gaoligong belt is same as the Fig. 5.

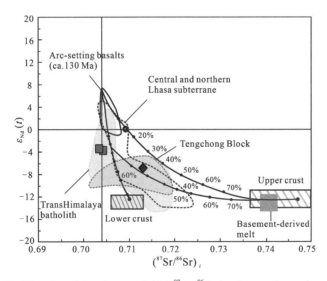

Fig. 7　Plot of $\varepsilon_{Nd}(t)$ values vs. initial $^{87}Sr/^{86}Sr$ for the syenites and granites
in the Tengchong Block, SW China.

Lower continental crust (Miller et al., 1999) and upper continental crust (Harris et al., 1988).

Arc-setting basalt was referenced by Chen et al. (2014). Symbols as in Fig. 5.

4. 4　Zircon chemical and Lu-Hf isotopic compositions

The zircons from dated samples were also analyzed for Lu-Hf isotopes on the same domains, and the results are listed in Table 5. Initial $^{176}Hf/^{177}Hf$ ratios and $\varepsilon_{Hf}(t)$ values of the zircons were calculated according to their LA-ICP-MS zircon U-Pb dates for the syenite. We selected 12 zircon grains from the syenite (DMJ13) for Lu-Hf isotopic analysis. All of these spots display concordant zircon U-Pb dating, which have various and relative enriched Hf isotopic compositions with $\varepsilon_{Hf}(t)$ values ranging from -9.11 to -0.27(but one is $+5.30$), with corresponding two-stage Hf model ages of 1. 18 Ga to 1. 75 Ga (but one is 0. 83 Ga). Above these various Hf compositions possibly indicate the more ancient continental crust (Yang et al. , 2007; Zheng et al. , 2007).

Furthermore, zircons from both syenite and moznogranite have moderate Th and U content and most of their Th/U ratios are distributed in the near the line of Th/U ratio = 1. 0, which plotted in the field of the zircon in the Lhasa Terrane records of continental crustal reworking during Mesozoic-Cenozoic magmatisms (Liu et al. , 2014) (Fig. 8). They also display the high U/Yb ratios and moderate Y content , suggesting the typical continental crust characters rather than the recycled oceanic crust (Grimes et al. , 2007) (Fig. 8).

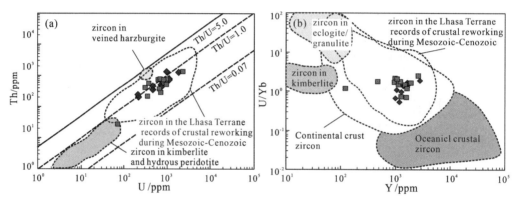

Fig. 8 Discriminant diagrams for zircon origin for the syenites
and granites in the Tengchong Block, SW China.
Modified from Liu et al. (2014) and Grimes et al. (2007). Symbols as in Fig. 5.

5 Discussion

5. 1 Petrogenesis of the syenite

Yang et al. (2012) has summarized several models to interpret the genesis of the syenite: extensive differentiation from mantle-derived basaltic magmas (Litvinovsky et al. , 2002; Yang et al. , 2005), magma mixing between basic and silicic melts with subsequent differentiation of the hybrid melts/liquids (Dorais, 1990; Sheppard, 1995; Riishuus et al. , 2005; Yang et al. , 2008), and partial melting of crustal rocks at high pressures (Huang and Wyllie, 1975; Lubala et al. , 1994; Yang et al. , 2012). Before any further discussions, it is important to distinguish the silica-saturated and silica-undersaturated syenites, as we will know whether it was derived from normal magmatic series or special alkaline series or not. The presence of the quartz can be up to 5%−10% and absence of nepheline-bearing in the syenites may infer that the Early Cretaceous syenites belong to the silica-saturated syenites (Fig. 2a, b), similar with the silica-saturated syenites in northern China Craton (Yang et al. , 2012). The syenites have been characterized by low abundances of MgO = 0. 74−0. 85 wt%, $Fe_2O_3^T$ = 3. 64−4. 31 wt%, TiO_2 = 0. 47−0. 51 wt%, and Cr (5. 71−17. 4 ppm) and Ni (2. 58−8. 33 ppm) (Table 2), which may preclude the model of differentiation from mantle-derived basaltic magmas but indicate a product from fractionation of alkaline magmas derived from an enriched mantle and/ or a continental crustal source. The absence of related early Cretaceous ultramafic to mafic rocks in the Tengchong Block (YBGMR, 1991) and associated mafic magmatic enclaves in the syenites (Fig. 2a) also preclude the conclusion of magma mixing between basic and felsic melts, although some enclaves are developed in the calc-alkaline granodiorite in the middle to southern Linaghe area which is far from our fields (Cong et al., 2011a; Li et al., 2012). The high initial $^{87}Sr/^{86}Sr$ = 0. 713 013, high $^{143}Nd/^{144}Nd$ ratios = 0. 512 121 with $\varepsilon_{Nd}(t)$ values of −6. 8 overlap with that of felsic granitoids in the Tengchong Block and intermediate to felsic granitoids and volcanic rocks in the Central and Northern Lhasa blocks, both of which are typical continental crustal sources (Yang et al., 2006; Zhu et al., 2009; Chen et al., 2014; Zhu et al., 2015b). Then the zircon Hf components are characterized by significant negative zircon $\varepsilon_{Hf}(t)$ values ranging from −9. 11 to −0. 27(although one is positive) with two-stage Hf model ages of 1. 18 Ga to 1. 75 Ga (Fig. 9), which also imply the mainly Paleo- and Meso-Proterozoic continental sources (Yang et al., 2007; Zheng et al., 2007). In addition, the low Nb and Ta content and Nb/Ta ratios range from 11. 1 to 11. 6, which are less than those from mantle-derived melts (17. 5 ± 2. 0) (Kamber and Collerson, 2000), also suggesting a crustal source. The notably high Na_2O (7. 04−7. 24 wt%) with high Na_2O/K_2O ratios = 2. 01−2. 10 indicate a sodium-rich source.

Experimental data show that syenitic magma could be formed by the melting at the base of continental crust in two conditions (Huang and Wyllie, 1975; Johannes and Holtz, 1990): ① partial melting of granite at pressure >10 kbar may produce the melts with low SiO_2 content in the water-undersaturated conditions, with the increasing normative feldspar; and ②partial melting of muscovite granite with 74 wt% SiO_2 under 15 kbar pressure and ca. 5 wt% water, which can produced the syenitic melts. But these experimental data are suited to haplogranitic system which does not contain mafic components, contrast to our syenites. And the juvenile crustal rocks have a significant Mg, Fe, and Ca along with the elevated pressure that would leave a residual clinopyroxene and garnet, then the granitic melts would be produced rather than syenitic melts (Litvinosky and Steele, 2000). Also, experimental petrological studies show that Na-rich intermediate to felsic melts are formed by either ①the differentiation of mantle-derived mafic magmas (Stern et al., 1989) or ②the high-temperature dehydration melting of mafic lower-crustal material (Rapp and Watson, 1995). However, the Early Cretaceous syenite share similar evolved whole-rock initial $^{87}Sr/^{86}Sr$ and $\varepsilon_{Nd}(t)$ and zircon Hf compositions with the Early Cretaceous granitoids near the study area in the Tengchong Block

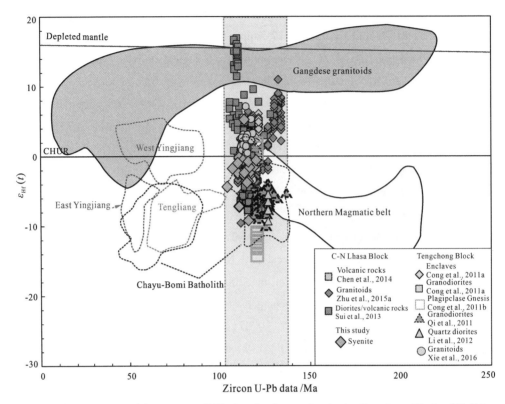

Fig. 9　Plot of zircon $\varepsilon_{Hf}(t)$ vs. zircon U-Pb data for the syenites in the Tengchong Block, SW China.

(Figs. 7 and 9) (Yang et al., 2006; Zhu et al., 2015b), indicating that their similar and/or common sources-melting of continental crustal materials, such as, mafic lower crustal and/or upper evolved crustal materials. Then the variable zircon Hf isotopic compositions also imply the mixing components (Kemp et al., 2007). The high Th and U abundances and moderate Y abundances and U/Yb ratios of the zircons (Fig. 8), as the robust mineral in the rocks, also show the continental crustal source. In addition, the low abundances of MgO, $Fe_2O_3^T$, TiO_2, Cr, Ni and depletion of P and Ti (Fig. 6a) imply the significant fractional crystallization of ferro-magnesian minerals such as pyroxene, amphibole, biotite and Fe-Ti oxides and so on. The significant negative Eu anomalies (Fig. 6b) also indicate the fractionation of plagioclase. Generally, the accessory minerals would be controlled much of the REE variation (Yang et al., 2012), the higher total REE than granites (Fig. 6b) suggests more accumulation of mineral such as titanite, zircon, apatite and monzonite, which may be a crucial accessory minerals in the rocks. In addition, the MgO-$\varepsilon_{Nd}(t)$, Ba/Th-La/Sm, Ba/Th-Th/Nb and Ba/La-Th/Nb diagram indicate that the sediments-derived fluids/melts play a key role in the genesis of the syenite (Fig. 10).

Based on the geochemical, whole-rock isotopic and zircon Hf isotopic data, we suggest

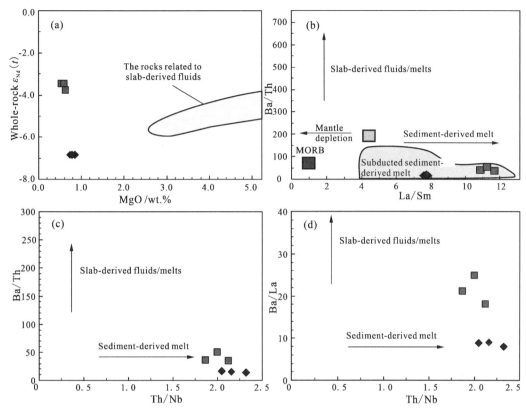

Fig. 10　Diagram of distinguishing the slab-derived and/or sediments-derived fluids/melts
(modified from Wang et al., 2014; Zhao et al., 2016; Elliott, 2003).
Symbols as in Fig. 5.

that the Early Cretaceous syenites were possibly derived from mixing of mafic lower crustal and evolved sediments-derived materials, and the fractionation of some ferro-magnesian mineral and plagioclase may occur during the evolution of magma.

5.2　Petrogenesis of the granite

In general, the granites are crust-derived materials mainly including metabasaltic rocks and/or metasedimentary rocks. Our granites are high SiO_2 content (71.35−74.41 wt%) and A/CNK (0.91−1.07) (Fig. 5a,b), belong to the metaluminous to peraluminous (Frost et al., 2001), similar with the metasedimentary materials (Chappell and White, 1992). The low MgO, $Fe_2O_3^T$, TiO_2, Cr and Ni indicate that they did not directly produced from differentiation of mantle-derived mafic magmas. On the primitive-mantle-normalized multi-element variation and REE diagrams (Fig. 6a), these granites show some features similar to that of upper crust with enrichment of LILEs (Rb, Th, U, K, and Pb) and depletion of HFSEs (Nb, Ta, P, and Ti), implying they were derived from crustal sources region (Harris et al., 1988), which was coupled with fractional crystallization to some extents during the evolution of magma. But the more differentiated pattern and depletion of HREE with high $(La/Yb)_N = 23.1-29.1$ than the upper crust (Fig. 6b) indicates the some HREE-rich minerals as the residue (such as garnet and so on) (Qu et al., 2012). In addition, the less pronounced or slightly positive Eu anomalies (Fig. 6b) may imply the negligible fractional crystallization of plagioclase. The relative low Rb/Sr and Rb/Ba (Fig. 11a) prove that the possible melting of metasedimentary-derived greywacke. The low $Al_2O_3/(MgO + FeO^T + TiO_2)$ and moderate $(Al_2O_3 + MgO + FeO^T + TiO_2)$ content (Fig. 11b) also infer that the significant involvement of metasedimentary-derived greywackes (Patiño-Douce et al., 1999). On the initial $^{87}Sr/^{86}Sr$ and $\varepsilon_{Nd}(t)$ diagram (Fig. 7), the samples are plotted into the fields between the lower crust and arc-setting basalts (Miller et al., 1999; Chen et al., 2014), similar with igneous rocks in the TransHimalaya batholith (Lin et al., 2012). However, the majority are still typical continental crustal sources confirmed by the crustal-like zircon chemical characters (Fig. 8).

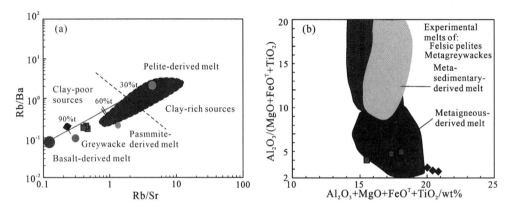

Fig. 11　(a) Rb/Ba vs. Rb/Sr diagrams (Janoussek et al., 2004) and (b) compositional field of experimental melts derived from melting of felsic pelites (muscovite schists), metagreywackes and amphibolites (Patiño Douce, 1999). Symbols as in Fig. 5.

In summary, together with its whole-rock chemical and isotopic data and zircon chemical data, we suggest that the granites were derived from the melting of predominated metasedimantary greywackes in the ancient continental crust.

5.3　Tectonic implications

Zircon U-Pb data show that the peraluminous, high-K, calc-alkaline granites and alkaline syenite in the Tengchong Block were formed ca. 115.3–115.7 Ma, which is consistent with abundant Early Cretaceous granitoids (of age range from 143 Ma to 115 Ma) in the Tengchong Block (Yang et al., 2006; Cong et al., 2011a, b; Li et al., 2012; Luo et al., 2012; Qi et al., 2011; Xie et al., 2010; Xu et al., 2012; Cao et al., 2014; Zhu et al., 2015b). The

Early Cretaceous magmatism in the Tengchong Block intensified from 135 Ma to ~110 Ma and generated intrusions with negative zircon $\varepsilon_{Hf}(t)$ (Fig. 9) and whole-rock $\varepsilon_{Nd}(t)$ compositions (Fig. 7), peraluminous characteristics (A/CNK > 1) (Zhu et al., 2015b), and dominantly Mesoproterozoic two-stage isotopic model ages (Yang et al., 2006; Cong et al., 2011a,b; Qi et al., 2011; Li et al., 2012; Xu et al., 2012; Zhu et al., 2015b; Xie et al., 2016). This led previous workers to suggest that the Early Cretaceous magmatisms in the Tengchong Block was dominated by magmas generated by the partial melting of ancient crustal material.

However, the geodynamical setting of the Early Cretaceous tectono-thermal magmatism in the Tengchong Block has remained controversial due to: ① the active continental margin is related to the eastward subduction of the Putao-Myitkyina paleo-oceanic slab as a branch of the Neo-Tethys Ocean (Cong et al., 2010a, 2011a, b); ② the southward subduction of the Bangong-Nujiang Tethyan oceanic slab occurred during the collision between the Lhasa-Tengchong and the Qiangtang-Baoshan blocks (Yang et al., 2006; Qi et al., 2011; Li et al., 2012; Luo et al., 2012; Cao et al., 2014; Zhu et al., 2015b); and ③ there is a post-collisional setting between the Lhasa and Qiangtang blocks being partly influenced by the far field of the Neo-Tethyan oceanic slab (Xu et al., 2012). In a recent study, Xie et al. (2016) pointed out the closely resemblance between the Early Cretaceous magmatism in the Tengchong Terrane and those in the central and northern Lhasa subterranes. The geochronological, geochemical and paleomagnetic data are consistent with magma generation associated with the subduction of the Tethyan Bangong-Nujiang Ocean lithosphere (Zhu et al., 2009, 2011, 2013, 2015a; Sui et al., 2013; Chen et al., 2014; Yan et al., 2016). Investigations based on petrology, stratigraphy and paleobiogeography (YNBGMR, 1991) also suggest that the Tengchong Terrane is the eastern extension of the Lhasa Terrane (Xie et al., 2016), both of which experienced similar tectono-magmatic histories. Generally, the Early Cretaceous magmatisms were regarded as the products of southward subduction of Bangong-Nujiang Meoso-Tethys (Zhu et al., 2009, 2011, 2015a; Sui et al., 2013; Chen et al., 2014; Zhu et al., 2015b; Xie et al., 2016). In recent, the slab break-off of subducting Bangong-Nujiang Tethyan oceanic lithosphere was proposed (Zhu et al., 2017b). In the slab break-off setting (Fig. 12), a hybrid basaltic magma that had affinity with arc-type and within-plate suites would be produced, the late Early Cretaceous Xainza basalts and bimodal volcanic rocks in central and northern Lhasa subterrane (Sui et al., 2013; Chen et al., 2014) (Fig. 12). Then the hybrid magma provides heat to induce partial melting of ancient lower crust, mature continental basement, and juvenile crust, which resulted in various melts (Fig. 12) including Xainza andesites and dacites (Chen et al., 2014), Daguo rhyolites (Sui et al., 2013), quartz diorite-granodiorite-monzogranite (Zhu et al., 2017b), and the granite-syenite (this study). In this study, both our whole-rock Sr-Nd isotopic and zircon Hf isotopic data share similar characteristics with the magmatism not only in the Tengchong Block but also in the central and northern Lhasa subterrane (both of which were belonging to Northern Magmatic belt which mainly contained the Early-Cretaceous magmatism from central and northern Lhasa to Tengchong Block (Xu et al., 1985; Zhu et al., 2011; Xu et al., 2012). Therefore, we infer that the Early Cretaceous syenites and granites are also the products that associated to the slab break-off of subducted Bangong-Nujiang Meso-Tethys oceanic lithosphere.

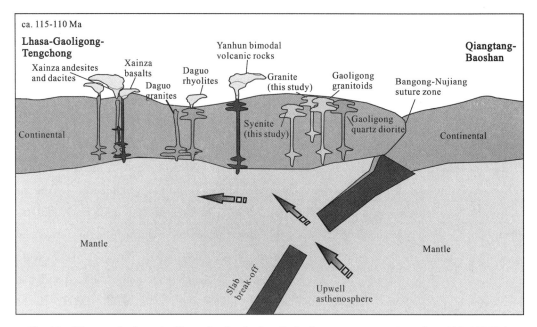

Fig. 12   Diagram of schematic illustration for the late Early-Cretaceous magmatisms (ca. 115−110 Ma) from Lhasa to Gaoligong (modified after Chen et al., 2014 and Zhu et al., 2017b).

6   Conclusions

(1) Zirocn LA-ICP-MS U-Pb data show that the syenites and granites were emplaced at ca.115. 3 ± 0. 9 Ma and ca. 115. 7 ± 0. 8 Ma.

(2) Geochemical, isotopic and in-situ zircon Hf isotopic data show that the syenites were possibly derived from mixing of mafic lower crustal and evolved sediments-derived materials, and the fractionation of some ferro-magnesian mineral and plagioclase may occur during the evolution of magma. The granite were derived from the melting of predominated metasedimantary greywackes in the ancient continental crust.

(3) Together with magmatic data in the literatures, we suggest the Early Cretaceous magmatism in the Tengchong Block was dominated by magmas generated by the partial melting of ancient crustal material. And ca. 115 Ma syenites and granites are also the products that associated to the slab break-off of subducted Bangong-Nujiang Meso-Tethys oceanic lithosphere.

Acknowledgements   This work was jointly supported by the National Natural Science Foundation of China (Grant Nos. 41421002, 41190072, 41372067 and 41102037) and the program for Changjiang Scholars and Innovative Research Team in University (Grant IRT1281). Support was also provided by the Foundation for the Author of National Excellent Doctoral Dissertation of PR China (201324), and the MOST Special Fund from the State Key Laboratory of Continental Dynamics, Northwest University, and Province Key Laboratory Construction Item (08JZ62).

References

Atherton, M.P., Petford, N., 1993. Generation of sodium-rich magmas from newly underplated basaltic crust. Nature, 362: 144-146.

Bonin, B., 1990. From orogenic to anorogenic settings: Evolution of granitoid suites after a major orogenesis. Geological Journal, 25: 261-270.

Brown, M., 2013. Granite: From genesis to emplacement. Geological Society of America Bulletin, 125(7-8): 1079-1113.

Cao, H.W., Zhang, S.T., Lin, J.Z., Zheng, L., Wu, J.D., Li, D., 2014. Geology, geochemistry and geochronology of the Jiaojiguanliangzi Fe-polymetallic deposit, Tengchong County, western Yunnan (China): Regional tectonic implications. Journal of Asian Earth Sciences, 81: 142-152.

Chappell, B.W., White, A.J.R., 1992. I- and S-type granites in the Lachlan Fold Belt. Geological Society of America Special Papers, 272: 1-26.

Chen, S.Z., Xing, G.F., Li,Y.N., Xi, W.W., Zhu, X.T., Zhang, X.D., 2017. Re-recognition of Tieshan "Syenite" and its geological significance in Zhenghe, Fujian Province. Acta Geologica Sinica (English Edition), 91(S1):72-73.

Chen, Y., Zhu, D. C., Zhao, Z. D., Meng, F. Y., Wang, Q., Santosh, M., Wang, L.Q., Dong, G.C., Mo, X.X., 2014. Slab breakoff triggered ca. 113 Ma magmatism around xainza area of the Lhasa terrane, Tibet. Gondwana Research, 26(2): 449-463.

Chu, Z.Y., Chen, F.K., Yang, Y.H., Guo, J.H., 2009. Precise determination of Sm, Nd concentrations and Nd isotopic compositions at the nanogram level in geological samples by thermal ionization mass spectrometry. Journal of Analytical Atomic Spectrometry, 24: 1534-1544.

Cong, F., Lin, S.L., Zou, G.F., Li, Z.H., Xie, T., Peng, Z.M., Liang, T., 2011a. Magma mixing of granites at Lianghe: In-situ zircon analysis for trace elements, U-Pb ages and Hf isotopes. Science China: Earth Sciences, 54: 1346-1359.

Cong, F., Lin, S.L., Zou, G.F., Xie, T., Li, Z.H., Tang, F.W., Peng, Z.M., 2011b. Geochronology and petrogenesis for the protolith of biotite plagioclase gneiss at Lianghe, western Yunnan. Acta Geologica Sinica (English Edition), 85(4): 870-880.

Dorais, M.J., 1990. Compositional variations in pyroxenes and amphiboles of the Belknap Mountain complex, New Hampshire: Evidence for origin of silica-saturated alkaline rocks. American Mineralogist, 75: 1092-1105.

Elliott, T., 2003. Tracers of the slab. In: Eiler, J. Inside the subduction factory, 138. American Geophysical Union, Geophysical Monograph, Washington DC: 23-45.

Frost, B.R., Barnes, C.G., Collins, W.J., Arculus, R.J., Ellis, D.J., Frost, C.D., 2001. A geochemical classification for granitic rocks. Journal of Petrology, 42(11): 2033-2048.

Goolaerts, A., Mattielli, N., de Jong, J., Weis, D., Scoates, J.S., 2004. Hf and Lu isotopic reference values for the zircon standard 91500 by MC-ICP-MS. Chemical Geology, 206: 1-9.

Grimes, C.B., John, B.E., Kelemen, P.B., Mazdab, F.K., Wooden, J.L., Cheadle, M.J., Hanghoj, K., Schwartz, J.J., 2007. Trace element chemistry of zircons from oceanic crust: A method for distinguishing detrital zircon provenance. Geology, 35: 643-646.

Hao,Z.G., Fei, H. Hao,Q.Q., 2015. China's largest granite-type gas field was discovered in Qinghai: The Inorganic theory has aroused attention again. Acta Geologica Sinica, 89(1):302-303.

Harris, N.B.W., Xu, R.H., Lewis, C.L., and Jin, C., 1988. Plutonic rocks of the 1985 Tibet Geotraverse: Lhasa to Golmud. Philosophical Transactions of the Royal Society London, A327: 145-168.

Hoskin, P.W.O., Schaltegger, U., 2003. The composition of zircon and igneous and metamorphic petrogenesis. Reviews in Mineralogy and Geochemistry, 53: 27-62.

Huang, W.L., Wyllie, P.J., 1975. Melting reaction in the system $NaAlSi_3O_8$-$KalSi_3O_8$-SiO_2 to 35 kilobars, dry and with excess water. Journal of Geology, 83: 737-748.

Janoušek, V., Finger, F., Roberts, M., Frýda, J., Pin, C., Dolejš, D., 2004. Deciphering the petrogenesis of deeply buried granites: Whole-rock geochemical constraints on the origin of largely undepleted felsic granulites from the Moldanubian Zone of the Bohemian Massif. Geological Society of America Special Papers, 389: 141-159.

Jin, X.C., 1996. Tectono-stratigraphic units in western Yunnan and their counterparts in southeast Asia. Continent Dynamics, 1: 123-133.

Johannes, W., Holtz, F., 1990. Formation and composition of H_2O-undersaturated granitic melts. In: Ashworth, J. R., Brown, M. High-temperature Metamorphism and Crustal Anatexis. London: Unwin Hyman: 87-104.

Kamber, B.S., Collerson, K.D., 2000. Role of hidden subducted slabs in mantle depletion. Chemical Geology, 166: 241-254.

Li, Z.H., Lin, S.L., Cong, F., Zou, G.F., Xie., T., 2012. U-Pb dating and Hf isotopic compositions of quartz diorite and monzonitic granite from the Tengchong-Lianghe Block, western Yunnan, and its geological implications. Acta Geology Sinica, 86(7):1047-1062 (in Chinese with English abstract).

Lin, I. J., Chung, S. L., Chu, C. H., Lee, H. Y., Gallet, S., Wu, G., et al., 2012. Geochemical and Sr-Nd isotopic characteristics of Cretaceous to Paleocene granitoids and volcanic rocks, SE, Tibet: Petrogenesis and tectonic Implications. Journal of Asian Earth Sciences, 53(2143): 131-150.

Litvinovsky, B.A., Jahn, B.M., Zanvilevich, A.N., Saunders, A., Poulain, S., Kuzmin, D.V., Reichow, M. K., Titov, A. V., 2002. Petrogenesis of syenite-granite suites from the Bryansky Complex (Transbaikalia, Russia): Implications for the origin of A-type granitoid magmas. Chemical Geology, 189: 105-133.

Liu, D., Zhao, Z., Zhu, D.C., Niu, Y., Harrison, T.M., 2014. Zircon xenocrysts in Tibetan ultrapotassic magmas: Imaging the deep crust through time. Geology, 42(1): 43-46.

Lubala, R.T., Frick, C., Roders, J.H., Walraven, F., 1994. Petrogenesis of syenites and granites of the Schiel Alkaline complex, Northern Transvaal, South Africa. Journal of Geology, 102: 307-309.

Ludwig, K.R., 2003. ISOPLOT 3.0: A Geochronological Toolkit for Microsoft Excel. Berkeley Geochronology Center Special Publications, 4.

Luo, G., Jia, X.C., Yang, X.J., Xiong, C.L., Bai, X.Z., Huang, B.X., Tan, X.L., 2012. LA-ICP-MS zircon U-Pb dating of southern Menglian granite in Tengchong area of western Yunnan Province and its tectonic implications. Geological Bulletin of China, 31: 287-296 (in Chinese with English abstract).

Metcalfe, I., 2013. Gondwana dispersion and Asian accretion: Tectonic and paleogeography evolution of eastern Tethys. Journal of Asian Earth Sciences, 66: 1-33.

Miller, C., Schuster, R., Kotzli, U., et al., 1999. Post-collisional potassic and ultrapotassic magmatism in SW Tibet: Geochemical and Sr-Nd-Pb-O isotopic constraints for mantle source characteristics and petrogenesis. Journal of Petrology, 40: 1399-1424.

Patiño Douce, A.E., 1999. What do experiments tell us about the relative contributions of crust and mantle to the origin of the granitic magmas. Geology Society of London, 168: 55-75.

Qi, L., Hu J., Gregoire, D.C., 2000. Determination of trace elements in granites by inductively coupled plasma mass spectrometry. Talanta, 51: 507-513.

Qi, X.X., Zhu, L.H., Hu, Z.C., Li, Z.Q., 2011. Zircon SHRIMP U-Pb dating and Lu-Hf isotopic composition for Early Cretaceous plutonic rocks in Tengchong Block, southeastern Tibet, and its tectonic implications. Acta Petrologica Sinica, 27: 3409-3421 (in Chinese with English abstract).

Qu, X.M., Wang, R.J., Xin, H.B., Jiang, J.H., Chen, H., 2012. Age and petrogenesis of A-type granites in the middle segment of the Bangonghu-Nujiang suture, Tibetan plateau. Lithos, 146: 264-275.

Riishuus, M.S., Peate, D.W., Tegner, C., Wilson, J.R., Brooks, C.K., Waight, T.E., 2005. Petrogenesis of syenites at a rifted continental margin: Origin, contamination and interaction of alkaline mafic and felsic magmas in the Astrophyllite Bay Complex, East Greenland. Contributions to Mineralogy and Petrology, 149: 350-371.

Sheppard, S., 1995. Hybridization of shoshonitic lamprophyre and calc-alkaline granite magma in the Early Proterozoic Mt. Bundey igneous suite, Northern Territory. Australian Journal of Earth Sciences, 42: 173-185.

Song, S.G., Niu, Y.L.,Wei, C.J., Ji, J.Q., Su, L., 2010. Metamorphism, anatexis, zircon ages and tectonic evolution of the Gongshan Block in the northern Indochina continent: An eastern extension of the Lhasa Block. Lithos, 120: 327-346.

Sui, Q. L., Wang, Q., Zhu, D. C., Zhao, Z. D., Chen, Y., Santosh, M., Hu, Z.C., Yuan, H.L., Mo, X. X., 2013. Compositional diversity of ca. 110 ma magmatism in the northern Lhasa Terrane, Tibet: Implications for the magmatic origin and crustal growth in a continent-continent collision zone. Lithos, 168-169(3): 144-159.

Sun, S.S., McDonough, W.F., 1989. Chemical and isotopic systematics of oceanic basalts: Implications or mantle composition and processes. Geological Society, London, Special Publications, 42(1): 313-345.

Sylvester, P.J.,1989. Post-collisional alkaline granites. Journal of Geology, 97: 261-280.

Sylvester, P.J., 1998. Postcollisional strongly peraluminous granites. Lithos, 45: 29-44.

Wang, E., Burchfiel, B., 1997. Interpretation of Cenozoic tectonics in the right-lateral accommodation zone between the Ailao Shan shear zone and the eastern Himalayan syntaxis. International Geology Review, 39: 191-219.

Wang, Y.J., Xing, X.W., Cawood, P.A., Lai, S.C., Xia, X.P., Fan, W.M., Liu, H.C., Zhang, F.F., 2013. Petrogenesis of early Paleozoic peraluminous granite in the Sibumasu Block of SW Yunnan and diachronous accretionary orogenesis along the northern margin of Gondwana. Lithos, 182: 67-85.

Wu, F.Y., Liu, Z.C., Liu, X.C., Ji, W.Q., 2015. Himalayan leucogranite: Petrogenesis and implications to orogenesis and plateau uplift. Acta Petrologica Sinica, 31(1): 1-36.

Xie, J.C., Zhu, D.C., Dong, G., Zhao, Z.D., Wang, Q. Mo, X., 2016. Linking the tengchong terrane in SW Yunnan with the Lhasa terrane in southern Tibet through magmatic correlation.Gondwana Research, 39: 217-229.

Xie, T., Lin, S.L., Cong, F., Li, Z.H., Zou, G.F., Li, J.M., Liang, T., 2010. LA-ICP-MS zircon U-Pb dating for K-feldspar granites in Lianghe region, western Yunnan and its geological significance. Geotectonica et Metallogenia, 34: 419-428 (in Chinese with English abstract).

Xu, R., Scharer, U., Allègre, C.J., 1985. Magmatism and metamorphism in the Lhasa Block(Tibet): A geochronological study. Journal of Geology, 93: 41-57.

Xu, Y.G., Yang, Q.J., Lan, J.B., Luo, Z.Y., Huang, X.L., Shi, Y.R., Xie, L.W., 2012. Temporal-spatial distribution and tectonic implications of the batholiths in the Gaoligong-Tengliang-Yingjiang area, western Yunnan: Constraints from zircon U-Pb ages and Hf isotopes. Journal of Asian Earth Sciences, 53: 151-175.

YBGMR(Yunnan Bureau Geological Mineral Resource), 1991. Regional geology of Yunnan Province. Beijing: Geological Publishing House: 1-729 (in Chinese with English abstract).

Yan M, Zhang D, Fang X, et al., 2016. Paleomagnetic data bearing on the Mesozoic deformation of the Qiangtang Block: Implications for the evolution of the Paleo- and Meso-Tethys. Gondwana Research, 39: 292-316.

Yang, J.H., Sun, J.F., Zhang, M., Wu, F.Y., Wilde, S.A., 2012. Petrogenesis of silica-saturated and silica-undersaturated syenites in the northern north China craton related to post-collisional and intraplate extension. Chemical Geology, 328(11): 149-167.

Yang, J.H., Wu, F.Y., Wilde, S.A., Xie, L.W., Yang, Y.H., Liu, X.M., 2007. Tracing magma mixing in granite genesis: In-situ U-Pb dating and Hf-isotope analysis of zircons. Contributions to Mineralogy and Petrology, 135: 177-190.

Yang, J.H., Chung, S.L., Wilde, S.A., Wu, F.Y., Chu, M.F., Lo, C.H., Fan, H.R., 2005. Petrogenesis of post-orogenic syenites in the Sulu Orogenic Belt, East China: Geochronological, geochemical and Nd-Sr isotopic evidence. Chemical Geology, 214: 99-125.

Yang, J.H., Wu, F.Y., Wilde, S.A., Chen, F., Liu, X.M., Xie, L.W., 2008a. Petrogenesis of an Alkali syenite-granite-rhyolite suite in the Yanshan Fold and Thrust Belt, Eastern North China Craton: Geochronological, geochemical and Nd-Sr-Hf isotopic evidence for lithospheric thinning. Journal of Petrology, 49: 315-351.

Yang, Q. J. , Xu, Y. G. , Huang, X. L. , Luo, Z. Y. , 2006. Geochronology and geochemistry granites in the Gaoligong tectonic belt, western Yunnan, tectonic implications. Acta Petrologica Sinica, 22: 817-834 (in Chinese with English abstract).

Yuan, H. L. , Gao, S. , Liu, X. M. , Li, H. M. , Gunther, D. , Wu, F. Y. , 2004. Accurate U-Pb age and trace element determinations of zircon by laser ablation-inductively coupled plasma mass spectrometry. Geostandard Newsletters, 28: 353-370.

Zhang, B. , Zhang, J. , Zhong, D. , Yang, L. , Yue, Y. , Yan, S. , 2012. Polystage deformation of the Gaoligong metamorphic zone: Structures, ^{40}Ar/^{39}Ar mica ages, and tectonic implications. Journal of Structural Geology, 37: 1-18.

Zhang, L. , Yang, J. , Zhang, J. , 2015. Geochronology and geochemistry of Zengga Mesozoic grantoids from East Gangdese Batholith, implications for the Remelting mechanism of granite Formation. Acta Geologica Sinica (English Edition), 89(A02): 113-114.

Zheng, Y. F. , Zhang, S. B. , Zhao, Z. F. , Wu, Y. B. , Li, X. , Li, Z. , Wu, F. Y. , 2007. Contrasting zircon Hf and O isotopes in the two episodes of Neoproterozoic granitoids in South China: Implications for growth and reworking of continental crust. Lithos, 96: 127-150.

Zhong, D. L, 1998. Paleo-Tethyan orogenic belt in the western parts of the Sichuan and Yunnan Provinces. Beijing: Science Press: 1-231.

Zhong, D. L. , 2000. Paleotethys Sides in West Yunnan and Sichuan, China. Beijing: Science Press: 1-248 (in Chinese with English abstract).

Zhu, D. C. , Li, S. M. , Cawood, P. A. , Wang, Q. , Zhao, Z. D. , Liu, S. A. , et al. , 2015a. Assembly of the lhasa and qiangtang terranes in central Tibet by divergent double subduction. Lithos, 245: 7-17.

Zhu, D. C. , Mo, X. X. , Niu, Y. L. , Zhao, Z. D. , Wang L. Q. , Liu Y. S. , Wu F. Y. , 2009. Geochemical investigation of Early Cretaceous igneous rocks along an east-west traverse throughout the central Lhasa Terrane, Tibet. Chemical Geology, 268: 298-312.

Zhu, D. C. , Zhao, Z. D. , Niu, Y. L. , Mo, X. X. , Chung, S. L. , Hou, Z. Q. , Wang, L. Q. , Wu, F. Y. , 2011. The Lhasa Terrane: Record of a microcontinent and its histories of drift and growth. Earth & Planetary Science Letters, 301: 241-255.

Zhu, D. C. , Zhao, Z. D. , Niu, Y. L. , Dilek, Y. , Hou, Z. Q. , Mo, X. X. , 2013. The origin and pre-Cenozoic evolution of the Tibetan Plateau. Gondwana Research, 23: 1429-1454.

Zhu, R. Z. , Lai, S. C. , Qin, J. F. , Zhao, S. W. , 2015b. Early Cretaceous highly fractionated I-type granites from the northern Tengchong Block, western Yunnan: Petrogenesis and tectonic implications. Journal of Asian Earth Sciences, 100: 145-163.

Zhu, R. Z. , Lai, S. C. , Santosh, M. , Qin, J. F. , Zhao, S. W. , 2017a. Early Cretaceous Na-rich granitoids and their enclaves in the Tengchong Block, SW China: Magmatism in relation to subduction of the Bangong-Nujiang Tethys ocean. Lithos, 286-287: 175-190.

Zhu, R. Z. , Lai, S. C. , Qin, J. F. , Zhao, S. W. , 2017b. Late Early-Cretaceous quartz diorite-granodiorite-monzogranite association from the Gaoligong belt, southeastern Tibet Plateau: Chemical variations and geodynamic implications. Lithos, 288-289: 311-325.

Geochemistry and zircon U-Pb-Hf isotopes of the 780 Ma I-type granites in the western Yangtze Block: Petrogenesis and crustal evolution[①]

Zhu Yu   Lai Shaocong[②]   Qin Jiangfeng   Zhu Renzhi   Zhang Fangyi   Zhang Zezhong

Abstract: Neoproterozoic I-type granites could provide vital insights into the crust-mantle interaction and the crustal evolution along the western Yangtze Block, South China. This paper presents new zircon U-Pb ages, bulk-rock geochemistry, and in situ zircon Lu-Hf isotope on the Dalu I-type granites from the southwestern Yangtze Block. Zircon U-Pb dating show the crystallization ages of 781.1 ± 2.8 Ma for granodiorites and 779.8 ± 2.0 Ma for granites, respectively. The Dalu granodiorites are Na-rich, calc-alkaline, metaluminous to slightly peraluminous (A/CNK = 0.94 − 1.08). Zircons from granodiorite have positive $\varepsilon_{Hf}(t)$ values (+2.16 to +7.39) with crustal model ages of 1.21−1.54 Ga, indicating juvenile mafic lower crust

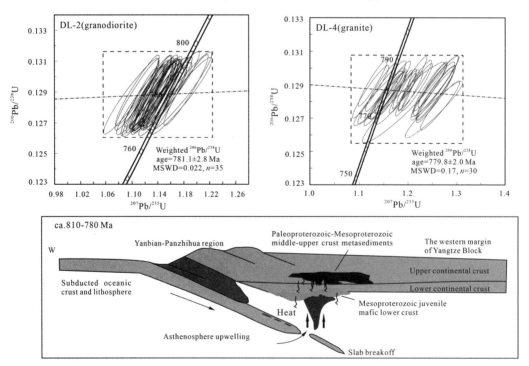

①   Published in *International Geology Review*, 2019, 61(10).

②   Corresponding author.

source. The Dalu granites are high-K calc-alkaline, peraluminous rocks. They have variable zircon $\varepsilon_{Hf}(t)$ values (-4.65 to $+5.80$) with crustal model ages of $1.31-1.97$ Ga, suggesting that they were derived from the mature metasediment-derived melts by the mixing of newly formed mafic lower crust-derived melts. The geochemical variations in Dalu pluton is dominated not only by the different source rocks but also by the different melting temperatures. Combining with the geochemistry and isotopic compositions of I-type granitoids and tectonic setting in the western Yangtze Block, we propose that the Dalu I-type granodiorites-granites associations are the magmatic response from different crustal levels, which were induced by the heat anomaly due to the asthenosphere upwelling in the subduction-related setting.

1 Introduction

Widespread Neoproterozoic mafic-felsic magmatism along the western periphery of Yangtze Block is the best media for investigating the tectonic evolution of South China with the assemblage and breakup of the supercontinent Rodinia (Li, 1999; Li et al., 1995, 1999, 2003a, 2006; Zhou et al., 2002, 2006a,b; Zheng et al., 2007, 2008; Zhu et al., 2008; Cawood et al., 2013; Lai et al., 2015a,b; Zhao et al., 2008a, 2010, 2011, 2017a). Different geodynamic models, including the slab-arc model (Zhou et al., 2002, 2006a,b; Sun et al., 2007, 2008; Sun and Zhou, 2008; Zhao and Zhou, 2008; Zhao et al., 2008a, 2010, 2011, 2017a; Munteanu et al., 2010), the mantle plume model (Li et al., 1995, 1999, 2003b, 2006, 2008), and the plate-rift model (Zheng et al., 2007, 2008), are proposed by various geochronology, geochemistry, and isotopic features from these mafic-felsic rocks. Although we cannot confirm that which geodynamic model is the most suitable mechanism for the tectonic evolution of South China, we could provide more constraints by clarifying the crustal evolution (growth or/and reworking) and the crust-mantle interaction in the western Yangtze Block during the Neoproterozoic.

I-type granites, as the most common continental granites types, are a significant "window" for understanding the crust-mantle interaction and crustal differentiation process (Hawkesworth and Kemp, 2006; Kemp et al., 2007; Liu and Zhao, 2018). They were commonly derived from different sources including purely crustal source or mantle sources (Depaolo et al., 1992; Rudnick, 1995; Collins, 1996; Hawkesworth and Kemp., 2006; Kemp et al., 2007; Clemens et al., 2016), the mixing of mantle-crust compositions (Kemp and Hawkesworth, 2014; Weidendorfer et al., 2014; Zheng et al., 2015; Liu et al., 2018), the ancient or juvenile igneous rocks (Zhao et al., 2007; Chappell et al., 2012; Lu et al., 2016, 2017). Furthermore, their geochemical variations are dictated by the various factors, such as magma mixing, fractional crystallization or partial melting (Chappell and White, 1992; Pitcher, 1997; Miller et al., 2003; Barbarin, 2005; Kemp et al., 2007; Yang et al.,

2007), peritectic assemblages entrainment (Clemens et al., 2010; Clemens and Stevens, 2012), partial melting conditions (Zhao et al., 2015; Lu et al., 2016, 2017), and distinct source rocks (Lai et al., 2015a; Qin et al., 2016; Zhu et al., 2018). Because of these enigmatic arguments, deciphering the petrogenesis and geochemistry diversity of I-type granites is very important to elucidate the crustal evolution and crust-mantle interaction.

Although previous studies have paid more attention to the mafic-ultramafic intrusions, intermediate rocks, and adakitic rocks in the Panzhihua-Yanbian region along the southwestern margin of Yangtze Block (Zhou et al., 2002, 2006a; Zhao and Zhou, 2007a,b; Zhu et al., 2008; Zhao et al., 2008a; Du et al., 2014), the I-type granitoids were rarely reported for investigating their petrogenesis and tectonic implications. In this study, we present new zircon U-Pb age data, whole rock major and trace elements, and zircon Lu-Hf isotopic data from the Dalu I-type granitoids pluton, which is located at the northeast of Yanbian Terrane. On the basis of other reported data and tectonic setting, this paper proposes that the ca. 780 Ma Dalu I-type granodiorite-granite associations are the different crustal response in depth under the subduction-related tectonic setting. Neoproterozoic extensive subduction-related crustal growth and reworking are important for understanding the tectonic evolution of the western Yangtze Block, South China.

2 Geological background

South China Block has unique geological tectonic position, which is divided into the Yangtze Block to the northwest and the Cathaysian Block to the southeast, by the NNE-trending, 1500-km-long Neoproterozoic Jiangnan orogenic belt (Zhao et al., 2011; Wang et al., 2013, 2014; Lai et al., 2015a). The Yangtze Block is separated from the North China Block by the Qinling-Dabie-Sulu high-ultrahigh pressure metamorphic belt to the north (Zheng et al., 2005b; Zhao and Zhou, 2008), from the Indochina Block by the Ailaoshan-Songma suture zone to the southwest, and from the Songpan-Ganzi terrane by the Longmenshan fault to the west (Gao et al., 1999; Zhang et al., 2005; Zhao and Cawood, 2012) (Fig. 1).

A 1 000-km long Neoproterozoic igneous and metamorphic belt, named Hannan-Panxi arc, was located around the western margin of the Yangtze Block (Zhou et al., 2002, 2006a, b; Sun et al., 2007; Sun and Zhou, 2008; Zhao and Zhou, 2008; Du et al., 2014) (Fig. 2a). Neoproterozoic sequences, such as the Yanbian Groups, Ebian Groups, Huangshuihe Groups, and Sinian strata (Geng et al., 2008; Zhao et al., 2010), are overlain by a thick sequence (> 9 km) of Neoproterozoic to Permian strata composed of clastic, carbonate, and metavolcanic rocks (Sun et al., 2007, 2008; Zhu et al., 2008). The Neoproterozoic granitoids are spatially preserved with the basement complexes including the high-grade metamorphic complexes, Mesoproterozoic to earliest Neoproterozoic metasedimentary strata, and mafic metavolcanic rocks (Li et al., 2003b, 2006; Zhou et al., 2006b; Zhao et

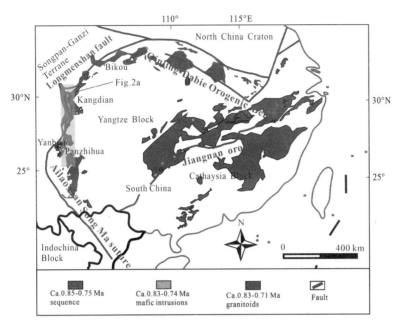

Fig. 1　Sketch geological map of the Yangtze Block, South China
(modified after Zhou et al., 2014; Gan et al., 2017).

al., 2008a, 2010). Moreover, Neoproterozoic volcanic-sedimentary sequences are known as the Yanbian, Suxiong, and Kaijianqiao Groups. Two large serpentinite blocks in Shimian area were considered as ophiolite which indicate an Andean-type arc system in Rodinia (Sun and Vuagnat, 1992; Zhao et al., 2017a).

In the Yanbian Terrane, the Yanbian Group is unconformably overlain by Sinian strata known as the Lieguliu Formation and consists of a sequence of metasediments, pillow lavas, and embedded basaltic sills, with a total thickness of > 6 km (Li et al., 2003a; Sun et al., 2007, 2008; Sun and Zhou, 2008). The Neoproterozoic Guandaoshan dioritic pluton, Tongde dioritic pluton, Gaojiacun and Lengshuiqing mafic-ultramafic intrusions intruded the Yanbian Group and were well preserved and profoundly studied for the tectonic evolution of South China (Li et al., 2003b; Zhou et al., 2006a; Zhu et al., 2008; Munteanu et al., 2010; Du et al., 2014). In the Panzhihua area, the adakitic plutons are spatially accompanied with high-grade metamorphic complexes and low-grade Mesoproterozoic strata (Zhao et al., 2008a). The Datian pluton, dated at 760 Ma by the SHRIMP zircon U-Pb technique (Zhao et al., 2007), is intruded by the Dadukou gabbroic complex to the northwest and Datian mafic-ultramafic dikes to the south (Zhao and Zhou, 2007a; Zhao et al., 2008a; Yang et al., 2016). The Dajianshan pluton, which is chiefly composed of the Neoproterozoic diorite and granodiorite, is situated in the northeast of the Datian pluton (Zhao and Zhou, 2007a). Further northeast, the Neoproterozoic Shuilu, Nanba, and Dalu granitic plutons are located at the northwest of the Puwei town about 10 km distance. These three Neoproterozoic granitoids plutons are rarely

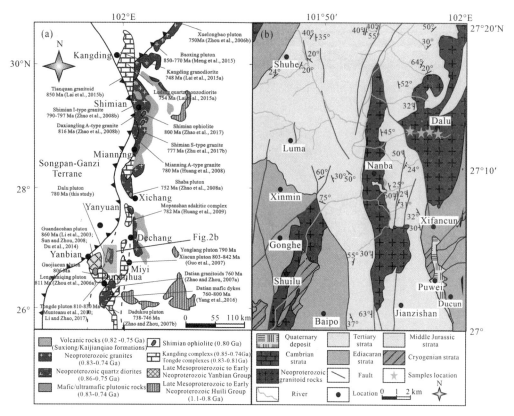

Fig. 2　Sketch geological map of the western margin of Yangtze Block (a) (modified after
Zhou et al., 2006a) and regional geological map of the Dalu pluton (b).

Modified after Yanbian 1∶200 000 geological map, Sichuan Provincial Bureau
of Geology and Mineral Resources(SPBGMR), 1972.

researched for their profound petrogenesis and tectonic implication. In the studied area,
bedding mainly dips 30°–60° with the fault of N-S trend. The Dalu pluton is varying from
northeast to east in strikes, which intruded the older strata, such as Paleoproterozoic
Lengguanzhu Formation, Tianbaoshan Formation, and Huangtian Formation. The Dalu pluton
crops out over 14. 84 km^2 and were overlain by the Sinian strata and Tertiary sediments strata
and underwent variable degrees of regional metamorphism (SPBGMR, 1972) (Fig. 2b).

The Dalu pluton is a complex massif which comprises granodiorite in its dominant facies
and granite along its margins. The granodiorites samples are mostly grey, medium grained
(Fig. 3a). They are composed of plagioclase (30%–35%), quartz (20%–25%), K-feldspar
(10% – 15%), hornblende (10% – 15%), biotite (5% – 10%), and accessory minerals
including magnetite and zircon. The plagioclases are zoned with idiomorphic plate and show the
polysynthetic twin (Fig. 3b,c). The K-feldspars show evidence of alteration in their surfaces.
The quartz exhibits xenomorphic granular among other minerals. Hornblendes are subhedral to
euhedral with some biotite clusters. The biotite displays subhedral shapes. The granites samples

are medium-fine grained (Fig. 3d), and mainly consist of K-feldspar (25% – 30%), plagioclase (22%–27%), quartz (30%–35%), hornblende (2%–5%), biotite (3%–5%), and accessory minerals including magnetite (0%–2%) and zircon. The K-feldspar in Dalu granites show the distinctive feature of the gridiron twining compared with the Dalu granodiorites (Fig. 3e,f). Biotite is filled between the quartz and K-feldspar grains. Analogous to granodiorites, the plagioclases show polysynthetic twin with fine grain.

Fig. 3   Field petrography (a,d) and microscope photographs of granodiorites (b,c) and granites (e,f) from Dalu pluton along the southwestern margin of the Yangtze Block.

Pl: plagioclase; Kfs: K-feldspar; Qtz: quartz; Hbl: hornblende; Mag: magnetite; Bt: biotite.

3   Analytical methods

In this paper, the major and trace elements analyses, zircon geochronology, and in situ zircon Lu-Hf isotopic analyses were carried out at the State Key Laboratory of Continental Dynamics, Northwest University, Xi'an, China (SKLCD).

3. 1　Zircon U-Pb dating

Zircon was separated from ~5 kg samples taken from sampling locations within the Dalu granitoids pluton so that we can select enough and representative zircons. Zircon grains from the sample were separated using conventional heavy liquid and magnetic techniques. Representative zircon grains were handpicked and mounted in epoxy resin disks, then polished and carbon coated. Internal morphology was examined with cathodoluminescence (CL) microscopy prior to U-Pb isotopic dating. Zircon Laser Ablation Inductively Coupled Plasma Mass Spectrometry (LA-ICP-MS) U-Pb analyses were conducted on Agilent 7500a ICP-MS equipped with a 193-nm laser, following the method of Yuan et al. (2004). The $^{207}Pb/^{235}U$ and $^{206}Pb/^{238}U$ ratios were calculated using the GLITTER program, which was corrected using the Harvard zircon 91500 as external calibration. Common Pb contents were subsequently evaluated using the method described in Andersen (2002). The age calculations and plotting of concordia diagrams were made using ISOPLOT (version 3.0; Ludwig, 2003).Uncertainties are quoted at 2σ level.

3. 2　Major and trace elements

Weathered surfaces of samples were removed and the fresh parts were then chipped and powdered to about a 200-mesh size using a tungsten carbide ball mill. Major and trace elements were analyzed by X-ray fluorescence (Rikagu RIX 2100) and inductively coupled plasma mass spectrometry (ICP-MS; Agilent 7500a), respectively. Analyses of USGS and Chinese national rock standards (BCR-2, GSR-1, and GSR-3) showed that the analytical precision and accuracy for the major elements were generally better than 5%. For the trace element analyses, sample powders were digested using an $HF + HNO_3$ mixture in high-pressure Teflon bombs at 190 ℃ for 48 h. For most trace elements, the analytical error was less than 2% and the precision was greater than 10% (Liu et al., 2007).

3. 3　In situ zircon Lu-Hf isotopic analyses

The in situ zircon Hf isotopic analyses were made using a Neptune MC-ICPS. The laser repetition rate was 6 Hz at 100 mJ and the spot sizes were 30 μm. The instrument information is obtainable from Bao et al. (2017). The detailed analytical technique is depicted by Yuan et al. (2008). During analyses, the measured values of well-characterized zircon standards (91500, GJ-1, and Monastery) were consistent with the recommended values within 2σ (Yuan et al., 2008). The obtained Hf isotopic compositions were 0.282 016 ± 20 ($2\sigma_n$, n = 84) for the GJ-1 standard and 0.282 735 ± 24 ($2\sigma_n$, n = 84) for the Monastery standard, respectively, consistent with the recommended values (Yuan et al., 2008) to within 2σ. The initial ^{176}Hf/^{177}Hf ratios and $\varepsilon_{Hf}(t)$ values were calculated with the reference to the chondritic reservoir (CHUR) at the time of zircon growth from the magmas. The decay constant for ^{176}Lu of 1.867×10^{-11} year^{-1} (Soderlund et al., 2004), the chondritic ^{176}Hf/^{177}Hf ratio of 0.282 785 and ^{176}Lu/^{177}Hf ratio of 0.033 6 were adopted. The depleted mantle model ages that were (T_{DM}) used for basic rocks were calculated with reference to the depleted mantle at the present-day ^{176}Hf/^{177}Hf ratio of 0.283 25, similar to that of the average MORB (Nowell et al., 1998) and ^{176}Lu/^{177}Hf = 0.038 4 (Griffin et al., 2000). For the zircons from felsic rocks, we also calculated the Hf isotope "crustal" model age (T_{DMC}) by assuming that its parental magma was derived from an average continental crust, with ^{176}Lu/^{177}Hf = 0.015, that originated from the depleted mantle source (Griffin et al., 2000). Our conclusions would not be affected even if other decay constants were used.

4　Results

4.1　Zircon LA-ICP-MS U-Pb analysis

4.1.1　Zircon trace elements

Zircon trace elements were analyzed simultaneously when the zircon U-Pb ages were determined. The analyzed results are shown in Supplementary Table 1. The rare-Earth elements (REE) pattern of the concordant zircon grains was displayed in the Supplementary Fig. 1.

In general, zircons from Dalu granodiorite (DL-2) and granite (DL-4) are characterized by recognizable fractionated REE. There is uniform REE trend that HREE (heavy RREs) enrichment, LREE (light REEs) depletion, and significant positive Ce anomalies, which indicates a typical magmatic origin (Hoskin and Schaltegger, 2003; Liu et al., 2009). The Dalu granodiorites have U concentrations from 126.21 ppm to 592.54 ppm, Th concentrations from 64.56 ppm to 636.15 ppm, and Th/U ratios from 0.36 to 1.87. For the Dalu granites, there are concordant 30 spots have U concentrations from 91.92 ppm to 594.10 ppm, Th concentrations from 62.44 ppm to 259.26 ppm, and Th/U ratios from 0.20 to 1.07. In addition, they have negative Eu anomalies in the range of 0.01−0.13 for Dalu granodiorite (DL-2) and 0.02−0.17 for Dalu granite (DL-4), respectively.

4.1.2　Zircon morphology and U-Pb dating

Zircon U-Pb concordia diagram and representative CL images of granodiorites and granites from the Dalu Pluton are presented in the Supplementary Table 2, Fig. 4.

(1) *Granodiorite* (*DL-2*). Zircons from Dalu granodiorites (DL-2) are dominantly colorless, transparent and euhedral prismatic crystals. They range in size from 50 μm to 200 μm with aspect ratios about 1 : 1 to 5 : 2. In CL images, these zircon grains are grey and

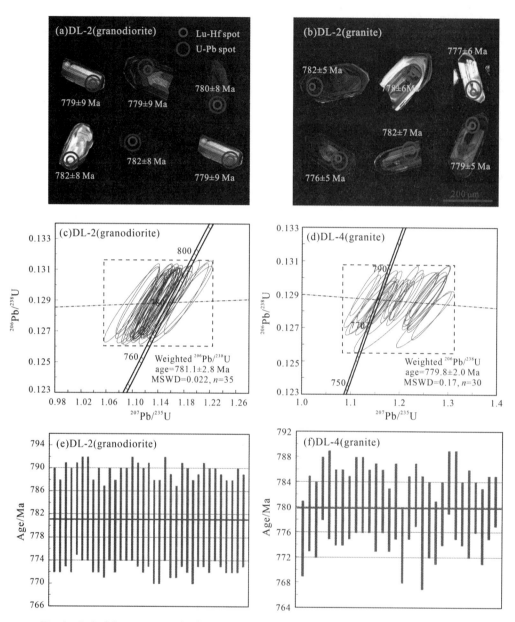

Fig. 4　Cathodoluminescence (CL) images of representative zircons (a, b), LA-ICP-MS U-Pb
zircon concordia diagrams for granodiorites and granites (c-f) from the Dalu pluton
along the southwestern margin of the Yangtze Block.

Ellipse dimensions are 1σ.

black (Fig. 4a). Most of the grains have well-developed oscillatory zoning, which is typical of magmatic origin. One out of 36 spots (3) yields discordant U-Pb age. In addition, these concordant 35 spots are characterized by $^{206}Pb/^{238}U$ ages ranging from 779 ± 8 Ma to 783 ± 9 Ma (weighted mean age of 781.1 ± 2.8 Ma, MSWD = 0.022) (Fig. 4c, e), which represent the crystallization age of the Dalu granodiorite.

（2）*Granite（DL*-4）. Zircons are mainly colorless, transparent and subhedral prismatic crystals, with aspect ratios ranging from 1 : 1 to 2 : 1 in the Dalu granites（DL-4）. In CL images, these zircon grains range $100-350$ μm and display grey or black（Fig. 4b）. Some grains normally show well-developed oscillatory zoning, indicating that they were also formed during the magmatic evolution. Three out of 36 spots（5, 24, 31）yield significantly younger ages（690 ± 4 Ma to 728 ± 8 Ma）, which may be caused by Pb loss in subsequent geological process. These concordant 30 spots are characterized by $^{206}Pb/^{238}U$ ages ranging from 774 ± 6 Ma to 784 ± 5 Ma（weighted mean age of $779. 8 \pm 2. 0$ Ma, MSWD = 0. 17）（Fig. 4d,f）.

4. 2　Major and trace elemental geochemistry

Major and trace element results for the Dalu granodiorite and granite samples are listed in Supplementary Table 3. In the Q-A-P-F diagram（Fig. 5a）, SiO_2 vs. $K_2O + Na_2O$ diagram（Fig. 5d）, and Rb-Ba-Sr diagram（Fig. 5e）, the samples from the Dalu pluton are dominantly plotted into the field of granodiorite and granite.

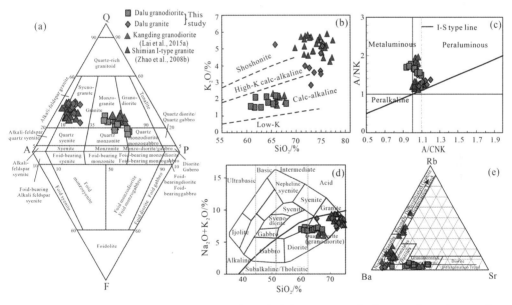

Fig. 5　Q-A-P-F diagram after Middlemost（1994）（a）; K_2O vs. SiO_2 diagram after Roberts and Clemens（1993）（b）; A/NK vs. A/CNK diagram after Frost et al.（2001）（c）; TAS diagram for classification of the rocks Middlemost（1994）（d）; Rb-Ba-Sr diagram（Wang et al., 2013）（e）for granodiorites and granites from the Dalu pluton along the southwestern margin of the Yangtze Block. The data of Kangding granodiorites from Lai et al.（2015a）, Shimian I-type granites from Zhao et al.（2008b）.

4. 2. 1　Dalu granodiorite

The Dalu granodiorites display intermediate SiO_2（$60. 88-68. 07$ wt%）, medium K_2O（$1. 47-2. 18$ wt%）with high Na_2O/K_2O ratio（$2. 27-3. 65$）which are similar with the Kangding granodiorites in the northwestern margin of Yangtze Block（Lai et al., 2015a）

(Fig. 5). All the granodiorite samples are calc-alkaline, metaluminous to weakly peraluminous rocks with A/CNK values of 0. 94–1. 08 (Fig. 5b,c). These samples have high Na_2O = 4. 95– 5. 49 wt%, medium $Fe_2O_3^T$ = 3. 83–5. 45 wt%, CaO = 1. 99–4. 69 wt%, and MgO = 1. 21– 2. 00 wt%, with $Mg^#$ values of 39. 94–47. 03. In the chondrite-normalized rare-earth element REE diagram (Fig. 6a), the Dalu granodiorite samples have fractionated REE patterns with total REE from 101. 19 ppm to 412. 17 ppm and show enrichment of LREE with $(La/Yb)_N$ = 4. 46– 13. 35 and flat HREE pattern with low $(Dy/Yb)_N$ = 4. 50–9. 20. In addition, they display negative Eu anomalies of 0. 13–0. 25. In the N-MORB-normalized trace element diagram (Fig. 6b), the Dalu granodiorites are enriched in Rb, Ba, and Pb, depleted in Nb, Ta, P and Sr.

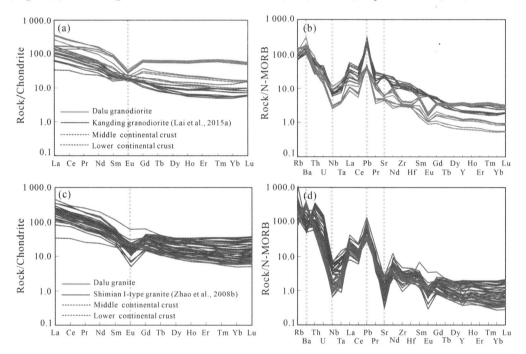

Fig. 6　Diagrams of chondrite-normalized REE patterns (a,c) and N-MORB-normalized trace element patterns (b,d) for granodiorites and granites from the Dalu pluton along the southwestern margin of the Yangtze Block.

The N-MORB and chondrite values are from Sun and McDonough (1989).

4. 2. 2　Dalu granite

The Dalu granites have high concentrations of SiO_2(71. 80–75. 38 wt%, except a 65. 74 wt%), Na_2O (3. 09–4. 42 wt%), elevated concentrations of K_2O (2. 85–5. 31 wt%), lower Na_2O/K_2O values of 0. 58–1. 51 than the Dalu granodiorite. They contain low concentrations of $Fe_2O_3^T$(0. 85–4. 31 wt%) and relatively high concentrations of Al_2O_3(12. 84–16. 49 wt%) with A/CNK ratios of 1. 05–1. 20. On the SiO_2-K_2O and A/CNK-A/NK diagrams (Fig. 5b,c), the Dalu granites are peraluminous, high-K calc-alkaline rocks, indicating the similar compositions with the Shimian I-type granites (Zhao et al., 2008b). The Dalu granites contain 118. 95 – 445. 96 ppm total REE and have $(La/Yb)_N$ values of 3. 78 – 12. 24 and $(Dy/Yb)_N$ values of 3. 40–13. 47. In the chondrite-normalized REE diagram (Fig. 6c), the Dalu granites show slightly negative Eu anomalies ($Eu/Eu^* = 0. 22–0. 31$). They have similar N-MORB-normalized trace element distribution patterns (Fig. 6d), which are more enriched in large ion lithophile elements (LILE) (e.g., Rb, Ba), but depleted in high field strength elements (HFSEs) (e.g., Nb, Ta) than the Dalu granodiorites.

4. 3　Zircon Lu-Hf isotopic compositions

We selected typical zircon grains from the granodiorites (DL-2) and granites (DL-4) for in situ Lu-Hf isotopic analysis. The zircons from two samples were also analyzed for Lu-Hf isotopes on the same domains, and the results are listed in Supplementary Table 4 and Fig. 7. Initial $^{176}Hf/^{177}Hf$ ratios and $\varepsilon_{Hf}(t)$ values were calculated according to their crystallization ages.

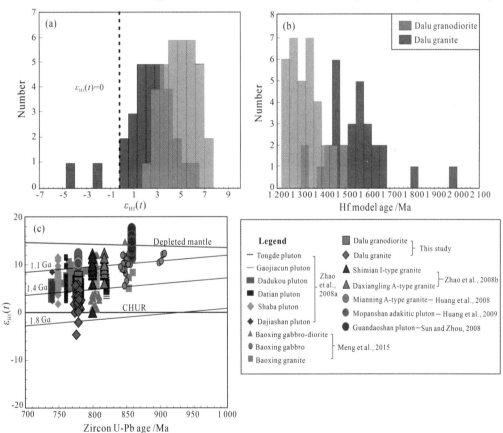

Fig. 7　Histograms of $\varepsilon_{Hf}(t)$ isotope ratios (a)and Hf model age (b)and zircon U-Pb age vs. $\varepsilon_{Hf}(t)$ (c) for granodiorites and granites from the Dalu pluton along the southwestern margin of the Yangtze Block. Data sources of Fig. 7c from Huang et al.(2008, 2009), Sun and Zhou (2008), Zhao et al.(2008a,b), and Meng et al.(2015).

4.3.1 Granodiorite (DL-2)

Zircon from Dalu granodiorites (DL-2) have low $^{176}Lu/^{177}Hf$ ratios range from 0.000 793 to 0.002 738 and present-day $^{176}Hf/^{177}Hf$ ratios from 0.282 397 to 0.282 560. The initial $^{176}Hf/^{177}Hf$ ratios ($^{176}Hf/^{177}Hf$)$_i$ vary from 0.282 371 to 0.282 527 with an average of 0.282 462. Zircon from granodiorite shows varied and positive $\varepsilon_{Hf}(t)$ values varying from +2.16 to +7.39 ($n=35$) (Fig. 7a), with corresponding single-stage Hf model age of 1 063–1 270 Ma and crustal Hf model ages of 1 214–1 544 Ma (Fig. 7b).

4.3.2 Granites (DL-4)

The 30 zircon grains from Dalu granites (DL-4) also share the various Hf isotopic compositions. Their $^{176}Lu/^{177}Hf$ and present-day $^{176}Hf/^{177}Hf$ ratios range from 0.000 985 to 0.002 578 and from 0.282 207 to 0.282 494, respectively. Initial $^{176}Hf/^{177}Hf$ ratios ($^{176}Hf/^{177}Hf$)$_i$ range from 0.282 182 to 0.282 473 with a mean of 0.282 375. Two of 30 $\varepsilon_{Hf}(t)$ values are negative (−4.65 to −2.03) with the single-stage Hf model age of 1 453–1 540 Ma (Fig. 7a), and 3 are 28 positive $\varepsilon_{Hf}(t)$ values from +0.11 to +5.80 with single-stage Hf model age of 1 116–1 338 Ma. Their crustal Hf model age varies from 1 310 Ma to 1 968 Ma (Fig. 7b).

5 Discussion

5.1 Genetic type of the Dalu pluton

Dalu pluton consists of the granodiorites and granites, which the geochronological data were rarely reported. Our new LA-ICP-MS zircon U-Pb dating yielded the crystallization age of 781.1 ± 2.8 Ma for the Dalu granodiorite, while the Dalu granite was formed at 779.8 ± 2.0 Ma. Therefore, the Dalu pluton was emplaced at ca. 780 Ma belonging to the Neoproterozoic magmatism. Their crystallization ages are same with the Mianning A-type granites and Mopanshan adakitic rocks in the west margin of Yangtze Block (Huang et al., 2008, 2009).

The Dalu granodiorites-granites associations have similar I-type trend with the typical I-type granitoids (Kangding granodiorites and Shimian granites) in the western Yangtze Block (Fig. 8) (Zhao et al., 2008b; Lai et al., 2015a). Concretely, the Dalu granodiorites have medium and variable SiO_2 contents (60.88–68.07 wt%) and are calc-alkaline, metaluminous to weakly peraluminous series (A/CNK = 0.94 − 1.08, but most being > 1.0). The Dalu granites are high-K calc-alkaline, peraluminous rocks (A/CNK = 1.05−1.20) with high SiO_2 contents (71.80–75.38 wt%, except a 65.74 wt%). Experimental petrology displayed that the apatite is highly soluble in peraluminous melts (Wolf and London, 1994). The low P_2O_5 contents (0.02–0.21 wt%), and negative correlation between SiO_2 and P_2O_5 for the Dalu

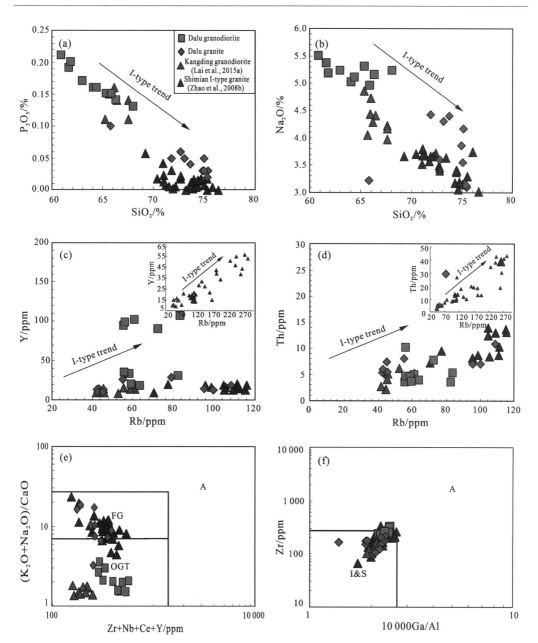

Fig. 8 SiO_2 vs. P_2O_5(a), SiO_2 vs. Na_2O (b), Rb vs. Y (c), Rb vs. Th (d) (Chappell and White, 1992), $(Zr+Nb+Ce+Y)$ vs. $(K_2O+Na_2O)/CaO$ (e), and 1 000 Ga/Al vs. Zr (f) (Whalen et al., 1987) diagrams for granodiorites and granites from the Dalu pluton along the southwestern margin of the Yangtze Block.

pluton indicate I-type granites trend and preclude the possibly of S-type granites that are strongly peraluminous with high P_2O_5 contents (Chappell and White, 1992; Clemens, 2003) (Fig. 8a). The obvious Na_2O decreases with increasing SiO_2 also show the I-type tendency (Fig. 8b) (Chappell and White, 1992). The presence of hornblende and biotite in the Dalu pluton is the significant mineral indicator of I-type granite (Chappell and White, 1992) (Fig. 3). Similarly to the typical I-type granites from the Tengchong Block (Zhu et al., 2015), the Dalu pluton also shows positive relation between Rb and Th with Y (Fig. 8c,d), displaying the I-type affinity (Chappell and White, 1992). Moreover, in the ($Zr+Nb+Ce+Y$) vs. (Na_2O+K_2O)/CaO and 10 000 Ga/Al vs. Zr diagrams (Fig. 8e,f), the samples plot into the field of I and S types. Thus, rocks from the Dalu pluton are considered to be I-type granites (Whalen et al., 1987).

5. 2　Nature of the magma source

5. 2. 1　Dalu granodiorites：Melts from juvenile mafic lower crust

The Dalu granodiorites are calc-alkaline and metaluminous to weakly peraluminous I-type series. Commonly, the calc-alkaline I-type granitic rocks could be formed by：① the partial melting of intermediate to mafic igneous sources (Roberts and Clements, 1993; Kemp et al., 2007; Clemens et al., 2009); ② magma mixing of basaltic and felsic compositions (Kemp et al., 2007); and ③ differentiation of mantle-derived mafic magmas (Soesoo, 2000).

The differentiation of mantle-derived mafic magmas would leave larger amounts of mafic-ultramafic cumulates (Clemens et al., 2011; Clemens and Stevens, 2012). As we can see from Fig. 2b, there are not any mafic-ultramafic intrusions around the Dalu pluton. Zircon can maintain its primary composition features due to the high stability (Valley, 2003; Zheng et al., 2004). The zircon trace elements of the Dalu granodiorites have medium Th and U contents, indicating the crustal affinity (Fig. 9a). Their low U/Yb ratios and high Y contents show the typical continental granitoids (Fig. 9b). The Dalu granodiorites contain variable SiO_2 (60. 88-68. 07 wt%), higher Na_2O contents (4. 95-5. 49 wt%), lower MgO (1. 21-2. 00 wt%) with $Mg^{\#}$ values (39. 94-47. 03), lower Cr (6. 08-11. 63 ppm) and Ni (2. 95-7. 62 ppm) contents than the typical mantle-derived intermediate-felsic rocks (Stern and Hanson, 1991; Smithies and Champion, 2000), which suggest that the mantle-derived source is unreasonable. Kemp et al. (2007) and Xiao et al. (2017) indicated that magma mixing could generate massive mafic enclaves and geochemical variations. Although a relatively wide range of geochemical compositions are detected from the Dalu granodiorites, our field observations and previous researchers did not discover the mafic microgranular enclaves. In the SiO_2 vs. $Mg^{\#}$ diagram (Fig. 10a), the Dalu granodiorites are not in conformity with the magma mixing trend. Additionally, an individual zircon sample for $\varepsilon_{Hf}(t)$ values displaying up to 10 epsilon units may show the magma mixing tendency (Kemp et al., 2007). Our zircon Hf isotopes from the Dalu granodiorite have positive $\varepsilon_{Hf}(t)$ values ranging from + 2. 16 to + 7. 39 (Fig. 7). Therefore, the scenario of magma mixing could be precluded for the petrogenesis of Dalu granodiorites.

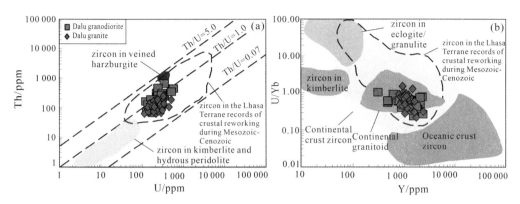

Fig. 9　Discriminant diagrams for zircon origin for the granodiorites and granites from the Dalu pluton
along the southwestern margin of the Yangtze Block.

U vs. Th diagram after Zhu et al. (2018); Y vs. U/Yb diagram after Grimes et al. (2007).

The calc-alkaline I-type granodiorites can also be formed by the partial melting of intermediate to mafic compositions (Beard and Lofgren, 1991; Wolf and Wyllie, 1994). The variable and positive zircon $\varepsilon_{Hf}(t)$ values (+2. 16 to +7. 39) (Fig. 7a,c), coupled with the crustal Hf model ages of 1. 21−1. 54 Ga (Fig. 7b), showing that the Dalu granodiorites were mainly derived from Mesoproterozoic juvenile lower crustal materials. The low Nb/Y (0. 20− 0. 30) and Rb/Y (0. 57−3. 61) ratios also indicate a lower crust source (Rudnick and Fountain, 1995) (Fig. 10b). The Na-rich Dalu granodiorites have high Na_2O (4. 95−5. 36 wt%) and Al_2O_3(15. 51−18. 34 wt%), low MgO (1. 21−1. 90 wt% <3 wt%) with $Mg^{\#}$ values (39. 94−47. 03, most of being <45), medium CaO (1. 99−4. 69 wt%), and so their parental melts are derived from the basaltic (mafic) rocks (Rapp and Watson, 1995). Their medium CaO/Na_2O (0. 38−0. 84) and Al_2O_3/TiO_2(28. 79−49. 20) ratios, low Rb/Ba (0. 03−0. 09) and Rb/Sr (0. 10−0. 20) ratios demonstrate the basalt-derived compositions (Fig. 10c,d). Rapp and Watson(1995) further proposed that ~20%−40% partial melting of basaltic rocks produce high Al_2O_3 and sodic granodioritic melts, which from the amphibole-out phase boundary. Voluminous experimental data affirm that dehydration melting of amphibolites in the lower crust can produce metaluminous to mildly peraluminous granodioritic melts (Beard and Lofgren, 1991; Wyllie and Wolf, 1993; Rapp and Watson, 1995). In the major elements feature diagrams (Fig. 10e,f), the Dalu granodiorites display the significant similarity with the experimental melts of amphibolite-bearing mafic rocks (Patiño Douce, 1999; Lu et al., 2016, 2017). Their high Yb (1. 57−10. 87 ppm, average value = 6. 14 ppm) and Y (17. 61−106. 80 ppm, average value = 62. 84 ppm) contents indicate that the source rocks were situated at the stability field of amphibole. Moreover, the depletion of HFSEs (e.g., Nb, Ta, and Hf) also give the constraint for the presence of residual amphibole in the source (Zhao et al., 2017b) (Fig. 6b).

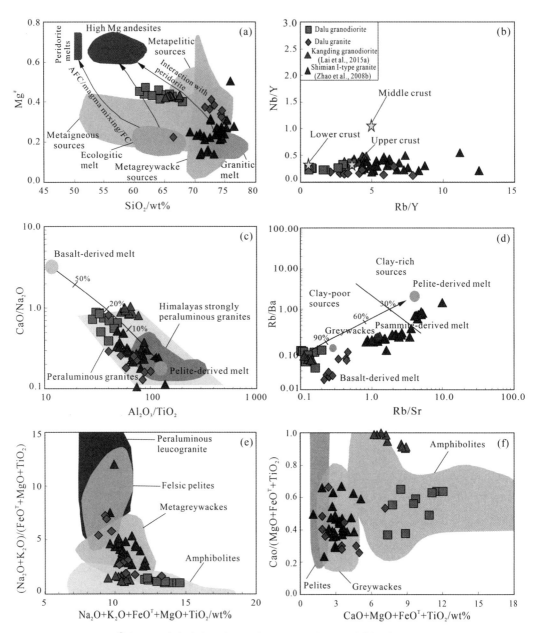

Fig. 10　SiO_2 vs. $Mg^\#$ diagram （a）（after Zhu et al., 2018; reference fields after Patiño Douce, 1999; Wolf and Wyllie, 1994）, Rb/Y vs. Nb/Y diagram （b）（the lower and middle crustal compositions are from Rudnick and Fountain, 1995; and the upper crustal compositions are from Taylor and McLennan, 1985）, Al_2O_3/TiO_2 vs. CaO/Na_2O （c）（Sylvester, 1998）, Rb/Sr vs. Rb/Ba （d）（Patiño Douce, 1999）, （$Na_2O+K_2O+FeO^T+MgO+TiO_2$） vs. （$Na_2O+K_2O$）/（$FeO^T+MgO+TiO_2$）（e）（Patiño Douce, 1999）, （$CaO+MgO+FeO^T+TiO_2$） vs. $CaO/$（$MgO+FeO^T+TiO_2$）（f）（Patiño Douce, 1999）, for the granodiorites and granites from the Dalu pluton along the southwestern margin of the Yangtze Block.

5.2.2 Dalu granites: Melts from mature metasediment by the mixing of newly formed mafic lower crust-derived melts

Compared with the Dalu granodiorites, the Dalu granites have higher SiO_2 contents (71.80−75.38 wt%), K_2O contents (2.85−5.31 wt%), lower CaO contents (0.45−2.58 wt%). They belong to the high-K calc-alkaline and peraluminous I-type granites. The partial melting of crustal source materials is a predominant explanation for the peraluminous granites (Chappell et al., 2012). On the Al_2O_3/TiO_2 vs. CaO/Na_2O and Rb/Sr vs. Rb/Ba diagrams (Fig. 10c,d), the Dalu granites show the affinity with the mixing sedimentary source of pelite-derived melts and greywacke melts. In the major elements features diagrams (Fig. 10e,f), most samples of Dalu granites are plotted in the field of metagreywacks melts, a few samples show the affinity with the experimental melts of pelites. Their low Nb/Y (0.12−0.29) and high Rb/Y (2.07−7.87) ratios show the approach to the upper crust source (Taylor and Mclennan, 1985) (Fig. 10b). As mentioned above, the Dalu granodiorites were derived from the partial melting of newly formed mafic lower crust. The zircon $\varepsilon_{Hf}(t)$ values of Dalu granites (−4.65 to +5.80) are lower than the Dalu granodiorites (+2.16 to +7.39) (Fig. 7a,c). Their crustal Hf model ages (1.31−1.97 Ga) are obviously older than the Dalu granodiorites (1.21−1.54 Ga) (Fig. 7b), which suggest the addition of the mature crustal components. The variation zircon $\varepsilon_{Hf}(t)$ units of Dalu granites may be caused by the fact that zircon crystallization commenced before the ingestion of supracrustal sedimentary (Kemp et al., 2007). Their positive $\varepsilon_{Hf}(t)$ values (+0.11 to +5.80) show the information of juvenile lower crust source, while the negative $\varepsilon_{Hf}(t)$ values (−4.65 to −2.03) indicate the injection of mature crustal materials. Actually, it is common that the source of I-type granites involves into the mature sedimentary materials. Kemp et al. (2007) suggested that classic hornblende-bearing I-type granites of eastern Australia form by the reworking of sedimentary materials by mantle-like magmas instead of by remelting ancient metamorphosed igneous rocks. Chappell et al. (2012) revealed that the incorporation of sedimentary components, either in a partial melt or through bulk assimilation, is an obvious mechanism by which the peraluminous I-type granites can be generated. Zhao et al. (2015) proposed that the sedimentary components in the form of hydrous melts would be conveyed to the I-type granites. Combining the geochronological, geochemical and isotopic data of Dalu granites, we therefore concluded that the Dalu granites were derived from the mature metasediment-derived melts by the mixing of newly formed mafic lower crust-derived melts.

In general, we concluded that the Dalu granodioritewas generated by the partial melting of juvenile mafic lower crust, while the Dalu granite was derived from the mature metasediment-derived melts by the mixing of newly formed mafic lower crust-derived melts.

5. 3 Geochemical variations of Dalu Pluton

In the major elements Harker diagrams, with increasing SiO_2 contents, the samples of Dalu pluton show decrease of major elements (Supplementary Fig. 2a-f). The geochemical variations could depend on many factors such as the source rocks (Lai et al., 2015a; Qin et al., 2016; Zhu et al., 2018), partial melting conditions (Zhao et al., 2015; Lu et al., 2016, 2017), entrainment peritectic minerals (Qin et al., 2016; Zhu et al., 2018), magma evolution processes (e.g., magma mixing, fractional crystallization, and assimilation of country rocks) (DePaolo, 1981; Collins, 1996; Healy et al., 2004; Kemp et al., 2007).

As discussed about the magma nature of the Dalu pluton, it is unlikely that the magma mixing model and AFC process are the main mechanisms in their magma evolution. The absence of peritectic minerals (e.g. garnet and so on) in the Dalu granites could preclude the possibly of peritectic connexion (Clemens et al., 2011; Zhu et al., 2018) (Fig. 3). The Dalu granodiorites were derived from the partial melting of mafic lower crustal compositions, whereas the Dalu granites were derived from the magma mixing of mature crustal metasedimentary and juvenile lower crust compositions. It is obvious that the different protolith is a foremost factor in their geochemical variations. In the Rb-Ba-Sr diagram (Fig. 5e), the Dalu granodiorites display slightly differentiation trend and the Dalu granites were plotted into the field of normal granites, not the strongly differentiated granites. The positive geochemical trend of Th and Th/Nd ratios with La_N and $(La/Sm)_N$ support that the partial melting trend dominantly dictates the geochemistry compositions, the fractional crystallization may play a minor role for the geochemical diversity (Schiano et al., 2010; Zhang et al., 2018) (Fig. 11). Besides the nature of protolith and the partial melting process, the partial melting conditions, especially the melting temperature (Zhao et al., 2015; Lu et al., 2016, 2017), also play a key role in dictating the composition variations in the Dalu pluton. The zircon saturation thermometer (T_{Zr}) is an effective parameter to estimate the partial melting temperature of granitoids (Hanchar and Watson, 2003; Miller et al., 2003). The calculated zircon saturation temperatures for the Dalu granodiorites span a range from 791 ℃ to 819 ℃, which are higher than the Dalu granites ranging of 763−811 ℃. The Dalu granodiorites display higher Rb/Ba ratios, ($Fe_2O_3^T + MgO$) contents, and P_2O_5 contents but lower Al_2O_3/TiO_2 ratios, Rb/Sr ratios, and Th contents than the Dalu granites (Supplementary Fig. 3a-f). The systematic differences in the geochemical parameters and zircon saturation temperatures from the Dalu granitoids samples indicate that melting temperature has affected the geochemical compositions.

In general, the geochemical compositions variations from Dalu granodiorites to granites are determined not only by the protolith nature but also by the melting temperature.

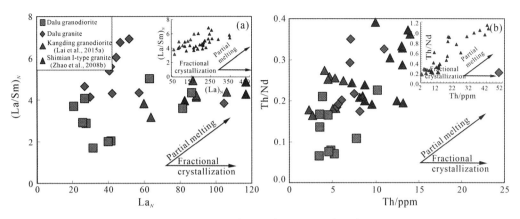

Fig. 11　The partial melting trend of La$_N$ vs. (La/Sm)$_N$ diagram (a) (after Zhang et al., 2018) and
Th vs. Th/Nd diagram (b) (after Schiano et al., 2010) for the granodiorites and granites
from the Dalu pluton along the southwestern margin of the Yangtze Block.

5.4　Geodynamic implication

5.4.1　Subduction-related tectonic setting

The Dalu granitoids was established at the Neoproterozoic with ages of ca. 780 Ma. Neoproterozoic magmatism along the western periphery of Yangtze Block includes not only felsic plutons but also some intermediate-basic volcanic rocks which have ages ranging from ca. 860 Ma to ca. 740 Ma (Zhou et al., 2002, 2006b; Sun et al., 2007, 2008; Zhao and Zhou, 2007a,b; Huang et al., 2008, 2009; Sun and Zhou, 2008; Zhao et al., 2008b, 2017a; Munteanu et al., 2010; Meng et al., 2015; Lai et al., 2015a,b; Zhu et al., 2017b; Li and Zhao, 2018) (Figs. 2a and 12, and Table 1). Although such Neoproterozoic magmatism is successive, the statistical data display that they have three main age peaks at ca. 860 Ma, ca. 780 – 800 Ma, and ca. 740 Ma (Fig. 12). Zhou et al. (2006b) suggested that the Neoproterozoic magmatism was subduction-related by the evidence of ca. 750 Ma Xuelongbao adakitic complex. They emphasized that the 1 000-km-long Panxi-Hannan arc was located at the periphery of Yangtze Block and the South China was placed at the margin of supercontinent Rodinia (Zhou et al., 2006a,b; Sun et al., 2008; Zhao et al., 2008a). Zhao et al. (2017a) proposed that the western margin of Yangtze Block is an Andean-type arc system by the ca. 800 Ma SSZ-type (supra-subduction zone-type) Shimian ophiolite. The ca. 748 Ma Kangding granodiorites, ca. 754 Ma Luding quartz monzodiorites, ca. 800 Ma Shimian I-type granites, and ca. 816 Ma Daxiangling A-type granites support the subduction-related proposal (Zhao et al., 2008b; Lai et al., 2015a). Massive geochemical and geochronological data showed that the Neoproterozoic mafic-felsic intrusions were formed at arc-related tectonic background (Zhou et al., 2002, 2006b; Sun et al., 2007, 2008; Zhao and Zhou, 2007a,b; Zhao et al., 2008b;

Du et al., 2014; Lai et al., 2015a, b). In the southwestern Yangtze Block, these Neoproterozoic intermediate-basaltic plutons contain the Guandaoshan pluton (granodiorites, diorites, gabbro-diorites) at ca. 860 Ma (Sun and Zhou, 2008; Du et al., 2014); Fangtian basaltic lavas at ca. 840 Ma (Li et al., 2006; Sun et al., 2007); the Tongde intrusions at ca. 810–830 Ma and high-Mg diorite dikes at ca. 833 Ma (Munteanu et al., 2010; Li and Zhou, 2018); the Gaojiacun and Lengshuiqing mafic intrusions at ca. 806–812 Ma (Zhou et al., 2006a); the Dadukou olivine gabbro plutons at ca. 746 Ma and hornblende olivine gabbro at ca. 738 Ma (Zhao and Zhou., 2007b). In addition, there are felsic magmatism including ca. 760 Ma Datian and Dajianshan adakitic pluton (Zhao and Zhou, 2007a; Zhao et al., 2008a). It is clear that the Neoproterozoic continuous arc-related magmatism was documented in the tectonic evolution of Yangtze Block, South China.

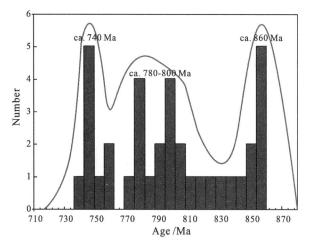

Fig. 12　The histogram of zircon U-Pb ages of Neoproterozoic magmatism in the western Yangtze Block, South China.

Data source from Table 1.

5.4.2　Widespread crustal evolution

Our study for Dalu granitoid pluton yields the zircon U-Pb ages of ca. 780 Ma which is older than the ca. 750 Ma Xuelongbao adakitic complex that derived from the partial melting of a subducted oceanic slab (Zhou et al., 2006b). The Dalu granodiorites are typical of calc-alkaline magmatism of active continental margins under the subduction-related tectonic background (Chen et al., 2015), while the Dalu granites are magmatic response on the middle-upper crust in the subduction-related setting. Their arc-like geochemical compositions are further characterized by enrichment of LILE and LREE and depletion of HREE (Zhao and Zhou, 2008; Cai et al., 2015; Chen et al., 2015) (Fig. 6). Neoproterozoic continuous crustal growth and reworking were widespread under the subduction setting along the periphery of western Yangtze Block (Zhao et al., 2008a; Huang et al., 2009) (Table 1). In the

Table 1 Isotopic ages of the Neoproterozoic magmatism along the western margin of the Yangtze Block, South China.

Location	Pluton name	Lithology	Age/Ma	Dating method	Isotopic	Model age/Ga	Reference
Kangding-Mianning region	Xuelongbao adakitic complex	Tonalite and granodiorite	748±7	SHRIMP	$\varepsilon_{Nd}(t)=+0.36$ to $+2.88$		Zhou et al., 2006b
	Baoxing	Gabbro	~850	SHRIMP	$\varepsilon_{Hf}(t)=+5.44$ to $+11.94$	$T_{DM(Hf)}=0.94-1.19$	Meng et al., 2015
		Gabbro-diorites	~800		$\varepsilon_{Hf}(t)=+3.99$ to $+14.62$	$T_{DM(Hf)}=0.89-1.21$	
					$\varepsilon_{Hf}(t)=+0.68$ to $+5.66$	$T_{DM(Hf)}=1.21-1.37$	
		Granites	~770		$\varepsilon_{Hf}(t)=+2.53$ to $+10.58$	$T_{DM2(Hf)}=1.00-1.51$	
	Kangding	Granodiorite	748±11	LA-ICP-MS	$\varepsilon_{Nd}(t)=+0.8$ to $+1.3$	$T_{DM2(Nd)}=1.50-1.51$	Lai et al., 2015a
	Luding	Quartz monzodiorite	754±10		$\varepsilon_{Nd}(t)=+2.4$ to $+4.8$	$T_{DM2(Nd)}=1.08-1.31$	
	Tianquan	Granite	851±15	LA-ICP-MS	$\varepsilon_{Nd}(t)=+4.4$ to $+8.3$	$T_{DM2(Nd)}=1.21-1.41$	Lai et al., 2015b
	Daxiangling	Granite (A-type)	816±10	SHRIMP	$\varepsilon_{Nd}(t)=+0.3$ to $+1.8$	$T_{DM2(Nd)}=1.7-1.8$	Zhao et al., 2008b
					$\varepsilon_{Hf}(t)=+6.0$ to $+12.4$	$T_{DM2(Hf)}=0.93-1.34$	
	Shimian	Granite (I-type)	797±22	SHRIMP	$\varepsilon_{Nd}(t)=+0.3$ to $+1.8$	$T_{DM2(Nd)}=1.25-1.56$	
			790±10	LA-ICP-MSP	$\varepsilon_{Hf}(t)=+2.9$ to $+8.2$	$T_{DM2(Hf)}=1.34-1.36$	
	Shimian ophiolite	Gabbro + Peridotite	~800	SHRIMP	$\varepsilon_{Hf}(t)=-2.23$ to -2.81		Zhao et al., 2017
	Shimian	Granite	777±4.8	LA-ICP-MS	$\varepsilon_{Nd}(t)=+0.5$ to $+3.3$	$T_{DM2(Nd)}=1.19-1.61$	Zhu et al., 2017b
	Mianning	Granite (A-Type)	~780	SHRIMP + LA-ICP-MS	$\varepsilon_{Nd}(t)=+2.97$ to $+4.71$	$T_{DM2(Nd)}=0.99-1.99$	Huang et al., 2008
					$\varepsilon_{Hf}(t)=+9.2$ to $+12.1$	$T_{DM(Hf)}=0.87-1.00$	
Xicalang-Miyi region	Shaba	Gabbro	748±12	SHRIMP	$\varepsilon_{Hf}(t)=+1.8$ to $+11.63$	$T_{DM2(Hf)}=0.93-1.56$	Zhao et al., 2008a
	Mopanshan adakitic complex	Tonalite and granodiorite	782±6	SHRIMP	$\varepsilon_{Nd}(t)=-2.06$ to -0.54	$T_{DM(Nd)}=1.28-1.46$	Huang et al., 2009
					$\varepsilon_{Hf}(t)=+1.91$ to $+7.07$	$T_{DM(Hf)}=1.06-1.27$	
	Xiacun	Mafic rock	842±14	SHRIMP	$\varepsilon_{Nd}(t)=-1.7$ to $+7.4$		Guo et al., 2007
		Granite	803±15		$\varepsilon_{Nd}(t)=-7.3$ to -2.9	$T_{DM2(Nd)}=1.8-2.4$	
	Yonglang	Granite	790±16				

Continued

Location	Pluton name	Lithology	Age/Ma	Dating method	Isotopic	Model age/Ga	Reference
Yanbian-Panzhihua region	Guandaoshan	Granodiorite	857±13	SHRIMP	$\varepsilon_{Nd}(t) = +3.3$ to $+5.2$	$T_{DM2(Nd)} = 1.1–1.3$	Li et al., 2003
		Quartz diorite	857±7	SHRIMP	$\varepsilon_{Nd}(t) = +4.8$ to $+5.2$	$T_{DM2(Nd)} = 1.09–1.11$	Du et al., 2014
		Gabbroic diorite	856±6				
		Gabbro	856±8				
		Diorite	858±7	SHRIMP	$\varepsilon_{Nd}(t) = +3.9$ to $+5.1$; $\varepsilon_{Hf}(t) = +11$ to $+17$	$T_{DM(Hf)} = 0.86$	Sun and Zhou, 2008
	Fangtian	Basaltic lavas	840		$\varepsilon_{Nd}(t) = +3.8$ to $+8.0$		Sun et al., 2007
	Tongde	Diorite	833±15; 825±7	SHRIMP	$\varepsilon_{Nd}(t) = +0.6$ to $+2.0$; $\varepsilon_{Hf}(t) = +2.69$ to $+11.68$	$T_{DM(Hf)} = 1.25–1.56$	Zhao et al., 2008a; Munteanu et al., 2010; Li and Zhao, 2017
	Lengshuiqing	Diorite	811.6±3	SHRIMP	$\varepsilon_{Nd}(t) = +1.5$ to $+6.0$		Zhou et al., 2006a;
	Gaojiacun	Diorite	806±4		$\varepsilon_{Hf}(t) = +4.33$ to $+9.31$	$T_{DM2(Hf)} = 1.12–1.44$	Zhao et al., 2008a
	Dadukou	Olivine gabbro	746±10	SHRIMP	$\varepsilon_{Nd}(t) = -0.93$ to -0.28		Zhao and Zhou, 2007b
		Hornblende gabbro	738±23		$\varepsilon_{Nd}(t) = -1.73$ to -0.64; $\varepsilon_{Hf}(t) = +2.96$ to $+6.44$	$T_{DM2(Hf)} = 1.26–1.48$	Zhao et al., 2008a
	Datian adakitic complex	Granodiorite	760±4	SHRIMP	$\varepsilon_{Nd}(t) = -0.92$ to $+0.66$; $\varepsilon_{Hf}(t) = +2.69$ to $+11.68$	$T_{DM2(Hf)} = 0.93–1.51$	Zhao and Zhou, 2007a; Zhao et al., 2008a
	Dajiashan		745Ma	SHRIMP	$\varepsilon_{Hf}(t) = +2.26$ to $+8.36$	$T_{DM2(Hf)} = 1.15–1.54$	Zhao et al., 2008a
	Datian mafic intrusion	Dolerite dykes	~760; ~800	SIMS	$\varepsilon_{Nd}(t) = +4.3$ to $+5.2$; $\varepsilon_{Nd}(t) = -3.3$ to $+1.1$		Yang et al., 2016
	Dalu	Granodiorite	781.1±2.8	LA-ICP-MS	$\varepsilon_{Hf}(t) = +2.16$ to $+7.36$	$T_{DM2(Hf)} = 1.21–1.54$	This study
		Granite	779.8±2.0		$\varepsilon_{Hf}(t) = +0.11$ to $+5.80$	$T_{DM2(Hf)} = 1.31–1.97$	

Kangding-Mianning region, the ca. 850 – 800 Ma depleted-mantle-derived Baoxing mafic intrusions supplied the enough mafic compositions for the crustal growth (Meng et al., 2015). The ca. 850 Ma Tianquan peraluminous granodiorite was derived from the partial melting of crust-derived greywackes (Lai et al., 2015b). The ca. 816 Ma Daxiangling A-type granite was reworked or recycled from juvenile crustal materials (Zhao et al., 2008b). Huang et al. (2008) proposed that the ca. 780 Ma Mianning highly fractionated A_2-type granite was formed by partial melting of the juvenile crust (ca. 1 000−900 Ma). They further suggested that the ca. 780 Ma Mopanshan adakites may be originated from a thickened lower crust (Huang et al., 2009). Moreover, the ca. 770 Ma Baoxing granite was ascribed to a juvenile crust source (Meng et al., 2015). The ca. 754 Ma Luding high-$Mg^\#$ quartz monzodiorite was generated by the ~40% partial melting of newly formed mafic lower crust, while the ca. 748 Ma Kangding granodiorite was derived from the partial melting of a plagioclase-rich crust source (Lai et al., 2015a). In the Xichang-Miyi region, the ca. 750 Ma Shaba gabbroic pluton and ca. 842 Ma Xiacun mafic rocks were related with the mantle sources and provided mantle-derived mafic magma for the juvenile crust (Guo et al., 2007; Zhao et al., 2008a). The ca. 803 Ma Xiacun granite and ca. 790 Ma Yonglang granite were generated by the reworking of ancient crustal materials (Guo et al., 2007). These ca. 850−740 Ma magmatism indicate that the extensive crustal growth or/and reworking are significant for the Neoproterozoic tectonic evolution. In the Panzhihua-Yanbian region, the early emplacement of mafic-ultramafic plutons (ca. 860 – 810 Ma), including the Guandaoshan, Tongde, Gaojiacun, and Lengshuiqing intrusions, were derived from a lithospheric mantle modified by oceanic slab components (fluids+melts). These massive mafic-ultramafic plutons imply that the crustal growth was involved in the extraction of mantle melts (Zhao et al., 2008a). The latter Datian (ca. 760 Ma) and Dajianshan adakites underwent the partial melting of the subduction slab, which suggest that the recycling of oceanic crust was a main contributor to continental crustal growth (Defant and Drummond, 1990; Smithies and Champion, 2000; Zhao and Zhou, 2007a, 2008; Zhao et al., 2008a). The ca. 780 Ma Dalu granitoids pluton was perched during the interval between the mafic-ultramafic intrusions and adakitic plutons, which imply that it may experience the partial melting of juvenile lower crust (Zhao and Zhou, 2008). The Dalu granodiorites were originated from the Mesoproterozoic juvenile mafic lower crust. Furthermore, the Dalu granites indicate the middle-upper mature crustal reworking. Taking into account of above massive evidences, we conclude that the Neoproterozoic extensive growth and/or reworking play a vital role for the voluminous igneous genesis and tectonic setting in the western margin of Yangtze Block, South China.

Zhao and Zhou (2008) proposed that the rigid and condensed oceanic slab broke off and induced the upwelling of the asthenosphere which heated the thickened crust, producing voluminous adakitic magmas in the margin of Yangtze Block. They further suggested that the heat anomaly from 820 Ma to 746 Ma may have resulted from roll-back and steepening of the subducted oceanic slab, causing slab break off (Aldanmaz et al., 2000; Zhao and Zhou, 2009). Therefore, we thought that the slab breakoff mechanism can explain the different melting temperature of lower crust and middle-upper crust in the Dalu pluton. The oceanic slab break-off may lead to the upwelling of asthenosphere. Large amounts of mantle-derived magma that carry massive heat ascend and triggered crustal anataxis at different levels in a short time (Liu and Zhao, 2018). When the hot mantle-derived magma rises to the middle-upper crust, due to the partial melting of mafic lower crust and heat loss, the discrepant melting temperature can occur in the lower crust and middle-upper crust.

In such a context, the Dalu pluton may be formed by the following processes (Fig. 13): The early subduction stage of the oceanic crust occurred continually from ca. 860 Ma to ca. 810 Ma, the mantle wedge was metasomatized by subduction slab components (fluids or melts) (Sun et al., 2007; Zhao and Zhou, 2008; Zhao et al., 2008a). This stage was accompanied by the partial melting of the modified mantle wedge and the upwelling of mantle-derived magmas. Consequently, these mantle-derived magmas emplaced in the middle-upper crust underwent strong differentiation and formed voluminous mafic intrusions in the Yanbian Terrane (Tongde, Gaojiacun, and Lengshuiqing) (Fig. 13a). Subsequently, at ca. 780 Ma, the oceanic slab break-off and upwelling of asthenosphere heated the pre-existing Mesoproterozoic

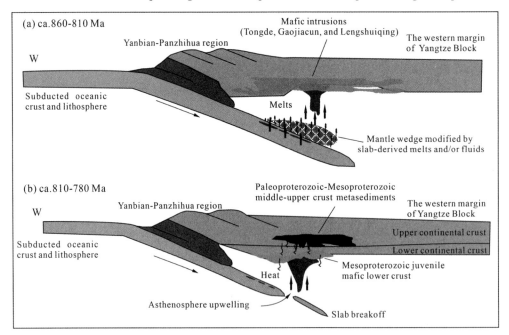

Fig. 13 Tectonic evolution of the Neoproterozoic magmatism ca. 860−780 Ma along the southwestern margin of the Yangtze Block, South China (after Sun et al., 2007, 2008; Zhao and Zhou, 2008, 2009).

juvenile mafic lower crust, producing the Dalu granodioritic melts (Zhao and Zhou, 2008; Zhao et al., 2008a). The juvenile mafic lower crust-derived melts further ascended and induced the Paleoproterozoic-Mesoproterozoic crustal reworking (metasedimentary materials) on the middle-upper crustal level and formed the Dalu granitic melts (Fig. 13b). We infer that the Dalu granodiorites-granites are the magmatic response from different crustal levels due to the asthenosphere upwelling in the subduction-related tectonic setting.

6　Conclusion

(1) Zircon U-Pb dating show ages for the Dalu granodiorites of 781.1 ± 2.8 Ma and for the Dalu granites of 779.8 ± 2.0 Ma.

(2) The Dalu Na-rich granodiorites are calc-alkaline and metaluminous to weakly peraluminous, which were derived from the partial melting of Mesoproterozoic juvenile mafic lower crust. The Dalu granites belong to the high-K calc-alkaline and peraluminous I-type series. They were derived from the mature metasediment-derived melts by the mixing of newly formed mafic lower crust-derived melts.

(3) The geochemical variations from granodiorites to granites in Dalu pluton were dominated by the different protolith and the different melting temperatures.

(4) Neoproterozoic continuous crustal growth and reworking were widely distributed and associated with the subduction process along the western periphery of Yangtze Block. The Dalu granodiorites-granites are the magmatic response from different crustal levels due to the heat anomaly from asthenosphere upwelling in the subduction-related tectonic setting.

Highlights

● Ca. 780 Ma I-type granitoids pluton in southwestern margin of Yangtze Block, South China.

● The Dalu granodiorites-granites are the different magmatic response in depth which were induced by the heat anomaly due to the asthenosphere upwelling under the subduction-related tectonic setting.

● Their geochemical variations depend on the distinct source rocks and the different melting temperatures.

● Neoproterozoic continuous crustal evolution was widely distributed and associated with the subduction process along the western periphery of Yangtze Block.

Acknowledgements　We are grateful to Dr. Robert J. Stern and two anonymous reviewers for their kind help and constructive reviews. This work was jointly supported by the National Natural Science Foundation of China (Grant Nos. 41421002 and 41772052) and the programme for Changjiang Scholars and Innovative Research Team in University (Grant No. IRT1281). Support was also provided by the Foundation for the Author of National Excellent

Doctoral Dissertation of China (Grant No. 201324) and independent innovation project of graduate students of Northwest University (Grant No. YZZ17192).

References

Aldanmaz, E., Pearce, J.A., Thirlwall, M.F., Mitchell, J.G., 2000. Petrogenetic evolution of Late Cenozoic, post-collision volcanism in western Anatolia, Turkey. Journal of Volcanology and Geothermal Research, 102, 67-95.

Andersen, T., 2002. Correction of common lead in U-Pb analyses that do not report [204]Pb. Chemical Geology, 192(1-2), 59-79.

Bao, Z.A., Chen, L., Zong, C.L., Yuan, H.L., Chen, K.Y., Dai, M.N., 2017. Development of pressed sulfide powder tablets for in situ, sulfur and lead isotope measurement using LA-MC-ICP-MS. International Journal of Mass Spectrometry, 421, 255-262.

Barbarin, B., 2005. Mafic magmatic enclaves and mafic rocks associated with some granitoids of the central Sierra Nevada batholith, California: Nature, origin, and relations with the hosts. Lithos, 80, 155-177.

Beard, J.S., Lofgren, G.E., 1991. Dehydration melting and water-saturated melting of basaltic and andesitic greenstones and amphibolites at 1, 3, and 6. 9 kb.. Journal of Petrology, 32(2), 365-401.

Cai, Y.F., Wang, Y.J., Cawood, P.A., Zhang, Y.Z., Zhang, A.M., 2015. Neoproterozoic crustal growth of the southern Yangtze block: Geochemical and zircon U-Pb geochronological and Lu-Hf isotopic evidence of Neoproterozoic diorite from the ailaoshan zone. Precambrian Research, 266, 137-149.

Cawood, P.A., Wang, Y.J., Xu, Y.G., Zhao, G.C., 2013. Locating South China in Rodinia and Gondwana: A fragment of greater India lithosphere? Geology, 41(8), 903-906.

Chappell, B., White, A., 1992. I- and S-type granites in the Lachlan Fold Belt. Transactions of the Royal Society of Edinburgh Earth Sciences, 83, 1-26.

Chappell, B.W., Bryant, C.J., Wyborn, D., 2012. Peraluminous I-type granites. Lithos, 153(8), 142-153.

Chen, Q., Sun, M., Long, X.P., Yuan, C., 2015. Petrogenesis of Neoproterozoic adakitic tonalites and high-K granites in the eastern Songpan-Ganze fold belt and implications for the tectonic evolution of the western Yangtze Block. Precambrian Research, 270, 181-203.

Clemens, J. D., 2003. S-type granitic magmas-Petrogenetic issues, models and evidence. Earth-Science Reviews, 61, 1-18.

Clemens, J.D., Darbyshire, D.P.F., Flinders, J., 2009. Sources of post-orogenic calcalkaline magmas: The arrochar and garabal hill-glen fyne complexes, Scotland. Lithos, 112, 524-542.

Clemens, J.D., Helps, P.A., Stevens, G., 2010. Chemical structure in granitic magmas: A signal from the source? Earth and Environmental Science Transactions of the Royal Society of Edinburgh, 100, 159-172.

Clemens, J.D., Regmi, K., Nicholls, I., Weinberg, R., Maas, R., 2016. The Tynong pluton, its mafic synplutonic sheets and igneous microgranular enclaves: The nature of the mantle connection in I-type granitic magmas. Contributions to Mineralogy and Petrology, 171, 1-17.

Clemens, J.D., Stevens, G., 2012. What controls chemical variation in granitic magmas? Lithos, 134-135, 317-329.

Clemens, J.D., Stevens, G., Farina, F., 2011. The enigmatic sources of I-type granites: The peritectic connexion. Lithos, 126(3), 174-181.

Collins, W., 1996. Lachlan Fold Belt granitoids. Products of three-component mixing. Special Paper, 315, 171-181.

Defant, M.J., Drummond, M.S., 1990. Derivation of some modern arc magmas by melting of young subducted lithosphere. Nature, 347, 662-665.

DePaolo, D. J., 1981. A neodymium and strontium isotopic study of the Mesozoic calcalkaline granitic batholiths of the Sierra Nevada and Peninsular Ranges, California. American Geophysical Union, 86, 10470-10488.

DePaolo, D.J., Perry, F.V., Baldridge, W.S., 1992. Crustal versus mantle sources of granitic magmas: A two-parameter model based on Nd isotopic studies. Transactions of the Royal Society of Edinburgh Earth Sciences, 83, 439-446.

Du, L.L., Guo, J.H., Nutman, A.P., Wyman, D., Geng, Y.S., Yang, C.H., Liu, F.L., Ren, L.D., Zhou, X.W., 2014. Implications for Rodinia reconstructions for the initiation of Neoproterozoic subduction at ~860 Ma on the western margin of the Yangtze Block: Evidence from the Guandaoshan Pluton. Lithos, 196-197, 67-82.

Frost, B.R., Barnes, C.G., Collins, W.J., Arculus, R.J., Ellis, D.J., Frost, C.D., 2001. A geochemical classification for granitic rocks. Journal of Petrology, 42, 2033-2048.

Gan, B.P., Lai, S.C., Qin, J.F., Zhu, R.Z., Zhao, S.W., Li, T., 2017. Neoproterozoic alkaline intrusive complex in the northwestern Yangtze Block, Micang Mountains region, South China: Petrogenesis and tectonic significance. International Geology Review, 59(3), 311-332.

Gao, S., Ling, W.L., Qiu, Y., Zhou, L., Hartmann, G., Simon, K., 1999. Contrasting geochemical and Sm-Nd isotopic compositions of Archean metasediments from the Kongling high-grade terrain of the Yangtze craton: Evidence for cratonic evolution and redistribution of REE during crustal anatexis. Geochimica et Cosmochimica Acta, 63(13-14), 2071-2088.

Geng, Y.S., Yang, C.H., Wang, X.S., Du, L.L., Ren, L.D., Zhou, X.W., 2008. Metamorphic basement evolution in western margin of Yangtze Block. Beijing: Geological Publishing House, 1-215 (in Chinese).

Griffin, W.L., Pearson, N.J., Belousova, E., Jackson, S.E., Van Achterbergh, E., O'Reilly, S.Y., Shee, S.R., 2000. The Hf isotope composition of cratonic mantle: LAM-MC-ICPMS analysis of zircon megacrysts in kimberlites. Geochimica et Cosmochimica Acta, 64(1), 133-147.

Grimes, C.B., John, B.E., Kelemen, P.B., Mazdab, F.K., Wooden, J.L., Cheadle, M.J., Hanghoj, K., Schwartz, J.J., 2007. Trace element chemistry of zircons from oceanic crust: A method for distinguishing detrital zircon provenance. Geology, 35, 643-646.

Guo, C.L., Wang, D.H., Chen, Y.C., Zhao, Z.G., Wang, Y.B., Fu, X.F., Fu, D.M., 2007. SHRIMP U-Pb zircon ages and major element, trace element and Nd-Sr isotope geochemical studies of a Neoproterozoic granitic complex in western Sichuan: Petrogenesis and tectonic significance. Acta Petrologica Sinica, 23(10), 2457-2470 (in Chinese with English abstract).

Hanchar, J. M., Watson, E. B., 2003. Zircon saturation thermometry. Reviews in Mineralogy and Geochemistry, 53, 89-112.

Hawkesworth, C.J., Kemp, A.I.S., 2006. Evolution of the continental crust. Nature, 443, 811-817.

Healy, B., Collins, W. J., Richards, S. W., 2004. A hybrid origin for Lachlan S-type granites: The Murrumbidgee Batholith example. Lithos, 78, 197-216.

Hoskin, P.W.O., Schaltegger, U., 2003. The composition of zircon and igneous and metamorphic Petrogenesis. Reviews in Mineralogy and Geochemistry, 53, 27-62.

Huang, X.L., Xu, Y.G., Lan, J.B., Yang, Q.J., Luo, Z.Y., 2009. Neoproterozoic adakitic rocks from Mopanshan in the Western Yangtze craton: Partial melts of a thickened lower crust. Lithos, 112(3), 367-381.

Huang, X.L., Xu, Y.G., Li, X.H., Li, W.X., Lan, J.B., Zhang, H.H., Liu, Y.S., Wang, Y.B., Li, H.Y., Luo, Z.Y., Yang, Q.J., 2008. Petrogenesis and tectonic implications of Neoproterozoic, highly fractionated A-type granites from Mianning, South China. Precambrian Research, 165(3-4), 190-204.

Kemp, A.I.S., Hawkesworth, C.J., 2014. Growth and differentiation of the continental crust from isotope studies of accessory minerals. Treatise on Geochemistry, 4, 379-421.

Kemp, A.I.S., Hawkesworth, C.J., Foster, G.L., Paterson, B.A., Woodhead, J.D., Hergt, J.M., Gray, C.M., Whitehouse, M.J., 2007. Magmatic and crustal differentiation history of granitic rocks from Hf-O isotopes in zircon. Science, 315(5814), 980-983.

Lai, S.C., Qin, J.F., Zhu, R.Z., Zhao, S.W., 2015a. Neoproterozoic quartz monzodiorite-Granodiorite association from the Luding-Kangding area: Implications for the interpretation of an active continental margin along the Yangtze Block (South China Block). Precambrian Research, 267(3-4), 196-208.

Lai, S.C., Qin, J.F., Zhu, R.Z., Zhao, S.W., 2015b. Petrogenesis and tectonic implication of Neoproterozoic peraluminous granitoids from the Tianquan area, western Yangtze Block, South China. Acta Petrologica Sinica, 31(8), 2245-2258 (in Chinese with English abstract).

Li, Q.W., Zhao, J.H., 2018. The neoproterozoic high-mg dioritic dikes in south china formed by high pressures fractional crystallization of hydrous basaltic melts. Precambrian research., 309, 198-211.

Li, X.H., 1999. U-Pb zircon ages of granites from the southern margin of the Yangtze Block: Timing of the Neoproterozoic Jinning Orogeny in SE China and implications for Rodinia assembly. Precambrian Research, 97, 43-57.

Li, X.H., Li, Z.X., Ge, W.C., Zhou, H.W., Li, W.X., Liu, Y., Wingate, M.T.D., 2003a. Neoproterozoic granitoids in South China: Crustal melting above a mantle plume at ca. 825 Ma? Precambrian Research, 122(s1-4), 45-83.

Li, X.H., Li, Z.X., Sinclair, J.A., Li, W.X., Carter, G., 2006. Revisiting the "Yanbian Terrane": Implications for Neoproterozoic tectonic evolution of the western Yangtze block, South China. Precambrian Research, 151(1-2), 14-30.

Li, X.H., Li, Z.X., Zhou, H.W., Liu, Y., Liang, X.R., Li, W.X., 2003b. SHRIMP U-Pb zircon age, geochemistry and Nd isotope of the Guandaoshan pluton in SW Sichuan: Petrogenesis and tectonic significance. Science in China: Series D, 46(1), 73-83.

Li, Z.X., Bogdanova, S.V., Collins, A.S., Davidson, A., De Waele, B., Ernst, R.E., Fitzsi-Mons, I.C.W., Fuck, R.A., Gladkochub, D.P., Jacobs, J., Karlstrom, K.E., Lu, S., Natapov, L.M., Pease, V., Pisarevsky, S.A., Thrane, K., Vernikovsky, V., 2008. Assembly, configuration, and break-up history of Rodinia: A synthesis. Precambrian Research, 160, 179-210.

Li, Z.X., Li, X.H., Kinny, P.D., Wang, J., 1999, The breakup of Rodinia: Did it start with a mantle plume beneath South China? Earth & Planetary Science Letters, 173(3), 171-181.

Li, Z.X., Zhang, L., Powell, C.M., 1995. South China in Rodinia: Part of the missing link between Australia-East Antarctica and Laurentia? Geology, 23(5), 407-410.

Liu, F.L., Xue, H.M., Liu, P.H., 2009. Partial melting time of ultrahigh-pressure metamorphic rocks in the Sulu UHP terrane: Constrained by zircon U-Pb ages, trace elements and Lu-Hf isotope compositions of biotite-bearing granite. Acta Petrologica Sinica, 25, 1039-1055 (in Chinese with English abstract).

Liu, H., Zhao, J.H., 2018. Neoproterozoic peraluminous granitoids in jiangnan fold belt: Implications for lithospheric differentiation and crustal growth. Precambrian Research., 309, 152-165

Liu, J.H., Xie, C.M., Li, C., Wang, M., Wu, H., Li, X.K., Liu, Y.M., Zhang, T.Y., 2018. Early Carboniferous adakite-like and I-type granites in central Qiangtang, northern Tibet: Implications for intra-oceanic subduction and back-arc basin formation within the Paleo-Tethys Ocean. Lithos, 296, 265-280.

Liu, Y., Liu, X.M., Hu, Z.C., Diwu, C.R., Yuan, H.L., Gao, S., 2007. Evaluation of accuracy and long-term stability of determination of 37 trace elements in geological samples by ICP-MS. Acta Petrologica Sinica, 23(5), 1203-1210 (in Chinese with English abstract).

Lu, Y.H., Zhao, Z.F., Zheng, Y.F., 2016. Geochemical constraints on the source nature and melting conditions of Triassic granites from South Qinling in central China. Lithos, 264, 141-157.

Lu, Y.H., Zhao, Z.F., Zheng, Y.F., 2017. Geochemical constraints on the nature of magma sources for Triassic granitoids from South Qinling in central China. Lithos, 284-285, 30-49.

Ludwig, K.R., 2003. ISOPLOT 3.0. A geochronological toolkit for Microsoft Excel. Berkeley Geochronology Center, Special Publication, 4.

Meng, E., Liu, F.L., Du, L.L., Liu, P.H., Liu, J.H., 2015. Petrogenesis and tectonic significance of the Baoxing granitic and mafic intrusions, southwestern china: Evidence from zircon U-Pb dating and Lu-Hf isotopes, and whole-rock geochemistry. Gondwana Research, 28(2), 800-815.

Middlemost, E.A.K., 1994. Naming materials in the magma/igneous rock system. Earth-Science Reviews, 37 (3-4), 215-224.

Miller, C.F., McDowell, S.M., Mapes, R.W., 2003. Hot and cold granites? Implications of zircon saturation temperatures and preservation of inheritance. Geology, 31, 529-532.

Munteanu, M., Wilson, A., Yao, Y., Harris, C., Chunnett, G., Luo, Y., 2010. The Tongde dioritic pluton (Sichuan, SW China) and its geotectonic setting: Regional implications of a local-scale study. Gondwana Research, 18, (2), 455-465.

Nowell, G.M., Kempton, P.D., Noble, S.R., Fitton, J.G., Saunders, A.D., Mahoney, J.J., Taylor, R.N., 1998. High precision Hf isotope measurements of MORB and OIB by thermal ionization mass spectrometry: Insights into the depleted mantle. Chemical Geology, 149(3-4), 211-233.

Patiño Douce, A.E., 1999. What do experiments tell us about the relative contributions of crust and mantle to the origin of the granitic magmas. Geological Society London Special Publications, 168(1), 55-75.

Pitcher, W.S., 1997. The Nature and Origin of Granite (2nd). London: Chapman and Hall, 387.

Qin, J.F., Lai, S.C., Li, Y.F., Ju, Y.J., Zhu, R.Z., Zhao, S.W., 2016. Early Jurassic monzogranite-Tonalite association from the southern Zhangguangcai Range: Implications for Paleo-Pacific plate subduction along northeastern China. Lithosphere, 8(4), 396-411.

Rapp, R. P., Watson, E. B., 1995. Dehydration melting of metabasalt at 8 – 32 kbar: Implications for continental growth and crust-mantle recycling. Journal of Petrology, 36(4), 891-931.

Roberts, M.P., Clemens, J.D., 1993. Origin of high-potassium, calc-alkaline, I-type granitoids. Geology, 21, 825-828.

Rudnick, R.L., 1995. Making continental crust. Nature, 378, 571-578.

Rudnick, R.L., Fountain, D.M., 1995. Nature and composition of the continental crust: A lower crustal perspective. Reviews of Geophysics, 33(3), 267-309.

Schiano, P., Monzier, M., Eissen, J.P., Martin, H., Koga, K.T., 2010. Simple mixing as the major control of the evolution of volcanic suites in the Ecuadorian Andes. Contributions to Mineralogy and Petrology, 160, 297-312.

Sichuan Provincial Bureau of Geology and Mineral Resources (SPBGMR), 1972. Regional geological survey of the People's Republic of China, the Yanbian Sheet map and report, scale 1∶200 000 (in Chinese).

Smithies, R.H., Champion, D.C., 2000. The Archaean high-Mg diorite suite: Links to tonalite-Trondhjemite-Granodiorite magmatism and implications for early archaean crustal growth. Journal of Petrology, 41(12), 1653-1671.

Soderlund, U., Patchett, P.J., Vervoort, J.D., Isachsen, C.E., 2004. The ^{176}Lu decay constant determined by Lu-Hf and U-Pb isotope systematics of Precambrian mafic intrusions. Earth & Planetary Science Letters, 219(3-4), 311-324.

Soesoo, A., 2000. Fractional crystallization of mantle-derived melts as a mechanism for some I-type granite petrogenesis: An example from Lachlan Fold Belt, Australia. Journal of the Geological Society, 157, 135-149.

Stern, R.A., Hanson, G.N., 1991, Archean high-Mg granodiorite-A derivative of light rare-eart element-enriched monzodiorite of mantle origin. Journal of Petrology, 32, 201-238.

Sun, C.M., Vuagnat, M., 1992. Proterozoic ophiolites from Yanbian and Shimian (Sichuan Province, China): Petrography, geochemistry, petrogenesis, and geotectonic environment. Schweizerische Mineralogischeund Petrographische Mitteilungen, 72(3), 389-413.

Sun, S.S., McDonough, W.F., 1989. Chemical and isotopic systematics of oceanic basalts; implications for mantle composition and processes. In: Saunders, A.D., Norry, M.J. Magmatism in the Ocean Basins. Geological Society, London, Special Publications, 42, 313-345.

Sun, W.H., Zhou, M.F., 2008. The ~860 Ma, cordilleran-type Guandaoshan dioritic pluton in the Yangtze Block, SW China: Implications for the origin of Neoproterozoic magmatism. Journal of Geology, 116(3), 238-253.

Sun, W.H., Zhou, M.F., Yan, D.P., Li, J.W., Ma, Y.X., 2008. Provenance and tectonic setting of the Neoproterozoic Yanbian Group, western Yangtze Block (SW China). Precambrian Research, 167(s1-2), 213-236.

Sun, W.H., Zhou, M.F., Zhao, J.H., 2007. Geochemistry and Tectonic Significance of Basaltic Lavas in the Neoproterozoic Yanbian Group, Southern Sichuan Province, Southwest China. International Geology Review, 49(6), 554-571.

Sylvester, P.J., 1998. Post-collisional strongly peraluminous granites. Lithos, 45, 29-44.

Taylor, S.R., Mclennan, S.M., 1985. The continental crust. Its composition and evolution, an examination of the geochemical record preserved in sedimentary rocks. Journal of Geology, 94(4). 632-633.

Valley, J.W., 2003. Oxygen isotopes in zircon. Reviews in Mineralogy and Geochemistry, 53, 343-385.

Wang, X.L., Zhou, J.C., Griffin, W.L., Zhao, G.C., Yu, J.H., Qiu, J.S., Zhang, Y.J., Xing, G.F., 2014. Geochemical zonation across a Neoproterozoic orogenic belt: Isotopic evidence from granitoids and metasedimentary rocks of the Jiangnan orogen, China. Precambrian Research, 242(2), 154-171.

Wang, X.L., Zhou, J.C., Wan, Y.S., Kitajima, K., Wang, D., Bonamici, C., Qiu, J.S., Sun, T., 2013. Magmatic evolution and crustal recycling for Neoproterozoic strongly peraluminous granitoids from southern China: Hf and O isotopes in zircon. Earth & Planetary Science Letters, 366(2), 71-82.

Weidendorfer, D., Mattsson, H.B., Ulmer, P., 2014. Dynamics of magma mixing in partially crystallized magma chambers: Textural and petrological constraints from the basal complex of the austurhorn intrusion (SE Iceland). Journal of Petrology, 55, 1865-1903.

Whalen, J.B., Currie, K.L., Chappell, B.W., 1987. A-type granites: Geochemical characteristics, discrimination and petrogenesis. Contributions to Mineralogy and Petrology, 95(4), 407-441.

Wolf, M.B., London, D., 1994. Apatite dissolution into peraluminous haplogranitic melts: An experimental study of solubilities and mechanisms. Geochimica et Cosmochimica Acta, 58, 4127-4145.

Wolf, M.B., Wyllie, P.J., 1994. Dehydration-melting of amphibolite at 10 kbar: The effects of temperature and time. Contributions to Mineralogy and Petrology, 115(4), 369-383.

Wyllie, P.J., Wolf, M.B., 1993. Amphibolite dehydration melting: Sorting out the solidus. Geological Society London Special Publications, 76, 405-416.

Xiao, B., Li, Q.G., He, S.Y., Chen, X., Liu, S.W., Wang, Z.Q., Xu, X.Y., Chen, J.L., 2017. Contrasting geochemical signatures between upper Triassic Mo-hosting and barren granitoids in the central segment of the South Qinling orogenic belt, central China: Implications for Mo exploration. Ore Geology Reviews, 81, 518-534.

Yang, J.H., Wu, F.Y., Wilde, S.A., Xie, L.W., Yang, Y.H., Liu, X.M., 2007. Tracing magma mixing in granite genesis: In situ U-Pb dating and Hf-isotope analysis of zircons. Contribution to Mineralogy and Petrology, 153, 177-190.

Yang, Y.J., Zhu, W.G., Bai, Z., Zhong, H., Ye, X.T., Fan, H.P., 2016. Petrogenesis and tectonic implications of the Neoproterozoic Datian mafic-Ultramafic dykes in the Panzhihua area, western Yangtze Block, SW China. International Journal of Earth Sciences, 106(1), 1-29.

Yuan, H.L., Gao, S., Dai, M.N., Zong, C.L., Gunther, D., Fontaine, G.H., Liu, X.M., Diwu, C.R., 2008. Simultaneous determinations of U-Pb age, Hf isotopes and trace element compositions of zircon by excimer laser-ablation quadrupole and multiple-collector ICPMS. Chemical Geology, 247(1-2), 100-118.

Yuan, H.L., Gao, S., Liu, X.M., Li, H.M., Gunther, D., Wu, F.Y., 2004. Accurate U-Pb age and trace element determinations of zircon by laser ablation-inductively coupled plasma mass spectrometry. Geostandards & Geoanalytical Research, 28(3), 353-370.

Zhang, J., Zhang, H.F., Li, L., 2018. Neoproterozoic tectonic transition in the south Qinling belt: New constraints from geochemistry and zircon U-Pb-Hf isotopes of diorites from the douling complex.

Precambrian Research., 306, 112-118.

Zhang, Z.J., Badal, B., Li, Y.K., Chen, Y., Yang, L.P., Teng, J.W., 2005. Crust-Upper mantle seismic velocity structure across Southeastern China. Tectonophysics, 395(1), 137-157.

Zhao, G.C., Cawood, P.A., 2012. Precambrian geology of china. Precambrian Research, 222-223, 13-54.

Zhao, J.H., Asimow, P.D., Zhou, M.F., Zhang, J., Yan, D.P., Zheng, J.P., 2017a. An Andean-type arc system in Rodinia constrained by the Neoproterozoic Shimian ophiolite in South China. Precambrian Research, 296, 93-111.

Zhao, J.H., Zhou, M.F., 2007a. Neoproterozoic Adakitic Plutons and Arc Magmatism along the western Margin of the Yangtze Block, South China. Journal of Geology, 115(6), 675-689.

Zhao, J.H., Zhou, M.F., 2007b. Geochemistry of Neoproterozoic mafic intrusions in the Panzhihua district (Sichuan Province, SW China): Implications for subduction-related metasomatism in the upper mantle. Precambrian Research, 152(1), 27-47.

Zhao, J.H., Zhou, M.F., 2008. Neoproterozoic adakitic plutons in the northern margin of the Yangtze Block, China: Partial melting of a thickened lower crust and implications for secular crustal evolution. Lithos, 104, 231-248.

Zhao, J.H., Zhou, M.F., 2009. Secular evolution of the Neoproterozoic lithospheric mantle underneath the northern margin of the Yangtze Block, South China. Lithos, 107(3), 152-168.

Zhao, J.H., Zhou, M.F., Yan, D.P., Yang, Y.H., Sun, M., 2008a. Zircon Lu-Hf isotopic constraints on Neoproterozoic subduction-related crustal growth along the western margin of the Yangtze Block, South China. Precambrian Research, 163(3), 189-209.

Zhao, J.H., Zhou, M.F., Yan, D.P., Zheng, J.P., Li, J.W., 2011. Reappraisal of the ages of Neoproterozoic strata in South China: No connection with the grenvillian orogeny. Geology, 39(4), 299-302.

Zhao, J.H., Zhou, M.F., Zheng, J.P., Fang, S.M., 2010. Neoproterozoic crustal growth and reworking of the Northwestern Yangtze Block: Constraints from the Xixiang dioritic intrusion, South China. Lithos, 120(3-4), 439-452.

Zhao, S.W., Lai, S.C., Qin, J.F., Zhu, R.Z., Wang, J.B., 2017b. Geochemical and geochronological characteristics of late cretaceous to early Paleocene granitoids in the Tengchong Block, Southwestern China: Implications for crustal anatexis and thickness variations along the eastern Neo-Tethys subduction zone. Tectonophysics, 694, 87-100.

Zhao, X.F., Zhou, M.F., Li, J.W., Wu, F.Y., 2008b. Association of Neoproterozoic A- and I-type granites in south china: Implications for generation of A-type granites in a subduction-related environment. Chemical Geology, 257(s1), 1-15.

Zhao, Z.F., Gao, P., Zheng, Y.F., 2015. The source of Mesozoic granitoids in South China: Integrated geochemical constraints from the Taoshan batholith in the Nanling range. Chemical Geology, 395, 11-26.

Zhao, Z.F., Zheng, Y.F., Wei, C.S., Wu, Y.B., 2007. Post-collisional granitoids from the Dabie orogen in China: Zircon U-Pb age, element and O isotope evidence for recycling of subducted continental crust. Lithos, 93, 248-272.

Zheng, Y.F., Chen, Y.X., Dai, L.Q., Zhao, Z.F., 2015. Developing plate tectonics theory from oceanic subduction zones to collisional orogens. Science China: Earth Sciences, 58, 1045-1069.

Zheng, Y.F., Wu, R.X., Wu, Y.B., Zhang, S.B., Yuan, H.L., Wu, F.Y., 2008. Rift melting of juvenile arc-derived crust: Geochemical evidence from Neoproterozoic volcanic and granitic rocks in the Jiangnan orogen, South China. Precambrian Research, 163(3-4), 351-383.

Zheng, Y.F., Wu, Y.B., Chen, F.K., Gong, B., Li, L., Zhao, Z.F., 2004. Zircon U-Pb and oxygen isotope evidence for a large-scale 18O depletion event in igneous rocks during the Neoproterozoic. Geochimica et Cosmochimica Acta, 68, 4145-4165.

Zheng, Y.F., Wu, Y.B., Zhao, Z.F., Zhang, S.B., Xu, P., Wu, F.Y., 2005a. Metamorphic effect on zircon Lu-Hf and U-Pb isotope systems in ultrahigh-pressure eclogite-facies meta-granite and metabasite. Earth & Planetary Science Letters, 240(2), 378-400.

Zheng, Y.F., Zhang, S.B., Zhao, Z.F., Wu, Y.B., Li, X.H., Li, Z.X., Wu, F.Y., 2007. Contrasting zircon Hf and O isotopes in the two episodes of Neoproterozoic granitoids in South China: Implications for growth and reworking of continental crust. Lithos, 96(12), 127-150.

Zheng, Y.F., Zhou, J.B., Wu, Y.B., Xie, Z., 2005b. Low-grade metamorphic rocks in the Dabie-Sulu orogenic belt: A passive-margin accretionary wedge deformed during continent subduction. International Geology Review, 47(8), 851-871.

Zhou, M.F., Ma, Y.X., Yan, D.P., Xia, X.P., Zhao, J.H., Sun, M., 2006a. The Yanbian Terrane (Southern Sichuan Province, SW China): A Neoproterozoic arc assemblage in the western margin of the Yangtze Block. Precambrian Research, 144(1-2), 19-38.

Zhou, M.F., Yan, D., Kennedy, A.K., Li, Y.Q., Ding, J., 2002. SHRIMP U-Pb zircon geochronological and geochemical evidence for Neoproterozoic arc-magmatism along the western margin of the Yangtze Block, South China. Earth & Planetary Science Letters, 196(1-2), 51-67.

Zhou, M.F., Yan, D.P., Wang, C.L., Qi, L., Kennedy, A., 2006b. Subduction-related origin of the 750 Ma Xuelongbao adakitic complex (Sichuan Province, China): Implications for the tectonic setting of the giant Neoproterozoic magmatic event in South China. Earth & Planetary Science Letters, 248(1-2), 286-300.

Zhou, M.F., Zhao, X.F., Chen, W.T., Li, X.C., Wang, W., Yan, D.P., Qiu, H.N., 2014. Proterozoic Fe-Cu metallogeny and supercontinental cycles of the southwestern Yangtze Block, southern China and northern Vietnam. Earth-Science Reviews, 139, 59-82.

Zhu, R.Z., Lai, S.C., Qin, J.F., Zhao, S.W., 2015. Early-cretaceous highly fractionated I-type granites from the northern Tengchong Block, Western Yunnan, SW China: Petrogenesis and tectonic implications. Journal of Asian Earth Sciences, 100, 145-163.

Zhu, R.Z., Lai, S.C., Qin, J.F., Zhao, S.W., 2018. Petrogenesis of late paleozoic-to-early mesozoic granitoids and metagabbroic rocks of the tengchong block, SW China: Implications for the evolution of the eastern paleo-tethys. International Journal of Earth Sciences, 107, 431-82.

Zhu, W.G., Zhong, H., Li, X.H., Deng, H.L., He, D.F., Wu, K.W., Bai, Z.J., 2008. SHRIMP zircon U-Pb geochronology, elemental, and Nd isotopic geochemistry of the Neoproterozoic mafic dykes in the Yanbian area, SW China. Precambrian Research, 164(1-2), 66-85.

Zhu, Y., Lai, S.C., Zhao, S.W., Zhang, Z.Z., Qin, J.F., 2017b. Geochemical characteristics and geological significance of the Neoproterozoic K-feldspar granites from the Anshunchang, Shimian area, Western Yangtze Block. Geological Review, 63(5), 1193-1208 (in Chinese with English abstract).

Compositional variations of granitic rocks in continental margin arc: Constraints from the petrogenesis of Eocene granitic rocks in the Tengchong Block, SW China [1]

Zhao Shaowei　Lai Shaocong[2]　Pei Xianzhi　Qin Jiangfeng
Zhu Renzhi　Tao Ni　Gao Liang

Abstract: In order to address the causes of compositional variations of crust-derived granitic rocks in continental margin arc, we focused on the petrogenesis of Eocene granitic host rocks with enclosed MMEs, associated metagabbro, fine-grained quartz-diorite in Xima-Tongbiguan area, the Tengchong Block, which is a continental margin arc related to the subduction of Neo-Tethys. Both the Eocene granitic and mafic rocks are characterized by gradual increment in enriched compositions, such as K, Rb, Th, LREE, initial $^{87}Sr/^{86}Sr$, $\varepsilon_{Nd}(t)$, and $\varepsilon_{Hf}(t)$, but decreasing Pb isotopic ratios from the west to the east of the Tengchong Block.

Zircon U-Pb age data of granitic host rocks, MMEs, metagabbro and fine-grained quartz-diorite reveal a similar age range of 55 – 50 Ma. The lithologies, zircon Hf compositions, geochemical and isotopic signatures of host granitic rocks show that the origin of these host rocks are mainly controlled by magma mixing between felsic and mafic magma. The products of the mafic magma are dominated by metagabbro, fine-grained quartz-diorite and MMEs enclosed in granitic rocks, which are all resulted from the reaction and mixture between different mafic magmatic products and felsic magma. The metagabbros could be the products of felsic magma mixing with early crystallized mafic magma, and the fine-grained quartz-diorites could be the products of felsic magma mixing with crystal-rich mafic magma, and the MMEs could be the products of felsic magma mixing with crystal-poor mafic magma or melt. The metagabbro has positive $\varepsilon_{Hf}(t)$ values of 1. 8 to 10. 9, which could provide the relative primitive Lu-Hf isotopic compositions of mafic magma, indicating that the mafic magma is derived from a depleted mantle. The host granitic rocks have negative $\varepsilon_{Nd}(t)$ values of −6. 5 to −5. 2, indicating that the felsic magma is derived from partial melting of an ancient mafic lower crust. Furthermore, there are excess reactions after magma mixing between the mafic and felsic magma, such as multi-stage reactions between solidified MMEs and felsic magma, and the disaggregation of MMEs, finally resulting in the basification of felsic magma. Magma mixing is therefore the main mechanism of the formation of granitic rocks in the Tengchong

---

①　Published in *Lithos*, 2019, 326-327.

②　Corresponding author.

Block. The compositional variation of intermediate to acid igneous rock in a continental margin arc is likely dominated by the characteristics of mafic magma because the intermediate to acid igneous rock is derived from mixing of crust-derived granitic and different mantle sources-derived mafic magma.

1 Introduction

The geochemical variations of igneous rocks related to subduction are noted in many arcs and subduction zones, e.g., east Sunda arc, volcanic zone in Chile, Central Andean (e.g., Elburg et al., 2002; Jacques et al., 2013; Mamani et al., 2010). Previous studies pay more attention to reasons for geochemical changes of mafic rocks that are derived from the mantle in subduction zones, which could be resulted from decreasing melt fractions induced by a decrease in the slab-derived fluids inputting to the mantle wedge (Stern et al., 1993), or progressively greater contributions from the continental crust and subcrustal lithosphere through assimilation processes (Hildreth and Moorbath, 1988; Mamani et al., 2010), or other complicated processes. However, there are a large volume of granitic rocks derived from the crustal materials in continental margin arc, and these granitic rocks also show notable geochemical variations through time and space (e.g., Zhao et al., 2016a). In order to address the effects on geochemical changes of the granitic rocks in continental margin arc, this study focuses on the petrogenesis of Eocene granitic rocks and associated metagabbro, MMEs and fine-grained quartz-diorite in the western Tengchong Block, SW China.

According to the igneous rocks comparison, the Gangdese magmatic arc, Tengchong Block and Mogok metamorphic belt have been structurally linked since the Early Cretaceous (Xu et al., 2015). The Tengchong Block could be southeastward extension of the Lhasa Block after the Early Cretaceous (Ma et al., 2014; Wang et al., 2014, 2015), and can be considered as a Andean-type active continental margin arc related to the subduction of Neo-Tethys prior to the collision between the Indian and Asian continents, similar to the Lhasa Block (Zhu et al., 2011). Eocene (~50 Ma) mafic and felsic plutonic rocks are abundant in the western Tengchong Block (Fig. 1a; Wang et al., 2014, 2015; Ma et al., 2014; Xu et al., 2012; Zhao et al., 2016a). The Eocene granitic rocks in western Tengchong Block show compositional variations in space with increasing values of enriched compositions (e.g., K_2O, LREE, and Th), $^{87}Sr/^{86}Sr$, $\varepsilon_{Nd}(t)$, and $\varepsilon_{Hf}(t)$, but decreasing Pb isotopic ratios from the west to the east of the western Tengchong Block (Fig. 2; Zhao et al., 2016a). The mafic rocks also have similar eastward compositional changes despite of the missing of the Pb isotopic data (Fig. 2; Wang et al., 2014), implying an eastward subduction of Neo-Tethys in SW Yunnan, Southwest China. The granitic rocks show a range from Na-rich to K-rich granitoids, and the mafic rocks show a transition from tholeiite and calc-alkaline through high-K calc-alkaline to

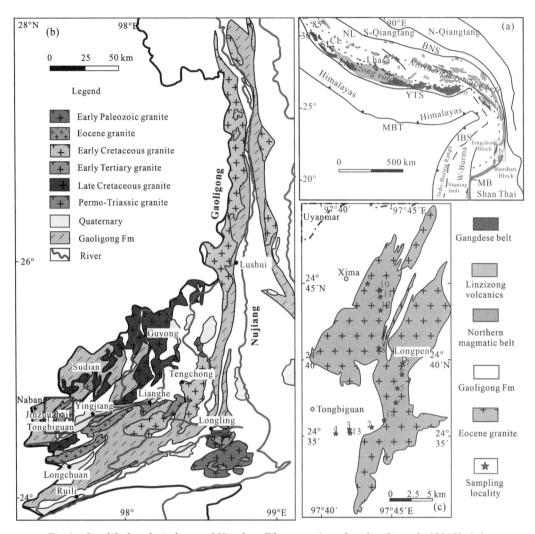

Fig. 1 Simplified geological map of Himalaya-Tibet tectonic realm after Qi et al. (2015) (a),
Tengchong Block and adjacent area after Xu et al. (2012) (b) and
the samples localities in the Xima-Tongbiguan pluton (c).

shoshonite series from west to east (Zhao et al., 2016a). The geochemical variations of granitic rocks are contradictory between Pb isotopes and other compositions, possibly indicating that the compositional changes of granitic rocks cannot be resulted from regular crustal sourced rocks. It is worth noting that the compositional variations of felsic rocks are spatially consistent with the coeval mafic rocks in western Tengchong Block (Fig. 2; Zhao et al., 2016a). The petrogenesis of mafic and felsic rocks there could therefore be closely related.

This study aims to clarify the relationship between the mafic and felsic rocks in a continental margin arc, and potential underlying magmatic evolution process. The granitoids, enclaves enclosed in the host rocks, and associated mafic rocks in the Xima-Tongbiguan pluton, the western Tengchong Block, are targeted to shed light on the geochemical changes of granitic rocks derived from thick crust in a continental margin arc.

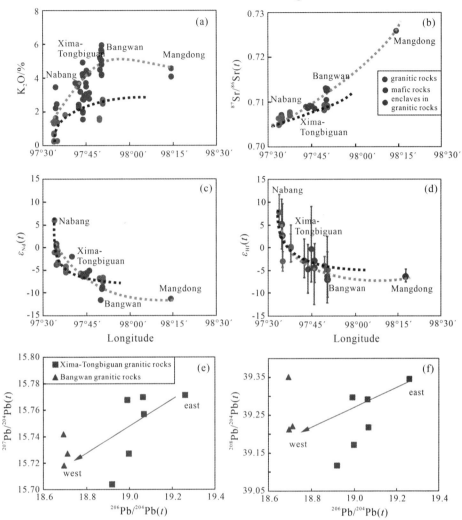

Fig. 2   Diagrams of the spatial varitions of K_2O, $^{87}Sr/^{86}Sr(t)$, $\varepsilon_{Nd}(t)$, $\varepsilon_{Hf}(t)$, and $^{207}Pb/^{204}Pb(t)$
vs. $^{206}Pb/^{204}Pb(t)$ and $^{208}Pb/^{204}Pb(t)$ vs. $^{206}Pb/^{204}Pb(t)$ for granitic rocks in Tengchong Block.
Data from Ma et al. (2014), Wang et al. (2014, 2015), Zhao et al. (2016a, 2017c).

2   Geological setting

The Tengchong Block is located in Southwest Yunnan Province, China, and bounded by the Gaoligong Belt from the Baoshan Block to the east, the Sagaing Fault from the west Burma Block to the west (Replumaz and Tapponnier, 2003; Fig. 1a). The Tengchong Block was once located in the northern margin of Gondwana during Early Paleozoic and accreted to the Eurasia in late Mesozoic (Morley et al., 2001; Zhao et al., 2017a-c). It is mainly comprised of pre-Mesozoic metamorphic rocks, Mesozoic-Cenozoic granitic and mafic rocks (Wang et al., 2015). The basement of Tengchong Block is named by Gaoligong Formation and exposed as

Mesoproterozoic metamorphic rocks, including metamorphosed volcano-sedimentary rocks, tonalite, and gabbro (Zhao et al., 2017b). The high-grade metamorphic rocks are exposed in the Nabang area, the China-Myanmar border, constituting a metamorphic zone named as the Nabang Metamorphic Zone, which extends southward adjoining the Mogok metamorphic belt in the east Burma Highland (Mitchell, 1993; Morley et al., 2001; Xu et al., 2012). Magmatic rocks in the Tengchong Block have been described in detail in Zhao et al. (2016a, b, 2017b, c). The Mesozoic-Cenozoic granitic and mafic rocks are considered as the southeastern continuation of the Gangdese batholith that has been dated at ca.70−50 Ma and related to the eastward subduction of Neo-Tethys (Ma et al., 2014; Qi et al., 2015; Wang et al., 2014, 2015; Xu et al., 2012, 2015; Zhao et al., 2016a, b). The Eocene igneous rock is abundant occurred at the western Tengchong Block (Fig. 1), which is the significant carrier to research the subduction of Neo-Tethys and collision of Indian-Asian continent in the southeast Yunnan.

3 Sampling petrography and description

3. 1 Observation from the field

The Xima-Tongbiguan Eocene granitic intrusion is located in the east of the Nabang town, the western Tengchong Block, SW China, with a NNE-SSW trend and approximately parallel to the Gaoligong Belt. The Xima-Tongbiguan pluton is mainly composed of granodiorite and quartz-diorite (Fig. 3). The abundant mafic microgranular enclaves (MMEs) are enclosed in the coarse-grained granodiorite and quartz-diorite but absent in the fine-grained quartz-diorite. The diameters of the MMEs are mostly 10−40 cm and appear as enclave swarms in the host granitoids on several outcrops (Fig. 4), some of which are round or ovoid and the rest are elongated. There are at least two different types of MMEs; the elongated MMEs with biotite-rich rind (type I) and the round/ovoid MMEs without biotite-rich rind (type II) (Fig. 4). The biotite-rich rinds are thin (0. 2−1 cm) and different from the interior of the amphibole-rich enclaves. The biotite-rich rinds might delaminate or break off from the enclave body. There is biotite-rich schlieren within the host granitoid (Fig. 4) that is characterized by a foliation fabric with preferentially aligned biotite. In many cases the schlieren could be derived directly from the biotite-rich rinds (Fig. 4). The variations of enclave and schlieren shapes indicate multiple stages in the strain history and the reactions between enclave and host granodiorites (Farner et al., 2014). In the southwest margin of Xima-Tongbiguan pluton, there is metagabbro occurring as a stock with 100 m in length. The metagabbro is massive and shows obvious igneous texture (Figs. 3d and 5a, b). The contact relationship between the metagabbro and the host granitic rock is unclear due to overlying plants. Sample locations are listed in Supplementary Table 1.

Fig. 3   Field and hand specimen photographs of the host granitic rocks, fine-grained
quartz-diorites and metagabbros in Xima-Tongbiguan area.

3.2 Petrography

The mineral assemblages of coarse-grained host granitic rocks, MMEs, fine-grained
quartz-diorites and metagabbros are mainly comprised of plagioclase, quartz, alkali feldspar,
biotite, amphibole, with accessory phases of apatite, ilmenite, magnetite, titanite and zircon.
Amphibole and biotite within MMEs, granitoid rocks and metagabbro are the most
ferromagnesian minerals in spite of various proportions.

Host granitic rocks: plagioclase (50% − 70%) is euhedral-subhedral, columnar or
granular. Most of the plagioclase grains have clear compositional zoning separated by resorption
surface and albite twinning. Alkali feldspar (3%−25%) is subhedral and anhedral, mainly
comprised of microcline and perthite. Alkali feldspar and quartz (7%−20%) also present as
interstitial phase. Biotite (4%−12%) occurs as subhedral flaky crystals and anhedral crystals
in clots. Amphibole (1%−10%) occurs in green euhedral to subhedral crystals.

Type Ⅰ *MMEs*: Plagioclase crystals (45% − 50%) are smaller than those in the host
rocks, though some euhedral feldspar megacrysts with compositional zoning are 2−5 mm in size
(Fig. 4). Biotites (20%−25%) are directionally orientated and anhedral crystals as interstitial

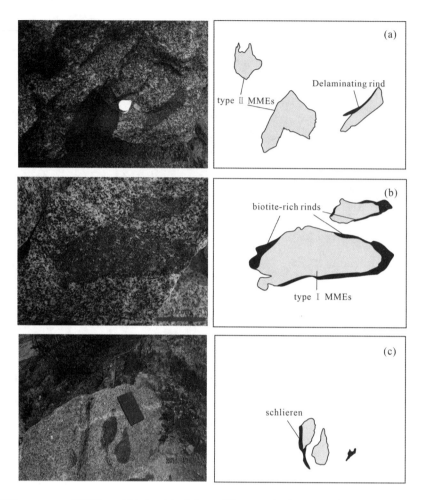

Fig. 4　Different types of MMEs and biotite-rich rind or schlieren in the Xima-Tongbiguan host granitic rocks,
type I MMEs with biotite-rich rinds and associated schlieren, type II MMEs without biotite-rich rinds.

phase or associated with amphibole. Amphibole (15%) commonly forms as clots of subhedral to anhedral green crystals. Quartz (10%–15%) and alkali feldspar (5%) occur as interstitial phase. Apatite, titanite, epidote and magnetite are present as accessory minerals.

　　Type II MMEs: Plagioclase (55%) in type II MMEs is similar to that in type I MMEs, but large euhedral feldspar megacrysts are much less (Fig. 4). Amphibole (30%) is more abundant than biotite in type II MMEs (10%), and both amphibole and biotite occur as euhedral to subhedral crystals. Quartz (2%) and alkali feldspar (3%) in type II are less than those in type I MMEs, and present as anhedral crystals and interstitial phase.

　　Fine-grained quartz-diorite: Plagioclase (65%) is euhedral crystal, and it has a compositional zoning with albite twining but lacks a resorption zone, and is different from plagioclase in the host granitic rocks (Fig. 5). Amphibole (20%) occurs as dark-greeneuhedral-subhedral crystals, or clot crystals surrounded or replaced by the biotite. Biotite

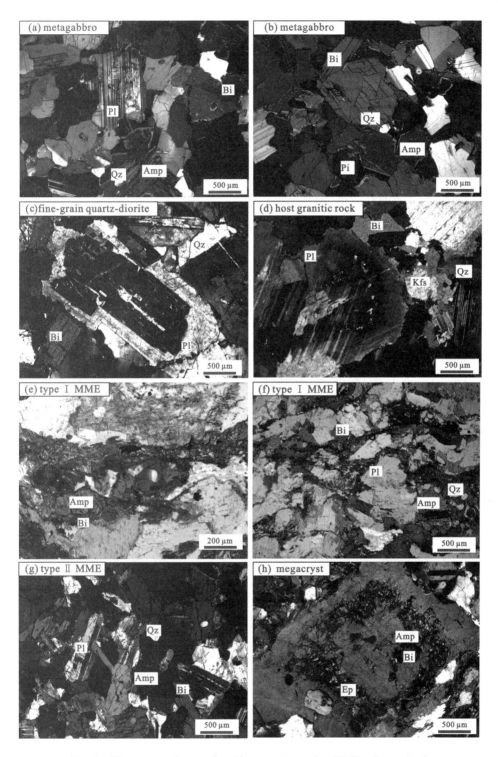

Fig. 5 Microscrope photographs of host granitic rocks, MMEs, fine-grained
quartz-diorites and metagabbros in Xima-Tongbiguan area.

（10%）occurs as yellow-brown subhedral-anhedral crystals. Quartz（5%）presents as interstitial phase. Alkali feldspar is absent in the fine-grained quartz-diorite.

Metagabbro：Mafic minerals in the metagabbro are mainly comprised of amphiboles and a small population of biotites, without pyroxene. The amphibole（40%）and plagioclase（55%）are subhedral granular or stumpy crystal, and many quartz and plagioclase crystals are enclosed in these amphiboles. The compositional zoning occurs in few plagioclases, but missing in most of them（Fig. 5a）. The plagioclase with compositional zoning lacks the resorption zone that is shown by plagioclase in the host granitic rock, similar to the plagioclase in fine-grained quartz-diorite. The biotite（3%）in the metagabbro intergrows with amphibole instead of occurring as individual crystal. The quartz（~2%）occurs as interstitial phase in the mineral pore around amphibole or enclosed in amphibole. The metagabbros do not contain alkali feldspar.

4　Results

4. 1　Zircon U-Pb dating

A total of six samples were selected for the zircon U-Pb isotopic analyses, two samples（NB13 and YJ53）from the host granitic rock, and two（NB70 and NB90）from the two different types of enclaves, one sample（NB09）from fine-grained quartz-diorite and one sample（XM05）from the metagabbro. Zircon U-Pb isotopic results and CL images are presented in Figs. 6 and 7, and U-Pb isotopic data are listed in Supplementary Table 2.

Coarse-grained host granitic rock：Zircons selected from the host granitic rocks are elongate prisms（150−250 μm in length）with aspect ratios of 1∶2 to 1∶3. They show well-defined oscillatory zoning on CL images, indicative of the magmatic origin. Thirty-one spots were analyzed for NB13 and yielded concordant $^{206}Pb/^{238}U$ ages of 57−51 Ma, with a weighted-mean age of 54 ± 1 Ma（$n = 31$, MSWD = 4. 3）. These spots have U contents of 170−1 720 ppm, and Th contents of 143−1 025 ppm, Th/U ratios of 0. 32−1. 02, except for the spots 27. Forty spots were analyzed for YJ53. Thirty-seven spot analyses gave $^{206}Pb/^{238}U$ ages of 56−48 Ma, with a weighted-mean age of 51 ± 1 Ma（$n = 37$, MSWD = 2. 7）, U contents of 253−1 353 ppm, Th contents of 71−764 ppm, and Th/U ratios of 0. 12−0. 92. The other three spots yielded concordant $^{206}Pb/^{238}U$ ages of 1 058−67 Ma, which are interpreted as the xenocrysts from the wall-rocks or inherited zircons from the source region.

Enclaves：

Type I：Zircon grains selected from the type I enclave（NB70）are granular or stubby prisms（100−150 μm long）with aspect ratios of 1∶1 to 1∶1. 5. On the CL images, the zircon grains show sector zoning without the typical magmatic oscillatory zoning. This internal structure indicates that these zircons could be crystallized at the fluids-rich condition. Thirty-two spots

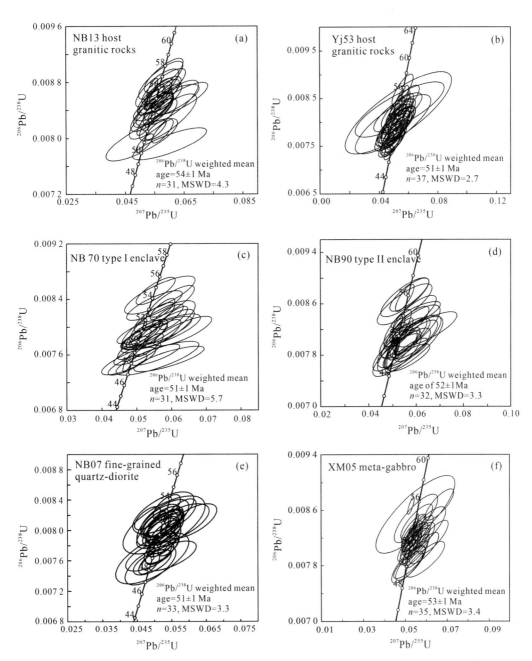

Fig. 6   Zircon U-Pb concordia diagrams for the host granitic rocks, MMEs, fine-grained quartz-diorites and metagabbros in Xima-Tongbiguan area.

were selected for analyses. Spot #12 yielded $^{206}Pb/^{238}U$ age of 799 Ma, interpreted as the xenocryst or inherited zircon, whereas all other thirty-one spots yielded concordant $^{206}Pb/^{238}U$ ages of 55−48 Ma, with a weighted-mean age of 51 ± 1 Ma ($n = 31$, MSWD = 5.7), U and Th contents of 284−2 446 ppm and 149−4 456 ppm, respectively, and Th/U ratios of 0.45−1.91.

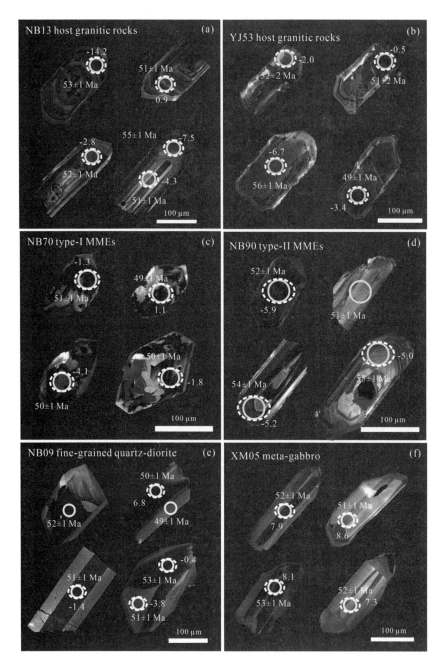

Fig. 7　CL images of representative zircons for the host granitic rocks, MMEs, fine-grained
quartz-diorites and metagabbros in Xima-Tongbiguan area.

Type II: Zircon grains selected from the type II enclave (NB90) are elongate prism,
100–200 μm in length, with aspect ratios of 1:1.5 to 1:2. They show well-defined oscillatory
zoning on the CL images. Thirty-two spots were analyzed and yielded concordant $^{206}Pb/^{238}U$
ages of 56–50 Ma, with a weighted mean age of 52±1 Ma (n=32, MSWD=3.3), U contents

of 224−2 420 ppm, Th contents of 41−4 253 ppm, and Th/U ratios of 0. 14−1. 76.

Fine-grained quartz-diorite: Zircons in the fine-grained quartz-diorite (NB09) are euhedral to subhedral crystals (150−200 μm long) with aspect ratios of 1∶1. 5 to 1∶2. These zircons display the wide compositional zoning, and differ from the zircons in the host rock and MMEs (Fig. 7), suggesting that the zircons are formed at the high temperature condition (Wu and Zheng, 2004). Thirty-three spots were analyzed and yielded concordant $^{206}Pb/^{238}U$ ages of 53−48 Ma, with a weighted-mean age of 51 ± 1 Ma ($n = 33$, MSWD = 3. 3) that is considered as the crystallization age of the quartz-diorite. The zircons have U contents of 133−865 ppm, Th contents of 65−1 137 ppm, and Th/U ratios of 0. 23−1. 33.

Metagabbro: Zircons in metagabbro (XM05) are euhedral with respect ratios of 1∶1. 5 to 1∶2 and show wide compositional zoning in CL images, indicating that these zircons are formed at the high temperature condition (Wu and Zheng, 2004), similar to zircons in the fine-grained quartz-diorite. Thirty-five spots were analyzed and yielded concordant $^{206}Pb/^{238}U$ ages of 56−50 Ma with a weighted-mean age of 53 ± 1 Ma ($n = 35$, MSWD = 3. 4), variable U contents of 134−13 277 ppm, Th contents of 53−971 ppm, and Th/U ratios of 0. 03−1. 14.

In summary, the MMEs are coeval with their hostgranitic rocks, and all of MMEs, host rocks and fine-grained quartz-diorite and metagabbro emplaced in the Early Eocene (ca. 55−50 Ma).

4. 2 Major elements

Thirty-four samples, including six metagabbros, one fine-grained quartz-diorite, eighteen host granitic rocks and nine MMEs, were analyzed for major and trace elements and the results are listed in Supplementary Table 3.

The metagabbros have low SiO_2 contents of 49. 41−51. 90 wt%, Na_2O contents of 2. 06− 2. 14 wt%, K_2O contents of 1. 12−1. 26 wt%, Al_2O_3 contents of 15. 11−15. 44 wt%, relatively high MgO contents of 7. 80 − 8. 46 wt%, $Fe_2O_3^T$ contents of 9. 60 − 10. 36 wt%. These metagabbros have high $Mg^#$ values of 65−66, alkaline contents of 3. 22−3. 38 wt%, belonging to subalkaline (Fig. 8a). The fine-granite quart-diorite has SiO_2 content of 54. 70 wt%, K_2O content of 1. 87 wt%, CaO content of 8. 36 wt%, MgO content of 2. 97 wt% with $Mg^#$ value of 50, and A/CNK ratios of 0. 90, characterized by the high-K calc-alkaline and metaluminous in geochemical signatures. The host granitic rocks have variable SiO_2 = 54. 08−70. 09 wt%, K_2O = 2. 24−4. 45 wt%, CaO = 2. 56−6. 16 wt%, MgO = 0. 84−3. 04 wt%, with $Mg^#$ values of 39−49. These host rocks are monzodiorite to granite in compositions (Fig. 8a), and metaluminous to weakly peraluminous with A/CNK values of 0. 92−1. 01 (Fig. 8c). They are all high-K calc-alkaline (Fig. 8b), and have trends of decreasing MgO, $Fe_2O_3^T$, CaO, Al_2O_3, TiO_2, and P_2O_5, and increasing K_2O with increasing silica content, but no obvious linear-relationship between Na_2O and SiO_2 (not shown). The type I MMEs have relatively high SiO_2 = 51. 49−

54. 50 wt%, $K_2O = 2.82 - 3.48$ wt%, low $MgO = 2.56 - 2.99$ wt%, with relatively low $Mg^{\#}$ value of 40. Comparing with type I MMEs, the type II MMEs have relatively low $SiO_2 = 48.90 - 52.65$ wt%, variable $K_2O = 1.53 - 4.89$ wt%, high $MgO = 4.51 - 5.07$ wt%, with relatively high $Mg^{\#}$ ratios of $51 - 54$. The type I MMEs are monzodiorite and type II are monzogabbro-monzodiorite in compositions (Fig. 8a), and they all belong to high-K calc-alkaline to shoshonitic series and metaluminous with A/CNK ratios of $0.96 - 0.98$ and $0.74 - 0.84$, respectively (Fig. 8b, c). MgO, $Fe_2O_3^T$, CaO, TiO_2 and P_2O_5 of host granitic rocks and MMEs decrease whereas Al_2O_3 and Na_2O scatter with increasing SiO_2. These major elements among host rocks, fine-grained quartz-diorite and metagabbro are discrete, except for $Fe_2O_3^T$ and K_2O, which show the linear relationship.

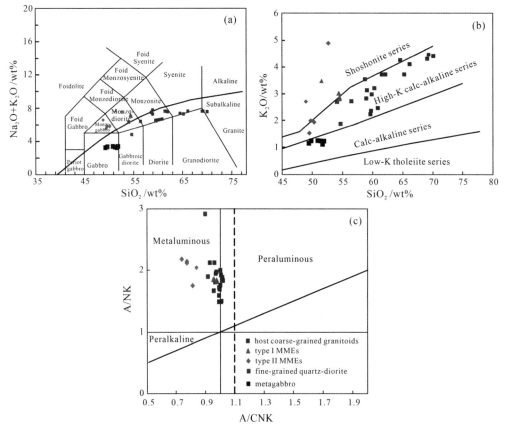

Fig. 8　Diagrams of SiO_2 vs. $K_2O + Na_2O$ (a), and SiO_2-K_2O (b), and A/CNK-A/NK (c) for the host granitic rocks, MMEs, fine-grained quartz-diorites and metagabbros after Middlemost (1994); Peccerillo and Taylor (1976), and Maniar and Piccoli (1989), respectively.

4. 3　Trace elements

The metagabbros and fine-grained quartz-diorite have different trace element pattern (Fig. 9a, b). Both of them are enriched in LILE and depleted in HFSE, but higher LILE in fine-grained quartz-diorite. Comparing with metagabbro and quartz-diorite, the host rocks have higher LILE and incompatible elements, depleted in HFSE. The MMEs have similar pattern to host rocks, but the type I MMEs have higher LILE than type II MMEs (Fig. 9c-f).

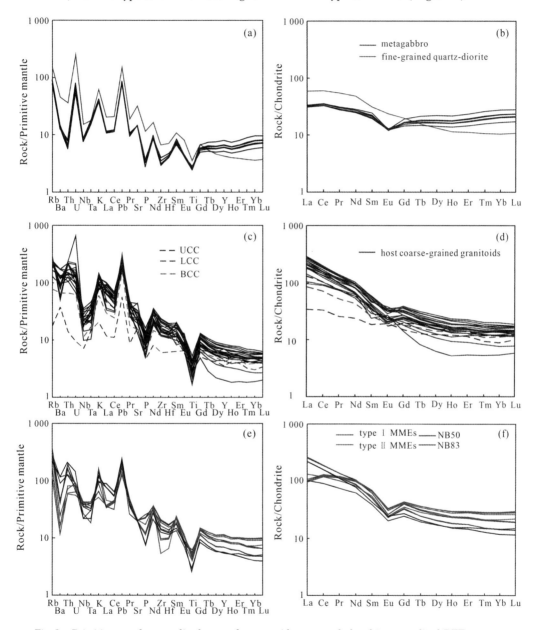

Fig. 9　Primitive mantle-normalized trace element spidergramsand chondrite-normalized REE pattern diagrams for the metagabbros and fine-grained quartz-diorites (a,b), host granitic rocks (c,d) and MMEs (e,f) in Xima-Tongbiguan area, the normalized data after Sun and McDonough (1989).

The REE patterns are almost flat for metagabbro with obvious negative Eu anomalies (δEu =0. 58- 0. 73) and total REE contents of 59－71 ppm, and gently right-dipping for fine-grained quartz-diorite that is slightly enriched in LREE and with weakly negative Eu anomaly (δEu = 0. 95) and total REE content of 97 ppm. The host granitic rocks are enriched in LREE and depleted in HREE with total REE contents of $132-281$ ppm, negative to positive Eu anomalies with δEu = 0. 65 - 1. 47, the positive Eu anomalies indicating accumulation of plagioclase in local area. The type I MMEs have similar REE patterns to host rocks, enriched in LREE and depleted in HREE, total REE contents of $231-276$ ppm, negative Eu anomalies with δEu = 0. 52-0. 57. Comparing with type I MMEs, type II MMEs have relatively low REE contents of $188-208$ ppm, more weakly enriched in LREE, and similar negative Eu anomalies with δEu = 0. 56 - 0. 61. The trace elements among these metagabbros, fine-grained quartz-diorite, host rocks, two types MMEs have no obvious linear relationship (not shown). Selected MMEs-host rocks pairs suggest partial equilibration, such as similar Rb , and different immobile elements Ti, Zr, Hf, Nb and Ta (Fig. 9e,f).

4. 4　Whole rock Sr-Nd-Pb isotopic data

The Sr, Nd and Pb isotopic compositions of the representative metagabbros, host rocks and two types MMEs are listed in Supplementary Table 4 and Figs. 10 and 11. Initial isotopic values were calculated at $t = 50$Ma on the basis of zircon U-Pb ages. The metagabbros have relatively low initial ^{87}Sr/^{86}Sr(t) ratios of 0. 706 319-0. 706 338, negative $\varepsilon_{Nd}(t)$ values of $-4. 2$ to $-3. 4$, with Nd two-stage model ages of 0. 97-1. 02 Ga, ^{206}Pb/^{204}Pb(t) = 19. 103- 19. 116, ^{207}Pb/^{204}Pb(t) = 15. 627 - 15. 644, ^{208}Pb/^{204}Pb(t) = 38. 621-38. 684. The host rocks have initial ^{87}Sr/^{86}Sr(t) ratios of 0. 708 502-0. 709 175, negative $\varepsilon_{Nd}(t)$ values of $-6. 5$ to $-5. 2$, Nd two-stage mode ages of 1. 10-1. 19 Ga, ^{206}Pb/^{204}Pb(t) = 18. 919-19. 260, ^{207}Pb/^{204}Pb(t) = 15. 704-15. 771, ^{208}Pb/^{204}Pb(t) = 39. 117-39. 345.

Type I MMEs have initial Sr ratio of 0. 709 082, $\varepsilon_{Nd}(t)$ value of $-5. 7$, Nd two-stage model age of 1. 13 Ga, ^{206}Pb/^{204}Pb(t) = 19. 016, ^{207}Pb/^{204}Pb(t) = 15. 733, ^{208}Pb/^{204}Pb(t) = 39. 181. Type II MMEs have initial Sr ratios of 0. 708 655-0. 708 660, $\varepsilon_{Nd}(t)$ values of $-6. 3$ to $-5. 9$, Nd two-stage model ages of 1. 14-1. 17Ga, ^{206}Pb/^{204}Pb(t) = 19. 173-19. 213, ^{207}Pb/^{204}Pb(t) = 15. 757-15. 776, ^{208}Pb/^{204}Pb(t) = 39. 208-39. 268.

4. 5　Zircon in-suit Lu-Hf isotopic compositions

The six zircon samples were analyzed for their in-suit Lu-Hf isotopes, and the result were presented in Supplementary Table 5 and Fig. 12. The metegabbros have positive $\varepsilon_{Hf}(t)$ values of 1. 08 to 10. 9, the corresponding single-stage model ages of 0. 29-0. 65 Ga. The fine-grained quartz-diorites have positive to negative $\varepsilon_{Hf}(t)$ values of $-4. 9$ to 6. 8, and eleven spots have single-stage model ages of 0. 48-0. 74 Ga with $\varepsilon_{Hf}(t)$ values of 0. 5 to 6. 8, while the other eleven spots have two-stage model ages of 1. 15-1. 43 Ga with $\varepsilon_{Hf}(t)$ values of $-4. 9$ to $-0. 4$.

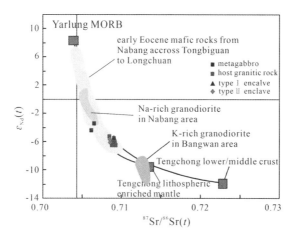

Fig. 10　Diagram of $^{87}Sr/^{86}Sr(t)$ vs. $\varepsilon_{Nd}(t)$ for the host granitic rocks,

MMEs and metagabbros in Xima-Tongbiguan area.

Data for mafic and granitic rocks in Nabang, Jinzhuzhai, Tongbiguan and Longchuan area from Wang et al. (2014,
2015), Ma et al. (2014). Data for Yarlung MORB from Xu and Castillo (2004). Data for lithospheric enriched
mantle from Wang et al. (2014).

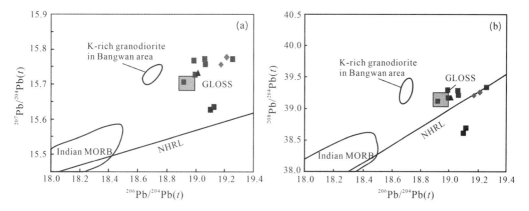

Fig. 11　Diagrams of $^{207}Pb/^{204}Pb(t)$ vs. $^{206}Pb/^{204}Pb(t)$ and $^{208}Pb/^{204}Pb(t)$ vs. $^{206}Pb/^{204}Pb(t)$

for the host granitic rocks, MMEs and metagabbros in Xima-Tongbiguan area.

Data for Indian MORB from Hofmann (1988).

The host granitic rocks have positive to negative of −7.5 to 1.9, and four spots have single-stage model ages of 0.66−0.73 Ga with $\varepsilon_{Hf}(t)$ values of 1.9 to 0.4, and the other spots have two-stage model ages of 1.15−1.60 Ga with $\varepsilon_{Hf}(t)$ values of −7.5 to −0.3. The type Ⅰ MMEs have positive to negative $\varepsilon_{Hf}(t)$ values of −7.7 to 9.0, and eight spots have single-stage model ages of 0.39−0.73 Ga, and the other spots have two-stage model ages of 1.14−1.61 Ga with $\varepsilon_{Hf}(t)$ values of −7.7 to −0.2. Type Ⅱ MMEs have positive to negative $\varepsilon_{Hf}(t)$ values of 0.6 to −10.2, and two spots have single-stage model ages of 0.71−0.73 Ga with $\varepsilon_{Hf}(t)$ values of 0.6, the other spots have two-stage model ages of 1.14−1.77 Ga with $\varepsilon_{Hf}(t)$ values of −10.2 to −0.4.

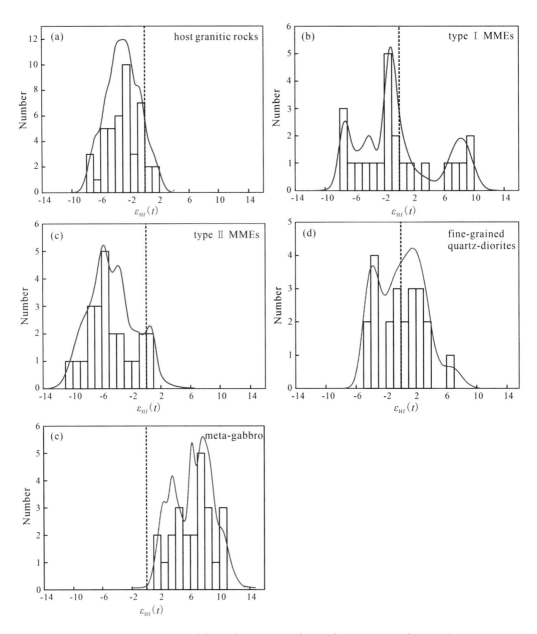

Fig. 12   Diagrams of $\varepsilon_{Hf}(t)$ for the Xima-Tongbiguan host granitic rocks, MMEs,
fine-grained quartz-diorites and metagabbros in Xima-Tongbiguan area.

5   Discussion

5. 1   Magma mixing for the petrogenesis of MMEs

The classification and petrogenesis of MMEs in the granitic rocks remains under debate. Generally, three hypotheses of the genesis of MMEs has been proposed: ① solid residues of partial melting or samples of unmelted, refractory materials from the source region (e. g., Chappell et al., 1987; White et al., 1999); ② autoliths segregated from the early crystallized layer of parent magma (e. g., Dahlquist, 2002; Dodge and Kistler, 1990; Donaire et al., 2005); and ③ mixing by injection of more mafic magma into cooler, partially crystalline felsic magma (e. g., Barbarin, 2005; Kocak, 2006; Kocak and Zedef, 2016; Kumar and Rino, 2006; Yang et al., 2007).

Metamorphic or residual sedimentary fabric and abundant inherited zircons were expected in the first model (Chappell et al., 1987; White et al., 1999). However, such fabrics and inherited zircons are absent in these MMEs from the Xima-Tongbiguan pluton that clearly point to an igneous texture and magmatic origin zircon (Figs. 3,5 and 7).

The hypotheses of "autoliths" and "mixing model" for generation of MMEs is difficult to discriminate according to the texture and geochemical and isotopic data, because these two opinions share many similar signatures. The linear correlation in chemical characteristics, similar mineral assemblage and identical isotopic features between MMEs and granitoids have been interpreted as autoliths because of crystallization fractionation from one parent magma (Dodge and Kistler, 1990; Donaire et al., 2005) or the result of magma mixing of two contrasted magma (Barbarin, 2005; Wiebe et al., 1997). Experimental studies indicate that diffusion processes induce chemical and isotopic equilibration of silicates of contrasted compositions (Lesher, 1990), and that isotopic equilibration is generally more easily achieved than chemical equilibration because isotopic exchanges proceed faster than chemical exchanges (Lesher, 1990). Therefore, the similar isotopic and linear geochemical correlation is not enough to distinguish the origin of MMEs in granitic rocks. The MMEs and host rocks in the Xima-Tongbiguan pluton have parallel variation in major elements of $Fe_2O_3^T$, MgO, CaO, and TiO_2 and some of trace elements, e.g. V and Cr vs. SiO_2(Fig. 13). In addition, the MMEs and host rocks have similar initial $^{87}Sr/^{86}Sr$ ratios of 0.708 660 − 0.709 082 and 0.708 502 − 0.709 175, respectively, and negative $\varepsilon_{Nd}(t)$ values of −6.3 to −5.7 and −6.5 to −5.2, respectively. This is also characterized by either autolith or chemical equilibration after magma mixing (Barbarin, 2005; Donaire et al., 2005). These geochemical signatures are insufficient to support either magma mixing model (e.g., Wiebe et al., 1997) or fractional crystallization process within closed system (Donaire et al., 2005).

Disequilibrium textures are also observed in the MMEs including fine-grained texture, acicular apatite crystals, and plagioclase megacrysts (Figs. 4 and 5). The fine-grained texture and acicular apatite crystals are interpreted as the rapid crystallization of the mafic magma undercooling in a more felsic and cooler granitic magma, representing magma mixing (Baxter and Feely, 2002). Nevertheless, experiments also indicate that, during the earliest crystallization phase of a granodioritic melt, mafic minerals should nucleate more quickly than

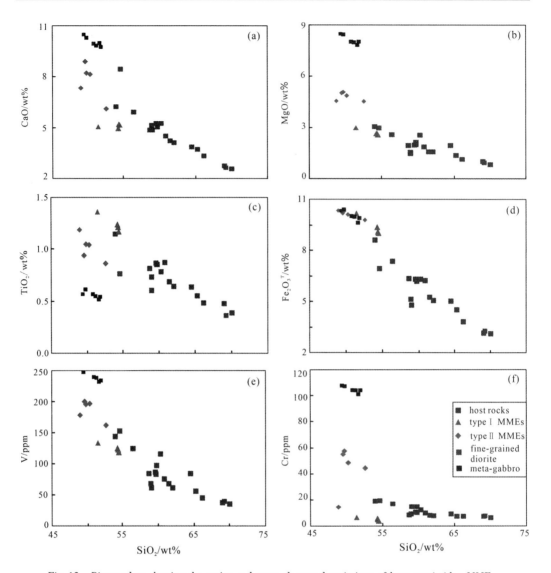

Fig. 13　Binary plots showing the major and trace elemental variations of host granitoids, MMEs,
fine-grained quartz-diorites and metagabbros from the Xima-Tongbiguan area.

the feldspars and quartz, and mafic minerals should be enriched in early crystallization products in the magma chamber margins (Castro, 2013; Holtz et al., 2001). In addition, apatite could also crystallize as acicular shape at high temperatures and low water concentration in the granitic magma system (García-Morene et al., 2006). The plagioclase megacrysts in MMEs are products of mechanical transfer from the granitic magma to enclaves during mingling (Kocak et al., 2011; Didier and Barbarin, 1991), which cannot resolve the petrogenesis of enclave yet.

　　The Hf isotope of zircon is critical for tracking its crystallization condition (Yang et al., 2007). The metagabbros have positive $\varepsilon_{Hf}(t)$ values of 1.8 to 10.9 with one peak value of 7.

The quartz-diorites have positive to negative $\varepsilon_{Hf}(t)$ values of -4.9 to 6.8 with two peak values of -4 and 2, respectively. Both of two types MMEs have various $\varepsilon_{Hf}(t)$ values ranging from negative to positive, 9.0 to -7.7 for type I MMEs and 0.6 to -10.2 for type II MMEs, which are characterized by gradually increasing degree of mixture of the inhomogeneous sources or different magma (Fig. 12; Yang et al., 2007; Kemp et al., 2007). The mafic rocks are abundant in western Tengchong Block (Wang et al., 2014, 2015), and there are metagabbros occurring at the southwest margin of Xima-Tongbiguan pluton. Thus, a plausible explanation for the formation of mixed zircon is magma mixing between two contrasting melts. During the mixing process, zircons crystallized in the mafic magma provide important information of the parent magma, but they are unable to survive in the high temperature intermediate magma ($> 1\ 000\ ℃$) that would exceed the closure temperature of zircon Lu-Hf isotopic system (Zheng et al., 2005), then zircons dissolve in or re-equilibrate with the high temperature intermediate magma. However, some zircons could have been occasionally enclosed in the refractory mafic minerals, i.e. pyroxene, and isolated from mixed intermediate melts or liquids, which could result in the preserve of primitive Lu-Hf isotopic composition of the mafic magma. Thus, the zircon Lu-Hf isotopic compositions in metagabbros could be represent the primitive information of mafic magma, and variable zircon Lu-Hf isotopic compositions in quartz-diorites and the two types of MMEs, could be resulted from mixing by two different sources.

To sum up, the MMEs in granitic rocks from Xima-Tongbiguan pluton are likely to be resulted from magma mixing based on its major and trace elemental composition, fine-grained igneous texture, apatite crystals in acicular shape, and plagioclase megacrysts, zircon Lu-Hf isotopic compositions.

5.2 Fractionation crystallization and Magma mixing for the host granitic rocks

The calc-alkaline granitic rocks occurred as large batholiths in the active continental margin are considered as responses to subduction and orogensis (e.g., Castro, 2013, 2014; Castro et al., 2010). Their petrogenesis are still unclear, although various models, based on experimental studies and natural research, have been proposed, including fractionation from the basaltic magma (e.g., Castro, 2013; Deering and Bachmann, 2010; Lee and Bachmann, 2014), magma mixing between felsic and mafic magma (e.g., Barbarin, 2005; Kent et al., 2010; Reubi and Blundy, 2009; Ruprecht et al., 2012; Yang et al., 2007), direct lithospheric mantle melting (e.g., Fallon and Danyushevsky, 2000), subducted young oceanic slab and sediments melting (e.g., Carter et al., 2015; Defant and Drumond, 1990; Mibe et al., 2011; Plank, 2005; Prouteau et al., 2001; Yogodzinski et al., 1995), or melting of pure mafic lower crust (Rapp et al., 1991; Rapp and Watson, 1995; Roberts and Clemens,

1993). These potential source regions and genetic mechanism could be applied to reveal origin of the intermediate to acidic rocks in active continental margin, but the given rocks have to been identified the required sources regions and magma evolved process.

The major and trace element data of these host granitic rocks signify fractional crystallization or magma mixing (e.g. Figs. 8 and 9; Chappell, 1996). The host granitic rocks show negative to positive Eu anomalies with $\delta Eu = 0.65 - 1.47$, but there is no obvious linear relationship between SiO_2 and δEu, indicating accumulating of plagioclase in local areas (Fig. 14). But, the linear relationship of host rocks between SiO_2 and Sr exhibits the fractional crystallization of plagioclase during the entire granitic magma evolution (Fig. 14). The amphibole fractionation would be indicated by increased La/Yb and decreased Dy/Yb ratios (Davidson et al., 2007; Macpherson, 2008), and this trend is also observed in host rocks, displaying increased La/Yb and decreased Dy/Yb with increasing SiO_2 (Fig. 14). These

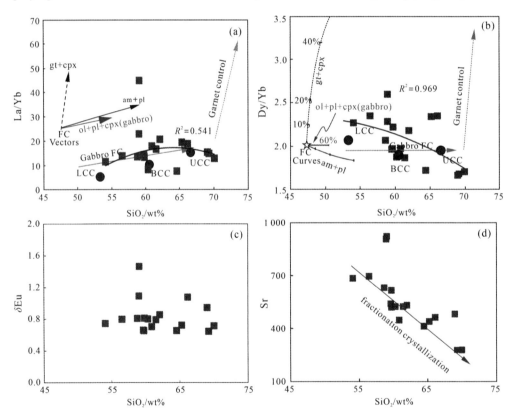

Fig. 14   Variation of (a) La/Y vs. SiO_2, (b) Dy/Yb vs. SiO_2, (c) δEu vs. SiO_2 and (d) Sr vs. SiO_2 for the host granitic rocks.

The calculation of correlation coefficient use all the data, except for deleting four samples including YJ43, NB20, NB21 and NB83 for La/Y vs. SiO_2. The trend of mineral fractionation refer to Davidson et al. (2007) and Macpherson (2008).

geochemical signatures reveal that the fractionation crystallization could take place in the granitic magma chamber, and plagioclase and amphibole are the major fractionated phases during granitic magma evolution.

The trace elements of Zr and P in magma are controlled by accessory minerals, e.g. zircon and apatite. Zircon and apatite would not crystalize when the magma is not saturated in Zr and P, and the Zr and P increase in liquid with progressive crystal segregation. Once the magma is saturated in Zr and P, zircon and apatite start to crystallize, and Zr and P decrease in liquid with further crystallization. The appearance of saturated for Zr and P are in accompany with SiO_2 contents at ca. 65 wt% and ca. 60 wt%, respectively (Gualda et al., 2012), which are also controlled by crystallization temperature, mineral assemblage, melt composition and H_2O content in magma system (Lee and Bachmann, 2014). Therefore, the crystal-liquid segregation leads to kinks in differentiation trend of Zr and P, while magma mixing would generate linear array in elemental variation diagrams (Fig. 15a,b; Lee and Bachman, 2014). The Zr and P contents against with SiO_2 of the host granitic rocks show the discrete and linear decreasing array, respectively, indicating magma mixing for the petrogenesis of host rocks

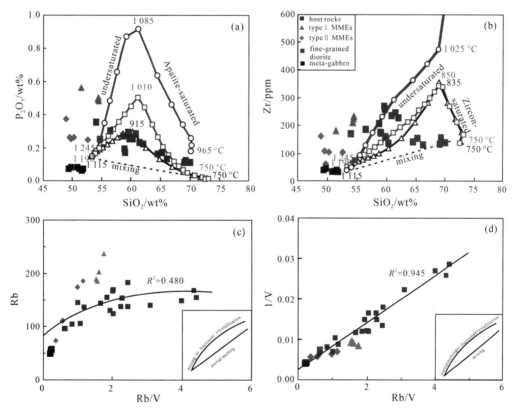

Fig. 15   Diagrams of SiO_2 vs. P_2O_5(a) and Zr(b), Rb/V vs. Rb (c) and 1/V (d) for the host granitic rocks in the Xima-Tongbiguan area after Lee and Bachman (2014) and Schiano et al. (2010).

(Fig. 15a, b). In addition, the host rocks show magma mixing trend in Rb-Rb/V and 1/V-Rb/V diagrams (Schiano et al., 2010; Fig. 15c, d). The zircon Lu-Hf isotopic compositions of host rocks show variable $\varepsilon_{Hf}(t)$ values of -7.5 to 1.9, with a peak of -3, indicating the magma mixing (Fig. 12; Kemp et al., 2007; Yang et al., 2007). The positive $\varepsilon_{Hf}(t)$ values could represent the information of mafic magma because the zircon crystallized from mafic magma could be enclosed in unresolved refractory minerals, such as pyroxene, amphibole and biotite, during magma mixing.

According to the lithologies, geochemical and zircon Lu-Hf compositional characteristics, the granitic rocks in Xima-Tongbiguan pluton were formed by magma mixing between felsic and mafic magma, and the plagioclase and amphibole fractionation also occurred during magma evolution.

5.3 Characteristics of felsic and mafic magma

The products associated with Eocene mafic magma are metagabbros, fine-grained quartz-diorites and two types of the MMEs in the Tengchong Block. According to the mineral petrography and zircon Lu-Hf isotopic compositions (Figs. 5 and 12), we proposed that the metagabbros could be the products of felsic magma mixing with early crystallized mafic magma, and the MMEs could be the products of felsic magma mixing with crystal-poor mafic magma or melt, and the fine-grained quartz-diorites are the products of felsic magma mixing with crystal-rich mafic magma. The metagabbros have a low degree reaction and mixing with felsic magma, thus the characteristics obtained from metagabbros could reveal primitive information of mafic magma. The metagabbros have positive $\varepsilon_{Hf}(t)$ values of 1.8 to 10.9, but negative $\varepsilon_{Nd}(t)$ value of -4.2 to -3.4 and high initial Sr ratios of $0.706\ 319-0.706\ 338$, which are characterized by decoupled Hf-Nd, depleted Hf and enriched Nd isotopes (Figs. 10 and 12). This is resulted from reaction and mixing between the metagabbros and felsic magma, and Nd isotopic compositions could be contaminated during reaction between the metagabbros and felsic magma, but zircons crystallized from mafic magma could be preserved and provide the primitive Lu-Hf isotopic compositions of mafic magma. Therefore, the mafic magma could be derived from the depleted mantle sources.

The host granitic rocks are characterized by variable SiO_2 contents of $54.08-70.09$ wt%, and mostly high-K calc-alkaline. They have high initial Sr ratios of $0.708\ 502-0.709\ 175$ and negative $\varepsilon_{Nd}(t)$ values of -6.5 to -5.2 and variable $\varepsilon_{Hf}(t)$ values of -7.5 to 1.9. These isotopic signatures preclude their derivation solely from pure mantle melts and indicate that the felsic magma could be mainly contributed by the ancient crustal rocks. The available experimental data indicate that the dehydration partial melting of mafic lower crust could generate metaluminous granitic melts, which could have relatively low $Mg^{\#}$ (<45) value and

MgO content, and are low-K calc-alkaline (e. g. , Rapp et al. , 1991; Rapp and Watson, 1995; Roberts and Clemens, 1993). It should be noted that the most basic granitic rocks in Xima-Tongbiguan (NB107), has the silica content of 54. 08 wt%, $Mg^{\#} = 45$, and is high-K calc-alkaline, which is inconsistent with the directly melting of mafic lower crust (Rapp and Watson, 1995). However, the K_2O and H_2O in the mafic magma could transfer into the felsic magma during the magma mixing (Lóperz et al. , 2005), and the mixed process could rise the $Mg^{\#}$ values in the felsic magma (Karsli et al. , 2010). Thus, the felsic melts could have been derived from ancient mafic lower crust materials and then mix with mafic magma derived from depleted mantle to form the host granitic rocks.

5. 4   Hybrid process during the evolution of the complex magmatic system

The MMEs are expected when the high temperature mafic magma freeze into the cooling silicic magma, which would result in a high viscosity contrast between the solidified enclaves and inhibit the mixing (Paterson et al. , 2004). The efficient mixing process requires high temperature in the magma system and is constrained by chaos dynamics (Perugini and Poli, 2012). The Xima-Tongbiguan area, in the western Tengchong Block, has a large number of Eocene mafic magmatism (Wang et al. , 2014, 2015), which could provide enough heat to support the melting of the crustal rocks and magma mixing. The classic two end-members mixing theory consider that the compositions of granitic rocks resulted from magma mixing could follow the equation of $C_m = C_a(1-x) + C_b x$ (Fourcade and Allegre, 1981), where C_m is hybrid composition, and C_a and C_b represent acid and mafic end-members, respectively, x is the weight proportion of acid end-member. In the Xima-Tongbiguan pluton, the type II MMEs must have mixed with the felsic magma, but they could represent the rapid cooling and crystallization of evolved mafic magma because of less reaction with granitic system than type I MMEs (Fig. 9e, f). Thus, their average composition (except NB87) could serve as the mafic end-member in this study, and the average compositions of most acid granitic rocks (YJ37, YJ38 and NB01) could serve as the acid end-member. The simulation results indicate that the host granitic rocks show a straight line, implying magma mixing trend, and the felsic magma contribution proportion varies from 0. 23 to 0. 84, whereas the R^2 values of granitic rocks decrease from acid to basic member (Fig. 16), suggesting that the other reaction process must exist in the formation of relatively basic granitic rocks during magma mixing. The low R^2 values of 0. 804 − 0. 753 and 0. 691 for type I MMEs and quartz-diorite, respectively, reveal that these rocks are similar to the relatively basic granitic rocks, which are not resulted from simple liquid mixture of two end-members where other process could have happened, such as mechanical transfer or peritectic mineral reaction (Clemens et al. , 2011).

The presences of feldspar megacrysts in the MMEs show that the felsic magma is

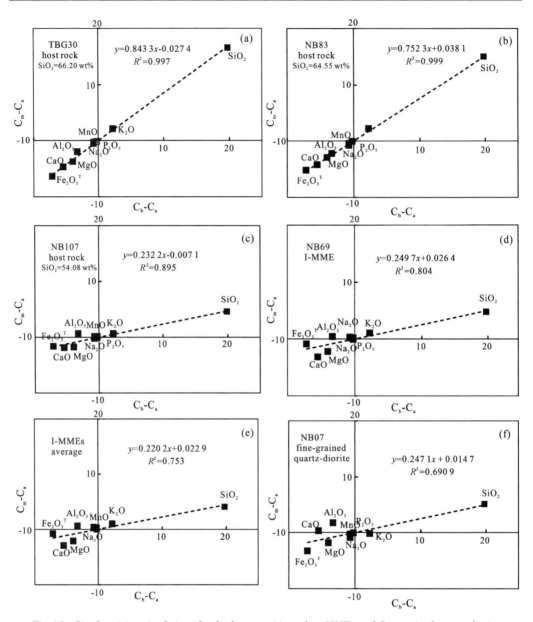

Fig. 16   Simple mixing simulations for the host granitic rocks, MMEs and fine-grained quartz diorites in the Xima-Tongbiguan area.

crystallizing during magma mixing. In addition, the absent of Qtz (quartz) megacrysts in the MMEs but as interstitial phase indicate that the felsic magma is initial crystallization because the plagioclase is normally generated in the earlier stage and Qtz in the later stage in the host granitoids (Castro, 2013). Therefore, the felsic magma is crystallizing during the injection of the mafic magma into the felsic magma chamber, and these mafic rocks could have different patterns, metagabbro being early crystallized products, fine-grained quartz-diorite representing crystal-rich mafic magma and MMEs representing crystal-poor mafic magma or melt before reaction and mixing with felsic magma, according to the petrography and lithology.

Actually, the mafic mineral in the metagabbros and fine-grained quartz-diorites is dominantly amphibole. This could be resulted from the chemical reaction as the following: opx + cpx + oxides + calc plagioclase + hydrous melt = amphibole and/or mica + quartz + sodic plagioclase (Beard et al., 2004, 2005). This reaction could be further evidenced by the fact that quartz and sodic plagioclase crystals was enclosed in the amphibole crystal or as interstitial phase around amphibole in metagabbros and fine-grained quartz-diorites (Fig. 5). This reaction could lead to entirely consumed pyroxene crystallized from mafic rocks if enough hydrous melt is provided. The MMEs are also absent of pyroxene, which could be also transferred to amphibole/mica during magma mixing. The reaction of pyroxene to amphibole requires hydrous melt. Generally, magmas containing even a trace amount of water will eventually become water saturated (Beard et al., 2004), then the anhydrous mineral previously crystallized could transform into hydrous mineral in their own magma system. In addition, the felsic magma could also provide hydrous melt during magma mixing. This could be further verified by mineral research.

The host granitic rocks have variable major elements, partial congruent trace elements and identical Sr-Nd-Pb isotopic composition (Figs. 8, 9 and 10). Comparing with type I MMEs, the type II MMEs have low SiO_2, high K_2O and MgO and similar Sr-Nd-Pb isotopic compositions (Figs. 8, 9 and 10). It is inferred that different diffusion rate among these compositions could induced the disequilibrium in major elements, incomplete equilibrium in trace elements and equilibrium in isotopic compositions in host rocks and MMEs (Dahlquist, 2002). The major elements in silicate melt are network-forming components and they are difficult diffusion and inhomogeneous; on the contrary, trace elements and associated isotopic system are non-network components and they are easily activated and rapid in diffusion to achieve homogeneity (Lesher, 1990). However, it is difficult to estimate the influence by magma mixing for granitic rocks composition and the scale of isotopic composition equilibration in granitic pluton.

The poorer Fe + Mg in experimental melt compared with natural granitic rocks has been noted by many scholars, which could be resulted from entrainment of peritectic minerals, such as pyroxene and garnet (Clemens et al., 2011), or fractionation from a more primitive liquid (Castro, 2013). However, during the mafic magma freezing into the felsic magma chamber and occurring as MMEs, this process also show that the mafic compositions contribute to the felsic rocks, including the dissolution of MMEs in a hydrous felsic magma and contributing to an increasing in the Mg + Fe content, finally resulting in the basification of felsic magma (Farner et al., 2014; García-Moreno et al., 2006; Miles et al., 2013). The MMEs in the Xima-Tongbiguan granitic rocks have at least two different types, including elongated type I MMEs with a biotite-rich rind and ovoid type II MMEs lack of a rind. The degree of reaction between type Ⅰ and corresponding host rocks is more intensive than that between type Ⅱ and the corresponding host rock, because of more quartz and alkali feldspar in type Ⅰ MMEs. In addition, more primitive MMEs (type Ⅱ) appears to have less reaction with their host granitoids because type Ⅱ MMEs have lower LREE, Rb and K contents and more amphibole than type Ⅰ MMEs (Fig. 9). The type Ⅰ MMEs have higher SiO_2 and K_2O, lower MgO, $Fe_2O_3^T$ and CaO contents than type Ⅱ MMEs, which further supports that the Fe + Mg + Ca could transfer from MMEs to felsic magma during reaction between MMEs and felsic magma. The biotite-rich rind around type Ⅰ MMEs is considered to form by successive reaction between hydrous residual liquids or fluids and the solidifying MMEs during the late stages of pluton crystallization (Farner et al., 2014). The biotite-rich rinds are rheologically weak and easy to break and delaminate from the parent MMEs (Fig. 5). These segregative biotite-rich rinds could occur in the crystallizing felsic magma as schlieren (Fig. 4) or the minerals distributed in the granitic rocks. The process during generation of biotite-rich rinds could follow the simplified reaction: amphibole + K_2O + Al_2O_3 = biotite + CaO + MgO + SiO_2 (Farner et al., 2014), where the amphiboles changes into biotites. The formation of biotite-rich rinds surrounded the enclaves is continuous after the delamination of old rinds until the entire disaggregation of enclaves or the entire solidarity of felsic magma. This process is controlled by the K_2O and H_2O and thermal limit of biotite stability in the residual liquids in the felsic magma (Farner et al., 2014), and can provide Fe + Mg + Ca to felsic melts, and further rise the content of these compositions in granitic rocks.

5.5 Implications for compositional variations of granitic rocks in continental margin arc

The whole Eocene mafic and granitic igneous rocks in the western Tengchong Block display a synchronous compositional variation from west to east (Fig. 2; Wang et al., 2014; Zhao et al., 2016b), revealing geochemical characteristics of magmatism in active continental margin arc. The abundant Eocene mafic rocks in the western Tengchong Block are derived from subcontinental lithospheric mantle and caused by the roll-back of Neo-Tethyan subducted slab (Wang et al., 2014, 2015; Xu et al., 2008; and reference therein), which could induce melting of subcontinental lithospheric mantle materials and produce the mafic magma. The spatial compositional variations of mafic rocks were controlled by the expected variations in the subducted slab/sediment-derived fluids and melts released into the sub-continental lithospheric mantle above the subduction zone, which were correlated with the gradual increase in the depth of the subduction zone from west to east. The characteristic of subducted slab/sediment-derived fluids/melts would be mainly aqueous fluids in the west (Nabang area, shallow zone), and more K and LILEs (Rb, Ba, and Sr) and lack Th and REEs during the dehydration of serpentine in the middle (Xima-Tongbiguan area, middle zone), and enriched in K, LILEs, Th, and REEs in the east (Bangwan area, deep zone), which is described in Zhao et al. (2016a). These fluids/melts would have metasomatized the overlying mantle material and induced its partial melting to produce the corresponding mafic magma, transitioning from tholeiite and calc-alkaline through high-K calc-alkaline to shoshonite series from west to east (Zhao et al., 2016a). The ascent and injection of mafic magma could have provided enough heat and material to induce the melting of crustal rocks and form deep crustal hot zones in the Tengchong Block (Annen et al., 2006). Together with the Eocene granitic rocks in Bangwan and Xima-Tongbiguan plutons (Zhao et al., 2016a), the magma mixing could happen and is the mainly genetic mechanism when the basaltic magma input into the crystallizing felsic magma, and the compositions of felsic magma could be remodified by the basaltic materials. Thus, the compositional variation of Eocene granitic rocks from west to east in Tengchong Block could be caused by the magma mixing between the corresponding mafic magma derived from different mantle sources and felsic magma (Fig. 17). Therefore, the compositional variations of crust-derived intermediate to acid igneous rocks in active continental margin arc are caused by mixing of different mantle sources-derived mafic magma and crust-derived granitic magma, and such variations are dominated by the composition of corresponding mafic magma.

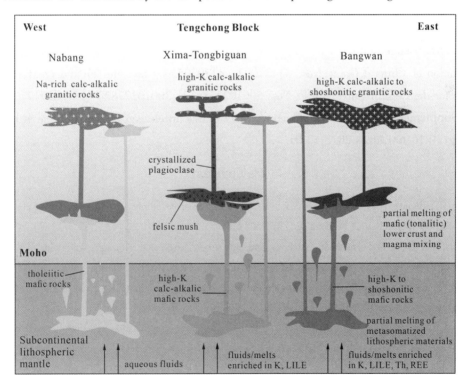

Fig. 17　Schematic cartoon illustrating the compositional variations and petrogenesis of Eocene granitic and mafic rocks from west to east in Tengchong Block.

6 Conclusions

(1) Zircon U-Pb dating results indicate the felsic and basaltic magmatisms in the Xima-Tongbiguan area are coeval at early Eocene with age of ca. 50 Ma.

(2) The field and petrological observations, geochemical and isotopic compositions indicate that metagabbros, fine-grained diorites and MMEs are the mainly products of mafic magma in Xima-Tongbiguan area. The metagabbros could be the products of felsic magma mixing with early crystallized mafic magma, and the fine-grained quartz-diorites could be felsic magma mixing with crystal-rich mafic magma, and the MMEs could be the products of felsic magma mixing with crystal-poor mafic magma or melt. The mafic magma could be derived from depleted mantle according to the Hf isotopic compositions of metagabbro. The host granitic rocks are derived from magma mixing between depleted mantle-derived mafic and mafic lower crust-derived felsic magma and then multi-stage reactions between solidified MMEs and felsic magma, and the disaggregation of MMEs finally result in the basification of felsic magma during the formation of host rocks.

(3) The compositional variations of crust-derived intermediate to acid igneous rocks in an active continental margin arc are resulted from mixing of different mantle sources-derived mafic magma and crust-derived granitic magma, and such variations are dominated by the composition of corresponding mafic magma.

Acknowledgements This work was jointly supported by the National Natural Science Foundation of China (Grant Nos. 41421002, 41703055), and the Fundamental Research Funds for the Central Universities (Nos. 300102278112, 300102278204, 310827172004).

Supplementary data Supplementary data to this article can be found online at https://doi.org/10.1016/j.lithos.2018.12.06.

References

Annen, C., Blundy, J.D., Sparks, R.S.J., 2006. The genesis of Intermediate and silicic magmas in deep crustal hot zones. Journal of Petrology, 47, 505-539.

Barbarin, B., 2005. Mafic magmatic enclaves and mafic rocks associated with some granitoids of the central Sierra Nevada batholith, California: Nature, origin, and relations with the hosts. Lithos, 80, 155-177.

Baxter, S., Feely, M., 2002. Magma mixing and mingling textures in granitoids: Examples from the Galway Granites, Connemara, Ireland. Mineralogy and Petrology, 76, 63-74.

Beard, J.S., Ragland, P.C., Rushmer, T., 2004. Hydration crystallization reaction between anhydrous minerals and hydrous melt to yield amphibole and biotite in igneous rocks: Description and implications. The Journal of Geology, 112, 617-621.

Beard, J.S., Ragland, P.C., Crawford, M.L., 2005. Reactive bulk assimilation: A model for crust mantle mixing in silicic magmas. Geology, 33, 681-684.

Carter, L.B., Skora, S., Blundy, J.D., De Hoog, J.C.M., Elliott, T., 2015. An experimental study of trace element fluxes from subducted oceanic crust. Journal of Petrology, 56, 1585-1606.

Castro, A., 2013. Tonalite-granodiorite suites as cotectic systems: A review of experimental studies with applications to granitoids petrogenesis. Earth-Science Reviews, 12, 68-95.

Castro, A., 2014. The off-crust origin of granite batholiths. Geoscience Frontiers, 5, 63-75.

Castro, A., Gerya, T., García-Casco, A., Fernández, C., DíazAlvarado, J., Moreno-Ventas, I., Loew, I., 2010. Melting relations of MORB-sediment mélanges in underplated mantle wedge plumes: Implications for the origin of cordilleran-type batholiths. Journal of Petrology, 51, 1267-1295.

Chappell, B.W., 1996. Magma mixing and the production of compositional variation within granite suites: Evidence from the granites of southeastern Australia. Journal of Petrology, 37, 449-470.

Chappell, B.W., White, A.J.R, Wyborn, D., 1987. The importance of residual source material (restite) in granite petrogenesis. Journal of Petrology, 28, 1111-1138.

Clemens, J.D., Stevens, G., Farina, F., 2011. The enigmatic sources of I-type granites: The peritectic connexion. Lithos, 126, 174-181.

Dahlquist, J.A., 2002. Mafic microgranular enclaves: Early segregation from metaluminousmagma (Sierra de Chepes), Pampean Ranges, NW Argentina. Journal of South American Earth Sciences, 15, 643-655.

Davidson, J., Turner, S., Handley, H., Macpherson, C., Dosseto, A., 2007. Amphibole "sponge" in arc crust? Geology, 35(9), 787-790.

Deering, C.D., Bachmann, O., 2010.Trace element indicators of crystal accumulation in silicic igneous rocks. Earth & Planetary Science Letters, 297, 324-331.

Defant, M.J., Drummond, M.S., 1990. Derivation of some modern arc magmas by melting of young subducted lithosphere. Nature, 347, 662-665.

Didier, J., Barbarin, B., 1991.Enclaves and Granite Petrology. Amsterdam: Elsevier,1-625.

Dodge, F.C.W., Kistler, R.W., 1990. Some additional observations on inclusions in the granitic rocks of Sierra Nevada. Journal of Geophysical Research, 95, 17841-17848.

Donaire, T., Pascual, E., Pin, C., Duthou, J.L., 2005. Microgranular enclaves as evidence of rapid cooling in granitoid rocks: The case of the Los Pedroches granodiorite, Iberian Massif, Spain. Contributions to Mineralogy and Petrology, 149(3), 247-265.

Elburg, M.A., Bergen, M.V., Hoogewerff, J., Foden, J., Vroon, P., Zulkarnain, I., Nasution, A., 2002. Geochemical trends across an arc-continent collision zone: Magma sources and slab-wedge transfer processes below the Pantar Strait volcanoes, Indonesia. Geochimica et Cosmochimica Acta, 66, 2771-2789.

Fallon, T.J., Danyushevsky, L.V., 2000. Melting of refractory mantle at 1.5, 2 and 2.5 GPa under anhydrous and H_2O-undersaturated conditions: Implications for the petrogenesis of high-Ca Boninites and the influence of subduction components on mantle melting. Journal of Petrology, 41, 257-283.

Farner, M.J., Lee, C.T.A., Putirka, K.D., 2014. Mafic-felsic magma mixing limited by reactive process: A case study of biotite-rich rinds on mafic enclaves. Earth & Planetary Science Letters, 393, 49-59.

Fourcade, S., Allegre, C.J., 1981. Trace elements behavior in granite genesis: A case study the calc-alkaline plutonic association from the Querigut complex (Pyrénées, France). Contributions to Mineralogy and

Petrology, 76, 177-195.

García-Moreno, O., Castro, A., Corretgé, L.G., El-Hmidi, H., 2006. Dissolution of tonalitic enclaves in ascending hydrous granitic magmas: An experimental study. Lithos, 89, 245-258.

Gualda, G.A.R., Ghiorso, M.S., Lemons, R.V., Carley, T.L., 2012. Rhyolite-Melts: A modified calibration of MELTS optimized for silica-rich, fluid-bearing magmatic systems. Journal of Petrology, 53, 875-890.

Hildreth, W., Moorbath, S., 1988. Crustal contributions to arc magmatism in the Andes of central Chile. Contributions to Mineralogy and Petrology, 98, 455-489.

Hofmann, A.W., 1988. Chemical differentiation of the Earth: The relationship between mantle, continental crust, and oceanic crust. Earth & Planetary Science Letters, 90, 279-314.

Holtz, F., Johannes, W., Tamic, N., Behrens, H., 2001. Maximum and minimum water contents of granitic melts generated in the crust: A reevaluation and implications. Lithos, 56, 1-14.

Jacques, G., Hoernle, K., Gill, J., Hauff, F., Wehrmann, H., Garbe-Schönberg, D., van den Bogaard, P., Bindeman, I., Lara, L.E., 2013. Across-arc geochemical variations in the Southern Volcanic Zone, Chile (34.5° - 38.0° S): Constraints on mantle wedge and slab input compositions. Geochimica et Cosmochimica Acta, 123, 218-243.

Karsli, O., Dokuz, A., Uysal, I., Aydin, F., Chen, B., Kandemir, R., Wijbrans, J., 2010. Relative contributions of crust and mantle to generation of Campanian high-K calc-alkaline I-type granitoids in a subduction setting, with special reference to the Harşit Pluton, Eastern Turkey. Contributions to Mineralogy and Petrology, 160, 467-487.

Kemp, A.I.S., Hawkesworth, C.J., Foster, G.L., Paterson, B.A., Woodhead, J.D., Hergt, J.M., Gray, C.M., Whitehouse, M.J., 2007. Magmatic and crustal differentiation history of granitic rocks from Hf-O isotopes in zircon. Science, 315, 980-983.

Kent, A.J.R., Darr, C., Koleszar, A.M., Salisbury, M.J., Cooper, K.M., 2010. Preferential eruption of andesitic magmas through recharge filtering. Nature Geoscience, 3, 631-636.

Kocak, K., 2006. Hybridization of mafic microgranular enclaves: Mineral and whole-rock chemistry evidence from the Karamadaz Granitoid, Central Turkey. International Journal of Earth Sciences, 95, 587-607.

Kocak, K., Zedef, V., 2016. Interaction of the lithospheric mantle and crustal melts for the generation of the Horoz pluton (Nigde, Turkey): Whole-rock geochemical and Sr-Nd-Pb isotopic evidence. Estonian Journal of Earth Science, 65, 138-160.

Kocak, K., Zedef, V., Kansun, G., 2011. Magma mixing/mingling in the Eocene Horoz (Nigde) granitoids, Central southern Turkey: Evidence from mafic microgranular enclaves. Mineralogy and Petrology, 103, 149-167.

Kumar, S., Rino, V., 2006. Mineralogy and geochemistry of microgranular enclaves in Palaeoproterozoic Malanjkhand granitoids, Central India: Evidence of magma mixing, mingling, and chemical equilibration. Contributions to Mineralogy and Petrology, 152, 591-609.

Lee, C.T.A., Bachmann, O., 2014. How important is the role of crystal fractionation in making intermediate magmas? Insights from Zr and P systematic. Earth & Planetary Science Letters, 393, 266-274.

Lesher, C.E., 1990. Decoupling of chemical and isotopic exchange during mixing. Nature, 344, 235-237.

Lóperz, S., Castro, A., García-Casco, A., 2005. Production of granodiorite melt by interaction between hydrous magma and tonalitic crust. Experimental constraints and implication for the generation of Archaean TTG complexes. Lithos, 79, 229-250.

Ma, L.Y., Wang, Y.J., Fan, W.M., Geng, H.Y., Cai, Y.F., Zhong, H., Liu, H.C., Xing, X.W., 2014. Petrogenesis of the early Eocene I-type granites in west Yingjiang (SW Yunnan) and its implications for the eastern extension of the Gangdese batholiths. Gondwana Research, 25, 401-419.

Macpherson, C.G., 2008. Lithosphere erosion and crustal growth in subduction zones: Insights from initial of the nascent East Philippine Arc. Geology, 36(4), 311-314.

Mamani, M., Wörner, G., Sempere, T., 2010. Geochemical variations in igneous rocks of the Central Andean orocline (13°S to 18°S): Tracing crustal thickening and magma generation through time and space. Geological Society of America Bulletin, 122, 162-182.

Maniar, P.D., Piccoli, P.M., 1989. Tectonic discrimination of granitoids. Geological Society of American Bulletin, 101, 635-643.

Mibe, K., Kawamoto, T., Matsukage, K.N., Fei, Y.W., Ono, S., 2011. Slab melting versus slab dehydration in subduction-zone magmatism. PANS, 108, 8177-8182.

Middlemonst, E.A.K., 1994. Naming materials in the magma/igneous rocks system. Earth-Science Reviews, 37, 215-224.

Miles, A.J., Graham, C.M., Hawkesworth, C.J., Gillespie, M.R., Hinton, R.W., EIMF, 2013. Evidence for distinct stages of magma history recorded by the compositions of accessory apatite and zircon. Contributions to Mineralogy and Petrology, 166, 1-19.

Mitchell, A.H.G., 1993. Cretaceous-Cenozoic tectonic events in the western Myanmar (Burma)-Assam region. Journal of the Geological Society, London, 150, 1089-1102.

Morley, C., Woganan, N., Sankumarn, N., Hoon, T., Alief, A., Simmons, M., 2001. Late Oligocene-recent stress evolution in rift basins of northern and central Thailand: Implications for escape tectonic. Tectonophysics, 334, 115-150.

Paterson, S.R., Pignotta, G.S., Vernon, R.H., 2004. The significance of microgranitoid enclave shapes and orientations. Journal of Structural Geology, 26, 1465-1481.

Peccerillo, A., Taylor, S.R., 1976. Geochemistry of Eocene calc-alkaline volcanic rocks from the Kastamonu area, northern Turkey. Contributions to Mineralogy and Petrology, 58, 63-81.

Perugini, D., Poli, G., 2012. The mixing of magmas in plutonic and volcanic environments: Analogies and differences. Lithos, 153, 261-277.

Plank, T., 2005. Constraints from Thorium/Lanthanum on sediment recycling at subduction zones and the evolution of the continents. Journal of Petrology, 46, 921-944.

Prouteau, G., Scaillet, B., Pichavant, M., Maury, R., 2001. Evidence for mantle metasomatism by hydrous silicic melts derived from subducted oceanic crust. Nature, 410, 197-200.

Qi, X.X., Zhu, L.H., Grimmer, J.C., Hu, Z.C., 2015. Tracing the Transhimalayan magmatic belt and the Lhasa block southward using zircon U-Pb, Lu-Hf isotopic and geochemical data: Cretaceous-Cenozoic granitoids in the Tengchong block, Yunnan, China. Journal of Asian Earth Sciences, 110, 170-188.

Rapp, R.P., Watson, E.B., 1995. Dehydration melting of metabasalt at 8-32 kbar: Implications for continental growth and crust-mantle recycling. Journal of Petrology, 36, 891-931.

Rapp, R.P., Watson, E.B., Miller, C.F., 1991. Partial melting of amphibolite/eclogite and the origin of Archean trondhjemites and tonalities. Precambrian Research, 51, 1-25.

Replumaz, A., Tapponnier, P., 2003. Reconstruction of deformed collision zone Between India and Asia by backward motion of lithospheric blocks. Journal of Geophysical Research, 108 (B6) 2285.

Reubi, O., Blundy, J., 2009. A dearth of intermediate melts at subduction zone volcanoes and the petrogensis of arc andesites. Nature, 461,1274-1296.

Roberts, M.P., Clemens, J.D., 1993. Origin of high-potassium, calc-alkaline, I-type granitoids. Geology, 21, 825-828.

Ruprecht, P., Bergantz, G.W., Cooper, K.M., Hildreth, W., 2012. The crustal magma storage system of volcán Quizapu, Chile, and the effects of magma mixing on magma diversity. Journal of Petrology, 53, 801-840.

Schiano, P., Monzier, M., Eissen, J.P., Martin, H., Koga, K.T., 2010. Simple mixing as the major control of the evolution of volcanic suites in the Ecuadorian Andes. Contributions to Mineralogy and Petrology, 160, 297-312.

Stern, R.J., Jackson, M.C., Fryer, P., Ito, E., 1993. O, Sr, Nd and Pb isotopic composition of Kasuga Cross-Chain in the Mariana Arc: A new perspective on the K-h relationship. Earth & Planetary Science Letters, 119, 459-475.

Sun, S.S., McDonough, W.F., 1989. Chemical and isotopic systematics of oceanic basalts: implications for mantle composition and processes. In: Saunders, A.D, Norry, M.J. Magmatism in the Ocean Basins, 42. Geological Society Special Publication, London, 313-345.

Wang, Y.J., Zhang, L.M., Cawood, P.A., Ma, L.Y., Fan, W.M., Zhang, A.M., Zhang, Y.Z., Bi, X.W., 2014. Eocene supra-subduction zone mafic magmatism in the Sibumasu block of SW Yunnan: Implications for Neotethyan subduction and India-Asia collision. Lithos, 206-207, 384-399.

Wang, Y. J., Li, S. B., Ma, L. Y., Fan, W. M., Cai, Y. F., Zhang, Y. H., Zhang, F. F., 2015. Geochronological and geochemical constraints on the petrogenesis of Early Eocene metagabbroic rocks in Nabang (SW Yunnan) and its implications on the Neotethyan slab subduction. Gondwana Research, 27, 1474-1486.

White, A. J. R., Chappell, B. W, Wyborn, D., 1999. Application of the restite model to the Deddick granodiorite and its enclaves-a reinterpretation of the observations and data of Maas et al. (1997). Journal of Petrology, 40(3), 413-421.

Wiebe, R.A., Smith, D., Sturn, M., King, E.M., 1997. Enclaves in the Cadillac mountain granite (Coastal Marine): Samples of hybrid magma from the base of the chamber. Journal of Petrology, 38, 393-426.

Wu, Y.B., Zheng, Y.F., 2004. Genesis of zircon and its constraints on interpretation of U-Pb age. Chinese Science Bulletin, 49, 1557-1569.

Xu, J.F., Castillo, P.R., 2004. Geochemical and Nd-Pb isotopic characteristics of the Tethyan asthenosphere: Implications for the origin of the Indian Ocean mantle domain. Tectonophysics, 393, 9-27.

Xu, Y.G., Lan, J.B., Yang, Q.J., Huang, X.L., Qiu, H.N., 2008. Eocene break-off of the Neo-Tethyan slab as inferred from intraplate-type mafic dykes in the Gaoligong orogenic belt, eastern Tibet. Chemical Geology, 255, 439-453.

Xu, Y.G., Yang, Q.J., Lan, J.B., Luo, Z.Y., Huang, X.L., Shi, Y.B., Xie, L.W., 2012. Temporal-spatial distribution and tectonic implications of the batholiths in the Gaoligong-Tengliang-Yingjiang area, western Yunnan: Constraints from zircon U-Pb ages and Hf isotopes. Journal of Asian Earth Sciences, 53, 151-175.

Xu, Z.Q., Wang, Q., Cai, Z.H., Dong, H.W., Li, H.Q., Chen, X.J., Duan, X.D., Cao, H., Li, J., Burg, J.P., 2015. Kinematics of the Tengchong Terrane in SE Tibet from the late Eocene to early Miocene: Insights from coeval mid-crustal detachments and strike-slip shear zones. Tectonophysics, 665, 127-148.

Yang, J.H., Wu, F.Y., Wilde, S.A., Xie, L.W., Yang, Y.H., Liu, X.M., 2007. Tracing magma mixing in granite genesis in situ U-Pb dating and Hf isotopes analysis of zircons. Contributions to Mineralogy and Petrology, 153, 177-190.

Yogodzinski, G.M., Kay, R.W., Volynets, O.N., Koloskov, A.V., Kay, S.M., 1995. Magnesian andesite in the western Aleutian Komandorsky region: Implications for slab melting and processes in the mantle wedge. Geological Society of American Bulletin, 107, 505-519.

Zhao, S.W., Lai, S.C., Qin, J.F., Zhu, R.Z., 2016a. Petrogenesis of Eocene granitoids and microgranular enclaves in the western Tengchong Block: Constraints on eastward subduction of the Neo-Tethys. Lithos, 264, 96-107.

Zhao, S.W., Lai, S.C., Qin, J.F., Zhu, R.Z., 2016b. Tectono-magmatic evolution of the Gaoligong belt, southeastern margin of the Tibetan plateau: Constraints from granitic gneisses and granitoid intrusions. Gondwana Research, 35, 238-256.

Zhao, S.W., Lai, S.C., Gao, L., Qin, J.F., Zhu, R.Z., 2017a. Evolution of the Proto-Tethys in the Baoshan block along the East Gondwana margin: Constraints from early Palaeozoic magmatism. International Geology Review, 59(1), 1-15.

Zhao, S.W., Lai, S.C., Qin, J.F., Zhu, R.Z., Gan, B.P., 2017b. The petrogenesis and implications of the Early Eocene granites in Lianghe area, Tengchong Block. Acta Petrologica Sinica, 33(1), 191-203.

Zhao, S.W., Lai, S.C., Qin, J.F., Zhu, R.Z., Wang, J.B., 2017c. Geochemical and geochronological characteristics of Late Cretaceous to Early Paleocene granitoids in the Tengchong Block, Southwestern China: Implications for crust anatexis and thickness variations along eastern Neo-Tethys subduction zone. Tectonophysics, 694, 87-100.

Zheng, Y.F., Wu, Y.B., Zhao, Z.F., Zhang, S.B., Xu, P., Wu, F.Y., 2005. Metamorphic effect on zircon Lu-Hf and U-Pb isotope systems in ultrahigh-pressure eclogite-facies metagranite and metabasite. Earth & Planetary Science Letters, 240, 378-400.

Zhu, D.C., Zhao, Z.D., Niu, Y.L., Mo, X.X., Chung, S.L., Hou, Z.Q., Wang, L.Q., Wu, F.Y., 2011. The Lhasa Terrane: Record of a microcontinent and its histories of drift and growth. Earth & Planetary Science Letters, 301, 241-255.

Middle Permian high Sr/Y monzogranites in central Inner Mongolia: Reworking of the juvenile lower crust of Bainaimiao arc belt during slab break-off of the Palaeo-Asian oceanic lithosphere[①]

Liu Min   Lai Shaocong[②]   Zhang Da   Zhu Renzhi   Qin Jiangfeng   Xiong Guangqiang

Abstract: The high Sr/Y geochemical feature of granitoids can be attributed to various mechanisms, and elucidating genesis of high Sr/Y granitoids provides insights into the material recycling and magmatic processes at depth. In southeastern Central Asian Orogenic Belt (CAOB), many Middle Permian granitoids exhibit high Sr/Y ratios, but their origins remain unclear, inhibiting a comprehensive understanding of the magmatic response to the final closure of the Palaeo-Asian ocean. Here we present new zircon U-Pb ages, Lu-Hf isotopes and whole-rock geochemical data for the Middle Permian high Sr/Y monzogranites from central Inner Mongolia, southeastern CAOB. LA-ICP-MS zircon U-Pb data shows that these high Sr/Y rocks were emplaced during 273 – 261 Ma. They are calc-alkaline, sodium-rich and metaluminous to weakly peraluminous, with enriched large-ion lithophile elements (Rb, Th, K and Pb) and depleted high field strength elements (Nb, Ta, P and Ti), suggesting a mafic lower crustal source rather than evolved potassic crustal materials. Their relatively low $(Gd/Yb)_N$(1.1–2.0), $(Dy/Yb)_N$(1.0–1.3), Nb/Ta (7.9–10.9) ratios and flat heavy rare earth element patterns are characteristics of derivation from a relatively shallow depth with amphibolite as dominant residue. They also have highly variable $\varepsilon_{Hf}(t)$ values (−8.2 to +10.0) and T_{DMC}(1 814–649 Ma), similar to those of the Early Palaeozoic high Sr/Y intrusions along the Bainaimiao arc belt. Combined with data from literatures, we suggest that the high Sr/Y monzogranites in this study were probably generated by reworking of the newly underplated juvenile high Sr/Y lower crust of the Bainaimiao arc belt. Moreover, taking into account the regional investigations, the sublinear distributed Middle Permian magmatic rocks in the southeastern CAOB were likely associated with the incipient slab break-off of the Palaeo-Asian oceanic lithosphere following initial collision between the North China craton and the South Mongolia terranes.

1   Introduction

Granitoids constitute an essential component in generation of the Earth's continental

---

①   Published in *International Geology Review*,2019,61(17).

②   Corresponding author.

crust, extraction and emplacement of granitoids with diversely geochemical characteristics are the principal processes by which continental crust has become differentiated (e. g., Hawkesworth and Kemp, 2006). Generally, the high Sr/Y granitoids are attributed to high-pressure magmatic processes, involving: ① partial melting of subducted oceanic slab (Defant and Drummond, 1990; Martin et al., 2005) or differentiation of basaltic magma (Macpherson et al., 2006; Castillo, 2012) in subduction zone; and ② partial melting of thickened or delaminated lower continental crust (e.g., Atherton and Petford, 1993; Xu et al., 2002; He et al., 2011). However, melting experiments and modellings have shown that high Sr/Y melts can also be generated by low-pressure partial melting of the lower crust without eclogitic residues (Qian and Hermann, 2013; Dai et al., 2017). In addition, numerous studies have demonstrated that source inheritance (e.g., Ma et al., 2012, 2015) and magma mixing (Streck et al., 2007) can also account for the generation of high Sr/Y magmas. Thus, deciphering the genesis of the high Sr/Y granitoids provides important insights into the material recycling and magmatic processes in the deep crust that caused by various geodynamic settings.

The southeastern Central Asian Orogenic Belt (CAOB; Fig. 1a) recorded significant information concerning the magmatic responses to the accretionary convergence between the South Mongolia terranes (SMT) and the North China craton (NCC) (Zhang et al., 2009a, 2016; Jian et al., 2010; Li et al., 2016b), driven by the subduction and closure of the Palaeo-Asian ocean since the Early Palaeozoic or Neoproterozoic (Xiao et al., 2003, 2015; Windley et al., 2007; Wilde, 2015). Previous work has documented that the Carboniferous to Early Permian magmatic rocks in southeastern CAOB represent the prolonged arc magmatism in response to the subduction of the Palaeo-Asian ocean (Xiao et al., 2003; Zhang et al., 2009a,b; Liu et al., 2013; Li et al., 2016b), and the Late Permian to Early-Middle Triassic magmatic rocks were mainly related to the terminal collision between the NCC and SMT along the Solonker suture zone (Xiao et al., 2009, 2015; Schulmann and Paterson, 2011; Eizenhöfer et al., 2014; Li et al., 2016a, 2017). Recently, many Middle Permian intermediate-felsic intrusions were also identified along the south side of the Solonker suture zone (Fig. 1; Table 1), which are considered to be generated by the magma mixing processes in MASH (melting, assimilation, storage, and homogenization) zone that induced by the southward subduction of the Palaeo-Asian ocean beneath the northern NCC (Zeng et al., 2011; Liu et al., 2014; Zhang and Zhao, 2017), or by the mixing of metasomatized mantle-derived with various amounts of crust-derived magma during the post-orogenic extension (Jiang et al., 2013; Luo et al., 2013; Zhao et al., 2016) or in relation to the tectonic switching from the oceanic subduction to the terminal collision between the NCC and SMT (Liu, 2010; Wang, 2014). However, we have known little about the reworking of the juvenile lower crust in response to the final closure of the Palaeo-Asian ocean, which may have played crucial role in the Middle Permian magmatism in southeastern CAOB.

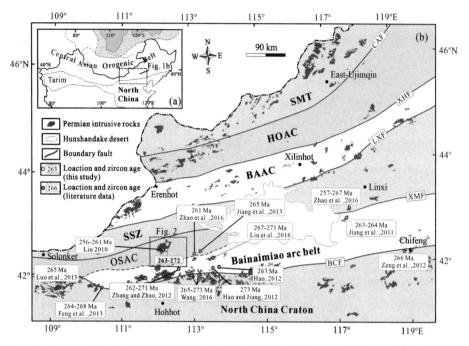

Fig. 1 (a) Simplified sketch map of the Central Asian Orogenic Belt (modified after Jahn, 2004); (b) Sketch geological map of the central-western Xing-Meng Orogenic Belt (modified after Jian et al., 2008; Xiao et al., 2015), and distributions of the Permian intrusive rocks (based on the Solon-HolinGola region 1 : 500 000 geological map compiled by Tianjin Institute of Geology and Mineral Resource).

SMT: South Mongolia terranes; HOAC: Hegenshan ophiolite-accretionary complex; BAAC: Baolidao arc-accretionary complex; SSZ: Solonker suture zone; OSAC: Ondor Sum subduction-accretionary complex; CAF: Chagan'aobao-Arongqi fault; XHF: Xilinhot fault; LXF: Linxi fault; XMF: Xar Moron fault; BCF: Bayan Obo-Chifeng fault. Literature data: Hao and Jiang (2010), Jiang et al. (2011), Liu (2010), Hao (2012), Zeng et al. (2011), Feng et al. (2013), Jiang et al. (2013), Luo et al. (2013), Liu et al. (2014), Wang (2014), Zhao et al. (2016), Zhang and Zhao (2017).

In recent studies, many Middle Permian granitoids exhibiting high Sr/Y geochemical signature have been identified from the southeastern CAOB (e.g., Liu, 2010; Zeng et al., 2011; Zhao et al., 2016). In this paper, we present new zircon U-Pb ages, Lu-Hf isotopes, major and trace element compositions for the Middle Permian Sr/Y monzogranites in the Bulitai area of central Inner Mongolia (Fig. 1b). Together with data from previous studies in adjacent areas, we have identified that these Sr/Y rocks were products of the reworking of the juvenile high Sr/Y lower crust of Bainaimiao arc belt in response to the slab break-off of the Palaeo-Asian oceanic lithosphere in southeastern CAOB.

Table 1 Summary of the Middle Permian intermediate-felsic intrusions in the southeastern CAOB.

Sample	Location	Rock type	Age/Ma	Method	$\varepsilon_{Hf}(t)$	T_{DMC}/Ga	Reference
C270	W. of Chifeng-Chehugou Mo deposit	Syenogranite porphyry	266 ± 4	SIMS	—	—	Zeng et al., 2011
130619-10	S. of Hexigten Qi-Guangxingyuan	K-feldspar granite	257 ± 3	LA-ICP-MS	+1.1 to +5.1	0.88−1.09	Zhao et al., 2016
110718-07	S. of Hexigten Qi-Guangxingyuan	Syenite	265 ± 2	LA-ICP-MS	−15.5 to +6.7	0.78−2.01	Zhao et al., 2016
110718-13	S. of Hexigten Qi-Guangxingyuan	Tonalite	267 ± 2	LA-ICP-MS	−12.4 to +4.7	0.9−1.86	Zhao et al., 2016
PM33TW1	S. of Hexigten Qi-Guangxingyuan	Granodiorite	263 ± 3	SHRIMP	—	—	Jiang et al., 2011
TW5532-1	S. of Hexigten Qi-Guangxingyuan	Quartz monzonite	264 ± 2	SHRIMP	—	—	Jiang et al., 2011
130610-01	W. of Xianghuang Qi-Durenwuliji	Quartz syenite	261 ± 4	LA-ICP-MS	+6.2 to +8.9	0.67−0.82	Zhao et al., 2016
HK3	W. of Xianghuang Qi-Hadamiao	Granite porphyry	271 ± 3	SHRIMP	—	—	Liu et al., 2014
H30-19	W. of Xianghuang Qi-Hadamiao	Diorite	267 ± 3	SHRIMP	—	—	Liu et al., 2014
XN09-203	W. of Xianghuang Qi-Houerdaogou	K-feldspar granite	263 ± 2	LA-ICP-MS	−2.7 to +2.4	0.69−1.22	Hao, 2012
H20	W. of Xianghuang Qi-Hadamiao	Quartz diorite	273 ± 2	LA-ICP-MS	—	—	Hao and Jiang, 2010
JN51	N. of Xianghuang Qi-Wulanhada	Alkali-feldspar granite	265 ± 2	LA-ICP-MS	—	—	Jiang et al., 2013
JN35	S. of Ondor Sum-Qianhushao	Quartz diorite	272 ± 2	LA-ICP-MS	—	—	Wang, 2014
P38b2-1	S. of Ondor Sum-Nanaobaotu	Granodiorite	271 ± 1	LA-ICP-MS	—	—	Wang, 2014
JN56	S. of Ondor Sum-Huaaobao	Granodiorite	269 ± 1	LA-ICP-MS	—	—	Wang, 2014
—	S. of Ondor Sum-Tumuertai	Monzogranite	273 ± 1	LA-ICP-MS	—	—	Wang, 2014
P02B16-2	S. of Ondor Sum-Tumuertai	Granite	267 ± 1	LA-ICP-MS	—	—	Wang, 2014
JN51-1	S. of Ondor Sum-Huaaobao	Alkali-feldspar granite	265 ± 2	LA-ICP-MS	—	—	Wang, 2014
WET19	N. of Bulitai-Dabusu	Monzogranite	256 ± 3	LA-ICP-MS	—	—	Liu, 2010
WET14	N. of Bulitai-Dabusu	Syenogranite	261 ± 2	LA-ICP-MS	—	—	Liu, 2010
Db8-b1	N. of Bulitai-Dabusu	Monzogranite	266 ± 1	LA-ICP-MS	−0.0 to +5.7	0.93−1.29	This study

Continued

Sample	Location	Rock type	Age/Ma	Method	$\varepsilon_{Hf}(t)$	T_{DMC}/Ga	Reference
Db9-b1	N. of Bulitai- Dabusu	Monzogranite	261±1	LA-ICP-MS	-0.0 to +5.8	0.93–1.30	This study
Db11-b1	N. of Bulitai- Dabusu	Monzogranite	263±1	LA-ICP-MS	-0.0 to +5.9	0.93–1.31	This study
Db15-b1	N. of Bulitai- Dabusu	Monzogranite	272±1	LA-ICP-MS	-0.0 to +5.10	0.93–1.32	This study
Db15-b10	N. of Bulitai- Dabusu	Monzogranite	272±1	LA-ICP-MS	-0.0 to +5.11	0.93–1.33	This study
08435-1	Xuniwusu	Porphyric quartz diorite	269±1	LA-ICP-MS	-0.0 to +5.7	0.93–1.29	Zhang and Zhao, 2017
08435-2	Xuniwusu	Porphyric diorite enclave	271±1	LA-ICP-MS	-0.4 to +6.3	0.89–1.32	Zhang and Zhao, 2017
08458-1	Hongge'er	Granodiorite	262±2	LA-ICP-MS	-9.3 to +8.4	0.76–1.88	Zhang and Zhao, 2017
08458-2	Hongge'er	Diorite enclave	264±2	LA-ICP-MS	-9.8 to +2.2	1.16–1.91	Zhang and Zhao, 2017
NM08-59	N. of Damao Qi	Syenogranite	268±2	LA-ICP-MS	-	-	Feng et al., 2013
NM08-061	N. of Damao Qi	Granite	264±2	LA-ICP-MS	-	-	Feng et al., 2013
BY06-16	W. of Bayan Obo- Bayinzhurihe	Quartz diorite	265±2	LA-ICP-MS	-17.2 to -13.0	2.38–2.12	Luo et al., 2013

2 Geological background and petrography

The southeastern CAOB is regarded as a complex collage of arcs, microcontinents, remnants of oceanic crust and associated volcanic-sedimentary sequences that were amalgamated between the NCC and the SMT during the evolution of the Palaeo-Asian ocean (e.g., Xiao et al., 2003, 2009, 2015). From south to north, the southeastern CAOB consists of the Bainaimiao arc belt abutting the northern NCC (Zhang et al., 2014), the Ondor Sum subduction-accretion complex (Xiao et al., 2003; De Jong et al., 2006; Jian et al., 2008), the Solonker suture zone (Jian et al., 2010; Fu et al., 2018), the Baolidao arc-accretionary complex (Xu et al., 2013; Li et al., 2014), the Hegenshan ophiolite-accretionary complex (Miao et al., 2008; Jian et al., 2012; Zhou et al., 2015) and the SMT (Fig. 1b). The Solonker suture zone is widely accepted to have marked the final closure of the Palaeo-Asian ocean in the southeastern CAOB (Xiao et al., 2003, 2015; Eizenhöfer et al., 2014, 2015; Wilde, 2015). Ages of the tectonic mélanges and ophiolitic fragments distributed along the Solonker suture zone are consistently Early Permian (Jian et al., 2010; Song et al., 2015). The Ondor Sum subduction-accretion complex is characterized by ophiolitic fragments or mélanges with ages of ca. 490−450 Ma (e.g., Xiao et al., 2003; Jian et al., 2008; Shi et al., 2013), and Early Palaeozoic high-pressure metamorphic rocks that in relation to the southward subduction of the Palaeo-Asian ocean (De Jong et al., 2006). The Early Palaeozoic Bainaimiao arc belt is composed of metamorphic sedimentary and volcanic rocks and plenty of felsic plutons (Zhang et al., 2014), which are unconformably overlain by the Late Silurian to Early Devonian Xibiehe Formation continental molasse or quasi-molasse deposition (e.g., Zhang et al., 2010).

Extensively Late Palaeozoic to Early Mesozoic magmatic activities occurred in the southeastern CAOB (Chen et al., 2009; Zhang et al., 2009a, 2016; Li et al., 2016a, 2017). The Carboniferous to Early Permian intermediate-acid intrusive and volcanic rocks are calc-alkaline or high-K calc-alkaline, metaluminous or weak peraluminous, representing the prolonged arc magmatism in response to the subduction of the Palaeo-Asian ocean in southeastern CAOB (e.g., Zhang et al., 2009a,b; Liu et al., 2013; Li et al., 2016b). The latest Permian to Early-Middle Triassic magmatism was sublinear distributed along the Solonker suture zone, and was considered as collision-related products by many recent studies, such as the ca. 255−250 Ma E-MORB dolerite, sanukitoid, high-Mg diorite and anorthosite in the Solonker ophiolitic mélanges (Jian et al., 2010), the ca. 255−251 Ma calc-alkaline granodiorites in the southeastern Xilinhot area (Li et al., 2016a), and the ca. 251−245 Ma high Sr/Y granitoids in the Linxi area (Li et al., 2017).

This study investigated the Dabusu pluton in the northern Bulitai, which is tectonically located at the Ondor Sum subduction-accretion complex (Fig. 1b). The Dabusu pluton with outcrop area over 500 km^2 intrudes into the Late Carboniferous Benbatu Formation carbonate rocks and the Early Palaeozoic Ondor Sum Group metamorphic strata in the field (Fig. 2). It is dominated by grey, medium- to fine-grained monzogranite, with fine-grained syenogranite occurring in the centre and coarse-grained hornblende syenite occurring in northern periphery (Liu, 2010). The central part and the north part of the Dabusu pluton are seriously intruded by a lot of NE-trending and NW-trending syenogranite veins, granite-porphyry veins and quartz veins, and are unconformably covered by the Jurassic-Quaternary sedimentary strata. Most surface outcrops of monzogranites, syenogranites and hornblende syenites show varying degrees of alteration. The monzogranite samples in this study were collected from the southwestern part of the Dabusu pluton, which processes relatively fresh outcrops and clear contacting relations (Fig. 3a). The studied monzogranites are porphyritic with phenocrysts of plagioclase (25% - 30%) and K-feldspar (15% - 20%), the groundmass (45% - 55%) is composed of plagioclase, quartz, biotite, K-feldspar and hornblende (Fig. 3b).

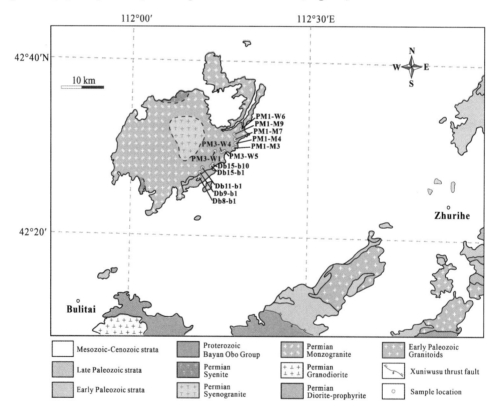

Fig. 2　Geological map of the northeastern Bulitai area.
Modified from the geological map (scale, 1 : 250 000) of Bulitai region (IGSCUGB, 2008).

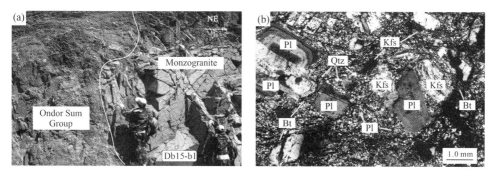

Fig. 3 Representative field photographs and micrographs of the studied monzogranite
from the Dabusu pluton.

Pl：plagioclase；Kfs：K-feldspar；Bt：biotite；Qtz：quartz.

3 Analytical methods

3. 1 Zircon U-Pb dating and Hf isotopes

Zircons from the studied monzogranites were selected by standard techniques of density and magnetic methods and further purified by hand-picking at the Langfang Diyan Mineral Separating Limited Company, Hebei Province, China. Cathodoluminescence (CL) images were taken at the Gaonian Navigation Technology Company (Beijing) to reveal the internal structures. Zircon U-Pb age determinations of the studied granitoid samples were conducted at the Tianjin Institute of Geology and Mineral Resources, Tianjin, China. The analyses were carried out by using a Finnigan Neptune ICP-MS equipped with a New Wave 193 nm excimer laser, and a 35 μm spot was used for the laser ablation of a single grain. Harvard zircon 91500 was used as a standard and NIST 610 was used to optimize the analytical results. Details of the analytical methodology are described by Li et al. (2009). The common Pb corrections were made following the method of Anderson (2002). Data processing was carried out using the Isoplot 3. 0 program (Ludwig, 2003).

In situ zircon Lu-Hf isotope measurements were conducted by using a Nu Plasma II MC-ICP-MS coupled with a RESOLution M-50 193 nm laser at the State Key Laboratory of Continental Dynamics, Northwest University, Xi'an, China. A 44 μm laser beam with a repetition rate of 6 Hz was used for the analysis. Detailed information of instrumental settings and analytical procedures were described in detail by Bao et al. (2017) and Yuan et al. (2008). The Harvard zircon 91500 and MT standards yielded weighted ^{176}Hf/^{177}Hf ratios of 0. 282 264 ±0. 000 030 and 0. 282 528 ±0. 000 019, respectively. The initial ^{176}Hf/^{177}Hf ratios and $\varepsilon_{Hf}(t)$ values were calculated based on the chondritic ^{176}Hf/^{177}Hf ratio of 0. 282 785 and ^{176}Lu/^{177}Hf ratio of 0. 033 6 (Bouvier et al., 2008). The depleted mantle model ages (T_{DM})

were calculated with reference to the depleted mantle source with ^{176}Hf/^{177}Hf ratio of 0. 283 25 and ^{176}Lu/^{177}Hf of 0. 038 4 (Griffin et al. , 2000). The crustal model ages (T_{DMC}) were calculated assuming a ^{176}Lu/^{177}Hf ratio of 0. 015 for the average continental crust (Griffin et al. , 2002).

3. 2　Major and trace elements

Fresh chips of whole rock samples were carefully crushed, and then powdered through a 200-mesh screen using a tungsten carbide ball mill. The major and trace element compositions were determined at the Analytical Laboratory of the Beijing Research Institute of Uranium Geology (BRIUG), China. Major elements were analysed using a Philips PW2404 X-ray fluorescence spectrometry with analytical errors of < 5%. The FeO content was determined using the classical wet chemical method. Trace elements were analysed using a Perkin-Elmer Sciex ELAN DRC-e inductively coupled plasma mass spectrometer (ICP-MS) and the analytical uncertainties were better than 5%. Detailed information of the analytical procedures and instrument parameters are described by Gao et al. (2007).

4　Results

4. 1　Zircon U-Pb ages

Five monzogranite samples from the Dabusu pluton were selected for zircon U-Pb dating. The results are given in Supplementary Table 1. Representative zircon CL images and corresponding U-Pb concordia diagrams are shown in Fig. 4. Zircons from all the samples are subhedral to euhedral, colourless or buff to transparent, with crystal length of $70-200$ μm. Most zircon grains exhibit clearly oscillatory zoning in the CL images (Fig. 4), which is typical feature of magmatic zircon. The studied zircons have varying Th and U abundances with Th/U ratios ranging from 0. 06 to 2. 17. These features are consistent with a magmatic origin (Hoskin and Schaltegger, 2003).

The analyses of 36 zircons from sample Db8-b1 yield ^{206}Pb/^{238}U ages ranging from 261 Ma to 273 Ma, with a weighted mean age of 266 ± 1 Ma (MSWD = 2. 3, $n = 36$) (Fig. 4a). For sample Db9-b1, excluding one analysis with an older age of ca. 1 729 Ma, the other 34 analyses gave a range of ^{206}Pb/^{238}U ages of $257-267$ Ma, with a weighted mean age of 261 ± 1 Ma (MSWD = 2. 5, $n = 34$) (Fig. 4b,c). Twenty-six analyses of zircons from sample Db11-b1 yield ^{206}Pb/^{238}U ages between 255 Ma and 269 Ma, with a weighted mean age of 263 ± 1 Ma (MSWD = 2. 5, $n = 26$) (Fig. 4d). Twenty-five analyses were obtained from sample Db15-b1 and yielded ^{206}Pb/^{238}U ages of $271-274$ Ma, with a weighted mean age of 273 ± 1 Ma (MSWD = 0. 1, $n = 25$) (Fig. 4e). A similar weighted mean age (272 ± 1 Ma) was also obtained from sample Db15-b10 (MSWD = 0.1, n = 25) (Fig. 4f), which yields $^{206}Pb/^{238}U$ ages ranging from 271 Ma to 273 Ma.

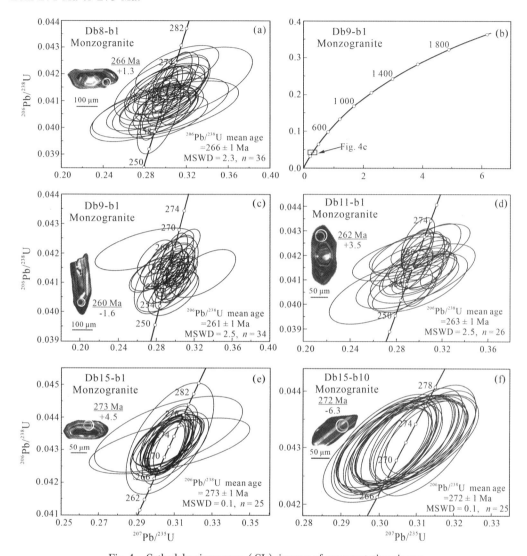

Fig. 4   Cathodoluminescence (CL) images of representative zircons
and concordia diagrams of the studied samples.

Red and yellow circles indicate the locations of U-Pb dating and Hf analyses, respectively.

4.2   Zircon Lu-Hf isotopic data

Representative zircons from the above five dated samples were also selected for in-situ Lu-Hf isotopic analysis, and the results are presented in Supplementary Table 2 and Fig. 5. The initial $^{176}Hf/^{177}Hf$ ratios and $\varepsilon_{Hf}(t)$ values were calculated according to their $^{206}Pb/^{238}U$ ages. Twenty-six analyses of zircons from sample Db8-b1 have $\varepsilon_{Hf}(t)$ values of −0.1 to +7.3 and T_{DMC} of 1 295 − 824 Ma. Seventeen analyses of zircons from sample Db9-b1 yielded $\varepsilon_{Hf}(t)$

values of -2.4 to $+10.0$, corresponding to a T_{DMC} of $1\,442-649$ Ma. Twenty analyses of zircons from sample Db11-b1 yielded $\varepsilon_{Hf}(t)$ values and T_{DMC} ranging from -3.5 to $+7.5$ and $1\,516$ Ma to 810 Ma, respectively. Nineteen analyses of zircons from sample Db15-b1 yielded $\varepsilon_{Hf}(t)$ values of $+1.3$ to $+8.1$, with T_{DMC} ranging from $1\,212$ Ma to 776 Ma. Fifteen analyses of zircons from sample Db15-b10 yielded $\varepsilon_{Hf}(t)$ values of -8.2 to $+9.0$, corresponding to a T_{DMC} from $1\,814$ Ma to 772 Ma.

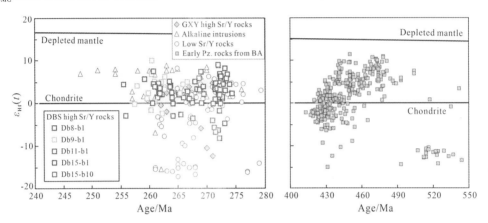

Fig. 5　Zircon $\varepsilon_{Hf}(t)$ vs. U-Pb age diagram for the high Sr/Y

monzogranites from the Dabusu (DBS) pluton.

Data source for the Middle Permian Guangxingyuan (GXY) high Sr/Y rocks and alkaline intrusions in the southeastern CAOB are from Zhao et al. (2016). Data source for the coeval low Sr/Y granitoids in the southeastern CAOB are from Zhang and Zhao (2017) and Luo et al. (2013). Data source for the Early Paleozoic (Early Pz.) magmatic rocks in the Bainaimiao arc (BA) are from Zhang et al. (2014) and Zhou et al. (2018).

4. 3　Major and trace elements

The studied monzogranite samples are characterized by relatively uniform major and trace element compositions, and assay $68.85-72.18$ wt% SiO_2 with $Mg^{\#}$ values $40.7-48.3$ (Supplementary Table 3). These rocks plot in the monzogranite field in a QAP diagram (Fig. 6). They are varying from granite to granodiorite in the TAS diagram (Fig. 7a) and belonging to the calc-alkaline series (Fig. 7b). The studied monzogranites also show sodium-rich affinity with high Na_2O/K_2O ratios ($2.07-4.39$) (Fig. 7c). They also display moderate Al_2O_3 concentrations ($14.42-15.38$ wt%), variable A/CNK values [molar $Al_2O_3/(CaO + Na_2O+K_2O)$] of $0.97-1.20$ (Fig. 7d) and are moderately fractionated as suggested by their DI (differentiation index) values ($76-87$).

The monzogranites have moderate total rare earth element (REE) contents ($39.4-74.0$ ppm), they are enriched in light REEs (LREEs) (Fig. 8a) with relatively low $(La/Yb)_N$ values ($4.0-11.8$), $(Gd/Yb)_N$ values ($1.1-2.0$) and slightly negative to slightly positive Eu anomalies ($\delta Eu = 0.85-1.28$). Their normal mid-ocean ridge basalt-normalized (N-

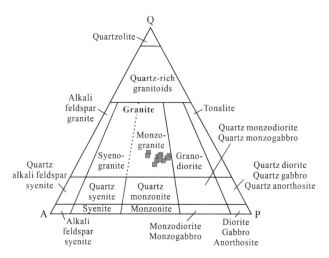

Fig. 6   QAP triangular classification diagram for the studied high Sr/Y rocks (Streckeisen, 1974).

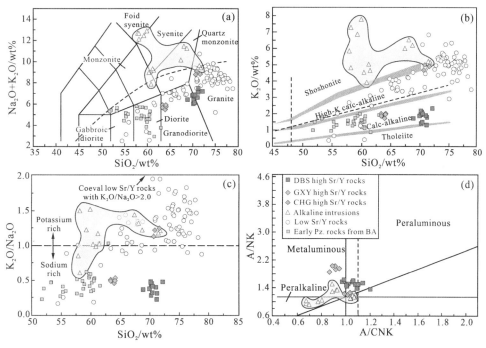

Fig. 7   (a) TAS diagram (Irvine and Baragar, 1971; Middlemost, 1994), (b) K_2O vs. SiO_2 diagram (Rickwood, 1989), (c) K_2O/Na_2O vs. SiO_2 diagram and (d) A/NK vs. A/CNK diagram(Peccerillo and Taylor, 1976) showing the compositional variation of the high Sr/Y monzogranites from the DBS pluton. Data source for the DBS monzogranites are from Xiong et al. (2013) and this study. Previously reported coeval GXY (Zhao et al., 2016) and Chehugou (CHG) (Zeng et al., 2011) high Sr/Y rocks, alkaline intrusions (Zhao et al., 2016) and low Sr/Y rocks in the southeastern CAOB (Hao, 2012; Jiang et al., 2013; Liu et al., 2014; Wang, 2014; Zhao et al., 2016; Zhang and Zhao, 2017), as well as the Early Paleozoic magmatic rocks from the BA (Zhang et al., 2014; Zhou et al., 2018) are plotted for comparison.

MORB-normalized) patterns are characterized by positive large-ion lithophile element (LILE) anomalies (e.g., Rb, Th, U, K and Pb) and negative high field strength element (HFSE) element anomalies (e.g., Nb, Ta, P and Ti) (Fig. 8b). Moreover, these monzogranite samples have relatively high Sr (285−413 ppm) contents and Sr/Y ratios (29−44), and low Y (8.94−12.80 ppm) and Yb (1.10−2.03 ppm) concentrations, which are comparable to those of the high Sr/Y or adakitic-like rocks (Fig. 9a,b).

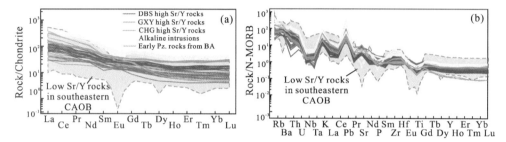

Fig. 8 (a) Chondrite-normalized REE diagram and (b) N-MORB-normalized trace element diagram. The values of chondrite and primitive mantle are from Boynton (1984) and Sun and McDonough (1989), respectively. Data sources are the same as in Fig. 6.

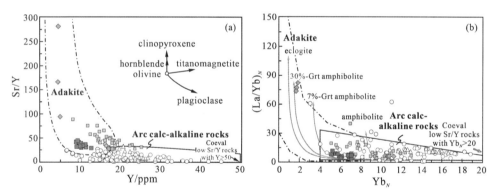

Fig. 9 Plots of Sr/Y vs. Y (a) and (La/Yb)$_N$ vs. Yb$_N$(b) for the studied monzogranite samples. Fields for adakite and arc calc-alkaline rocks are based on Defant and Drummond (1990) and Petford and Atherton (1996). The crystal fractionation paths of the primary minerals in Fig. 9a are from Castillo et al. (1999). The batch partial melting trends in Fig. 9b are based on Petford and Atherton (1996).

5 Discussion

5.1 Petrogenesis

5.1.1 Origin of the Sr/Y geochemical signature

The monzogranites from the DBS pluton have U-Pb ages ranging from 272 Ma to 261 Ma, share similar geochemistry and zircon Lu-Hf isotopic compositions (Figs. 5−8), and show geochemical features of high Sr/Y-type granitoids, which are generally characterized by high

Sr, low Y and Yb with high Sr/Y ratios (Fig. 9). As stated above, different petrogenetic models have been proposed for the origin of high Sr/Y rocks, including slab melting (e. g. Defant and Drummond, 1990), high-pressure partial melting of thickened or delaminated lower crust (Atherton and Petford, 1993; He et al., 2011), magma mixing (Streck et al., 2007), assimilation and fractional crystallization (AFC) processes of parental basaltic magmas (Castillo et al., 1999; Castillo, 2012), and partial melting of an intrinsically high Sr/Y source (Qian and Hermann, 2013; Ma et al., 2015).

High Sr/Y granitoids formed by AFC processes of basaltic magmas generally contain clinopyroxenes and amphiboles with complex compositional zonation and show systematic geochemical correlations (e. g. MgO, Cr, Ni concentrations) (Castillo et al., 1999; Macpherson et al., 2006). However, the high SiO_2 contents (68. 85 – 72. 1 wt%), the relatively uniform geochemistry, as well as the absence of clinopyroxenes in the high Sr/Y monzogranites in this study, preclude their derivation from primary basaltic magma by AFC processes. Moreover, the lack of mafic xenoliths/enclaves, mingling textures, together with the absence of contemporaneous basaltic rocks in the study area, are inconsistent with the mixing model between felsic and basaltic magmas. In water-rich ($\geqslant 2$ wt% H_2O) magmas, high H_2O contents would suppress fractional crystallization of plagioclase until after mafic minerals (hornblende), resulting in relatively high Sr/Y ratios (Moore and Carmichael, 1998; Richards and Kerrich, 2007; Blatter et al., 2013). However, hornblende fractionation usually produces U-shaped REE patterns due to the high partition coefficients of middle REEs (MREEs) in horblende (e.g. Garrison and Davidson, 2003), which was not observed in the high Sr/Y rocks in this study. Generally, high Sr/Y granitoids generated by high-pressure (>45 km) partial melting of the lower continental crust are characterized by high $(Gd/Yb)_N$ ratios (up to 5. 8) due to the significantly higher partition coefficients of heavy REEs (HREEs) in garnet (Moyen, 2009; Huang and He, 2010), but the studied high Sr/Y monzogranites have low $(Gd/Yb)_N$ ratios of 1. 1–2. 0, indicating that they are unlikely to have been formed by high-pressure partial melting of a lower crust source. Besides, their relatively low Nb/Ta ratios, high Zr/Sm ratios (Fig. 10a), and flat HREEs patterns (Fig. 8a) also support an amphibolite melting, rather than the eclogite melting (Martin et al., 2005; He et al., 2011). Oceanic slab-derived high Sr/Y rocks usually exhibit high $Mg^{\#}$, Cr and Ni values because of the inevitable interaction between the primary magmas and the overlying mantle peridotite during magma ascent (e.g., Martin et al., 2005). The high Sr/Y monzogranites in this study process relatively low $Mg^{\#}$ values (40. 7–48. 3), MgO (0. 82–1. 55 wt%), TiO_2 (0. 33 – 0. 52 wt%), Cr (20. 60 – 61. 60 ppm) and Ni (3. 00 – 24. 10 ppm) contents, precluding a direct oceanic slab melting mechanism in their petrogenesis. Besides, their SiO_2 contents (68. 85–72. 1 wt%) and Rb/Sr ratios (0. 1–0. 3) are also much higher than those of the typical slab-derived high Sr/Y magma (Fig. 10b-f).

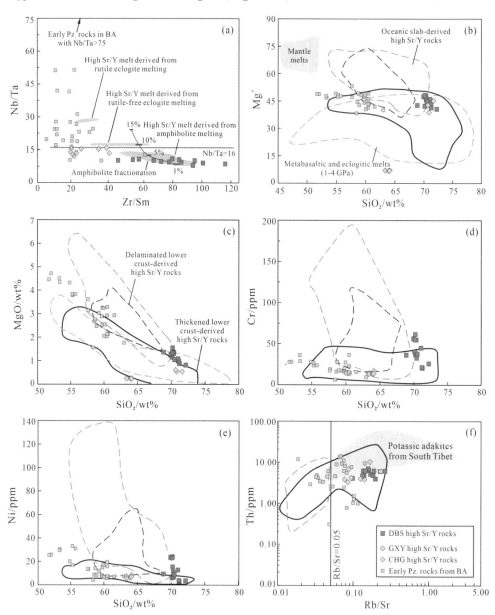

Fig. 10   Discrimination diagrams of the high Sr/Y rocks.

Note that Nb/Ta vs. Zr/Sm diagram (a) is from Martin et al. (2005); Mg# vs. SiO_2, MgO vs. SiO_2, Cr vs. SiO_2 and Ni vs. SiO_2 diagrams are modified after Long et al. (2015, and references therein); Th vs. Rb/Sr diagram is from Huang et al. (2009, and references therein).

It is also crucial to examine whether the high Sr/Y geochemical signature of the rock samples are directly inherited from their source rocks (He et al., 2011; Ma et al., 2015). The high Sr/Y monzogranites in this study are calc-alkaline and sodium-rich with low K_2O/Na_2O

ratios (0.23−0.48) (Fig. 7c) and high $Mg^{\#}$ values (40.7−48.3), these geochemical features are similar to those of the normal-arc series rocks generated in subduction zones (Rushmer and Jackson, 2006; Zhu et al., 2017). The arc-related affinity is also indicated by their trace element characteristics, such as enrichment in LILEs (e.g., Cs, Rb, Ba, Th, and U) and depletion in HFSEs (e.g., Nb, Ta, and Ti) (Fig. 8b). It is noteworthy that the Early Palaeozoic (mainly Ordovician to Silurian) magmatic rocks in the Bainaimiao arc belt, which constitute a major component of the juvenile lower crust source of the Bainaimiao arc belt, are also characterized by high contents of Sr and low contents of Y and Yb with high Sr/Y ratios (up to 88.7) (Fig. 9a) (Zhang et al., 2014; Zhou et al., 2018). These high Sr/Y rocks are composed of gabbroic diorites and diorites, and belong to the calc-alkaline, sodium-rich, metaluminous to weak peraluminous series (Fig. 7). More importantly, a large majority of these Early Palaeozoic magmatic rocks have variable Hf isotopic compositions similar to those of the Middle Permian high Sr/Y monzogranites in this study (Fig. 5). Partial melting of the lower crust with high Sr/Y ratios with normal crustal thickness generally generates high Sr/Y melts, and the lower the degree of partial melting of this source, the higher the Sr/Y ratios would be produced in the melt (Moyen, 2009; Ren et al., 2018). Thus, we may suggest that the high Sr/Y geochemical signature of the high Sr/Y monzogranites in this study might be inherited from an intrinsically high Sr/Y crustal source, which was likely dominated by the Early Palaeozoic newly underplated magma beneath the Bainaimiao arc belt. A more detailed petrogenesis is described in the following section.

5.1.2 Petrogenetic model

The high Sr/Y monzogranites in this study have slightly negative to slightly positive Eu anomalies ($\delta Eu = 0.85 - 1.28$), suggesting that plagioclase is minor or absent in their source residual. As pointed out above, these rocks were likely generated at relatively low pressures with amphibolite as dominant residual phases (Fig. 10a). Their low $(Dy/Yb)_N$ ratios (1.0−1.3) also agree with a low-pressure melting condition, because melting of high Sr/Y crustal materials at low pressures with amphibole as the dominant residual phases would produce magmas with substantially low $(Dy/Yb)_N$ ratios (Macpherson et al., 2006; S.B. Zhang et al., 2009; He et al., 2011). In the $(La/Yb)_N$ vs. Yb_N diagram (Fig. 9b), the high Sr/Y monzogranites in this study plot closed to the 7%-Grt amphibolite melting curve, further indicating a relatively shallow melting depth (around 40 km; Rapp et al., 1999) for their generation.

According to Zhang et al. (2014), most Early Palaeozoic high Sr/Y magmatic rocks in the Bainaimiao arc belt are characterized by low initial $^{87}Sr/^{86}Sr$ ratios (0.704 92−0.707 65), and variable $\varepsilon_{Nd}(t)$ values (−4.9 to +5.2) and Nd isotopic model ages (0.78−2.39 Ga). Combined with their adakitic-like geochemical characteristics, high Nb/Ta and low Zr/Sm ratios (Fig. 10a), as well as variable $\varepsilon_{Hf}(t)$ values and T_{DMC}, these high Sr/Y rocks were suggested to be derived from the partial melting of the subducted slab of the Paleo-Asian ocean with involvements of the overlying mantle wedge and the ancient crustal materials (Zhang et al., 2014; Zhou et al., 2018). This means that the lower crust of the Bainaimiao arc belt were obviously subduction-modified and relatively heterogeneous during the Early Paleozoic. Although the Middle Permian high Sr/Y rocks in this study exhibit similar major and trace element compositions, their relatively large variations in $\varepsilon_{Hf}(t)$ values (-8.2 to $+10.0$) and T_{DMC}(1 814 $-$ 649 Ma) (especially the early Middle Permian samples) might be indicative of a relatively heterogeneous magma source. Thus, we argue that the variations in Hf isotopic composition of these high Sr/Y rocks were likely in relation to the intrinsical heterogeneity of the subduction-modified juvenile high Sr/Y lower crust beneath the Bainaimiao arc belt.

Partial melting of crustal materials at relatively shallow depth would inevitably require an external heat source. In southeastern CAOB, the final closure of the Palaeo-Asian ocean was suggested to be started with the initial collision/amalgamation between the NCC and the SMT in the Middle Permian, which led to cessation of the prolonged arc magmatism, disappearance of the marine sedimentation, and formation of a regional angular unconformity in central Inner Mongolia (Jian et al., 2010; Eizenhöfer et al., 2014; Xiao et al., 2015; Li et al., 2014, 2016a,b, 2017). Combined with the presence of the Middle Permian peralkaline-alkaline intrusions in the southeastern CAOB (Zhao et al., 2016), an extensional regime related to the ocean closure should be taken into account for the petrogenesis of the high Sr/Y monzogranites in this study. We suggest that slab break-off of the Palaeo-Asian oceanic lithosphere following the collision between the NCC and the SMT was a plausible mechanism. Because the tensile stresses between the buoyant continental lithosphere and the previously subducted oceanic lithosphere would lead to the detachment of the oceanic slab and formation of a slab window soon after the collision, which would enable upwelling of the asthenosphere through the slab window (Davies and von Blanckenburg, 1995; Van Hunen and Allen, 2011). Then, the heat from the underplated magma in response to the asthenosphere upwelling would trigger a linear zone of magmatism with limited spatial distribution (Li et al., 2016a), which is consistent with the linear distributed Middle Permian magmatism in southeastern CAOB (Fig. 1b). Thus, we propose that the partial melting of the source materials of the high Sr/Y rocks in this study was likely triggered by the asthenosphere upwelling during the slab break-off of the subducted Palaeo-Asian oceanic lithosphere following the final ocean closure.

5.2 Tectonic implications

5.2.1 Diversified sources in generation of the Middle Permian magmatism

In southeastern CAOB, the Middle Permian was previously regarded as a short magmatic hiatus caused by the initial collision between the NCC and the SMT (e. g., Li et al., 2016a, b). However, more and more Middle Permian magmatic rocks were identified along the south side of the Solonker suture zone in recent years, they are composed of high Sr/Y calc-alkaline granitoids, low Sr/Y high-K calc-alkaline rocks and peralkaline-alkaline intrusions (Fig. 1; Table 1).

As discussed in the preceding sections, generation of the Middle Permian high Sr/Y monzogranites in this study were likely related to the reworking of the juvenile lower crust of the Bainaimiao arc belt. In comparison, the coeval high Sr/Y tonalites from the Guangxingyuan area (Zhao et al., 2016) have lower $Mg^{\#}$ values, MgO, Cr, Ni contents (Fig. 10b-e), as well as lower Na_2O/K_2O ratios and higher Yb contents than those of the high Sr/Y rocks in this study (Fig. 9b). We suggest that these geochemical variations might be caused by the heterogeneous nature of the juvenile lower crust of the Bainaimiao arc belt. The high Sr/Y rocks from the Chehukou deposit in Chifeng area (Zeng et al., 2012) are dominantly high-K calc-alkaline and potassic-rich, with prominently enriched LREEs, high Sr/Y and $(La/Yb)_N$ ratios, low Y and Yb_N values (Figs. 7-9). Considering that the Chehukou high Sr/Y rocks are exposed at the northern margin of the NCC, we propose that they were probably generated in relation to the partial melting of the ancient lower crust of the NCC at deeper crustal levels or higher pressures. The Middle Permian peralkaline-alkaline intrusions in southeastern CAOB were suggested to be formed though the partial melting of the subduction-modified lithospheric mantle with mixing of different amount of the ancient crust-derived magma during the asthenosphere upwelling (Zhao et al., 2016). The Middle Permian low Sr/Y magmatic rocks in southeastern CAOB are dominantly high-K calc-alkaline, potassic-rich, showing large variations in major and trace element compositions, as well as Hf isotopic compositions (Figs. 5, 7 and 8). It seems plausible that variable amounts of the ancient crust materials beneath the Bainaimiao arc belt might have been involved in the formation of those low Sr/Y magmatic rocks. The above inferences suggest that the magma source of the Middle Permian magmatism in southeastern CAOB is relatively complex, materials from the juvenile lower crust, ancient lower crust components and subduction-modified lithospheric mantle were all likely to be involved in the magma generation.

5. 2. 2　Closure of the Palaeo-Asian ocean in southeastern CAOB

Slab break-off of the subducted oceanic lithosphere is believed to be a natural consequence after the ocean closure due to the attempted subduction of continental lithosphere (Wortel and Spakman, 2000; Atherton and Ghani, 2002). Continental collision is not an instantaneous process, the evolution from initial collision to complete detachment of the oceanic slab may cost 5-40 Ma (e. g., Royden, 1993; Andrews and Billen, 2009; van Hunen and Allen, 2011). In southeastern CAOB, after the long-lived subduction of the Palaeo-Asian ocean since the Neoproterozoic (Xiao et al., 2003, 2015; Wilde, 2015), the NCC and the SMT eventually collided following the ocean closure in the Middle Permian (Jian et al., 2010; Xiao et al., 2015; Li et al., 2016a, 2017). The available paleomagnetic data also suggest that the NCC and SMT were very close during Early Permian (Zhang et al., 2014c). The collision between the NCC and SMT led to the cessation of the arc magmatism and marine sedimentation, accompanied by a short magmatic hiatus and a regional angular unconformity in the XMOB (e.g., Eizenhöfer et al., 2014; Li et al., 2014, 2016b). Furthermore, Li et al. (2016a, 2017) proposed that the collision between the NCC and SMT can be further divided into three stages, including the initial collision during Middle Permian, the slab break-off during Late Permian and the intracontinental contraction during Early-Middle Triassic.

The Middle Permian high Sr/Y monzogranites in this study were likely derived from the partial melting of the juvenile lower crust beneath the Bainaimiao arc belt during slab break-off of the Palaeo-Asian oceanic lithosphere. The presence of this slab break-off process is also supported by many recent studies, for example, Jian et al. (2010) argue that the Late Permian to earliest Triassic igneous rocks (255–250 Ma) in the Mandula forearc mélange were derived from the decompression melting of the mantle peridotite as the result of the asthenosphere upwelling during the slab break-off of the Palaeo-Asian oceanic lithosphere. Li et al. (2016a) also suggested that the latest Permian magmatic flare-up in the southeastern Xilinhot area was formed in response to the slab break-off. Slab break-off generally has three evolutionary stages, including: ① lithospheric tearing; ② slow sinking of the detaching root associated with downward dragging of the overriding lithosphere; and ③ complete detachment with faster slab sinking and the overriding lithosphere rebounding (e.g., Ferrari, 2004; Yuan et al., 2010). In this case, the Middle Permian high Sr/Y monzogranites in this study, as well as the contemporaneous magmatic rocks in southeastern CAOB, were more likely corresponding to the early stage of slab break-off (lithospheric tearing) (Fig. 11), which indicates that slab break-off of the Palaeo-Asian oceanic lithosphere might have already started in Middle Permian. The more intense Late Permian magmatism in southeastern CAOB might be related to the late stage of the slab break-off (Jian et al., 2010; Li et al., 2016a). Afterwards, a sublinear distributed Early-Middle Triassic high Sr/Y magma belt was formed probably in response to the moderately thick stacking of crustal rocks due to the subsequently ongoing weak compression between the NCC and the SMT (Wang et al., 2015; Li et al., 2017), which marks the termination of Palaeo-Asian ocean. The regionally post-collisional extension probably occurred until the Early Mesozoic, as evidenced by the Triassic metamorphic core complex in the Sonid Zuoqi area (Davis et al., 2004), and the widespread Mesozoic A-type granitic rocks and alkaline complexes along the northern NCC (Wu et al., 2011; Zhang et al., 2012, 2014b).

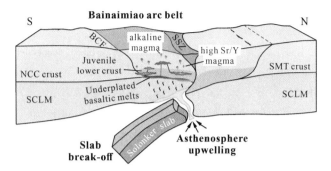

Fig. 11   Tentative scenario illustrating the Middle Permian evolution and
magmatic responses in southeastern CAOB.

6   Conclusions

(1) Zircon U-Pb dating indicates that the high Sr/Y monzogranites from the Dabusu pluton in central Inner Mongolia were emplaced at 273−261 Ma.

(2) They are calc-alkaline, sodium-rich and LILEs enriched with variable $\varepsilon_{Hf}(t)$ values (−8.2 to +10.0), and were likely formed by reworking of the juvenile lower crust of Bainaimiao arc belt at a relatively shallow melting depth.

(3) These high Sr/Y rocks was associated with the slab break-off of the Palaeo-Asian oceanic lithosphere following the collision between NCC and SMT.

Acknowledgements   We are grateful to Haoran Wang, Zhen Wang, and Zhong Wang for their help with the field investigation. We thank Yu Zhu, Fangyi Zhang and Zezhong Zhang for their laboratory assistance. We also appreciate Yuan Yuan and Yaoyao Zhang for their insightful suggestions. Critical and constructive comments from the Editor and the anonymous reviewers significantly improved the quality of this manuscript. This work was jointly supported by the China Geological Survey (1212011085490) and National Natural Science Foundation of China (41421002).

References

Andersen, T., 2002. Correction of common lead in U-Pb analyses that do not report [204]Pb. Chemical Geology, 192(1-2), 59-79.

Andrews, E.R., Billen, M.I., 2009. Rheologic controls on the dynamics of slab detachment. Tectonophysics, 464, 60-69.

Atherton, M.P., Ghani, A.A., 2002. Slab breakoff: A model for Caledonian, Late Granite syn-collisional magmatism in the orthotectonic (metamorphic) zone of Scotland and Donegal, Ireland. Lithos, 62, 65-85.

Atherton, M.P., Petford, N., 1993. Generation of sodium-rich magmas from newly underplated basaltic crust. Nature, 362, 144-146.

Bao, Z.A., Chen, L., Zong, C.L., Yuan, H.L., Chen, K.Y., Dai, M.N., 2017. Development of pressed sulfide powder tablets for in situ sulfur and lead isotope measurement using LA-MC-ICP-MS. International Journal of Mass Spectrometry, 421, 255-262.

Blatter, D., Sisson, T., Hankins, W.B., 2013. Crystallization of oxidized, moderately hydrous arc basalt at mid- to lower-crustal pressures: Implications for andesite genesis. Contributions to Mineralogy and Petrology, 166, 861-886.

Bouvier, A., Vervoort, J.D., Patchett, P.J., 2008. The Lu-Hf and Sm-Nd isotopic composition of CHUR: Constraints from unequilibrated chondrites and implications for the bulk composition of terrestrial planets. Earth & Planetary Science Letters, 273, 48-57.

Boynton, W.V., 1984. Cosmochemistry of the rare earth elements: Meteorite studies. In: Henderson, P.E. Rare Earth Element Geochemistry. Amsterdam: Elsevier, 63-144.

Castillo, P.R., 2012. Adakite petrogenesis. Lithos, 134-135, 304-316.

Castillo, P.R., Janney, P.E., Solidum, R.U., 1999. Petrology and geochemistry of Camiguin Island, southern Philippines: Insights to the source of adakites and other lavas in a complex arc setting. Contributions to Mineralogy and Petrology, 134, 33-51.

Chen, B., Jahn, B.M., Tian, W., 2009. Evolution of the Solonker Suture Zone: Constraints from Zircon U-Pb Ages, Hf Isotopic Ratios and Whole-rock Nd-Sr Isotope Compositions of Subduction- and Collision-related Magmas and Forearc Sediments. Journal of Asian Earth Sciences, 34, 245-257.

Dai, H.K., Zheng, J.P., Zhou, X., Griffin, W.L., 2017. Generation of continental adakitic rocks: Crystallization modeling with variable bulk partition coefficients. Lithos, 272-273, 222-231.

Davies, J.H., von Blanckenburg, F., 1995. Slab breakoff: A model of lithospheric detachment and its test in the magmatism and deformation of collisional orogens. Earth & Planetary Science Letters, 129, 85-102.

Davis, G.A., Xu, B., Zheng, Y.D., Zhang, W.J., 2004. Indosinian extension in the Solonker suture zone: The Sonid Zuoqi metamorphic core complex, Inner Mongolia, China. Earth Science Frontiers, 11, 135-143.

De Jong, K., Xiao, W.J., Windley, B.F., Masago, H., Lo, C.H., 2006. Ordovician $^{40}Ar/^{39}Ar$ phengite ages from the blueschist-facies Ondor Sum subduction-accretion complex (Inner Mongolia) and implications for the early Paleozoic history of continental blocks in China and adjacent areas. American Journal of Science, 306, 799-845.

Defant, M.J., Drummond, M.S., 1990. Derivation of some modern arc magmas by melting of young subducted lithosphere. Nature, 347, 662-665.

Eizenhöfer, P.R., Zhao, G., Zhang, J., Han, Y., Hou, W., Liu, D., Wang, B., 2015. Geochemical characteristics of the Permian basins and their provenances across the Solonker Suture Zone: Assessment of net crustal growth during the closure of the Palaeo-Asian Ocean. Lithos, 224-225, 240-255.

Eizenhöfer, P.R., Zhao, G.C., Zhang, J., Sun, M., 2014. Final closure of the Paleo-Asian Ocean along the Solonker suture zone: Constraints from geochronological and geochemical data of Permian volcanic and sedimentary rocks. Tectonics, 33, 441-463.

Feng, L.X., Zhang, Z.C., Han, B.F., Ren, R., Li, J.F., Su, L., 2013. LA-ICP-MS zircon U-Pb ages of granitoids in Darhan Muminggan Joint Banner, Inner Mongolia, and their geological significance. Geological Bulletin of China, 11, 1737-1748 (in Chinese with English abstract).

Ferrari, L., 2004. Slab detachment control on mafic volcanic pulse and mantle heterogeneity in central Mexico. Geology, 32, 77-80.

Fu, D., Huang, B., Kusky, T.M., Li, G.Z., Wilde, S.A., Zhou, W.X., Yu, Y., 2018. A Middle Permian Ophiolitic Mélange Belt in the Solonker Suture Zone, Western Inner Mongolia, China: Implications for the Evolution of the Paleo-Asian Ocean. Tectonics, 37, 1292-1320.

Gao, S., Liu, X.M., Yuan, H.L., Hattendorf, B., Günther, D., Chen, L., 2007. Determination of forty two major and trace elements in USGS and NIST SRM glasses by laser ablation-inductively coupled plasma-mass spectrometry. Geostandards and Geoanalytical Research, 26, 181-196.

Garrison, J.M., Davidson, J.P., 2003. Dubious case for slab melting in the Northern volcaniczone of the Andes. Geology, 31, 565-568.

Griffin, W.L., Pearson, N.J., Belousova, E., Jackson, S.E., van Achterbergh, E., O'Reilly, S.Y., Shee, S.R., 2000. The Hf isotope composition of cratonic mantle: LA-MC-ICP-MS analysis of zircon megacrysts in kimberlites. Geochimica et Cosmochimica Acta, 64, 133-147.

Griffin, W.L., Wang, X., Jackson, S.E., Pearson, N.J., O'Reilly, S.Y., Xu, X., Zhou, X., 2002. Zircon chemistry and magma mixing, SE China: In-situ analysis of Hf isotopes, Tonglu and Pingtan igneous complexes. Lithos, 61, 237-269.

Hao, B.W., 2012. Discovery, genesis and tectonic significance of the Late-Paleozoic miarolitic K-feldspar granite in southern Xianghuangqi, Inner Mongolia. Journal of Jilin University: Earth Science Edition, 42 (S2), 269-284 (in Chinese with English abstract).

Hao, B.W., Jiang, J., 2010. Chronology, geochemistry of the hadamiao complex related to gold deposits in xianghuang banner, Inner Mongolia. Acta Petrologica et Mineralogica, 29, 750-762 (in Chinese with English abstract).

Hawkesworth, C.J., Kemp, A., 2006. Evolution of the continental crust. Nature, 443, 811-817.

He, Y.S., Li, S.G., Hoefs, J., Huang, F., Liu, S.A., Hou, Z.H., 2011. Post-collisional granitoids from the Dabie orogen: New evidence for partial melting of a thickened continental crust. Geochimica et Cosmochimica Acta, 75, 3815-3838.

Hoskin, P.W.O., Schaltegger, U., 2003. The composition of zircon and igneous and metamorphic petrogenesis. Reviews in Mineralogy and Geochemistry, 53, 27-62.

Huang, F., He, Y.S., 2010. Partial melting of the dry mafic continental crust: Implications for petrogenesis of C-type adakites. Chinese Science Bulletin, 55, 2428-2439.

Huang, X.L., Xu, Y.G., Lan, J.B., Yang, Q.J., Luo, Z.Y., 2009. Neoproterozoic adakitic rocks from Mopanshan in the western Yangtze Craton: Partial melts of a thickened lower crust. Lithos, 112, 367-381.

Jahn, B.M., 2004. The Central Asian Orogenic Belt and Growth of the continental crust in the Phanerozoic. Geological Society, London, Special Publications, 226, 73-100.

Jian, P., Kröner, A., Windley, B.F., Shi, Y.R., Zhang, W., Zhang, L.B., Yang, W.R., 2012. Carboniferous and Cretaceous mafic-ultramafic massifs in Inner Mongolia (China): A SHRIMP zircon and geochemical study of the previously presumed integral "Hegenshan ophiolite". Lithos, 142-143, 48-66.

Jian, P., Liu, D.Y., Kröner, A., Windley, B.F., Shi, Y.R., Zhang, F.Q., Shi, G.H., Miao, L.C., Zhang, W., Zhang, Q., Zhang, L.Q., Ren, J.S., 2008. Time scale of an early to mid-Paleozoic orogenic cycle of the long-lived Central Asian Orogenic Belt, Inner Mongolia of China: Implications for continental growth. Lithos, 101, 233-259.

Jian, P., Liu, D.Y., Kröner, A., Windley, B.F., Shi, Y.R., Zhang, W., Zhang, F.Q., Miao, L.C., Zhang, L.Q., Tomurhuu, D., 2010. Evolution of a Permian intraoceanic arc-trench system in the Solonker suture zone, Central Asian Orogenic Belt, China and Mongolia. Lithos, 118, 169-190.

Jiang, X.J., Liu, Y.Q., Peng, N., Shi, Y.R., Xu, H., Wei, W.T., Liu, Z.X., Zhao, H.P., Yao, B.G., 2011. Geochemistry and SHRIMP U-Pb Dating of the Guangxingyuan Composite Pluton in Hexigten Qi, Inner Mongolia and Its Geological Implication. Acta Geologica Sinica, 85, 114-128 (in Chinese with English abstract).

Jiang, X.J., Liu, Z.H., Xu, Z.Y., Wang, W.Q., Wang, X.A., Zhang, C., 2013. LA-ICP-MS zircon U-Pb dating of Wulanhada middle Permian alkali-feldspar granites in Xianghuang Banner, central Inner Mongolia, and its geochemical characteristics. Geological Bulletin of China, 32, 1760-1768 (in Chinese with English abstract).

Li, H.K., Geng, J.Z., Hao, S., Zhang, Y.Q., Li, H.M., 2009. The study of zircon U-Pb dating by means LA-MC-ICPMS. Bulletin of Mineralogy, Petrology and Geochemistry, 28 (suppl.), 77 (in Chinese).

Li, S., Chung, S.L., Wilde, S.A., Jahn, B.M., Xiao, W.J., Wang, T., Guo, Q.Q., 2017. Early-Middle Triassic high Sr/Y granitoids in the southern Central Asian Orogenic Belt: Implications for ocean closure in accretionary orogens. Journal of Geophysical Research: Solid Earth, 122, 2291-2309.

Li, S., Chung, S.L., Wilde, S.A., Wang, T., Xiao, W.J., 2016a. Linking magmatism with collision in an accretionary orogen. Scientific Reports, 6, 25751.

Li, S., Wilde, S.A., Wang, T., Xiao, W.J., Guo, Q.Q., 2016b. Latest Early Permian granitic magmatism in southern Inner Mongolia, China: Implications for the tectonic evolution of the southeastern Central Asian Orogenic Belt. Gondwana Research, 29, 168-180.

Li, Y.L., Zhou, H.W., Brouwer, F.M., Xiao, W.J., Wijbrans, J.R., Zhong, Z.Q., 2014. Early Paleozoic to Middle Triassic bivergent accretion in the Central Asian Orogenic Belt: Insights from zircon U-Pb dating of ductile shear zones in central Inner Mongolia, China. Lithos, 205, 84-111.

Liu, C.F., 2010. Paleozoic-Early Mesozoic magmatic belt and tectonic significance in Siziwangqi area, Inner Mongolia [Ph.D thesis]. Beijing: China University of Geosciences, 132 (in Chinese with English abstract).

Liu, J., Wu, G., Li, T.G., Wang, G.R., Wu, H., 2014. SHRIMP zircon U-Pb dating, geochemistry, Sr-Nd isotopic analysis of the Late Paleozoic intermediate-acidic intrusive rocks in the Hadamiao area, Xianghuang Banner, Inner Mongolia and its geological significances. Acta Petrologica Sinica, 30, 95-108 (in Chinese with English abstract).

Liu, J.F., Li, J.Y., Chi, X.G., Qu, J.F., Hu, Z.C., Fang, S., Zhang, Z., 2013a. A late-Carboniferous to early early-Permian subduction-accretion complex in Daqing pasture, southeastern Inner Mongolia: Evidence of northward subduction beneath the Siberian paleoplate southern margin. Lithos, 177, 285-296.

Long, X.P., Wilde, S.A., Wang, Q., Yuan, C., Wang, X.C., Li, J., Jiang, Z.Q., Dan, W., 2015. Partial melting of thickened continental crust in central Tibet: Evidence from geochemistry and geochronology of Eocene adakitic rhyolites in the northern Qiangtang Terrane. Earth & Planetary Science Letters, 414,

30-44.

Ludwig, K.R., 2003. Isoplot 3.00: A geochronological toolkit for Microsoft Excel. Berkeley Geochronology Center, Special Publication, 4, 70.

Luo, H.L., Wu, T.R., Zhao, L., He, Y.K., Jing, X., 2013. Geochemicalcharacteristics of Bayinzhurihe pluton and its tectonic significance, Bayan Obo, Inner Mongolia. Geological Journal of China Universities, 19, 123-132 (in Chinese with English abstract).

Ma, Q., Zheng, J.P., Griffin, W.L., Zhang, M., Tangm, H.Y., Su, Y.P., Ping, X.Q., 2012. Triassic "adakitic" rocks in an extensional setting (North China): Melts from the cratonic lower crust. Lithos, 149, 159-173.

Ma, Q., Zheng, J.P., Xu, Y.G., Griffin, W.L., Zhang, R.S., 2015. Are continental "adakites" derived from thickened or foundered lower crust. Earth & Planetary Science Letters, 419, 125-133.

Macpherson, C.G., Dreher, S.T., Thirlwall, M.F., 2006. Adakites without slab melting: High pressure differentiation of island arc magma, Mindanao, the Philippines. Earth & Planetary Science Letters, 243, 581-593.

Martin, H., Smith, R.H., Rapp, R., Moyen, J.F., Champion, D., 2005. An overview of adakite, tonalite-trondhjemite-granodiotite (TTG), and sanitoid: Relationship and some implications for crustal evolution. Lithos, 79, 1-24.

Miao, L.C., Fan, W.M., Liu, D.Y., Zhang, F.Q., Shi, Y.R., Guo, F., 2008. Geochronology and geochemistry of the Hegenshan ophiolitic complex: Implications for late-stage tectonic evolution of the Inner Mongolia-Daxinganling Orogenic Belt, China. Journal of Asian Earth Sciences, 32, 348-370.

Moore, G.M., Carmichael, I.S.E., 1998. The hydrous phase equilibria (to 3 kbar) of an andesite and basaltic andesite from western Mexico: Constraints on water content and conditions of phenocryst growth. Contributions to Mineralogy and Petrology, 130, 304-319.

Moyen, J.F., 2009. High Sr/Y and La/Yb ratios: The meaning of the "adakitic signature". Lithos, 112, 556-574.

Petford, N., Atherton, M., 1996. Na-rich partial melts from newly underplated basaltic crust: The Cordillera Blanca Batholith, Peru. Journal of Petrology, 37, 1491-1521.

Qian, Q., Hermann, J., 2013. Partial melting of lower crust at 10-15 kbar: Constraints on adakite and TTG formation. Contributions to Mineralogy and Petrology, 165, 1195-1224.

Rapp, R.P., Shimizu, N., Norman, M.D., Applegate, G.S., 1999. Reaction between slab-derived melts and peridotite in the mantle wedge: Experimental constraints at 3.8 GPa. Chemical Geology, 160, 335-356.

Ren, L., Liang, H.Y., Bao, Z.W., Zhang, J., Li, K.X., Huang, W.T., 2018. Genesis of the high Sr/Y rocks in Qinling orogenic belt, central China. Lithos, 314-315, 337-349.

Richards, J.P., Kerrich, R., 2007. Adakite-like rocks: Their diverse origins and questionable role in metallogenesis. Economic Geology, 102, 537-576.

Royden, L.H., 1993. Evolution of retreating subduction boundaries formed during continental collision. Tectonics, 12, 629-638.

Rushmer, T., Jackson, M., 2006. Impact of melt segregation on tonalite-trondhjemite-granodiorite (TTG) petrogenesis. Earth and Environmental Science Transactions of the Royal Society of Edinburgh, 97,

325-336.

Schulmann, K., Paterson, S., 2011. Geodynamics: Asian continental growth. Nature Geoscience, 4, 827-829.

Shi, G.Z., Faure, M., Xu, B., Zhao, P., Chen, Y., 2013. Structural and kinematic analysis of the Early Paleozoic Ondor Sum-Hongqi mélange belt, eastern part of the Altaids (CAOB) in Inner Mongolia, China. Journal of Asian Earth Sciences, 66, 123-139.

Song, S.G., Wang, M.M., Xu, X., Wang, C., Niu, Y.L., Allen, M.B., Su, L., 2015. Ophiolites in the Xing'an-Inner Mongolia accretionary belt of the CAOB: Implications for two cycles of seafloor spreading and accretionary orogenic events. Tectonics, 34, 2221-2248.

Streck, M.J., Leeman, W.P., Chesley, J., 2007. High-magnesian andesite from Mount Shasta: A product of magma mixing and contamination, not a primitive mantle melt. Geology, 35, 351-354.

Streckeisen, A.L., 1974. Classification and nomenclature of plutonic rocks. Recommendations of the IUGS subcommission on the systematics of igneous rocks. Geologische Rundschau. Internationale Zeitschrift für Geologie, Stuttgart, 63, 773-786.

Sun, S.S., McDonough, W.F., 1989. Chemical and isotopic systematics of oceanic basalts: Implications for mantle composition and processes. Geol. Soc., Lond., Spec. Publ. 42, 313-345.

van Hunen, J., Allen, M.B., 2011. Continental collision and slab break-off: A comparison of 3-D numerical models with observations. Earth & Planetary Science Letters, 302, 27-37.

Wang, W.Q., 2014. Late Paleozoic tectonic evolution of the central-northern margin of the North China Craton: Constraints from zircon U-Pb ages and geochemistry of igneous rocks in Ondor Sum-Jining area [Ph.D thesis]. Changchun: Jilin University, 156 (in Chinese with English abstract).

Wang, Z.J., Xu, W.L., Pei, F.P., Wang, Z.W., Li, Y., Cao, H.H., 2015. Geochronology and geochemistry of middle Permian-Middle Triassic intrusive rocks from central-eastern Jilin Province, NE China: Constraints on the tectonic evolution of the eastern segment of the Paleo-Asian Ocean. Lithos, 238, 13-25.

Wilde, S.A., 2015. Final amalgamation of the Central Asian Orogenic Belt in NE China: Paleo-Asian Ocean closure versus Paleo-Pacific plate subduction-A review of the evidence. Tectonophysics, 662, 345-362.

Windley, B.F., Alexeiev, D., Xiao, W.J., Kröner, A., Badarch, G., 2007. Tectonic models for accretion of the Central Asian Orogenic Belt. Journal of the Geological Society, 164, 31-47.

Wortel, M.J.R., Spakman, W., 2000. Subduction and slab detachment in the Mediterranean-Carpathian Region. Science, 290, 1910-1917.

Wu, F.Y., Sun, D.Y., Ge, W.C., Zhang, Y.B., Grant, M.L., Wilde, S.A., Jahn, B.M., 2011. Geochronology of the Phanerozoic granitoids in northeastern China. Journal of Asian Earth Sciences, 41, 1-30.

Xiao, W.J., Windley, B.F., Hao, J., Zhai, M.G., 2003. Accretion leading to collision and the Permian Solonker suture, Inner Mongolia, China: Termination of the central Asian orogenic belt. Tectonics, 22, 1-8.

Xiao, W.J., Windley, B.F., Huang, B.C., Han, C.M., Yuan, C., Chen, H.L., Sun, M., Sun, S., Li, J. L., 2009. End-Permian to Early-Triassic termination of the accretionary processes of the southern Altaids: Implications for the geodynamic evolution, Phanerozoic continental growth, and metallogeny of Central

Asia. International Journal of Earth Sciences, 98, 1189-1217.

Xiao, W.J., Windley, B.F., Sun, S., Li, J.L., Huang, B.C., Han, C.M., Yuan, C., Sun, M., Chen, H. L., 2015. A Tale of Amalgamation of Three Permo-Triassic Collage Systems in Central Asia: Oroclines, Sutures, and Terminal Accretion. Annual Review of Earth and Planetary Sciences, 43, 477-507.

Xiong, G.Q., Zhao, H.T., Liu, M., Zhang, D., Wang, H.R., Wang, Z., 2013. Geochronology and geochemistry of the Heinaobao pluton in Siziwangqi, Inner Mongolia and its tectonic evolution. Journal of Geomechanics, 19, 162-177 (in Chinese with English abstract).

Xu, B., Charvet, J., Chen, Y., Zhao, P., Shi, G.Z., 2013. Middle paleozoic convergent orogenic belts in western Inner Mongolia (China): Framework, kinematics, geochronology and implications for tectonic evolution of the central asian orogenic belt. Gondwana Research, 23, 1342-1364.

Xu, J.F., Shinjo, R., Defant, M.J., Wang, Q., Rapp, R.P., 2002. Origin of Mesozoic adakitic intrusive rocks in the Ningzhen area of east China: Partial melting of delaminated lower continental crust? Geology, 30, 1111-1114.

Yuan, C., Sun, M., Wilde, S., Xiao, W.J., Xu, Y.G., Long, X.P., Zhao, G.C., 2010. Post-collisional plutons in the Balikun area, East Chinese Tianshan: Evolving magmatism in response to extension and slab break-off. Lithos, 119, 269-288.

Yuan, H.L., Gao, S., Dai, M.N., Zong, C.L., Gunther, D., Fontaine, G.H., Liu, X.M., Diwu, C.R., 2008. Simultaneous determinations of U-Pb age, Hf isotopes and trace element compositions of zircon by excimer laser-ablation quadrupole and multiple-collector ICP-MS. Chemical Geology, 247, 100-118.

Zeng, Q.D., Yang, J.H., Liu, J.M., Chu, S.X., Duan, X.X., Zhang, Z.L., Zhang, W.Q., Zhang, S., 2011. Genesis of the Chehugou Mo-bearing granitic complex on the northern margin of the North China Craton: Geochemistry, zircon U-Pb age and Sr-Nd-Pb isotopes. Geological Magazine, 149, 753-767.

Zhang, S., Zhao, Y., 2017. Cogenetic origin of mafic microgranular enclaves in calc-alkaline granitoids: The Permian plutons in the northern North China Block. Geosphere, 13, 482-517.

Zhang, S.B., Zheng, Y.F., Zhao, Z.F., Wu, Y.B., Yuan, H.L., Wu, F.Y., 2009. Origin of TTG-like rocks from anatexis of ancient lower crust: Geochemical evidence from Neoproterozoic granitoids in South China. Lithos, 113, 347-368.

Zhang, S.H, Gao, R., Li, H.Y., Hou, H.S., Wu, H.C., Li, Q.S., Yang, K., Li, C., Li, W.H., Zhang, J.S., Yang, T.S., Keller, G.R., Liu, M., 2014c. Crustal structures revealed from a deep seismic reflection profile across the Solonker suture zone of the Central Asian Orogenic Belt, northern China: An integrated interpretation. Tectonophysics, 612-613, 26-39.

Zhang, S.H., Zhao, Y., Davis, G.A., Ye, H., Wu, F., 2014b. Temporal and spatial variations of Mesozoic magmatism and deformation in the North China Craton: Implications for lithospheric thinning and decratonization. Earth-Science Reviews, 131, 49-87.

Zhang, S.H., Zhao, Y., Liu, J.M., Hu, Z.C., 2016. Different sources involved in generation of continental arc volcanism: The Carboniferous-Permian volcanic rocks in the northern margin of the North China Block. Lithos, 240-243, 382-401.

Zhang, S.H., Zhao, Y., Liu, X.C., Liu, D.Y., Chen, F., Xie, L.W., Chen, H.H., 2009b. Late Paleozoic to Early Mesozoic mafic-ultramafic complexes from the northern North China Block: Constraints on the composition and evolution of the lithospheric mantle. Lithos, 110, 229-246.

Zhang, S.H., Zhao, Y., Song, B., Hu, J.M., Liu, S.W., Yang, Y.H., Chen, F.K., Liu, X.M., Liu, J., 2009a. Contrasting Late Carboniferous and Late Permian-Middle Triassic intrusive suites from the northern margin of the North China Craton: Geochronology, petrogenesis, and tectonic implications. Geological Society of America Bulletin, 121, 181-200.

Zhang, S.H., Zhao, Y., Ye, H., Hou, K.J., Li, C.F., 2012. Early Mesozoic alkaline complexes in the northern North China Craton: Implications for cratonic lithospheric destruction. Lithos, 155, 1-18.

Zhang, S.H., Zhao, Y., Ye, H., Liu, J.M., Hu, Z.C., 2014a. Origin and evolution of the Bainaimiao arc belt: Implications for crustal growth in the southern Central Asian orogenic belt. Geological Society of America Bulletin, 126, 1275-1300.

Zhang, Y.P., Su, Y.Z., Li, J.C., 2010. Regional tectonic significance of the Late Silurian Xibiehe Formation in central Inner Mongolia, China. Geological Bulletin of China, 29, 1599-1605 (in Chinese with English abstract).

Zhao, P., Jahn, B.M., Xu, B., Liao, W., Wang, Y.Y., 2016. Geochemistry, geochronology and zircon Hf isotopic study of peralkaline-alkaline intrusions along the northern margin of the North China Craton and its tectonic implication for the southeastern Central Asian Orogenic Belt. Lithos, 261, 92-108.

Zhou, H., Zhao, G.C., Han, Y.G., Wang, B., 2018. Geochemistry and zircon U-Pb-Hf isotopes of Paleozoic intrusive rocks in the Damao area in Inner Mongolia, northern China: Implications for the tectonic evolution of the Bainaimiao arc. Lithos, 314-315, 119-139.

Zhou, J.B., Han, J., Zhao, G.C., Zhang, X.Z., Cao, J.L., Wang, B., Pei, S.H., 2015. The emplacement time of the Hegenshan ophiolite: Constraints from the unconformably overlying Paleozoic strata. Tectonophysics, 662, 398-415.

Zhu, R.Z., Lai, S.C., Santosh, M., Qin, J.F., Zhao, S.W., 2017. Early Cretaceous Na-rich granitoids and their enclaves in the Tengchong Block, SW China: Magmatism in relation to subduction of the Bangong-Nujiang Tethys ocean. Lithos, 286-287, 175-190.

Petrogenesis of high-K calc-alkaline granodiorite and its enclaves from the SE Lhasa Block, Tibet (SW China): Implications for recycled subducted sediments[①]

Zhu Renzhi   Lai Shaocong[②]   Qin Jiangfeng   Zhao Shaowei   M Santosh

Abstract: The genesis of high-K calc-alkaline rocks is important in evaluating subduction-related magmatism, particularly with regard to melts derived from recycled subducted sediments, which require unusually steep geothermal gradients. Here, we present high-precision laser ablation-inductively coupled plasma-mass spectrometry (LA-ICP-MS) zircon U-Pb, bulk-rock geochemical, Sr and Nd isotopic, mineral chemical, and in situ zircon Hf isotope data for high-K calc-alkaline granodiorites and associated enclaves in the SE Lhasa Block, Tibet. The LA-ICP-MS zircon U-Pb data show that the granodiorites were emplaced at ca. 124.1–123.8 Ma, and the associated enclaves formed at ca. 132.7–129.6 Ma. The granodiorites are high-K, calc-alkaline, and slightly peraluminous with A/CNK [= molar ratio of $Al_2O_3/(CaO + Na_2O + K_2O)$] values in the range 0.96–1.11. The high Th/La and La/Sm ratios, Th and Nb content, and low Nb/Ta, Ba/Th, U/Th, and Zr/Nb ratios are similar to rocks formed from sediment-derived melts. The uniformly high initial $^{87}Sr/^{86}Sr$ values (0.710 277–0.712 581), low $\varepsilon_{Nd}(t)$ values (−11.2 to −8.3), and negative and variable zircon $\varepsilon_{Hf}(t)$ values (−11.71 to −0.96) imply continental crustal sources. The enclaves are basaltic to andesitic and have high MgO (3.82–5.48 wt%), and their zircon Hf isotopic compositions and presence of magnesian-hornblende and plagioclase with labradorite-bytownite composition indicate a possible mantle connection. In conjunction with regional geological features, we infer that these rocks were derived from melting of recycled subducted sedimentary materials in the mantle wedge during the southward subduction of Bangong-Nujiang Tethyan Ocean. Our new data, together with those from the Bangong-Nujiang suture zone and its flanks, suggest that magmatic activity associated with the subduction of Bangong-Nujiang Tethyan oceanic lithosphere occurred from the Middle Triassic to Early Cretaceous.

Introduction

Voluminous Early Cretaceous magmatic suites mainly consisting of granitoids and volcanic

---

① Published in *Geological Society of America Bulletin*, 2019, 131(7-8).

② Corresponding author.

rocks are widely distributed within the Bangong-Nujiang suture zone and its flanks and are considered to have formed in response to the subduction and closure process of Bangong-Nujiang Tethyan ocean (Zhu et al., 2009a,b, 2011, 2013, 2016; Chiu et al., 2009; Sui et al., 2013; Qu et al., 2012; Lin et al., 2013; Chen et al., 2014; S.S. Chen et al., 2017; J. X. Li et al., 2014; S.M. Li et al., 2014, 2016; Y. Li et al., 2016a,b; Hu et al., 2017). Recent studies have proposed enhanced mantle contributions during the subduction of Bangong-Nujiang Ocean lithosphere, as inferred from granitoids and andesites, minor basalts and rhyolites, and the presence of dioritic enclaves and basaltic dykes (Sui et al., 2013; Chen et al., 2014; Zhu et al., 2011, 2016). Middle Triassic to Early Cretaceous suites formed from hybrid basaltic and sediment-derived melts also occur on both sides of the Bangong-Nujiang suture zone (Chen et al., 2016a,b; Zhu et al., 2016; S.M. Li et al., 2016). However, no detailed studies have been done on the role of recycled crustal materials in the mantle sources in this region. Previous studies have proposed the recycling of crustal material into the mantle sources through delaminated of lower continental crust (Gao et al., 2004; Xu et al., 2008; Chen et al., 2015), a subducted oceanic slab (Yogodzinski et al., 1995; Rapp et al., 1999; Martin et al., 2005), subducted sediments (Shimoda et al., 1998; Johnson and Plank, 1999; Tatsumi, 2001; Schmidt et al., 2004; Hanyu et al., 2006; Wang et al., 2011; Behn et al., 2011), or subducted continental crust (Qin et al., 2010). For example, subduction of pelagic and terrestrial sediment plays an important role in influencing the compositional variations of both subduction-related magmas and heterogeneity in mantle (Shimoda et al., 1998; Tatsumi, 2001; Hanyu et al., 2006; Wang et al., 2011; Behn et al., 2011; Guo et al., 2014). Partial melting of sediments requires an unusually steep geothermal gradient under a specific geodynamic environment (Johnson and Plank, 1999; Schmidt et al., 2004; Behn et al., 2011). Generally, incorporation of subducted sediments would generate high-K calc-alkaline andesitic to rhyolitic volcanic or intrusive rocks with high Th/La and La/Sm ratios and low Nb/Ta, Ba/Th, U/Th, and Zr/Nb ratios (Shimoda et al., 1998; Tatsumi, 2001; Hanyu et al., 2006; Wang et al., 2011; Guo et al., 2014).

Our recent studies led to the identification of typical high-K calc-alkaline granodiorites and associated enclaves in SE Lhasa terrane (SW China), which are important features from which to gain insights on source characteristics, magma petrogenesis, and tectonic settings. In this contribution, we present petrological, geochronological, bulk-rock geochemical, Sr-Nd isotopic, mineral chemical, and in-situ zircon Hf isotopic data for Early Cretaceous granodiorites and associated enclaves in the SE Lhasa terrane. We attempt to evaluate the genesis of the granodiorite and the role of sediment recycling associated with the subduction of Bangong-Nujiang Tethyan Ocean.

Geological background and petrography

The Tibet Plateau incorporates the Songpan-Garze flysch complex and several tectonic blocks, including Northern Qiangtang, Southern Qiangtang, Lhasa, and the Himalaya blocks, separated by the Jinshajiang, Longmucuo-Shuanghu-Changning-Menglian, Bangong-Nujiang, and Yarlung-Zangbo suture zones from north to south. The Lhasa terrane is bordered by the Bangong-Nujiang suture in the north and the Yarlung-Zangbo suture in the south (Fig. 1a; Chiu et al., 2009), and it is considered to have dispersed from Gondwana supercontinent during the Permian to Triassic, drifting northward to finally collide with the Qiangtang terrane during the Late Jurassic to Early Cretaceous (J. L. D. Kapp et al., 2005; P. Kapp et al., 2005; Zhu et al., 2011). Three distinct magmatic belts are identified in the Lhasa terranes (Fig. 1a; Zhu et al., 2011): the widespread Early Jurassic to Early Tertiary Gangdese batholiths; the Linzizong volcanic successions, which are regarded as the products of northward subducted Neo-Tethyan oceanic lithosphere in the southern Lhasa subterrane (Coulon et al., 1986; Harris et al., 1990; Lee et al., 2009; Mo et al., 2005, 2007; Wen et al., 2008; Guo et al., 2013; Kang et al., 2014); and the voluminous Mesozoic magmatic rocks occurring in in the central and northern Lhasa subterranes (Coulon et al., 1986; Harris et al., 1990; Chiu et al., 2009; Chu et al., 2006; Guynn et al., 2006; Zhu et al., 2009a, b, 2011). These Mesozoic magmatic suites constitute the so-called northern magmatic belt, which has been correlated to the closure of the Bangong-Nujiang Tethyan Ocean in recent studies (Zhu et al., 2009a, b, 2011, 2013, 2015; Chiu et al., 2009; Chen et al., 2014; Sui et al., 2013).

The northern Lhasa subterrane is overlain by the Middle Triassic to Cretaceous sedimentary rocks with widespread Early Cretaceous volcanic and intrusive rocks (Pan et al., 2004; Zhu et al., 2009b, 2011; Sui et al., 2013), Zircon Hf isotopic data from these rocks suggest that the northern Lhasa subterranes are mainly comprise juvenile basement (Zhu et al., 2011). The exposed Amdo Neoproterozoic to early Paleozoic basement (Guynn et al., 2006) likely represents an isolated microcontinent accreted to the northern part of the Lhasa subterrane (Zhu et al., 2013). The central Lhasa subterrane has been considered as a microcontinent with Precambrian basement rocks (Zhu et al., 2011), and it is dominated by a Carboniferous-Permian metasedimentary sequence, Lower Cretaceous volcano-sedimentary sequence, and associated Mesozoic granitoids (J. L. D. Kapp et al., 2005; P. Kapp et al., 2005, 2007; Volkmer et al., 2007; Zhu et al., 2009a, b; Cao et al., 2016).

In the eastern part, extensive Cretaceous granitoids and minor Ordovician, Jurassic, and Cenozoic granitoids are exposed in a NW-SE belt to the southeast of the Bangong-Nujiang suture zone (Pan et al., 2004; Zhu et al., 2008, 2009b; Liang et al., 2008; Chiu et al., 2009), and they occur within Carboniferous-Permian, Devonian, and Proterozoic metamorphic

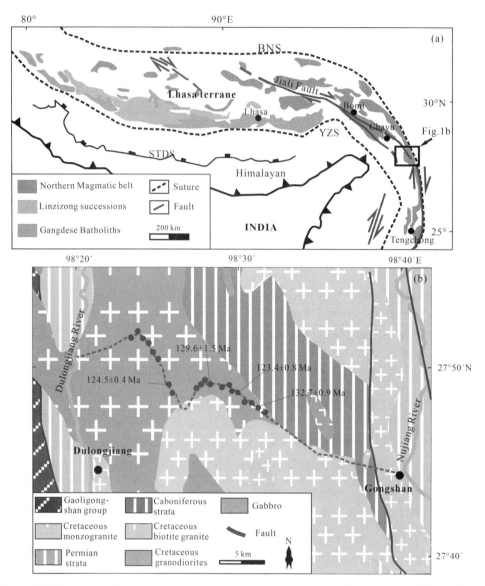

Fig. 1 (a) Distribution of principal magmatic units and strata in the central and southern Tibetan Plateau (modified after Chiu et al., 2009). (b) Study area showing the Dulongjiang pluton along the southeastern extension of the Bomi-Chayu batholith in the SE Lhasa block (modified after YNBGMR, 1991).

BNS: Bangong-Nujiang suture; YZS: Yarlu-Zangpo suture; STDS: South Tibet Detachment system.

rocks. These Cretaceous granitoids are distributed mostly as batholiths, mainly in the Bomi, Basu, Ranwu, and Chayu area (so-called Bomi-Chayu batholiths) They are mainly composed of monzogranites and granodiorites with minor dioritic veins and dioritic enclaves (Pan et al., 2004; Liang et al., 2008; Chiu et al., 2009; Zhu et al., 2009b; Lin et al., 2012).

This study focuses on the Dulongjiang pluton (Fig. 1b), a NW-SE-elongated intrusion, parallel to the long axis of the southeastern part of Bomi-Chayu batholith, which is located in between the Jiali fault and the Bangong-Nujiang suture (Fig. 1a,b; Zhu et al., 2009b; Chiu et al., 2009). The pluton is exposed in the area between the town of Dulongjiang and Gongshan County, with its southern part dominated by hornblende-bearing granitic rocks.

The Dulongjiang pluton is mainly represented by medium-grained granodiorites and associated enclaves (Fig. 2a,b) that occur scattered throughout the pluton (Fig. 2a,b). The enclaves are mainly composed of quartz (<8%), alkali feldspar (8%-10%), calcic plagioclase (40%-45%), hornblende (30%), biotite (~8%), and accessory apatite, zircon, magnetite and other Fe-Ti oxides (Fig. 2c). The granodiorites mainly contains quartz (18% - 20%), alkaline feldspar (~15%), plagioclase (35%-40%), euhedral hornblende (15%-20%), biotite (~8%), and accessory apatite, zircon, magnetite and other Fe-Ti oxides (Fig. 2d).

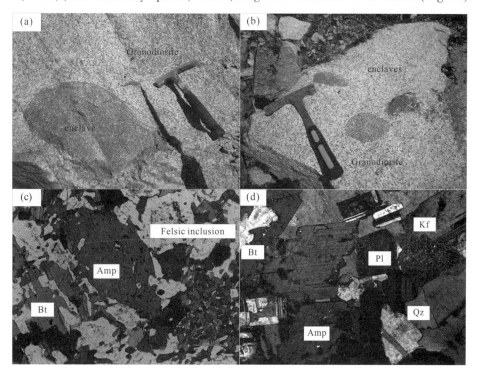

Fig. 2   Field photos and photomicrographs of granodiorite and associated enclaves
from the Dulongjiang pluton, SE Lhasa Block, SW China.
Amp: amphibole; Bt: biotite; Qz: quartz; Pl: plagioclase; Kf: K-feldspar.

Analytical methods

Major and trace elements

Whole-rock samples were trimmed to remove weathered surfaces, cleaned with deionized water, crushed, and then powdered through a 200 mesh screen using a tungsten carbide ball mill. Major elements were analyzed using an X-ray fluorescence (XRF) spectrometer (Rikagu RIX 2100) at the Guizhou Tuopu Resource and Environmental Analysis Center, Institute of

Geochemistry, Chinese Academy of Sciences, Guiyang, China. Analyses of U.S. Geological Survey (USGS) and Chinese national rock standards BCR-2 ($SiO_2 = 51.3$ wt%, $Na_2O = 0.4$ wt%, $K_2O = 0.31$ wt%, $MgO = 3.22$ wt%, $Al_2O_3 = 13.2$ wt%, $P_2O_5 = 0.29$ wt%, $CaO = 6.81$ wt%, $TiO_2 = 1.95$ wt%, $MnO = 0.122$ wt%, and $Fe_2O_3^T = 12.43$ wt%), GSR-1 ($SiO_2 = 72.06$ wt%, $Na_2O = 2.83$ wt%, $K_2O = 4.85$ wt%, $MgO = 0.36$ wt%, $Al_2O_3 = 13.3$ wt%, $P_2O_5 = 0.08$ wt%, $CaO = 1.46$ wt%, $TiO_2 = 0.172$ wt%, $MnO = 0.05$ wt%, and $Fe_2O_3^T = 1.95$ wt%), and GSR-3 ($SiO_2 = 43.8$ wt%, $Na_2O = 2.866$ wt%, $K_2O = 2.04$ wt%, $MgO = 6.68$ wt%, $Al_2O_3 = 0.14$ wt%, $P_2O_5 = 0.09$ wt%, $CaO = 8.25$ wt%, $TiO_2 = 1.42$ wt%, $MnO = 0.14$ wt%, and $Fe_2O_3^T = 13.06$ wt%) were conducted for comparison. Comparison with the standards indicated that the analytical precision and accuracy for major elements were generally better than 5%. Trace elements were determined by using a Bruker Aurora M90 inductively coupled plasma-mass spectrometry (ICP-MS) at the Guizhou Tuopu Resource and Environmental Analysis Center, Institute of Geochemistry, Chinese Academy of Sciences, Guiyang, China, following the method of Qi et al. (2000). Sample powders were dissolved using an $HF + HNO_3$ mixture in a high-pressure polytetrafluoroethylene bomb at 185 ℃ for 36 h. The accuracy of the ICP-MS analyses was estimated to be better than ±5%−10% (relative) for most elements.

Sr-Nd isotopic analyses

Whole-rock Sr-Nd isotopic data were obtained by using a Neptune Plasma high-resolution (HR) multicollector (MC) mass spectrometer at the Guizhou Tuopu Resource and Environmental Analysis Center, Institute of Geochemistry, Chinese Academy of Sciences, Guiyang, China. The Sr and Nd isotopes were determined by using a method similar to that of Chu et al. (2009). Sr and Nd isotopic fractionation was corrected to $^{87}Sr/^{86}Sr = 0.119\ 4$ and $^{146}Nd/^{144}Nd = 0.721\ 9$, respectively. The average $^{143}Nd/^{144}Nd$ ratio of the La Jolla standard measured during the sample runs is $0.511\ 862 \pm 5$ (2σ), and the average $^{87}Sr/^{86}Sr$ ratio of the NBS987 standard is $0.710\ 236 \pm 16$ (2σ). The total procedural Sr and Nd blanks are b1 ng and b50 pg, respectively. NIST SRM-987 and JMC-Nd were used as certified reference standard solutions for $^{87}Sr/^{86}Sr$ and $^{143}Nd/^{144}Nd$ isotopic ratios, respectively. BCR-1 and BHVO-1 standards were $0.705\ 014 \pm 3$ (2σ) and $0.703\ 477 \pm 20$ (2σ), respectively. The average $^{146}Nd/^{144}Nd$ ratios of the BCR-1 and BHVO-1 standards were $0.512\ 615 \pm 12$ (2σ) and $0.512\ 987 \pm 23$ (2σ), respectively, which were used as the reference materials.

Zircon U-Pb and Hf isotopic analyses

Zircon was separated from four ~5 kg rock samples taken from various sampling locations within the Dulongjiang pluton. The zircon grains were separated by using conventional heavy liquid and magnetic techniques. Representative zircon grains were handpicked and mounted in epoxy resin disks and then polished and coated with carbon. Internal morphology was examined using cathodoluminescent (CL) prior to U-Pb analyses.

Laser-ablation ICP-MS zircon U-Pb analyses were conducted on an Agilent 7500a ICP-MS equipped with a 193-nm laser, which is housed at the State Key Laboratory of Continental Dynamics, Northwest University, Xi'an, China. The $^{207}Pb/^{206}Pb$ and $^{206}Pb/^{238}U$ ratios were calculated using the GLITTER program and corrected using the Harvard zircon 91500 as external calibration. These correction factors were then applied to each sample to correct for both instrumental mass bias and depth-dependent elemental and isotopic fractionation. The detailed analytical technique was described in Yuan et al. (2004). Common Pb contents were evaluated by using the method described in Andersen (2002). The age calculations and plotting of concordia diagrams were made using Isoplot (version 3.0; Ludwig, 2003). The errors quoted in tables and figures are at the 2σ level.

In situ zircon Hf isotopic analyses were conducted using a Neptune MC-ICPMS, equipped with a 193-nm laser. During analyses, a laser repetition rate of 10 Hz at 100 mJ was used, and spot sizes were 32 μm. The detailed analytical technique was described by Yuan et al. (2008). Zircon standards (91500, GJ-1, and Monastery) were simultaneously analyzed, which yielded the recommended values (cf. Yuan et al., 2008) within 2σ. The obtained Hf isotopic compositions were $0.282\,016 \pm 20$ ($2\sigma_n$, $n = 84$) for the GJ-1 standard and $0.282\,735 \pm 24$ ($2\sigma_n$, $n = 84$) for the Monastery standard, respectively, consistent with the recommended values (cf. Yuan et al., 2008) within 2σ. The initial $^{176}Hf/^{177}Hf$ ratios and $\varepsilon_{Hf}(t)$ values were calculated with reference to the chondritic uniform reservoir (CHUR) at the time of zircon growth from the magmas. A decay constant for ^{176}Lu of 1.867×10^{-11} year^{-1} (Soderlund et al., 2004), chondritic $^{176}Hf/^{177}Hf$ ratio of $0.282\,785 \pm 7$, and $^{176}Lu/^{177}Hf$ ratio of $0.033\,6$ were adopted. The depleted mantle model ages (T_{DM}) for basic rocks were calculated with reference to the depleted mantle at the present-day $^{176}Hf/^{177}Hf$ ratio of $0.283\,25 \pm 20$, similar to that of average mid-ocean-ridge basalt (MORB; Nowell et al., 1998) and $^{176}Lu/^{177}Hf = 0.038\,4$ (Griffin et al., 2000). For the zircons from felsic rocks, we also calculated the Hf isotope "crustal" model age (T_{DMC}) by assuming that the parental magma was derived from an average continental crust, with $^{176}Lu/^{177}Hf = 0.015$, which originated from the depleted mantle source (Griffinet al., 2000). Our conclusions are not affected even if other decay constants are used.

Electron microprobe analyses

The compositions of rock-forming minerals were analyzed using a JXA-8230 electron microprobe at the State Key Laboratory of Continental Dynamics, Northwest University, Xi'an, China. The microprobe analyses were carried out with an accelerating voltage of 15 kV, a sample current of 10 nA, and a beam diameter of 1 μm. Microprobe standards of natural and synthetic phases were supplied by SPI Company: jadeite for Si, Al, and Na, diopside for Ca, olivine for Mg, sanidine for K, hematite for Fe, rhodonite for Mn, and rutile for Ti.

Results

Zircon LA-ICP-MS U-Pb dating

We chose two typical samples from each rock type of U-Pb geochronodogy, and the analytical results are listed in the GSA Data Repository Supplementary Dataset.[①] The sampling locations, lithology, CL images, and data are summarized in Figs. 1–4.

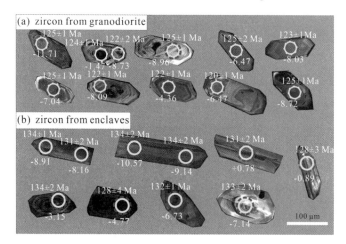

Fig. 3　Cathodoluminescence (CL) images of representative zircon grains
for the Dulongjiang pluton, SW China.

The white numbers at the top of each grain represent the U-Pb age of zircon grains, and the white numbers below each grain represent zircon $\varepsilon_{Hf}(t)$ values.

Granodiorites

Zircon grains from granodiorites are subhedral to euhedral, measuring up to 150–300 μm in length, with length: width ratios of 2:1 to 3:1. The prismatic crystals are colorless to light brown and transparent to translucent, and most grains display weak oscillatory zoning in CL images (Fig. 3a), typical of a magmatic origin (Hoskin and Schaltegger, 2003). The 21 data

---

① GSA Data Repository item 2019034, Figure DR1: Diagram of classification of amphiboles (a and b) (Leak et al., 2003; Hawthorne and Oberti, 2007); Figure DR2: The Ab-An-Or triangular diagram of classification (Parsons, 2010); Table DR1: Major- (wt%) and trace-element (ppm) analysis of Dulongjiang pluton; Table DR2: Whole-rock Rb-Sr and Sm-Nd isotopic data for Dulongjiang pluton; Table DR3: Zircon Hf isotopic data for Dulongjiang pluton; Table DR4: Representative major-element analyses (wt%) of amphibole and plagioclase from the Dulongjiang pluton; Supplementary Dataset: Results of zircon LA-ICP-MS U-Pb analysis for granitic rocks in Dulongjiang, SW China, is available at http://www.geosociety.org/datarepository/2019 or by request to editing @ geosociety.org.

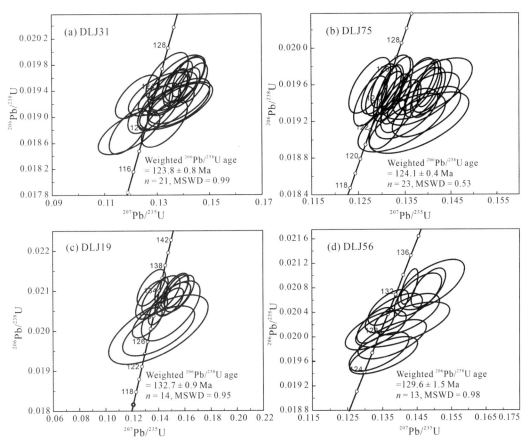

Fig. 4　Laser ablation-in-ductively coupled plasma-mass spectrometry (LA-ICP-MS) U-Pb zircon concordia diagrams of representative zircon grains for the Dulongjiang pluton, SW China.

MSWD: mean square of weighted deviates.

points from DLJ31 show high Th (199-2 074 ppm) and U (283-1 419 ppm) contents, with Th/U ratios of 0. 48-1. 72. Their yielded $^{206}Pb/^{238}U$ ages ranging from 120. 0±4. 0 Ma to 125. 0±2. 0 Ma, with a weighted mean age of 123. 3±0. 8 Ma (mean square of weighted deviates MSWD=0. 99, $n=21$, 95% confidence, probability 0. 99, 2σ; Fig. 4a). The 23 data points from DLJ75 show high Th (255-867 ppm) and U (272-1 405 ppm) contents, with moderate to high Th/U ratios (0. 41-1. 39). The analytical data yielded $^{206}Pb/^{238}U$ ages in the range of 122. 0±1. 0 Ma to 125. 0±1. 0 Ma, with a weighted mean age of 124. 1±0. 4 Ma (MSWD=0. 53, $n=23$, 95% confidence, probability 0. 99, 2σ; Fig. 4b).

Enclaves

Zircons grains from the enclaves are subhedral to anhedral. Some are prismatic grains with weak oscillatory zones, and they show higher length: width ratios than their host rocks (Fig. 3b). The 14 data points from DLJ19 show a range of Th (214-883 ppm) and U (190-1 517 ppm) contents, and Th/U ratios of 0. 28-1. 87 (most >0. 40). The data yielded

^{206}Pb/^{238}U ages from 128.0±4.0 Ma to 134.0±4.0 Ma, with a weighted mean age of 132.7±0.9 Ma (MSWD=0.94, n=14, 95% confidence, probability 0.99, 2σ; Fig. 4c). Zircons in sample DLJ56 also have a wide range of Th (148−2 418 ppm) and U (212−3 113 ppm) contents, with Th/U ratios of 0.26−1.16 (mostly >0.40). The data yielded ^{206}Pb/^{238}U ages from 126.0±1.0 Ma to 133.0±2.0 Ma, with a weighted mean age of 129.6±1.5 Ma (MSWD =0.98, n=13, 95% confidence, probability 0.99, 2σ; Fig. 4d).

Major- and trace-elemental geochemistry

Major- and trace-element analyses of granodiorites and associated enclaves from the Dulongjiang pluton are listed in Table DR1 (see footnote 1).

Granodiorite

The granodiorites show low SiO_2 = 59.88−65.86 wt%, various K_2O/Na_2O ratio = 0.81−1.26, high CaO = 4.18−5.47 wt% and $Fe_2O_3^T$ = 4.70−6.77 wt%, and low MgO = 2.04−2.91 wt%, with $Mg^\#$ = 49.8−54.9. The high Al_2O_3 contents of 15.71−18.66 wt% with A/CNK [molar Al_2O_3/ (CaO+Na_2O+K_2O)] ratios of 0.96−1.11 (most <1.10) classify the rocks as weakly peralumious. They are plotted as high-K calc-alkaline and low-Fe series (Fig. 5a-c). The primitive mantle-normalized trace-element diagram (Fig. 6a) shows enrichment of large ion lithophile elements (LILEs), such as positive Pb, Rb, Th, U contents, and a slight K anomaly, and depletions of high field strength elements (HFSEs), such as Nb, Ta, P, Zr, and Ti. The chondrite-normalized rare-earth element (REE) diagram (Fig. 6b) shows enrichment in light (L)REEs with $(La/Yb)_N$ = 8.20−23.52 and $(Gd/Yb)_N$ = 1.43−1.82, and small to moderate negative Eu anomalies (δEu = 0.61−0.82, where δEu = Eu_N · $\sqrt{Sm_N \cdot Gd_N}$).

Enclave

The enclaves in host granodiorites contained variable SiO_2 = 51.23−58.54 wt% and Na_2O/K_2O = 0.93−1.74, high CaO = 6.07−7.31 wt% and $Fe_2O_3^T$ = 7.38−10.53 wt%, and higher MgO = 3.82−5.48 wt% and $Mg^\#$ = 53.1−54.8 than in the host granodiorite. The primitive mantle-normalized trace-element diagram (Fig. 6c) shows positive Rb, U, and Pb, and slight K, Nd, and Sm anomalies, and depletions in Nb, Ta, P, Zr, Hf, and Ti. The chondrite-normalized REE diagram (Fig. 6d) shows relative weakly fractionated light to heavy REE (LREE/HREE) ratios of 4.82−7.83 and significant negative Eu anomalies (δEu = 0.58−0.69).

Whole-rock Sr-Nd isotopes

Whole-rock Sr-Nd isotopic data for the granodiorites and associated enclaves from the

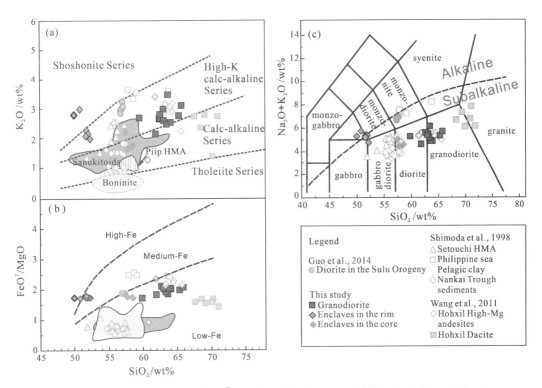

Fig. 5　(a) K_2O vs. SiO_2 diagram. (b) FeO^T/MgO vs. SiO_2 diagram. (c) ($Na_2O + K_2O$) vs. SiO_2 diagram.
HMA：high-Mg andesites.

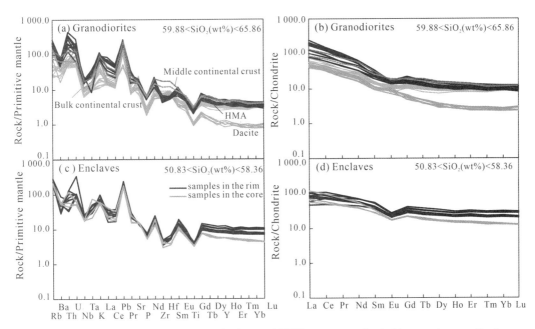

Fig. 6　Chondrite-normalized rare earth element (REE) patterns and primitive mantle-normalized
trace-element spider diagrams (a,b) for granodiorite and (c,d) enclave from Dulongjiang pluton.
The primitive mantle and chondrite values are from Sun and McDonough (1989). HMA：high-Mg andesites.

Dulongjiang pluton are listed in Table DR2 (see footnote 1) and the diagram of $\varepsilon_{Nd}(t)$ vs. initial $^{87}Sr/^{86}Sr$ (Fig. 7). All the initial $^{87}Sr/^{86}Sr$ isotopic ratios (I_{Sr}) and $\varepsilon_{Nd}(t)$ values were based on the time of magma crystallization. The host granodiorites (sample DLJ39, DLJ49, DLJ71, and DLJ100) have I_{Sr} ratios ranging from 0.710 277 to 0.711 824, $\varepsilon_{Nd}(t)$ values

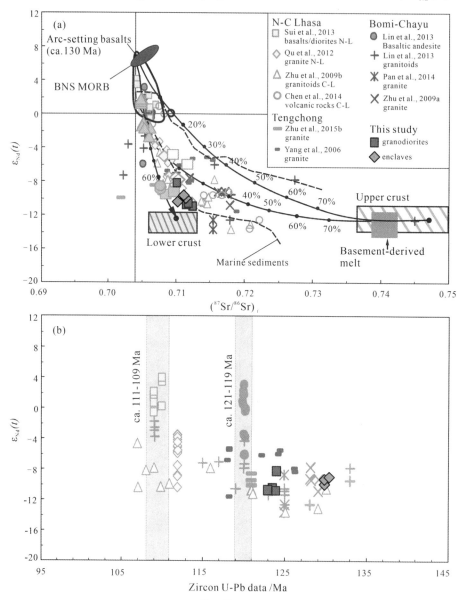

Fig. 7 (a) $\varepsilon_{Nd}(t)$ values vs. initial $^{87}Sr/^{86}Sr$. (b) $\varepsilon_{Nd}(t)$ values vs. zircon U-Pb age.

The data for juvenile lower crust and back-arc basalts are from Chen et al. (2014), which also cites Zhu et al. and Wang et al. (unpublished data). Basement-derived melts are from Zhu et al. (2011). Lower continental crust is from Miller et al. (1999) and upper continental crust is from Harris et al. (1988). Symbols are as in Fig. 5. HMA: high-Mg andesites; BNS MORB: Bangong-Nujiang suture mid-ocean-ridge basalt; N-C: north central; N-L: Northern Lhasa; C-L: Central Lhasa.

ranging from -11.2 to -8.3, and T_{DM2} values of $1.37-1.57$ Ga. The enclaves (sample DLJ23, DLJ58, DLJ59, and DLJ89) have I_{Sr} ratios ranged from $0.710\,482$ to $0.711\,389$, and $\varepsilon_{Nd}(t)$ values ranged from -10.8 to -9.8, and T_{DM2} values of $1.48-1.55$ Ga.

Zircon Hf isotopic composition

Zircon grains from three dated samples were also analyzed for in situ Lu-Hf isotopes on the same domains, and the results are listed in Table DR3 (see footnote 1). Initial $^{176}Hf/^{177}Hf$ ratios and $\varepsilon_{Hf}(t)$ values of the zircon grains were calculated based on their in situ zircon U-Pb age. Fig. 8 shows $\varepsilon_{Hf}(t)$ plots where zircon grains from both the granodiorites and associated enclave show variable and evolved Hf isotopic composition, with $\varepsilon_{Hf}(t)$ values of -11.71 to -0.96 for granodiorites (16 zircons), and $\varepsilon_{Hf}(t)$ values of -10.76 to $+0.78$ for enclaves (14 zircons). They data suggest crustal Hf model ages of $1.24-1.92$ Ga for granodiorites, and $1.13-1.86$ Ga for enclaves.

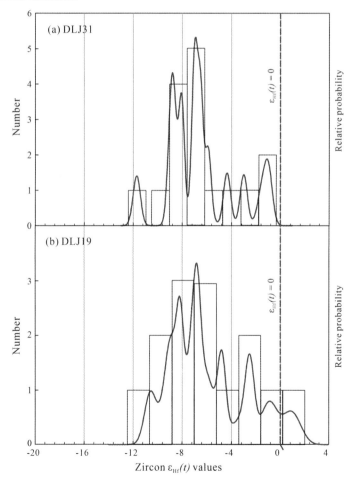

Fig. 8　Histograms of initial zircon Hf isotope ratios for: (a) granodiorite and (b) enclaves from the Dulongjiang pluton, SW China.

Mineral chemistry

The analytical data for plagioclase and amphibole are listed in Table DR4 (see footnote 1). The amphibole from the Dulongjiang pluton is mainly magnesian-hornblende, and the plagioclase is high andesine to bytownite, according to the classification of Leake et al. (1997) and Parsons (2010) [see Data Repository Figs. DR1 and DR2 (footnote 1)].

Amphibole and plagioclase in the granodiorites

Amphibole (DLJ46-AMP) is subhedral and ~1.10 mm×1.50 mm in size. It shows high MgO content (10.87 - 12.85 wt%) with high $Mg^{\#}$ of 57.5 - 64.8, and low Al_2O_3 content (6.15 - 9.63 wt%). An inclusion in amphibole (sample DLJ46-AMP) was confirmed as carbonate, and another inclusion was variously enriched in K or Ca. Plagioclase (DLJ46-Pl) is euhedral and ~0.50 mm×0.85 mm in size. It shows a compositional range of $Ab_{11-46}An_{53-89}Or_{0-1}$ and high SrO content of 0.14 - 0.32 wt%.

Amphibole and plagioclase in the enclaves

Amphibole (DLJ21-AMP) in the enclaves is euhedral and ~0.45mm × 0.85mm in size. It shows high MgO content (10.52 - 11.62 wt%) with high $Mg^{\#}$ of 55.7 - 60.1, and low Al_2O_3 content (7.28 - 8.84 wt%). A plagioclase inclusion was An-rich, with $Ab_{28-35}An_{65-72}Or_{0-1}$. Amphibole from sample DLJ61-AMP is subeuhedral and ~0.45 mm × 0.75 mm in size. It has higher MgO content (11.50 - 13.49 wt%) with high $Mg^{\#}$ of 60.1 - 67.3, and lower Al_2O_3 content (6.85 - 8.07 wt%) than those in sample DLJ21-AMP. The plagioclase inclusion had a compositions in the range $Ab1_{3-37}An_{62-87}Or_{0-1}$. Plagioclase (DLJ61-Pl) is euhedral and ~0.40 mm × 0.80 mm in size with a compositional range of $Ab_{15-39}An_{60-85}Or_{0-1}$ and high SrO content of 0.15 - 0.29 wt%.

Pressure of crystallization

Pressures of crystallization were estimated from the composition of amphibole, using the Al-in-hornblende geobarometer (Schmidt, 1992). This barometer requires that a buffering mineral assemblage is present consisting of quartz, K-feldspar, plagioclase, biotite, titanite and magnetite. All above mentioned minerals were observed in the samples, including DLJ21, DLJ46, and DLJ61. The granodiorite and its enclave have low Al content and yielded an average pressure of ~3.6 kbar [granodiorites; Table DR4 (see footnote 1)].

Discussion

Petrogenesis of the high-K calc-alkaline granodiorite

Origin of enclaves

The wide range of zircon Hf isotopic compositions (-10.76 to +0.78) of the enclaves precludes a simple, common evolution by closed-system fractionation process (Yang et al., 2007). Kemp et al. (2007) suggested that such variations with a spectrum of zircon $\varepsilon_{Hf}(t)$ values up to 10ε unites within a single sample can only be reconciled by the operation of open-system processes (Fig. 9). Processes such as magma mixing, wall-rock assimilation, or restile separation can most likely explain such variations (Kemp et al., 2007; Yang et al., 2007). The absence of inherited zircons and metamorphic texture in the enclaves (Figs. 2 and 3) precludes the model of restite separation (Chappell et al., 1987; White et al., 1999; Chappell et al., 2000). Yang et al. (2007) suggested that some enclave and monzogranite samples that have inherited zircons might potentially indicate wall-rock contamination, although there is no evidence for this process in our rocks. In addition, the absence of residue information from wall rock and partially disaggregated metasedimentary enclaves in the plutons do not support assimilation process (Kemp et al., 2007). There is increasing evidence to suggest that many magma chambers are open systems and may be fed with more primitive magma, resulting in magma mixing and mingling and generation of mafic microgranular enclaves (Vernon et al., 1988; Maas et al., 1997; Yang et al., 2007; Santosh et al., 2017). In this context, the following features in our study provide some clues: ① Both enclaves and host granodiorites have significant variations of zircon Hf isotopic composition [Fig. 9; Table DR3 (see footnote 1)]. ② Enclaves are characterized by typical quenched igneous textures, including lath-shaped amphibole and acicular apatite, and absence of cumulate textures (Fig. 2). ③ The shape of the enclaves varies from rounded to ovoid, with sharp interfaces with host granodiorites (Fig. 2). ④ The size of the enclaves is larger in the rim of the pluton than those in the core (Fig. 2a, b), with lower MgO and $Fe_2O_3^T$ and higher SiO_2 contents for those in the core than in the rim (Fig. 5). Accordingly, the enclaves were likely formed as mafic magma blobs injected

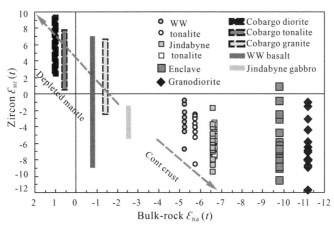

Fig. 9 Zircon Hf isotopic composition vs. whole-rock Nd isotopic composition of samples from Dulongjiang pluton, SW China (modified after Kemp et al., 2007).

WW tonalite: Why Worry tonalites.

into the intermediate to felsic magma chambers, similar to the top-to-down model (Castro et al., 2008). The contrasting rheological properties between mafic and felsic magmas may have inhibited physical mixing (Sparks and Marshall, 1986), whereas interdiffusion of elements did occur (Lesher, 1990), particularly mobile elements such as Na, K, Rb, and Sr between the enclaves and host granitoids. However, the highly charged elements, including U, Th, Nb, Ta, Zr, Hf, and REEs, were slow to interact and were easily retained in the minerals (Klein et al., 1997).

Generation of host granodiorite

As mentioned above, an important feature of both host granodiorites and enclaves is the marked variation in the zircon Hf isotopic data (Figs. 8–9). A similar feature was observed in the enclaves and host granites of the Gudaoling batholiths, where varying degrees of hybridization between magmas derived from pre-existing crustal materials and mantle-derived components have been proposed (Yang et al., 2007). Also, these zircon Hf isotopic features are similar to those in examples where granitic magmas form by reworking of sedimentary materials through input of mantle magmas, such as the Why Worry tonalites in the Lachlan I-type granitic belt (Fig. 9; Kemp et al., 2007). The enclaves in host granodiorites are basaltic to andesitic ($SiO_2 = 51.23 - 58.54$ wt%) and have higher MgO (3.82–5.48 wt%) and $Mg^#$ values (53.1–54.8) than melts derived from metabasalts (e.g., MgO < 3 wt% and $Mg^# < 45$; Fig. 10a, b; Rapp and Watson, 1995), and the parental melts were most likely of mantle origin (Guo et al., 2014). Unlike typical high-Mg andesites, however, the enclaves have $Mg^#$ values that are not high enough to reflect equilibration with the mantle peridotites ($Mg^# > 60$; Fig. 10a, b). This is also supported by correspondingly low concentrations of Cr and Ni. In addition, the model of fractional crystallization of primitive melts can be excluded based on the following lines of evidence: ① In the case of fractional crystallization of primitive melts, the zircon Hf and whole-rock Sr-Nd isotopic compositions will be consistent with those of the primitive melts. However, the rocks in this study have variable zircon $\varepsilon_{Hf}(t)$ values (Fig. 8 and 9) that cannot be produced by simple fractional crystallization (Kemp et al., 2007). Also, the high initial $^{87}Sr/^{86}Sr$ (0.710 277–0.712 581) and low $\varepsilon_{Nd}(t)$ values (−11.2 to −8.3) of whole rocks imply an evolved continental crust source rather than primitive melts. ② Typical petrological features for fractional crystallization, such as cumulate texture in the pluton (Fig. 2a, b) and early-crystallization minerals surrounded by later ones (Fig. 2c, d), were not observed. ③ Mafic magmatic enclaves of variable size range from the margin to the core of the granodiorite intrusion, which also suggests that magma mixing, rather than fractional crystallization, played an important role. Although most the zircon $\varepsilon_{Hf}(t)$ values are negative, some values are positive (up to + 0.78), suggesting the involvement of juvenile and/or depleted mantle material. Furthermore, the enclaves are coeval with the arc-related

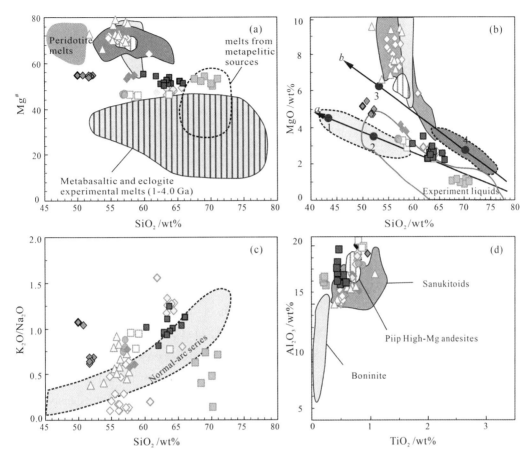

Fig. 10 （a）Plots of SiO_2 vs. $Mg^{\#}$. Marked fields outline experimental melts by Rap and Watson （1995）
and Rapp etal. （1999）; （b）MgO vs. SiO_2 systematic diagram after Moyen and Martin （2012）.
Trend "a" is a differentiation trend （irrespective of the actual process generating it）. Sample 1
has higher MgO than sample 2. Interactions of magma 2 with the mantle will shift it to posi-tion 3;
from this point onwards, the magma will define a new trend （"b"）. Yet, samples 3 and 4 have
experienced interactions with the mantle. If the two dark gray and light gray fields （close to trends
a and b）represent two sample sets, it is not immediately intuitive that the mantle-contaminated set
is group "b". Experimental liquids are after Martin and Moyen （2002）. （c）SiO_2 vs. K_2O/Na_2O
diagram after Moyen and Martin （2012）. （d）TiO_2 vs. Al_2O_3 diagram after Wang et al. （2011）.
Symbols are as in Fig. 5.

basalts （ca. 130 Ma） and Bangong-Nujiang MORB （Fig. 7; Sui et al., 2013; Chen et al.,
2014）. Amphiboles from the granodiorites and enclaves are magnesio-hornblende ［Data
Repository Fig. DR1A, B （footnote 1）］, similar to those from calc-alkaline, arc magmatic
rocks in subduction settings （Martin, 2007; Coltorti et al., 2007）, and values also compare
with the hornblendes in calc-alkaline basaltic magmas with mantle-derived features （Rock,
1987）. Plagioclase in the granodiorites and enclaves show $An_{>50}$ with high CaO, FeO^T, and
TiO_2 contents ［Data Repository Fig. DR2 （footnote 1）］, suggesting crystallization from basic/

mafic melts (Singer, 1995; Ginibre et al., 2002; Chen et al., 2008). Therefore, the basaltic to andesitic enclaves may represent a mantle-like magmatic component and significant involvement of mantle-derived materials. The negative zircon $\varepsilon_{Hf}(t)$ values (-11.83 to -0.72; Fig. 8a) and bulk-rock Sr-Nd isotopic compositions [I_{Sr} ratios ranging from 0.710 3 to 0.711 8 and $\varepsilon_{Nd}(t)$ values ranging from -11.2 to -8.3] of granodiorites (Fig. 7) would most likely suggest preexisting and evolved crustal components (Yang et al., 2004, 2007; Wu et al., 2006; Kemp et al., 2007). Recycling of these crustal materials into the mantle sources would require lithospheric delamination, subducted continental crust, and/or subducted oceanic sediments. A delamination model, however, is not supported by the geological record and geochemical characters. The delamination of a thickened lithosphere would lead to rapid uplift along the Bangong-Nujiang suture zone during the Early Cretaceous (ca. 130 Ma), evidence for which is lacking during. Lithospheric delamination would also induce asthenospheric upwelling, but there is no evidence for contemporaneous asthenospheric-derived magmatism. Furthermore, magmas that are derived from delaminated lower crust typically display adakitic affinity, with high Sr/Y ratios (>20) and low Y content attributable to residual garnet (Gao et al., 2004), which is also absent in the present case. Subduction of continental lithosphere would also generate high-pressure or untrahigh-pressure rocks such as eclogites, crustal xenoliths in the lavas, and high-K_2O (at least >3 wt%) magmas in the overlying plate, including late-orogenic, postorogenic, and possibly anorogenic granitoids (Brueckner, 2009; Qin et al., 2010). All these features are absent in our study area.

The rocks in our study display relatively high K_2O/Na_2O ratios, high Al_2O_3 content, and low TiO_2 content, which are similar to Philippine Sea pelagic clay and Nankai Trough sediments, as well as the Piip high-Mg andesites, sanukitoids, diorites, and high-Mg andesites and dacites in northern Hohxil (Fig. 10c, d). These high-Mg andesites and diorites were derived through significant addition of melts from subducted oceanic sediments (Shimoda, 1998; Wang et al., 2011; Guo et al., 2014). The relatively low Ba/Th, U/Th, and Zr/Nb ratios and high La/Sm ratios also correspond to sediment-derived melts (Fig. 11a, b, and e; Stern et al., 2006). The high Th/La ratios and high Th and Nb contents, and low Nb/Ta ratios (Fig. 11c, d) are similar to those of modern marine sediments and/or upper continental crust (Wang et al., 2011). This is supported by the elevated Th/Yb (~12.0) and low Ba/La (~13.8) values, since sediments have high Th/Yb and low Ba/La ratios (Hanyu et al., 2006). The possible addition of fluid from the subducted slab into the source region is unlikely, since Ba is more soluble in aqueous fluids than REEs, and Th and Yb are immobile in aqueous fluids. These features coupled with the negative zircon $\varepsilon_{Hf}(t)$ values (Fig. 8) and evolved bulk-rock Sr-Nd isotopic compositions (Fig. 7) are best explained by sediment contribution (Hanyu et al., 2006). Island-arc magmatism derived from subducted sediments is

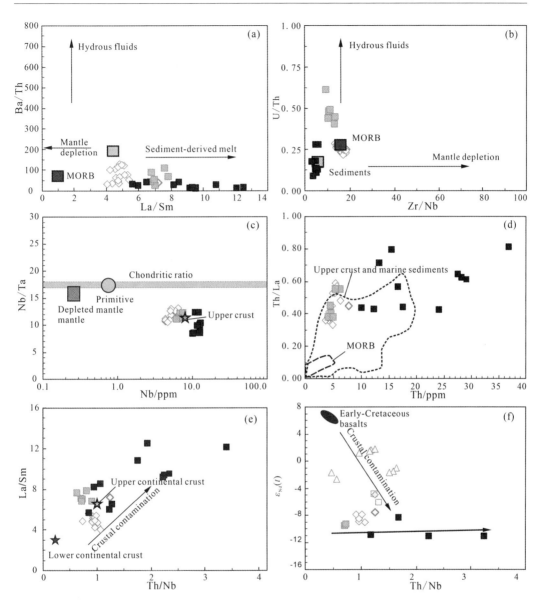

Fig. 11   Behavior of diagnostic incompatible trace elements in the host granodiorites and their enclaves.
(a) Ba/Th vs. La/Sm (Elliott, 2003). (b) U/Th vs. Zr/Nb (Stern, 2006). (c) Nb/Ta vs. Nb
(Rudnick et al., 2000); (d) Th/La vs. Th (Plank, 2005). (e) La/Sm vs. Th/Nb
(Wang et al., 2011). (f) $\varepsilon_{Nd}(t)$ vs. Th/Nb(Wang et al., 2011).

Marine sediments are from Plank (2005) and Plank and Langmuir (1998). MORB data are from Hofmann (1988)
and Niu and Batiza (1997). Data for LCC (lower continental crust) and UCC (upper continental crust) are from
Rudnick and Gao (2003). Composition of sediment-melts from Hochstaedter et al. (1996). Early Cretaceous basalts
are the same as Fig. 7. Symbols are as Fig. 5.

characterized by Eu depletion, with typical Eu anomalies of ~ 0.65 (Mclennan and Taylor,
1981), which is similar to our rocks. Recently, Zhu et al. (2016) also reported rocks negative
zircon $\varepsilon_{Hf}(t)$ values (−8.5 to −6.7) from rocks in the northern Lhasa terrane, which were

attributed to partial melting of subducted Bangong-Nujiang oceanic lithosphere and sediments. Therefore, we suggest that the granodiorites were derived from reworking of recycled subducted oceanic sedimentary materials by mantle-like magmas.

Petrotectonic implications

The melting of subducted sediments played a significant role in the production of the granodiorites in our study area, similar to the Choshi rocks (Hanyu et al., 2006). Based on this inference, some constraints can be derived for the temperature and pressure regime in the subduction zone (Johnson and Plank, 1999; Hanyu et al., 2006). Although the sodius for sediment-derived melts is relatively lower than that of the underlying oceanic crust, partial melting of sediment still requires an unusually steep geothermal gradient. Previous studies have indicated that sediments on top of the subducting oceanic crust have solidus ($H_2O + Cl$ fluid-saturated) at 775 ± 25 ℃ at 2 GPa, 810 ± 15 ℃ at 3 GPa, and $1\,025 \pm 25$ ℃ at 4 GPa, saturated and undersaturated, and 600 ℃ and 800 ℃ at 1.5GPa (Nichols et al., 1996; Johnson and Plank, 1999). Stable hydrous phases at 2 GPa include phengite (dehydration solidus is at $810 - 825$ ℃), amphibole, and biotite (which breaks down between 850 ℃ and 900 ℃; Johnson and Plank, 1999). Schmidt et al. (2004) also suggested that the minimum temperature required for fluid-saturated sediment melting in the subducting crust is 825 ℃ at 4 GPa. Thus, normal geothermal models cannot produce a slab hot enough to melt the sediments. The temperatures at the top of the slab must be hot enough (200 ℃ higher than normal slabs) to melt the recycled subducting sediments at the subarc depth, but a typical subducted slab is always below the solidus at any given depth. Several mechanisms can be considered to account for the abnormal temperature conditions, including the thermal age of the subducting plate and overriding plate, subduction velocity, mantle flow and rheology, and shear and viscous heating (Johnson and Plank, 1999). These may be achieved by slow subduction of the slab, a low angle of slab penetration, very young slab subduction, and extremely high temperatures in the mantle wedge (Hanyu et al., 2006). Indeed, young (< 45 Ma), slowly and obliquely subducting slabs would provide sufficient time for reheating and result in the formation of adakites (e.g., Defant and Drummond, 1990) rather than granodiorite. There is no robust evidence to support a young, slow, or low-angle subduction of the Bangong-Nujiang Ocean during the Early Cretaceous (Kapp et al., 2007; Zhu et al., 2011, 2013, 2016; Sui et al., 2013). In addition, none of the paths of shear heating leads to slab temperatures that approach the experimentally determined sediment solidus (> 750 ℃; Johnson and Plank, 1999). The other heat source to the slab is the mantle, which is supported by the mantle-like origin of enclaves in the host granodiorite. In addition to significant viscous heating, the viscosity structure of the mantle wedge that induces the convective flow pattern can greatly affect slab temperatures (Kincaid and Sacks, 1997; Behn et al., 2011). However, instead of the sediment melting at the slab surface, the subducted sediments could detach from the downgoing slab to form buoyant diapirs and subsequently release the trace elements through the overlying hot mantle wedge, which would later impart the sediment melt signature in the erupted lavas. Such a process would produce peraluminous granitic melts, which is consistent with the characteristics observed in the granodiorite of present study (A/CNK = 0. 96 − 1. 11). Bulk partition coefficients demonstrate that incompatible Th and Be cannot be transferred from the sediment pile by an aqueous fluid, but instead they require the presence of a silicate melt above the solidus (Johnson and Plank, 1999), which is consistent with the observed high La/Sm and low Ba/Th ratios in the samples (Fig. 11a). Even if amphibolite melting took place, the sediment-derived components were still predominant in the slab-derived partial melt (Shimoda et al., 1998). Sediment melting models have primarily been applied to arc settings involving subduction of oceanic crust containing sediments (Johnson and Plank, 1999; Plank, 2005; Wang et al., 2011, reference therein). Compared with regional magmatism along the Bangong-Nujiang suture, a typical arc-arc model at ca. 131 − 120 Ma has been suggested to explain calc-alkaline to high-K calc-alkaline melts with arc signature, including the 131 − 120 Ma Yanhu volcanic rocks, a ca. 120 Ma mafic dike, and 134 − 130 Ma metaluminous and 125 − 120 Ma peraluminous granitoids in the northern Lhasa subterrane (Zhu et al., 2016), which are related to the subducting and closure of the Bangong-Nujiang Ocean (Zhu et al., 2011, 2013, 2016). However, the timing of Lhasa-Qiangtang collision and the closure of the Bangong-Nujiang Tethyan Ocean remains debated, with several models proposed, as summarized below.

P.Kapp et al. (2005) suggested that the red beds that lie unconformably on Jurassic flysch in the Gaize area delimit the initiation of non-marine sedimentation during the Early Cretaceous. However, the age of nonmarine red beds is in the range of ca. 116 − 107 Ma., Kapp et al. (2007) focused on the Nima area along the Bangong-Nujiang suture in central Tibet, where Jurassic to Lower Cretaceous (≤ 125 Ma) marine sedimentary rocks were transposed, intruded by granitoids, and uplifted above sea level by ca. 118 Ma. Subsequently, some ca. 110 − 106 Ma nonmarine Cretaceous rocks were also formed. Both of these suggest that a marine environment existed at least until 125 − 118 Ma (P. Kapp et al., 2005, 2007). Stratigraphic and structural geology studies suggest that ophiolitic mélange and Jurassic flysch in the western Bangong-Nujiang suture zone were obducted southward onto the northern margin of the Lhasa terrane (Kapp et al., 2003; Shi et al., 2012). The Middle Triassic to Early Cretaceous magmatic rocks on the southern margin of the western Qiangtang subterrane were the products of northward subduction of Bangong-Nujiang Tethyan Ocean lithosphere (Kapp et al., 2005; Shi et al., 2008; J.X. Li et al., 2014; Chen et al., 2016a,b; S.M. Li et al., 2014, 2016; Y. Li et al.,

2016a,b). The Lhasa-Qiangtang collision occurred prior to ca. 110 Ma (Kapp et al., 2007; Raterman et al., 2014).

Paleomagnetic data indicate tha the northern Lhasa terrane and southern Qiangtang terrane shared similar paleolatitude at ca. 110 – 104 Ma (Chen et al., 2017; Hu et al., 2017, references therein). The ca. 114 – 100 Ma A2-type magmatism in the northern Lhasa terrane and Bangong-Nujiang suture zone also indicates a post-collisional setting (Eby, 1992; Qu et al., 2012), However, Y. Li et al. (2016b) proposed that the Late Jurassic to Early Cretaceous (> 100 Ma) magmatism in the southern Qiangtang terrane was related to the northward subduction of Bangong-Nujiang oceanic crust, and the ca. 100 Ma magmatism was the product of slab breakoff. J.X. Li et al. (2014) also suggested that the ca. 170 – 110 Ma intermediate-felsic intrusions in the southern Qiangtang terrane were related to the northward subduction of the Bangong-Nujiang Tethyan Ocean in a continental setting. The 120 – 108 Ma basaltic rocks and interbedded limestones within the Bangong-Nujiang suture zone were interpreted to indicate the presence of oceanic islands, implying that the Bangong-Nujiang Tethyan Ocean was still open until 108 Ma (Fan et al., 2015). However, alternate models consider that the basaltic rocks may not correspond to the presence of oceanic crust (Zhu et al., 2016; Hu et al., 2017). The abundant late Jurassic to Early Cretaceous magmatic rocks in the central and northern Lhasa subterrane were interpreted to be the products of southward subduction of the Bangong-Nujiang Tethyan Ocean (Zhu et al., 2009a,b, 2011, 2013, 2016; Sui et al., 2013; Chen et al., 2014).

Chen et al. (2017) proposed that the Late Early Cretaceous (ca. 110 – 105 Ma) high-K calc-alkaline to shoshonitic volcanic rocks from the central and northern Lhasa subterranes were related to the recycled fluids derived from the subducted slab associated with the subduction of the Bangong-Nujiang Tethyan Ocean, athough they were formed in a syncollision setting. The absence of Early Cretaceous high-grade metamorphic rocks in the Lhasa-Qiangtang collision zone indicates that the Bangong-Nujiang ocean closed via divergent double-sided subduction resulting in soft collision of opposing arcs rather than continent-continent "hard" collision (Soesoo et al., 1997; Zhu et al., 2016). According to available stratigraphic, structural, metamorphic, and magmatic features of the region, Zhu et al. (2016) proposed subduction prior to 140 Ma, which continued into divergent double subduction ultimately leading to the closure of the Bangong-Nujiang ocean ca. 140 – 130 Ma. The cold and dense Bangong-Nujiang oceanic lithosphere below the "soft" arc-arc collision zone ruptured, with roll-back and detachment occurring through gravitational instability (130 – 120 Ma). Continued sinking of the Bangong-Nujiang oceanic lithosphere resulted in the rupture propagating from south to north and/or breakoff (120 – 110 Ma).

A synopsis of the inferences from the above models is as follows: ① The influence of subduction of Bangong-Nujiang Tethyan oceanic lithosphere occurred prior to 125–118 Ma, and continued to 110 Ma, which is supported by the abundant subduction-related melts and/or fluids before ca. 110 Ma in both northern Lhasa and southern Qiangtang terranes. ② The Lhasa-Qiangtang collision occurred from Late Jurassic to Early Cretaceous, but the Bangong-Nujiang Tethyan Ocean closed via divergent double-sided subduction resulting in soft collision of opposing arcs. Thus, two distinct types of magmatic events occurred during the closure of the Bangong-Nujiang Tethyan Ocean and collision between the Lhasa and Qiangtang terranes: One was related to the soft collision between the Lhasa and Qiangtang terranes, and another was associated with the rupture, roll-back, and slab breakoff of the subducting and sinking Bangong-Nujiang Tethyan oceanic lithosphere. The Dulongjiang granodiorites and associated enclaves represent the melts of recycled subducted sediments by mantle-like magmas in the mantle wedge (Fig. 12). This model is consistent with the paleogeographic and tectonic reconstruction of SE Asia during the Early Cretaceous (ca. 120 Ma; Metcalfe, 2006, 2013), including subduction and closure of the Bangong-Nujiang Tethyan Ocean (Fig. 12a).

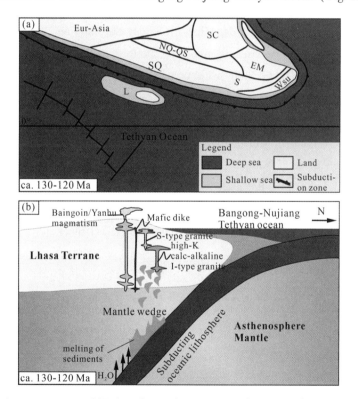

Fig. 12 Tectonic reconstruction of SE Asia during the Cretaceous (ca. 120 Ma) showing (a) formation of an extensive continental arc (after Metcalfe, 2006, 2013; J.X. Li et al., 2014, reference therein) and (b) geodynamic and petrogenetic models for the ca. 130–120 Ma magmatism in the northern Lhasa terrane.

SC: South China; NQ-QS: North Qiangtang-Qamdo-Simao; SQ: South Qiangtang;

S: Sibumasu; EM: East Malaya; Wsu: West Sumatra; L: Lhasa.

Conclusions

LA-ICP-MS zircon U-Pb geochronology indicates that the granodiorites and associated enclaves formed at ca.124. 1-123. 8 Ma and ca. 132. 7-129. 6 Ma.

The bulk-rock geochemistry, Sr and Nd isotopic compositions, mineral chemistry, and in-situ zircon Hf isotope compositions suggest that the granodiorites were formed by melting of recycled subducted sediments above the southward-dipping Bangong-Nujiang Tethyan Ocean slab by mantle-like magmas in the mantle wedge.

Our new data, integrated with those from published works from both sides and within Bangong-Nujiang suture zone, suggest that the Lhasa-Qiangtang collision might have started from the Late Jurassic to Early Cretaceous. However, the Bangong-Nujiang Tethyan Ocean closed via divergent double-sided subduction resulting in soft collision of opposing arcs. The influence of subduction of Bangong-Nujiang Tethyan oceanic lithosphere was still active prior to 125-118 Ma, and even to 110 Ma.

Acknowledgements　We thank Prof. Aaron J. Cavosie for his valuable editorial comments and Prof. Haibo Zou, Zhidan Zhao, Rendeng Shi, and one anonymous reviewer for constructive reviews, which significantly improved the focus and quality of this manuscript. We also thank Mike Fowler for his help with polishing the manuscript. This work was jointly supported by the National Natural Science Foundation of China (grant Nos. 41802054, 41421002, 41372067, and 41772052) and the program for Changjiang Scholars and Innovative Research Team in University (grant IRT1281). Support was also provided bythe Foundation for the Author of National Excellent Doctoral Dissertation of China (201324) and the MOST Special Fund from the State Key Laboratory of Continental Dynamics, Northwest University, and Province Key Laboratory Construction Item (08JZ62). We also thank Di-cheng Zhu for providing zircon U-Pb and Hf isotopic data from Lhasa Block.

References

Andersen, T., 2002. Correction of common lead in U-Pb analyses that do not report [204]Pb. Chemical Geology, 192(1-2), 59-79.

Behn, M.D., Kelemen, P.B., Hirth, G., Hacker, B.R., Massonne, H.J., 2011. Diapirs as the source of the sediment signature in arc lavas. Nature Geoscience, 4(9), 641-646.

Brueckner, H. H., 2009. Subduction of continental crust, the origin of post-orogenic granitoids (and anorthosites?) and the evolution of Fennoscandia. Journal of the Geological Society, London, 166, 753-762.

Cao, M.J., Qin, K.Z., Li, G.M., Li, J.X., Zhao, J.X., Evans, N.J., Hollings, P., 2016. Tectono-magmatic evolution of Late Jurassic to Early Cretaceous granitoids in the west central Lhasa subterrane, Tibet. Gondwana Research, 39, 386-400.

Castro, A., Martino, R., Vujovich, G., Otamendi, J., Pinotti, L., D'Eramo, F., et al., 2008. Top-down structures of mafic enclaves within the Valle Fértil magmatic complex (Early Ordovician, San Juan, Argentina). Geologica Acta, 6(3), 217-229.

Chappell, B.W., White, A.J.R., Wyborn, D., 1987. The importance of residual source material (restite) in granite petrogenesis. Journal of Petrology, 28, 1111-1138.

Chappell, B.W., White, A.J.R., Williams, I.S., Wyborn, D., Wyborn, L.A.I., 2000. Lachlan fold belt granites revisited: High- and low-temperature granites and their implications. Australian Journal of Earth Sciences, 47, 123-138.

Chen, B., Tian, W., Jahn, B.M., Chen, Z.C., 2008. Zircon SHRIMP U-Pb ages and in-situ Hf isotopic analysis for the Mesozoic intrusions in South Taihang, North China craton: Evidence for hybridization between mantle-derived magmas and crustal components. Lithos, 102, 118-137.

Chen, J.L., Xua, J.F., Yua, H.X., Wang, B.D., Wu, J.B., Feng, Y.X., 2015. Late Cretaceous high-Mg$^{\#}$ granitoids in southern Tibet: Implications for the early crustal thickening and tectonic evolution of the Tibetan Plateau? Lithos, 232, 12-22.

Chen, S.S., Shi, R.D., Yi, G.D., Zou, H.B., 2016a. Middle Triassic volcanic rocks in the northern Qiangtang (central Tibet): Geochronology, petrogenesis, and tectonic implications. Tectonophysics, 666, 90-102.

Chen, S.S., Shi, R.D., Fan, W.M., Zou, H.B., Liu, D.L., Huang, Q.S. Gong, X.H., Yi, G.D., Wu, K., 2016b. Middle Triassic ultrapotassic rhyolites from the Tanggula Pass, southern Qiangtang, China: A previously unrecognized stage of silicic magmatism. Lithos, 264, 258-276.

Chen, S.S., Shi, R.D., Gong, X.H., Liu, D.L., Huang, Q.S., Yi, G.D., Wu, K., Zou, H.B., 2017. A syn-collisional model for Early Cretaceous magmatism in the northern and central Lhasa subterranes. Gondwana Research, 41, 93-109.

Chen, Y., Zhu, D.C., Zhao, Z.D., Meng, F.Y., Wang, Q., Santosh, M., Wang, L.Q., Dong, G.C., Mo, X.X., 2014. Slab breakoff triggered ca. 113 Ma magmatism around Xainza area of the Lhasa terrane, Tibet. Gondwana Research, 26(2), 449-463.

Chiu, H.Y., Chung, S.L., Wu, F.Y., Liu, D.Y., Liang, Y.H., Lin, I.J., Lizuka, Y., Xie, L.W., Wang, Y.B., Chu, M.F., 2009. Zircon U-Pb and Hf isotopic constraints from eastern Transhimalayan batholiths on the precollisional magmatic and tectonic evolution in southern Tibet. Tectonophysics, 477, 3-19.

Chu, M.F., Chung, S.L., Song, B., Liu, D.Y., O'Reilly, S.Y., Pearson, N.J., Ji, J.Q., Wen, D.J., 2006. Zircon U-Pb and Hf isotope constraints on the Mesozoic tectonics and crustal evolution of southern Tibet. Geology, 34, 745-748.

Chu, Z.Y., Chen, F.K., Yang, Y.H., Guo, J.H., 2009. Precise determination of Sm, Nd concentrations and Nd isotopic compositions at the nanogram level in geological samples by thermal ionization mass spectrometry. Journal of Analytical Atomic Spectrometry, 24, 1534-1544.

Coltorti, M., Bonadiman, C., Faccini, B., Grégoire, M., O'Reilly, S.Y., Powell, W., 2007. Amphiboles from suprasubduction and intraplate lithospheric mantle. Lithos, 99(1-2), 68-84.

Cong, F., Lin, S.L., Zou, G.F., Li, Z.H., Xie, T., Peng, Z.M., Liang, T., 2011. Magma mixing of granites at Lianghe: In-situ zircon analysis for trace elements, U-Pb ages and Hf isotopes. Science China:

Earth Sciences, 54, 1346-1359.

Coulon, C., Maluski, H., Bollinger, C., Wang, S., 1986. Mesozoic and Cenozoic volcanic rocks from central and southern Tibet: ^{39}Ar-^{40}Ar dating, petrological characteristics and geodynamical significance. Earth & Planetary Science Letters, 79, 281-302.

Defant, M.J., Drummond, M.S., 1990. Derivation of some modern arc magmas by melting of young subducted lithosphere. Nature, 347(6294), 662-665.

Eby, G.N., 1992. Chemical subdivision of the A-type granitoids: Petrogenetic and tectonic implications. Geology, 20, 641-644.

Elliott, T., 2003. Tracers of the slab. In: Eiler, J. Inside the Subduction Factory. American Geophysical Union, Geophysical Monograph, 138, 23-45

Fan, J.J., Li, C., Xie, C.M., Wang, M., Chen, J.W., 2015. Petrology and U/Pb zircon geochronology of bimodal volcanic rocks from the Maierze Group, northern Tibet: Constraints on the timing of closure of the Banggong-Nujiang Ocean. Lithos, 227, 148-160.

Gao, S., Rudnick, R.L., Yuan, H.L., Liu, X.M., Liu, Y.S., Xu, W.L., Ling, W.L., Ayers, J., Wang, X.C., Wang, Q.H., 2004. Recycling lower continental crust in the North China craton. Nature, 432, 892-897.

Ginibre, C., Wörner, G., Kronz, A., 2002. Minor- and trace-element zoning in plagioclase: Implications for magma chamber processes at parinacota volcano, northern Chile. Contributions to Mineralogy and Petrology, 143(3), 300-315.

Griffin, W.L., Pearson, N.J., Belousova, E., Jackson, S.E., van Achterbergh, E., O'Reilly, S.Y., Shee, S.R., 2000. The Hf isotope composition of cratonic mantle: LAM-MC-ICP-MS analysis of zircon megacrysts in kimberlites. Geochimica et Cosmochimica Acta, 64, 133-147.

Guo, F., Fan, W., Li, C., Wang, C.Y., Li, H., Zhao, L., et al., 2014. Hf-Nd-O isotopic evidence for melting of recycled sediments beneath the Sulu orogen, North China. Chemical Geology, 381, 243-258.

Guo, L., Liu, Y., Liu, S., Cawood, P.A., Wang, Z., Liu, H., 2013. Petrogenesis of Early to Middle Jurassic granitoid rocks from the Gangdese belt, southern Tibet: Implications for early history of the Neo-Tethys. Lithos, 179(5), 320-333.

Guynn, J.H, Kapp, P., Pullen, A., et al., 2006. Tibetan basement rocks near Amdo reveal "missing" Mesozoic tectonism along the Bangong suture, central Tibet. Geology, 34, 505-508

Hanyu, T., Tatsumi, Y., Nakai, S., Chang, Q., Miyazaki, T., Sato, K., et al., 2006. Contribution of slab melting and slab dehydration to magmatism in the Japan arc for the last 25 Myr: Constraints from geochemistry. Geochemistry, Geophysics, Geosystems, 7(8), 1-29.

Harris, N.B.W., Inger, S., Xu, R., 1990. Cretaceous plutonism in Central Tibet: An example of post-collision magmatism? Journal of Volcanology and Geothermal Research, 44, 21-32.

Harris, N.B.W., Xu, R.H., Lewis, C.L., Jin, C., 1988. Plutonic rocks of the 1985 Tibet Geotraverse: Lhasa to Golmud. Philosophical Transactions of the Royal Society of London, A327, 145-168.

Hawthorne, F.C., Oberti, R., 2007. Classification of the Amphiboles. Reviews in Mineralogy and Geochemistry, 67, 55-88.

Hochstaedter, A.G., Kepezhinskas, P., Defant, M., Drummond, M., Koloskov, A., 1996. Insights into the volcanic arc mantle wedge from magnesian lavas from the Kamchatka arc. Journal of Geophysical Research, 101(B1), 697-712

Hofmann, A.W., 1988. Chemical differentiation of the Earth: The relationship between mantle, continental crust, and oceanic crust. Earth & Planetary Science Letters, 90, 297-314.

Hoskin, P. W. O., Schaltegger, U., 2003. The Composition of Zircon and Igneous and Metamorphic Petrogenesis. Reviews in Mineralogy and Geochemistry, 53, 27-62.

Hu, P.Y., Zhai, Q.G., Jahn, B.M., Wang, J., Li, C., Chung, S.L., et al., 2017. Late Early Cretaceous magmatic rocks (118 - 113 Ma) in the middle segment of the Bangong-Nujiang suture zone, Tibetan Plateau: Evidence of lithospheric delamination. Gondwana Research, 44, 116-138.

Johnson, M.C., Plank, T., 1999. Dehydration and melting experiments constrain the fate of subducted sediments. Geochemistry, Geophysics, Geosystems, 1(12), 597-597.

Kang, Z.Q., Xu, J.F., Wilde, S. A., Feng, Z.H., Chen, J.L., Wang, B.D., et al., 2014. Geochronology and geochemistry of the Sangri group volcanic rocks, southern Lhasa terrane: Implications for the early subduction history of the Neo-Tethys and Gangdese magmatic arc. Lithos, 200-201(1), 157-168.

Kapp, J.L.D., Harrison, T.M., Grove, M., Lovera, O.M., Lin, D., 2005a. Nyainqentanglha Shan: A window into the tectonic, thermal, and geochemical evolution of the Lhasa block, southern Tibet. Journal of Geophysical Research: Solid Earth (1978-2012), 110 (B8).

Kapp, P., Murphy, M.A., Yin, A., Harrison, T.M., Ding, L., Guo, J.H., 2003. Mesozoic and Cenozoic tectonic evolution of the Shiquanhe area of western Tibet. Tectonics, 22, 1029

Kapp, P., Peter, G.D., George, E., Gehrels., Matthew, H., Lin, D., 2007. Geological records of the Lhasa-Qiangtang and Indo-Asian collisions in the Nima area of central Tibet. Geological Society of America Bulletin, 119(7-8), 917-993.

Kemp, A.I.S., Hawkesworth, C.J., Foster, G.L., Paterson, B.A., Woodhead, J.D., Hergt, J.M., Whitehouse, M.J., 2007. Magmatic and crustal differentiation history of granitic rocks from Hf-O isotopes in zircon. Science, 315 (5814), 980-983.

Kincaid, C., Sacks, I.S., 1997. Thermal and dynamical evolution of the upper mantle in subduction zones. Journal of Geophysical Research: Solid Earth, 102(B6), 295-315.

Klein, M., Stosch, H.G., Seck, H.A., 1997. Partitioning of high field strength and rare-earth elements between amphibole, and quartz dioritic to tonalitic melts: An experimental study. Chemical Geology, 138, 257-271.

Leake, B.E., Woolley, A.R., Arps, C.E.S. et al., 1997. Nomenclature of amphiboles: Report of the subcommittee on amphiboles of the international mineralogical association, commission on new minerals and minerals names. Canadian Mineralogist, 35, 219-246.

Lee, H.Y., Chung, S.L., Lo, C.H., Ji, J., Lee, T.Y., Qian, Q., Zhang, Q., 2009. Eocene Neo-tethyan slab breakoff in southern Tibet inferred from the Linzizong volcanic record. Tectonophysics, 477, 20-35.

Lesher, C.E., 1990. Decoupling of chemical and isotopic exchange during magma mixing. Nature, 344, 235-237.

Li, J. X., Qin, K. Z., Li, G. M., Richards, J. P., Zhao, J. X., Cao, M. J., 2014. Geochronology, geochemistry, and zircon Hf isotopic compositions of Mesozoic intermediate-felsic intrusions in central

Tibet: Petrogenetic and tectonic implications. Lithos, 198-199, 77-91.

Li, S.M., Zhu, D.C., Wang, Q., et al., 2014. Northward subduction of Bangong-Nujiang Tethys: Insight from Late Jurassic intrusive rocks from Bangong Tso in western Tibet. Lithos, 205(9), 284-297.

Li, S.M., Zhu, D.C., Wang, Q., Zhao, Z., Zhang, L.L., Liu, S.A., et al., 2016. Slab-derived adakites and subslab asthenosphere-derived OIB-type rocks at 156 ± 2 Ma from the north of Gerze, central Tibet: Records of the Bangong-Nujiang oceanic ridge subduction during the Late Jurassic. Lithos, 262, 456-469.

Li, Y., He, J., Han, Z., Wang, C., Ma, P., Zhou, A., et al., 2016a. Late Jurassic sodium-rich adakitic intrusive rocks in the southern Qiangtang Terrane, central Tibet, and their implications for the Bangong-Nujiang ocean subduction. Lithos, 245, 34-46.

Li, Y., He, H., Wang, C., Wei, Y., Chen, X., He, J., et al., 2016b. Early Cretaceous (ca. 100 Ma) magmatism in the southern Qiangtang subterrane, central Tibet: Product of slab break-off? International Journal of Earth Sciences, 106(4), 1289-1310.

Liang, Y.H., Chung, S.L., Liu, D., Xu, Y., Wu, F.Y., Yang, J.H., Wang, Y., Lo, C.H., 2008. Detrital zircon evidence from Burma for reorganization of the eastern Himalayan river system. American Journal of Science, 308, 618-638.

Lin, I.J., Chung, S.L., Chu, C.H., Lee, H.Y., Gallet, S., Wu, G., Ji, J., Zhang, Y., 2013. Geochemical and Sr-Nd isotopic characteristics of Cretaceous to Paleocene granitoids and volcanic rocks, SE Tibet: Petrogenesis and tectonic implications. Journal of Asian Earth Sciences, 53, 131-150.

Ludwig, K.R., 2003. ISOPLOT 3.0: A geochronological toolkit for Microsoft Excel. Special Publication, 4, Berkeley Geochronology Center.

Maas, R., Nicholls, I.A., Legg, C., 1997. Igneous and metamorphic encalves in the S-type Deddick Granodiorite, Lachlan Fold Belt, SE Australia: Petrographic, geochemical and Nd-Sr isotopic evidence for crustal melting and magma mixing. Journal of Petrology, 38, 815-841.

Martin, H., Moyen, J.F., 2002. Secular changes in TTG composition as markers of the progressive cooling of the Earth. Geology, 30, 319-322.

Martin, H., Smithies, R.H., Rapp, R.P., Moyen, J.F., Champion, D.C., 2005. An overview of adakite, tonalite-trondhjemite-granodiorite (TTG) and sanukitoid: Relationships and some implications for crustal evolution. Lithos, 79, 1-24.

Martin, R.F., 2007. Amphiboles in the igneous environment. Reviews in Mineralogy and Geochemistry, 67(1), 323-358.

McLenan, S.M., Taylor, S.R., 1981. Role of subducted sediments in island-arc magmatism: Constraints from REE patterns. Earth & Planetary Science Letters, 54(3), 423-430.

Metcalfe, I., 2006. Paleozoic and Mesozoic tectonic evolution and palaeogeography of East Asian crustal fragments: The Korean Peninsula in context. Gondwana Research, 9, 24-46.

Metcalfe, I., 2013. Gondwana dispersion and Asian accretion: Tectonic and palaeogeographic evolution of eastern Tethys. Journal of Asian Earth Sciences, 66, 1-33.

Miller, C., Schuster, R., Kotzli, U., Frank, W., Purtscheller, F., 1999. Post-collisional potassic and ultrapotassic magmatism in SW Tibet: Geochemical and Sr-Nd-Pb-O isotopic constraints for mantle source characteristics and petrogenesis. Journal of Petrology, 40, 1399-1424.

Mo, X.X., Dong, G., Zhao, Z., Zhou, S., Wang, L., Qiu, R., Zhang, F., 2005. Spatial and temporal distribution and characteristics of granitoids in the Gangdese, Tibet and implication for crustal growth and evolution. Geology Journal of Chinese University, 11, 281-190.

Mo, X.X., Hou, Z., Niu, Y., Dong, G., Qu, X., Zhao, Z., Yang, Z., 2007. Mantle contributions to crustal thickening during continental collision: Evidence from Cenozoic igneous rocks in southern Tibet. Lithos, 96, 225-242.

Moyen, J.F., Martin, H., 2012. Forty years of TTG research. Lithos, 148(148), 312-336.

Niu, Y., Batiza, R., 1997. Trace element evidence from seamounts for recycled oceanic crust in the eastern equatorial Pacific mantle. Earth & Planetary Science Letters, 148, 471-484.

Nichols, G.T., Wyllie, P.J., Stern, C.R., 1996. Experimental melting of pelagic sediment, constraints relevant to subduction. In: Bebout, G.E., Scholl, E.W., Kirby, S.H., Platt, J.P. Subduction Top to Bottom. American Geophysical Union, Geophysical Monograph, 96, 293-298.

Niu, Y., Batiza, R., 1997. Trace element evidence from seamounts for recycled oceanic crust in the eastern equatorial Pacific mantle. Earth & Planetary Science Letters, 148, 471-483.

Nowell, G.M., Kempton, P.D., Noble, S.R., Fitton, J.G., Saunders, A.D., Mahoney, J.J., Taylor, R.N., 1998. High precision Hf isotope measurements of MORB and OIB by thermal ionisation mass spectrometry: Insights into the depleted mantle. Chemical Geology, 149, 211-233.

Pan, G., Ding, J., Yao, D., Wang, L., 2004. Guidebook of 1:1 500 000 geologic map of the Qinghai-Xizang (Tibet) plateau and adjacent areas. Chengdu: Chengdu Cartographic Publishing House, 48.

Pan, F.B., Zhang, H.F., Xu, W.C., Guo, L., Wang, S., Luo, B.J., 2014. U-Pb zircon chronology, geochemical and Sr-Nd isotopic composition of Mesozoic-Cenozoic granitoids in the SE Lhasa terrane: Petrogenesis and tectonic implications. Lithos, 192, 142-157.

Parsons, I., 2010. Feldspars defined and described: A pair of posters published by the Mineralogical Society. Sources and supporting information. Mineralogical Magazine, 74, 529-551.

Plank, T., 2005. Constraints from Thorium/Lanthanumon sediment recycling at subduction zones and the evolution of the continents. Journal of Petrology, 46(5), 921-944.

Plank, T., Langmuir, C.H., 1998. The chemical composition of subducting sediment and its consequences for the crust and mantle. Chemical Geology, 145, 325-394.

Qi, L., Hu, J., Gregoire, D.C., 2000. Determination of trace elements in granites by inductively coupled plasma mass spectrometry. Talanta, 51, 507-513.

Qi, X.X., Zhu, L.H., Hu, Z.C., Li, Z.Q., 2011. Zircon SHRIMP U-Pb dating and Lu-Hf isotopic composition for Early Cretaceous plutonic rocks in Tengchong block, southeastern Tibet, and its tectonic implications. Acta Petrologica Sinica, 27, 3409-3421 (in Chinese with English abstract).

Qin, J.F., Lai, S.C., Grapes, R., Diwu, C.R., Ju, Y.J., Li, Y.F., 2010. Origin of late Triassic high-Mg adakitic granitoid rocks from the Dongjiangkou area, Qinling orogen, central China: Implications for subduction of continental crust. Lithos, 120(3-4), 347-367.

Qu, X.M., Wang, R.J., Xin, H.B., Jiang, J.H., Chen, H., 2012. Age and petrogenesis of A-type granites in the middle segment of the Bangonghu-Nujiang suture, Tibetan plateau. Lithos, 146-147, 264-275.

Rapp, R.P., Shimizu, N., Norman, M.D., Applegate, G.S., 1999. Reaction between slab-derived melts and peridotite in the mantle wedge: Experimental constraints at 3. 8 GPa. Chemical Geology, 160, 335-356.

Rapp, R.P., Watson, E.B., 1995. Dehydration melting of metabasalt at 8-32 kbar: Implications for continental growth and crust-mantle recycling. Journal of Petrology, 36, 891-931.

Raterman, N.S., Robinson, A.C., Cowgill, E.S., 2014. Structure and detrital zircon geochronology of the Domar fold-thrust belt: Evidence of pre-Cenozoic crustal thickening of the western Tibetan Plateau. Geological Society of America Special Papers, 507, 89-114.

Rock, N.M.S, 1987. The nature and origin of lamprophyres: an overview. Geological Society of London Special Publications, 30(1) , 191-226.

Rudnick, R. L. , Barth, M. G., Horn, I., McDonough, W. F., 2000. Rutile-bearing refractory eclogites: Missing link between continents and depleted mantle. Science, 287, 278-281.

Rudnick, R.L., Gao, S., 2003. Composition of the continental crust. In: Holland, H.D., Turekian, K. K. Treatise on Geochemistry, 3. The Crust. Oxford: Elsevier-Pergamon, 1-64.

Santosh, M., Hu, C. N., He, X. F., Li, S. S., Tsunogae, T., Shaji, E., Indu, G., 2017. Neoproterozoic arc magmatism in the southern Madurai block, India: Subduction, relamination, continental outbuilding, and the growth of Gondwana. Gondwana Research, 45, 1-42.

Schmidt, M. W., 1992. Amphibole composition in tonalite as a function of pressure: An experimental calibration of the Al-in-hornblende barometer. Contributions to Mineralogy and Petrology, 110, 304-310.

Schmidt, M.W., Vielzeuf, D., Auzanneau, E., 2004. Melting and dissolution of subducting crust at high pressures: The key role of white mica. Earth & Planetary Science Letters, 228, 65-84.

Shi, R.D., Yang, J.S., Xu, Z.Q., Qi, X.X., 2008. The Bangong Lake ophiolite (NW Tibet) and its bearing on the tectonic evolution of the Bangong-Nujiang suture zone. Journal of Asian Earth Sciences, 32, 438-457.

Shi, R.D., Griffin, W.L., O'Reilly, S.Y., Huang, Q.S., Zhang, X.R., Liu, D.L., Zhi, X.C., Xia, Q.X., Ding, L., 2012. Melt/mantle mixing produces podiform chromite deposits in ophiolites: Implications of Re-Os systematics in the Dongqiao Neo-tethyan ophiolite, northern Tibet. Gondwana Research, 21, 194-206.

Shimoda, G., Tatsumi, Y., Nohda, S., Ishizaka, K., Jahn, B. M., 1998. Setouchi high-Mg andesites revisited: Geochemical evidence for melting of subducting sediments. Earth & Planetary Science Letters, 160(3-4) , 479-492.

Singer, B.S., Dungan, M. A., Layne, G.D., 1995. Textures and Sr, Ba, Mg, Fe, K, and Ti compositional profiles in volcanic plagioclase: Clues to the dynamics of calc-alkaline magma chambers. American Mineralogist, 80(7) , 776-798.

Soderlund, U., Patchett, P.J., Vervoort, J.D., Isachsen, C.E., 2004.The ^{176}Lu decay constantdetermined by Lu-Hf and U-Pb isotope systematics of Precambrian mafic intrusions. Earth & Planetary Science Letters, 219, 311-324.

Sparks, R.S.J., Marshall, L.A., 1986. Thermal and mechanical constraints on mixing between mafic and silicic magmas. Journal of Volcanology and Geothermal Research, 29, 99-129.

Stern, R.J., Kohut, E., Bloomer, S.H., Leybourne, M., Fouch, M., Vervoort, J., 2006. Subduction factory processes beneath the Guguan cross-chain, Mariana arc: No role for sediments, are serpentinites important? Contributions to Mineralogy and Petrology, 151(2) , 202-221.

Sui, Q. L., Wang, Q., Zhu, D. C., Zhao, Z. D., Chen, Y., Santosh, M., Hu, Z.C., Yuan, H.L., Mo, X.

X., 2013. Compositional diversity of ca. 110 Ma magmatism in the northern Lhasa Terrane, Tibet: Implications for the magmatic origin and crustal growth in a continent-continent collision zone. Lithos, 168-169(3), 144-159.

Sun, S.S., McDonough, W.F., 1989. Chemical and isotopic systematics of oceanic basalts: Implications or mantle composition and processes. Geological Society, London, Special Publications, 42(1), 313-345.

Tatsumi, Y., 2001. Geochemical modeling of partial melting of subducting sediments and subsequent melt-mantle interaction: Generation of high-Mg andesites in the Setouchi volcanic belt, southwest Japan. Geology, 29(4), 323-326.

Vernon, R.H., Etheridge, M.E., Wall, V.J., 1988. Shape and microstructure of microgranitoid enclaves: Indicators of magma mingling and flow. Lithos, 22, 1-11.

Volkmer, J.E., Kapp, P., Guynn, J.H., Lai, Q., 2007. Cretaceous-Tertiary structural evolution of the north central Lhasa terrane, Tibet. Tectonics 26, TC6007.

Wang, Q., Li, Z.X., Chung, S.L., Wyman, D.A., Sun, Y.L., Zhao, Z.H.,Zhu, Y.T., Qiu, H.N., 2011. Late Triassic high-Mg andesite/dacite suites from northern Hohxil, north Tibet: Geochronology, geochemical characteristics, petrogenetic processes and tectonic implications. Lithos, 126(1-2), 54-67.

Wen, D.R., Liu, D., Chung, S.L., Chu, M.F., Ji, J., Zhang, Q., Song, B., Lee, T.Y., Yeh, M.W., Lo, C.H., 2008. Zircon SHRIMP U-Pb ages of the Gangdese Batholith and implications for Neo-Tethyan subduction in southern Tibet. Chemical Geology, 252, 191-201.

White, A.J.R., Chappell, B.W., Wyborn, D., 1999. Application of the restite model to the Deddick granodiorite and its enclaves: A reinterpretation of the observations and data of Mass et al. (1997). Journal of Petrology, 40, 413-421.

Wu, Y.B., Zheng, Y.F., Zhao, Z.F., Gong, B., Liu, X., Wu., F.Y., 2006. U-Pb, Hf and O isotope evidence for two episodes of fluid-assisted zircon growth in marble-hosted eclogites from the Dabie Orogen. Geochimica et Cosmochimica Acta, 70, 3743-3761.

Xu, W., Hergt, J.M., Gao, S., Pei, F., Wang, W., Yang, D., 2008. Interaction of adakitic melt peridotite: Implications for the high-$Mg^{\#}$ signature of Mesozoic adakitic rocks in the eastern North China Craton. Earth & Planetary Science Letters, 265, (1-2), 123-137.

Xu, Y.G., Yang, Q.J., Lan, J.B., Luo, Z.Y., Huang, X.L., Shi, Y.R., Xie, L.W., 2012. Temporal-spatial distribution and tectonic implications of the batholiths in the Gaoligong-Tengliang-Yingjiang area, western Yunnan: Constraints from zircon U-Pb ages and Hf isotopes. Journal of Asian Earth Sciences, 53, 151-175.

Yang, J.H., Wu, F.Y., Chung, S.L., Wilde, S.A., Chu, M.F., 2004. Multiple sources for the origin of granites: Geochemical and Nd/Sr isotopic evidence from the Gudaoling granite and its mafic enclaves, northeast China. Geochimica et Cosmochimica Acta, 68(21), 4469-4483.

Yang, J.H., Wu, F.Y., Wilde, S.A., Xie, L.W., Yang, Y.H., Liu, X.M., 2007. Tracing magma mixing in granite genesis: In-situ U-Pb dating and Hf-isotope analysis of zircons. Contributions to Mineralogy and Petrology, 135, 177-190.

Yang, Q.J., Xu, Y.G., Huang, X.L., Luo, Z.Y., 2006. Geochronology and geochemistry granites in the Gaoligong tectonic belt, western Yunnan, tectonic implications. Acta Petrologica Sinica, 22, 817-834 (in

Chinese with English abstract).

Yogodzinski, G.M., Kay, R.W., Volynets, O.N., Koloskov, A.V. Kay, S.M., 1995. Magnesian andesite in the western Aleutian Komandorsky region: Implications for slab melting and processes in the mantle wedge. Geological Society of America Bulletin, 107, 505-519.

Yuan, H.L., Gao, S., Dai, M.N., Zong, C.L., Gunther, D., Fontaine, G.H., Liu, X.M., Diwu, C.R., 2008. Simultaneous determinations of U-Pb age, Hf isotopes and trace element compositions of zircon by excimer laser-ablation quadrupole and multiple-collector ICP-MS. Chemical Geology, 247, 100-118.

Yuan, H.L., Gao, S., Liu, X.M., Li, H.M., Gunther, D., Wu, F.Y., 2004. Accurate U-Pb age and trace element determinations of zircon by laser ablation-inductively coupled plasma mass spectrometry. Geostandard Newsletters, 28, 353-370.

Yunnan Bureau Geological Mineral Resource (YBGMR.), 1991. Regional geology of Yunnan Province. Beijing: Geological Publishing House, 1-729 (in Chinese with English abstract).

Zhu, D.C., Li, S. M., Cawood, P. A., Wang, Q., Zhao, Z. D., Liu, S. A., et al., 2016. Assembly of the Lhasa and Qiangtang terranes in central Tibet by divergent double subduction. Lithos, 245, 7-17.

Zhu, D.C., Mo, X.X., Wang L.Q., Zhao, Z.D., Niu, Y.L., Zhou, C.Y., Yang, Y.H., 2009a. Petrogenesis of highly fractionated I-type granites in the Zayu area of eastern Gangdese, Tibet: Constraints from zircon U-Pb geochronology, geochemistry and Sr-Nd-Hf isotopes. Science in China: Series D, Earth Sciences, 52 (9), 1223-1239.

Zhu, D.C., Mo, X.X., Niu, Y.L., Zhao, Z.D., Wang L.Q., Liu Y.S., Wu F.Y., 2009b. Geochemical investigation of Early Cretaceous igneous rocks along an east-west traverse throughout the central Lhasa Terrane, Tibet. Chemical Geology, 268, 298-312.

Zhu, D.C., Pan, G., Chung, S.L., Liao, Z., Wang, L., Li, G., 2008. SHRIMP zircon age and geochemical constraints on the origin of lower Jurassic volcanic rocks from the Yeba Formation, southern Gangdese, south Tibet. Internatinal Geology Review, 50, 442-471.

Zhu, D.C., Zhao, Z.D., Niu, Y.L., Dilek, Y., Hou, Z.Q., Mo, X.X., 2013. The origin and pre-Cenozoic evolution of the Tibetan Plateau. Gondwana Research, 23, 1429-1454.

Zhu, D.C., Zhao, Z.D., Niu, Y.L., Mo, X.X., Chung, S.L., Hou, Z.Q., Wang, L.Q., Wu, F.Y., 2011. The Lhasa Terrane: Record of a microcontinent and its histories of drift and growth. Earth & Planetary Science Letters, 301, 241-255.

Zhu, R.Z., Lai, S.C., Qin, J.F., Zhao, S.W., 2015. Early-cretaceous highly fractionated I-type granites from the northern Tengchong block, western Yunnan, SW China: Petrogenesis and tectonic implications. Journal of Asian Earth Sciences, 100, 145-163.

Late Triassic biotite monzogranite from the western Litang area, Yidun terrane, SW China: Petrogenesis and tectonic implications[①]

Zhu Yu Lai Shaocong[②] Qin Jiangfeng Zhang Zezhong Zhang Fangyi

Abstract: The Late Triassic igneous rocks in the Yidun terrane can provide vital insights into the evolution of Plaeo-Tethys in western China. We present new zircon U-Pb, whole-rock geochemistry, and Sr-Nd-Pb-Hf isotopic data for the Litang biotite monzogranites, Yidun terrane. The biotite monzogranites have zircon U-Pb age of 206.1 ± 1.0 Ma ($MSWD = 1.9$, $n = 30$), which indicates Late Triassic magmatism. The biotite monzogranites display I-type affinity, high Na_2O ($3.38-3.60$ wt%) contents, medium SiO_2 ($67.12-69.13$ wt%), and low P_2O_5 contents ($0.10-0.12$ wt%). They are enriched in Rb, Th, and Ba and depleted in Nb and Ta, with negative Eu anomalies ($Eu/Eu^* = 0.74-0.81$). They have evolved Sr-Nd-Pb-Hf isotopic composition, i.e., $(^{87}Sr/^{86}Sr)_i = 0.714225-0.714763$, negative $\varepsilon_{Nd}(t)$ values of -2.6 to -2.0 with two-stage Nd model ages ranging from 1.01 Ga to 1.05 Ga, negative $\varepsilon_{Hf}(t)$ values of -4.1 to -3.4 with two-stage Hf model ages of $1.85-1.88$ Ga, suggesting a matured crustal sources. Their low Al_2O_3/TiO_2 ratios and medium CaO/Na_2O ratios, medium $Mg^{\#}$ and SiO_2 contents, low [molar $Al_2O_3/(MgO + FeO^T)$] values, and high [molar $CaO/(MgO + FeO^T)$] values indicate that the Litang biotite monzogranite was formed by partial melting of metabasaltic rocks. Based on the previous studies, we propose that the Litang biotite monzogranite was derived from the westward subduction and closure of the Ganzi-Litang ocean during the Late Triassic. The mantle wedge-derived mafic melts provided sufficient heat for partial melting of ancient metabasalt protolith within the middle-lower crust.

1 Introduction

The eastern Tibetan Plateau locates in the junction zone among Songpan-Ganzi, Yidun, Qiangtang and eastern Lhasa terranes (Peng et al., 2014). The Yidun terrane locates between the Qiangtang and the Songpan-Ganzi blocks (Fig. 1b), which attracted wide interest. Voluminous Triassic arc-related granitic igneous rocks in the eastern Yidun terrane were considered to be related to the closure of Paleo-Tethys in this region (Roger et al., 2004; Zhang et al., 2006, 2007; Xiao et al., 2007; Zhao Yongjiu, 2007; Yuan et al., 2010).

---

① Published in *Acta Geologica Sinica* (*English Edition*), 2019, 93(2).

② Corresponding author.

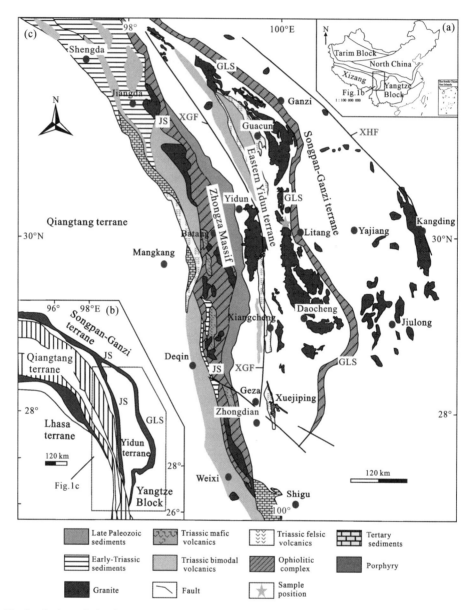

Fig. 1   Geological sketch map of the Yidun terrane and strata distribution (after Peng et al., 2014)
(China basemap is from China National Bureau of Surveying and Mapping Geographical Information).
CAOB: the central Asia orogenic belt; SGT: Songpan-Ganzi terrane; GLS: Ganzi-Litang suture; JS: Jinsha suture;
BNS: Bangong-Nujiang suture; XGF: Xiangcheng-Geza fault; XHF: Xianshuihe fault.

These NNW-trending granites distributed more than 300 km (Fig. 1c), most occur in the southern to middle segment of the belt (Qu et al., 2002).

The NNW-trending Yidun arc is composed of Triassic calc-alkaline volcanic rocks with intercalated flysch deposits, and overlie variably metamorphosed Paleozoic rocks (Reid et al., 2007). Two Triassic deformation events have been identified in this area (Reid et al., 2005).

The Early Triassic deformation was associated with the closure of the Jinsha Paleo-Tethyan Ocean and the late-phase deformation likely was a consequence of closure of the Ganzi-Litang Palaeo-Tethyan ocean (Reid et al., 2005). The volcanic rocks in back-arc basins in the northern and the middle part of the eastern Yidun terrane provide a great deal of information to elucidate the interactions between the subduction slab and mantle wedge (Floyd et al., 1991; Todd et al., 2011). Voluminous Late Triassic porphyry- or skarn-type Cu-polymetallic ore deposits occur in the southern part of Yidun terrane (Hou Zengqian et al., 2001, 2003; Wang et al., 2011; Leng et al., 2012; He et al., 2013; Peng et al., 2014), such as the Zhongdian and Pulang porphyry Cu ore deposits (Leng et al., 2012). Late Triassic bimodal volcanic rocks host multiple sulfide deposits (Hou Zengqian et al., 2003), including the Gacun and Luchun volcanogenic massive sulfide (VMS) ore deposits (Hou Zengqian et al., 2001, 2003), were also discovered.

The Ganzi-Litang suture locates in the eastern Tibetan Plateau, it is still unclear that it is an in-stiu suture zone marking relict of a subducted Paleo-Tethyan oceanic crust or a failed intracontinental rift (Li et al., 2017). Triassic granites recorded the information about the subduction of Paleo-Tethyan oceanic lithosphere (BGMRSP, 1991; Hou, 1993; Hou Zengqian, 2004). Previous researches proposed that the arc volcanic rocks and granites formed during the westward subduction of Ganzi-Litang ocean in Late Triassic (Hou Zengqian and Mo Xuanxue, 1991; Hou, 1993). Available ages for the granites of the Yidun terrane show that there was more than one phase of granite intrusion (Hou Zengqian et al., 2001). However, it has always been a matter whether these granites were genetically related to Triassic westward or eastward subduction of the Yidun terrane (Hou Zengqian, 1993; Roger et al., 2010). The coetaneous I-type granites in the middle part of the Litang area could reveal significant geological significance. We aim to clarify the geochemical characteristics and relationships of these granites in different tectonic stages. This contribution presents new geochemical data, zircon U-Pb age data from laser ablation inductively coupled plasma mass spectrometry (LA-ICP-MS), and Sr-Nd-Pb-Hf isotopic compositions of the biotite monzogranite from the western Litang area. The study may provide reasonable and serviceable information about the tectonic stage of the Yidun terrane.

2 Regional geology

2.1 Geological background and field geology

The Yidun terrane is located between the Qiangtang and the Songpan-Ganzi terranes, separated by two Paleo-Tethyan oceanic subduction zones: The Jinsha suture to the west and the Ganzi-Litang suture to the east (Roger et al., 2008, 2010; Wang et al., 2011). It adjoins the Yangtze block, separated by the Longmenshan-Jinhe fault. The Songpan-Ganzi Triassic turbidite complex is one of the largest flysch turbiditic basins on Earth, which contains a succession with an average thickness of up to 10 km accumulated during Ladinian through Norian times (~230–203 Ma) (Zhang et al., 2014). Traditionally, the Songpan-Ganzi terrane has been believed to be underlain by an oceanic basement (Roger et al., 2003), however, Zhang (2001) proposed that it could be underlain by a South China-type Precambrian continental basement. The Ganzi-Litang suture in the eastern Tibetan Plateau, trending NNW-SSE, separates the Songpan-Ganzi terrane to the northeast and the Yidun terrane to the southwest (Li et al., 2017). Some of the mafic fragments in Ganzi-Litang suture were identified with affinity to the mid-ocean-ridge basalts (MORB) and thus the Ganzi-Litang suture was believed to be a Palaeo-Tethyan suture zone (Yao Xueliang and Lan Yan, 2001; Li et al., 2017). In addition, the Ganzi-Litang suture merges into the Jinsha suture toward the northern end of the Yidun terrane, but the southern segment of the suture is poorly preserved (Yin and Harrison, 2000; Li et al., 2011; Wang et al., 2013a). The Jinsha suture represents the remnants of Palaeo-Tethyan branches that closed approximately during the Late Triassic to the earliest Jurassic (Zhang et al., 2014). The Jinsha and Ganzi-Litang oceans were consumed by Permian-Triassic subductions (Mo Xuanxue et al., 1994; Chang Chengfa, 1996). The collision between the Yidun terrane and the Qiangtang Block was occurred in the Early-Middle Triassic (Zhu et al., 2011).

The strata of the Yidun terrane vary across the north-south-trending Xiangcheng-Geza fault. The oldest rocks within the Yidun terrane are palaeozoic metasedimentary rocks in the Zhongza Massif (Leng et al., 2014; Li et al., 2017). The western part of the Yidun terrane, also known as the Zhongza Massif, is composed of weakly metamorphosed Paleozoic carbonate, clastic rocks and minor mafic volcanic rocks, with a Neoproterozoic basement composed of granitic gneisses and meta-volcanic rocks (BGMRSP, 1991; Chang et al., 2000; Reid et al., 2005). The Zhongza Massif is stratigraphically similar to the Yangtze Block (BGMRSP, 1991) and was previously considered to be a microcontinent that rifted from the western Yangtze Block during the opening of the Ganzi-Litang ocean in the Late Permian (Hou Zengqian, 1993). The eastern Yidun terrane consists mainly of Middle-Upper Triassic volcanic rocks intercalated with flysch deposits which were deformed and metamorphosed at a low metamorphic grade (BGMRSP, 1991; Hou, 1993; Reid et al., 2005). In detail, the northern part of the Yidun terrane (the "Changtai" arc) is dominated by Upper Triassic volcanic rocks, which are bimodal and include both major rhyolites and minor interlayered basalts (Hou Zengqian et al., 2003). The southern part of the Yidun terrane (the "Zhongdian" arc) is composed on a small quantity of porphyries, felsic volcanics, bimodal volcanics, and granites (Ren Jiangbo et al., 2011) (Fig. 1c). Additionally, small volumes of Middle-Late Triassic intermediate-felsic porphyries intruded into the Upper Triassic Tumugou Group composed of volcanic and sedimentary rocks (Wang et al., 2011; Leng et al., 2012; Peng et al., 2014).

The Triassic successions were extensively deformed by a single generation of upright folding and thrust faulting (Reid et al., 2005). From the base upward, the Yidun Group is subdivided into the Lieyi, Qugasi, Tumugou and Lanashan Formations (BGMRSP, 1991; Wang et al., 2013a). The entire Lanashan Formation consists predominantly of black or gray slate and sandstone, whereas the Qugasi and Tumugou Formations have variable amounts of mafic to felsic volcanic rocks and tuffs accompanied by gray slate and sandstone (Wang et al., 2013b). Voluminous Late Triassic granitic and granodioritic plutons intruded deformed Paleozoic and Middle-Late Triassic volcano-sedimentary sequences across the Yidun terrane, such as the immense Dongcuo and Shengmu batholiths (Peng et al., 2014). Granitic plutons were emplaced into the sedimentary pile during and after Indosinian deformation across the Yidun terrane. Since the collision of India with Asia, the Yidun terrane has been incorporated into the eastern Tibetan Plateau (Reid et al., 2007).

2.2 Field geology and petrography

Six biotite monzogranite samples were collected in the western Litang area (Fig. 2). Field outcrops show that these rocks are medium- to fine-grained, and consist mainly of quartz (20–25 vol%), plagioclase (25–35 vol%), K-feldspar (20–30 vol%), and biotite (5–10 vol%), accessory minerals include zircon, apatite, and magnetite. The quartz crystals are xenomorphic to granular. The K-feldspar crystals are subhedral. The plagioclase crystals are subhedral platy (> 2 mm) and exhibit well-developed polysynthetic twinning. Some plagioclase (andesine) crystals show apparent zonal textures. The biotite consists of reddish brown euhedral to subhedral crystal (< 2 mm).

Fig. 2   Field photograph (a) and microscope photograph (b) of the biotite monzogranite from the western Litang area.

Kfs: potassium feldspar; Qtz: quartz; Bi: biotite; Pl: plagioclase.

3 Analytical methods

In this study, major and trace elemental geochemistry, whole-rock Sr-Nd-Pb-Hf isotopic geochemistry, and zircon geochronology were analyzed at the State Key Laboratory of Continental Dynamics, Northwest University, Xi'an, China.

3. 1 Zircon U-Pb dating

Zircon grains were separated from sample using conventional heavy liquid and magnetic techniques. Representative zircon grains were handpicked, mounted in epoxy resin disks, polished, and carbon- coated. Internal morphology was examined with cathodoluminescence (CL) microscopy prior to U-Pb isotopic dating. Zircon LA-ICP-MS U-Pb analyses were conducted with an Agilent 7500a ICP-MS equipped with a 193-nm laser following the method of Yuan et al. (2004). The $^{207}Pb/^{235}U$ and $^{206}Pb/^{238}U$ ratios were calculated using the GLITTER program and corrected using the Harvard zircon 91500 standard for external calibration. Common Pb contents were subsequently valuated using the method described by Andersen (2002). Ages were calculated and concordia diagrams were plotted using ISOPLOT (version 3. 0; Ludwig, 2003).

3. 2 Major and trace elements

The weathered surfaces of the samples were removed, and the fresh parts were then chipped and powdered to about a 200 mesh size using a tungsten carbide ball mill. Major and trace elements were analyzed using X-ray fluorescence (XRF; Rikagu RIX 2100) and inductively coupled plasma mass spectrometry (ICP-MS; Agilent 7500a), respectively. Analyses of United States Geological Survey and Chinese national rock standards (BCR-2, GSR-1, and GSR-3) showed that the analytical precision and accuracy for major elements were generally better than 5%. For trace element analyses, sample powders were digested using a $HF + HNO_3$ mixture in high-pressure Teflon bombs at 190 ℃ for 48 h. For most trace elements, analytical error was less than 2% and the precision was greater than 10% (Liu Ye et al., 2007).

3. 3 Whole-rock Sr-Nd-Pb-Hf isotopes

Whole-rock Sr-Nd-Pb isotopic data were obtained using a Nu Plasma HR multi-collector (MC) mass spectrometer. Sr and Nd isotopic fractionations were corrected to $^{87}Sr/^{86}Sr = 0.119\ 4$ and $^{146}Nd/^{144}Nd = 0.721\ 9$, respectively. During the analysis period, the NIST SRM 987 standard gave an average value of $^{87}Sr/^{86}Sr = 0.710\ 250 \pm 12$ (2σ, $n = 15$) and the La Jolla standard gave an average value of $^{146}Nd/^{144}Nd = 0.511\ 859 \pm 6$ (2σ, $n = 20$). Whole-rock

Pb was separated through anion exchange in HCl-Br columns, and Pb isotopic fractionation was corrected to $^{205}Tl/^{203}Tl = 2.3875$. Within the period of analysis, 30 measurements of NBS981 yielded average values of $^{206}Pb/^{204}Pb = 16.937 \pm 1$ (2σ), $^{207}Pb/^{204}Pb = 15.491 \pm 1$ (2σ), and $^{208}Pb/^{204}Pb = 36.696 \pm 1$ (2σ). BCR-2 standard yielded average values of $^{206}Pb/^{204}Pb = 18.742 \pm 1$ (2σ), $^{207}Pb/^{204}Pb = 15.620 \pm 1$ (2σ), and $^{208}Pb/^{204}Pb = 38.705 \pm 1$ (2σ) (Yuan et al., 2008). Total procedural Pb blanks were in the range 0.1−0.3 ng. Whole-rock Hf was also separated using single-anion exchange columns. In the course of the analysis, 22 measurements of JCM 475 yielded an average of $^{176}Hf/^{177}Hf = 0.2821613 \pm 0.0000013$ (2σ) (Yuan Honglin et al., 2007).

4　Results

4.1　Zircon LA-ICP-MS U-Pb age

Zircon U-Pb concordia diagrams and CL images of the biotite monzogranite (LTW-05) from the western Litang area are presented in Fig. 3, and the results of the analyses are listed in Table 1.

Zircons from the biotite monzogranite (LTW-05) are fawn to colorless, euhedral, long prismatic crystals (200−300 μm in length), with aspect ratios ranging from 2:1 to 3:1. Most of the grains are gray and display well-developed oscillatory zoning in CL images (Fig. 3a). Spots #7, #18, #22, #29 and #30 yielded relatively younger $^{206}Pb/^{238}U$ ages (186 ± 2 Ma to 198 ± 2 Ma), which are attributed to Pb loss. The other 30 spots have U contents from 229 ppm to 980 ppm, Th contents from 100 ppm to 695 ppm, and Th/U ratios from 0.35 to 0.78. These 30 spots have concordant $^{206}Pb/^{238}U$ ages from 202 ± 2 Ma to 213 ± 2 Ma (weighted mean age of 206.1 ± 1.0 Ma, MSWD = 1.9, $n = 30$), which represent the crystallization age of the biotite monzogranite.

4.2　Major and trace element chemistry

Analytical data for major (wt%) and trace element (ppm) analyses of the biotite monzogranite samples are listed in Table 2.

The whole-rock geochemical results show $SiO_2 = 67.12-69.13$ wt%, $K_2O = 2.54-2.95$ wt%, $Na_2O = 3.38-3.60$ wt%, and $CaO = 3.56-3.71$ wt%. These rocks belong to the calc-alkaline series (Fig. 4a). The TiO_2, Al_2O_3, and MgO contents of these samples are 0.44−0.50 wt%, 14.94−15.96 wt%, and 1.02−1.17 wt%, respectively. These rocks are metaluminous with A/CNK values of 0.99−1.01 (Fig. 4b). The samples have $Fe_2O_3^T$ contents from 3.82 wt% to 4.81 wt%, low P_2O_5 contents from 0.10 wt% to 0.12 wt%.

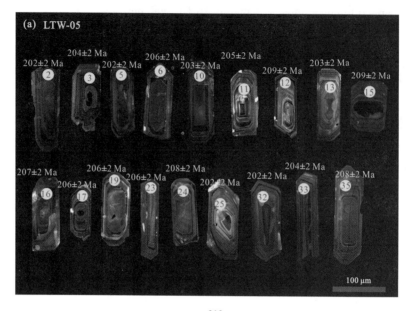

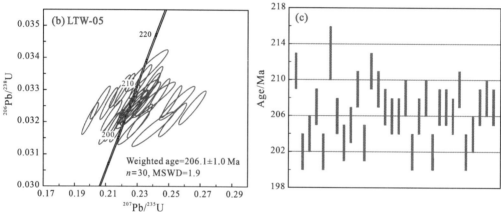

Fig. 3　Representative zircons cathodoluminescene (CL) (a) images and zircon U-Pb concordia diagrams (b,c) of the biotite monzogranite from the western Litang area.

The total REE ($\sum$ REE) values of the biotite monzogranite samples range from about 122.9 ppm to 157.7 ppm. The LREE/HREE ratios are within the range of 8.1–11.0. These samples have relatively low Rb (102–107 ppm) and high Sr (185–197 ppm) contents, with low Rb/Sr ratios of 0.53–0.57. In the chondrite-normalized REE patterns diagrams (Fig. 5a), the biotite monzogranites enriched LREEs, and flat HREEs. They have variable $(La/Yb)_N$ and $(Gd/Yb)_N$ ratios of 9.73–14.00 and 1.59–1.74, respectively, with insignificant negative Eu anomalies ($Eu/Eu^* = 0.74 - 0.81$). The primitive mantle-normalized spider diagrams (Fig. 5b) show that the biotite monzogranite samples have the enrichment of Rb, Th, Pb, the depletion in Nb, Ta, Sr, and Ba.

Table 1　Results of zircon LA-ICP-MS U-Pb ages for the biotite monzogranite from the western Litang area.

Analysis	Content/ppm			Ratios								Age/Ma							
	Th	U	Th/U	$^{207}Pb/^{206}Pb$	1σ	$^{207}Pb/^{235}U$	1σ	$^{206}Pb/^{238}U$	1σ	$^{208}Pb/^{232}Th$	1σ	$^{207}Pb/^{206}Pb$	1σ	$^{207}Pb/^{235}U$	1σ	$^{206}Pb/^{238}U$	1σ	$^{208}Pb/^{232}Th$	1σ
LTW-05-01	376	725	0.52	0.052 14	0.001 37	0.239 08	0.004 67	0.033 22	0.000 34	0.010 31	0.000 13	292	26	218	4	211	2	207	3
LTW-05-02	184	529	0.35	0.051 89	0.001 64	0.227 64	0.006 76	0.031 81	0.000 34	0.010 00	0.000 09	281	74	208	6	202	2	201	2
LTW-05-03	440	980	0.45	0.048 09	0.001 28	0.213 51	0.004 26	0.321 80	0.000 33	0.009 26	0.000 12	104	28	196	4	204	2	186	2
LTW-05-04	175	391	0.45	0.053 27	0.002 12	0.240 13	0.009 17	0.032 69	0.000 37	0.010 24	0.000 10	340	92	219	8	207	2	206	2
LTW-05-05	334	612	0.55	0.054 74	0.002 10	0.239 85	0.008 79	0.031 78	0.000 36	0.009 92	0.000 09	402	88	218	7	202	2	200	2
LTW-05-06	100	229	0.44	0.050 88	0.002 16	0.235 81	0.008 99	0.033 60	0.000 43	0.011 79	0.000 26	235	64	215	7	213	3	237	5
LTW-05-09	258	546	0.47	0.051 03	0.001 64	0.228 35	0.006 08	0.032 44	0.000 36	0.010 09	0.000 17	242	41	209	5	206	2	203	3
LTW-05-10	288	640	0.45	0.052 42	0.001 53	0.231 56	0.005 40	0.032 03	0.000 35	0.009 99	0.000 15	304	33	211	4	203	2	201	3
LTW-05-11	391	879	0.44	0.050 94	0.001 65	0.227 00	0.006 92	0.032 32	0.000 35	0.010 18	0.000 09	238	76	208	6	205	2	205	3
LTW-05-12	243	566	0.43	0.048 14	0.001 28	0.218 74	0.004 43	0.032 95	0.000 34	0.009 83	0.000 13	106	29	201	4	209	2	198	3
LTW-05-13	238	394	0.60	0.057 13	0.001 66	0.252 49	0.005 85	0.032 07	0.000 35	0.010 20	0.000 14	497	32	229	5	203	2	205	3
LTW-05-14	251	560	0.45	0.050 63	0.001 36	0.232 06	0.004 78	0.033 26	0.000 35	0.009 95	0.000 13	224	28	212	4	211	2	200	3
LTW-05-15	275	667	0.41	0.052 26	0.001 34	0.237 72	0.004 55	0.033 01	0.000 34	0.011 07	0.000 14	297	25	217	4	209	2	223	3
LTW-05-16	695	889	0.78	0.056 00	0.001 34	0.251 82	0.004 26	0.032 64	0.000 33	0.010 23	0.000 10	452	20	228	3	207	2	206	2
LTW-05-17	279	635	0.44	0.050 47	0.001 57	0.225 80	0.005 80	0.032 47	0.000 36	0.011 01	0.000 17	217	39	207	5	206	2	221	3
LTW-05-19	315	687	0.46	0.050 21	0.001 20	0.224 50	0.003 82	0.032 46	0.000 33	0.010 20	0.000 11	205	21	206	3	206	2	205	2
LTW-05-20	238	487	0.49	0.051 23	0.001 85	0.231 41	0.007 95	0.032 76	0.000 37	0.010 31	0.000 09	251	85	211	7	208	2	207	2
LTW-05-21	239	522	0.46	0.048 50	0.001 29	0.212 36	0.004 36	0.031 8	0.000 33	0.009 59	0.000 13	124	29	196	4	202	2	193	3
LTW-05-23	278	652	0.43	0.050 81	0.001 44	0.227 64	0.005 99	0.032 49	0.000 34	0.010 24	0.000 09	232	67	208	5	206	2	206	2
LTW-05-24	236	571	0.41	0.053 92	0.001 36	0.243 72	0.004 57	0.032 84	0.000 34	0.011 41	0.000 14	368	24	221	4	208	2	229	3
LTW-05-25	244	497	0.49	0.049 03	0.001 21	0.215 16	0.003 95	0.031 89	0.000 33	0.009 48	0.000 11	149	24	198	3	202	2	191	2
LTW-05-26	336	779	0.43	0.047 28	0.001 11	0.211 84	0.003 56	0.032 57	0.000 33	0.009 47	0.000 11	63	22	195	3	207	2	191	2
LTW-05-27	226	436	0.52	0.050 08	0.001 29	0.224 89	0.004 43	0.032 64	0.000 34	0.009 98	0.000 12	199	27	206	4	207	2	201	2
LTW-05-28	117	399	0.29	0.059 36	0.001 62	0.264 94	0.005 72	0.032 45	0.000 35	0.013 84	0.000 21	580	28	239	5	206	2	278	4
LTW-05-31	271	632	0.43	0.048 86	0.001 26	0.221 69	0.004 39	0.033 00	0.000 34	0.010 32	0.000 13	141	27	203	4	209	2	208	3
LTW-05-32	381	838	0.45	0.049 80	0.001 19	0.218 22	0.003 79	0.031 87	0.000 33	0.009 60	0.000 11	186	22	200	3	202	2	193	2
LTW-05-33	491	858	0.57	0.045 38	0.001 09	0.200 20	0.003 56	0.032 10	0.000 33	0.008 77	0.000 10	185	22	185	3	204	2	176	2

Continued

Analysis	Content/ppm			Ratios								Age/Ma							
	Th	U	Th/U	$\frac{207Pb}{206Pb}$	1σ	$\frac{207Pb}{235U}$	1σ	$\frac{206Pb}{238U}$	1σ	$\frac{208Pb}{232Th}$	1σ	$\frac{207Pb}{206Pb}$	1σ	$\frac{207Pb}{235U}$	1σ	$\frac{206Pb}{238U}$	1σ	$\frac{208Pb}{232Th}$	1σ
LTW-05-34	424	620	0.68	0.056 83	0.001 77	0.255 35	0.007 46	0.032 59	0.000 35	0.010 13	0.000 09	485	70	231	6	207	2	204	2
LTW-05-35	288	653	0.44	0.052 66	0.001 46	0.236 82	0.005 28	0.032 72	0.000 35	0.011	0.000 16	314	31	216	4	208	2	221	3
LTW-05-36	268	645	0.41	0.046 1	0.001 19	0.206 51	0.004 14	0.032 6	0.000 34	0.009 7	0.000 13	3	24	191	3	207	2	195	3

Table 2 Analytical results of major (wt%) and trace (ppm) elements of the biotite monzogranite from the western Litang area.

Sample	SiO_2	TiO_2	Al_2O_3	$Fe_2O_3^T$	MgO	CaO	Na_2O	K_2O	P_2O_5	MnO	LOI	Total	A/CNK	$Mg^\#$
LTW-01	69.12	0.45	14.94	3.86	1.06	3.59	3.38	2.62	0.1	0.06	0.43	99.61	1	44
LTW-03	68.42	0.44	15.59	3.85	1.02	3.7	3.56	2.73	0.1	0.06	0.45	99.92	1	45
LTW-04	68.85	0.46	15.23	4.01	1.07	3.56	3.57	2.84	0.11	0.06	0.43	100.19	0.99	21
LTW-08	67.12	0.5	15.96	4.26	1.14	3.71	3.6	2.95	0.12	0.07	0.48	99.91	1.01	19
LTW-09	69.13	0.45	14.98	3.82	1.03	3.6	3.41	2.54	0.1	0.06	0.42	99.54	1.01	21
LTW-10	68.9	0.49	15.15	4.31	1.17	3.68	3.39	2.81	0.11	0.07	0.39	100.47	0.99	20

Sample	Li	Be	Sc	V	Cr	Co	Ni	Cu	Zn	Ga	Ge	Rb	Sr	Y	Zr	Nb	Cs	Ba
LTW-01	26.9	2.05	10.8	44.2	9.53	106	5.31	2.94	51.9	17.6	1.19	106	185	22.1	168	11.1	2.62	624
LTW-03	26.7	2.16	10.9	43.4	8.54	90	4.15	3.22	52.3	18.2	1.23	106	192	23	193	10.9	2.58	622
LTW-04	28.4	1.88	9	45.9	10.4	117	5.03	2.95	55.5	17.9	1.2	102	188	20.1	195	11.2	2.69	769
LTW-08	29.2	1.98	8.92	46.6	9.67	76.5	4.84	2.78	55.4	18.7	1.21	105	197	20.4	189	11.7	3.1	792
LTW-09	27	2.08	10.2	44.1	10.5	124	5.36	3.06	51.6	17.6	1.2	104	186	21.7	181	11	2.57	603
LTW-10	29.3	1.89	10.7	49.7	10.9	102	4.91	3.42	57.5	17.9	1.25	107	187	22.9	216	11.8	2.71	743

Sample	La	Ce	Pr	Nd	Sm	Eu	Gd	Tb	Dy	Ho	Er	Tm	Yb	Lu	Hf	Ta	Pb	Th	U	$(La/Yb)_N$	$(Gd/Yb)_N$	Eu/Eu^*	ΣREE
LTW-01	29.3	54.9	5.84	21.6	4.19	1.03	3.96	0.6	3.65	0.74	2.1	0.31	2.03	0.31	4.09	1.07	13.9	9.12	1.71	10.35	1.61	0.77	130.6
LTW-03	35.6	67.1	7.01	25.5	4.62	1.07	4.31	0.63	3.8	0.76	2.18	0.33	2.1	0.33	4.75	0.92	15.4	12.5	2.14	12.19	1.7	0.74	155.3
LTW-04	36.1	67.4	6.91	24.8	4.29	1.06	3.93	0.56	3.32	0.66	1.92	0.29	1.87	0.28	4.67	0.89	14.3	11.4	1.73	13.81	1.74	0.79	153.3
LTW-08	37.2	69.2	7.14	25.5	4.44	1.12	4.02	0.57	3.4	0.68	1.95	0.29	1.91	0.28	4.64	0.91	14.7	11.6	1.84	14.00	1.74	0.81	157.7
LTW-09	27.1	51.1	5.48	20.6	4.08	1.04	3.89	0.59	3.6	0.73	2.09	0.3	2	0.3	4.44	0.99	13.8	10.2	2.03	9.73	1.61	0.79	122.9
LTW-10	31.3	59.2	6.28	23.1	4.41	1.06	4.16	0.63	3.79	0.76	2.23	0.33	2.17	0.32	5.23	0.93	13.9	9.69	1.88	10.37	1.59	0.76	139.8

The results were analyzed by XRF and ICP-MS in the State Key Laboratory of Continental Dynamics, Northwest University, Xi'an, China.

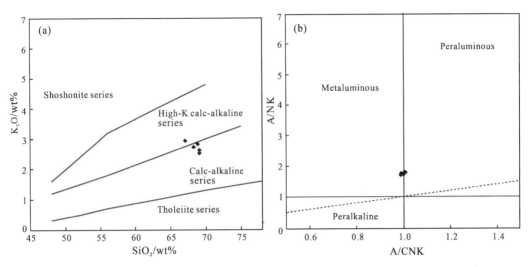

Fig. 4　SiO$_2$-K$_2$O (4a) (after Maniar and Piccoli, 1989) and A/NK [Al$_2$O$_3$/(Na$_2$O+K$_2$O)] vs. A/CNK [molar ratio Al$_2$O$_3$/(CaO+Na$_2$O+K$_2$O)] (b) (after Rollinson, 1993) diagrams for the biotite monzogranite from the western Litang area.

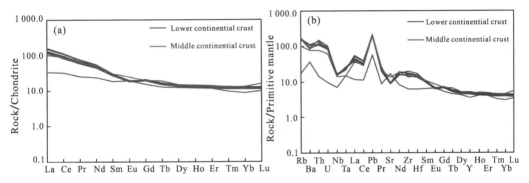

Fig. 5　Chondrite-normalized REE patterns (a) and primitive mantle (PM) normalized trace element spider diagrams (b) for the biotite monzogranite from the western Litang area. The normalized values are from Sun and McDonough (1989).

4.3　Whole-rock Sr-Nd-Pb-Hf isotopes

Whole-rock Sr-Nd-Pb-Hf isotopic data for the biotite monzogranite from the western Litang area are listed in Tables 3, 4 and 5. All the initial ^{87}Sr/^{86}Sr isotopic ratios (I_{Sr}), $\varepsilon_{Nd}(t)$, and $\varepsilon_{Hf}(t)$ values were calculated for the time of 210 Ma based on the LA-ICP-MS U-Pb zircon ages for the Litang biotite monzogranite.

Three samples (LTW-03, LTW-04, and LTW-09) have high (^{87}Sr/^{86}Sr)$_i$(I_{Sr}) ratios of 0.714 225−0.714 763, negative $\varepsilon_{Nd}(t)$ values of −2.0 to −2.6, with two-stage model ages of 1.01−1.05 Ga. These samples are characterized by Sr = 186−192 ppm and Rb = 102−106 ppm. The I_{Sr} vs. $\varepsilon_{Nd}(t)$ diagram (Fig. 6a) shows similar features to matured continental crustal source.

Table 3 Whole-rock Rb-Sr and Sr-Nd isotopic compositions of the biotite monzogranite from the western Litang area

Sample	$^{87}Sr/^{86}Sr$	2σ	Sr /ppm	Rb /ppm	$^{143}Nd/^{144}Nd$	2σ	Nd /ppm	Sm /ppm	T_{DM} /Ga	T_{DM2} /Ga	$\varepsilon_{Nd}(t)$ ($t=210$ Ma)	I_{Sr}
LTW-03	0.714 763	4	192	106	0.512 233	7	25.5	4.62	0.65	1.05	−2.6	0.714 763
LTW-04	0.714 588	12	188	102	0.512 247	4	24.8	4.29	0.64	1.03	−2.4	0.714 588
LYW-09	0.714 225	5	186	104	0.512 265	4	20.6	4.08	0.63	1.01	−2.0	0.714 225

$^{87}Rb/^{86}Sr$ and $^{147}Sm/^{144}Nd$ ratios were calculated using Rb, Sr, Nd and Sm contents analyzed by ICP-MS.

T_{DM2} values represent the two-stage model age and were calculated using present-day $(^{147}Sm/^{144}Nd)_{DM} = 0.213\ 7$, $(^{143}Nd/^{144}Nd)_{DM} = 0.513\ 15$, $(^{147}Sm/^{144}Nd)_{crust} = 0.101\ 2$ $(^{147}Sm/^{144}Nd)_{CHUR} = 0.196\ 7$ and $(^{143}Nd/^{144}Nd)_{CHUR} = 0.512\ 638$.

$$\varepsilon_{Nd}(t) = \left[(^{143}Nd/^{144}Nd)_S(t)/(^{143}Nd/^{144}Nd)_{CHUR}(t) - 1 \right] \times 10^4.$$

$$T_{DM} = \frac{1}{\lambda}\left\{ 1 + \left[(^{143}Nd/^{144}Nd)_S - ((^{147}Sm/^{144}Nd)_S - (^{147}Sm/^{144}Nd)_{crust})(e^{\lambda t} - 1) - (^{143}Nd/^{144}Nd)_{DM} \right] / \left[(^{147}Sm/^{144}Nd)_{crust} - (^{147}Sm/^{144}Nd)_{DM} \right] \right\}, \lambda = 6.54 \times 10^{-12}/a.$$

Table 4 Whole-rock Pb isotopic analysis results of the biotite monzogranite from the western Litang area.

Sample	U /ppm	Th /ppm	Pb /ppm	$^{206}Pb/^{204}Pb$	$^{207}Pb/^{204}Pb$	$^{208}Pb/^{204}Pb$	$^{238}U/^{204}Pb$	$^{232}Th/^{204}Pb$	$(^{206}Pb/^{204}Pb)_i$	$(^{207}Pb/^{204}Pb)_i$	$(^{208}Pb/^{204}Pb)_i$
LTW-03	2.14	12.5	15.4	18.894 4	15.744 8	39.515 1	8.952	53.640	18.598 0	15.729 8	38.951 8
LTW-04	1.73	11.4	14.3	18.827 1	15.745 6	39.467 9	7.781	52.603	18.569 4	15.732 6	38.915 5
LYW-09	2.03	10.2	13.8	18.942 0	15.747 4	39.477 2	9.478	48.853	18.628 2	15.731 6	38.964 2

U, Th, Pb concentrations were analysed by ICP-MS.

Table 5 Whole-rock Lu-Hf isotopic analysis results of the biotite monzogranite from the western Litang area.

Sample	Lu/ppm	Hf/ppm	$^{176}Hf/^{177}Hf$	$^{176}Lu/^{177}Hf$	2σ	$\varepsilon_{Hf}(t)$ ($t=210$ Ma)	$f_{Lu/Hf}$	T_{DM2}/Ga
LTW-03	0.31	4.75	0.282 583	0.009 266	5	−3.4	−0.72	1.85
LTW-04	0.28	4.67	0.282 560	0.008 513	5	−4.1	−0.74	1.87
LYW-09	0.30	4.44	0.282 581	0.009 594	4	−3.5	−0.71	1.88

Lu and Hf concentrations were analysed by ICP-MS. T_{DM2} was calculated using present-day $(^{176}Lu/^{177}Hf)_{DM} = 0.038\ 4$ and $(^{176}Hf/^{177}Hf)_{DM} = 0.283\ 25$, $f_{CC} = -0.55$, and $f_{DM} = 0.16$. f_S is the sample's $f_{Lu/Hf}$.

$\varepsilon_{Hf}(t)$ values were calculated using present-day $(^{176}Lu/^{177}Hf)_{CHUR} = 0.033\ 2$ and $(^{176}Hf/^{177}Hf)_{CHUR} = 0.282\ 772$.

$$T_{DM1} = \frac{1}{\lambda}\ln\left\{ 1 + \left[(^{176}Hf/^{177}Hf)_S - (^{176}Hf/^{177}Hf)_{DM} \right] / \left[(^{176}Lu/^{177}Hf)_S - (^{176}Lu/^{177}Hf)_{DM} \right] \right\}.$$

$$T_{DM2} = T_{DM1} - (T_{DM1} - t) \cdot \left[(f_{CC} - f_S)/(f_{CC} - f_{DM}) \right], \lambda = 1.876 \times 10^{-11}/a.$$

The biotite monzogranite samples have Pb contents of 13.8−15.4 ppm. In the diagram of the $(^{206}Pb/^{204}Pb)_i$-$(^{208}Pb/^{204}Pb)_i$ and $(^{206}Pb/^{204}Pb)_i$-$(^{207}Pb/^{204}Pb)_i$ (Fig. 7), the three samples plot in the range of MORB and lower continental crust, which indicate a depleted source region. The initial $^{206}Pb/^{204}Pb$, $^{207}Pb/^{204}Pb$, and $^{208}Pb/^{204}Pb$ ratios of the biotite monzogranite samples from the western Litang area are 18. 569 4 - 18. 628 2, 15. 729 8 - 15. 732 6, and 38. 915 5-38. 964 2, respectively.

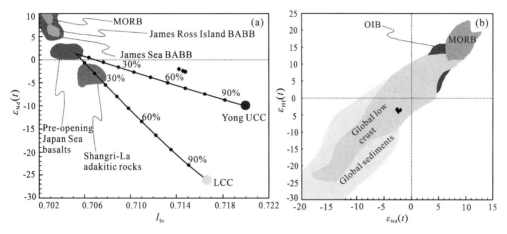

Fig. 6 I_{Sr} vs. $\varepsilon_{Nd}(t)$ (a) (after Wang et al., 2013a) and $\varepsilon_{Nd}(t)$ vs. $\varepsilon_{Hf}(t)$ (b) (after Dobosi et al., 2003) diagram of the biotite monzogranite from the western Litang area.

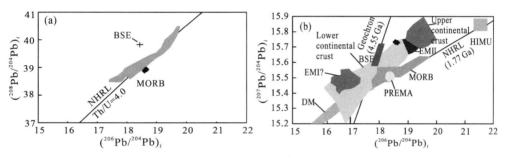

Fig. 7 Diagrams of $(^{208}Pb/^{204}Pb)_i$ vs. $(^{206}Pb/^{204}Pb)_i$(a) and $(^{208}Pb/^{204}Pb)_i$ vs. $(^{206}Pb/^{204}Pb)_i$(b) for the biotite monzogranite from the western Litang area (after Hugh, 1993).
NHRL: northern hemisphere reference line (Th/U = 0. 4).

These samples (LTW-03, LTW-04, and LTW-09) have Lu contents of 0. 28-0. 31 ppm and Hf contents of 4. 44-4. 75 ppm. They have the $^{176}Hf/^{177}Hf$ ratios ranging from 0. 282 560 to 0. 282 583 and $^{176}Lu/^{177}Hf$ ratios from 0. 008 513 to 0. 009 594. The $\varepsilon_{Hf}(t)$ values are -4. 1 to -3. 4 with the two-stage model ages of 1. 85-1. 88 Ga. In the $\varepsilon_{Nd}(t)$ vs. $\varepsilon_{Hf}(t)$ diagram (Fig. 6b), the Litang biotite monzogranite samples indicate a lower crust source.

5　Discussion

5. 1　Magmatic type of the Litang biotite monzogranite

The Late Triassic biotite monzogranite from the western Litang area has SiO_2 (67. 12-69. 13 wt%), medium K_2O (2. 54 - 2. 95 wt%), calc-alkalinity, A/CNK = 0. 99 - 1. 01, relatively high Na$_2$O contents (3.38-3.60 wt%) and low P$_2$O$_5$(0.10-0.12 wt%) contents. The A/CNK values are regarded as the criteria for I-type and S-type granites (Chappell and White, 1974, 2001; Wu Fuyuan et al., 2007; Wang Dezi et al., 1993). Wang Dezi et al. (1993) proposed that K and Rb are enriched in matured crust, whereas Sr is enriched in immature continental crust. The low Rb/Sr ratios (Rb/Sr = 0.53-0.57 < 0.9) of the Litang biotite monzogranite suggest an I-type granite affinity. The CIPW normative corundum values of the studied samples, < 1 wt% (0.073-0.386 wt%), show the features of I-type granites (Chappell and White, 1974). Moreover, in the (Na$_2$O+K$_2$O)/CaO vs. 10 000 Ga/Al and Y vs. 10 000 Ga/Al diagrams (Fig. 8a,b), all samples plot within the fields of I and S types. The Litang biotite monzogranite samples display a positive correlation between the Y and Rb (Fig. 8c), which indicates a typical I-type trend. Furthermore, in the K$_2$O-Na$_2$O diagram (Fig. 8d), the biotite monzogranite samples also plot in the I-type granite region.

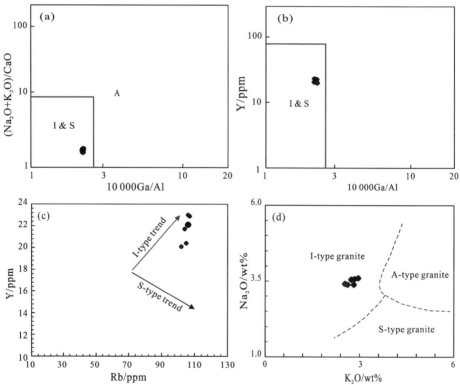

Fig. 8 (Na$_2$O+K$_2$O)/CaO vs. 10 000 Ga/Al (a) and Y vs. 10 000 Ga/Al (after Whalen et al., 1987) (b), Y-Rb (c) (after Li et al., 2007) and Na$_2$O-K$_2$O (d) (after Collins et al., 1982) diagrams for the biotite monzogranite from the western Litang area.

5.2 Characteristics of sources

The biotite monzogranite samples in this study have SiO$_2$(67.12-69.13 wt%), with Na$_2$O/

K_2O ratios of 1.20−1.34, similar to granitic rocks generated by the dehydration partial melting of basaltic rocks (Rapp and Watson, 1995). The low Al_2O_3/TiO_2 and medium Cao/Na_2O ratios indicate that the samples approached the source of basalt-derived melt (Fig.9a). In the $Mg^{\#}$ vs. SiO_2 diagram (Fig.9b), these rocks also show the features of melts from meta-igneous sources. Furthermore, in the [molar $Al_2O_3/(MgO + FeO^T)$] vs. [molar $CaO/(MgO + FeO^T)$] diagram (Fig.9c), the biotite monzogranite samples plot with partial melts from metabasaltic sources. The studied biotite monzogranite samples have negative $\varepsilon_{Nd}(t)$ values of −2.6 to −2.0 with the two-stage model ages of 1.01−1.05 Ga, and corresponding negative $\varepsilon_{Hf}(t)$ values of −4.1 to −3.4 with two-stage model ages from 1.85 Ga to 1.88 Ga, which indicate that the protolith was from the ancient continental crust. The Rb/Y-Nb/Y diagram of the biotite monzogranites also shows the feature of a crustal source (Fig.9d). The Nb/Ta ratios (10.37 − 12.86) are considerably lower than those of mantle-derived rock (17.5±2) and closely approximate those of

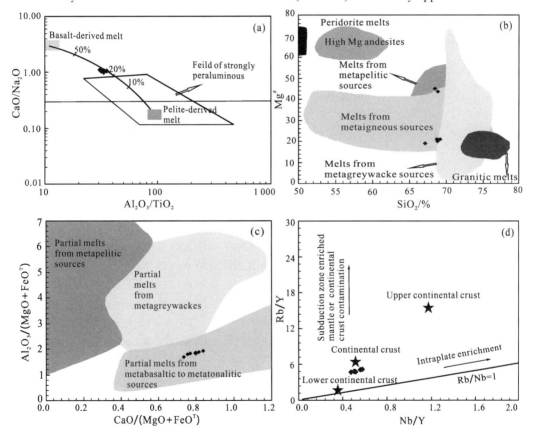

Fig.9 Cao/Na$_2$O vs. Al$_2$O$_3$/TiO$_2$ diagram (a) (after Sylvester, 1998), Mg$^{\#}$ vs. SiO$_2$ diagram (b) (after Zhu et al., 2017), [molar Al$_2$O$_3$/(MgO+FeO^T)] vs. [molar Cao/(MgO+FeO^T)] (c) (after Altherr et al., 2000) and Rb/Y-Nb/Y (d) (after Jahn et al., 1999) diagrams for the biotite monzogranite from the western Litang area.

crustal rock (~11) (Taylor and McLennan, 1985). In summary, the biotite monzogranite was mainly derived from partial melting of the basalt in the ancient crustal sources.

5.3 Tectonic implications

Voluminous Late-Triassic granitoids and volcanics have been documented in the Yidun terrane. The chronology results for these igneous rocks are summarized in the Table 6 and Fig. 10.

The crystallization age of the biotite monzogranite from the western Litang area is ~206.1 Ma. The geochemical characteristics indicate that these biotite monzogranites are derived from Paleoproterozoic to Mesoproterozoic mafic crustal sources. Indeed, the ~225–206 Ma

Table 6 Summary of chronology results for Triassic igneous rocks in the Yidun terrane (after Hou et al., 2001; Wang et al.,2013a; Peng et al., 2014).

Region	Pluton/ Volcanic rocks	Lithology	Methology	Age /Ma	Source
Changtai	Guojiaoma	Granite	SHRIMP	215.1 ± 2.0	Weislogel, 2008)
		Granite	SHRIMP	224.5 ± 2.0	
	Ganzi	Granodiorite	LA-ICPMS	218.5 ± 3.0	Reid et al., 2007
		Granodiorite	LA-ICPMS	~213.6	Lai et al., 2007
	Litang	Granite	SHRIMP	220.3 ± 3.2	Wang et al., 2013a from Zhou Meifu
	The west of Litang	Biotite monzogranite	LA-ICPMS	206.1 ± 1.0	This study
	Litang(Yongjie)	Granodiorite	K-Ar	214	
	Litang(Chajiu)	Granodiorite	K-Ar	219	Lu Boxi et al., 1993
	Litang(Rongse)	Monzogranite	K-Ar	219	
	Tumugou	Rhyolite	LA-ICPMS	219.3 ± 1.4	Yan, 2016
	Gacun	Dacite	LA-ICPMS	230.6 ± 1.3	Wang et al., 2013a
	Miange	Dacite	LA-ICPMS	230.0 ± 2.5	
Xiangcheng	Daocheng	Granodiorite	LA-ICPMS	215.3 ± 3.4	Reid et al., 2007
		Granodiorite	LA-ICPMS	~201.8	Lai et al., 2007
		Granite	LA-ICPMS	217.4 ± 1.1	Wang Nan et al., 2016
	Daocheng (Dongcuo)	Monzogranite	Rb-Sr	208	Lu Boxi et al., 1993
	Daocheng (Bangretang)	Monzogranite	K-Ar	213	
	Xiangcheng	Monzogranite	SHRIMP	224.0 ± 3.0	Lin et al., 2006
		Diorite	SHRIMP	222.0 ± 3.0	
	Xiangcheng	Dacite	LA-ICPMS	227.9 ± 1.5	Wang et al., 2013a
	Xiangcheng (Maxionggou)	K-feldspar granite	K-Ar	205	Lu Boxi et al., 1993

Continued

Region	Pluton/ Volcanic rocks	Lithology	Methology	Age /Ma	Source
Shangri-La	Pulang	Quartz monzonite	SHRIMP	228.0 ± 3.0	Wang Shouxu et al., 2008
		Quartz monzonite		226.3 ± 2.8	
		Quartz monzonite		226.3 ± 3.0	
		Quartz diorite	TIMS	221.0 ± 1.0	Pang Zhenshan et al., 2009
		Quartz monzonite		211.8 ± 0.5	
		Granodiorite		206.3 ± 0.7	
		Quartz diorite	LA-ICPMS	217.9 ± 1.8	Wang et al., 2011
		Quartz diorite		224.2 ± 1.7	
	Qiansui	Quartz diorite	LA-ICPMS	217.1 ± 1.5	Ren Jiangbo et al., 2011
		Quartz diorite	SHRIMP	215.3 ± 2.3	Lin et al., 2006
	Xuejiping	Quartz diorite	SHRIMP	215.2 ± 1.9	Cao Dianhua et al., 2009
		Quartz diorite	LA-ICPMS	213.4 ± 1.5	Ren Jiangbo et al., 2011
	Pingdong	Quartz diorite	LA-ICPMS	230.3 ± 1.7	Wang et al., 2011
	Disuga	Quartz diorite	LA-ICPMS	219.8 ± 3.0	Wang et al., 2011
		Andesite	LA-ICPMS	220.7 ± 2.0	Wang et al.. 2011
	Chundu	Quartz monzonite	SIMS	219.7 ± 1.8	Zhang Xingchun et al., 2009
	Songnuo	Quartz monzonite	SHRIMP	220.9 ± 3.5	Leng Chengbiao et al, 2008

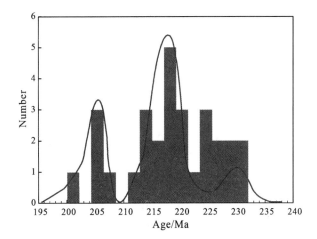

Fig. 10 The summary of chronology results for Triassic igneous rocks in the Yidun terrane.

granitic rock of the Yidun terrane has I-type granite affinity (Hou Zengqian et al., 2001), which is explained as the products of partial melting of meta-igneous rocks (Chappell and White, 2001). In the Rb vs. (Y + Nb) and Rb/30-Hf-3Ta diagrams (Fig. 11), all of the samples fall into the field of VAG (volcanic arc granites). The Litang biotite monzogranite is typical calc-alkaline magmatism of subduction-related tectonic background (Chen et al., 2015). Based on the geochemical characteristics and tectonic discrimination diagrams, the Litang biotite monzogranite was most likely dominated by an arc-continental collision setting.

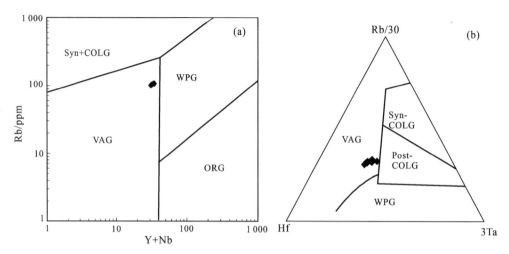

Fig. 11   The Rb vs. (Y + Nb) (a) (after Pearce, 1984) and Rb/30-Hf-3Ta (b) (after Harris et al.,
1986) diagrams of the biotite monzogranite from the western Litang area.
Syn-COLG: syn-collisional granite; WPG: within-plate granite; VAG: volcanic arc granites;
ORG: oceanic ridge granite; Post-COLG: post-collisional granite.

There are three main tectonic models for Middle-Late Triassic volcanism in the Yidun
terrane (Hou Zengqian, 1993; Roger et al., 2010; Wang et al., 2011; Peng et al., 2014).

(1) Eastward-dipping subduction of the Jinsha oceanic lithosphere under the Zhongza
Massif (Reid et al., 2005; Roger et al., 2008, 2010). They proposed that the east dipping
subduction zone of Jinsha ocean generated the ~245–229 Ma granites in the western Yidun
terrane (Roger et al., 2010). These ~245–229 Ma granitoids are inconsistent with the main
age range of ~225–206 Ma (Hou zengqian et al., 2001; Roger et al., 2010). The Litang
biotite monzogranite was formed at ~206 Ma and located at the eastern Yidun terrane
(Fig. 1c). Although they have the typical subduction-related magmatism feature, the
crystallization age is later than the main stage of Jinsha ocean eastward subduction which they
proposed. Moreover, this assumption cannot account for the spatial distribution of the arc
granitic plutons and the back-arc volcanic rocks in the northern part of Yidun terrane.
Actually, the dominant granitic rocks are located at the eastern part of the Yidun terrane and
the back-arc volcanic rocks in the Changtai region of the northern part of the Yidun terrane
(Fig. 1c) (Wang et al., 2013a). In fact, there is no associated magmatic type ore deposits
which were found to the east of the Ganzi-Litang suture (i.e., the Songpan-Ganzi fold belt) so
far, but many plutons in the eastern Yidun terrane to the west of Ganzi-Litang suture (Peng et
al., 2014). In addition, Wang et al. (2013b) pointed out that the eastward subduction of the
Jinsha Ocean was implausible in terms of the source recycling feature. If the Jinsha oceanic slab
subducted eastward, the Zhongza Massif would be an active continental margin and the Yidun
Group would receive massive detritus in such an active continental margin. However, the Yidun

Group displays the source features of a passive margin by the geochemical compositions of sedimentary materials (Wang et al., 2013b). We therefore conclude that the eastward subduction of the Jinsha ocean was implausible.

(2) Orogenic collapse (Peng et al., 2014). They argued that the coeval occurrence of the Late Triassic intrusive rocks in the two sides of the Ganzi-Litang suture formed after Triassic crustal thickening of the Yidun and Songpan-Ganzi terranes caused by convergence between the Yangtze, North China and Qiangtang blocks (Xiao et al., 2007; Zhang et al., 2007; Peng et al., 2014). They explained that the enriched lithospheric mantle and lower crust are metamorphosed up to the eclogite facies and delaminated downwards into asthenosphere after Triassic crustal thickening (Peng et al., 2014). Subsequently, the delamination would cause the rise of hot asthenosphere to reach the lower crust and crustal extension (Sacks and Secor, 1990). These basaltic magmas which were formed by the partial melting of the upwelling asthenosphere lea to large-scale partial fusion of middle-lower crust to form felsic magma (Stel et al., 1993). Although this scenario can give a reasonable interpretation of the formation of Late Triassic granitic batholith in the Yidun terrane, this model neglects the discovery of crustal thickening and cannot explain the existence of the Ganzi-Litang and Jinsha ophiolite sutures which were considered as the remants of the subduction slab in the Zhongza Massif and eastern Yidun terrane. In fact, no basement rocks have been reported in the eastern Yidun terrane because of thick coverage by Upper Triassic strata. It has been suggested that the Yidun arc was built on thinned continental crust, which may have consisted of Proterozoic basement derived from the Yangtze block (Hou, 1993; Chang Chengfa, 1996; Reid et al., 2007).

(3) Westward subduction of the Ganzi-Litang oceanic lithosphere (Hou Zengqian et al., 2001, 2003, 2007; Leng et al., 2012; Wang et al., 2013a). Hou Zengqian et al. (2001, 2003, 2007) pointed out that the differences of magmatic type are due to different subduction dip angles based on the differences of magmatic type, rock assemblages and their host deposits in the northern and southern parts of eastern Yidun terrane. Wang et al (2013a) further proposed that the high silica adakites of the sourthern Yidun terrane (Xiangcheng and Shangri-La regions) were formed by early-stage subduction (~230–224 Ma) of Ganzi-Litang ocean. The transition from HSA (high silica adakitic rocks) to LSA (low silica adakitic rocks) at ~218–215 Ma in the Shangri-La region suggests partial melting of the metasomatized mantle wedge (Wang et al., 2011). Those authors also proposed that the back-arc volcanic rocks in the northern part of Yidun terrane (Changtai region) were caused by later-stage westward subduction (~224 Ma to the end of the Triassic) of Ganzi-Litang oceanic lithosphere, leading to the formation of arc granitic plutons. The ~216 Ma Daocheng granites in the south of Litang area, which are high-K, calc-alkaline and I-type granites, were derived from the partial melting of mafic-intermediate lower crust with a variably minor addition (<20%) of depleted mantle-derived magma (He et al., 2013). Combined with the results of previous studies on mafic to intermediate volcanic rocks (Leng Chengbiao, 2009), the assemblage of fossils from the Tumugou formation (Huang Jianguo and Zhang Liuqing, 2005), and the diorites and granites in the eastern Yidun terrane (Hou Zengqian et al., 2001, 2004; Reid et al., 2007; Leng et al., 2012; Leng Chengbiao, 2009), He et al. (2013) further proposed that the Daocheng granites was generated in a syn-collisional tectonic setting during the westward subduction and closure of Ganzi-Litang paleo-ocean. In fact, subduction orogeny is the basic cause of the Yidun arc (Mabi Awei et al., 2015). The arc-related granitoids or volcanics have been dated to have formed mostly between ~225 Ma and 206 Ma (Hou Zengqian et al., 2001, 2003; Lin Qingcha et al., 2006; Lai et al., 2007; Reid et al., 2007; Weislogel, 2008). Even the Cu (-Mo) mineralization in the southern Yidun arc was likely related to the westward subduction of the Ganzi-Litang oceanic plate during the Late Triassic (Dong et al., 2014). The Litang biotite monzogranites were formed at ~206 Ma which were accompanied by the voluminous arc-related magmatism in the eastern Yidun terrane during the ~225 – 206 Ma. The trace elements feature displays the obvious enrichment of LILE (e.g. Rb, Ba) and LREE and depletion of the HREE which are typical arc-like geochemical compositions (Fig. 5). The isotopic signatures show the lower crust source (Fig. 6, Fig. 7). The geochemical discriminant diagrams show the similar sources with the experimental melts of the meta-igneous (Fig. 9). The Litang biotite granites were located at the same tectonic region with those typical subduction-related magmatism (e. g. Shangri-La adakitic rocks and Daocheng I-type granites) and were the magmatism respond in the middle-lower crust under the tectonic setting of the westward subduction of Ganzi-Litang ocean in the eastern Yidun terrane. Thus, we could deduce that the formation of Yidun arc-basin system also occurred within this time span.

Based on the evidence presented in this study and previous studies, we propose that the mechanism of the Litang biotite monzogranite may be as follows (Fig. 12).

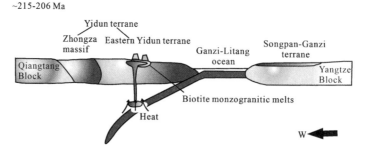

Fig. 12 Geodynamic model for the biotite monzogranite from the western Litang area
(after Wang et al., 2011).

In the Late Permian, the Ganzi-Litang ocean opened because of upwelling of the mantle plume, which formed the Emeishan large igneous province on the western margin of the Yangtze block at ~ 260 Ma (Tan, 1987; Song et al., 2004). During the Early-Middle Triassic, the westward subduction of the Jinsha oceanic lithosphere ceased and the Yidun terrane collided with the Qiangtang block (Hou Zengqian and Mo Xuanxue, 1991; Pullen et al., 2008; Zhu et al., 2011). At the beginning of the Late Triassic, the closure and westward subduction of Ganzi-Litang oceanic lithosphere started (Wang et al., 2011; He et al., 2013; Leng et al., 2014). The subduction triggered asthenosphere upwelling, which resulted in the upwelling of hot mantle-derived melt into the base of the crust and provided a large amount of heat required for partial melting of the crustal rocks (He et al., 2013; Wang et al., 2013a). The mantle-derived basaltic magma in the Litang area, similar with the presence of coeval mafic to intermediate volcanic rocks in the Zhongdian area (Leng Chengbiao, 2009) and the Daocheng I-type granites in the south of the Litang area (He et al., 2013), could have provided the heat source for the partial melting of Paleoproterozoic to Mesoproterozoic middle-lower crust (Leng Chengbiao, 2009; He et al., 2013). Actually, such massive granites (~5 200 km^2) along the Ganzi-Litang suture zone in eastern Yidun terrane (Fig. 1c) required the upwelling of much more voluminous mantle-derived magmas, which supplied sufficient heat to melt the middle-lower crust (He et al., 2013). As subduction proceeded, these basaltic melts provided the heat source for the partial melting of the basalt protolith within the middle-lower crust and formed the biotite monzogranite magma in the western of Litang area.

6 Conclusions

Zircon U-Pb geochronology indicates that the biotite monzogranite from the western Litang area was emplaced at ~206. 1 Ma, and thus belongs to the Late Triassic granitic rocks. The Litang biotite monzogranite has the similar features, with medium SiO$_2$(67. 12−69. 13 wt%), medium K (K$_2$O = 2. 54−2. 95 wt%), and relatively high sodium (Na$_2$O = 3. 38−3. 60 wt%) contents. These rocks are metaluminous, calc-alkaline series, and I-type granites. The initial ^{87}Sr/^{86}Sr isotopic ratios (I_{Sr}) were high (0. 714 225−0. 714 763), and $\varepsilon_{Nd}(t)$ values were negative (−2. 0 to −2. 6), with T_{DM2} from 1. 01 Ga to 1. 05 Ga; $\varepsilon_{Hf}(t)$ values were negative (−3. 4 to −4. 1), with T_{DM2} ranging from 1. 85 Ga to 1. 88 Ga. Moreover, the initial ^{206}Pb/^{204}Pb, ^{207}Pb/^{204}Pb, and ^{208}Pb/^{204}Pb ratios of 18. 569 4−18. 628 2, 15. 729 8−15. 732 6, and 38. 915 5 − 38. 964 2, respectively, suggest that the parent magma was derived from matured continental crust sources.

Based on previous researches, geochemical characteristics, and tectonic evolution, the Litang biotite monzogranite was derived from the partial melting of the metabasaltic rocks within the middle-lower crust under the tectonic background of the closure and westward subduction of the Ganzi-Litang oceanic lithosphere during the Late Triassic, where mantle-derived basaltic melts provided sufficient heat for the partial melting of basalt protolith within the ancient crustal.

Acknowledgements We gratefully acknowledge the editor Prof. Hongfu Zhang for his work about this study. We thank the anonymous reviewers for through revision of the manuscript. This work was supported by the National Natural Science Foundation of China (Grant Nos. 41421002, 41772052, and 41372067), and independent innovation project of graduate students of Northwest University (YZZ17192).

References

Altherr, R., Holl, A., Hegner, E., Langer, C., and Kreuzer, H., 2000. High-potassium, calc-alkaline I-type plutonismin the European Variscides: Northern Vosges (France) and northern Schwarzwald (Germany). Lithos, 50(1): 51-73.

Andersen, T., 2002. Correction of common lead in U-Pb analyses that do not report ^{204}Pb. Chemical Geology, 192(1-2): 59-79.

BGMRSP (Bureau of Geology and Mineral Resources of Sichuan Province), 1991.Regional Geology of Sichuan Province. Beijing: Geological Publishing House (in Chinese).

Cao Dianhua, Wang Anjian, Li Wenchang, Wang Gaoshang, Li Ruiping, and Li Yike, 2009. Magma mixing in the Pulang porphyry copper deposit: Evidence from petrology and element geochemistry. Acta Geologica Sinica, 83(2): 166-175 (in Chinese with English abstract).

Chang Chengfa, 1996.Geology and Tectonics of Qinghai-Xizang Plateau. Beijing: Science Press (in Chinese).

Chang, E.Z., 2000. Geology and tectonics of the Songpan-Ganzi fold belt, southwestern China. International Geology Review, 42(9): 813-831.

Chappell, B.W., and White, A.J.R., 1974. Two contrasting granite types. Pacific Geology, 8(2): 173-174.

Chappell, B.W., and White, A.J.R., 2001. Two contrasting granite types: 25 years later. Australian Journal of Earth Sciences, 48(4): 489-499.

Chen, Q., Sun, M., Long, X.P., and Yuan, C., 2015. Petrogenesis of Neoproterozoic adakitic tonalites and high-k granites in the eastern Songpan-Ganze fold belt and implications for the tectonic evolution of the western Yangtze block.Precambrian Research, 270: 181-203.

Collins, W.J., Beams, S.D., White, A.J.R., and Chappell, B.W., 1982. Nature and origin of A-type granites with particular reference to Southeastern Australia. Contributions to Mineralogy and Petrology, 80(2): 189-200.

Dong Guochen, Zhang Yuling, Wang Peng, Li Xuefeng, and Zeng Yang., 2014. The Triassic Intrusions and Cu (-Mo) mineralization in the Southern Yidun arc, China: Implications for metallogenesis and geodynamic setting. Acta Geologica Sinica (English Edition), 88(Supp. 2): 898-899.

Floyd, P.A., Kelling, G., Gökçen, S.L., and Gökçen, N., 1991. Geochemistry and tectonic environment of basaltic rocks from the Misis ophiolitic mélange, south Turkey. Chemical Geology, 89(3-4): 263-280.

Harris, N.B.W., Pearce, J.A., and Tindle, A.G., 1986. Geochemical characteristics of collision-zone magmatism.In: Coward, M.P., and Reis, A.C. Collision Tectonics. Geological Society, London, Special Publications, 19(1): 67-81.

He Defeng, Zhu Weiguang, Zhong Hong, Ren Tao, Bai Zhongjie, and Fan Hongpeng, 2013. Zircon U-Pb geochronology and elemental and Sr-Nd-Pb isotopic geochemistry of the daocheng granitic pluton from the Yidun arc, SW China. Journal of Asian Earth Sciences, s67-68(1): 1-17.

Hou Zengqian, and Mo Xuanxue, 1991. The evolution of Yidun island arc and implications in the exploration of Kuroko-type volcanogenic massive sulphide deposits in Sanjiang area, China. Earth Science: Journal of China University of Geosciences, 16: 153-164 (in Chinese with English abstract).

Hou Zengqian, 1993. Tectono-magmatic evolution of the Yidun island-arc and geodynamic setting of Kuroko-type sulfide deposits in Sanjiang region, SW China. Resource Geology Special Issue, 17: 336-350.

Hou Zengqian, Qu Xiaoming, Zhou Jirong, Yang Yueqing, Huang Dianhao, Lv Qingtian, Tang Shaohua, Yu Jinjie, Wang Haiping, and Zhao Jinhua., 2001. Collision orogeny in the Yidun arc: Evidence from granites in the Sanjiang region, China. Acta Geologica Sinica, 75(4): 484-497 (in Chinese with English abstract).

Hou Zengqian, Yang Yueqing, Wang Haiping, Qu Xiaoming, Lv Qingtian, Huang Dianhao, Tang Shaohua, and Zhao Jinhua, 2003. Collision-Orogenic Progress and Mineralization System of Yidun arc. Beijing: Geological Publishing House: 345 (in Chinese).

Hou Zengqian, Yang Yueqing, Qu Xiaoming, Huang Dianhao, Lv Qingtian, Wang Haiping, Yu Jinjie, and Tang Shaohua, 2004. Tectonic evolution and mineralization systems of the Yidun arc orogen in Sanjiang region, China. Acta Geologica Sinica, 78(1): 109-120 (in Chinese with English abstract).

Huang Jianguo, and Zhang Liuqing, 2005. The petrochemistry and tectonics of Late Triassic Tumugou Formation in Zhongdian. Yunnan Geology, 24(2): 186-192 (in Chinese with English Abstract).

Hugh, R.R., 1993. Using Geochemical Data.Singapore: Longman Singapore Publishers, 234-240.

Jahn, B.M., Wu, F.Y., Lo, C.H., and Tsai, C.H., 1999. Crust-mantle interaction induced by deep subduction of the continental crust: Geochemical and Sr-Nd isotopic evidence from post-collisional mafic-ultramafic intrusions of the northern Dabie complex, central China. Chemical Geology, 157(1-2): 119-146.

Lai Qingzhou, Ding Lin, Wang Hongwei, Yue Yahui, and Cai Fulong, 2007. Constraining the stepwise migration of the eastern Tibetan Plateau margin by apatite fission track thermochronology. Science in China: Series D, Earth Sciences, 50: 172-183.

Leng Chengbiao, Zhang Xingchun, Wang Shouxu, Qin Chaojian, Gou Tizhong, and Wang Waiquan, 2008. SHRIMP zircon U-Pb dating of the Songnuo ore-hosted porphyry, Zhongdian, Northwest Yunnan, China and its geological implication. Geotectonica et Metallogenia, 32: 124-130 (in Chinese with English abstract).

Leng Chengbiao, 2009.Ore Deposit Geochemistry and Regional Geological Setting of the Xuejiping Porphyry Copper Deposit, Northwest Yunnan, China. Doctor Paper of the Graduate University of Chinese Academy of Sciences (in Chinese with English abstract).

Leng Chengbiao, Zhang Xingchun, Hu Ruizhong, Wang Shouxu, Zhong Hong, Wang Waiquan, and Bi Xianwu, 2012. Zircon U-Pb and molybdenite Re-Os geochronology and Sr-Nd-Pb-Hf isotopic constraints on the genesis of the Xuejiping porphyry copper deposit in Zhongdian, Northwest Yunnan, China. Journal of Asian Earth Sciences, 60: 31-48.

Leng Chengbiao, Huang Qiuyue, Zhang Xingchun, Wang Shouxu, Zhong Hong, Hu Ruizhong, Bi Xianwu,

Zhu Jingjing, and Wang Xinsong, 2014. Petrogenesis of the Late Triassic volcanic rocks in the Southern Yidun Arc, SW China: Constraints from the geochronology, geochemistry, and Sr-Nd-Pb-Hf isotopes. Lithos, 190-191(2), 363-382.

Li Qiuhuan, Zhang Yuxiu, Zhang Kaijun, Yan Lilong, Zeng Lu, Jin Xin, Sun Jinfeng, Zhou Xiaoyao, Tang Xianchun, and Lu Lu, 2017. Garnet amphibolites from the Ganzi-Litang fault zone, eastern Tibetan Plateau: Mineralogy, geochemistry, and implications for evolution of the eastern Palaeo-Tethys Realm. International Geology Review, 1-14.

Li Wenchang, Zeng Pusheng, Hou Zengqian, and White, N.C., 2011. The Pulang porphyry copper deposit and associated felsic intrusions in Yunnan Province, southwest China. Economic Geology, 106: 79-92.

Li Xianhua, Li Zhengxiang, Li Wuxian, Liu Ying, Yuan Chao, Wei Gangjian, and Qi Changshi, 2007. U-Pb zircon, geochemical and Sr-Nd-Hf isotopic constraints on age and origin of Jurassic I- and A-type granites from central Guangdong, SE China: A major igneous event in response to foundering of a subducted flat-slab? Lithos, 96(1): 186-204.

Liu Ye, Liu Xiaoming, Hu Zhaochu, Diwu Chunrong, Yuan Honglin, and Gao Shan, 2007. Evaluation of accuracy and long-term stability of determination of 37 trace elements in geological samples by ICP-MS. Acta Petrologica Sinica, 23: 1203-1210 (in Chinese with English abstract).

Lin Qingcha, Xia Bin, and Zhang Yuquan, 2006. Zircon SHRIMP U-Pb dating of the syncollisional Xuejiping quartz diorite pophyrite in Zhongdian, Yunnan, China, and its geological implications. Geological Bulletin of China, 25: 133-137 (in Chinese with English abstract).

Ludwig, K.R., 2003. ISOPLOT 3.0: A Geochronological Toolkit for Microsoft Excel. Berkeley Geochronology Center, Special Publication, 4.

Lu Boxi, Wang Zeng, and Zhang Nengde, 1993. Granitoids in the Sanjiang Region and their metallogenic specialization. Beijing: Geological Publishing House (in Chinese).

Mabi Awei, Muhetaer Zari, Wen Dengkui, and Zhang Mingchun., 2015. Geochemical characteristics of southern Genie granite in the eastern Tibet and its geological significance. Acta Petrologica Sinica, 89(2): 305-318 (in Chinese with English abstract).

Maniar, P.D, and Piccoli, P.M., 1989. Tectonic discrimination of granitoids. Geological Society of America Bulletin, 101: 635-643.

Mo Xuanxue, Deng Jinfu, and Lu Fengxiang, 1994. Volcanism and the evolution of Tethys in Sanjiang area, southeastern China. Journal of Southeast Asian Earth Sciences, 9: 325-333.

Pang Zhenshan, Du Yangsong, Wang Gongwen, Guo Xin, Cao Yi, and Li Qing, 2009. Single-grain zircon U-Pb isotopic ages, geochemistry and implication of the Pulang complex in Yunnan Province, China. Acta Petrologica Sinica, 25(1): 159-165.

Pearce, J.A., Harris, N.B.W., and Tindle, A.G., 1984. Trace element discrimination diagrams for the tectonic interpretation of granitic rocks. Journal of Petrology, 25(4): 956-983.

Peng Touping, Zhao Guochun, Fan Weiming, Peng Bingxia, and Mao Yongsheng, 2014. Zircon geochronology and Hf isotopes of Mesozoic intrusive rocks from the Yidun terrane, eastern Tibetan Plateau: Petrogenesis and their bearings with Cu mineralization. Journal of Asian Earth Sciences, 80(2): 18-33.

Pullen, A., Kapp, P., Gehrels, G.E., Vervoort, J.D., and Ding L., 2008. Triassic continental subduction in central Tibet and Mediterranean-style closure of the Paleo-Tethys Ocean. Geology, 36: 351-354.

Qu Xiaoming, Hou Zengqian, and Zhou Shugui, 2002. Geochemical and Nd, Sr isotopic study of the post-orogenic granites in the Yidun arc belt of northern sanjiang region, southwestern china. Resource Geology, 52(2): 163-172.

Rapp, R.P., and Watson, E.B., 1995. Dehydration melting of metabasalt at 8−32 kbar: Implications for continental growth and crust-mantle recycling. Journal of Petrology, 36(4):891-931

Reid, A.J, Wilson, C.J.L., and Liu, S., 2005. Structural evidence for the Permo-Triassic tectonic evolution of the Yidun arc, eastern Tibetan Plateau. Journal of Structural Geology, 27(1): 119-137.

Reid, A.J., Wilson, C.J.L., Liu, S, Pearson N., and Belousova, E., 2007. Mesozoic plutons of the Yidun arc, SW China: U/Pb geochronology and Hf isotopic signature. Ore Geology Reviews, 31(1-4): 88-106.

Ren Jiangbo, Xu Jifeng, and Chen Jianlin., 2011. LA-ICP-MS zircon U-Pb geochronology of porphyries in Zhongdian arc. Acta Petrologica Sinica, 27(9): 2591-2599 (in Chinese with English abstract).

Roger, F., Arnaud, N., Gilder, S., Tapponnier, P., Jolivet, M., Brunel, M., Malavieille, J., Xu, Zhiqin, and Yang, J.S., 2003. Geochronological and geochemical constraints on Mesozoic suturing in East Central Tibet. Tectonics, 22(4): 1037.

Roger, F., Malavieille, J., Leloup, P.H., Calassou S., and Xu, Z., 2004. Timing of granite emplacement and cooling in the Songpan-Ganzi fold belt (eastern Tibetan plateau) with tectonic implications. Journal of Asian Earth Sciences, 22(5): 465-481.

Roger, F., Jolivet, M., and Malavieille J., 2008. Tectonic evolution of the Triassic fold belts of Tibet. Comptes Rendus Geoscience, 340(2): 180-189.

Roger, F., Jolivet, M., and Malavieille J., 2010. The tectonic evolution of the Songpan-Garze (North Tibet) and adjacent areas from Proterozoic to Present: A synthesis. Journal of Asian Earth Sciences, 39: 254-269.

Rollinson, H., 1993. Using geochemical data: Evaluation, presentation, interpretation. London: Longman Scientific and Technical.

Sacks, P.E., and Secor, D.T., 1990. Delamination in collisional orogens. Geology, 18, 999-1002.

Song Xieyan, Zhou Meifu, Cao Zhimin, and Robinson, P.T., 2004. Late Permian rifting of the South China Craton caused by the Emeishan mantle plume? Journal of the Geological Society, 161: 773-781.

Stel, H., Cloetingh, S., Heeremans, M., and Beek, P.V.D., 1993. Anorogenic granites, magmatic underplating and the origin of intracratonic basins in a nonextensional setting. Tectonophysics, 226: 285-299.

Sun, S.S., and McDonough, W.F., 1989. Chemical and isotopic systematics of oceanic basalts: implications for mantle composition and processes. In: Saunders, A.D., and Norry, M.J. Magmatism in the ocean Basins. Geological Society Special Publication, London, 42: 313-345.

Sylvester, P.J., 1998. Post collisional strongly peraluminous granites. Lithos, 45(1-4): 29-44.

Tan, T.K., 1987. Geodynamics and tectonic evolution of the Panxi rift.Tectonophysics, 133: 287-304.

Taylor, S.R., and Mclennan, S.M., 1985. The continental crust: Its composition and evolution, an examination of the geochemical record preserved in sedimentary rocks. Journal of Geology, 94(4):632-633.

Todd, E., Gill, J.B., Wysoczanski, R.J., Hergt, J., Wright, I.C., Leybourne, M.I., and Mortimer, N., 2011. Hf isotopic evidence for small-scale heterogeneity in the mode of mantle wedge enrichment: Southern

Havre Trough and South Fiji Basin back arcs. Geochemistry, Geophysics, Geosystems, 12(9): 1525-2027.

Wang Baiqiu, Zhou Meifu, Li Jianwei, and Yan Danping, 2011. Late Triassic porphyritic intrusions and associated volcanic rocks from the Shangri-La region, Yidun terrane, eastern Tibetan Plateau: Adakitic magmatism and porphyry copper mineralization. Lithos, 127: 24-38.

Wang Baiqiu, Zhou Meifu, Chen, W.T, Gao Jianfeng, and Yan Danping., 2013a. Petrogenesis and tectonic implications of the Triassic volcanic rocks in the northern Yidun terrane, eastern Tibet. Lithos, s175-176 (8): 285-301.

Wang Baiqiu, Wang Wei, and Zhou Meifu., 2013b. Provenance and tectonic setting of the Triassic Yidun Group, the Yidun terrane, Tibet. Geoscience Frontiers, 4(6):765-777.

Wang Dezi, Liu Changshi, Shen Weizhou, and Chen Fanrong, 1993. The contrast between Tonglu I-type and Xiangshan S-type clastoporphyritic lava. Acta Petrologica Sinica, 9(1): 44-53 (in Chinese with English abstract).

Wang Nan, Wu Cailai, Qin Haipeng, Lei Min, Guo Wenfeng, Zhang Xin, and Chen Hongjie, 2016. Zircon U-Pb Geochronology and Hf isotopic characteristics of the Daocheng Granite and Haizishan Granite in the Yidun arc, Western Sichuan, and their geological significance. Acta Petrologica Sinica, 90(11): 3227-3245 (in Chinese with English abstract).

Wang Shouxu, Zhang Xingchun, Leng Chengbiao, Qin Chaojian, Ma Deyun, and Wang Waiqua, 2008. Zircon SHRIMP U-Pb dating of the Pulang porphyry copper deposit, northwestern Yunnan, China: The ore-forming time limitation and geological significance. Acta Petrologica Sinica, 25(1): 2313-2321 (in Chinese with English abstract).

Weislogel, A.L., 2008. Tectonostratigraphic and geochronologic constraints on evolution of the northeast Paleotethys from the Songpan-Ganzi complex, central China. Tectonophysics, 451: 331-345.

Whalen, J.B., Currie, K.L., and Chappell, B.W., 1987. A-type granites: Geochemical characteristics, discrimination and petrogenesis. Contributions to Mineralogy and Petrology, 95: 407-419.

Wu Fuyuan, Li Xianhua, Yang Jinhui, and Zheng Yongfei., 2007. Discussions on the petrogenesis of granites. Acta Petrologica Sinica, 23(6): 1217-1238 (in Chinese with English abstract).

Xiao Long, Zhang Hongfei, Clemens, J.D., Wang, Q.W., Kan, Z.Z., Wang, K.M., Ni, P.Z., and Liu Xiaoming, 2007. Late Triassic granitoids of the eastern margin of the Tibetan Plateau: Geochronology, petrogenesis and implications for tectonic evolution. Lithos, 96: 436-452.

Yao Xueliang, and Lan Yan, 2001. N-type oceanic ridge basalt in the Garge-Litang ophiolite mélange zone. Acta Geologica Sichuan, 21: 138-140 (in Chinese with English abstract).

Yin, An, and Harrison, T, M., 2000. Geologic evolution of the Himalayan-Tibetan orogen. Annual Review of Earth and Planetary Sciences, 28: 211-280.

Yuan Chao, Zhou Meifu, Sun Min, Zhao Yongjiu, Wilde, A.S, Long Xiaoping, and Yan Danping, 2010. Triassic granitoids in the eastern Songpan Ganzi Fold Belt, SW China: Magmatic response to geodynamics of the deep lithosphere. Earth & Planetary Science Letters, 290(3-4): 481-492.

Yuan Honglin, Gao Shan, Liu Xiaoming, Li Huiming, Gunther, D., and Wu Fuyuan, 2004. Accurate U-Pb age and trace element determinations of zircon by laser ablation-inductively coupled plasma mass spectrometry. Geostandard and Geoanalytical Research, 28(3): 353-370.

Yuan Honglin, Gao Shan, Luo Yan, Zong Chunlei, Dai Mengning, Liu Xiaoming, and Diwu Chunrong, 2007. Study of Lu-Hf geochronology: A case study of eclogite from Dabie UHP Belt. Acta Petrologica Sinica, 23 (2): 233-239 (in Chinese with English abstract).

Yuan Honglin, Gao Shan, Dai Mengning, Zong Chunlei, Gunther, D., Fontaine, G.H., Liu Xiaoming, and Diwu Chunrong, 2008. Simultaneous determinations of U-Pb age, Hf isotopes and trace element compositions of zircon by excimer laser-ablation quadrupole and multiple-collector ICP-MS. Chemical Geology, 247(1-2): 100-118.

Zhang Hongfei., Zhang Li, Harris, N., Jin Lanlan, and Yuan Honglin, 2006. U-Pb zircon ages, geochemical and isotopic compositions of granitoids in Songpan-Garze fold belt, eastern Tibetan Plateau: Constraints on petrogenesis and tectonic evolution of the basement. Contributions to Mineralogy and Petrology, 152(1): 75-88.

Zhang Hongfei, Parrish, R., Zhang Li, Xu Wangchun, Yuan Honglin, Gao Shan, and Crowley, Q.G., 2007. A type granite and adakitic magmatism association in Songpan-Ganzi fold belt, eastern Tibetan Plateau: Implication for lithospheric delamination. Lithos, 97: 323-335.

Zhang Kaijun, 2001. Is the Songpan-Ganzi terrane (central China) really underlain by oceanic crust? Journal of the Geological Society of India, 57(3): 223-230.

Zhang Xingchun, Leng Chengbiao, Yang Chaozhi, Wang Waiquan, and Qin Chaojian, 2009. The SIMS U-Pb zircon age and its geological significance of porphyry copper ore-bearing porphyry of Chundou area in Northwest Yunnan, China. Acta Mineralogica Sinica, 29(Suppl.): 359-360 (in Chinese).

Zhang Yuxiu, Tang Xianchun, Zhang Kaijun, Zeng Lu, and Gao Changliang, 2014. U-Pb and Lu-Hf isotope systematics of detrital zircons from the Songpan-Ganzi Triassic flysch, NE Tibetan Plateau: Implications for provenance and crustal growth. International Geology Review, 56(1): 29-56.

Zhao Yongjiu, 2007. Mesozoic Granitoids in eastern Songpan-Ganzi: Geochemistry, Petrogenesis and Tectonic Implications. Dissertation for the Doctoral Degree. Guangzhou: Guangzhou Institute of Geochemistry, Chinese Academy of Sciences: 1-101.

Zhu Jingjing, Hu Ruizhong, Bi Xianwu, Zhong Hong, and Chen Heng, 2011. Zircon U-Pb ages, Hf-O isotopes and whole-rock Sr-Nd-Pb isotopic geochemistry of granitoids in the Jinshajiang suture zone, SW China: Constraints on petrogenesis and tectonic evolution of the Paleo-Tethys ocean. Lithos, 126(3): 248-264.

Zhu Renzhi, Lai Shaocong, Qin Jiangfeng, and Zhao Shaowei, 2017. Petrogenesis of late Paleozoic-to-early Mesozoic granitoids and metagabbroic rocks of the Tengchong Block, SW China: Implications for the evolution of the eastern Paleo-Tethys. International Journal of Earth Sciences, (3): 1-27.

Early-Middle Triassic intrusions in western Inner Mongolia, China: Implications for the final orogenic evolution in southwestern Xing-Meng orogenic belt[①]

Liu Min　Lai Shaocong[②]　Zhang Da　Zhu Renzhi　Qin Jiangfeng　Di Yongjun

Abstract: The end-Permian to Early-Middle Triassic magmatic rocks in Inner Mongolia can provide valuable insights into the relationships between the collisional processes and the magmatic responses during the final orogenic evolution of Xing-Meng orogenic belt (XMOB). This paper presents zircon U-Pb ages and Hf isotopes, whole rock geochemical and Sr-Nd-Pb isotopic data for the Early-Middle Triassic diabases and monzogranites from the Langshan area, southwestern XMOB. Our results suggest that the studied diabases and monzogranites were respectively formed during Early Triassic and Middle Triassic. The Early Triassic diabases are characterized by "arc-like" geochemical signatures, including enrichment in Rb, U and K, and depletion in Nb, Ta, P and Ti. They have negative to weak positive $\varepsilon_{Nd}(t)$ values (-3.1 to $+1.5$) and relatively high initial ratios of $^{208}Pb/^{204}Pb$ ($35.968-37.346$), $^{207}Pb/^{204}Pb$ ($15.448-15.508$) and $^{206}Pb/^{204}Pb$ ($16.280-17.492$), indicating a subduction-metasomatized enriched lithospheric mantle source. Their low Ba/Rb ($2.72-6.56$), Ce/Y ($0.97-1.39$) and $(Tb/Yb)_N$ ratios ($1.31-1.45$) suggest that the parental magma was likely originated from low degree partial melting of the phlogopite-bearing lherzolite in a spinel-stability field. The Middle Triassic monzogranites show high Sr/Y ratios, low MgO, Cr and Ni contents, high Zr/Sm ratios ($40-64$), negative zircon $\varepsilon_{Hf}(t)$ values (-25.8 to -8.8), as well as relatively flat heavy rare earth element patterns. They were likely derived from low degree partial melting of a moderately thickened ancient lower crust. The diabases and the slightly postdated high Sr/Y granites in this study represent the magmatic responses to the final orogenic evolution in the southwestern XMOB. Together with regional works, we propose that the slab break-off of the Paleo-Asian oceanic lithosphere following the terminal collision between the North China Craton and the South Mongolia terranes triggered asthenospheric upwelling, and the ongoing convergence further initiated moderately crustal thickening and uplift in the XMOB.

1　Introduction

Accretionary orogens formed at sites of subduction of the oceanic lithosphere are ultimately

①　Published in *Journal of Earth Science*, 2019, 30(5).
②　Corresponding author.

involved in a collisional phase when plate subduction ceases (Cawood et al., 2009), leading to continental amalgamation/collision, slab break-off, crustal thickening and subsequent lithospheric delamination, accompanied by extensive and diverse magmatism (e.g., Condie et al., 2009; Ferrari, 2004). Consequently, identification of magmatic records in response to these processes can help us better understand the final orogenic evolution in an accretionary orogen.

The Central Asian orogenic belt (CAOB) (Fig. 1a) is one of the world's largest and best-preserved accretionary orogens (Wilde, 2015; Xiao et al., 2003, 2015; Windley et al., 2007; Jahn et al., 2000). The southeastern segment of the CAOB, which is also known as Xing-Meng orogenic belt (XMOB), was mainly constructed by the convergence between the South Mongolia terranes (SMT) to the north and the North China Craton (NCC) to the south, driven by the evolution of the Paleo-Asian Ocean from Neoproterozoic to Late Paleozoic (Zhao et al., 2018; Ma et al., 2017; Li et al., 2016a; Wilde, 2015; Xiao et al., 2003, 2015; Xu et al., 2013; Jian et al., 2010; Windley et al., 2007). The Solonker suture zone is widely considered to have marked the terminal closure of the Paleo-Asian Ocean during the end-Permian to Early-Middle Triassic (Li et al., 2016a, 2017; Xiao et al., 2003, 2015; Eizenhöfer et al., 2014; Schulmann and Paterson, 2011). However, unlike the well

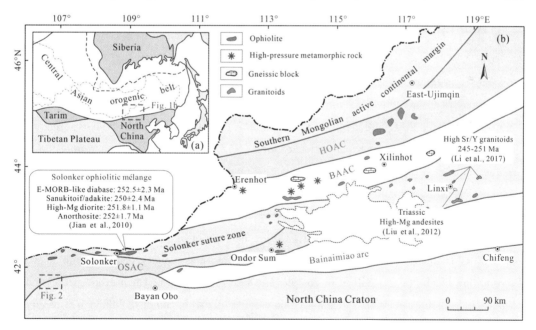

Fig. 1 (a) Simplified sketch map of the Central Asian orogenic belt showing major tectonic sub-divisions of Central Asia (modified after Jahn et al., 2000). (b) Geological sketch map of the central-western Xing-Meng orogenic belt (modified after Xiao et al., 2015; Jian et al., 2008).

OSAC: Ondor Sum subduction-accretionary complex; BAAC: Baolidao arc-accretionary complex; HOAC: Hegenshan ophiolite-accretionary complex.

documented Paleozoic arc magmatism and oceanic subduction (e. g. , Li et al. , 2016b; Eizenhöfer et al. , 2014, 2015; Xiao et al. , 2015; Xu et al. , 2013; Jian et al. , 2008, 2010; Zhang et al. , 2009a, b), origin and petrogenesis of the end-Permian to Early-Middle Triassic collision-related magmatism in the XMOB are still enigmatic. Recent studies propose that the end-Permian to Early Triassic magmatic rocks in Mandula and southeastern Xilinhot were generated in relation to the slab break-off following the terminal collision (Li et al. , 2016a; Jian et al. , 2010), and the Early-Middle Triassic high Sr/Y granitoids in Linxi and central Jilin were magmatic response to the subsequent crustal thickening (Li et al. , 2017; Wang et al. , 2015a). However all those works were centered on the Solonker suture zone, the end-Permian to Early-Middle Triassic magmatism in southwestern XMOB receives much less attention.

In this study, we present new zircon U-Pb ages and Lu-Hf isotopes, whole rock geochemical and Sr-Nd-Pb isotopic data for the Early-Middle Triassic diabases and monzogranites in the Langshan area, southwestern XMOB (Fig. 1b). Our results, in combination with regional studies, enable us to propose that the studied diabases and the monzogranites were respectively generated in response to the slab break-off and the following crustal thickening during the final orogenic evolution in southwestern XMOB.

2 Geological background and samples

The XMOB has long been regarded as a complex collage of arcs, microcontinents, remnants of oceanic crust and associated volcanic-sedimentary sequences that amalgamated between the NCC and the SMT during the evolution and closure of the Paleo-Asian ocean (e.g. , Xiao et al. , 2003, 2015; Jahn et al. , 2000). The main part of the XMOB, from south to north, consists of the southern accretionary zone (SAZ) between the north margin of the NCC and the Solonker suture zone, the Solonker suture zone, the northern accretionary zone (NAZ) and the Hegenshan ophiolite-accretionary complex (Xiao et al. , 2015; Jian et al. , 2010). The SAZ includes the Early Paleozoic Bainaimiao arc (Zhang et al. , 2014c) and the Ondor Sum subduction-accretion complex (Jian et al. , 2008; de Jong et al. , 2006; Xiao et al. , 2003), the NAZ is characterized by the Paleozoic Baolidao arc-accretionary complex with accreted Precambrian blocks (e.g. , Xilin Gol complex) (Xiao et al. , 2015; Chen et al. , 2009). The Hegenshan ophiolite-accretionary complex possesses a relatively complete lithological sequence (Miao et al. , 2008), which is suggested be formed in a subduction-related tectonic setting, such as back-arc or island arc-marginal basin, that related to the Early Carboniferous north-dipping subduction of the Paleo-Asian ocean (Eizenhöfer et al. , 2015; Miao et al. , 2008; Robinson et al. , 1999).

The Solonker suture zone, which strikes from the Solonker area to the Linxi area

(Fig. 1b) (Xiao et al.,2015; Jian et al., 2008, 2010), marks the final closure of the Paleo-Asian Ocean in the southeastern CAOB (e.g., Li et al., 2017, 2016a; Zhang et al., 2017; Eizenhöfer et al., 2014, 2015; Wilde, 2015; Xiao et al., 2003, 2015). The ages of the tectonic mélanges and ophiolitic fragments that discontinuously distributed along the Solonker suture zone are consistently Early Permian (Song et al., 2015; Jian et al., 2010). The Middle Permian Zhesi Formation volcaniclastic and turbiditic strata in the Solonker suture zone are unconformably overlain by the Late Permian-Early Triassic Linxi Formation clastic rocks, which are considered to be accumulated in a terrestrial setting (Eizenhöfer et al., 2015; Li et al., 2014). Zhang et al. (2016a) suggests that the Solonker suture zone underwent intermediate-to-low P-T type metamorphism during the Triassic. Several end-Permian to Early-Middle Triassic collision-related granitoids also have been identified in the Linxi area (Li et al., 2016a, 2017).

The Langshan area in the southwestern XMOB is approximately 100 km south of the Solonker suture zone (Fig. 1b). The Paleozoic tectonic affinity of the Langshan area was disputed because of the controversial relationship between the NCC and the Alxa block (Zhang et al., 2016b; Yuan and Yang, 2015). Many recent studies have revealed that the Alxa Block had collided with the NCC before the Late Paleozoic (e.g., Wang et al., 2016; Zhang et al., 2016b), although the accurate timing and location of this amalgamation are still in debate. The eldest rocks in the Langshan area is the Archean high-grade metamorphic complex (Fig. 2), which are dominated by amphibolite, quartz schist and schistose to magmatic gneiss with granitic gneiss of 2 563 Ma and 2 619 Ma (Liu, 2012). The Proterozoic basement rocks in the Langshan area mainly consist of greenschist-amphibolite facies metamorphosed carbonates, clastic rocks, with some Neoproterozoic meta-volcanic interlayers (Hu et al., 2014; Peng et al., 2010). Outcrops of the Paleozoic stratigraphy are only limited to the west of the Langshan area. They are mainly composed of some Carboniferous-Permian marine sedimentary and volcanic rocks, and are uncomfortably covered by the Jurassic-Quaternary sedimentary strata (Fig. 2). The Langshan area is also characterized by a complicated framework of Mesozoic multi-stage strike-slip faults and thrusts, which are interpreted to be related to the intraplate deformation phases across the southern CAOB (Zhang et al., 2013, 2016b; Darby and Ritts, 2007).

Igneous rocks in the Langshan area are mainly composed of the Late Paleozoic and Early Mesozoic felsic intrusions (Fig. 2) (e.g., Liu et al., 2016; Wang et al., 2015b; Lin et al., 2014; Peng et al., 2013), such as the ~338 Ma Baoribu quartz diorite with $\varepsilon_{Hf}(t)$ values of −6.9 to +2.0 in the north Chaoge Ondor (Liu et al., 2016), the ~281 Ma deformed granitic porphyry in the Urad Houqi area (Lin et al., 2014), and the ~259 Ma granodiorite with $\varepsilon_{Nd}(t)$ values of −2.7 to −1.6 close to the Huogeqi ore district (Wang et al., 2015b).

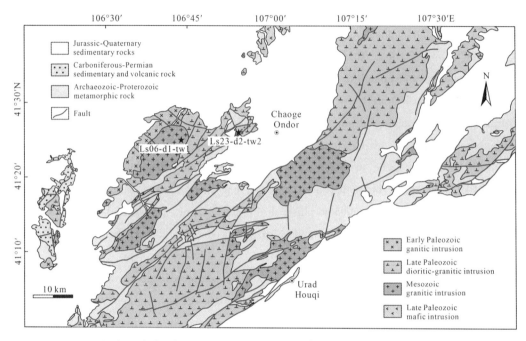

Fig. 2   Geological sketch map of the Langshan area (modified after Liu et al., 2016).

Additionally, some small outcrops of the Early Paleozoic granitoids and the Late Paleozoic mafic intrusions are also observed (e.g., Liu et al., 2016; Wang et al., 2015b, 2016).

The investigated diabases are located immediately to the east of the Chaoge Ondor (Fig. 2), most of them are vertical and NE-SW trending. They intruded into the Late Paleozoic granitic rocks (Fig. 3a) and the Proterozoic greenschist-amphibolite facies metamorphic clastic rocks (Zhaertai Group), but did not penetrate the Mesozoic granitoids in the field. The intrusive contacts are very clear and thermal recrystallization can be observed near the boundary of most dykes. In contrast to their country rocks, the diabases are rarely subjected to metamorphism and deformation. Individual diabase dykes in the Langshan area typically have widths of 0.8 – 3.0 m and can be traced for distances of 0.2 – 8 km. The investigated monzogranites are located approximately 15 km west of Chaoge Ondor, with surface outcrops over 60 km^2(Figs. 2 and 3b). They mainly intruded into the Early Carboniferous quartz diorite in the field, and are devoid of any effects of metamorphism or deformation. The diabase displays ophitic or sub-ophitic textures, with similar mineral compositions of lath-shaped plagioclase (50–60 vol%) and pyroxene (30–40 vol%), with minor hornblende (~5 vol%) and Fe-Ti oxides (Fig. 3c). The monzogranite is composed of plagioclase (25–30 vol%), K-feldspar (30–35 vol%), quartz (25–30 vol%), biotite (3–10 vol%) and minor accessory minerals (e.g., magnetite, zircon and apatite) (Fig. 3d).

Fig. 3　Representative field photographs and micrographs of the studied diabase and monzogranite from the Langshan area.

Py: pyroxene; Hbl: hornblende; Pl: plagioclase; Kf: K-feldspar; Bt: biotite; Qtz: quartz.

3　Analytical methods

Zircon grains were separated from the whole-rock samples by standard techniques of density and magnetic methods and further purified by hand-picking using a binocular microscope. The internal structures of each grains were revealed by Cathodoluminescence (CL) imaging prior to the U-Pb analysis. Zircon U-Pb analysis for sample Ls23-d2-tw2 was carried out at the State Key Laboratory of Continental Dynamics, Northwest University, Xi'an, China, by using an Agilent 7500a ICP-MS equipped with a 193 nm laser, the spot size is 35 μm. Zircon U-Pb age analysis for sample Ls06-d1-tw1 was carried out on a Finnigan Neptune ICP-MS that equipped with New Wave 193 nm excimer lasers, at the Tianjin Institute of Geology and Mineral Resources, Tianjin, China. A 35 μm diameter spot was used for the laser ablation of a single zircon grain. The $^{206}Pb/^{238}U$ and $^{207}Pb/^{206}Pb$ ratios were calculated by the GLITTER program and corrections were applied using Harvard zircon 91500 as an external calibration. Details of the analytical methodology are described by Li et al. (2009) and Yuan et al. (2004). The common Pb corrections were made following the method of Andersen (2002). Isoplot 3.0 program (Ludwig, 2003) was used for data processing.

Whole-rock geochemical compositions were determined at the Analytical Laboratory of the Beijing Research Institute of Uranium Geology (BRIUG), China. Major elements were analysed on fused glass discs by a Philips PW2404 X-ray fluorescence spectrometry with analytical errors better than 1%−5%, FeO contents were determined using the classical wet chemical method. Trace elements were determined by using a Perkin-Elmer Sciex ELAN DRC-e inductively coupled plasma mass spectrometer (ICP-MS) and the analytical uncertainties were mostly better than 5%. Analytical procedures and instrument parameters were similar to those described by Gao et al. (2002).

The whole rock Sr-Nd and Pb isotopic analysis were performed on triton thermal ionization mass spectrometer (TIMS) and MAT261 TIMS, respectively, at the Central-South China Supervision and Inspection Center of Mineral Resources, Ministry of Land and Resources, Wuhan, China. The mass fractionation of Sr isotopic ratios has been exponentially corrected to the $^{86}Sr/^{88}Sr = 0.119\ 4$ and Nd isotopic ratios to $^{146}Nd/^{144}Nd = 0.721\ 9$. The values for the NBS 987 Sr-standard and the BCR-2 Nd-standard measured during the data acuisition were $^{87}Sr/^{86}Sr = 0.710\ 31 \pm 4$ (2σ) and $^{143}Nd/^{144}Nd = 0.512\ 62 \pm 5$ (2σ), respectively. Repeated analysis yielded the $^{208}Pb/^{206}Pb$, $^{207}Pb/^{206}Pb$ and $^{204}Pb/^{206}Pb$ ratios of the Standard NBS981 in this study were $2.164\ 91 \pm 13$ (2σ), $0.914\ 37 \pm 5$ (2σ) and $0.059\ 104 \pm 5$ (2σ), respectively. Detailed descriptions of the analytical techniques are given by Zeng et al. (2013).

In situ zircon Lu-Hf isotopic analysis was conducted by using a Nu Plasma II MC-ICP-MS coupled with a Resolution M-50 193 nm laser at the State Key Laboratory of Continental Dynamics, Northwest University, Xi'an, China. A 44 μm laser beam with a repetition rate of 6 Hz was used for the analysis. Detailed information of these instruments and analytical technique can be found in Bao et al. (2017) and Yuan et al. (2008). The 91500 and MT standards yielded weighted $^{176}Hf/^{177}Hf$ ratios of $0.282\ 264 \pm 0.000\ 030$ and $0.282\ 528 \pm 0.000\ 019$, respectively.

4　Results

4.1　Zircon U-Pb ages

Zircons selected from diabase sample Ls23-d2-tw2 and monzogranite sample Ls06-d1-tw1 are mostly subhedral to euhedral, colourless or buff to transparent, with lengths of 50−200 μm. The majority of zircon grains exhibit clearly oscillatory zoning in the CL images (Fig. 4), suggesting their magmatic origin (Hoskin and Schaltegger, 2003). The Th/U ratios of zircons from Ls23-d2-tw2 and Ls06-d1-tw1 rang from 0.04 to 1.34 and 0.06 to 0.42, respectively. Detailed zircon U-Pb isotopic data are presented in Table 1. Twenty-one spots were conducted on zircons from sample Ls23-d2-tw2, and most of them are located below the concordia line (Fig. 5a). However, four relatively younger zircon grains have concordant ages that range from 248 Ma to 255 Ma and yield a weighted mean $^{206}Pb/^{238}U$ age of 251.4 ± 3.6 Ma, and another five grains have concordant ages ranging from 271 Ma to 273 Ma, yielding a concordant age of 272.1 ± 1.6 Ma (Fig. 5b). A total of twenty-five spots of zircons from the monzogranite sample Ls06-d1-tw1 show nearly concordant ages and yield a well-weighted mean $^{206}Pb/^{238}U$ age of

245. 5 ± 0. 7 Ma（Fig. 5c,d）, which means that emplacement of the studied monzogranite took place in the Middle Triassic.

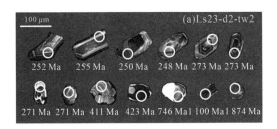 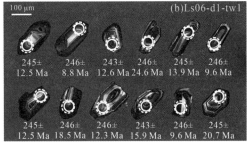

Fig. 4   CL images of representative zircons from the studied diabases and monzogranites.
The solid and dotted circles denote the spots for U-Pb dating and Lu-Hf, respectively.

Table 1   LA-ICP-MS U-Pb isotopic data for the zircons from the studied diabase and monzogranite samples.

| No. | Th/U | Isotopic ratios | | | | | | Age/Ma | | | | | |
		$^{206}Pb/^{238}U$	1σ	$^{207}Pb/^{235}U$	1σ	$^{207}Pb/^{206}Pb$	1σ	$^{206}Pb/^{238}U$	1σ	$^{207}Pb/^{235}U$	1σ	$^{207}Pb/^{206}Pb$	1σ
Ls23-d2-tw2（N41°25′46″/E106°54′02″）													
1	0. 68	0. 039 80	0. 000 50	0. 279 70	0. 013 40	0. 050 90	0. 002 50	252	3	250	11	238	117
2	1. 26	0. 043 00	0. 000 80	0. 308 50	0. 020 20	0. 052 10	0. 003 60	271	5	273	16	288	117
3	0. 91	0. 040 30	0. 000 60	0. 277 20	0. 012 10	0. 049 90	0. 002 40	255	4	248	10	188	74
4	0. 73	0. 039 60	0. 000 60	0. 282 90	0. 021 80	0. 051 90	0. 004 10	250	4	253	17	280	180
5	0. 29	0. 328 60	0. 003 70	5. 492 70	0. 067 90	0. 121 20	0. 002 60	1 832	18	1 899	11	1 974	10
6	0. 71	0. 065 90	0. 000 90	0. 528 10	0. 018 50	0. 058 20	0. 002 30	411	6	431	12	536	52
7	0. 45	0. 043 20	0. 000 60	0. 315 60	0. 010 30	0. 053 00	0. 002 00	273	3	279	8	330	50
8	0. 68	0. 039 20	0. 000 60	0. 296 00	0. 015 20	0. 054 70	0. 003 00	248	4	263	12	401	86
9	1. 34	0. 122 70	0. 001 70	1. 140 80	0. 033 80	0. 067 50	0. 002 30	746	10	773	16	852	39
10	0. 04	0. 090 90	0. 001 00	0. 938 50	0. 014 90	0. 074 90	0. 001 50	561	6	672	8	1 065	40
11	0. 68	0. 043 30	0. 000 60	0. 355 60	0. 010 80	0. 059 60	0. 002 10	273	3	309	8	591	43
12	0. 60	0. 067 80	0. 000 80	0. 524 50	0. 010 80	0. 056 10	0. 001 50	423	5	428	7	456	26
13	0. 72	0. 071 20	0. 000 80	0. 592 80	0. 010 40	0. 060 40	0. 001 50	444	5	473	7	617	20
14	1. 08	0. 197 50	0. 002 20	2. 696 00	0. 035 20	0. 099 00	0. 002 20	1 162	12	1 327	10	1 606	11
15	0. 63	0. 212 00	0. 002 80	2. 646 30	0. 081 30	0. 090 50	0. 003 00	1 240	15	1 314	23	1 437	65
16	0. 51	0. 043 00	0. 000 70	0. 334 10	0. 016 80	0. 056 40	0. 003 10	271	4	293	13	467	83
17	0. 14	0. 095 40	0. 001 10	1. 015 80	0. 019 80	0. 077 20	0. 002 00	588	7	712	10	1 127	21
18	0. 71	0. 043 00	0. 000 60	0. 326 80	0. 015 20	0. 055 10	0. 002 70	271	4	287	12	416	111
19	0. 44	0. 186 10	0. 002 20	2. 095 90	0. 046 20	0. 081 70	0. 002 00	1 100	12	1 147	15	1 238	50
20	0. 68	0. 337 40	0. 004 30	5. 418 60	0. 097 20	0. 116 50	0. 002 90	1 874	21	1 888	15	1 903	16
Ls06-d1-tw1（N41°26′16″/E106°44′30″）													
1	0. 91	0. 038 77	0. 000 36	0. 280 13	0. 009 80	0. 052 41	0. 001 88	245	2	251	9	303	82

Continued

No.	Th/U	Isotopic ratios						Age/Ma					
		$^{206}Pb/^{238}U$	1σ	$^{207}Pb/^{235}U$	1σ	$^{207}Pb/^{206}Pb$	1σ	$^{206}Pb/^{238}U$	1σ	$^{207}Pb/^{235}U$	1σ	$^{207}Pb/^{206}Pb$	1σ
2	0.73	0.038 82	0.000 36	0.273 59	0.005 48	0.051 12	0.000 89	246	2	246	5	246	40
3	0.29	0.038 93	0.000 31	0.276 62	0.008 70	0.051 53	0.001 62	246	2	248	8	265	72
4	0.71	0.038 78	0.000 40	0.273 56	0.019 43	0.051 16	0.003 77	245	3	246	17	248	170
5	0.45	0.038 84	0.000 37	0.275 39	0.020 03	0.051 42	0.003 68	246	2	247	18	260	164
6	0.68	0.039 14	0.000 30	0.277 27	0.005 75	0.051 38	0.001 00	248	2	248	5	258	45
7	1.34	0.038 43	0.000 23	0.454 56	0.009 99	0.085 79	0.001 80	243	1	380	8	1 333	41
8	0.04	0.039 42	0.000 24	0.275 38	0.005 73	0.050 66	0.000 99	249	2	247	5	226	45
9	0.68	0.038 24	0.000 28	0.280 37	0.009 66	0.053 17	0.001 72	242	2	251	9	336	73
10	0.60	0.038 95	0.000 23	0.279 60	0.006 37	0.052 07	0.001 17	246	1	250	6	288	51
11	0.72	0.038 98	0.000 23	0.279 26	0.006 01	0.051 96	0.001 10	246	1	250	5	284	49
12	1.08	0.038 76	0.000 24	0.278 24	0.005 40	0.052 07	0.000 99	245	2	249	5	288	43
13	0.63	0.039 05	0.000 24	0.280 64	0.007 87	0.052 12	0.001 43	247	2	251	7	291	63
14	0.51	0.038 82	0.000 25	0.274 83	0.012 22	0.051 34	0.002 25	246	2	247	11	256	101
15	0.14	0.038 92	0.000 22	0.275 04	0.004 93	0.051 25	0.000 92	246	1	247	4	252	41
16	0.71	0.038 66	0.000 24	0.273 00	0.014 41	0.051 21	0.002 68	245	2	245	13	250	121
17	0.44	0.038 96	0.000 22	0.282 89	0.003 31	0.052 66	0.000 60	246	1	253	3	314	26
18	0.35	0.038 87	0.000 23	0.273 73	0.006 22	0.051 08	0.001 13	246	1	246	6	244	51
19	0.32	0.038 91	0.000 26	0.278 55	0.016 33	0.051 92	0.003 02	246	2	250	15	282	133
20	0.28	0.038 90	0.000 22	0.280 12	0.006 36	0.052 23	0.001 16	246	1	251	6	295	51
21	0.22	0.038 34	0.000 22	0.282 48	0.006 86	0.053 44	0.001 27	243	1	253	6	348	54
22	0.06	0.038 69	0.000 25	0.275 32	0.010 41	0.051 62	0.001 93	245	2	247	9	268	86
23	0.31	0.038 34	0.000 22	0.413 30	0.012 61	0.078 18	0.002 32	243	1	351	11	1 151	59
24	0.40	0.038 93	0.000 26	0.273 36	0.016 08	0.050 93	0.002 95	246	2	245	14	237	133
25	0.42	0.039 02	0.000 23	0.279 43	0.007 54	0.051 94	0.001 39	247	1	250	7	283	61

4.2　Major and trace elements

The major and trace element concentrations of studied diabase and monzogranite samples are listed in Table 2. The relatively low LOI (loss on ignition) values of most samples indicate a limited degree of hydrothermal alteration.

The diabase samples have SiO_2 contents of 51.13 − 54.55 wt%, K_2O contents of 0.76 − 1.84 wt% and Na_2O contents of 2.56 − 4.23 wt%. They plot in the compositional fields of gabbro, gabbroic diorite, monzogabbro and monzodiorite (Fig. 6a) in the TAS diagram, and belong to the calc-alkaline series (Fig. 6b-d). These characteristics are similar to those of the ~250 Ma E-MOEB-like diabase from the Solonker ophiolitic mélanges (Jian et al., 2010). In the Harker diagrams, they exhibit regular trends of decreasing Al_2O_3, Na_2O, K_2O, TiO_2 and P_2O_5 contents with increasing MgO contents (Fig. 7). All the studied diabase samples show

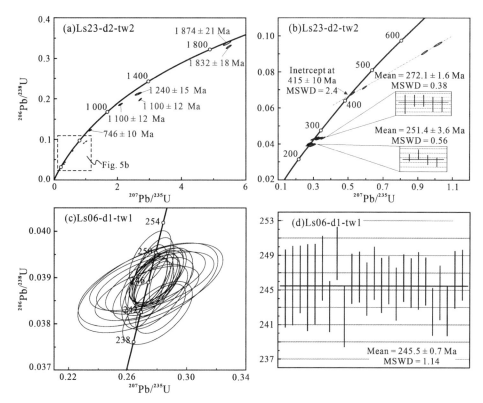

Fig. 5　(a,b) U-Pb concordia diagrams of zircon grains from the studied diabase sample Ls23-d2-tw2; (c,d) monzogranite sample Ls06-d1-tw1.

Fig. 5b shows the results of zircons with $^{206}Pb/^{238}U$ ages less than 600 Ma for Ls23-d2-tw2.

similar rare earth element (REE) compositions, with total REE contents ranging from 74. 7 ppm to 135. 0 ppm. In the chondrite-normalized REE diagram (Fig. 8a), they are characterized by light REE (LREE) enrichment and slightly negative to positive Eu anomalies, with $(La/Yb)_N$ of 2. 99−4. 14 and Eu/Eu^* of 0. 88−1. 17. Their primitive mantle-normalized trace element patterns (Fig. 8b) are generally characterized by enrichment in large ion lithophile elements (LILE; e.g., Rb, U, K, Pb) and depletion in high field strength elements (HFSE; e.g., Nb, Ta, P, Ti), which are similar to those of the "arc-like" magmatic rocks from the subduction zone (Ma et al., 2014).

The monzogranite samples exhibit relatively high SiO_2 (70. 09 − 72. 63 wt%), Na_2O (3. 98−4. 08 wt%), K_2O (3. 76−4. 67 wt%) and low $Fe_2O_3^T$ (1. 54 − 2. 27 wt%), MgO (0. 40−0. 87 wt%), TiO_2 (0. 25−0. 33 wt%) contents and $Mg^#$ [$100MgO/(MgO + FeO^T)$] values (31. 6−40. 5). These rocks belong to high-K calc-alkaline series, and are metaluminous to weakly peraluminous with A/CNK ratios of 0. 99 − 1. 01. They also have relatively high concentrations of Al_2O_3 (14. 10 − 15. 07 wt%), Sr (310 − 380 ppm) but low Y (7. 16 − 610. 50 ppm) and Yb (0. 60−1. 15 ppm) contents, with high Sr/Y ratios (30−45) (Fig. 9). The

Table 2 Major oxide and trace elements composition of the studied diabases and monzogranites.

Sample	Ls01-d1-q2	Ls01-d1-q10	Ls01-d1-q17	Ls01-d2-q7	Ls01-d2-q10	Ls01-d3-q5	Ls01-d1-q8	Ls01-d1-q14	Ls03-d6-q1	Ls06-d1-q1	Ls06-d3-q1	PM007D4-1GS	PM007D4-1W
Rock type	Diabase	Diabase	Diabase	Diabase	Diabase	Diabase	Diabase	Diabase	Diabase	Monzogranite	Monzogranite	Monzogranite	Monzogranite
Location	N41°25'29" E106°54'15"	N41°25'32" E106°54'10"	N41°25'39" E106°54'14"	N41°25'41" E106°53'52"	N41°25'41" E106°53'50"	N41°26'01" E106°53'42"	N41°25'31" E106°54'11"	N41°25'35" E106°54'07"	N41°26'18" E106°54'36"	N41°26'16" E106°44'30"	N41°25'56" E106°44'01"	N41°25'32" E106°44'25"	N41°25'32" E106°44'25"
Major element/wt%													
SiO_2	52.14	52.38	54.55	51.45	51.13	52.92	53.21	53.13	52.22	70.09	70.27	72.55	72.63
TiO_2	1.35	1.14	1.32	1.38	1.27	1.40	1.45	1.51	1.86	0.32	0.33	0.25	0.26
Al_2O_3	17.41	15.01	15.53	18.20	17.03	17.06	20.83	21.57	18.42	15.07	14.79	14.17	14.10
Fe_2O_3	7.90	2.61	2.53	2.89	2.39	2.51	2.06	2.40	8.18	1.30	1.39	0.99	1.00
FeO	1.00	5.68	5.15	5.42	5.79	5.57	4.60	3.15	1.06	1.08	1.02	0.65	0.70
MnO	0.12	0.12	0.11	0.12	0.12	0.13	0.09	0.07	0.12	0.05	0.05	0.02	0.03
MgO	5.75	9.66	7.53	5.58	6.66	5.26	3.01	2.27	3.78	0.86	0.87	0.40	0.42
CaO	8.00	7.81	6.63	8.88	8.72	7.62	7.91	7.65	7.02	2.26	2.28	1.37	1.40
Na_2O	3.82	2.56	3.34	3.36	3.55	4.09	4.11	4.07	4.23	4.08	3.98	4.07	4.06
K_2O	1.20	1.31	1.06	0.94	0.76	1.54	1.84	1.75	1.58	3.76	3.96	4.67	4.58
P_2O_5	0.18	0.17	0.18	0.18	0.16	0.20	0.24	0.24	0.31	0.10	0.11	0.07	0.08
LOI	1.82	1.77	2.27	1.83	2.55	1.91	0.84	2.34	2.01	1.38	1.22	0.81	0.77
Trace element/ppm													
Sc	22.1	22.0	20.7	25.9	25.7	27.0	15.1	15.2	19.8	4.1	4.6	2.7	2.8
V	128.0	145.0	132.0	121.0	149.0	161.0	102.0	100.0	128.0	23.0	31.5	36.3	37.7
Cr	59.2	494.0	268.0	82.4	158.0	139.0	62.4	53.6	66.3	4.7	8.3	5.7	5.8
Co	29.7	33.2	25.1	31.3	32.9	24.3	21.1	16.0	25.8	3.1	4.1	2.1	2.1
Ni	30.8	123.0	88.0	26.9	57.6	31.3	30.6	24.5	36.2	3.6	6.4	1.2	1.9
Cu	30.7	29.4	50.8	40.3	39.5	45.8	49.4	51.2	64.4	7.5	7.8	12.3	13.2
Zn	70.5	66.6	63.4	77.2	75.3	74.0	69.4	61.9	89.1	32.9	41.6	47.6	49.9
Ga	16.8	14.5	15.8	19.2	18.5	18.1	20.7	19.7	20.7	13.8	16.2	21.3	21.4
Cs	4.3	4.0	4.2	2.8	2.2	1.4	5.7	5.9	2.5	5.5	4.9	4.9	5.0
La	11.8	11.3	14.5	12.0	11.5	12.4	17.9	16.7	21.4	19.7	20.1	34.3	39.7

Continued

Sample	Ls01-d1-q2	Ls01-d1-q10	Ls01-d1-q17	Ls01-d2-q7	Ls01-d2-q10	Ls01-d3-q5	Ls01-d1-q8	Ls01-d1-q14	Ls03-d6-q1	Ls06-d1-q1	Ls06-d3-q1	PM007D4-1GS	PM007D4-1W
Rock type	Diabase	Diabase	Diabase	Diabase	Diabase	Diabase	Diabase	Diabase	Diabase	Monzogranite	Monzogranite	Monzogranite	Monzogranite
Location	N41°25'29" E106°54'15"	N41°25'32" E106°54'10"	N41°25'39" E106°54'14"	N41°25'41" E106°53'52"	N41°25'41" E106°53'50"	N41°26'01" E106°53'42"	N41°25'31" E106°54'11"	N41°25'35" E106°54'07"	N41°26'18" E106°54'36"	N41°26'16" E106°44'30"	N41°25'56" E106°44'01"	N41°25'32" E106°44'25"	N41°25'32" E106°44'25"
Ce	26.2	24.8	30.8	26.7	26.0	28.3	37.0	36.0	46.6	35.1	37.6	58.2	67.5
Pr	3.62	3.44	4.12	3.75	3.50	3.92	5.24	5.02	6.21	3.81	3.89	5.97	7.01
Nd	17.2	16.0	18.3	17.4	16.0	18.2	23.5	22.7	28.1	13.7	13.9	20.0	22.2
Sm	3.99	3.76	4.13	4.47	4.05	4.71	5.41	5.44	6.71	2.43	2.21	2.97	3.24
Eu	1.50	1.24	1.33	1.57	1.44	1.47	1.90	1.85	1.83	0.49	0.54	0.72	0.72
Gd	3.88	3.47	3.69	4.11	3.76	4.23	5.18	4.98	6.04	2.07	1.93	2.14	2.40
Tb	0.82	0.69	0.79	0.83	0.78	0.87	1.03	1.06	1.25	0.36	0.31	0.30	0.35
Dy	4.63	3.91	4.31	4.92	4.54	5.10	5.79	5.96	6.79	1.84	1.48	1.49	1.80
Ho	0.93	0.79	0.85	0.97	0.91	1.02	1.16	1.19	1.34	0.35	0.29	0.25	0.26
Er	2.60	2.26	2.49	2.77	2.55	2.86	3.27	3.37	3.78	1.07	0.85	0.68	0.76
Tm	0.44	0.37	0.40	0.45	0.44	0.46	0.52	0.55	0.60	0.16	0.15	0.10	0.10
Yb	2.72	2.31	2.51	2.81	2.72	2.97	3.22	3.34	3.87	1.15	1.01	0.60	0.70
Lu	0.39	0.33	0.35	0.38	0.36	0.40	0.45	0.46	0.50	0.17	0.16	0.09	0.10
Rb	35.5	48.1	44.1	40.4	29.9	73.2	83.2	71.7	72.2	109.0	128.0	169.0	173.0
Sr	339	287	343	413	352	405	438	391	398	310	383	320	324
Y	23.10	21.00	22.20	27.40	25.00	27.80	32.40	32.80	37.50	10.50	9.26	7.16	8.03
Zr	159	148	160	179	165	181	212	209	285	116	142	120	142
Nb	4.24	4.10	5.61	4.99	4.40	4.65	5.77	6.10	7.01	8.44	9.22	9.13	9.60
Ba	233	131	163	133	116	223	229	195	255	658	991	785	770
Hf	4.39	3.93	4.33	4.53	4.32	4.92	5.57	5.43	7.22	3.73	3.88	3.13	3.86
Ta	0.39	0.36	0.49	0.38	0.37	0.36	0.49	0.51	0.44	0.87	0.87	1.07	1.13
Pb	5.69	5.61	7.79	5.47	6.44	6.16	9.02	9.72	9.56	22.00	27.50	31.70	31.70
Th	2.77	2.41	3.48	2.62	2.74	2.98	4.27	4.28	5.60	8.69	8.87	22.20	25.60
U	0.74	0.62	0.92	1.24	0.78	0.83	1.90	1.27	1.75	1.92	1.50	1.45	1.76

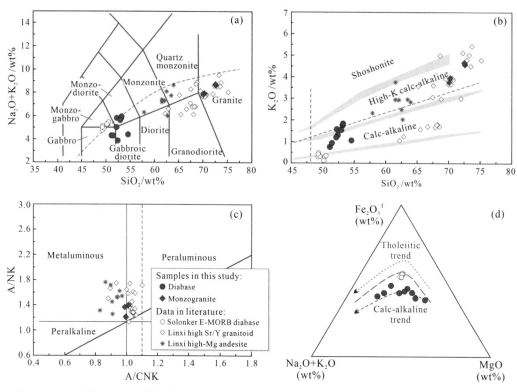

Fig. 6 (a) TAS diagram (Middlemost, 1994; Irvine and Baragar, 1971); (b) K₂O vs. SiO₂ diagram
(Rickwood, 1989); (c) A/NK vs. A/CNK diagram (Peccerillo and Taylor, 1976);
(d) AFM diagram (Irvine and Baragar, 1971).

The previously reported ~250 Ma Solonker E-MORB like diabases (Jian et al., 2010), the Linxi Triassic high-Mg
adakitic andesites (Liu et al., 2012) and the Early-Middle Triassic high Sr/Y granitoids (Li et al., 2017) are
plotted for comparison.

monzogranites show notable enrichment in LREE $[(La/Yb)_N = 12.29-40.80]$ and slightly
negative Eu anomalies ($Eu/Eu^* = 0.66 - 0.88$) in the chondrite-normalized diagram
(Fig. 8a). Their primitive mantle-normalized trace element diagrams are generally
characterized by enrichment in LILE, such as Rb, U and K, and depletion in Nb, Ta, Ti,
and P (Fig. 8b). Overall, the monzogranites in this study display geochemical characteristics
similar to those of the coeval adakitic rocks along the Solonker suture zone, such as the Early-
Middle Triassic high Sr/Y granitoids in the Linxi area (Li et al., 2017) and the Early Triassic
adakitic granitoids in the Central Jilin area (Wang et al., 2015a).

4.3 Sr-Nd-Pb isotopes

Detailed results of whole-rock Rb-Sr, Sm-Nd and Pb isotopes for the studied diabases
are given in Table 3. Initial values of all the samples have been calculated based on their
youngest concordant zircon U-Pb age of ~251 Ma. As shown in Fig. 10, the diabase samples

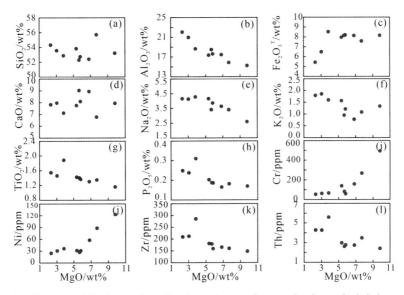

Fig. 7   Harker diagrams of MgO vs. selected major and trace elements for the studied diabase samples.

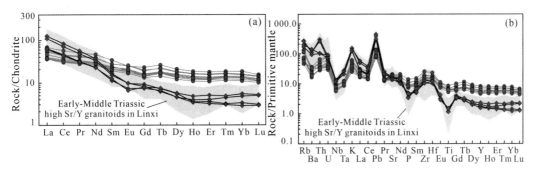

Fig. 8   (a) Chondrite-normalized REE diagram and (b) primitive mantle-normalized trace element diagram
for the studied diabase and monzogranite samples.

The values of chondrite and primitive mantle are from Boynton (1984) and Sun and McDonough (1989),
respectively. Data for the Early-Middle Triassic high Sr/Y granitoids in Linxi area are from Li et al. (2017).

have relatively variable initial $^{87}Sr/^{86}Sr$ values of 0. 705 86 − 0. 708 87 and $\varepsilon_{Nd}(t)$ values of −3. 1 to + 1. 5, with Nd model ages ranging from 1 473 Ma to 1 034 Ma, these characteristics are apparently different from the end-Permain mafic rocks in the Solonker area (Luo et al., 2016) and the Early Mesozoic ultramafic-mafic complexes along the northern NCC (Zhang et al., 2012), but similar to those of the Late Paleozoic enriched mantle-derived mafic intrusions in the region (Wang et al., 2015b). As for the Pb isotopic compositions, the initial ratios of $^{208}Pb/^{204}Pb$, $^{207}Pb/^{204}Pb$ and $^{206}Pb/^{204}Pb$ of the diabase samples vary from 35. 968 to 37. 346, 15. 448 to 15. 508 and 16. 280 to 17. 492, respectively, they plot above the North Hemisphere Reference Line (NHRL) and close to the compositional field of EM1 (Fig. 11).

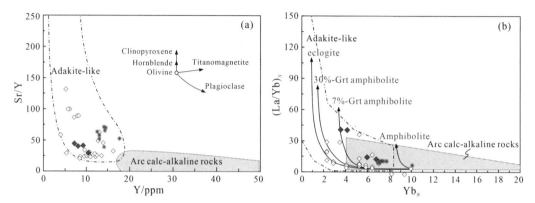

Fig. 9　Adakite discrimination diagrams (a) Sr/Y vs. Y and (b) (La/Yb)$_N$ vs. Yb$_N$

for the studied monzogranite samples.

Fields for adakite and arc calc-alkaline rocks are based on Petford and Atherton (1996), and Defant and Drummond (1990). The crystal fractionation paths of the primary minerals in Fig. 9a are from Castillo et al. (1999). The batch partial melting trends in Fig. 9b are based on Petford and Atherton (1996).

Table 3　Whole rocks Sr-Nd-Pb isotopic data for the studied diabase samples.

Sample	Ls01-d1-2	Ls01-d1-8	Ls01-d1-17	Ls01-d2-7	Ls03-d6-1
Rb/(μg/g)	39.46	86.87	48.36	39.51	75.61
Sr/(μg/g)	376.9	452.3	385	412.1	414.1
$^{87}Rb/^{86}Sr$	0.301 8	0.553 9	0.362 2	0.276 3	0.526 5
$^{87}Sr/^{86}Sr$	0.707 56	0.709 84	0.710 16	0.706 85	0.710 26
1δ	0.000 003	0.000 003	0.000 005	0.000 003	0.000 003
$(^{87}Sr/^{86}Sr)_i$	0.706 483	0.707 863	0.708 867	0.705 864	0.708 381
Sm/(μg/g)	5.13	5.474	4.319	4.163	6.707
Nd/(μg/g)	20.3	23.24	18.58	16.6	28.42
$^{147}Sm/^{144}Nd$	0.152 9	0.142 5	0.140 6	0.151 7	0.142 8
$^{143}Nd/^{144}Nd$	0.512 668	0.512 556	0.512 48	0.512 717	0.512 618
1δ	0.000 005	0.000 005	0.000 004	0.000 004	0.000 004
$(^{143}Nd/^{144}Nd)_i$	0.512 421	0.512 326	0.512 253	0.512 472	0.512 388
$\varepsilon_{Nd}(t)$	2.06	0.19	−1.23	3.05	1.4
T_{DM}/Ma	1 076	1 158	1 283	1 473	1 034
Pb/(μg/g)	5.47	1.54	3	1.38	3.12
U/(μg/g)	0.97	0.58	0.8	0.55	1.55
Th/(μg/g)	2.89	2.17	2.79	2.21	5.09
$^{206}Pb/^{204}Pb$	18.508	18.524	18.48	18.591	18.587
$^{207}Pb/^{204}Pb$	15.55	15.533	15.52	15.523	15.545
$^{208}Pb/^{204}Pb$	38.523	38.549	38.438	38.429	38.603
$(^{206}Pb/^{204}Pb)_i$	17.492	16.841	17.363	16.544	16.28
$(^{207}Pb/^{204}Pb)_i$	15.508	15.453	15.476	15.451	15.448
$(^{208}Pb/^{204}Pb)_i$	37.346	36.042	36.825	35.628	35.968

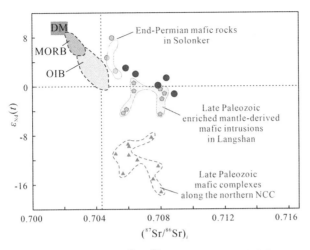

Fig. 10　Plot of $\varepsilon_{Nd}(t)$ vs. $({}^{87}Sr/{}^{86}Sr)_i$ for the studied diabase samples.

The fields for DM (depleted mantle), MORB (middle ocean ridge basalt) and OIB (ocean island basalt) are from Zindler and Hart (1986). Data for the Late Paleozoic ultramafic-mafic complexes from the northern NCC are from Zhang et al. (2009b). Data for the Late Paleozoic enriched mantle-derived mafic intrusions in Langshan area and the end-Permian mafic rocks in the Solonker area are from Wang et al. (2015b) and Luo et al. (2016), respectively.

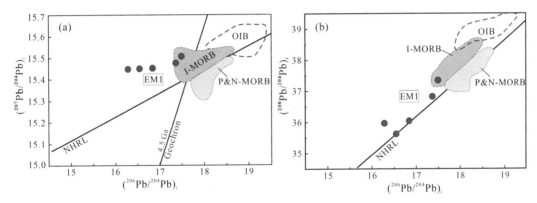

Fig. 11　(a) $({}^{206}Pb/{}^{204}Pb)_i$ vs. $({}^{207}Pb/{}^{204}Pb)_i$ and (b) $({}^{206}Pb/{}^{204}Pb)_i$ vs. $({}^{208}Pb/{}^{204}Pb)_i$ diagrams for the studied diabase samples.

The fields for I-MORB (Indian MORB), P&N-MORB (Pacific and North Atlantic MORB), OIB, EM1, NHLR and the 4.5 Ga geochron are from Zou et al. (2000), Zindler and Hart (1986) and Hart (1984).

4.4　Zircon Lu-Hf isotopes

The zircon grains of monzogranite sample Ls06-d1-tw1 which were previously used for U-Pb age dating were also selected for Lu-Hf isotopes on the same spot. Individual Lu-Hf isotopic data, as well as the $^{206}Pb/^{238}U$ ages of the corresponding spots, are given in Table 4. Twenty analyses on zircons from sample Ls06-d1-tw1 yielded initial $^{176}Hf/^{177}Hf$ ratios of 0.281 896 – 0.282 373, their $\varepsilon_{Hf}(t)$ values and the corresponding T_{DM2} range from −25.8 to −8.8 (Fig. 12) and 2 900 Ma to 1 833 Ma, respectively.

Table 4　Zircon Lu-Hf isotopic compositions of the studied monzogranite sample Ls06-d1-tw1.

Sample spot	Age /Ma	^{176}Hf/^{177}Hf	2σ	^{176}Lu/^{177}Hf	2σ	^{176}Yb/^{177}Hf	2σ	$\varepsilon_{Hf}(t)$	T_{DM1} /Ma	T_{DM2} /Ma	$f_{Lu/Hf}$
Ls06-d1-tw1. 01	245	0. 282 271	0. 000 024	0. 020 840	0. 000 081	0. 000 859	0. 000 003	−12. 5	1 379	2 064	−0. 97
Ls06-d1-tw1. 02	246	0. 282 227	0. 000 019	0. 024 826	0. 000 150	0. 000 958	0. 000 006	−14. 0	1 444	2 163	−0. 97
Ls06-d1-tw1. 03	246	0. 282 351	0. 000 019	0. 015 781	0. 000 135	0. 000 623	0. 000 005	−9. 6	1 260	1 883	−0. 98
Ls06-d1-tw1. 04	246	0. 282 373	0. 000 017	0. 013 352	0. 000 114	0. 000 550	0. 000 005	−8. 8	1 227	1 833	−0. 98
Ls06-d1-tw1. 05	248	0. 281 896	0. 000 020	0. 034 959	0. 000 328	0. 001 350	0. 000 012	−25. 8	1 923	2 900	−0. 96
Ls06-d1-tw1. 06	243	0. 282 267	0. 000 018	0. 015 929	0. 000 048	0. 000 665	0. 000 002	−12. 6	1 378	2 073	−0. 98
Ls06-d1-tw1. 07	242	0. 282 247	0. 000 020	0. 017 003	0. 000 428	0. 000 685	0. 000 016	−13. 4	1 406	2 118	−0. 98
Ls06-d1-tw1. 08	246	0. 281 930	0. 000 020	0. 036 772	0. 000 395	0. 001 405	0. 000 014	−24. 6	1 879	2 827	−0. 96
Ls06-d1-tw1. 09	246	0. 282 232	0. 000 016	0. 018 090	0. 000 126	0. 000 729	0. 000 005	−13. 8	1 428	2 148	−0. 98
Ls06-d1-tw1. 10	245	0. 282 229	0. 000 017	0. 017 079	0. 000 140	0. 000 683	0. 000 005	−13. 9	1 431	2 156	−0. 98
Ls06-d1-tw1. 11	247	0. 282 232	0. 000 020	0. 016 948	0. 000 131	0. 000 691	0. 000 005	−13. 8	1 427	2 148	−0. 98
Ls06-d1-tw1. 12	246	0. 282 228	0. 000 020	0. 015 719	0. 000 280	0. 000 638	0. 000 011	−13. 9	1 430	2 157	−0. 98
Ls06-d1-tw1. 13	245	0. 282 306	0. 000 020	0. 013 909	0. 000 099	0. 000 565	0. 000 004	−11. 2	1 320	1 983	−0. 98
Ls06-d1-tw1. 14	246	0. 282 100	0. 000 019	0. 021 041	0. 000 066	0. 000 838	0. 000 003	−18. 5	1 616	2 445	−0. 97
Ls06-d1-tw1. 15	246	0. 282 351	0. 000 018	0. 013 326	0. 000 214	0. 000 527	0. 000 008	−9. 6	1 257	1 883	−0. 98
Ls06-d1-tw1. 16	246	0. 282 275	0. 000 022	0. 014 781	0. 000 150	0. 000 585	0. 000 005	−12. 3	1 364	2 053	−0. 98
Ls06-d1-tw1. 17	246	0. 282 155	0. 000 021	0. 018 748	0. 000 121	0. 000 748	0. 000 004	−16. 5	1 535	2 320	−0. 98
Ls06-d1-tw1. 18	243	0. 282 175	0. 000 019	0. 016 144	0. 000 044	0. 000 668	0. 000 002	−15. 9	1 504	2 277	−0. 98
Ls06-d1-tw1. 19	245	0. 282 040	0. 000 019	0. 024 647	0. 000 176	0. 000 986	0. 000 007	−20. 7	1 704	2 578	−0. 97
Ls06-d1-tw1. 20	246	0. 282 239	0. 000 019	0. 016 306	0. 000 111	0. 000 650	0. 000 004	−13. 6	1 416	2 134	−0. 98

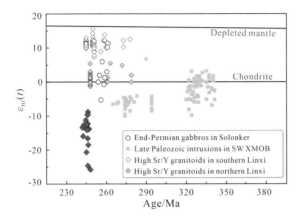

Fig. 12　Zircon $\varepsilon_{Hf}(t)$ vs. U-Pb age diagram for the studied monzogranite sample Ls06-d1-tw1.

Data source for the Triassic high Sr/Y granitoids in northern and southern Linxi area are from Li et al. (2017). The end-Permian gabbros in Solonker (Luo et al., 2016) and the Late Paleozoic intrusions in SW XMOB (Liu et al., 2016; Peng et al., 2013; Pi et al., 2010) are also shown for comparison.

5　Discussion

5. 1　Petrogenesis of the diabases

5. 1. 1　Formation age

　　Since the diabase dykes in this study are mostly hosted by the Late Paleozoic granitoids and unconformably covered by the Early Cretaceous Guyang Formation in the field, their formation ages can be roughly restricted to Late Paleozoic to Early Mesozoic. Besides, the pre-Mesozoic geological units in the Langshan area are pervasively deformed (Zhang et al., 2013; Darby and Ritts, 2007), whereas the studied diabases are undeformed and unmetamorphosed. Thus, the timing of the youngest deformation in the region can provide constraint on the upper age limit of the studied diabase. As reported by Lin et al. (2014), the Early Permian (~290 Ma) granitic porphyries in Urad Houqi are generally characterized by NE-striking and N-vergent orientations and structures, and they are the youngest deformed intrusions in the Langshan area, which implies that the diabase dykes in this study must have been formed after Early Permian.

　　Although zircons are volumetrically minor in mafic-ultramafic magma and zircon grains obtained from the mafic-ultramafic rocks are mostly captured, previous studies have shown that high-precision zircon U-Pb dating remains useful in placing constraints on the formation ages of mafic-ultramafic rocks (e.g., Yu et al., 2017; Zhang et al., 2009b). Our U-Pb dating results indicate that the zircon $^{206}Pb/^{238}U$ ages acquired from diabase sample Ls23-d2-tw2 are varying from 248 Ma to 1 874 Ma. The older ages may be xenocrysts that potentially captured during the magma ascent, but nine relatively younger zircons yielded two weighted mean $^{206}Pb/^{238}U$ ages of ~272 Ma and ~251 Ma (Fig. 5b). Considering that the Early Permian igneous rocks are widely distributed in the southwestern XMOB (e. g. , Wang et al. , 2015b; Lin et al. , 2014; Peng et al. , 2013; Zhang et al. , 2009a), we propose that the weighted mean age of ~272 Ma was more likely to be in relation to the Early Permian magmatism in the region. The four zircons with ages of ~251 Ma are relatively larger in size and exhibit developed oscillatory zoning in the CL images (Fig. 4), these features are more akin to those of zircons from the intermediate-felsic magmatic rocks instead of the mafic magma, suggesting that the studied diabase probably formed after ~251 Ma. Moreover, previous studies show that extensive Middle-Late Triassic magmatism (including the Middle Triassic high Sr/Y monzogranite in this study) was developed in southwestern XMOB (e. g. , Zhang et al. , 2014a and references therein). However, no Middle-Late Triassic zircon is observed in sample Ls23-d2-tw2, which means that the studied diabase is unlikely to be formed after Middle-Late Triassic. Consequently, we assume that Early Triassic, which is slightly older than the studied monzogranite (245.5 ± 0.7 Ma), might be the best estimate of the formation age of the diabases. The Langshan Early Triassic diabase, together with the recently identified coeval mafic rocks in Wengunshan (Lin et al. , 2014), Mandula area (Jian et al. , 2010) and Solonker area (Luo et al. , 2016; Jian et al. , 2010), indicate the presence of the end-Permian to Early Triassic magmatic flare-up in southwestern XMOB.

5. 1. 2　Crustal contamination and fractional crystallization

Presence of the Archean and Proterozoic zircon grains (Fig. 5a), together with the positive Zr-Hf-Pb and negative Nb-Ta anomalies in the primitive mantle-normalized diagram of the diabase samples (Fig. 8b), indicate that the parental magma may have undergone crustal contamination during magma ascent. However, all the diabases are mafic with low concentrations of SiO_2(51. 13 – 54. 55 wt%) and relatively high $Mg^{\#}$ values (43. 2 – 68. 2), suggesting that bulk crustal contamination is unlikely. Their low Lu/Yb ratios (0. 13 – 0. 14), narrow range of (Th/Yb)$_N$ values (5. 41 – 8. 39), consistent LREE and LILE distribution patterns (Fig. 8), as well as relatively high $\varepsilon_{Nd}(t)$ values (– 3. 1 to + 1. 5), are also inconsistent with significant crustal contamination. Besides, significant crustal contamination would result in linear correlations between $Mg^{\#}$ values and $\varepsilon_{Nd}(t)$ values, (Th/Nb)$_N$ and Nb/La ratios, but such characteristics are not observed the diabase samples. Thus, crustal contamination, although it cannot be ruled out, does not appear to be a significant role in the petrogenesis of the studied diabases.

The diabase samples have variable MgO (2. 27 – 9. 66 wt%), Ni (24. 5 – 123 ppm) and Cr (53. 6 – 494 ppm) contents, indicating that fractional crystallization might have occurred prior to their emplacement. In the Harker diagram (Fig. 7), the positive correlations between MgO and $Fe_2O_3^T$, Cr and Ni and the negative correlations between MgO and SiO_2 indicate considerable degree of fractionation of olivine and/or clinopyroxene. The increasing TiO_2 with decreasing MgO, together with the relatively consistent CaO contents imply that the parental magmas only underwent a minor degree of clinopyroxene fractionation (Fig. 7d, g). Although Al_2O_3 is negatively correlated with MgO (Fig. 7b), the negligible Sr and Eu anomalies (Fig. 8) suggest that plagioclase fractionation was insignificant. Moreover, the roughly negative correlations between MgO and TiO_2, P_2O_5 (Fig. 7), along with the negligible Ti anomalies (Fig. 8b), argue against significant fractionation of the Fe-Ti oxides and apatite. Therefore, olivine with a minor amount of clinopyroxene are the major fractionated minerals of the studied diabases.

5. 1. 3 Nature of mantle sources

Overall, all the diabase samples have similar trace element and Sr-Nd-Pb isotopic characteristics, so their derivation can be ascribed to a common magma source. The low SiO_2 contents (51. 13–54. 55 wt%) and relatively high MgO, Cr, Ni contents suggest that they were likely originated from mantle-derived components rather than a crustal source (e.g., Rapp et al., 2003). Isotopically, these rocks display relatively high initial ratios of $^{87}Sr/^{86}Sr$ (0. 705 9–0. 708 9), with $\varepsilon_{Nd}(t)$ values and Nd model ages respectively ranging from −3. 1 to +1. 5 (Fig. 10) and 1 473 Ma to 1 034 Ma, implying an enriched lithospheric mantle source. This is also consistent with their relatively high initial ratios of $^{208}Pb/^{204}Pb$ (35. 968–37. 346), $^{207}Pb/^{204}Pb$ (15. 448–15. 508) and $^{206}Pb/^{204}Pb$ (16. 280–17. 492).

Arc-like geochemical signatures, such as enrichment in LREE and LILE relative to heavy REE (HREE) and HFSE, are generally related to either extensively crustal contamination or partial melting of an enriched mantle source (e.g., Zhang et al., 2011). As significant crustal contamination can be excluded, the "arc-like" trace element patterns (Fig. 8) and the enriched isotopic signatures (Fig. 10) of the studied diabase were therefore mainly inherited from an enriched lithospheric mantle source. Depletion of HFSE in the primitive mantle normalized diagrams is generally caused by fluid-related and/or melt-related metasomatism in subduction process (e.g., Ma et al., 2014; Duggen et al., 2005; Thirlwall et al., 1994). Thus, we propose that the mantle source of the studied diabase may have been modified by subduction-related fluids and/or melt. Furthermore, the LILE enriched geochemical features are calling for a LILE-enriched mantle source, which means that volatile-bearing minerals (such as phlogopite and amphibole) might be the major host phases in the mantle source origin (e.g., Ionov et al., 1997; Foley et al., 1996). Since melts in equilibrium with amphibole generally have significantly lower Rb/Sr (< 0. 1) and higher Ba/Rb (> 20) ratios, whereas melts of phlogopite-bearing source are expected to show significantly lower Ba/Rb ratios (Ma et al., 2014; Furman and Graham, 1999). The relatively low Ba/Rb ratios (2. 72–6. 56) and variable Rb/Sr ratios (0. 08 – 0. 19) of the diabases in this study indicate predominance of phlogopite in their mantle source (Fig. 13a), implying that metasomatism by volatile-rich melts must have occurred prior to the partial melting. Moreover, partial melting in the spinel-stability field generally has low Dy/Yb ratios (<1.5), while melting in the garnet-stability field shows much higher Dy/Yb ratios (>2.5) (e.g., Ma et al., 2014). The diabase samples in this study have Dy/Yb ratios ranging from 1.67 to 1.80, indicating that the partial melting was probably operated within the spinel-garnet transition zone lherzolite (Ma et al., 2014; Jiang et al., 2010; Duggen et al., 2005). However, the relatively low Ce/Y (0.97−1.39) and (Tb/Yb)$_N$ ratios (1.31−1.45) of the diabase samples indicate that the melts were more likely to be derived from a spinel-stability mantle source (McKenzie and Bickle, 1988). In the Sm/Yb vs. La/Sm and Lu/Hf vs. Sm/Nd diagrams, all the diabase samples plot closed to the 1%−5% partial melting curve of spinel facies (Fig. 13b,c), indicating that parental magma of the diabases were likely derived from the low degree partial melting of the phlogopite-bearing lherzolite in the spinel-stability field. Compared with the coeval enriched lithospheric mantle derived mafic rocks in the Solonker area (Jian et al., 2010), the diabases in this study have relatively lower $\varepsilon_{Nd}(t)$ values (−3.1 to +1.5), older Nd model ages (1 473−1 034 Ma) and variable initial $^{87}Sr/^{86}Sr$ ratios (0.705 9−0.708 9) (Fig. 10). Such isotopic characteristics

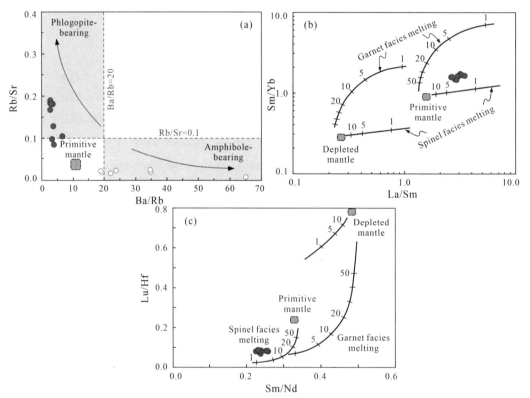

Fig. 13 (a) Ba/Rb vs. Rb/Sr (Furman and Graham, 1999), (b) La/Sm vs. Sm/Yb, and (c) Sm/Nd vs. Lu/Hf diagrams for the studied diabase samples.
The melting curves in Fig. 13b,c are from Zhao and Asimow (2014).

might point to a relatively heterogeneous SCLM (subcontinental lithospheric mantle) source beneath the southwestern XMOB.

5. 2　Petrogenesis of the monzogranites

The monzogranite samples in this study are characterized by high Sr/Y (30−45) and La/Yb ratios (Fig. 9), with relatively high Al_2O_3 (14. 15−15. 98 wt%) and Sr (310−380 ppm) contents and low Y (7. 16−10. 50 ppm) and Yb (0. 60−1. 15 ppm) contents, indicating geochemical characteristics of adakite-like rocks (Martin et al., 2005; Defant and Drummond, 1990). Generally, adakite or adakite-like rocks can be generated by the partial melting of subducted oceanic slab (e. g., Defant and Drummond, 1990), the partial melting of delaminated lower crust (Wang et al., 2006; Kay and Kay, 1993), the partial melting of thickened lower crust (Chung et al., 2003; Atherton and Petford, 1993) and the differentiation of parental basaltic magmas (Castillo, 2012; Prouteau and Scaillet, 2003).

The studied monzogranites possess high SiO_2 (70. 09−72. 63 wt%), low MgO (0. 40−0. 87 wt%), Cr (4. 74−8. 31 ppm) and Ni (1. 15−6. 35 ppm) contents, these features are more akin to the thickened lower crust-derived adakites and metabasaltic and eclogite experimental melts (Fig. 14) (Wang et al., 2006), rather than the subducted oceanic slab-derived or delaminated lower crust-derived adakites (Martin et al., 2005). Their uniform geochemistry, together with the absence of mafic minerals (e. g., amphibole and clinopyroxene), do not support an origin from primary basaltic magma by crustal assimilation and fractional crystallization processes (Castillo et al., 1999). However, the relatively flat HREE patterns (Fig. 8a), as well as the high Zr/Sm ratios (40−64) of these monzogranite

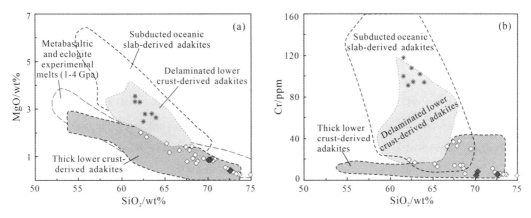

Fig. 14　(a) MgO vs. SiO_2 and (b) Cr vs. SiO_2 diagrams for the studied monzogranites.
Fields for the subducted oceanic slab-derived adakites, delaminated lower crust-derived adakites, thick lower crust-derived adakites and metabasaltic and eclogite experimental melts (1−4 GPa) are after Wang et al. (2006). The Triassic high-Mg adakitic andesites (Liu et al., 2012) and Early-Middle Triassic high Sr/Y granitoids (Li et al., 2017) in the Linxi area are also plotted for comparison.

samples suggest amphibole-dominated residual phases in the source region, which is consistent with the coeval high Sr/Y granitoids in the Linxi area (Li et al., 2017). Thus, these rocks are unlikely to be originated from a significantly thickened lower crust. Recent partial melting experiments shows that adakite-like magmas can also be produced by the low degree partial melting (10% – 40%) of a slightly thickened mafic lower crust with P-T condition of 1 – 1.25 GPa and 800 – 950 ℃ (corresponding to a depth of 30 – 40 km; Qian and Hermann, 2013). Considering that only moderate thickening of the crust (40 – 45 km) occurred during the final orogenic evolution in the southern CAOB (Li et al., 2016a, 2017; Zhang et al., 2014b), we propose that the low degree partial melting of the moderately thickened lower crust is likely to be a plausible explanation for the genesis of the studied high Sr/Y monzogranites.

As shown in the $\varepsilon_{Hf}(t)$ vs. U-Pb age diagram (Fig. 12), the Early-Middle high Sr/Y granitoids in northern Linxi have positive zircon $\varepsilon_{Hf}(t)$ values of +7.3 to +15.6, while those in southern Linxi have weak negative to positive zircon $\varepsilon_{Hf}(t)$ values of −1.3 to +5.4 (Li et al., 2017). These characteristics are apparently different from the Middle Triassic high Sr/Y monzogranites in this study, which are characterized by negative zircon $\varepsilon_{Hf}(t)$ values (−25.8 to −8.8) and old T_{DM2} model ages (2 900 – 1 833 Ma). This kind of spatial distribution may indicate that the lower crust source region of the Early-Middle Triassic high Sr/Y rocks in the southwestern XMOB was dominated by ancient crustal materials, the source region south of the Solonker suture zone got a greater contribution from juvenile crustal materials, whereas the area north of the Solonker suture zone was dominated by juvenile components.

5.3　Tectonic implications

The final closure of the Paleo-Asian ocean and the final continental amalgamation/collision were suggested to begin with the initial collision between the SMT and the NCC during the Middle Permian (e.g., Li et al., 2014, 2016a, 2017; Xiao et al., 2015; Jian et al., 2010), leading to the cessation of the arc magmatism and marine sedimentation, accompanied by a short magmatic hiatus and a regional angular unconformity in the XMOB (Li et al., 2014, 2016b; Eizenhöfer et al., 2014). Afterwards, the terminal collision triggered a regional greenschist-blueschist facies metamorphism along the Solonker suture zone in the Late Permian (Li et al., 2016a, 2017; Jian et al., 2010), which marks the onset of the final episode of orogenic evolution in the XMOB.

As stated above, the Early Triassic diabases in this study were derived from the enriched lithospheric mantle source beneath the southwestern XMOB. Considering that subduction of the Paleo-Asian Ocean had ceased since the Middle-Late Permian, we assume that the slab break-off following the terminal collision was a plausible mechanism for the generation of the studied diabases. Slab break-off can significantly increase the temperature of the overlying lithosphere due to the upwelling of the hot asthenosphere through the slab window, leading to partial melting of the overlying lithospheric mantle, as well as crustal anatexis (e.g., van de Zedde and Wortel, 2001). Likewise, Jian et al. (2010) suggested that the end-Permian to Early Triassic igneous rocks (255−250 Ma) in the Mandula forearc mélange were derived from the decompression melting of mantle peridotite by heat from the upwelling asthenosphere during the slab break-off. Li et al. (2016a) also proposed that the end-Permian to Early Triassic magmatic flare-up in the southeastern Xilinhot area was in response to the final slab break-off in the XMOB.

The Middle Triassic high Sr/Y monzogranites in this study, combined with the Early-Middle Triassic high Sr/Y granitoids in Linxi area (Li et al., 2017) and central Jilin (Wang et al., 2015a), indicate moderate thickening and shortening of the crust. Thus, considerable contraction must have occurred after the slab break-off, which might be in relation to the ongoing convergence between the NCC and the SMT. The heat from the underplated basaltic magma may lead to: ① partial melting of the moderately thickened ancient lower crust beneath the southwestern XMOB to generate high Sr/Y magma with negative $\varepsilon_{Hf}(t)$ values, and ② partial melting of the moderately thickened juvenile lower crust beneath the Solonker suture zone to produce high Sr/Y magma with positive $\varepsilon_{Hf}(t)$ values. In addition, recent studies revealed that the aqueous fluids released from an residual oceanic slab in the upper mantle also had participated in the generation of the Early-Middle Triassic high Sr/Y magmatism along the Solonker suture zone (Li et al., 2017; Song et al., 2015; Zhang et al., 2014b; Liu et al., 2012; Jian et al., 2010).

The Paleozoic tectonic development of the XMOB was overall dominated by the evolution of the Paleo-Asian Ocean between the NCC and the SMT (e.g., Eizenhöfer et al., 2014, 2015; Xiao et al., 2003, 2015; Windley et al., 2007). During Carboniferous to Early Permian, the progressively double-sided subduction of the Paleo-Asian Ocean enabled an Andean-style continental margins to develop on the northern NCC and the SMT (Liu et al., 2013; Chen et al., 2009; Zhang et al., 2009a; Xiao et al., 2003), accompanied by extensive arc magmatism (e.g., Li et al., 2016b; Chen et al., 2009; Zhang et al., 2009a). The initial collision between the NCC and SMT in Middle Permian prevented further oceanic subduction and terminated the arc-related magmatism. The Paleo-Asian Ocean in the XMOB ultimately closed during the Late Permian, leading to the formation of the Solonker suture zone (Li et al., 2017; Xiao et al., 2015; Jian et al., 2010; Zhang et al., 2009a). The slab break-off of the Paleo-Asian oceanic lithosphere switched the regionally tectonic regime from the prolonged compression to a temporary extension, and then resulted in the magmatic flare-up during end-Permian to Early Triassic (Fig. 15a), such as the Langshan diabase dykes in this study, the E-MORB diabase, sanukitoid, high-Mg diorite and anorthosite in the Solonker ophiolitic mélanges (Jian et al., 2010), as well as those collision-related calc-alkaline granitoids in the southeastern Xilinhot area (Li et al., 2016a). Subsequently, the ongoing weak convergence between the NCC and the SMT initiated moderately crustal thickening and uplift, generating the Early-Middle Triassic high Sr/Y magmas and the high-Mg adakitic andesites (Fig. 15b) (Li et al., 2017; Liu et al., 2012). The post-orogenic extension in the XMOB might bestarted in the Late Triassic, as evidenced by the 224−208 Ma metamorphic core complex in the Sonid Zuoqi area (Davis et al., 2004), as well as the widespread A-type granitic rocks and alkaline complexes along the northern NCC (Zhang et al., 2012, 2014a). By the end of Early Mesozoic, tectonic evolution of the XMOB was overall controlled by the Mongol-Okhotsk tectonic regime to the northwest and the Paleo-Pacific tectonic regime to the east (e.g., Zhou and Wilde, 2013).

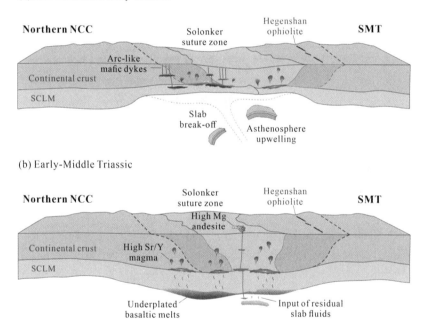

Fig. 15　Schematic illustrations showing the genesis model and geodynamic environment of end-Permian to Early-Middle Triassic magmatism during the final orogenic evolution of the XMOB.
SCLM: sub-continental lithospheric mantle.

6　Conclusions

(1)The studied diabases and monzogranites were respectively formed during Early Triassic and Middle Triassic, corresponding to the end-Permian to Early-Middle Triassic magmatic events identified in the southwestern XMOB.

(2)The Early Triassic diabases were derived from the low degree partial melting of a subduction-metasomatized lithospheric mantle followed by fractionation of olivine and minor clinopyroxene, the Middle Triassic monzogranites were generated by the low degree partial melting of a moderately thickened ancient lower crust.

(3) The diabases were resulted from the slab break-off of the Paleo-Asian oceanic lithosphere following the terminal collision between the NCC and the SMT in the southwestern XMOB, while the monzogranites were resulted from the subsequently crustal thickening and uplift.

Acknowledgements This study was jointly supported by the Geological Survey of China (No.1212011085490) and the National Natural Science Foundation of China (No.41421002). We sincerely thank the editors and the anonymous reviewers for their critical and constructive comments. We are grateful to Guangqiang Xiong, Hongtao Zhao, Quanliu Chen, and Zhong Wang for their help with the field work. We thank Yu Zhu, Fangyi Zhang and Zezhong Zhang for their laboratory assistance. We also appreciate Yuan Yuan and Yaoyao Zhang for their insightful suggestions. The final publication is available at Springer via https://doi. org/ 10. 1007/s12583-019-1015-5.

References

Andersen, T., 2002. Correction of common lead in U-Pb analyses that do not report ^{204}Pb. Chemical Geology, 192(1/2): 59-79.

Atherton, M.P., Petford, N., 1993. Generation of sodium-rich magmas from newly underplated basaltic crust. Nature, 362(6416): 144-146.

Bao, Z.A., Chen, L., Zong, C.L., et al., 2017. Development of pressed sulfide powder tablets for in situ sulfur and lead isotope measurement using LA-MC-ICP-MS. International Journal of Mass Spectrometry, 421: 255-262.

Boynton, W.V., 1984. Cosmochemistry of the rare earth elements: Meteorite studies. In: Henderson, P.E. Rare Earth Element Geochemistry. Amsterdam: Elsevier: 63-114.

Castillo, P.R., 2012. Adakite petrogenesis.Lithos, 134-135(3): 304-316.

Castillo, P.R., Janney, P.E., Solidum, R.U., 1999. Petrology and geochemistry of Camiguin Island, southern Philippines: Insights to the source of adakites and other lavas in a complex arc setting. Contributions to Mineralogy and Petrology, 134(1): 33-51.

Cawood, P.A., Kröner, A., Collins, W.J., et al., 2009. Accretionary orogens through Earth history. Geological Society of London, Special Publications, 318(1): 1-36.

Chen, B., Ma, X.H., Liu, A.K., et al., 2009. Zircon U-Pb ages of the Xilinhot metamorphic complex and blueschist and implications for tectonic evolution of the Solonker suture.Acta Petrologica Sinica, 25(12): 3123-3129 (in Chinese with English Abstract).

Chung, S.L., Liu, D., Ji, J., et al., 2003. Adakites from continental collision zones: Melting of thickened lower crust beneath southern Tibet.Geology, 31(11): 1021-1024.

Condie, K.C., Belousova, E., Griffin, W.L., et al., 2009. Granitoid events in space and time: Constraints from igneous and detrital zircon age spectra. Gondwana Research, 15(3): 228-242.

Darby, B. J, Ritts, B. D., 2007. Mesozoic structural architecture of the Lang Shan, North-Central China: Intraplate contraction, extension, and synorogenic sedimentation. Journal of Structural Geology, 29(12): 2006-2016.

Davis, G.A., Xu, B., Zheng, Y.D., et al., 2004. Indosinian extension in the Solonker suture zone: The Sonid Zuoqi metamorphic core complex, Inner Mongolia, China. Earth Science Frontiers, 11(3): 135-143.

de Jong, K., Xiao, W.J., Windley, B.F., et al., 2006. Ordovician ^{40}Ar/^{39}Ar phengite ages from the blueschist-facies Ondor Sum subduction-accretion complex (Inner Mongolia) and implications for the early Paleozoic history of continental blocks in China and adjacent areas. American Journal of Science, 306 (10): 799-845.

Defant, M.J., Drummond, M.S., 1990. Derivation of some modern arc magmas by melting of young subducted lithosphere. Nature, 347(6294): 662-665.

Duggen, S., Hoernle, K., Bogaard, P., et al., 2005. Post-collisional transition from subduction- to intraplate-type magmatism in the westernmost Mediterranean: Evidence for continental-edge delamination of subcontinental lithosphere. Journal of Petrology, 46(6): 1155-1201.

Eizenhöfer, P.R., Zhao, G.C, Zhang, J., et al., 2015. Geochemical characteristics of the Permian basins and their provenances across the Solonker Suture Zone: Assessment of net crustal growth during the closure of the Palaeo-Asian Ocean. Lithos, 224-225: 240-255.

Eizenhöfer, P.R., Zhao, G.C., Zhang, J., et al., 2014. Final closure of the Paleo-Asian Ocean along the Solonker suture zone: Constraints from geochronological and geochemical data of Permian volcanic and sedimentary rocks. Tectonics, 33(4): 441-463.

Ferrari, L., 2004. Slab detachment control on mafic volcanic pulse and mantle heterogeneity in central Mexico. Geology, 32(1): 77-80.

Foley, S.F., Jackson, S.E., Fryer, B.J., et al., 1996. Trace element partition coefficients for clinopyroxene and phlogopite in an alkaline lamprophyre from Newfoundland by LAM-ICP-MS. Geochimica et Cosmochimica Acta, 60(4): 629-638.

Furman, T., Graham, D., 1999. Erosion of lithospheric mantle beneath the East African Rift system: Geochemical evidence from the Kivu volcanic province. Developments in Geotectonics, 24: 237-262.

Gao, S., Liu, X.M., Yuan, H.L., et al., 2002. Determination of Forty Two Major and Trace Elements in USGS and NIST SRM Glasses by Laser Ablation-Inductively Coupled Plasma-Mass Spectrometry. Geostandards and Geoanalytical Research, 26: 181-196.

Hart, S.R., 1984. A large-scale isotope anomaly in the southern hemisphere mantle. Nature, 309(5971): 753-757.

Hoskin, P.W.O., Schaltegger, U., 2003. The composition of zircon and igneous and metamorphic petrogenesis. Reviews in Mineralogy and Geochemistry, 53(1): 27-62.

Hu, J.M., Gong, W., Wu, S., et al., 2014. LA-ICP-MS zircon U-Pb dating of the Langshan Group in the northeast margin of the Alxa block, with tectonic implications. Precambrian Research, 255: 756-770.

Ionov, D.A., Griffin, W.L., O'Reilly, S.Y., 1997. Volatile-bearing minerals and lithophile trace elements in the upper mantle. Chemical Geology, 141(3-4): 153-184.

Irvine, T. N., Baragar, W. R. A., 1971. A guide to the chemical classification of common volcanic rocks. Canadian Journal of Earth Sciences, 8(5): 523-548.

Jahn, B.M., Wu, F.Y., Chen, B., 2000. Granitoids of the Central Asian Orogenic Belt and continental growth in the Phanerozoic. Transactions of the Royal Society of Edinburgh: Earth Sciences, 91(1-2): 181-193.

Jian, P., Liu, D.Y., Kröner, A., et al., 2010. Evolution of a Permian Intraoceanic arc-trench system in the Solonker suture zone, Central Asian Orogenic Belt, China and Mongolia. Lithos, 118(1): 169-190.

Jian, P., Liu, D.Y., Kröner, A., et al., 2008. Time scale of an early to mid-Paleozoic orogenic cycle of the long-lived Central Asian Orogenic Belt, Inner Mongolia of China: Implications for continental growth. Lithos, 101(3): 233-259.

Jiang, Y.H., Jiang, S.Y., Ling, H.F., et al., 2010. Petrogenesis and tectonic implications of Late Jurassic shoshonitic lamprophyre dikes from the Liaodong Peninsula, NE China. Mineralogy and Petrology, 100(3-4): 127-151.

Kay, R.W., Kay, S.M., 1993. Delamination and delamination magmatism. Tectonophysics, 219: 177-189.

Li, H.K., Geng, J.Z., Hao, S., et al., 2009. The study of zircon U-Pb dating by means LA-MC-ICP-MS. Bulletin of Mineralogy, Petrology and Geochemistry, 28 (Suppl.): 77 (in Chinese).

Li, S., Chung, S.L., Wilde, S.A., et al., 2017. Early-Middle Triassic high Sr/Y granitoids in the southern Central Asian Orogenic Belt: Implications for ocean closure in accretionary orogens. Journal of Geophysical Research: Solid Earth, 122(6): 2291-2309.

Li, S., Chung, S.L., Wilde, S.A., et al., 2016a. Linking magmatism with collision in an accretionary orogen. Scientific Reports, 6: 25751.

Li, S., Wilde, S.A., He, Z.H., et al., 2014. Triassic sedimentation and postaccretionary crustal evolution along the Solonker suture zone in Inner Mongolia, China. Tectonics, 33(6): 960-981.

Li, S., Wilde, S.A., Wang, T., et al., 2016b. Latest Early Permian granitic magmatism in southern Inner Mongolia, China: Implications for the tectonic evolution of the southeastern Central Asian Orogenic Belt. Gondwana Research, 29(1): 168-180.

Lin, L.N., Xiao, W.J., Wan, B., et al., 2014. Geochronologic and geochemical evidence for persistence of south-dipping subduction to late Permian time, Langshan area, Inner Mongolia (China): Significance for termination of accretionary orogenesis in the southernAltaids. American Journal of Science, 314 (2): 679-703.

Liu, J.F., Li, J.Y., Chi, X.G., et al., 2013. A late-Carboniferous to early early-Permian subduction-accretion complex in Daqing pasture, southeastern Inner Mongolia: Evidence of northward subduction beneath the Siberian paleoplate southern margin. Lithos, 177(2): 285-296.

Liu, M., Zhang, D., Xiong, G. Q., et al., 2016. Zircon U-Pb age, Hf isotope and geochemistry of Carboniferous intrusions from the Langshan area, Inner Mongolia: Petrogenesis and tectonic implications. Journal of Asian Earth Sciences, 120: 139-158.

Liu, Y.S., Wang, X.H., Wang, D.B., et al., 2012. Triassic high-Mg adakitic andesites from Linxi, Inner Mongolia: Insights into the fate of the Paleo-Asian ocean crust and fossil slab-derived melt-peridotite interaction. Chemical Geology, 328(11): 89-108.

Liu, Y., 2012. Geochemical and Chronological Characteristics of the Granitic Gneisses and Intrusive Rocks from Dongshengmiao Region, Inner Mongolia and Their Tectonic Implications [Dissertation]. Lanzhou: Lanzhou University: 21-46 (in Chinese).

Ludwig, K.R., 2003. ISOPLOT 3.0: A Geochronological Toolkit for Microsoft Excel. Geochronology Center: Special Publication, Berkeley, 4.

Luo, Z.W., Xu, B., Shi, G.Z., et al., 2016. Solonker ophiolite in Inner Mongolia, China: A late Permian continental margin-type ophiolite. Lithos, 261: 72-91.

Ma, L., Jiang, S.Y., Hofmann, A.W., et al., 2014. Lithospheric and asthenospheric sources of lamprophyres in theJiaodong Peninsula: A consequence of rapid lithospheric thinning beneath the North China Craton? Geochimica et Cosmochimica Acta, 124(1): 250-271.

Ma, S.W., Liu, C.F., Xu, Z.Q., et al., 2017. Geochronology, geochemistry and tectonic significance of the early Carboniferous gabbro and diorite plutons in West Ujimqin, Inner Mongolia. Journal of Earth Science, 28(2): 249-264.

Martin, H., Smithies, R.H., Rapp, R., et al., 2005. An overview of adakite, tonalite-trondhjemite-granodiorite (TTG), and sanukitoid: Relationships and some implications for crustal evolution. Lithos, 79(1): 1-24.

McKenzie, D.P., Bickle, M.J., 1988. The volume and composition of melt generated by extension of the lithosphere. Journal of Petrology, 29(3): 625-679.

Miao, L.C., Fan, W.M., Liu, D.Y., et al., 2008. Geochronology and geochemistry of the Hegenshan ophiolitic complex: Implications for late-stage tectonic evolution of the Inner Mongolia-Daxinganling Orogenic Belt, China. Journal of Asian Earth Sciences, 32(5): 348-370.

Middlemost, E.A.K., 1994. Naming materials in the magma/igneous rock system.Earth-Science Reviews, 37(3-4): 215-224.

Peccerillo, A., Taylor, S.R., 1976. Geochemistry of Eocene calc-alkaline volcanic rocks from the Kastamonu area, Northern Turkey. Contributions to Mineralogy and Petrology, 58(1): 63-81.

Peng, R.M., Zhai, Y.S., Li, C.S., et al., 2013. The Erbutu Ni-Cu deposit in the Central Asian Orogenic Belt: A Permian magmatic sulfide deposit related to boninitic magmatism in an arc setting. Economic Geology, 108(8): 1879-1888.

Peng, R.M., Zhai, Y.S., Wang, J.P., et al., 2010. Discovery of Neoproterozoic acid volcanic rock in the south-western section of Langshan, Inner Mongolia.Chinese Science Bulletin, 55(26): 2611-2620 (in Chinese with English abstract).

Petford, N., Atherton, M., 1996. Na-rich partial melts from newly underplated basaltic crust: The Cordillera Blanca Batholith, Peru. Journal of Petrology, 37(6): 1491-1521.

Pi, Q.H., Liu, C.Z., Chen, Y.L., et al., 2010. Formation epoch and genesis of intrusive rocks in Huogeqi ore field of Inner Mongolia and their relationship with copper mineralization. Mineral Deposits, 29(3): 437-451 (in Chinese with English abstract).

Prouteau, G., Scaillet, B., 2003. Experimental constraints on the origin of the Pinatubo dacite. Journal of Petrology, 44(12): 2203-2241.

Qian, Q., Hermann, J., 2013. Partial melting of lower crust at 10−15 kbar: constraints on adakite and TTG formation. Contributions to Mineralogy and Petrology, 165(6): 1195-1224.

Rapp, R.P., Shimizu, N., Norman, M.D., 2003. Growth of early continental crust by partial melting of eclogite. Nature, 425(6958): 605-609.

Rickwood, P.C., 1989. Boundary lines within petrologic diagrams which use oxides of major and minor elements. Lithos, 22(4): 247-263.

Robinson, P.T., Zhou, M.F., Hu, X.F., et al., 1999. Geochemical constraints on the origin of the Hegenshan Ophiolite, Inner Mongolia, China. Journal of Asian Earth Sciences, 17(4): 423-442.

Schulmann, K., Paterson, S., 2011. Geodynamics: Asian continental growth. Nature Geoscience, 4(12): 827-829.

Song, S.G., Wang, M.M., Xu, X., et al., 2015. Ophiolites in the Xing'an-Inner Mongolia accretionary belt of the CAOB: Implications for two cycles of seafloor spreading and accretionary orogenic events. Tectonics, 34(10): 2221-2248.

Sun, S.S., McDonough, W.F., 1989. Chemical and isotopic systematics of oceanic basalts: Implications for mantle composition and processes. Geological Society, London, Special Publications, 42(1): 313-345.

Thirlwall, M.F., Smith, T.E., Graham, A.M., et al., 1994. High field strength element anomalies in arc lavas: Source or process? Journal of Petrology, 35(3): 819-838.

van de Zedde, D.M.A., Wortel, M.R.J., 2001. Shallow slab detachment as a transient source of heat at midlithospheric levels. Tectonics, 20(6): 868-882.

Wang, Q., Xu, J.F., Jian, P., et al., 2006. Petrogenesis of adakitic porphyries in an extensional tectonic setting, Dexing, South China: Implications for the genesis of porphyry copper mineralization. Journal of Petrology, 47(1): 119-144.

Wang, Z.J., Xu, W.L., Pei, F.P., et al., 2015a. Geochronology and geochemistry of middle Permian-Middle Triassic intrusive rocks from central-eastern Jilin Province, NE China: Constraints on the tectonic evolution of the eastern segment of the Paleo-Asian Ocean. Lithos, 238: 13-25.

Wang, Z.Z., Han, B.F., Feng, L.X., et al., 2015b. Geochronology, geochemistry and origins of the Paleozoic-Triassic plutons in the Langshan area, western Inner Mongolia, China. Journal of Asian Earth Sciences, 97: 337-351.

Wang, Z.Z., Han, B.F., Feng, L.X., et al., 2016. Tectonic attribution of the Langshan area in western Inner Mongolia and implications for the Neoarchean-Paleoproterozoic evolution of the Western North China Craton: Evidence from LA-ICP-MS zircon U-Pb dating of the Langshan basement. Lithos, 261: 278-295.

Wilde, S.A., 2015. Final amalgamation of the Central Asian Orogenic Belt in NE China: Paleo-Asian Ocean closure versus Paleo-Pacific plate subduction-A review of the evidence. Tectonophysics, 662: 345-362.

Windley, B.F., Alexeiev, D., Xiao, W.J., et al., 2007. Tectonic models for accretion of the Central Asian Orogenic Belt. Journal of the Geological Society, 164(12): 31-47.

Xiao, W.J., Windley, B.F., Hao, J., et al., 2003. Accretion leading to collision and the Permian Solonker suture, Inner Mongolia, China: Termination of the central Asian orogenic belt. Tectonics, 22(6): 1-8.

Xiao, W.J., Windley, B.F., Sun, S., et al., 2015. A Tale of Amalgamation of ThreePermo-Triassic Collage Systems in Central Asia: Oroclines, Sutures, and Terminal Accretion. Annual Review of Earth and Planetary Sciences, 43(1): 477-507.

Xu, B., Charvet, J., Chen, Y., et al., 2013. Middle Paleozoic Convergent Orogenic Belts in Western Inner

Mongolia (China): Framework, Kinematics, Geochronology and Implications for Tectonic Evolution of the Central Asian Orogenic Belt. Gondwana Research, 23(4): 1342-1364.

Yu, Y., Sun, M., Huang, X.L., et al., 2017. Sr-Nd-Hf-Pb isotopic evidence for modification of the Devonian lithospheric mantle beneath the Chinese Altai. Lithos, 284-285: 207-221.

Yuan, H.L., Gao, S., Dai, M.N., et al., 2008. Simultaneous determinations of U-Pb age, Hf isotopes and trace element compositions of zircon by excimer laser-ablation quadrupole and multiple-collector ICP-MS. Chemical Geology, 247(1): 100-118.

Yuan, H.L., Gao, S., Liu, X.M., et al., 2004. Accurate U-Pb age and trace element determinations of zircon by laser ablation-inductively coupled plasma-mass spectrometry. Geostandards and Geoanalytical Research, 28(3): 353-370.

Yuan, W., Yang, Z.Y., 2015. The Alashan terrane was not part of North China by the late Devonian: Evidence from detrital zircon U-Pb geochronology and Hf isotopes. Gondwana Research, 27: 1270-1282.

Zeng, Q.D., Yang, J.H., Zhang, Z.L., et al., 2013. Petrogenesis of the Yangchang Mobearing granite in the Xilamulun metallogenic belt, NE China: Geochemistry, zircon U-Pb ages and Sr-Nd-Pb isotopes. Geological Journal, 49(1): 1-14.

Zhang, J.R, Wei, C.J., Chu, H., et al., 2016a. Mesozoic metamorphism and its tectonic implication along the Solonker suture zone in central Inner Mongolia, China. Lithos, 261: 262-277.

Zhang, J., Zhang, B.H., Zhao, H., 2016b. Timing of amalgamation of the Alxa Block and the North China Block: Constraints based on detrital zircon U-Pb ages and sedimentologic and structural evidence. Tectonophysics, 668-669: 65-81.

Zhang, J., Li, J.Y., Xiao, W.J., et al., 2013. Kinematics and geochronology of multi-stage ductile deformation along the eastern Alxa block, NW China: New constraints on the relationship between the NCP and the Alxa block. Journal of Structural Geology, 57: 38-57.

Zhang, S.H., Zhao, Y., Davis, G.A., et al., 2014a. Temporal and spatial variations of Mesozoic magmatism and deformation in the North China Craton: Implications for lithospheric thinning and decratonization. Earth-Science Reviews, 131(4): 49-87.

Zhang, S.H., Gao, R., Li, H.Y., et al., 2014b. Crustal Structures Revealed from a Deep Seismic Reflection Profile across the Solonker Suture Zone of the Central Asian Orogenic Belt, Northern China: An Integrated Interpretation. Tectonophysics, 612-613(3): 26-39.

Zhang, S.H., Zhao, Y., Ye, H., et al., 2014c. Origin and evolution of the Bainaimiao arc belt: Implications for crustal growth in the southern Central Asian orogenic belt. Geological Society of America Bulletin, 126: 1275-1300.

Zhang, S.H., Zhao, Y., Song, B., et al., 2009a. Contrasting Late Carboniferous and Late Permian-Middle Triassic intrusive suites from the northern margin of the North China craton: Geochronology, petrogenesis, and tectonic implications. Geological Society of America Bulletin, 121: 181-200.

Zhang, S.H., Zhao, Y., Liu, X.C., et al., 2009b. Late Paleozoic to Early Mesozoic mafic-ultramafic complexes from the northern North China Block: Constraints on the composition and evolution of the lithospheric mantle. Lithos, 110: 229-246.

Zhang, S.H., Zhao, Y., Ye, H., et al., 2012. Early Mesozoic alkaline complexes in the northern North China

Craton: Implications for cratonic lithospheric destruction. Lithos, 155(2): 1-18.

Zhang, X.B., Wang, K.Y., Wang, C.Y., et al., 2017. Age, Genesis, and Tectonic Setting of the Mo-W Mineralized Dongshanwan Granite Porphyry from the Xilamulun Metallogenic Belt, NE China. Journal of Earth Science, 28(3): 433-446.

Zhang, X.H., Mao, Q., Zhang, H.F., et al., 2011. Mafic and felsic magma interaction during the construction of high-K calc-alkaline plutons within a metacratonic passive margin: The early Permian Guyang batholith from the northern North China Craton. Lithos, 125(9-10): 569-591.

Zhao, J.H., Asimow, P.D., 2014. Neoproterozoicboninite-series rocks in South China: A depleted mantle source modified by sediment-derived melt. Chemical Geology, 388: 98-111.

Zhao, X.C., Zhou, W.X., Fu, D., et al., 2018. Isotope Chronology and Geochemistry of the Lower Carboniferous Granite in Xilinhot, Inner Mongolia, China. Journal of Earth Science, 29(2): 280-294.

Zhou, J.B., Wilde, S.A., 2013. The crustal accretion history and tectonic evolution of the NE China segment of the Central Asian Orogenic Belt. Gondwana Research, 23(4): 1365-1377.

Zindler, A., Hart, S.R., 1986. Chemical geodynamics. Annual Review of Earth and Planetary Sciences, 14(1): 493-571.

Zou, H.B., Zindler, A., Xu, X.S., et al., 2000. Major, trace element, and Nd, Sr and Pb isotope studies of Cenozoic basalts in SE China: Mantle sources, regional variations, and tectonic significance. Chemical Geology, 171(1): 33-47.

Neoproterozoic peraluminous granites in the western margin of the Yangtze Block, South China: Implications for the reworking of mature continental crust[①]

Zhu Yu   Lai Shaocong[②]   Qin Jiangfeng   Zhu Renzhi

Zhang Fangyi   Zhang Zezhong   Zhao Shaowei

Abstract: Significant widespread melting of the mantle and juvenile mafic lower crust is known to have occurred along the western margin of the Yangtze Block during the Neoproterozoic, but melting of the mature continental crust remains poorly understood. Peraluminous granites can provide vital insights on the reworking of mature continental crustal materials. We present zircon U-Pb-Hf isotopic, whole-rock geochemical, and Sr-Nd isotopic data from the Neoproterozoic peraluminous granites in the western Yangtze Block, South China. Zircon U-Pb dating displays concordant crystallization ages of ca. 840 Ma for the Kuanyu granites and ca. 835 Ma for the Cida granites. These peraluminous granites have high SiO_2(66.9-75.6 wt%), K_2O (4.61-7.29 wt%) contents, as well as high K_2O/Na_2O (1.44-3.25) and A/CNK [molar ratio of $Al_2O_3/(CaO + Na_2O + K_2O)$] (1.04-1.18) ratios. The samples are enriched in Rb, K, Th, U, and Pb, and depleted in Nb, Ta, Ba, Sr, and Ti, indicating a middle-upper crustal affinity. They are also characterized by high $(^{87}Sr/^{86}Sr)_i$(0.709 9-0.721 7) and negative $\varepsilon_{Nd}(t)$ values (-5.1 to -2.9), which resemble the isotopic features of an evolved continental crust source. Furthermore, these peraluminous granites possess variable CaO/Na_2O (0.09-0.65) and Al_2O_3/TiO_2(25.3-88.4) ratios, moderate Rb/Ba (1.68-3.86) and Rb/Sr (0.32-0.85) ratios, as well as high molar $Al_2O_3/(MgO+FeO^T)$ (2.04-5.23) and low molar $CaO/(MgO+FeO^T)$ (0.15-0.48), implying that they predominantly originate from heterogeneous metasedimentary sources (metapelites + metagreywackes). Scarce evidence of hybridization processes between crust- and mantle-derived components indicates that their heterogeneous zircon Hf isotopic compositions [$\varepsilon_{Hf}(t) = -7.75$ to +3.31] may be caused by disequilibrium partial melting of heterogeneous metasedimentary sources, similar to peraluminous granites from the Altai area and Jiangnan orogen in China. In combination with previously reported results, we suggest that the Kuanyu and Cida peraluminous granites represent melting of mature crustal material in an evolved middle-upper crust source during the early stages of the Neoproterozoic subduction process. The western margin of the Yangtze Block

---

① Published in *Precambrian Research*, 2019, 333.

② Corresponding author.

underwent not only melting of the juvenile mafic lower crust but also reworking of the mature continental crust during the Neoproterozoic.

1　Introduction

The South China Block (SCB) is composed of the Yangtze Block to the northwest and the Cathaysia Block to the southeast (Wang et al., 2013; Zhao and Cawood, 2012; Zhao et al., 2018), and preserves extensive Neoproterozoic magmatism generated during the assembly and breakup of the Rodinia supercontinent (Li et al., 2003, 2006; Zhou et al., 2002, 2006a,b; Zhao et al., 2013, 2017, 2018). The Neoproterozoic magmatism is characterized by voluminous felsic intrusions and some mafic-ultramafic intrusions around the Yangtze Block (Zhou et al., 2002, 2006a,b; Zhao et al., 2008a, 2013, 2018). Previous studies mainly considered these intrusions to have resulted from either a mantle plume upwelling (Li et al., 2003, 2006; Yang et al., 2016), lithospheric extension in response to plate-rift process (Zheng et al., 2007), or persistent subduction process (Zhou et al., 2002, 2006a,b; Zhao et al., 2018 and references therein). On the western margin of the Yangtze Block, Neoproterozoic massive mafic to intermediate, adakitic, and Na-rich granitic rocks have been investigated to evaluating the crustal evolution, mantle melting and fractionation during the last decade years (Du et al., 2014; Lai et al., 2015a,b; Li and Zhao, 2018; Sun et al., 2007; Zhao et al., 2008a,b, 2010; Zhao and Cawood, 2012; Zhou et al., 2002, 2006a,b; Zhu et al., 2019a, b). The ca. 860−740 Ma mafic-intermediate rocks are dominantly considered to be from a mantle source metasomatized by subduction-related fluids and/or melts (Du et al., 2014; Sun and Zhou, 2008; Zhao et al., 2008a; Zhao and Zhou, 2007a; Zhou et al., 2006b). The ca. 800− 750 Ma adakitic rocks have been explained as the products of a subducted oceanic slab (Zhao and Zhou, 2007b; Zhou et al., 2006a) or thickened continental lower crust (Huang et al., 2009; Zhu et al., 2019b). In addition, the ca. 800−750 Ma Na-rich granitoids are thought to have been produced by partial melting of juvenile mafic lower crust (Lai et al., 2015a; Zhao et al., 2008b; Zhu et al., 2019a). However, few detailed studies have addressed partial melting of mature continental crust along the western margin of the Yangtze Block.

Peraluminous granites are universal in various tectonic setting and geochemically characterized by a high aluminous saturation index [ASI = molar $Al_2O_3/(CaO + Na_2O + K_2O)$] (A/CNK >1.0) (Chappell and White, 1992; Chen et al., 2014; Kemp et al., 2007; Patiño Douce, 1995; Zhao et al., 2013). Most peraluminous granites are thought to be the products of mature metasedimentary (e.g., metapelites and metagreywackes) in a relatively evolved crust source induced by mantle-derived magma (Clemens, 2003; Clemens et al., 2016; Chappell et al., 2012). Although some peraluminous granites originated from metaluminous igneous protoliths (e. g., basaltic to andesitic rocks) (Chappell and White, 1992, 2001; Chappell et al., 2012; Clemens, 2003), mature sedimentary components were also significantly incorporated into their magmatic evolution (Clemens, 2018; Chappell et al., 2012; Kemp et al., 2007; Zhao et al., 2015). Identification of the magma source of the peraluminous granites can therefore provide important insights on the melting of mature continental crustal materials.

We recently identified two peraluminous granitic plutons, including the Kuanyu and Cida granites in the Miyi region along the western margin of the Yangtze Block, South China. In this study, we present new zircon U-Pb ages, whole-rock major and trace element compositions, Sr-Nd isotopic data, and in-situ zircon Lu-Hf isotopic data of these peraluminous granites. The objectives of our work are to ① investigate the petrogenesis of Neoproterozoic peraluminous granites and ② identify the Neoproterozoic mature continental crustal magmatism in the western margin of the Yangtze Block that responded to early subduction process along the Rodinia supercontinent.

2 Regional geology and sample

South China is divided into two large blocks by the NNE-trending, 1 500-km-long Neoproterozoic Jiangnan orogenic belt (Li et al., 2008; Wang et al., 2013, 2014), which resulted from the assembly of the Yangtze and Cathaysia blocks around 830 Ma (Zhao et al., 2011) (Fig. 1). The Jiangnan orogenic belt is mainly composed of Neoproterozoic massive, undeformed granitoids and low-grade greenschist facies metamorphosed sedimentary rocks (Wang et al., 2014). The Yangtze Block is bounded by the Indochina block to the southwest and Songpan-Ganzi terrane to the west (Gao et al., 1999; Zhao and Cawood, 2012), and separated from the North China Block by the Qinling-Dabie-Sulu orogenic belt, which was

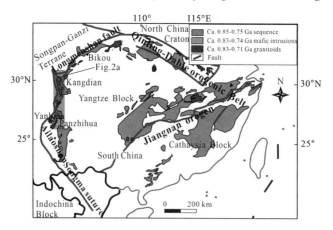

Fig. 1　Sketch geological map of the Yangtze Block, South China
(modified after Gan et al., 2017; Zhao et al., 2017).

generated by the Triassic collision between the North China and Yangtze blocks (Harker et al., 2006; Zhao et al., 2008a, 2013; Zheng et al., 2005) (Fig. 1).

The Yangtze Block consists of Archean to Proterozoic basement overlain by Neoproterozoic (Sinian) to Cenozoic cover sequences (Li and Zhao, 2018; Zhou et al., 2002, 2006b). The Archean to Early Neoproterozoic basement complexes are composed of metamorphosed arenaceous to argillaceaous sedimentary strata (Zhao et al., 2010). The Kongling TTG (tonalite-trondhjemite-granodiorite) suite is the only Archean terrane exposed in the Yangtze Block (Gao et al., 1999), but zircon U-Pb-Hf isotopic evidence from Paleozoic granulite xenoliths suggest that Archean crust most probably exists extensively beneath the Yangtze Block (Zheng et al., 2006). Voluminous Neoproterozoic intrusive and extrusive igneous rocks were emplaced along the western margin of the Yangtze Block (Fig. 2a) (Du et al., 2014; Li et al., 2006; Munteanu et al., 2010; Zhao et al., 2010; Zhou et al., 2002, 2006a,b; Zhu et al., 2008). Neoproterozoic metamorphic complexes are well preserved and include, from north to south, the Kangding, Miyi and Yuanmou complexes whose igneous rocks have concordant crystallization ages ranging from ca. 860 Ma to 750 Ma (Zhou et al., 2002). Neoproterozoic sequences, such as the Yanbian, Ebian, Huangshuihe Groups, and Sinian strata (Geng et al., 2008), are overlain by a thick sequence (>9 km) of Neoproterozoic to Permian strata composed of clastic, carbonate and meta-volcanic rocks (Zhu et al., 2008). Moreover, Neoproterozoic volcanic-sedimentary rocks are well-known as the Yanbian, Suxiong, and Kaijianqiao groups (Sun et al., 2008). Two large serpentinite blocks associated with gabbros and pillow basalts are traditionally considered to be ophiolites in the Shimian area. These ophiolite sequences have been dated to be Late Mesoproterozoic and Early Neoproterozoic, and are geochemically similar to the SSZ (supra-subduction zone)-type ophiolites (Hu et al., 2017; Zhao et al., 2017). In the western Yangtze Block, massive Neoproterozoic granites are spatially associated with high-grade metamorphic complexes and low-grade metamorphic Mesoproterozoic strata (Zhao and Zhou, 2007a).

The Miyi region is located in the southern segment of the western margin of the Yangtze Block, South China (Fig. 2a). In this area, the Late Mesoproterozoic to Neoproterozoic Huili Group is widespread and consists of meta-clastic and meta-carbonate rocks interbedded with meta-volcanic rocks of about 10 km in thickness (Zheng et al., 2007; Zhu et al., 2016). From the base upward, the Huili Group contains the Yinmin, Luoxue, Heishan, Qinglongshan, Limahe, Fengshan, and Tianbaoshan formations (SPBGMR, 1966; Zhu et al., 2016). The Huili Group was intruded by later Neoproterozoic and Triassic granitic magmatism around the Kuanyu and Cida areas. In addition, there have been reports of Neoproterozoic mafic and felsic magmatism including the ca. 803−790 Ma granites and ca. 840 Ma mafic rocks from the Xiacun and Mosuoying areas (Guo et al., 2007).

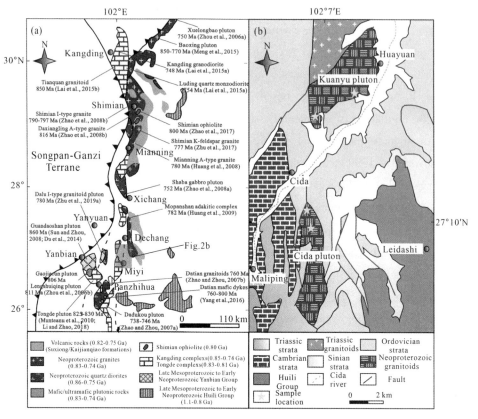

Fig. 2   Geological map of the western margin of the Yangtze Block (a) (modified after
Zhou et al., 2006b) and regional geological map of the Kuanyu and Cida pluton (b)
(revised from the Miyi 1 : 200 000 geological map, SPBGMR, 1966).

In this study, we investigate the Kuanyu and Cida granitic plutons situated near the town
of Huayuan, which is located at the northwest part of the Miyi County. These plutons are
located on both sides of the Cida River, and tectonically bounded by several faults (Fig. 2b).
The Kuanyu and Cida plutons are mainly composed of medium- to medium-coarse grain biotite
granites (Fig. 3a,b). Samples from the Kuanyu granites consist of K-feldspar (20−30 vol%),
plagioclase (20−25 vol%), quartz (20−25 vol%), biotite (10−15 vol%), and accessory
minerals including magnetite (0−3 vol%) and zircon (Fig. 3c,e). The plagioclase typically
shows polysynthetic twinning and the K-feldspar displays the distinctive crossed twinning. The
biotite is subhedral shape and the quartz occurs as xenomorphs to subhedral crystals. Some
biotite minerals were partly modified by later magmatism. The Cida granites mainly consist of
K-feldspar (15−20 vol%), perthite (0−10 vol%), plagioclase (20−25 vol%), quartz (30−
35 vol%), biotite (0−5 vol%), magnetite, and zircon (Fig. 3d,f). The plagioclase displays
the well-developed polysynthetic twins and sodium-compound twins. The K-feldspar is
subhedral and partly altered to kaolinite. Some K-feldspar minerals also show crossed twinning.

The biotite is inlaid between the K-feldspar and quartz.

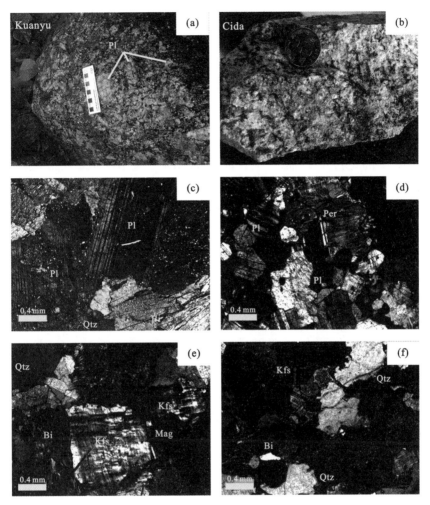

Fig. 3　Field and hand specimen photographs as well as representative microscope photographs of the Kuanyu (a,c,e) and Cida (b,d,f) granites from the western margin of the Yangtze Block, South China.
Pl: plagioclase; Kfs: K-feldspar; Per: perthite; Qtz: quartz; Bi: biotite; Mag: magnetite.

3　Analytical methods

We analyzed major and trace elements analyses, zircon geochronology, and in-situ zircon Lu-Hf isotopic at the State Key Laboratory of Continental Dynamics, Northwest University, Xi'an, China (SKLCD). Whole-rock Sr-Nd isotopic data were obtained at the Guizhou Tuopu Resource and Environment Analysis Center.

3.1　Zircon U-Pb dating

Zircons were separated from ~ 5 kg samples (KY-2, CD-1, and CD-2) collected from different sampling locations within the Kuanyu and Cida granitic plutons so that sufficient and representative zircons could be selected. Zircon grains from the samples were separated using conventional heavy liquid and magnetic techniques. Representative zircon grains were handpicked and mounted in epoxy resin disks and then polished and carbon coated. The internal morphology was examined using cathodoluminescence (CL) microscopy prior to U-Pb isotopic dating.

Laser ablation inductively-coupled plasma mass spectrometry (LA-ICP-MS) U-Pb analyses were conducted following the method of Yuan et al. (2004) on an Agilent 7500a ICP-MS equipped with a 193-nm laser. The laser ablation spot size was approximately 32 μm. $^{207}Pb/^{206}Pb$, $^{206}Pb/^{238}U$, $^{237}Pb/^{235}U$ and $^{208}Pb/^{232}Th$ ratios were calculated using GLITTER 4.0 (Macquarie University), and then corrected using the Harvard zircon 91500 as an external standard with a recommended $^{206}Pb/^{238}U$ isotopic age of 1 065.4±0.6 Ma (Wiedenbeck et al., 2004). The GJ-1 is also a standard sample with a recommended $^{206}Pb/^{238}U$ isotopic age of 603.2±2.4 Ma (Liu et al., 2007a,b). Detailed analytical techniques are described in Yuan et al. (2004). Common Pb contents were subsequently valuated using the method described in Andersen (2002). The age calculations and plotting of concordia diagrams were made using ISOPLOT (version 3.0; Ludwig, 2003). Uncertainties are quoted at the 2σ level.

3.2   Major and trace elements

Weathered surfaces of the whole-rock samples were removed and fresh parts were chipped and powdered to a ~200 mesh using a tungsten carbide ball mill. Major and trace elements were analyzed by X-ray fluorescence (XRF; Rikagu RIX 2100) and inductively coupled plasma mass spectrometry (ICP-MS; Agilent 7500a), respectively. Analyses of USGS and Chinese national rock standards (BCR-2, GSR-1, and GSR-3) showed that the analytical precision and accuracy for the major elements were generally better than 5%. For the trace element analyses, sample powders were digested using a HF+HNO$_3$ mixture in a high-pressure Teflon bombs at 190 ℃ for 48 h., The analytical error was less than 2% for most trace elements and the precision was greater than 10% (Liu et al., 2007a,b).

3.3   Whole-rock Sr-Nd isotopes

Whole-rock Sr-Nd isotopic data were obtained using a Neptune Plasma high-resolution (HR) multi-collector (MC) mass spectrometer. Sr and Nd isotopes were determined using a method similar to that of Chu et al. (2009). Sr and Nd isotopic fractionation was corrected to $^{87}Sr/^{86}Sr=0.119~4$ and $^{146}Nd/^{144}Nd=0.721~9$. During the sample runs, the La Jolla standard yielded an average value of $^{143}Nd/^{144}Nd=0.511~862±5$ (2σ), and the NBS987 standard yielded an average value of $^{87}Sr/^{86}Sr=0.710~236±16$ (2σ). The total procedural Sr and Nd blanks are b1 ng and b50 pg, respectively. NIST SRM-987 and JMC-Nd were used as certified reference standard solutions for $^{87}Sr/^{86}Sr$ and $^{143}Nd/^{144}Nd$ isotopic ratios, respectively. The BCR-1 and BHVO-1 standards yielded an average of $^{87}Sr/^{86}Sr$ ratio are 0.705 014 $\pm$ 3 (2σ) and 0.703 477 $\pm$ 20 (2σ), respectively. The BCR-1 and BHVO-1 standards yielded an average of $^{146}Nd/^{144}Nd$ ratio are 0.512 615 $\pm$ 12 (2σ) and 0.512 987 $\pm$ 23 (2σ), respectively.

3.4　In-situ zircon Lu-Hf isotopic analyses

The in-situ zircon Hf isotopes were analyzed using a Neptune MC-ICP-MS. The laser repetition rate was 6 Hz at 100 mJ and the spot size was 30 μm. Instrument information is available in Bao et al. (2017). The detailed analytical technique is depicted by Yuan et al. (2008). During the analyses, the measured values of well-characterized zircon standards (91500, GJ-1, and Monastery) were consistent with the recommended values within 2σ (Yuan et al., 2008). The obtained Hf isotopic compositions were 0.282 016 $\pm$ 20 ($2\sigma_n$, $n = $ 84) for the GJ-1 standard and 0.282 735 $\pm$ 24 ($2\sigma_n$, $n = $ 84) for the Monastery standard, consistent to within 2σ of the recommended values (Yuan et al., 2008). The initial $^{176}Hf/^{177}Hf$ ratios and $\varepsilon_{Hf}(t)$ were calculated with reference to the chondritic reservoir (CHUR) at the time of zircon growth from the magmas. The decay constant for ^{176}Lu of 1.867 $\times$ 10^{-11} year^{-1} (Soderlund et al., 2004), chondritic $^{176}Hf/^{177}Hf$ ratio of 0.282 785, and $^{176}Lu/^{177}Hf$ ratio of 0.033 6 were adopted. The depleted mantle model ages (T_{DM}) used for basic rocks were calculated with reference to the present-day depleted mantle $^{176}Hf/^{177}Hf$ ratio of 0.283 25, similar to that of the average MORB (Nowell et al., 1998) and $^{176}Lu/^{177}Hf = 0.038 4$ (Griffin et al., 2000). For the zircons from felsic rocks, we also calculated the Hf isotope "crustal" model age (T_{DMC}) by assuming that its parental magma was derived from an average continental crust, with $^{176}Lu/^{177}Hf = 0.015$, that originated from the depleted mantle source (Griffin et al., 2000). Our conclusions remain unaffected even if other decay constants were used.

4　Results

4.1　Zircon U-Pb geochronology

Zircon grains from three typical peraluminous granites samples, including one sample (KY-2) from the Kuanyu pluton and two samples (CD-1, CD-2) from the Cida pluton, were collected and analyzed. The zircon U-Pb isotopic data are presented in the Table 1 and illustrated in a U-Pb concordia diagram (Fig. 4).

4.1.1　Kuanyu granites

Zircon grains from sample KY-2 in the Kuanyu granites are dominantly colorless, transparent,

Table 1　Concordant results of zircon U-Pb dating for the Neoproterozoic Kuanyu and Cida peraluminous granites from the western margin of the Yangtze Block, South China.

| Analysis | Content/ppm | | Th/U | ratios | | | | | | | | Age/Ma | | | | | | | |
	Th	U		$^{207}Pb/^{206}Pb$	2σ	$^{207}Pb/^{235}U$	2σ	$^{206}Pb/^{238}U$	2σ	$^{208}Pb/^{232}Th$	2σ	$^{207}Pb/^{206}Pb$	2σ	$^{207}Pb/^{235}U$	2σ	$^{206}Pb/^{238}U$ UP	2σ	$^{208}Pb/^{232}Th$	2σ
KY-2																			
KY-2-01	49.48	686.15	0.07	0.065 99	0.001 44	1.273 45	0.017 01	0.139 99	0.001 51	0.046 15	0.000 49	806	45	834	8	845	9	912	10
KY-2-02	99.49	975.09	0.10	0.065 85	0.001 42	1.262 78	0.016 32	0.139 10	0.001 49	0.039 71	0.000 58	802	45	829	7	840	8	787	11
KY-2-03	101.59	395.18	0.26	0.066 07	0.001 55	1.257 22	0.019 89	0.138 03	0.001 53	0.043 36	0.000 72	809	48	827	9	834	9	858	14
KY-2-04	113.60	1 908.49	0.06	0.066 35	0.001 38	1.269 93	0.014 63	0.138 83	0.001 47	0.043 75	0.000 51	818	43	832	7	838	8	866	10
KY-2-05	115.23	513.23	0.22	0.065 87	0.001 52	1.268 43	0.019 27	0.139 69	0.001 54	0.042 54	0.000 52	802	48	832	9	843	9	842	10
KY-2-07	116.05	747.83	0.16	0.067 66	0.001 50	1.306 99	0.018 13	0.140 12	0.001 52	0.042 99	0.000 65	858	45	849	8	845	9	851	13
KY-2-08	117.45	414.96	0.28	0.066 15	0.001 51	1.280 48	0.018 95	0.140 42	0.001 54	0.043 34	0.000 58	811	47	837	8	847	9	858	11
KY-2-09	123.50	338.12	0.37	0.065 06	0.001 55	1.252 7	0.020 38	0.139 66	0.001 56	0.041 34	0.000 55	776	49	825	9	843	9	819	11
KY-2-10	137.83	838.62	0.16	0.067 12	0.001 45	1.289 17	0.016 72	0.139 33	0.001 50	0.041 06	0.000 42	841	44	841	7	841	8	813	8
KY-2-11	140.74	959.04	0.15	0.066 41	0.001 43	1.278 04	0.016 44	0.139 60	0.001 50	0.044 97	0.000 61	819	44	836	7	842	8	889	12
KY-2-12	145.32	717.85	0.20	0.066 01	0.001 46	1.261 14	0.017 24	0.138 58	0.001 50	0.041 38	0.000 46	807	45	828	8	837	8	820	9
KY-2-13	147.37	567.66	0.26	0.066 01	0.001 46	1.258 12	0.017 37	0.138 24	0.001 50	0.046 67	0.000 56	807	46	827	8	835	8	922	11
KY-2-14	147.58	512.79	0.29	0.066 62	0.001 47	1.286 47	0.017 54	0.140 07	0.001 51	0.042 53	0.000 48	826	45	840	8	845	8	842	9
KY-2-15	163.27	1 710.68	0.10	0.065 67	0.001 37	1.262 35	0.014 61	0.139 43	0.001 47	0.040 42	0.000 49	796	43	829	7	841	8	801	9
KY-2-16	171.79	496.82	0.35	0.065 58	0.001 68	1.252 73	0.023 57	0.138 55	0.001 60	0.039 44	0.000 63	793	53	825	11	837	9	782	12
KY-2-17	172.67	2 153.67	0.08	0.065 9	0.001 39	1.260 38	0.015 24	0.138 73	0.001 47	0.040 43	0.000 48	803	44	828	7	838	8	801	9
KY-2-18	172.79	4 307.67	0.04	0.065 6	0.001 35	1.255 06	0.013 91	0.138 77	0.001 46	0.037 37	0.000 36	794	42	826	6	838	8	742	7
KY-2-20	179.49	979.67	0.18	0.066 29	0.001 42	1.274 89	0.016 11	0.139 51	0.001 49	0.044 44	0.000 57	815	44	835	7	842	8	879	11
KY-2-21	188.71	584.59	0.32	0.066 48	0.001 55	1.275 15	0.019 72	0.139 13	0.001 53	0.041 19	0.000 56	822	48	835	9	840	9	816	11

Continued

Analysis	Content/ppm		Th/U	ratios								Age/Ma							
	Th	U		$\frac{207Pb}{206Pb}$	2σ	$\frac{207Pb}{235U}$	2σ	$\frac{206Pb}{238U}$	2σ	$\frac{208Pb}{232Th}$	2σ	$\frac{207Pb}{206Pb}$	2σ	$\frac{207Pb}{235U}$	2σ	$\frac{206Pb}{238}UP$	2σ	$\frac{208Pb}{232Th}$	2σ
KY-2-22	211.06	670.06	0.31	0.066 65	0.001 55	1.279 45	0.019 82	0.139 23	0.001 54	0.040 45	0.000 53	827	48	837	9	840	9	801	10
KY-2-23	213.41	532.30	0.40	0.066 99	0.001 55	1.285 50	0.019 67	0.139 18	0.001 53	0.042 12	0.000 56	838	48	839	9	840	9	834	11
KY-2-25	220.48	367.37	0.63	0.067 01	0.001 64	1.287 23	0.022 20	0.139 34	0.001 57	0.041 79	0.000 60	838	50	840	10	841	9	827	12
KY-2-26	220.75	481.82	0.21	0.067 09	0.001 62	1.289 43	0.021 46	0.139 40	0.001 56	0.056 77	0.000 86	841	49	841	10	841	9	1 116	16
KY-2-27	223.73	922.68	0.32	0.066 31	0.001 42	1.273 44	0.016 09	0.139 30	0.001 49	0.038 14	0.000 43	816	44	834	7	841	8	757	8
KY-2-28	231.02	627.06	0.08	0.065 80	0.001 48	1.251 67	0.017 81	0.137 98	0.001 50	0.039 94	0.000 59	800	46	824	8	833	8	792	11
KY-2-29	294.15	471.58	0.30	0.066 29	0.001 52	1.279 50	0.019 20	0.140 00	0.001 53	0.041 92	0.000 56	816	47	837	9	845	9	830	11
KY-2-30	294.63	3 516.92	0.06	0.066 01	0.001 37	1.261 56	0.014 43	0.138 61	0.001 46	0.043 51	0.000 53	807	43	829	6	837	8	861	10
KY-2-31	307.09	664.54	0.46	0.067 41	0.001 54	1.298 60	0.019 26	0.139 73	0.001 53	0.045 90	0.000 77	850	47	845	9	843	9	907	15
KY-2-32	347.20	799.31	0.37	0.066 64	0.001 48	1.277 75	0.017 65	0.139 07	0.001 50	0.042 89	0.000 61	827	46	836	8	839	9	849	12
KY-2-33	351.54	1 748.38	0.10	0.066 95	0.001 42	1.283 81	0.015 72	0.139 09	0.001 48	0.045 03	0.000 59	836	44	839	7	840	8	890	11
KY-2-34	359.49	479.83	0.29	0.065 53	0.001 49	1.251 31	0.018 37	0.138 50	0.001 51	0.049 98	0.000 62	791	47	824	8	836	9	986	12
KY-2-35	541.28	453.68	0.47	0.066 64	0.001 57	1.271 34	0.020 02	0.138 37	0.001 53	0.042 34	0.000 51	827	48	833	9	835	9	838	10
KY-2-36	1 723.10	610.81	0.36	0.066 82	0.001 53	1.285 43	0.019 00	0.139 53	0.001 52	0.042 53	0.000 52	832	47	839	8	842	9	842	10
CD-1																			
CD-1-01	151.00	240.01	0.63	0.067 15	0.001 63	1.294 50	0.021 66	0.139 83	0.001 57	0.042 80	0.000 53	842	50	843	10	844	9	847	10
CD-1-02	448.65	526.01	0.85	0.067 80	0.001 51	1.292 75	0.018 05	0.138 30	0.001 51	0.042 31	0.000 44	862	46	843	8	835	9	838	9
CD-1-03	399.22	563.71	0.71	0.065 15	0.001 51	1.237 30	0.018 75	0.137 76	0.001 52	0.043 11	0.000 48	779	48	818	9	832	9	853	9
CD-1-04	341.29	466.04	0.73	0.066 22	0.001 50	1.262 17	0.018 30	0.138 24	0.001 51	0.043 04	0.000 47	813	47	829	8	835	9	852	9
CD-1-05	456.74	693.40	0.66	0.066 05	0.001 49	1.262 52	0.018 05	0.138 64	0.001 51	0.043 05	0.000 47	808	46	829	8	837	9	852	9
CD-1-06	336.73	626.80	0.54	0.067 29	0.001 52	1.298 80	0.018 67	0.139 99	0.001 53	0.043 71	0.000 49	847	46	845	8	845	9	865	10

Continued

Analysis	Content/ppm			ratios								Age/Ma							
	Th	U	Th/U	$\frac{207\text{Pb}}{206\text{Pb}}$	2σ	$\frac{207\text{Pb}}{235\text{U}}$	2σ	$\frac{206\text{Pb}}{238\text{U}}$	2σ	$\frac{208\text{Pb}}{232\text{Th}}$	2σ	$\frac{207\text{Pb}}{206\text{Pb}}$	2σ	$\frac{207\text{Pb}}{235\text{U}}$	2σ	$\frac{206\text{Pb}}{238\text{UP}}$	2σ	$\frac{208\text{Pb}}{232\text{Th}}$	2σ
CD-1-07	538.78	1 001.27	0.54	0.065 23	0.001 41	1.243 68	0.015 79	0.138 29	0.001 48	0.045 07	0.000 46	782	45	821	7	835	8	891	9
CD-1-08	574.13	931.01	0.62	0.064 82	0.001 44	1.230 13	0.016 99	0.137 65	0.001 49	0.041 77	0.000 45	769	46	814	8	831	8	827	9
CD-1-10	362.55	650.15	0.56	0.067 10	0.001 49	1.274 95	0.017 51	0.137 82	0.001 50	0.043 67	0.000 47	841	46	835	8	832	8	864	9
CD-1-12	406.81	676.51	0.60	0.066 55	0.001 58	1.262 52	0.020 17	0.137 60	0.001 53	0.041 72	0.000 51	824	49	829	9	831	9	826	10
CD-1-16	332.67	786.75	0.42	0.065 41	0.001 46	1.249 23	0.017 39	0.138 53	0.001 51	0.042 88	0.000 48	788	46	823	8	836	9	849	9
CD-1-17	373.75	502.83	0.74	0.065 68	0.001 83	1.253 17	0.026 96	0.138 39	0.001 66	0.041 16	0.000 58	796	57	825	12	836	9	815	11
CD-1-18	1 105.70	915.47	1.21	0.064 93	0.001 45	1.236 10	0.017 21	0.138 08	0.001 50	0.040 04	0.000 40	772	46	817	8	834	9	794	8
CD-1-21	294.33	367.92	0.80	0.068 03	0.001 61	1.297 12	0.020 73	0.138 30	0.001 54	0.043 07	0.000 49	869	48	844	9	835	9	852	9
CD-1-22	282.98	439.70	0.64	0.067 02	0.001 77	1.281 07	0.025 18	0.138 64	0.001 62	0.044 62	0.000 60	838	54	837	11	837	9	882	12
CD-1-24	243.76	407.78	0.60	0.066 67	0.001 61	1.270 34	0.021 11	0.138 20	0.001 55	0.044 62	0.000 54	828	50	833	9	835	9	882	10
CD-1-26	369.36	794.21	0.47	0.065 65	0.001 54	1.250 60	0.019 44	0.138 17	0.001 53	0.042 87	0.000 52	795	48	824	9	834	9	848	10
CD-1-27	200.67	312.30	0.64	0.066 41	0.001 56	1.265 36	0.019 74	0.138 20	0.001 53	0.042 94	0.000 49	819	48	830	9	835	9	850	10
CD-1-28	617.64	620.32	1.00	0.067 04	0.001 51	1.278 80	0.018 17	0.138 36	0.001 51	0.040 30	0.000 42	839	46	836	8	835	9	799	8
CD-1-29	229.44	504.54	0.45	0.069 31	0.001 61	1.330 44	0.020 12	0.139 22	0.001 54	0.040 79	0.000 49	908	47	859	9	840	9	808	10
CD-1-30	720.19	719.70	1.00	0.065 48	0.001 50	1.250 65	0.018 28	0.138 53	0.001 52	0.040 69	0.000 42	790	47	824	8	836	9	806	8
CD-1-31	1 044.07	1 068.27	0.98	0.068 07	0.001 80	1.290 06	0.025 45	0.137 46	0.001 61	0.044 93	0.000 62	871	54	841	11	830	9	888	12
CD-1-32	540.48	747.67	0.72	0.066 13	0.001 45	1.260 71	0.016 74	0.138 26	0.001 49	0.041 50	0.000 42	811	45	828	8	835	8	822	8
CD-1-33	432.87	514.55	0.84	0.066 45	0.001 53	1.267 24	0.018 9	0.138 32	0.001 52	0.040 62	0.000 44	821	47	831	8	835	9	805	8
CD-1-34	942.23	835.64	1.13	0.065 68	0.001 44	1.253 81	0.016 57	0.138 46	0.001 49	0.041 94	0.000 41	796	45	825	7	836	8	830	8
CD-1-35	365.92	704.67	0.52	0.066 88	0.001 47	1.268 58	0.016 88	0.137 56	0.001 49	0.042 99	0.000 45	834	45	832	8	831	8	851	9
CD-1-38	111.38	175.34	0.64	0.065 99	0.002 09	1.258 36	0.032 80	0.138 31	0.001 77	0.041 99	0.000 71	806	65	827	15	835	10	832	14

Continued

Analysis	Content/ppm		Th/U	ratios								Age/Ma							
	Th	U		$\frac{207Pb}{206Pb}$	2σ	$\frac{207Pb}{235U}$	2σ	$\frac{206Pb}{238U}$	2σ	$\frac{208Pb}{232Th}$	2σ	$\frac{207Pb}{206Pb}$	2σ	$\frac{207Pb}{235U}$	2σ	$\frac{206Pb}{238U}$	2σ	$\frac{208Pb}{232Th}$	2σ
CD-1-39	106.55	185.55	0.57	0.065 58	0.001 83	1.249 62	0.026 86	0.138 21	0.001 65	0.042 91	0.000 64	793	57	823	12	835	9	849	12
CD-2																			
CD-2-02	353.13	707.69	0.50	0.067 56	0.001 57	1.288 59	0.019 19	0.138 32	0.001 51	0.040 56	0.000 48	855	48	841	9	835	9	804	9
CD-2-03	370.25	499.51	0.62	0.067 27	0.001 57	1.289 46	0.019 37	0.139 02	0.001 52	0.040 58	0.000 45	846	48	841	9	839	9	804	9
CD-2-07	182.68	286.70	0.64	0.066 78	0.001 60	1.268 98	0.020 40	0.137 82	0.001 52	0.041 48	0.000 49	831	49	832	9	832	9	822	9
CD-2-08	774.91	1 049.09	0.64	0.066 11	0.001 43	1.257 62	0.015 66	0.137 98	0.001 46	0.040 52	0.000 40	810	45	827	7	833	8	803	8
CD-2-09	681.34	945.32	0.69	0.065 39	0.001 48	1.247 76	0.017 55	0.138 40	0.001 49	0.042 11	0.000 45	787	47	822	8	836	8	834	9
CD-2-11	358.63	380.42	0.70	0.065 94	0.001 52	1.260 35	0.018 84	0.138 64	0.001 51	0.043 24	0.000 46	804	48	828	8	837	9	856	9
CD-2-12	821.14	901.12	0.72	0.066 15	0.001 44	1.265 69	0.016 37	0.138 77	0.001 48	0.040 66	0.000 40	811	45	831	7	838	8	806	8
CD-2-13	887.34	1 439.34	0.73	0.065 63	0.001 40	1.241 97	0.015 26	0.137 27	0.001 45	0.042 09	0.000 42	795	44	820	7	829	8	833	8
CD-2-14	448.00	699.77	0.74	0.065 45	0.001 49	1.245 52	0.018 14	0.138 03	0.001 49	0.042 50	0.000 47	789	47	821	8	834	8	841	9
CD-2-15	675.92	925.90	0.74	0.066 25	0.001 45	1.264 24	0.016 53	0.138 42	0.001 47	0.039 86	0.000 41	814	45	830	7	836	8	790	8
CD-2-19	700.24	1 009.08	0.85	0.065 48	0.001 53	1.248 34	0.019 27	0.138 29	0.001 51	0.043 32	0.000 51	790	48	823	9	835	9	857	10
CD-2-20	1 081.39	1 197.91	0.90	0.064 72	0.001 52	1.228 89	0.019 35	0.137 72	0.001 51	0.042 39	0.000 46	765	49	814	9	832	9	839	9
CD-2-21	867.46	1 232.49	0.91	0.065 17	0.001 45	1.242 79	0.017 15	0.138 32	0.001 48	0.042 16	0.000 46	780	46	820	8	835	8	835	9
CD-2-22	1 592.62	1 325.90	0.94	0.065 57	0.001 44	1.245 34	0.016 87	0.137 76	0.001 47	0.039 89	0.000 40	793	46	821	8	832	8	791	8
CD-2-27	1 229.46	1 446.69	1.20	0.065 52	0.001 44	1.252 96	0.016 98	0.138 71	0.001 48	0.041 16	0.000 44	791	45	825	8	837	8	815	9
CD-2-30	1 659.58	1 375.52	1.21	0.064 20	0.001 38	1.225 01	0.015 86	0.138 39	0.001 47	0.041 69	0.000 42	748	45	812	7	836	7	826	8
CD-2-31	2 795.03	2 046.03	1.37	0.064 17	0.001 36	1.226 30	0.015 13	0.138 60	0.001 46	0.040 27	0.000 40	747	44	813	7	837	8	798	8
CD-2-32	1 736.61	875.82	1.98	0.065 78	0.001 62	1.255 45	0.022 11	0.138 43	0.001 55	0.043 25	0.000 51	799	51	826	10	836	9	856	8

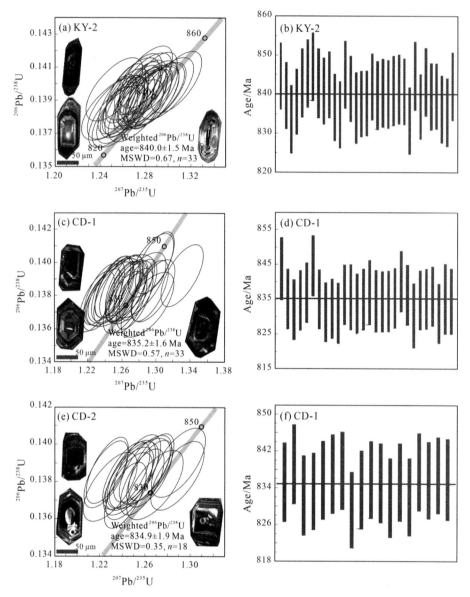

Fig. 4   LA-ICP-MS U-Pb zircon concordia diagrams for the Kuanyu (a,b) and Cida (c-f) granites from the western margin of the Yangtze Block, South China.

and euhedral prismatic crystals. The crystal length ranges from 50 μm to 200 μm with aspect ratios around 1:1 to 2:1. Most grains from this sample have well-developed oscillatory zoning, which indicates a magmatic origin (Hoskin and Schaltegger, 2003). The 33 concordant analyses spots are characterized by $^{206}Pb/^{238}U$ ages ranging from 833 ± 8 Ma to 847 ± 9 Ma with a weighted mean age of 840.0 ± 1.5 Ma (MSWD = 0.67, $n = 33$) (Fig. 4a,b), interpreted to be the crystallization age of the Kuanyu granites.

4.1.2 Cida granites

Zircons from sample CD-1 are transparent and subhedral short-prismatic crystals with aspect ratios from 1 : 1 to 3 : 2. These zircon grains range from 50 μm to 100 μm long and also show the well-developed oscillatory zoning, a typical indicator of an igneous origin. The zircons from sample CD-1 have a wide range of Th contents (107−1 101 ppm), U contents (175−1 068 ppm), and Th/U ratios (0.42−1.21). Their 28 concordant analytical spots yield $^{206}Pb/^{238}U$ ages varying from 830 ± 9 Ma to 845 ± 9 Ma and produce a weighted mean age of 835.2 ± 1.6 Ma (MSWD = 0.57, n = 33) (Fig. 4c,d).

Zircons from sample CD-2 are also subhedral stubby crystals with aspect ratios around 1 : 1. The crystals are furvous and range from 40 μm to 100 μm long. The zircon grains from the sample CD-2 contain variable U (287−2 046 ppm), Th (83−2 795 ppm), with Th/U ratios of 0.50−1.98. Among the 18 concordant analyses plots, these samples yield a concordant weighted mean age of 834.9 ± 1.9 Ma (MSWD = 0.35, n = 18) (Fig. 4e,f).

4.2 Major and trace element compositions

Major and trace element data measured from the Kuanyu and Cida peraluminous granites are listed in Table 2 and Figs. 5−7. The Neoproterozoic juvenile mafic lower crust-derived granitoids are shown for comparison (Lai et al., 2015a; Zhu et al., 2019a). All samples in the A/NK versus A/CNK diagram (Fig. 5a) plot in the peraluminous granites field, and mostly in the alkali-calcic field on the ($Na_2O + K_2O - CaO$) vs. SiO_2 diagram (Fig. 5b). The Kuanyu and Cida peraluminous granites are both a potassium-rich series with high K_2O/Na_2O ratios (1.44−3.25) (Fig. 5c). They also show the similar Mg# [molar 100MgO/(MgO + FeO^T)] values (17−33) to those pure crustal melts under continental crust pressure and temperature (P-T) conditions (Fig. 5d).

4.2.1 Kuanyu granites

The Kuanyu granites show a range of major elements compositions, particularly with regards to SiO_2 contents (66.9−75.0 wt%), total alkali contents ($K_2O + Na_2O$ = 7.43−9.53 wt%), K_2O/Na_2O ratios (1.44−3.25), and Al_2O_3 contents (12.3−15.8 wt%), with A/CNK values of 1.05−1.18. FeO^T, CaO, Al_2O_3, MgO, and TiO_2 all decrease with increasing SiO_2 contents (Fig. 6). The samples also contain moderate P_2O_5(0.06−0.22 wt%) and Mg# values (17−33). On the chondrite-normalized REE pattern diagram (Fig. 7a), the Kuanyu granites display similar rare earth elements (REEs) distribution patterns, enrichment in light rare earth elements (LREEs) with $(La/Yb)_N$ = 8.7−19.6, relatively flat heavy rare earth elements (HREEs) patterns with $(Dy/Yb)_N$ = 1.22−1.80. They have high total REE contents of 231−336 ppm and clearly display negative Eu anomalies (Eu/Eu^* = 0.23−0.57). On the primitive-mantle normalized diagram (Fig. 7b), the Kuanyu granites exhibit the enrichment of

Table 2　Major(wt%) and trace(ppm) elements for the Neoproterozoic Kuanyu and Cida peraluminous granites in the western Yangtze Block, South China.

Sample	Kuanyu pluton 27°15′20″N 102°10′33″E								Cida pluton 27°10′50″N 102°5′19″E												
	KY-1-1	KY-1-2	KY-1-3	KY-2-2	KY-2-3	KY-2-4	KY-2-5	KY-2-6	CD-1-1	CD-1-2	CD-1-4	CD-1-5	CD-1-6	CD-2-1	CD-2-2	CD-2-3	CD-2-4	CD-2-5	CD-2-6	CD-2-7	CD-2-8
Major element /wt%																					
SiO_2	74.98	73.30	73.44	69.62	71.19	71.44	68.21	66.88	75.56	73.13	73.17	74.76	73.55	75.12	73.94	74.27	75.21	75.47	73.56	73.94	73.48
TiO_2	0.20	0.20	0.20	0.57	0.33	0.37	0.40	0.54	0.14	0.21	0.20	0.17	0.21	0.24	0.21	0.19	0.23	0.24	0.23	0.21	0.20
Al_2O_3	12.30	13.10	13.15	14.44	14.74	14.00	15.80	15.77	12.38	13.37	13.81	12.64	13.29	12.53	13.29	13.14	12.43	12.28	13.26	13.19	13.36
$Fe_2O_3^T$	2.41	2.46	2.48	4.06	2.07	3.13	3.04	4.00	1.54	2.07	2.07	1.76	2.13	2.12	1.96	2.03	2.06	2.13	2.06	2.22	1.86
MnO	0.02	0.02	0.02	0.04	0.02	0.04	0.03	0.04	0.02	0.02	0.02	0.02	0.02	0.02	0.02	0.02	0.02	0.02	0.02	0.02	0.02
MgO	0.23	0.22	0.22	0.78	0.43	0.54	0.59	0.75	0.17	0.22	0.25	0.19	0.23	0.23	0.21	0.19	0.20	0.22	0.22	0.21	0.21
CaO	0.72	0.79	0.79	1.75	0.97	1.01	1.11	1.74	0.49	0.57	0.56	0.50	0.59	0.28	0.36	0.39	0.25	0.26	0.31	0.28	0.33
Na_2O	3.21	3.26	3.22	2.70	2.73	2.53	2.24	2.79	3.09	3.06	3.22	3.00	3.14	3.01	3.24	3.19	2.88	2.92	3.18	2.99	3.29
K_2O	4.61	5.26	5.26	4.73	5.83	5.42	7.29	5.98	5.46	6.00	5.99	5.68	5.84	5.23	5.54	5.46	5.32	5.20	5.58	5.81	5.66
P_2O_5	0.06	0.06	0.06	0.15	0.09	0.22	0.11	0.17	0.02	0.03	0.03	0.03	0.04	0.03	0.03	0.03	0.03	0.04	0.03	0.02	0.03
LOI	0.93	0.98	1.00	0.76	1.10	1.19	1.04	0.85	0.83	0.87	0.82	0.76	0.77	0.94	1.00	1.18	1.21	1.02	1.10	1.19	1.14
Total	99.67	99.65	99.84	99.60	99.50	99.89	99.86	99.51	99.70	99.55	100.14	99.51	99.81	99.75	99.80	100.09	99.84	99.80	99.55	100.08	99.58
K_2O/Na_2O	1.44	1.61	1.63	1.75	2.14	2.14	3.25	2.14	1.77	1.96	1.86	1.89	1.86	1.74	1.71	1.71	1.85	1.78	1.75	1.94	1.72
K_2O+Na_2O	7.82	8.52	8.48	7.43	8.56	7.95	9.53	8.77	8.55	9.06	9.21	8.68	8.98	8.24	8.78	8.65	8.20	8.12	8.76	8.80	8.95
$Mg^\#$	18	17	17	31	33	29	31	30	20	20	22	20	20	20	20	18	18	19	20	18	21
A/CNK	1.06	1.05	1.06	1.13	1.17	1.18	1.16	1.11	1.04	1.06	1.08	1.05	1.06	1.13	1.11	1.11	1.14	1.13	1.12	1.13	1.10
Trace element /ppm																					
Li	12.3	13.1	12.6	42.4	17.7	25.2	37.4	42.4	2.20	2.38	3.64	2.50	3.14	2.11	2.09	2.20	2.77	1.20	1.12	2.02	2.19
Be	2.40	2.97	2.95	2.15	3.64	3.48	2.32	2.07	1.51	1.75	2.17	1.63	1.79	1.72	1.78	1.78	1.68	1.59	1.78	1.76	1.77

Continued

Sample	Kuanyu pluton 27°15′20″N 102°10′33″E								Cida pluton 27°10′50″N 102°5′19″E												
	KY-1-1	KY-1-2	KY-1-3	KY-2-2	KY-2-3	KY-2-4	KY-2-5	KY-2-6	CD-1-1	CD-1-2	CD-1-4	CD-1-5	CD-1-6	CD-2-1	CD-2-2	CD-2-3	CD-2-4	CD-2-5	CD-2-6	CD-2-7	CD-2-8
Sc	5.91	5.93	5.97	8.36	3.89	6.60	6.38	8.13	3.12	3.57	4.23	3.52	4.35	4.35	4.46	4.06	4.05	3.61	3.55	3.72	3.83
V	5.13	4.95	4.90	32.0	11.2	18.4	23.9	31.4	4.66	6.47	6.42	5.43	6.44	6.66	6.42	6.13	6.99	6.92	6.39	7.09	5.67
Cr	3.75	3.75	7.84	15.7	7.59	10.3	12.8	16.8	3.95	3.82	3.65	4.22	4.09	3.98	4.09	3.83	4.31	4.08	4.35	4.03	4.05
Co	203	186	187	181	192	169	189	161	242	229	187	251	202	220	213	225	244	237	228	241	245
Ni	1.32	1.43	4.22	6.05	2.34	3.42	4.90	6.12	1.32	1.24	1.41	1.47	1.65	1.45	1.24	1.52	1.66	1.64	1.58	1.51	1.43
Cu	1.33	1.84	1.92	6.70	1.65	2.37	6.72	6.08	0.86	1.06	2.04	0.91	1.64	2.93	5.37	13.0	4.92	2.04	2.84	18.1	2.03
Zn	26.6	26.4	26.3	59.1	26.1	43.8	42.5	58.7	15.7	18.1	17.9	15.4	19.5	24.1	21.8	21.1	22.1	25.4	24.3	21.6	21.3
Ga	19.2	20.3	20.4	24.5	23.3	21.5	21.4	25.0	14.6	16.6	17.2	15.7	17.9	16.2	17.9	17.8	16.3	15.0	16.2	17.0	17.3
Ge	1.54	1.68	1.68	1.81	1.68	1.96	1.76	1.96	1.20	1.46	1.32	1.31	1.49	1.27	1.45	1.46	1.31	1.26	1.33	1.44	1.35
Rb	198	218	218	268	309	305	306	305	223	246	206	222	240	211	219	218	212	206	225	226	227
Sr	57.6	56.7	56.4	106	83.9	96.5	132	123	74.9	120	108	119	125	103	126	130	103	102	119	111	128
Y	34.4	43.1	42.8	31.2	34.4	44.8	36.2	36.7	20.0	22.1	20.2	16.5	23.2	17.9	22.6	19.8	14.7	13.6	17.5	17.4	16.7
Zr	181	181	170	304	232	191	212	296	159	205	204	171	212	235	221	207	261	231	252	212	211
Nb	10.9	10.9	10.7	19.1	19.1	16.4	12.5	18.3	6.54	7.76	8.51	7.25	10.0	9.50	8.99	8.10	9.37	6.93	9.36	8.19	8.50
Cs	1.80	1.92	1.93	7.27	3.76	10.1	7.34	7.22	0.98	0.94	0.99	0.89	0.94	1.02	1.09	1.21	1.21	0.93	0.98	1.21	1.04
Ba	420	453	460	571	514	359	956	777	372	422	434	411	418	382	403	385	384	369	407	417	405
La	52.3	53.4	54.2	67.9	60.2	44.8	51.9	68.8	46.3	59.3	46.7	40.9	64.5	56.6	70.1	68.6	57.6	51.1	56.5	72.2	53.1
Ce	116	117	118	146	128	92.9	107	146	97.6	124	95.8	83.6	135	115	152	143	128	109	126	166	110
Pr	13.8	14.2	14.5	17.2	15.1	12.0	13.0	17.4	11.4	14.8	11.5	9.64	16.1	14.2	17.8	16.8	14.5	12.6	14.5	17.8	13.3
Nd	49.1	50.5	51.3	63.5	52.3	42.3	45.5	64.2	39.2	50.7	50.6	34.7	55.8	48.4	63.8	57.3	49.2	42.9	49.8	60.6	45.6
Sm	10.1	10.6	10.7	11.5	9.47	9.34	8.74	11.7	7.09	9.12	8.72	6.36	10.0	8.45	10.7	9.79	8.36	7.35	8.83	10.1	7.92

Continued

Sample	Kuanyu pluton 27°15′20″N 102°10′33″E								Gida pluton 27°10′50″N 102°5′19″E												
	KY-1-1	KY-1-2	KY-1-3	KY-2-2	KY-2-3	KY-2-4	KY-2-5	KY-2-6	CD-1-1	CD-1-2	CD-1-4	CD-1-5	CD-1-6	CD-2-1	CD-2-2	CD-2-3	CD-2-4	CD-2-5	CD-2-6	CD-2-7	CD-2-8
Eu	0.77	0.78	0.78	1.21	0.88	0.74	1.52	1.42	0.55	0.62	0.68	0.59	0.64	0.62	0.66	0.63	0.57	0.57	0.63	0.62	0.64
Gd	8.85	9.56	9.62	9.47	7.40	8.56	7.60	9.94	5.64	7.26	6.04	5.11	7.90	6.39	8.05	7.34	6.09	5.41	6.56	7.39	5.90
Tb	1.31	1.47	1.47	1.25	1.05	1.39	1.10	1.35	0.77	0.94	0.80	0.68	1.02	0.82	1.02	0.92	0.72	0.67	0.82	0.88	0.74
Dy	7.17	8.52	8.54	6.42	6.02	8.21	6.44	7.19	4.22	4.79	4.16	3.45	5.15	4.14	5.14	4.57	3.49	3.34	4.14	4.23	3.76
Ho	1.30	1.63	1.63	1.11	1.18	1.56	1.27	1.30	0.77	0.83	0.73	0.61	0.87	0.71	0.89	0.78	0.58	0.56	0.71	0.70	0.65
Er	3.41	4.41	4.43	2.98	3.51	4.28	3.71	3.55	2.10	2.10	1.90	1.57	2.19	1.82	2.27	2.00	1.52	1.47	1.83	1.78	1.70
Tm	0.46	0.60	0.60	0.40	0.52	0.61	0.53	0.49	0.29	0.26	0.24	0.20	0.27	0.23	0.29	0.25	0.20	0.19	0.23	0.22	0.22
Yb	2.67	3.51	3.47	2.49	3.31	3.71	3.30	2.98	1.72	1.50	1.39	1.14	1.52	1.36	1.65	1.47	1.17	1.10	1.33	1.31	1.29
Lu	0.37	0.48	0.48	0.36	0.48	0.51	0.47	0.43	0.23	0.21	0.20	0.17	0.21	0.19	0.23	0.21	0.18	0.16	0.19	0.19	0.19
Hf	5.60	5.65	5.25	8.43	6.49	5.67	5.92	8.21	5.00	6.17	6.05	5.15	6.41	7.03	6.69	6.26	7.96	6.96	7.57	6.45	6.25
Ta	1.21	1.31	1.31	1.73	2.45	2.11	1.21	1.67	0.58	0.50	0.58	0.65	1.00	0.75	0.59	0.59	0.72	0.28	0.70	0.53	0.74
Pb	18.5	21.1	21.4	33.4	27.3	42.9	47.8	41.7	18.5	16.2	15.7	15.5	17.3	14.7	14.7	13.8	14.0	14.5	15.6	14.3	14.9
Th	41.0	43.1	43.3	42.9	42.5	30.4	33.2	43.7	27.0	40.7	33.2	30.0	44.8	35.6	48.7	43.1	48.9	35.2	38.7	55.9	34.1
U	5.5	7.2	7.5	5.8	4.1	4.8	4.1	4.3	3.6	4.8	3.5	3.1	4.9	3.7	4.8	3.9	4.6	3.9	3.7	4.7	3.3
T_{Zr}/℃	803	799	795	851	831	814	817	841	790	812	812	797	814	834	824	818	845	833	837	822	819
Eu/Eu*	0.25	0.24	0.23	0.35	0.32	0.25	0.57	0.40	0.27	0.23	0.31	0.32	0.22	0.26	0.22	0.23	0.24	0.27	0.25	0.22	0.29
$(La/Yb)_N$	14.1	10.9	11.2	19.6	13.0	8.7	11.3	16.5	19.3	28.3	24.1	25.6	30.5	29.8	30.5	33.6	35.2	33.4	30.4	39.5	29.5
$(Dy/Yb)_N$	1.80	1.62	1.65	1.73	1.22	1.48	1.31	1.61	1.64	2.13	2.00	2.02	2.27	2.03	2.09	2.09	1.99	2.04	2.08	2.16	1.96
REE	267.41	276.25	279.76	331.32	289.19	230.92	252.39	336.39	217.98	276.20	217.20	188.71	301.26	259.10	334.84	313.36	272.08	236.47	271.73	344.01	244.94

Mg# = molar $100MgO/(MgO+FeO^T)$; FeO^T = All Fe calculated as FeO; A/CNK = $Al_2O_3/(CaO+K_2O+Na_2O)$ molar ratio; $\delta Eu = Eu_N/(Sm_N \cdot Gd_N)^{1/2}$.

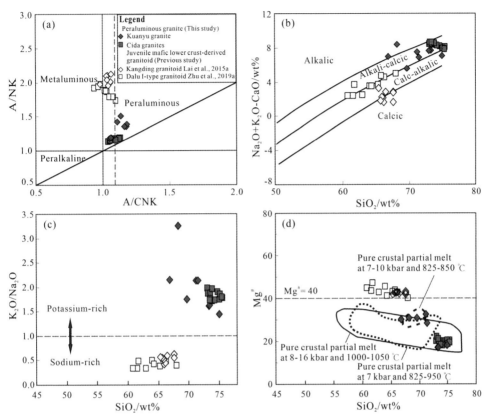

Fig. 5　The A/NK vs. A/CNK diagram (a), (Na$_2$O+K$_2$O−CaO) vs. SiO$_2$(b) (Frost et al., 2001),

K$_2$O/Na$_2$O vs. SiO$_2$(c) (Moyen and Martin, 2012), and Mg$^{\#}$ vs. SiO$_2$(d) diagrams

for the Kuanyu and Cida peraluminous granites

from western margin of the Yangtze Block, South China.

The Neoproterozoic juvenile mafic lower crust-derived granitoids were shown for comparison. The Kangding granitoids are from Lai et al. (2015a). The Dalu I-type granitoids are from Zhu et al. (2019a). In Fig. 5d, also shown are the fields of pure crustal partial melts determined in experimental studies on dehydration melting of two-mica schist at 7−10 kbar, 825−850 ℃ (Patiño Douce and Johnston, 1991), of dehydration melting of low-K basaltic rocks at 8−16 kbar, 1 000−1 050 ℃ (Rapp and Watson, 1995), and of moderately hydrous (1. 7−2. 3 wt% H$_2$O) medium- to high-K basaltic rocks at 7 kbar, 825−950 ℃ (Sisson et al., 2005).

large ion lithophile elements (LILEs) (e.g., Rb, Th, U, and K) and Pb, and depletion of high field strength elements (HFSEs) (e.g., Nb, Ta, Zr, and Ti), resembling trends of the middle-upper crust (Rudnick and Gao, 2003). According to the whole-rock zircon saturation thermometry from Watson and Harrison (1983), the Kuanyu granites yield high crystallization temperature varying from 795 ℃ to 851 ℃ with an average of 819 ℃.

4. 2. 2　Cida granites

All the Cida granites samples show high and concentrated SiO$_2$ contents ranging from 73. 1 wt% to 75. 6 wt%, K$_2$O contents from 5. 20 wt% to 6. 00 wt%, Na$_2$O contents from 2. 88 wt%

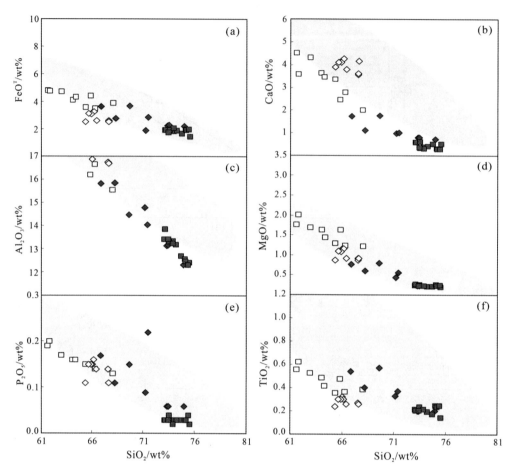

Fig. 6   Harker diagrams of major elements for the Kuanyu and Cida peraluminous granites
from the western Yangtze Block, South China.

The shaded areas denote the experimental melts at crustal conditions (referred from Guo et al., 2012). Symbols as Fig. 5.

to 3. 29 wt%, K_2O/Na_2O ratios from 1. 71 to 1. 96, and A/CNK values from 1. 04 to 1. 14. They contain relatively low CaO (0. 25 - 0. 57 wt%), Al_2O_3(12. 4 - 13. 8 wt%), MgO (0. 17 - 0. 25 wt%), P_2O_5(0. 02 - 0. 04 wt%), and TiO_2(0. 14 - 0. 24 wt%) (Fig. 6). They also have low $Fe_2O_3^T$(1. 54 - 2. 22 wt%) and $Mg^\#$ values (18 - 20). In the chondrite-normalized REE pattern diagram (Fig. 7a), the Cida granites display more fractionated REE patterns with $(La/Yb)_N$ ratios of 19. 3 - 39. 5, $(Dy/Yb)_N$ ratios of 1. 64 - 2. 27, and pronouncedly negative Eu anomalies ($Eu/Eu^* = 0. 22 - 0. 32$). On the primitive-mantle normalized diagram (Fig. 7b), the enrichment of Rb, Th, U, K, and Pb and the depletion of Nb, Ta, Sr, and Ti are also observed. All the Cida granites have high zircon saturation temperature ranging from 790 ℃ to 845 ℃ with an average of 820 ℃.

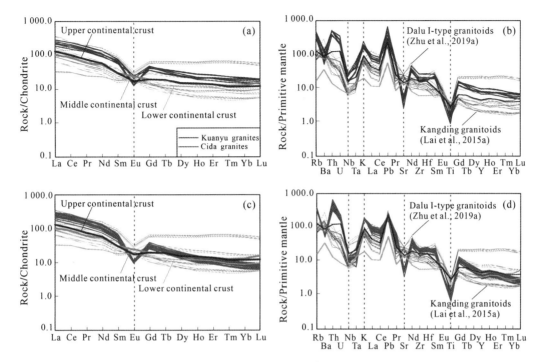

Fig. 7   Diagrams of chondrite-normalized REE patterns (a,c) and primitive mantle-normalized

trace element patterns (b,d) for the Kuanyu and Cida peraluminous granites

from the western margin of the Yangtze Block, South China.

The primitive mantle and chondrite values are from Sun and McDonough (1989). The lower, middle, and upper

continental crust references are from Rudnick and Gao (2003). The Kangding granitoids are from Lai et al. (2015a).

The Dalu I-type granitoids are from Zhu et al. (2019a).

4. 3   Whole-rock Sr-Nd isotopic compositions

The whole-rock Sr-Nd isotopic data for the Kuanyu and Cida peraluminous granites are listed in Table 3, and a diagram of $\varepsilon_{Nd}(t)$ vs. initial $^{87}Sr/^{86}Sr$ ratios (I_{Sr}) is shown in Fig. 8. The $\varepsilon_{Nd}(t)$ and initial $^{87}Sr/^{86}Sr$ ratios of all samples were calculated according to concordant crystallization ages and using the model of Depaolo (1981).

The Kuanyu granites have variable initial $^{87}Sr/^{86}Sr$ ratios (I_{Sr}) from 0.709 9 to 0.721 7. They are characterized by enriched Nd isotopic compositions with $\varepsilon_{Nd}(t)$ values of −5.1 to −4.9. They yield the ancient two-stage Nd model ages ranging from 1.74 Ga to 1.75 Ga.

The Cida granites yield relatively constant initial $^{87}Sr/^{86}Sr$ ratios (I_{Sr}) varying from 0.712 8 to 0.714 8. They also show the negative $\varepsilon_{Nd}(t)$ values of −3.5 to −2.9 and old two-stage Nd model ages of 1.60−1.64 Ga.

4. 4   Zircon Lu-Hf isotopic compositions

Zircon grains from there peraluminous granites samples (KY-2, CD-1, and CD-2) were

Table 3 Whole-rock Sr-Nd isotopes data for the Neoproterozoic Kuanyu and Cida peraluminous granites in the western Yangtze Block, South China.

Sample	$^{87}Sr/^{86}Sr$	2SE	Sr /ppm	Rb /ppm	$^{143}Nd/^{144}Nd$	2SE	Nd /ppm	Sm /ppm	T_{DM2} /Ga	$\varepsilon_{Nd}(t)$	$(^{87}Sr/^{86}Sr)_i$ (I_{Sr})
Kuanyu pluton											
KY-1-1	0.829 926	0.000 005	57.6	198	0.511 983	0.000 005	49.1	10.1	1.75	−5.1	0.709 9
KY-2-2	0.809 826	0.000 003	106	268	0.511 908	0.000 004	63.5	11.5	1.74	−4.9	0.721 7
Cida pluton											
CD-1-4	0.781 161	0.000 005	108	206	0.511 984	0.000 004	50.6	8.72	1.60	−2.9	0.714 8
CD-2-4	0.784 435	0.000 004	103	212	0.511 943	0.000 004	49.2	8.36	1.64	−3.5	0.712 8

$^{87}Rb/^{86}Sr$ and $^{147}Sm/^{144}Nd$ ratios were calculated using Rb, Sr, Sm and Nd contents analyzed by ICP-MS.

T_{DM2} represents the two-stage model age and were calculated using present-day $(^{147}Sm/^{144}Nd)_{DM} = 0.213\ 7$, $(^{147}Sm/^{144}Nd)_{DM} = 0.513\ 15$ and $(^{147}Sm/^{144}Nd)_{crust} = 0.101\ 2$. $\varepsilon_{Nd}(t)$ values were calculated using present-day $(^{147}Sm/^{144}Nd)_{CHUR} = 0.196\ 7$ and $(^{147}Sm/^{144}Nd)_{CHUR} = 0.512\ 638$.

$$\varepsilon_{Nd}(t) = [(^{143}Nd/^{144}Nd)_S(t)/(^{143}Nd/^{144}Nd)_{CHUR}(t) - 1] \times 10^4.$$

$$T_{DM2} = \frac{1}{\lambda}\{1 + [(^{143}Nd/^{144}Nd)_S - ((^{147}Sm/^{144}Nd)_S - (^{147}Sm/^{144}Nd)_{crust})(e^{\lambda t} - 1) - (^{143}Nd/^{144}Nd)_{DM}] /$$
$$[(^{147}Sm/_{144}Nd)_{crust} - (^{147}Sm/^{144}Nd)_{DM})]\}.$$

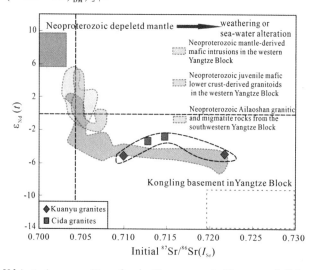

Fig. 8 Initial Sr-Nd isotopic compositions for the Neoproterozoic Kuanyu and Cida peraluminous granites in the western Yangtze Block, South China (revised after Wang et al., 2016).

The Neoproterozoic mantle-derived mafic rocks in the western Yangtze Block are from Zhao et al. (2008a); Zhou et al. (2006b); Zhao and Zhou (2007a). The Neoproterozoic juvenile mafic lower crust-derived granitoids in the western Yangtze Block are from Huang et al. (2008, 2009); Lai et al. (2015a); Zhao and Zhou (2007b); Zhao et al. (2008a,b); Zhou et al. (2006a); Zhu et al. (2019a). The Neoproterozoic Ailaoshan granitic and migmatite rocks in the southwestern Yangtze Block are from Wang et al. (2016). The Neoproterozoic depleted mantle is from Zimmer et al. (1995). The Kongling basement in the Yangtze Block is from Gao et al. (1999).

also analyzed for in-situ Lu-Hf isotopes on the same domains with U-Pb dating. The results are shown in Table 4, and Figs. 9 and 10. The initial $^{176}Hf/^{177}Hf$, $\varepsilon_{Hf}(t)$, and two-stage model ages were calculated according to their corresponding zircon crystallization ages.

Table 4　Zircon Lu-Hf isotopic data for the Neoproterozoic Kuanyu and Cida peraluminous granites along the western margin of the Yangtze Block, South China.

Grain spot	Age /Ma	$^{176}Yb/^{177}Hf$	2SE	$^{176}Lu/^{177}Hf$	2SE	$^{176}Hf/^{177}Hf$	2SE	$(^{176}Hf/^{177}Hf)_i$	$f_{Lu/Hf}$	$\varepsilon_{Hf}(t)$	2SE	T_{DM1} /Ma	T_{DM2} /Ma
KY-2													
KY-2-1	840	0.043 258	0.000 573	0.001 365	0.000 019	0.282 305	0.000 017	0.282 283	-0.96	0.51	0.59	1 382	1 693
KY-2-2	840	0.050 549	0.000 525	0.001 565	0.000 014	0.282 220	0.000 012	0.282 195	-0.95	-2.70	0.41	1 514	1 895
KY-2-3	840	0.033 467	0.000 719	0.001 144	0.000 027	0.282 320	0.000 014	0.282 302	-0.97	1.31	0.50	1 347	1 643
KY-2-4	840	0.045 416	0.000 147	0.001 285	0.000 004	0.282 251	0.000 014	0.282 230	-0.96	-1.29	0.51	1 453	1 807
KY-2-5	840	0.034 377	0.000 125	0.001 038	0.000 003	0.282 225	0.000 014	0.282 209	-0.97	-1.94	0.50	1 474	1 847
KY-2-6	840	0.030 490	0.000 179	0.000 962	0.000 006	0.282 183	0.000 011	0.282 168	-0.97	-3.34	0.40	1 527	1 935
KY-2-7	840	0.030 637	0.000 187	0.000 994	0.000 005	0.282 236	0.000 018	0.282 221	-0.97	-1.50	0.65	1 456	1 819
KY-2-8	840	0.027 283	0.000 309	0.000 865	0.000 010	0.282 196	0.000 015	0.282 182	-0.97	-2.80	0.51	1 504	1 901
KY-2-9	840	0.039 709	0.000 569	0.001 230	0.000 016	0.282 186	0.000 016	0.282 166	-0.96	-3.56	0.56	1 541	1 948
KY-2-10	840	0.029 163	0.000 168	0.000 941	0.000 009	0.282 234	0.000 014	0.282 219	-0.97	-1.51	0.50	1 456	1 821
KY-2-11	840	0.026 102	0.000 420	0.000 807	0.000 015	0.282 266	0.000 016	0.282 253	-0.98	-0.23	0.57	1 403	1 740
KY-2-12	840	0.030 977	0.000 193	0.000 951	0.000 007	0.282 279	0.000 013	0.282 263	-0.97	0.02	0.45	1 394	1 723
KY-2-13	840	0.034 476	0.000 252	0.001 068	0.000 009	0.282 226	0.000 015	0.282 209	-0.97	-1.98	0.52	1 475	1 849
KY-2-14	840	0.043 320	0.001 200	0.001 344	0.000 037	0.282 215	0.000 015	0.282 194	-0.96	-2.63	0.54	1 507	1 891
KY-2-15	840	0.030 038	0.000 174	0.000 937	0.000 005	0.282 221	0.000 013	0.282 206	-0.97	-1.98	0.46	1 474	1 850
KY-2-16	840	0.030 447	0.000 432	0.000 939	0.000 012	0.282 235	0.000 012	0.282 220	-0.97	-1.50	0.41	1 455	1 819
KY-2-17	840	0.078 538	0.000 388	0.002 468	0.000 010	0.282 243	0.000 014	0.282 204	-0.93	-2.90	0.50	1 538	1 907
KY-2-18	840	0.031 873	0.000 093	0.001 008	0.000 002	0.282 206	0.000 014	0.282 190	-0.97	-2.61	0.49	1 499	1 888
KY-2-19	840	0.020 377	0.000 059	0.000 625	0.000 003	0.282 283	0.000 012	0.282 273	-0.98	0.54	0.41	1 369	1 691
KY-2-20	840	0.026 911	0.000 026	0.000 859	0.000 001	0.282 275	0.000 013	0.282 261	-0.97	0.00	0.46	1 394	1 725
KY-2-21	840	0.035 141	0.000 156	0.001 117	0.000 006	0.282 200	0.000 015	0.282 182	-0.97	-2.95	0.52	1 514	1 910
KY-2-22	840	0.037 578	0.000 189	0.001 149	0.000 007	0.282 261	0.000 016	0.282 243	-0.97	-0.81	0.55	1 431	1 776
KY-2-23	840	0.039 546	0.000 441	0.001 270	0.000 012	0.282 229	0.000 015	0.282 209	-0.96	-2.07	0.51	1 483	1 855
KY-2-24	840	0.023 154	0.000 580	0.000 693	0.000 017	0.282 224	0.000 017	0.282 213	-0.98	-1.55	0.53	1 455	1 824
KY-2-25	840	0.013 304	0.000 191	0.000 387	0.000 007	0.282 209	0.000 012	0.282 202	-0.99	-1.80	0.42	1 458	1 838
KY-2-26	840	0.034 983	0.000 128	0.001 112	0.000 007	0.282 064	0.000 014	0.282 047	-0.97	-7.75	0.51	1 703	2 210

Continued

Grain spot	Age /Ma	$^{176}\mathrm{Yb}/^{177}\mathrm{Hf}$	2SE	$^{176}\mathrm{Lu}/^{177}\mathrm{Hf}$	2SE	$^{176}\mathrm{Hf}/^{177}\mathrm{Hf}$	2SE	$(^{176}\mathrm{Hf}/^{177}\mathrm{Hf})_i$	$f_{\mathrm{Lu/Hf}}$	$\varepsilon_{\mathrm{Hf}}(t)$	2SE	T_{DM1} /Ma	T_{DM2} /Ma
KY-2-27	840	0.047 581	0.001 470	0.001 458	0.000 012	0.282 248	0.000 042	0.282 225	-0.96	-1.58	0.42	1 468	1 825
KY-2-28	840	0.051 591	0.001 260	0.001 583	0.000 014	0.282 232	0.000 033	0.282 207	-0.95	-2.31	0.50	1 498	1 870
KY-2-29	840	0.043 175	0.000 138	0.001 319	0.000 014	0.282 295	0.000 003	0.282 274	-0.96	0.20	0.49	1 394	1 713
KY-2-30	840	0.070 099	0.001 200	0.002 202	0.000 013	0.282 249	0.000 041	0.282 214	-0.93	-2.43	0.46	1 513	1 877
KY-2-31	840	0.025 763	0.000 133	0.000 781	0.000 014	0.282 246	0.000 005	0.282 234	-0.98	-0.95	0.51	1 429	1 784
KY-2-32	840	0.034 935	0.000 271	0.001 098	0.000 017	0.282 286	0.000 009	0.282 268	-0.97	0.13	0.61	1 393	1 717
KY-2-33	840	0.037 344	0.000 067	0.001 156	0.000 015	0.282 232	0.000 001	0.282 214	-0.97	-1.81	0.53	1 471	1 840
CD-1													
CD-1-01	835	0.044 827	0.000 059	0.001 387	0.000 030	0.282 247	0.000 005	0.282 225	-0.96	-1.64	1.06	1 465	1 826
CD-1-02	835	0.023 330	0.000 236	0.000 761	0.000 023	0.282 301	0.000 006	0.282 289	-0.98	0.97	0.80	1 352	1 662
CD-1-03	835	0.025 763	0.000 254	0.000 817	0.000 022	0.282 294	0.000 006	0.282 281	-0.98	0.65	0.77	1 365	1 681
CD-1-04	835	0.043 642	0.000 108	0.001 334	0.000 028	0.282 309	0.000 004	0.282 288	-0.96	0.57	0.99	1 374	1 685
CD-1-05	835	0.046 547	0.000 322	0.001 444	0.000 024	0.282 329	0.000 007	0.282 306	-0.96	1.18	0.85	1 352	1 648
CD-1-06	835	0.028 904	0.000 160	0.000 912	0.000 023	0.282 278	0.000 004	0.282 264	-0.97	-0.05	0.80	1 392	1 724
CD-1-07	835	0.041 728	0.000 122	0.001 338	0.000 019	0.282 384	0.000 005	0.282 362	-0.96	3.21	0.66	1 269	1 519
CD-1-08	835	0.045 957	0.000 067	0.001 370	0.000 022	0.282 307	0.000 004	0.282 286	-0.96	0.49	0.78	1 378	1 691
CD-1-09	835	0.036 242	0.000 068	0.001 127	0.000 022	0.282 323	0.000 002	0.282 305	-0.97	1.31	0.78	1 343	1 640
CD-1-10	835	0.037 529	0.000 614	0.001 150	0.000 021	0.282 273	0.000 017	0.282 255	-0.97	-0.47	0.74	1 413	1 751
CD-1-11	835	0.051 225	0.000 225	0.001 546	0.000 022	0.282 294	0.000 005	0.282 270	-0.95	-0.18	0.79	1 407	1 732
CD-1-12	835	0.030 516	0.000 165	0.000 930	0.000 026	0.282 253	0.000 005	0.282 238	-0.97	-0.96	0.90	1 428	1 781
CD-1-13	835	0.076 121	0.000 221	0.002 319	0.000 024	0.282 312	0.000 009	0.282 275	-0.93	-0.39	0.86	1 430	1 747
CD-1-14	835	0.039 162	0.000 105	0.001 181	0.000 014	0.282 279	0.000 002	0.282 260	-0.96	-0.28	0.51	1 407	1 740
CD-1-15	835	0.031 628	0.000 043	0.001 086	0.000 028	0.282 264	0.000 004	0.282 247	-0.97	-0.71	1.00	1 422	1 766
CD-1-16	835	0.052 256	0.000 463	0.001 729	0.000 043	0.282 201	0.000 011	0.282 174	-0.95	-3.71	1.54	1 551	1 953
CD-1-17	835	0.039 020	0.000 281	0.001 197	0.000 021	0.282 288	0.000 007	0.282 269	-0.96	-0.02	0.73	1 396	1 723
CD-1-18	835	0.031 997	0.000 119	0.000 995	0.000 018	0.282 334	0.000 005	0.282 318	-0.97	1.83	0.62	1 319	1 606
CD-1-19	835	0.032 707	0.000 349	0.001 028	0.000 020	0.282 230	0.000 011	0.282 213	-0.97	-1.92	0.72	1 467	1 840
CD-1-20	835	0.041 519	0.000 292	0.001 281	0.000 024	0.282 237	0.000 008	0.282 217	-0.96	-1.89	0.86	1 472	1 840

Continued

Grain spot	Age /Ma	$^{176}\mathrm{Yb}/^{177}\mathrm{Hf}$	2SE	$^{176}\mathrm{Lu}/^{177}\mathrm{Hf}$	2SE	$^{176}\mathrm{Hf}/^{177}\mathrm{Hf}$	2SE	$(^{176}\mathrm{Hf}/^{177}\mathrm{Hf})_i$	$f_{\mathrm{Lu/Hf}}$	$\varepsilon_{\mathrm{Hf}}(t)$	2SE	T_{DM1} /Ma	T_{DM2} /Ma
CD-1-21	835	0.041 500	0.000 172	0.001 283	0.000 005	0.282 204	0.000 029	0.282 184	-0.96	-3.11	1.04	1 519	1 915
CD-1-22	835	0.035 231	0.000 073	0.001 078	0.000 003	0.282 196	0.000 022	0.282 179	-0.97	-3.16	0.78	1 518	1 919
CD-1-23	835	0.044 314	0.000 192	0.001 368	0.000 006	0.282 312	0.000 023	0.282 290	-0.96	0.63	0.80	1 372	1 681
CD-1-24	835	0.048 728	0.000 177	0.001 487	0.000 005	0.282 351	0.000 027	0.282 327	-0.96	1.86	0.95	1 324	1 603
CD-1-25	835	0.042 244	0.000 455	0.001 372	0.000 013	0.282 350	0.000 016	0.282 329	-0.96	1.99	0.56	1 318	1 596
CD-1-26	835	0.043 384	0.000 496	0.001 298	0.000 015	0.282 244	0.000 029	0.282 224	-0.96	-1.66	1.02	1 463	1 826
CD-1-27	835	0.033 887	0.000 193	0.001 061	0.000 006	0.282 311	0.000 025	0.282 295	-0.97	0.96	0.87	1 355	1 661
CD-1-28	835	0.036 010	0.000 096	0.001 105	0.000 003	0.282 243	0.000 017	0.282 226	-0.97	-1.49	0.61	1 453	1 815
CD-2													
CD-2-01	835	0.047 357	0.000 574	0.001 434	0.000 015	0.282 275	0.000 020	0.282 253	-0.96	-0.72	0.72	1 428	1 767
CD-2-02	835	0.085 741	0.000 334	0.002 599	0.000 011	0.282 323	0.000 029	0.282 282	-0.92	-0.29	1.02	1 430	1 740
CD-2-03	835	0.028 067	0.000 170	0.000 961	0.000 004	0.282 325	0.000 015	0.282 310	-0.97	1.59	0.55	1 330	1 623
CD-2-04	835	0.040 418	0.000 239	0.001 249	0.000 008	0.282 301	0.000 014	0.282 281	-0.96	0.33	0.50	1 380	1 698
CD-2-05	835	0.041 037	0.000 270	0.001 269	0.000 007	0.282 296	0.000 014	0.282 276	-0.96	0.18	0.51	1 388	1 709
CD-2-06	835	0.040 791	0.000 496	0.001 347	0.000 017	0.282 378	0.000 018	0.282 357	-0.96	3.06	0.63	1 277	1 530
CD-2-07	835	0.050 442	0.000 391	0.001 493	0.000 009	0.282 256	0.000 017	0.282 232	-0.96	-1.52	0.61	1 459	1 815
CD-2-08	835	0.045 227	0.000 600	0.001 505	0.000 020	0.282 352	0.000 017	0.282 328	-0.95	1.93	0.61	1 323	1 601
CD-2-09	835	0.028 394	0.000 095	0.000 904	0.000 003	0.282 251	0.000 015	0.282 237	-0.97	-1.01	0.53	1 429	1 783
CD-2-10	835	0.044 653	0.000 216	0.001 547	0.000 007	0.282 378	0.000 020	0.282 353	-0.95	2.75	0.72	1 289	1 547
CD-2-11	835	0.041 337	0.000 314	0.001 251	0.000 008	0.282 297	0.000 019	0.282 277	-0.96	0.22	0.67	1 386	1 706
CD-2-12	835	0.029 799	0.000 253	0.000 939	0.000 006	0.282 374	0.000 015	0.282 360	-0.97	3.31	0.52	1 259	1 512
CD-2-13	835	0.029 876	0.000 080	0.000 922	0.000 002	0.282 277	0.000 015	0.282 263	-0.97	-0.05	0.53	1 394	1 725
CD-2-14	835	0.053 500	0.000 379	0.001 675	0.000 014	0.282 262	0.000 019	0.282 236	-0.95	-1.46	0.67	1 460	1 812
CD-2-15	835	0.038 353	0.000 221	0.001 181	0.000 006	0.282 266	0.000 016	0.282 247	-0.96	-0.78	0.57	1 425	1 770
CD-2-16	835	0.042 838	0.000 133	0.001 420	0.000 004	0.282 252	0.000 021	0.282 230	-0.96	-1.54	0.75	1 459	1 817
CD-2-17	835	0.043 756	0.000 130	0.001 347	0.000 004	0.282 288	0.000 016	0.282 267	-0.96	-0.22	0.57	1 405	1 734
CD-2-18	835	0.063 272	0.000 140	0.001 912	0.000 004	0.282 251	0.000 018	0.282 221	-0.94	-2.15	0.64	1 491	1 854

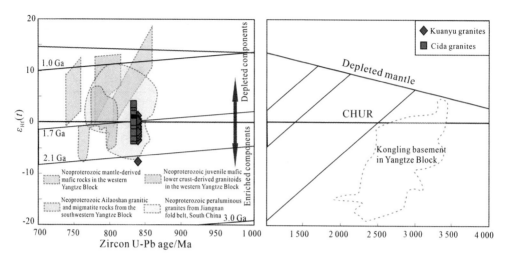

Fig. 9　The $\varepsilon_{Hf}(t)$ values vs. Zircon U-Pb age diagram for the Neoproterozoic Kuanyu and Cida peraluminous granites from the Yangtze Block, South China (revised after Zhao et al., 2013).

Zircon Hf isotopic data of Neoproterozoic mantle-derived mafic rocks in the western Yangtze Block are from Sun and Zhou (2008); Zhao et al. (2008a). Neoproterozoic juvenile mafic lower crust-derived granitoids in the western Yangtze Block are from Huang et al. (2008, 2009); Zhao et al. (2008b); Zhu et al. (2019a). Neoproterozoic Ailaoshan granitic and migmatite rocks in the southwestern Yangtze Block are from Wang et al. (2016). Neoproterozoic peraluminous granites in Jiangnan fold belt, South China are from Zhao et al. (2013). The Kongling basement is from Gao et al. (2011).

Zircons from sample KY-2 show initial $^{176}Hf/^{177}Hf$ ratios ($^{176}Hf/^{177}Hf$)$_i$ ranging from 0.282 047 to 0.282 302. They yield variable $\varepsilon_{Hf}(t)$ values of −7.75 to +1.31 and two-stage Hf model ages of 1 643−2 210 Ma (a mean of 1 831 Ma) (Figs. 9 and 10a,b).

Zircon grains from sample CD-1 share high initial $^{176}Hf/^{177}Hf$ ratios ($^{176}Hf/^{177}Hf$)$_i$ from 0.282 196 to 0.282 384. The $\varepsilon_{Hf}(t)$ values are distributed from −3.71 to +3.21 with corresponding two-stage Hf model ages of 1 519−1 953 Ma (a mean of 1 735 Ma) (Figs. 9 and 10c,d).

The zircons from sample CD-2 exhibit initial $^{176}Hf/^{177}Hf$ ratios ($^{176}Hf/^{177}Hf$)$_i$ varying from 0.282 221 to 0.282 360. Their $\varepsilon_{Hf}(t)$ values and two-stage Hf model ages are in the range of −2.15 to +3.31 and 1 512 Ma to 1 854 Ma, respectively (Figs. 9 and 10e,f).

5　Discussion

5.1　Petrogenesis of the Neoproterozoic peraluminous granites

5.1.1　Magma source

The Kuanyu and Cida granites formed at ca. 840 − 835 Ma. The samples are all geochemically characterized by the peraluminous to strongly peraluminous features with high A/CNK values ranging from 1.04 to 1.18. They display a geochemical consistent with

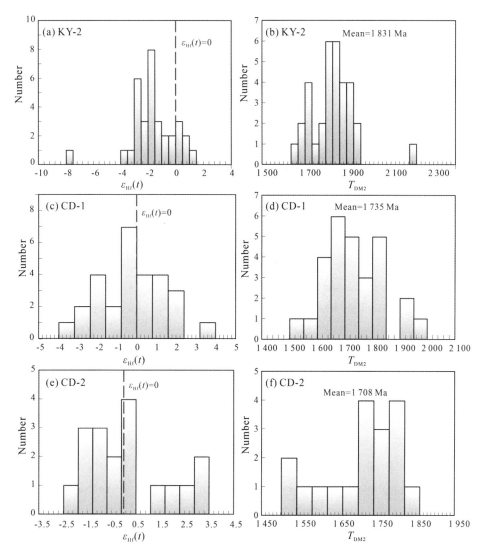

Fig. 10 Histograms of $\varepsilon_{Hf}(t)$ isotope ratios and Hf model age for the Kuanyu (a,b) and Cida (c-f) peraluminous granites from the western margin of the Yangtze Block, South China.

experimental melts under crustal conditions (Figs. 6 and 7) (Guo et al., 2012). Experimental petrology has revealed that peraluminous intermediate to felsic melts can form by the partial melting of metaluminous basaltic to andesitic rocks under crustal conditions (Chappell, 1999; Chappell et al., 2012; Rapp and Watson, 1995; Sission et al., 2005; Sylvester, 1998). Nevertheless, such peraluminous rocks usually contain low K_2O contents with K_2O/Na_2O ratios <1, which significantly contrasts with the results presented here of peraluminous granites with particularly high K_2O contents (4.61−7.29 wt%) and K_2O/Na_2O ratios (1.44−3.25 > 1) (Fig. 5c). We can therefore preclude metabasaltic rocks in the magma source. Previous voluminous studies also suggested that the peraluminous, silica-rich rocks are mainly generated by the partial melting of metasedimentary protoliths in a mature continental crust source including the clay-rich metapelites and clay-poor metagreywackes (psammites) (Clemens, 2003; Sylvester, 1998). Several lines of evidence suggest that the Kuanyu and Cida peraluminous granites predominantly originate from a metasedimentary source under middle-upper crust conditions.

(1) The Kuanyu and Cida peraluminous granites contain high and variable concentrations of SiO_2 (66. 9 – 75. 6 wt%), K_2O (4. 61 – 7. 29 wt%), and A/CNK ratios (1. 04 – 1. 18), suggesting that their sources are dominated by metasedimentary rocks rather than meta-igneous protoliths (Cai et al., 2011; Zhu et al., 2018).

(2) The weakly fractionated HREE patterns, low $(Gd/Yb)_N$ values (1. 85 – 4. 66) and Sr/Y ratios (1. 31 – 7. 62) indicate that these peraluminous granites were derived from a relatively shallow crustal source above the garnet stability depth (Patiño Douce, 1996; Rossi et al., 2002). In addition, the samples display the pronounced Th, U, K, and Pb peaks and Nb, Ta, Ti, Ba, Sr, and Eu troughs (Fig. 7), which resemble geochemical characteristics of middle-upper crust-derived melts (Chen et al., 2014; Rudnick and Gao, 2003).

(3) On the diagrams of CaO/Al_2O_3 vs. $CaO+Al_2O_3$ and $Al_2O_3/(Fe_2O_3^T+MgO+TiO_2)$ vs. $Al_2O_3+Fe_2O_3^T+MgO+TiO_2$ (Patiño Douce, 1999) (Fig. 11a,b), the compositions of the peraluminous granites are comparable with experimental melts of various metasediments under relatively low pressure (<5 kbar), suggesting a shallow crustal level (Cai et al., 2011). This is in agreement with CIPW-normative Qz-Ab-Or phase compositions (Fig. 11c), which also reflect precursor magmas that were derived from the relatively low pressure conditions (1 – 5 kbar) in a shallow crustal source. They also display relatively higher whole-rock Zr saturation temperatures than juvenile mafic lower crust-derived granitoids (Fig. 11d) (Lai et al., 2015a; Zhu et al., 2019a).

(4) The enriched Sr-Nd isotopic compositions [e.g., $(^{87}Sr/^{86}Sr)_i$ = 0. 709 9 – 0. 721 7, $\varepsilon_{Nd}(t)$ = −5. 1 to −2. 9] together with ancient two-stage Nd model ages (1. 60 – 1. 75 Ga) suggest an evolved and ancient crustal source of these peraluminous granites (Zhu et al., 2018), distinct from the Sr-Nd isotopic features of previously reported Neoproterozoic mantle-derived mafic-intermediate rocks and juvenile mafic lower crust-derived granitoids in the western Yangtze Block (Fig. 8).

Sylvester (1998) proposed that peraluminous granites can inherit different CaO/Na_2O ratios by the partial melting of various protoliths. Metapelite-derived melts usually contain lower CaO/Na_2O ratios (<0. 3) than the metagreywacke-derived counterparts (CaO/Na_2O >0. 3) (Sylvester, 1998). On the CaO/Na_2O vs. Al_2O_3/TiO_2 diagram (Fig. 12a), the majority of the Kuanyu peraluminous granite samples display high and variable CaO/Na_2O ratios (0. 22 – 0. 65, mostly >0. 3) and low Al_2O_3/TiO_2 ratios (25. 3 – 65. 8), suggesting a dominant origin

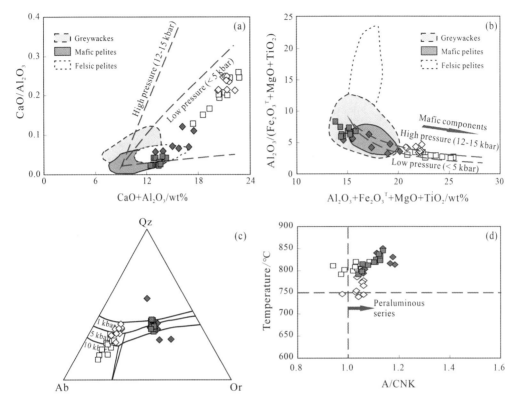

Fig. 11　The CaO/Al_2O_3 vs. $CaO+Al_2O_3$(a)(Patiño Douce, 1999), $Al_2O_3/(Fe_2O_3^T+MgO+TiO_2)$ vs. $Al_2O_3+Fe_2O_3^T+MgO+TiO_2$(b) (Patiño Douce, 1999), CIPW-normative Qz-Ab-Or (c) (Johannes and Holtz, 1996), and zircon saturation temperature vs. A/CNK (d) diagrams for the Kuanyu and Cida peraluminous granites from the western margin of the Yangtze Block, South China. Symbols as Fig. 5.

of psammites (metagreywackes). All of the Cida peraluminous granite samples yield low CaO/Na_2O ratios (0. 09−0. 19 <0. 3) and moderate Al_2O_3/TiO_2 ratios (51. 2−88. 4), indicating a metapelite source (Fig. 12a). In term of the Rb/Ba versus Rb/Sr diagram (Fig. 12b), the Kuanyu and Cida peraluminous granites suggest an affinity to a heterogeneous source of metapelite- and metagreywacke-derived melts, as indicated by moderate Rb/Ba (1. 68−3. 86) and Rb/Sr (0. 32−0. 85) ratios. Furthermore, differing from juvenile mafic lower crust-derived granitoids along the western periphery of the Yangtze Block (Lai et al., 2015a; Zhu et al., 2019a), the peraluminous granites in this study contain high molar $Al_2O_3/(MgO+FeO^T)$ values of 2. 04−5. 23 and low molar $CaO/(MgO+FeO^T)$ values of 0. 15−0. 48 (Fig. 12c) (Altherr et al., 2000), which are scattered in the metapelites and metagreywackes fields. These geochemical features all support a heterogeneous metasedimentary source for the peraluminous granites studied here. Previous studies have interpreted the wide range of isotopic variation in peraluminous granites to reflect isotopic heterogeneity inherited from a heterogenous

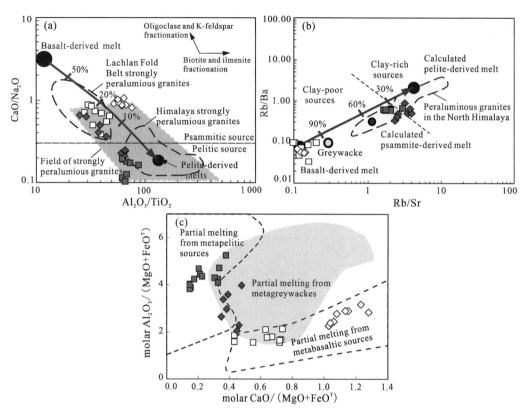

Fig. 12   The CaO/Na$_2$O vs. Al$_2$O$_3$/TiO$_2$(a) (Sylvester, 1998), Rb/Ba vs. Rb/Sr (b)
(Patiño Douce, 1999), and molar Al$_2$O$_3$/(MgO + FeO^T) vs. molar CaO/(MgO + FeO^T) (c)
(Altherr et al., 2000) diagrams for the Kuanyu and Cida peraluminous granites
from the western Yangtze Block, South China.
Symbols as Fig. 5.

source (Champion and Bultitude, 2013; Huang et al., 2019; Villaros et al., 2012; Wang et al., 2018; Zhu et al., 2018). Source heterogeneity can be transferred to a granitic magma by the formation of discrete magma batches (Villaros et al., 2012). The Kuanyu and Cida peraluminous granites in this study contain a wide range of zircon $\varepsilon_{Hf}(t)$ values varying from −7. 75 to +1. 31 and −3. 71 to +3. 31, respectively (Figs. 9 and 10), further demonstrating the heterogeneous metasedimentary source (metapelites + metagreywackes) (Zhu et al., 2018). In addition, the ancient two-stage Hf model ages of 1 643−2 210 Ma for the Kuanyu granites and 1 512−1 953 Ma for the Cida granites can also support an evolved continental crustal source (Fig. 10) (Zhu et al., 2018).

In summary, our results suggest that the Kuanyu and Cida peraluminous granites in the western Yangtze Block predominantly formed by the partial melting of heterogeneous metasedimentary protoliths, including variable proportions of metagreywackes and metapelites from an evolved continental crust source (i.e., middle-upper crust level).

5. 1. 2 Hybridization process or disequilibrium melting?

As mentioned above, the peraluminous granites investigated in this study were derived from heterogeneous metasedimentary sources (metapelites + metagreywackes). It is notable that these peraluminous granites partly display a depleted Hf isotopic compositions [$\varepsilon_{Hf}(t)$ values up to +3.31] (Figs. 9 and 10). This phenomenon is likely caused by the hybridization process between mantle- and crust-derived inputs (Brown, 2013; Chen et al., 2014; Clemens, 2003; Kemp et al., 2007) or disequilibrium partial melting of zircon-bearing crustal rocks (Dou et al., 2019; Farinia et al., 2014; Flowerdew et al., 2006; Iles et al., 2018; Kong et al., 2019; Tang et al., 2014; Wang et al., 2018).

There is a general knowledge that mantle-derived magmas may provide necessary heat and materials for crustal melting, and peraluminous granites can crystallize from both mantle- and crust-derived inputs (Brown, 2013; Clemens, 2003; Jiang and Zhu, 2017; Sylvester, 1998). The Kuanyu and Cida peraluminous granites in this study yield positive and negative zircon $\varepsilon_{Hf}(t)$ values ranging from −7.75 to +3.31 (Figs. 9 and 10), suggesting an open system and mixing source (Kemp et al., 2007). Their heterogeneous zircon $\varepsilon_{Hf}(t)$ values differ from the zircon Hf isotopic compositions of Neoproterozoic juvenile mafic lower crust-derived granitoids in the western Yangtze Block, but are partly consistent with Neoproterozoic peraluminous granites from the Jiangnan fold belt (Jiang and Zhu, 2017; Wang et al., 2013, 2014; Zhao et al., 2013), further indicating the possibility of magma mixing between the mantle- and crust-derived components (Belousova et al., 2006). However, the following hints lend limited support to the hybridization process between crustal melts and mantle-derived magmas. First, the significant microgranular mafic enclaves and disequilibrium mineral pairs, which are strong evidences for magma mixing (Tang et al., 2014; Vernon, 1984), are rarely observed in the field and micrographs (Fig. 3). Second, magma mixing of voluminous mantle- and crust-derived melts elevates the $Mg^{\#}$ values of the resultant melts (Jiang and Zhu, 2017), but the high observed SiO_2 contents (66.9−75.6 wt%) and low $Mg^{\#}$ values (17−33) measured here are similar to pure crustal partial melts (Fig. 5d), which weakens the hybridization process. There is no clear positive correlation between La/Sm and Th/Sc ratios for our granites samples (Fig. 13a), further ruling out the possibility of magma mixing (Huang et al., 2019). Third, Dickin (2018) suggested that a positive correlation between 1/Sr and initial $^{87}Sr/^{86}Sr$ values is expected for magma mixing between mafic and felsic magmas. However, the negative correlation of our peraluminous granites argues against the hybridization process (Fig. 13b). In addition, a negative correlation of SiO_2 and $\varepsilon_{Nd}(t)$ values would develop during mixing (Jiang and Zhu, 2017), which is not observed in the samples studied here (Fig. 13c). In summary, the comprehensive evidence leads us to support that the hybridization of crustal melts with mantle-derived magmas played a minor role in the genesis of the peraluminous granites.

Underplating of mantle-derived magma merely acted as a heat source that triggered the partial melting of a middle-upper curst source.

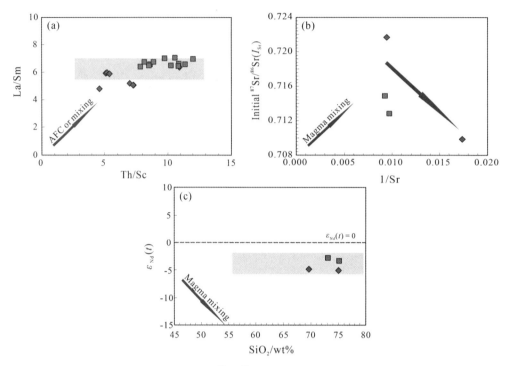

Fig. 13　The La/Sm vs. Th/Sc, initial ^{87}Sr/^{86}Sr (I_{Sr}) vs. 1/Sr, and $\varepsilon_{Nd}(t)$ vs. SiO$_2$ diagrams for the Kuanyu and Cida peraluminous granites from the western Yangtze Block, South China. Symbols as Fig. 5.

Alternatively, the zircon Hf isotopic heterogeneity in our peraluminous granites may result from disequilibrium partial melting (IIes et al., 2018; Kong et al., 2019; Tang et al., 2014; Wang et al., 2018). As the major carrier of Hf, zircon controls the Hf budget at the source, and its dissolution may dictate Hf isotopic evolution in the melt (Tang et al., 2014). Flowerdew et al. (2006) proposed that the varying zircon dissolution rate can lead to heterogeneous zircon $\varepsilon_{Hf}(t)$ values between different melt batches at a single source and give rise to a decoupling of the Hf isotope system from other radiogenic isotope systems. In this study, the existence of zircons xenocrysts in our peraluminous granites suggests that some residual zircons from the source were entrained by the melt during melt loss (Fig. 4) (Kong et al., 2019). These residual zircons would retain the Hf relative to Nd, which leads to relatively high Nd/Hf (6. 17 – 9. 76) ratios and low Hf (5. 00 – 8. 43 ppm) concentrations in our peraluminous granites. Accordingly, the undissolved residual zircons may retain a significant amount of ^{177}Hf at the source, elevating the ^{177}Hf/^{176}Hf ratios of the melts and decoupling them from the ^{143}Nd/^{144}Nd ratios (Kong et al., 2019; Tang et al., 2014). Nd-Hf isotopic decoupling was observed in our peraluminous granites, which show that enriched Nd isotopes $[\varepsilon_{Nd}(t) = -5.1$ to $-2.9]$ and heterogeneous Hf isotopes $[\varepsilon_{Hf}(t) = -7.75$ to $+3.31]$. Zirconium is abundant in the continental crust with average concentrations ranging from 68 ppm in the lower crust to 193 ppm in the upper curst (Rudnick and Gao, 2003). This means that the zircon effect may be common during the crustal melting because of the slow disequilibrium melting of zircon (Tang et al., 2014). Because high Zr concentrations at the source can easily saturate the melt with zircon and zircon dissolution will cease until more melt is generated to dilute the remaining Zr at the source, the Zr concentration at the source is a significant factor for controlling the zircon dissolution rate (Kong et al., 2019; Tang et al., 2014). As mentioned in Section 5.1.1, our peraluminous granites were derived from a middle-upper crustal source. The presence of residual zircon grains and high Zr concentrations (159−304 ppm) in our peraluminous granites indicate the high initial Zr concentrations. The model presented by Tang et al. (2014) showed that when the initial Zr concentration at the source is sufficiently high (e.g., >100 ppm), the inter-batch Hf varies from low concentration and highly radiogenic to high concentration and less radiogenic than the bulk protolith, which may produce a melt extracted from a single source that evolves from a mantle-like source at the early stage to a crust-like source after extensive melting. They emphasized that the mixing of magma batches from individual crustal sources may mimic the Hf isotopic features from the hybridization of crust- and mantle-derived magmas (Tang et al., 2014), which is similar to that observed in our peraluminous granites with positive and negative zircon $\varepsilon_{Hf}(t)$ values (−7.75 to +3.31). The high Zr content in the source along with disequilibrium melting can therefore explain the isotopic heterogeneity in our peraluminous granites. In addition, variable Sr isotopic compositions may also reflect significant disequilibrium melting process during crustal anatexis (Farina et al., 2014; Mcleod et al., 2012; Zeng et al., 2005a,b), which can be displayed in our peraluminous granites with variable $(^{87}Sr/^{86}Sr)_i$ values from 0.709 9 to 0.721 7 (Fig. 8).

Therefore, the integrated evidences leads to the conclusion that the zircon Hf isotopic heterogeneity in our peraluminous granites may be attributed to disequilibrium partial melting of a heterogeneous metasedimentary source, similar to that of peraluminous granites from the Altai area and Jiangnan orogen (Kong et al., 2019; Tang et al., 2014). Multi-sourced magma mixing between crust- and mantle-derived components is therefore insignificant for our peraluminous granites in the western margin of the Yangtze Block.

5.2 Petrological and geological implications

5.2.1 Formation of Neoproterozoic peraluminous granites

There are several main mechanisms that account for the generation of peraluminous granites, such as crustal reworking during ridge subduction, crustal thickening and melting during continental collision, post-collisional collapse and melting of pre-existed sediments in the back-arc basin, or continental crust reworking in the early stage of subduction (Barbarin, 1999; Cai et al., 2011; Chen et al., 2014; Collision, 2002; Collins and Richard, 2008; Liu and Zhao, 2018; Sylvester, 1998). Ridge subduction can trigger the partial melting of continental crust and generate peraluminous granites (Cai et al., 2011) but it will cause the high-T and low-P metamorphism (Jiang et al., 2010), which remain unreported from Neoproterozoic rocks of the western Yangtze Block. We can thus preclude the ridge subduction model. The zircon U-Pb ages (ca. 840–835 Ma) of the Kuanyu and Cida peraluminous granites are obviously older than the formation age of thickened crust-derived adakites (ca. 800–780 Ma) along the margins of Yangtze Block (Huang et al., 2009; Zhao et al., 2010; Zhu et al., 2019b). There is also no geological evidence for the presence of an early Neoproterozoic back-arc basin around the Miyi region. Therefore, the continental collisional-related models are therefore unreasonable, and continental crust reworking in the early stage of subduction setting might account for the genesis of the Kuanyu and Cida peraluminous granites in the western Yangtze Block. Chen et al. (2014) suggested that the Cambrian Chaidanuo peraluminous granites in the North Qilian suture zone were formed by the re-melting of crustal materials during the early onset of subduction initiation. Similarly, a subduction setting is also proposed in the western margin of the Yangtze Block during the dominant interval of ca. 860–750 Ma (Lai et al., 2015a,b; Li et al., 2018; Sun et al., 2007, 2008; J. H. Zhao et al., 2008; X. F. Zhao et al., 2008; Zhao et al., 2017, 2018; Zhou et al., 2002, 2006a,b; Zhu et al., 2017, 2019a,b). Although a mantle plume setting is a controversial model for the widespread Neoproterozoic magmatism in the western Yangtze Block (Li et al., 2003, 2006, 2008), an increasing number of sedimentary, geophysical, and igneous studies of the region strengthen the viewpoint of subduction process (Gao et al., 2016; Sun et al., 2008, 2009; Zhao et al., 2017, 2018). Sun et al. (2009) concretely proposed that detrital zircons in Precambrian strata show a vital period of juvenile magmatism at ca. 1 000–740 Ma, which is consistent with the interval of long-lasting Neoproterozoic subduction along the western Yangtze Block. Gao et al. (2016) proposed that the multichannel seismic reflection profiles across the Sichuan Basin are similar to the geometry of ancient subduction-related mantle, which has been explained as the remnant mantle of Neoproterozoic subduction process in the Yangtze Block. Zhao et al. (2017) suggested that the western margin of the Yangtze Block is an Andean-type arc system by the evidence of the ca. 800 Ma SSZ-type Shimian ophiolite assemblage. They further summarized that Neoproterozoic magmatism in the western and northern margins of the Yangtze Block was controlled by continuous slab subduction and subduction-transform-edge-propagator system (Zhao et al., 2018 and references therein). We therefore support that the massive Neoproterozoic magmatism in the western Yangtze Block, including the Kuanyu and

Cida peraluminous granites in this study, were formed under persistent subduction process.

As mentioned above, the crystallization ages of our studied peraluminous granites (ca. 840–835 Ma) approach the age of the Guandaoshan pluton (ca. 860 Ma) that is considered to be the products of initial subduction process in the western Yangtze Block (Du et al., 2014). The age constraint implies that these peraluminous granites were generated in the early stage of subduction. Combining with the regional tectonic background, we suggest that the Kuanyu and Cida peraluminous granites may have formed by the following process (Fig. 14a). During the

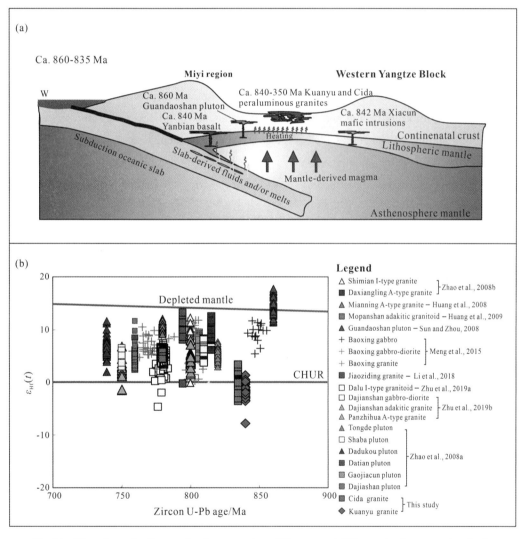

Fig. 14　The schematic diagram for the generation of Kuanyu and Cida peraluminous granites in the western margin of the Yangtze Block, South China (modified after Chen et al., 2014) (a), and the detailed comparison on the zircon U-Pb age and Hf isotopes of the Neoproterozoic magmatism from the western Yangtze Block (b).

early subduction of an oceanic slab beneath the western margin of the Yangtze Block, strong slab rollback towards the vertical direction increased the subduction rate (Niu et al., 2003) and lead to intensive seafloor spreading (Gerya, 2011; Zhu et al., 2009). Trench retreat and extension of the overriding plate would subsequently be expected in the fore-arc region (Chen et al., 2014). The overriding plate would be rheologically weakened by arc magmatism, therefore resulting in a regionally thinning of the overriding plate (Gerya and Meilick, 2011). This process will also lead to decompression melting of the asthenosphere mantle (Gerya et al., 2008) and simultaneously generation of basaltic magmatism along the western Yangtze Block, i.e., the parental magma of the ca. 860 Ma Guandaoshan pluton, the ca. 842 Ma Xiacun mafic intrusions, and the ca. 840 Ma MORB-type Yanbian basalt (Du et al., 2014; Guo et al., 2007; Sun et al., 2007). In the meantime, the overlaying continental crust in the Miyi region may have been heated by the strong upwelling of voluminous mantle-derived magma and undergone disequilibrium partial melting of heterogeneous metasediments when the temperature reached the solidus (Chen et al., 2014), producing the Kuanyu and Cida peraluminous granites in the Miyi region. In summary, the Kuanyu and Cida peraluminous granites represent partial melting of mature continental crust under early subduction stages during the Neoproterozoic.

5.2.2　Widespread Neoproterozoic juvenile and mature crustal melting

The enriched Sr-Nd isotopes and negative-dominantly zircon $\varepsilon_{Hf}(t)$ values, together with the whole-rock geochemical evidences, suggest that the Kuanyu and Cida peraluminous granites in the western Yangtze Block represent mature continental crust-derived metasediment melts. Actually, Neoproterozoic large-scale crustal growth and melting occurred along the western periphery of the Yangtze Block in a subduction setting (Huang et al., 2009; Li et al., 2018; Zhao et al., 2008; Zhu et al., 2019a,b). Zhao et al. (2008a) proposed that the voluminous mafic-intermediate plutons, which display depleted zircon Hf isotopic compositions (Fig. 14b), were derived from a depleted mantle source modified by subduction components. The extraction of voluminous mantle melts was a significant process involved in crustal growth. They also suggested that several adakitic granitoid plutons formed by the partial melting of a subduction oceanic slab, and the recycling of oceanic crust was also a main contributor for crustal growth (Zhao et al., 2008a). Moreover, there was widespread juvenile and ancient crustal melting along the western Yangtze Block during the Neoproterozoic (Zhu et al., 2019a) (Fig. 14b).

In the Kangding-Shimian region (Fig. 2a), the ca. 850 Ma Tianquan peraluminous granites were thought to generate from the partial melting of crust-derived basalts and greywackes under H_2O saturation and a high geothermal gradient (Lai et al., 2015b). The ca. 816 Ma Daxiangling A-type granites have been considered from recycled juvenile crust-derived rocks, and the ca. 800 Ma Shimian I-type granites were derived from an ancient TTG source as a result of underplating of mantle-derived mafic magmas (Zhao et al., 2008b). In addition, the ca. 777 Ma Shimian K-feldspar granites formed by the re-melting of middle-upper crust-derivation metagreywackes (Zhu et al., 2017), and the ca. 770 Ma Baoxing granites have been explained by the partial melting of the juvenile crust according to the positive zircon $\varepsilon_{Hf}(t)$ values of +2.53 to +10.58 (Fig. 14b) (Meng et al., 2015). Moreover, Lai et al. (2015a) proposed that relatively younger ca. 754 Ma Luding high-Mg$^{\#}$ quartz monzodiorites were generated by ~40% partial melting of newly formed mafic lower crust. They also suggested that the ca. 748 Ma Kangding granodiorite was produced by the partial melting of a plagioclase-rich crustal source (Lai et al., 2015a). In the Mianning region (Fig. 2a), the ca. 780 Ma Mianning highly fractionated A$_2$-type granites were ascribed from a juvenile crust source owing to the depleted whole-rock Nd isotopes [$\varepsilon_{Nd}(t)$= +2.97 to +5.24] and zircon Hf isotopes compositions [$\varepsilon_{Hf}(t)$= +9.2 to +12.1] (Fig. 14b) (Huang et al., 2008). In the Miyi region, our identification is that the ca. 840–835 Ma Kuanyu and Cida peraluminous granites were mainly formed by the reworking of heterogeneous metasediments in the evolved middle-upper crust source. Additionally, Guo et al. (2009) proposed that the ca. 803 Ma Xiacun granite and ca. 790 Ma Yonglang granite were derived from the ancient crust-derived materials. In the Panzhihua-Yanbian region (Fig. 2a), the recently identified ca. 780 Ma Dalu I-type granodiorites-granites association yielded positive-dominantly zircon $\varepsilon_{Hf}(t)$ values (−4.65 to +7.39) (Fig. 14b), and were considered to be from the juvenile mafic lower crust-derived melts and mixing of juvenile curst-derived melts and mature metasediments melts (Zhu et al., 2019a). These ca. 850–740 Ma granitoids magmatism correspond with an interval of a Neoproterozoic subduction setting. In general, the western Yangtze Block underwent long-lasting crustal re-melting during the Neoproterozoic subduction process, including not only the juvenile mafic lower crust but also mature continental crustal compositions.

6　Conclusions

(1)Zircon U-Pb ages show that the Kuanyu and Cida peraluminous granites were formed at ca. 840–835 Ma, coeval with the early Neoproterozoic magmatism in the western margin of the Yangtze Block.

(2)The Kuanyu and Cida peraluminous granites were mainly produced by the partial melting of mature and heterogeneous metasediments in the middle-upper crust source. The absence of sufficient evidence for magma mixing between mantle-derived magmas and crust-derived melts, the heterogeneous zircon Hf [$\varepsilon_{Hf}(t)$= −7.75 to +3.31] isotopic compositions can be ascribed to disequilibrium partial melting.

(3)Compared with the mantle-derived mafic-intermediate magmatism and juvenile mafic lower crust-derived granitoid magmatism in the western Yangtze Block, we suggest that the Kuanyu and Cida peraluminous granites represent melting of mature continental crust during the early stage of Neoproterozoic subduction. Widespread Neoproterozoic crustal melting, including juvenile mafic lower crust and mature crustal sources, occurred along the western margin of the Yangtze Block.

Acknowledgements　Thanks so much for the help and constructive comments from Chief editor Prof. Guochun Zhao and other anonymous reviewers, which significantly improved the shape, language, and discussion of this manuscript. We also grateful to the further English polish from Esther Posner, PhD. This work was jointly supported by the National Natural Science Foundation of China (Grant Nos. 41421002, 41802054, and 41772052) and the program for Changjiang Scholars and Innovative Research Team in University (Grant IRT1281). Support was also provided by the Foundation for the Author of National Excellent Doctoral Dissertation of China (201324) and independent innovation project of graduate students of Northwest University (YZZ17192).

Appendix A. Supplementary data　Supplementary data to this article can be found online at https://doi.org/10. 1016/j.precamres.2019. 105443.

References

Altherr, R., Holl, A., Hegner, E., Langer, C., Kreuzer, H., 2000. High-potassium, calc-alkaline I-type plutonismin the European Variscides: Northern Vosges (France) and northern Schwarzwald (Germany). Lithos, 50, 51-73.

Andersen, T., 2002. Correction of common lead in U-Pb analyses that do not report [204]Pb. Chemical Geology, 192(1-2), 59-79.

Bao, Z.A., Chen, L., Zong, C.L., Yuan, H.L., Chen, K.Y., Dai, M.N., 2017. Development of pressed sulfide powder tablets for in situ, sulfur and lead isotope measurement using LA-MC-ICP-MS. International Journal of Mass Spectrometry, 421, 255-262.

Barbarin, B., 1999. A review of the relationships between granitoid types, their origins and their geodynamic environments. Lithos, 46(3), 605-626.

Belousova, E.A., Griffin, W.L., O'Reilly, S.Y., 2006. Zircon crystal morphology, trace element signatures and Hf isotope composition as a tool for petrogenetic modelling: Examples from eastern Australian granitoids. Journal of Petrology, 47, 329-353.

Brown, M., 2013. Granites: From genesis to emplacement. Geological Society of America Bulletin, 125, 1079-1113.

Cai, K.D., Sun, M., Yuan, C., Zhao, G.C., Xiao, W.J., Long, X.P., Wu, F.Y., 2011. Geochronology, petrogenesis and tectonic significance of peraluminous granites from the Chinese Altai, NW China. Lithos, 127(1-2), 261-281.

Champion, D.C., Bultitude, R.J., 2013. The geochemical and Sr, Nd isotopic characteristics of Paleozoic fractionated S-types granites of north Queensland: Implications for S-type granite petrogenesis. Lithos, 162-

163(2), 37-56.

Chappell, B.W., White, A.J.R., 1992. I- and S-type granites in the Lachlan fold belt. Transactions of the Royal Society of Edinburgh: Earth Sciences, 83.

Chappell, B.W., 1999. Aluminium saturation in I- and S-type granites and the characterization of fractionated haplogranites. Lithos, 46(3), 535-551.

Chappell, B.W., White, A.J.R., 2001. Two contrasting granite types: 25 years later. Australian Journal of Earth Sciences, 48(4), 489-499.

Chappell, B.W., Bryant, C.J., Wyborn, D., 2012. Peraluminous I-type granites. Lithos, 153(8), 142-153.

Chen, Y.X., Song, S.G., Niu, Y.L., Wei, C.J., 2014. Melting of continental crust during subduction initiation: A case study from the Chaidanuo peraluminous granite in the north Qilian suture zone. Geochimica et Cosmochimica Acta, 132, 311-336.

Chu, Z.Y., Chen, F.K., Yang, Y.H., Guo, J.H., 2009. Precise determination of Sm, Nd concentrations and Nd isotopic compositions at the nanogram level in geological samples by thermal ionization mass spectrometry. Journal of Analytical Atomic Spectrometry, 24, 1534-1544.

Clemens, J.D., 2003. S-type granitic magmas-petrogenetic issues, models and evidence. Earth-Science Reviews, 61, 1-18.

Clemens, J.D., 2018. Granitic magmas with I-type affinities, from mainly metasedimentary sources: The Harcourt batholith of southeastern Australia. Contributions to Mineralogy and Petrology, 173(11), 93.

Clemens, J.D., Regmi, K., Nicholls, I.A., Weinberg, R., Maas, R., 2016. The Tynong pluton, its mafic synplutonic sheets and igneous microgranular enclaves: The nature of the mantle connection in I-type granitic magmas. Contributions to Mineralogy and Petrology, 171(4), 35.

Collins, W.J., 2002. Hot orogens, tectonic switching, and creation of continental crust. Geology, 30, 535-538.

Collins, W.J., Richards, S.W., 2008. Geodynamic significance of S-type granites in circum-Pacific orogens. Geology, 36, 559-562.

DePaolo, D.J., 1981. A Neodymium and Strontium Isotopic Study of the Mesozoic Calc-Alkaline Granitic Batholiths of the Sierra Nevada and Peninsular Ranges, California Granites and Rhyolites. Journal of Geophysical Research, 86(B11), 10470-10488.

Dickin, A.P., 2018. Radiogenic Isotope Geology, Third Edition. Cambridge: Cambridge University Press, 171-172.

Dou, J.Z., Sibel, W., He, J.F., Chen, F.K., 2019. Different melting conditions and petrogenesis of peraluminous granites in western Qinling, China, and tectonic implications. Lithos, 336-337, 97-111.

Du, L.L., Guo, J.H., Nutman, A.P., Wyman, D., Geng, Y.S., Yang, C.H., Liu, F.L., Ren, L.D., Zhou, X.W., 2014.Implications for Rodinia reconstructions for the initiation of Neoproterozoic subduction at ~860 Ma on the western margin of the Yangtze Block: Evidence from the Guandaoshan Pluton. Lithos, s196-197, 67-82.

Farina, F., Dini, A., Rocchi, S., Stevens, G., 2014. Extreme mineral-scale sr isotope heterogeneity in granites by disequilibrium melting of the crust. Earth & Planetary Science Letters, 399, 103-115.

Flowerdew, M.J., Millar, I.L., Vaughan, A.P.M., Horstwood, M.S.A., Fanning, C.M., 2006. The source of granitic gneisses and migmatites in the Antarctic Peninsula: A combined U-Pb SHRIMP and laser ablation Hf isotope study of complex zircons. Contributions to Mineralogy and Petrology, 151, 751-768.

Frost, B.R., Barnes, C.G., Collins, W.J., Arculus, R.J., Ellis, D.J., Frost, C.D., 2001. A geochemical classification for granitic rocks. Journal of Petrology, 42, 2033-2048.

Gan, B.P., Lai, S.C., Qin, J.F., Zhu, R.Z., Zhao, S.W., Li, T., 2017. Neoproterozoic alkaline intrusive complex in the northwestern Yangtze Block, Micang Mountains region, South China: Petrogenesis and tectonic significance. International Geology Review, 59(3), 311-332.

Gao, R., Chen, C., Wang, H.Y., Lu, Z.W., Brown, L., Dong, S.W., Feng, S.Y., Li, Q.S., Li, W.H., Wen, Z.P., Li, F., 2016. SINOPROBE deep reflection profile reveals a Neo-proterozoic subduction zone beneath Sichuan basin. Earth & Planetary Science Letters, 454, 86-91.

Gao, S., Ling, W.L., Qiu, Y., Zhou, L., Hartmann, G., Simon, K., 1999.Contrasting geochemical and Sm-Nd isotopic compositions of Archean metasediments from the Kongling high-grade terrain of the Yangtze Craton: Evidence for cratonic evolution and redistribution of REE during crustal anatexis. Geochimica et Cosmochimica Acta, 63(13-14), 2071-2088.

Geng, Y.S., Yang, C.H., Wang, X.S., Du, L.L., Ren, L.D., Zhou, X.W., 2008. Metamorphic basement evolution in western margin of Yangtze Block. Beijing: Geological Publishing House, 1-215 (in Chinese).

Gerya, T.V., 2011. Future directions in subduction modeling. Journal of Geodynamic, 52, 344-378.

Gerya, T.V., Meilick, F.I., 2011. Geodynamic regimes of subduction under an active margin: Effects of rheological weakening by fluids and melts. Journal of Metamorphic Geology, 29, 7-31.

Gerya, T.V., Connolly, J.A.D., Yuen, D.A., 2008. Why is terrestrial subduction one-sided? Geology, 36, 43-46.

Griffin, W.L., Pearson, N.J., Belousova, E., Jackson, S.E., van Achterbergh, E., O'Reilly, S.Y., Shee, S.R., 2000. The Hf isotope composition of cratonic mantle: LAM-MC-ICP-MS analysis of zircon megacrysts in kimberlites. Geochimica et Cosmochimica Acta, 64(1), 133-147.

Guo, C.L., Wang, D.H., Chen, Y.C., Zhao, Z.G., Wang, Y.B., Fu, X.F., Fu, D.M., 2007. SHRIMP U-Pb zircon ages and major element, trace element and Nd-Sr isotope geochemical studies of a Neoproterozoic granitic complex in western Sichuan: Petrogenesis and tectonic significance. Acta Petrologica Sinica, 23 (10), 2457-2470 (in Chinese with English abstract).

Guo, F., Fan, W.M., Li, C.W., Zhao, L., Li, H.X., Yang, J.H., 2012. Multi-stage crust-mantle interaction in SE China: Temporal, thermal and compositional constraints from the Mesozoic felsic volcanic rocks in eastern Guangdong-Fujian provinces. Lithos, 150, 62-84.

Hacker., B.R., Wallis., S.R., Ratschbacher, L., Grove, M., Gehrels, G., 2006. High-temperature geochronology constraints on the tectonic history and architecture of the ultrahigh-pressure Dabie-Sulu Orogen. Tectonics, 25, TC5006,

Hoskin, P.W.O., Schaltegger, U., 2003. The composition of zircon and igneous and metamorphic petrogenesis. Reviews in Mineralogy and Geochemistry, 53,27-62.

Hu, P.Y., Zhai, Q.G., Wang, J., Tang, Y., Ren, G.M., 2017. The Shimian ophiolite in the western Yangtze Block, SW China: Zircon SHRIMP U-Pb ages, geochemical and Hf-O isotopic characteristics, and tectonic implications. Precambrian Research, 298, 107-122.

Huang, S.F., Wang, W., Pandit, M.K., Zhao, J.H., Lu, G.M., Xue, E.K., 2019. Neoproterozoic S-type granites in the western Jiangnan Orogenic Belt, South China: Implications for petrogenesis and geodynamic significance. Lithos, 342-343, 45-58.

Huang, X.L., Xu, Y.G., Lan, J.B., Yang, Q.J., Luo, Z.Y., 2009. Neoproterozoic adakitic rocks from Mopanshan in the Western Yangtze Craton: Partial melts of a thickened lower crust. Lithos, 112 (3), 367-381.

Huang, X.L., Xu, Y.G., Li, X.H., Li, W.X., Lan, J.B., Zhang, H.H., Liu, Y.S., Wang, Y.B., Li, H. Y., Luo. Z. Y., Yang, Q. J., 2008. Petrogenesis and tectonic implications of Neoproterozoic, highly fractionated A-type granites from Mianning, South China. Precambrian Research, 165(3-4), 190-204.

IIes, K.A., Hergt, J.M., Woodhead, J.D., 2018. Modelling isotopic responses to disequilibrium melting in granitic systems. Journal of Petrology, 59(1), 87-114.

Jiang, Y.D., Sun, M., Zhao, G.C., Yuan, C., Xiao, W.J., Xia, X.P., Long, X.P., Wu, F.Y., 2010. The 390 Ma high-T metamorphism in the Chinese Altai: consequence of ridge-subduction ? American Journal of Sciences, 310 (10), 1421-1452.

Jiang, Y.H., Zhu, S.Q., 2017. Petrogenesis of the Late Jurassic peraluminous biotite granites and muscovite-bearing granites in SE China: Geochronological, elemental and Sr-Nd-O-Hf isotopic constraints. Contributions to Mineral and Petrology, 172, 1-27.

Johannes, W., Holtz, F., 1996. Petrogenesis and Experimental petrology of Granitic rocks. Berlin: Springer, 1-335.

Kemp, A.I.S., Hawkesworth, C.J., Foster, G.L., Paterson, B.A., Woodhead, J.D., Hergt, J.M., Gray, C. M., Whitehouse, M.J., 2007. Magmatic and crustal differentiation history of granitic rocks from Hf-O isotopes in zircon. Science, 315, 980-983.

Kong, X.Y., Zhang, C., Liu, D.D., Jiang, S., Luo, Q., Zeng, J.H., Liu, L.F., Luo, L., Shao, H.B., Liu, D., Liu, X.Y., Wang, X.P., 2019. Disequilibrium partial melting of metasediments in subduction zones: Evidence from O-Nd-Hf isotopes and trace elements in S-type granites of the Chinese Altai. Lithosphere, 11(1), 149-168.

Lai, S.C., Qin, J.F., Zhu, R.Z., Zhao, S.W., 2015a. Neoproterozoic quartz monzodiorite-granodiorite association from the Luding-Kangding area: Implications for the interpretation of an active continental margin along the Yangtze Block (South China Block). Precambrian Research, 267(3-4), 196-208.

Lai, S.C., Qin, J.F., Zhu, R.Z., Zhao, S.W., 2015b. Petrogenesis and tectonic implication of Neoproterozoic peraluminous granitoids from the Tianquan area, western Yangtze Block, South China. Acta Petrologica Sinica, 31(8), 2245-2258 (in Chinese with English abstract).

Li, J.Y., Wang, X.L., Gu, Z.D., 2018. Petrogenesis of the Jiaoziding granitoids and associated basaltic porphyries: Implications for extensive early Neoproterozoic arc magmatism in western Yangtze Block. Lithos, s296-299, 547-562.

Li, Q.W., Zhao, J.H., 2018. The Neoproterozoic high-Mg dioritic dikes in south china formed by high pressures fractional crystallization of hydrous basaltic melts. Precambrian Research, 309, 198-211.

Li, X.H., Li, Z.X., Ge, W.C., Zhou, H.W., Li, W.X., Liu, Y., Wingate, M.T.D., 2003. Neoproterozoic granitoids in South China: Crustal melting above a mantle plume at ca. 825 Ma? Precambrian Research, 122

（1-4）, 45-83.

Li, X.H., Li, Z.X., Sinclair, J.A., Li, W.X., Carter, G., 2006. Revisiting the "Yanbian Terrane": Implications for Neoproterozoic tectonic evolution of the western Yangtze Block, South China. Precambrian Research, 151(1-2),14-30.

Li, Z.X., Bogdanova, S.V., Collins, A.S., Davidson, A., De Waele, B., Ernst, R.E., Fitzsi-mons, I.C. W., Fuck, R.A., Gladkochub, D.P., Jacobs, J., Karlstrom, K.E., Lu, S., Natapov, L.M., Pease, V., Pisarevsky, S.A., Thrane, K., Vernikovsky, V., 2008. Assembly, configuration, and break-up history of Rodinia: A synthesis. Precambrian Research, 160, 179-210.

Liu, X.M., Gao, S., Diwu, C.R., Yuan, H.L., Hu, Z.C., 2007. Simultaneous in-situ determination of U-Pb age and trace elements in zircon by LA-ICP-MS in 20 μm spot size. Chinese Science Bulletin, 52(9), 1257-1264.

Liu, Y., Liu, X.M., Hu, Z.C., Diwu, C.R., Yuan, H.L., Gao, S., 2007. Evaluation of accuracy and long-term stability of determination of 37 trace elements in geological samples by ICP-MS. Acta Petrologica Sinica, 23(5), 1203-1210 (in Chinese with English abstract).

Liu, H., Zhao, J.H., 2018. Neoproterozoic peraluminous granitoids in the Jiangnan Fold Belt: Implications for lithospheric differentiation and crustal growth. Precambrian Research, 309, 152-165.

Ludwig, K.R., 2003. ISOPLOT 3. 0: A geochronological toolkit for Microsoft Excel. Berkeley Geochronology Center, Special Publication, 4.

Mcleod, C.L., Davidson, J.P., Nowell, G.M., De Silva, S.L., 2012. Disequilibrium melting during crustal anatexis and implications for modeling open magmatic systems. Geology, 40(5), 435-438.

Meng, E., Liu, F.L., Du, L.L., Liu, P.H., Liu, J. H., 2015. Petrogenesis and tectonic significance of the Baoxing granitic and mafic intrusions, southwestern China: Evidence from zircon U-Pb dating and Lu-Hf isotopes, and whole-rock geochemistry. Gondwana Research, 28(2), 800-815.

Moyen, J.F., Martin, H., 2012. Forty years of TTG research. Lithos, 148, 312-336.

Munteanu, M., Wilson, A., Yao, Y., Harris, C., Chunnett, G., Luo, Y., 2010. The Tongde dioritic pluton (Sichuan, SW China) and its geotectonic setting: Regional implications of a local-scale study. Gondwana Research, 18(2), 455-465.

Niu, Y.L., O'Hara M.J., Pearce J.A., 2003. Initiation of subduction zones as a consequence of lateral compositional buoyancy contrast within the lithosphere: A petrological perspective. Journal of Petrology, 44, 851-866.

Nowell, G.M., Kempton, P.D., Noble, S.R., Fitton, J.G., Saunders, A.D., Mahoney, J.J., Taylor, R.N., 1998. High precision Hf isotope measurements of MORB and OIB by thermal ionisation mass spectrometry: Insights into the depleted mantle. Chemical Geology, 149(3-4), 211-233.

Patiño Douce, A.E., 1995. Experimental generation of hybrid silicic melts by reaction of high-Al basalt with metamorphic rocks. Journal of Geophysical Research Solid Earth, 100, 15623-15639.

Patiño Douce, A.E., 1996. Effects of pressure and H_2O content on the compositions of primary crustal melts. Transaction of the Royal Society of Edinburgh: Earth Science, 87, 11-21.

Patiño Douce, A.E., 1999. What do experiments tell us about the relative contributions of crust and mantle to the origin of the granitic magmas. Geological Society, London, Special Publications, 168(1), 55-75.

Patiño Douce, A.E., Johnston, A.D., 1991. Phase equilibria and melt productivity in the pelitic system: Implications for the origin of peraluminous granitoids and aluminous granulites. Contributions to Mineralogy and Petrology, 107, 202-218.

Rapp, R.P., Watson, E.B., 1995.Dehydration melting of metabasalt at 8–32 kbar: Implications for continental growth and crust-mantle recycling. Journal of Petrology, 36(4), 891-931.

Rossi, J.N., Toselli, A.J., Saavedra, J., Sial, A.N., Pellitero, E., Ferreira, V.P., 2002. Common crustal sources for contrasting peraluminous facies in the early Paleozoic Capillitas Batholith, NW Argentina. Gondwana Research, 5, 325-337.

Rudnick, R.L., Gao, S., 2003. Composition of the continental crust. Treatise Geochem, 3, 1-64.

Sichuan Provincial Bureau of Geology and Mineral Resources (SPBGMR), 1966. Regional geological survey of the People's Republic of China, the Miyi Sheet map and report, scale 1:200 000 (in Chinese).

Sisson, T.W., Ratajeski, K., Hankins, W.B., Glazner, A.F., 2005. Voluminous granitic magmas from common basaltic sources. Contributions to Mineralogy and Petrology, 148, 635-661.

Soderlund, U., Patchett, P.J., Vervoort, J.D., Isachsen, C.E., 2004.The ^{176}Lu decay constant determined by Lu-Hf and U-Pb isotope systematics of Precambrian mafic intrusions. Earth & Planetary Science Letters, 219(3-4), 311-324.

Sun, S.S., McDonough, W.F., 1989. Chemical and isotopic systematics of oceanic basalts: Implications for mantle composition and processes. In: Saunders, A.D., Norry, M.J. Magmatism in the Ocean Basins. Geological Society, London, Special Publications, 42, 313-345.

Sun, W.H., Zhou, M.F., 2008. The ~860 Ma, cordilleran-type Guandaoshan dioritic pluton in the Yangtze Block, SW China: Implications for the origin of Neoproterozoic magmatism. Journal of Geology, 116(3), 238-253.

Sun, W.H., Zhou, M.F., Gao, J.F., Yang, Y.H., Zhao, X.F., Zhao, J.H., 2009. Detrital zircon U-Pb geochronological and Lu-Hf isotopic constraints on the Precambrian magmatic and crustal evolution of the western Yangtze Block, SW China. Precambrian Research, 172(1), 99-126.

Sun, W.H., Zhou, M.F., Yan, D.P., Li, J.W., Ma, Y.X., 2008. Provenance and tectonic setting of the Neoproterozoic Yanbian Group, western Yangtze Block (SW China). Precambrian Research, 167(s1-2), 213-236.

Sun, W.H., Zhou, M.F., Zhao, J.H., 2007. Geochemistry and Tectonic Significance of Basaltic Lavas in the Neoproterozoic Yanbian Group, Southern Sichuan Province, Southwest China. International Geology Review, 49(6), 554-571.

Sylvester, P.J., 1998. Post-collisional strongly peraluminous granites. Lithos, 45, 29-44.

Tang, M., Wang, X.L., Shu, X.J., Wang, D., Yang, T., Gopon, P., 2014. Hafnium isotopic heterogeneity in zircons from granitic rocks: Geochemical evaluation and modeling of "zircon effect" in crustal anatexis. Earth & Planetary Science Letters, 389, 188-199.

Vernon, R.H., 1984. Microgranitoid enclaves in granites: Globules of hybrid magma quenched in a plutonic environment. Nature, 309, 438-439.

Villaros, A., Buick, I.S., Stevens, G., 2012. Isotopic variations in S-type granites: An inheritance from a heterogeneous source? Contributions to Mineralogy and Petrology, 163(2), 243-257.

Wang, D., Wang, X.L., Cai, Y., Goldstein, S.L., Yang, T., 2018. Do Hf isotopes in magmatic zircons represent those of their host rocks? Journal of Asian Earth Sciences, 154, 202-212.

Wang, X.L., Zhou, J.C., Griffin, W.L., Zhao, G.C., Yu, J.H., Qiu, J.S., Zhang, Y.J., Xing, G.F., 2014. Geochemical zonation across a Neoproterozoic orogenic belt: Isotopic evidence from granitoids and metasedimentary rocks of the Jiangnan orogen, China. Precambrian Research, 242(2), 154-171.

Wang, X.L., Zhou, J.C., Wan, Y.S., Kitajima, K., Wang, D., Bonamici, C., Qiu, J.S., Sun, T., 2013. Magmatic evolution and crustal recycling for Neoproterozoic strongly peraluminous granitoids from Southern China: Hf and O isotopes in zircon. Earth & Planetary Science Letters, 366(2), 71-82.

Wang, Y.J., Zhou, Y.Z., Cai, Y.F., Liu, H.C., Zhang, Y.Z., Fan, W.M., 2016. Geochronological and geochemical constraints on the petrogenesis of the Ailaoshan granitic and migmatite rocks and its implications on Neoproterozoic subduction along the SW Yangtze Block. Precambrian Research, 283, 106-124.

Watson, E.B., Harrison, T.M., 1983. Zircon saturation revisited: Temperature and composition effects in a variety of crustal magma types. Earth & Planetary Science Letters, 64, 295-304.

Wiedenbeck, M., Hanchar, J.M., Peck, W.H., Sylvester, P., Valley, J., Whitehouse, M., Kronz, A., Morishita, Y., et al., 2004. Further characterisation of the 91500 zircon crystal. Geostandards and Geoanalytical Research, 28(1), 9-39.

Yang, Y.J., Zhu, W.G., Bai, Z., Zhong, H., Ye, X.T., Fan, H.P., 2016. Petrogenesis and tectonic implications of the Neoproterozoic Datian mafic-Ultramafic dykes in the Panzhihua area, western Yangtze Block, SW China. International Journal of Earth Sciences, 106 (1), 1-29.

Yuan, H.L., Gao, S., Liu, X.M., Li, H.M., Gunther, D., Wu, F.Y., 2004. Accurate U-Pb age and trace element determinations of zircon by laser ablation-inductively coupled plasma mass spectrometry. Geostandards & Geoanalytical Research, 28(3), 353-370.

Yuan, H.L., Gao, S., Dai, M.N., Zong, C.L., Gunther, D., Fontaine, G.H., Liu, X.M., Diwu, C.R., 2008. Simultaneous determinations of U-Pb age, Hf isotopes and trace element compositions of zircon by excimer laser-ablation quadrupole and multiple-collector ICPMS. Chemical Geology, 247(1-2), 100-118.

Zeng, L.S., Asimow, P.D., Saleeby, J.B., 2005a. Coupling of anatectic reactions and dissolution of accessory phases and the Sr and Nd isotope systematics of anatectic melts from a metasedimentory source. Geochimica et Cosmochimica Acta, 69, 3671-3682.

Zeng, L.S., Saleeby, J.B., Asimow, P., 2005b. Nd isotopic disequilibrium during crustal anatexis: A record from the Goat Ranch migmatite complex, southern Sierra Nevada batholith, California. Geology, 33, 53-56.

Zhao, G.C., Cawood, P.A., 2012. Precambrian geology of china. Precambrian Research, s 222-223, 13-54.

Zhao, J.H., Asimow, P.D., Zhou, M.F., Zhang, J., Yan, D.P., Zheng, J.P., 2017. An Andean-type arc system in Rodinia constrained by the Neoproterozoic Shimian ophiolite in South China. Precambrian Research, 296, 93-111.

Zhao, J.H., Zhou, M.F., 2007a. Geochemistry of Neoproterozoic mafic intrusions in the Panzhihua district (Sichuan Province, SW China): Implications for subduction-related metasomatism in the upper mantle. Precambrian Research, 152(1), 27-47.

Zhao, J.H., Zhou, M.F., 2007b. Neoproterozoic adakitic plutons and arc magmatism along the western margin of the Yangtze Block, South China. Journal of Geology, 115(6), 675-689.

Zhao, J.H., Zhou, M.F., Yan, D.P., Yang, Y.H., Sun, M., 2008a. Zircon Lu-Hf isotopic constraints on Neoproterozoic subduction-related crustal growth along the western margin of the Yangtze Block, South China. Precambrian Research, 163(3), 189-209.

Zhao, J. H., Zhou, M. F., Yan, D. P., Zheng, J. P., Li, J. W., 2011. Reappraisal of the ages of Neoproterozoic strata in South China: No connection with the Grenvillian orogeny. Geology, 39 (4), 299-302.

Zhao, J. H., Zhou, M. F., Zheng, J. P., 2013. Constraints from zircon U-Pb ages, O and Hf isotopic compositions on the origin of Neoproterozoic peraluminous granitoids from the Jiangnan fold belt, South China. Contributions to Mineralogy and Petrology, 166(5), 1505-1519.

Zhao, J.H., Zhou, M.F., Zheng, J.P., Fang, S.M., 2010. Neoproterozoic crustal growth and reworking of the Northwestern Yangtze Block: Constraints from the Xixiang dioritic intrusion, South China. Lithos, 120(3-4), 439-452.

Zhao, J.H., Li, Q.W., Liu, H., Wang, W., 2018. Neoproterozoic magmatism in the western and northern margins of the Yangtze Block (South China) controlled by slab subduction and subduction-transform-edge-propagator. Earth-Science Reviews, 187, 1-18.

Zhao, X.F., Zhou, M.F., Li, J.W., Wu, F.Y., 2008b. Association of Neoproterozoic A- and I-type granites in south china: Implications for generation of A-type granites in a subduction-related environment. Chemical Geology, 257(s1), 1-15.

Zhao, Z.F., Gao, P., Zheng, Y.F., 2015. The source of Mesozoic granitoids in South China: Integrated geochemical constraints from the Taoshan batholith in the Nanling Range. Chemical Geology, 395, 11-26.

Zheng, J.P., Griffin, W.L., O'Reilly, S.Y., Zhang, M., Pearson, N., Pan, Y.M., 2006. Widespread Archean basement beneath the Yangtze Craton. Geology, 34, 417-420.

Zheng, Y.F., Wu, Y.B., Zhao, Z.F., Zhang, S.B., Xu, P., Wu, F.Y., 2005. Metamorphic effect on zircon Lu-Hf and U-Pb isotope systems in ultrahigh-pressure eclogite-facies metagranite and metabasite. Earth & Planetary Science Letters, 240(2), 378-400.

Zheng, Y.F., Zhang, S.B., Zhao, Z.F., Wu, Y.B., Li, X.H., Li, Z.X., Wu, F.Y., 2007. Contrasting zircon Hf and O isotopes in the two episodes of Neoproterozoic granitoids in South China: Implications for growth and reworking of continental crust. Lithos, 96, 127-150.

Zhou, M.F., Ma, Y.X., Yan, D.P., Xia, X.P., Zhao, J.H., S, M., 2006b. The Yanbian Terrane (Southern Sichuan Province, SW China): A Neoproterozoic arc assemblage in the western margin of the Yangtze Block. Precambrian Research, 144(1-2), 19-38.

Zhou, M.F., Yan, D., Kennedy, A.K., Li, Y.Q., Ding, J., 2002. SHRIMP U-Pb zircon geochronological and geochemical evidence for Neoproterozoic arc-magmatism along the western margin of the Yangtze Block, South China. Earth & Planetary Science Letters, 196(1-2), 51-67.

Zhou, M.F., Yan, D.P., Wang, C.L., Qi, L., Kennedy, A., 2006a. Subduction-related origin of the 750 Ma Xuelongbao adakitic complex (Sichuan Province, China): Implications for the tectonic setting of the giant Neoproterozoic magmatic event in South China. Earth & Planetary Science Letters, 248, (1-2), 286-300.

Zhu, G.Z., Gerya, T.V., Yuen, D.A., Honda, S., Yoshida, T., Connolly, J.A.D., 2009. Three-dimensional dynamics of hydrous thermal-chemical plumes in oceanic subduction zones. Geochemistry, Geophysics,

Geosystems, 10, Q11006.

Zhu, R.Z., Lai, S.C., Qin, J.F., Zhao, S.W., Santosh, M., 2018. Strongly peraluminous fractionated S-type granites in the Baoshan Block, SW China: Implications for two-stage melting of fertile continental materials following the closure of Bangong-Nujiang Tethys. Lithos, 316-317, 178-198.

Zhu, W.G., Zhong, H., Li, X.H., Deng, H.L., He, D.F., Wu, K.W., Bai, Z.J., 2008. SHRIMP zircon U-Pb geochronology, elemental, and Nd isotopic geochemistry of the Neoproterozoic mafic dykes in the Yanbian area, SW China. Precambrian Research, 164(1-2), 66-85.

Zhu, W.G., Zhong, H., Li, Z.X., Bai, Z.J., Yang, Y.J., 2016. SIMS zircon U-Pb ages, geochemistry and Nd-Hf isotopes of ca. 1.0 Ga mafic dykes and volcanic rocks in the Huili area, SW China: Origin and tectonic significance. Precambrian Research, 273, 67-89.

Zhu, Y., Lai, S.C., Zhao, S.W., Zhang, Z.Z., Qin, J.F., 2017. Geochemical characteristics and geological significance of the Neoproterozoic K-feldspar granites from the Anshunchang, Shimian area, Western Yangtze Block. Geological Review, 63(5), 1193-1208 (in Chinese with English abstract).

Zhu, Y., Lai, S.C., Qin, J.F., Zhu, R.Z., Zhang, F.Y., Zhang, Z.Z., 2019a. Geochemistry and zircon U-Pb-Hf isotopes of the 780 Ma I-type granites in the western Yangtze Block: Petrogenesis and crustal evolution. International Geology Review, 61(10), 1222-1243.

Zhu, Y., Lai, S.C., Qin, J.F., Zhu, R.Z., Zhang, F.Y., Zhang, Z.Z., Gan, B.P., 2019b. Petrogenesis and geodynamic implications of Neoproterozoic gabbro-diorites, adakitic granites, and A-type granites in the southwestern margin of the Yangtze Block, South China. Journal of Asian Earth Sciences.

Zimmer, M., Kroner, A., Jochum, K.P., Reischmann, T., Todt, W., 1995. The Gabal Gerf complex: A precambrian N-MORB ophiolite in the Nubian Shield, NE Africa. Chemical Geology, 123, 29-51.

Petrogenesis and geodynamic implications of Neoproterozoic gabbro-diorites, adakitic granites, and A-type granites in the southwestern margin of the Yangtze Block, South China[①]

Zhu Yu   Lai Shaocong[②]   Qin Jiangfeng   Zhu Renzhi

Zhang Fangyi   Zhang Zezhong   Gan Baoping

Abstract: The occurrence of A-type granites following adakitic granites can provide unique insights into the tectonic evolution of subduction zone. This paper presents a comprehensive study of zircon U-Pb-Hf isotopes and whole-rock geochemistry on the Neoproterozoic gabbro-diorites, biotite granites, and K-feldspar granites in the southwestern margin of the Yangtze Block, South China. Zircon U-Pb dating of three distinctive intrusions yield concordant ages of 810.4 ± 2.0 Ma for gabbro-diorites, 800.3 ± 2.1 Ma for biotite granites, and 749.1 ± 2.0 Ma for K-feldspar granites. The gabbro-diorites are sodic and calc-alkaline rocks with low SiO_2 ($52.62-53.87$ wt%), medium MgO ($2.67-3.41$ wt%), high $Fe_2O_3^T$ ($7.18-7.49$ wt%) and CaO ($5.68-7.50$ wt%) contents. They display high Th/Zr and Rb/Y ratios but low Nb/Zr and Nb/Y ratios. Together with their positive whole-rock $\varepsilon_{Nd}(t)$ ($+1.0$ to $+1.5$) and zircon $\varepsilon_{Hf}(t)$ values ($+3.66$ to $+8.18$), we suggest that these gabbro-diorites were derived from the partial melting of depleted lithospheric mantle modified by subduction-related fluids. The biotite granites display relatively high Sr contents ($335-395$ ppm), Sr/Y ratios ($38.9-54.3$), low Y ($7.04-9.71$ ppm) and Yb contents ($0.78-1.08$ ppm), resembling the adakitic granites. Their low $Mg^{\#}$ values ($36-41$), Cr ($2.94-3.59$ ppm) and Ni ($1.32-1.55$ ppm) contents as well as positive $\varepsilon_{Nd}(t)$ ($+0.5$ to $+0.6$) and $\varepsilon_{Hf}(t)$ values ($+1.62$ to $+8.07$) indicate that they were formed by partial melting of thickened juvenile mafic lower crust, probably within the existence of garnet. The K-feldspar granites have high SiO_2 ($76.61-77.14$ wt%) and alkalis ($Na_2O + K_2O = 8.55-9.69$ wt%) contents, high 10 000 Ga/Al ($2.56-2.80$) ratios and differentiation index ($95.3-96.9$), indicating the affinity to highly fractionated A-type granites. We suggest that these A-type granites were mainly generated by the partial melting of felsic crustal rocks under low pressure condition and subsequently extensive fractional crystallization is significant. Taking into account previous studies from the western Yangtze Block, South China, we propose that the gabbro-diorites, adakitic granites and A-type granites in this study were formed in persistent subduction process during the Neoproterozoic.

---

① Published in *Journal of Asian Earth Sciences*, 2019, 183.

② Corresponding author.

1　Introduction

The South China Block, which is largely characterized by the Neoproterozoic long-term igneous activity, has been considered as an important part of supercontinent Rodinia (Li et al., 2003; Wang et al., 2013; Zhao and Cawood, 2012; Zhang et al., 2012; Zhou et al., 2002, 2006a,b), and thus it is a key area for understanding Neoproterozoic mantle nature, crust-mantle interaction, and crustal evolution during the assembly and breakup of Rodinia (Li et al., 2003; Li and Zhao, 2018; Wang et al., 2018; Zhang et al., 2012, 2015; Zhao et al., 2008a, 2019; Zhao and Zhou, 2007a,b, 2008, 2009). In the past few decades, massive mafic to felsic intrusions along the western and northern margins of the Yangtze Block have been studied (Table 1), and three kinds of geodynamic mechanisms were proposed, including the mantle plume model (Li et al., 2003, 2006), subduction model (Zhou et al., 2002, 2006a,b; Zhao et al., 2008a, 2010, 2018, 2019), and plate rifting model (Zheng et al., 2007, 2008). The different understanding of tectonic models might be caused by distinct interpretation for single lithology. Therefore, deciphering the petrogenesis of various rock association is essential for us to further understand Neoproterozoic tectonic evolution of the Yangtze Block, South China.

The adakitic magma is geochemically characterized by high Sr, Sr/Y, and La/Yb ratios, as well as low Y and Yb contents (Defant and Drummond, 1990; Martin et al., 2005), which can be selectively thought to source from the thickened continental lower crust (Huang et al., 2009; Wang et al., 2006, 2012; Zhang et al., 2018). On the other hand, the A-type (alkaline, anhydrous, and anorogenic) granites usually contain high SiO_2, $Na_2O + K_2O$ contents and show high Ga/Al ratios (Whalen et al., 1987), which are generally linked to extensional tectonism in the continental rift setting, subduction zone, or post-collisional setting (Eby, 1992). Therefore, the occurrence of A-type granites following adakitic granites can provide a window for probing crustal evolution from regionally thickening to thinning.

Although voluminous studies have been devoted to Neoproterozoic igneous rocks in the southwestern periphery of the Yangtze Block (Panzhihua-Yanbian region) (Fig. 2), most of them were predominantly focused on the ultramafic-intermediate rocks that were related to the mantle-derived melts modified by subduction-related fluids and/or melts, such as the ca. 860 Ma Guandaoshan pluton (Du et al., 2014; Sun and Zhou, 2008), ca. 812 Ma Gaojiacun and ca. 806 Ma Lengshuiqing mafic intrusions (Zhao et al., 2008a; Zhou et al., 2006b), as well as ca. 746 Ma olivine gabbro and ca. 738 Ma hornblende gabbro in Panzhihua district (Zhao et al., 2008a; Zhao and Zhou, 2007b). However, Neoproterozoic felsic intrusions in this region, especially adakitic granites and A-type granites, were little systematically examined but can provide significant information about the geodynamic evolution. In this paper, we provided

Table 1 Summary of the Neoproterozoic magmatism in the western Yangtze Block, South China.

Pluton location	Lithology	Age /Ma	SiO_2 /%	A/CNK	Zircon Hf isotopes $\varepsilon_{Hf}(t)$	Whole-rock Sr-Nd isotopes $(^{87}Sr/^{86}Sr)_i$	$\varepsilon_{Nd}(t)$	Major mineral assemblage	Reference
Kangding-Mianning region									
Tianquan	Granodiorite	851±15	64.48–65.82	0.96–1.18			+0.6 to +0.9	Pl(40%–50%)+Kfs(20%–30%)+Qtz(10%–20%)+Hbl(10%)+Tit+Apa+Zr	Lai et al., 2015b
	Granite		73.31–74.93	1.01–1.33			+4.4 to +8.3	Pl(20%–25%)+Kfs(40%–50%)+Qtz(20%–25%)+Bi(5%)+Hbl(1%–2%)+Tit+Apa+Zr	
Baoxing	Gabbro	ca. 850	49.59–50.94		+5.44 to +11.94			Pl(40%–45%)+Hbl(35%–40%)+Pyroxene(5%–10%)+Bi+Epi+Mag	Meng et al., 2015
	Gabbro-diorites	ca. 800	47.67–57.62		+3.99 to +14.62 +0.68 to +5.66			Pl(60%–65%)+Hbl(20%–25%)+Qtz(5%–10%)+Bi(5%)+Chl+Epi+Mag	
	Granites	ca. 770	71.58–76.26		+2.53 to +10.58			Qtz(25%–30%)+Pl(40%–50%)+alkali feldspar(20%–30%)+Bi+Mus	
Daxiangling	Granite (A-type)	816±10	76.3–79.3	1.00–1.12	+6.0 to +12.4	0.659 38–0.727 60	+0.3 to +1.8	Qtz(30%–50%)+Per(50%–70%)+Albite+Bi	Zhao et al., 2008b
Shimian	Granite (I-type)	797±22 790±10	69.3–76.26		+2.9 to +8.2			Qtz(20%–40%)+Hbl+Bi+Fe-Ti oxide+Zir+K-feldspar	Zhao et al., 2017
Shimian ophiolite	Gabbro Peridotite	ca. 800	37.31–51.18 37.65–41.95		-1.83 to +10.69			Pl+Hbl+Cpx+Fe-Ti oxide	Zhao et al., 2017
Mianning	Granite (A-Type)	ca. 780	75.4–77.96	1.00–1.26	+9.2 to +12.1		+2.97 to +4.71	Alkali-feldspar(65%–70%)+Qtz(20%–30%)+Hbl(<3%)+Pl(<5%)+Mag	Huang et al., 2008
Shimian (Anshunchang)	K-feldspar granites	777±4.8	72.64–76.27	1.06–1.24			+0.5 to +3.3	Kfs(45%–50%)+Pl(25%–30%)+Qtz(20%–25%)+Bi(5%)+Zr+Mag	Zhu et al., 2017

Continued

Pluton location	Lithology	Age /Ma	SiO$_2$ /%	A/CNK	Zircon Hf isotopes $\varepsilon_{Hf}(t)$	Whole-rock Sr-Nd isotopes (^{87}Sr/^{86}Sr)$_i$	$\varepsilon_{Nd}(t)$	Major mineral assemblage	Reference
Kangding	Granodiorite	748±11	65.32–67.59	0.98–1.06		0.704 071–0.705 414	+0.8 to +1.3	Pl(50%–55%)+Alkali feldspar(30%–40%)+Bi+Amp	Lai et al., 2015a
Luding	Quartz monzodiorite	754±10	60.76–63.78	0.90–0.99		0.703 513–0.704 519	+2.4 to +4.8	Pl(40%–45%)+Amp(20%–25%)+Alkali feldspar(5%–10%)+Bi(10%–15%)+Qtz+Zr+Apa+Mag	
Xuelongbao	Tonalite+granodiorite	748±7	62.0–74.8			0.703 3–0.705 4	+0.30 to +2.88	Pl(50%–70%)+K-feldspar(<10%)+Qtz(20%–30%)+Bi(5%–10%)	Zhou et al., 2006b
Xichang-Miyi region									
Xiacun	Mafic rock	842±14	43.03–49.7				–1.7 to +7.4	Pl(45%–55%)+Hbl(35%–45%)+Qtz+Bi+Fe-Ti oxide	Guo et al., 2007
Yonglang	Granite Granite	803±15 790±16	65.02–76.44	1.03–2.21			–7.3 to –2.9	Kfs+Pl(65%–70%)+Qtz(25%–30%)+Bi(5%–10%)+Mus	
Mopanshan	Tonalite+granodiorite	782±6	63.92–69.69	0.92–1.14	+1.91 to +7.07	0.704 275–0.705 649	–2.06 to –0.54	Pl(65%–75%)+Hbl(5%–10%)+Qtz(5%–25%)+Bi(5%–15%)+Tit+Mag	Huang et al., 2009
Shaba	Gabbro	748±12			+1.8 to +11.63				Zhao et al., 2008a
Yanbian-Panzhihua region									
Guandaoshan	Granodiorite	857±13	64.03–65.96	0.98–1.04			+3.3 to +5.2	Pl+Hbl+Qtz+Hbl+Bi	Li et al., 2003
	Quartz diorite	857±7	47.02–67.66			0.703 0–0.703 3	+4.8 to +5.2		Du et al., 2014
	Gabbroic diorite	856±6						Pl(40%–60%)+Hbl(25%–40%)+K-feldspar(0%–7%)+Qtz(5%–10%)+Mag(3%–5%)	
	Gabbro	856±8						Pl(50%–55%)+Hbl(10%–20%)+K-feldspar(7%–10%)+Qtz(10%–20%)+Mag(1%–2%)	

Continued

Pluton location	Lithology	Age /Ma	SiO$_2$ /%	A/CNK	Zircon Hf isotopes $\varepsilon_{Hf}(t)$	Whole-rock Sr-Nd isotopes $(^{87}Sr/^{86}Sr)_i$	$\varepsilon_{Nd}(t)$	Major mineral assemblage	Reference
Guandaoshan	Quartz diorite	858±7	52.8-63.8		+11 to +17	0.702 8-0.703 3	+3.9 to +5.1	Pl(45%-55%)+K-feldspar(5%-10%)+Amp(10%-25%)+Qtz(5%-20%)+Mag(2%-5%)	Sun and Zhou, 2008
	Diorites							Pl(50%-65%)+Amp(25%-40%)+Bi(2%-5%)+Mag(2%-4%)+Qtz(<5%)	
Fangtian	Basaltic lavas	ca.840	45.3-50.7			0.703 0-0.704 0	+3.8 to +8.0	Pl+Cpx+Fe-Ti oxide	Sun et al., 2007
Tongde	Quartz diorite+Diorite+Gabbro	825±7	47-59					Pl+Hbl+Mag+Opx+Cpx+Bi	Munteanu et al., 2010
	Diorite	ca.830	50.98-61.01		+4.33 to +9.31	0.706 842-0.706 320	+0.6 to +2.0	Hbl(30%-75%)+Pl(25%-60%)+Mag+Apa	Li and Zhao, 2018
Leng-shuiqing	Diorite	811.6±3	40.1-50.0			0.700 5-0.706 0	-2.06 to -0.54	Pl+Cpx+Qtz+Zr	Zhou et al., 2006a;
Gaojiacun	Diorite	806±4						Pl+Cpx+Qtz	Zhao et al., 2008a
Dajianshan	Gabbro-diorite	810.4±2.0	52.62-53.87	0.89-1.02	+3.64 to +8.14	0.705 184-0.705 392	+1.0 to +1.5	Pl(45%-50%)+Hbl(25%-35%)+Cpx(~10%)+Qtz(<5%)+Mag(1%-2%)+Zr	This study
Dajianshan	Biotite granite (adakite)	800.3±2.1	74.08-74.82	1.04-1.13	+1.66 to +8.10	0.705 262-0.705 324	+0.5 to +0.6	Pl(40%-50%)+Qtz(20%-30%)+K-feldspar(15%-20%)+Bi(5%-10%)+Zr	This study
Yanbian	Hornblende gabbro	792±13	48.6-51.6				+5.8 to +7.2	Pl(45%-65%)+Cpx(8%-40%)+Hbl(8%-25%)+Bi(1%)+Fe-Ti oxide(1%-2%)+Apa+Zr	Zhu et al., 2008
	Gabbro+Dolerite	761±14	44.7-50.2				+4.5 to +7.2	Cpx(48%-60%)+Pl(40%-50%)+Hbl(<1%)+Bi(<1%)+Fe-Ti oxide(<1%)	

Continued

Pluton location	Lithology	Age /Ma	SiO$_2$ /%	A/CNK	Zircon Hf isotopes $\varepsilon_{Hf}(t)$	Whole-rock Sr-Nd isotopes ($^{87}Sr/^{86}Sr$)$_i$	$\varepsilon_{Nd}(t)$	Major mineral assemblage	Reference
Datian	Dolerite	ca. 800	47.3-53.3				-3.3 to +1.1	Pl(40%-55%)+Cpx(10%-40%)+Hbl(5%-25%)+Bi(1%-2%)+Fe-Ti oxide(1%-2%)	Yang et al., 2016
		ca. 760	47.8-50.1				+4.3 to +5.2	Pl(40%-50%)+Hbl(20%-40%)+Cpx(5%-10%)+Bi(1%-2%)+Fe-Ti oxide(1%-2%)	
Dalu	Granodiorite	781.1±2.8	60.88-68.70	0.94-1.08	+2.16 to +7.36			Pl(30%-35%)+Qtz(20%-25%)+K-feldspar(10%-15%)+Hbl(10%-15%)+Bi(5%-10%)+Mag+Zr	Zhu et al., 2019
	Granite	779.8±2.0	65.74-75.38	1.05-1.20	-4.65 to +5.80			K-feldspar(25%-30%)+Pl(22%-27%)+Qtz(30%-35%)+Hbl(2%-5%)Bi(3%-5%)+Mag(0%-2%)+Zr	
Datian	Granodiorite	760±4	50.96-73.40	0.69-1.10	+2.69 to +11.68	0.704 308-0.705 068	-0.92 to -0.01	Pl(35%-45%)+Qtz(20%-30%)+Amp(10%-15%)+Bi(10%-15%)+K-feldspar(<10%)	Zhao and Zhou,2007a;
Dajianshan	Diorite				+2.22 to +8.36	0.704 613-0.704 688	+0.45 to +0.66	Amp(40%-50%)+Pl(20%-30%)+Qtz(10%-20%)+Bi+K-feldspar	Zhao et al., 2008a
Dadukou	Olivine gabbro	746±10	48.0-51.6		+2.96 to +6.44	0.707 0-0.707 5	-0.64 to -1.73	Ol+Cpx(20%-30%)+Pl(40%-60%)+Hbl+Fe-Ti oxide(5%-10%)	Zhao and Zhou,2007b;
	Hornblende gabbro	738±23	46.1-52.5			0.704 5-0.707 0	-0.12 to -0.93	Hbl(50%-70%)+Pl(30%-50%)+Fe-Ti oxide	Zhao et al., 2008a
Panzhihua	K-feldspar gr-anite(A-type)	749.1±2.0	76.61-77.14	0.85-1.02	-1.85 to +6.74		-1.2 to -1.6	Pl(20%-30%)+Qtz(20%-30%)+K-feldspar(30%-40%)+Per(5%-10%)+Bi+Zr+Apa	This study

systemic analyses of zircon U-Pb-Hf isotopes and whole-rock geochemistry for gabbro-diorites, adakitic granites, and A-type granites along the southern part of the western Yangtze Block. The objectives of this paper are: ① to elucidate their magma source; and ② to provide reasonable insights for tectonic evolution of the western Yangtze Block during the Neoproterozoic.

2 Geological background, regional geology, and petrography

2. 1 Geological background

The South China Block was divided into the Yangtze Block in the northwest and the Cathaysia Block in the southeast by the Jiangnan Fold Belt (also known as the Jiangnan Orogen or Sibao Orogen) (Wang et al., 2013; Zhao et al., 2011; Zhao and Cawood, 2012) (Fig. 1a). The Jiangnan Fold Belt is a NE-trending tectonic zone, geographically along the northeastern Guangxi, southeastern Guizhou, western Hunan and northwest Jiangxi Provinces (Zhang et al., 2015). It is mainly composed of the early Neoproterozoic strata which underwent greenschist facies metamorphism and strongly deformation (Zhao et al., 2018). The

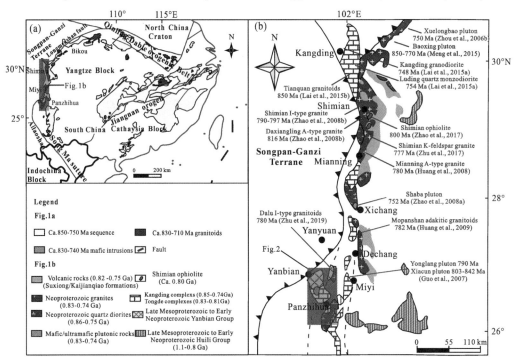

Fig. 1   Sketch geological map of South China (a) (after Gan et al., 2017) and
the western margin of the Yangtze Block (b) (after Zhou et al., 2006b).
The Neoproterozoic igneous rocks data are from Huang et al., 2008, 2009; Guo et al., 2007; Lai et al., 2015a,b;
Meng et al., 2015; Zhao et al., 2008a,b, 2017; Zhou et al., 2002, 2006a,b; Zhu et al., 2017, 2019.

Yangtze Block is surrounded by the Qinling-Dabie-Sulu high-ultrahigh pressure metamorphic belt to the north (Zhao et al., 2008a; Zhou et al., 2006b), the Longmenshan fault to the west, and the Ailaoshan-Songma suture to the southwest. (Gao et al., 1999; Zhao and Cawood, 2012) (Fig. 1a). The Kongling metamorphic complex, consisting of the Archean to Paleoproterozoic high-grade metamorphic TTG gneisses, metasedimentary rocks and amphibolites, covers an area of 360 km^2 and is the oldest exposed rock unit in the Yangtze Block (Gao et al., 1999). In addition, the Paleoproterozoic to Mesoproterozoic low-grade greenschist facies volcanic-sedimentary sequences are preserved along the northern and western margins of the Yangtze Block (Li and Zhao, 2018). Neoproterozoic sedimentary-volcanic rocks (ca. 850 − 750 Ma), granitoids (ca. 830 − 710 Ma), mafic intrusions (ca. 830 − 740 Ma), granitic gneisses, and associated calc-alkaline volcanic rocks are discontinuously distributed along the margins of the Yangtze Block (Li et al., 2003; Gan et al., 2017) (Fig. 1a).

2. 2 Regional geology

A major N-S trending 1 000-km long Neoproterozoic igneous and metamorphic belt occurs along the western periphery of the Yangtze Block (Zhou et al., 2002; Du et al., 2014) (Fig. 1b). The Neoproterozoic granitoids are spatially associated with the basement complexes including the high-grade metamorphic complexes and Mesoproterozoic strata as well as its equivalents (Li et al., 2006; Zhou et al., 2006a; Zhao et al., 2008a). Neoproterozoic volcanic-sedimentary sequences are well preserved and contain the Yanbian, Suxiong, Kaijianqiao Groups, and two large serpentinite blocks on the western margin of Yangtze Block (Geng et al., 2008; Zhou et al., 2002). The Suxiong Group is mainly composed of silicic volcanic rocks, whereas the Yanbian Group consists of flysch-type sedimentary sequence (Zhou et al., 2006a). The Neoproterozoic Shimian ophiolite sequence consists of the gabbro and serpentinized peridotite intruded by mafic dikes (Zhao et al., 2017).

The Panzhihua-Yanbian region is located at the southern segment of the western margin of the Yangtze Block (Fig. 1b). The Yanbian Terrane, with a total area of ~300 km^2 (Fig. 2a), is situated at the northwestern corner of the Panzhihua city, Sichuan province. The Yanbian Group is strongly deformed and metamorphosed to low-grade greenschist facies and unconformably overlain by Sinian strata known as the Lieguliu Formation (Du et al., 2014; Sun et al., 2007). There are three kinds of Precambrian rock units in the east-western trending Yanbian Terrane: ①the Yanbian Group; ②Neoproterozoic igneous bodies that intruded the Yanbian Group; and ③the Upper Neoproterozoic (Sinian) successions and some igneous intrusions (Li et al., 2006). The Yanbian Group is composed of lower ~ 1 500-m-thick sequence of basaltic lava and upper ~3 500-m-thick flysch deposit that intruded by the ca. 860 Ma Guandaoshan pluton (Li et al., 2006; Sun and Zhou, 2008). They can be subdivided into

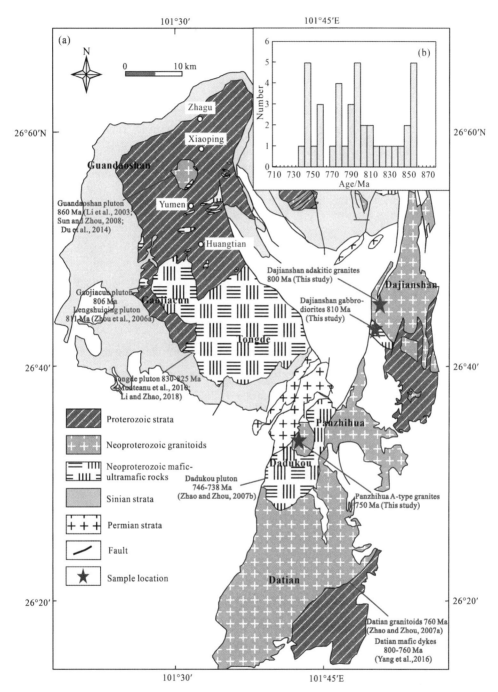

Fig. 2　Sketch geological map of the Panzhihua-Yanbian region along the southwestern Yangtze Block (a)
(after Zhao et al., 2008a) and the summary of the Neoproterozoic magmatism
in the western Yangtze Block (b) (Data source is from Table 1).

the Huangtian, Yumen, Xiaoping, and Zagu Formations from bottom to top (Du et al., 2014;
Zhou et al., 2006a). Moreover, Neoproterozoic intermediate and mafic-ultramafic intrusions

have been identified, i.e. the ca. 830−825 Ma Tongde pluton and dikes (Li and Zhao, 2018; Munteanu et al., 2010), ca. 810 Ma Gaojiacun and Lengshuiqing pluton (Zhou et al., 2006a) (Fig. 2a). In the Panzhihua city, voluminous granitoids are spatially accompanied with the high-grade metamorphic complex and low-grade metamorphic Mesoproterozoic strata. The Datian and Dajianshan granitoid plutons intruded a metamorphic complex of fine-grained amphibolite and amphibole-bearing gneiss (Zhao et al., 2008a). The ca. 760 Ma Datian granitic pluton is intruded by the ca. 740 Ma Dadukou gabbroic complex in the northwest and associated with ca. 800−760 Ma Datian mafic-ultramafic dikes in the southeast (Yang et al., 2016; Zhao et al., 2008a; Zhao and Zhou, 2007a,b).

2.3 Petrography

In this study, three groups of intrusive rocks were recognized and collected in the Panzhihua-Yanbian region along the southwestern margin of the Yangtze Block, South China (Fig. 2a), i.e. gabbro-diorite (7 samples, DJS-1-1 ~ DJS-1-7), biotite granite (7 samples, DJS-2-1 ~ DJS-2-7), and K-feldspar granite (6 samples, PZH-01 ~ PZH-06)

In the Dajianshan area, the igneous rocks intrude Mesoproterozoic foliated metasedimentary rocks in the south and are in fault contact with Mesozoic strata to the north. They are overlain unconformably by Sinian strata (Zhao et al., 2008a). In this study, the gabbro-diorites around the Dajianshan area are grey and medium-grained granular texture and massive structure (Fig. 3a). They are composed of plagioclase (45−50 vol%), hornblende (25−35 vol%), partly altered clinopyroxene (~10 vol%), quartz (<5 vol%), magnetite (1−2 vol%), some accessory minerals (e.g., zircon), and secondary minerals (e.g., chlorite) (Fig. 3b,c). The plagioclases are subhedral and weakly altered on the surface. The hornblendes are subhedral to euhedral and some of them are partly altered into the biotite.

The studied biotite granites are situated at the north part of the gabbro-diorites. The collected biotite granites are medium- to fine-grained and mainly consist of plagioclase (40−50 vol%), quartz (20−30 vol%), K-feldspar (15−20 vol%), biotite (5−10 vol%), minor zircon (Fig. 3d). Some plagioclases display the typical feature of polysynthetic twin (Fig. 3e,f). Some plagioclases which show the concentric zonation were partly altered to sericite on the surface (Fig. 3f). The brownish biotite is subhedral and filled into the crevice among plagioclase and quartz.

The studied K-feldspar granitic pluton is solely located at the vicinity of Panzhihua University in the Panzhihua city. These K-feldspar granites are flesh red, medium grained and massive structure (Fig. 3g). They are mainly composed of plagioclase (20−30 vol%), quartz (25−30 vol%), K-feldspar (30−40 vol%), perthite (5−10 vol%), minor biotite, as well as some accessory minerals including zircon and apatite. The plagioclase shows the polysynthetic

Fig. 3   Field petrography (a,d,g) and microscope photographs of the gabbro-diorites (b,c),
biotite granites (e,f), and K-feldspar granites (g,h) from the southwestern margin
of the Yangtze Block, South China.

Pl: plagioclase; Kfs: K-feldspar; Qtz: quartz; Hbl: hornblende;

Mag: magnetite; Bi: biotite; Cpx: Clinopyxene; Zr: zircon.

twin and the K-feldspar displays distinctive feature of the gridiron twining (Fig. 3h). Some K-feldspars are partly altered to kaolinite. The quartz exhibits xenomorphic grains among other minerals. They also display vermicular structure at the margin of the feldspar minerals (Fig. 3i).

3   Analytical methods

In this paper, the in-situ zircon U-Pb-Hf isotopic analyses and whole-rock major and trace elements analyses were carried out at the State Key Laboratory of Continental Dynamics, Northwest University, Xi'an, China. The whole-rock Sr-Nd isotopic data were conducted at the Guizhou Tuopu Resource and Environmental Analysis Center.

3.1   Zircon U-Pb dating

Zircons were separated from ~5kg samples taken from sampling locations for the studied gabbro-diorites (DJS-1) and granites (DJS-2, PZH-09) so that we can select enough and representative zircons. Zircon grains from the studied samples were separated using conventional heavy-liquids and magnetic methods. Representative zircon grains were handpicked and mounted in epoxy resin disks, then polished and carbon coated. Internal morphology of the zircon grains was shown with cathodoluminescence (CL) microscopy before the U-Pb isotopic dating. Zircon Laser Ablation Inductively Coupled Plasma Mass Spectrometry (LA-ICP-MS) U-Pb analyses were conducted on Agilent 7500a ICP-MS equipped with a 193 nm laser following the analytical method of Yuan et al. (2004). Helium was used as the carrier gas to ensure efficient aerosol delivery to the troch and a beam diameter of 32 μm with a laser pulse width of 15ns was adopted throughout analysis processes. The $^{207}Pb/^{235}U$ and $^{206}Pb/^{238}U$ ratios were calculated using the GLITTER program, which was corrected using the Harvard zircon 91500 as external calibration with a recommended $^{206}Pb/^{238}U$ isotopic age of 1 065. 4 ± 0. 6 Ma (Wiedenbeck et al., 2004). The GJ-1 is also a standard sample with a recommended $^{206}Pb/^{238}U$ isotopic age of 603. 2 ± 2. 4 Ma (Liu et al., 2007a). Common Pb contents were subsequently valuated using the method described in Andersen (2002). The age calculations and plotting of concordia diagrams were made using ISOPLOT (version 3. 0; Ludwig, 2003). Uncertainties are quoted at the 2σ level.

3. 2　Major and trace elements

The fresh parts were powdered to about 200 mesh using a tungsten carbide ball mill. Major and trace elements were analyzed by X-ray fluorescence (XRF; Rikagu RIX 2100) and inductively coupled plasma mass spectrometry (ICP-MS; Agilent 7500a), respectively. Analyses of USGS and Chinese national rock standards (BCR-2, GSR-1, and GSR-3) showed that the analytical precision and accuracy for the major elements were generally better than 5%. For the trace element analyses, sample powders were digested using an $HF + HNO_3$ mixture in high-pressure Teflon bombs at 190 ℃ for 48 h. For most trace elements, the analytical error was less than 2% and the precision was greater than 10% (Y. Liu et al., 2007b).

3. 3　Whole-rock Sr-Nd isotopes

Whole-rock Sr-Nd isotopic data were obtained using a Neptune Plasma high-resolution (HR) multicollector (MC) mass spectrometer. The Sr and Nd isotopes were determined using a method similar to that of Chu et al. (2009). Sr and Nd isotopic fractionation was corrected to $^{87}Sr/^{86}Sr = 0. 119 4$ and $^{146}Nd/^{144}Nd = 0. 721 9$, respectively. During the sample runs, the La Jolla standard yielded an average value of $^{143}Nd/^{144}Nd = 0. 511 862 \pm 5$ (2σ), and the NBS987 standard yielded an average value of $^{87}Sr/^{86}Sr = 0. 710 236 \pm 16$ (2σ). The total procedural Sr and Nd blanks are b1 ng and b50 pg, and NIST SRM-987 and JMC-Nd were used as certified reference standard solutions for $^{87}Sr/^{86}Sr$ and $^{143}Nd/^{144}Nd$ isotopic ratios, respectively. The BCR-1 and BHVO-1 standards yielded an average of $^{87}Sr/^{86}Sr$ ratio are 0. 705 014 ± 3 (2σ) and 0. 703 477 ± 20 (2σ), respectively. The BCR-1 and BHVO-1 standards yielded an average of $^{146}Nd/^{144}Nd$ ratio are 0. 512 615 ± 12 (2σ) and 0. 512 987 ± 23 (2σ), respectively.

3. 4　In-situ zircon Lu-Hf isotopic analyses

The in-situ zircon Lu-Hf isotopic analyses were made using a Neptune MC-ICP-MS. The laser repetition rate was 10 Hz at 100 mJ and the spot sizes were 32 μm. The instrument information is obtainable from Bao et al. (2017). The detailed analytical technique is depicted by Yuan et al. (2008). During the analyses process, the beam diameter of 44 μm was used for all the analytical spots. The repetition rate was 5 Hz and the laser energy was 6 mJ/cm^2. Raw count rates for ^{172}Yb, ^{173}Yb, ^{175}Lu, ^{176}(Hf + Yb + Lu), ^{177}Hf, ^{178}Hf, ^{179}Hf and ^{180}Hf were collected simultaneously. Typical ablation time was 40 s, resulting in ablation depths of pits ca. 20–30 μm. The measured values of well-characterized zircon standards (91500, GJ-1, and Monastery) were consistent with the recommended values within 2σ (Yuan et al., 2008). The obtained Hf isotopic compositions were 0. 282 016±20 (2σ, $n=84$) for the GJ-1 standard and 0. 282 735 ± 24 (2σ, $n=84$) for the Monastery standard, respectively, which are consistent with the recommended values (Yuan et al., 2008) within 2σ. The initial ^{176}Hf/^{177}Hf ratios and $\varepsilon_{Hf}(t)$ values were calculated with the chondritic reservoir (CHUR) at the time of zircon growth from the magmas. The decay constant for ^{176}Lu of 1. 867 × 10^{-11} year^{-1}, the chondritic ^{176}Hf/^{177}Hf ratio of 0. 282 785 and ^{176}Lu/^{177}Hf ratio of 0. 033 6 were adopted (Soderlund et al., 2004). The depleted mantle model ages (T_{DM}) for basic rocks were calculated with reference to the depleted mantle at the present-day ^{176}Hf/^{177}Hf ratio of 0. 283 25, similar to that of the average MORB (Nowell et al., 1998) and ^{176}Lu/^{177}Hf = 0. 038 4 (Griffin et al., 2000). For the zircons from felsic rocks, we also calculated the "crustal" Hf model age (T_{DMC}) by assuming that its parental magma was derived from an average continental crust with ^{176}Lu/^{177}Hf = 0. 015, which was originated from the depleted mantle source (Griffin et al., 2000). Our conclusions would not be affected even if other decay constants were used.

4　Results

4. 1　Zircon LA-ICP-MS U-Pb dating

Representative zircon CL images, zircon U-Pb data and concordia diagrams of the studied samples [gabbro-diorite (DJS-1), biotite granite (DJS-2), and K-feldspar granite (PZH-09)] from the southwestern Yangtze Block are summarized and shown in Table 2 and Fig. 4.

4. 1. 1　Gabbro-diorite (DJS-1)

Zircons from gabbro-diorite sample (DJS-1) are mainly transparent and subhedral to euhedral prismatic crystals, with aspect ratios ranging from 1:1 to 3:2. In CL images, these zircon grains range from 100 μm to 150 μm in length. The 29 reliable analytical data from

Table 2　Concordant results of zircon U-Pb ages for the Neoproterozoic gabbro-diorites, biotite granites, and K-feldspar granites from the southwestern Yangtze Block, South China.

Analysis	Content/ppm			Ratios								Age/Ma							
	Th	U	Th/U	$\frac{^{207}Pb}{^{206}Pb}$	1σ	$\frac{^{207}Pb}{^{235}U}$	1σ	$\frac{^{206}Pb}{^{238}U}$	1σ	$\frac{^{208}Pb}{^{232}Th}$	1σ	$\frac{^{207}Pb}{^{206}Pb}$	1σ	$\frac{^{207}Pb}{^{235}U}$	1σ	$\frac{^{206}Pb}{^{238}U}$	1σ	$\frac{^{208}Pb}{^{232}Th}$	1σ
DJS-1 (Gabbro diorite)																			
DJS-1-01	54.43	48.24	1.13	0.064 66	0.002 80	1.193 67	0.046 72	0.133 87	0.002 03	0.040 53	0.000 78	763	89	798	22	810	12	803	15
DJS-1-02	71.42	105.24	0.68	0.064 78	0.001 91	1.194 36	0.027 96	0.133 70	0.001 62	0.041 32	0.000 63	767	61	798	13	809	9	819	12
DJS-1-03	47.65	61.88	0.77	0.064 41	0.002 15	1.195 70	0.033 65	0.134 60	0.001 75	0.042 42	0.000 76	755	69	799	16	814	10	840	15
DJS-1-04	44.89	46.39	0.97	0.065 46	0.002 65	1.220 86	0.043 98	0.135 24	0.001 97	0.040 91	0.000 79	789	83	810	20	818	11	810	15
DJS-1-05	36.03	54.47	0.66	0.065 80	0.002 59	1.204 98	0.041 99	0.132 80	0.001 90	0.038 17	0.000 81	800	80	803	19	804	11	757	16
DJS-1-06	37.73	43.91	0.86	0.066 26	0.002 58	1.224 77	0.042 07	0.134 03	0.001 92	0.043 09	0.000 83	815	79	812	19	811	11	853	16
DJS-1-07	39.36	56.90	0.69	0.066 50	0.002 35	1.246 43	0.037 81	0.135 92	0.001 82	0.041 24	0.000 78	822	72	822	17	822	10	817	15
DJS-1-08	10.09	9.34	1.08	0.065 19	0.005 88	1.209 09	0.105 86	0.134 50	0.003 23	0.043 64	0.001 51	780	179	805	49	814	18	863	29
DJS-1-09	27.96	29.63	0.94	0.071 25	0.006 06	1.324 88	0.108 60	0.134 83	0.003 66	0.041 13	0.001 97	965	165	857	47	815	21	815	38
DJS-1-10	30.23	39.22	0.77	0.068 11	0.002 73	1.256 81	0.044 77	0.133 81	0.001 97	0.042 26	0.000 90	872	81	827	20	810	11	837	17
DJS-1-11	40.79	47.52	0.86	0.072 31	0.002 64	1.349 83	0.042 55	0.135 37	0.001 88	0.042 55	0.000 79	995	72	868	18	818	11	842	15
DJS-1-12	38.50	47.48	0.81	0.063 76	0.002 44	1.192 10	0.040 04	0.135 59	0.001 90	0.041 97	0.000 80	734	79	797	19	820	11	831	16
DJS-1-13	53.69	86.65	0.62	0.063 09	0.002 02	1.178 38	0.031 21	0.135 46	0.001 71	0.039 06	0.000 72	711	67	791	15	819	10	775	14
DJS-1-15	43.47	47.85	0.91	0.065 82	0.002 73	1.211 84	0.044 88	0.133 53	0.001 97	0.041 43	0.000 83	801	84	806	21	808	11	821	16
DJS-1-17	28.75	52.01	0.55	0.065 67	0.003 63	1.206 86	0.062 36	0.133 29	0.002 44	0.037 74	0.001 52	796	112	804	29	807	14	749	30
DJS-1-18	50.93	56.84	0.90	0.068 02	0.002 65	1.251 85	0.043 07	0.133 48	0.001 90	0.041 54	0.000 79	869	79	824	19	808	11	823	15
DJS-1-19	96.08	74.73	1.29	0.060 66	0.002 31	1.120 33	0.037 48	0.133 95	0.001 83	0.039 17	0.000 63	627	80	763	18	810	10	777	12
DJS-1-20	51.51	55.65	0.93	0.063 03	0.002 86	1.161 22	0.048 04	0.133 62	0.002 08	0.040 26	0.000 85	709	94	783	23	809	12	798	16
DJS-1-22	55.01	58.47	0.94	0.069 78	0.002 58	1.289 35	0.041 38	0.134 01	0.001 85	0.043 75	0.000 77	922	74	841	18	811	11	866	15
DJS-1-23	35.23	56.28	0.63	0.064 00	0.002 15	1.177 46	0.033 28	0.133 43	0.001 71	0.040 71	0.000 77	742	69	790	16	807	10	807	15
DJS-1-24	36.10	39.50	0.91	0.065 16	0.002 75	1.200 11	0.045 40	0.133 58	0.001 98	0.041 15	0.000 83	780	86	801	21	808	11	815	16
DJS-1-25	28.08	24.04	1.17	0.065 85	0.004 13	1.205 79	0.071 55	0.132 81	0.002 71	0.038 23	0.001 00	802	126	803	33	804	15	758	20

Continued

Analysis	Content/ppm			Ratios								Age/Ma							
	Th	U	Th/U	$^{207}Pb/^{206}Pb$	1σ	$^{207}Pb/^{235}U$	1σ	$^{206}Pb/^{238}U$	1σ	$^{208}Pb/^{232}Th$	1σ	$^{207}Pb/^{206}Pb$	1σ	$^{207}Pb/^{235}U$	1σ	$^{206}Pb/^{238}U$	1σ	$^{208}Pb/^{232}Th$	1σ
DJS-1-26	91.36	128.26	0.71	0.062 67	0.002 12	1.145 45	0.032 68	0.132 57	0.001 71	0.042 07	0.000 72	697	70	775	15	803	10	833	14
DJS-1-29	28.25	32.70	0.86	0.064 52	0.003 42	1.185 56	0.058 43	0.133 28	0.002 32	0.041 06	0.001 11	759	108	794	27	807	13	813	22
DJS-1-30	50.15	53.92	0.93	0.064 20	0.002 64	1.180 89	0.043 42	0.133 41	0.001 94	0.039 08	0.000 77	748	85	792	20	807	11	775	15
DJS-1-33	57.08	60.46	0.94	0.066 66	0.002 87	1.218 54	0.047 27	0.132 59	0.002 01	0.041 62	0.000 84	827	87	809	22	803	11	824	16
DJS-1-34	82.82	74.25	1.12	0.062 90	0.003 28	1.147 94	0.055 67	0.132 39	0.002 30	0.042 03	0.000 99	705	107	776	26	802	13	832	19
DJS-1-35	56.91	58.65	0.97	0.065 23	0.002 73	1.204 79	0.045 13	0.133 98	0.001 97	0.039 03	0.000 78	782	86	803	21	811	11	774	15
DJS-1-36	45.71	57.74	0.79	0.070 31	0.002 30	1.301 52	0.035 31	0.134 27	0.001 71	0.041 63	0.000 70	938	66	846	16	812	10	824	14
DJS-2（Biotite granite）																			
DJS-2-01	48.14	46.67	1.03	0.068 88	0.002 61	1.262 89	0.041 78	0.132 99	0.001 83	0.042 61	0.000 72	895	76	829	19	805	10	843	14
DJS-2-03	50.71	39.39	1.29	0.065 50	0.003 64	1.191 97	0.061 97	0.132 00	0.002 39	0.038 18	0.000 91	791	112	797	29	799	14	757	18
DJS-2-05	32.84	39.62	0.83	0.064 13	0.002 90	1.173 04	0.048 13	0.132 68	0.002 05	0.040 33	0.000 92	746	93	788	22	803	12	799	18
DJS-2-06	70.44	46.63	1.51	0.069 86	0.004 36	1.260 16	0.074 32	0.130 85	0.002 70	0.041 36	0.001 07	924	123	828	33	793	15	819	21
DJS-2-07	45.00	46.15	0.98	0.069 87	0.003 21	1.266 10	0.053 06	0.131 44	0.002 09	0.040 17	0.000 85	925	92	831	24	796	12	796	17
DJS-2-08	39.44	43.42	0.91	0.066 19	0.003 34	1.217 78	0.056 80	0.133 45	0.002 29	0.041 18	0.001 01	813	102	809	26	808	13	816	20
DJS-2-09	23.15	35.97	0.64	0.067 09	0.003 40	1.234 51	0.057 93	0.133 47	0.002 28	0.038 95	0.001 17	841	102	816	26	808	13	772	23
DJS-2-10	125.15	189.56	0.66	0.066 10	0.002 01	1.203 60	0.029 52	0.132 07	0.001 62	0.042 05	0.000 68	810	62	802	14	800	9	833	13
DJS-2-11	145.41	163.62	0.89	0.067 60	0.002 05	1.241 02	0.030 26	0.133 15	0.001 63	0.035 83	0.000 63	857	62	819	14	806	9	712	12
DJS-2-12	64.28	59.69	1.08	0.068 76	0.002 89	1.248 94	0.046 99	0.131 75	0.001 98	0.038 73	0.000 72	892	84	823	21	798	11	768	14
DJS-2-13	68.36	50.44	1.36	0.062 72	0.002 44	1.131 07	0.038 81	0.130 80	0.001 80	0.040 87	0.000 63	699	81	768	18	792	10	810	12
DJS-2-15	59.06	45.40	1.30	0.071 88	0.003 04	1.305 98	0.049 48	0.131 79	0.002 01	0.041 72	0.000 77	983	84	848	22	798	11	826	15
DJS-2-17	101.08	208.13	0.49	0.067 01	0.001 80	1.232 37	0.024 80	0.133 40	0.001 53	0.040 05	0.000 64	838	55	815	11	807	9	794	12
DJS-2-18	33.06	41.15	0.80	0.065 17	0.002 74	1.183 83	0.044 70	0.131 76	0.001 93	0.040 31	0.000 85	780	86	793	21	798	11	799	16
DJS-2-21	34.23	33.46	1.02	0.067 84	0.002 99	1.241 83	0.049 50	0.132 77	0.002 04	0.040 19	0.000 83	864	89	820	22	804	12	796	16
DJS-2-23	83.94	153.13	0.55	0.066 46	0.002 05	1.199 95	0.029 99	0.130 95	0.001 61	0.038 40	0.000 64	821	63	801	14	793	9	762	12

Continued

Analysis	Content/ppm			Ratios								Age/Ma							
	Th	U	Th/U	$^{207}Pb/^{206}Pb$	1σ	$^{207}Pb/^{235}U$	1σ	$^{206}Pb/^{238}U$	1σ	$^{208}Pb/^{232}Th$	1σ	$^{207}Pb/^{206}Pb$	1σ	$^{207}Pb/^{235}U$	1σ	$^{206}Pb/^{238}U$	1σ	$^{208}Pb/^{232}Th$	1σ
DJS-2-24	128.66	176.24	0.73	0.067 14	0.001 82	1.220 55	0.024 78	0.131 85	0.001 52	0.041 87	0.000 55	842	55	810	11	798	9	829	11
DJS-2-25	60.93	47.70	1.28	0.066 66	0.002 62	1.213 16	0.042 02	0.132 01	0.001 87	0.040 38	0.000 68	827	80	807	19	799	11	800	13
DJS-2-27	155.66	272.40	0.57	0.065 10	0.001 82	1.179 47	0.025 39	0.131 40	0.001 53	0.039 92	0.000 60	778	58	791	12	796	9	791	12
DJS-2-28	71.87	55.93	1.28	0.064 62	0.003 30	1.171 73	0.055 37	0.131 52	0.002 24	0.039 18	0.000 85	762	104	788	26	797	13	777	16
DJS-2-29	7.97	8.83	0.90	0.069 69	0.007 48	1.272 55	0.132 97	0.132 45	0.004 11	0.042 18	0.002 16	919	206	834	59	802	23	835	42
DJS-2-32	128.92	138.64	0.93	0.066 85	0.001 87	1.222 91	0.026 26	0.132 69	0.001 55	0.042 27	0.000 55	833	57	811	12	803	9	837	11
DJS-2-34	106.39	152.99	0.70	0.065 84	0.001 76	1.203 63	0.024 07	0.132 60	0.001 51	0.040 85	0.000 55	801	55	802	11	803	9	809	11
DJS-2-36	60.58	61.66	0.98	0.068 71	0.002 47	1.251 91	0.038 81	0.132 14	0.001 78	0.040 33	0.000 69	890	73	824	18	800	10	799	13
PZH-09（K-feldspar granite）																			
PZH-09-01	307.11	556.92	0.55	0.064 23	0.001 40	1.090 03	0.015 39	0.123 11	0.001 38	0.039 36	0.000 42	749	45	749	7	748	8	780	8
PZH-09-02	402.53	782.49	0.51	0.064 88	0.001 40	1.114 68	0.015 57	0.124 63	0.001 39	0.039 87	0.000 43	770	45	760	7	757	8	790	8
PZH-09-04	415.11	762.27	0.54	0.065 19	0.001 43	1.099 24	0.015 88	0.122 30	0.001 37	0.040 20	0.000 44	780	46	753	8	744	8	797	9
PZH-09-05	263.15	670.20	0.39	0.064 90	0.001 43	1.108 55	0.015 96	0.123 87	0.001 38	0.041 96	0.000 48	771	46	758	8	753	8	831	9
PZH-09-07	342.42	670.69	0.51	0.066 34	0.001 46	1.123 27	0.015 98	0.122 77	0.001 36	0.042 58	0.000 46	817	45	765	8	747	8	843	9
PZH-09-08	357.68	726.04	0.49	0.067 74	0.001 52	1.149 45	0.017 16	0.123 04	0.001 37	0.044 29	0.000 50	861	46	777	8	748	8	876	10
PZH-09-09	551.34	956.87	0.58	0.065 45	0.001 50	1.111 94	0.017 33	0.123 18	0.001 38	0.040 37	0.000 46	789	47	759	8	749	8	800	9
PZH-09-10	348.90	634.55	0.55	0.066 14	0.001 44	1.130 10	0.015 62	0.123 88	0.001 36	0.041 31	0.000 43	811	45	768	7	753	8	818	8
PZH-09-12	329.85	605.94	0.54	0.065 57	0.001 46	1.119 13	0.016 33	0.123 74	0.001 37	0.040 84	0.000 44	793	46	763	8	752	8	809	9
PZH-09-15	341.68	726.89	0.47	0.066 91	0.001 80	1.143 49	0.023 51	0.123 89	0.001 47	0.038 69	0.000 58	835	55	774	11	753	8	767	11
PZH-09-16	127.56	146.99	0.87	0.064 87	0.002 44	1.103 52	0.036 75	0.123 39	0.001 78	0.037 36	0.000 84	770	77	755	18	750	10	742	16
PZH-09-18	342.07	616.39	0.55	0.066 51	0.001 44	1.133 72	0.015 07	0.123 57	0.001 33	0.041 08	0.000 42	822	45	770	7	751	7	814	8
PZH-09-24	318.95	669.54	0.48	0.069 05	0.001 55	1.164 74	0.016 75	0.122 28	0.001 33	0.041 37	0.000 44	900	46	784	8	744	8	819	9
PZH-09-26	403.63	652.61	0.62	0.069 20	0.001 76	1.167 60	0.021 16	0.122 42	0.001 36	0.040 82	0.000 51	905	52	786	10	745	8	809	10
PZH-09-35	253.32	505.97	0.50	0.068 37	0.001 59	1.162 06	0.017 29	0.123 33	0.001 31	0.041 82	0.000 47	880	47	783	8	750	8	828	9
PZH-09-36	468.55	793.34	0.59	0.063 52	0.002 39	1.069 45	0.035 50	0.122 10	0.001 75	0.037 76	0.000 85	726	78	739	17	743	10	749	17

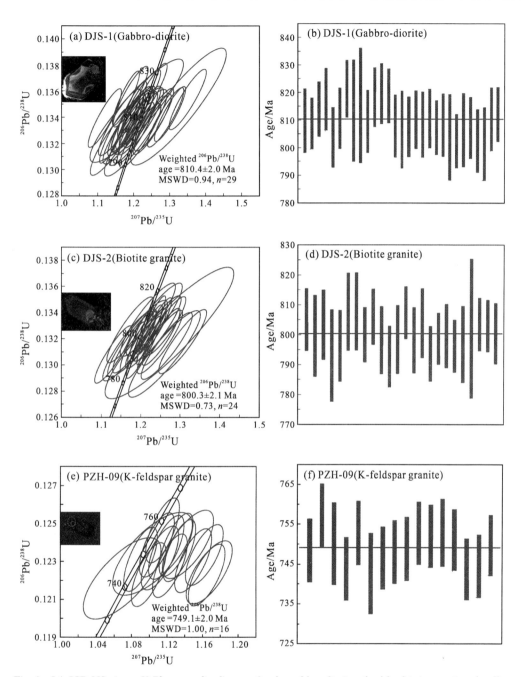

Fig. 4 LA-ICP-MS zircon U-Pb concordia diagram for the gabbro-diorites (a,b), biotite granites (c,d), and K-feldspar granites (e,f) from the southwestern margin of the Yangtze Block, South China.

gabbro-diorite samples have Th (10.1−96.1 ppm) and U (9.34−128 ppm) contents with Th/U ratios of 0.55−1.29, indicating the magmatic origin. These reliable twenty-nine analyses spots yield a weighted mean age of 810.4±2.0 Ma (MSWD=0.94, n=29) (Fig. 4a,b).

4.1.2 Biotite granite (DJS-2)

Zircon grains in biotite granite sample DJS-2 are mainly idiomorphic, and short or long prismatic crystals, measuring up to 50–200 μm with length to width ratios about 1:1 to 4:1. Most of the zircon grains exhibit well-defined oscillatory zoning in the CL images. Twenty-four concordant analytical spots have variable Th (7.97–156 ppm) and U (8.83–272 ppm) contents with Th/U ratios of 0.49–1.51. They yield the $^{206}Pb/^{238}U$ ages from 792 ± 10 Ma to 808 ± 13 Ma with a weighted mean age of 800.3 ± 2.1 Ma (MSWD = 0.73, $n = 24$) (Fig. 4c,d).

4.1.3 K-feldspar granite (PZH-09)

Zircons from K-feldspar granite sample (PZH-09) in the Panzhihua city are dominantly colorless, transparent and subhedral prismatic crystals, with aspect ratios varying from 1:1 to 2:1. In CL images, these zircon grains range from 100 μm to 350 μm in length. The 16 analytical spots have U concentrations from 147 ppm to 957 ppm, Th concentrations from 128 ppm to 551 ppm, and Th/U ratios from 0.39 to 0.87. These concordant 16 analytical spots are characterized by $^{206}Pb/^{238}U$ ages ranging from 743 ± 10 Ma to 757 ± 8 Ma with a weighted mean age of 749.1 ± 2.0 Ma (MSWD = 1.00, $n = 16$), which can represent the crystallization age of the K-feldspar granites (Fig. 4e,f).

4.2 Major and trace elemental geochemistry

Whole-rock major and trace geochemical data for the representative samples of studied gabbro-diorites, biotite granites, and K-feldspar granites are shown in the Table 3, and Figs. 5 and 6.

4.2.1 Gabbro-diorites

The gabbro-diorite samples are characterized by low SiO_2 (52.62–53.87 wt%), moderate MgO (2.67–3.41 wt%) contents with $Mg^{\#}$ values of 46–52, high Al_2O_3 (19.98–20.63 wt%), $Fe_2O_3^T$ (7.18–7.49 wt%), and CaO (5.68–7.50 wt%) contents. They have low loss on ignition (LOI) values with a range of 2.39–2.80 wt%, which suggest minimal weathering secondary processes and hydrothermal alteration (Karsli et al., 2007). These gabbro-diorite samples are plotted into the field of medium-K, calc-alkaline series in the SiO_2 vs. K_2O diagram (Fig. 5b), and are metaluminous to weakly peraluminous rocks with A/CNK [molar $Al_2O_3/(CaO + K_2O + Na_2O)$] ratios of 0.89–1.02 (Fig. 5c). Moreover, they show uniform chondrite-normalized rare earth element (REE) patterns [$(La/Yb)_N = 2.61–7.87$] with slightly negative Eu anomalies ($Eu/Eu^* = 0.84–0.96$) (Fig. 6a). In the primitive-mantle-normalized diagram (Fig. 6b), these gabbro-diorite samples are characterized by the moderately enrichment in light rare earth elements (LREEs) and the depletion in Nb, Ta, P, and Ti. They also contain medium amounts of Cr (20.7–26.4 ppm), Co (45.9–60.1 ppm), and Ni (8.86–11.1 ppm).

Table 3 Major(wt%) and trace(ppm) element analysis results for the Neoproterozoic gabbro-diorites, biotite granites, and K-feldspar granites from the southwestern Yangtze Block, South China.

Sample	DJS-1-1	DJS-1-2	DJS-1-3	DJS-1-4	DJS-1-5	DJS-1-6	DJS-1-7	DJS-2-1	DJS-2-2	DJS-2-3	DJS-2-4	DJS-2-5	DJS-2-6	DJS-2-7	PZH-01	PZH-02	PZH-03	PZH-04	PZH-05	PZH-06
Lithology	Gabbro-diorite N: 26°43′9″ E: 101°51′1″							Biotite granite N: 26°45′10″ E: 101°50′1″							K-feldspar granite N: 26°34′2″ E: 101°43′44″					
SiO_2	52.62	53.25	52.72	53.87	53.55	52.66	52.74	74.62	74.73	74.08	74.60	74.82	74.32	74.46	77.10	76.80	77.02	77.14	76.67	76.61
TiO_2	0.53	0.54	0.51	0.53	0.56	0.53	0.54	0.10	0.11	0.09	0.10	0.11	0.10	0.13	0.05	0.04	0.04	0.05	0.04	0.04
Al_2O_3	20.15	20.19	20.10	20.14	19.98	20.63	20.21	13.85	14.32	14.70	14.52	13.97	14.25	14.11	11.71	11.72	11.79	11.91	11.43	11.86
$Fe_2O_3^{T}$	7.49	7.45	7.45	7.18	7.26	7.46	7.45	1.04	1.10	0.86	1.12	1.08	0.98	1.16	1.76	1.45	1.20	1.22	1.19	1.16
MnO	0.13	0.15	0.14	0.13	0.13	0.13	0.14	0.04	0.05	0.03	0.04	0.04	0.04	0.05	0.02	0.01	0.01	0.01	0.01	0.01
MgO	3.04	3.09	3.41	3.38	2.67	3.40	3.17	0.28	0.29	0.25	0.27	0.29	0.29	0.30	0.05	0.06	0.06	0.02	0.10	0.06
CaO	7.50	6.86	6.32	5.68	7.36	6.18	6.89	1.49	1.66	1.79	1.41	1.48	1.50	1.53	0.25	0.34	0.34	0.37	0.27	0.25
Na_2O	4.85	4.84	4.99	4.80	4.83	4.83	4.85	4.76	5.16	5.06	5.50	5.60	5.07	4.92	3.23	4.15	3.86	3.13	4.34	3.75
K_2O	0.84	1.04	1.21	1.47	0.92	1.34	1.06	2.39	1.82	2.31	1.13	1.04	2.36	2.19	5.32	4.83	4.82	5.68	5.35	5.60
P_2O_5	0.21	0.21	0.23	0.20	0.21	0.22	0.21	0.04	0.09	0.03	0.04	0.04	0.04	0.05	0.01	0.01	0.01	0.01	0.01	0.01
LOI	2.47	2.51	2.69	2.61	2.39	2.80	2.54	1.06	0.84	1.06	0.97	1.15	0.85	0.84	0.33	0.45	0.39	0.29	0.20	0.36
Total	99.83	100.13	99.77	99.99	99.86	100.18	99.80	99.67	100.17	100.26	99.70	99.62	99.80	99.74	99.82	99.85	99.53	99.82	99.60	99.70
$Mg^{\#}$	49	49	52	52	46	52	48	39	38	40	36	38	41	38	6	9	10	4	16	11
A/CNK	0.89	0.94	0.96	1.02	0.90	1.00	0.93	1.06	1.06	1.04	1.13	1.07	1.05	1.07	1.02	0.92	0.97	1.00	0.85	0.93
Differentiation index	–	–	–	–	–	–	–	89.2	88.29	88.24	88.65	89.06	89.36	88.67	95.74	96.21	96.92	96.32	95.3	96.85
Li	1.63	4.36	5.69	5.55	4.84	6.52	5.51	4.85	4.64	3.91	4.90	4.83	4.52	5.46	0.29	0.32	0.32	0.20	0.99	0.23
Be	1.38	1.54	1.69	1.99	1.34	1.75	1.71	1.65	1.83	1.74	1.57	1.60	1.71	1.82	1.54	2.53	2.58	1.70	1.40	1.51
Sc	15.80	19.04	20.98	19.63	16.23	15.55	18.99	2.14	2.03	1.64	1.94	2.06	2.04	2.30	0.94	1.05	1.03	0.89	0.76	0.84
V	124.10	127.55	125.04	123.29	133.17	114.92	122.81	5.30	5.12	4.25	5.24	5.26	4.96	5.91	0.84	3.28	2.35	0.62	0.81	0.69
Cr	22.69	24.44	23.20	26.38	23.11	20.72	24.73	3.52	3.36	2.94	3.00	3.27	3.16	3.59	7.13	3.87	3.61	3.79	4.60	3.25

Continued

Sample	DJS-1-1	DJS-1-2	DJS-1-3	DJS-1-4	DJS-1-5	DJS-1-6	DJS-1-7	DJS-2-1	DJS-2-2	DJS-2-3	DJS-2-4	DJS-2-5	DJS-2-6	DJS-2-7	PZH-01	PZH-02	PZH-03	PZH-04	PZH-05	PZH-06
Lithology	Gabbro-diorite N: 26°43'9" E: 101°51'1"							Biotite granite N: 26°45'10" E: 101°50'1"							K-feldspar granite N: 26°34'2" E: 101°43'44"					
Co	53.37	51.40	45.86	47.88	60.57	46.90	50.44	230.57	228.18	214.42	222.72	243.29	251.16	251.26	157.48	253.80	157.96	187.53	188.99	218.82
Ni	11.11	9.78	9.58	8.86	11.11	11.14	9.69	1.39	1.55	1.36	1.32	1.46	1.35	1.53	2.98	1.59	1.35	1.88	2.35	1.54
Cu	18.36	12.78	13.58	13.95	34.41	21.64	17.33	2.73	2.07	1.83	2.43	2.68	2.25	3.36	10.90	22.69	17.79	13.07	8.96	8.36
Zn	74.91	77.87	83.12	84.28	67.49	81.60	80.99	28.23	25.34	21.65	24.30	25.75	25.88	31.50	108.18	127.25	107.85	193.25	101.81	96.67
Ga	21.03	21.61	21.41	21.41	21.96	20.60	21.61	12.86	13.83	13.33	13.57	13.24	13.38	13.95	16.70	17.00	17.46	16.58	16.28	16.09
Ge	1.53	1.50	1.41	1.38	1.51	1.29	1.41	0.94	0.83	0.80	0.89	0.98	1.08	1.14	1.63	1.77	1.77	1.45	1.41	1.63
Rb	13.84	18.21	21.35	26.39	13.53	18.18	20.92	29.89	26.37	28.62	20.49	18.95	29.95	30.21	198.52	158.26	166.43	175.31	181.13	199.70
Sr	796.43	724.21	686.69	631.22	787.69	672.47	693.78	346.93	395.26	382.34	341.29	334.94	371.49	377.76	13.31	14.22	11.36	7.89	9.61	10.72
Y	24.13	27.98	29.47	27.92	22.03	22.47	25.91	8.26	8.62	7.04	7.13	7.90	7.97	9.71	37.28	48.24	49.95	50.29	38.57	39.15
Zr	145.24	191.60	173.87	187.46	164.53	156.52	158.49	98.95	98.48	87.95	89.96	93.87	90.28	104.77	172.08	171.65	161.26	181.24	170.03	154.19
Nb	3.99	5.07	4.52	4.96	4.16	3.85	4.73	4.69	5.61	3.84	4.36	4.52	4.32	5.79	6.50	6.25	12.15	6.71	5.61	5.72
Cs	0.10	0.11	0.14	0.17	0.11	0.15	0.14	0.27	0.28	0.24	0.22	0.22	0.24	0.28	0.59	0.39	0.40	0.34	0.41	0.54
Ba	445.21	510.97	698.43	448.00	677.06	613.55	1 727.35	1 271.04	1 505.19	572.34	456.20	1 586.17	1 553.78		47.61	41.23	35.54	32.29	34.13	38.88
La	18.58	9.90	15.26	25.21	24.21	11.75	21.80	14.02	17.52	14.49	16.19	15.90	17.57	16.24	25.51	28.46	26.10	39.91	20.93	24.40
Ce	42.08	35.04	54.53	51.20	29.04	47.85	26.01	32.64	26.43	30.00	29.57	32.63	30.71		64.61	78.24	62.28	98.28	55.43	63.21
Pr	5.38	3.59	6.70	6.10	3.91	5.98	2.71	3.39	2.74	3.12	3.11	3.41	3.21		7.29	8.37	7.82	11.19	6.18	7.07
Nd	22.84	17.13	27.50	24.29	17.69	24.93	9.47	11.83	9.54	10.92	10.82	11.93	11.17		28.60	33.07	31.67	41.79	25.02	27.67
Sm	4.81	4.41	5.64	4.59	4.14	5.16	1.59	1.94	1.55	1.76	1.78	1.95	1.87		6.72	7.85	7.95	9.31	6.16	6.65
Eu	1.35	1.30	1.50	1.35	1.28	1.43	0.66	0.71	0.68	0.62	0.58	0.71	0.69		0.09	0.11	0.09	0.09	0.08	0.08
Gd	4.54	4.47	5.31	4.27	4.02	4.88	1.45	1.69	1.38	1.49	1.53	1.65	1.68		6.31	7.53	7.70	8.46	6.06	6.35
Tb	0.68	0.72	0.80	0.62	0.62	0.73	0.21	0.25	0.19	0.21	0.22	0.23	0.25		1.11	1.33	1.37	1.45	1.08	1.13

Continued

Sample	DJS-1-1	DJS-1-2	DJS-1-3	DJS-1-4	DJS-1-5	DJS-1-6	DJS-1-7	DJS-2-1	DJS-2-2	DJS-2-3	DJS-2-4	DJS-2-5	DJS-2-6	DJS-2-7	PZH-01	PZH-02	PZH-03	PZH-04	PZH-05	PZH-06
Lithology	Gabbro-diorite N: 26°43'9" E: 101°51'1"							Biotite granite N: 26°45'10" E: 101°50'1"							K-feldspar granite N: 26°34'2" E: 101°43'44"					
Dy	4.16	4.46	5.01	4.88	3.73	3.78	4.48	1.32	1.47	1.17	1.21	1.32	1.34	1.56	7.14	8.68	8.99	9.06	7.03	7.30
Ho	0.85	0.93	1.03	0.99	0.77	0.79	0.92	0.28	0.30	0.24	0.25	0.27	0.27	0.33	1.52	1.84	1.91	1.87	1.51	1.57
Er	2.47	2.72	3.00	2.88	2.23	2.30	2.68	0.83	0.88	0.72	0.73	0.81	0.81	0.99	4.54	5.61	5.76	5.52	4.55	4.67
Tm	0.37	0.41	0.44	0.42	0.33	0.34	0.40	0.13	0.14	0.11	0.11	0.12	0.13	0.16	0.70	0.87	0.89	0.84	0.69	0.70
Yb	2.41	2.72	2.95	2.77	2.21	2.29	2.60	0.92	0.95	0.78	0.81	0.88	0.87	1.08	4.62	5.86	5.91	5.46	4.57	4.62
Lu	0.37	0.42	0.45	0.42	0.34	0.36	0.40	0.14	0.15	0.12	0.13	0.14	0.14	0.17	0.68	0.86	0.86	0.77	0.66	0.65
Hf	3.48	4.49	4.17	4.46	3.82	3.74	3.76	2.52	2.47	2.21	2.29	2.35	2.29	2.65	7.51	7.86	7.22	8.02	7.79	6.96
Ta	0.18	0.25	0.20	0.28	0.19	0.17	0.25	0.60	0.71	0.54	0.56	0.59	0.58	0.71	0.34	0.25	0.51	0.38	0.27	0.29
Pb	14.49	13.59	12.42	10.60	14.61	11.65	12.44	25.49	12.98	14.79	8.75	11.64	15.92	15.89	19.59	37.04	30.09	26.86	18.20	18.74
Th	3.04	1.68	2.40	4.76	4.02	2.02	3.87	1.91	2.30	1.64	2.07	2.09	2.24	2.15	21.03	26.57	26.84	25.00	23.12	24.84
U	0.41	0.37	0.41	0.50	0.44	0.35	0.43	0.32	0.40	0.30	0.30	0.30	0.36	0.32	2.52	3.33	3.34	2.23	2.71	2.52
REE	111	78.5	102	140	126	82.3	124	59.8	73.9	60.1	67.6	67.1	73.6	70.1	159	189	169	234	140	156
$(La/Yb)_N$	5.54	2.61	3.71	6.52	7.87	3.67	6.01	10.93	13.28	13.40	14.32	13.00	14.52	10.78	3.96	3.48	3.17	5.24	3.28	3.79
Eu/Eu^*	0.89	0.90	0.85	0.84	0.93	0.96	0.87	1.33	1.20	1.41	1.17	1.07	1.21	1.19	0.04	0.04	0.04	0.03	0.04	0.04
Nb/Ta	22.7	20.5	22.2	18.0	21.4	23.2	18.9	7.83	7.88	7.11	7.75	7.61	7.41	8.17	19.2	25.3	23.7	17.9	20.8	19.6
Sr/Y	33.0	25.9	23.3	22.6	35.8	29.9	26.8	42.0	45.9	54.3	47.9	42.4	46.6	38.9	0.36	0.29	0.23	0.16	0.25	0.27
Ta/La	0.01	0.03	0.01	0.01	0.01	0.01	0.01	0.04	0.04	0.04	0.03	0.04	0.03	0.04	0.01	0.01	0.02	0.01	0.01	0.01
Nb/La	0.21	0.51	0.30	0.20	0.17	0.33	0.22	0.33	0.32	0.27	0.27	0.28	0.25	0.36	0.25	0.22	0.47	0.17	0.27	0.23
Nb/U	9.66	13.6	11.1	9.99	9.54	10.9	10.9	14.5	14.2	13.0	14.4	15.1	11.9	18.1	2.57	1.88	3.64	3.01	2.07	2.27

$Mg^\# =$ molar $100MgO/(MgO+FeO^T)$; Differentiation index (DI) $= Qz+Or+Ab+Ne+Lc+Kp$; $A/CNK = Al_2O_3/(CaO+K_2O+Na_2O)$ molar ratio; subscript $N =$ chondrite-normalized value; $Eu/Eu^* = Eu_N/(Sm_N \cdot Gd_N)^{1/2}$.

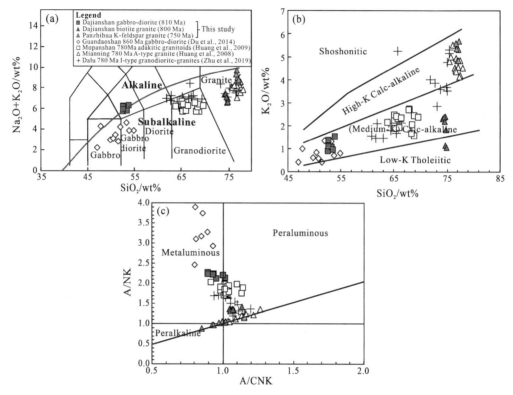

Fig. 5   Plots of total alkali vs. SiO$_2$(TAS)(a)(Middlemost, 1994), SiO$_2$ vs. K$_2$O(b)(Roberts and Clemens, 1993), and A/CNK vs. A/NK (c)(Frost et al., 2001) for the gabbro-diorites, biotite granites, and K-feldspar granites from the southwestern margin of the Yangtze Block, South China.

4.2.2 Biotite granites

The biotite granites have high contents of SiO$_2$ (74.08 – 74.82 wt%), Al$_2$O$_3$ (13.85 – 14.70 wt%), and Na$_2$O (4.76 – 5.60 wt%), low contents of K$_2$O (1.04 – 2.39 wt%) and MgO (0.25 – 0.30 wt%) with Mg$^{\#}$ values of 36−41, belonging to the peraluminous (A/CNK = 1.04 – 1.13) and medium-K calc-alkaline to low-K series (Fig. 5b, c). On the chondrite-normalized REE patterns (Fig. 6c), they contain low total REE (59.8 – 73.8 ppm) contents and are enriched in LREEs [(La/Yb)$_N$ = 10.8 – 14.5] with negligible Eu anomalies (Eu/Eu* = 1.07 – 1.41). On the primitive-mantle-normalized diagram (Fig. 6d), our biotite granite samples exhibit the remarkably positive Ba, K, Pb, and Sr anomalies, and negative Nb, Ta, P, and Ti anomalies. Moreover, they have relatively high Sr (335 – 395 ppm), and obviously low Y (7.04 – 9.71 ppm) and Yb (0.78 – 1.08 ppm) contents, with high Sr/Y ratios (38.9 – 54.3). They also have low concentrations of Cr = 2.94 – 3.59 ppm and Ni = 1.32 – 1.55 ppm.

4.2.3 K-feldspar granites

The K-feldspar granites in this study contain high SiO$_2$ (76.61 – 77.14 wt%), Na$_2$O (3.13 – 4.34 wt%), and K$_2$O (4.82 – 5.68 wt%) contents. They have obviously low Al$_2$O$_3$

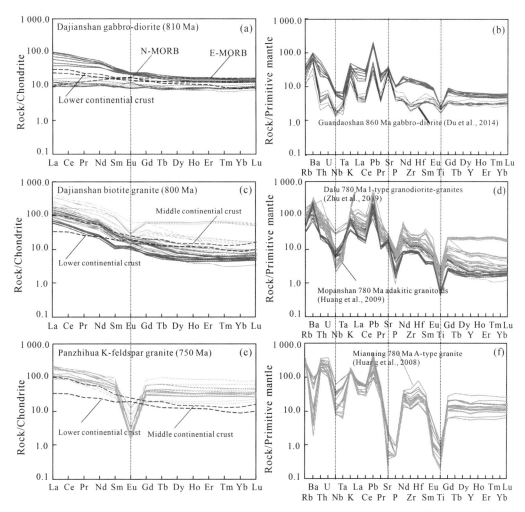

Fig. 6 Diagrams of chondrite-normalized REE patterns (a,c,e) and primitive mantle-normalized
trace element patterns (b,d,f) for the gabbro-diorites, biotite granites, and K-feldspar granites
from the southwestern margin of the Yangtze Block, South China.

The primitive mantle and chondrite values are from Sun and McDonough (1989). The N-MORB,
E-MORB, middle crust, and lower crust reference from Rudnick and Gao (2003).

(11. 43−11. 86 wt%), CaO (0. 25−0. 37 wt%), and MgO (0. 02−0. 10 wt%) contents. They contain high total REE contents (140−189 ppm) and show the relatively enrichment in LREEs $[(La/Yb)_N = 3.28−5.24]$, flat heavy rare earth elements (HREEs), and pronouncedly negative Eu anomalies with Eu/Eu* values of 0. 03−0. 04 (Fig. 6e). In the primitive-mantle-normalized diagram (Fig. 6f), these K-feldspar granites are characterized by the significant enrichment of Rb, Th, U, K, and Pb, and the depletion of Ba, Nb, Ta, Sr, Zr, and Ti.

4. 3 Whole-rock Sr-Nd isotopic data

Whole-rock Sr-Nd isotopic data for the studied rocks are listed in the Table 4. All the

$(^{87}Sr/^{86}Sr)_i$ (I_{Sr}) isotopic ratios and $\varepsilon_{Nd}(t)$ values are calculated at the time of magma crystallization.

Table 4 **Whole-rock Sr-Nd isotopes data for the Neoproterozoic gabbro-diorites, biotite granites, and K-feldspar granites from the southwestern Yangtze Block, South China.**

Sample	$^{87}Sr/^{86}Sr$	2SE	Sr /ppm	Rb /ppm	$^{143}Nd/^{144}Nd$	2SE	Nd /ppm	Sm /ppm	T_{DM2} /Ga	$\varepsilon_{Nd}(t)$	$(^{87}Sr/^{86}Sr)_i$ (I_{Sr})
Gabbro-diorite											
DJS-1-1	0.705 973	0.000 003	796	13.8	0.512 345	0.000 004	22.8	4.81	1.28	+1.5	0.705 392
DJS-1-2	0.706 025	0.000 005	724	18.2	0.512 472	0.000 003	17.1	4.41	1.31	+1.0	0.705 184
Biotite granite											
DJS-2-2	0.707 529	0.000 006	395	26.4	0.512 157	0.000 004	11.8	1.94	1.33	+0.6	0.705 324
DJS-2-3	0.707 736	0.000 004	382	28.6	0.512 145	0.000 003	9.54	1.55	1.34	+0.5	0.705 262
K-feldspar granite											
PZH-01	–	–	–	–	0.512 308	0.000 002	28.6	6.72	1.41	−1.2	–
PZH-02	–	–	–	–	0.512 296	0.000 002	33.1	7.85	1.44	−1.6	–

$^{87}Rb/^{86}Sr$ and $^{147}Sm/^{144}Nd$ ratios were calculated using Rb, Sr, Sm and Nd contents analyzed by ICP-MS. T_{DM2} represent the two-stage model age and were calculated using present-day $(^{147}Sm/^{144}Nd)_{DM} = 0.213\,7$, $(^{147}Sm/^{144}Nd)_{DM} = 0.513\,15$ and $(^{147}Sm/^{144}Nd)_{crust} = 0.101\,2$.

$\varepsilon_{Nd}(t)$ values were calculated using present-day $(^{147}Sm/^{144}Nd)_{CHUR} = 0.196\,7$ and $(^{147}Sm/^{144}Nd)_{CHUR} = 0.512\,638$. $\varepsilon_{Nd}(t) = [(^{143}Nd/^{144}Nd)_S(t)/(^{143}Nd/^{144}Nd)_{CHUR}(t) - 1] \times 10^4$.

$$T_{DM2} = \frac{1}{\lambda}\left\{1 + \left[(^{143}Nd/^{144}Nd)_S - ((^{147}Sm/^{144}Nd)_S - (^{147}Sm/^{144}Nd)_{crust})(e^{\lambda t} - 1) - (^{143}Nd/^{144}Nd)_{DM}\right] / \left[(^{147}Sm/^{144}Nd)_{crust} - (^{147}Sm/^{144}Nd)_{DM}\right]\right\}.$$

4.3.1 Gabbro-diorites

The gabbro-diorites (DJS-1-1 and DJS-1-2) contain the $(^{87}Sr/^{86}Sr)_i$ values ranging from 0.705 184 to 0.705 392. They show the depleted Nd isotopes compositions with the $\varepsilon_{Nd}(t)$ values ranging from +1.0 to +1.5.

4.3.2 Biotite granites

The biotite granites (DJS-2-2 and DJS-2-3) display the similar $(^{87}Sr/^{86}Sr)_i$ values from 0.705 262 to 0.705 324. They yield the positive $\varepsilon_{Nd}(t)$ values (+0.5 to +0.6) and old two-stage model ages (T_{DM2}) (1.33−1.34 Ga).

4.3.3 K-feldspar granites

According to the extremely low Sr contents (7.89−14.2 ppm) of K-feldspar granites, we just obtained the Nd isotopic data. The studied K-feldspar granites (PZH-01 and PZH-02) have negative $\varepsilon_{Nd}(t)$ values from −1.6 to −1.2 and old T_{DM2} values from 1.41 Ga to 1.44 Ga.

4.4 Zircon Lu-Hf isotopic compositions

The zircons from three studied igneous samples (DJS-1, DJS-2, and PZH-09) were analyzed for Lu-Hf isotopes on the same sites as those used for U-Pb dating, and the results are shown in Table 5, and Figs. 7 and 8. The initial ^{176}Hf/^{177}Hf ratios were calculated according to their corresponding LA-ICP-MS zircon U-Pb ages.

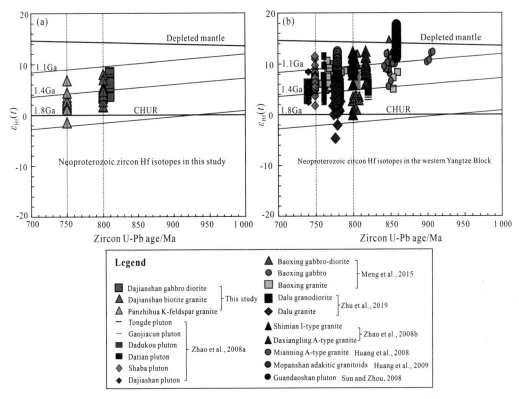

Fig. 7　Zircon U-Pb age vs. $\varepsilon_{Hf}(t)$ for the gabbro-diorites, biotite granites, and K-feldspar granites from the southwestern margin of the Yangtze Block, South China.

4.4.1　Gabbro-diorite (DJS-1)

The Lu-Hf isotopic compositions of the 29 zircons from gabbro-diorite sample (DJS-1) yield the initial^{176}Hf/^{177}Hf ratios 0.282 389−0.282 509 with an average value of 0.282 455. These analytical spots yield positive $\varepsilon_{Hf}(t)$ values of +3.66 to +8.18 and the single-stage model age from 1 045 Ma to 1 229 Ma, respectively (Figs. 7a and 8a,b).

4.4.2　Biotite granite (DJS-2)

A total of 24 spots from biotite granite sample (DJS-2) share the initial ^{176}Hf/^{177}Hf ratios from 0.282 346 to 0.282 518 with a mean of 0.282 418. The zircon $\varepsilon_{Hf}(t)$ values are positive ranging from +1.62 to +8.07 with single-stage Hf model age of 1 043−1 309 Ma (Fig. 7a). Their crustal Hf model age varies from 1 186 Ma to 1 592 Ma (Fig. 8c,d).

4.4.3　K-feldspar granite (PZH-09)

For the K-feldspar granite sample (PZH-09), the initial^{176}Hf/^{177}Hf ratios of 15 reliable analytical spots span a range from 0.282 291 to 0.282 527. The two $\varepsilon_{Hf}(t)$ values are negative

Table 5 Zircon Lu-Hf isotopic data for the Neoproterozoic gabbro-diorites, biotite granites, and K-feldspar granites from the southwestern Yangtze Block, South China.

Grain spot	Age /Ma	$^{176}Yb/^{177}Hf$	2SE	$^{176}Lu/^{177}Hf$	2SE	$^{176}Hf/^{177}Hf$	2SE	$(^{176}Hf/^{177}Hf)_i$	$f_{Lu/Hf}$	$\varepsilon_{Hf}(t)$	2SE	T_{DM1} /Ma	T_{DMC} /Ma
Gabbro diorite (DJS-1)													
DJS1-1	810	0.021 981	0.000 342	0.000 759	0.000 011	0.282 521	0.000 019	0.282 509	−0.98	8.18	0.67	1 045	1 187
DJS1-2	810	0.013 462	0.000 046	0.000 486	0.000 001	0.282 505	0.000 015	0.282 497	−0.99	7.90	0.54	1 054	1 204
DJS1-3	810	0.016 782	0.000 331	0.000 611	0.000 009	0.282 468	0.000 017	0.282 459	−0.98	6.48	0.61	1 111	1 294
DJS1-4	810	0.022 973	0.000 146	0.000 778	0.000 005	0.282 468	0.000 017	0.282 456	−0.98	6.29	0.60	1 120	1 306
DJS1-5	810	0.014 931	0.000 232	0.000 528	0.000 008	0.282 503	0.000 016	0.282 495	−0.98	7.80	0.57	1 059	1 211
DJS1-6	810	0.014 392	0.000 061	0.000 501	0.000 002	0.282 446	0.000 018	0.282 438	−0.98	5.81	0.63	1 136	1 336
DJS1-7	810	0.018 272	0.000 252	0.000 656	0.000 009	0.282 470	0.000 017	0.282 460	−0.98	6.49	0.59	1 111	1 293
DJS1-8	810	0.031 225	0.000 147	0.001 143	0.000 005	0.282 442	0.000 019	0.282 424	−0.97	4.97	0.69	1 175	1 389
DJS1-9	810	0.013 498	0.000 096	0.000 462	0.000 004	0.282 465	0.000 015	0.282 457	−0.99	6.51	0.51	1 109	1 292
DJS1-10	810	0.035 754	0.000 405	0.001 187	0.000 014	0.282 484	0.000 016	0.282 466	−0.96	6.42	0.56	1 118	1 298
DJS1-11	810	0.034 595	0.000 358	0.001 165	0.000 014	0.282 455	0.000 017	0.282 437	−0.96	5.40	0.61	1 158	1 362
DJS1-12	810	0.018 387	0.000 090	0.000 625	0.000 003	0.282 461	0.000 016	0.282 451	−0.98	6.20	0.56	1 122	1 311
DJS1-13	810	0.015 110	0.000 122	0.000 536	0.000 004	0.282 462	0.000 015	0.282 454	−0.98	6.34	0.53	1 116	1 303
DJS1-14	810	0.050 802	0.000 205	0.001 661	0.000 007	0.282 476	0.000 020	0.282 451	−0.95	5.63	0.71	1 154	1 348
DJS1-15	810	0.008 178	0.000 119	0.000 318	0.000 004	0.282 477	0.000 014	0.282 472	−0.99	7.12	0.50	1 084	1 254
DJS1-16	810	0.039 810	0.000 035	0.001 344	0.000 002	0.282 435	0.000 019	0.282 415	−0.96	4.52	0.68	1 195	1 417
DJS1-17	810	0.021 407	0.000 352	0.000 738	0.000 011	0.282 419	0.000 018	0.282 407	−0.98	4.59	0.62	1 186	1 413
DJS1-18	810	0.029 351	0.000 192	0.000 992	0.000 007	0.282 478	0.000 016	0.282 463	−0.97	6.43	0.57	1 116	1 297
DJS1-19	810	0.037 605	0.000 628	0.001 252	0.000 022	0.282 450	0.000 023	0.282 431	−0.96	5.15	0.82	1 169	1 378
DJS1-20	810	0.015 139	0.000 013	0.000 532	0.000 001	0.282 470	0.000 019	0.282 462	−0.98	6.62	0.68	1 105	1 285
DJS1-21	810	0.019 739	0.000 046	0.000 681	0.000 002	0.282 464	0.000 017	0.282 454	−0.98	6.26	0.60	1 120	1 308
DJS1-22	810	0.034 230	0.000 277	0.001 257	0.000 010	0.282 408	0.000 019	0.282 389	−0.96	3.66	0.67	1 229	1 472
DJS1-23	810	0.013 032	0.000 029	0.000 528	0.000 001	0.282 478	0.000 017	0.282 470	−0.98	6.90	0.60	1 094	1 267

Continued

Grain spot	Age /Ma	$^{176}Yb/^{177}Hf$	2SE	$^{176}Lu/^{177}Hf$	2SE	$^{176}Hf/^{177}Hf$	2SE	$(^{176}Hf/^{177}Hf)_i$	$f_{Lu/Hf}$	$\varepsilon_{Hf}(t)$	2SE	T_{DM1} /Ma	T_{DMC} /Ma
DJS1-24	810	0.020 869	0.000 254	0.000 734	0.000 009	0.282 514	0.000 019	0.282 503	-0.98	7.98	0.68	1 053	1 199
DJS1-25	810	0.028 253	0.000 145	0.000 969	0.000 005	0.282 490	0.000 019	0.282 476	-0.97	6.88	0.68	1 098	1 269
DJS1-26	810	0.031 720	0.000 367	0.001 051	0.000 013	0.282 479	0.000 018	0.282 463	-0.97	6.40	0.63	1 118	1 299
DJS1-27	810	0.027 142	0.000 237	0.000 936	0.000 007	0.282 444	0.000 018	0.282 429	-0.97	5.26	0.63	1 162	1 371
DJS1-28	810	0.031 305	0.000 138	0.001 087	0.000 006	0.282 455	0.000 020	0.282 439	-0.97	5.50	0.72	1 154	1 356
DJS1-29	810	0.027 853	0.000 257	0.000 934	0.000 008	0.282 469	0.000 015	0.282 455	-0.97	6.16	0.53	1 126	1 314
Granite (DJS-2)													
DJS2-1	800	0.031 271	0.000 178	0.001 108	0.000 005	0.282 534	0.000 017	0.282 518	-0.97	8.07	0.58	1 043	1 186
DJS2-2	800	0.046 562	0.000 469	0.001 671	0.000 019	0.282 470	0.000 022	0.282 445	-0.95	5.21	0.77	1 162	1 366
DJS2-3	800	0.036 813	0.000 136	0.001 345	0.000 006	0.282 430	0.000 020	0.282 409	-0.96	4.11	0.72	1 203	1 435
DJS2-4	800	0.070 869	0.000 260	0.002 240	0.000 009	0.282 489	0.000 022	0.282 455	-0.93	5.27	0.79	1 165	1 362
DJS2-5	800	0.032 548	0.000 129	0.001 141	0.000 006	0.282 460	0.000 021	0.282 442	-0.97	5.40	0.74	1 150	1 355
DJS2-6	800	0.033 897	0.000 087	0.001 193	0.000 003	0.282 411	0.000 019	0.282 393	-0.96	3.63	0.66	1 220	1 466
DJS2-7	800	0.026 908	0.000 109	0.000 931	0.000 004	0.282 430	0.000 017	0.282 416	-0.97	4.58	0.61	1 180	1 406
DJS2-8	800	0.051 382	0.000 988	0.001 822	0.000 036	0.282 373	0.000 028	0.282 346	-0.95	1.62	0.98	1 309	1 592
DJS2-9	800	0.061 319	0.000 282	0.001 909	0.000 008	0.282 384	0.000 020	0.282 355	-0.94	1.91	0.69	1 299	1 574
DJS2-10	800	0.036 007	0.000 129	0.001 298	0.000 006	0.282 447	0.000 020	0.282 428	-0.96	4.79	0.70	1 175	1 393
DJS2-11	800	0.061 119	0.000 298	0.001 990	0.000 009	0.282 432	0.000 020	0.282 402	-0.94	3.53	0.71	1 234	1 472
DJS2-12	800	0.055 801	0.000 319	0.001 999	0.000 009	0.282 476	0.000 022	0.282 445	-0.94	5.05	0.79	1 172	1 377
DJS2-13	800	0.024 941	0.000 124	0.000 869	0.000 003	0.282 396	0.000 014	0.282 383	-0.97	3.44	0.49	1 224	1 478
DJS2-14	800	0.052 439	0.000 195	0.001 860	0.000 008	0.282 464	0.000 020	0.282 436	-0.94	4.77	0.70	1 182	1 394
DJS2-15	800	0.029 585	0.000 174	0.001 025	0.000 006	0.282 410	0.000 019	0.282 395	-0.97	3.76	0.68	1 213	1 457
DJS2-16	800	0.024 414	0.000 347	0.000 841	0.000 012	0.282 419	0.000 017	0.282 407	-0.97	4.29	0.61	1 191	1 425
DJS2-17	800	0.035 372	0.000 123	0.001 244	0.000 005	0.282 387	0.000 019	0.282 369	-0.96	2.73	0.67	1 257	1 523
DJS2-18	800	0.046 972	0.000 087	0.001 679	0.000 004	0.282 523	0.000 023	0.282 498	-0.95	7.08	0.83	1 087	1 249

Continued

Grain spot	Age /Ma	$^{176}\mathrm{Yb}/^{177}\mathrm{Hf}$	2SE	$^{176}\mathrm{Lu}/^{177}\mathrm{Hf}$	2SE	$^{176}\mathrm{Hf}/^{177}\mathrm{Hf}$	2SE	$(^{176}\mathrm{Hf}/^{177}\mathrm{Hf})_i$	$f_{\mathrm{Lu/Hf}}$	$\varepsilon_{\mathrm{Hf}}(t)$	2SE	T_{DM1} /Ma	T_{DMC} /Ma
DJS2-19	800	0.028 718	0.000 333	0.000 941	0.000 009	0.282 430	0.000 019	0.282 416	−0.97	4.57	0.66	1 181	1 407
DJS2-20	800	0.057 188	0.000 081	0.002 018	0.000 003	0.282 445	0.000 023	0.282 414	−0.94	3.93	0.81	1 218	1 447
DJS2-21	800	0.022 838	0.000 096	0.000 866	0.000 004	0.282 409	0.000 021	0.282 396	−0.97	3.91	0.75	1 206	1 448
DJS2-22	800	0.038 764	0.000 562	0.001 246	0.000 018	0.282 454	0.000 016	0.282 436	−0.96	5.10	0.56	1 163	1 373
DJS2-23	800	0.023 426	0.000 137	0.000 894	0.000 006	0.282 411	0.000 016	0.282 398	−0.97	3.95	0.56	1 205	1 446
DJS2-24	800	0.029 992	0.000 222	0.001 049	0.000 009	0.282 440	0.000 017	0.282 424	−0.97	4.79	0.61	1 173	1 393
Granite (PZH-09)													
PZH-09-1	750	0.078 315	0.000 534	0.002 403	0.000 013	0.282 452	0.000 030	0.282 418	−0.93	2.83	1.07	1 225	1 477
PZH-09-2	750	0.063 557	0.000 725	0.001 824	0.000 021	0.282 441	0.000 025	0.282 415	−0.95	3.01	0.90	1 210	1 466
PZH-09-3	750	0.057 335	0.000 572	0.001 698	0.000 013	0.282 435	0.000 027	0.282 411	−0.95	2.94	0.94	1 212	1 471
PZH-09-4	750	0.047 033	0.000 329	0.001 439	0.000 007	0.282 406	0.000 021	0.282 386	−0.96	2.17	0.73	1 239	1 519
PZH-09-5	750	0.047 932	0.000 189	0.001 515	0.000 004	0.282 390	0.000 028	0.282 368	−0.95	1.51	0.99	1 266	1 561
PZH-09-6	750	0.065 285	0.000 260	0.002 056	0.000 006	0.282 320	0.000 035	0.282 291	−0.94	−1.50	1.25	1 397	1 750
PZH-09-7	750	0.078 698	0.000 638	0.002 305	0.000 014	0.282 471	0.000 031	0.282 438	−0.93	3.60	1.09	1 192	1 429
PZH-09-8	750	0.060 324	0.000 223	0.001 754	0.000 005	0.282 406	0.000 024	0.282 381	−0.95	1.84	0.85	1 257	1 540
PZH-09-9	750	0.062 039	0.000 159	0.001 972	0.000 013	0.282 428	0.000 026	0.282 400	−0.94	2.39	0.91	1 237	1 505
PZH-09-10	750	0.065 940	0.000 359	0.002 188	0.000 015	0.282 558	0.000 036	0.282 527	−0.93	6.78	1.28	1 060	1 228
PZH-09-11	750	0.069 897	0.000 868	0.002 231	0.000 026	0.282 442	0.000 049	0.282 411	−0.93	2.66	1.73	1 230	1 488
PZH-09-12	750	0.053 196	0.000 391	0.001 643	0.000 006	0.282 399	0.000 023	0.282 376	−0.95	1.71	0.81	1 260	1 548
PZH-09-13	750	0.054 143	0.000 235	0.001 750	0.000 007	0.282 387	0.000 031	0.282 363	−0.95	1.19	1.10	1 283	1 581
PZH-09-14	750	0.044 574	0.000 327	0.001 362	0.000 008	0.282 470	0.000 024	0.282 451	−0.96	4.50	0.86	1 145	1 372
PZH-09-15	750	0.061 006	0.000 501	0.001 958	0.000 022	0.282 321	0.000 032	0.282 293	−0.94	−1.38	1.14	1 390	1 743

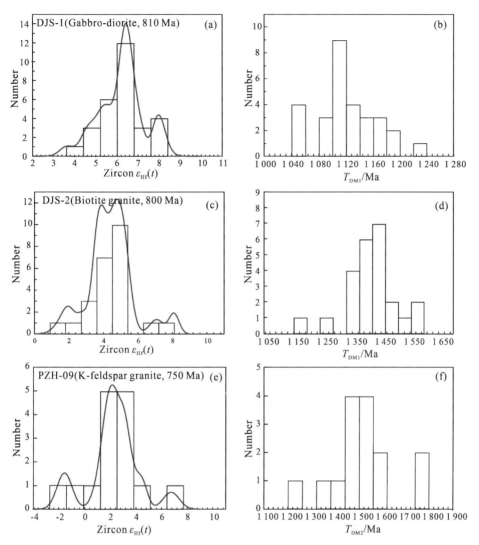

Fig. 8　Histograms of $\varepsilon_{Hf}(t)$ isotope data (a,c,e) and Hf model age (b,d,f) for the gabbro-diorites, biotite granites, and K-feldspar granites from the southwestern margin of the Yangtze Block, South China.

(−1.50 to −1.38) with the single-stage Hf model age of 1 390 − 1 397 Ma and crustal Hf model age of 1 743 − 1 750 Ma (Fig. 7a). There are 13 positive $\varepsilon_{Hf}(t)$ values from + 1.19 to + 6.78 with single-stage Hf model age of 1 060 − 1 283 Ma. Their crustal Hf model age varies from 1 228 Ma to 1 581 Ma (Fig. 8e,f).

5　Discussion

5.1　Magmatic type of the Neoproterozoic granites

5.1.1　Biotite granites: The affinity of the adakitic granites

　　Defant and Drummond (1990) firstly use the term adakite to describe Cenozoic arc magma in Adak Island that were derived from the young and hot slab melts. These volcanic or intrusions are geochemically characterized by high concentrations of SiO_2($\geqslant 56$ wt%), Al_2O_3 ($\geqslant 15$ wt%, rarely lower), and Sr (>400 ppm), as well as low concentrations of MgO (<3 wt%), Y ($\leqslant 18$ ppm), and Yb ($\leqslant 1.9$ ppm). The later scholars concluded that the adakites have high Na_2O (>3.5 wt%) contents, Sr/Y (>40) and La/Yb (>20) [or (La/Yb) $_N>10$] ratios (Martin et al., 2005; Moyen et al., 2009).

In this study, the ca. 800 Ma biotite granites have high SiO_2(74.08 – 74.82 wt%), Na_2O (4.76 – 5.60 wt%) contents, and Sr/Y (38.9 – 54.3) ratios, low MgO (0.25 – 0.30 wt%), Y (7.04 – 9.71 ppm), and Yb (0.78 – 1.08 ppm) contents, resembling the adakitic signatures (Castillo, 2012; Martin et al., 2005; Moyen et al., 2009). On the (La/Yb) $_N$ vs. Yb_N and Sr/Y vs. Y diagrams (Fig. 9a,b), these biotite granites are different from arc volcanic rocks and chiefly plotted into the field of adakitic rocks. Although the Sr (335 – 395 ppm <400 ppm) and Al_2O_3(13.85 – 14.70 wt% <15 wt%) contents of our biotite granites are slightly lower than typical adakitic rocks, this phenomenon might be caused by the fractional crystallization of

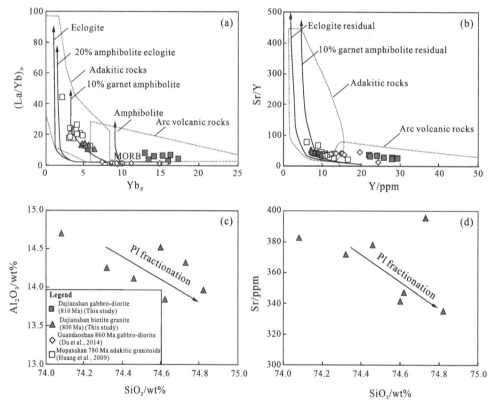

Fig. 9 (La/Yb) $_N$ vs. Yb_N(a), Sr/Y vs. Y (b) (Defant and Drummond, 1990; Martin et al., 2005), Al_2O_3 vs. SiO_2(c), and Sr vs. SiO_2(d) diagrams for the gabbro-diorites and biotite granites from the southwestern margin of the Yangtze Block, South China.

plagioclase minerals during the late magmatic evolution (Huang et al., 2009; Liu et al., 2018). The fractionation of plagioclase can be evidenced by the existence of euhedral and zoned plagioclase in our samples (Huang et al., 2009) (Fig. 3f). The decrease in Sr and Al_2O_3 contents with increasing SiO_2 contents is also the indicative of plagioclase fractionation (Liu et al., 2018) (Fig. 9c, d). Moreover, these biotite granites have high SiO_2 (74.08 - 74.82 wt%) contents and differentiation index (88.2-89.3), further supporting the fractional crystallization during magma evolution. Thus, we suggest that the ca. 810 Ma biotite granites in this study can be classified as the adakitic granites.

5.1.2 K-feldspar granites: Highly fractionated A_2-type granites

The ca. 750 Ma K-feldspar granites in the Panzhihua city are characterized by high SiO_2 (76.61 - 77.14 wt%), alkaline ($Na_2O + K_2O$ = 8.55 - 9.69 wt%) contents, and K_2O/Na_2O (1.16 - 1.81) ratios, as well as low concentrations of Al_2O_3 (11.43 - 11.86 wt%), CaO (0.25 - 0.37 wt%), and MgO (0.02 - 0.10 wt%). All these geochemical signatures are comparable to those A-type granites (Bonin, 2007; Karsli et al., 2018; King et al., 1997; Sami et al., 2018; Whalen et al., 1987). As the primitive-mantle-normalized diagram shows (Fig. 6f), these samples display similar trend with Neoproterozoic Mianning A-type granites, which show the strong depletion of Ba, Sr, P, and Eu, and the enrichment of Th, U, and K (Huang et al., 2008). Although the highly fractionated I-type granites may share similar geochemical characteristics with high-Si A-type granites (King et al., 1997; Whalen et al., 1987), some samples of our K-feldspar granites display the peralkaline nature with low A/NK ratios (0.88-0.97) (Fig. 5c), which are distinct from the highly fractionated I-type granites that are usually metaluminous to weakly peraluminous (Jahn et al., 2001; Luo et al., 2018). Their high 10 000 Ga/Al (2.56-2.80) ratios and $Fe^{\#}$ values [$FeO^T/(FeO^T + MgO)$ = 0.91-0.97] are obviously different from those fractionated I-type granites (Frost et al., 2001; Wu et al., 2002; Liao et al., 2019; Sami et al., 2018). In addition, the existence of perthite and interstitial biotite in our K-feldspar granites can also imply an A-type affinity (Bonin, 2007; Huang et al., 2008; Zhao et al., 2008b). In the granite classification diagrams defined by Whalen et al. (1987) (Fig. 10a-c), the K-feldspar granites fall into the field of A-type granites, but not the I-, S-, and M-types. Notably, the K-feldspar granites in this study contain extremely high SiO_2 (76.61 - 77.14 wt%) contents and differentiation index (95.3 - 96.9), indicating that they are highly evolved and have experienced a notably fractional crystallization. The highly evolved granites usually underwent the fractional crystallization of accessory minerals, which results in the reduction of high field strength elements (e.g., Zr, Nb, Ce, and Y) (Chappell, 1999). Similar to those highly fractionated A-type granites in the north Arabian-Nubian Shield, Egypt (Sami et al., 2018), the lower contents of (Zr+Nb+Ce+ Y) (262 - 337 ppm) than traditional A-type granites (> 350 ppm) may be attributed to extensive fractional crystallization during the late magma evolution. In conclusion, these K-feldspar granites belong to the highly fractionated A-type series.

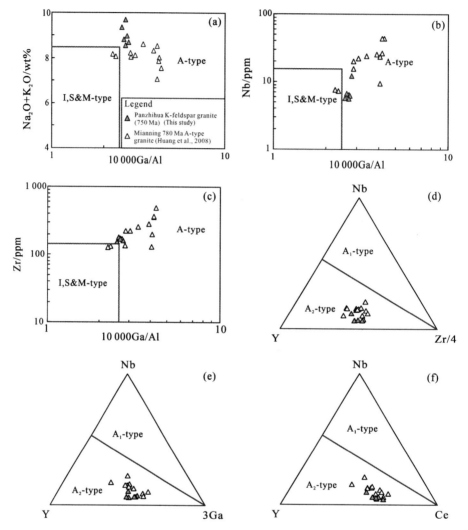

Fig. 10 Plots of (Na₂O+K₂O) vs. 10 000 Ga/Al, Nb vs. 10 000 Ga/Al, Zr vs. 10 000 Ga/Al (a-c)
discrimination diagrams (Whalen et al., 1987), and Nb-Y-Zr/4, Nb-Y-3Ga, and
Nb-Y-Ce (f-h) discrimination diagrams (Eby, 1992) for the K-feldspar granites
from the southwestern margin of the Yangtze Block, South China.

Eby (1992) subdivided A-type granites into A_1 and A_2 groups by different geochemical characteristics and tectonic environments. The A_1-type granites represent OIB-like (ocean island basalt-like) magma sources in continental rifts or intraplate magmatism, whereas the A_2-type granites are originated from continental crust or island-arc origins which are generated at convergent plate margin. On the Nb-Y-Zr/4, Nb-Y-3 Ga, and Nb-Y-Ce diagrams (Fig. 10e, f), all the K-feldspar granites in this study are plotted into the A_2-type subgroup. As a result, the ca. 750 Ma Panzhihua K-feldspar granites are classified as the highly fractionated A_2-type granites.

5.2 Petrogenesis of the Neoproterozoic intrusive rocks

5.2.1 Gabbro-diorites: Partial melting of depleted lithospheric mantle modified by subduction slab fluids

The ca. 810 Ma gabbro-diorites in this study have low loss on ignition (LOI) (2.39–2.80 wt%) values, suggesting minor or no alteration. We therefore can use their whole-rock geochemical compositions to constrain the magma source (Karsli et al., 2017). These gabbro-diorites are characterized by high Sr (631–796 ppm), Y (22.0–29.5 ppm), HREE contents (e.g., Yb = 2.21–2.95 ppm) and low (La/Yb)$_N$ (2.61–7.87) values, suggesting the geochemical features of normal arc volcanic rocks (Defant and Drummond, 1990) (Fig. 10a,b). They are enriched in large ion lithophile elements (LILEs, e.g., Rb, Cs, Sr, and Ba) and depleted in high field strength elements (HFSEs, e.g., Nb, Ta, and Ti), showing the arc magma characteristics from subduction zone (Sun and McDonough, 1989) (Fig. 6b).

These gabbro-diorite samples display positive $\varepsilon_{Nd}(t)$ values (+1.0 to +1.5) and low (^{87}Sr/^{86}Sr)$_i$ values (0.705 184–0.705 392) (Fig. 11a), indicating the insignificant crustal contamination during late stage of magma evolution (Zhao et al., 2018). Their positive whole-rock $\varepsilon_{Nd}(t)$ (+1.0 to +1.5) and zircon $\varepsilon_{Hf}(t)$ values (+3.64 to +8.14) suggest the relatively depleted mantle source. Furthermore, the low Nb/La ratios (0.17–0.51) of these gabbro-diorites are partly similar to those from the lithospheric mantle (0.3–0.4) rather than asthenospheric mantle (>1), suggesting that they were likely originated from the depleted lithospheric mantle (Smith et al., 1999). Their Zr/Nb ratios (33.5–40.7) are obviously higher than the OIB-like mantle melts (5.83), further ruling out the asthenospheric mantle source (Meng et al., 2015). In addition, these gabbro-diorites show high Th/Yb (0.62–1.82) and Nb/Yb (1.53–1.88) ratios, plotting near the field between mantle array and subduction component (Fig. 11b). Moreover, our gabbro-diorites display high Th/Zr and Rb/Y ratios but low Nb/Zr and Nb/Y ratios (Figs. 11c,d), implying that the mantle source has been metasomatized by subduction-related fluids prior to its partial melting (Kepezhinskas et al., 1997). This is in agreement with the presence of coeval mafic rocks (e.g., ca. 812 Ma Gaojiacun and ca. 806 Ma Lengshuiqing plutons) in the western Yangtze Block, which were considered from the depleted lithospheric mantle modified by slab components (Zhao et al., 2008a; Zhou et al., 2006a). The existence of metasomatized lithospheric mantle can also be supported by recent summary that Neoproterozoic mantle sources were strongly modified by subduction slab fluids during ca. 850–800 Ma interval (Zhao et al., 2018). Accordingly, we

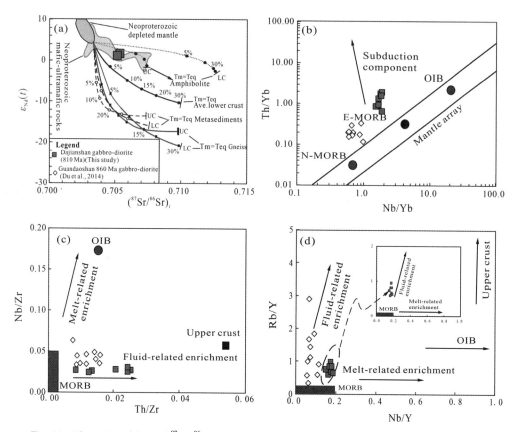

Fig. 11　Plots of $\varepsilon_{Nd}(t)$ vs. $(^{87}Sr/^{86}Sr)_i(I_{Sr})$ (a) (Zhao et al., 2018), Th/Yb vs. Nb/Yb (b)
(Pearce, 2008), Nb/Zr vs. Th/Zr, and Rb/Y vs. Nb/Y (c, d) (Kepezhinskas et al., 1997) diagrams
for the gabbro-diorites along the southwestern margin of the Yangtze Block, South China.
Fig. 11a display the EC-AFC modeling results that show the evolution of the primitive basaltic melts with
contamination by the different crustal materials (Zhao et al., 2013; Zhao and Asimow, 2018). The Neoproterozoic
depleted mantle is from Zimmer et al. (1995).

propose that the gabbro-diorites in this study were mainly formed by the partial melting of depleted lithospheric mantle modified by subduction-related fluids.

5. 2. 2　Adakitic granites: Partial melting of thickened juvenile mafic lower crust

　　Previous massive researches proposed that the adakitic magma can be formed by the following petrogenetic models: ① the fractional crystallization and assimilation of mantle-derived basaltic magma (Macpherson et al., 2006); ② the magma mixing between mantle- and crust- derived melts (Li et al., 2018; Streck et al., 2007); ③ the partial melting of subducted young and hot oceanic slab (Defant and Drummond, 1990; Martin et al., 2005); and ④ the partial melting of delaminated or thickened mafic lower crust (Hou et al., 2004; Huang et al., 2009; Martin et al., 2005; Smithies, 2000; Wang et al., 2006, 2007).

　　Adakitic rocks formed via the direct fractional crystallization of mantle-derived basaltic magma were generally intermediate to mafic rocks, but the adakitic granites in this study display more felsic compositions with high SiO_2 contents (74.08−74.82 wt%). Most adakites originated from mantle source have high $Mg^\#$ values (>60) (Rapp et al., 1999), which are higher than our adakitic granites ($Mg^\# = 36−41$). In addition, adakitic rocks formed by the fractional crystallization are typically accompanied with abundant mafic and ultramafic rocks (Macpherson et al., 2006). However, the adakitic granites in this study are not associated with coeval massive mafic magma. We therefore exclude the possibility that our adakitic granites were generated by the fractional crystallization process acting on the basaltic magma.

The uniform geochemical compositions and the absence of maficmicrogranular enclaves (MMEs) in the Dajianshan adakitic granites do not record the obvious hints of magma mixing between mantle- and crust-derived melts (Yu et al., 2015). The magma mixing usually forms intermediate and high Mg adakitic rocks (e.g., high-Mg andesites) (Streck et al., 2007; Wang et al., 2012; Yu et al., 2015), which is inconsistent with our adakitic granites with high SiO_2 (74.08 − 74.82 wt%) and low MgO (0.25 − 0.30 wt%) contents. The low concentrations of Cr (2.94−3.59 ppm) and Ni (1.32−1.55 ppm) indicate that mixing of mantle-derived magma were limited in their formation (Martin, 1999; Wang et al., 2012). Their positive whole-rock Nd isotopies [$\varepsilon_{Nd}(t) = +0.5$ to $+0.6$] can also preclude the magma mixing between mantle-derived mafic and crust-derived felsic magmas.

The adakitic melts derived from subducted slab or delaminated lower crust contain high $Mg^\#$, Cr, and Ni contents due to interaction with mantle components during subsequent magma ascent (Martin et al., 2005; Smithies, 2000). However, our adakitic granites have lower $Mg^\#$ (36−41) values, Cr (2.94−3.59 ppm), and Ni (1.32−1.55 ppm) contents than subduction slab-derived adakites (e.g., $Mg^\# = \sim48$; Cr = ~36 ppm; Ni = ~24 ppm) and delaminated lower crust-related adakites (e.g., $Mg^\# > 50$) (Fig. 12a-d) (Defant and Drummond, 1990; Huang et al., 2009). Zircon can provide reliable provenance information due to their refractory and resistant after the multiple episodes of sedimentation, magmatism, and metamorphism (Grimes et al., 2007; Valley, 2003). Average U/Yb ratios for zircons from different sources are disparate, and increase from oceanic source (0.18) to continental granitoids (1.07) and kimberlites source (2.1) (Grimes et al., 2007). The majority of zircon grains from our adakitic granites are plotted into the field of continental zircon and the overlapped area between continental zircon and oceanic zircon (Fig. 12e). Hou et al. (2004) suggested that the adakites derived from slab melts generally have low Rb/Sr ratios ranging from 0.01 to 0.04. In contrast, the high Rb/Sr ratios (0.06−0.09 >0.04) of our adakitic granites also indicate the crustal source (Hou et al., 2004; Huang et al., 2009) (Fig. 12f). Therefore, we suggest that the subduction slab and delaminated lower crust sources are also unreasonable for the genesis of our adakitic granites.

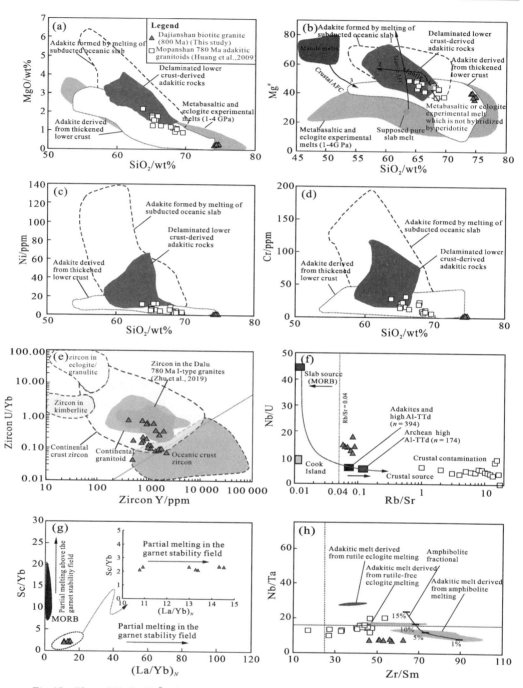

Fig. 12　Plots of MgO, Mg$^#$ values, Ni, and Cr vs. SiO$_2$(a-d), as well zircon U/Yb vs. Y (e)

(Grimes et al., 2007), Nb/U vs. Rb/Sr (f) (Hou et al., 2004), Sc/Yb vs. (La/Yb)$_N$(g),

and Nb/Ta vs. Zr/Sm (h) (Long et al., 2015 and references therein) diagrams

for the biotites granites from the southwestern margin of the Yangtze Block, South China.

Field of metabasaltic and eclogitic melts (dehydration melting) (1-4. 0 GPa), subducted oceanic

crust- and delaminated lower crust-derived adakites are after Wang et al. (2007) and references therein.

Therefore, we infer that the adakitic granites in this study were derived from the partial melting of thickened mafic lower crust (Fig. 12a-d). The positive whole-rock $\varepsilon_{Nd}(t)$ values (+0.5 to +0.6) and zircon $\varepsilon_{Hf}(t)$ values (+1.66 to +8.10) indicate that they were sourced from the juvenile mafic lower crust source. As mentioned in the Part 5.1.1, our adakitic granites might have experienced the fractional crystallization of plagioclase minerals during the late stage of magma evolution, but they have no Eu and Sr negative anomalies, implying that the original magma was richer in Sr and no plagioclase residual in source (Liu et al., 2018). The plagioclase becomes unstable and releases Sr when the pressure increases at over 1.2 GPa (corresponding to ~40 km depth) (Moyen, 2009; Wang et al., 2012). Experimental results suggested that the partial melting of mafic materials under high pressure (>1.2 GPa) can produce adakitic melts with a residual phase containing garnet but no plagioclase minerals (Rapp and Watson, 1995). Our adakitic granites display high Sr/Y (average value = 45.4 >40) and $(La/Yb)_N$ (10.8 – 14.5 >10) ratios, showing the presence of garnet residual in the source after melt extraction (Fig. 9a,b; Fig. 12g) (Defant and Drummond, 1990; Rapp and Watson, 1995). The garnet has high partition coefficients of HREEs, and amphibole has high partition coefficients of MREEs (Hanson, 1978). Therefore, if garnet is the only residual phase, the equilibrium magmas usually display a progressive decrease trend from Gd to Lu in REE pattern. On the contrary, if amphibole is the dominant residual phase, a concave-upward pattern between the MREEs and HREEs can be shown. In this study, the obvious concave-upward REE patterns [$(Dy/Yb)_N$ = 0.99–1.04] of our adakitic granites (Fig. 6c) indicate the dominantly residual of amphiboles in the source, consistent with their low Nb/Ta (7.11–8.17) and high Zr/Sm (46.2–62.3) ratios (Fig. 12h). In conclusion, we proposed that the studied adakitic granites in the Dajianshan area were derived from the partial melting of relatively thickened juvenile mafic lower crust, probably with the amphibole-dominant, garnet-present, and plagioclase-absent residual. Zhao et al. (2008a) proposed that the ca. 860 – 810 Ma mantle-derived magma formed widespread mafic rocks and led to Neoproterozoic crustal growth in the western Yangtze Block. This imply the presence of a relatively thickened mafic lower crust (Zhao et al., 2008a).

5.2.3 A-type granites: Partial melting of felsic crustal rocks under low-pressure condition

In this study, the Panzhihua K-feldspar granites belong to the highly fractionated A-type granites with high SiO_2 contents (76.61–77.14 wt%) and differentiation index (DI) (95.3 – 96.9), which suggest that they have experienced extensive fractional crystallization after the partial melting. The single-stage partial melting could not produce granitic melts with extremely low Sr (7.89 – 14.2 ppm) and relatively high Rb (158 – 200 ppm) contents (Karsli et al., 2018; Sami et al., 2018), implying that the subsequent fractional crystallization occurred. The strong depletion of Eu, Ba, and Sr in the trace elements spider diagrams (Fig. 6e,f) can be attributed to the fractionation of plagioclase and K-feldspar minerals. In the diagrams of Rb vs. Sr and Ba (Fig. 13a,b), the predominantly fractionation of plagioclase and K-feldspar were

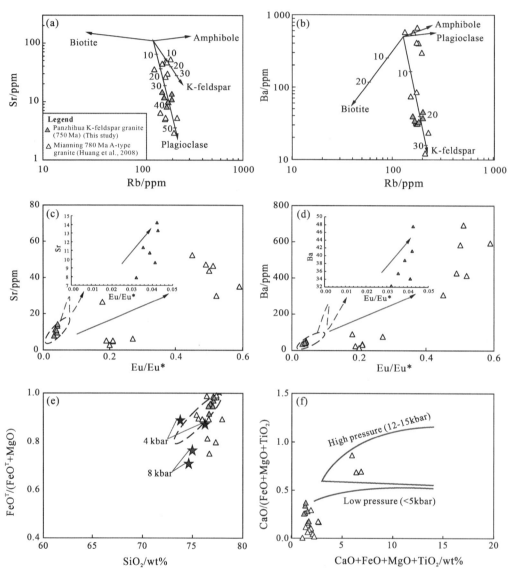

Fig. 13 Plots of Sr vs. Rb (a), Ba vs. Rb (b) (Sami et al., 2018), Sr vs. Eu/Eu* (c),

Ba vs. Eu* (d), $FeO^T/(FeO^T+MgO)$ vs. SiO_2(e) (Patiño Douce, 1997), and CaO/

$(FeO+MgO+TiO_2)$ vs. $(CaO+FeO+MgO+TiO_2)$ (f) (Patiño Douce, 1999) for the K-feldspar

granites from the southwestern margin of the Yangtze Block, South China.

Fig. 13a-d show that the fractional crystallization of plagioclase and K-feldspar played an important role in the generation of Panzhihua K-feldspar granites. The Rayleigh fractionation vectors are cited after Sami et al. (2018). The star points in Fig. 13e represent the experimental melt of TTGs at 4 kbar and 8 kbar from Patiño Douce (1997) and Frost et al. (2001). The area between the high-pressure (12–15 kbar) and low-pressure (≤5 kbar) curves in Fig. 13f encompasses the range of depths at which mantle-crust interaction takes place (Patiño Douce, 1999).

shown. This can also be demonstrated by the positive correlation between Sr and Ba with Eu/Eu* (Fig. 13c,d). In addition, the fractional crystallization of zircon can account for low Zr concentration (154 – 181 ppm) of Panzhihua A-type granites (King et al., 1997). The negative P and Ti anomalies are ascribed to the fractionation of apatite and Fe-Ti oxides. Therefore, the Panzhihua A-type granites experienced extensive fractional crystallization during the late stage of magma evolution.

Generally, the A-type granites can be generated by the fractionation of mantle-derived magmas or partial melting of various crustal rocks (Collins et al., 1982; Eby, 1992; Frost and Frost, 2011; Patiño Douce, 1997; Turner et al., 1992). The generation of A-type granites with extremely high concentrations of SiO_2 (76. 61 – 77. 14 wt% > 72 wt%) by the persistent fractionation from mafic through intermediate to felsic compositions, require a significant quantity of mafic magma (Turner et al., 1992). In such occurrences, the voluminous mafic residues can be found (Karsli et al., 2018). However, there are no massive mafic minerals (amphibole and clinopyroxene) in our samples and coetaneous mafic-ultramafic compositions around our A-type granites. Furthermore, the extremely low MgO (0. 02 – 0. 10 wt%), Cr (3. 25 – 7. 13 ppm), and Ni (1. 35 – 2. 98 ppm) concentrations of Panzhihua A-type granites indicate minor contribution of mantle-derived materials (Karsli et al., 2018; Sami et al., 2018). Their negative whole-rock $\varepsilon_{Nd}(t)$ values (-1. 6 to -1. 2) support a crustal source. Considering the crustal source, our A-type granites can be produced by several petrogenetic mechanisms, including: ① the partial melting of granulite facies metasedimentary rocks (Collins et al., 1982; Huang et al., 2011); ② the partial melting of anhydrous lower crustal granulitic residue which a granitoid melt was previously extracted (Clemens et al., 1986; Whalen et al., 1987); and ③ the dehydration melting of quartzofeldspathic crustal rocks (tonalities or granodiorites) in the shallow depth (Creaser et al., 1991; Frost and Frost, 2011; Patiño Douce, 1997).

The partial melts of metasedimentary and granulite residues may generate magmas with low alkaline contents relative to alumina (Huang et al., 2011; Patiño Douce, 1997; Sylvester, 1998). However, the studied A-type granites have high alkalis [($K_2O + Na_2O$) = 8. 55 – 9. 69 wt%] contents and low Al_2O_3 contents (11. 43 – 11. 91 wt%). The re-melting of granulitic residues could produce the A-type granites that are depleted in TiO_2 relative to MgO (Creaser et al., 1991; Patiño Douce, 1997). Our samples display low MgO (0. 02 – 0. 10 wt%) contents and high TiO_2/MgO (0. 40 – 2. 50) ratios, indicating the inconsistent signatures with granulitic residues source (Karsli et al., 2018). Moreover, the enrichment of LILEs (e.g., Rb, Th, and U) (Fig. 6f) further preclude the source of depleted granulitic residue (Bi et al., 2016; Sami et al., 2018). Therefore, the metasedimentary source and granulite residues can be precluded.

Thus, the Panzhihua A-type granites may be formed by the dehydration melting of quartzofeldspathic rocks in the shallow crust. Patiño Douce(1997) suggested that low-pressure (4 kbar) dehydration melting of tonalites or granodiorites would produce the A-type granitic magma. Our A-type granites contain high SiO_2 contents (76. 61−77. 14 wt%) and TiO_2/MgO ratios (0. 40−2. 50), as well low $CaO/(FeO+MgO+TiO_2)$ ratios (0. 17−0. 37) and ($CaO+FeO+MgO+TiO_2$) contents (1. 24−1. 70 wt%), indicating the relatively low pressure partial melting in the shallow crust (Fig. 13e,f). They possess the obviously low CaO (0. 25−0. 37 wt%), MgO (0. 02−0. 10 wt%) contents, and remarkably negative Eu anomalies, which coincides with the residual mineral assemblage of calcic plagioclase + orthopyroxene at low pressure condition (Patiño Douce, 1997). Furthermore, the plagioclase-dominated low pressure residual assemblage can produce the melts that are enriched in Ga relative to Al (Patiño Douce, 1997). This is consistent with the fact that our A-type granites yield high Ga/Al ratios (10 000Ga/Al=2. 56−2. 80). In summary, these geochemical fingerprints imply that the Panzhihua A-type granites were mainly sourced from the partial melting of felsic crustal rocks under low pressure condition. It is notable that these A-type granites show the positive and negative $\varepsilon_{Hf}(t)$ (−1. 50 to +6. 78) values, which suggest that the mantle- and/or juvenile crust- derived magma might provide necessary heat and materials for the partial melting of source rocks.

5. 3　Tectonic implication

5. 3. 1　Neoproterozoic subduction tectonic setting

In this study, the ca. 810 Ma Dajianshan gabbro-diorites were derived from the partial melting of depleted lithospheric mantle that have been modified by subduction slab fluids. They are typically calc-alkaline magmatism and enriched in LREEs and Rb, Ba, Sr, K, as well depleted in Nb, Ta, and Ti, indicative of the subduction-related magma (Wilson, 1989) (Fig. 6b). In combination with the existence of ca. 810 Ma arc-signature Gaojiacun and Lengshuiqing mafic intrusions (Zhou et al., 2006a), they may have experienced a common evolution history during the Neoproterozoic subduction process.

The ca. 800 Ma adakitic granites in the Dajianshan area wereg enerated by the partial melting of relatively thickened juvenile crust within the garnet stability field. They are enriched in LREEs and LILEs relative to HREEs (Fig. 6d), consistent with origination of active continental margin (Tang et al., 2016; Wang et al., 2012). Such adakites can be formed in the thickened continental crust source from subduction zone (Defant and Drummond, 1990). On the western margin of the Yangtze Block, the long-term extraction of Neoproterozoic mantle-derived magma is a major process involved in the generation of new continental crust, leading to formation of thickened continental crust (Zhao et al., 2008a). The existence of thickened lower crust along the western and northern margins of the Yangtze Block can be evidenced by the Mopanshan adakitic complex and Wudumen and Erliba adakitic plutons (Dong et al., 2012; Huang et al., 2008; Zhao and Zhou, 2008).

Moreover, the ca. 750 Ma Panzhihua A-type granites were mainly formed by the partial melting of felsic crustal rocks induced by underplating of mantle- and/or juvenile mafic lower crust-derived magma. These A-type granites were produced under a lower pressure condition, indicating an extensional environment (Eby, 1992; Whalen et al., 1987). These A-type granites belong to the A_2-type sub-group (Fig. 10d-f), which are associated with the waning stages of arc magmatism (Eby, 1992). We can thus rule out an origin related to a hot spot, plume or continental rift zone located in anorogenic settings (Jiang et al., 2011). The shallow origin of our A-type granites is in turn a consequence of their setting in noncompressive tectonic regimes, where the crust tends to be thin and magmatic advection of heat can approach the Earth's surface (Patiño Douce, 1997). It is notable that the ca. 738−746 Ma Dadukou mafic intrusions in Panzhihua district display obviously negative Nb-Ta anomalies and have arc-like geochemical characteristics (Zhao and Zhou, 2007b), which prove the regional extension under subduction setting. The underplating of coeval mantle-derived magmas formed mafic rocks and triggered the partial melting of felsic crustal rocks in the shallow depth, generating the A-type granites in this study. In addition, the combined occurrence of late Neoproterozoic Tiechuanshan, Huangguan, and Mianning A-type granites manifests regionally extensional environment under subduction background along the margins of the Yangtze Block (Eby, 1992; Luo et al., 2018; Huang et al., 2008).

During the last few years, Neoproterozoic voluminous seismic, sedimentary, and igneous viewpoints have also provided constraints for persistent subduction process along the periphery of the Yangtze Block during the evolution of supercontinent Rodinia. Gao et al. (2016) provided a new multichannel seismic reflection profile collected across the Sichuan Basin by SINOPROBE images prominent reflectors. They proposed that these mantle reflectors are similar to those observed on other deep reflection profiles that have been interpreted as relicts of ancient subduction. Therefore, these newly discovered reflectors were considered as the remnants of Neoproterozoic subduction zone along the Yangtze Block (Gao et al., 2016). Sun et al. (2008) proposed that the geochemical features of the sandstone and mudstones from Yanbian Group indicate an arc signature and intermediate-felsic volcanic source. They further studied the detrital zircon from Precambrian strata and suggested an important period of juvenile magmatism at 1 000−740 Ma, which is consistent with the interval of continuous subduction along the western Yangtze Block (Sun et al., 2009). In addition, massive igneous researches have also been investigated to support Neoproterozoic subduction setting in the western Yangtze Block. The ca. 860 Ma Guandaoshan pluton was considered from a depleted mantle source influenced by subduction-related fluids (Du et al., 2014; Sun and Zhou, 2008). Sun et al. (2007) suggested that the ca. 840 Ma Fangtian basaltic lavas had an N-MORB-type mantle affinity modified by slab-derived fluids. The ca. 830 Ma Tongde high-Mg diorite dikes were explained as the products of high pressure fractional crystallization of hydrous basaltic melts in an active continental arc setting (Li and Zhao, 2018). The ophiolite is considered part of continental margin arc (Ishiwatari and Tsujimori, 2003). Zhao et al. (2017) proposed that the geochemical features of ca. 800 Ma Shimian ophiolite is similar to the SSZ-type ophiolite, which suggested that Neoproterozoic voluminous arc-affinity igneous rocks were formed in a giant Andean-type arc system. Moreover, the widespread Neoproterozoic granitoids, such as the ca. 800 Ma A-type granites and I-type granites in Daxiangling-Shimian region (Zhao et al., 2008b), ca. 780 Ma I-type granodiorites-granites in Dalu area (Zhu et al., 2019), and ca. 750 Ma quartz monzodiorites-granodiorites in Kangding-Luding region (Lai et al., 2015a), were also considered from the arc continental crust source during the subduction process.

In summary, we are inclined to favor that the long duration of Neoproterozoic magmatism was generated in persistent subduction process. The ca. 810 Ma gabbro-diorites suggest the partial melting of depleted lithospheric mantle metasomatized by subduction-related fluids. The occurrence from ca. 800 Ma adakitic granites to ca. 750 Ma A-type granites indicate the geodynamic transition from regionally crustal thickening to thinning under subduction background.

5. 3. 2　Geodynamic evolution

The special igneous evidences in this study, combined with previous geochronological and geochemical data, imply a reasonable geodynamic evolution process in the western Yangtze Block, South China (Fig. 14). The early eastward subduction of oceanic crust beneath the continental margin of the Yangtze Block occurred during the period of ca. 860 – 810 Ma and accompanied by the upwelling of mantle-derived magma (Zhao and Zhou, 2008; Zhu et al., 2019), which were modified by the subduction slab fluids or melts (Zhao et al., 2018, 2019). The persistent mantle-derived magmas emplaced into the middle-upper crust underwent strong differentiation and formed voluminous mafic-intermediate rocks, including the Dajianshan gabbro-diorites in this study, the Guandaoshan pluton, the Tongde dioritic pluton and dikes, the Gaojiacun and Lengshuiqing pluton (Fig. 14a). The continuous mantle-derived magma also led to Neoproterozoic vertical crustal growth along the margin of Yangtze Block (Zhao et al., 2008a; Zhao and Zhou, 2008), bringing about the relatively new lower crust and thickening the lower crust. At ca. 800 Ma, the oceanic slab break-off and the upwelling of asthenosphere heated juvenile thickened continental lower crust (Zhao and Zhou, 2008), producing the adakitic granites in the Dajianshan area (Fig. 14b). Melting of the relatively thickened continental crust marked the beginning of continental crustal thinning (Zhao et al.,

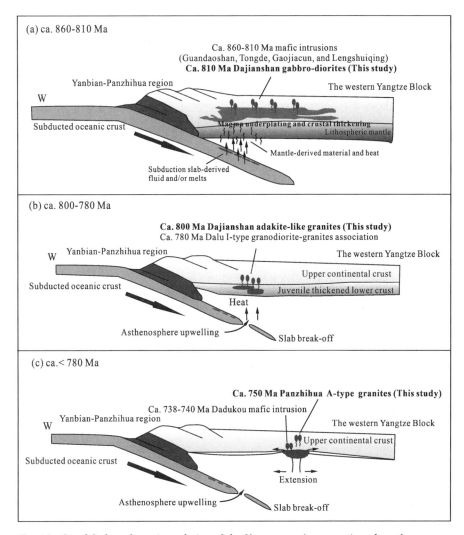

Fig. 14　Simplified geodynamic evolution of the Neoproterozoic magmatism along the western Yangtze Block（Zhao and Zhou，2008，2009；Zhu et al.，2019）.

2010；Zhao and Zhou，2008）. As the subduction process continues，the underplating of mantle- and/or juvenile mafic lower crust-derived magma penetrated the shallower crust and induced the partial melting of felsic crustal rocks under the regionally extensional background，forming the A-type granites at ca. 750 Ma in the Panzhihua district（Fig. 14c）.

6　Conclusion

（1）The Neoproterozoic gabbro-diorites，adakitic granites，and A-type granites were emplaced at ca. 810 Ma，800 Ma，and 750 Ma，respectively.

（2）The gabbro-diorites were derived from the partial melting of depleted lithospheric mantle metasomatized by subduction-related fluids. The adakitic granites were formed by the partial melting of the relatively thickened juvenile mafic lower crust within the garnet stability field. The A-type granites were mainly generated by the partial melting of felsic crustal rocks under low pressure condition and experienced extensive fractional crystallization. The underplating of mantle- and/or juvenile mafic lower crust-derived magma play a significant role for the genesis of these A-type granites.

(3) The Neoproterozoic gabbro-diorites, adakitic granites, and A-type granites in this study were formed during a persistent subduction process along the western Yangtze Block, South China.

Acknowledgements We thank Prof. Junhong Zhao for his editorial working and valuable comments as well as other constructive reviews from anonymous reviewers, which significantly improved the shape and content of this manuscript. This work was jointly supported by the National Natural Science Foundation of China (Grant Nos. 41421002 and 41772052) and the program for Changjiang Scholars and Innovative Research Team in University (Grant IRT1281). Support was also provided by the Foundation for the Author of National Excellent Doctoral Dissertation of China (201324) and independent innovation project of graduate students of Northwest University (YZZ17192).

Appendix A. Supplementary material Supplementary data to this article can be found online at https://doi.org/10. 1016/j.jseaes.2019. 103977.

References

Andersen, T., 2002. Correction of common lead in U-Pb analyses that do not report ^{204}Pb. Chemical Geology, 192(1-2), 59-79.

Bao, Z.A., Chen, L., Zong, C.L., Yuan, H.L., Chen, K.Y., Dai, M.N., 2017. Development of pressed sulfide powder tablets for in situ, sulfur and lead isotope measurement using LA-MC-ICP-MS. International Journal of Mass Spectrometry, 421, 255-262.

Bi, J.H., Ge, W.C., Yang, H., Wang, Z.H., Xu, W.L., Yang, J.H., 2016. Geochronology and geochemistry of late Carboniferous-middle Permian I- and A-type granites and gabbro-diorites in the eastern Jiamusi Massif, NE China: Implications for petrogenesis and tectonic setting. Lithos, 266-267, 213-232.

Bonin, B., 2007. A-type granites and related rocks: Evolution of a concept problems and prospects. Lithos, 97, 1-29.

Castillo, P.R., 2012. Adakite petrogenesis. Lithos, 134-135(3), 304-316.

Chappell, B.W., 1999. Aluminium saturation in I- and S-type granites and the characterization of fractionated haplogranites. Lithos, 46(3), 535-551.

Chu, Z.Y., Chen, F.K., Yang, Y.H., Guo, J.H., 2009. Precise determination of Sm, Nd concentrations and Nd isotopic compositions at the nanogram level in geological samples by thermal ionization mass spectrometry. Journal of Analytical Atomic Spectrometry, 24, 1534-1544.

Clemens, J.D., Holloway, J.R., White, A.J.R., 1986. Origin of A-type granite: Experimental constraints.

American Mineralogist, 71, 317-324.

Collins, W.J., Beams, S.D., White, A.J.R., Chappell, B.W., 1982. Nature and origin of A-type granites with particular reference to southeastern Australia. Contributions to Mineralogy and Petrology, 80, 189-200.

Creaser, R.A., Price, R.C., Wormald, R.J., 1991. A-type granites revisited: Assessment of a residual-source model. Geology, 19, 163-166.

Defant, M.J., Drummond, M.S., 1990. Derivation of some modern arc magmas by melting of young subducted lithosphere. Nature, 347, 662-665.

Dong, Y.P., Liu, X.M., Santosh, M., Chen, Q., Zhang, X.N., Li, W., He, D.F., Zhang, G.W., 2012. Neoproterozoic accretionary tectonics along the northwestern margin of the Yangtze Block, China: Constraints from zircon U-Pb geochronology and geochemistry. Precambrian Research, 196-197(1), 247-274.

Du, L.L., Guo, J.H., Nutman, A.P., Wyman, D., Geng, Y.S., Yang, C.H., Liu, F.L., Ren, L.D., Zhou, X.W., 2014. Implications for Rodinia reconstructions for the initiation of Neoproterozoic subduction at ~860 Ma on the western margin of the Yangtze Block: Evidence from the Guandaoshan Pluton. Lithos, 196-197, 67-82.

Eby, G.N., 1992. Chemical subdivision of the A-type granitoids: Petrogenesis and tectonic implications. Geology, 20, 641-644.

Frost, B.R., Barnes, C.G., Collins, W.J., Arculus, R.J., Ellis, D.J., Frost, C.D., 2001. A geochemical classification for granitic rocks. Journal of Petrology, 42(11), 2033-2048.

Frost, C.D., Frost, B.R., 2011. On ferroan (A-type) granitoids: Their compositional variability and modes of origin. Journal of Petrology, 52(1), 39-53.

Gan, B.P., Lai, S.C., Qin, J.F., Zhu, R.Z., Zhao, S.W., Li, T., 2017. Neoproterozoic alkaline intrusive complex in the northwestern Yangtze Block, Micang Mountains region, South China: Petrogenesis and tectonic significance. International Geology Review, 59(3), 311-332.

Gao, R., Chen, C., Wang, H.Y., Lu, Z.W., Brown, L., Dong, S.W., Feng, S.Y., Li, Q.S., Li, W.H., Wen, Z.P., Li, F., 2016. SINOPROBE deep reflection profile reveals a Neo-proterozoic subduction zone beneath Sichuan basin. Earth & Planetary Science Letters, 454, 86-91.

Gao, S., Ling, W.L., Qiu, Y., Zhou, L., Hartmann, G., Simon, K., 1999. Contrasting geochemical and Sm-Nd isotopic compositions of Archean metasediments from the Kongling high-grade terrain of the Yangtze craton: Evidence for cratonic evolution and redistribution of REE during crustal anatexis. Geochimica et Cosmochimica Acta, 63(13-14), 2071-2088.

Geng, Y.S., Yang, C.H., Wang, X.S., Du, L.L., Ren, L.D., Zhou, X.W., 2008. Metamorphic basement evolution in western margin of Yangtze Block. Beijing: Geological Publishing House, 1-215 (in Chinese).

Griffin, W.L., Pearson, N.J., Belousova, E., Jackson, S.E., van Achterbergh, E., O'Reilly, S.Y., Shee, S. R., 2000. The Hf isotope composition of cratonic mantle: LAM-MC-ICP-MS analysis of zircon megacrysts in kimberlites. Geochimica et Cosmochimica Acta, 64(1), 133-147.

Grimes, C.B., John, B.E., Kelemen, P.B., Mazdab, F.K., Wooden, J.L., Cheadle, M.J., Hanghoj, K., Schwartz, J.J., 2007. Trace element chemistry of zircons from oceanic crust: A method for distinguishing detrital zircon provenance. Geology, 35, 643-646.

Guo, C.L., Wang, D.H., Chen, Y.C., Zhao, Z.G., Wang, Y.B., Fu, X.F., Fu, D.M., 2007. SHRIMP U-Pb zircon ages and major element, trace element and Nd-Sr isotope geochemical studies of a Neoproterozoic granitic complex in western Sichuan: Petrogenesis and tectonic significance. Acta Petrologica Sinica, 23 (10), 2457-2470 (in Chinese with English abstract).

Hanson, G. N., 1978. The application of trace elements to the petrogenesis of igneous rocks of granitic composition. Earth & Planetary Science Letters, 38(1), 26-43.

Hou, Z.Q., Gao, Y.F., Qu, X.M., Rui, Z.Y., Mo, X.X., 2004. Origin of adakitic intrusives generated during mid-Miocene east-west extension in southern Tibet. Earth & Planetary Science Letters, 220(1-2), 139-155.

Huang, H.Q., Li, X.H., Li, W.X., Li, Z.X., 2011. Formation of high δ^{18}O fayalite-bearing A-type granite by high temperature melting of granulitic metasedimentary rocks, southern China. Geology, 39, 903-906.

Huang, X.L., Xu, Y.G., Li, X.H., Li, W.X., Lan, J.B., Zhang, H.H., Liu, Y.S., Wang, Y.B., Li, H. Y., Luo. Z. Y., Yang, Q.J., 2008. Petrogenesis and tectonic implications of Neoproterozoic, highly fractionated A-type granites from Mianning, South China. Precambrian Research, 165(3-4), 190-204.

Huang, X.L., Xu, Y.G., Lan, J.B., Yang, Q.J., Luo, Z.Y., 2009. Neoproterozoic adakitic rocks from Mopanshan in the Western Yangtze craton: Partial melts of a thickened lower crust. Lithos, 112(3), 367-381.

Ishiwatari, A., Tsujimori, T., 2003. Paleozoic ophiolites and blueschists in Japan and Russian Primorye in the tectonic framework of East Asia: A synthesis. Island Arc, 12, 190-206.

Jahn, B.M., Wu, F.Y., Capdevila, R., Martineau, F., Zhao, Z.H., Wang, Y.X., 2001. Highly evolved juvenile granites with tetrad REE patterns: The Woduhe and Baerzhe granites from the great Xing'an Mountains in NE China. Lithos, 59(4), 171-198.

Jiang, Y.H., Zhao, P., Zhou, Q., Liao, S.Y., Jin, G.D., 2011. Petrogenesis and tectonic implications of Early Cretaceous S- and A-type granites in the northwest of the Gan-Hang rift, SE China. Lithos, 121, 55-73.

Karsli, O., Chen, B., Aydin, F., Şen, C., 2007. Geochemical and Sr-Nd-Pb isotopic compositions of the Eocene Dölek and Sariçiçek Plutons, Eastern Turkey: Implications for magma interaction in the genesis of high-K calc-alkaline granitoids in a post-collision extensional setting. Lithos, 98, 67-96.

Karsli, O., Dokuz, A., Kandemir, R., 2017. Zircon Lu-Hf isotope systematics and U-Pb geochronology, whole-rock Sr-Nd isotopes and geochemistry of the early Jurassic Gokcedere pluton, Sakarya zone-NE Turkey: A magmatic response to roll-back of the Paleo-Tethyan oceanic lithosphere. Contributions to Mineralogy and Petrology, 172(5), 31.

Karsli, O., Aydin, F., Uysal, I., Dokuz, A., Kumral, M., Kandemir, R., Budakoglu, M., Ketenci, M., 2018. Latest cretaceous "A_2-type" granites in the Sakarya Zone, NE Turkey: Partial melting of mafic lower crust in response to roll-back of neo-tethyan oceanic lithosphere. Lithos, 302-303, 312-328.

Kepezhinskas, P., McDermott, F., Defant, M.J., Hochstaedter, A., Drummond, M.S., 1997. Trace element and Sr-Nd-Pb isotopic constraints on a three-component model of Kamchatka Arc petrogenesis. Geochimica et Cosmochimica Acta, 61(3), 577-600.

King, P.L., White, A.J.R., Chappell, B.W., Allen, C.M., 1997. Characterization and origin of aluminous A-type granites from the Lachlan Fold Belt, Southeastern Australia. Journal of Petrology, 38, 371-391.

Lai, S.C., Qin, J.F., Zhu, R.Z., Zhao, S.W., 2015a. Neoproterozoic quartz monzodiorite-granodiorite association from the Luding-Kangding area: Implications for the interpretation of an active continental margin along the Yangtze Block (South China Block). Precambrian Research, 267(3-4), 196-208.

Lai, S.C., Qin, J.F., Zhu, R.Z., Zhao, S.W., 2015b. Petrogenesis and tectonic implication of Neoproterozoic peraluminous granitoids from the Tianquan area, western Yangtze Block, South China. Acta Petrologica Sinica, 31(8), 2245-2258 (in Chinese with English abstract).

Li, Q.W., Zhao, J.H., 2018. The Neoproterozoic high-Mg dioritic dikes in South China formed by high pressures fractional crystallization of hydrous basaltic melts. Precambrian Research, 309, 198-211.

Li, X.H., Li, Z.X., Zhou, H.W., Liu, Y., Liang, X.R., Li, W.X., 2003. SHRIMP U-Pb zircon age, geochemistry and Nd isotope of the Guandaoshan pluton in SW Sichuan: Petrogenesis and tectonic significance. Science in China: Series D, 46(1), 73-83.

Li, X.H., Li, Z.X., Sinclair, J.A., Li, W.X., Carter, G., 2006. Revisiting the "Yanbian Terrane": Implications for Neoproterozoic tectonic evolution of the western Yangtze Block, South China. Precambrian Research, 151(1-2), 14-30.

Li, Y., Xu, W.L., Tang, J., Pei, F.P., Wang, F., Sun, C.Y., 2018. Geochronology and geochemistry of Mesozoic intrusive rocks in the Xing'an Massif of NE China: Implications for the evolution and spatial extent of the Mongol-Okhotsk tectonic regime. Lithos, 304-307, 57-73.

Liao, X.D., Sun, S., Chi, H.Z., Jia, D.Y., Nan, Z.Y., Zhou, W.N., 2019. The Late Permian highly fractionated I-type granites from Sishijia pluton in southestern Inner Mongolia, North China: A post-collisional magmatism record and its implication for the closure of Paleo-Asian Ocean. Lithos, 328-329, 262-275.

Liu, J.H., Xie, C.M., Li, C., Wang, M., Wu, H., Li, X.K., Liu, Y.M., Zhang, T.Y., 2018. Early Carboniferous adakite-like and I-type granites in central Qiangtang, northern Tibet: Implications for intra-oceanic subduction and back-arc basin formation within the Paleo-Tethys Ocean. Lithos, 296, 265-280.

Liu, X.M., Gao, S., Diwu, C.R., Yuan, H.L., Hu, Z.C., 2007. Simultaneous in-situ determination of U-Pb age and trace elements in zircon by LA-ICP-MS in 20 μm spot size. Chinese Science Bulletin, 52(9), 1257-1264.

Liu, Y., Liu, X.M., Hu, Z.C., Diwu, C.R., Yuan, H.L., Gao, S., 2007. Evaluation of accuracy and long-term stability of determination of 37 trace elements in geological samples by ICP-MS. Acta Petrological Sinica, 23(5), 1203-1210 (in Chinese with English abstract).

Long, X.P., Wilde, S.A., Wang, Q., Yuan, C., Wang, X.C., Li, J., Jiang, Z.Q., Dan, W., 2015. Partial melting of thickened continental crust in central Tibet: Evidence from geochemistry and geochronology of Eocene adakitic rhyolites in the northern Qiangtang terrane. Earth & Planetary Science Letters, 414, 30-44.

Luo, B.J., Liu, R., Zhang, H.F., Zhao, J.H., Yang, H., Xu, W.C., Guo, L., Zhang, L.Q., Tao, L., Pan, F.B., Wang, W., Gao, H., Shao, H., 2018. Neoproterozoic continental back-arc rift development in the Northwestern Yangtze Block: Evidence from the Hannan intrusive magmatism. Gondwana Research, 59, 27-42.

Ludwig, K.R., 2003. ISOPLOT 3.0: A geochronological toolkit for Microsoft Excel. Berkeley Geochronology Center, Special Publication, 4.

Macpherson, C.G., Dreher, S.T., Thirlwall, M.F., 2006. Adakites without slab melting: High pressure differentiation of island arc magma, Mindanao, the Philippines. Earth & Planetary Science Letters, 243, 581-593.

Martin, H., 1999. Adakitic magmas: Modern analogues of Archaean granitoids. Lithos, 46, 411-429.

Martin, H., Smithies, R.H., Rapp, R., Moyen, J.F., Champion, D., 2005. An overview of adakite, tonalite-trondhjemite-granodiorite (TTG), and sanukitoid: Relationships and some implications for crustal evolution. Lithos, 79, 1-24.

Meng, E., Liu, F.L., Du, L.L., Liu, P.H., Liu, J. H., 2015. Petrogenesis and tectonic significance of the Baoxing granitic and mafic intrusions, southwestern china: Evidence from zircon U-Pb dating and Lu-Hf isotopes, and whole-rock geochemistry. Gondwana Research, 28(2), 800-815.

Middlemost, E.A.K., 1994. Naming materials in the magma/igneous rock system. Earth-Science Reviews, 37, 215-224.

Moyen, J. F., 2009. High Sr/Y and La/Yb ratios: The meaning of the "adakitic signature". Lithos, 112, 556-574

Munteanu, M., Wilson, A., Yao, Y., Harris, C., Chunnett, G., Luo, Y., 2010.The Tongde dioritic pluton (Sichuan, SW China) and its geotectonic setting: Regional implications of a local-scale study. Gondwana Research, 18(2), 455-465.

Nowell, G.M., Kempton, P.D., Noble, S.R., Fitton, J.G., Saunders, A.D., Mahoney, J.J., Taylor, R.N., 1998. High precision Hf isotope measurements of MORB and OIB by thermal ionisation mass spectrometry: Insights into the depleted mantle. Chemical Geology, 149(3-4), 211-233.

Patiño Douce, A.E., 1997. Generation of metaluminous A-type granites by low-pressure melting of calc-alkaline granitoids. Geology, 25,743-746.

Patiño Douce, A.E., 1999. What do experiments tell us about the relative contributions of crust and mantle to the origin of granitic magmas? Geological Society, London, Special Publications, 168,55-75.

Pearce, J.A., 2008. Geochemical fingerprinting of oceanic basalts with applications to ophiolite classification and the search for Archean oceanic crust. Lithos, 100, 14-48.

Rapp, R.P., Watson, E.B., 1995. Dehydration melting of metabasalt at 8 − 32 kbar: Implications for continental growth and crust-mantle recycling. Journal of Petrology, 36, 891-931.

Rapp, R.P., Shimizu, N., Norman, M.D., Applegate, G.S., 1999. Reaction between slab derived melts and peridotite in the mantle wedge: Experimental constraints at 3. 8 GPa. Chemical Geology, 160, 335-356.

Roberts, M.P., Clemens, J.D., 1993. Origin of high-potassium, calc-alkaline, I-type granitoids. Geology, 21, 825-828.

Rudnick, R.L., Gao, S., 2003. The composition of the continental crust. In: Rudnick, R.L. The Crust. Oxford: Elsevier-Pergamon, 1-64.

Sami, M., Ntaflos, T., Farahat, E.S., Mohamed, H.A., Hauzenberger, C., Ahmed, A.F., 2018. Petrogenesis and geodynamic implications of Ediacaran highly fractionated A-type granitoids in the north Arabian-Nubian shield (Egypt): Constraints from whole-rock geochemistry and Sr-Nd isotopes. Lithos,

304-307, 329-346.

Smith, E.I., Sánchez, Alexander, Walker, J.D., Wang, K., 1999. Geochemistry of mafic magmas in the hurricane volcanic field, Utah: Implications for small- and large-scale chemical variability of the lithospheric mantle. Journal of Geology, 107(4), 433-448.

Smithies, R.H., 2000. The Archean tonalite-trondhjemite-granodiorite (TTG) series is not an analogue of Cenozoic adakite. Earth & Planetary Science Letters, 182, 115-125.

Soderlund, U., Patchett, P.J., Vervoort, J.D., Isachsen, C.E., 2004.The ^{176}Lu decay constant determined by Lu-Hf and U-Pb isotope systematics of Precambrian mafic intrusions. Earth & Planetary Science Letters, 219(3-4), 311-324.

Streck, M.J., Leeman, W.P., Chesley, J., 2007. High-magnesian andesite from Mount Shasta: A product of magma mixing and contamination, not a primitive mantle melt. Geology, 35, 351-354.

Sun, S.S., McDonough, W.F., 1989.Chemical and isotopic systematics of oceanic basalts: Implications for mantle composition and processes. In: Saunders, A.D., Norry, M.J. Magmatism in the Ocean Basins. Geological Society, London, Special Publications, 42, 313-345.

Sun, W.H., Zhou, M.F., Zhao, J.H., 2007. Geochemistry and Tectonic Significance of Basaltic Lavas in the Neoproterozoic Yanbian Group, Southern Sichuan Province, Southwest China. International Geology Review, 49 (6), 554-571.

Sun, W.H., Zhou, M.F., 2008. The ~860 Ma, Cordilleran-type Guandaoshan dioritic pluton in the Yangtze Block, SW China: Implications for the origin of Neoproterozoic magmatism. Journal of Geology, 116(3), 238-253.

Sun, W.H., Zhou, M.F., Yan, D.P., Li, J.W., Ma, Y.X., 2008. Provenance and tectonic setting of the Neoproterozoic Yanbian group, western Yangtze Block (SW China). Precambrian Research, 167(1), 213-236.

Sun, W.H., Zhou, M.F., Gao, J.F., Yang, Y.H., Zhao, X.F., Zhao, J.H., 2009. Detrital zircon U-Pb geochronological and Lu-Hf isotopic constraints on the Precambrian magmatic and crustal evolution of the western Yangtze Block, SW China. Precambrian Research, 172(1), 99-126.

Sylvester, P.J., 1998. Post-collisional strongly peraluminous granites. Lithos, 45, 29-44.

Tang, J., Xu, W.L., Niu, Y.L., Wang, F., Ge, W.C., Sorokin, A.A., Chekryzhov, I.Y., 2016. Geochronology and geochemistry of Late Cretaceous-Paleocene granitoids in the Sikhote-Alin Orogenic Belt: Petrogenesis and implications for the oblique subduction of the paleo-Pacific plate. Lithos, 266-267, 202-212.

Turner, S.P., Foden, J.D., Morrison, R.S., 1992. Derivation of some A-type magmas by fractionation of basaltic magma: An example from the Padthaway Ridge, South Australia. Lithos, 28, 151-179.

Valley, J.W., 2003. Oxygen isotopes in zircon. Reviews in Mineralogy and Geochemistry, 53, 343-385.

Wang, F., Xu, W.L., Meng, E., Cao, H.H., Gao, F.H., 2012. Early Paleozoic amalgamation of the Songnen-Zhangguangcai Range and Jiamusi massifs in the eastern segment of the Central Asian Orogenic Belt: Geochronological and geochemical evidence from granitoids and rhyolites. Journal of Asian Earth Sciences, 49, 234-248.

Wang, Q., Xu, J.F., Jian, P., Bao, Z.W., Zhao, Z.H., Li, C.F., Xiong, X.L., Ma, J.L., 2006.

Petrogenesis of adakitic porphyries in an extensional tectonic setting, Dexing, South China: Implications for the genesis of porphyry copper mineralization. Journal of Petrology, 47, 119-144.

Wang, Q., Wyman, D.A., Xu, J.F., Jian, P., Zhao, Z.H., Li, C.F., Xu, W., Ma, J.L., He, B., 2007. Early Cretaceous adakitic granites in the Northern Dabie Complex, central China: Implications for partial melting and delamination of thickened lower crust. Geochimica et Cosmochimica Acta, 71, 2609-2636.

Wang, Q., Li, X.H., Jia, X.H., Wyman, D., Tang, G.J., Li, Z.X., Ma, L., Yang, Y.H., Jiang, Z.Q., Gou, G.N., 2012. Late Early Cretaceous adakitic granitoids and associated magnesian and potassium-rich mafic enclaves and dikes in the Tunchang-Fengmu area, Hainan Province (South China): Partial melting of lower crust and mantle, and magma hybridization. Chemical Geology, 328, 222-243.

Wang, X.L., Zhou, J.C., Wan, Y.S., Kitajima, K., Wang, D., Bonamici, C., Qiu, J.S., Sun, T., 2013. Magmatic evolution and crustal recycling for Neoproterozoic strongly peraluminous granitoids from southern China: Hf and O isotopes in zircon. Earth & Planetary Science Letters, 366 (2), 71-82.

Wang, Y.J., Gan, C.S., Tan, Q.L., Zhang, Y.Z., He, H.Y., Xin, Q., Zhang, Y.H., 2018. Early Neoproterozoic (~840 Ma) slab window in South China: Key magmatic records in the Chencai Complex. Precambrian Research, 314, 434-451.

Whalen, J. B., Currie, K. L., Chappell, B. W., 1987. A-type granites: Geochemical characteristics, discrimination and petrogenesis. Contributions to Mineralogy and Petrology, 95, 407-419.

Wilson, M., 1989. Igneous Petrogenesis. Boston, Sydney, Wellington, London: Unwin Hyman.

Wiedenbeck, M., Hanchar, J.M., Peck, W.H., Sylvester, P., Valley, J., Whitehouse, M., Kronz, A., Morishita, Y., et al., 2004. Further characterisation of the 91500 zircon crystal. Geostandards and Geoanalytical Research, 28 (1), 9-39.

Wu, F.Y., Sun, D.Y., Li, H., Jahn, B.M., Wilde, S., 2002. A-type granites in northeastern China: Age and geochemical constraints on their petrogenesis. Chemical Geology, 187, 143-173.

Yang, Y.J., Zhu, W.G., Bai, Z., Zhong, H., Ye, X.T., Fan, H.P., 2016. Petrogenesis and tectonic implications of the Neoproterozoic Datian mafic-ultramafic dykes in the Panzhihua area, western Yangtze Block, SW China. International Journal of Earth Sciences, 106(1), 1-29.

Yu S.Y., Zhang J.X., Qin, H.P., Sun D.Y., Zhao, X.L., Cong, F., Li, Y.S., 2015. Petrogenesis of the early Paleozoic low-Mg and high-Mg adakitic rocks in the North Qilian orogenic belt, NW China: Implications for transition from crustal thickening to extension thinning. Journal of Asian Earth Sciences, 107, 122-139.

Yuan, H.L., Gao, S., Liu, X.M., Li, H.M., Gunther, D., Wu, F.Y., 2004. Accurate U-Pb age and trace element determinations of zircon by laser ablation-inductively coupled plasma mass spectrometry. Geostandards & Geoanalytical Research, 28(3), 353-370.

Yuan, H.L., Gao, S., Dai, M.N., Zong, C.L., Gunther, D., Fontaine, G.H., Liu, X.M., Diwu, C.R., 2008. Simultaneous determinations of U-Pb age, Hf isotopes and trace element compositions of zircon by excimer laser-ablation quadrupole and multiple-collector ICPMS. Chemical Geology, 247(1-2), 100-118.

Zhao, G.C., Cawood, P.A., 2012. Precambrian geology of China. Precambrian Research, s222-223, 13-54.

Zhao, J.H., Zhou, M.F., 2007a. Neoproterozoic Adakitic Plutons and Arc Magmatism along the western Margin of the Yangtze Block, South China. Journal of Geology, 115(6), 675-689.

Zhao, J.H., Zhou, M.F., 2007b. Geochemistry of Neoproterozoic mafic intrusions in the Panzhihua district (Sichuan Province, SW China): Implications for subduction-related metasomatism in the upper mantle. Precambrian Research, 152(1), 27-47.

Zhao, J.H., Zhou, M.F., 2008. Neoproterozoic adakitic plutons in the northern margin of the Yangtze Block, China: Partial melting of a thickened lower crust and implications for secular crustal evolution. Lithos, 104, 231-248.

Zhao, J.H., Zhou, M.F., 2009. Secular evolution of the Neoproterozoic lithospheric mantle underneath the northern margin of the Yangtze Block, South China. Lithos, 107(3), 152-168.

Zhao, J.H., Zhou, M.F., Yan, D.P., Yang, Y.H., Sun, M., 2008a. Zircon Lu-Hf isotopic constraints on Neoproterozoic subduction-related crustal growth along the western margin of the Yangtze Block, South China. Precambrian Research, 163 (3), 189-209.

Zhao, J.H., Zhou, M.F., Zheng, J.P., Fang, S.M., 2010. Neoproterozoic crustal growth and reworking of the Northwestern Yangtze Block: Constraints from the Xixiang dioritic intrusion, South China. Lithos, 120(3-4), 439-452.

Zhao, J. H., Zhou, M. F., Yan, D. P., Zheng, J. P., Li, J. W., 2011. Reappraisal of the ages of Neoproterozoic strata in South China: No connection with the Grenvillian orogeny. Geology, 39 (4), 299-302.

Zhao, J.H., Zhou, M.F., Zheng, J.P., Griffin, W.L., 2013b. Neoproterozoic tonalite and trondhjemite in the Huangling complex, South China: Crustal growth and reworking in a continental arc environment. American Journal of Science, 313, 540-583.

Zhao, J.H., Asimow, P.D., Zhou, M.F., Zhang, J., Yan, D.P., Zheng, J.P., 2017. An Andean-type arc system in Rodinia constrained by the Neoproterozoic Shimian ophiolite in South China. Precambrian Research, 296, 93-111.

Zhao, J.H., Asimow, P.D., 2018. Formation and evolution of a magmatic system in a rifting continental margin: The Neoproterozoic arc- and MORB-like dike swarms in South China. Journal of Petrology, 59 (9), 1811-1844.

Zhao, J.H., Li, Q.W., Liu, H., Wang, W., 2018. Neoproterozoic magmatism in the western and northern margins of the Yangtze Block (South China) controlled by slab subduction and subduction-transform-edge-propagator. Earth-Science Reviews, 187, 1-18.

Zhao, J.H., Zhou, M.F., Wu, Y.B., Zheng, J.P., Wang, W., 2019. Coupled evolution of Neoproterozoic arc mafic magmatism and mantle wedge in the western margin of the South China Craton. Contributions to Mineralogy and Petrology, 174, 36.

Zhao, X.F., Zhou, M.F., Li, J.W., Wu, F.Y., 2008b. Association of Neoproterozoic A- and I-type granites in south china: Implications for generation of A-type granites in a subduction-related environment. Chemical Geology, 257(s1), 1-15.

Zhang, W.X., Zhu, L.Q., Wang, H., Wu, Y.B., 2018. Generation of post-collisional normal calc-alkaline and adakitic granites in the Tongbai orogen, central China. Lithos, s296-299, 513-531.

Zhang, Y.Z., Wang, Y.J., Fan, W.M., Zhang, A.M., Ma, L.Y., 2012. Geochronological and geochemical constraints on the metasomatised source for the Neoproterozoic (~825 Ma) high-Mg volcanic rocks from the

Cangshuipu area (Hunan Province) along the Jiangnan domain and their tectonic implications. Precambrian Research, 220-221, 139-157.

Zhang, Y.Z., Wang, Y.J., Zhang, Y.H., Zhang, A.M., 2015. Neoproterozoic assembly of the Yangtze and Cathaysia blocks: Evidence from the Cangshuipu Group and associated rocks along the Central Jiangnan Orogen, South China. Precambrian Research, 269, 18-30.

Zheng, Y.F., Zhang, S.B., Zhao, Z.F., Wu, Y.B., Li, X.H., Li, Z.X., Wu, F.Y., 2007. Contrasting zircon Hf and O isotopes in the two episodes of Neoproterozoic granitoids in South China: Implications for growth and reworking of continental crust. Lithos, 96(12), 127-150.

Zheng, Y.F., Wu, R.X., Wu, Y.B., Zhang, S.B., Yuan, H.L., Wu, F.Y., 2008. Rift melting of juvenile arc-derived crust: Geochemical evidence from Neoproterozoic volcanic and granitic rocks in the Jiangnan orogen, South China. Precambrian Research, 163(3-4), 351-383.

Zhou, M.F., Yan, D., Kennedy, A.K., Li, Y.Q., Ding, J., 2002. SHRIMP U-Pb zircon geochronological and geochemical evidence for Neoproterozoic arc-magmatism along the western margin of the Yangtze Block, South China. Earth & Planetary Science Letters, 196(1-2), 51-67.

Zhou, M.F., Ma, Y.X., Yan, D.P., Xia, X.P., Zhao, J.H., Sun, M., 2006a. The Yanbian Terrane (Southern Sichuan Province, SW China): A Neoproterozoic arc assemblage in the western margin of the Yangtze Block. Precambrian Research, 144(1-2), 19-38.

Zhou, M.F., Yan, D.P., Wang, C.L., Qi, L., Kennedy, A., 2006b. Subduction-related origin of the 750 Ma Xuelongbao adakitic complex (Sichuan Province, China): Implications for the tectonic setting of the giant Neoproterozoic magmatic event in South China. Earth & Planetary Science Letters, 248(1-2), 286-300.

Zhu, W.G., Zhong, H., Li, X.H., Deng, H.L., He, D.F., Wu, K.W., Bai, Z.J., 2008. SHRIMP zircon U-Pb geochronology, elemental, and Nd isotopic geochemistry of the Neoproterozoic mafic dykes in the Yanbian area, SW China. Precambrian Research, 164(1), 66-85.

Zhu, Y., Lai, S.C., Zhao, S.W., Zhang, Z.Z., Qin, J.F., 2017. Geochemical characteristics and geological significance of the Neoproterozoic K-feldspar granites from the Anshunchang, Shimian area, Western Yangtze Block. Geological Review, 63(5), 1193-1208 (in Chinese with English abstract).

Zhu, Y., Lai, S.C., Qin, J.F., Zhu, R.Z., Zhang, F.Y., Zhang, Z.Z., 2019. Geochemistry and zircon U-Pb-Hf isotopes of the 780 Ma I-type granites in the western Yangtze Block: Petrogenesis and crustal evolution. International Geology Review, 61(10), 1222-1243.

Zimmer, M., Kroner, A., Jochum, K.P., Reischmann, T., Todt, W., 1995. The Gabal Gerf complex: A Precambrian N-MORB ophiolite in the Nubian Shield, NE Africa. Chemical Geology, 123, 29-51.

Genesis of high-potassium calc-alkaline peraluminous I-type granite: New insights from the Gaoligong belt granites in southeastern Tibet Plateau[①]

Zhu Renzhi[②]   Lai Shaocong[②]   Qin Jiangfeng   M Santosh   Zhao Shaowei

Zhang Encai   Zong Chunlei   Zhang Xiaoli   Xue Yuze

Abstract: Silicic magmas, primarily felsic I- and S-type granite (sensu lato), play a key role in understanding the compositional evolution and differentiation of the continental crust. However, the genetic models of felsic I-type granites remain controversial. Herein we investigate the Gaoligong granite (sensu lato) belt in southern Tibet and compare its features with an extensive dataset that we compiled on high-potassium calc-alkaline peraluminous I-type granite from typical granitic belts around the word with a view to gain insights on the genesis of felsic I-type granites. The Early-Cretaceous granites from Gaoligong belt mostly belong to high-K calc-alkaline peraluminous I-type series. Zircon Lu-Hf $[\varepsilon_{Hf}(t) = -16.3$ to $+0.9]$ and whole-rock Sr-Nd isotopic data $[\varepsilon_{Nd}(t) = -14.6$ to $+3.0]$ on these rocks show highly variable and evolved nature, suggesting that the granites were generated by partial melting of lower crustal materials with possible involvement of upper crustal materials, followed by mixing, leading to marked variation in source compositions.

Compilation of the dataset and comparison with felsic I- and S-type granites show the following salient features: ① possibly distinct trends of Al_2O_3, P_2O_5, K_2O, Rb, Sr, Pb, Th, U and Ga vs. $MgO + FeO^T$ between the felsic I-type granites and those S-types; and ② there are notably higher and variable zircon Hf $[\varepsilon_{Hf}(t) > 10\varepsilon$ units$]$ and whole-rock Sr-Nd $[\varepsilon_{Nd}(t) > 5\varepsilon$ units$]$ isotopic compositions of I-type granites than those of the S-types. Comparison with experimental liquids and other typical cases around the world, indicate that the felsic I-type granites from Gaoligong are similar to those secondary I-type granites, where mixing processes and extensive partial melting of the heterogeneous lower crust by a mantle-like magma control the variable components. The range of compositional diversity is significantly higher for the studied felsic I-type granites than those typical S-type granites from north Queensland. Our study provides further insights into the application of geochemical and isotopic proxies to differentiate felsic I- and S-type granites.

1　Introduction

Granite (sensu lato), including the most differentiated types, are the fundamental

---

①　Published in *Lithos*, 2020, 354−355.

②　Corresponding authors.

building blocks of continental crust (Clemens, 2018; Condie and Puetz., 2019; Lee and Morton, 2015). High-K calc-alkaline peraluminous I- and S-type granites represent two distinctive suites among the evolved and felsic rocks in the upper continental crust. The former is generally thought to be formed by partial melting of a metaigneous precursor, although recent studies show that melting of metagreywackes might also generate I-type granite (Clemens et al., 2011). Most Cordillera-type granites have chemical features resembling I-type granites (Castro, 2019). Thus, the genesis of I-type granites remains debated since magmas with I-type affinity can be formed by different processes and by melting of different sources (Castro, 2019; Chappell and Stephens, 1988). More complex is the genesis of peraluminous I-type granites with relatively high silica.

In case of evaluated genesis of granites, several processes and features are proposed: the melts of the peraluminous I-type granites are considered to have formed at pressures less than the garnet stability field at lower temperatures (Chappell et al., 2012), and melting is induced by water flux from hot magmas (Collins et al., 2016); felsic I- and S-type granites have been evaluated based on major and trace elements vs. mafic content (Mg + Fe and/or MgO + total FeO) (Champion and Bultitude, 2013; Clemens et al., 2011; Stevens et al., 2007; Villaros et al., 2009; Zhu et al., 2018); and Chappell and White (1992) proposed that magmas that are transitional between S- and I-type are not common, although many slightly peraluminous I-type granite form from chemically immature metasedimentary source (Clemens, 2018; Clemens et al., 2011). Therefore, diverse models have explained the petrogenesis of felsic I-type granites, which include the following: ① Experimental studies suggested that peraluminous leucogranites represent the products of melts formed through dehydration melting of muscovite and/or biotite (Patiño Douce, 1999), including leucogranites and high-silica granites, both of which form by fluid-absent reactions involving dehydration of biotite and muscovite in clastic aluminous sediments (Patiño Douce, 1999; Clemens and Stevens, 2012; Lee and Morton, 2015; Gao et al., 2016). ② P_2O_5 was regarded as a proxy for distinguishing between peraluminous I- and S-type granites because the solubility of apatite is higher in more peraluminous melts (Chappell, 1999; Wolf and London, 1994). ③ I-type granites show a gradual change from metaluminous to peraluminous composition with increasing SiO_2 whereas S-type granites display either constant values or slight decrease with increasing SiO_2, ascribed to entrainment of various peritectic minerals (Clemens and Stevens, 2012; Stevens et al., 2007). Recently, ④ Castro (2019) suggested the dual origin of I-type granites, including primary and secondary I-type, which was supported by the experimental data using granulite xenoliths. The former is directly fractionated liquids from broadly andesitic composition, but the later is formed by water-fluxed crustal melts from older subduction-related rocks in the lower crust.

In this study, we use a compilation of geochemical data to evaluate the genesis of high-potassium calc-alkaline peraluminous I-type granites. We focus on the whole-rock geochemistry and zircon U-Pb and Lu-Hf isotopes of high-K calc-alkaline peraluminous granites from the Gaoligong belt in Tibet, and compare these with those of typical high-K calc-alkaline peraluminous I- and S-type from north Queensland in eastern Australia. The combination of geochemical and isotopic data helps us to evaluate the petrogenesis of the felsic (high-K calc-alkaline peraluminous) granites of the Gaoligong belt in the southeastern Tibet, and also provide insights into the formation of felsic I- and S-type granites in general.

2 Geological setting and petrography

The Tibet Plateau is broadly composed of the Songpan-Garze flysch complex, and several tectonic entities including Northern Qiangtang, Southern Qiangtang, Lhasa, and Himalaya blocks, which are separated by the Jinshajiang, Longmucuo-Shuanghu-Changning-Menglian, Bangong-Nujiang, and Yarlung-Zangbo suture from north to south. The Lhasa terrane is bordered by the Bangong-Nujiang suture in the north and the Yarlung-Zangbo suture in the south (Zhu et al., 2009; Zhu et al., 2011) (Fig. 1a), and is considered to have dispersed from the Gondwana supercontinent during Permian to Triassic, subsequently drifting northward and welding with the Qiangtang block during the Late Triassic to Early Cretaceous (Kapp et al., 2005a; Kapp et al., 2007; Metcalfe, 2013; Zhu et al., 2011, 2016; Zhu et al., 2019). Three major magmatic belts are identified in the Lhasa terrane (Fig. 1a) (Zhu et al., 2011). The widespread Cretaceous-early Tertiary Gangdese batholiths and Linzizong volcanic successions occur in the southern Lhasa sub-terranes (Coulon et al., 1986; Harris et al., 1990; Lee et al., 2009; Zhu et al., 2009; Wang et al., 2018), and were formed by northward subduction of the Neo-Tethyan oceanic lithosphere. The next group is represented by the abundant Mesozoic magmatic rocks in the central and northern Lhasa sub-terranes (Chiu et al., 2009; Chu et al., 2006; Coulon et al., 1986; Guynn et al., 2006; Harris et al., 1990; Zhu et al., 2011), which are correlated to the closure of the Bangong-Nujiang Tethyan ocean (Zhu et al., 2009; Zhu et al., 2011; Zhu et al., 2016; Chiu et al., 2009; Chen et al., 2014; Sui et al., 2013; Qu et al., 2012; Li et al., 2014, 2016).

The central and northern Lhasa sub-terranes are covered by widespread Carboniferous metasedimentary rocks and Jurassic-Early Cretaceous volcano-sedimentary rocks, together with minor Ordovician, Silurian, and Triassic limestones (Pan et al., 2004) as well as voluminous Mesozoic volcanic rocks composed of andesites, dacite, rhyolite, and associated volcaniclastic rocks (Kapp et al., 2005a,b, 2007; Zhu et al., 2009; Chen et al., 2014; Sui et al., 2013). The Mesozoic intrusions in the sub-terranes were mainly emplaced within Pre-Ordovician to Carboniferous-Permian metasedimentary successions, during early Cretaceous (Harris et al., 1990; Guynn et al., 2006). Abundant dioritic enclaves occur within the Early Cretaceous

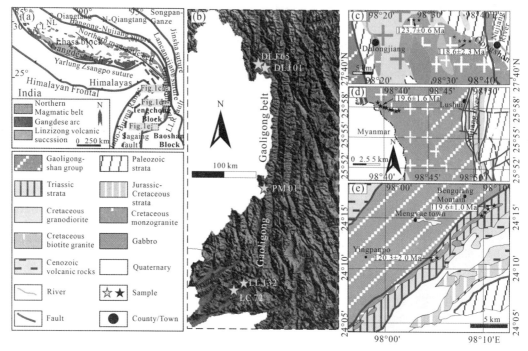

Fig. 1   Distribution of the major continental blocks of SE Asia (a) (Xu et al., 2012). Geological
map of the northern (b), central (c) and southern (d) Gaoligong belt in southeastern Tibet
(YNBGMR, 1991 and 1 : 250 000 regional geological map).
LLR: Lushui-Longling-Ruili.

granodiorite-granite association (Zhu et al., 2009, 2011, 2016; Sui et al., 2013; Li et al.,
2014, 2016). In the southeastern extension, abundant Cretaceous granitoids and minor
Ordovician, Jurassic, and Cenozoic granitoids are developed along a NW-SE belt to the
southwest of the Bangong-Nujiang suture (Pan et al., 2004). The Cretaceous granodiorite-
granite association is batholithic and is mainly distributed in Bomi, Basu, Ranwu, and Chayu
areas (so-called Bomi-Chayu batholiths) and consists mainly of monzogranites and
granodiorites with minor dioritic veins andenclaves (Pan et al., 2004; Chiu et al., 2009; Zhu
et al., 2009; Lin et al., 2012).

In the Gaoligong belt (orocline, Chiu et al., 2018), the right-lateral Gaoligong shear zone
has been regarded as one of the main Cenozoic strike-slip fault systems associated with the
clockwise rotation around the Eastern Himalayan syntaxis at ca. 40 Ma (Kornfeld et al.,
2014), which is located within the ca. 20 km wide, ca. 400 km long area between the
Tengchong and Baoshan blocks. The north branch of the Gaoligong orocline trends N-S but it
bends to trend NE-SW as it splays to the south (Chiu et al., 2018). It is mostly composed of
gneiss, granodiorite and granite, with four-stages of crystallization ages as ca. 495−487 Ma,
121 Ma, 89 Ma, and 70−63 Ma (Zhao et al., 2016), that intruded the Paleoproterozoic

metavolcanic and metasedimentary rocks. Generally, the Cretaceous magmatic rocks and associated deformation events were closely related to the closure of Meso- and Neo-Tethyan oceans (Xu et al., 2012; Zhu et al., 2017b, 2018, 2019). The Early-Cretaceous events are regarded as the southeastern extension of the northern magmatic belt of Lhasa terrane, which are generated at the setting of the closure of Bangong-Nujiang Tethyan ocean (Xu et al., 2012; Zhao et al., 2016; Zhu et al., 2017a,b; Zhu et al., 2018; Zhu et al., 2019). The Late-Cretaceous granites are closely related to the setting of subduction of Neo-Tethys (Xu et al., 2012; Zhao et al., 2016; Zhu et al., 2018).

This study is focused on the Early-Cretaceous granies in the Gaoligong belt (Fig. 1a,b). The NW-SE belt runs parallel to the long axis of the southeastern part of Bomi-Chayu batholith that is located between the Longchuanjaing fault and the Bangong-Nujiang suture (Xu et al., 2012; Yang et al., 2006; Zhu et al., 2017b). The plutons are mostly hornblende-bearing monzogranite, and are distributed in the area from Gongshan county, Pianma town, to Longling-Luxi (Fig. 1c-e). Our study focuses on the monzogranites (Fig. 2). The fine- to medium-grained monzogranites from the northern Gaoligong belt are dominantly composed of quartz (26%–32%), alkaline feldspar (18%–25%), albitic plagioclase (30%–38%), hornblende (~3%), biotite (~8%–10%), and others accessory minerals including apatite, zircon, magnetite and Fe-Ti oxides. The monzogranites in the Pianma area from central Gaoligong belt have higher modal content of quartz (27%–35%), together with alkali feldspar (24%–35%), albitic plagioclase (30%–34%), hornblende (~1%–2%), biotite (~5%–8%), and accessory minerals including apatite, zircon and titanite. The hornblende-bearing granite from the southern Gaoligong belt have less quartz (15%–28%), with alkali feldspar (10%–20%), and higher abundance of plagioclase (40%–50%), together with hornblende (~1%–3%), biotite (~2%–5%), and accessory minerals including apatite, zircon and titanite and Fe-Ti oxides.

3 Analytical methods

3.1 Whole-rock geochemistry and Sr-Nd isotopes

Whole rock analyses were performed at the State Key Laboratory of Continental Dynamics, Northwest University, Xi'an, China. Fresh chips of whole-rock samples (ca. 1 kg) were powdered to 200 mesh using a tungsten carbide ball mill. Major elements were analyzed using an X-ray fluorescence (Rikagu RIX 2100). Analyses of U.S. Geological Survey and Chinese national rock standards BCR-2, GSR-1, and GSR-3, indicating that both analytical precision and accuracy for major elements are generally better than 5%. Trace elements analyses were determined by using a Bruker Aurora M90 inductively coupled plasma mass spectrometer (ICP-

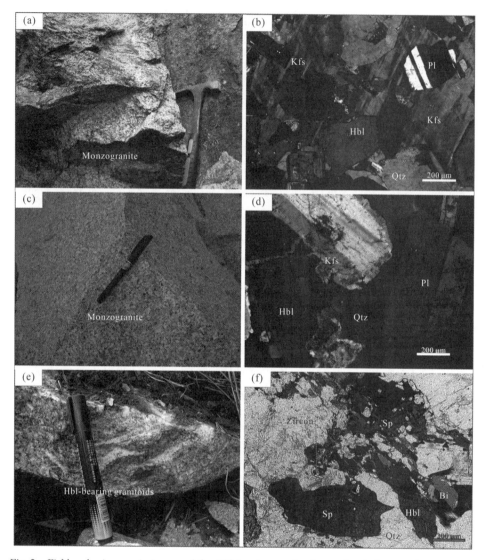

Fig. 2 Field and microscopic images of the Hbl-bearing granite (sensu lato) from northern, central and southern Gaoligong belt, both (b) and (d) are in crossed nicols and (f) is in paralle nicol.

Kfs: K-feldspar; Hbl: Hornblende; Pl: Plagioclase; Qtz: Quartz; Sp: Titanite.

MS) following the method of Qi et al. (2000). Sample powders were dissolved using an HF + HNO$_3$ mixture in a high-pressure PTFE bomb at 190 ℃ for 48 h. The analytical precision was better than ±5%−10% (relative) for most of the trace elements.

Whole-rock Sr-Nd isotopic data were obtained using a Nu Plasma HR multi-collector (MC) mass spectrometer. The Sr and Nd isotopic fractionation was corrected to $^{87}Sr/^{86}Sr = 0.119\ 4$ and $^{146}Nd/^{144}Nd = 0.721\ 9$, respectively (Chu et al., 2009). During the sample runs, the La Jolla standard yielded an average of $^{143}Nd/^{144}Nd = 0.511\ 862 ± 5$ (2σ), and the NBS987 standard yielded an average of $^{87}Sr/^{86}Sr$ ratio = $0.710\ 236 ± 16$ (2σ). The total procedural Sr and Nd blanks were about 1 ng and 50 pg. NIST SRM-987 and JMC-Nd were used as reference solutions for $^{87}Sr/^{86}Sr$ and $^{143}Nd/^{144}Nd$ isotopic ratios, respectively. The BCR-1 and BHVO-1 standards yielded an average of $^{87}Sr/^{86}Sr$ ratio are 0.705 014±3 (2σ) and 0.703 477±20 (2σ, $n=9$), The BCR-1 and BHVO-1 standards yielded an average of $^{146}Nd/^{144}Nd$ ratio are 0.512 615±12 (2σ) and 0.512 987±23 (2σ).

3.2 Zircon U-Pb and Hf isotopic analyses

Zircon grains were separated using conventional heavy liquid and magnetic techniques from approximately 5 kg of each sample. Representative grains were hand-picked and mounted on epoxy resin discs, polished and carbon coated. Internal morphology was examined using cathodoluminescent (CL) prior to U-Pb and Lu-Hf isotopic analyses. Laser ablation (LA) ICP-MS zircon U-Pb analyses were conducted on an Agilent 7500a ICP-MS equipped with a 193-nm laser, following method of Bao et al. (2017) at the State Key Laboratory of Continental Dynamics, Northwest University, Xi'an, China. Common Pb contents were evaluated using the method described by Andersen (2002). The age calculations and plotting of concordia diagrams were performed using ISOPLOT (version 3.0; Ludwig, 2003). The $^{176}Yb/^{172}Yb$ value of 0.588 7 and mean Yb value obtained during Hf analysis on the same spot were applied for the interference correction of ^{176}Yb on ^{176}Hf (Iizuka et al., 2005). The obtained Hf isotopic compositions were 0.282 016±20 (2σ, $n=84$) for the GJ-1 standard and 0.282 735±24 (2σ, $n=84$) for the Monastery standard, both of which are consistent with the recommended values (Yuan et al., 2008). The decay constant for ^{176}Lu of $1.867×10^{-11}$ year^{-1} (Soderlund et al., 2004), the chondritic $^{176}Hf/^{177}Hf$ ratio of 0.282 785±7 and $^{176}Lu/^{177}Hf$ ratio of 0.033 6 were adopted.

4 Results

The analytical data on major and trace elements in representative samples from the Gaoligong belt are listed in Supplementary Table 1. The whole-rock major and trace elements on volatile free basis for granites from the Gaoligong belt and north Queensland are given in Supplementary Table 2. The bulk Sr-Nd isotopic compositions are presented in Supplementary Table 3. The zircon Lu-Hf isotopic data are given in Supplementary Table 4, and zircon U-Pb concordia diagrams together with representative CL images are shown in Figs. 3 and 4.

4.1 Major and trace elemental geochemistry

In the SiO_2 vs. K_2O (Fig. 5a), SiO_2 vs. $Na_2O + K_2O$ (Fig. 5b), and SiO_2 vs. ASI (Fig. 5c), most of the granite (sensu lato) classify as high-K, calc-alkaline and peraluminous.

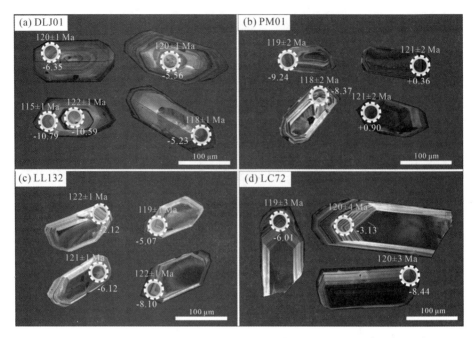

Fig. 3　Representative zircon CL images for the Hbl-bearing granite (sensu lato) from northern, central and southern Gaoligong belt.

4.1.1　Monzogranites from northern Gaoligong belt

The monzogranites in the Gongshan area show relatively low $SiO_2 = 67.72 - 69.67$ wt%, $K_2O = 3.09 - 4.28$ wt%, higher Al_2O_3 contents of $15.27 - 15.90$ wt%, and A/CNK ratios between 1.00 and 1.11, $CaO = 2.91 - 3.15$ wt%, $Fe_2O_3^T = 2.82 - 3.33$ wt%, and $MgO = 1.18 - 1.56$ wt% with $Mg^\# = 45.5 - 54.7$. The primitive mantle-normalized trace element spider diagram (Fig. 6a) shows positive Rb, Th, U, and Pb anomalies, enrichment of large ion lithophile elements and sharp depletions of P and Ti. The chondrite-normalized rare-earth elements (REE) diagram (Fig. 6b) shows notably more enrichment of LREE with high $(La/Yb)_N = 7.46 - 15.31$ and notably negative Eu anomalies ($\delta Eu = 0.57 - 0.67$).

4.1.2　Monzogranites from central Gaoligong belt

The monzogranites in the Pianma area have higher $SiO_2 = 68.51 - 74.76$ wt%, $K_2O = 3.95 - 6.05$ wt%, Al_2O_3 contents of $13.43 - 15.22$ wt%, and A/CNK ratios between 1.01 and 1.09, and lower $CaO = 1.40 - 2.75$ wt%, $Fe_2O_3^T = 1.07 - 2.97$ wt%, and $MgO = 0.30 - 1.04$ wt% with $Mg^\# = 31.6 - 45.5$ than those from the northern Gaoligong Belt. The primitive mantle-normalized trace element spider diagram (Fig. 6c) of these rocks shows positive Rb, Th, U, and Pb anomaly, enrichment of large ion lithophile elements and marked depletion in HFSE including P and Ti. The depletion is more pronounced than for rocks in north Gaoligong. The chondrite-normalized rare-earth element (REE) diagram (Fig. 6d) shows enrichment of LREE with high $(La/Yb)_N = 5.26 - 23.8$ and variable Eu anomalies ($\delta Eu = 0.38 - 1.04$).

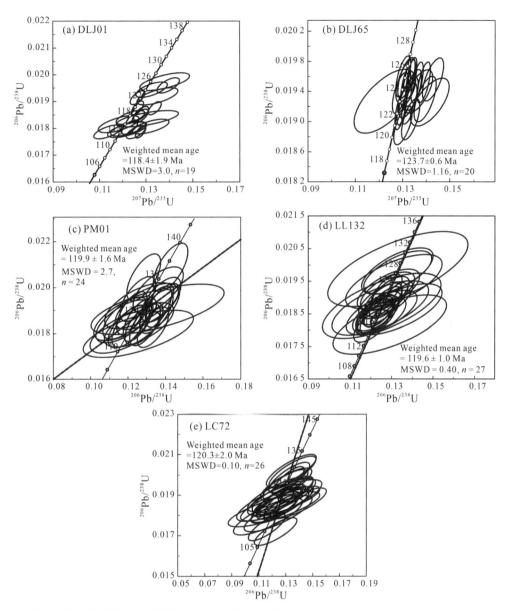

Fig. 4   LA-ICP-MS zircon U-Pb concordia diagram of representative granite (sensu lato) samples from northern, central and southern Gaoligong belt.

4.1.3   Monzogranites from southern Gaoligong belt

The granite in the Longling-Longchuan area of southern Gaoligong belt have variable SiO_2 = 59.85−70.58 wt%, lower K_2O = 1.58−3.55 wt%, higher Al_2O_3 contents of 15.36−18.64 wt% with A/CNK ratios between 0.94 and 1.06, CaO = 2.59−5.68 wt%, $Fe_2O_3^T$ = 2.71− 5.44 wt%, and MgO = 0.85 − 2.19 wt% with $Mg^{\#}$ = 41.6 − 48.9. The primitive mantle-normalized trace element spider diagram (Fig. 6e) shows positive Rb, Th, U, and Pb anomalies, enrichment of large ion lithophile elements and sharp depletion in HFSE including

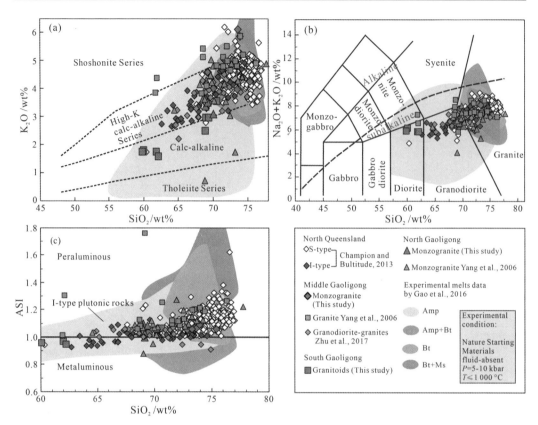

Fig. 5　SiO_2 vs. K_2O (a), SiO_2 vs. $Na_2O + K_2O$ (b) and SiO_2 vs. ASI (c) diagram (Frost et al., 2001) for the dataset from north Queensland I- and S-type granites (Champion and Bultitude, 2013), Gaoligong felsic granites (Yang et al., 2006; Zhu et al., 2017b; this study) and experimental melts data under normal crustal conditions (Gao et al., 2016). The full data are presented in Supplement Table 2.

Nb and Ta. The chondrite-normalized rare-earth element (REE) diagram (Fig. 6f) shows notably enrichment of LREE with $(La/Yb)_N = 4.39 - 17.9$, lower total HREE than the granites in the central and northern Gaoligong belt. They also show variable Eu anomalies ($\delta Eu = 0.60 - 1.02$).

4. 2　Whole-rock Sr-Nd isotopes

Whole-rock Sr-Nd isotopic data for the granite from the Gaoligong belt are listed in Supplementary Table 3. All the initial $^{87}Sr/^{86}Sr$ isotopic ratios (I_{Sr}) and $\varepsilon_{Nd}(t)$ values are calculated for the time of magma crystallization. The samples from northern Gaoligong belt have I_{Sr} ratios ranging from 0.710 34 to 0.716 15 and $\varepsilon_{Nd}(t)$ values ranging from -11.7 to -9.7, and T_{DM2} values of 1.46 - 1.60 Ga. The samples from central Gaoligong belt show relative lower I_{Sr} ratios ranging from 0.707 699 to 0.709 637 and $\varepsilon_{Nd}(t)$ values ranging from -8.8 to -8.7, and T_{DM2} values of 1.40 - 1.41 Ga. The samples from southern Gaoligong belt display similar

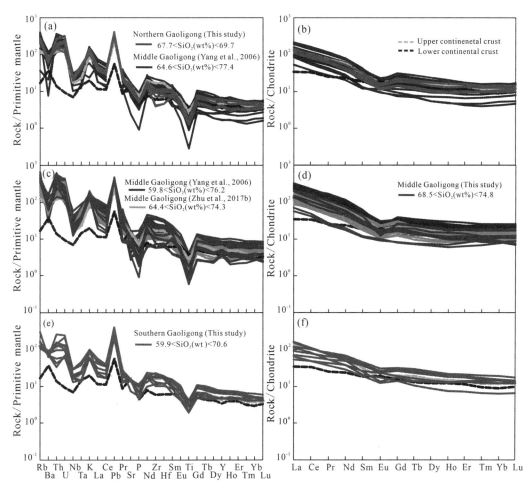

Fig. 6　Chondrite-normalized REE patterns and primitive-mantle-normalized trace element spider diagram
for Hbl-bearing granite (sensu lato) from northern, middle and southern Gaoligong belt.

The primitive mantle and chondrite values are from Sun and McDonough (1989).

Sr-Nd isotopic compositions with northern one, I_{Sr} ratios ranging from 0. 711 856 to 0. 716 634 and $\varepsilon_{Nd}(t)$ values are −10. 1, and T_{DM2} values of 1. 49−1. 50 Ga.

4. 3　Zircon Lu-Hf isotopic compositions

Zircon grains from the four dated samples were also analyzed for Lu-Hf isotopes on the same domains, and the results are listed in Supplementary Table 4. Initial $^{176}Hf/^{177}Hf$ ratios and $\varepsilon_{Hf}(t)$ values of the early Cretaceous zircons were calculated according to their crystallization age. The data show variable zircon $\varepsilon_{Hf}(t)$ compositions (Fig. 7). Zircon crystals from the granite show a significantly evolved and variable Hf isotopic composition, with $\varepsilon_{Hf}(t)$ values between −16. 3 and −1. 29 (Fig. 7a) with crustal model ages are range from 1 032 Ma to 1 808 Ma for hornblende-bearing monzogranites (DLJ01) in the northern Gaoligong belt (19 zircon grains), −16. 2 to +0. 90 with crustal model ages are range from 1 117 Ma to 2 193 Ma (Fig. 7b) for hornblende-bearing monzogranites (PM01) in the central Gaoligong belt (25 grains), and −8. 10 to −2. 12 with crustal model ages are range from 1 311 Ma to 1 689 Ma (Fig. 7c) (25 grains) (LC72) and −12. 3 to −3. 13 with crustal model ages are range from 1 374 Ma to 1 594 Ma (Fig. 7d) (26 grains) (LL132) for monzogranites in the southern Gaoligong belt.

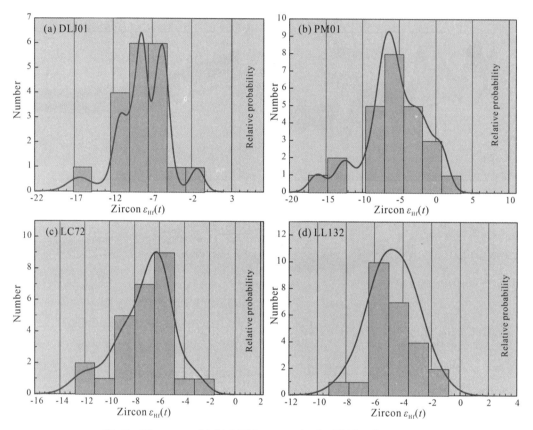

Fig. 7　Histograms of initial Hf isotope ratios for Hbl-bearing granites from northern, middle and southern Gaoligong belt.

5　Discussion

5. 1　Petrogenesis of the Gaoligong high-K calc-alkaline peraluminous I-type granite (sensu lato): Insights from geochemistry and isotopic signature

The granites (sensu lato) from northern, central and southern Gaoligong belt show similar geochemical characteristics including high-potassic, calc-alkaline and peraluminous nature with only a few samples plotting in the medium potassium and metaluminous fields (Fig. 5a-c). The major mineralogy is composed of hornblende, biotite, plagioclase, k-feldspar, quartz and accessory minerals including magnesite, titanite and zircon, in the absence of aluminous minerals such as garnet, cordierite, muscovite and accessory minerals including ilmenite. Various models have been proposed for the origin of high-K calc-alkaline peraluminous I-type granites including the following: ① Partial melting of hydrous calc-alkaline to high-K calc-alkaline, mafic to intermediate crustal lithologies (Roberts and Clemens, 1993). ② Melting and/or crystallization of medium- to high-K basaltic source rocks (Sisson et al., 2005). ③ Partial melting of crustal rocks with heat and water input from basaltic magmas emplaced beneath the lower crust (Annen et al., 2006; Collins et al., 2016), followed by mixing between crustal melts and residual melts after crystallization of basalts, leading to isotopic and trace element heterogeneity (Annen et al., 2006). ④ Mixing of mantle- and crust-derived components (Keay et al., 1997; Kemp et al., 2007). And ⑤ partial melting from intermediate igneous rocks (Clemens et al., 2011), or fractional crystallization from intermediate igneous rocks (Castro, 2013). Besides, in recently, the high-K calc-alkaline I-type granites in our study have A/CNK values from 0.94 to 1.11 (Supplemental Table 1), similar with the felsic peraluminous I-type granites in the Lachlan Fold belt and Hodgkinson (Fig. 5) (Champion and Bultitude, 2013; Chappell et al., 2012). Experimental data show that the melts generated by partial melting of basaltic to andesitic rocks under crustal conditions are mostly peraluminous (~1.0-1.10). Chappell et al. (2012) suggested that the excess Al is easily incorporated into the felsic liquids which resulted in the generation of peraluminous melts. During dehydration melting of metaigneous source at pressures below the garnet stability field, amphibole and biotite melt incongruently to yield pyroxenes. This process is considered to generate broadly felsic peraluminous granodioritic-monzogranitic batholiths (Chappell et al., 2012), a case that is similar to that of the Gaoligong granites. Compared with experimental data from dehydration melting at $P = 5-10$ kbar and $T \leqslant 1\,000$ ℃ and typical I- and S-type granites from northern Queensland (Supplemental Table 2, Fig. 5) (Champion and Bultitude, 2013; Gao et al., 2016), these samples show similar trend with I-types from northern Queensland. They also display significant overlap with melts derived from dehydration melting of amphibole + biotite-, biotite-, and biotite + muscovite-bearing source rocks. On the primitive mantle-normalized spider and chondrite-normalized REE diagram (Fig. 6), the granite (sensu lato) from northern, central and southern Gaoligong belt display affinity with continental crust and show enriched LILEs and depleted HFSEs. The trace element concentration in granites are generally controlled by both sources and magmatic processes such as partial melting and fractional crystallization. In this case, the whole-rock Rb-Sr and Sm-Nd isotopic components, and in particular, zircon Lu-Hf can preserve primary chemical and isotopic composition and residence time in the crust (Kemp et al., 2007; Zhu et al., 2011; Zhu et al., 2017a). The whole-rock Sr-Nd isotopic compositions of granite (sensu lato) from Gaoligong belt show enriched and evolved features including the relative low initial $^{87}Sr/^{86}Sr = 0.707\,7 - 0.709\,6$ and $\varepsilon_{Nd}(t) = -8.8$ to -8.7 from central Gaoligong belt, the relative high initial $^{87}Sr/^{86}Sr = 0.711\,9 - 0.716\,6$ and $\varepsilon_{Nd}(t) = -10.1$ from southern Gaoligong belt, and moderate values from northern Gaoligong belt, which are similar to those reported by Yang et al. (2006) and Zhu et al. (2017b), and compare with the value of lower crust and trend between MORB and lower to upper crust (Fig. 8a). The rocks also display coupled whole-rock Sr-Nd and zircon Hf isotopic composition. The zircon Lu-Hf isotopic compositions from Gaoligong belt also show highly evolved and variable features including $\varepsilon_{Hf}(t)$ values in the range of -16.3 to -1.29 for those in the northern Gaoligong belt (Fig. 7a), -16.2 to $+0.90$ for those in the central

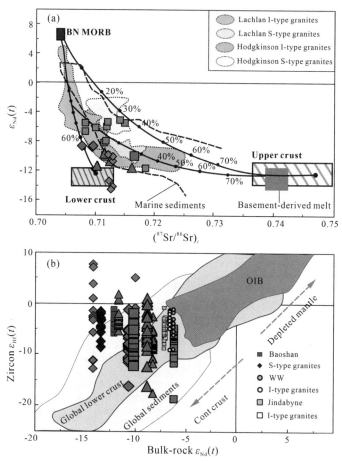

Fig. 8　Whole-rock $\varepsilon_{Nd}(t)$ values vs. initial $^{87}Sr/^{86}Sr$ ratios (a). The data for BN (Bangong-Nujiang) MORB are from Chen et al. (2014). Basement-derived melts (Zhu et al., 2011). Lower continental crust (Miller et al., 1999) and upper continental crust (Harris et al., 1988). Marine sediments (Wang et al., 2011). Zircon $\varepsilon_{Hf}(t)$ vs. zircon U-Pb data (b) for the various I- and S-type granites from Gaoligong belt. Global sediments referenced from Vervoort and Blichert-Toft (1999), Global lower crust value from Dobosi et al. (2003), and OIB from Kempton and McGill (2002).
WW: Why Worry. Symbols as the Fig. 5.

Gaoligong belt (Fig. 7b) , and −8. 10 to −2. 12 and −12. 3 to −3. 13 (Fig. 7c ,d) for those in the southern Gaoligong belt. Such variations of up to 10ε units within a single sample and sieve textures in the plagioclase core suggest an open system processes such as mixed sources (Kemp et al. , 2007; Słaby and Gotze, 2004; Słaby and Martin, 2008; Zhu et al. , 2017b). On the zircon $\varepsilon_{Hf}(t)$ vs. whole-rock $\varepsilon_{Nd}(t)$ diagram, the rocks plot in the field of global lower crust and sediments (Fig. 8b) , comparable with the granodiorite-monzogranite I-type associations which formed by mixed sources (Zhu et al. , 2017b). Castro (2019) proposed that intermediate magmas with sediment signatures can be emplaced at the lower crust and can be after crystallization the source of further granite magmas that inherit the early geochemical hybrid (mantel-crust) features, unless, sediments must be found at the source of magmas in a supra subduction mantle (subducted mélanges). As mentioned above, Zhu et al. (2019) reported that sediments were playing the key role in the genesis of Dulongjiang granodiorites (ca. 130 Ma) , which is earlier than granites (ca. 120 Ma) in this study. Therefore, the sediments could be another significant contribution in the sources. In addition, the high-K calc-alkaline peraluminous granites in this study share very similar characteristics those granitic rocks from Harcourt batholith in southeastern Australia (Clemens, 2018) , which also have I-type affinity with mildly peraluminous and remarkably radiogenic isotope compositions, including variable initial $^{87}Sr/^{86}Sr$ range from 0. 708 07 to 0. 714 21, and $\varepsilon_{Nd}(t)$ values in the range −5. 6 to −4. 3. Although the granites from Harcourt batholith have radiogenic whole-rock Sr-Nd isotope signatures, the features with brown biotite and accessory titanite that are normally considered characteristic of I-type granites (Clemens, 2018). Using the modelling parameters of Nd-Sr isotope components, Clemens (2018) suggested that the granitic rocks from the batholith could be produced by a source that comprised ca. 28% of a metamafic rock and ca. 72% of a metasedimentary rock. The high proportion of metasedimentary materials in the sources could trigger the elevated radiogenic Sr values, which is still consistent with overall I-type but transitional affinities of the product magmas (Clemens et al. , 2011). In a word, Clemens (2018) thought that the I-type Harcourt batholith with its peraluminous chemistry and evolved Sr-Nd isotopic components were rather variable in composition and lie in the lower crust, below 20 km depth, where significant volumes of mildly aluminous metasedimentary rocks must be present. This case is well consistent with the granites in the Gaoligong, with mildly peraluminous and radiogenic and variable isotopes, both of us suggested that there are a heterogeneous source rocks in the lower crust (Clemens, 2018, and this study) , but sediments should be present (Zhu et al. , 2019 and this study). As mentioned above, herein ,we suggest that the high-K calc-alkaline peraluminous I-type granites from Gaoligong belt were derived from partial melting of the heterogenous source rocks in the lower crust, with possible involvement of sediments in the source region (Zhu et al. , 2019).

5. 2 Implications for the genesis of high-K calc-alkaline peraluminous I-type granites

As discussed above, there is significant overlap between the high-K calc-alkaline peraluminous I-type granites in this study and high-K calc-alkaline peraluminous S-type granites, and Amp-, Amp + Bi-, Bi-, Bi + Mus-bearing source rocks such as those of northern Queensland (Fig. 5). In the Harker diagram, both high-K calc-alkaline peraluminous I- and S-type granites and those produced through dehydration melting of various sources at crustal conditions from experimental studies show significant overlap, although different trends also exist. There is only limited overlap between high-K calc-alkaline peraluminous I-type granites and the Lachlan and Hodgkinson S-type granites, especially, the SiO_2 range of $\geqslant 70$ wt%. Chappell et al. (2000) suggested that both the S-type, and most of the I-type granites in the Lachlan belt formed at relatively lower magmatic temperatures and the variation within these suites might represent restite fractionation. The former was derived from more feldspathic metasedimentary rocks, and the later were produced through partial melting pre-existing quartz-feldspathic igneous crust. Alternatively, at higher temperature conditions, Ca and other components of clinopyroxene are easily to dissolve in the melts which resulted in their evolution to more metaluminous melts (Chappell et al., 2012). Clemens et al. (2011) considered that the S-I dichotomy in granite typology reflects the nature of peritectic minerals entrained by the ascending granitic melts. The S- and I-type magmas are generally spatially and temporally separated from each other in terms of tectonic setting. Also, the evolved meta-pelitic sources including biotite + sillimanite assemblages occur earlier than the hornblende + biotite assemblages during crustal heating. The former is typical of S-type sources whereas the latter represents I-type sources. Harker plots are also sometimes ambiguous. In addition, evidence for peritectic melting is difficult to identify from field and petrological features.

Several studies have shown that the variations in normative mafic components and/or MgO + FeO^T is an important indicator for the genesis of I- and S-type granites. Frost et al. (2001) and Stevens et al. (2007) used Mg + Fe vs. A/CNK, K, $Mg^{\#}$, Si, Ca, Ti and Dy + Ho + Er diagram to distinguish the peritectic garnet entrainment as the origin of geochemical diversity of S-type granites, and noted that much of the compositional variation in the granites reflects the primary signature. Villaros et al. (2009) suggested that the trace element concentrations in S-type granites of the Peninsular pluton increases with maficity, which was correlated to the trace element disequilibrium melting model. Clemens et al. (2011) used the K, Zr, Ti, A/CNK, $Mg^{\#}$ and SiO_2 vs. Mg + Fe relations to identify mixing of mantle- and crust-derived materials as sources of I-types, although granitic magmas have a purely crustal origin. Clemens and Stevens (2012) also used the various elements vs. maficity to help understand the entrainable peritectic assemblage. The geochemical variations of Al_2O_3, P_2O_5, ASI, Rb, Sr and Zr vs. $MgO + FeO^T$ in the north Queensland I- and S-type granites provide crucial evidence for the nature of sources that control the behavior of granitic melts (Champion and Bultitude, 2013).

The I-type rocks contain minerals such as hornblende, magnetite and titanite whereas the S-type rocks are enriched biotite, muscovite, and monzonite. Especially in the felsic peraluminous melts, the different fractionation of K-feldspar and plagioclase within the I- and S-type granites will result in distinct Rb, Sr and Eu content. The higher content of P in S-type rocks with increasing SiO_2 in the peraluminous melt is controlled by apatite (Wolf and London, 1994). These minerals also influence the concentration of REE, Pb and Th. Therefore, as described above, the trends of major elements including Al_2O_3, P_2O_5 and K_2O and trace elements including Rb, Sr, Pb, Th, U and Ga could be used to distinguish the felsic I-type granites with those S-types. In addition to the variable zircon and whole-rock isotopic compositions, significant differences are also noted between the two rock types in their zircon Hf isotopic compositions. The I-type granites from Lachlan belt and Hodgkinson as well as those from Gaoligong belt in the southeastern Tibet show highly variable zircon $\varepsilon_{Hf}(t)$ values of $>10\varepsilon$ units (Fig. 8). However, those S-type granites from Baoshan in the southeastern Tibet display limited variations $< 10\varepsilon$ units (Fig. 8b). With regard to the whole-rock Sr-Nd isotopic data (Figs. 8a and 9), the variations in the whole-rock $\varepsilon_{Nd}(t)$ values from ca. -10 to $+5.0$ for

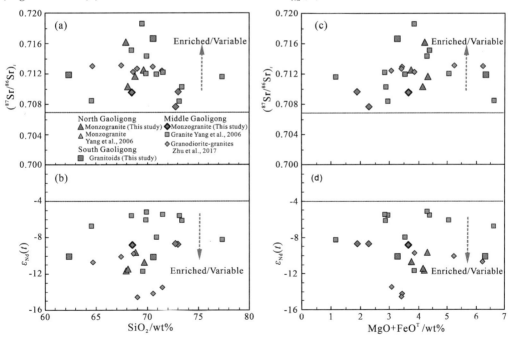

Fig. 9   Initial $^{87}Sr/^{86}Sr$ ratios and whole-rock $\varepsilon_{Nd}(t)$ values vs. $SiO_2(a,b)$; Initial $^{87}Sr/^{86}Sr$ ratios and whole-rock $\varepsilon_{Nd}(t)$ vs. $MgO + FeO^T(c,d)$.

Symbols as the Fig. 5.

Lachlan I-type, ca. -12.0 to -4.0 for Hodgkinson I-type and ca. -14.6 to -5.6 for Gaoligong I-type, however, the variations of S-type granites from the Lachlan, Hodgkinson and Baoshan are lower than 5ε units (Champion and Bultitudes, 2013; Zhu et al., 2018a). The highly variable initial $^{87}Sr/^{86}Sr$ for I-type granites from various parts of globe and those in this study are similar to those enriched S-type granites (Figs. 8a and 9). In general, the source of S-type granites is considered to be crustal metasedimentary materials including metagraywacke and meta-pelitic rocks. However, the source of I-type generally involves multiple components, including meta-igneous rocks, mixing, assimilation and/or contamination with other components including metasediments (Kemp et al., 2007; Zhu et al., 2017a,b; Zhu et al., 2019).

In this Gaoligong case, similar to typical phenomenon that peraluminous granites with modal Hbl seems to be incompatible but it is not uncommon, it has also occurred: the Hbl-bearing granodiorite-monzogranite association in the Pianma area has high-K calc-alkaline, weakly peraluminous characteristics that were resulted from mixed basaltic and metasedimentary source rocks (Zhu et al., 2017b). Zhu et al. (2019) also suggested that recycled sediments had played a significant role in the genesis of these peraluminous Hbl-bearing granodiorites. In this study, similarly, the almost Hbl-bearing granites from northern, middle and southern Gaoligong belt are high-K calc-alkaline and weakly peraluminous, although minor amounts are metaluminous and even to slightly strongly peraluminous. According to their geochemical data, we suggested that they are a mixed source of lower crustal materials and sediments. Both of these cases in the Gaoligong belt are well consistent with the results from experimental data (Castro et al., 1999, 2019; Patiño Douce, 1995a,b), that mixing processes in the source can be a charge of the Hbl-bearing peraluminous granodiorite-granite association. In consideration of the dual origin of I-type granites, these granites are similar to secondary I-type granites (Castro, 2019), in this instance, most experimental liquids from metaluminous, calc-alkaline reaction systems are peraluminous over a wide silica interval from 60 wt% to 75 wt% SiO_2. These peraluminous compositions include not only granites but also granodiorites and tonalites. In this experimental study (Castro, 2019), rehydration melting reactions entail the formation of peritectic Amphibole and a peraluminous liquid that departs considerably from the composition of the batholiths. Also, experimental liquids co-existing in equilibrium with amphibole at the condition of pressures >5 kbar are peraluminous (Zen, 1986). The presence of peraluminous leucosomes in amphibolitic migmatized rocks had been reported in metamorphic complexes (Lázaro and García-Casco, 2008). The values of aluminosity over a wide silica interval, supporting the implication of small fractions of entrained resites in the bulk chemical composition of I-type granites. As mentioned above, Chappell et al. (2012) also focused on this issue and suggested that high-temperature plutons in the Lachlan Fold Belt are difficult to produce large quantities of peraluminous melt by removal of amphibole, then chemical available and peraluminous I-type granodioritic-monzogranitic batholiths were partial melting of more mafic magmas below the garnet stability field, where amphibole and biotite melts incongruently to yield pyroxenes (Chappell et al., 2012). The observed gradation from peraluminous felsic granites to metaluminous compositions could be generated when the excess Al has been incorporated into the felsic liquid (Chappell et al., 2012). Recently experimental data also shown that peraluminous granodioritic liquid in equilibrium with Amp and Pl can up to ca. 50% of at temperatures at ca. 1 000 ℃ in the condition of the granulitic lower crust with a contribution of water and potassium (Castro, 2019). Comparison with experimental liquids, Castro (2019) suggested that restite entrainment can be necessary to explain an increase in maficity and decreasing in peraluminosity of magmas. Among them, Pl and Amp are of relevance, plagioclase calcic cores are common and Amp and Px are present over a wide range of temperature in the I-type granites. Of pure Amp could be produced in a peritectic reaction, for example, the Caledonian "Newer" granites have abundant Amp-rich clots which were regarded as restites from the sources. The similarity in Gaoligong case, the Amp and Pl are also widely present in the rocks as clots and/or equilibrium peritectic minerals.

6　Conclusions

LA-ICP-MS zircon U-Pb and Lu-Hf isotopic data and whole-rock geochemical data presented in this study indicate that the Early-Cretaceous high-K calc-alkaline peraluminous granite (sensu lato) from northern, central, and southern Gaoligong belt in the southeastern Tibet, represent typical I-type granites that were formed by partial melting of lower crustal materials with the possible involvement of sediments, followed by mixing of melts resulting variable source compositions.

A detailed comparison of the high-K calc-alkaline peraluminous I-type granites from the Gaoligong belt in the southeastern Tibet with the I- and S-types from north Queensland in eastern Australia, indicate that the notably higher and variable zircon Hf and whole-rock Sr-Nd isotopic compositions of I-type granites compared to those for S-type granites could also be a significant indicator to identify the difference between felsic I- and S-type granites.

Comparison with experimental liquids and other typical I-type granites around the world, the felsic I-type granites from Gaoligong display patterns similar to those secondary I-type granites, with multi-components in the source region where mixing processes and extensive partial melting of the heterogenous lower crust by a mantle-like magma could control the variable components. Similar to experimental results and Caledonian case, the amphibole is also widely present in the high-K calc-alkaline peraluminous Gaoligong granite, although it is partly incompatible but not uncommon.

Acknowledgements We sincerely thank handle editing of Prof. Michael Roden, constructive comments from Prof. Antonio Castro's and one anonymous reviewer, and Prof. David Champion for kindly providing us with the dataset of the felsic I- and S-type granites from Queensland, eastern Australia. This work was supported by the National Natural Science Foundation of China (Grant Nos. 41802054, 41421002, 41902046, and 41772052), China Postdoctoral Science Special Foundation Grant. 2019T120937 and Foundation Grant. 2018M643713, Natural Science Foundation of Shaanxi Grant. 2019JQ-719 and Shaanxi Postdoctoral Science Foundation.

Supplementary data Supplementary data to this article can be found online at https://doi.org/10.1016/j.lithos.2019.105343.

References

Andersen, T., 2002. Correction of common lead in U-Pb analyses that do not report ^{204}Pb. Chemical Geology, 192(1-2), 59-79.

Annen, C., Blundy, J.D., Sparks, R.S.J., 2006. The genesis of intermediate and silicicmagmas in deep crustal hot zones. Journal of Petrology, 47, 505-539.

Bao, Z., Chen, L., Zong, C., Yuan, H., Chen, K., Dai, M., 2017. Development of pressed sulfide powder tablets for in situ, Sulfur and Lead isotope measurement using LA-MC-ICP-MS. International Journal of Mass Spectrometry, 421.

Castro, A., 2013. Tonalite-granodiorite suites as cotectic systems: A review of experimental studies with applications to granite (sensu lato) petrogenesis. Earth-Science Reviews, 124, 68-95.

Castro, A., 2019. The Dual origin of I-type granites: The contribution fromexperiments. Post-Archean Granitic Rocks: Petrogenetic Processes and Tectonic Environments. Geological Society, London, Special Publications, 491.

Champion, D.C., Bultitude, R.J., 2013. The geochemical and Sr, Nd isotopic characteristics of Paleozoic fractionated S-types granites of north Queensland: Implications for S-type granite petrogenesis. Lithos, 162-163(2), 37-56.

Chappell, B.W., 1999. Aluminium saturation in I- and S-type granites and the characterization of fractionated haplogranites. Lithos, 46, 535-551

Chappell, B.W., Stephens, W.E., 1988. Origin of infracrustal(I-type) granite magmas. Transactions of the Royal Society of Edinburgh: Earth Sciences, 79, 71-86.

Chappell, B.W., White, A.J.R., 1992. I- and S-type granites in the lachlan fold belt. Transactions of the Royal Society of Edinburgh: Earth Sciences, 83 (1-2), 1-26.

Chappell, B.W., White, A.J.R., Williams, I.S., Wyborn, D., Wyborn, L.A.I., 2000. Lachlan fold belt granites revisited: High and low temperature granites and their implications. Journal of the Geological Society of Australia, 47(1), 123-138.

Chappell, B.W., Bryant, C.J., Wyborn, D., 2012. Peraluminous I-type granites.Lithos, 153, 142-153.

Chen, Y., Zhu, D.C., Zhao, Z.D., Meng, F.Y., Wang, Q., Santosh, M., Wang, L.Q., Dong, G.C., Mo,

X.X., 2014. Slab breakoff triggered ca. 113 Ma magmatism around Xainza area of the Lhasa terrane, Tibet. Gondwana Research, 26(2),449-463.

Chiu, H.Y., Chung, S.L., Wu, F.Y., Liu, D.Y., Liang, Y.H., Lin, I.J., Iizuka, Y., Xie, L.W., Wang, Y.B., Chu, M.F., 2009. Zircon U-Pb and Hf isotopic constraints from eastern Transhimalayan batholiths on the precollisional magmatic and tectonic evolution in southern Tibet. Tectonophysics, 477,3-19.

Chiu, Y.P., Yeh, M.W., Wu, K.H., Lee, T.Y., Lo, C.H., Chung, S.L., Iizuka, Y., 2018. Transition from extrusion to flow tectonism around the Eastern Himalaya syntaxis. Geological Society of America Bulletin, 130, 1675-1696.

Chu, M.F., Chung, S.L., Song, B., Liu, D.Y., O'Reilly, S.Y., Pearson, N.J., Ji, J.Q., Wen, D.J., 2006. Zircon U-Pb and Hf isotope constraints on the Mesozoic tectonics and crustal evolution of southern Tibet. Geology, 34,745-748.

Chu, Z.Y., Chen, F.K., Yang, Y.H., Guo, J.H., 2009. Precise determination of Sm, Nd concentrations and Nd isotopic compositions at the nanogram level in geological samples by thermal ionization mass spectrometry. Journal of Analytical Atomic Spectrometry, 24,1534-1544.

Clemens, J.D., 2018. Granitic magmas with I-type affinities, from mainly metasedimentary sources: The Harcourt batholith of southeastern Australia. Contributions to Mineralogy and Petrology, 173, 93.

Clemens, J.D., Stevens, G., 2012. What controls chemical variation in granitic magmas? Lithos, 134-135, 317-329.

Clemens, J.D., Stevens, G., Farina, F., 2011. The enigmatic sources of I-type granites: The peritectic connexion. Lithos, 126, 174-181.

Collins, W.J., Huang, H.Q., Jiang, X.Y., 2016. Water-fluxed crustal melting produces Cordilleran batholiths. Geology, 44, 143-146.

Condie, K.C., Puetz, S.J., 2019. Time series analysis of mantle cycles Part II: The geologic record in zircons, large igneous provinces and mantle lithosphere. Geoscience Frontiers (in press).

Coulon, C., Maluski, H., Bollinger, C., Wang, S., 1986. Mesozoic and Cenozoic volcanic rocks from central and southern Tibet: ^{39}Ar-^{40}Ar dating, petrological characteristics and geodynamical significance. Earth & Planetary Science Letters, 79, 281-302.

Dobosi, G., Kempton, P.D., Downes, H., Embey-Isztin, A., Thirlwall, M., Greenwood, P., 2003. Lower crustal granulite xenoliths from the Pannonian Basin, Hungary, Part 2: Sr-Nd-Pb-Hf and O isotope evidence for formation of continental lower crust by tectonic emplacement of oceanic crust. Contrib. Mineral. Petrol., 144 (6), 671-683.

Frost, B.R., Barnes, C.G., Collins, W.J., Arculus, R.J., Ellis, D.J., Frost, C.D., 2001. A geochemical classification for granitic rocks. Journal of Petrology, 42, 2033-2048

Gao, P., Zheng, Y., Zhao, Z., 2016. Experimental melts from crustal rocks: A lithochemical constraint on granite petrogenesis. Lithos, 266-267, 133-157.

Guynn, J.H, Kapp, P., Pullen, A., et al., 2006. Tibetan basement rocks near Amdo reveal "missing" Mesozoic tectonism along the Bangong suture, central Tibet. Geology, 34,505-508

Harris, N.B.W., Xu, R.H., Lewis, C.L., Jin, C., 1988. Plutonic rocks of the 1985 Tibet Geotraverse: Lhasa to Golmud. Philosophical Transactions of the Royal Society of London, A327, 145-168.

Harris, N.B.W., Inger, S., Xu, R., 1990. Cretaceous plutonism in Central Tibet: An example of post-collision magmatism? Journal of Volcanology and Geothermal Research, 44,21-32.

Iizuka, T., Hirata, T., Komiya, T., Rino, S., Katayama, I., Motoki, A., et al., 2005. U-Pb and Lu-Hf isotope systematics of zircons from the Mississippi river sand: Implications for reworking and growth of continental crust. Geology, 33(6), 485-488.

Kapp,J.L.D., Harrison, T.M., Grove, M., Lovera, O.M., Lin, D., 2005a. Nyainqentanglha Shan: A window into the tectonic, thermal, and geochemical evolution of the Lhasa block, southern Tibet. Journal of Geophysical Research: Solid Earth, 110(B8), 1978-2012.

Kapp, P., Peter, G.D., George, E., Gehrels., Matthew, H., Lin, D., 2007. Geological records of the Lhasa-Qiangtang and Indo-Asian collisions in the Nima area of central Tibet. Geological Society of America Bulletin, 9, 917-993.

Keay, S., Collins, W.J., McCulloch, M.T., 1997. A three-component Sr-Nd isotopic mixing model for granitoid genesis, Lachlan fold belt, eastern Australia. Geology, 25, 307-310.

Kemp, A.I., Hawkesworth, C.J., Foster, G.L., Paterson, B.A.,Woodhead, J.D., Hergt, J.M., Gray,C.M., Whitehouse, M.J., 2007. Magmatic and crustal differentiation history of granitic rocks from Hf-O isotopes in zircon. Science, 315, 980-983.

Kempton, P.D., McGill, R., 2002. Procedures for the analysis of common lead at the NERC Isotope Geosciences Laboratory and an assessment of data quality. NIGL Rep. Ser., 178.

Kornfeld, D., Eckert, S., Appel, E., Ratschbacher, L., Sonntag, B., Pfander, J., et al., 2014. Cenozoic clockwise rotation of the Tengchong Block, southeastern Tibetan Plateau: A paleomagnetic and geochronologic study, 628, 105-122.

Lázaro, C., García-Casco, A., 2008. Geochemical and Sr-Nd isotope signatures of pristine slab melts and their residues (Sierra del Convento mélange, eastern Cuba). Chemical Geology, 255, 120-133.

Lee, C.T.A., Morton, D.M., 2015. High silica granites: Terminal porosity and crystal settling in shallow magma chambers. Earth & Planetary Science Letters, 409, 23-31.

Lee, H.Y., Chung, S.L., Lo, C.H., Ji, J., Lee, T.Y., Qian, Q., Zhang, Q.,2009. Eocene Neo-tethyan slab breakoff in southern Tibet inferred from the Linzizong volcanic record. Tectonophysics, 477, 20-35.

Li, S.M., Zhu, D.C., Wang, Q., et al., 2014. Northward subduction of Bangong-Nujiang Tethys: Insight from Late Jurassic intrusive rocks from Bangong Tso in western Tibet. Lithos, 205(9), 284-297.

Li, S.M., Zhu, D.C., Wang, Q., Zhao, Z., Zhang, L.L., Liu, S.A., et al., 2016. Slab-derived adakites and subslab asthenosphere-derived OIB-type rocks at 156 ± 2 Ma from the north of Gerze, central Tibet: Records of the Bangong-Nujiang oceanic ridge subduction during the Late Jurassic. Lithos, 262, 456-469.

Lin, I.J., Chung, S.L., Chu, C.H., Lee, H.Y., Gallet, S., Wu, G., Ji, J., Zhang, Y., 2012. Geochemical and Sr-Nd isotopic characteristics of Cretaceous to Paleocene granitoids and volcanic rocks, SE Tibet: Petrogenesis and tectonic implications. Journal of Asian Earth Sciences, 53,131-150.

Ludwig, K.R., 2003. ISOPLOT 3.0: A geochronological toolkit for Microsoft Excel. Special Publication, 4, Berkeley Geochronology Center.

Metcalfe, I., 2013. Gondwana dispersion and Asian accretion: Tectonic and palaeogeographic evolution of eastern Tethys. Journal of Asian Earth Sciences, 66, 1-33.

Miller, C., Schuster, R., Kotzli, U., et al., 1999. Post-collisional potassic and ultrapotassic magmatism in SW Tibet: Geochemical and Sr-Nd-Pb-O isotopic constraints for mantle source characteristics and petrogenesis. Journal of Petrology, 40, 1399-1424.

Pan, G., Ding, J., Yao, D., Wang, L., 2004. Guidebook of 1:1 500 000 geologic map of the Qinghai-Xizang (Tibet) Plateau and adjacent areas. Chengdu: Chengdu Cartographic Publishing House, 48.

Patiño Douce, A.E., 1999. What do experiments tell us about the relative contributions of crust and mantle to the origin of granitic magmas? Geological Society Special Publication, 168, 55-75.

Qi, L., Hu, J., Gregoire, D.C., 2000. Determination of trace elements in granites by inductively coupled plasma mass spectrometry. Talanta, 51, 507-513.

Qu, X.M., Wang, R.J., Xin, H.B., Jiang, J.H., Chen, H., 2012. Age and petrogenesis of A-type granites in the middle segment of the Bangonghu-Nujiang suture, Tibetan plateau. Lithos, 146-147,264-275.

Roberts, M.P., Clemens, J.D., 1993. Origin of high-potassium, calc-alkaline, I-type granitoids. Geology, 21, 825-828.

Sisson, T.W., Ratajeski, K., Hankins, W.B., Glazner, A.F., 2005. Voluminous graniticmagmas from common basaltic sources. Contributions to Mineralogy and Petrology, 148,635-661.

Słaby, E., Gotze, J., 2004. Feldspar crystallization under magma mixing conditions shown by cathodoluminescence and geochemical modeling: A case study from the Karkonosze pluton(SW Poland). Mineralogical Magazine, 68, 541-557.

Słaby, E., Martin, H., 2008. Mafic and felsic magma interaction in granites: The Hercynian Karkonosze Pluton (Sudetes, Bohemian Massif). Journal of Petrology, 49, 353-391.

Soderlund, U., Patchett, P.J., Vervoort, J.D., Isachsen, C.E., 2004. The ^{176}Lu decay constant determined by Lu-Hf and U-Pb isotope systematics of Precambrian mafic intrusions.Earth & Planetary Science Letters, 219,311-324.

Stevens, G., Villaros, A., Moyen, J.F., 2007. Selective peritectic garnet entrainment as the origin of geochemical diversity in S-type granites. Geology, 35, 9-12.

Sui, Q.L., Wang, Q., Zhu, D.C., Zhao, Z.D., Chen, Y., Santosh, M.,Hu, Z.C., Yuan, H.L., Mo, X.X., 2013. Compositional diversity of ca. 110 Ma magmatism in the northern Lhasa Terrane, Tibet: Implications for the magmatic origin and crustal growth in a continent-continent collision zone. Lithos, 168-169, 144-159.

Sun, S.S., McDonough, W.F., 1989. Chemical and isotopic systematics of oceanic basalts: Implications or mantle composition and processes. Geological Society, London, Special Publications, 42(1),313-345.

Vervoort, J.D., Blichert-Toft, J., 1999. Evolution of the depleted mantle: Hf isotope evidence from juvenile rocks through time. Geochimica et Cosmochimica Acta, 63(3-4), 533-556.

Villaros, A., Stevens, G., Moyen, J.F., Buick, I.S., 2009. The trace element compositions of S-type granites: Evidence for disequilibrium melting and accessory phase entrainment in the source. Contributions to Mineralogy and Petrology, 158, 543-561.

Wang, Q., Li, Z.X., Chung, S.L., Wyman, D.A., Sun, Y.L., Zhao, Z.H., Zhu, Y.T., Qiu, H.N., 2011. Late Triassic high-Mg andesite/dacite suites from northern Hohxil, north Tibet: Geochronology, geochemical characteristics, petrogenetic processes and tectonic implications. Lithos, 126(1-2),54-67.

Wang, H.Q., Ding, L., Kapp, P., Cai, F.L., et al., 2018. Earliest Cretaceous accretion of Neo-Tethys oceanic subduction along the Yarlung Zangbo Suture Zone, Sangsang area, southern Tibet. Tectonophysics, 744, 373-389.

Wolf, M.B., London, D., 1994. Apatite dissolution into peraluminous haplogranitic melts: An experimental study of solubilities and mechanisms. Geochimica et Cosmochimica Acta, 58, 4127-4145.

Xu, Y.G., Yang, Q.J., Lan, J.B., Luo, Z.Y., Huang, X.L., Shi, Y.R., Xie, L.W., 2012. Temporal-spatial distribution and tectonic implications of the batholiths in the Gaoligong-Tengliang-Yingjiang area, western Yunnan: Constraints from zircon U-Pb ages and Hf isotopes. J. Asian Earth Sci., 53, 151-175.

Yang, Q.J., Xu, Y.G., Huang, X.L., Luo, Z.Y., 2006. Geochronology and geochemistry granites in the Gaoligong tectonic belt, western Yunnan, tectonic implications. Acta Petrol. Sinica, 22, 817-834 (in Chinese with English abstract).

Yuan, H.L., Gao, S., Dai, M.N., Zong, C.L., Gunther, D., Fontaine, G.H., Liu, X.M., Diwu, C.R., 2008. Simultaneous determinations of U-Pb age, Hf isotopes and trace element compositions of zircon by excimer laser-ablation quadrupole and multiple-collector ICPMS. Chemical Geology, 247, 100-118.

Zen, E.A., 1986. Aluminum enrichment in silicate melts by fractional crystallization: Some mineralogic and petrographic constraints. Journal of Petrology, 27, 1095-1117.

Zhao, S.W., Lai, S.C., Qin, J.F., Zhu, R.Z., 2016. Tectono-magmatic evolution of the Gaoligong belt, southeastern margin of the Tibetan Plateau: Constraints from granitic gneisses and granitoid intrusions. Gondwana Research, 35 (1), 238-256.

Zhu, R.Z., Lai, S.C., Santosh, M., Qin, J.F., Zhao, S.W., 2017a. Early Cretaceous Na-rich granitoids and their enclaves in the Tengchong Block, SW China: Magmatism in relation to subduction of the Bangong-Nujiang Tethys ocean. Lithos, 286-287, 175-190.

Zhu, D.C., Mo, X.X., Niu, Y.L., Zhao, Z.D., Wang, L.Q., Liu, Y.S., Wu, F.Y., 2009. Geochemical investigation of early cretaceous igneous rocks along an east-west traverse through out the central Lhasa Terrane, Tibet. Chemical Geology, 268, 298-312.

Zhu, D.C., Zhao, Z.D., Niu, Y.L., Mo, X.X., Chung, S.L., Hou, Z.Q., Wang, L.Q., Wu, F.Y., 2011. The Lhasa Terrane: Record of a microcontinent and its histories of drift and growth. Earth Planet. Sci. Lett., 301, 241-255.

Zhu, D.C., Li, S.M., Cawood, P.A., Wang, Q., Zhao, Z.D., Liu, S.A., et al., 2016. Assembly of the Lhasa and Qiangtang terranes in Central Tibet by divergent double subduction. Lithos, 245, 7-17.

Zhu, R.Z., Lai, S.C., Qin, J.F., Zhao, S.W., Wang, J.B., 2017b. Late Early-cretaceous quartz diorite-granodiorite-monzogranite association from the Gaoligong belt, southeastern Tibet Plateau: Chemical variations and geodynamic implications. Lithos, 288-289, 311-325.

Zhu, R.Z., Lai, S.C., Qin, J.F., Zhao, S.W., Santosh, M., 2019. Petrogenesis of high-K calc alkaline granodiorite and its enclaves from the SE Lhasa block, Tibet (SW China): Implications for recycled subducted sediments. Geological Society of America Bulletin, 131 (7-8), 1224-1238.

Zhu, R.Z., Lai, S.C., Qin, J.F., Zhao, S.W., Santosh, M., 2018. Strongly peraluminous fractionated S-type granites in the Baoshan Block, SW China: Implications for two-stage melting of fertile continental materials following the closure of Bangong-Nujiang Tethys. Lithos, 316-317, 178-198.

Petrogenesis and geochemical diversity of Late Mesoproterozoic S-type granites in the western Yangtze Block, South China: Co-entrainment of peritectic selective phases and accessory minerals[①]

Zhu Yu   Lai Shaocong[②]   Qin Jiangfeng   Zhu Renzhi   Zhang Fangyi   Zhang Zezhong

Abstract: Deciphering the geochemical diversity of S-type granites is crucial for obtaining more profound insight into their petrogenesis. We therefore undertook an integrated study of whole-rock geochemistry, Sr-Nd isotopes, and zircon U-Pb-Hf isotopes for newly recognized Late Mesoproterozoic S-type granites, including two-mica-, biotite-, garnet-bearing two-mica granites, from the western Yangtze Block, South China. The crystallization ages of these granites are ca. 1 040 Ma (weighted mean $^{206}Pb/^{238}U$ age = 1 036. 4 ± 4. 5 Ma to 1 042. 2 ± 1. 1 Ma). They are peraluminous to strongly peraluminous [A/CNK = molar ratio of $Al_2O_3/(CaO + Na_2O + K_2O)$ = 1. 02−1. 67], high-K calc-alkaline rocks, and display high concentrations of normative-corundum (0. 54−7. 04 wt%) as well as positive correlations of A/CNK and $FeO^T + MgO$ values, which are characteristics of S-type granites. These S-type granites are characterized by enriched in Rb, Th, K, and Pb, depleted in Ba, Sr, Ti, and Eu, with negative whole-rock $\varepsilon_{Nd}(t)$ (−6. 8 to −3. 0) and predominantly negative zircon $\varepsilon_{Hf}(t)$ values (−8. 09 to +5. 70), indicating the affinity of middle-upper crustal trends and a heterogeneous metasedimentary source. Compared with the geochemical diversity of S-type granites around the world, the variably negative $\varepsilon_{Nd}(t)$ values as well as positive and negative $\varepsilon_{Hf}(t)$ values of our S-type granites may be caused by source heterogeneity and disequilibrium melting processes. More importantly, similar to typical more mafic S-type granites from the Cape Granite Suite (South Africa) and north Queensland (Australia), the high and variable $FeO^T + MgO$ contents (2. 21−6. 64 wt%) are significantly attributed to coupled co-entrainment of peritectic and accessory minerals (e.g., garnet, ilmenite, zircon, and monazite), evidenced by positive relationships between $FeO^T + MgO$ and TiO_2, CaO, Zr, Th, Hf, Y, Yb, light rare earth elements (LREEs). In conjunction with the existence of coeval rift-related A-type igneous rocks, depleted asthenosphere-derived mafic rocks, marine sedimentary sequences, and subsequent extensive Neoproterozoic arc magmatism along the western Yangtze Block, we infer that the ca. 1 040 Ma S-type granites studied here were produced in a continental rifting basin that was transformed into a compression setting during the Neoproterozoic.

---

① Published in *Lithos*, 2020, 352-353.
② Corresponding author.

1　Introduction

S-type granites are widespread in Earth's upper crust under various geodynamic conditions (Barbarin, 1999; Chappell and White, 1992; Collins and Richards, 2008), and are usually characterized by a peraluminous to strongly peraluminous composition and the existence of aluminum-rich mineral phases (e. g., muscovite, cordierite, garnet, andalusite, and tourmaline) (Champion and Bultitude, 2013; Chappell, 1999; Huang et al., 2019; Stevens et al., 2007; Sylvester, 1998). Traditionally, S-type granites are believed to originate from partial melting of supracrustal components, especially metasedimentary protoliths (Barbarin, 1999; Clemens, 2003; Sylvester, 1998). Nevertheless, an increasing number of studies have argued that S-type granites originate from a multi-component mixing source, including the mantle-derived, infracrustal, and supracrustal components (Appleby et al., 2010; Healy et al., 2004; Wang et al., 2013). More importantly, different magmatic mechanisms, including source heterogeneity, disequilibrium melting, and co-entrainment of selective peritectic phases and accessary minerals, have been adapted to interpret the geochemical diversity of S-type granitic melts around the world. These include more mafic features (i.e., high $FeO^T + MgO$ contents) and variable zircon $\varepsilon_{Hf}(t)$ values in S-type granites from the Cape Granite Suite in South Africa (Stevens et al., 2007; Villaros et al., 2009, 2012), heterogeneous Sr-Nd isotopes and more ferromagnesian contents in strongly fractionated S-type granites from the northern Queensland in Australia (Champion and Bultitude, 2013), significant single-zircon Hf-isotopes variations in S-type granites from southern China (Tang et al., 2014; Wang et al., 2018), Nd-Hf-O isotopic decoupling in peraluminous S-type granites from the Chinese Altai areas (Kong et al., 2019), and heterogeneous zircon Hf isotopes in strongly peraluminous fractionated S-type granites from the southwestern China (Zhu et al., 2018). In these cases, there is indeed a strong link between magma origination and geochemical variations of S-type granites (Stevens et al., 2007). More rigorous investigation into the geochemical diversity of S-type granites can thus shed important light on their petrogenesis.

The South China Block consists of the Yangtze Block to the northwest and the Cathaysia Block to the southeast (Fig. 1) (Wang et al., 2013; Zhao et al., 2011; Zhao and Cawood, 2012). The western Yangtze Block records voluminous Late Mesoproterozoic magmatism, and their generations is considered to be related to the assembly of the Rodinia supercontinent (Zhao and Cawood, 2012). It has been documented that these magmatic rocks contain alkaline basalts, tholeiitic basalts, and A-type magmatic rocks. The alkaline and tholeiitic basalts have been explained to be originated from the depleted asthenospheric mantle (Chen et al., 2014; Greentree et al., 2006). The A-type magmatic rocks are considered to be from granulite-facies lower crust (Zhu et al., 2016), juvenile lower crust (Chen et al., 2018), and ancient

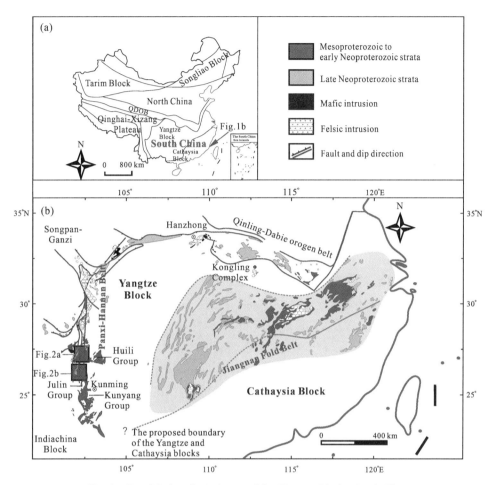

Fig. 1   Simplified geological map of the Yangtze Block, South China
(modified after Zhao and Cawood, 2012; Zhao et al., 2018).

continental crust sources (Chen et al., 2018; Wang et al., 2019). However, sparse research has studied S-type granites in detail. Here, we provide combined data of zircon U-Pb-Hf isotopes, whole-rock major- and trace-elements, and Sr-Nd isotopes from newly identified Late Mesoproterozoic S-type granites in the western Yangtze Block to decipher their petrogenesis and evaluate their geochemical diversity.

2 Regional geology and petrology

The South China Block comprises the Yangtze and Cathaysia blocks, which amalgamated along the ca. 1 500 km-long Neoproterozoic Jiangnan Fold Belt (Fig. 1b) (Wang et al., 2013; Zhao et al., 2011, 2018). The Yangtze Block is separated from the North China Craton by the Triassic Qinling-Dabie orogenic belt to the north and from the Indochina Block by the Ailaoshan-Songma suture zone to the southwest; it is bounded by the Songpan-Ganze terrane of the Tibetan

Plateau to the west (Lai et al., 2015; Wang et al., 2016; Zhao and Cawood, 2012; Zhu et al., 2019a) (Fig. 1). The Yangtze Block comprises Archean and Mesoproterozoic basement complexes, which are overlain by Neoproterozoic to Cenozoic cover (Zhao et al., 2018). The exposed Archean Kongling basement rocks, composed of felsic gneisses and metasedimentary rocks associated with minor amphibolite and mafic granulite (Gao et al., 1999, 2011), are sporadically located in the northern Yangtze Block (Zhao et al., 2018). The unexposed Archean basement is widely distributed beneath the Yangtze Block according to zircons from felsic granulite xenoliths in Mesozoic volcanic rocks (Zheng et al., 2006).

The western Yangtze Block is characterized by widespread Late Mesoproterozoic to Early Neoproterozoic volcanic-sedimentary sequences, including the Upper Kunyang, Upper Huili, Julin, and Yanbian groups (Chen et al., 2014; Greentree et al., 2006; Zhu et al., 2016). Apart from the Julin Group metamorphosed to lower amphibolite facies (Deng, 2000), the other groups have undergone only lower greenschist facies metamorphism (Li et al., 1988). The Kunyang Group is > 10 km in thickness and mainly consists of terrigenous clastic rocks, carbonates, and volcanic rocks. A tuff layer within the Heishantou Formation of the Kunyang Group yielded zircon U-Pb ages of 1 032 ± 9 Ma and 995 ± 15 Ma (Greentree et al., 2006; Zhang et al., 2007). Detrital zircons constrained the depositional age of the Upper Kunyang Group as young as ca. 960 Ma (Greentree et al., 2006). The Huili Group, with a total thickness > 10 km, is predominantly composed of meta-clastic and meta-carbonate rocks associated with meta-volcanic rocks (Chen et al., 2018). The Upper Huili Group is divided into the Limahe, Fengshan, and Tianbaoshan formations from the base upward (Wang et al., 2019). The U-Pb in zircons from Tianbaoshan volcanic rocks have been dated to 985 ± 16 Ma via thermal ionization mass spectrometry (TIMS) (Mou et al., 2003), 1 028 ± 9 Ma by sensitive high resolution ion microprobe (SHRIMP) (Geng et al., 2007), and 1 025±13 Ma and 1 021±6. 4 Ma by secondary ion mass spectrometry (SIMS) (Zhu et al., 2016). The mafic dykes that intruded into the lower Tianbaoshan Formation were dated to 1 023 ± 6. 7 Ma (Zhu et al., 2016). The Huidong granites, which intruded into the Upper Huili Group, have SIMS zircon U-Pb ages of 1 048. 5 ± 4. 9 Ma and 1 043. 1 ± 5. 1 Ma (Wang et al., 2019). The Julin Group, ~20 km southeast of the Huili Group, is a > 3 560 m-thick sequence of slate, gneiss, schist, sandstone, marble, and dolomite with subordinate meta-volcanic rocks (Chen et al., 2018; Wang et al., 2019). The Julin Group is composed of the Pudeng, Lugumo, Fenghuangshan, and Haizishao formations upwards from the base (Wang et al., 2019). The meta-basalts of the Pudeng Formation in the Julin Group have laser ablation-inductively coupled-mass spectrometry (LA-ICP-MS) zircon U-Pb ages of 1 043 ± 19 Ma and 1 050 ± 14 Ma (Chen et al., 2014). The recently identified Yuanmou granites were emplaced into the Julin Group and yielded LA-ICP-MS U-Pb ages of 1 041 ± 12 Ma (Wang et al., 2019). In addition, the A-type magmatic rocks, including the dacites and rhyolites in the Huili Group and granites in the Julin Group, were formed at ca. 1 050 Ma (Chen et al., 2018). The Yanbian Group is one of the earliest Neoproterozoic volcanic-sedimentary sequences and composed of volcanic and clastic rocks that were strongly deformed and variably metamorphosed (Zhu et al., 2016). Detrital zircons from the Yanbian Group yielded U-Pb ages ranging from ca. 1 000 Ma to 865 Ma with two peaks at ca. 920 Ma and 900 Ma (Sun et al., 2008).

The Late Mesoproterozoic granites in this study, including the two-mica granites, biotite granites, and garnet-bearing two-mica granites, were collected from the Mosuoying and Salian areas along the western margin of the Yangtze Block (southern China). In the Mosuoying area (Fig. 2a), the Late Mesoproterozoic to Neoproterozoic granites co-exist with the Neoproterozoic strata and Late Mesoproterozoic Tianbaoshan meta-sediments (Zhu et al., 2016). There have been reports of ca. 790−803 Ma granites from the Xiacun and Yonglang areas (Zhu et al., 2019a and reference therein). Late Mesoproterozoic S-type granites have also been reported in the study region and formed at ca. 1 050 Ma (Mabi et al., 2018). In this study, the two-mica granites, collected from the Mosuoying area, are medium to coarse-grained with granitic texture and composed of 25%−30% quartz, 20%−25% K-feldspar, 5%−10% perthite, 20%−25% plagioclase, 5%−7% muscovite, 5%−8% biotite, 0%−5% Fe-Ti oxides, and accessary minerals including zircon and ilmenite. The perthite is >0.5 cm in length. Quartz crystals are subhedral and intergrown with biotite and muscovite; some biotites are slightly chloritized (Fig. 3a-c). In

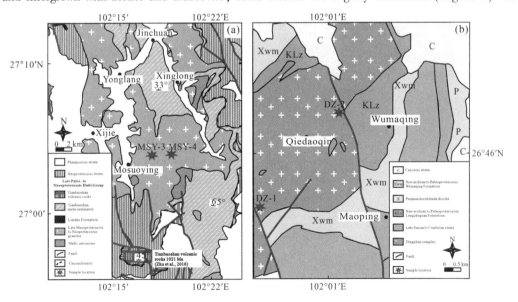

Fig. 2　Geological map showing disposition of studied granites in the Mosuoying (a) and Salian areas (b) from the western Yangtze Block, South China.

Fig. 2a was revised after Zhu et al. (2016). Fig. 2b was modified after the Salian 1：50 000 geological map, SPBGMR (2001).

the Salian area (Fig. 2b), the biotite granites and garnet-bearing two-mica granites were sampled from the Dingzhen complex that intruded into the Neo-archean to Paleoproterozoic strata (SPBGMR, 2001). The biotite granites are medium-grained texture and consist of 30%－35% quartz, 20%－30% K-feldspar, 15%－20% plagioclase, 5%－10% biotite, 0%－5% Fe-Ti oxides, zircon, and apatite. Some K-feldspars are subhedral and partly altered to kaolinite on the surface and other K-feldspar grains display gridiron twining. Plagioclases are subhedral to euhedral and slightly altered to sericite. Biotites occur between quartz and feldspar crystals. The Fe-Ti oxides infill the biotite grains (Fig. 3d-f). The medium-grained garnet-bearing two-mica granites are predominantly composed of 35%－40% quartz, 20%－30% K-feldspar, 10%－20% plagioclase,

Fig. 3　Field and hand specimen photographs as well as representative photomicrographs
of the two-mica granites(a-c), biotite granites(d-f), and garnet-bearing two-mica granites(g-l)
in the western Yangtze Block, South China.

Pl: plagioclase; Kfs: K-feldspar; Pth: perthite; Qtz: quartz; Bi: biotite; Mus: muscovite;
Grt: garnet. "＋": crossed polarized photograph; "-": single polarized photograph.

~5% garnet, 5% – 10% biotite, ~ 5% muscovite, and Fe-Ti oxides, zircon, apatite, and ilmenite. Muscovite and biotite are filled between the quartz and K-feldspar grains. The K-feldspars also show the distinct feature of crossed twinning and are partly altered to kaolinite (Fig. 3g-l).

3 Analytical methods

In this paper, zircon U-Pb-Hf isotopic analyses and whole-rock major- and trace-elemental analyses were processed at the State Key Laboratory of Continental Dynamics, Northwest University, Xi'an, China (SKLCD). Whole-rock Sr-Nd isotopes were performed in the Guizhou Tuopu Resource and Environmental Analysis Center in China.

3.1 Zircon U-Pb-Hf isotopes analysis

Zircon grains (samples MSY-3, MSY-4, DZ-2, and DZ-1) were separated using conventional heavy liquid and magnetic techniques. Representative zircon grains were handpicked and mounted in epoxy resin disks and then polished and carbon coated. The internal morphology of zircon grains was shown with cathodoluminescence (CL) microscopy before U-Pb isotopic dating. Laser ablation inductively-coupled plasma mass spectrometry (LA-ICP-MS) U-Pb analyses for four samples were conducted on an Agilent 7500a ICP-MS equipped with a 193 nm laser following the analytical method of Yuan et al. (2004). The $^{207}Pb/^{235}U$ and $^{206}Pb/^{238}U$ ratios were calculated using GLITTER program (Macquarie University), which was corrected using the Harvard zircon 91500 as an external calibration with a recommended $^{206}Pb/^{238}U$ isotopic age of 1 065.4 ± 0.6 Ma (Wiedenbeck et al., 2004). The GJ-1 is also a standard sample with a recommended $^{206}Pb/^{238}U$ isotopic age of 603.2 ± 2.4 Ma (Liu et al., 2007). Common Pb contents were subsequently evaluated using the method described in Andersen (2002). The age calculations and plotting of concordia diagrams were made using ISOPLOT (version 3.0; Ludwig, 2003). Uncertainties are quoted at the 2σ level.

In-situ zircon Lu-Hf isotopic analyses were performed on a Neptune MC-ICP-MS. The laser repetition rate was 10 Hz at 100 mJ and the spot size was 32 μm. Instrument information is available in Bao et al. (2017). The detailed analytical technique is depicted by Yuan et al. (2008). During the analyses, the measured values of well-characterized zircon standards (91500, GJ-1, and Monastery) were consistent with the recommended values within 2σ (Yuan et al., 2008). The obtained Hf isotopic compositions [GJ-1 standard = 0.282 016 ± 20 (2σ, $n = 84$), Monastery standard = 0.282 735 ± 24 (2σ, $n = 84$)] are also consistent with the recommended values (Yuan et al., 2008) within 2σ. The initial $^{176}Hf/^{177}Hf$ ratios and $\varepsilon_{Hf}(t)$ values were calculated with the chondritic reservoir (CHUR) at the time of zircon growth from the magmas. The decay constant for ^{176}Lu of 1.867×10^{-11} year^{-1}, chondritic

^{176}Hf/^{177}Hf ratio of 0.282 785 and ^{176}Lu/^{177}Hf ratio of 0.033 6 were adopted (Soderlund et al., 2004). The depleted mantle model ages (T_{DM}) used for basic rocks were calculated with reference to present-day depleted mantle ^{176}Hf/^{177}Hf ratio of 0.283 25, similar to that of the average MORB (Nowell et al., 1998) and ^{176}Lu/^{177}Hf = 0.038 4 (Griffin et al., 2000). For the zircons from felsic rocks, we also calculated the "crustal" Hf model age (T_{DMC}) by assuming that its parental magma was derived from an average continental crust with ^{176}Lu/^{177}Hf = 0.015, which was originated from the depleted mantle source (Griffin et al., 2000). Our conclusions would not be affected even if other decay constants were used.

3. 2   Whole-rock geochemistry and Sr-Nd isotopes

Weathered surfaces of samples were removed and the fresh parts were chipped and powdered to ~200 mesh using a tungsten carbide ball mill. Major and trace elements were examined by X-ray fluorescence (XRF; Rikagu RIX 2100) and inductively coupled plasma mass spectrometry (ICP-MS; Agilent 7500a), respectively. Analyses of USGS and Chinese national rock standards (BCR-2, GSR-1, and GSR-3) showed that the analytical precision and accuracy for the major elements were generally better than 5%. For the trace element analyses, sample powders were digested using an HF + HNO$_3$ mixture in high-pressure Teflon bombs at 190 ℃ for 48 h. For most trace elements, the analytical error was < 2% and the precision was > 10% (Liu et al., 2007b).

Sr and Nd isotopic fractionation was corrected to ^{87}Sr/^{86}Sr = 0.119 4 and ^{146}Nd/^{144}Nd = 0.721 9, respectively (Chu et al., 2009). During the analysis process, a Neptune multi-collector ICP-MS was used to measure the ^{87}Sr/^{86}Sr and ^{143}Nd/^{144}Nd isotope ratios. NIST SRM-987 and JMC-Nd were used as certified reference standard solutions for ^{87}Sr/^{86}Sr and ^{143}Nd/^{144}Nd isotopic ratios, respectively. BCR-1 and BHVO-1 were used as the reference materials.

4   Results

4. 1   LA-ICP-MS zircon U-Pb dating

Zircon CL images and LA-ICP-MS zircon U-Pb dating results for four granite samples, including two two-mica granites (MSY-3 and MSY-4), one biotite granite (DZ-2), and one garnet-bearing two-mica granite (DZ-1), are shown in Table 1 and illustrated in Figs. 4 and 5.

Sample MSY-3 is a two-mica granite from the Mosuoying area. Zircon grains from this sample were mostly euhedral long-prismatic crystals, with aspect ratios from 1∶1 to 3∶1. In the CL images (Fig. 4a), these zircon grains ranged from 30 μm to 200 μm and displayed gray and black. Most grains normally showed clear oscillatory zoning, indicating that they formed during magmatic evolution (Hoskin and Schaltegger, 2003). Zircons from MSY-3 contained low and

Table 1 LA-ICP-MS zircon U-Pb dating results for the Late Mesoproterozoic S-type granite in the western Yangtze Block, South China.

Analysis	Th	U	Th/U	$^{207}Pb/^{206}Pb$	1σ	$^{207}Pb/^{235}U$	1σ	$^{206}Pb/^{238}U$	1σ	$^{208}Pb/^{232}Th$	1σ	$^{207}Pb/^{206}Pb$	1σ	$^{207}Pb/^{235}U$	1σ	$^{206}Pb/^{238}U$	1σ	$^{208}Pb/^{232}Th$	1σ
MSY-3 (Two-mica granite, 27°4′43″N 102°16′31″E)																		
MSY-3-01	33.67	272.08	0.12	0.073 57	0.001 65	1.791 81	0.024 50	0.176 65	0.001 84	0.050 70	0.000 81	1 030	13	1 042	9	1 049	10	1 000	16
MSY-3-02	91.47	403.04	0.23	0.073 26	0.001 61	1.767 84	0.022 84	0.175 03	0.001 80	0.047 54	0.000 58	1 021	12	1 034	8	1 040	10	939	11
MSY-3-03	49.17	413.31	0.12	0.074 15	0.001 70	1.804 98	0.026 17	0.176 57	0.001 86	0.050 61	0.000 89	1 046	14	1 047	9	1 048	10	998	17
MSY-3-04	32.51	348.51	0.09	0.071 66	0.001 63	1.730 22	0.024 45	0.175 12	0.001 83	0.052 59	0.000 98	976	14	1 020	9	1 040	10	1 036	19
MSY-3-07	103.54	359.35	0.29	0.076 00	0.001 72	1.825 11	0.036 49	0.174 18	0.001 86	0.052 37	0.000 56	1 095	46	1 055	13	1 035	10	1 032	11
MSY-3-08	246.49	446.64	0.55	0.077 96	0.001 72	1.867 09	0.024 20	0.173 70	0.001 79	0.051 81	0.000 52	1 146	12	1 070	9	1 032	10	1 021	10
MSY-3-09	30.49	187.48	0.16	0.073 16	0.001 92	1.781 08	0.034 06	0.176 56	0.002 00	0.049 20	0.001 07	1 018	21	1 039	12	1 048	11	971	21
MSY-3-10	42.22	284.72	0.15	0.074 33	0.001 92	1.776 01	0.032 97	0.173 29	0.001 95	0.045 39	0.001 04	1 050	20	1 037	12	1 030	11	897	20
MSY-3-11	53.10	256.98	0.21	0.074 33	0.001 87	1.794 07	0.031 50	0.175 07	0.001 94	0.056 19	0.000 98	1 050	19	1 043	11	1 040	11	1 105	19
MSY-3-14	37.03	229.73	0.16	0.074 04	0.001 73	1.777 53	0.026 39	0.174 12	0.001 84	0.049 26	0.000 81	1 043	15	1 037	10	1 035	10	972	16
MSY-3-15	58.07	132.95	0.44	0.076 07	0.002 31	1.827 80	0.044 37	0.174 26	0.002 17	0.049 44	0.000 97	1 097	29	1 055	16	1 036	12	975	19
MSY-3-16	132.33	357.96	0.37	0.077 69	0.001 73	1.894 13	0.024 96	0.176 81	0.001 82	0.053 59	0.000 59	1 139	12	1 079	9	1 050	10	1 055	11
MSY-3-18	332.56	604.27	0.55	0.074 36	0.001 65	1.811 87	0.023 36	0.176 71	0.001 81	0.059 22	0.000 62	1 051	12	1 050	8	1 049	10	1 163	12
MSY-3-19	48.81	292	0.17	0.076 94	0.001 83	1.871 89	0.028 94	0.176 44	0.001 88	0.047 15	0.000 80	1 120	15	1 071	10	1 047	10	931	15
MSY-3-22	95.98	315.61	0.30	0.073 61	0.001 70	1.769 51	0.025 30	0.174 33	0.001 82	0.051 71	0.000 68	1 031	14	1 034	9	1 036	10	1 019	13
MSY-3-23	116.04	309.30	0.38	0.074 94	0.001 71	1.802 94	0.025 02	0.174 48	0.001 81	0.055 64	0.000 63	1 067	13	1 047	9	1 037	10	1 094	12
MSY-3-24	140.87	465.75	0.30	0.076 07	0.001 71	1.851 57	0.024 57	0.176 52	0.001 82	0.062 40	0.000 68	1 097	12	1 064	9	1 048	10	1 223	13
MSY-3-25	22.62	131.19	0.17	0.073 62	0.002 10	1.764 37	0.038 60	0.173 80	0.002 05	0.046 40	0.001 17	1 031	26	1 032	14	1 033	11	917	23
MSY-3-26	119.66	217.48	0.55	0.074 35	0.002 05	1.797 69	0.036 81	0.175 35	0.002 02	0.052 33	0.000 84	1 051	23	1 045	13	1 042	11	1 031	16
MSY-3-27	62.04	197	0.31	0.077 20	0.001 89	1.876 92	0.030 37	0.176 32	0.001 90	0.053 91	0.000 78	1 126	16	1 073	11	1 047	10	1 061	15
MSY-3-28	30.83	297.83	0.10	0.072 79	0.002 02	1.741 54	0.036 10	0.173 50	0.002 01	0.052 38	0.001 40	1 008	24	1 024	13	1 031	11	1 032	27

Continued

Analysis	Content/ppm			ratios								Age/Ma							
	Th	U	Th/U	$\frac{^{207}Pb}{^{206}Pb}$	1σ	$\frac{^{207}Pb}{^{235}U}$	1σ	$\frac{^{206}Pb}{^{238}U}$	1σ	$\frac{^{208}Pb}{^{232}Th}$	1σ	$\frac{^{207}Pb}{^{206}Pb}$	1σ	$\frac{^{207}Pb}{^{235}U}$	1σ	$\frac{^{206}Pb}{^{238}U}$	1σ	$\frac{^{208}Pb}{^{232}Th}$	1σ
MSY-3-29	41.33	230.72	0.18	0.075 18	0.001 70	1.827 30	0.036 10	0.176 28	0.001 92	0.053 07	0.000 54	1 073	46	1 055	13	1 047	11	1 045	10
MSY-3-30	44.46	229.93	0.19	0.075 59	0.001 81	1.816 47	0.028 01	0.174 28	0.001 85	0.052 88	0.000 85	1 084	15	1 051	10	1 036	10	1 042	16
MSY-3-33	273.28	566.09	0.48	0.073 56	0.001 73	1.762 21	0.036 91	0.173 74	0.001 84	0.052 43	0.000 51	1 029	49	1 032	14	1 033	10	1 033	10
MSY-3-34	167.72	182.33	0.92	0.074 09	0.003 13	1.777 67	0.071 42	0.174 01	0.002 27	0.052 47	0.000 56	1 044	87	1 037	26	1 034	12	1 034	11
MSY-3-36	194.42	394.34	0.49	0.074 34	0.002 21	1.778 53	0.041 07	0.173 49	0.002 09	0.052 83	0.000 93	1 051	27	1 038	15	1 031	11	1 041	18
MSY-3-05	225.21	591.31	0.38	0.133 42	0.002 78	4.765 36	0.052 33	0.259 07	0.002 62	0.073 82	0.000 69	2 143	9	1 779	9	1 485	13	1 440	13
MSY-3-12	271.89	414.23	0.66	0.096 45	0.002 22	3.411 58	0.069 04	0.256 53	0.002 81	0.075 19	0.000 77	1 557	44	1 507	16	1 472	14	1 465	14
MSY-3-20	58.39	409.77	0.14	0.088 18	0.001 93	2.770 86	0.034 34	0.227 89	0.002 34	0.067 49	0.000 91	1 386	11	1 348	9	1 323	12	1 320	17
MSY-3-21	54.93	1 194.19	0.05	0.097 10	0.001 43	3.523 72	0.038 72	0.263 20	0.002 58	0.077 09	0.000 78	1 569	28	1 533	9	1 506	13	1 501	15
MSY-4 (Two-mica granite, 27°4′46″N 102°16′41″E)																			
MSY-4-01	39.95	440.64	0.09	0.075 00	0.001 81	1.796 88	0.028 93	0.174 17	0.001 88	0.057 25	0.001 33	1 069	16	1 044	11	1 035	10	1 125	25
MSY-4-02	38.12	343.92	0.11	0.073 15	0.001 68	1.764 09	0.025 46	0.175 27	0.001 84	0.048 90	0.000 88	1 018	14	1 032	9	1 041	10	965	17
MSY-4-04	45.52	310.36	0.15	0.074 56	0.001 52	1.783 70	0.030 90	0.173 50	0.001 84	0.052 28	0.000 53	1 057	42	1 040	11	1 031	10	1 030	10
MSY-4-05	71.19	927.58	0.08	0.076 04	0.001 55	1.811 61	0.031 35	0.172 80	0.001 85	0.051 96	0.000 65	1 096	42	1 050	11	1 028	10	1 024	13
MSY-4-06	37.63	331.14	0.11	0.074 41	0.001 55	1.775 97	0.031 90	0.173 09	0.001 84	0.052 17	0.000 70	1 053	43	1 037	12	1 029	10	1 028	13
MSY-4-07	40.33	277.52	0.15	0.074 48	0.001 72	1.814 35	0.026 91	0.176 83	0.001 88	0.059 40	0.000 98	1 055	14	1 051	10	1 050	10	1 166	19
MSY-4-08	49.07	331.08	0.15	0.074 36	0.001 65	1.806 76	0.024 39	0.176 35	0.001 84	0.055 48	0.000 81	1 051	13	1 048	9	1 047	10	1 091	16
MSY-4-13	28.80	558.61	0.05	0.072 35	0.001 63	1.713 50	0.033 66	0.171 76	0.001 90	0.051 93	0.001 21	996	47	1 014	13	1 022	10	1 023	23
MSY-4-15	37.43	334.16	0.11	0.074 96	0.002 37	1.796 30	0.047 36	0.173 66	0.002 29	0.056 29	0.002 16	1 067	32	1 044	17	1 032	13	1 107	41
MSY-4-16	38.84	323.21	0.12	0.071 32	0.001 61	1.712 53	0.025 41	0.173 98	0.001 88	0.049 86	0.000 91	967	15	1 013	10	1 034	10	983	18
MSY-4-18	44.74	196.31	0.23	0.071 78	0.001 76	1.732 91	0.030 57	0.174 89	0.001 98	0.055 14	0.001 04	980	19	1 021	11	1 039	11	1 085	20
MSY-4-19	39.50	277.87	0.14	0.072 08	0.001 68	1.752 45	0.028 14	0.176 07	0.001 95	0.068 36	0.001 08	988	16	1 028	10	1 045	11	1 337	20

Continued

Analysis	Content/ppm			ratios								Age/Ma							
	Th	U	Th/U	$\frac{207Pb}{206Pb}$	1σ	$\frac{207Pb}{235U}$	1σ	$\frac{206Pb}{238U}$	1σ	$\frac{208Pb}{232Th}$	1σ	$\frac{207Pb}{206Pb}$	1σ	$\frac{207Pb}{235U}$	1σ	$\frac{206Pb}{238U}$	1σ	$\frac{208Pb}{232Th}$	1σ
MSY-4-20	34.19	258.58	0.13	0.070 54	0.002 11	1.696 71	0.041 54	0.174 18	0.002 21	0.047 17	0.001 57	944	30	1 007	16	1 035	12	932	30
MSY-4-22	38.89	274.64	0.14	0.075 02	0.001 63	1.824 42	0.025 17	0.176 08	0.001 89	0.064 63	0.000 97	1 069	13	1 054	9	1 046	10	1 266	18
MSY-4-24	33.63	440.94	0.08	0.074 01	0.001 56	1.810 01	0.023 45	0.177 06	0.001 88	0.071 73	0.001 14	1 042	12	1 049	8	1 051	10	1 400	21
MSY-4-25	41.89	289.78	0.14	0.068 36	0.001 59	1.662 49	0.027 13	0.176 03	0.001 97	0.049 87	0.000 97	879	17	994	10	1 045	11	984	19
MSY-4-26	55.86	375.95	0.15	0.067 27	0.001 53	1.638 44	0.025 57	0.176 28	0.001 95	0.048 74	0.000 90	846	16	985	10	1 047	11	962	17
MSY-4-27	27.85	264.35	0.11	0.069 10	0.001 52	1.671 11	0.024 39	0.175 04	0.001 92	0.050 17	0.001 10	902	14	998	9	1 040	11	989	21
MSY-4-28	43.68	303.4	0.14	0.067 78	0.001 63	1.636 24	0.028 60	0.174 71	0.001 99	0.049 31	0.000 94	862	19	984	11	1 038	11	973	18
MSY-4-29	56.34	273.91	0.21	0.069 00	0.001 53	1.672 58	0.024 98	0.175 44	0.001 93	0.050 77	0.000 78	899	15	998	9	1 042	11	1 001	15
MSY-4-30	53.44	215.22	0.25	0.076 28	0.001 67	1.836 18	0.026 86	0.174 21	0.001 92	0.048 60	0.000 71	1 102	14	1 058	10	1 035	11	959	14
MSY-4-31	45.53	493.39	0.09	0.067 96	0.001 40	1.648 21	0.021 14	0.175 51	0.001 89	0.055 61	0.000 89	867	12	989	8	1 042	10	1 094	17
MSY-4-32	136.33	423.23	0.32	0.067 93	0.001 48	1.641 04	0.023 81	0.174 83	0.001 92	0.052 23	0.000 78	866	14	986	9	1 039	11	1 029	15
MSY-4-34	45.22	348.12	0.13	0.078 52	0.001 89	1.904 93	0.034 04	0.175 58	0.002 06	0.116 55	0.002 56	1 160	18	1 083	12	1 043	11	2 228	46
MSY-4-10	78.45	222.84	0.35	0.085 39	0.001 94	2.831 71	0.041 09	0.240 59	0.002 60	0.075 44	0.001 01	1 324	13	1 364	13	1 390	14	1 470	19
MSY-4-11	44.4	363.08	0.12	0.088 06	0.001 48	2.675 60	0.035 63	0.220 37	0.002 26	0.065 22	0.000 66	1 384	33	1 322	10	1 284	12	1 277	13
MSY-4-14	53.88	341.08	0.16	0.081 87	0.001 76	2.163 01	0.028 29	0.191 47	0.002 02	0.051 43	0.000 74	1 242	12	1 169	9	1 129	11	1 014	14
MSY-4-21	98.31	282.29	0.35	0.075 60	0.001 63	2.047 42	0.027 67	0.196 11	0.002 10	0.057 62	0.000 69	1 084	12	1 131	9	1 154	11	1 132	13
MSY-4-33	89.34	364.98	0.24	0.099 03	0.002 45	3.427 17	0.075 05	0.250 99	0.002 91	0.073 37	0.001 30	1 606	47	1 511	17	1 444	15	1 431	24
MSY-4-35	41.34	241.78	0.17	0.081 02	0.001 79	2.221 73	0.033 61	0.198 45	0.002 24	0.082 00	0.001 25	1 222	14	1 188	11	1 167	12	1 593	23
MSY-4-36	159.51	262.92	0.61	0.134 08	0.002 60	7.270 88	0.081 23	0.392 48	0.004 25	0.107 25	0.001 06	2 152	9	2 145	10	2 134	20	2 059	19
DZ-2（Biotite granite，26°47′24″N 102°1′9″E）																			
DZ-2-03	92.08	520.93	0.18	0.077 17	0.001 40	1.850 47	0.027 80	0.173 92	0.001 78	0.052 21	0.000 52	1 125	37	1 064	10	1 034	10	1 029	10
DZ-2-10	72.88	484.92	0.15	0.073 80	0.001 96	1.773 15	0.035 22	0.174 25	0.002 01	0.051 98	0.001 33	1 036	22	1 036	13	1 035	11	1 024	26

Continued

Analysis	Content/ppm			ratios								Age/Ma							
	Th	U	Th/U	$^{207}Pb/^{206}Pb$	1σ	$^{207}Pb/^{235}U$	1σ	$^{206}Pb/^{238}U$	1σ	$^{208}Pb/^{232}Th$	1σ	$^{207}Pb/^{206}Pb$	1σ	$^{207}Pb/^{235}U$	1σ	$^{206}Pb/^{238}U$	1σ	$^{208}Pb/^{232}Th$	1σ
DZ-2-11	32.72	266.89	0.12	0.074 23	0.001 67	1.795 62	0.025 31	0.175 44	0.001 84	0.052 86	0.000 90	1 048	13	1 044	9	1 042	10	1 041	17
DZ-2-12	66.04	273.31	0.24	0.074 30	0.001 83	1.799 54	0.030 74	0.175 65	0.001 93	0.048 88	0.000 85	1 050	18	1 045	11	1 043	11	965	16
DZ-2-19	44.29	328.87	0.13	0.072 82	0.001 71	1.745 90	0.027 06	0.173 88	0.001 86	0.052 42	0.000 96	1 009	16	1 026	10	1 033	10	1 033	18
DZ-2-20	568.51	649.22	0.88	0.075 09	0.002 89	1.782 11	0.065 15	0.172 13	0.002 07	0.051 82	0.000 53	1 071	79	1 039	24	1 024	11	1 021	10
DZ-2-23	28.52	271	0.11	0.072 81	0.001 66	1.766 39	0.025 36	0.175 95	0.001 85	0.051 06	0.000 94	1 009	14	1 033	9	1 045	10	1 007	18
DZ-2-24	37.75	345.24	0.11	0.072 98	0.001 60	1.775 89	0.023 22	0.176 50	0.001 82	0.056 55	0.000 87	1 013	12	1 037	8	1 048	10	1 112	17
DZ-2-26	93.8	179.61	0.52	0.077 76	0.002 27	1.844 41	0.049 56	0.172 03	0.001 95	0.051 60	0.000 55	1 141	59	1 061	18	1 023	11	1 017	11
DZ-2-29	54.72	286.64	0.19	0.074 02	0.001 40	1.791 51	0.028 42	0.175 55	0.001 81	0.052 94	0.000 53	1 042	39	1 042	10	1 043	10	1 043	10
DZ-2-31	212.11	445.25	0.48	0.074 21	0.001 88	1.770 48	0.040 73	0.173 04	0.001 86	0.052 17	0.000 54	1 047	52	1 035	15	1 029	10	1 028	10
DZ-2-33	350.39	799.59	0.44	0.076 45	0.001 71	1.834 28	0.036 12	0.174 01	0.001 83	0.052 29	0.000 52	1 107	46	1 058	13	1 034	10	1 030	10
DZ-2-35	93.38	219.04	0.43	0.074 37	0.002 18	1.776 25	0.047 82	0.173 23	0.002 02	0.052 21	0.000 55	1 051	61	1 037	17	1 030	11	1 029	11
DZ-2-36	167.73	338.71	0.50	0.077 67	0.001 78	1.879 39	0.038 08	0.175 51	0.001 86	0.052 65	0.000 52	1 138	47	1 074	13	1 042	10	1 037	10
DZ-2-02	320.08	388.68	0.82	0.100 33	0.002 13	3.299 12	0.039 12	0.238 47	0.002 46	0.066 20	0.000 63	1 630	10	1 481	9	1 379	13	1 296	12
DZ-2-18	655.55	583.75	1.12	0.090 98	0.001 90	3.120 32	0.035 05	0.248 75	0.002 52	0.074 58	0.000 65	1 446	10	1 438	9	1 432	13	1 454	12
DZ-1 (Garnet-bearing two-mica granite, 26°45′16″N 101°59′35″E)																			
DZ-1-01	133.71	176.67	0.76	0.074 08	0.001 31	1.795 41	0.026 95	0.175 79	0.001 20	0.050 85	0.000 51	1 044	35	1 044	7	1 044	10	1 003	10
DZ-1-02	174.02	192.59	0.90	0.073 60	0.001 21	1.776 81	0.024 05	0.175 10	0.001 13	0.049 50	0.000 42	1 031	33	1 040	6	1 037	9	977	8
DZ-1-03	102.15	157.8	0.65	0.074 64	0.001 46	1.800 16	0.030 86	0.174 94	0.001 29	0.050 11	0.000 62	1 058	39	1 039	7	1 046	11	988	12
DZ-1-04	268.39	308.77	0.87	0.075 53	0.001 04	1.823 17	0.018 61	0.175 08	0.001 00	0.049 35	0.000 35	1 083	27	1 040	5	1 054	7	974	7
DZ-1-05	146.57	200.49	0.73	0.074 81	0.001 15	1.806 84	0.022 22	0.175 19	0.001 08	0.049 82	0.000 43	1 063	31	1 041	6	1 048	8	983	8
DZ-1-06	72.49	117.83	0.62	0.075 41	0.001 58	1.835 99	0.034 53	0.176 60	0.001 39	0.052 28	0.000 71	1 079	42	1 048	8	1 058	12	1 030	14
DZ-1-07	161.21	181.67	0.89	0.077 47	0.001 26	1.885 38	0.025 28	0.176 54	0.001 14	0.051 85	0.000 46	1 133	32	1 048	6	1 076	9	1 022	9

Continued

Analysis	Content/ppm			ratios								Age/Ma							
	Th	U	Th/U	$\frac{^{207}Pb}{^{206}Pb}$	1σ	$\frac{^{207}Pb}{^{235}U}$	1σ	$\frac{^{206}Pb}{^{238}U}$	1σ	$\frac{^{208}Pb}{^{232}Th}$	1σ	$\frac{^{207}Pb}{^{206}Pb}$	1σ	$\frac{^{207}Pb}{^{235}U}$	1σ	$\frac{^{206}Pb}{^{238}U}$	1σ	$\frac{^{208}Pb}{^{232}Th}$	1σ
DZ-1-08	101.62	159.73	0.64	0.073 67	0.001 17	1.790 33	0.023 33	0.176 28	0.001 12	0.048 33	0.000 47	1 032	32	1 047	6	1 042	8	954	9
DZ-1-09	189.56	235.12	0.81	0.074 08	0.001 08	1.801 29	0.020 51	0.176 38	0.001 05	0.050 95	0.000 40	1 044	29	1 047	6	1 046	7	1 005	8
DZ-1-10	214.15	283.01	0.76	0.072 79	0.001 06	1.766 06	0.020 14	0.176 00	0.001 04	0.050 99	0.000 41	1 008	29	1 045	6	1 033	7	1 005	8
DZ-1-11	86.31	134.82	0.64	0.075 27	0.001 32	1.823 65	0.027 27	0.175 74	0.001 20	0.050 54	0.000 55	1 076	35	1 044	7	1 054	10	997	11
DZ-1-12	78.43	121.89	0.64	0.074 41	0.001 33	1.797 96	0.027 65	0.175 27	0.001 21	0.050 74	0.000 56	1 053	36	1 041	7	1 045	10	1 000	11
DZ-1-14	127.5	181.1	0.70	0.073 19	0.001 18	1.768 66	0.023 42	0.175 29	0.001 11	0.050 43	0.000 47	1 019	32	1 041	6	1 034	9	994	9
DZ-1-15	205.12	240.94	0.85	0.073 47	0.001 14	1.774 26	0.022 29	0.175 19	0.001 09	0.049 84	0.000 42	1 027	31	1 041	6	1 036	8	983	8
DZ-1-16	180.08	225.27	0.80	0.077 17	0.001 22	1.871 80	0.024 16	0.175 95	0.001 12	0.051 74	0.000 46	1 126	31	1 045	6	1 071	9	1 020	9
DZ-1-17	294.15	325.35	0.90	0.073 79	0.001 03	1.784 25	0.019 03	0.175 40	0.001 01	0.050 62	0.000 37	1 036	28	1 042	6	1 040	7	998	7
DZ-1-18	158.45	222.17	0.71	0.073 77	0.001 16	1.786 74	0.022 92	0.175 70	0.001 10	0.053 67	0.000 49	1 035	31	1 043	6	1 041	8	1 057	9
DZ-1-19	142.3	189.7	0.75	0.072 61	0.001 16	1.755 28	0.023 02	0.175 35	0.001 11	0.051 42	0.000 47	1 003	32	1 042	6	1 029	8	1 014	9
DZ-1-20	133.25	199.56	0.67	0.072 90	0.001 17	1.753 27	0.023 29	0.174 46	0.001 11	0.050 13	0.000 48	1 011	32	1 037	6	1 028	9	989	9
DZ-1-22	173.04	216.75	0.80	0.073 88	0.001 37	1.783 45	0.028 73	0.175 12	0.001 25	0.050 66	0.000 53	1 038	37	1 040	7	1 039	10	999	10
DZ-1-23	144.65	192.88	0.75	0.074 03	0.001 38	1.783 38	0.028 98	0.174 76	0.001 25	0.049 54	0.000 55	1 042	37	1 038	7	1 039	11	977	11
DZ-1-25	140.48	174.33	0.81	0.071 49	0.001 13	1.721 43	0.022 40	0.174 67	0.001 09	0.052 39	0.000 47	972	32	1 038	6	1 017	8	1 032	9
DZ-1-27	130.52	185.48	0.70	0.074 64	0.001 27	1.801 58	0.025 86	0.175 09	0.001 17	0.051 18	0.000 53	1 059	34	1 040	6	1 046	9	1 009	10
DZ-1-28	131.62	184.75	0.71	0.073 46	0.001 17	1.780 21	0.023 43	0.175 79	0.001 12	0.051 47	0.000 49	1 027	32	1 044	6	1 038	9	1 015	9
DZ-1-29	179.96	243.87	0.74	0.072 57	0.001 03	1.752 88	0.019 27	0.175 23	0.001 02	0.052 56	0.000 43	1 002	29	1 041	6	1 028	7	1 035	8
DZ-1-30	186.92	245.46	0.76	0.073 07	0.001 07	1.769 10	0.020 42	0.175 63	0.001 05	0.050 86	0.000 43	1 016	29	1 043	6	1 034	7	1 003	8
DZ-1-34	166.87	201.91	0.83	0.073 41	0.001 19	1.780 49	0.023 97	0.175 93	0.001 13	0.051 16	0.000 48	1 025	32	1 045	6	1 038	9	1 009	9
DZ-1-36	236.69	289.71	0.82	0.072 72	0.001 06	1.756 45	0.020 02	0.175 21	0.001 04	0.050 46	0.000 42	1 006	29	1 041	6	1 030	7	995	8

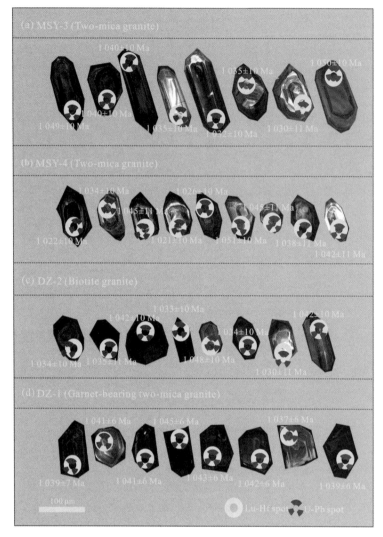

Fig. 4　Zircon cathodoluminescence (CL) images of representative grains for the two-mica granites (a,b), biotite granites (c), and garnet-bearing two-mica granites (d) in the western Yangtze Block, South China.

variable Th (22.6–333 ppm) and U contents (131–604 ppm), and low Th/U ratios (0.09–0.92). Twenty-six concordant spots yielded two subgroups of ^{206}Pb/^{238}U ages at ca. 1 050 Ma and ca. 1 030 Ma, characterized by a weighted mean age of 1 040.0±2.8 Ma (MSWD=1.7, n=26) (Fig. 5a,b). In addition, the four zircon grains displayed distinctly older U-Pb ages (1 323–1 506 Ma) (Table 1) that were probably inherited from the source region (Huang et al., 2019).

　　Zircon grains, separated from the two-mica granite sample (MSY-4), were subhedral to euhedral short-prismatic crystals, with aspect ratios around 1∶1–2∶1. In CL images (Fig. 4b), they were furvous and ranged from 50 μm to 100 μm. The sample had variable U (196–928 ppm) and Th (27.9–136 ppm) contents, and low Th/U ratios (0.05–0.32). There were

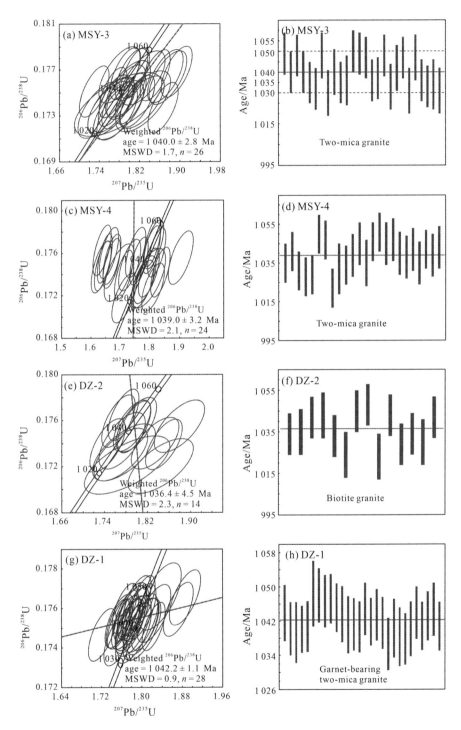

Fig. 5　Laser ablation-inductively coupled plasma-mass spectrometry (LA-ICP-MS) U-Pb zircon concordia diagrams for the Late Mesoproterozoic two-mica granites (a,d), biotite granites (e,f), and garnet-bearing two-mica granites (g,h) in the western Yangtze Block, South China.

seven inherited that showed variable and ancient U-Pb ages of 1 129- 2 134 Ma (Table 1). Among the other 24 concordant analyses plots, they displayed a concordant weighted mean age of 1 039. 0 ± 3. 2 Ma (MSWD = 2. 1, n = 24) (Fig. 5c,d).

Sample DZ-2 is a biotite granite collected from the Salian area. Zircon grains were subhedral to euhedral short-prismatic crystals. They ranged in size from 50 μm to 150 μm with length to width ratios of 1∶1-3∶1. In CL images (Fig. 4c), these zircon grains were colorless to gray. Fourteen analytical spots showed variable Th (28. 5-569 ppm) and U (180-800 ppm) contents as well as low Th/U ratios (0. 11-0. 52). These concordant spots were plotted in a group with a weighted mean $^{206}Pb/^{238}U$ age of 1 036. 4 ± 4. 5 Ma (MSWD = 2. 3, n = 14) (Fig. 5e,f). Two inherited grains had relatively older ages of 1 379 ± 13 Ma and 1 432 ± 13 Ma (Table 1).

Sample DZ-1 is a garnet-bearing two-mica granite from the Salian area. Zircon grains were euhedral to subhedral crystals and varied from 60 μm to 150 μm in length with aspect ratios of 1∶1 to 3∶2 (Fig. 4d). They displayed variable Th (72. 5-294 ppm) and U (118-325 ppm) contents as well as relatively high Th/U ratios (0. 62 - 0. 90). Twenty-eight concordant analyses were conducted and plotted in a group showing a weighted mean age of 1 042. 2 ± 1. 1 Ma (MSWD = 0. 9, n = 28) (Fig. 5g,h), which represents their emplacement time.

4. 2　Whole-rock major and trace elemental geochemistry

Whole-rock major and trace elements of 13 two-mica granites, 4 biotite granites, and 5 garnet-bearing two-mica granite samples from the Mosuoying and Salian areas along the western Yangtze Block are presented in Table 2, and Figs. 6 and 7.

All studied samples displayed moderate to high contents of SiO_2(65. 0-76. 9 wt%), total alkalis ($Na_2O + K_2O$ = 5. 88-8. 17 wt%), and K_2O/Na_2O ratios (0. 96-2. 48), plotting in the high-K calc-alkaline series (Fig. 6b). They contained highly variable Al_2O_3(11. 1-17. 0 wt%) and CaO (0. 74-2. 08 wt%) contents, and molar ratios of $Al_2O_3/(CaO + Na_2O + K_2O)$ (A/CNK) of 1. 02-1. 67, indicating that they classify as peraluminous to strongly peraluminous rocks (Fig. 6c). They also have high $FeO^T + MgO$ contents (2. 21-6. 64 wt%) and variable $FeO^T/(FeO^T + MgO)$ ratios (0. 76-0. 97) (Fig. 6d) as well as low MgO (0. 11-1. 45 wt%) contents and $Mg^\#$ values (7. 03-39. 6). In addition, these samples possessed variable contents of P_2O_5(0. 01-0. 14 wt%), TiO_2(0. 18-0. 73 wt%), and MnO (0. 02-0. 07 wt%).

The studied granites have high and variable total rare earth element (Σ REE) concentrations ranging from 168 ppm to 506 ppm. In the chondrite-normalized REE pattern diagram (Fig. 7a,c), they showed light rare earth elements (LREEs) enrichment relative to heavy rare earth elements (HREEs), yielding highly variable $(La/Yb)_N$(4. 55 - 44. 2), $(La/Sm)_N$(2. 05-5. 15), and $(Gd/Yb)_N$(1. 44-5. 10) ratios as well as moderately negative

Table 2 Major (wt%) and trace (ppm) element analysis results for the Late Mesoproterozoic S-type granites from the western Yangtze Block, South China.

Sample	MSY-3-1	MSY-3-2	MSY-3-3	MSY-3-4	MSY-3-5	MSY-3-6	MSY-4-1	MSY-4-2	MSY-4-3	MSY-4-4	MSY-4-5	MSY-4-6	MSY-4-7	DZ-2-1	DZ-2-2	DZ-2-3	DZ-2-4	DZ-1-1	DZ-1-2	DZ-1-3	DZ-1-4	DZ-1-5
Lithology	Two-mica granite													Biotite granite				Garnet-bearing two-mica granite				
SiO_2	67.1	66.1	65.0	65.7	67.8	66.0	66.2	66.7	68.0	66.3	65.6	66.6	65.6	71.0	71.5	71.6	71.2	76.9	76.5	76.6	76.6	76.6
TiO_2	0.71	0.68	0.64	0.68	0.65	0.70	0.69	0.61	0.63	0.56	0.73	0.59	0.57	0.35	0.33	0.35	0.37	0.20	0.20	0.18	0.19	0.21
Al_2O_3	15.7	16.4	17.0	16.2	15.6	16.3	16.3	15.8	15.0	16.5	16.1	16.1	16.6	14.9	14.7	14.6	14.7	11.1	11.5	11.4	11.5	11.2
$Fe_2O_3^T$	5.59	5.23	5.48	5.77	5.10	5.74	5.77	4.85	4.94	4.73	5.69	4.72	4.74	2.18	2.10	1.91	2.16	3.18	3.39	3.17	3.21	3.21
MnO	0.06	0.05	0.06	0.06	0.05	0.06	0.06	0.05	0.05	0.04	0.06	0.05	0.05	0.03	0.03	0.02	0.02	0.05	0.07	0.07	0.06	0.04
MgO	1.33	1.31	1.44	1.41	1.28	1.45	1.45	1.21	1.22	1.33	1.42	1.19	1.17	0.48	0.47	0.49	0.51	0.14	0.11	0.11	0.11	0.11
CaO	1.28	1.31	1.58	1.45	1.03	1.16	1.18	1.37	1.37	1.31	1.44	1.26	1.53	1.84	1.83	2.08	2.01	0.83	0.88	0.78	0.74	0.79
Na_2O	1.74	1.87	2.09	1.84	1.98	1.92	1.95	2.09	2.06	2.22	1.96	2.24	2.39	3.84	3.99	3.90	3.76	2.43	2.50	2.46	2.44	2.46
K_2O	4.16	4.64	4.10	4.25	3.90	4.16	4.17	4.49	3.90	4.38	3.99	4.24	4.54	4.33	4.05	3.76	3.99	4.23	4.14	4.40	4.46	4.31
P_2O_5	0.14	0.12	0.12	0.13	0.13	0.13	0.13	0.13	0.13	0.13	0.15	0.13	0.12	0.08	0.08	0.08	0.08	0.02	0.01	0.02	0.02	0.02
LOI	2.29	1.91	2.03	2.28	2.05	2.05	2.05	2.43	2.31	2.50	2.40	2.40	2.40	0.66	0.85	0.98	1.11	0.68	0.68	0.60	0.66	0.71
Total	100.1	99.6	99.5	99.8	99.5	99.7	100.0	99.7	99.6	100.0	99.5	99.5	99.7	99.7	99.9	99.7	99.9	99.8	100.0	99.8	100.0	99.6
K_2O/Na_2O	2.39	2.48	1.96	2.31	1.97	2.17	2.14	2.15	1.89	1.97	2.04	1.89	1.90	1.13	1.02	0.96	1.06	1.74	1.66	1.79	1.83	1.75
$Mg^\#$	35.7	36.9	38.0	36.3	36.9	37.1	36.9	36.8	36.5	39.6	36.8	37.0	36.5	33.9	34.3	37.4	35.5	9.31	7.03	7.48	7.40	7.40
A/CNK	1.62	1.56	1.58	1.58	1.67	1.67	1.66	1.46	1.49	1.53	1.59	1.52	1.42	1.04	1.03	1.02	1.04	1.10	1.13	1.11	1.13	1.10
FeO^T+MgO	6.36	6.02	6.37	6.60	5.87	6.61	6.64	5.57	5.67	5.59	6.54	5.44	5.44	2.44	2.36	2.21	2.45	3.00	3.16	2.96	3.00	3.00
Li	39.2	36.5	42.7	34.1	40.8	41.1	13.4	12.4	34.8	11.8	14.4	11.8	11.3	13.7	11.3	12.4	11.1	8.12	7.28	8.21	6.60	6.12
Be	3.70	4.06	4.06	3.44	4.22	4.20	3.53	3.48	3.89	3.47	3.57	3.57	3.48	5.06	4.84	4.11	4.42	3.64	3.83	3.48	3.45	3.57
Sc	10.9	10.7	10.7	10.0	11.5	11.5	9.24	9.62	10.6	8.99	11.9	8.84	9.22	4.16	3.92	4.36	4.07	4.68	4.80	4.97	4.20	4.12
V	65.5	64.3	64.0	60.9	69.5	69.9	54.6	56.3	64.8	52.3	67.8	51.7	53.5	20.3	19.1	21.4	20.0	2.11	2.19	2.24	2.16	2.17

Continued

Sample	MSY-3-1	MSY-3-2	MSY-3-3	MSY-3-4	MSY-3-5	MSY-3-6	MSY-4-1	MSY-4-2	MSY-4-3	MSY-4-4	MSY-4-5	MSY-4-6	MSY-4-7	DZ-2-1	DZ-2-2	DZ-2-3	DZ-2-4	DZ-1-1	DZ-1-2	DZ-1-3	DZ-1-4	DZ-1-5
Lithology	Two-mica granite													Biotite granite				Garnet-bearing two-mica granite				
Cr	37.5	37.3	36.2	33.9	42.6	40.5	30.8	31.6	36.5	28.7	37.2	28.3	29.3	3.45	3.38	3.84	3.92	1.22	2.62	1.36	1.24	1.37
Co	165	139	134	175	154	156	94.7	95.7	157	104	116	124	124	38.1	61.5	64.3	56.2	38.5	33.3	51.9	45.6	46.6
Ni	20.0	14.6	14.1	16.2	18.9	15.4	12.9	13.3	15.5	12.5	17.3	16.9	14.7	2.18	2.92	2.80	2.62	1.21	1.85	1.65	1.55	1.47
Cu	16.8	15.9	16.7	16.2	20.5	20.1	12.8	12.7	13.9	11.3	14.6	16.7	13.8	1.63	1.77	1.56	1.49	1.22	1.91	1.20	1.36	1.24
Zn	61.6	75.3	65.2	65.1	72.9	73.6	58.6	58.9	71.6	43.0	70.6	52.4	63.0	51.8	47.7	46.1	36.0	63.6	72.6	66.9	65.0	69.8
Ga	23.4	25.7	24.7	23.0	24.7	24.7	22.5	22.4	23.5	23.2	25.4	23.2	23.6	24.7	24.2	24.4	23.6	22.2	24.0	23.4	22.3	22.2
Ge	1.98	2.21	1.97	2.10	2.16	2.17	2.09	1.99	2.10	2.07	2.21	2.02	2.07	1.49	1.45	1.37	1.28	1.86	2.20	2.03	2.00	1.86
Rb	192	191	199	177	202	202	175	157	194	153	176	155	169	214	199	204	184	177	169	177	179	176
Sr	122	119	131	79.3	86.5	86.7	107	94.8	111	105	94.3	110	124	169	174	165	165	45.2	50.3	46.7	48.7	47.5
Y	34.3	38.1	34.6	33.6	40.8	38.2	27.1	28.7	33.0	24.9	41.7	28.7	23.1	15.8	14.8	14.4	12.0	90.5	110	110	99.0	88.9
Zr	276	288	275	273	304	272	259	261	275	256	287	253	248	201	193	214	201	332	363	339	329	346
Nb	18.2	17.1	18.2	16.8	18.7	18.8	17.0	17.1	17.6	15.6	20.3	16.2	16.1	13.6	12.1	13.1	11.8	28.5	30.1	28.7	28.0	30.2
Cs	7.15	7.56	8.03	7.45	9.70	9.63	6.26	5.79	7.86	4.61	6.75	4.93	5.25	5.47	4.19	5.55	4.27	2.80	2.44	2.65	2.38	2.58
Ba	801	845	785	749	810	806	806	669	958	803	613	768	850	515	525	563	534	845	825	873	872	839
La	59.9	63.5	59.3	56.1	55.2	57.6	54.9	59.4	55.8	53.2	63.3	54.8	53.8	46.4	51.8	54.1	39.2	48.4	89.2	77.2	73.5	43.8
Ce	127	133	126	119	116	122	116	126	119	113	135	116	114	89.8	100	105	76.8	105	198	167	162	98.8
Pr	15.0	15.8	14.9	14.1	13.7	14.5	13.7	14.9	14.2	13.4	16.2	13.8	13.6	9.41	10.5	11.2	8.43	13.3	24.2	20.4	19.9	12.8
Nd	55.0	57.8	54.4	51.5	50.7	53.2	50.2	54.8	51.2	47.9	59.7	49.7	48.6	33.1	37.2	38.4	28.2	53.7	97.4	82.6	82.0	53.7
Sm	10.6	11.1	10.5	9.95	9.91	10.4	9.81	10.7	9.96	9.42	11.6	9.77	9.44	5.96	6.70	6.77	5.00	13.4	20.3	18.1	17.8	13.8
Eu	1.67	1.87	1.74	1.77	1.56	1.60	1.69	1.58	1.70	1.77	1.59	1.64	1.77	0.96	0.94	0.96	0.92	2.12	2.35	2.38	2.27	2.05

Continued

Sample	MSY-3-1	MSY-3-2	MSY-3-3	MSY-3-4	MSY-3-5	MSY-3-6	MSY-4-1	MSY-4-2	MSY-4-3	MSY-4-4	MSY-4-5	MSY-4-6	MSY-4-7	DZ-2-1	DZ-2-2	DZ-2-3	DZ-2-4	DZ-1-1	DZ-1-2	DZ-1-3	DZ-1-4	DZ-1-5
Lithology	Two-mica granite													Biotite granite				Garnet-bearing two-mica granite				
Gd	8.72	9.21	8.69	8.28	8.60	8.86	8.11	8.81	8.43	7.74	10.0	8.15	7.69	4.89	5.44	5.41	4.15	14.7	20.1	18.7	18.3	15.1
Tb	1.19	1.28	1.19	1.14	1.26	1.26	1.09	1.17	1.12	0.98	1.38	1.08	0.97	0.65	0.70	0.68	0.53	2.62	3.30	3.22	3.09	2.65
Dy	6.37	7.02	6.45	6.19	7.26	7.05	5.52	5.94	6.10	4.96	7.69	5.66	4.78	3.28	3.35	3.28	2.58	16.5	20.3	20.3	19.1	16.6
Ho	1.26	1.40	1.29	1.23	1.51	1.42	1.01	1.08	1.20	0.89	1.53	1.05	0.84	0.53	0.52	0.50	0.42	3.38	4.11	4.12	3.72	3.26
Er	3.75	4.19	3.84	3.66	4.63	4.28	2.89	3.07	3.53	2.54	4.61	3.09	2.33	1.35	1.26	1.22	1.04	9.44	12.1	11.8	10.2	8.65
Tm	0.56	0.64	0.58	0.56	0.72	0.66	0.44	0.46	0.52	0.37	0.70	0.46	0.34	0.18	0.15	0.15	0.13	1.28	1.77	1.67	1.34	1.08
Yb	3.55	4.06	3.69	3.58	4.62	4.25	2.79	3.00	3.31	2.43	4.59	2.98	2.18	1.09	0.92	0.88	0.77	7.63	11.5	10.4	7.87	5.88
Lu	0.53	0.61	0.54	0.54	0.70	0.64	0.41	0.45	0.49	0.36	0.68	0.44	0.33	0.15	0.13	0.12	0.11	1.07	1.69	1.49	1.08	0.77
Hf	7.67	8.04	7.69	7.67	8.62	7.72	7.41	7.42	7.51	7.01	7.91	6.99	6.78	5.44	5.26	5.90	5.64	10.7	12.4	11.0	10.5	11.0
Ta	1.52	1.36	1.48	1.41	1.51	1.50	1.41	1.40	1.39	1.22	1.60	1.32	1.28	1.48	1.30	1.34	1.19	1.53	1.67	1.55	1.34	1.58
Pb	28.0	27.8	26.9	18.4	18.9	19.0	27.6	20.8	32.2	23.2	21.5	21.8	26.2	27.9	27.5	26.6	26.2	18.2	19.5	18.3	18.5	18.7
Th	27.0	28.8	26.8	27.5	27.4	28.1	27.2	28.2	28.3	24.8	31.2	26.4	25.7	27.4	32.1	34.9	27.0	22.9	29.6	25.8	25.3	23.9
U	2.12	3.51	2.41	3.10	2.69	2.70	2.88	2.98	2.34	3.37	4.27	2.47	2.56	6.01	8.19	7.05	7.67	3.84	4.91	4.21	3.87	4.17
REE	295	312	293	278	276	288	269	292	277	259	319	269	261	198	220	228	168	292	506	439	422	279
Eu/Eu*	0.53	0.57	0.56	0.60	0.52	0.51	0.58	0.50	0.57	0.63	0.45	0.56	0.63	0.54	0.48	0.48	0.62	0.46	0.36	0.40	0.39	0.44

$Mg^{\#} = $ molar $100MgO/(MgO+FeO^T)$; $A/CNK = Al_2O_3/(CaO+K_2O+Na_2O)$ molar ratio; $Eu/Eu^* = Eu_N/(Sm_N \cdot Gd_N)^{1/2}$.

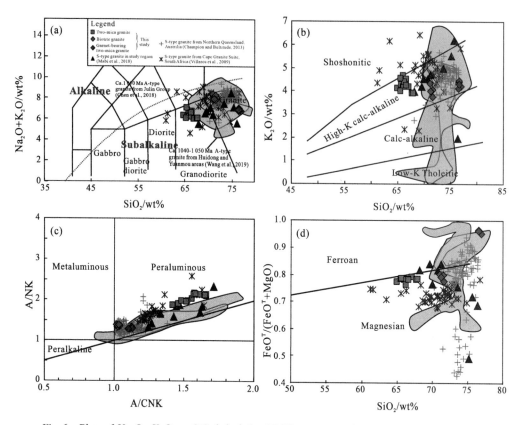

Fig. 6　Plots of $Na_2O + K_2O$ vs. SiO_2(a) (after Middlemost, 1994), K_2O vs. SiO_2(b) (after Roberts and Clemens, 1993), A/NK vs. A/CNK(c) (after Frost et al., 2001), $FeO^T/(FeO^T + MgO)$ vs. SiO_2(d) (after Frost et al., 2001) for the Late Mesoproterozoic two-mica granites, biotite granites, and garnet-bearing two-mica granites in the western Yangtze Block, South China.

The S-type granites in study region were from Mabi et al. (2018). The S-type granites in Northern Queensland, Australia were from Champion and Bultitude (2013). The S-type granites in Cape Granite Suite, South Africa were from Villaros et al. (2009). The ca. 1 050 Ma Julin A-type granites were from Chen et al. (2018). The ca. 1 040–1 050 Ma Huidong and Yuanmou A-type granites were from Wang et al. (2019).

Eu anomalies ($Eu/Eu^* = 0.36 - 0.63$). Similar to trace elemental patterns of middle-upper continental crustal compositions (Rudnick and Gao, 2003), they were enriched in Rb, Th, U, K, and Pb, and displayed distinctly negative Ba, Sr, Nb, Ta, P, and Ti anomalies, as shown by the primitive mantle-normalized diagram (Fig. 7b,d).

4.3　Whole-rock Sr-Nd isotopes

Whole-rock Sr-Nd isotopic data of the studied Late Mesoproterozoic granites are shown in Table 3 and Fig. 8. All of the initial $^{87}Sr/^{86}Sr$ isotopic (I_{Sr}) ratios and $\varepsilon_{Nd}(t)$ values were calculated using the LA-ICP-MS zircon U-Pb age (ca. 1 040 Ma).

A total of eight samples were analyzed for whole-rock Sr-Nd isotopes, including four two-

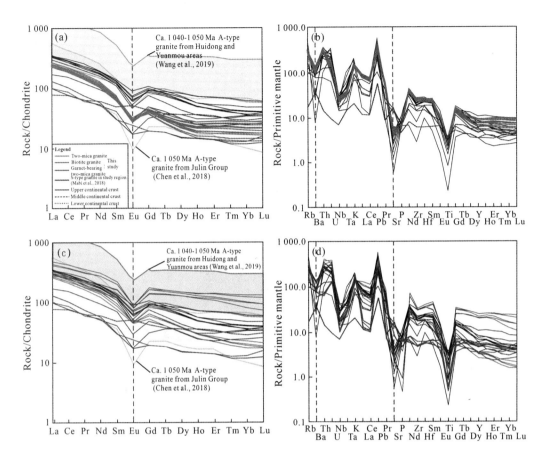

Fig. 7 Diagrams of chondrite-normalized REE patterns (a,c) and primitive mantle-normalized trace element patterns (b,d) for the Late Mesoproterozoic granites in the western Yangtze Block, South China. The primitive mantle and chondrite values are from Sun and McDonough (1989). The lower, middle, and upper continental crust references are from Rudnick and Gao (2003). The S-type granites in study region were from Mabi et al. (2018). The ca. 1 050 Ma Julin A-type granites were from Chen et al. (2018). The ca. 1 040−1 050 Ma Huidong and Yuanmou A-type granites were from Wang et al. (2019).

mica granite samples from the Mosuoying area, two biotite granite samples, and two garnet-bearing two-mica granite samples from the Salian area. The two-mica granite samples had relatively homogeneous $^{87}Rb/^{86}Sr$ (4.77 − 6.83) and $^{87}Sr/^{86}Sr$ (0.786 974 − 0.818 424) ratios, corresponding to high initial $^{87}Sr/^{86}Sr$ ratios ranging from 0.713 6 to 0.718 8. They exhibited $^{143}Nd/^{144}Nd$ ratios varying from 0.511 812 to 0.511 854 and negative $\varepsilon_{Nd}(t)$ values from −5.7 to −4.8. Their two-stage Nd model ages were calculated as ranging from 1.90 Ga to 1.96 Ga. The two biotite granite and two garnet-bearing two-mica granite samples displayed highly variable $^{87}Rb/^{86}Sr$ (3.32 − 11.5) and $^{87}Sr/^{86}Sr$ (0.753 727 − 0.882 115) ratios, yielding variable initial $^{87}Sr/^{86}Sr$ ratios (0.701 5 − 0.714 6). They are characterized by negative $\varepsilon_{Nd}(t)$ (−6.8 to −0.3) and ancient two-stage Nd model ages (1.60−2.04 Ga).

Table 3　Whole-rock Sr-Nd isotopic compositions of the Late Mesoproterozoic S-type granites in the western Yangtze Block, South China.

Sample	$^{87}Rb/^{86}Sr$	$^{87}Sr/^{86}Sr$	$2SE$	Sr /ppm	Rb /ppm	$^{143}Nd/^{144}Nd$	$2SE$	Nd/ppm	Sm/ppm	T_{DM2} /Ga	$\varepsilon_{Nd}(t)$	I_{Sr}
Two-mica granite												
MSY-3-5	6.83	0.815 153	0.000 007	86.5	202	0.511 818	0.000 002	50.7	9.91	1.95	-5.5	0.713 6
MSY-3-6	6.82	0.818 424	0.000 004	86.7	202	0.511 854	0.000 004	53.2	10.4	1.90	-4.8	0.717 0
MSY-4-1	4.77	0.786 974	0.000 006	107	175	0.511 812	0.000 002	50.2	9.81	1.96	-5.7	0.716 0
MSY-4-2	4.82	0.790 506	0.000 005	94.8	157	0.511 847	0.000 004	54.8	10.7	1.92	-5.0	0.718 8
Biotite granite												
DZ-2-1	3.68	0.756 277	0.000 006	169	214	0.511 850	0.000 002	82.6	18.1	2.04	-6.8	0.701 5
DZ-2-2	3.32	0.753 727	0.000 007	174	199	0.511 866	0.000 002	82.0	17.8	2.01	-6.3	0.704 3
Garnet-bearing two-mica granite												
DZ-1-1	11.5	0.882 115	0.000 008	45.2	177	0.512 259	0.000 002	53.7	13.4	1.66	-1.3	0.710 7
DZ-1-2	9.87	0.861 358	0.000 007	50.3	169	0.512 139	0.000 002	97.4	20.3	1.60	-0.3	0.714 6

$^{87}Rb/^{86}Sr$ and $^{147}Sm/^{144}Nd$ ratios were calculated using Rb, Sr, Sm and Nd contents analyzed by ICP-MS, referenced from DePaolo (1981).

T_{DM2} represent the two-stage model age and were calculated using present-day $(^{147}Sm/^{144}Nd)_{CHUR} = 0.196\ 7$ and $(^{147}Sm/^{144}Nd)_{DM} = 0.213\ 7$, $(^{147}Sm/^{144}Nd)_{DM} = 0.513\ 15$ and $(^{147}Sm/^{144}Nd)_{crust} = 0.101\ 2$.

$\varepsilon_{Nd}(t)$ values were calculated using present-day $(^{147}Sm/^{144}Nd)_{CHUR} = 0.196\ 7$ and $(^{147}Sm/^{144}Nd)_{CHUR} = 0.512\ 638$.

$$\varepsilon_{Nd}(t) = \left[(^{143}Nd/^{144}Nd)_S(t)/(^{143}Nd/^{144}Nd)_{CHUR}(t) - 1 \right] \times 10^4.$$

$$T_{DM2} = \frac{1}{\lambda}\left\{ 1 + \left[(^{143}Nd/^{144}Nd)_S - ((^{147}Sm/^{144}Nd)_S - (^{147}Sm/^{144}Nd)_{crust})(e^{\lambda t} - 1) - (^{143}Nd/^{144}Nd)_{DM} \right]/\left[(^{147}Sm/^{144}Nd)_{crust} - (^{147}Sm/^{144}Nd)_{DM} \right] \right\}.$$

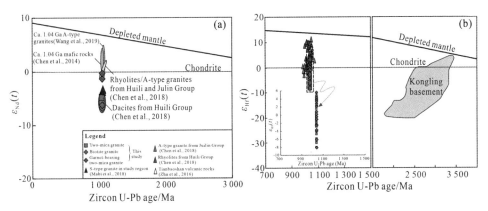

Fig. 8　Whole-rock $\varepsilon_{Nd}(t)$ values vs. zircon U-Pb ages (a) (after Chen et al., 2018) and zircon $\varepsilon_{Hf}(t)$ values vs. zircon U-Pb ages (b) (after Chen et al., 2018) diagrams for the Late Mesoproterozoic granites in the western Yangtze Block, South China.

The S-type granites in study region was from Mabi et al. (2018). The ca. 1 050 Ma Julin A-type granites and rhyolites were from Chen et al. (2018). The Tianbaoshan volcanic rocks were from Zhu et al. (2016). The ca. 1 040 Ma mafic rocks and A-type granites were from Chen et al. (2014) and Wang et al. (2019), respectively.

4.4　Zircon Lu-Hf isotopic compositions

In situ Lu-Hf isotopic analysis for four dating samples was undertaken on the same domains as the U-Pb spots. Initial $^{176}Hf/^{177}Hf$ ratios, $\varepsilon_{Hf}(t)$ values and two-stage depleted mantle model ages (T_{DM2}) were calculated according to zircon crystallization ages. The results are shown in Table 4, and Figs. 8 and 9.

Zircon grains from the two-mica granite sample MSY-3 have predominantly negative $\varepsilon_{Hf}(t)$ values, which are distributed from −5.95 to +1.55 (n=25) (Figs. 8b and 9a). Their two-stage Hf model ages varied from 1 781 Ma to 2 249 Ma (Fig. 9b). Zircon grains from sample MSY-4 (two-mica granite) also contained positive and negative $\varepsilon_{Hf}(t)$ values ranging from −8.02 to +0.46 (n=24) (Figs. 8b and 9c). The 24 analysis spots had two-stage Hf model ages of 1 849−2 378 Ma (Fig. 9d). The zircon $\varepsilon_{Hf}(t)$ values and two-stage Hf model ages from biotite granite sample DZ-2 were −8.09 to +0.80 (n=14) and 1 827 Ma to 2 382 Ma, respectively (Figs. 8b and 9e, f). The garnet-bearing two-mica granites possessed slightly positive zircon $\varepsilon_{Hf}(t)$ values of +2.05 to +5.70 (n=28) and two-stage Hf model ages of 1 521−1 749 Ma (Figs. 8b and 9g, h).

5　Discussion

5.1　Magmatic type and source

In this study, the studied Late Mesoproterozoic granites showed high and variable SiO_2 (65.0−76.9 wt%), K_2O (3.76−4.64 wt%), A/CNK values (1.02−1.67), and normative-CIPW

Table 4　Zircon Lu-Hf isotopic data for the Late Mesoproterozoic S-type granites in the western Yangtze Block, South China.

Grain spot	Age/Ma	$^{176}Yb/^{177}Hf$	2SE	$^{176}Lu/^{177}Hf$	2SE	$^{176}Hf/^{177}Hf$	2SE	$(^{176}Hf/^{177}Hf)_i$	$f_{Lu/Hf}$	$\varepsilon_{Hf}(t)$	2SE	T_{DM1}/Ma	T_{DMC}/Ma
Two-mica granite (MSY-3)													
MSY-3-1	1 040	0.009 844	0.000 061	0.000 299	0.000 002	0.282 07	0.000 015	0.282 060	-0.99	-2.4	0.51	1 649	2 028
MSY-3-2	1 040	0.013 733	0.000 512	0.000 403	0.000 015	0.282 13	0.000 012	0.282 118	-0.99	-0.43	0.43	1 575	1 904
MSY-3-3	1 040	0.010 028	0.000 064	0.000 293	0.000 001	0.282 070	0.000 012	0.282 059	-0.99	-2.42	0.44	1 650	2 029
MSY-3-4	1 040	0.011 562	0.000 154	0.000 372	0.000 006	0.282 14	0.000 012	0.282 132	-0.99	0.1	0.43	1 554	1 872
MSY-3-5	1 040	0.039 681	0.000 584	0.001 199	0.000 014	0.282 14	0.000 018	0.282 117	-0.96	-1.02	0.63	1 609	1 941
MSY-3-6	1 040	0.009 368	0.000 156	0.000 288	0.000 006	0.282 08	0.000 014	0.282 070	-0.99	-2.02	0.49	1 634	2 004
MSY-3-7	1 040	0.006 813	0.000 027	0.000 188	0.000 001	0.282 13	0.000 014	0.282 124	-0.99	-0.04	0.51	1 557	1 880
MSY-3-8	1 040	0.011 640	0.000 220	0.000 358	0.000 007	0.282 15	0.000 013	0.282 147	-0.99	0.64	0.46	1 533	1 838
MSY-3-9	1 040	0.006 904	0.000 020	0.000 193	0.000 001	0.282 09	0.000 013	0.282 087	-0.99	-1.37	0.45	1 608	1 963
MSY-3-10	1 040	0.024 585	0.000 117	0.000 789	0.000 003	0.282 10	0.000 014	0.282 087	-0.98	-1.8	0.49	1 634	1 990
MSY-3-11	1 040	0.006 122	0.000 208	0.000 166	0.000 007	0.282 10	0.000 012	0.282 097	-0.99	-1	0.41	1 593	1 940
MSY-3-12	1 040	0.059 797	0.002 440	0.001 801	0.000 071	0.282 08	0.000 015	0.282 042	-0.95	-4.07	0.53	1 741	2 131
MSY-3-13	1 040	0.018 873	0.000 076	0.000 618	0.000 003	0.281 98	0.000 014	0.281 966	-0.98	-5.95	0.50	1 792	2 249
MSY-3-14	1 040	0.013 560	0.000 289	0.000 403	0.000 009	0.282 13	0.000 013	0.282 127	-0.99	-0.1	0.46	1 562	1 884
MSY-3-15	1 040	0.017 471	0.000 570	0.000 521	0.000 017	0.282 11	0.000 015	0.282 095	-0.98	-1.3	0.54	1 610	1 959
MSY-3-16	1 040	0.038 198	0.001 240	0.001 263	0.000 042	0.282 17	0.000 022	0.282 145	-0.96	-0.04	0.78	1 572	1 880
MSY-3-17	1 040	0.004 972	0.000 066	0.000 133	0.000 002	0.282 15	0.000 011	0.282 146	-1	0.78	0.38	1 524	1 829
MSY-3-18	1 040	0.045 510	0.000 554	0.001 402	0.000 018	0.282 13	0.000 017	0.282 103	-0.96	-1.65	0.60	1 638	1 981
MSY-3-19	1 040	0.026 481	0.000 737	0.000 879	0.000 025	0.282 15	0.000 019	0.282 129	-0.97	-0.34	0.68	1 578	1 899
MSY-3-20	1 040	0.005 327	0.000 027	0.000 140	0.000 001	0.282 12	0.000 013	0.282 121	-1	-0.11	0.46	1 559	1 884
MSY-3-21	1 040	0.007 459	0.000 075	0.000 210	0.000 003	0.282 07	0.000 013	0.282 064	-0.99	-2.2	0.44	1 640	2 015
MSY-3-22	1 040	0.014 041	0.000 190	0.000 463	0.000 006	0.282 10	0.000 020	0.282 091	-0.99	-1.41	0.71	1 614	1 966
MSY-3-23	1 040	0.021 876	0.000 251	0.000 672	0.000 010	0.282 19	0.000 014	0.282 179	-0.98	1.55	0.49	1 501	1 781

Continued

Grain spot	Age/Ma	$^{176}Yb/^{177}Hf$	2SE	$^{176}Lu/^{177}Hf$	2SE	$^{176}Hf/^{177}Hf$	2SE	$(^{176}Hf/^{177}Hf)_i$	$f_{Lu/Hf}$	$\varepsilon_{Hf}(t)$	2SE	T_{DM1}/Ma	T_{DMC}/Ma
MSY-3-24	1 040	0.018 166	0.000 473	0.000 522	0.000 013	0.282 14	0.000 012	0.282 128	-0.98	-0.13	0.44	1 565	1 886
MSY-3-25	1 040	0.055 577	0.000 140	0.001 751	0.000 007	0.282 16	0.000 016	0.282 126	-0.95	-1.08	0.57	1 620	1 945
Two-mica granite (MSY-4)													
MSY-4-1	1 040	0.015 651	0.000 226	0.000 511	0.000 008	0.282 15	0.000 014	0.282 141	-0.98	0.33	0.51	1 547	1 857
MSY-4-2	1 040	0.020 623	0.000 023	0.000 661	0.000 001	0.282 16	0.000 012	0.282 148	-0.98	0.46	0.41	1 544	1 849
MSY-4-3	1 040	0.034 159	0.000 299	0.001 200	0.000 014	0.282 09	0.000 022	0.282 064	-0.96	-2.89	0.77	1 683	2 058
MSY-4-4	1 040	0.017 183	0.000 515	0.000 546	0.000 017	0.282 10	0.000 014	0.282 085	-0.98	-1.7	0.50	1 626	1 984
MSY-4-5	1 040	0.013 618	0.000 102	0.000 435	0.000 004	0.282 07	0.000 015	0.282 060	-0.99	-2.5	0.53	1 655	2 034
MSY-4-6	1 040	0.007 819	0.000 099	0.000 241	0.000 003	0.282 10	0.000 013	0.282 100	-0.99	-0.93	0.47	1 592	1 936
MSY-4-7	1 040	0.018 135	0.000 133	0.000 618	0.000 004	0.282 12	0.000 02	0.282 103	-0.98	-1.1	0.71	1 604	1 946
MSY-4-8	1 040	0.011 546	0.000 299	0.000 367	0.000 011	0.282 13	0.000 013	0.282 125	-0.99	-0.13	0.45	1 563	1 886
MSY-4-9	1 040	0.018 154	0.000 132	0.000 607	0.000 005	0.282 09	0.000 022	0.282 081	-0.98	-1.86	0.77	1 633	1 994
MSY-4-10	1 040	0.040 581	0.000 879	0.001 404	0.000 032	0.281 98	0.000 018	0.281 953	-0.96	-6.96	0.64	1 848	2 312
MSY-4-11	1 040	0.014 431	0.000 263	0.000 477	0.000 010	0.282 07	0.000 014	0.282 063	-0.99	-2.41	0.49	1 653	2 028
MSY-4-12	1 040	0.012 320	0.000 031	0.000 378	0.000 001	0.282 00	0.000 012	0.281 994	-0.99	-4.8	0.43	1 743	2 177
MSY-4-13	1 040	0.015 327	0.000 101	0.000 506	0.000 003	0.282 11	0.000 012	0.282 104	-0.98	-0.99	0.44	1 598	1 940
MSY-4-14	1 040	0.013 399	0.000 206	0.000 419	0.000 007	0.282 13	0.000 011	0.282 124	-0.99	-0.22	0.39	1 567	1 892
MSY-4-15	1 040	0.018 145	0.000 082	0.000 617	0.000 004	0.282 08	0.000 015	0.282 066	-0.98	-2.4	0.51	1 654	2 028
MSY-4-16	1 040	0.035 151	0.000 187	0.001 234	0.000 008	0.281 99	0.000 018	0.281 967	-0.96	-6.34	0.64	1 820	2 273
MSY-4-17	1 040	0.010 134	0.000 138	0.000 333	0.000 006	0.282 07	0.000 018	0.282 066	-0.99	-2.22	0.64	1 643	2 016
MSY-4-18	1 040	0.008 802	0.000 148	0.000 271	0.000 006	0.282 08	0.000 012	0.282 074	-0.99	-1.87	0.44	1 628	1 995
MSY-4-19	1 040	0.017 617	0.000 168	0.000 577	0.000 007	0.282 10	0.000 014	0.282 085	-0.98	-1.7	0.51	1 626	1 984
MSY-4-20	1 040	0.006 609	0.000 075	0.000 205	0.000 003	0.281 90	0.000 040	0.281 899	-0.99	-8.02	1.41	1 863	2 378
MSY-4-21	1 040	0.007 873	0.000 082	0.000 250	0.000 003	0.282 09	0.000 026	0.282 086	-0.99	-1.44	0.91	1 612	1 968

Continued

Grain spot	Age/Ma	^{176}Yb/^{177}Hf	2SE	^{176}Lu/^{177}Hf	2SE	^{176}Hf/^{177}Hf	2SE	$(^{176}$Hf/^{177}Hf$)_i$	$f_{Lu/Hf}$	$\varepsilon_{Hf}(t)$	2SE	T_{DM1}/Ma	T_{DMC}/Ma
MSY-4-22	1 040	0.016 267	0.000 071	0.000 559	0.000 003	0.282 110	0.000 017	0.282 095	-0.98	-1.35	0.61	1 613	1 962
MSY-4-23	1 040	0.023 778	0.000 377	0.000 845	0.000 015	0.282 160	0.000 021	0.282 140	-0.97	0.05	0.75	1 562	1 875
MSY-4-24	1 040	0.017 044	0.000 268	0.000 554	0.000 008	0.282 060	0.000 016	0.282 054	-0.98	-2.79	0.56	1 668	2 052
Biotite granite (DZ-2)													
DZ-2-1	1 040	0.028 860	0.000 474	0.000 945	0.000 012	0.282 050	0.000 025	0.282 029	-0.97	-3.96	0.90	1 721	2 125
DZ-2-2	1 040	0.038 575	0.000 253	0.001 266	0.000 007	0.281 940	0.000 018	0.281 918	-0.96	-8.09	0.63	1 889	2 382
DZ-2-3	1 040	0.026 494	0.000 420	0.000 858	0.000 014	0.282 080	0.000 022	0.282 068	-0.97	-2.51	0.76	1 662	2 034
DZ-2-4	1 040	0.056 863	0.000 200	0.001 867	0.000 008	0.282 090	0.000 024	0.282 054	-0.94	-3.7	0.84	1 727	2 108
DZ-2-5	1 040	0.013 067	0.000 264	0.000 404	0.000 010	0.282 110	0.000 018	0.282 099	-0.99	-1.08	0.65	1 600	1 945
DZ-2-6	1 040	0.037 416	0.000 434	0.001 257	0.000 011	0.282 000	0.000 037	0.281 976	-0.96	-6.03	1.33	1 808	2 254
DZ-2-7	1 040	0.019 716	0.000 415	0.000 657	0.000 015	0.282 140	0.000 031	0.282 123	-0.98	-0.4	1.11	1 577	1 902
DZ-2-8	1 040	0.033 761	0.001 210	0.001 161	0.000 043	0.282 190	0.000 036	0.282 167	-0.97	0.8	1.28	1 537	1 827
DZ-2-9	1 040	0.034 836	0.000 661	0.001 136	0.000 022	0.282 170	0.000 019	0.282 150	-0.97	0.21	0.67	1 560	1 864
DZ-2-10	1 040	0.022 997	0.000 375	0.000 703	0.000 012	0.282 120	0.000 018	0.282 106	-0.98	-1.04	0.63	1 603	1 943
DZ-2-11	1 040	0.033 314	0.000 106	0.001 184	0.000 006	0.282 070	0.000 018	0.282 042	-0.96	-3.65	0.63	1 713	2 106
DZ-2-12	1 040	0.028 765	0.000 265	0.001 036	0.000 009	0.282 090	0.000 029	0.282 074	-0.97	-2.39	1.02	1 661	2 027
DZ-2-13	1 040	0.041 554	0.000 703	0.001 447	0.000 023	0.282 100	0.000 018	0.282 075	-0.96	-2.68	0.64	1 679	2 045
DZ-2-14	1 040	0.034 176	0.000 791	0.001 205	0.000 026	0.282 000	0.000 020	0.281 973	-0.96	-6.11	0.71	1 810	2 258
Garnet-bearing two-mica granite (DZ-1)													
DZ-1-1	1 040	0.046 652	0.000 397	0.001 561	0.000 016	0.282 292	0.000 016	0.282 262	-0.95	3.88	0.55	1 420	1 635
DZ-1-2	1 040	0.032 063	0.000 385	0.001 077	0.000 014	0.282 291	0.000 018	0.282 270	-0.97	4.52	0.63	1 389	1 595
DZ-1-3	1 040	0.029 950	0.000 190	0.000 985	0.000 004	0.282 284	0.000 017	0.282 265	-0.97	4.40	0.60	1 393	1 602
DZ-1-4	1 040	0.060 566	0.000 767	0.001 960	0.000 026	0.282 309	0.000 016	0.282 271	-0.94	3.92	0.55	1 422	1 632
DZ-1-5	1 040	0.036 663	0.000 144	0.001 303	0.000 004	0.282 326	0.000 016	0.282 301	-0.96	5.44	0.55	1 355	1 537

Continued

Grain spot	Age/Ma	^{176}Yb/^{177}Hf	2SE	^{176}Lu/^{177}Hf	2SE	^{176}Hf/^{177}Hf	2SE	(^{176}Hf/^{177}Hf)$_i$	$f_{Lu/Hf}$	$\varepsilon_{Hf}(t)$	2SE	T_{DM1}/Ma	T_{DMC}/Ma
DZ-1-6	1 040	0.034 115	0.000 061	0.001 135	0.000 003	0.282 258	0.000 015	0.282 236	−0.97	3.25	0.54	1 440	1 674
DZ-1-7	1 040	0.043 703	0.000 506	0.001 514	0.000 017	0.282 297	0.000 021	0.282 268	−0.95	4.12	0.73	1 409	1 620
DZ-1-8	1 040	0.021 160	0.000 108	0.000 722	0.000 002	0.282 263	0.000 016	0.282 248	−0.98	3.99	0.58	1 407	1 628
DZ-1-9	1 040	0.044 529	0.000 233	0.001 442	0.000 004	0.282 254	0.000 014	0.282 226	−0.96	2.69	0.51	1 466	1 709
DZ-1-10	1 040	0.044 159	0.000 147	0.001 502	0.000 003	0.282 293	0.000 016	0.282 264	−0.95	3.98	0.58	1 415	1 628
DZ-1-11	1 040	0.031 002	0.000 108	0.001 045	0.000 004	0.282 276	0.000 016	0.282 255	−0.97	4.01	0.57	1 409	1 626
DZ-1-12	1 040	0.034 556	0.000 067	0.001 146	0.000 002	0.282 280	0.000 015	0.282 258	−0.97	4.02	0.52	1 410	1 626
DZ-1-13	1 040	0.041 175	0.000 103	0.001 471	0.000 001	0.282 243	0.000 018	0.282 214	−0.96	2.24	0.63	1 484	1 738
DZ-1-14	1 040	0.036 343	0.000 108	0.001 241	0.000 002	0.282 237	0.000 015	0.282 213	−0.96	2.37	0.54	1 476	1 729
DZ-1-15	1 040	0.035 973	0.000 168	0.001 227	0.000 006	0.282 228	0.000 018	0.282 204	−0.96	2.05	0.62	1 488	1 749
DZ-1-16	1 040	0.041 429	0.000 065	0.001 408	0.000 004	0.282 316	0.000 017	0.282 288	−0.96	4.93	0.59	1 376	1 569
DZ-1-17	1 040	0.043 334	0.000 037	0.001 457	0.000 003	0.282 277	0.000 016	0.282 249	−0.96	3.49	0.56	1 434	1 659
DZ-1-18	1 040	0.036 055	0.000 172	0.001 242	0.000 008	0.282 294	0.000 015	0.282 270	−0.96	4.38	0.54	1 396	1 603
DZ-1-19	1 040	0.034 300	0.000 052	0.001 186	0.000 004	0.282 248	0.000 015	0.282 225	−0.96	2.82	0.54	1 458	1 701
DZ-1-20	1 040	0.040 286	0.000 340	0.001 437	0.000 012	0.282 292	0.000 016	0.282 264	−0.96	4.03	0.55	1 412	1 625
DZ-1-21	1 040	0.034 817	0.000 125	0.001 210	0.000 004	0.282 330	0.000 016	0.282 306	−0.96	5.70	0.57	1 344	1 521
DZ-1-22	1 040	0.037 814	0.000 257	0.001 257	0.000 010	0.282 273	0.000 015	0.282 248	−0.96	3.60	0.51	1 427	1 652
DZ-1-23	1 040	0.062 134	0.000 146	0.002 032	0.000 008	0.282 279	0.000 018	0.282 239	−0.94	2.74	0.62	1 471	1 706
DZ-1-24	1 040	0.038 864	0.000 137	0.001 296	0.000 003	0.282 275	0.000 016	0.282 250	−0.96	3.63	0.56	1 427	1 650
DZ-1-25	1 040	0.046 011	0.000 287	0.001 511	0.000 008	0.282 304	0.000 014	0.282 274	−0.95	4.35	0.50	1 400	1 605
DZ-1-26	1 040	0.046 659	0.000 121	0.001 521	0.000 003	0.282 274	0.000 016	0.282 244	−0.95	3.29	0.55	1 442	1 671
DZ-1-27	1 040	0.038 383	0.000 525	0.001 277	0.000 015	0.282 278	0.000 015	0.282 253	−0.96	3.76	0.54	1 421	1 642
DZ-1-28	1 040	0.048 378	0.000 159	0.001 616	0.000 006	0.282 278	0.000 018	0.282 247	−0.95	3.30	0.64	1 443	1 671

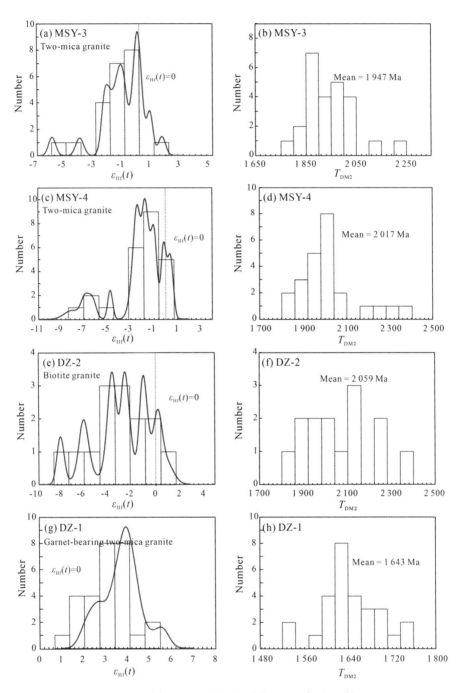

Fig. 9　Histograms of zircon $\varepsilon_{Hf}(t)$ values and Hf model age for the Late Mesoproterozoic granites
in the western Yangtze Block, South China.

corundum contents (0.54 - 7.04 wt%), which lead us to attribute them to S-type granites rather than I-type granites that are mostly Na-rich metaluminous rocks and usually contain amphiboles (Chappell et al., 2012; Clemens, 2003; Chappell and White, 2001). The studied granites contained relatively low FeO^T/MgO ratios ($0.78-0.97$, majority between $0.78-0.80$), mainly plotting into the magnesian fields (Fig. 6d), which differ from the ferroan (A-type) granitoids as described in Frost and Frost. (2011). Their low ($Na_2O + K_2O$)/CaO ratios ($3.68-9.32$, mainly $3.68-5.71$) are also distinct from A-type granites (Eby, 1990; Whalen et al., 1987). Zhang (2012) stated that extremely negative Eu anomalies ($Eu/Eu^* < 0.30$) can be considered a significant index of A-type granites. However, our granites displayed higher Eu/Eu^* values ($0.36-0.63$) than coeval Julin A-type granites ($Eu/Eu^* = 0.09-0.25$) (Fig. 7a,c) (Chen et al., 2018). Furthermore, our granites yielded relatively lower whole-rock Zr saturation temperatures ($799-887$ ℃) than the A-type granites from the Julin Group ($799-985$ ℃) as well as the Huidong and Yuanmou areas ($816-1\ 164$ ℃) (Fig. 10a) (Chen et al., 2018; Wang et al., 2019). The presence of aluminum-rich minerals (e. g., muscovite and garnet) also substantiates a S-type affinity (Fig. 3) (Chappell and White, 1992, 2001). In addition, our samples are comparable to S-type granites from northern Queensland in Australia and the Cape Granite Suite in South Africa (Champion and Bultitude, 2013; Villaros et al., 2009), which plotted within S-type fields (Fig. 10b). Also, the initial melts of S-type granites are felsic and gradually become more

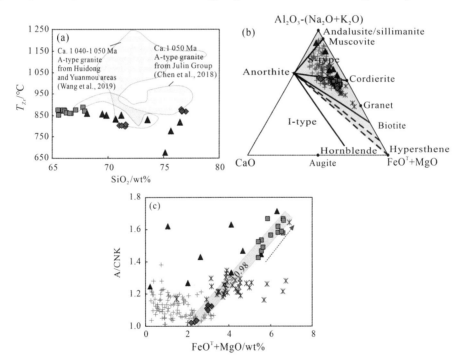

Fig. 10   The diagrams of whole-rock T_{Zr} vs. SiO_2(a), [$Al_2O_3-(Na_2O+K_2O)$]-CaO-(FeO^T+MgO) (b),

A/CNK vs. FeO^T+MgO (c) for the Late Mesoproterozoic granites

in the western Yangtze Block, South China.

Symbols as Fig. 6.

peraluminous with increasing maficity due to entrainment of the Grt + Ilm ± Opx peritectic assemblage (Clemens and Stevens, 2012). The granites studied here displayed a positive correlation between $FeO^T + MgO$ and A/CNK values (Fig. 10c), which fits the characteristic of S-type granites with peritectic assemblage entrainment (Champion and Bultitude, 2013; Huang et al., 2019; Stevens et al., 2007; Villaros et al., 2009). In summary, all of the mineral and geochemical evidences support the interpretation that the Late Mesoproterozoic granites in the western Yangtze Block belong to the S-type granites.

The studied S-type granites contained high Al_2O_3 contents (11.1 – 17.0 wt%) and A/CNK values (1.02 – 1.67). Traditionally, peraluminous melts can be produced by the fractional crystallization of hornblende (Chappell et al., 2012; Zen, 1986) and partial melting of metaluminous rocks (e.g., basaltic to andesitic protoliths) (Rapp and Watson, 1995). However, these two mechanisms usually generate metaluminous or weakly peraluminous melts with Na and Sr-rich features (Gaudemer et al., 1988), which significantly contrast with our S-type granites that are peraluminous to strongly peraluminous (Fig. 6c) and contain high K_2O/Na_2O ratios (0.96–2.48) and low Sr concentrations (45.2–174 ppm). Jiang et al. (2017) proposed that a positive correlation of SiO_2 and A/CNK values can be expected by the fractional crystallization of hornblende. Nevertheless, the S-type granites investigated here show a negative relationship (Fig. 11a). Furthermore, their enriched $\varepsilon_{Nd}(t)$ values (−6.8 to −0.3) differ from the coeval depleted asthenospheric mantle-derived basalts (−1.4 to +4.0) in the Julin Group (Fig. 8a) (Chen et al., 2014), which indicates a distinct magma source and thus diminishes the possibility of mafic metaluminous magma fractionation (Jiang et al., 2017). Their high SiO_2(65.0–76.9 wt%) and low MgO (0.11–1.45 wt%) contents as well as low $Mg^{\#}$ values (7.03–38.0) can also weakens the possibility of partial melting of basaltic to andesitic protoliths (Jiang et al., 2017; Zhu et al., 2018). Collectively, these two above mechanisms are unrealistic for the genesis of our S-type granites that likely originate from an aluminum-rich metasedimentary source (e.g., metapelites and metagreywackes) under crustal conditions (Clemens, 2003; Champion and Bultitude, 2013; Huang and Jiang, 2014; Sylvester, 1998; Zhu et al., 2018).

Our S-type granites displayed high SiO_2(65.0–76.9 wt%) and K_2O (3.76–4.64 wt%) contents, and A/CNK values (1.02–1.67), indicating that their sources are dominated by Al-rich metasedimentary rather than metaigneous protoliths (Clemens, 2003; Huang et al., 2019; Zhu et al., 2018). They show an obviously depletion of Eu, Sr, and Ba (Fig. 7b, d), resembling middle to upper crustal patterns (Rudnick and Gao, 2003). Both two-mica, biotite, and garnet-bearing two-mica granites display remarkably enriched whole-rock Nd isotopic compositions [$\varepsilon_{Nd}(t)$ = −6.8 to −0.3] and ancient two-stage Nd model ages (1.60–2.04 Ga), supporting an evolved and ancient crustal source, not a juvenile mafic crust

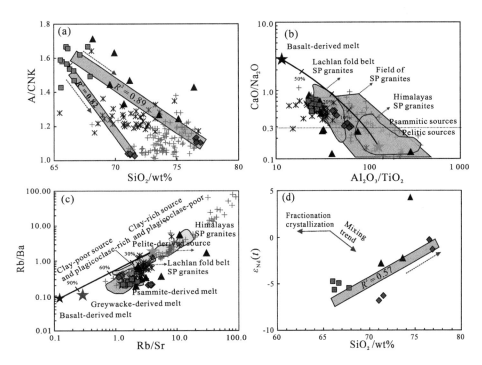

Fig. 11 The A/CNK vs. SiO₂(a), CaO/Na₂O vs. Al₂O₃/TiO₂(b) (Sylvester, 1998), Rb/Ba vs. Rb/Sr (c) (Patiño Douce, 1999), whole-rock $\varepsilon_{Nd}(t)$ vs. SiO₂(d) diagrams for the Late Mesoproterozoic S-type granites in the western Yangtze Block, South China. Symbols as Fig. 6.

(Huang et al., 2019; Zhu et al., 2018). In addition, the predominantly negative zircon $\varepsilon_{Hf}(t)$ values and old two-stage Hf model ages indicate that these S-type granites originated from a mature continental crust source (Figs. 8b and 9). Furthermore, our S-type granites data, combined with published S-type granite data in the study area (Mabi et al., 2018), exhibit high CaO/Na₂O and low Al₂O₃/TiO₂ as well as moderate Rb/Ba and Rb/Sr values (Fig. 11b, c), suggesting that their precursor magmas were predominantly sourced from heterogeneous metasedimentary protoliths, including metapelites and metagreywackes (Sylvester, 1998). In addition, the inherited zircons observed in our S-type granites display a wide range of ancient U-Pb ages (1 129−2 134 Ma) (Table 1), which can reflect variable metasedimentary sources (Huang and Jiang, 2014; Huang et al., 2019). In summary, it is inferred that the Late Mesoproterozoic S-type granites in this study were derived from heterogeneous metasedimentary sources.

5. 2 Geochemical diversity

To obtain more profound knowledge regarding the magmatic nature of our S-type granites, it is necessary to evaluate their geochemical diversity (Clemens, 2003; Clemens and Stevens,

2012; Champion and Bultitude, 2013 and reference therein), such as high and variable FeO^T + MgO contents, large variation in Sr-Nd isotopes, and heterogeneous zircon Hf isotopes. Several mainstream mechanisms have been invoked to explain the chemical heterogeneity exhibited by S-type granitic melts, such as magma mixing or assimilation and fractional crystallization (Castro et al., 1999; Clemens, 2003; Healy et al., 2004), source heterogeneity (Clemens and Stevens, 2012; Villaros et al., 2012; Zhu et al., 2018), disequilibrium melting (Kong et al., 2019; Tang et al., 2014; Villaros et al., 2009), and co-entrainment of selective peritectic assemblage and accessary mineral phases (Champion and Bultitude, 2013; Clemens and Stevens, 2012; Stevens et al., 2007; Villaros et al., 2009).

5.2.1　Magma mixing: An unimportant mechanism

Given the potential geochemical complexity, multiple processes might have been involved in our S-type granites. However, the only process we can readily rule out is magma mixing, for which there is minimal evidence in our samples.

Magma mixing between mantle magmas and crustal melts usually produce microdioritic enclaves and/or disequilibrium mineral pairs, and thus elevate the $Mg^{\#}$ values of resultant melts (Champion and Bultitude, 2013; Jiang et al., 2017; Tang et al., 2014; Vernon, 1984). However, a scarcity of microdioritic enclaves and disequilibrium mineral pairs (Fig. 3), together with low $Mg^{\#}$ values (7.03 − 38.0) in our samples, dramatically diminishes the possibility of magma mixing. Moreover, a negative correlation of SiO_2 vs. $\varepsilon_{Nd}(t)$ would develop during hybridization with both mantle and crustal inputs (Jiang and Zhu, 2017). This is contrary to our S-type granites that show a positive relationship between SiO_2 contents and $\varepsilon_{Nd}(t)$ values (Fig. 11d). Therefore, magma mixing is an unimportant mechanism for our S-type granites.

5.2.2　Source heterogeneity: Heterogeneous whole-rock geochemistry and variable Nd-Hf isotopes

Because protoliths (or magma source rocks) exert primary control on chemical features of granitic magmas (Clemens and Stevens, 2012; Stevens et al., 2007), source heterogeneity is a vital factor for influencing geochemistry of our S-type granites (Champion and Bultitude, 2013; Villaros et al., 2012; Zhu et al., 2018).

The S-type granites studied here were mainly derived from heterogeneous metasedimentary sources. Villaros et al. (2009) suggested that some compatible elements (e.g., Rb, Sr, Ba, and Eu) in S-type granites are concentrated within reactant minerals during incongruent melting processes, the considerable variation of these elements can thus reflect source heterogeneity. As shown in Fig. 12, our S-type granites display relatively scattered Rb, Sr, Ba, Eu/Eu* at given FeO^T + MgO contents, indicating that their sources have heterogeneous compositions with variable minerals and trace elements (Villaros et al., 2009). Furthermore, Villaros et al. (2012) proposed that zircon Hf isotopic heterogeneity observed in the Cape

Granite Suite S-type granites is related to heterogeneous sources rather than mixing between mantle magmas and crustal melts. Source heterogeneity can transfer to the granitic melts via the genesis of discrete magma batches. Champion and Bultitude (2013) summarized that S-type granites from northern Queensland in Australia yield wide range of whole-rock $\varepsilon_{Nd}(t)$ values (>4 epsilon units), indicating the involvement of multiple metasedimentary components. Likewise, our S-type granites showed large variability of single-grain zircon Hf isotopes (>3-8 epsilon units) (Fig. 9) and whole-rock Nd isotopes (>4 epsilon units) (Fig. 8a), which can demonstrates that their metasedimentary sources are compositionally variable. Consequently, the heterogeneity of metasedimentary sources may pose important significance for the S-type granites in this study.

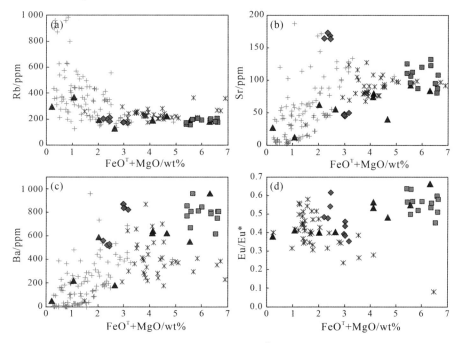

Fig. 12　The scattered Rb, Sr, Ba, and Eu at given FeO^T+MgO values for the Late Mesoproterozoic S-type granites in the western Yangtze Block, South China.

Symbols as Fig. 6.

5.2.3　Disequilibrium melting: Heterogeneous Hf isotopes

Disequilibrium melting controls Nd-Hf isotopic decoupling as well as positive and negative Hf isotopes in peraluminous to strongly peraluminous S-type granites, such as those from the Cape Granite Suite (Villaros et al., 2009) and Chinese Altai area (Kong et al., 2019) as well as southern China (Tang et al., 2014; Zhu et al., 2019c).

Previous studies proposed that a varying zircon dissolution rate at a single magma source can lead to varying Hf isotopic compositions between different batches of crustal melts and may result in the decoupling of Hf isotopes from other radiogenic isotopes (Flowerdew et al. 2006;

Tang et al., 2014). In this study, the positive and negative zircon $\varepsilon_{Hf}(t)$ values (-8.09 to $+5.70$) of our S-type granites are decoupled from variable whole-rock $\varepsilon_{Nd}(t)$ values (-0.3 to -6.8). Zirconium concentrations in the source are vital for influencing the zircon dissolution rate during disequilibrium partial melting (Tang et al., 2014; Villaros et al., 2009). Inherited zircons can be detected in our samples (Table 1), indicating that some residual zircons from the source were entrained by the melt during melt loss (Huang et al., 2019; Kong et al., 2019). As suggested by Stevens et al. (2007) and Villaros et al. (2009), a positive relationship of $FeO^T + MgO$ and Zr in S-type granites might be attributed to the entrainment of residual zircons from the source, which is well defined in our samples (Fig. 13a). Moreover, our samples possess obviously higher whole-rock Zr contents (193–363 ppm) than the average continental crust (e.g., lower crust = 68 ppm; upper crust = 193 ppm) and typical crustal rocks ($\sim 100 - 200$ ppm) (Miller et al., 2003 and references therein; Rudnick and Gao, 2003) (Fig. 13a). It is notable that a substantial fraction of the zirconium budget is contained within inherited zircons (Clemens and Stevens, 2012), the existence of residual zircons, and high whole-rocks Zr contents in our S-type granites indicate high initial Zr contents in the source. Because high initial Zr concentrations at the source can easily saturate the melt with zircon, and zircon dissolution will cease until more melt is produced to dilute the remaining Zr at the source, the initial Zr concentration in the source can significantly affect zircon dissolution during crustal melting (Tang et al., 2014). Tang et al. (2014) have modeled that when the

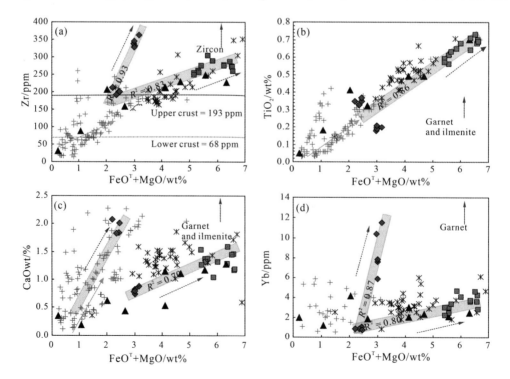

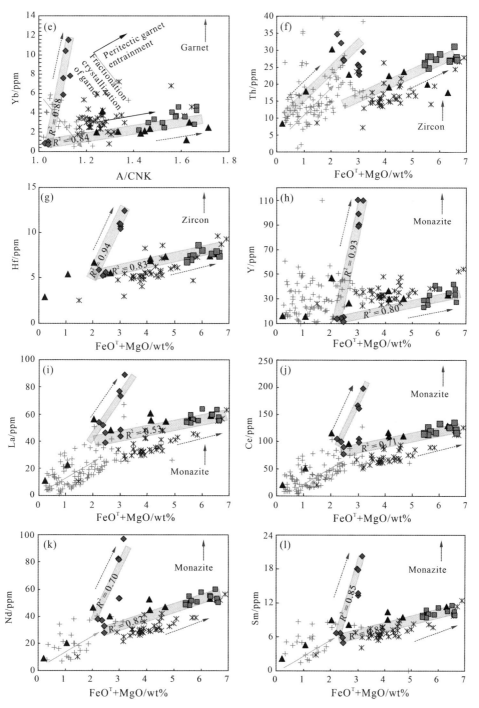

Fig. 13　The correlations between Zr(a), TiO$_2$(b), CaO(c), Yb(d), Th(f), Hf(g), Y(h), La(i),

Ce(j), Nd(k), Sm(l) and FeO^T+MgO values as well as Yb vs. A/CNK values (e) for the

Late Mesoproterozoic S-type granites in the western Yangtze Block, South China.

Symbols as Fig. 6.

initial Zr concentrations at a source are sufficiently high (> 100 ppm), the inter-batch Hf varies from low concentrations and highly radiogenic to high concentrations and less radiogenic than bulk protolith. This can generate a melt extracted from a single crustal source that evolves from a mantle-like source at an early stage to a crust-like source after extensive melting (Kong et al., 2019; Tang et al., 2014; Zhu et al., 2019c). Mixing of these crust-derived melts from different batches may mimic that of crust and mantle-derived magmas in terms of Hf isotopes (Tang et al., 2014), which is observed in our S-type granites that display positive and negative zircon $\varepsilon_{Hf}(t)$ values (-8.09 to $+5.70$) (Figs. 8b and 9). The high initial Zr in the source along with disequilibrium melting can therefore account for heterogeneous zircon Hf isotopes [i.e., positive and negative $\varepsilon_{Hf}(t)$ values] in our S-type granites.

5. 2. 4　Co-entrainment of peritectic selective phases and accessory minerals: High and variable FeO^T+MgO contents

Although conventional studies have suggested that S-type granitic melts have more felsic compositions with low Fe + Mg or Fe + Mg + Ti contents at suitable crustal pressures and temperatures (Stevens et al., 1997), more mafic S-type granites with Mg- and Fe-rich compositions have been investigated in detail. They were found not to derive from simple sedimentary melts and/or mixing of mafic magmas, but rather represent the existence of peritectic materials from heterogeneous sources (Champion and Bultitude, 2013; Clemens and Stevens, 2012; Stevens et al., 2007; Villaros et al., 2009).

Champion and Bultitude (2013) suggested that some North Queensland S-type granites with high FeO^T + MgO contents (> 2. 5 – 3. 0 wt%) reflect the existence of restitic/peritectic materials and accessory phases. Clemens and Stevens (2012) and Villaros et al. (2009) also maintained that co-entrainment of both a peritectic assemblage entrainment and accessary minerals is responsible for those low solubility elements (e. g., Fe, Mg, and Ti) in S-type granites from the Cape Granite Suite. Similarly, our S-type granites contain high and variable FeO^T + MgO (2. 21 – 6. 64 wt%) contents, and display decreasing A/CNK values with increasing differentiation (decreasing maficity) (Fig. 10b). This is interpreted to be a consequence of co-entrained peritectic/restitic phases and accessory minerals from the source (Stevens et al., 2007; Villaros et al., 2009), not fractionation, for which the fractional crystallization is commonly characterized by increasing peraluminosity accompanied with decreasing maficity (Champion and Bultitude, 2013). Like the Cape Granite Suite S-type granites (Stevens et al., 2007; Villaros et al., 2009), positive correlations between TiO$_2$, CaO, and FeO^T+MgO in our S-type granites were observed, and explained by entrainment of garnet and ilmenite produced by biotite incongruent melting (Fig. 13b,c). A positive relationship of FeO^T+MgO vs. Yb can also reflect garnet entrainment (Fig. 13d) (Stevens et al., 2007; Villaros et al., 2009). Furthermore, a fractionated REE pattern with variable $(La/Yb)_N$ (4.55-44.2) and $(Gd/Yb)_N$ (1.44-5.10) ratios in our S-type granites can demonstrate the presence of residual garnet in the source (Fig. 7) (Wang et al., 2007). As the HREE (e.g., Yb) concentrations and A/CNK values are essentially controlled by garnet, a negative correlation between Yb and A/CNK develops during crystal fractionation (Champion and Bultitude, 2013), while a positive correlation is expected by the garnet entrainment model (Stevens et al., 2007). In contrast to the North Queensland S-type granites (Champion and Bultitude, 2013), a positive correlation defined by A/CNK vs. Yb in our S-type granites delineates the peritectic garnet entrainment from the source (Fig. 13e).

In addition, it has been documented that selective entrainment of peritectic garnet in more mafic S-type granites is usually accompanied by entrainment of available accessory minerals (e.g., zircon and monazite) (Clemens and Stevens, 2012; Stevens et al., 2007; Villaros et al., 2009). The presence of entrained zircons in our S-type granites was recognized by the occurrence of inherited zircons and positive correlation of $FeO^T + MgO$ vs. Zr (Fig. 13a). The positive correlations that exist between Th, Hf, and $FeO^T + MgO$ in our samples also suggest that zircon is a co-entrained accessary phase during partial melting (Fig. 13f,g) (Clemens and Stevens, 2012; Villaros et al., 2009). Moreover, the LREEs and Y are concentrated within the accessary phase of monazite; thus, the variation in Y and LREEs concentrations is tightly related to the behavior of monazite (Villaros et al., 2009). For our S-type granites, monazite is involved in co-entrainment of accessary minerals because of positive correlations between Y, La, Ce, Nd, Sm, and $FeO^T + MgO$ (Fig. 13h-l) (Stevens et al., 2007; Villaros et al., 2009). In summary, the more mafic features of S-type granites can be explained by co-entrainment of peritectic and accessary minerals (e. g., garnet, ilmenite, zircon, and monazite), similar to other typical ferromagnesian-rich S-type granites (e.g., Cape Granite Suite and North Queensland) (Champion and Bultitude, 2013; Stevens et al., 2007; Villaros et al., 2009).

The integrated lines of evidence therefore lead us to the conclusion that the primary geochemical diversity of our S-type granites is principally inherited from heterogeneous metasedimentary sources. Disequilibrium melting plays a vital role in variable zircon $\varepsilon_{Hf}(t)$ values. The more mafic characteristics (i. e., high and variable $FeO^T + MgO$ contents) are ascribed to co-entrainment of selective peritectic phases (garnet + ilmenite) and accessary minerals (zircon + monazite) from the residue.

5. 3 Petrological and geodynamic implications

5. 3. 1 Petrological implications

It has been documented that S-type granites can exhibit geochemical characteristics that overlap with typical A-type granites (Huang and Jiang, 2014; Whalen, 1987). The S-type granites studied here displayed relatively high 10 000Ga/Al (2. 61-3. 95) ratios and Zr + Nb + Ce + Y (301-701 ppm) contents analogous to an A-type affinity. These elements, such as Ga, Zr, Ce, and Y, are predominantly concentrated within the garnet and refractory accessary phases (e.g., zircon and monazite) (Villaros et al., 2009). Our discussion has implied that the geochemical diversity of our S-type granites is significantly attributed to coupled co-entrainment of selective peritectic phases and accessary minerals. The presence of these co-entrained minerals can account for high concentrations of Ga, Zr, Ce, and Y in resultant granitic melts. We therefore suggest that S-type granites that show overlapping trace elemental compositions with A-type granites, especially high 10 000Ga/Al ratios and Zr + Nb + Ce + Y values, may be caused by the entrainment of garnet, zircon, and monazite from the source during partial melting.

5. 3. 2 Tectonic setting

Our investigation demonstrated that the ca. 1 040 Ma S-type granites in the western Yangtze Block were generated by disequilibrium partial melting of heterogenous metasedimentary rocks under evolved crustal conditions. S-type granites are common in the syn-collision and post-collision settings (Barbarin, 1999; Chappell and White, 2001; Sylvester, 1998).

In this study, the formation ages of our S-type granites correspond with global Grenvillian magmatism, which was produced by the collisional events during the assembly of the Rodinia supercontinent (Chen et al., 2018; Jacobs et al., 2003). Although the Jiangnan orogenic belt (Sibao orogenic belt) was traditionally considered to represent the suturing of the Yangtze and Cathaysia blocks in response to the Grenvillian orogenic events (Greentree et al., 2006), new geochronological and geochemical studies of sedimentary and igneous rocks from the Jiangnan orogen belt indicated that it likely formed during the Neoproterozoic, postdating the global Grenvillian orogenesis (Huang et al., 2019; Wang et al., 2013; Zhao et al., 2011). The identification of coetaneous within-plate A-type felsic rocks, alkaline basalts, mafic rocks, and marine sedimentary sequences in the western Yangtze Block suggests that the southern part of the western Yangtze Block was located in an intra-plate rifting setting rather than a collisional setting linked to the Grenvillian orogenic events (Chen et al., 2014, 2018; Wang et al., 2019; Zhu et al., 2016). The absence of Late Mesoproterozoic high grade metamorphic rocks and crustal thickening events in the southwestern Yangtze Block also indicate that a continent-continent "hard" collision is unreasonable (Sylvester, 1998; Zhu et al., 2018 and reference therein). In summary, the Jiangnan orogenic belt is not a part of the global Greenville orogenic belt and the southern part of the western Yangtze Block is not the southwestward extension of the Jiangnan orogenic belt in the Late Mesoproterozoic (Chen et al., 2018; Zhao et al.,

2011). Therefore, it seems unlikely that our S-type granites are the products of a syn-collision responding to the Grenvillian orogenic events.

Furthermore, the coeval basalts and mafic dykes in the western Yangtze Block are tholeiitic in composition without pronounced negative Nb-Ta anomalies (Chen et al., 2014; Zhu et al., 2016), distinct from back-arc, impactogen, or post-orogenic basalts that display obvious depletion of Nb-Ta (Chen et al., 2014). The western Yangtze Block preserves the Late Mesoproterozoic within-plated magmatism (Chen et al., 2014, 2018; Wang et al., 2019; Zhu et al., 2016) and subsequently prolonged Neoproterozoic arc magmatism (Zhao et al., 2018 and reference therein; Zhu et al., 2019a-c), which indicates that a tectonic switch, from an intra-continental rifting in a passive margin to oceanic subduction in a positive margin, is reasonable for these special igneous rocks during ca. 1 050 Ma to 730 Ma (Chen et al., 2014, 2018). In addition, the basaltic layers in the Pudeng Formation were followed by deposition of massive sandstones in the Lugumo Formation and subsequently massive marbles or dolomites in the Fenghuangshan Formation (Chen et al., 2014). These sedimentary sequences are related to a shallow marine environment in a rifting basin (Deng, 2000). Consequently, both sedimentary and igneous sequences suggested that subduction beneath the western margin of the Yangtze Block occurred at least after ca. 1.0 Ga when the passive margin rifting evolved to compression setting in a convergent plate margin during the Late Mesoproterozoic to Early Neoproterozoic (Chen et al., 2018). We concluded that the southern part of the western Yangtze Block was located in an intra-continental rifting basin in a passive continental margin during the Late Mesoproterozoic (Chen et al., 2014, 2018).

Therefore, we suggest that our S-type granites were not produced in syn-collision and post-collision settings, but in a rifting basin formed in a passive margin during the Late Mesoproterozoic. The ca. 1 050 Ma metabasalts in the Julin Group have predominantly positive whole-rock $\varepsilon_{Nd}(t)$ (-1.4 to $+4.0$) and zircon $\varepsilon_{Hf}(t)$ (-1.7 to $+11.0$) values, indicating a depleted asthenospheric mantle source at a rifting basin in a passive margin (Chen et al., 2014). The ca. 1 020 Ma mafic dykes in the Tianbaoshan Formation with depleted Nd-Hf isotopic compositions [e.g., $\varepsilon_{Nd}(t) = +0.41$ to $+1.6$; $\varepsilon_{Hf}(t) = +7.0$ to $+10.3$] were also generated by the partial melting of depleted asthenospheric mantle under an extensional background within an intra-plate rifting setting (Zhu et al., 2016). The voluminous advective heat supplied from underplating asthenospheric mantle-derived magmas heated the newly formed crust and overlying ancient continental crust (Chen et al., 2014, 2018; Wang et al., 2019; Zhu et al., 2016), leading to the formation of massive felsic igneous rocks. These include the ca. 1 020 Ma A-type Tianbaoshan intermediate-felsic volcanic rocks (Zhu et al., 2016), ca. 1 050 Ma A-type Huili dacites and rhyolites as well as Julin granites (Chen et al., 2018), ca. 1 040-1 050 Ma Huidong and Yuanmou A-type granites (Wang et al., 2019), and ca. 1 040 Ma S-type granites in the Mosuoying and Salian areas (Mabi et al. , 2018 and this study). Therefore, the ca. 1 040 Ma S-type granites in the western Yangtze Block were formed in a rifting basin in a passive margin that was gradually transformed into a compressional setting during the Neoproterozoic. They originated from ancient and heterogeneous metasedimentary sources induced by the upwelling of voluminous mantle-derived magmas.

6 Conclusions

(1)The newly identified S-type granites in the western Yangtze Block, including two-mica granites, biotite granites, and garnet-bearing two-mica granites, were emplaced at ca. 1 040 Ma.

(2)They originated from heterogeneous metasedimentary protoliths.

(3)The heterogeneous Nd-Hf isotopes of our S-type granites were mainly attributed to source heterogeneity and disequilibrium melting. Similar to typical more mafic S-type granites from the Cape Granite Suite (South Africa) and northern Queensland (Australia), their high and variable FeO^T+MgO contents, coupled with positive relationships between FeO^T+MgO and TiO_2, CaO, Zr, Th, Hf, Y, Yb, LREEs, might be caused by coupled co-entrainment of peritectic and accessory minerals (e.g., garnet, ilmenite, zircon, and monazite).

(4)Considering the regional geology, we propose that the ca. 1 040 Ma S-type granites from the western Yangtze Block were produced in a continental rifting basin that was transformed into a compressional setting during the Neoproterozoic.

Acknowledgements We thank Prof. Michael Roden for valuable editorial comments and two anonymous reviewer for constructive comments. We also appreciate Kara Bogus, PHD, for further English polish. This work was jointly supported by the National Natural Science Foundation of China (Grant Nos. 41421002 and 41772052) and the program for Changjiang Scholars and Innovative Research Team in University (Grant IRT1281). Support was also provided by independent innovation project of graduate students of Northwest University (YZZ17192).

References

Andersen, T., 2002. Correction of common lead in U-Pb analyses that do not report [204]Pb. Chemical Geology, 192(1-2), 59-79.

Appleby, S.K., Gillespie, M.R., Graham, C.M., Hinton, R.W., Oliver, G.J.H., Kelly, N.M., 2010. Do S-type granites commonly sample infracrustal sources? New results from an integrated O, U-Pb and Hf isotope study of zircon. Contributions to Mineralogy and Petrology, 160(1), 115-132.

Bao, Z.A., Chen, L., Zong, C.L., Yuan, H.L., Chen, K.Y., Dai, M.N., 2017. Development of pressed sulfide powder tablets for in situ, sulfur and lead isotope measurement using LA-MC-ICP-MS. International

Journal of Mass Spectrometry, 421, 255-262.

Barbarin, B., 1999. A review of the relationships between granitoid types, their origins and their geodynamic environments. Lithos, 46(3), 605-626.

Champion, D.C., Bultitude, R.J., 2013. The geochemical and Sr, Nd isotopic characteristics of Paleozoic fractionated S-types granites of north Queensland: Implications for S-type granite petrogenesis. Lithos, 162-163(2), 37-56.

Chappell, B.W., White, A.J.R., 1992. I- and S-type granites in the Lachlan fold belt. Transactions of the Royal Society of Edinburgh: Earth Sciences, 83.

Chappell, B.W., 1999. Aluminium saturation in I- and S-type granites and the characterization of fractionated haplogranites. Lithos, 46(3), 535-551.

Chappell, B.W., White, A.J.R., 2001. Two contrasting granite types: 25 years later. Australian Journal of Earth Sciences, 48(4), 489-499.

Chappell, B.W., Bryant, C.J., Wyborn D., 2012. Peraluminous I-type granites. Lithos, 153(8), 142-153.

Chen, W.T., Sun, W.H., Wang, W., Zhao, J.H., Zhou, M.F., 2014. "Grenvillian" intra-plate mafic magmatism in the southwestern Yangtze Block SW China. Precambrian Research, 242, 138-153.

Chen, W.T., Sun, W.H., Zhou, M.F., Wang, W., 2018. Ca. 1 050 Ma intra-continental rift related A-type felsic rocks in the southwestern Yangtze Block, South China. Precambrian Research, 309, 22-44.

Chu, Z.Y., Chen, F.K., Yang, Y.H., Guo, J.H., 2009, Precise determination of Sm, Nd concentrations and Nd isotopic compositions at the nanogram level in geological samples by thermal ionization mass spectrometry. Journal of Analytical Atomic Spectrometry, 24, 1534-1544.

Clemens, J.D., 2003. S-type granitic magmas-petrogenetic issues, models and evidence. Earth-Science Reviews, 61, 1-18.

Clemens, J.D., Stevens, G., 2012. What controls chemical variation in granitic magmas? Lithos, 134-135, 317-329.

Collins W.J., Richards, S.W., 2008. Geodynamic significance of S-type granites in circum-Pacific orogens. Geology, 36, 559-562.

Deng, S.X., 2000. The evolution of metamorphism and geochemistry for the Cangshan and Julin Groups in Central Yunnan, China. Unpublished Ph. D thesis. Guangzhou: Guangzhou Institute of Geochemistry, Chinese Academy of Sciences, 41-49 (in Chinese with English abstract).

DePaolo, D.J., 1981. A Neodymium and Strontium Isotopic Study of the Mesozoic Calc-Alkaline Granitic Batholiths of the Sierra Nevada and Peninsular Ranges, California Granites and Rhyolites. Journal of Geophysical Research, 86(B11), 10470-10488.

Eby, G.N., 1992. Chemical subdivision of the A-type granitoids: Petrogenesis and tectonic implications. Geology, 20, 641-644.

Flowerdew, M.J., Millar, I.L., Vaughan, A.P.M., Horstwood, M.S.A., Fanning, C.M., 2006. The source of granitic gneisses and migmatites in the Antarctic Peninsula: A combined U-Pb SHRIMP and laser ablation Hf isotope study of complex zircons. Contributions to Mineralogy and Petrology, 151, 751-768.

Frost, B.R., Barnes, C.G., Collins, W.J., Arculus, R.J., Ellis, D.J., Frost, C.D., 2001. A Geochemical Classification for Granitic Rocks. Journal of Petrology, 42, 2033-2048.

Frost, C.D., Frost, B.R., 2011. On ferroan (A-type) granitoids: Their compositional variability and modes of origin. Journal of Petrology, 52(1), 39-53.

Gao, S., Ling, W.L., Qiu, Y., Zhou, L., Hartmann, G., Simon, K., 1999.Contrasting geochemical and Sm-Nd isotopic compositions of Archean metasediments from the Kongling high-grade terrain of the Yangtze craton: Evidence for cratonic evolution and redistribution of REE during crustal anatexis. Geochimica et Cosmochimica Acta, 63(13-14), 2071-2088.

Gao, S., Yang, J., Zhou, L., Li, M., Hu, Z., Guo, J., Yuan, H., Gong, H., Xiao, G., Wei, J., 2011. Age and growth of the Archean Kongling terrain, South China, with emphasis on 3.3 Ga granitoid gneisses. American Journal of Science, 311, 153-182.

Gaudemer, Y., Jaupart, C., Tapponnier, P., 1988. Thermal control on post-orogenic extension in collision belts. Earth & Planetary Science Letters, 89, 48-62.

Greentree, M.R., Li, Z.X., Li, X.H., Wu, H., 2006. Latest Mesoproterozoic to earliest Neoproterozoic basin record of the Sibao orogenesis in western South China and relationship to the assembly of Rodinia. Precambrian Research, 151, 79-100.

Geng, Y.S., Yang, C.H., Du, L.L., Wang, X.S., Ren, L.D., Zhou, X.W., 2007. Chronology and tectonic environment of the Tianbaoshan Formation: New evidence from zircon SHRIMP U-Pb age and geochemistry. Geological Review, 53, 556-563 (in Chinese with English abstract).

Griffin, W.L., Pearson, N.J., Belousova, E., Jackson, S.E., van Achterbergh, E., O'Reilly, S.Y., Shee, S.R., 2000. The Hf isotope composition of cratonic mantle: LAM-MC-ICP-MS analysis of zircon megacrysts in kimberlites. Geochimica et Cosmochimica Acta, 64(1), 133-147.

Healy, B., Collins, W.J., Richards, S.W., 2004. A hybrid origin for Lachlan S-type granites: The Murrumbidgee batholith example. Lithos, 79, 197-216.

Hoskin, P.W.O., Schaltegger, U., 2003. The composition of zircon and igneous and metamorphic petrogenesis. Reviews in Mineralogy and Geochemistry, 53,27-62.

Huang, L.C., Jiang, S.Y., 2014. Highly fractionated S-type granites from the giant Dahutang Tungsten deposit in Jiangnan orogen, southeast China: Geochronology, petrogenesis and their relationship with W-mineralization. Lithos, 202-203 (4), 207-226.

Huang, S.F., Wang, W., Pandit, M.K., Zhao, J.H., Lu, G.M., Xue, E.K., 2019. Neoproterozoic S-type granites in the western Jiangnan Orogenic Belt, South China: Implications for petrogenesis and geodynamic significance. Lithos, 342-343, 45-58.

Jacobs, J., Fanning, C.M., Bauer, W., 2003. Timing of Grenville-age vs. Pan-African medium- to high grade metamorphism in western Dronning Maud Land (East Antarctica) and significance for correlations in Rodinia and Gondwana. Precambrian Research, 125, 1-20.

Jiang, Y.H., Zhu, S.Q., 2017. Petrogenesis of the Late Jurassic peraluminous biotite granites and muscovite-bearing granites in SE China: Geochronological, elemental and Sr-Nd-O-Hf isotopic constraints. Contributions to Mineralogy and Petrology, 172, 1-27.

Kong, X.Y., Zhang, C., Liu, D.D., Jiang, S., Luo, Q., Zeng, J.H., Liu, L.F., Luo, L., Shao, H.B., Liu, D., Liu, X.Y., Wang, X.P., 2019. Disequilibrium partial melting of metasediments in subduction zones: Evidence from O-Nd-Hf isotopes and trace elements in S-type granites of the Chinese Altai.

Lithosphere, 11(1), 149-168.

Lai, S.C., Qin, J.F., Zhu, R.Z., Zhao, S.W., 2015. Neoproterozoic quartz monzodiorite-granodiorite association from the Luding-Kangding area: Implications for the interpretation of an active continental margin along the Yangtze Block (South China Block). Precambrian Research, 267(3-4), 196-208.

Li, F.H., Tan, J.M., Shen, Y.L., Yu, F.X., Zhou, G.F., Pan, X.N., Li, X.Z., 1988. The Pre-Sinian in the Kangdian area. Chongqing: Chongqing Publishing House, 396 (in Chinese with English abstract).

Liu, X.M., Gao, S., Diwu, C.R., Yuan, H.L., Hu, Z.C., 2007. Simultaneous in-situ determination of U-Pb age and trace elements in zircon by LA-ICP-MS in 20 μm spot size. Chinese Science Bulletin, 52(9), 1257-1264.

Liu, Y., Liu, X.M., Hu, Z.C., Diwu, C.R., Yuan, H.L., Gao, S., 2007. Evaluation of accuracy and long-term stability of determination of 37 trace elements in geological samples by ICP-MS. Acta Petrologica Sinica, 23(5), 1203-1210 (in Chinese with English abstract).

Ludwig, K.R., 2003. ISOPLOT 3.0: A geochronological toolkit for Microsoft Excel. Berkeley Geochronology Center, Special Publication, 4.

Mabi, A.W., Yang, Z.X., Zhang, M.C., Wen, D.K., Li, Y.L., Liu, X.Y., 2018. Two types of granites in the western Yangtze block and their implications for regional tectonic evolution: Constraints from geochemistry and isotopic data. Acta Geologica Sinica: English Edition, 92(1), 89-105.

Middlemost, E.A.K., 1994. Naming materials in the magma/igneous rocks system. Earth-Science Reviews, 37, 215-224.

Miller, C.F., McDowell, S.M., Mapes, R.W., 2003. Hot and cold granites? Implications of zircon saturation temperatures and preservation of inheritance. Geology, 31, 529-532.

Mou, C.L., Lin, S.L., Yu, Q., 2003. The U-Pb ages of the volcanic rock of the Tianbaoshan formation, Huili, Sichuan Province. Journal of Stratigraphy, 27, 216-219 (in Chinese with English abstract).

Nowell, G.M., Kempton, P.D., Noble, S.R., Fitton, J.G., Saunders, A.D., Mahoney, J.J., Taylor, R.N., 1998. High precision Hf isotope measurements of MORB and OIB by thermal ionisation mass spectrometry: Insights into the depleted mantle. Chemical Geology, 149(3-4), 211-233.

Patiño Douce, A.E., 1999. What do experiments tell us about the relative contributions of crust and mantle to the origin of the granitic magmas. Geological Society London Special Publications, 168(1), 55-75.

Rapp, R.P., Watson, E.B., 1995.Dehydration melting of metabasalt at 8−32 kbar: Implications for continental growth and crust-mantle recycling. Journal of Petrology, 36(4), 891-931.

Roberts, M.P., Clemens, J.D., 1993. Origin of high-potassium, calc-alkaline, I-type granitoids. Geology, 21 (9), 825.

Rudnick, R.L., Gao, S., 2003. Composition of the continental crust. Treatise Geochem, 3, 1-64.

Sichuan Provincial Bureau of Geology and Mineral Resources (SPBGMR), 2001. Regional geological survey of the People's Republic of China, the Salian Sheet map and report, scale 1:50 000 (in Chinese).

Soderlund, U., Patchett, P.J., Vervoort, J.D., Isachsen, C.E., 2004.The [176]Lu decay constant determined by Lu-Hf and U-Pb isotope systematics of Precambrian mafic intrusions. Earth & Planetary Science Letters, 219(3-4), 311-324.

Stevens, G., Clemens, J.D., Droop, G.T.R., 1997. Melt production during granulite-facies anatexis:

Experimental data from "primitive" metasedimentary protoliths. Contributions to Mineralogy and Petrology, 128, 352-370.

Stevens, G., Villaros, A., Moyen, J. F., 2007. Selective peritectic garnet entrainment as the origin of geochemical diversity in S-type granites. Geology, 35(1), 9-12.

Sun, S.S., McDonough, W.F., 1989. Chemical and isotopic systematics of oceanic basalts: Implications for mantle composition and processes. In: Saunders, A.D., Norry, M.J. Magmatism in the Ocean Basins. Geological Society, London, Special Publications, 42, 313-345.

Sun, W.H., Zhou, M.F., Yan, D.P., Li, J.W., Ma, Y.X., 2008. Provenance and tectonic setting of the Neoproterozoic Yanbian Group, western Yangtze Block (SW China). Precambrian Research, 167(s1-2), 213-236.

Sylvester, P.J., 1998. Post-collisional strongly peraluminous granites. Lithos, 45, 29-44.

Tang, M., Wang, X.L., Shu, X.J., Wang, D., Yang, T., Gopon, P., 2014. Hafnium isotopic heterogeneity in zircons from granitic rocks: Geochemical evaluation and modeling of "zircon effect" in crustal anatexis. Earth & Planetary Science Letters, 389, 188-199.

Vernon, R.H., 1984. Microgranitoid enclaves in granites: Globules of hybrid magma quenched in a plutonic environment. Nature, 309, 438-439.

Villaros, A., Stevens, G., Moyen, J.F., Buick, I.S., 2009. The trace element compositions of S-type granites: Evidence for disequilibrium melting and accessory phase entrainment in the source. Contributions to Mineralogy and Petrology, 158(4), 543-561.

Villaros, A., Buick, I.S., Stevens, G., 2012. Isotopic variations in S-type granites: An inheritance from a heterogeneous source? Contributions to Mineralogy and Petrology, 163(2), 243-257.

Wang, D., Wang, X.L., Cai, Y., Goldstein, S.L., Yang, T., 2018. Do Hf isotopes in magmatic zircons represent those of their host rocks? Journal of Asian Earth Sciences, 154, 202-212.

Wang, X.L., Zhou, J.C., Wan, Y.S., Kitajima, K., Wang, D., Bonamici, C., Qiu, J.S., Sun, T., 2013. Magmatic evolution and crustal recycling for Neoproterozoic strongly peraluminous granitoids from Southern China: Hf and O isotopes in zircon. Earth & Planetary Science Letters, 366(2), 71-82.

Wang, Y.J., Fan, W.M., Sun, M., Liang, X.Q., Zhang, Y.H., Peng, T.P., 2007. Geochronological, geochemical and geothermal constraints on petrogenesis of the Indosinian peraluminous granites in the South China Block: A case study in the Hunan province. Lithos, 96(3-4), 475-502.

Wang, Y.J., Zhou, Y.Z., Cai, Y.F., Liu, H.C., Zhang, Y.Z., Fan, W.M., 2016. Geochronological and geochemical constraints on the petrogenesis of the Ailaoshan granitic and migmatite rocks and its implications on Neoproterozoic subduction along the SW Yangtze Block. Precambrian Research, 283, 106-124.

Wang, Y.J., Zhu, W.G., Huang, H.Q., Zhong, H., Bai, Z.J., Fan, H.P., Yang, Y.J., 2019. Ca. 1.04 Ga hot Grenville granites in the western Yangtze block, southwest China. Precambrian Research, 328, 217-234.

Whalen, J. B., Currie, K. L., Chappell, B. W., 1987. A-type granites: Geochemical characteristics, discrimination and petrogenesis. Contributions to Mineralogy and Petrology, 95, 407-419.

Wiedenbeck, M., Hanchar, J.M., Peck, W.H., Sylvester, P., Valley, J., Whitehouse, M., Kronz, A., Morishita, Y., et al., 2004. Further characterisation of the 91500 zircon crystal. Geostandards and

Geoanalytical Research, 28(1), 9-39.

Yuan, H.L., Gao, S., Liu, X.M., Li, H.M., Gunther, D., Wu, F.Y., 2004. Accurate U-Pb age and trace element determinations of zircon by laser ablation-inductively coupled plasma mass spectrometry. Geostandards & Geoanalytical Research, 28(3), 353-370.

Yuan, H.L., Gao, S., Dai, M.N., Zong, C.L., Gunther, D., Fontaine, G.H., Liu, X.M., Diwu, C.R., 2008. Simultaneous determinations of U-Pb age, Hf isotopes and trace element compositions of zircon by excimerlaser-ablation quadrupole and multiple-collector ICPMS. Chemical Geology, 247(1-2), 100-118.

Zen, E., 1986. Aluminum enrichment in silicate melts by fractional crystallization: Some mineralogic and petrographic constraints. Journal of Petrology, 27, 1095-1117.

Zhang, Q., Ran, H., Li, C.D., 2012. A-type granite: What is the essence? Acta Petrologica et Mineralogical, 31(4), 621-626 (in Chinese with English abstract).

Zhao, G.C., Cawood, P.A., 2012. Precambrian geology of China. Precambrian Research, s 222-223, 13-54.

Zhao, J.H., Zhou, M.F., Yan, D.P., Zheng, J.P., Li, J.W., 2011. Reappraisal of the ages of Neoproterozoic strata in South China: No connection with the Grenvillian orogeny. Geology, 39(4), 299-302.

Zhao, J.H., Li, Q.W., Liu, H., Wang, W., 2018. Neoproterozoic magmatism in the western and northern margins of the Yangtze Block (South China) controlled by slab subduction and subduction-transform-edge-propagator. Earth-Science Reviews, 187, 1-18.

Zheng, J.P., Griffin, W.L., O'Reilly, S.Y., Zhang, M., Pearson, N., Pan, Y.M., 2006. Widespread Archean basement beneath the Yangtze craton. Geology, 34, 417-420.

Zhu, R.Z., Lai, S.C., Qin, J.F., Zhao, S.W., Santosh, M., 2018. Strongly peraluminous fractionated S-type granites in the Baoshan Block, SW China: Implications for two-stage melting of fertile continental materials following the closure of Bangong-Nujiang Tethys. Lithos, 316-317, 178-198.

Zhu, W.G., Zhong, H., Li, Z.X., Bai, Z.J., Yang, Y.J., 2016. SIMS zircon U-Pb ages, geochemistry and Nd-Hf isotopes of ca. 1.0 Ga mafic dykes and volcanic rocks in the Huili area, SW China: Origin and tectonic significance. Precambrian Research, 273, 67-89.

Zhu, Y., Lai, S.C., Qin, J.F., Zhu, R.Z., Zhang, F.Y., Zhang, Z.Z., 2019a.Geochemistry and zircon U-Pb-Hf isotopes of the 780 Ma I-type granites in the western Yangtze Block: Petrogenesis and crustal evolution. International Geology Review, 61(10), 1222-1243.

Zhu, Y., Lai, S.C., Qin, J.F., Zhu, R.Z., Zhang, F.Y., Zhang, Z.Z., Gan, B.P., 2019b. Petrogenesis and geodynamic implications of Neoproterozoic gabbro-diorites, adakitic granites, and A-type granites in the southwestern margin of the Yangtze Block, South China. Journal of Asian Earth Sciences, 183, 103977.

Zhu, Y., Lai, S.C., Qin, J.F., Zhu, R.Z., Zhang, F.Y., Zhang, Z.Z., Zhao, S.W., 2019c. Neoproterozoic peraluminous granites in the western margin of the Yangtze Block, South China: Implications for the reworking of mature continental crust. Precambrian Research, 333, 105443.

Constructing the latest Neoproterozoic to Early Paleozoic multiple crust-mantle interactions in western Bainaimiao arc terrane, southeastern Central Asian Orogenic Belt[①]

Liu Min　Lai Shaocong[②]　Zhang Da　Zhu Renzhi

Qin Jiangfeng　Xiong Guangqiang　Wang Haoran

Abstract: Identifying the crust-mantle interactions in association with the evolution of the Precambrian microcontinents provides critical constraints on the accretionary evolution in the Central Asian Orogenic Belt (CAOB). The Bainaimiao arc terrane (BAT) is one of the most important Precambrian microcontinents in southeastern CAOB, however, few studies have paid attention to the types and the evolving processes of the crust-mantle interactions that occurred before its final accretion onto the northern North China Craton. This study presents an integrated study of geochronology, zircon Hf isotope and whole-rock geochemistry on the latest Neoproterozoic diabases and the Early Paleozoic arc intrusions in the western BAT. The latest Neoproterozoic (ca. 546 Ma) diabases display low SiO_2(46.52−49.24 wt%) with high MgO (8.23−14.41 wt%), Cr (66−542 ppm) and Ni (50−129 ppm), consisting with mantle origin. Their highly negative zircon $\varepsilon_{Hf}(t)$ (−24.7 to −12.0) and high Fe/Mn ratios (62.1−81.7) further indicate a significantly enriched mantle source. Considering that the BAT maybe initially separated from the Tarim Craton with a thickened crustal root, we propose that these diabases were generated through partial melting of an enriched lithospheric mantle source that had been hybridized by lower-crustal eclogites during foundering of the BAT lower crust. The Early Paleozoic (ca. 475−417 Ma) arc intrusions in western BAT can be divided into Periods Ⅰ and Ⅱ at approximately 450 Ma. The Period Ⅰ (> 450 Ma) intrusions contain abundant mafic minerals like hornblende and pyroxene, and show positive zircon $\varepsilon_{Hf}(t)$ (+1.5 to +10.9). They are predominantly medium-K calc-alkaline with broad correlations of SiO_2 versus various major and trace elements, which correlate well with the experimental melts produced by the fractional crystallization of primitive hydrous arc magmas at 7 kbar. We assume they were formed through mid-crustal differentiation of the mantle wedge-derived hydrous basaltic melts. By contrast, the Period Ⅱ (≤ 450 Ma) intrusions are characterized by variable zircon $\varepsilon_{Hf}(t)$ (−15.0 to +11.5) with irregular variations in most major and trace elements, which are more akin to the arc magmas generated in an open system. The general occurrence of elder inherited zircons,

---

①　Published in *Geoscience Frontiers*, 2020, 11.

②　Corresponding author.

along with the relatively high Mg# (> 45) of some samples, call upon a derivation from the reworking of the previously subduction-modified BAT lower crust with the input of mantle-derived mafic components. In combination with the Early Paleozoic tectonic mélanges flanking western BAT, we infer that the compositional transition from Period Ⅰ to Ⅱ can be attributed to the tectonic transition from south-dipping subduction of Solonker ocean to north-dipping subduction of South Bainaimiao ocean in southeastern CAOB. The above results shed light not only on the latest Neoproterozoic to Early Paleozoic multiple crust-mantle interactions in western BAT, but also on the associated crustal construction processes before the final arc-continent accretion.

1 Introduction

Magmatic records preserved in accretionary orogens often show intense flare-ups with significant compositional variations (Condie, 2007; Collins et al., 2011; Xiao and Santosh, 2014; Tang et al., 2017), which are considered to be general consequences of distinct crust-mantle interactions during subduction and accretion of different tectonic domains (e. g., Cawood et al., 2009; Xiao et al., 2015). In this respect, for a given accretionary orogen, linking variations of magma composition with changes of crust-mantle interaction with time in different orogenic components is critical in dissecting its architecture and amalgamation history.

Microcontinents comprising ancient basements as well as ensuing continental arcs constitute major components of the Central Asian Orogenic Belt (CAOB), such as the Xing'an, Bureya and Xingkai blocks in the eastern CAOB, the Junggar terrane in the southwestern CAOB and the Tuva-Mongolia block in the central CAOB (Fig. 1a; Kozakov et al., 2007; Kröner et al., 2013; He et al., 2018; Zhou et al., 2018a; Yuan et al., 2019). These microcontinents were either accreted onto the neighbouring blocks or assemblages as coherent allochthonous terranes through ancient subduction systems, or acting as nuclei for accretionary growth (Kröner et al., 2013; Zhou et al., 2018a). Identifying the crust-mantle interactions and deep dynamic processes associated with the evolution of the microcontinents can provide critical insights into the crustal construction processes of the microcontinents and the accretionary history in the CAOB (Kröner et al., 2013; He et al., 2018; Zhou et al., 2018a).

The Bainaimiao arc terrane (BAT), as a prominent part of the southeastern CAOB, was recently recognized as an exotic Precambrian microcontinent with development of extensive Early Paleozoic arc magmatism (Fig. 1b; Zhang et al., 2014; Eizenhöfer and Zhao, 2018; Zhou et al., 2018b,c; Ma et al., 2019). It was accreted onto the northern North China Craton (NCC) through the latest Silurian arc-continent collision (Tang, 1990; Zhang et al., 2010, 2014). However, previous works have primarily focused on the basement affinity, the geochronological framework of the Early Paleozoic arc magmatism, and the postcollisional

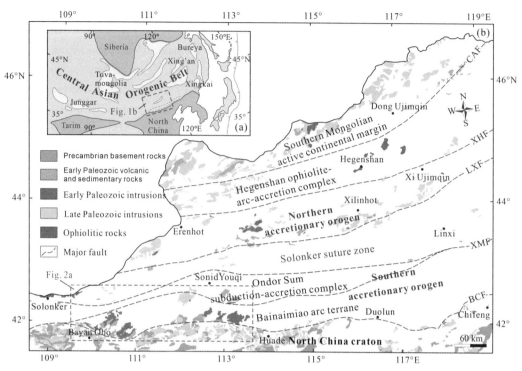

Fig. 1 (a) Sketch map of the Central Asian Orogenic Belt (CAOB) showing the main tectonic subdivisions (modified after Zhou et al., 2018a; Yuan et al., 2019). (b) Simplified geological map of the southeastern CAOB (modified after Jian et al., 2008; Xiao et al., 2015).

CAF: Chagan'aobao-Arongqi fault; XHF: Xilinhot fault; LXF: Linxi fault;

XMF: Xar Moron fault; BCF: Bayan Obo-Chifeng fault.

sedimentation in the BAT (Jian et al., 2008; Zhang et al., 2014; Wu et al., 2016; Eizenhöfer and Zhao, 2018; Zhou et al., 2018c; Ma et al., 2019), few attention has been paid to the compositional variations of the Late Neoproterozoic to Early Paleozoic magmas, which may have recorded significant changing of crust-mantle interactions and associated deep mantle dynamics that dominated the evolution of the BAT.

This contribution presents geochronological, whole-rock geochemical and in-situ zircon Hf isotopic data of the latest Neoproterozoic diabases and the Early Paleozoic granodiorite porphyries from the western BAT. Our results, together with the literature data of the Early Paleozoic arc intrusions in adjacent areas (Fig. 2a; Table 1), lead us to propose that multiple crust-mantle interactions have occurred beneath the western BAT before its final accretion onto the northern NCC, which were triggered by the Late Neoproterozoic foundering of the BAT lower crust, and the Early Paleozoic transition from south-dipping subduction of the Solonker ocean to north-dipping subduction of the South Bainaimiao ocean in southeastern CAOB.

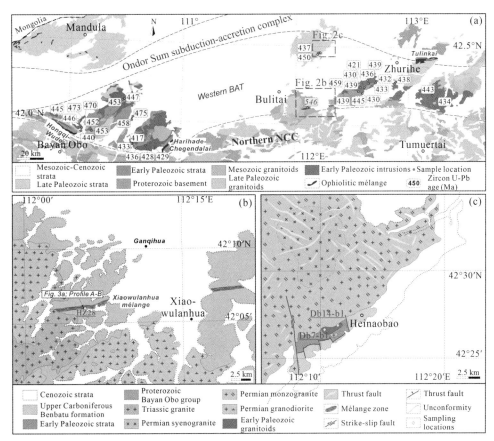

Fig. 2 (a) Simplified geological map of the western BAT (modified from 1:500 000 scale geological map of the Solonker-Xilin Gol region compiled by Tianjin Institute of Geology and Mineral Resource), showing the spatial and temporal distribution of the Early Paleozoic intrusions in western BAT. References for the literature age data are listed in Table 1. (b) Geological map of the northeastern Bulitai area (IGSCUGB, 2008). (c) Geological map of the southeastern Bulitai area (IGSCUGB, 2008).

2 Geological background

The southeastern CAOB (also known as the Xing-Meng orogenic belt) preserves five typical orogenic units: the Southern Mongolian active continental margin, the Hegenshan ophiolite-arc-accretion complex, the northern accretionary orogen, the Solonker suture zone and the southern accretionary orogen (Fig. 1b; Jian et al., 2008; Xiao et al., 2015). It is generally accepted that the southeastern CAOB represents a complicated college of arcs, fragments of Paleo-Asian ocean (PAO) crust and microcontinents that were amalgamated between the northern NCC and the South Mongolia terranes (e.g., Xiao et al., 2003, 2015). Moreover, recent studies have suggested that the southeastern CAOB eventually evolved into the collisional phase during the end-Permian to Early-Middle Triassic (Li et al., 2016, 2017), after the prolonged subduction of PAO since the Neoproterozoic (e.g., Xiao et al., 2015).

Table 1　Summary of zircon U-Pb ages of the Early Paleozoic arc intrusions in the western BAT.

Location	Coordinate	Sample	Rock type	Age/Ma	$\varepsilon_{Hf}(t)$	T_{DMC}/Ga	Reference
Northern Damaoqi	N41°50′42.0″, E110°41′58.9″	DM132	Diorite	428±2	−6.1 to −0.9	1.47–1.79	Zhou et al., 2018c
	N42°00′09.2″, E110°29′51.1″	DM210	Gabbroic diorite	458±2	+2.1 to +6.7	1.01–1.30	
	N42°05′31.4″, E110°29′48.8″	DM232	Gabbroic diorite	475±4	+1.5 to +6.7	1.00–1.33	
Baiyanhua	N41°55′06.2″, E110°07′15.3″	BYH01	Granodiorite	440±2	–	–	Zhang and Jian, 2008
	N41°55′54.0″, E110°07′20.7″	BYH02	Diorite	452±3	–	–	
	N41°59′14.5″, E110°06′58.7″	BYH03	Quartz diorite	446±2	–	–	
Chaganhushao	N41°50′16.1″, E110°34′11.4″	DARB01	Tonalite	417±2	–	–	
Bateaobao	–	–	Diorite	472	–	–	Xu et al., 2003
	–	–	Granodiorite	450	–	–	
	–	–	Tonalite	427	–	–	
Bailiutu	N42°07′00.1″, E110°25′48.0″	NM01-13	Diorite	453±3	–	–	Li et al., 2010
	N42°07′00.3″, E110°25′58.0″	NM08-17	Diorite	447±5	–	–	
Bayan Obo-Damaoqi	N41°53′47.4″, E110°12′23.3″	07130-1	Quartz diorite	453±2	+2.3 to +7.0	0.99–1.29	Zhang et al., 2014
	N41°48′40.1″, E110°36′35.4″	08480-1	Quartz diorite	436±4	−4.6 to −0.5	1.45–1.72	
	N41°49′17.5″, E110°47′22.9″	08487-1	Alkali feldspar granite	429±4	−6.4 to +0.5	1.39–1.82	
	N41°59′12.8″, E110°07′04.6″	08500-1	Tonalite	473±2	+6.8 to +10.9	0.76–1.02	
	N41°59′52.5″, E110°07′12.8″	08502-1	Tonalite	470±3	+4.0 to +8.7	0.89–1.19	
	N41°50′16.1″, E110°34′11.5″	08515-2	Tonalite	433±6	−15.0 to −2.6	1.59–2.36	
	N41°59′53.8″, E109°54′11.8″	08557-1	Quartz diorite	445±3	+5.4 to +11.4	0.70–1.09	

Continued

Location	Coordinate	Sample	Rock type	Age/Ma	$\varepsilon_{Hf}(t)$	T_{DMC}/Ga	Reference
Bainaimiao	N42°15′27.9″, E112°34′37.2″	07813-7	Quartz diorite	421±2	−4.4 to −0.9	1.47−1.69	Zhang et al., 2014
	N42°16′49.6″, E112°43′15.1″	08404-1	Quartz diorite	439±3	+1.7 to +7.1	0.97−1.32	
	N42°17′34.4″, E112°43′19.9″	08406-1	Muscovite granite	432±5	−3.9 to +6.9	1.04−1.66	
	N42°12′47.4″, E112°36′36.0″	08409-1	Quartz diorite	430±3	+0.3 to +2.8	1.24−1.40	
	N42°13′24.1″, E112°31′56.2″	08413-1	Tonalite	439±5	−10.3 to +1.9	1.31−2.08	
	N42°14′16.3″, E112°33′31.7″	08417-1	Tonalite	430±3	−0.8 to +2.6	1.30−1.47	
	N42°16′13.0″, E112°37′48.4″	08429-1	Tonalite	433±3	−1.3 to +4.5	1.14−1.50	
	N42°17′05.9″, E112°37′27.6″	08432-1	Diorite	436±4	−14.4 to +7.1	1.00−2.38	
Bainaimiao	N42°12′42.0″, E112°25′40.2″	BNM-1	Diorite porphyrite	439±2	−	−	Wu et al., 2016
Baiyinduxi	−	BNM08	Metadiorite	438±2	−	−	Zhang et al., 2013a
Taigushengmiao	−	JN81	Tonalite	443±2	−	−	Bai et al., 2015
	−	JN82	Quartz diorite	434±2	−	−	
Southern Zhurihe	N42°18′13.3″, E113°00′06.4″	NM13-40	Granite	433±3	−1.2 to +11.5	0.68−1.49	Qian et al., 2017
	N42°17′59.2″, E113°00′15.4″	NM13-42	Granite	428±2	+3.3 to +10.3	0.76−1.20	
Bainaimiao deposit	−	BN-VIII-08	Quartz diorite	459±6	−	−	Zhou et al., 2017
	−	BNM-3	Granodiorite porphyry	445±6	−	−	Li et al., 2012
	−	99-11	Quartz diorite	459±3	−	−	Tong et al., 2010
Heinaobao	N42°26′16.1″, E112°10′32.1″	DB7-b1	Granodiorite porphyry	450±1	−10.1 to +3.9	1.18−2.06	This study
	N42°26′15.5″, E112°10′33.7″	DB14-b1	Granodiorite porphyry	437±2	−	−	

The Solonker suture zone marked the final closure of the PAO and as well as the termination of the subduction-accretion processes in the southeastern CAOB (e. g. , Xiao et al. , 2015; Eizenhöfer and Zhao, 2018; Fu et al. , 2018).

The southern accretionary orogen can be further divided into the BAT and the Ondor Sum subduction-accretion complex (OSC) (Fig. 1b). The OSC is characterized by the ca. 490 – 450 Ma mélanges/ophiolitic fragments and the 453 – 449 Ma high-pressure metamorphic rocks, which were juxtaposed in a southward thrust system (de Jong et al. , 2006; Jian et al. , 2008; Xiao et al. , 2015). The BAT, which extends eastward from the northern Bayan Obo to the Yanbian area (e. g. , Zhang et al. , 2014; Zhou et al. , 2018b; Ma et al. , 2019), was traditionally regarded as an island arc or a continental arc with NCC affinity that was developed in relation to the Early Paleozoic subduction of the PAO (e.g. , Jia et al. , 2003; Xiao et al. , 2003; Zhang et al. , 2014). However, recent works demonstrate that the BAT was actually developed on an exotic Precambrian microcontinent, which shows a tectonic affinity to the Tarim Craton (e.g. , Zhang et al. , 2014; Eizenhöfer and Zhao, 2018). As suggested by Zhang et al. (2014), the BAT was likely separated from the NCC by the South Bainaimiao ocean during the Early Paleozoic, and was accreted to the passive northern margin of the NCC through the latest Silurian arc-continent collision event. Moreover, the ophiolitic mélanges discontinuously exposed along tectonic boundary between the BAT and the northern NCC, including the Hongqi-Wude mélange and the Harihada-Chegendalai mélange (Fig. 2a), may be formed during the subduction of the South Bainaimiao ocean (Zhang et al. , 2014; Zhou et al. , 2018c).

The western BAT is relatively well-exposed and preserves the majority of the Early Paleozoic geological records of the BAT (Eizenhöfer and Zhao, 2018). It is characterized by the presence of extensive Early Paleozoic intrusive rocks (Fig. 1b), which comprise diorites, quartz diorites, tonalites, granodiorites, granites and minor gabbro diorites (Fig. 2a; Table 1). They generally exhibit arc-related geochemical signatures, including enriched light rare-earth elements (LREE) and large-ion lithophile elements (LILE) and depleted high field strength elements (HFSE). These intrusive rocks are widely attributed be to the partial melting of oceanic crust and overlying mantle wedge, and the interactions/mixing of melts derived from the ancient crust beneath the BAT and the mantle wedge during the Early Paleozoic oceanic subduction (e.g. , Zhang et al. , 2014; Zhou et al. , 2018c). Some outcrops of the Early to Middle Ordovician arc-related volcanic rocks, including dacites, rhyolites and andesites (Zhang et al. , 2014; Zhou et al. , 2017), are also identified from the Early Paleozoic metamorphic volcanic-sedimentary units in the western BAT (Fig. 2a). Nevertheless, distinct subduction polarities have been proposed for generation of the above Early Paleozoic magmatism: ① the north-dipping subduction of the South Bainaimiao oceanic lithosphere (Zhang et al. , 2014; Zhou et al. , 2018c; Ma et al. , 2019); ② the south-dipping subduction of the Solonker oceanic lithosphere (Jian et al., 2008; Eizenhöfer and Zhao, 2018; Zhou et al., 2018a). Termination of the Early Paleozoic arc magmatism may have occurred before the Early Devonian, as inferred from the emplacement of an undeformed pegmatite dike (ca. 411 Ma) into the Early Paleozoic metamorphic volcanic-sedimentary units in the Bainaimiao area (Zhang et al., 2013a).

3　Petrography and sample descriptions

The investigated latest Neoproterozoic diabases and Early Paleozoic granodiorite porphyries in this study were respectively collected from the Xiaowulanhua tectonic mélange in the southwestern Bulitai area and the Heinaobao pluton in the northeastern Bulitai area (Fig. 2b,c).

The Xiaowulanhua mélange, which is considered to be the tectonic boundary line of the BAT and the northern NCC in this region [Institute of Geological Survey of China University of Geosciences, Beijing (IGSCUGB), 2008], discontinuously outcrops in the Proterozoic Bayan Obo group with a width of 100−500 m. It is characterized by a ductile shear zone comprising sericite-quartz schists, foliated quartzites, granitic mylonites and foliated limestones, with tectonic breccias and exotic blocks of mafic-ultramafic rocks (Fig. 3). Our field observations

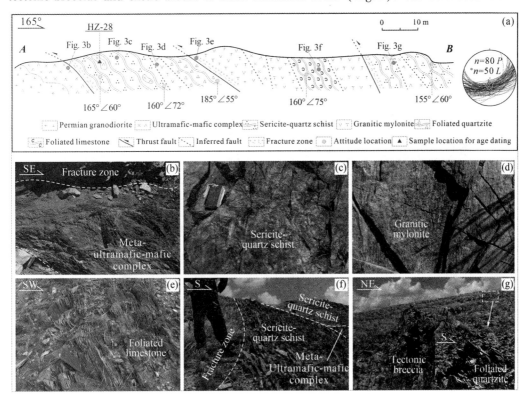

Fig. 3　(a) Cross section A-B showing the structural relationships between different lithostructural units and lower hemisphere projection of the foliation and lineation for the Xiaowulanhua mélange.

(b-g) Field photographs of the main lithostructural units of the Xiaowulanhua mélange.

suggest that the rocks within the ductile shear zone have a pervasive E-W or NE-SW striking medium- to high-angle foliation, as well as a well-developed medium- to high-angle, SSE to SSW plunging stretching lineation (Fig. 3a). Kinematic indicators, including asymmetric tectonic lenses, small scale asymmetric folds, asymmetric rotational augens, S-C fabrics and mica fishes, σ-type K-feldspar porphyroclasts and rotated quartz grains with undulose extinction, indicate a dominant SW-NE dextral shearing (Supplementary Fig. 1). The investigated Xiaowulanhua diabases in this study were mostly selected from the mafic-ultramafic complex within the sericite-quartz schists (Fig. 4a). They mainly contain plagioclase (50% - 55%) and pyroxene (35% - 40%), with minor Fe-Ti oxides and hornblende (Fig. 4b). The diabasic texture can hardly be identified due to alteration. Most pyroxene crystals were replaced by chlorite, and most plagioclase crystals were altered to clay. Some samples were also variably carbonated and contain opaque minerals like magnetite and ilmenite.

Fig. 4   Field photographs and representative photomicrographs of the Xiaowulanhua diabases (a,b) and the Heinaobao granodiorite porphyries (c,d).
amp: amphibole; cal: calcite; chl: chlorite; mt: magnetite; px: pyroxene; qtz: quartz; pl: plagioclase; Kf: K-feldspar; bt: biotite.

The Heinaobao pluton is mainly exposed at the Heinaobao iron deposit with a surface outcrop of about 60 km^2(Fig. 2c). It is characterized by an irregular NW-SE-trending body dominated by grey, medium-grained granodiorite porphyry. The granodiorite porphyries emplaced into the Early Paleozoic Ondor Sum group metamorphic strata (Fig. 4c), and were uncomfortably covered by the Upper Carboniferous Benbatu formation in the field. Most surface outcrops of the Heinaobao pluton are deformed and affected by varying degrees of alteration. The investigated granodiorite porphyry samples in this study are relatively fresh and possess a porphyritic texture. The phenocrysts is dominantly composed of plagioclase (20%−30%) and K-feldspar (10% − 15%), whereas the groundmass (50% − 60%) mainly contains plagioclase (~20%), quartz (~15%), biotite (~10%) with minor K-feldspar (~5%) and amphibole (~5%) (Fig. 4d). Magnetite, zircon, titanite and apatite occur as accessory phases in the granodiorite porphyries.

4 Analytical methods

The separation of zircons grains from the studied diabase and granodiorite porphyry samples carried out at the Langfang Diyan-separation Company (China) by the traditional density and magnetic methods. The internal structure of the zircons grains was revealed by cathodoluminescence (CL) imaging at the Beijing Gaonian-navigation Technology Company (China). Zircon U-Pb analysis was performed on a Finnigan Neptune ICP-MS, which was equipped with a 193 nm New Wave laser and a 35 μm beam at the Tianjin Institute of Geology and Mineral Resources (China). Harvard zircon 91500 and NIST 610 were simultaneously measured as external calibrations during the analysis. The detailed analytical techniques and the method of common Pb corrections were described in Andersen (2002) and Yuan et al. (2004), respectively. The age calculation and concordia plotting were based on the Isoplot 3.0 program (Ludwig, 2003).

Whole-rock major and trace elements of the studied diabase and granodiorite porphyry were measured at the Beijing Research Institute of Uranium Geology (China). Determination of the major element concentrations was carried out on a Philips PW2404 X-ray fluorescence spectrometry, and the data accuracy was better than 5%. Concentration of FeO was further measured by the conventional wet chemical procedure. Determination of the trace element compositions was carried out on a PerkinElmer Sciex ELAN DRC-e ICP-MS, with data accuracy better than 5%−10%. The instrument parameters and detailed analytical techniques are given in Gao et al. (2007).

Measurements of the in-situ zircon Lu-Hf compositions for the studied diabase and granodiorite porphyry were carried out on a Nu Plasma II MC-ICP-MS equipped with a 193 nm RESOLution M-50 laser and a 44 μm beam at the State Key Laboratory of Continental Dynamics, Northwest University, Xi'an, China. The detailed analytical techniques and instrument information can be found in Yuan et al. (2008). The MT and 91500 standards were simultaneously measured during the analysis, which respectively yielded weighted $^{176}Hf/^{177}Hf$ ratios of 0.282 528 ± 0.000 019 and 0.282 264 ± 0.000 030, consisting with the recommended values (Yuan et al., 2008). The calculations of initial $^{176}Hf/^{177}Hf$ ratios and $\varepsilon_{Hf}(t)$ values were referenced to the chondritic uniform reservoir (Bouvier et al., 2008), which typically has a ^{176}Hf/^{177}Hf ratio of 0. 282 785 and a ^{176}Lu/^{177}Hf ratio of 0. 033 6. The calculations of the depleted mantle model (T_{DM}) age and the crustal model (T_{DMC}) age were respectively based on the present-day depleted mantle with ^{176}Lu/^{177}Hf ratio of 0. 038 4 and ^{176}Hf/^{177}Hf ratio of 0. 283 25, and the average continental crust with ^{176}Lu/^{177}Hf of 0. 015 (Griffin et al., 2000).

5 Results

5. 1 Zircon U-Pb ages

Detailed zircon U-Pb data for the diabase sample HZ28 (N42°06′16. 0″, E112°05′12. 3″) from the Xiaowulanhua tectonic mélange, and the granodiorite porphyry samples Db7-b1 (N42°26′16. 1″, E112°10′32. 1″) and Db14-b1 (N42°26′15. 5″, E112°10′33. 7″) from the Heinaobao pluton are listed in Supplementary Table 1.

The majority of zircons from the diabase sample HZ28 are euhedral to subhedral, 50–300 μm in length, and show homogeneous or faint, broad zoning (Fig. 5a), in accordance with magmatic origin (Hoskin and Schaltegger, 2003). Twenty-two dating spots of the sample HZ28 show variable Th/U ratios of 0. 1–2. 1 and concordant ^{206}Pb/^{238}U age clusters of 544–548 Ma, which yield a weighted mean age of 546 ± 2 Ma (n = 22, MSWD = 0. 1) (Fig. 5a).

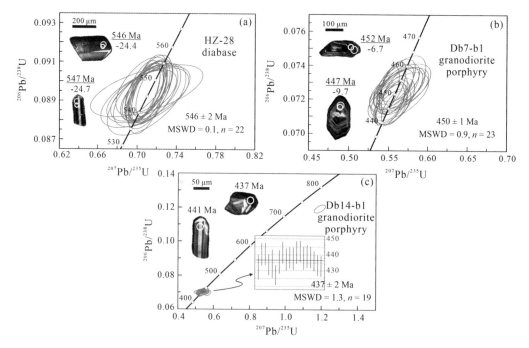

Fig. 5　CL images of representative zircons and U-Pb concordia diagrams of diabase sample HZ-28 (a) and granodiorite porphyry samples Db7-b1 (b) and Db14-b1 (c).

White and yellow circles indicate the spots of U-Pb dating and Hf analyses, respectively.

Zircons from the granodiorite porphyry samples Db7-b1 and Db14-b1 are also euhedral to subhedral, 80-200 μm in length and show clear oscillatory zonings (Fig. 5b,c). They have Th/U ratios varying from 0. 1 to 1. 1. Twenty-three dating spots of the sample Db7-b1 show concordant $^{206}Pb/^{238}U$ age clusters of 445-455 Ma and yield a weighted mean age of 450 ± 1 Ma ($n = 23$, MSWD = 0. 9) (Fig. 5b). Twenty dating spots of the sample Db14-b1 show a concordant $^{206}Pb/^{238}U$ age cluster of 427-442 Ma and an elder inherited age of 727 Ma, yielding a weighted mean age of 437 ± 2 Ma ($n = 19$, MSWD = 1. 3) (Fig. 5c).

5. 2 Major and trace element geochemistry

Detailed major and trace element data for the studied diabases and granodiorite porphyries are given in Supplementary Table 2. As the studied Xiaowulanhua diabases and Heinaobao granodiorite porphyries have varying LOI (loss on ignition) values (0. 61 - 3. 23 wt% and 0. 38-1. 45 wt%, respectively), all major element data used in the following discussion were normalized to 100% on an anhydrous basis.

The Xiaowulanhua diabases are characterized by high MgO (8. 23-14. 41 wt%), FeO^T (13. 07-17. 56 wt%) and CaO (8. 10-12. 81 wt%), low SiO_2 (46. 52-49. 24 wt%), Na_2O (1. 43-2. 22 wt%) and K_2O (0. 11-0. 57 wt%) (Fig. 6a), plotting into the tholeiitic field in the AFM diagram (Fig. 6b) and straddling the medium-K to low-K boundary in the K_2O vs. SiO_2 plot (Fig. 6c). The Aluminium Saturation Index [ASI = $Al_2O_3/(Na_2O + K_2O)$ molar units] of the Xiaowulanhua diabases are ranging from 0. 48 to 0. 66, showing a metaluminous affinity (Fig. 6d). In the chondrite-normalized REE diagram, the Xiaowulanhua diabases generally have slight enriched LREE with $(La/Yb)_N$ ratios of 0. 97-2. 11 and slightly negative to positive Eu anomalies with Eu/Eu^* ratios of 0. 97 - 1. 47 (Fig. 7a). The Xiaowulanhua diabases are overall enriched in Rb, Th, U, and Pb, depleted in Zr and P, and lacking of remarkable negative Nb, Ta and Ti anomalies in the primitive mantle-normalized trace element diagram (Fig. 7b).

The Heinaobao granodiorite porphyries display relatively high SiO_2 (64. 81-71. 39 wt%), K_2O (1. 42-4. 19 wt%) and Na_2O (3. 05-4. 62 wt%) (Fig. 6a), low MgO (0. 61-2. 36 wt%) and TiO_2 contents (0. 41-0. 95 wt%), straddling the high-K to medium-K calc-alkaline series (Fig. 6b,c). Moreover, the varying ASI of 0. 89-1. 29 indicates that the Heinaobao granodiorite porphyries are mainly metaluminous to weakly peraluminous, consisting with the Early Paleozoic arc intrusions in the western BAT (Fig. 6d). The granodiorite porphyries also show arc-related geochemical features in their normalized trace element patterns (Fig. 7), including enriched LREE with relatively high $(La/Yb)_N$ of 9. 83-35. 82, negative to slightly positive Eu anomalies with Eu/Eu^* of 0. 72-1. 08, depleted HFSE (such as P, Nb, Ta and Ti) and enriched LILE (such as Rb, Th, U and Pb).

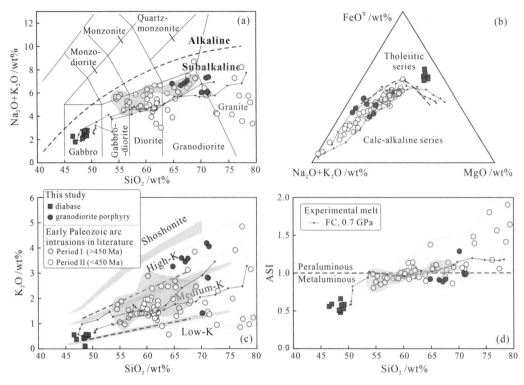

Fig. 6　Geochemical classification diagrams.

（a）TAS diagram（Middlemost，1994）.（b）AFM diagram with discriminatory line of Irvine and Barager（1971）.（c）K₂O vs. SiO₂ diagram（Peccerillo and Taylor，1976）.（d）ASI vs. SiO₂ diagram. Data sources for the Early Paleozoic arc-related intrusions are from Xu et al.（2003），Jian et al.（2008），Zhang and Jian（2008），Zhang et al.（2013a，2014），Bai et al.（2015），Wu et al.（2016），Qian et al.（2017）and Zhou et al.（2018b）. Experimental primitive，hydrous arc melts from fractional crystallization experiments at 0. 7 GPa（corresponding to mid-crustal level）are from Nandedkar et al.（2014）.

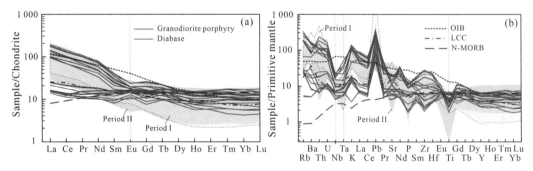

Fig. 7　（a）Chondrite-normalized REE patterns and（b）primitive mantle-normalized trace element patterns. Chondrite values are from Boynton（1984），primitive mantle，OIB and N-MORB values are from Sun and McDonough（1989），LCC（lower continental crust）values are from Rudnick and Gao（2003）. The Period Ⅰ and Period Ⅱ arc intrusions in western BAT（Jian et al.，2008；Zhang and Jian，2008；Zhang et al.，2014；Wu et al.，2016；Qian et al.，2017；Zhou et al.，2018c）are shown for comparison.

5.3 Zircon Hf isotopic composition

The diabase sample HZ28 and the granodiorite porphyry sample Db7-b1 were further chosen for in-situ zircon Hf isotopic analyses. Detailed results are presented in Supplementary Table 3. The initial ^{176}Hf/^{177}Hf ratios and corresponding $\varepsilon_{Hf}(t)$ values and T_{DMC} ages of the samples HZ28 and Db7-b1 were calculated based on their weighted mean U-Pb ages.

Ten analyzed zircons from the diabase sample HZ28 yield initial ^{176}Hf/^{177}Hf ratios of 0.281 735−0.282 092, corresponding to highly negative $\varepsilon_{Hf}(t)$ values of −24.7 to −12.0 and elder T_{DMC} ages of 2.26−3.05 Ga (Fig. 8a). Eighteen analyzed zircons from the granodiorite porphyry sample Db7-b1 yield initial ^{176}Hf/^{177} Hf ratios of 0.282 209 − 0.282 602, corresponding to negative to slightly positive $\varepsilon_{Hf}(t)$ values of −10.1 to +3.9 as well as varying T_{DMC} ages of 1.18−2.06 Ga (Fig. 8a).

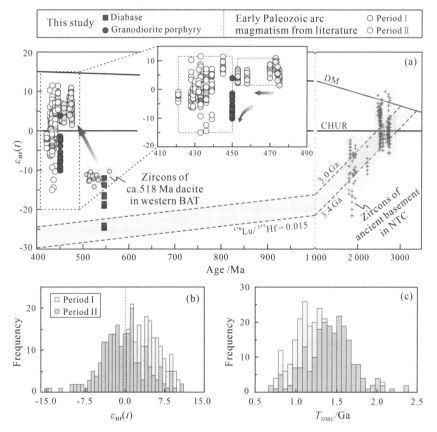

Fig. 8 (a) Plot of zircon $\varepsilon_{Hf}(t)$ vs. U-Pb age. (b,c) Histogram of zircon $\varepsilon_{Hf}(t)$ values and T_{DMC} ages of the Early Paleozoic arc-related rocks in western BAT.

Zircon $\varepsilon_{Hf}(t)$ values of the ancient basement rocks in northern Tarim Craton (NTC) are from Long et al.(2010), Ge et al.(2014), Wang et al.(2017) and Cai et al.(2018). Data sources in Fig. 8b,c are from Zhang et al. (2014), Qian et al.(2017), Zhou et al.(2018c) and this study.

6　Discussion

6. 1　Petrogenesis of the Xiaowulanhua diabases: Melting of the enriched mantle source hybridized by foundered BAT lower crust

The variable LOI values (0. 88 – 3. 23 wt%), together with the presence of alteration minerals like chlorites and sericites (Fig. 4b), indicate that the Xiaowulanhua diabases have experienced varying degrees of alteration after the emplacement. Therefore, this study mainly focuses on the HFSE, REE and compatible elements (e. g., Cr, Co and Ni) in their petrogenetic discussion.

6. 1. 1　Crustal assimilation and fractional crystallization

Crustal assimilation is an inevitable course for most mantle-derived melts during the ascent (Castillo et al., 1999). The Xiaowulanhua diabases are characterized by low SiO_2 (46. 52 – 49. 24 wt%) contents, high MgO (8. 23 – 14. 41 wt%), Cr (66 – 542 ppm), Ni (50 – 129 ppm) contents and high $Mg^#$ values (47 – 66), consisting with a mantle-derived origin (e.g., Rapp et al., 2003). Meanwhile, they also show highly negative $\varepsilon_{Hf}(t)$ values (−24. 7 to −12. 0) and old T_{DMC} (2. 26 – 3. 05 Ga) (Fig. 8a), which are comparable to those of the ancient crust-derived dacites (ca. 518 Ma) in the northern Bayan Obo (Zhang et al., 2014). The above features suggest that crustal assimilation may have contributed to the generation of the Xiaowulanhua diabases.

However, the Xiaowulanhua diabases show low Th (0. 5 – 2. 1 ppm) and U (0. 2 – 0. 5 ppm) contents, which are much lower than the Th and U concentrations of the average middle-upper continental crust (Rudnick and Gao, 2003), implying insignificant crustal assimilation during the magma ascent. Their relatively low Th/Nb (0. 11 – 0. 31), high Nb/La (0. 65 – 1. 24) and uniform Ta/La (0. 04 – 0. 07) ratios are also in agreement with insignificant crustal assimilation (e.g., Song et al., 2006). Moreover, the general lack of negative Nb, Ta and Ti anomalies (Fig. 7b), together with the absence of crustal xenoliths or xenocrysts in all the diabase samples, are also inconsistent with significant contribution from crustal assimilation. Thus, we tend to propose that crustal assimilation played a limited role in the petrogenesis of the Xiaowulanhua diabases.

The Xiaowulanhua diabases are not mantle-equilibrated primitive magmas, and probably have undergone substantial fractional crystallization of clinopyroxene and/or combined olivine and spinel as evidenced by their variable MgO (8. 23 – 14. 41 wt%), Cr (66 – 542 ppm), Ni (50 – 129 ppm) and $Mg^#$ values (47 – 66) (e.g., Pfänder et al., 2002). The less varied TiO_2 and FeO^T contents, along with the negligible Ti anomalies (Fig. 7b), indicate that Fe-Ti oxides was not an important fractionating phase. In addition, although the general absence of negligible Eu anomalies and the slightly positive Sr anomalies are inconsistent with significant plagioclase fractionation (Fig. 7), the roughly decreasing Eu/Eu^* with decreasing $Mg^\#$ in most samples indicate that considerable separation of plagioclase may have occurred during the magma evolution.

6. 1. 2   Nature of the mantle source

Since crustal assimilation was an insignificant factor during the formation of the Xiaowulanhua diabases, we propose that the high MgO, Cr, Ni, $Mg^\#$ values, together with the highly negative zircon $\varepsilon_{Hf}(t)$ values are jointly indicative of partial melting of an enriched mantle source. Moreover, the Xiaowulanhua diabases show low Dy/Yb ratios of 1. 4−1. 8 ppm, which are more likely consistent with partial melting within a spinel-stability field (Duggen et al., 2005; Ma et al., 2014), because melting within a spinel-stability field should have yielded higher Dy/Yb ratios (over 2. 5). In addition, the flat chondrite-normalized REE patterns as well as the relatively high heavy REE abundances also indicate a garnet-free source for the magma generation (Fig. 7a).

Partial melting experiments demonstrate that mantle-derived melts produced through partial melting of dry peridotite generally have low Fe/Mn ratios (<60), whereas magmas in equilibrium with pyroxenite in mantle show higher Fe/Mn ratios (>60), because the $D_{Fe/Mn}$ values for olivine are >1 but are <1 for pyroxene (e.g., Pertermann and Hirschmann, 2003; Kogiso et al., 2004). The Xiaowulanhua diabases typically have high Fe/Mn ratios of 62. 1− 81. 7 with high MnO contents of 0. 20−0. 25 wt%, which are in line with partial melting of a pyroxenite-dominated mantle source (Xu et al., 2012). Moreover, the Nb/Ta ratios and Zr/Hf ratios of mantle-derived magmas are usually unaffected by the magmatic evolution process because of the similar geochemical behaviors of the the Nb-Ta and Zr-Hf pairs, unless an unusual metasomatic assemblage occurred (e.g., Jochum et al., 1989; Xiong et al., 2005). Generally, low-Mg amphibole is the major phase to form melts with low Nb/Ta and high Zr/Hf ratios, whereas rutile is the dominant phase to generate melts with super-chondritic Nb/Ta and Zr/Hf ratios (e.g., Jörg et al., 2007). The Xiaowulanhua diabases are characterized by low Nb/Ta ratios of 13. 3−18. 8 (19. 9 for chondrite) and sub-chondritic to super-chondritic Zr/Hf ratios of 20. 3 − 37. 0 (34. 3 for chondrite), which are more likely indicative of a low-Mg amphibole-dominated metasomatism of mantle source region (Münker et al., 2003). Those samples with relatively low Zr/Hf ratios might be caused by the presence of an unstable rutile phase in the mantle source given their relatively high MgO, Cr and Ni concentrations (Xiong et al., 2005).

6. 1. 3   Petrogenetic mechanism

The above discussions suggest that Xiaowulanhua diabases were likely formed through partial melting of an enriched, pyroxene-dominated, low-Mg amphibole-metasomatized lithospheric mantle source within a spinel-stability field, and then underwent fractionation of olivine+clinopyroxene±plagioclase. However, the Xiaowulanhua diabases typically show highly negative zircon $\varepsilon_{Hf}(t)$ values (-24.7 to -12.0) and old T_{DMC} ages ($2.26-3.05$ Ga). Specifically, the maximum zircon $\varepsilon_{Hf}(t)$ values of the Xiaowulanhua diabases are equal to the ca. 518 Ma ancient crust-derived dacites in northern Bayan Obo (Zhang et al., 2014), and the minimum values are comparable to the ancient basements beneath northern Tarim craton (Fig. 8a. e.g., Long et al., 2010; Ge et al., 2014; Wang et al., 2017; Cai et al., 2018). This strongly "crustal-like" Hf isotopic feature may be indicative of significant input of the crustal components in the formation of the mantle source. Thus, the way by which the crustal components had been entrained into the mantle source region of the Xiaowulanhua diabases needs to be investigated.

Recent studies suggest that the BAT was initiated as a Precambrian microcontinent, which was separated from the Tarim craton, and situated in the PAO between Siberia and East Asian blocks (Tarim, Alax, NCC and South China) before being accreted onto the northern NCC (Zhang et al., 2014; Eizenhöfer and Zhao, 2018; Zhao et al., 2018). So, there are two likely mechanisms for the above mantle enrichment process: ① subduction of an ancient oceanic crust and ② foundering of the lower crust of BAT. Given the strongly "crustal-like" Hf isotopic composition, a significant digestion of the subducted materials into the overlying mantle must be needed. However, all the Xiaowulanhua diabase samples typically lack geochemical characteristics of subduction-related magmas (especially the negative Nb, Ta and Ti anomalies), as well as hydrous minerals like amphibole and subordinate biotite, so significant flux of slab fluids/melts into their mantle source in response to oceanic subduction is unlikely. In addition, the general absence of Precambrian subduction-related magmatism in the western BAT is also inconsistent with an oceanic subduction setting. Thus, foundering of the BAT lower crust into the underlying lithospheric mantle seems to be a more plausible explanation.

Generally, a thickened crustal root is required to accomplish the foundering of lower crust during tectonic collapse (e.g., Lustrino, 2005). Previous works suggest that a thickened crust (>50 km) was likely existed beneath the northern Tarim craton during the Early Neoproterozoic (ca. 820-750 Ma) (e.g., Long et al., 2011). Besides, extensive bimodal volcanic series/complex with U-Pb ages of ca. 740-735 Ma were identified from the northern margin of the Tarim craton (e.g., Zhang et al., 2013b). Taking into account the Tarim affinity of the BAT basements (Eizenhöfer and Zhao, 2018), we infer that the BAT may have inherited a thick lower crust from the northern Tarim carton by the Middle Neoproterozoic detachment. Due to the higher density and lower melting temperature (Anderson, 2006; Sobolev et al., 2007), the lower-crustal eclogites beneath the BAT can be removed and recycled into the underlying hot lithospheric mantle by foundering, and can produce silicic fluids/melts to progressively hybridize the ambient mantle peridotite (Xu et al., 2012). Meanwhile, the loss of the thick crustal root would further trigger lithospheric thinning and asthenospheric upwelling in the region, which may have resulted in decompressional melting of the hybridized lithospheric mantle to form the parental magmas of the Xiaowulanhua diabases. Asthenospheric upwelling may also have induced a radical increase of heat flow into the overlying crust (e.g., Menzies et al., 2007; Xu et al., 2012), triggering extensive partial melting of the lower-crustal materials to generate the ca. 518 Ma dacites in northern Bayan Obo (Zhang et al., 2014). However, more evidences, including coeval magmatic, metamorphic and metallogenic records, are still needed to further constrain the above lower crust foundering mechanism.

6. 2 Petrogenesis of the Early Paleozoic arc intrusions in western BAT

As stated before, extensive Early Paleozoic arc intrusions have been identified in the western BAT (Fig. 2a), which primarily consist of calc-alkaline, metaluminous to peraluminous quartz diorites, diorites, tonalites, granodiorites, granites and minor gabbroic diorites (e.g., Zhang and Jian, 2008; Li et al., 2010; Zhang et al., 2014; Qian et al., 2017; Zhou et al., 2018c). The new ages of the Heinaobao granodiorite porphyries, along with the literature data of other Early Paleozoic arc intrusions in adjacent areas (Table 1), indicate that the Early Paleozoic arc intrusions in western BAT were gradually emplaced during ca. 475–417 Ma. Moreover, recent studies suggest that diversified petrogenetic processes were involved in the generation of the Early Paleozoic arc intrusions in western BAT, including partial melting of the subducted oceanic crust and the overlying mantle wedge, as well as mixing of the ancient crust-derived melts with the mantle wedge and slab-derived melts (e.g., Zhang et al., 2014; Zhou et al., 2018c).

6. 2. 1 Composition of the Early Paleozoic arc intrusions transited at ca. 450 Ma

As illustrated in Fig. 8a, the Early Paleozoic arc intrusions with zircon U-Pb age >450 Ma have consistently positive zircon $\varepsilon_{Hf}(t)$ values (+1.5 to +10.9), but the ones with age ≤450 Ma are show large variations in the $\varepsilon_{Hf}(t)$ values (−15.0 to +11.5) (Table 1). This observation may suggest that a significant changing of arc magma composition occurred at ca. 450 Ma. Thus, we further divided the Early Paleozoic arc intrusions in western BAT into the Period I (>450 Ma) and the Period II (≤450 Ma) intrusions.

Considerable differences can also be observed in the whole-rock geochemical compositions of Period I and Period II intrusions. For instance, the Period I and Period II intrusions are respectively ranging from gabbro-diorite to granodiorite, and from diorite to granite in composition (Fig. 6a); the Period I intrusions are mainly medium-K calc-alkaline and metaluminous to weakly peraluminous, whereas the Period II intrusions are varying from high-

K to low-K calc-alkaline and from metaluminous to strongly peraluminous (Fig. 6b-d) ; the Period II intrusions overall have more elevated (La/Yb)$_N$ ratios and more enriched LREE than that of the Period I intrusions (Fig. 7a). The investigated Heinaobao granodiorite porphyries were formed during ca. 450–437 Ma (Fig. 5c,d) , and show zircon Hf isotopic and whole-rock geochemical characteristics similar to the Period II intrusions (Figs. 6 – 8) , indicating that these rocks may share a common genesis.

6. 2. 2 Period I intrusions: Mid-crustal level differentiation of an enriched mantle-derived hydrous basaltic magma

It should be noted that same samples of the Period I and II intrusions have relatively high Sr/Y and (La/Yb)$_N$ ratios, and fall into the adakite fields in the discrimination diagrams (Fig. 9a,b) , indicating geochemical signatures of adakite-like rocks. However, their the adakite-like signatures cannot be simply attributed to the partial melting of oceanic crust, thickened or foundered lower continental crust, as these high-pressure melting processes would have given rise to adakite-like signatures in all the Period I and II intrusions due to the inevitable interaction with the overlying mantle peridotites (e. g. , Defant and Drummond, 1990; Xu et al. , 2002; Chung et al. , 2003; Ma et al. , 2015). Besides, the relatively low MgO, Cr, Ni and Mg$^{\#}$ values of the Period I and Period II intrusions are also inconsistent with partial melting of the foundered lower crust or the oceanic slab (Fig. 9c-f).

The Period I intrusions typically display broad correlations of SiO$_2$ with various major and trace elements, including Al$_2$O$_3$, P$_2$O$_5$, FeO^T, TiO$_2$, Sr/Y, Sr, Y, (Dy/Yb)$_N$ and total REE (Fig. 10a-i). Besides, they also contain abundant mafic minerals like hornblende and pyroxene (Zhang et al. , 2014; Zhou et al. , 2018c). These observations are comparable to the adakite-like rocks generated by fractionation of the hydrous basaltic melts (Castillo et al. , 1999; Macpherson et al. , 2006). Furthermore, the Period I intrusions have positive zircon $\varepsilon_{Hf}(t)$ values of +1. 5 to +10. 9 (Fig. 8a; Table 1) , and show enriched LILE and LREE and depleted HFSE (Fig. 7) , which are consistent with an origination from partial melting of the subduction-modified lithospheric mantle (e. g. , Hawkesworth et al. , 1997). Therefore, we propose that the Period I intrusions were possibly formed through fractional crystallization within an evolving mantle-derived hydrous basaltic magma in subduction zone (Castillo et al. , 1999; Macpherson et al. , 2006).

Within the compositional range of the Period I intrusions, the decreases of Y and (Dy/Yb)$_N$ with increasing SiO$_2$ are consistent with dominant hornblende fractionation (Fig. 10g,h) , since hornblende preferentially incorporates middle REE rather than HREE (Davidson et al. , 2007). The negative correlations of SiO$_2$ with P$_2$O$_5$ and total REE indicate that fractionation of accessory apatite (Fig. 10b,i) , because apatite is an important factor in controlling the P$_2$O$_5$ content in hydrous arc melts, and has high partition coefficients for REE (Lee and Bachmann,

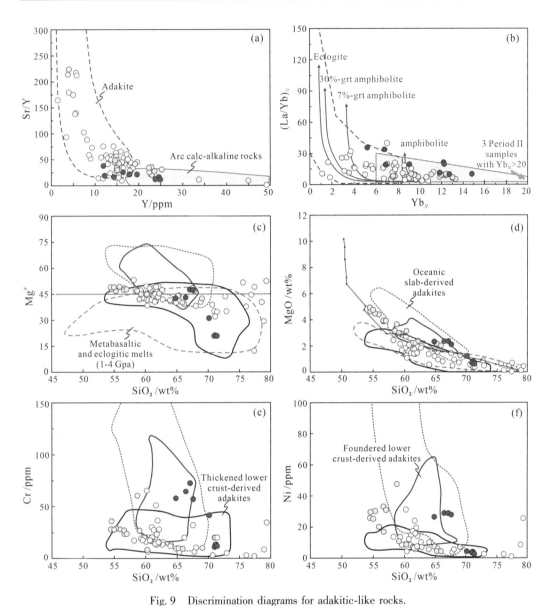

Fig. 9　Discrimination diagrams for adakitic-like rocks.

(a,b) Sr/Y vs. Y and (La/Yb)$_N$ vs. Yb$_N$ diagrams based on Defant and Drummond (1990) and Petford and Atherton (1996). (c-f) Mg$^{\#}$, MgO, Cr and Ni vs. SiO$_2$ diagrams based on Long et al. (2015) and references therein.

2014). The negative correlations of SiO$_2$ with FeO^T and TiO$_2$ indicate that Fe-Ti oxide fractionation was insignificant (Fig. 10c,d). Moreover, plagioclase fractionation was likely involved in the late magmatic evolution, as evidenced by the decreases of Al$_2$O$_3$ and Sr at higher SiO$_2$ contents (Fig. 10a,f). We infer that the fractionation of hornblende and apatite removed Y from the residual, whereas the early suppression of plagioclase fractionation kept Sr in the melt (Castillo et al., 1999; Richards and Kerrich, 2007; Xu et al., 2019). These processes jointly caused the increase of Sr/Y ratios with the increase of SiO$_2$(Fig. 10e), and

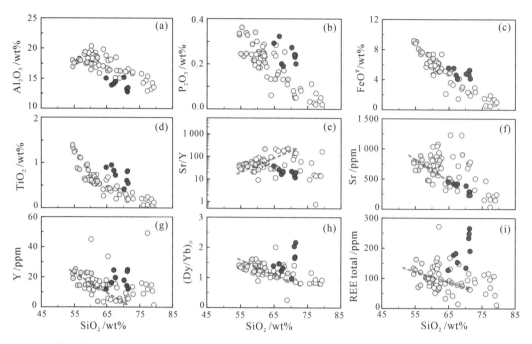

Fig. 10　Variation diagrams for selected major elements, trace elements and trace element
ratios vs. SiO$_2$ contents for the Early Paleozoic arc intrusions in western BAT.

gave rise to the adakite-like signatures of the Period I intrusions.

Experimental studies suggest that garnet would remain stable under pressure condition over 10 kbar in a hydrous basaltic melt, and would crystallize after the pyroxene or hornblende fractionation (e.g., Müntener et al., 2001; Ulmer et al., 2018). As discussed before, the lack of positive correlation between SiO$_2$ and (Dy/Yb)$_N$, along with the general absence of garnet in the Period I intrusions (e.g., Zhang et al., 2014; Zhou et al., 2018c), are inconsistent with fractionation of magmatic garnet at depth, implying that the crystallization pressure shall not exceed 10 kbar. Moreover, the occurrence of magmatic epidote in the ca. 470 Ma tonalites in northern Bayan Obo gives a minimum pressure condition of 5 kbar (Schmidt and Poli, 2004; Zhang et al., 2014; Xu et al., 2019). Thus, crystallization of the Period I intrusions most likely occurred between 5 kbar and 10 kbar. When compared with a recent fractional crystallization experiment that conducted on a near-primitive hydrous mantle-derived olivine tholeiite at 7 kbar (corresponding to the mid-crustal level) (Nandedkar et al., 2014), the Period I intrusions show well agreement with the obtained experimental data in most major element trends (Figs. 6 and 9d). Thus, we infer that the differentiation of the mantle-derived basaltic precursor of Period I intrusions probably took place at mid-crustal depth.

6.2.3　Period II intrusions: Reworking of subduction-modified BAT lower crust and mixing with mantle wedge-derived melts at MASH zone

Unlike the Period I intrusions, the Period II intrusions are characterized by more variable zircon $\varepsilon_{Hf}(t)$ values of -15.0 to $+11.5$ (Fig. 8; Table 1). More importantly, their negative-end zircon $\varepsilon_{Hf}(t)$ values are comparable to the ca. 518 Ma dacites in northern Bayan Obo (Fig. 8a), which were formed through partial melting of the lower-crustal materials beneath the western BAT (Zhang et al., 2014). The Period Ⅱ intrusions overall have low MgO, Cr and Ni (Fig. 8d-f), high SiO$_2$ and ASI (Fig. 6a,d). Moreover, their normalized trace element patterns are characterized by enriched LREE and LILE with depleted HFSE (Fig. 7). The above geochemical features, along with the occurrence of abundant elder inherited zircons in most samples (Zhang et al., 2014; Zhou et al., 2018c), suggest that the Period Ⅱ intrusions was possibly formed through reworking of the previously subduction-modified lower-crustal materials beneath the western BAT.

Magmas equilibrated with or originated from lower crust are generally barren and dry, and direct melting of a lower-crustal source can not generate melts with Mg$^{\#}$ values over 45 (e.g., Rapp and Watson, 1995). However, although most Period Ⅱ intrusions have low Mg$^{\#}$ values, there are still a number of samples with Mg$^{\#}$ values exceeding 45 (Fig. 9c), which indicates considerable input of mantle-derived mafic components in their generation. In addition, given that the positive-end zircon $\varepsilon_{Hf}(t)$ values of Period Ⅱ intrusions are just equal to the zircon $\varepsilon_{Hf}(t)$ values of the Period Ⅰ intrusions (Fig. 8a), it is likely that an enriched mantle-derived hydrous melt, which was similar to the precursor of the Period Ⅰ intrusions, had contributed to the generation of the Period Ⅱ intrusions. Thus, we assume that the Period Ⅱ intrusions were probably produced by reworking of the subduction-modified BAT lower crust and mixing with the enriched mantle wedge-derived hydrous melts at a lower-crustal MASH zone (melting, assimilation, storage and homogenization; Hildreth and Moorbath, 1988). The compositional diversity observed in the Period Ⅱ intrusions can be ascribed either to the different input of the mantle-derived melts during the magma mixing, or to the heterogeneity of the subduction-modified lower crust beneath the BAT.

Recent studies suggest that source inheritance can also account for the generation of adakite-like signatures in igneous rocks, even with normal crustal thickness (e.g., Moyen, 2009; Ma et al., 2015). Prior to the emplacement of the Period Ⅱ intrusions, the lower crust beneath the western BAT had already been significantly modified by the Period Ⅰ oceanic subduction as discussed before. Thus, abundant juvenile components with adakite-like signatures may have been entrained into the lower crust of western BAT, including the melts generated by partial melting of the subducted slab, as well as the subduction-metasomatized mantle wedge peridotites under hydrous conditions (Defant and Drummond, 1990; Calmus et al., 2003; Macpherson et al., 2006; Ma et al., 2015). Thus, we infer that the adakite-like signatures of the Period Ⅱ intrusions were possibly inherited from the previously subduction-modified lower crust of western BAT.

6. 3   Geodynamic implications

6. 3. 1   Two Early Paleozoic arc systems flanking the western BAT

The Early Paleozoic arc intrusions are covering a broad region with maximum N-S width over 50 km in the western BAT (Fig. 2a). In the previous sections, we have shown that the composition of these Early Paleozoic arc intrusions significantly transited at ca. 450 Ma, which implies that a transition of deep dynamic process may have occurred at approximately 450 Ma in the western BAT. However, as stated before, most previous studies simply ascribed them to either the south-dipping subduction of the Solonker ocean beneath northern BAT (Jian et al., 2008; Eizenhöfer and Zhao et al., 2018; Zhou et al., 2018a), or the north-dipping subduction of the South Bainaimiao ocean beneath southern BAT (e.g., Zhang et al., 2014; Ma et al., 2019). Thus, the Early Paleozoic arc magmatic evolution of western BAT and the corresponding oceanic subduction polarity need to be revisited.

The OSC abutting the northern BAT is widely considered to have recorded the Early Paleozoic south-dipping subduction process of the Solonker ocean (Fig. 2a) (e.g., Xiao et al., 2003,2015; Jian et al., 2008). Nevertheless, in the well-exposed Tulinkai ophiolite in the central OSC, the ophiolitic units (mainly metagabbros and tonalites) were mainly formed at ca. 497–477 Ma, and the incorporated felsic rock blocks (trondhjemites, quartz diorites and dacites) were mostly formed at ca. 472 – 454 Ma, which suggest that the south-dipping subduction of the Solonker oceanic lithosphere was likely initiated during ca. 497–477 Ma, and was carried on until 454 Ma, approximately (Jian et al., 2008). Moreover, the thrusted quartzite mylonites from the OSC have phengite $^{40}Ar/^{39}Ar$ plateau ages ranging from 453 Ma to 449 Ma (de Jong et al., 2006), indicating that the last metamorphism caused by the south-dipping subduction of the Solonker oceanic lithosphere probably occurred at ca. 453–449 Ma. Thus, the formation of the Early Paleozoic remnants in the OSC was likely in accordance with the emplacement of the Period Ⅰ intrusions (ca. 475–450 Ma), which indicates that the Period Ⅰ intrusions and the Early Paleozoic OSC in the western BAT were likely developed in a common arc system in relation to the south-dipping subduction of the Solonker oceanic lithosphere.

Similarly,several Early Paleozoic tectonic mélanges have also been recognized from the southern margin of the western BAT, such as the Hongqi-Wude mélange in northern Bayan Obo and the Harihada-Chegendalai mélange in northeastern Damaoqi (Shao, 1991; Jia et al., 2003; Shi et al., 2013; Zhang et al., 2014). They were more likely to have recorded the Early Paleozoic north-dipping subduction process of the South Bainaimiao oceanic lithosphere, considering that the northern continental margin of the NCC has remained passive during this period (Zhang et al., 2014; Wu et al., 2016). The Xiaowulanhua mélange in this study is also tectonically outcropped close to boundary between the western BAT and northern NCC [Institute of Geological Survey of China University of Geosciences, Beijing (IGSCUGB), 2008] (Fig. 2a). Besides, the well-developed E-W to NE-SW striking high-angle foliation and SSE-SSW plunging stretching lineation in the Xiaowulanhua mélange are similar to those of the Harihada-Chegendalai mélange (Shao, 1991; Zhang et al., 2014). Thus, the Xiaowulanhua mélange might be the east extension of the Hongqi-Wude and Harihada-Chegendalai mélanges in western BAT. More importantly, the occurrence of deformed Late Silurian fossiliferous limestone blocks in the Hongqi-Wude mélange indicates that the north-dipping subduction of the South Bainaimiao ocean probably had remained active during the Late Silurian (Shi et al., 2013), corresponding to the emplacement of the Period Ⅱ intrusions (ca. 450–417 Ma) in western BAT. Therefore, we infer that the Period Ⅱ intrusions and the Early Paleozoic mélanges along the southern margin of western BAT were jointly formed in the same arc system during the north-dipping subduction of the South Bainaimiao ocean.

6.3.2 Elucidating the multiple crust-mantle interactions in western BAT

Based on the detrital zircon age spectra of Early Paleozoic metasedimentary rocks and the Early Paleozoic paleo-geographic reconstructions of the southeastern CAOB, the BAT was suggested to be initially developed as an independent Precambrian microcontinent detached from the Tarim craton (Zhang et al., 2014; Wang et al., 2016; Eizenhöfer and Zhao, 2018; Zhou et al., 2018b,c). Final accretion of the BAT onto the northern NCC was likely occurred during latest Silurian (de Jong et al., 2006; Zhang et al., 2014; Ma et al., 2019), which was accompanied by the deposition of the Xibiehe formation molasse or quasi-molasse (e.g., Tang, 1990; Zhang et al., 2010). Our new petrogenetic observations on the latest Neoproterozoic diabases and the Early Paleozoic arc intrusions in this study suggest that a series of crust-mantle interactions coupled with distinct deep dynamic processes had occurred beneath the western BAT before the final accretion. Together with previous studies on the regional tectonic-magmatic history (Jian et al., 2008; Zhang et al., 2014; Eizenhöfer and Zhao, 2018; Zhou et al., 2018c; Ma et al., 2019), we employed the following model to clarify the multiple crust-mantle interactions in western BAT (Fig. 11).

In the Late Neoproterozoic, foundering of the thick eclogitic lower crust led to a significant hybridization of the underlying mantle peridotite beneath the western BAT. Meanwhile, the foundering further triggered lithospheric thinning and asthenospheric upwelling, which successively induced the melting of the hybridized lithospheric mantle to produce the parental magmas for the ca. 546 Ma Xiaowulanhua diabases (Fig. 11a,b), and the melting of the residual lower-crustal materials to generate the Early Cambrian magmatism in the western BAT. It is noteworthy that an obvious increasing of zircon $\varepsilon_{Hf}(t)$ values can be observed from the Early Cambrian magmatism to the Period Ⅰ intrusions (Fig. 8a), which indicates that

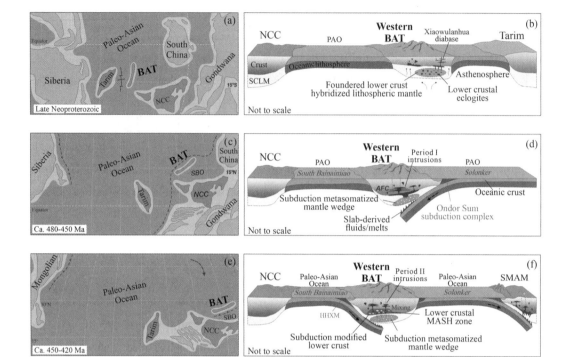

Fig. 11　Schematic illustrations showing the geodynamic evolution and associated crust-mantle interactions in western BAT before the latest Silurian arc-continent collision.

The paleo-geographic reconstructions of Siberia, NCC, Tarim, South China and Gondwana are modified from Eizenhöfer and Zhao (2018) and Zhao et al. (2018). SBO: South Bainaimiao ocean; SCLM: subcontinental lithospheric mantle; HHXM: Hongqi-Wude, Harihada-Chegendalai and Xiaowulanhua mélanges.

abundant depleted components may have eventually been entrained into the lithospheric mantle by the asthenospheric upwelling.

After the NCC changed its drift northwards in the Early Ordovician (Eizenhöfer and Zhao, 2018; Zhao et al., 2018), the south-dipping subduction of the Solonker oceanic lithosphere beneath the BAT was initiated, leading to the development of the Period Ⅰ arc system in western BAT (Fig. 11c). The descending Solonker oceanic slab dehydrated and released hydrous fluids/melts into the overlying mantle wedge, triggering partial melting of the enriched mantle wedge peridotite to generate primitive hydrous basaltic magmas (Castillo et al., 1999). The hydrous basaltic magmas then intruded into the mid-crustal level, where they experienced AFC processes to form the Period Ⅰ intrusions in the western BAT (Fig. 11d), or further erupted to form the Early to Middle Ordovician medium-K calc-alkaline volcanic rocks (Zhang et al., 2014; Zhou et al., 2017).

A transition from south-dipping subduction of the Solonker oceanic lithosphere to north-dipping subduction of the South Bainaimiao oceanic lithosphere was likely occurred at ca. 450 Ma, which initiated the development of the Period Ⅱ arc system in the western BAT

(Fig. 11e). In this case, the heat and volatiles provided by the mantle-derived melts lowered the melting temperature (Collins et al., 2016), inducing partial melting of the lower-crustal materials beneath the western BAT. The melts produced by reworking of the subduction-modified BAT lower crust further mixed with the mantle wedge-derived hydrous melts at the lower-crustal MASH zone, giving rise to an intense arc magmatic flare-up as represented by the Period Ⅱ intrusions in western BAT (Fig. 11f).

In brief, this model highlights that multiple crust-mantle interactions, including: ①recycling of eclogitic lower crust into the lithospheric mantle and following enhanced input of depleted components into the lithospheric mantle triggered by lower crust foundering, ② mid-crustal differentiation of an enriched mantle-derived hydrous basaltic magma during the south-dipping subduction of Solonker ocean, and ③ mixing between the subduction-modified BAT lower crust-derived magmas and the mantle wedge-derived magmas during the north-dipping subduction of South Bainaimiao ocean, had successively occurred in the western BAT. The documentation of the above multiple crust-mantle interactions not only sheds light on the mechanisms of mass/energy transfer at depth associated with the evolution of the BAT, but also provides an example to probe into the compositional modification and crustal construction of the Precambrian microcontinents during the long-term evolution of the CAOB.

7　Conclusions

(1) The latest Neoproterozoic diabases (ca. 546 Ma) in western BAT display low SiO_2, high MgO and highly negative zircon $\varepsilon_{Hf}(t)$ values, they were likely formed through partial melting of a hybridized lithospheric mantle during the foundering of the BAT lower crust.

(2) The Early Paleozoic arc intrusions (ca. 475−417 Ma) in western BAT can be divided into Period Ⅰ and Ⅱ at ca. 450 Ma. The Period Ⅰ intrusions (> 450 Ma) with consistently positive zircon $\varepsilon_{Hf}(t)$ were formed by mid-crustal differentiation of an enriched mantle-derived hydrous magma, whereas the Period Ⅱ intrusions (≤ 450 Ma) with variable zircon $\varepsilon_{Hf}(t)$ were resulted from reworking of the subduction-modified BAT lower crust and mixing with the mantle wedge-derived melts.

(3) The compositional changing from Period Ⅰ to Ⅱ arc intrusions at ca. 450 Ma in western BAT can be attributed to the tectonic transition from the south-dipping subduction of Solonker ocean to the north-dipping subduction of South Bainaimiao ocean in southeastern CAOB.

Acknowledgements　This research was financially supported by the China Geological Survey (1212011085490 and 1212011220465) and the National Natural Science Foundation of China (41421002). We gratefully acknowledge the guest Editor and the reviewers for their critical and constructive comments. We also sincerely thank Yongjun Di, Zhen Wang and

Zhong Wang for their insightful suggestions.

Appendix A. Supplementary data　Supplementary data to this article can be found online at https://doi.org/10. 1016/j.gsf.2020. 01. 012.

References

Andersen, T., 2002. Correction of common lead in U-Pb analyses that do not report ^{204}Pb. Chemical Geology, 192, 59-79.

Anderson, D.A., 2006. Speculations on the nature and cause of mantle heterogeneity. Tectonophysics, 146, 7-22.

Bai, X.H., Xu, Z.Y., Liu, Z.H., Xin, H.T., Wang, W.Q., Wang, X., Lei, C.C., 2015. Zircon U-Pb dating, geochemistry and geological significance of the Early Silurian plutons from the southeastern margin of the Central Asian Orogenic Belt. Acta Petrologica Sinica, 31, 67-79 (in Chinese with English abstract).

Bouvier, A., Vervoort, J.D., Patchett, P.J., 2008. The Lu-Hf and Sm-Nd isotopic composition of CHUR: Constraints from unequilibrated chondrites and implications for the bulk composition of terrestrial planets. Earth & Planetary Science Letters, 273, 48-57.

Boynton, W.V., 1984. Cosmochemistry of the rare earth elements: Meteorite studies. In: Henderson, P.E. Rare Earth Element Geochemistry. Amsterdam: Elsevier, 63-144.

Cai, Z.H., Xu, Z.Q., Yu, S.Y., Li, S.Z., He, B.Z., Ma, X.X., Chen, X.J., Xu, X.Y., 2018. Neoarchean magmatism and implications for crustal growth and evolution of the Kuluketage region, northeastern Tarim Craton. Precambrian Research, 304, 156-170.

Calmus, T., Aguillon-Robles, A., Maury, R.C., Bellon, H., Benoit, M., Cotten, J., Bourgois, J., Michaud, F., 2003. Spatial and temporal evolution of basalts and magnesian andesites (bajaites) from Baja California, Mexico: The role of slab melts. Lithos, 66, 77-105.

Castillo, P.R., Janney, P.E., Solidum, R.U., 1999. Petrology and geochemistry of Camiguin Island, southern Philippines: Insights to the source of adakites and other lavas in a complex arc setting. Contributions to Mineralogy and Petrology, 134, 33-51.

Cawood, P.A., Kroner, A., Collins, W.J., Kusky, T.M., Mooney, W.D., Windley, B.F., 2009. Accretionary orogens through earth history. Geological Society, London, Special Publications, 318, 1-36.

Chung, S.L., Liu, D.Y., Ji, J.Q., Chu, M.F., Lee, H.Y., Wen, D.J., Lo, C.H., Lee, T.Y., Qian, Q., Zhang, Q., 2003. Adakites from continental collision zones: Melting of thickened lower crust beneath southern Tibet. Geology, 31, 1021-1024.

Collins, W.J., Belousova, E.A., Kemp, A.I.S., Murphy, J.B., 2011. Two contrasting Phanerozoic orogenic systems revealed by hafnium isotope data. Nature Geoscience, 4, 333-337.

Collins, W.J., Huang, H.Q., Jiang, X.Y., 2016. Water-fluxed crustal melting produces Cordilleran batholiths. Geology, 44, 143-146.

Condie, K.C., 2007. Accretionary orogens in space and time. Memoir of the Geological Society of America, 200, 145-158.

Davidson, J., Turner, S., Handley, H., Macpherson, C., Dosseto, A., 2007. Amphibole "sponge" in arc crust? Geology, 35, 787-790.

de Jong, K., Xiao, W.J., Windley, B.F., Masago, H., Lo, C.H., 2006. Ordovician ^{40}Ar/^{39}Ar phengite ages from the blueschist-facies Ondor Sum subduction-accretion complex (Inner Mongolia) and implications for the early Paleozoic history of continental blocks in China and adjacent areas. American Journal of Science, 306, 799-845.

Defant, M.J., Drummond, M.S., 1990. Derivation of some modern arc magmas by melting of young subducted lithosphere. Nature, 347, 662-665.

Duggen, S., Hoernle, K., Bogaard, P., Garbe-Schonberg, D., 2005. Post-collisional transition from subduction- to intraplate-type magmatism in the westernmost Mediterranean: Evidence for continental-edge delamination of subcontinental lithosphere. Journal of Petrology, 46, 1155-1201.

Eizenhöfer, P.R., Zhao, G.C., 2018. Solonker Suture in East Asia and its bearing on the final closure of the eastern segment of the Palaeo-Asian Ocean. Earth-Science Reviews, 186, 153-172.

Fu, D., Huang, B., Kusky, T.M., Li, G.Z., Wilde, S.A., Zhou, W.X., Yu, Y., 2018. A middle Permian Ophiolitic Mélange Belt in the Solonker Suture Zone, Western Inner Mongolia, China: Implications for the Evolution of the Paleo-Asian Ocean. Tectonics, 37, 1292-1320.

Gao, S., Liu, X.M., Yuan, H.L., Hattendorf, B., Günther, D., Chen, L., 2007. Analysis of forty-two major and trace elements in USGS and NIST SRM glasses by LA-ICP-MS. Geostandards and Geoanalytical Research, 26, 181-196.

Ge, R.F., Zhu, W.B., Wilde, S.A., Wu, H.L., He, J.W., Zheng, B.H., 2014. Archean magmatism and crustal evolution in the northern Tarim Craton: Insights from zircon U-Pb-Hf-O isotopes and geochemistry of ~2.7 Ga orthogneiss and amphibolite in the Korla Complex. Precambrian Research, 252, 145-165.

Griffin, W.L., Pearson, N.J., Belousova, E., Jackson, S.E., van Achterbergh, E., O'Reilly, S.Y., Shee, S.R., 2000. The Hf isotope composition of cratonic mantle: LAM-MC-ICP-MS analysis of zircon megacrysts in kimberlites. Geochimica et Cosmochimica Acta, 64, 133-147.

Hawkesworth, C.J., Turner, S.P., McDermott, F., Peate, D.W., Calsteren, P.V., 1997. U-Th isotopes in arc magmas: Implications for element transfer from the subducted crust. Science, 276, 551-555.

He, Z.Y., Klemd, R., Yan, L.L., Zhang, Z.M., 2018. The origin and crustal evolution of microcontinents in the Beishan orogen of the southern Central Asian Orogenic Belt. Earth-Science Reviews, 185, 1-14.

Hildreth, W., Moorbath, S., 1988. Crustal contributions to arc magmatism in the Andes of central Chile. Contributions to Mineralogy and Petrology, 98, 455-489.

Hoskin, P.W.O., Schaltegger, U., 2003. The composition of zircon and igneous and metamorphic petrogenesis. Reviews in Mineralogy and Geochemistry, 53, 27-62.

Institute of Geological Survey of China University of Geosciences, Beijing (IGSCUGB), 2008. Geological map of Bulitai Region (K49C002003), Scale 1:250 000 (in Chinese).

Irvine, T.N., Baragar, W.R.A., 1971. A guide to the chemical classification of common volcanic rocks. Canadian Journal of Earth Sciences, 8, 523-548.

Jia, H.Y., Baoyin, W.L.J., Zhang, Y.Q., 2003. Characteristics and tectonic significance of the Wude suture zone in northern Damaoqi, Inner Mongolia. Journal of Chengdu University of Technology (Science & Technology Edition), 30, 30-34 (in Chinese with English abstract).

Jian, P., Liu, D.Y., Kröner, A., Windley, B.F., Shi, Y.R., Zhang, F.Q., Shi, G.H., Miao, L.C.,

Zhang, W., Zhang, Q., Zhang, L.Q., Ren, J.S., 2008. Time scale of an early to mid-Paleozoic orogenic cycle of the long-lived Central Asian Orogenic Belt, Inner Mongolia of China: Implications for continental growth. Lithos, 101, 233-259.

Jochum, K. P., McDonough, W. F., Palme, H., Spettel, B., 1989. Compositional constraints on the continental lithospheric mantle from trace elements in spinel peridotite xenoliths. Nature, 340, 548-550.

Jörg, A., Pfänder, J.A., Münker, C., Stracke, A., Mezger, K., 2007. Nb/Ta and Zr/Hf in ocean island basalts: Implications for crust-mantle differentiation and the fate of Niobium. Earth & Planetary Science Letters, 254, 158-172.

Kogiso, T., Hirschmann, M.M., Pertermann, M., 2004. High pressure partial melting of mafic lithologies in the mantle. Journal of Petrology, 45, 2407-2422.

Kozakov, I.K., Sal'nikova, E.B., Wang, T., Didenko, A.N., Plotkina, Y.V., Podkovyrov, V.N., 2007. Early Precambrian crystalline complexes of the central Asian microcontinent: Age, sources, tectonic position. Stratigraphy and Geological Correlation, 15, 121-140.

Kröner, A., Alexeiev, D. V., Rojas-Agramonte, Y., Hegner, E., Wong, J., Xia, X., Belousova, E., Mikolaichuk, A., Seltmann, R., Liu, D., Kisilev, V., 2013. Mesoproterozoic (Grenvilleage) terranes in the Kyrgyz North Tianshan: Zircon ages and Nd-Hf isotopic constraints on the origin and evolution of basement blocks in the southern Central Asian Orogen. Gondwana Research, 23, 272-295.

Lee, C.T.A., Bachmann, O., 2014. How important is the role of crystal fractionation in making intermediate magmas? Insights from Zr and P systematics. Earth & Planetary Science Letters, 393, 266-274.

Li, J.F., Zhang, Z.C., Han, B.F., 2010. Ar-Ar and zircon SHRIMP geochronology of hornblendite and diorite in northern Darhan Muminggan Joint Banner, Inner Mongolia, and its geological significance. Acta Petrologica et Mineralogica, 29, 732-740 (in Chinese with English abstract).

Li, S., Chung, S.L., Wilde, S.A., Jahn, B.M., Xiao, W.J., Wang, T., Guo, Q.Q., 2017. Early-Middle Triassic high Sr/Y granitoids in the southern Central Asian Orogenic Belt: Implications for ocean closure in accretionary orogens. Journal of Geophysical Research: Solid Earth, 122, 2291-2309.

Li, S., Chung, S.L., Wilde, S.A., Wang, T., Xiao, W.J., 2016. Linking magmatism with collision in an accretionary orogen. Scientific Reports, 6, 25751.

Li, W.B., Zhong, R.C., Xu, C., Song, B., Qu, W.J., 2012. U-Pb and Re-Os geochronology of the Bainaimiao Cu-Mo-Au deposit, on the northern margin of the North China Craton, Central Asia Orogenic Belt: Implications for ore genesis and geodynamic setting. Ore Geology Reviews, 48, 139-150.

Long, X.P., Wilde, S.A., Wang, Q., Yuan, C., Wang, X.C., Li, J., Jiang, Z.Q., Dan, W., 2015. Partial melting of thickened continental crust in central Tibet: Evidence from geochemistry and geochronology of Eocene adakitic rhyolites in the northern Qiangtang Terrane. Earth & Planetary Science Letters, 414, 30-44.

Long, X.P., Yuan, C., Sun, M., Kröner, A., Zhao, G.C., Wilde, S., Hu, A.Q., 2011. Reworking of the Tarim Craton by underplating of mantle plume-derived magmas: Evidence from Neoproterozoic granitoids in the Kuluketage area, NW China. Precambrian Research, 187, 1-14.

Long, X.P., Yuan, C., Sun, M., Zhao, G.C., Xiao, W.J., Wang, Y.J., Yang, Y.H., Hu, A.Q., 2010. Achean crustal evolution of the northern Tarim Craton, NW China: Zircon U-Pb and Hf isotopic constraints. Precambrian Research, 180, 272-284.

Ludwig, K.R., 2003. Isoplot 3. 00: A Geochronological Toolkit for Microsoft Excel. Berkeley Geochronology Center Special Publication, Berkeley, 4.

Lustrino, M., 2005. How the delamination and detachment of lower crust can influence basaltic magmatism. Earth-Science Reviews, 72, 21-38.

Ma, L., Jiang, S.Y., Hofmann, A.W., Dai, B.Z., Hou, M.L., Zhao, K.D., Chen, L.H., Li, J.W., Jiang, Y.H., 2014. Lithospheric and asthenospheric sources of lamprophyres in the Jiaodong Peninsula: A consequence of rapid lithospheric thinning beneath the North China Craton? Geochimica et Cosmochimica Acta, 124, 250-271.

Ma, Q., Zheng, J.P., Xu, Y.G., Griffin, W.L., Zhang, R.S., 2015. Are continental "adakites" derived from thickened or foundered lower crust. Earth & Planetary Science Letters, 419, 125-133.

Ma, S.X., Wang, Z.Q., Zhang, Y.L., Sun, J.X., 2019. Bainaimiao arc as an exotic terrane along the northern margin of the North China Craton: Evidences from petrography, zircon U-Pb dating, and geochemistry of the early devonian deposits. Tectonic, 38.

Macpherson, C.G., Dreher, S.T., Thirlwall, M.F., 2006. Adakites without slab melting: High pressure differentiation of island arc magma, Mindanao, the Philippines. Earth & Planetary Science Letters, 243 (3-4), 581-593.

Menzies, M.A., Xu, Y., Zhang, H., Fan, W., 2007. Integration of geology, geophysics and geochemistry: A key to understanding the North China Craton. Lithos, 96, 1-21.

Middlemost, E.A.K., 1994. Naming materials in the magma/Igneous rock system. Earth-Science Reviews, 37, 215-224.

Moyen, J.F., 2009. High Sr/Y and La/Yb ratios: The meaning of the "adakitic signature". Lithos, 112, 556-574.

Münker, C., Pfänder, J.A., Weyer, S., Büchl, A., Kleine, T., Mezger, K., 2003. Evolution of planetary cores and the Earth-Moon system from Nb/Ta systematics. Science, 301, 84-87.

Müntener, O., Kelemen, P.B., Grove, T.L., 2001. The role of H_2O during crystallization of primitive arc magmas under uppermost mantle conditions and genesis of igneous pyroxenites: An experimental study. Contributions to Mineralogy and Petrology, 141, 643-658.

Nandedkar, R.H., Ulmer, P., Müntener, O., 2014. Fractional crystallization of primitive hydrous arc magmas: An experimental study at 0. 7 GPa. Contributions to Mineralogy and Petrology, 167, 1015.

Peccerillo, A., Taylor, S.R., 1976. Geochemistry of Eocene calc-alkaline volcanic rocks from the Kastamonu area, northern Turkey. Contributions to Mineralogy and Petrology, 58, 63-81.

Pertermann, M., Hirschmann, M.M., 2003. Anhydrous partial melting experiment on MORB-like eclogites phase relations, phase composition and mineral-melt partitioning of major elements at 2−3 GPa. Journal of Petrology, 44, 2173-2202.

Petford, N., Atherton, M., 1996. Na-rich partial melts from newly underplated basaltic crust: The Cordillera Blanca Batholith, Peru. Journal of Petrology, 37, 1491-1521.

Pfänder, J.A., Jochum, K.P., Kozakov, I., Kröner, A., Todt, W., 2002. Coupled evolution of back-arc and island arc-like mafic crust in the late-Neoproterozoic Agardagh Tes-Chem ophiolite, Central Asia: Evidence from trace element and Sr-Nd-Pb isotope data. Contributions to Mineralogy and Petrology, 143, 154-174.

Qian, X.Y., Zhang, Z.C., Chen, Y., Yu, H.F., Luo, Z.W., Yang, J.F., 2017. Geochronology and geochemistry of Early Paleozoic igneous rocks in Zhurihe area, Inner Mongolia and their tectonic significance. Earth Science, 42, 1472-1494 (in Chinese with English abstract).

Rapp, R.P., Shimizu, N., Norman, M.D., 2003. Growth of early continental crust by partial melting of eclogite. Nature, 425, 605-609.

Rapp, R.P., Watson, E.B., 1995. Dehydration melting of metabasalt at 8 − 32 kbar: Implications for continental growth and crust-mantle recycling. Journal of Petrology, 36, 891-931.

Richards, J.P., Kerrich, R., 2007. Special Paper: Adakite-like rocks: their diverse origins and questionable role in metallogenesis. Economic Geology, 102, 537-576.

Rudnick, R.L., Gao, S., 2003. Composition of the Continental Crust. In: Holland, H.D., Turekian, K.K., Rudnick, R.L. Treatise on the Geochemistry Vol. 3: The Crust. Oxford: Elsevier-Pergamon, 1-64.

Schmidt, M., Poli, S., 2004. Magmatic epidote. Reviews in Mineralogy and Geochemistry, 56, 399-430.

Shao, J.A., 1991. Crustal Evolution in the Middle Part of the Northern Margin of the Sino-Korean Plate. Beijing: Peking University Press (in Chinese with English abstract).

Shi, G.Z., Faure, M., Xu, B., Zhao, P., Chen, Y., 2013. Structural and kinematic analysis of the early Paleozoic Ondor Sum-Hongqi mélange belt, eastern part of the Altaids (CAOB) in Inner Mongolia, China. Journal of Asian Earth Sciences, 66, 123-139.

Sobolev, A.V., et al., 2007. The amount of recycled crust in sources of mantle-derived melts. Science, 316, 412-417.

Song, X.Y., Zhou, M.F., Keays, R.R., Cao, Z., Sun, M., Qi, L., 2006. Geochemistry of the Emeishan flood basalts at Yangliuping, Sichuan, SW China: Implications for sulphide segregation. Contributions to Mineralogy and Petrology, 152, 53-74.

Sun, S.S., McDonough, W.F., 1989. Chemical and isotopic systematics of oceanic basalts: Implications for mantle composition and processes. Geological Society, London, Special Publications, 42, 313-345.

Tang, K.D., 1990. Tectonic development of Paleozoic foldbelts at the north margin of the Sino-Korean craton. Tectonics, 9, 249-260.

Tang, G.J., Chung, S.L., Hawkesworth, C.J., Cawood, P.A., Wang, Q., Wyman, D.A., Xu, Y.G., Zhao, Z.H., 2017. Short episodes of crust generation during protracted accretionary processes: Evidence from Central Asian Orogenic Belt, NW China. Earth & Planetary Science Letters, 464, 142-154.

Tong, Y., Hong, D.W., Wang, T., Shi, X.J., Zhang, J.J., Zeng, T., 2010. Spatial and temporal distribution of granitoids in the middle segment of the Sino-Mongolian border and its tectonic and metallogenic implications. Acta Geoscientica Sinica, 31, 395-412 (in Chinese with English abstract).

Ulmer, P., Kaegi, R., Müntener, O., 2018. Experimentally derived intermediate to silica-rich arc magmas by fractional and equilibrium crystallization at 10 GPa: An evaluation of phase relationships, compositions, liquid lines of descent and oxygen fugacity. Journal of Petrology, 59, 11-58.

Wang, X.S., Gao, J., Klemd, R., Jiang, T., Li, J.L., Zhang, X., Xue, S.C., 2017. The Central Tianshan block: A microcontinent with a Neoarchean-Paleoproterozoic basement in the southwestern Central Asian Orogenic Belt. Precambrian Research, 295, 130-150.

Wang, Z.W., Pei, F.P., Xu, W.L., Cao, H.H., Wang, Z.J., Zhang, Y., 2016. Tectonic evolution of the eastern Central Asian Orogenic Belt: Evidence from zircon U-Pb-Hf isotopes and geochemistry of early Paleozoic rocks in Yanbian region, NE China. Gondwana Research, 38, 334-350.

Wu, C., Liu, C.F., Zhu, Y., Zhou, Z.G., Jiang, T., Liu, W.C., Li, H.Y., Wu, C., Ye, B.Y., 2016. Early Paleozoic magmatic history of central Inner Mongolia, China: Implications for the tectonic evolution of the Southeast Central Asian Orogenic Belt. International Journal of Earth Sciences, 105, 1307-1327.

Xiao, W.J., Santosh, M., 2014. The western Central Asian Orogenic Belt: A window to accretionary orogenesis and continental growth. Gondwana Research, 25, 1429-1444.

Xiao, W.J., Windley, B.F., Hao, J., Zhai, M.G., 2003. Accretion leading to collision and the Permian Solonker suture, Inner Mongolia, China: Termination of the central Asian orogenic belt. Tectonics, 22, 1-8.

Xiao, W.J., Windley, B.F., Sun, S., Li, J.L,, Huang, B.C., Han, C.M., Yuan, C., Sun, M., Chen, H. L., 2015. A tale of amalgamation of three Permo-Triassic collage systems in Central Asia: Oroclines, sutures, and terminal accretion. Annual Review of Earth and Planetary Sciences, 43, 477-507.

Xiong, X.L., Adam, T.J., Green, T.H., 2005. Rutile stability and rutile/melt HFSE partitioning during partial melting of hydrous basalt: Implications for TTG genesis. Chemical Geology, 218, 339-359.

Xu, H.J., Ma, C.Q., Song, Y.R., Zhang, J.F., Ye, K., 2012. Early Cretaceous intermediate-mafic dykes in the Dabie orogen, eastern China: Petrogenesis and implications for crust-mantle interaction. Lithos, 154, 83-99.

Xu, J.F., Shinjo, R., Defant, M.J., Wang, Q., Rapp, R.P., 2002. Origin of Mesozoic adakitic intrusive rocks in the Ningzhen area of east China: Partial melting of delaminated lower continental crust? Geology, 30, 1111-1114.

Xu, L.Q., Deng, J.F., Chen, Z.Y., Tao, J.X., 2003. The identification of Ordovician adakites and its significance in northern Damao, Inner Mongolia. Geoscience, 17, 428-434 (in Chinese with English abstract).

Xu, W., Zhu, D.C., Wang, Q., Weinberg, R.F., Wang, R., Li, S.M., Zhang, L.L., Zhao, Z.D., 2019. Constructing the early Mesozoic Gangdese crust in southern Tibet by hornblende-dominated magmatic differentiation. Journal of Petrology, 60, 515-552.

Yuan, H.L., Gao, S., Dai, M.N., Zong, C.L., Gunther, D., Fontaine, G.H., Liu, X.M., Diwu, C.R., 2008. Simultaneous determinations of U-Pb age, Hf isotopes and trace element compositions of zircon by excimer laser-ablation quadrupole and multiple-collector ICP-MS. Chemical Geology, 247, 100-118.

Yuan, H.L., Gao, S., Liu, X.M., Li, H.M., Gunther, D., Wu, F.Y., 2004. Accurate U-Pb age and trace element determinations of zircon by laser ablation-inductively coupled plasma mass spectrometry. Geostandards Newsletter, 28, 353-370.

Yuan, Y., Zong, K.Q., Cawood, P.A., Cheng, H., Yu, Y.Y., Guo, J.L., Liu, Y.S., Hu, Z.C., Zhang, W., Li, M., 2019. Implication of Mesoproterozoic (~1.4 Ga) magmatism within microcontinents along the southern central Asian orogenic belt. Precambrian Research, 327, 314-326.

Zhang, C.L., Zou, H.B., Li, H.K., Wang, H.Y., 2013b. Tectonic framework and evolution of the Tarim block in NW China. Precambrian Research, 23, 1306-1315.

Zhang, S.H., Zhao, Y., Ye, H., Liu, J.M., Hu, Z.C., 2014. Origin and evolution of the Bainaimiao arc belt: Implications for crustal growth in the southern Central Asian orogenic belt. Geological Society of America Bulletin, 126, 1275-1300.

Zhang, W., Jian, P., 2008. SHRIMP dating of early Paleozoic granites from north Damaoqi, Inner Mongolia. Acta Geologica Sinica, 82, 778-787 (in Chinese with English abstract).

Zhang, W., Jian, P., Kröner, A., Shi, Y.R., 2013a. Magmatic and metamorphic development of an early to mid-Paleozoic continental margin arc in the southernmost Central Asian Orogenic Belt, Inner Mongolia, China. Journal of Asian Earth Sciences, 72, 63-74.

Zhang, Y.P., Su, Y.Z., Li, J.C., 2010.Regional tectonic significance of the late Silurian Xibiehe formation in central Inner Mongolia, China. Geological Bulletin of China, 29, 1599-1605 (in Chinese with English abstract).

Zhao, G.C., Wang, Y.J., Huang, B.C., Dong, Y.P., Li, S.Z., Zhang, G.W., Yu, S., 2018. Geological reconstructions of the east Asian blocks: From the breakup of Rodinia to the assembly of Pangea. Earth-Science Reviews, 186, 262-286.

Zhou, H., Pei, F.P., Zhang, Y., Zhou, Z.B., Xu, W.L., Wang, Z.W., Cao, H.H., Yang, C., 2018b. Origin and tectonic evolution of early Paleozoic arc terranes abutting the northern margin of North China Craton. International Journal of Earth Sciences, 107, 1911-1933.

Zhou, H., Zhao, G.C., Han, Y.G., Wang, B., 2018c. Geochemistry and zircon U-Pb-Hf isotopes of Paleozoic intrusive rocks in the Damao area in Inner Mongolia, northern China: Implications for the tectonic evolution of the Bainaimiao arc. Lithos, 314-315, 119-139.

Zhou, J.B., Wilde, S.A., Zhao, G.C., Han, J., 2018a. Nature and assembly of microcontinental blocks within the Paleo-Asian Ocean. Earth-Science Reviews, 186, 76-93.

Zhou, Z.H. Mao, J.W., Ma, X.H., Che, H.W., Ouyang, H.G., Gao, X., 2017. Geochronological framework of the early Paleozoic Bainaimiao Cu-Mo-Au deposit, NE China, and its tectonic implications. Journal of Asian Earth Sciences, 144, 323-338.

Genesis of ca. 850−835 Ma high-Mg# diorites in the western Yangtze Block, South China: Implications for mantle metasomatism under the subduction process[①]

Zhu Yu   Lai Shaocong[②]   Qin Jiangfeng   Zhu Renzhi
Liu Min   Zhang Fangyi   Zhang Zezhong   Yang Hang

Abstract: High-Mg# [molar $100Mg/(Mg+Fe)$] diorites can provide significant insights on mantle metasomatism under a subduction zone. Here we investigate the genesis of the Neoproterozoic Shuilu high-Mg# diorites in the western Yangtze Block to evaluate the role of subduction-related fluids and sediment melts acting on mantle sources during the subduction process. Zircon U-Pb dating results display new weighted mean $^{206}Pb/^{238}U$ ages of 850.1 ± 1.7 Ma, 840.9 ± 2.4 Ma, and 836.6 ± 1.9 Ma for these high-Mg# diorites. They are metaluminous and calc-alkaline rocks, and characterized by moderate SiO_2 contents($57.08−61.12$ wt%) and high MgO contents ($3.36−4.30$ wt%) and Mg# values ($56−60$). The relatively low initial $^{87}Sr/^{86}Sr$ ratios ($0.703\ 406−0.704\ 157$) and highly positive whole-rock $\varepsilon_{Nd}(t)$ ($+3.26$ to $+4.26$) and zircon $\varepsilon_{Hf}(t)$ values ($+8.43$ to $+13.6$) imply that these rocks were predominantly sourced from depleted lithospheric mantle. These high-Mg# diorites also show enrichment of light rare earth elements and large ion lithophile elements (e.g., Rb, Ba, K, and Sr) as well as depletion of high field strength elements (e.g., Nb, Ta, Zr, and Hf), resembling a typical arc magma affinity. The highly variable Ba contents and Rb/Y, Th/Ce, Th/Sm, Ba/La, and Th/Yb ratios indicate significant incorporation of subduction-related fluids and sediment-derived melts into the primary mantle source. We therefore propose that the ca. 850−835 Ma high-Mg# diorites investigated in this study were formed by partial melting of a metasomatized mantle source enriched by subduction fluids and sediment melts. Our new data, in conjunction with numerous studies of metasomatized mantle magmatism from the western Yangtze Block, suggest that the Neoproterozoic mantle sources beneath the western Yangtze Block were metasomatized by subduction-related compositions involving slab fluids, sediment melts, and oceanic slab melts during the subduction process.

1   Introduction

High-Mg diorites, as the intrusive equivalents of high-Mg andesites, are geochemically

---

① Published in *Precambrian Research*, 2020, 343.

② Corresponding author.

characterized by moderate SiO_2 contents and high MgO contents and/or $Mg^#$ [molar 100Mg/ (Mg + Fe)] numbers (Kelemen et al., 2003, 2014; Qian and Hermann, 2010; Tatsumi, 2008). High-Mg intermediate rocks are crucial for evaluating continental crustal evolution because they compositionally approximate the average continental crust (Qian and Herman, 2010; Smithies and Champion, 2000; Tatsumi, 2006). They are also of significant interest because of their particular geochemical characteristics that show both crust- and mantle-derived inputs (Kamei et al., 2004; Qian and Hermann, 2010; Tatsumi, 2006). They usually contain elevated mantle-like indices (e.g., $Mg^#$ and MgO, Cr, and Ni), indicating an affinity to primitive mantle-derived magmas (Smithies and Champion, 2000; Tatsumi, 2006). However, they also show the obvious enrichment in large ion lithophile elements (LILEs; e.g., Rb, Ba, K, and Sr) and the depletion of high field strength elements (HFSEs; e.g., Nb, Ta, and Ti), which are distinctive trace elemental characteristics of mafic lower crust-derived melts (Martin et al., 2005; Qian and Hermann, 2010; Smithies and Champion, 2000). High-Mg diorites and andesites are commonly classified into four types: boninites, bajaites, adakites, and sanukitoids (Defant and Drummond, 1990; Kamei et al., 2004; Kuroda et al., 1978; Rogers et al., 1985; Shimoda et al., 1998; Tatsumi and Ishizaka, 1981). These high-Mg intermediate rocks (e.g., the Setouchi high-Mg andesites in Japan, the Pilbara high-Mg diorite suite in Australia, and the Dongma high-Mg diorites of the Jiangnan Orogen in South China) have mainly been interpreted as representing an equilibrium between overlying mantle melts and subduction slab-related fluids, sediment melts, or oceanic slab melts (Chen et al., 2014a; Hanyu et al., 2006; Shimoda et al., 1998; Shirey and Hanson, 1984; Smithies and Champion, 2000; Stern et al., 1989; Tang et al., 2010; Tatsumi et al., 2002, 2003; Tatsumi, 2006, 2008; Wang et al., 2007, 2009). They are predominantly associated with subduction-related tectonic settings, requiring significant melting of subducted oceanic slab and/or subduction component-modified mantle (Hanyu et al., 2006; Martin et al., 2005; Shimoda et al., 1998; Stern and Kilian, 1996; Tatsumi, 2006). However, a considerable number of studies have proposed that high-Mg intermediate rocks can also be generated by other mechanisms without significant contributions of subduction-related components. For example, the Mesozoic Jinling-Tietongou high-Mg adakitic diorites (North China Craton) were formed by magma mixing between siliceous crustal melts and basaltic magma derived from metasomatized mantle in a post-kinematic setting (Chen et al., 2013). The Early Cretaceous Han-Xing high-Mg diorites (Central North China Block) have been interpreted as derived from the assimilation of mantle peridotite by felsic monzodioritic magma at crustal depth in an intra-continental setting (Qian and Hermann, 2010). Moreover, some high-Mg adakites and sanukitoids have been predominantly interpreted as products of the partial melting of delaminated or thickened lower continental crust (Gao et al., 2004; Hou et al., 2004; Huang et al., 2008a; Moyen et al., 2003; Rapp et al., 2010; Xu et al., 2002, 2008). Considering the unique geochemical features and particular tectonic implications for the subduction setting, it is thus crucial to perform rigorous investigations of the genesis of high-Mg intermediate rocks.

The South China Craton is one of the largest cratonic blocks in eastern Asia and is characterized by abundant Neoproterozoic igneous rocks with direct bearings on the assembly and dispersal of the supercontinent Rodinia (Lai et al., 2015; Li et al., 2003; Zhao et al., 2018 and reference therein; Zhao and Cawood, 2012; Zhou et al., 2002, 2006a, b). Widespread Neoproterozoic magmatism along the western margin of the Yangtze Block has been proposed to be associated with controversial tectonic settings, including a mantle plume setting (Li et al., 2003, 2006; Wang et al., 2008; Wu et al., 2019), an intra-continental rift setting (Huang et al., 2008b, 2009), and an oceanic slab subduction setting (Munteanu et al., 2010; Zhao et al., 2017, 2018, 2019; Zhao and Zhou, 2007a, b; Zhou et al., 2002, 2006a, b; Zhu et al., 2017, 2019a-c). Although the voluminous mafic-ultramafic and intermediate intrusions have been attributed to high-pressure fractional crystallization of hydrous basaltic melts (Li and Zhao, 2018) or the partial melts of mantle sources metasomatized by subducted fluids and oceanic slab melts (Du et al., 2014; Munteanu et al., 2010; Sun and Zhou, 2008; Zhao et al., 2008; Zhao and Zhou, 2007b; Zhou et al., 2006a; Zhu et al., 2019a), the metasomatized mantle-derived magmatism, which has been correlated with the subducted slab fluids and sediment melts, has rarely been studied in detail. We therefore present our new identification of high-$Mg^{\#}$ diorites in the Shuilu area along the western Yangtze Block, South China, to probe the genesis of these high-$Mg^{\#}$ diorites and evaluate the role of slab fluids and sediment melts acting on the mantle source associated with Neoproterozoic subduction processes.

2 Geological background and petrography

The South China Block, as a vital part of the supercontinent Rodinia, is composed of two principal Precambrian blocks: the Yangtze Block to the northwest and the Cathaysia Block to the southeast (Fig. 1a) (Wang et al., 2013, 2014; Zhao and Cawood, 2012; Zhao et al., 2011, 2018; Zhao and Asimow, 2018). The Yangtze Block is bounded by the Longmenshan thrust belt in the northwest, separated from the North China Block by the Qinling-Dabie-Sulu orogenic belt in the north, and surrounded by the Ailaoshan-Songma suture to the southwest (Gao et al., 1999; Zhao and Cawood, 2012). The ca. 1 500-km-long Neoproterozoic Jiangnan orogenic belt is located in the middle part of the South China Block, and the ca. 1 000-km-long Neoproterozoic igneous and metamorphic belt is embedded at the western margin of the Yangtze Block (Du et al., 2014; Lai et al., 2015; Wang et al., 2013; Zhao et al., 2011, 2018) (Fig. 1a).

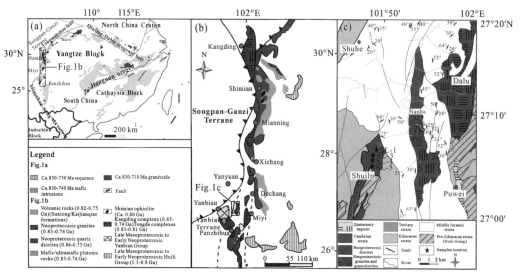

Fig. 1　Simplified geological sketch map of South China (a) (Zhu et al., 2019a), the western margin
of the Yangtze Block (b) (Zhu et al., 2019a), and the regional geological map
in the Shuilu area (c) (SPBGMR, 1972; Zhu et al., 2019b).

The Yangtze Block consists of Archean to Early Neoproterozoic basement complexes overlain by Late Neoproterozoic to Cenozoic sequences (Li and Zhao, 2018; Zhao and Asimow, 2018). The Archean Kongling complex, as old as ca. 3.45 Ga (Guo et al., 2014), crops out in the northern Yangtze Block and consists of felsic gneisses and metasedimentary rocks as well as minor amphibolites and mafic granulites (Gao et al., 1999, 2011; Guo et al., 2014). The widespread Late Mesoproterozoic to Early Neoproterozoic volcanic-sedimentary sequences, including the Upper Kunyang, Upper Huili, Julin, and Yanbian groups (Chen et al., 2014b; Greentree et al., 2006; Zhao et al., 2018; Zhu et al., 2016), are distributed at the southwestern margin of the Yangtze Block. The Kunyang Group, with a total thickness of >10 km, is mainly composed of terrigenous clastic rocks, carbonates, and volcanic rocks (Greentree et al., 2006). The Huili Group is >10 km in thickness and predominantly composed of meta-clastic and meta-carbonate rocks as well as meta-volcanic rocks (Chen et al., 2018). The Julin Group is a >3 560-m-thick sequence located ~20 km southeast of the Huili Group. It is composed of slate, gneiss, schist, sandstone, marble, and dolomite, as well as subordinate meta-volcanic rocks (Chen et al., 2018; Wang et al., 2019). The Yanbian Group, as one of the earliest Neoproterozoic volcanic-sedimentary sequences, consists of volcanic and clastic rocks (Zhu et al., 2016). In addition, various Late Mesoproterozoic magmatic rocks, including A-type volcanic rocks and granites as well as S-type granites, have been documented in the southwestern Yangtze Block (Zhu et al., 2020 and reference therein). At higher levels, Neoproterozoic sequences were overlain by thick Late Neoproterozoic to Cenozoic strata (Zhao and Asimow, 2018). Widespread Neoproterozoic igneous and sedimentary sequences are well preserved across the western Yangtze Block, including the Kangding, Miyi, Tongde, Datian, and Yuanmou complexes from north to south with dominant ages of ca. 860－750 Ma (Fig. 1b) (Sun et al., 2008, 2009; Zhou et al., 2002, 2006a; Zhu et al., 2019a-c). The Neoproterozoic granitoids are spatially associated with high-grade metamorphic complexes and low-grade metamorphic Mesoproterozoic strata along the western Yangtze Block (Zhao et al., 2008; Zhou et al., 2006a).

The Yanbian Terrane consists of the Yanbian Group and is located at the southern segment of the western Yangtze Block with a total area of 300 km^2. The Yanbian Group can be subdivided into the Fangtian, Yumen, Xiaoping, and Zagu formations (Zhou et al., 2006a; Zhu et al., 2008). They are dominantly flysch-type volcanic-sedimentary sequences consisting of basalt, slate, sandstone, limestone, and carbonaceous slate (Du et al., 2014). Four Neoproterozoic mafic-ultramafic and intermediate plutons intruded into the Yanbian Group, including the ca. 860 Ma Guandaoshan diorites and gabbro-diorites, the ca. 830－825 Ma Tongde diorites, the ca. 810 Ma Gaojiacun and Lengshuiqing mafic rocks, and the ca. 740 Ma Dadukou mafic-ultramafic intrusions (Du et al., 2014; Munteanu et al., 2010; Sun and Zhou, 2008; Zhao and Zhou, 2007b; Zhou et al., 2006a). These large-scale intermediate-felsic intrusions are situated in the northeast part of the Yanbian Group (Fig. 1c). In this region, the Neoproterozoic Shuilu, Nanba, and Dalu granitoid plutons are well preserved, but their geochronology, petrogenesis, and tectonic implications have rarely been investigated (SPBGMR, 1972). These Neoproterozoic intermediate-felsic rocks are spatially accompanied by the Ediacaran strata and pre-Ediacaran strata (Huili Group) (SPBGMR, 1972; Zhu et al., 2019b). Zhu et al. (2019b) documented that the Dalu I-type granitic pluton was composed of granodiorites and granites and formed ca. 780 Ma. They were originated from the mafic lower crust with the addition of mature metasedimentary components (Zhu et al., 2019b).

This study focused on the Shuilu dioritic pluton (Fig. 1c), a nearly N-S-elongated intrusion that covers an area of ca. 40 km^2 (SPBGMR, 1972). It is tectonically cut by several faults along the northwestern margin and by the river in the interior (Fig. 1c) (SPBGMR, 1972). The southern part of the Shuilu pluton intrudes the Late Mesoproterozoic Huili Group, which mainly consists of slate and meta-sandstone (Fig. 1c) (SPBGMR, 1972). The diorite samples collected from the Shuilu pluton are dark gray and have a medium- to fined-grained texture (Fig. 2a-c). They are predominantly composed of plagioclase (40%－45%), amphibole (20%－35%), quartz (10%－20%), K-feldspar (10%－15%), with minor biotite, clinopyroxene, magnetite, and zircon (Fig. 2d,e). Plagioclases are subhedral to euhedral grains and show a typical polysynthetic twinning texture. Some of the plagioclase grains have been slightly altered to the sericite. Amphibole occurs as subhedral to euhedral crystals. Some K-feldspar crystals display crossed twinning.

Fig. 2   Field petrography (a-c) and microphotographs of the Shuilu diorites (d,e)
in the western Yangtze Block, South China.

Pl: plagioclase; Kfs: K-feldspar; Qtz: quartz; Amp: amphibole; Bi: biotite; Mag: magnetite.

3   Analytical methods

3. 1   Zircon U-Pb dating and Lu-Hf isotopes analysis

Zircon grains were separated from diorite samples taken from different sampling locations in the Shuilu pluton from the Panzhihua-Yanbian region along the western margin of the Yangtze Block (Fig. 1). Zircons were separated using conventional heavy liquid and magnetic techniques. Representative grains were mounted in epoxy resin discs, which were then polished and carbon coated. Internal morphology was examined at the State Key Laboratory of Continental Dynamics, Northwest University, Xi'an, China with cathodoluminescence (CL) microscopy prior to U-Pb and Lu-Hf isotopic analyses. Laser Ablation Inductively Coupled Plasma Mass Spectrometry (LA-ICP-MS) U-Pb analyses were conducted on Agilent 7500a ICP-MS equipped with a 193-nm laser, following the method of Yuan et al. (2004). The $^{207}Pb/^{206}Pb$ and $^{206}Pb/^{238}U$ ratios were calculated using the GLITTER program and corrected using the Harvard zircon 91500 as an external calibration with a recommended $^{206}Pb/^{238}U$ isotopic age of 1 065. 4±0. 6 Ma (Wiedenbeck et al., 2004). The detailed analytical technique is described in Yuan et al. (2004). Common Pb contents were subsequently evaluated using the method described in Andersen (2002). The age calculations and plotting of concordia diagrams were made using ISOPLOT (version 3. 0; Ludwig, 2003). Uncertainties are quoted at the 1σ level.

The in-situ zircon Lu-Hf isotopic analyses were conducted using a Neptune multiconductor IPC-MS (MC-ICP-MS) equipped with a 193-nm laser in the State Key Laboratory of

Continental Dynamics, Northwest University. During analyses, a laser repetition rate of 6 Hz at 100 mJ was used and the spot sizes were 30 μm. The instrument information is obtainable from Bao et al. (2017). The detailed analytical procedures are depicted by Yuan et al. (2008). Zircon standards (91500, GJ-1, and Monastery) were simultaneously analyzed, which yielded the recommended values within 2σ (Yuan et al., 2008). The obtained Hf isotopic compositions were 0.282 016 ± 20 (2σ, $n=84$) for the GJ-1 standard and 0.282 735 ± 24 (2σ, $n=84$) for the Monastery standard, respectively, consistent with the recommended values with 2σ (Yuan et al., 2008). The initial $^{176}Hf/^{177}Hf$ ratios and $\varepsilon_{Hf}(t)$ values were calculated with the reference to the chondritic reservoir (CHUR) at the time of zircon growth from the magmas. A decay constant for ^{176}Lu of 1.867×10^{-11} year^{-1} (Söderlund et al., 2004), the chondritic $^{176}Hf/^{177}Hf$ ratio of 0.282 785 and $^{176}Lu/^{177}Hf$ ratio of 0.033 6 were adopted. The depleted mantle model ages for basic rocks were calculated with reference to the depleted mantle at present-day $^{176}Hf/^{177}Hf$ ratio of 0.283 25 ± 20, similar to that of average mid-ocean-ridge basalt (MORB; Nowell et al., 1998) and $^{176}Lu/^{177}Hf=0.038\ 4$ (Griffin et al., 2000).

3.2 Whole-rock major and trace elements and Sr-Nd isotopes

Weathered surfaces of diorite samples were removed and the fresh parts were then chipped and powdered to about a ~200 mesh size using a tungsten carbide ball mill. Major and trace elements were analyzed in the State Key Laboratory of Continental Dynamics, Northwest University, Xi'an by X-ray fluorescence (XRF; Rikagu RIX 2100) and inductively coupled plasma mass spectrometry (ICP-MS; Agilent 7500a), respectively. Analyses of USGS and Chinese national rock standards (BCR-2, GSR-1, and GSR-3) indicate that both analytical precision and accuracy for the major elements were generally better than 5%. For the trace element analyses, sample powders were digested using an $HF+HNO_3$ mixture in high-pressure Teflon bombs at 190 ℃ for 48 h. The analytical error was less than 2% and the precision was greater than 10% (Liu et al., 2007).

Whole-rock Sr-Nd isotopic data were obtained using a Neptune Plasma HR multi-collector mass spectrometer at the Guizhou Tuopu Resource and Environmental Analysis Center in China. Sr and Nd isotopic fractionation was corrected to $^{87}Sr/^{86}Sr=0.119\ 4$ and $^{146}Nd/^{144}Nd=0.721\ 9$, respectively. During the samples run, the La Jolla standard yielded an average value of $^{143}Nd/^{144}Nd=0.511\ 862 \pm 5$ (2σ), and the NBS987 standard yielded an average value of $^{87}Sr/^{86}Sr=0.710\ 236 \pm 16$ (2σ). The total procedural Sr and Nd blanks are b1 ng and b50 pg, and NIST SRM-987 and JMC-Nd were used as certified reference standard solutions for $^{87}Sr/^{86}Sr$ and $^{143}Nd/^{144}Nd$ isotopic ratios, respectively. The average $^{87}Sr/^{86}Sr$ ratios of the BCR-1 and BHVO-1 standards are 0.705 014 ± 3 (2σ) and 0.703 477 ± 20 (2σ), respectively. The average $^{146}Nd/^{144}Nd$ ratios of the BCR-1 and BHVO-1 are 0.512 615 ± 12 (2σ) and 0.512 987

± 23 (2σ), which were used as the reference values.

4   Results

4.1   Zircon LA-ICP-MS U-Pb ages

Zircon U-Pb data, representative zircon CL images, and concordia diagrams of three high-Mg# diorite samples (SL-3, SL-5, and SL-1) from the Shuilu pluton in the western Yangtze Block are shown in Table 1 and Fig. 3. The separated zircons from three samples are mainly subhedral to euhedral prismatic crystals, with aspect ratios around 1 : 1 to 3 : 2. In the CL images (Fig. 3a, d and g), these zircon grains show the size of 50–250 μm, and some grains display weak oscillatory zoning.

For the sample SL-3, 35 reliable analytical spots have variable Th (23.4–189 ppm) and U (32.7–326 ppm) contents and Th/U ratios of 0.26–1.17. These 35 concordant spots yield $^{206}Pb/^{238}U$ ages from 842±11 Ma to 859±11 Ma, with a weighted mean age of 850.1±1.7 Ma (MSWD=0.93, n=35) (Fig. 3b, c). For the sample SL-5, 22 analytical spots show the low Th contents of 28.4–122 ppm, U contents of 43.4–99.5 ppm, and Th/U ratios of 0.57–1.22. The 22 concordant spots are characterized by a weighted mean age of 840.9±2.4 Ma (MSWD =0.81, n=22) (Fig. 3e, f). In addition, sample SL-1 also has relatively low Th contents of 17.5–113 ppm, U contents of 25.9–136 ppm, and Th/U ratios of 0.43–0.95. The 36 concordant spots yield $^{206}Pb/^{238}U$ ages ranging from 822±14 Ma to 846±12 Ma, with a weighted mean age of 836.6±1.9 Ma (MSWD=0.81, n=36) (Fig. 3h, i).

4.2   Whole-rock major and trace elemental geochemistry

In this study, the whole-rock major and trace element compositions of the ca. 850–835 Ma Shuilu high-Mg# diorites are given in Table 2, and Figs. 4 and 5. The values for the Neoproterozoic metasomatized mantle-derived magmatism, including the ca. 860 Ma Guandaoshan pluton, ca. 825 Ma Tongde pluton, and ca. 810 Ma Dajianshan gabbro-diorites, are shown for comparison (Du et al., 2014; Munteanu et al., 2010; Zhu et al., 2019a).

The diorites studied here are dominant of the calc-alkaline series with moderate SiO_2 contents of 57.08–61.12 wt%, low K_2O contents of 1.16–2.20 wt%, and high Na_2O contents of 3.87–4.57 wt% (Fig. 4a, b). The moderate Al_2O_3 contents of 15.36–16.81 wt% and A/CNK [molar $Al_2O_3/(CaO+Na_2O+K_2O)$] ratios of 0.78–1.01 classify these diorites mainly as the metaluminous series (Fig. 4c). They possess high contents of MgO = 3.36–4.30 wt% and Mg# values=56–60 (>50), plotting into the field of high-Mg diorite suites (Kelemen et al., 2014; Smithies and Champion, 2000) (Fig. 4d). They also contain high CaO (4.35–5.94 wt%) and $Fe_2O_3^T$(6.02–7.03 wt%) contents.

Table 1　Concordant results of zircon U-Pb ages for Neoproterozoic Shuilu high-Mg# diorites in western Yangtze Block, South China.

Analysis	Content/ppm								Ratios								Age/Ma							
	Th	U	Th/U	Nb	Yb	Hf	10 000 Nb/Yb	U/Yb	$^{207}Pb/^{206}Pb$	1σ	$^{207}Pb/^{235}U$	1σ	$^{206}Pb/^{238}U$	1σ	$^{208}Pb/^{232}Th$	1σ	$^{207}Pb/^{206}Pb$	1σ	$^{207}Pb/^{235}U$	1σ	$^{206}Pb/^{238}U$	1σ	$^{208}Pb/^{232}Th$	1σ
Diorite SL-3 (N 27°6′44″, E 101°49′27″)																								
SL-3-01	79.29	83.45	0.95	0.83	200	8 411	41.5	0.42	0.067 14	0.002 18	1.319 43	0.035 90	0.142 57	0.001 87	0.045 18	0.000 68	842	35	854	16	859	11	893	13
SL-3-02	74.79	67.42	1.11	0.48	214	7 573	22.6	0.32	0.065 18	0.001 99	1.275 09	0.031 73	0.141 94	0.001 78	0.043 77	0.000 59	780	32	835	14	856	10	866	11
SL-3-03	43.92	48.41	0.91	0.60	162	7 673	37.0	0.30	0.066 54	0.002 66	1.283 52	0.045 78	0.139 94	0.002 07	0.044 64	0.000 89	823	50	838	20	844	12	883	17
SL-3-04	91.41	131.20	0.70	0.90	159	10 348	56.9	0.83	0.066 21	0.001 69	1.288 32	0.024 11	0.141 17	0.001 62	0.042 81	0.000 54	813	21	841	11	851	9	847	10
SL-3-05	56.83	57.62	0.99	0.47	176	7 924	26.8	0.33	0.069 03	0.002 43	1.344 27	0.040 82	0.141 29	0.001 95	0.044 13	0.000 76	900	40	865	18	852	11	873	15
SL-3-06	64.59	61.77	1.05	0.61	342	7 763	17.7	0.18	0.064 04	0.002 16	1.233 21	0.035 24	0.139 69	0.001 85	0.040 96	0.000 62	743	38	816	16	843	10	811	12
SL-3-07	60.42	91.40	0.66	0.78	159	9 077	48.9	0.57	0.066 65	0.002 12	1.284 32	0.033 99	0.139 78	0.001 80	0.045 17	0.000 77	827	34	839	15	843	10	893	15
SL-3-08	57.55	55.63	1.03	0.71	192	8 025	37.2	0.29	0.067 71	0.002 12	1.317 98	0.033 97	0.141 21	0.001 80	0.043 38	0.000 62	860	33	854	15	852	10	858	12
SL-3-09	103.95	88.47	1.17	0.54	265	7 925	20.4	0.33	0.068 92	0.001 94	1.327 83	0.029 28	0.139 76	0.001 69	0.041 71	0.000 52	896	26	858	13	843	10	826	10
SL-3-10	34.04	49.86	0.68	0.64	117	7 583	54.6	0.43	0.071 97	0.003 45	1.398 50	0.061 59	0.140 96	0.002 42	0.044 88	0.001 08	985	62	888	26	850	14	887	21
SL-3-11	58.58	53.89	1.09	0.62	189	7 294	32.9	0.28	0.069 09	0.002 42	1.339 81	0.040 40	0.140 67	0.001 93	0.043 20	0.000 70	901	40	863	18	848	11	855	14
SL-3-12	65.52	73.85	0.89	0.54	205	8 014	26.3	0.36	0.067 68	0.002 40	1.316 18	0.040 22	0.141 07	0.001 94	0.046 43	0.000 84	859	41	853	18	851	11	917	16
SL-3-13	29.48	42.61	0.69	0.68	151	7 924	44.9	0.28	0.067 40	0.002 29	1.321 72	0.038 25	0.142 25	0.001 90	0.043 42	0.000 78	850	38	855	17	857	11	859	15
SL-3-14	45.10	50.60	0.89	0.72	182	8 821	39.8	0.28	0.068 6	0.002 33	1.322 79	0.038 15	0.139 86	0.001 87	0.041 94	0.000 69	887	38	856	17	844	11	830	13
SL-3-15	65.22	78.24	0.83	0.75	206	8 807	36.3	0.38	0.066 03	0.001 85	1.287 74	0.028 10	0.141 46	0.001 69	0.042 44	0.000 57	807	26	840	12	853	10	840	11
SL-3-16	72.20	63.81	1.13	0.50	210	6 394	23.6	0.30	0.068 59	0.002 54	1.346 62	0.043 55	0.142 40	0.002 02	0.044 17	0.000 72	886	43	866	19	858	11	874	14
SL-3-17	44.28	54.48	0.81	0.79	205	9 141	38.3	0.27	0.065 17	0.002 49	1.253 07	0.042 24	0.139 45	0.001 99	0.043 48	0.000 82	780	47	825	19	842	11	860	16
SL-3-18	187.00	326.20	0.57	1.46	212	11 731	68.6	1.54	0.071 42	0.001 68	1.394 37	0.021 94	0.141 60	0.001 57	0.047 02	0.000 55	969	16	887	9	854	9	929	11
SL-3-19	79.55	99.62	0.80	0.59	175	8 516	33.5	0.57	0.071 03	0.002 47	1.379 58	0.041 00	0.140 86	0.001 93	0.041 91	0.001 02	958	38	880	17	850	10	830	20
SL-3-20	90.35	120.82	0.75	0.95	159	9 158	59.5	0.76	0.066 50	0.002 43	1.293 66	0.044 59	0.141 09	0.001 73	0.043 06	0.000 45	822	78	843	20	851	10	852	9
SL-3-21	47.11	179.46	0.26	0.51	132	7 545	38.9	1.36	0.065 88	0.001 91	1.292 26	0.029 70	0.142 27	0.001 73	0.042 14	0.000 79	803	28	842	13	857	10	834	15
SL-3-22	45.15	80.02	0.56	0.62	109	9 302	56.8	0.73	0.067 49	0.002 35	1.312 31	0.039 05	0.141 03	0.001 91	0.044 96	0.000 89	853	39	851	17	850	11	889	17
SL-3-23	52.47	61.49	0.85	0.62	139	8 283	44.6	0.44	0.069 90	0.002 44	1.351 99	0.040 49	0.140 27	0.001 92	0.043 7	0.000 76	925	39	868	17	846	11	865	15

Continued

Analysis	Content/ppm						10 000 Nb/Yb	U/Yb	Ratios								Age/Ma							
	Th	U	Th/U	Nb	Yb	Hf			$\frac{^{207}Pb}{^{206}Pb}$	1σ	$\frac{^{207}Pb}{^{235}U}$	1σ	$\frac{^{206}Pb}{^{238}U}$	1σ	$\frac{^{208}Pb}{^{232}Th}$	1σ	$\frac{^{207}Pb}{^{206}Pb}$	1σ	$\frac{^{207}Pb}{^{235}U}$	1σ	$\frac{^{206}Pb}{^{238}U}$	1σ	$\frac{^{208}Pb}{^{232}Th}$	1σ
SL-3-24	58.20	88.40	0.66	0.79	152	9 632	52.0	0.58	0.068 36	0.001 91	1.335 04	0.028 81	0.141 64	0.001 69	0.042 37	0.000 62	879	25	861	13	854	10	839	12
SL-3-25	88.98	86.15	1.03	0.70	209	8 540	33.8	0.41	0.068 75	0.002 05	1.328 06	0.031 78	0.140 09	0.001 73	0.042 49	0.000 58	891	29	858	14	845	10	841	11
SL-3-26	46.09	53.27	0.87	0.70	133	6 889	52.4	0.40	0.066 85	0.002 69	1.293 50	0.046 30	0.140 32	0.002 08	0.043 74	0.000 91	833	50	843	20	846	12	865	18
SL-3-27	42.94	56.27	0.76	0.70	132	8 289	53.3	0.43	0.066 76	0.002 54	1.304 94	0.043 65	0.141 75	0.002 02	0.044 59	0.000 87	830	46	848	19	855	11	882	17
SL-3-28	57.96	59.15	0.98	0.64	330	6 926	19.3	0.18	0.070 22	0.002 42	1.370 17	0.040 24	0.141 50	0.001 91	0.043 71	0.000 71	935	38	876	17	853	11	865	14
SL-3-29	188.71	320.67	0.59	1.40	209	11 557	66.9	1.53	0.068 34	0.001 59	1.339 55	0.020 50	0.142 14	0.001 55	0.044 70	0.000 50	879	15	863	9	857	9	884	10
SL-3-30	54.05	63.90	0.85	0.83	172	8 973	48.0	0.37	0.069 85	0.002 49	1.343 52	0.041 28	0.139 48	0.001 92	0.041 93	0.000 75	924	41	865	18	842	11	830	15
SL-3-31	75.87	66.59	1.14	0.58	254	8 088	23.0	0.26	0.067 86	0.002 22	1.320 17	0.036 23	0.141 08	0.001 84	0.043 15	0.000 62	864	35	855	16	851	10	854	12
SL-3-32	66.86	60.70	1.10	0.47	204	8 045	22.9	0.30	0.070 49	0.002 09	1.366 51	0.032 50	0.140 57	0.001 73	0.042 62	0.000 56	943	29	875	14	848	10	844	11
SL-3-33	62.38	89.51	0.70	0.77	135	8 534	56.8	0.66	0.067 40	0.003 28	1.300 77	0.060 67	0.139 97	0.001 97	0.042 65	0.000 49	850	104	846	27	844	11	844	9
SL-3-34	60.63	94.89	0.64	0.79	143	9 894	55.2	0.66	0.066 18	0.001 78	1.286 92	0.025 96	0.141 02	0.001 64	0.042 48	0.000 57	812	23	840	12	850	9	841	11
SL-3-36	23.42	32.71	0.72	0.50	163	7 268	30.6	0.20	0.065 40	0.003 15	1.266 34	0.056 10	0.140 42	0.002 34	0.041 52	0.001 06	787	65	831	25	847	13	822	21
Diorite SL-5 (N 27°6'13", E 101°48'57")																								
SL-5-01	63.42	66.62	0.95	0.63	273	9 048	23.1	0.24	0.065 73	0.002 29	1.249 22	0.037 20	0.137 83	0.001 81	0.040 72	0.000 66	798	40	823	17	832	10	807	13
SL-5-02	56.41	70.05	0.81	0.60	232	10 191	26.0	0.30	0.067 11	0.002 74	1.281 79	0.046 70	0.138 53	0.002 04	0.039 88	0.000 84	841	51	838	21	836	12	790	16
SL-5-04	61.35	62.55	0.98	0.73	197	8 287	37.2	0.32	0.066 61	0.002 41	1.277 30	0.040 00	0.139 07	0.001 87	0.040 74	0.000 68	826	43	836	18	839	11	807	13
SL-5-05	35.68	43.38	0.82	0.60	167	8 082	36.3	0.26	0.067 36	0.002 41	1.295 89	0.040 08	0.139 53	0.001 88	0.041 50	0.000 75	849	42	844	18	842	11	822	15
SL-5-06	48.26	52.43	0.92	0.57	256	8 669	22.1	0.20	0.068 30	0.003 29	1.311 39	0.058 08	0.139 26	0.002 35	0.039 29	0.000 90	878	64	851	26	840	13	779	18
SL-5-08	50.68	56.30	0.90	0.67	223	9 062	30.2	0.25	0.062 88	0.003 04	1.192 55	0.053 20	0.137 55	0.002 27	0.038 41	0.000 91	704	67	797	25	831	13	762	18
SL-5-09	70.65	66.37	1.06	0.53	282	8 119	18.8	0.24	0.065 43	0.002 18	1.252 55	0.035 26	0.138 83	0.001 78	0.040 54	0.000 61	788	38	825	16	838	10	803	12
SL-5-14	121.66	99.49	1.22	0.75	389	7 608	19.1	0.26	0.068 22	0.002 39	1.326 03	0.039 92	0.140 97	0.001 90	0.040 28	0.000 62	875	40	857	17	850	11	798	12
SL-5-15	55.27	61.01	0.91	0.62	271	8 313	22.7	0.23	0.066 31	0.002 50	1.272 49	0.042 11	0.139 17	0.001 94	0.039 63	0.000 71	816	46	834	19	840	11	786	14
SL-5-16	43.56	50.94	0.86	0.52	190	8 615	27.3	0.27	0.070 68	0.003 29	1.370 10	0.058 32	0.140 59	0.002 33	0.041 30	0.001 00	948	60	876	25	848	13	818	19
SL-5-17	86.23	79.23	1.09	0.57	276	8 883	20.7	0.29	0.064 90	0.002 02	1.252 75	0.031 99	0.140 00	0.001 73	0.040 45	0.000 56	771	33	825	14	845	10	801	11

Continued

Analysis	Content/ppm								Ratios								Age/Ma							
	Th	U	Th/U	Nb	Yb	Hf	10 000 Nb/Yb	U/Yb	207Pb/206Pb	1σ	207Pb/235U	1σ	206Pb/238U	1σ	208Pb/232Th	1σ	207Pb/206Pb	1σ	207Pb/235U	1σ	206Pb/238U	1σ	208Pb/232Th	1σ
SL-5-18	65.82	79.35	0.83	0.71	278	9 162	25.5	0.29	0.066 26	0.002 00	1.269 13	0.030 91	0.138 91	0.001 69	0.040 23	0.000 59	815	31	832	14	838	10	797	11
SL-5-20	45.93	48.48	0.95	0.48	170	8 454	28.4	0.29	0.066 01	0.002 45	1.277 79	0.041 38	0.140 4	0.001 94	0.040 99	0.000 72	807	45	836	18	847	11	812	14
SL-5-22	60.33	68.18	0.88	0.84	223	9 433	37.4	0.31	0.067 52	0.002 48	1.306 80	0.041 81	0.140 37	0.001 94	0.042 37	0.000 76	854	43	849	18	847	11	839	15
SL-5-24	34.01	43.78	0.78	0.55	160	8 983	34.1	0.27	0.071 44	0.003 77	1.371 59	0.067 28	0.139 24	0.002 56	0.041 95	0.001 39	970	70	877	29	840	14	831	27
SL-5-25	63.27	65.09	0.97	0.69	241	9 821	28.6	0.27	0.065 27	0.004 78	1.248 57	0.087 65	0.138 74	0.003 27	0.038 17	0.001 51	783	108	823	40	838	19	757	29
SL-5-26	38.12	66.98	0.57	0.79	144	10 522	55.0	0.46	0.067 13	0.002 49	1.290 57	0.041 69	0.139 42	0.001 94	0.040 17	0.000 89	842	44	842	18	841	11	796	17
SL-5-27	46.21	52.55	0.88	0.75	198	8 350	37.8	0.26	0.066 10	0.002 51	1.279 33	0.042 70	0.140 38	0.001 98	0.040 78	0.000 77	810	46	837	19	847	11	808	15
SL-5-28	45.13	46.97	0.96	0.70	227	8 011	31.0	0.21	0.065 31	0.003 75	1.240 31	0.066 88	0.137 73	0.002 62	0.039 60	0.001 07	784	81	819	30	832	15	785	21
SL-5-30	40.45	63.25	0.64	0.73	150	10 044	48.5	0.42	0.070 42	0.002 97	1.349 10	0.053 96	0.138 95	0.001 87	0.042 13	0.000 47	941	89	867	23	839	11	834	9
SL-5-32	48.82	65.93	0.74	0.83	169	8 905	48.8	0.39	0.068 54	0.002 33	1.317 55	0.037 96	0.139 41	0.001 85	0.040 71	0.000 73	885	38	853	17	841	10	807	14
SL-5-36	28.44	48.43	0.59	0.57	125	10 627	45.5	0.39	0.065 43	0.002 58	1.258 26	0.043 87	0.139 47	0.002 01	0.039 20	0.000 90	788	49	827	20	842	11	777	18
Diorite SL-1 (N 27°7′44″, E 101°49′27″)																								
SL-1-01	56.12	64.10	0.88	0.73	299	8 427	24.3	0.21	0.071 50	0.003 28	1.359 57	0.056 80	0.137 85	0.002 26	0.040 88	0.001 00	972	58	872	24	832	13	810	19
SL-1-02	49.17	51.66	0.95	0.62	263	8 990	23.7	0.20	0.060 98	0.005 02	1.150 04	0.091 38	0.136 73	0.003 47	0.039 14	0.001 46	639	128	777	43	826	20	776	28
SL-1-03	46.15	58.11	0.79	0.81	197	7 931	41.4	0.30	0.065 96	0.002 20	1.258 70	0.035 18	0.138 35	0.001 76	0.041 18	0.000 68	805	37	827	16	835	10	816	13
SL-1-04	22.27	35.85	0.62	0.57	136	8 815	41.9	0.26	0.065 93	0.002 64	1.251 00	0.044 36	0.137 56	0.001 96	0.039 57	0.000 90	804	50	824	20	831	11	784	17
SL-1-05	17.49	25.91	0.68	0.48	123	7 822	38.7	0.21	0.066 91	0.003 09	1.286 85	0.054 23	0.139 43	0.002 23	0.042 73	0.001 10	835	61	840	24	841	13	846	21
SL-1-06	33.29	46.41	0.72	0.66	159	8 179	41.3	0.29	0.070 15	0.002 60	1.339 57	0.043 10	0.138 45	0.001 93	0.043 11	0.000 87	933	43	863	19	836	11	853	17
SL-1-07	81.21	94.23	0.86	0.54	248	8 898	21.9	0.38	0.066 47	0.001 89	1.268 80	0.028 09	0.138 39	0.001 62	0.039 37	0.000 54	821	27	832	13	836	9	780	11
SL-1-08	38.24	50.66	0.75	0.71	185	8 184	38.3	0.27	0.066 77	0.002 50	1.284 67	0.041 97	0.139 48	0.001 94	0.040 39	0.000 81	831	45	839	19	842	11	800	16
SL-1-09	56.28	66.33	0.85	0.78	249	8 528	31.3	0.27	0.068 12	0.002 35	1.299 60	0.038 02	0.138 31	0.001 83	0.039 90	0.000 67	872	39	846	17	835	10	791	13
SL-1-10	19.70	34.48	0.57	0.51	118	8 337	43.0	0.29	0.070 62	0.002 71	1.363 82	0.045 99	0.140 01	0.002 01	0.040 69	0.000 95	946	46	874	20	845	11	806	18
SL-1-11	45.98	63.47	0.72	0.88	294	7 814	30.0	0.22	0.065 21	0.002 49	1.255 08	0.042 09	0.139 55	0.001 96	0.042 45	0.000 83	781	47	826	19	842	11	840	16
SL-1-12	31.20	50.36	0.62	0.64	169	9 077	38.2	0.30	0.068 18	0.002 01	1.310 00	0.030 69	0.139 31	0.001 68	0.041 63	0.000 65	874	29	850	13	841	10	824	13

Continued

Analysis	Content/ppm								Ratios								Age/Ma							
	Th	U	Th/U	Nb	Yb	Hf	$\frac{10\,000}{Nb/Yb}$	U/Yb	$\frac{^{207}Pb}{^{206}Pb}$	1σ	$\frac{^{207}Pb}{^{235}U}$	1σ	$\frac{^{206}Pb}{^{238}U}$	1σ	$\frac{^{208}Pb}{^{232}Th}$	1σ	$\frac{^{207}Pb}{^{206}Pb}$	1σ	$\frac{^{207}Pb}{^{235}U}$	1σ	$\frac{^{206}Pb}{^{238}U}$	1σ	$\frac{^{208}Pb}{^{232}Th}$	1σ
SL-1-13	20.64	47.68	0.43	0.51	168	8 264	30.2	0.28	0.075 78	0.004 98	1.439 73	0.089 83	0.137 76	0.003 11	0.042 79	0.001 89	1 089	89	906	37	832	18	847	37
SL-1-14	18.49	32.38	0.57	0.51	129	8 475	39.4	0.25	0.063 03	0.002 78	1.209 97	0.048 50	0.139 19	0.002 13	0.048 85	0.001 17	709	59	805	22	840	12	964	23
SL-1-15	25.98	35.54	0.73	0.55	199	8 069	27.8	0.18	0.068 64	0.002 80	1.296 92	0.047 21	0.136 99	0.002 04	0.040 96	0.000 90	888	51	844	21	828	12	811	17
SL-1-16	20.66	37.68	0.55	0.58	126	8 778	46.1	0.30	0.068 39	0.002 75	1.313 71	0.047 08	0.139 29	0.002 05	0.041 91	0.001 02	880	50	852	21	841	12	830	20
SL-1-17	25.58	37.83	0.68	0.48	159	8 652	30.2	0.24	0.067 63	0.002 67	1.299 53	0.045 52	0.139 34	0.002 03	0.041 14	0.000 88	857	48	846	20	841	11	815	17
SL-1-18	23.83	37.78	0.63	0.53	183	8 092	29.1	0.21	0.067 47	0.002 88	1.304 31	0.050 19	0.140 18	0.002 15	0.042 57	0.001 06	852	54	848	22	846	12	843	21
SL-1-19	23.97	29.98	0.80	0.51	168	7 294	30.5	0.18	0.064 59	0.003 12	1.240 50	0.055 31	0.139 28	0.002 32	0.041 29	0.001 04	761	66	819	25	841	13	818	20
SL-1-20	53.38	65.20	0.82	0.77	303	7 520	25.4	0.22	0.065 96	0.002 16	1.257 88	0.034 54	0.138 29	0.001 78	0.040 93	0.000 66	805	36	827	16	835	10	811	13
SL-1-21	47.00	58.29	0.81	0.74	225	7 876	32.8	0.26	0.066 70	0.003 10	1.256 60	0.053 42	0.136 63	0.002 24	0.040 23	0.000 95	828	61	826	24	826	13	797	18
SL-1-22	29.94	39.07	0.77	0.55	227	7 651	24.1	0.17	0.065 59	0.002 76	1.252 23	0.047 41	0.138 46	0.002 11	0.041 16	0.000 86	793	54	824	21	836	12	815	17
SL-1-23	39.97	64.20	0.62	0.94	277	7 456	34.0	0.23	0.067 90	0.002 97	1.298 88	0.051 47	0.138 73	0.002 19	0.040 41	0.001 04	866	56	845	23	837	12	801	20
SL-1-24	52.89	63.70	0.83	0.85	291	7 634	29.3	0.22	0.064 10	0.002 52	1.222 33	0.042 68	0.138 30	0.002 00	0.042 50	0.000 86	745	49	811	20	835	11	841	17
SL-1-25	62.08	67.81	0.92	0.65	276	8 402	23.4	0.25	0.064 52	0.002 04	1.235 10	0.032 38	0.138 86	0.001 75	0.040 51	0.000 61	759	34	817	15	838	10	803	12
SL-1-26	32.04	44.16	0.73	0.60	189	8 280	31.6	0.23	0.066 58	0.002 53	1.261 36	0.042 30	0.137 43	0.001 95	0.040 71	0.000 82	825	46	829	19	830	11	807	16
SL-1-27	18.33	31.88	0.57	0.49	116	8 269	42.1	0.27	0.061 63	0.002 66	1.171 75	0.045 80	0.137 92	0.002 08	0.042 79	0.001 03	661	58	787	21	833	12	847	20
SL-1-28	19.24	33.51	0.57	0.49	106	8 623	46.5	0.32	0.065 77	0.003 57	1.260 86	0.064 14	0.139 07	0.002 54	0.042 29	0.001 41	799	76	828	29	839	14	837	27
SL-1-29	47.31	61.13	0.77	0.73	262	8 038	28.0	0.23	0.065 46	0.002 14	1.254 34	0.034 56	0.139 05	0.001 80	0.040 36	0.000 69	789	36	825	16	839	10	800	13
SL-1-30	46.38	57.71	0.80	0.78	197	8 135	39.7	0.29	0.061 47	0.002 45	1.161 83	0.041 44	0.137 15	0.001 99	0.041 34	0.000 83	656	52	783	19	829	11	819	16
SL-1-31	42.22	53.35	0.79	0.79	240	8 106	32.9	0.22	0.066 35	0.002 57	1.264 01	0.043 48	0.138 26	0.002 01	0.041 80	0.000 82	817	48	830	20	835	11	828	16
SL-1-32	47.34	57.32	0.83	0.73	195	8 122	37.4	0.29	0.066 94	0.002 65	1.283 98	0.045 31	0.139 22	0.002 06	0.043 80	0.000 88	836	49	839	20	840	12	866	17
SL-1-33	113.43	135.90	0.83	1.73	577	16 210	30.0	0.24	0.069 72	0.002 73	1.338 21	0.046 55	0.139 33	0.002 07	0.043 67	0.000 90	920	47	862	20	841	12	864	17
SL-1-34	48.46	53.85	0.90	0.65	227	7 140	28.6	0.24	0.071 06	0.004 98	1.332 30	0.090 17	0.135 98	0.002 45	0.041 19	0.000 57	959	147	860	39	822	14	816	11
SL-1-35	26.11	39.96	0.65	0.57	150	8 831	38.1	0.27	0.067 66	0.003 12	1.290 59	0.054 56	0.138 48	0.002 30	0.045 54	0.001 24	858	60	842	24	836	13	900	24
SL-1-36	33.39	37.80	0.88	0.56	210	7 070	26.5	0.18	0.059 85	0.002 80	1.151 15	0.049 56	0.139 64	0.002 24	0.042 74	0.000 97	598	65	778	23	843	13	846	19

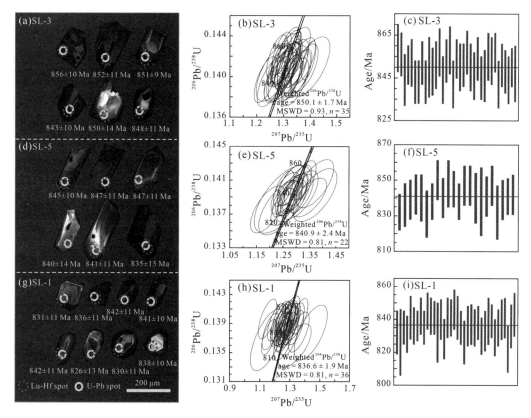

Fig. 3　Cathodoluminescence (CL) images of representative zircon grains and LA-ICP-MS zircon U-Pb Concordia diagrams for the Shuilu high-Mg$^{\#}$ diorites from the western Yangtze Block, South China.

In the chondrite-normalized rare earth element (REE) patterns (Fig. 5a), the Shuilu diorites display uniform REE patterns with moderate total REE contents (REE = 85. 8 – 120 ppm). They are enriched in light rare earth elements (LREEs) [(La/Yb)$_N$ = 6. 26 – 13. 5], relatively flat in heavy rare earth elements (HREEs), and negligible Eu anomalies (Eu/Eu* = 0. 80 – 1. 03) (Figs. 4f and 5a). In the primitive mantle-normalized trace-element diagram (Fig. 5b), the samples show enrichment of large ion lithophile elements (LILEs) (e.g., Rb, Ba, K, and Sr) and the depletion of high field strength elements (HFSEs) (e.g., Nb, Ta, Zr, and Hf). In addition, these diorites have moderate Sr (470 – 606 ppm) and Y (16. 2 – 20. 5 ppm) concentrations, yielding relatively low Sr/Y (26. 6 – 32. 3) ratios compared with typical adakites (Defant and Drummond, 1990) (Fig. 4e). They also contain high concentrations of compatible elements (e.g., Cr = 60. 2 – 107 ppm, Ni = 27. 0 – 47. 6 ppm).

4. 3　Whole-rock Sr-Nd isotopes

The whole-rock Sr-Nd isotopic data of the Neoproterozoic high-Mg$^{\#}$ diorites in this study are displayed in Table 3 and Fig. 6. All the initial $^{87}Sr/^{86}Sr$ isotopic [($^{87}Sr/^{86}Sr$)$_i$] ratios and

Table 2 Major(wt%) and trace(ppm) elements analysis data for Neoproterozoic Shuilu high-Mg# diorites in the western Yangtze Block, South China.

Sample	SL-1-1	SL-1-2	SL-1-3	SL-1-4	SL-1-5	SL-1-6	SL-3-1	SL-3-2	SL-4-1	SL-4-2	SL-5-1	SL-5-2	SL-5-3	SL-5-4	SL-5-5
			N27°7'44"				N 27°6'44"		N 27°5'35"				N 27°6'13"		
			E 101°49'27"				E 101°49'27"		E 101°48'39"				E 101°48'57"		
SiO_2	59.33	58.45	57.08	57.76	57.69	57.32	61.12	61.05	59.23	59.65	60.56	60.48	59.29	59.14	59.46
TiO_2	0.65	0.6	0.71	0.68	0.67	0.72	0.68	0.67	0.78	0.77	0.62	0.62	0.69	0.69	0.68
Al_2O_3	15.36	16.45	16.58	16.23	16.34	16.11	16.06	16.16	16.81	16.29	15.76	15.71	16.33	16.25	16.32
$Fe_2O_3{}^T$	6.23	6.02	6.84	6.86	6.86	7.03	6.1	6.15	6.69	6.67	6.05	6.04	6.49	6.51	6.41
MnO	0.12	0.1	0.12	0.13	0.12	0.13	0.09	0.1	0.11	0.11	0.1	0.1	0.11	0.11	0.11
MgO	3.69	3.36	3.89	4.1	4.12	4.3	3.38	3.42	3.58	3.8	3.92	3.87	3.81	3.92	3.83
CaO	5.94	5.74	5.82	5.41	5.44	5.44	4.86	4.9	4.98	4.35	5.31	5.3	5.7	5.74	5.62
Na_2O	4.57	4.55	4.46	4.14	4.06	3.87	4.16	4.18	4.34	4.06	4.18	4.21	4.18	4.14	4.12
K_2O	1.16	1.93	1.57	2.2	2.18	2.07	1.76	1.77	1.22	1.43	1.92	1.77	1.85	1.74	1.9
P_2O_5	0.16	0.18	0.18	0.18	0.18	0.17	0.16	0.16	0.19	0.18	0.16	0.16	0.18	0.18	0.18
LOI	2.62	2.55	3	1.88	2.2	2.63	1.23	1.31	2.14	2.44	1.03	1.37	1.27	1.37	1.16
Total	99.83	99.93	100.25	99.57	99.86	99.79	99.6	99.87	100.07	99.75	99.61	99.63	99.9	99.79	99.79
Na_2O/K_2O	3.94	2.36	2.84	1.88	1.86	1.87	2.36	2.36	3.56	2.84	2.18	2.38	2.26	2.38	2.17
Mg#	58	56.5	57	58.2	58.3	58.8	56.4	56.4	55.5	57	60.2	59.9	57.8	58.4	58.2
A/CNK	0.78	0.82	0.85	0.85	0.86	0.87	0.91	0.91	0.96	1.01	0.85	0.85	0.85	0.85	0.86
Li	19.3	18.9	26.6	19.6	21	23.1	14	14.3	25.9	28.9	10.7	11.7	14.9	15.6	14.5
Be	1.01	1.14	1.15	1.21	1.22	1.22	1.46	1.43	1.39	1.4	1.47	1.44	1.42	1.41	1.4
Sc	17.8	14.5	16.7	18.3	18.6	19.9	15.4	15.3	16.4	16.4	16.7	16.4	16.7	16.9	16.4
V	135	131	144	144	146	156	136	133	145	137	122	126	131	134	130

Continued

Sample	SL-1-1	SL-1-2	SL-1-3	SL-1-4	SL-1-5	SL-1-6	SL-3-1	SL-3-2	SL-4-1	SL-4-2	SL-5-1	SL-5-2	SL-5-3	SL-5-4	SL-5-5
			N27°7'44" E 101°49'27"				N 27°6'44" E 101°49'27"		N 27°5'35" E 101°48'39"		N 27°6'13" E 101°48'57"				
Cr	74.2	60.2	72.7	78.5	79.7	86.8	71.1	71.9	63	79.4	105	107	89.8	103	88.2
Co	73.1	72.4	66.9	45.2	46	72.3	131	124	93.9	89.4	106	116	96.7	97.9	96
Ni	35.9	29.8	34.2	34.4	35.1	38.2	27	27.3	26.2	28.6	47.6	46.2	38.5	43.8	38.4
Cu	16.1	37.6	42.2	20.7	21.3	18.8	20.2	19.7	10.9	12.8	5.56	3.79	9.19	8.5	9.29
Zn	67.8	64.8	75.7	77.2	78.6	80.1	72.1	71.5	74.3	81	70.2	70.1	77.5	78.2	76.6
Ga	16.6	18.1	18.7	17.9	18	18.1	19.4	19.4	19.8	18.8	18.7	18.6	19.2	19.4	19.3
Ge	1.22	1.23	1.31	1.2	1.19	1.24	1.14	1.14	1.16	1.11	1.17	1.15	1.17	1.18	1.17
Rb	25.1	39.6	35.5	46.5	46.3	43.5	46.4	46.5	36	39.8	45.6	39.8	47.8	45	50.7
Sr	500	508	542	474	470	541	578	577	606	550	545	548	581	582	588
Y	18.3	16.2	17.9	19.6	19.8	20.5	18	17.8	18.8	18.9	20.5	20.5	19.7	19.8	19.7
Zr	100	125	154	134	99.3	122	136	155	119	113	167	142	154	155	146
Nb	3.32	3.24	3.85	3.57	3.73	3.9	4.3	4.28	4.39	4.22	4.3	4.38	4.29	4.41	4.31
Cs	1.07	1.34	1.49	1.7	1.71	1.65	2.48	2.46	3.68	4.27	2.21	1.8	1.88	1.6	2.06
Ba	476	919	792	1 000	1 000	918	671	664	441	599	689	643	608	595	622
La	15.3	25.8	19.3	25.7	18.6	37.5	20	20.2	16.8	17.5	18.8	20.9	20.2	19.7	20.1
Ce	32.5	45.3	38.7	49.1	37.3	62.8	41.3	42.1	35.7	37.8	40.8	44.4	42.4	41.6	42.3
Pr	3.97	4.78	4.51	5.48	4.45	6.33	4.91	4.97	4.49	4.65	5.04	5.39	5.1	5.06	5.1
Nd	17.1	18.4	18.3	21.6	18.6	23.5	20	20.2	19.5	19.9	21.6	22.5	21.5	21.3	21.3
Sm	3.81	3.55	3.74	4.23	4.05	4.5	4.19	4.17	4.26	4.29	4.64	4.66	4.51	4.53	4.52
Eu	1.08	1.1	1.21	1.22	1.2	1.24	1.13	1.14	1.19	1.2	1.17	1.15	1.22	1.24	1.22

Continued

Sample	SL-1-1	SL-1-2	SL-1-3	SL-1-4	SL-1-5	SL-1-6	SL-3-1	SL-3-2	SL-4-1	SL-4-2	SL-5-1	SL-5-2	SL-5-3	SL-5-4	SL-5-5
			N27°7'44" E 101°49'27"				N 27°6'44" E 101°49'27"		N 27°5'35" E 101°48'39"		N 27°6'13" E 101°48'57"				
Gd	3.57	3.29	3.48	3.94	3.82	4.22	3.72	3.72	3.97	3.98	4.23	4.22	4.16	4.19	4.17
Tb	0.54	0.48	0.52	0.58	0.58	0.61	0.55	0.54	0.58	0.59	0.62	0.62	0.61	0.62	0.61
Dy	3.27	2.8	3.09	3.42	3.44	3.62	3.18	3.19	3.4	3.4	3.64	3.64	3.55	3.61	3.6
Ho	0.65	0.56	0.62	0.68	0.68	0.72	0.63	0.63	0.66	0.66	0.72	0.72	0.69	0.7	0.69
Er	1.84	1.61	1.78	1.97	1.98	2.07	1.79	1.78	1.85	1.84	2.03	2.05	1.94	1.97	1.95
Tm	0.27	0.24	0.27	0.29	0.29	0.31	0.27	0.26	0.27	0.27	0.3	0.3	0.28	0.29	0.28
Yb	1.75	1.59	1.77	1.92	1.92	1.99	1.71	1.7	1.68	1.69	1.91	1.92	1.79	1.81	1.79
Lu	0.27	0.25	0.27	0.29	0.3	0.3	0.26	0.26	0.26	0.25	0.29	0.28	0.27	0.27	0.27
Hf	2.74	3.3	3.9	3.6	2.82	3.45	3.72	4.23	3.23	3.1	4.5	3.91	4.16	4.24	4.02
Ta	0.3	0.29	0.32	0.27	0.3	0.35	0.42	0.41	0.39	0.38	0.39	0.39	0.4	0.4	0.4
Pb	9.08	8.24	8.23	7.18	7.01	10.1	12.2	12.1	8.81	9.24	12.7	11.6	11.9	11.5	12.5
Th	4.55	10.2	6.37	8.03	6.81	17.1	6.55	6.55	3.83	4.16	6.15	3.43	6.32	5.95	6.61
U	0.65	0.62	0.76	0.68	0.63	0.97	1.38	1.39	0.87	1.41	0.8	0.52	1.56	1.53	1.62
Sr/Y	27.31	31.34	30.26	24.25	23.75	26.37	32.18	32.34	32.14	29.1	26.6	26.7	29.42	29.38	29.8
Nb/Ta	11.2	11	12	13	12.4	11	10.2	10.5	11.4	11.1	11.1	11.2	10.8	11.1	10.8
Eu/Eu*	0.89	0.98	1.03	0.91	0.93	0.87	0.88	0.88	0.89	0.89	0.81	0.8	0.86	0.87	0.86
$(La/Yb)_N$	6.26	11.62	7.83	9.59	6.96	13.49	8.41	8.56	7.15	7.42	7.08	7.81	8.09	7.81	8.03
REE	85.84	109.67	97.61	120.36	97.28	149.73	103.58	104.91	94.58	98.08	105.79	112.75	108.14	106.96	107.91

$Mg^{\#}$ = molar $100Mg/(Mg+Fe)$; A/CNK = $Al_2O_3/(CaO+K_2O+Na_2O)$ molar ratio; $Eu/Eu^* = Eu_N/(Sm_N \cdot Gd_N)^{1/2}$.

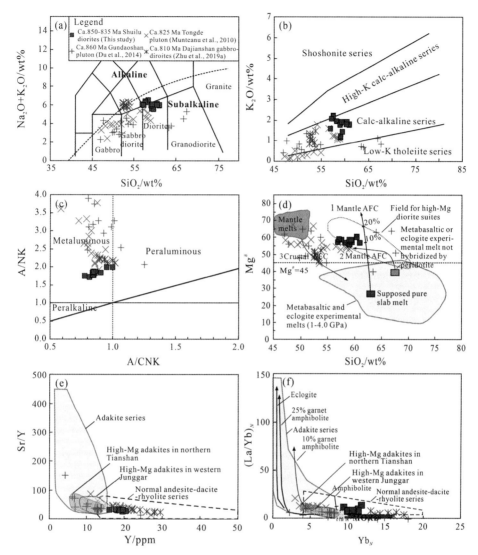

Fig. 4　Plots of total alkali vs. SiO₂(TAS) (a) (Middlemost, 1994), SiO₂ vs. K₂O (b) (Roberts and Clemens, 1993), A/CNK vs. A/NK (c) (Frost et al., 2001), SiO₂ vs. Mg#(d), Y vs. Sr/Y (e), and Yb$_N$ vs. (La/Yb)$_N$(f) (Defant and Drummond, 1990) diagrams for the Shuilu high-Mg# diorites along the western Yangtze Block, South China.

In Fig. 4d, the fields of metabasaltic and eclogite experimental melts (1.0-4.0 GPa) are from Rapp et al. (1999) and references therein. The mantle AFC curves are after Stern and Kilian (1996) (curve 1) and Rapp et al. (1999) (curve 2). The crustal AFC is after Stern and Kilian (1996). The supposed pure slab melt is from Stern and Kilian (1996). The metabasaltic or eclogite experimental melt not hybridized by peridotite is from Rapp et al. (1999). The field of high-Mg diorite suites is after Smithies and Champion (2000). In Fig. 4e, f, the slab-derived high-Mg diorites in northern Tianshan and western Junggar are from Wang et al. (2007) and Tang et al. (2010). The Neoproterozoic metasomatized mantle-derived magmatism, including the ca. 860 Ma Guandaoshan pluton (Du et al., 2014), ca. 825 Ma Tongde pluton (Munteanu et al., 2010), and ca. 810 Ma Dajianshan gabbro-diorites (Zhu et al., 2019a), are shown for comparison.

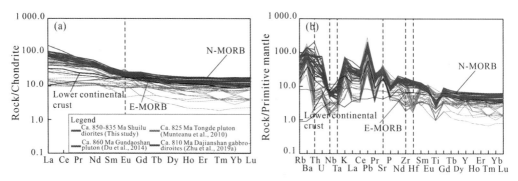

Fig. 5   Diagrams of chondrite-normalized REE patterns (a) and primitive mantle-normalized trace-element patterns (b) for the Shuilu high-Mg# diorites in the western Yangtze Block, South China.

The primitive mantle and chondrite values are from Sun and McDonough (1989).

The reference values of lower crust are from Rudnick and Gao (2003).

$\varepsilon_{Nd}(t)$ values are calculated based on the crystallization ages as show in Table 3. Eight high-Mg# diorite samples from Shuilu area yield relatively constant $(^{87}Sr/^{86}Sr)_i$ ratios ranging from 0. 703 406 to 0. 704 157 and positive $\varepsilon_{Nd}(t)$ values from +3. 26 to +4. 26 (Fig. 6).

4. 4   Zircon Lu-Hf isotopic compositions

The in-situ zircon Lu-Hf isotopic analysis results for the three dated samples are presented in Table 4, and Figs. 7 and 8. The initial $^{176}Hf/^{177}Hf$ ratios and $\varepsilon_{Hf}(t)$ values were calculated according to their crystallization ages. Zircons from diorite sample SL-3 yield positive zircon $\varepsilon_{Hf}(t)$ values of +8. 43 to +13. 4 [average $\varepsilon_{Hf}(t)$ = +10. 8, n = 15] (Figs. 7 and 8a), single-stage model ages of 812−1 069 Ma, and two-stage model ages of 885−1 202 Ma. The 32 analyzed spots of sample SL-5 display depleted zircon Hf compositions with $\varepsilon_{Hf}(t)$ values of +10. 2 to +13. 5 [average $\varepsilon_{Hf}(t)$ = +11. 7, n = 22] (Figs. 7 and 8b). They yield relatively young single-stage model ages ranging from 862 Ma to 993 Ma and two-stage model ages from 876 Ma to 1 085 Ma. Zircons of sample SL-1 have positive zircon $\varepsilon_{Hf}(t)$ values of +9. 41 to +13. 6 [average $\varepsilon_{Hf}(t)$ = +11. 8, n = 25] (Figs. 7 and 8c). The 25 analyzed spots possess single-stage model ages of 854−1 018 Ma and two-stage model ages of 866−1 128 Ma.

5   Discussion

5. 1   Petrogenesis of Shuilu high-Mg# diorites

5. 1. 1   Insignificant crustal contamination

The Shuilu high-Mg# diorites in this study show enrichment of LREEs, LILEs (e.g., Rb, Ba, Sr, and K), and Pb, and depletion of Nb, Ta, and Ti (Fig. 5d), indicating enriched "crust-like" features (Wilson, 1989). There are two main mechanisms to account for this

Table 3 Whole-rock Sr-Nd isotopic data for Neoproterozoic Shuilu high-Mg# diorites in the western Yangtze Block, South China.

Sample	$^{87}Sr/^{86}Sr$	$2SE$	Sr/ppm	Rb/ppm	$^{143}Nd/^{144}Nd$	$2SE$	Nd/ppm	Sm/ppm	Age/Ma	T_{DM2}/Ga	$\varepsilon_{Nd}(t)$	$I_{Sr}[(^{87}Sr/^{86}Sr)_i]$
SL-1-1	0.705 888	0.000 003	500	25.1	0.512 512	0.000 003	17.1	3.81	835	1.11	4.19	0.704 157
SL-1-3	0.706 053	0.000 003	542	35.5	0.512 454	0.000 003	18.3	3.74	835	1.11	4.26	0.703 796
SL-3-1	0.706 226	0.000 005	578	46.4	0.512 439	0.000 003	20.0	4.19	850	1.16	3.73	0.703 406
SL-3-2	0.706 380	0.000 003	577	46.5	0.512 427	0.000 002	20.2	4.17	850	1.16	3.72	0.703 549
SL-4-1	0.705 798	0.000 009	606	36	0.512 476	0.000 003	19.5	4.26	850	1.15	3.87	0.703 711
SL-4-2	0.706 164	0.000 005	550	39.8	0.512 435	0.000 006	19.9	4.29	850	1.19	3.26	0.703 622
SL-5-2	0.705 942	0.000 004	548	39.8	0.512 462	0.000 003	22.5	4.66	840	1.11	4.26	0.703 421
SL-5-3	0.706 358	0.000 004	581	47.8	0.512 470	0.000 003	21.5	4.51	840	1.11	4.24	0.703 502

^{87}Rb/^{86}Sr and ^{147}Sm/^{144}Nd ratios were calculated using Rb, Sr, Sm and Nd contents analyzed by ICP-MS.

T_{DM2} represent the two-stage model age and were calculated using present-day $(^{147}Sm/^{144}Nd)_{DM} = 0.213\ 7$, $(^{147}Sm/^{144}Nd)_{DM} = 0.513\ 15$ and $(^{147}Sm/^{144}Nd)_{crust} = 0.101\ 2$.

$\varepsilon_{Nd}(t)$ values were calculated using present-day $(^{147}Sm/^{144}Nd)_{CHUR} = 0.196\ 7$ and $(^{147}Sm/^{144}Nd)_{CHUR} = 0.512\ 638$.

$\varepsilon_{Nd}(t) = [(^{143}Nd/^{144}Nd)_S(t)/(^{143}Nd/^{144}Nd)_{CHUR}(t) - 1] \times 10^4$.

$T_{DM2} = \dfrac{1}{\lambda}\left\{1 + \left[(^{143}Nd/^{144}Nd)_S - (^{147}Sm/^{144}Nd)_S(e^{\lambda t}-1) - (^{143}Nd/^{144}Nd)_{DM}\right] / \left[(^{147}Sm/^{144}Nd)_{crust} - (^{147}Sm/^{144}Nd)_{DM}\right]\right\}$.

Table 4 Zircon Lu-Hf isotopic data for Neoproterozoic Shuilu high-Mg# diorites in the western Yangtze Block, South China.

Grain spot	Age/Ma	$^{176}Yb/^{177}Hf$	$2SE$	$^{176}Lu/^{177}Hf$	$2SE$	$^{176}Hf/^{177}Hf$	$2SE$	$(^{176}Hf/^{177}Hf)_i$	$f_{Lu/Hf}$	$\varepsilon_{Hf}(t)$	$2SE$	T_{DM1}/Ma	T_{DMC}/Ma
Diorite (SL-3)													
SL-3-1	850	0.019 315	0.000 093	0.000 626	0.000 004	0.282 525	0.000 022	0.282 514	-0.98	9.32	0.79	1 034	1 146
SL-3-2	850	0.025 314	0.000 021	0.000 821	0.000 001	0.282 606	0.000 025	0.282 592	-0.98	11.97	0.90	930	978
SL-3-3	850	0.021 308	0.000 119	0.000 682	0.000 003	0.282 505	0.000 026	0.282 494	-0.98	8.57	0.91	1 064	1 193
SL-3-4	850	0.011 411	0.000 020	0.000 393	0.000 001	0.282 545	0.000 024	0.282 539	-0.99	10.30	0.84	994	1 084
SL-3-5	850	0.011 018	0.000 182	0.000 372	0.000 006	0.282 547	0.000 022	0.282 541	-0.99	10.39	0.78	991	1 078
SL-3-6	850	0.016 133	0.000 062	0.000 537	0.000 002	0.282 538	0.000 021	0.282 530	-0.98	9.90	0.75	1 011	1 109
SL-3-7	850	0.019 714	0.000 046	0.000 640	0.000 001	0.282 500	0.000 026	0.282 490	-0.98	8.43	0.92	1 069	1 202
SL-3-8	850	0.021 671	0.000 140	0.000 701	0.000 004	0.282 625	0.000 029	0.282 614	-0.98	12.81	1.02	897	925

Continued

Grain spot	Age/Ma	$^{176}Yb/^{177}Hf$	2SE	$^{176}Lu/^{177}Hf$	2SE	$^{176}Hf/^{177}Hf$	2SE	$(^{176}Hf/^{177}Hf)_i$	$f_{Lu/Hf}$	$\varepsilon_{Hf}(t)$	2SE	T_{DM1}/Ma	T_{DMC}/Ma
SL-3-9	850	0.015 939	0.000 125	0.000 535	0.000 004	0.282 594	0.000 065	0.282 585	−0.98	11.88	2.30	933	984
SL-3-10	850	0.015 717	0.000 075	0.000 523	0.000 002	0.282 559	0.000 027	0.282 551	−0.98	10.66	0.95	981	1 061
SL-3-11	850	0.014 838	0.000 046	0.000 500	0.000 001	0.282 579	0.000 025	0.282 571	−0.98	11.40	0.89	952	1 014
SL-3-12	850	0.020 885	0.000 049	0.000 697	0.000 002	0.282 643	0.000 029	0.282 632	−0.98	13.44	1.04	872	885
SL-3-13	850	0.020 989	0.000 307	0.000 684	0.000 009	0.282 534	0.000 028	0.282 523	−0.98	9.59	1.00	1 024	1 128
SL-3-14	850	0.021 489	0.000 312	0.000 700	0.000 008	0.282 639	0.000 042	0.282 628	−0.98	13.30	1.48	877	894
SL-3-15	850	0.012 857	0.000 021	0.000 455	0.000 001	0.282 538	0.000 021	0.282 531	−0.99	10.00	0.74	1 006	1 102
Diorite (SL-5)													
SL-5-01	840	0.019 466	0.000 113	0.000 644	0.000 004	0.282 569	0.000 012	0.282 559	−0.98	10.67	0.44	972	1 053
SL-5-02	840	0.015 142	0.000 068	0.000 487	0.000 002	0.282 621	0.000 013	0.282 613	−0.99	12.66	0.46	894	927
SL-5-03	840	0.025 253	0.000 057	0.000 810	0.000 003	0.282 610	0.000 014	0.282 597	−0.98	11.92	0.50	923	973
SL-5-04	840	0.024 001	0.000 086	0.000 780	0.000 003	0.282 615	0.000 014	0.282 602	−0.98	12.13	0.50	915	960
SL-5-05	840	0.023 268	0.000 184	0.000 755	0.000 007	0.282 640	0.000 014	0.282 628	−0.98	13.05	0.51	879	902
SL-5-06	840	0.015 847	0.000 189	0.000 520	0.000 006	0.282 580	0.000 014	0.282 571	−0.98	11.18	0.49	952	1 020
SL-5-07	840	0.026 596	0.000 082	0.000 838	0.000 003	0.282 561	0.000 016	0.282 548	−0.97	10.15	0.57	993	1 085
SL-5-08	840	0.026 172	0.000 039	0.000 802	0.000 002	0.282 585	0.000 013	0.282 572	−0.98	11.04	0.47	958	1 029
SL-5-09	840	0.021 064	0.000 023	0.000 679	0.000 001	0.282 649	0.000 012	0.282 638	−0.98	13.46	0.41	862	876
SL-5-10	840	0.023 626	0.000 197	0.000 748	0.000 005	0.282 617	0.000 016	0.282 605	−0.98	12.23	0.56	911	954
SL-5-11	840	0.014 516	0.000 101	0.000 468	0.000 003	0.282 590	0.000 013	0.282 583	−0.99	11.61	0.47	935	993
SL-5-12	840	0.018 885	0.000 080	0.000 608	0.000 003	0.282 575	0.000 015	0.282 566	−0.98	10.92	0.53	962	1 037
SL-5-13	840	0.018 453	0.000 118	0.000 603	0.000 003	0.282 555	0.000 015	0.282 546	−0.98	10.22	0.52	990	1 081
SL-5-14	840	0.027 168	0.000 153	0.000 834	0.000 004	0.282 601	0.000 014	0.282 588	−0.97	11.57	0.51	937	995
SL-5-15	840	0.035 074	0.000 394	0.001 118	0.000 015	0.282 607	0.000 016	0.282 590	−0.97	11.48	0.55	942	1 001
SL-5-16	840	0.018 180	0.000 072	0.000 588	0.000 002	0.282 589	0.000 013	0.282 580	−0.98	11.44	0.44	942	1 004
SL-5-17	840	0.021 544	0.000 060	0.000 684	0.000 002	0.282 614	0.000 013	0.282 603	−0.98	12.22	0.45	911	955
SL-5-18	840	0.024 875	0.000 202	0.000 783	0.000 006	0.282 628	0.000 013	0.282 616	−0.98	12.59	0.47	897	931
SL-5-19	840	0.024 247	0.000 061	0.000 782	0.000 001	0.282 623	0.000 014	0.282 611	−0.98	12.43	0.48	903	941

Continued

Grain spot	Age/Ma	^{176}Yb/^{177}Hf	2SE	^{176}Lu/^{177}Hf	2SE	^{176}Hf/^{177}Hf	2SE	$(^{176}$Hf/^{177}Hf$)_i$	$f_{Lu/Hf}$	$\varepsilon_{Hf}(t)$	2SE	T_{DM1}/Ma	T_{DMC}/Ma
SL-5-20	840	0. 029 005	0. 000 064	0. 000 904	0. 000 001	0. 282 609	0. 000 013	0. 282 595	−0. 97	11. 78	0. 47	929	982
SL-5-21	840	0. 020 733	0. 000 447	0. 000 652	0. 000 014	0. 282 627	0. 000 013	0. 282 617	−0. 98	12. 71	0. 46	892	923
SL-5-22	840	0. 019 439	0. 000 245	0. 000 628	0. 000 008	0. 282 556	0. 000 013	0. 282 546	−0. 98	10. 21	0. 45	990	1 082
Diorite (SL-1)													
SL-1-01	835	0. 024 767	0. 000 031	0. 000 802	0. 000 001	0. 282 621	0. 000 017	0. 282 609	−0. 98	12. 22	0. 58	907	950
SL-1-02	835	0. 022 973	0. 000 044	0. 000 719	0. 000 002	0. 282 627	0. 000 021	0. 282 616	−0. 98	12. 52	0. 73	895	932
SL-1-03	835	0. 019 536	0. 000 032	0. 000 619	0. 000 002	0. 282 638	0. 000 019	0. 282 628	−0. 98	13. 01	0. 68	876	900
SL-1-04	835	0. 015 560	0. 000 032	0. 000 509	0. 000 001	0. 282 598	0. 000 017	0. 282 590	−0. 98	11. 73	0. 60	926	982
SL-1-05	835	0. 018 717	0. 000 209	0. 000 671	0. 000 006	0. 282 588	0. 000 013	0. 282 577	−0. 98	11. 20	0. 45	947	1 015
SL-1-06	835	0. 012 031	0. 000 040	0. 000 416	0. 000 001	0. 282 633	0. 000 011	0. 282 626	−0. 99	13. 06	0. 40	874	898
SL-1-07	835	0. 018 405	0. 000 223	0. 000 593	0. 000 007	0. 282 608	0. 000 013	0. 282 599	−0. 98	12. 00	0. 47	916	965
SL-1-08	835	0. 015 443	0. 000 032	0. 000 508	0. 000 001	0. 282 631	0. 000 013	0. 282 623	−0. 98	12. 89	0. 46	881	908
SL-1-09	835	0. 026 403	0. 000 147	0. 000 854	0. 000 005	0. 282 591	0. 000 013	0. 282 577	−0. 97	11. 09	0. 46	952	1 022
SL-1-10	835	0. 017 710	0. 000 045	0. 000 588	0. 000 001	0. 282 602	0. 000 013	0. 282 593	−0. 98	11. 79	0. 45	924	978
SL-1-11	835	0. 019 812	0. 000 177	0. 000 634	0. 000 006	0. 282 597	0. 000 013	0. 282 587	−0. 98	11. 55	0. 47	933	993
SL-1-12	835	0. 022 737	0. 000 019	0. 000 709	0. 000 002	0. 282 587	0. 000 015	0. 282 576	−0. 98	11. 12	0. 53	950	1 020
SL-1-13	835	0. 013 819	0. 000 048	0. 000 445	0. 000 001	0. 282 566	0. 000 014	0. 282 559	−0. 99	10. 67	0. 48	967	1 049
SL-1-14	835	0. 017 272	0. 000 172	0. 000 577	0. 000 005	0. 282 640	0. 000 013	0. 282 631	−0. 98	13. 15	0. 45	870	892
SL-1-15	835	0. 017 898	0. 000 110	0. 000 590	0. 000 004	0. 282 596	0. 000 014	0. 282 587	−0. 98	11. 56	0. 49	933	992
SL-1-16	835	0. 015 304	0. 000 039	0. 000 503	0. 000 001	0. 282 650	0. 000 014	0. 282 642	−0. 98	13. 56	0. 49	854	866
SL-1-17	835	0. 023 659	0. 000 159	0. 000 765	0. 000 006	0. 282 598	0. 000 014	0. 282 586	−0. 98	11. 44	0. 50	938	1 000
SL-1-18	835	0. 022 996	0. 000 018	0. 000 733	0. 000 001	0. 282 577	0. 000 013	0. 282 565	−0. 98	10. 73	0. 46	966	1 045
SL-1-19	835	0. 026 406	0. 000 237	0. 000 834	0. 000 005	0. 282 560	0. 000 015	0. 282 547	−0. 97	10. 02	0. 54	994	1 090
SL-1-20	835	0. 025 076	0. 000 045	0. 000 847	0. 000 002	0. 282 608	0. 000 015	0. 282 595	−0. 97	11. 70	0. 53	928	983
SL-1-21	835	0. 029 621	0. 000 071	0. 000 949	0. 000 003	0. 282 613	0. 000 015	0. 282 598	−0. 97	11. 76	0. 53	926	979
SL-1-22	835	0. 020 899	0. 000 099	0. 000 696	0. 000 002	0. 282 597	0. 000 012	0. 282 586	−0. 98	11. 47	0. 42	937	998
SL-1-23	835	0. 021 281	0. 000 059	0. 000 688	0. 000 003	0. 282 538	0. 000 015	0. 282 527	−0. 98	9. 41	0. 54	1 018	1 128
SL-1-24	835	0. 018 033	0. 000 203	0. 000 585	0. 000 007	0. 282 617	0. 000 013	0. 282 608	−0. 98	12. 31	0. 45	903	945
SL-1-25	835	0. 011 521	0. 000 057	0. 000 393	0. 000 001	0. 282 638	0. 000 015	0. 282 632	−0. 99	13. 27	0. 52	865	884

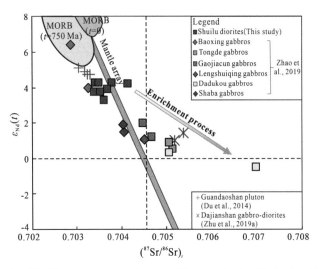

Fig. 6 Whole-rock $(^{87}Sr/^{86}Sr)_i$ vs. $\varepsilon_{Nd}(t)$ diagram (Zhao et al., 2008) for the Shuilu high-Mg$^{\#}$ diorites from the western Yangtze Block, South China.

The Sr-Nd isotopes of Neoproterozoic metasomatized mantle magmatism are from Du et al. (2014), Zhao et al. (2019), and Zhu et al. (2019a).

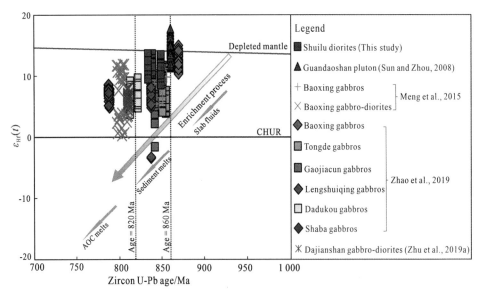

Fig. 7 Zircon U-Pb age vs. $\varepsilon_{Hf}(t)$ for the Shuilu high-Mg$^{\#}$ diorites from the western Yangtze Block, South China.

The Hf isotopes for the Neoproterozoic metasomatized mantle-derived igneous are from Sun and Zhou (2008), Meng et al. (2015), Zhao et al. (2019), and Zhu et al. (2019a).

phenomenon, including crustal contamination during late magma evolution (Chen et al., 2014a; van de Flierdt et al., 2003; Jung et al., 2002, 2009) and metasomatism of a depleted mantle source prior to parental magma generation (Chen et al., 2014a; Jung et al., 2015; Karsli et al., 2017; Zhang et al., 2019).

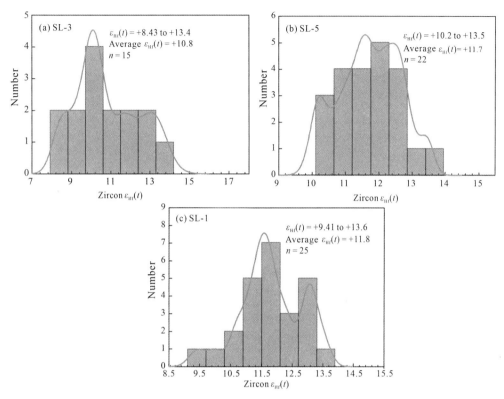

Fig. 8　Histograms of zircon $\varepsilon_{Hf}(t)$ values for the Shuilu high-Mg$^{\#}$ diorites
from the western Yangtze Block, South China.

Karsli et al. (2017) proposed that ancient zircon populations represent the obvious crustal contamination during magma evolution. However, our high-Mg$^{\#}$ diorites display completely concordant ages (Fig. 3), and zircon populations of older age were not observed during in-situ analysis. It is notable that continental crustal contamination may lead to negative Nb, Ta, and Ti anomalies as well as slightly positive Zr and Hf anomalies (Du et al., 2014; Sun and McDonough, 1989; Zhou et al., 2006a). Nevertheless, the diorite samples studied here show depletion of Nb, Ta, and Ti, but no evident enrichment of Zr and Hf, compared with adjacent Nd and Sm contents (Fig. 5b), reflecting the negligible influence of crustal contamination. Crustal contamination is usually characterized by an increase of whole-rock $(^{87}Sr/^{86}Sr)_i$ values and a decrease of $\varepsilon_{Nd}(t)$ values following magmatic evolution (Guo et al., 2016; Zhang et al., 2019). The narrow variations of $(^{87}Sr/^{86}Sr)_i$ (0.703 406−0.704 157) and $\varepsilon_{Nd}(t)$ (+3.26 to +4.26) values at given SiO$_2$ and Mg$^{\#}$ values in our high-Mg$^{\#}$ diorites therefore diminish the possibility of crustal contamination (Fig. 9). In summary, these lines of evidence argue against obvious crustal contamination en route for the genesis of the high-Mg$^{\#}$ diorites in this study. Their enriched "crust-like" characteristics may be caused by mantle metasomatism prior to the generation of the parental magma (Detailed discussion as shown in part 5.1.3).

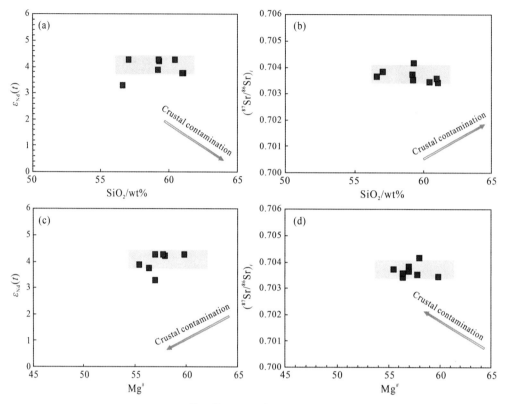

Fig. 9   The SiO_2 vs. $\varepsilon_{Nd}(t)$, SiO_2 vs. $(^{87}Sr/^{86}Sr)_i$, $Mg^{\#}$ vs. $\varepsilon_{Nd}(t)$, and $Mg^{\#}$ vs. $(^{87}Sr/^{86}Sr)_i$ diagrams

for the Shuilu high-$Mg^{\#}$ diorites from the western Yangtze Block, South China.

5. 1. 2   Magma source

The Shuilu high-$Mg^{\#}$ diorites in this study were formed at ca. 850–835 Ma. They have moderate Sr (470–606 ppm) and Y contents (16. 2–20. 5 ppm) and low Sr/Y (26. 6–32. 3) ratios, as well as low $(La/Yb)_N$ (6. 26–13. 5) and high Yb_N (9. 36–11. 7) values, suggesting that they belong to the andesite-dacite-rhyolite series (Fig. 4e,f), rather than the typical adakites that originate from subduction slab melts along high thermal gradients (Defant and Drummond, 1990). Their trace elemental characteristics are also different from those of subducting oceanic slab-derived high-Mg adakitic diorites in northwestern Tianshan and western Junggar (Wang et al., 2007; Tang et al., 2010) (Fig. 4e,f). We thus can preclude a direct subduction oceanic slab source for the Shuilu high-$Mg^{\#}$ diorites. The homogeneous whole-rock Nd isotopes [$\varepsilon_{Nd}(t)$ = +3. 26 to +4. 26] (~1ε Nd units) and zircon Hf isotopes [$\varepsilon_{Hf}(t)$ = +8. 43 to +13. 6] (~5 εHf units) (Figs. 6 and 8), together with the absence of mafic micro-granular enclaves and disequilibrium mineral pairs in these diorites (Fig. 2), indicate that magma mixing between mafic and felsic magmas is not a plausible mechanism (Karsli et al., 2017; Kemp et al., 2007). Experimental studies have demonstrated that dehydration melting of mafic lower crust usually produces melts with low MgO contents and $Mg^{\#}$ values (<45) regardless of the melting degrees (Rapp and Watson, 1995; Rapp et al., 1999). In contrast, the Shuilu diorites studied here possess distinctly higher MgO contents (3. 36−4. 30 wt%) and $Mg^{\#}$ values (56−60) than experimental melts from metabasalts and eclogites (Fig. 4d) (Patiño Douce, 1999; Rapp and Watson, 1995; Rapp et al., 1999). Therefore, they were likely derived from a mantle source rather than a juvenile mafic lower crust source. Moreover, they have high concentrations of Cr = 60. 2−107 ppm, Co = 45. 2−131 ppm, and Ni = 27. 0−47. 6 ppm, which also point to a mantle origin (Cao et al., 2018; Karsli et al., 2017). Their low Nb/La ratios (0. 10−0. 26) are close to the lithospheric mantle (0. 3−0. 4) rather than OIB-like asthenospheric mantle (>1) (Smith et al., 1999). The relatively low $(^{87}Sr/^{86}Sr)_i$ values (0. 703 406−0. 704 157) and highly positive whole-rock $\varepsilon_{Nd}(t)$ (+3. 26 to +4. 26) and zircon $\varepsilon_{Hf}(t)$ (+8. 43 to +13. 6) values are comparable to those of Neoproterozoic metasomatized mantle-derived mafic-intermediate rocks from the western Yangtze Block (Du et al., 2014; Meng et al., 2015; Sun and Zhou, 2008; Zhao et al., 2019; Zhu et al., 2019a), further supporting interpretation as a depleted lithosphere mantle source (Figs. 6 and 7) (Griffin et al., 2000; Karsli et al., 2017). In general, all the geochemical features indicate that the parental melts of the ca. 850−835 Ma Shuilu high-$Mg^{\#}$ diorites were originated from depleted lithospheric mantle.

5. 1. 3 Mantle metasomatism induced by slab fluids and sediments melts

Geochemical and isotopic evidence support the inference that the Shuilu high-$Mg^{\#}$ diorites were sourced from depleted lithospheric mantle. Nevertheless, they show distinctive "crust-like" features with enriched LREEs, LILEs, and Pb, and depleted Nb, Ta, and Ti (Fig. 5d). Considering the insignificant crustal contamination as discussed above (part 5. 1. 1), the enriched "crust-like" characteristics of these high-$Mg^{\#}$ diorites may be caused by mantle metasomatism prior to the generation of the parental magma. Numerous studies have suggested that a depleted mantle source can been modified by enriched compositions, such as LILEs-rich fluids, subducted sediment melts, and altered oceanic crust (AOC) melts, under an arc magma system (Eiler et al., 2000, 2005; Guo et al., 2015; Hanyu et al., 2002, 2006; Karsli et al., 2017; Shimoda et al., 2003; Tatsumi, 2006; Zhao et al., 2018, 2019). The Shuilu high-$Mg^{\#}$ diorites in this study display enrichment of LILEs and LREEs and depletion of HFSEs (Fig. 5), which are typically indicative of subduction-related magmas (Sun and McDonough, 1989). Experimental petrology has established that hydrous melting of upper mantle peridotite is a reasonable mechanism for high-magnesium magma production (Tatsumi, 2006). The LILE-rich aqueous fluids released from subducting lithosphere are likely a reasonable agent during the partial melting of the mantle wedge (Crawford et al., 1989; Hanyu et al., 2006). In this study, the calc-alkaline affinity (Fig. 4b), together with the existence of voluminous hydrous minerals (e. g., amphibole) in our high-Mg$^{\#}$ diorites (Fig. 2), confirms that the mantle sources were hydrous (Grove et al., 2002; Smith et al., 2009). LILEs (e.g., Rb and Ba) are mobile in aqueous fluids, whereas Y is controlled by the oceanic slab melts (Elliott et al., 1997; Kepezhinskas et al., 1997; Hanyu et al., 2006). Therefore, the incorporation of oceanic slab melts into the mantle sources could elevate the Nb/Y ratios, as slab melts are depleted in Y because of residual garnet in the slab (Defant and Drummond, 1990; Rapp and Watson, 1995), whereas fluids-related enrichment will drive the resultant melts toward high Rb/Y ratios and Ba concentrations (Kepezhinskas et al., 1997). Similar to the Golovin and Belaya arc lavas from the Kamchatka arc in Russia (Kepezhinskas et al., 1997), the Shuilu high-Mg$^{\#}$ diorites display strikingly high enrichment of Rb/Y ratios (1. 37-2. 60) and Ba contents (441-1 000 ppm) but a narrow range of Nb/Y ratios (0. 18-0. 24) (Fig. 10a,b), also suggesting that the primary mantle sources were influenced by fluid-related enrichment, not oceanic slab melt (adakite)-related enrichment (Kepezhinskas et al., 1997).

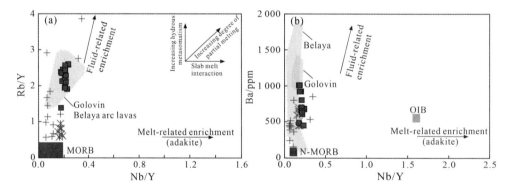

Fig. 10　Plots of Nb/Y vs. Rb/Y (a) and Nb/Y vs. Ba (b) (Kepezhinskas et al., 1997)

for the Shuilu high-Mg$^{\#}$ diorites from the western Yangtze Block, South China.

The field of Golovin and Belaya arc lavas is from Kepezhinskas et al. (1997). Symbols as Fig. 4.

Previous studies have proposed that melting of subducting sediments could occur accompanied with dehydration of subducted oceanic slab in a specific thermal regime during the subduction process (Nichols et al., 1994; Shimoda et al., 1998, 2003). It is noteworthy that the whole-rock Nd and zircon Hf isotopes of our high-Mg$^{\#}$ diorites are less radiogenic than that of depleted mantle (Figs. 6 and 7). Considering that slab-derived aqueous fluids generally possess low partition coefficients of Sm, Nd, Lu, and Hf (Bau, 1991, Hanyu et al., 2006), the relatively low whole-rock $\varepsilon_{Nd}(t)$ and zircon $\varepsilon_{Hf}(t)$ values (relative to depleted mantle) thus cannot be explained by the involvement of slab fluids alone (Chen et al., 2019; Guo et al., 2015; Zhao et al., 2019). Accordingly, subducting sediments is a suitable contributor of enriched component for the mantle wedge above the subduction zone, and will dramatically change the Hf-Nd isotopic compositions of mantle source (Guo et al., 2015; Hanyu et al., 2006; Shimoda et al., 1998, 2003; Tatsumi et al., 2006; van de Flierdt et al., 2003; Zhao et al., 2018, 2019). The Shuilu high-Mg$^#$ diorites studied here possess slightly high and positive whole-rock $\varepsilon_{Nd}(t)$ and zircon $\varepsilon_{Hf}(t)$ values, and plot above the mantle-crustal array in the $\varepsilon_{Hf}(t)$-$\varepsilon_{Nd}(t)$ diagram (Fig. 11a), indicating that subducted sediment melts might have been incorporated into the mantle sources (Guo et al., 2015; Zhao et al., 2018, 2019). Such a Hf-Nd isotopic decoupling phenomenon in our samples is widely observed in the Neoproterozoic mafic-intermediate rocks along the western Yangtze Block, in which subducted sediment melts have been explained to be incorporated into the mantle wedge prior to its partial melting (Zhao et al., 2008, 2018, 2019). Thorium is immobile by aqueous fluids but it can be transferred by subducted sediment melts from the oceanic slab to mantle wedge (Johnson and Plank, 1999; Hawkesworth et al., 1997; Hanyu et al., 2006; Woodhead et al., 2001; Xu et al., 2014), so the Th-rich geochemical features can favor the contribution of sediment melts. On the primitive mantle-normalized trace-element diagram (Fig. 5b), the Shuilu high-Mg$^#$ diorites display obvious enrichment of Th, which is different from those of only slab fluid-enriched mantle magmatism depleted in Th (e.g., the Guandaoshan mafic-intermediate rocks and Dajianshan gabbro-diorites) (Du et al., 2014; Zhu et al., 2019a). Our samples display obviously higher Th/Ce (0.08-0.27) ratios than those of average N-MORB (Th/Ce = 0.016) (Sun and McDonough, 1989) and global subducting sediment components (Th/Ce = 0.12) (Plank and Langmuir, 1998), reflecting the incorporation of subducting sediment melts into primary mantle sources. Their Th/Yb (1.79-8.59) ratios are obviously higher than that of N-MORB (Th/Yb = 0.04) (Sun and McDonough, 1989), further indicating the involvement of sediment-derived melts (Zeng et al., 2018). As illustrated in the plots of Th/Sm vs. Th/Ce (Fig. 11b), the elevated Th/Ce (0.08-0.27) and Th/Sm (0.90-3.80) ratios in our high-Mg$^#$ diorites can also demonstrate the input of sediment melts (Guo et al., 2015; Zhang et al., 2019). Zhao et al. (2019) proposed that the ca. 850-840 Ma gabbros in the western Yangtze Block possess high and constant zircon $\delta^{18}O$ values as well as variable εHf values, which were attributed to the partial melting of mantle source modified by $\delta^{18}O$-enriched sediment melts. This conclusion further demonstrates that the mantle sources beneath the western Yangtze Block underwent subducted sediment-related enrichment during the interval of ca. 850-840 Ma, which partly overlapped with the emplacement ages of the Shuilu high-Mg$^#$ diorites (ca. 850-835 Ma) studied here. In summary, the sediment melts, apart from the aqueous fluids, were also incorporated into the primary mantle source of our high-Mg$^#$ diorites. As mentioned above, Ba is enriched in the aqueous phase and Th is mobile during the transfer of sediment melts (Johnson and Plank, 1999; Hawkesworth et al., 1997; Hanyu et al., 2006; Woodhead et al., 2001). The variable

Th/Yb (1.79−8.59) and Ba/La (24.5−53.6) ratios of our high-Mg$^#$ diorites depict a dual-evolutionary trend (Fig. 11c), which also argue for the enrichment of both subduction-related fluids and sediment melts (Hanyu et al., 2006; Zeng et al., 2018).

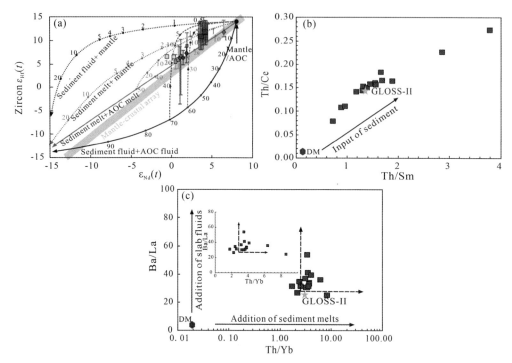

Fig. 11　Whole-rock $\varepsilon_{Nd}(t)$ vs. zircon $\varepsilon_{Hf}(t)$ (a) (Zhao et al., 2019 and reference therein), Th/Sm vs. Th/Ce (b) (Guo et al., 2015; Zhang et al., 2019), and Th/Yb vs. Ba/La (c) (Hanyu et al., 2006) diagrams for the Shuilu high-Mg$^#$ diorites in the western Yangtze Block, South China. Symbols as Fig. 7. In Fig. 11a, the compositions of mixing members are from Hanyu et al. (2006). The isotope compositions of AOC melts show $\varepsilon_{Hf}(t) = +14$ and $\varepsilon_{Nd}(t) = +8$; sediment melts show $\varepsilon_{Hf}(t) = -13.7$ and $\varepsilon_{Nd}(t) = -15.54$.

Actually, it is common that the subduction-related fluids and sediment melts were simultaneously introduced into mantle sources, leading to the metasomatism of overlying mantle wedge and generating the arc magmatism in other subduction zones, such as the Neoproterozoic boninitic lavas in the southeastern Gorny Altai terrane (Chen et al., 2019), the Neoproterozoic boninite-series rocks in the eastern Yangtze Block, South China (Zhao and Asimow, 2014), the Permian Luobei mafic intrusions in NE China (Yang et al., 2019), the Early Cretaceous gabbros from the Asa ophiolite in Tibet (Zeng et al., 2018), the Early Jurassic Tumen mafic intrusions in NE China (Guo et al., 2015), and Miocene to Quaternary volcanic rocks in NE Japan (Hanyu et al., 2006). We thus conclude that the Shuilu high-Mg$^#$ diorites in this study were formed by the partial melting of depleted lithospheric mantle previously metasomatized by subduction-related fluids and sediment melts.

5.2 Neoproterozoic subduction setting and mantle metasomatism

5.2.1 Implications for the Neoproterozoic subduction tectonic setting

High-Mg intermediate igneous rocks (andesites and diorites) are commonly related to metasomatized mantle melts influenced by subduction slab components (Hanyu et al., 2006; Kamei et al., 2004; Martin et al., 2005; Tatsumi, 2006). Therefore, these high-Mg magmatic rocks can provide important constraints on the subduction-related tectonic setting. In this study, the Shuilu high-$Mg^\#$ diorites display concordant crystallization ages of ca. 850 – 835 Ma, corresponding to the early stage of Neoproterozoic magmatism. They belong to the calc-alkaline series (Fig. 4b), and are enriched in LREEs and LILEs, and are depleted in HFSEs (Fig. 5). These geochemical characteristics are predominantly similar to those of Neoproterozoic ca. 860–810 Ma arc-affinity mafic to intermediate rocks (Figs. 4b and 5), which are thought to be derived from subduction-modified lithospheric mantle sources (Du et al., 2014; Munteanu et al., 2010; Zhu et al., 2019a). Such kinds of geochemical fingerprints are also consistent with the typical arc magmatic affinity described by Stern and Killian (1996). In addition, the zircons of the Shuilu high-$Mg^\#$ diorites in this study possess Th/U (0.26–1.17) ratios of igneous zircon (0.2 – 4) (Hoskin and Schalteger, 2003), indicting insignificant late alteration. They yield moderate Hf concentrations (6 393–11 731 ppm, except one spot with 16 210 ppm) as well as high U/Yb (0.17–1.54) and 10 000Nb/Yb (17.7–68.6) ratios (Table 1). Since U/Yb ratios in zircon is a reflection of parental melt compositions, magmatic settings characterized by different U/Yb ratios can be discriminated (Grimes et al., 2015). Almost all zircon analytic spots in this study are characterized by high U/Yb (0.17–1.54) ratios and compiled into the field of continental arc compilation (U/Yb = 0.1–4) (Fig. 12), supporting that these magmatic zircons in the Shuilu high-$Mg^\#$ diorites were originated from LILE-enriched, hydrous melts (Grimes et al., 2015). Grimes et al. (2015) have defined a reference "mantle-zircon array" that encompasses massive analyses on mid-ocean ridge (MOR) and oceanic island zircons based on the U/Yb-Nb/Yb proxy (Fig. 12b). Different from those zircon compilations from mantle plume settings (e.g., Hawaii and Iceland), the zircon spots in our samples are plotted above the mantle-zircon array and geochemically similar to those of Neoproterozoic gabbros in the western Yangtze Block (Fig. 12), which were interpreted to source from mantle wedge enriched by subducted components (e.g., slab fluids, sediment melts, and/or oceanic slab melts) (Zhao et al., 2019). Accordingly, our high-$Mg^\#$ diorites were generated under the subduction zone, not a mantle plume setting. As mentioned above (part 5.1), we have concluded that the Shuilu high-$Mg^\#$ diorites were generated by the partial melting of a metasomatized mantle source enriched by subduction-related fluids and sediment melts, as envisaged in Fig. 13a. This can also provide robust and direct evidence for the subduction tectonic setting along the western Yangtze Block during the Neoproterozoic. In combination with previous comprehensive evidence of detrital zircons, igneous geochemistry, and geophysics (Gao et al., 2016; Sun et al., 2008, 2009; Zhao et al., 2018 and reference therein; Zhu et al., 2017, 2019a-c), our newly identification of the ca. 850–835 Ma Shuilu high-Mg# diorites suggests that the western Yangtze Block underwent persistent subduction process during the Neoproterozoic. We also support that the western margin of the Yangtze Block is part of an arc system responding to the evolution of the supercontinent Rodinia.

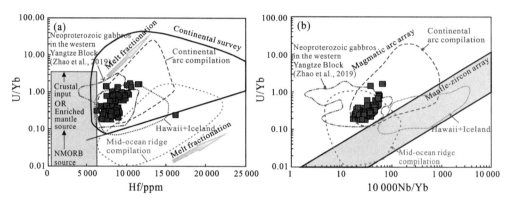

Fig. 12　Plots of zircon Hf vs. U/Yb (a) and 10 000 Nb/Yb vs. U/Yb (b) (Grimes et al., 2015; Zhao et al., 2019) from the Shuilu high-Mg# diorites in the western Yangtze Block, South China.

5.2.2　Insights on the subduction-related mantle metasomatism

The subduction zone magmatism is characterized by metasomatism of the mantle source influenced by subduction-related fluids, sediments melts, and oceanic slab melts (Defant and Drummond, 1990; Eiler et al., 2000, 2005; Hanyu et al., 2006; Shimoda et al., 1998; Tatsumi, 2006; Zhao et al., 2018, 2019). On the western margin of the Yangtze Block, Neoproterozoic subducted fluids-enriched metasomatized mantle-derived magmatism have been documented (Fig. 13a-c). For instance, the ca. 870 Ma Baoxing gabbros show normal mantle-like zircon $\delta^{18}O$ (5.13‰–6.07‰) values and MORB-like whole-rock $\varepsilon_{Nd}(t)$ (+4.00 to +6.42) and zircon $\varepsilon_{Hf}(t)$ values (+10 to +15), indicating that they originated from a depleted mantle source solely metasomatized by slab fluids (Zhao et al., 2019). The ca. 860 Ma Guandaoshan pluton was thought to be derived from a depleted mantle source modified by slab fluids and experienced obvious fractional crystallization (Du et al., 2014; Sun and Zhou, 2008). In addition, the ca. 810 Ma Gaojiacun, Lengshuiqing, and Dajianshan mafic rocks were attributed to sourcing from the depleted, subduction fluids-modified mantle wedge above the subduction zone (Zhou et al., 2006a; Zhu et al., 2019a). The ca. 740 Ma hornblende gabbros in Panzhihua district also came from a metasomatized mantle source enriched by slab fluids, based on their low K/Rb and Nb/Zr ratios as well as high Th/Zr and Rb/Y ratios (Zhao and Zhou, 2007b). More importantly, Zhao et al. (2018) have summarized that the trace elements of the ca. 850–800 Ma mafic-ultramafic rocks in the Yangtze Block display increasingly Rb/Y, Th/La, and Th/Zr ratios and these indexes maintain high levels during the interval of ca. 800–740 Ma, thus reflecting significant fluid-related mantle metasomatism. In general, it is evident that the subducted slab fluids were widely incorporated into the overlying mantle wedge associated with Neoproterozoic subduction process (Fig. 13).

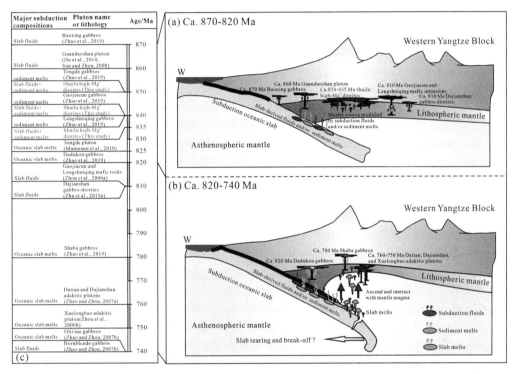

Fig. 13 A sketch map (a, b) and a summary (c) for the Neoproterozoic metasomatized mantle magmatism in the subduction zone, which display that Neoproterozoic mantle source was progressively metasomatized by the subduction-related compositions from slab fluids, sediment melts, to oceanic slab melts during persistent subduction process.

Recently, high zircon $\delta^{18}O$ (5.61‰ – 7.42‰) values observed in ca. 850 – 840 Ma gabbros (e.g., the Tongde, Gaojiacun, and Lengshuiqing gabbros) suggest that their mantle sources were modified by ^{18}O-rich subducted sediment melts (Zhao et al., 2019). The subducted sediment melts might have been gradually incorporated into the mantle source in the western Yangtze Block after ca. 860 Ma (Zhao et al., 2019). In this study, the wide range of Th/Ce, Th/Sm, and Th/Yb ratios, together with the slight Nd-Hf isotopic decoupling in the ca. 850–835 Ma Shuilu high-$Mg^{\#}$ diorites, lead us to support the inference that the subducted sediment melts were significantly introduced into their magma sources. All these geochemical evidences therefore imply that the subducted sediment melt is also a major metasomatic agent acting on the mantle source during the Neoproterozoic (Fig. 13).

In addition, the subduction oceanic slab melts and the slab melt-enriched mantle magmatism have also been presented during the Late Neoproterozoic (Fig. 13b,c). The ca. 750 Ma Xuelongbao and ca. 760 Datian and Dajianshan adakitic rocks were thought to have been generated by the partial melting of subduction oceanic slab (Zhao and Zhou, 2007a; Zhou et al., 2006b). The occurrence of these oceanic slab-derived adakitic magmatism indicate that the mantle source might have been hybridized by oceanic slab melts (Zhao et al., 2008, 2019). Zhao et al. (2019) reported that the ca. 820-780 Ma Dadukou and Shaba gabbros display low zircon $\delta^{18}O$ values (4.22‰ - 5.49‰), which match the potential metasomatized agent of altered oceanic crustal melts incorporated into the mantle source (Bindeman et al., 2005). Considering the high Nb/Ta, Nb/Zr, and Nb/Y ratios as well as relatively low Rb/Y and Th/Zr ratios, the ca. 740 Ma olivine gabbros in the Panzhihua district were also explained from a mantle source metasomatized by subducted slab melts (Zhao and Zhou, 2007b). Therefore, the overlying mantle sources in the western Yangtze Block were modified by subducted oceanic slab melts during the Late Neoproterozoic.

In summary, our identification of ca. 850 - 835 Ma high-Mg$^{\#}$ diorites, coupled with numerous studies of metasomatized mantle magmatism (Fig. 13c), lead us to support that the Neoproterozoic mantle sources beneath the western Yangtze Block were modified by subducted slab fluids, sediment melts, and oceanic slab melts (Zhao et al., 2018, 2019).

6　Conclusions

The geochronological results suggest that the newly recognized Shuilu high-Mg$^{\#}$ diorites in the western Yangtze Block were emplaced at ca. 850-835 Ma. Geochemical and isotopic data indicate that these high-Mg$^{\#}$ diorites were produced by partial melting of depleted lithospheric mantle, which have been metasomatized by subduction-related fluids and sediment-derived melts. Our new data, integrated with numerous previous studies of metasomatized mantle magmatism, support that the Neoproterozoic mantle sources in the western Yangtze Block were metasomatized by subduction slab-related compositions involving slab fluids, sediment melts, and oceanic slab melts during the subduction process.

Acknowledgements　We are grateful to Editor-in Chief Professor Guochun Zhao and two anonymous reviewers for their kind help and constructive comments, which significantly improve the discussion and language of this manuscript. We thank Sara J. Mason for further English polish. We also appreciate Prof. Junhong Zhao for providing whole-rock Sr-Nd isotopic and average zircon Hf-O isotopic data of Neoproterozoic gabbros in the western Yangtze Block. This work was jointly supported by the National Natural Science Foundation of China (Grant Nos. 41421002 and 41772052) and the program for Changjiang Scholars and Innovative Research Team in University (Grant IRT1281). Support was also provided by the Foundation for the Author of National Excellent Doctoral Dissertation of China (201324).

Appendix A. Supplementary data   Supplementary data to this article can be found online at https://doi.org/10. 1016/j.precamres.2020. 105738.

References

Andersen, T., 2002. Correction of common lead in U-Pb analyses that do not report [204]Pb. Chemical Geology, 192(1-2), 59-79.

Bao, Z.A., Chen, L., Zong, C.L., Yuan, H.L., Chen, K.Y., Dai, M.N., 2017. Development of pressed sulfide powder tablets for in situ, sulfur and lead isotope measurement using LA-MC-ICP-MS. International Journal of Mass Spectrometry, 421, 255-262.

Bindeman, I.N., Eiler, J.M., Yogodzinski, G.M., Tatsumi, Y., Stern, C.R., Grove, T.L., Portnyagin, M., Hoernle, K., Danyushevsky, L.V., 2005. Oxygen isotope evidence for slab melting in modern and ancient subduction zones. Earth & Planetary Science Letters, 235, 480-496.

Bau, M., 1991. Rare-earth element mobility during hydrothermal and metamorphic fluid-rock interaction and the significance of the oxidation state of europium. Chemical Geology, 93 (3-4), 219-230.

Cao, K., Yang, Z.M., Xu, J.F., Fu, B., Li, W.K., Sun, M.Y., 2018. Origin of dioritic magma and its contribution to porphyry Cu-Au mineralization at Pulang in the Yidun arc, eastern Tibet. Lithos, 304-307, 436-449.

Chen, B., Jahn, B.M., Suzuki, K., 2013. Petrological and Nd-Sr-Os isotopic constraints on the origin of high-Mg adakitic rocks from the North China Craton: Tectonic implications. Geology, 41(1), 91-94.

Chen, X., Wang, D., Wang, X.L., Gao, J.F., Shu, X.J., Zhou, J. C., Qi, L., 2014a. Neoproterozoic chromite-bearing high-Mg diorites in the western part of the Jiangnan orogen, southern China: Geochemistry, petrogenesis and tectonic implications. Lithos, 200-201, 35-48.

Chen, W.T., Sun, W.H., Wang, W., Zhao, J.H., Zhou, M.F., 2014b. "Grenvillian" intra-plate mafic magmatism in the southwestern Yangtze Block, SW China. Precambrian Research, 242 138-153.

Chen, W.T., Sun, W.H., Zhou, M.F., Wang, W., 2018. Ca. 1 050 Ma intra-continental rift related A-type felsic rocks in the southwestern Yangtze Block, South China. Precambrian Research, 309, 22-44.

Chen, M., Sun, M., Buslov, M., Cai, K.D., Jiang, Y.D., Kulikova, A.V., Zheng, J.P., Xia, X,P., 2019. Variable slab-mantle interaction in a nascent Neoproterozoic arc-back-arc system generating boninitic-tholeiitic lavas and magnesian andesites. The Geological Society of America, 130, 9-10, 1562-1581.

Crawford A J, 1989. Boninites and Related Rocks. London: Unwin Hyman.

Defant, M.J., Drummond, M.S., 1990. Derivation of some modern arc magmas by melting of young subducted lithosphere. Nature, 347, 662-665.

Du, L.L., Guo, J.H., Nutman, A.P., Wyman, D., Geng, Y.S., Yang, C.H., Liu, F.L., Ren, L.D., Zhou, X.W., 2014.Implications for Rodinia reconstructions for the initiation of Neoproterozoic subduction at ~860 Ma on the western margin of the Yangtze Block: Evidence from the Guandaoshan Pluton. Lithos, 196-197, 67-82.

Elliott, T., Plank, T., Zindler, A., White, W., Bourdon, B., 1997. Element transport from slab to volcanic front at the Mariana arc. Journal of Geophysical Research-Solid Earth, 102(B7), 14991-15019.

Eiler, J.M., Crawford, A., Elliott, T., Farley, K.A., Valley, J.W., Stolper, E.M., 2000. Oxygen isotope geochemistry of oceanic-arc lavas. Journal of Petrology, 41, 229-256.

Eiler, J.M., Carr, M.J., Reagan, M., Stolper, E., 2005. Oxygen isotope constraints on the sources of central american arc lavas. Geochemistry, Geophysics, Geosystems, 6, Q07007.

Frost, B.R., Barnes, C.G., Collins, W.J., Arculus, R.J., Ellis, D.J., Frost, C.D., 2001. A geochemical classification for granitic rocks.Journal of Petrology, 42 (11), 2033-2048.

Gao, R., Chen, C., Wang, H.Y., Lu, Z.W., Brown, L., Dong, S.W., Feng, S.Y., Li, Q.S., Li, W.H., Wen, Z.P., Li, F., 2016. SINOPROBE deep reflection profile reveals a Neo-proterozoic subduction zone beneath Sichuan basin. Earth & Planetary Science Letters, 454, 86-91.

Gao, S., Ling, W.L., Qiu, Y., Zhou, L., Hartmann, G., Simon, K., 1999.Contrasting geochemical and Sm-Nd isotopic compositions of Archean metasediments from the Kongling high-grade terrain of the Yangtze craton: Evidence for cratonic evolution and redistribution of REE during crustal anatexis. Geochimica et Cosmochimica Acta, 63(13-14), 2071-2088.

Gao, S., Rudnick, R.L., Yuan, H.L., Liu, X.M., Liu, Y.S., Xu, W.L., Ling, W.L., Ayers, J., Wang, X.C., Wang, Q.H., 2004. Recycling lower continental crust in the North China Craton. Nature, 432, 892-897.

Gao, S., Jie, Y., Lian, Z., Li, M., Hu, Z.C., Guo, J.L., Yuan, H.L., Gong, H.J., Xiao, G.Q., Wei, J.Q., 2011.Age and growth of the Archean Kongling terrain, South China, with emphasis on 3. 3 Ga granitoid gneisses. Geochimica et Cosmochimica Acta Supplement, 72(12),153-182.

Greentree, M.R., Li, Z.X., Li, X.H., Wu, H., 2006. Latest Mesoproterozoic to earliest Neoproterozoic basin record of the Sibao orogenesis in western South China and relationship to the assembly of Rodinia. Precambrian Research, 151, 79-100.

Griffin, W.L., Pearson, N.J., Belousova, E., Jackson, S.E., van Achterbergh, E., O'Reilly, S.Y., Shee, S.R., 2000. The Hf isotope composition of cratonic mantle: LAM-MC-ICP-MS analysis of zircon megacrysts in kimberlites. Geochimica et Cosmochimica Acta, 64, 133-147.

Grimes, C. B., Wooden, J. L., Cheadle, M. J., John, B. E., 2015. "Fingerprinting" tectono-magmatic provenance using trace elements in igneous zircon. Contributions to Mineralogy and Petrology, 170, 46.

Grove, T., Parman, S., Bowring, S., Price, R., Baker, M., 2002. The role of an H_2O-rich fluid component in the generation of primitive basaltic andesites and andesites from the Mt. Shasta region, N California. Contributions to Mineralogy and Petrology, 142, 375-396.

Guo, F., Li, H.X., Fan, W.M., Li, J.Y., Zhao, L., Huang, M.W., Xu, W.L., 2015. Early Jurassic subduction of the Paleo-Pacific Ocean in NE China: Petrologic and geochemical evidence from the Tumen mafic intrusive complex. Lithos, 224-225, 46-60.

Guo, F., Li, H.X., Fan, W.M., Li, J.Y., Zhao, L., Huang, M.W., 2016. Variable sediment flux in generation of Permian subduction-related mafic intrusions from the Yanbian region, NE China. Lithos, 261, 195-215.

Guo, J.L., Gao, S., Wu, Y.B., Li, M., Chen, K., Hu, Z.C., Liang, Z.W., Liu, Y.S., Zhou, L., Zong, K.Q., Zhang, W., Chen, H.H., 2014. 3. 45 Ga granitic gneisses from the Yangtze Craton, South China: Implications for Early Archean crustal growth. Precambrian Research, 242, 82-95.

Hanyu, T., Tatsumi, Y., Nakai, S., 2002. A contribution of slab-melts to the formation of high-Mg andesite magmas; Hf isotopic evidence from SW Japan. Geophysical Research Letters, 29(8),1-4.

Hanyu, T., Tatsumi, Y., Nakai, S., Chang, Q., Miyazaki, T., Sato, K., Tani, K., Shibata, T., Yoshida, T., 2006.Contribution of slab melting and slab dehydration to magmatism in the ne japan arc for the last 25 Myr: Constraints from geochemistry. Geochemistry, Geophysics, Geosystems, 7(8).

Hawkesworth, C.J., Turner, S.P., McDermott, F., Peate, D.W, van Calsteren, P., 1997. U-Th isotopes in arc magmas: Implications for element transfer from the subducted crust. Science, 276, 551-555.

Hoskin, P.O, Schaltegger, U., 2003. The composition of zircon and igneous and metamorphic petrogenesis. In: Hanchar, J.M., Hoskin, P.W.O. Zircon, Reviews in Mineralogy and Geochemistry, 53. Mineralogical Society of America, 27-62.

Hou, Z.Q., Gao, Y.F., Qu, X.M., Rui, Z.Y., Mo, X.X., 2004. Origin of adakitic intrusives generated during mid-Miocene east-west extension in southern Tibet. Earth & Planetary Science Letters, 220 (1-2), 139-155.

Huang, F., Li, S.G., Dong, F., He, Y.S., Chen, F.K., 2008a. High-Mg adakitic rocks in the Dabie orogen, central China: Implications for foundering mechanism of lower continental crust. Chemical Geology, 255(1-2), 0-13.

Huang, X.L., Xu, Y.G., Li, X.H., Li, W.X., Lan, J.B., Zhang, H.H., Liu, Y.S., Wang, Y.B., Li, H.Y., Luo. Z.Y., Yang, Q.J., 2008b. Petrogenesis and tectonic implications of Neoproterozoic, highly fractionated A-type granites from Mianning, South China. Precambrian Research, 165(3-4), 190-204.

Huang, X.L., Xu, Y.G., Lan, J.B., Yang, Q.J., Luo, Z.Y., 2009. Neoproterozoic adakitic rocks from Mopanshan in the Western Yangtze craton: Partial melts of a thickened lower crust. Lithos, 112(3), 367-381.

Johnson, M.C., Plank, T., 1999. Dehydration and melting experiments constrain the fate of subducted sediments. Geochemistry, Geophysics, Geosystems, 1, 1007.

Jung, S., Hoernes, S., Mezger, K., 2002. Synorogenic melting of mafic lower crust: Constraints from geochronology, petrology and Sr, Nd and O isotope geochemistry of quartz diorites (Damara orogen, Namibia). Contributions to Mineralogy and Petrology, 143, 551-566.

Jung, S., Masberg, P., Mihm, D., Hoernes, S., 2009. Partial melting of diverse crustal sources-constraints from Sr-Nd-O isotope compositions of quartz diorite-granodiorite-leucogranite associations (Kaoko Belt, Namibia). Lithos, 111(3), 236-251.

Jung, S., Kröner, A., Hauff, F., Masberg, P., 2015. Petrogenesis of synorogenic diorite-granodiorite-granite complexes in the Damara belt, Namibia: Constraints from U-Pb zircon ages and Sr-Nd-Pb isotopes. Journal of African Earth Sciences, 101, 253-265.

Kamei, A., Owada, M., Nagao, T., Shiraki, K., 2004. High-Mg diorites derived from sanukitic HMA magmas, Kyushu Island, southwest Japan arc: Evidence from clinopyroxene and whole rock compositions. Lithos, 75, 359-371.

Karsli, O., Dokuz, A., Kandemir, R., 2017. Zircon Lu-Hf isotope systematics and U-Pb geochronology, whole-rock Sr-Nd isotopes and geochemistry of the early Jurassic Gokcedere pluton, Sakarya zone-NE Turkey: A magmatic response to roll-back of the Paleo-Tethyan oceanic lithosphere. Contributions to Mineralogy and Petrology, 172(5), 31.

Kelemen, P. B., Yogodzinski, G. M., Scholl, D.W., 2003. Along-strike variation in the Aleutian Island arc: Genesis of high Mg# andesite and implications for continental crust. In: Eiler, J. Inside the subduction factory. American Geophysical Union, Washington, DC, 223-276.

Kelemen, P. B., Hanghøj, K., Greene, A. R., 2014. One view of the geochemistry of subduction-related magmatic arcs, with an emphasis on primitive andesite and lower crust. In: Turekian, H.D.H.K. Treatise on geochemistry. Oxford: Elsevier, 749-806.

Kemp, A.I.S., Hawkesworth, C.J., Foster, G.L., Paterson, B.A., Woodhead, J.D., Hergt, J.M., Gray, C. M., Whitehouse, M.J., 2007. Magmatic and crustal differentiation history of granitic rocks from Hf-O isotopes in zircon. Science, 315, 980-983.

Kepezhinskas, P., McDermott, F., Defant, M.J., Hochstaedter, A., Drummond, M.S., 1997. Trace element and Sr-Nd-Pb isotopic constraints on a three-component model of Kamchatka Arc petrogenesis. Geochimica et Cosmochimica Acta, 61(3), 577-600.

Kuroda, N., Shiraki, K., Urano, H., 1978. Boninite as a possible calc-alkalic primary magma. Bulletin of Volcanology, 41, 563-575.

Lai, S.C., Qin, J.F., Zhu, R.Z., Zhao, S.W., 2015. Neoproterozoic quartz monzodiorite-granodiorite association from the Luding-Kangding area: Implications for the interpretation of an active continental margin along the Yangtze Block (South China Block). Precambrian Research, 267(3-4), 196-208.

Li, Q.W., Zhao, J.H., 2018. The Neoproterozoic high-Mg dioritic dikes in south China formed by high pressures fractional crystallization of hydrous basaltic melts. Precambrian Research, 309, 198-211.

Li, X.H., Li, Z.X., Zhou, H.W., Liu, Y., Liang, X.R., Li, W.X., 2003. SHRIMP U-Pb zircon age, geochemistry and Nd isotope of the Guandaoshan pluton in SW Sichuan: Petrogenesis and tectonic significance. Science in China: Series D, 46(1), 73-83.

Li, X.H., Li, Z.X., Sinclair, J.A., Li, W.X., Carter, G., 2006. Revisiting the "Yanbian Terrane": Implications for Neoproterozoic tectonic evolution of the western Yangtze Block, South China. Precambrian Research, 151(1-2), 14-30.

Liu, Y., Liu, X.M., Hu, Z.C., Diwu, C.R., Yuan, H.L., Gao, S., 2007. Evaluation of accuracy and long-term stability of determination of 37 trace elements in geological samples by ICP-MS. Acta Petrological Sinica, 23(5), 1203-1210 (in Chinese with English abstract).

Ludwig, K.R., 2003. ISOPLOT 3.0: A geochronological toolkit for Microsoft Excel. Berkeley Geochronology Center, Special Publication, 4.

Martin, H., Smithies, R.H., Rapp, R., Moyen, J.F., Champion, D., 2005. An overview of adakite, tonalite-trondhjemite-granodiorite (TTG), and sanukitoid: Relationships and some implications for crustal evolution. Lithos, 79, 1-24.

Meng, E., Liu, F.L., Du, L.L., Liu, P.H., Liu, J.H., 2015. Petrogenesis and tectonic significance of the Baoxing granitic and mafic intrusions, southwestern china: Evidence from zircon U-Pb dating and Lu-Hf isotopes, and whole-rock geochemistry.Gondwana Research, 28(2), 800-815.

Middlemost, E.A.K., 1994. Naming materials in the magma/igneous rock system. Earth-Science Reviews, 37, 215-224.

Moyen, J.F., Martin, H., Jayananda, M., Auvray, B., 2003. Late Archaean granites: A typology based on the Dharwar Craton (India). Precambrian Research, 127, 103-123.

Munteanu, M., Wilson, A., Yao, Y., Harris, C., Chunnett, G., Luo, Y., 2010. The Tongde dioritic pluton (Sichuan, SW China) and its geotectonic setting: Regional implications of a local-scale study. Gondwana Research, 18(2), 455-465.

Nichols, G.T., Wyllie, P.J., Stern, C.R., 1994. Subduction zone melting of pelagic sediments constrained by melting experiments. Nature, 371, 785-788.

Nowell, G.M., Kempton, P.D., Noble, S.R., Fitton, J.G., Saunders, A.D., Mahoney, J.J., Taylor, R.N., 1998. High precision Hf isotope measurements of MORB and OIB by thermal ionisation mass spectrometry: Insights into the depleted mantle. Chemical Geology, 149, 211-233.

Patiño Douce, A.E., 1999. What do experiments tell us about the relative contributions of crust and mantle to the origin of the granitic magmas. Geol. Soc. London Spec. Publ., 168 (1), 55-75.

Plank, T., Langmuir, C.H., 1998. The chemical composition of subducting sediment and its consequences for the crust and mantle. Chemical Geology, 145 (3-4), 325-394.

Qian, Q., Hermann, J., 2010. Formation of high-Mg diorites through assimilation of peridotite by monzodiorite magma at crustal depths. Journal of Petrology, 51(7), 1381-1416.

Rapp, R.P., Shimizu, N., Norman, M.D., Applegate, G.S., 1999. Reaction between slab-derived melts and peridotite in the mantle wedge: Experimental constraints at 3.8 GPa. Chemical Geology, 160, 335-356.

Rapp, R.P., Watson, E.B., 1995. Dehydration melting of metabasalt at 8-32 kbar: Implications for continental growth and crust-mantle recycling. Journal of Petrology, 36, 891-931.

Rapp, R.P., Norman, M.D., Laporte, D., Yaxley, G.M., Martin, H., Foley, S.F., 2010. Continental formation in the Archean and chemical evolution of the cratonic lithosphere: Melt-rock reaction experiments at 3-4GPa and petrogenesis of Archean Mg-diorites. Journal of Petrology, 52 (6), 1237-1266.

Roberts, M.P., and Clemens, J.D., 1993. Origin of high-potassium, calc-alkaline, I-type granitoids. Geology, 21, 825-828.

Rogers, N.W., Hawkesworth, C.J., Parker, R.J., Marsh, J.S., 1985. The geochemistry of potassic lavas from Vulsini, central Italy and implications for mantle enrichment processes beneath the Roman region. Contributions to Mineralogy and Petrology, 90, 244-257.

Rudnick, R.L., Gao, S., 2003. The composition of the continental crust. In: Rudnick, R.L. The Crust. Oxford: Elsevier-Pergamon, 1-64.

Shimoda, G., Tatsumi, Y., Nohda, S., Ishizaka, K., Jahn, B.M., 1998. Setouchi high-Mg andesites revisited: Geochemical evidence for melting of subducting sediments. Earth & Planetary Science Letters, 160, 479-492.

Shimoda, G., Tatsumi, Y., Morishita, Y., 2003. Behavior of subducting sediments beneath an arc under high geothermal gradient: Constraints from the Miocene SW Japan arc. Geochemical Journal, 37, 503-518.

Shirey, S.B., Hanson, G.N., 1984. Mantle-derived Archean monozodiorites and trachyandesites. Nature, 310, 222-224.

Smithies, R.H., Champion, D.C., 2000. The Archaean high-Mg diorite suite: Links to tonalite-trondhjemite-granodiorite magmatism and implications for Early Archaean crustal growth. Journal of Petrology, 41, 1653-1671.

Smith, E.I., Sánchez, Alexander, Walker, J.D., Wang, K., 1999. Geochemistry of mafic magmas in the hurricane volcanic field, Utah: Implications for small- and large-scale chemical variability of the lithospheric mantle. Journal of Geology, 107(4), 433-448.

Smith, D.J., Petterson, M.G., Saunders, A.D., Millar, I.L., Jenkin, G.R.T., Toba, T., Naden, J., Cook, J.M., 2009. The petrogenesis of sodic island arc magmas at Savo volcano, Solomon islands. Contributions to Mineralogy and Petrology, 158, 785-801.

Sichuan Provincial Bureau of Geology and Mineral Resources (SPBGMR), 1972. Regional geological survey of the People's Republic of China, the Yanbian Sheet map and report, scale 1:200 000. (in Chinese).

Söderlund, U., Patchett, P.J., Vervoort, J.D., Isachsen, C.E., 2004. The ^{176}Lu decay constant determined by Lu-Hf and U-Pb isotope systematics of Precambrian mafic intrusions. Earth & Planetary Science Letters, 219(3-4), 311-324.

Sun, S.S., McDonough, W.F., 1989. Chemical and isotopic systematics of oceanic basalts: Implications for mantle composition and processes. In: Saunders, A.D., Norry, M.J. Magmatism in the Ocean Basins. Geological Social Lond Special Publication, 42, 313-345.

Sun, W.H., Zhou, M.F., 2008. The ~860 Ma, cordilleran-type Guandaoshan dioritic pluton in the Yangtze Block, SW China: Implications for the origin of Neoproterozoic magmatism. Journal of Geology, 116(3), 238-253.

Sun, W.H., Zhou, M.F., Yan, D.P., Li, J.W., Ma, Y.X., 2008. Provenance and tectonic setting of the Neoproterozoic Yanbian group, western Yangtze Block (SW China). Precambrian Research, 167(1), 213-236.

Sun, W.H., Zhou, M.F., Gao, J.F., Yang, Y.H., Zhao, X.F., Zhao, J.H., 2009. Detrital zircon U-Pb geochronological and Lu-Hf isotopic constraints on the Precambrian magmatic and crustal evolution of the western Yangtze Block, SW China. Precambrian Research, 172(1), 99-126.

Stern, R.A., Hanson, G.N., Shirey, S.B., 1989. Petrogenesis of mantle-derived, LILE-enriched Archean monzodiorites and trachyandesites (sanukitoids) in southwestern Superior Province. Canadian Journal of Earth Science, 26, 1688-1712.

Stern, C.R., Kilian R., 1996. Role of the subducted slab, mantle wedge and continental crust in the generation of adakites from the Austral Volcanic Zone. Contributions to Mineralogy and Petrology, 123, 263-281.

Tang, G.J., Wang, Q., Wyman, D.A., Li, Z.X., Zhao, Z.H., Jia, X.H., Jiang, Z.Q., 2010. Ridge subduction and crustal growth in the Central Asian orogenic belt: Evidence from Late Carboniferous adakites and high-Mg diorites in the western Junggar region, northern Xinjiang (West China). Chemical Geology, 277(3-4), 1-300.

Tatsumi, Y., 2006. High-Mg andesites in the Setouchi volcanic belt, southwestern Japan: Analogy to Archean magmatism and continental crust formation? Annual Review of Earth and Planetary Science, 34, 467-499.

Tatsumi, Y., 2008. Making continental crust: The sanukitoid connection. Science Bulletin, 53 (11), 1620-1633.

Tatsumi, Y., Ishizaka, K., 1981. Existence of andesitic primary magma: An example from Southwest Japan. Earth & Planetary Science Letters, 53, 124-130.

Tatsumi, Y., Nakashima, T., Tamura, Y., 2002. The petrology and geochemistry of calc-alkaline andesites on

Shodo-Shima Island, SW Japan. Journal of Petrology, 43, 3-16.

Tatsumi, Y., Shukuno, H., Sato, K., Shibata, T., Yoshikawa, M., 2003. The petrology and geochemistry of high-magnesium andesites at the western tip of the Setouchi volcanic belt, SW Japan. Journal of Petrology, 44, 1561-1578.

van de Flierdt, T., Hoernes, S., Jung, S., Masberg, P., Hoffer, E., Schaltegger, U., Friedrichsen, H., 2003. Lower crustal melting and the role of open-system processes in the genesis of syn-orogenic quartz diorite-granite-leucogranite associations: Constraints from Sr-Nd-O isotopes from the Bandombaai Complex, Namibia. Lithos, 67(3), 205-226.

Wang, Q., Wyman, D.A., Zhao, Z.H., Xu, J.F., Bai, Z.H., Xiong, X.L., Dai, T.M., Li, C.F., Chu, Z. Y., 2007. Petrogenesis of carboniferous adakites and Nb-enriched arc basalts in the Alataw area, northern Tianshan range (western china): Implications for phanerozoic crustal growth in the Central Asia orogenic belt. Chemical Geology, 236(1-2), 0-64.

Wang, X.C., Li, X.H., Li, W.X., Li, Z.X., Tu, X.L., 2008. The bikou basalts in the northwestern Yangtze Block, South China: Remnants of 820−810 Ma continental flood basalts? Geological Society of America Bulletin, 120(11).

Wang, X.L., Zhou, J.C., Wan, Y.S., Kitajima, K., Wang, D., Bonamici, C., Qiu, J.S., Sun, T., 2013. Magmatic evolution and crustal recycling for Neoproterozoic strongly peraluminous granitoids from southern China: Hf and O isotopes in zircon. Earth & Planetary Science Letters, 366 (2), 71-82.

Wang, X.L., Zhou, J.C., Griffin, W.L., Zhao, G.C., Yu, J.H., Qiu, J.S., Zhang, Y.J., Xing, G.F., 2014. Geochemical zonation across a Neoproterozoic orogenic belt: Isotopic evidence from granitoids and metasedimentary rocks of the Jiangnan orogen, China. Precambrian Research, 242 (2), 154-171.

Wang, Y.J., Zhang, Y.Z., Zhao, G.C., Fan, W.M., Xia, X.P., Zhang, F.F., Zhang, A.M., 2009. Zircon U-Pb geochronological and geochemical constraints on the petrogenesis of the Taishan sanukitoids (Shandong): Implications for Neoarchean subduction in the Eastern Block, North China Craton. Precambrian Research, 174(3-4), 0-286.

Wang, Y.J., Zhu, W.G., Huang, H.Q., Zhong, H., Bai, Z.J., Fan, H.P., Yang, Y.J., 2019. Ca. 1.04 Ga hot Grenville granites in the western Yangtze Block, southwest China. Precambrian Research, 328, 217-234.

Wiedenbeck, M., Hanchar, J.M., Peck, W.H., Sylvester, P., Valley, J., Whitehouse, M., Kronz, A., Morishita, Y., et al., 2004. Further characterisation of the 91500 zircon crystal. Geostandards & Geoanalytical Research, 28 (1), 9-39.

Wilson, M., 1989. Review of igneous petrogenesis: A global tectonic approach. Terra Nova, 1(2), 218-222.

Woodhead, J.D., Hergt, J.M., Davidson, J.P., Eggins, S.M., 2001. Hafnium isotope evidence for "conservative" element mobility during subduction zone processes. Earth & Planetary Science Letters, 192, 331-346.

Wu, T., Wang, X.C., Li, W.X., Wilde, S., Tian, L.Y., 2019. Petrogenesis of the ca. 820−810 Ma felsic volcanic rocks in the Bikou Group: Implications for the tectonic setting of the western margin of the Yangtze Block. Precambrian Research, 331.

Xu, J.F., Shinjo, R., Defant, M.J., Wang, Q., Rapp, R.P., 2002. Origin of Mesozoic adakitic intrusive rocks in the Ningzhen area of east China: Partial melting of delaminated lower continental crust. Geology,

30, 1111-1114.

Xu, M.J., Li, C., Xu, W., Xie, C., Hu, P., Wang, M., 2014. Petrology, geochemistry and geochronology of gabbros from the Zhongcang ophiolitic mélange, Central Tibet: Implications for an intra-oceanic subduction zone within the Neo-Tethys Ocean. Journal of Earth Sciences, 25, 224-240.

Xu, W.L., Hergt, J.A., Gao, S., Pei, F.P., Wang, W., Yang, D.B., 2008. Interaction of adakitic melt-peridotite: Implications for the high-Mg$^{\#}$ signature of Mesozoic adakitic rocks in the eastern North China Craton. Earth & Planetary Science Letters, 265, 123-137.

Yang, H., Ge, W.C., Dong, Y., Bi, J.H., Ji, Z., He, Y., Jing, Y., Xu, W.L., 2019. Permian subduction of the Paleo-Pacific (Panthalassic) oceanic lithosphere beneath the Jiamusi Block: Geochronological and geochemical evidence from the Luobei mafic intrusions in Northeast China. Lithos, 332-333, 207-225.

Yuan, H.L., Gao, S., Liu, X.M., Li, H.M., Gunther, D., Wu, F.Y., 2004. Accurate U-Pb age and trace element determinations of zircon by laser ablation-inductively coupled plasma mass spectrometry. Geostandards & Geoanalytical Research, 28 (3), 353-370.

Yuan, H.L., Gao, S., Dai, M.N., Zong, C.L., Gunther, D., Fontaine, G.H., Liu, X.M., Diwu, C.R., 2008. Simultaneous determinations of U-Pb age, Hf isotopes and trace element compositions of zircon by excimer laser-ablation quadrupole and multiple-collector ICPMS. Chemical Geology, 247(1-2), 100-118.

Zeng, X.W., Wang, M., Fan, J.J., Li, C., Xie, C.M., Liu, Y.M., Zhang, T.Y., 2018. Geochemistry and geochronology of gabbros from the Asa Ophiolite, Tibet: Implications for the early Cretaceous evolution of the Meso-Tethys Ocean. Lithos, 320-321, 192-206.

Zhang, B., Guo, F., Zhang, X.B., Wu, Y.M., Wang, G.Q., Zhao, L., 2019. Early Cretaceous subduction of Paleo-Pacific Ocean in the coastal region of SE China: Petrological and geochemical constraints from the mafic intrusions. Lithos, 334-334, 8-24.

Zhao, G.C., Cawood, P.A., 2012. Precambrian geology of China. Precambrian Research, 222-223, 13-54.

Zhao, J.H., Zhou, M.F., 2007a. Neoproterozoic Adakitic Plutons and Arc Magmatism along the western Margin of the Yangtze Block, South China. Journal of Geology, 115(6), 675-689.

Zhao, J.H., Zhou, M.F., 2007b. Geochemistry of Neoproterozoic mafic intrusions in the Panzhihua district (Sichuan Province, SW China): Implications for subduction-related metasomatism in the upper mantle. Precambrian Research, 152(1), 27-47.

Zhao, J.H., Zhou, M.F., Yan, D.P., Yang, Y.H., Sun, M., 2008. Zircon Lu-Hf isotopic constraints on Neoproterozoic subduction-related crustal growth along the western margin of the Yangtze Block, South China. Precambrian Research, 163(3), 189-209.

Zhao, J.H., Zhou, M.F., Yan, D.P., Zheng, J.P., Li, J.W., 2011. Reappraisal of the ages of Neoproterozoic strata in South China: No connection with the Grenvillian orogeny. Geology, 39 (4), 299-302.

Zhao, J.H., Asimow, P.D., 2014. Neoproterozoic boninite-series rocks in South China: a depleted mantle source modified by sediment-derived melt. Chemical Geology, 388, 98-111.

Zhao, J.H., Asimow, P.D., Zhou, M.F., Zhang, J., Yan, D.P., Zheng, J.P., 2017. An Andean-type arc system in Rodinia constrained by the Neoproterozoic Shimian ophiolite in South China. Precambrian Research, 296, 93-111.

Zhao, J.H., Li, Q.W., Liu, H., Wang, W., 2018. Neoproterozoic magmatism in the western and northern margins of the Yangtze Block (South China) controlled by slab subduction and subduction-transform-edge-propagator. Earth-Science Reviews, 187, 1-18.

Zhao, J.H., Asimow, P.D., 2018. Formation and evolution of a magmatic system in a rifting continental margin: The Neoproterozoic arc- and MORB-like dike swarms in South China. Journal of Petrology, 59 (9), 1811-1844.

Zhao, J.H., Zhou, M.F., Wu, Y.B., Zheng, J.P., Wang, W., 2019. Coupled evolution of Neoproterozoic arc mafic magmatism and mantle wedge in the western margin of the South China Craton. Contributions to Mineralogy and Petrology, 174, 36.

Zhou, M.F., Yan, D., Kennedy, A.K., Li, Y.Q., Ding, J., 2002. SHRIMP U-Pb zircon geochronological and geochemical evidence for Neoproterozoic arc-magmatism along the western margin of the Yangtze Block, South China. Earth & Planetary Science Letters, 196(1-2), 51-67.

Zhou, M.F., Ma, Y.X., Yan, D.P., Xia, X.P., Zhao, J.H., Sun, M., 2006a. The Yanbian Terrane (Southern Sichuan Province, SW China): A Neoproterozoic arc assemblage in the western margin of the Yangtze Block. Precambrian Research, 144(1-2), 19-38.

Zhou, M.F., Yan, D.P., Wang, C.L., Qi, L., Kennedy, A., 2006b. Subduction-related origin of the 750 Ma Xuelongbao adakitic complex (Sichuan Province, China): Implications for the tectonic setting of the giant Neoproterozoic magmatic event in South China. Earth & Planetary Science Letters, 248(1-2), 286-300.

Zhu, W.G., Zhong, H., Li, X.H., Deng, H.L., Bai, Z.J., 2008. SHRIMP zircon U-Pb geochronology, elemental, and Nd isotopic geochemistry of the Neoproterozoic mafic dykes in the Yanbian area, SW China. Precambrian Research, 164(1), 66-85.

Zhu, W.G., Zhong, H., Li, Z.X., Bai, Z.J., Yang, Y.J., 2016. SIMS zircon U-Pb ages, geochemistry and Nd-Hf isotopes of ca. 1.0 Ga mafic dykes and volcanic rocks in the Huili area, SW China: Origin and tectonic significance. Precambrian Research, 273, 67-89.

Zhu, Y., Lai, S.C., Zhao, S.W., Zhang, Z.Z., Qin, J.F., 2017. Geochemical characteristics and geological significance of the Neoproterozoic K-feldspar granites from the Anshunchang, Shimian area, Western Yangtze Block. Geological Review, 63(5), 1193-1208 (in Chinese with English abstract).

Zhu, Y., Lai, S.C., Qin, J.F., Zhu, R.Z., Zhang, F.Y., Zhang, Z.Z., Gan, B.P., 2019a. Petrogenesis and geodynamic implications of Neoproterozoic gabbro-diorites, adakitic granites, and A-type granites in the southwestern margin of the Yangtze Block, South China. Journal of Asian Earth Science, 183, 103977.

Zhu, Y., Lai, S.C., Qin, J.F., Zhu, R.Z., Zhang, F.Y., Zhang, Z.Z., 2019b. Geochemistry and zircon U-Pb-Hf isotopes of the 780 Ma I-type granites in the western Yangtze Block: Petrogenesis and crustal evolution. International Geology Review, 61 (10), 1222-1243.

Zhu, Y., Lai, S.C., Qin, J.F., Zhu, R.Z., Zhang, F.Y., Zhang, Z.Z., Zhao, S.W., 2019c. Neoproterozoic peraluminous granites in the western margin of the Yangtze Block, South China: Implications for the reworking of mature continental crust. Precambrian Research, 333, 105443.

Zhu, Y., Lai, S.C., Qin, J.F., Zhu, R.Z., Zhang, F.Y., Zhang, Z.Z., 2020. Petrogenesis and geochemical diversity of late Mesoproterozoic S-type granites in the western Yangtze Block, South China: Co-entrainment of peritectic selective phases and accessory minerals. Lithos, 352-353, 105326.

Vein-plus-wall rock melting model for the origin of Early Paleozoic alkali diabases in the South Qinling Belt, central China[①]

Zhang Fangyi　Lai Shaocong[②]　Qin Jiangfeng

Zhu Renzhi　Zhao Shaowei　Zhu Yu　Yang Hang

Abstract: Early Paleozoic mafic dykes are widespread in the South Qinling Belt, central China. In this study, we present new major and trace element, zircon U-Pb age and Sr-Nd-Hf isotopic results of Early Paleozoic diabases dykes in the South Qinling Belt to explore the nature of their mantle source. The zircon U-Pb dating yielded ages of 455.9 ± 1.5 Ma and 446.2 ± 1.1 Ma. The South Qinling Belt diabases had low SiO_2 (42.1–49.5 wt%), high TiO_2 (2.89–5.17 wt%) and variable MgO (4.0–9.4 wt%) contents. In primitive mantle normalized multi-element diagrams, all samples were strongly enriched in the majority of incompatible trace elements but showed systematic depletion in Rb, K, Pb, Zr and Hf. The negative K and Rb anomalies, together with high TiO_2 and high Na_2O/K_2O, suggests magma was derived from a source rich in amphibole. Partial melting modeling indicated that 20%–36% partial melting of amphibole-clinopyroxene-phlogopite veins with subsequent dissolution of ~30% orthopyroxene from the wall-rock peridotite within the spinel stability field can produce the observed diabase compositions. Additionally, the South Qinling Belt diabases were characterized by moderately depleted Nd [$\varepsilon_{Nd}(t) = +2.2$ to $+3.3$] and Hf [$\varepsilon_{Hf}(t) = +6.2$ to $+7.2$] isotopic compositions without pronounced isotope decoupling, indicating mantle metasomatism occurred prior to Early Paleozoic magmatism. We propose that low-degree silicate melts released from the asthenosphere infiltrated and solidified within the lithospheric mantle, forming non-peridotitic lithologies rich in amphibole, clinopyroxene and phlogopite. Subsequent lithospheric extension caused the melting of the most easily fusible material in the lithosphere, which gave rise to the Early Paleozoic alkaline magmatism in South Qinling.

1　Introduction

The petrogenesis of alkaline mafic rocks provides important information about the chemical composition of the Earth's upper mantle (Jung et al., 2006; Mayer et al., 2013). However, the nature of the source region and the mantle melting processes that generate alkaline magmas

---

①　Published in *Lithos*, 2020, 370-371.

②　Corresponding author.

are still contentious (Pilet et al., 2008; Zeng et al., 2010). Alkaline rocks are generally characterized by highly enriched incompatible trace element and radiogenic isotope compositions (Tappe et al., 2016). There is increasing evidence that highly enriched alkaline magmas which occur in a variety of intraplate tectonic settings (e.g., ocean islands and continental rifts), cannot be produced by partial melting of homogenous depleted mantle peridotite (Nelson et al., 2019; Pilet, 2015). Therefore, more enriched non-peridotite or volatile-bearing mantle sources are required to explain the magmatic major and trace element compositions (Foley, 1992; Pilet et al., 2011; Prytulak and Elliott, 2007). The source enrichment has been attributed to the recycling of subducted oceanic crust and variable sediment into the convecting mantle (Hofmann and White, 1982; Stracke, 2012). Alternatively, the enrichment can also be explained by mantle metasomatism (Niu and O'Hara, 2003; Pilet et al., 2008).

It has long been considered that mantle metasomatism caused either by hydrous or carbonate fluids/melts could be a common precursor to mafic alkaline magmatism (Fitzpayne et al., 2019). The interaction between metasomatic fluids/melts and peridotitic lithospheric mantle will result in the formation of metasomatic phases (amphibole and phlogopite) disseminated in peridotite, or veins dominated by those hydrous phases (Pilet et al., 2011). Significant amounts of incompatible trace elements (e.g., large ion lithophile elements (LILEs), high field strength elements (HFSEs), light rare earth elements (LREEs), and middle rare earth elements (MREEs)) are stored within these minerals, and their breakdown during melting processes would enhance the contents of these elements in melts (Ionov and Hofmann, 1995; LaTourrette et al., 1995; Mayer et al., 2014). In particular, given the low melting points of metasomatic phases (Foley et al., 1999; Médard et al., 2006; Pilet et al., 2008), their partial melting can readily be achieved by limited thermobaric perturbation of the lithosphere or lithospheric stretching (Rooney et al., 2017). Melting experiments and numerical modeling results indicate that the major and trace element compositions of alkaline mafic magmas can be reproduced by a high degree of melting of metasomatic veins (Pilet et al., 2008, 2011). Therefore, metasomatized lithospheric mantle is a potentially important source of alkaline mafic rocks.

The Early Paleozoicigneous rocks in the South Qinling Belt (central China) are characterized by voluminous alkaline intrusions that span a wide compositional range from alkali diabase to syenite and minor volcanic rocks (alkali basalt and trachyte; Chen et al., 2010; Wang et al., 2017; Zhang et al., 2002). Additionally, petrological studies of mantle xenoliths hosted in alkali basalt showed that the underlying lithospheric mantle was metasomatized and contained large amounts of hydrous mineral phases such as amphibole and phlogopite (Huang, 1993; Xu et al., 1997). These observations provide a unique opportunity to examine the potential role of lithospheric mantle in the genesis of alkaline rocks. In this study, we present combined zircon U-Pb ages, whole-rock major and trace element data and Sr-Nd-Hf isotope data for alkali diabases in the South Qinling Belt. These data were used to constrain the source composition and melting processes for alkali diabases, as well as to unravel the evolution of the primary magma. The results confirmed that diabases from the South Qinling Belt were derived from a metasomatized lithospheric mantle source.

2 Geological setting and petrography

The Qinling Orogenic Belt is a composite collision orogenic belt formed by the collision between the North China and South China Blocks along the Shangdan suture and Mianlue suture during the Paleozoic and Mesozoic, respectively (Dong and Santosh, 2016, and references therein). The collision zone links the Dabie-Sulu Orogenic Belt in the east and the Qilian-Kunlun Orogenic Belt in the west, forming the Central China Orogenic Belt (Xu et al., 2002). The Shangdan suture in the north and the Mianlue suture in the south have divided the Qinling orogen into the South Qinling and North Qinling belts (Fig. 1).

The South Qinling Belt (SQB) has been considered as the northern part of the South China Block before the Late Paleozoic (Dong and Santosh, 2016), and mainly consists of Precambrian crystalline basement and overlying Sinian to Triassic strata (Meng and Zhang, 2000). The crystalline basement is represented by the Paleoproterozoic Douling complex, which is predominantly composed of biotite gneiss, amphibolite and schist (Hu et al., 2013; Wang et al., 2013). The Neoproterozoic basement complexes in the South Qinling Belt mainly consist of the Wudang and Yaolinghe groups (Dong et al., 2017; Zhang et al., 2001). The Wudang and Yaolinghe groups are the most widely exposed basement strata in the Ziyang-Langao area, and consist of metabasalt and basaltic andesite with minor rhyolite or dacite. The felsic volcanic rocks have zircon U-Pb ages of around 680 Ma (Ling et al., 2008). The Paleoproterozoic basement rocks and metamorphosed Neoproterozoic clastic-volcanic rocks are unconformably overlain by a continuous succession of latest Neoproterozoic to Early Paleozoic strata (Dong et al., 2017). The thick cover of Sinian to Ordovician sedimentary rocks mainly consists of carbonate, shale, and sandstone (Gao et al., 1995); Silurian sedimentary rocks mainly consist of carbonaceous shale, carbonaceous slate, sericite phyllite and sandstone. In particular, there are abundant graptolite fossils in the sedimentary deposits (Luo and Duanmu, 2001).

The Early Paleozoic magmatism in the SQB is mainly composed of mafic and intermediate alkaline intrusions, alkali basalts, trachytes and a few carbonatite-syenite complexes (Chen et al., 2010; Wang et al., 2017; Xu et al., 2008; Zhang et al., 2002, 2007). The mafic intrusions include diabases, gabbros and pyroxenites. Most of them were concordantly intruded into surrounding Cambrian to Silurian strata and emplaced as sills, with several to hundreds of

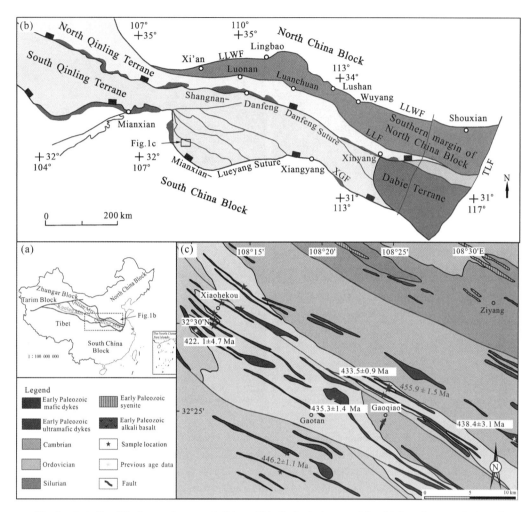

Fig. 1 (a) Simplified tectonic map of China. (b) Geological map of the Qinling Orogenic Belt and adjacent areas (modified after Dong and Santosh, 2016). (c) Simplified geological map showing the distribution of Early Paleozoic alkali diabases and syenites in the South Qinling Belt (modified after the 1 : 200 000 geological map of the Ziyang sheet). Sample locations are marked by stars. The age data are from Chen et al. (2014), Wang et al. (2015), and Zhang et al. (2020).

meters in width, and hundreds to thousands of meters in length (Fig. 2a). Samples used in this study were collected from the Xiaohekou and Gaoqiao areas (Fig. 1c).

The diabases in the SQB are medium to fine-grained with ophitic textures (Fig. 2c). The rocks dominantly consist of plagioclase (50 − 60 vol%), clinopyroxene (30 − 40 vol%), amphibole (~5 vol%) and opaque minerals (Ti-magnetite, <5 vol%) with accessory apatite, titanite and zircon. The plagioclase occurs as euhedral tabular grains (0.2 − 1.5 mm). Clinopyroxene grains are subhedral to anhedral and filled into the feldspar framework. Some well-developed plagioclase laths are enveloped by larger clinopyroxenes, forming a typical

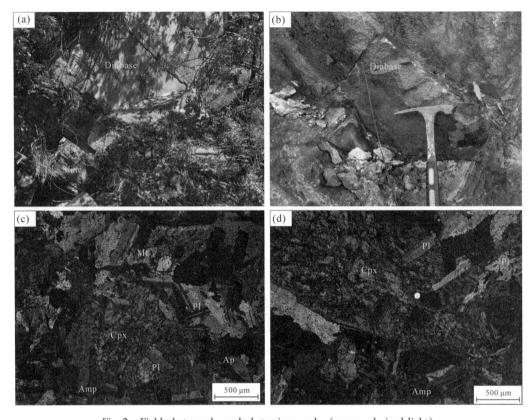

Fig. 2　Field photographs and photomicrographs (cross-polarized light)
of representative diabase samples from the South Qinling Belt.

(a) Diabases concordantly intruded into Silurian carbonaceous slate. (b) Field occurrence. (c) Clinopyroxenes partially replaced by amphiboles. (d) Euhedral plagioclase enclosed in a large clinopyroxene, forming an ophitic texture. Pl: plagioclase; Cpx: clinopyroxene; Amp: amphibole; Mt: magnetite.

diabasic texture (Fig. 2d). The grain sizes range from 0.2 mm to 1 mm. Amphibole occurs commonly as patches at the edges of clinopyroxene grains, indicating the high H_2O content of the magma (Fig. 2c). Some samples are slightly altered; secondary minerals include chlorite and clay minerals.

3　Analytical methods

Twenty diabase samples with minimal visible secondary alteration were selected for whole-rock major and trace element analyses at the State Key Laboratory of Continental Dynamics, Northwest University in Xi'an, China. Samples were powdered to 200 mesh size using a tungsten carbide ball mill. Major elements were determined on lithium borate fusion disks using X-ray fluorescence spectrometry (XRF). Analyses of USGS and Chinese national rock standards (BCR-2, GSR-1 and GSR-3) indicate that both the analytical precision and accuracy for major elements were generally better than 5%. Trace element contents were determined by inductively coupled plasma mass spectrometry (ICP-MS); sample powders were digested using an HF + HNO$_3$ mixture in high-pressure Teflon bombs at 190 ℃ for 48 h. The detailed sample-digesting procedure for ICP-MS analysis was described by Liu et al. (2007). Precision and accuracy were monitored by measurements of replicates and the standard reference materials BHVO-2, AGV-2, BCR-2 and GSP-2 (Table S1). Analytical precision for most trace elements was better than 10%.

Sr-Nd-Hf isotopic data were obtained using a Nu Plasma HR multi-collector mass spectrometer at the State Key Laboratory of Continental Dynamics, Northwest University. The isotopic fractionation of Sr and Nd was corrected to $^{87}Sr/^{86}Sr = 0.119\ 4$ and $^{146}Nd/^{144}Nd = 0.721\ 9$, respectively. During the analysis, the NIST SRM 987 standard yielded an average value of $^{87}Sr/^{86}Sr = 0.710\ 250 \pm 12$ (2σ, $n = 15$) and the La Jolla standard gave an average of $^{143}Nd/^{144}Nd = 0.511\ 859 \pm 6$ (2σ, $n = 20$). Full details of analytical methods were present in Gao et al. (2008). Whole-rock Hf was separated by single anion exchange columns, and Hf isotope data were normalized to $^{179}Hf/^{177}Hf = 0.732\ 5$. During analysis, 22 measurements of the JCM 475 standard yielded an average $^{176}Hf/^{177}Hf = 0.282\ 161\ 3 \pm 0.000\ 001\ 3$ (2σ) (Yuan et al., 2007).

Zircon grains from the diabases were separated using conventional heavy liquid and magnetic techniques. Representative zircon grains were handpicked, mounted in epoxy resin disks, and then polished and coated with carbon. Internal morphology was examined by cathodoluminescence (CL) prior to U-Pb isotopic analyses. Laser ablation (LA) ICP-MS zircon U-Pb analyses were conducted on an Agilent 7500a ICP-MS equipped with a 193 nm laser at the State Key Laboratory of Continental Dynamics at Northwest University. The detailed analytical technique is described by Yuan et al. (2004). Common Pb contents were evaluated using the method described by Andersen (2002). Age calculations and concordia diagrams were completed using the ISOPLOT version 3.0 software (Ludwig, 2003). The errors quoted in tables and figures are 2σ.

Major-element compositions of the minerals were determined using an electron microprobe (JXA-8230) at the State Key Laboratory of Continental Dynamics, Northwest University. The operating conditions included an acceleration voltage of 15 kV, a beam current of 10 nA, and a beam diameter of 1 μm. Natural and synthetic microprobe standards were supplied by SPI, and included olivine for Mg, jadeite for Si, Al, and Na, diopside for Ca, sanidine for K, rhodonite for Mn, hematite for Fe, and rutile for Ti.

4　Results

4.1　Zircon U-Pb dating

Two diabase samples were selected for zircon U-Pb analysis (GQ-4-1 and GT-10-1).

Zircon CL images and U-Pb isotopic results of the SQB diabases are presented in Fig. 3, and the isotope data are listed in Supplementary Table S2. Zircon grains from the diabase were mainly broken prismatic crystals, 50 – 200 μm long with aspect ratios of 1 : 1 – 1 : 5. In CL images, most grains showed dark and unzoned textures; only a few zircons had weak banded zoning (Fig. 3a,b). High Th/U ratios (1.20 – 2.79) in zircon grains indicated a magmatic origin. Nine analyses from sample GQ-4-1 yielded a weighted mean $^{206}Pb/^{238}U$ age of 455.9 ± 1.5 Ma (MSWD = 0.77, 2σ; Fig. 3c). Seventeen analyses of sample GT-10-1 yielded a weighted mean $^{206}Pb/^{238}U$ age of 446.2 ± 1.1 Ma (MSWD = 0.9, 2σ; Fig. 3d). These ages show that the SQB diabases crystallized at ~450 Ma.

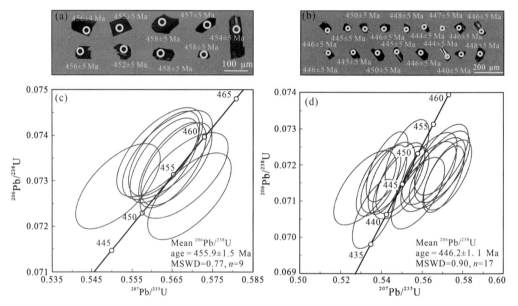

Fig. 3 Cathodoluminescence (CL) images of representative zircons from diabase dykes in the South Qinling Belt(SQB), (a) GQ-4-1 and (b) GT-10-1. U-Pb zircon concordia diagram for SQB diabases, (c) GQ-4-1 and (d) GT-10-1.

4.2 Major and trace element compositions

Major and trace element data for the SQB diabases are presented in Table 1. The samples showed low SiO_2 contents (42.1 – 49.5 wt%) and high TiO_2 contents (2.89 – 5.17 wt%). MgO contents varied between 4.0 wt% and 9.4 wt%. All of the samples were sodic with Na_2O = 2.45 – 4.59 wt%, K_2O = 0.24 – 1.55 wt% and Na_2O/K_2O = 2.5 – 11.9 (Fig. 4). The majority of these basalts fall in the domain of foidgabbro and monzogabbro in the total alkalis versus SiO_2 (TAS) diagram (Fig. 5). Al_2O_3 contents were between 11.72 wt% and 14.25 wt%, CaO contents between 6.64 wt% and 11.73 wt% and CaO/Al_2O_3 ratios between 0.49 and 0.91, which correlated positively with MgO contents(See Table 1).

Table 1 Major(wt%) and trace(ppm) element and whole-rock Sr-Nd-Hf isotopic compositions of diabase samples from the South Qinling Belt.

Sample	GQ-1-1	GQ-3-1	GQ-3-1*	GQ-4-1	GQ-4-2	XHK-2-1	XHK-2-2	XHK-4-1	XHK-5-1	XHK-7-1	XHK-8-1
Latitude(N)	32°21′48″	32°26′6″	32°26′6″	32°26′15″	32°26′15″	32°30′11″	32°30′11″	32°30′8″	32°30′10″	32°30′15″	32°30′23″
Longitude(E)	108°17′1″	108°24′34″	108°24′34″	108°24′45″	108°24′45″	108°26′40″	108°12′19″	108°12′26″	108°12′34″	108°12′40″	108°12′35″
SiO_2	44.73	42.65	42.56	48.71	49.02	44.20	44.16	42.70	42.10	45.61	43.61
TiO_2	4.97	3.46	3.45	3.16	3.16	4.91	5.17	3.87	4.03	4.57	4.41
Al_2O_3	12.81	12.06	11.99	14.23	14.21	12.66	12.46	11.72	14.25	13.44	11.80
$Fe_2O_3^T$	15.56	14.58	14.53	12.78	12.81	15.11	15.64	16.38	15.40	13.85	16.19
MnO	0.22	0.21	0.21	0.27	0.27	0.20	0.20	0.21	0.18	0.28	0.25
MgO	5.55	9.29	9.26	4.02	3.98	5.69	5.93	9.38	5.57	5.65	5.48
CaO	8.78	10.24	10.20	7.41	6.96	9.14	9.07	8.51	11.73	6.64	9.64
Na_2O	3.55	2.55	2.52	4.43	4.59	3.55	3.59	2.45	2.86	4.07	2.95
K_2O	0.58	0.72	0.71	1.41	1.42	0.69	0.55	0.54	0.24	1.25	0.78
P_2O_5	0.45	0.51	0.51	1.36	1.34	0.85	0.86	0.77	0.44	1.17	2.33
LOI	2.30	3.64	3.61	1.98	1.95	2.69	2.29	3.25	2.96	2.97	2.48
Total	99.50	99.91	99.55	99.76	99.71	99.69	99.92	99.78	99.76	99.50	99.92
$Mg^{\#}$	45	60	60	42	42	47	47	57	46	49	44
T_p/°C	–	1 261	1 258	–	–	–	–	1 300	–	–	–
P/GPa	–	1.39	1.37	–	–	–	–	1.50	–	–	–
Li	27.29	36.98	37.10	19.56	20.33	20.05	18.66	39.01	20.15	24.55	17.33
Be	0.89	0.96	0.94	1.92	1.77	1.05	0.99	0.69	1.03	0.94	1.02
Sc	28.51	22.43	22.37	12.66	12.49	25.22	26.09	20.45	25.43	21.94	21.38
V	434.59	326.82	325.14	189.44	184.47	385.12	394.79	347.61	460.36	335.72	338.16
Cr	4.00	56.81	56.83	2.68	2.42	44.93	47.67	303.32	49.99	6.14	1.80
Co	47.89	61.40	60.49	31.67	29.59	54.12	57.01	62.74	60.94	47.81	49.41

Continued

Sample	GQ-1-1	GQ-3-1	GQ-3-1*	GQ-4-1	GQ-4-2	XHK-2-1	XHK-2-2	XHK-4-1	XHK-5-1	XHK-7-1	XHK-8-1
Latitude(N)	32°21'48"	32°26'6"	32°26'6"	32°26'15"	32°26'15"	32°30'11"	32°30'11"	32°30'8"	32°30'10"	32°30'15"	32°30'23"
Longitude(E)	108°17' 1"	108°24' 34"	108°24' 34"	108°24' 45"	108°24' 45"	108°26' 40"	108°12' 19"	108°12' 26"	108°12' 34"	108°12' 40"	108°12' 35"
Ni	3.05	132.42	130.41	1.39	1.06	60.49	57.95	173.88	73.51	31.43	1.02
Cu	17.45	50.21	49.82	7.55	6.92	76.79	70.35	32.84	104.97	41.82	11.36
Zn	142.30	118.10	116.58	186.78	178.77	143.64	136.13	144.78	129.53	132.99	158.84
Ga	22.07	19.20	19.02	25.30	24.63	21.77	20.80	20.06	23.20	23.05	23.42
Ge	1.83	1.35	1.34	1.84	1.78	1.47	1.55	1.63	1.56	1.64	1.85
Rb	17.65	27.43	27.54	35.70	35.39	13.97	11.03	11.52	5.54	23.45	18.81
Sr	792.85	569.58	563.07	456.19	449.54	1 077.06	997.05	459.41	1 240.97	689.45	1 137.65
Y	26.55	23.63	23.39	45.42	44.39	30.22	30.61	26.64	23.34	36.94	42.49
Zr	165.27	154.41	152.45	283.69	285.87	170.85	176.60	151.66	152.47	190.86	180.81
Nb	26.11	26.26	26.11	53.01	51.55	27.91	29.15	24.84	24.51	31.43	31.07
Cs	4.59	2.92	2.89	2.37	2.64	2.00	2.23	0.92	0.35	0.45	5.55
Ba	513.60	436.48	430.70	765.01	855.46	1 021.17	816.95	500.44	205.04	1 281.83	2 358.34
La	25.66	24.68	24.34	52.71	51.68	30.42	30.19	26.59	22.47	39.26	41.34
Ce	57.62	55.20	54.28	119.03	117.16	69.89	69.82	61.12	49.87	92.42	100.86
Pr	7.78	7.19	7.13	15.75	15.53	9.72	9.61	8.41	6.38	12.80	14.41
Nd	35.91	32.24	31.88	69.72	68.57	43.59	43.64	37.86	28.73	60.13	70.46
Sm	7.92	7.10	7.01	14.39	14.18	9.56	9.59	8.27	6.40	12.64	15.06
Eu	3.18	2.67	2.62	5.19	5.05	3.45	3.35	2.88	2.32	4.62	5.52
Gd	7.71	6.78	6.71	13.50	13.29	9.05	9.08	7.88	6.20	11.85	14.31
Tb	1.06	0.94	0.93	1.84	1.82	1.22	1.24	1.07	0.88	1.58	1.87
Dy	5.79	5.11	5.06	9.83	9.71	6.47	6.61	5.70	4.83	8.30	9.63

Continued

Sample	GQ-1-1	GQ-3-1	GQ-3-1*	GQ-4-1	GQ-4-2	XHK-2-1	XHK-2-2	XHK-4-1	XHK-5-1	XHK-7-1	XHK-8-1
Latitude(N)	32°21'48"	32°26'6"	32°26'6"	32°26'15"	32°26'15"	32°30'11"	32°30'11"	32°30'8"	32°30'10"	32°30'15"	32°30'23"
Longitude(E)	108°17'1"	108°24'34"	108°24'34"	108°24'45"	108°24'45"	108°26'40"	108°12'19"	108°12'26"	108°12'34"	108°12'40"	108°12'35"
Ho	1.04	0.91	0.91	1.76	1.74	1.15	1.16	1.01	0.87	1.43	1.65
Er	2.59	2.29	2.25	4.40	4.34	2.86	2.92	2.53	2.21	3.55	4.02
Tm	0.33	0.30	0.29	0.57	0.56	0.37	0.37	0.32	0.29	0.45	0.49
Yb	1.94	1.71	1.70	3.28	3.22	2.13	2.14	1.87	1.66	2.52	2.69
Lu	0.27	0.24	0.23	0.45	0.45	0.30	0.30	0.26	0.23	0.35	0.37
Hf	4.20	3.95	3.89	6.83	6.83	4.34	4.45	3.84	3.78	4.85	4.59
Ta	1.80	1.75	1.72	3.43	3.36	1.89	1.98	1.67	1.57	2.20	2.16
Pb	1.46	2.53	3.08	3.11	2.79	4.63	4.74	2.08	1.96	2.19	3.13
Th	1.81	2.21	2.20	4.75	4.61	2.47	2.39	2.18	2.06	2.69	2.55
U	0.46	0.53	0.53	1.11	1.08	0.57	0.59	0.53	0.52	0.66	0.61
$^{87}\mathrm{Sr}/^{86}\mathrm{Sr}$	0.705 363(7)	–	–	0.706 851(5)	–	–	–	–	–	–	0.704 746(7)
$(^{87}\mathrm{Sr}/^{86}\mathrm{Sr})_i$	0.704 945	–	–	0.705 382	–	–	–	–	–	–	0.704 435
$^{143}\mathrm{Nd}/^{144}\mathrm{Nd}$	0.512 629(6)	–	–	0.512 560(7)	–	–	–	–	–	–	0.512 558(8)
$(^{143}\mathrm{Nd}/^{144}\mathrm{Nd})_i$	0.512 231	–	–	0.512 188	–	–	–	–	–	–	0.512 173
$\varepsilon_{\mathrm{Nd}}(t)$	3.5	–	–	2.7	–	–	–	–	–	–	2.4
T_{DMNd}	992	–	–	1 014	–	–	–	–	–	–	1 071
$^{176}\mathrm{Hf}/^{177}\mathrm{Hf}$	0.282 761(9)	–	–	0.282 769(10)	–	–	–	–	–	–	0.282 771(10)
$(^{176}\mathrm{Hf}/^{177}\mathrm{Hf})_i$	0.282 684	–	–	0.282 689	–	–	–	–	–	–	0.282 675
$\varepsilon_{\mathrm{Hf}}(t)$	6.9	–	–	7.1	–	–	–	–	–	–	6.6
T_{DMHf}	885	–	–	881	–	–	–	–	–	–	938

Continued

Sample	XHK-8-2	XHK-10-1	XHK-10-2	XHK-10-2*	GT-10-1	GT-10-2	GT-11-1	GT-11-2
Latitude(N)	32°30'23"	32°31'28"	32°31'28"	32°31'28"	32°22'46"	32°22'46"	32°21'48"	32°21'48"
Longitude(E)	108°12'35"	108°13'52"	108°13'52"	108°13'52"	108°17'51"	108°17'51"	108°16'60"	108°16'60"
SiO_2	43.57	49.32	49.52	49.22	42.81	38.24	42.78	42.95
TiO_2	4.40	2.94	2.89	2.90	5.59	6.73	4.68	4.66
Al_2O_3	12.06	13.63	13.68	13.64	13.32	13.07	12.25	12.36
$Fe_2O_3^T$	16.18	12.74	12.74	12.73	16.36	20.31	15.68	15.71
MnO	0.24	0.32	0.32	0.32	0.20	0.22	0.24	0.23
MgO	5.41	4.62	4.52	4.52	5.51	4.91	4.92	4.85
CaO	9.56	6.97	7.02	6.97	9.60	10.19	11.17	10.88
Na_2O	2.93	3.89	3.93	3.95	3.07	2.51	3.23	3.09
K_2O	0.78	1.55	1.54	1.55	0.70	0.29	0.78	0.81
P_2O_5	2.41	1.03	1.04	1.02	0.48	0.62	1.79	1.90
LOI	2.58	2.56	2.70	2.68	2.18	2.50	2.43	2.52
Total	100.12	99.57	99.90	99.50	99.82	99.59	99.95	99.96
$Mg^{\#}$	44	46	45	45	44	36	42	42
$T_p/℃$	-	-	-	-	-	-	-	-
P/GPa	-	-	-	-	-	-	-	-
Li	17.95	20.03	20.02	19.84	16.48	11.65	14.37	14.51
Be	0.96	1.56	1.53	1.53	0.92	0.79	1.20	1.15
Sc	20.67	18.28	18.00	18.25	24.69	20.74	19.28	19.16
V	335.19	181.21	178.67	181.35	437.08	583.08	374.32	373.44
Cr	1.78	1.38	1.39	1.43	4.60	4.65	3.43	3.64
Co	40.85	26.72	26.57	26.54	54.10	62.27	46.73	43.83
Ni	1.00	0.58	0.60	0.65	4.69	5.69	2.59	2.65

Continued

Sample	XHK-8-2	XHK-10-1	XHK-10-2	XHK-10-2*	GT-10-1	GT-10-2	GT-11-1	GT-11-2
Latitude(N)	32°30'23"	32°31'28"	32°31'28"	32°31'28"	32°22'46"	32°22'46"	32°21'48"	32°21'48"
Longitude(E)	108°12'35"	108°13'52"	108°13'52"	108°13'52"	108°17'51"	108°17'51"	108°16'60"	108°16'60"
Cu	11.35	6.20	6.19	6.30	19.96	27.02	16.50	15.53
Zn	151.19	168.39	167.97	168.99	154.06	186.84	177.23	170.63
Ga	23.78	26.14	26.01	26.09	22.09	23.70	22.75	23.17
Ge	1.84	1.81	1.82	1.83	1.42	1.34	1.75	1.73
Rb	18.49	34.28	34.13	34.44	14.92	7.36	14.82	15.78
Sr	1 167.41	902.31	897.71	904.17	738.39	857.12	561.95	581.04
Y	43.03	50.09	50.08	50.11	23.97	21.17	37.77	38.90
Zr	188.50	278.32	271.24	275.90	152.11	130.81	190.05	190.35
Nb	31.21	43.47	42.95	43.40	28.21	27.89	35.28	35.80
Cs	5.20	3.10	3.11	3.12	0.46	0.81	0.64	0.64
Ba	2 343.31	1 699.75	1 700.86	1 707.98	471.07	209.19	264.01	279.90
La	42.55	51.19	51.91	51.37	23.43	21.12	37.26	38.29
Ce	103.21	121.29	122.66	121.48	52.29	47.40	87.70	90.41
Pr	14.72	16.76	16.87	16.86	6.92	6.32	12.26	12.65
Nd	71.97	79.11	79.20	78.85	31.28	28.84	56.48	58.42
Sm	15.32	16.61	16.65	16.59	7.01	6.41	12.31	12.72
Eu	5.66	6.33	6.35	6.35	2.61	2.46	4.41	4.50
Gd	14.57	15.62	15.52	15.51	6.78	6.20	11.81	12.19
Tb	1.89	2.11	2.10	2.11	0.94	0.85	1.58	1.63
Dy	9.73	11.17	11.10	11.12	5.11	4.53	8.38	8.58
Ho	1.66	1.96	1.94	1.96	0.92	0.81	1.46	1.50
Er	3.98	4.84	4.83	4.83	2.29	2.00	3.55	3.67

Continued

Sample	XHK-8-2	XHK-10-1	XHK-10-2	XHK-10-2*	GT-10-1	GT-10-2	GT-11-1	GT-11-2
Latitude(N)	32°30′23″	32°31′28″	32°31′28″	32°31′28″	32°22′46″	32°22′46″	32°21′48″	32°21′48″
Longitude(E)	108°12′35″	108°13′52″	108°13′52″	108°13′52″	108°17′51″	108°17′51″	108°16′60″	108°16′60″
Tm	0.48	0.61	0.61	0.61	0.30	0.25	0.44	0.45
Yb	2.69	3.49	3.45	3.48	1.70	1.44	2.46	2.55
Lu	0.36	0.48	0.47	0.47	0.24	0.20	0.34	0.34
Hf	4.70	6.72	6.58	6.68	3.92	3.36	4.70	4.73
Ta	2.13	2.92	2.88	2.90	1.95	1.93	2.36	2.38
Pb	1.31	2.52	2.43	2.21	1.35	1.85	1.20	1.14
Th	2.57	3.76	3.68	3.70	2.00	1.65	2.70	2.73
U	0.64	0.92	0.91	0.91	0.48	0.41	0.64	0.65
$^{87}Sr/^{86}Sr$	–	–	0.705 104(10)	0.705 156(5)	–	–	0.704 996(5)	–
$(^{87}Sr/^{86}Sr)_i$	–	–	0.704 390	0.704 443	–	–	0.704 501	–
$^{143}Nd/^{144}Nd$	–	–	0.512 570(8)	0.512 604(6)	–	–	0.512 575(6)	–
$(^{143}Nd/^{144}Nd)_i$	–	–	0.512 191	0.512 225	–	–	0.512 182	–
$\varepsilon_{Nd}(t)$	–	–	2.7	3.4	–	–	2.5	–
T_{DMNd}	–	–	1023	964	–	–	1073	–
$^{176}Hf/^{177}Hf$	–	–	0.282 749(12)	0.282 767(8)	–	–	0.282 768(14)	–
$(^{176}Hf/^{177}Hf)_i$	–	–	0.282 663	0.282 680	–	–	0.282 681	–
$\varepsilon_{Hf}(t)$	–	–	6.2	6.8	–	–	6.8	–
T_{DMHf}	–	–	942	909	–	–	906	–

$Fe_2O_3^T$, total Fe as ferric iron; LOI, loss on ignition.

* Replicate analyses.

Mantle potential temperature (T_P) and melting pressure (P) was estimated using the approach from Putirka (2016).

Rb, Sr, Sm, Nd, Lu and Hf concentrations were analyzed by ICP-MS.

$(^{87}Sr/^{86}Sr)_i$, $\varepsilon_{Nd}(t)$, $\varepsilon_{Hf}(t)$ were calculated for the emplacement age of 455 Ma.

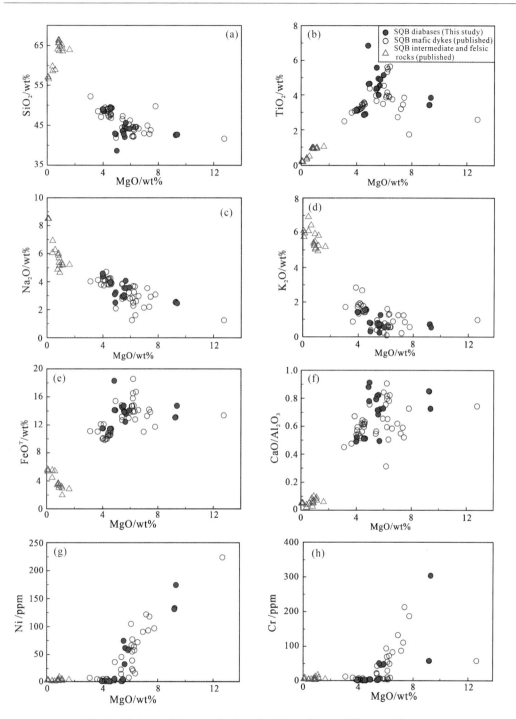

Fig. 4 Variation diagrams of selected major and compatible trace elements
for South Qinling Belt (SQB) diabases.

Also shown for comparison are previous data for Early Paleozoic mafic dykes in the SQB (Chen et al., 2014; Wang et al., 2015; Wang et al., 2017; Xiang et al., 2016; Zhang et al., 2007), and intermediate and felsic rocks in the SQB (Wang et al., 2017; Xiang et al., 2016).

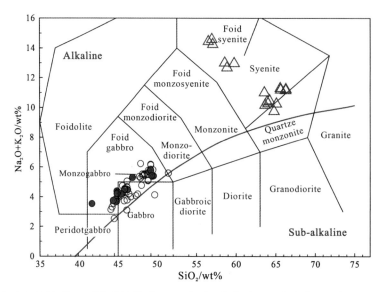

Fig. 5　Total alkali vs. SiO$_2$(TAS) diagram for South Qinling Belt diabases (Middlemost, 1994).

The concentrations of the compatible trace elements Ni and Cr were 0. 6 – 174 ppm and 1. 39 – 303 ppm, respectively. Ni and Cr concentrations were positively correlated with MgO. Chondrite-normalized rare earth element (REE) patterns are given in Fig. 6a. All samples were highly enriched in LREEs relative to heavy (H) REEs, with 90 to 220 times chondritic values for La and 10 to 20 times chondritic values for Yb. Chondrite normalized (La/Yb)$_N$ ranged from 9. 5 to 11. 5. All of the samples had slightly positive Eu anomalies (Eu/Eu* = 1. 09 – 1. 24). The primitive mantle (PM)-normalized incompatible trace element patterns are shown in Fig. 6b. All the samples displayed strong enrichment of most of the incompatible elements relative to the primitive mantle, including Ba, Sr, Nb and Ta. However, they also showed significant depletions in Rb, K, Pb, Zr and Hf compared with trace elements with similar incompatibility.

4. 3　Sr-Nd-Hf isotopes

Sr-Nd-Hf isotope data are reported in Table 1. The initial isotopic ratios of the samples were calculated for an emplacement age of 455 Ma. The SQB diabases displayed variable initial ^{87}Sr/^{86}Sr ratios from 0. 704 39 to 0. 705 38, in contrast with the narrow range of moderately depleted initial Nd isotope compositions [$\varepsilon_{Nd}(t)$ of + 2. 4 and + 3. 5; Fig. 7a]. The samples also exhibited moderately depleted $\varepsilon_{Hf}(t)$ values of + 6. 1 to + 7. 1 (Fig. 7b). It should be noted that the initial Nd isotopic compositions of diabases overlapped with the compositional fields of early Paleozoic hydrous xenoliths but had more radiogenic Sr isotope compositions.

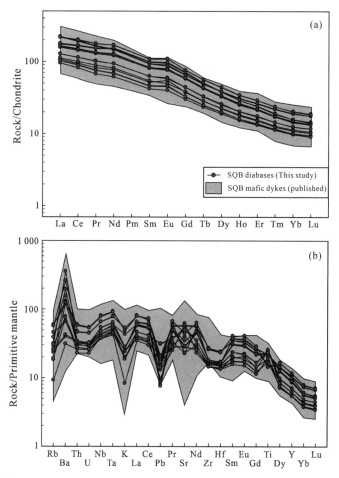

Fig. 6　(a) Chondrite-normalized rare earth element (REE) patterns and (b) primitive-mantle normalized trace element distribution patterns for SQB diabases.

The chondrite and primitive mantle data are from Sun and McDonough (1989) and McDonough and Sun (1995), respectively.

4. 4　Mineral chemistry

Clinopyroxenewas a ubiquitous phase observed in all the SQB diabase samples. According to the classification of Morimoto et al. (1988), all of the clinopyroxene was augite (En_{38-44} $Fs_{13-17}Wo_{41-45}$; Supplementary Table S3). The compositions of clinopyroxene were rich in MgO (12. 9-15. 1 wt%) and CaO (19. 3-21. 3 wt%) but poor in Cr_2O_3(<0. 1 wt%) and NiO (<0. 02 wt%). No evident compositional zoning was observed in the clinopyroxene.

The compositions of amphibolewere were rich in FeO (13. 7-15. 2 wt%), MgO (11. 9-12. 3 wt%), CaO (10. 5-10. 8 wt%) and TiO_2(3. 5-3. 9 wt%), with minor amounts of Na_2O (2. 2 - 2. 4 wt%) and K_2O (0. 6 - 0. 7 wt%; Supplementary Table S4). Based on the recommendations of the International Mineralogical Association Commission on New Minerals,

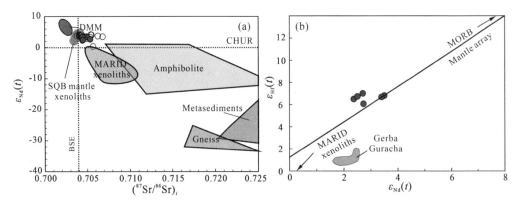

Fig. 7　The (a) $\varepsilon_{Nd}(t)$ vs. $({}^{87}Sr/{}^{86}Sr)_i$ and (b) $\varepsilon_{Hf}(t)$ vs. $\varepsilon_{Nd}(t)$ composition
for South Qinling Belt (SQB) diabases.

Depleted MORB mantle (DMM) is from Workman and Hart (2005). Hydrous mantle xenoliths in the SQB are from Xu (1998) and mica-amphibole-rutile-ilmenite-diopside (MARID) xenoliths are from Fitzpayne et al. (2019). Fields for Gerba Guracha lavas in the East African Rift are shown for comparison. Crustal rock from the Kongling Complex consists of amphibolite, gneiss and metasediment (Gao et al., 1999). Mantle array after Chauvel et al. (2008).

Nomenclature and Classification subcommittee on amphiboles (Hawthorne et al., 2012), the amphiboles were classified as Ti-rich magnesio-hastingsite to Ti-rich pargasite. The crystallization temperature (T), pressures (P) and H_2O content of the melt in equilibrium with amphibole were calculated using the algorithm of Ridolfi et al. (2010). The results indicate that the amphiboles crystallized at pressures from 196 MPa to 230 MPa and temperatures of 887−909 ℃, with H_2O contents of 4.4−4.7 wt%.

5　Discussion

5.1　Sample alteration

　　The SQB diabases showed minor alteration with chlorite replacement of clinopyroxene and clay minerals replacing plagioclase. Loss on ignition (LOI) values (1.95 − 3.64 wt%) reflected the presence of hydrous minerals (amphibole and biotite) or secondary minerals (chlorite and calcite), indicating that alteration or metamorphism had affected the rocks to various degrees. LILEs, such as Ba, Rb and K, are regarded as alteration-sensitive elements, whereas HFSEs (such as Zr, Hf, Nb, Ta and Th) are immobile during alteration. Thus, the correlations of trace elements versus Zr are useful to test potential shifts caused by alteration (Polat et al., 2002). The SQB diabases showed good correlations of K, Rb, REEs and HFSEs versus Zr, suggesting that these elements were not modified significantly during alteration (Fig. 8). However, Ba and Sr were probably affected by alteration, because there were no correlations between Ba, Sr and Zr (Fig. 8). Therefore, Ba and Sr are not used in the subsequent discussion of source characteristics.

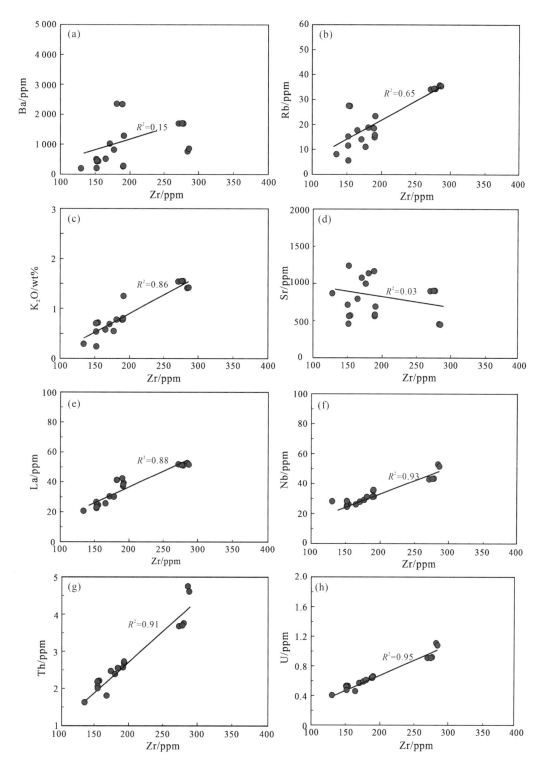

Fig. 8 Variation of selected trace elements vs. Zr for South Qinling Belt diabases.

5. 2　Fractional crystallization and crustal contamination

The samples from this study showed variable MgO contents (4. 0–9. 4 wt%) and variable Ni and Cr concentrations (0. 6–174 ppm and 1. 38–303 ppm, respectively; Fig. 4), indicating that they were not primary melts but experienced fractional crystallization. The negative correlations between MgO and SiO_2, CaO, CaO/Al_2O_3, TiO_2 and FeO^T coupled with the occurrence of minerals in rocks indicate fractionation of clinopyroxene, amphibole and Fe-Ti oxides from the parental magmas. To deduce the source characteristics from the most primitive samples, only samples with <45 wt% SiO_2 and >5 wt% MgO ($Mg^{\#}$>44) were used to identify mantle source characteristics.

Mantle-derived mafic dikes may be contaminated by crustal melts during ascent or emplacement into the shallow crust. Some trace element ratios, such as Ce/Pb and Nb/U, can effectively assess the influence of crustal assimilation because they are not fractionated during mantle melting and therefore reflect the composition of the Earth's mantle (Hofmann et al., 1986). Ocean island basalts (OIBs) have high Ce/Pb (25±5) and Nb/U (47±10), whereas continental crust has low Ce/Pb (< 5) and Nb/U (< 10; Rudnick and Gao, 2003). Substantial assimilation of continental crust lowers these ratios in mantle-derived magmas. The majority of the SQB diabase samples plot within the range for undifferentiated oceanic and continental intraplate basalts with high Ce/Pb (> 20) and Nb/U (> 40), except three samples, GQ-3-1, XHK-2-1 and XHK-2-2, which exhibit Ce/Pb < 20, indicating crustal contamination (Fig. 9); thus, we do not consider these samples further.

5. 3　Mantle source of SQB diabases

5. 3. 1　The nature and mineralogy of the enriched component

The high TiO_2 and low SiO_2 contents coupled with enriched incompatible elements in the SQB diabases cannot be explained solely by the melting of a homogenous peridotite source (Prytulak and Elliott, 2007; Tappe et al., 2006). The concentration of highly incompatible elements such as Th, U and LREEs in the SQB diabases can be explained by extremely low degrees of partial melting of peridotite sources, because the concentrations of those elements are controlled by that process (Pilet, 2015). However, the concentrations of intermediately incompatible elements such as Ti, Y and HREEs are controlled by the mineralogical and geochemical composition of the source region (Pilet, 2015; Rooney et al., 2017). Peridotite with a primitive mantle or depleted mantle composition contains insufficient TiO_2 to produce the high TiO_2 alkaline melts (Prytulak and Elliott, 2007). Therefore, an additional TiO_2 enriched component such as pyroxenite/eclogite or hornblendite, is required to produce the observed compositional character.

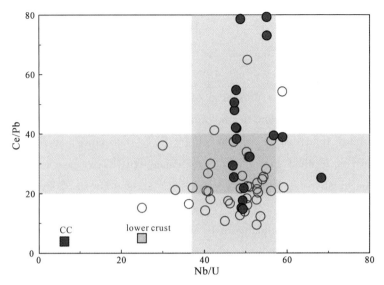

Fig. 9　Ce/Pb vs. Nb/U for South Qinling Belt diabases.

The light gray bars denote the range of ocean island basalt (OIB) magmas (Hofmann et al., 1986).

Average continental crust (CC) and lower continental crust (LC) are from Rudnick and Gao (2003).

The presence of hydrous K-bearing phases (amphibole and/or phlogopite) in the mantle source is supported by negative K anomalies in PM-normalized trace element patterns of the SQB diabases (Jung et al., 2006; Mayer et al., 2013). Because K is a stoichiometric component in amphibole and phlogopite, it will be retained in the mantle source as long as these minerals are residual during partial melting processes (Panter et al., 2018; Tappe et al., 2006). Amphibole is expected to be the dominant phase in melting assemblages as a result of its potential to produce silica-undersaturated melts with $Na_2O/K_2O > 1$ (2.5−11.9 for SQB diabases). In contrast, partial melting of phlogopite dominated lithologies will produce potassic and ultrapotassic melts with $Na_2O/K_2O < 1$ (Condamine and Médard, 2014; Foley et al., 1999; Médard et al., 2006; Pilet et al., 2008). Furman and Graham (1999) suggested that melts in equilibrium with amphibole should have lower Rb/Sr (< 0.1) and higher Ba/Rb (> 10) than melts coexisting with phlogopite; the SQB diabases generally had low Rb/Sr (0.004−0.078) and high Ba/Rb (16−127), suggesting the existence of amphibole in their mantle source. It should be noted that minor residual phlogopite is required in melting assemblages because K is moderately incompatible in amphibole with the partition coefficients $^{amp/melt}D_K = 0.17−1.36$, whereas K is compatible in phlogopite ($^{phl/melt}D_K = 3−4$), which can fractionate K more efficiently than amphibole (Adam and Green, 2006; Dalpe and Baker, 1994; LaTourrette et al., 1995; Tiepolo et al., 2007). Additionally, melting experiments have demonstrated that amphibole is completely consumed only 50 ℃ above the solidus, whereas phlogopite persists to comparatively higher temperatures above the solidus with the potential to be the residual mineral (Condamine and Médard, 2014; Foley et al., 1999). The primitive SQB diabases have low CaO and CaO/Al_2O_3 values, indicating that residual clinopyroxene exerts significant control on CaO content. Specifically, clinopyroxene is the product of incongruent amphibole melting (Foley et al., 1999; Pilet et al., 2008).

REE systematics can be used as proxies to estimate melting depth, because REEs carry information about source mineralogy and degree of melting. Whether the melting occurred in the spinel or garnet stability field can be determined by REE systematics, such as La/Yb vs. Dy/Yb. The presence of residual garnet would fractionate HREEs because they are strongly partitioned into garnet relative to LREEs and MREEs (Adam and Green, 2006; Hauri et al., 1994). It is noteworthy that MREE/HREE (e.g., Dy/Yb and Tb/Yb) ratios have been regarded as a better indicator for residual garnet, whereas LREE/HREE (e.g., La/Yb) ratios are primarily a function of the degree of melting (Pfänder et al., 2018). The SQB diabases showed relatively high La/Yb (11.1 – 19.5) and Dy/Yb (3.0 – 3.6) ratios (Fig. 10a), indicating that garnet was present in their mantle source. The effect of residual garnet is also supported by the elevated Fe/Mn values in primitive diabases (> 60; Davis et al., 2013). However, a low-degree of melting of metasomatized harzburgite or a high-degree of melting of hornblendite within spinel-facies mantle would also produce melts with high La/Yb and Dy/Yb signatures (Pilet et al., 2008; Rooney et al., 2017). Thus, whether garnet was present in the mantle source of the SQB diabases requires further confirmation.

Metasomatized harzburgite is unlikely to have been the mantle source for the SQB diabases, because magmas in equilibrium with metasomatized harzburgite are characterized by high SiO_2 but low CaO and Al_2O_3 contents (Prelević and Foley, 2007). The parameter Dy/Dy^* can be used to quantify the curvature of the REE pattern (Davidson et al., 2013). In the mantle, amphibole and clinopyroxene are major minerals that preferentially sequester MREEs with respect to LREEs and HREEs (Adam and Green, 2006; Tiepolo et al., 2007). Partial melting of amphibole-rich veins or amphibole-bearing peridotite will produce melts with lower Dy/Yb and Dy/Dy^* values (Davidson et al., 2013). In contrast, partial melting with residual garnet in the source will not change the curvature of the melts, but will drastically increase Dy/Yb ratios (Fig. 10b). Despite elevated Dy/Yb values, the SQB diabases displayed a trend toward lower Dy/Yb and Dy/Dy^* values, suggesting the presence of residual amphibole and/or clinopyroxene as well as the absence of garnet in the mantle source.

The geochemical data indicate that the SQB diabases were derived from a metasomatized mantle assemblage rich in amphibole, clinopyroxene and phlogopite. Additional evidence that amphibole-rich lithologies were involved in the generation of the SQB diabases comes from xenoliths hosted in Early Paleozoic alkali basalts in the Langao area of the SQB (Huang, 1993; Xu et al., 1997). Various stages of mantle metasomatism in the SQB have been recorded

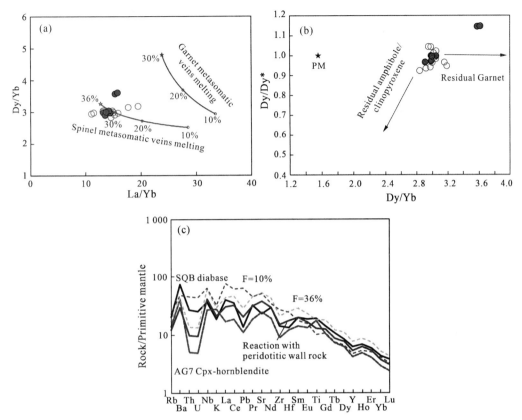

Fig. 10 (a) Dy/Yb vs. La/Yb for South Qinling Belt (SQB) diabases. Continuous curves represent incongruent batch melting models for spinel and garnet amphibole-clinopyroxene-phlogopite vein sources using the trace element composition of AG7 cpx-hornblendite xenolith from the French Pyrenees as the source composition (Pilet et al., 2008). Model parameters and partition coefficients are given in Table 2. The asterisk in the melting curve represents the complete consumption of amphibole with Ol + Cpx + Gt/Sp as residual minerals. The number along the model curves indicate the degree of melting. The vein-plus-wall rock melting result is not shown in rare earth element (REE) systematics because melt reactions with wall rock will dilute the REE concentrations but not change the REE ratios unless the amount of orthopyroxene dissolved in the melts is unrealistically high (>80%). (b) Dy/Dy* vs. Dy/Yb values were used to evaluate the residual minerals in the mantle (Davidson et al., 2013). (c) Primitive-mantle normalized multi-element diagrams depicting the model results of vein-plus-wall rock melting. The effect of vein-plus-wall rock melting was quantitatively simulated by dissolving orthopyroxene from the peridotitic wall-rock in alkaline melts. The orthopyroxene composition was taken from Su et al. (2012). A representative SQB diabase pattern (XHK-4-1) is shown for comparison. The deep and light blue dashed lines represent 10% and 36% partial melting of amphibole-clinopyroxene-phlogopite veins, respectively. The purple line represents 36% partial melting of amphibole-clinopyroxene-phlogopite veins with subsequent dissolution of ~30% orthopyroxene from the wall-rock peridotite within the spinel stability field, which resembles the composition of the SQB diabases.

For interpretation of the references to colour in this figure legend, the reader is referred to the web version of this article.

Table 2 Partition coefficients, source composition, source modes and melting reactions used in the partial melting modeling.

	Partition coefficient						Abundance /ppm	
	Ol	Cpx	Gt	Sp	Amp	Phl	AG7 metasome	Opx
Rb	0.000 18	0.000 7	0.000 7	0.000 01	0.1	3	8.1	0.18
Ba	0.000 3	0.000 68	0.000 7	0.000 01	0.36	4	275	1.07
Th	0.000 01	0.000 8	0.001 5	0.000 01	0.018	0.01	0.4	0.02
U	0.000 4	0.000 8	0.005	0.000 01	0.016	0.01	0.1	0.01
K	0.000 001	0.007 2	0.000 01	0.000 01	0.434	3.67	6 641	0
Nb	0.005	0.007 7	0.02	0.01	0.53	0.6	18	0.22
La	0.000 4	0.053 6	0.01	0.000 6	0.23	0.05	11.2	0.08
Ce	0.000 5	0.085 8	0.021	0.000 6	0.4	0.055	32.1	0.21
Pb	0.000 01	0.01	0.000 5	0.000 01	0.096	0.2	1.7	0.23
Pr	0.001	0.1	0.045	0.000 6	0.63	0.06	4.8	0.03
Sr	0.000 19	0.13	0.006	0.000 01	0.69	0.3	500	1.72
Nd	0.001	0.19	0.087	0.000 6	0.91	0.065	24.2	0.13
Sm	0.001	0.29	0.217	0.000 6	1.39	0.07	5.8	0.05
Zr	0.01	0.12	0.32	0.07	0.43	0.08	96	3.49
Hf	0.005	0.26	0.32	0.07	0.72	0.08	3.5	0.09
Eu	0.002	0.47	0.4	0.000 6	1.44	0.09	2.1	0.02
Ti	0.02	0.38	0.2	0.15	2.5	3	21 758	719
Gd	0.002	0.48	0.498	0.000 6	1.65	0.095	5.7	0.04
Tb	0.002	0.48	0.75	0.000 6	1.62	0.1	0.71	0.01
Dy	0.002	0.44	1.06	0.001 5	1.58	0.1	4.2	0.05
Ho	0.002	0.42	1.53	0.002 3	1.54	0.1	0.73	0.02
Y	0.005	0.4	2.11	0.004 5	1.48	0.1	18	0.43
Er	0.002	0.39	3	0.003	1.34	0.1	1.79	0.07
Yb	0.001 5	0.43	4.03	0.004 5	1.1	0.1	1.27	0.1
Lu	0.001 5	0.43	5.5	0.005 3	1.1	0.1	0.16	0.02

Modes and melting reaction						
	Ol	Cpx	Gt	Sp	Amp	Phl
Spinel metasome						
Source mode	0	0.25	0	0	0.6	0.15
Melt mode	−0.07	−0.3	0	−0.03	1	0
Garnet metasome						
Source mode	0	0.25	0	0	0.6	0.15
Melt mode	0	−0.34	−0.16	0	1	

Partition coefficients for Ol, Cpx, Gt and Amp are from Pilet et al. (2011 and reference therein), for Sp are from Kelemen et al. (2003). Source mode and melt mode reactions of spinel and garnet hornblendite are from Pilet et al. (2008) and Ma et al. (2011), respectively. OPX composition are from Su et al. (2012).

in these xenoliths, in which mantle peridotite has been completely replaced by amphibole pyroxenites, hornblendite or glimmerites. Similar phenomena have been observed in East African Rift localities such as Uganda and northern Kenya (Davies and Lloyd, 1989). Notably, the SQB diabases had initial Nd isotopic compositions overlapping with the compositional fields of Early Paleozoic hydrous xenoliths (Fig. 7a). The isotopic similarity between the diabases and hydrous mantle xenoliths confirms the interpretation that amphibole-rich metasomes are a potential mantle source for the SQB diabases.

The metasomatic vein model satisfies the majority of geochemical characteristics of the SQB diabases, such as high TiO_2 content and incompatible trace element enrichment, as well as the REE pattern. This includes the low MgO content and contents of the compatible elements Ni and Cr in the SQB diabases, which were not solely caused by fractional crystallization processes but also consistent with the involvement of amphibole-rich non-peridotitic veins (Mayer et al., 2013). However, there is one problem in using a model where melts derived from amphibole-rich veins are highly silica-undersaturated ($SiO_2 < 40$ wt%; Foley et al., 1999; Pilet et al., 2008); it differs from primitive SQB diabase samples (42-45 wt% SiO_2). To reconcile this discrepancy, a vein-plus-wall rock melting model was invoked to explain the petrogenesis of the SQB diabases (Foley, 1992).

5.3.2 Melting mechanism and condition

Metasomatic veins have lower solidus temperatures than volatile-free peridotite as a result of the abundance of volatiles and incompatible elements accommodated in hydrous and accessory phases (Foley, 1992; Foley et al.,1999; Pilet et al., 2008). During partial melting processes, amphibole-rich metasomatic veins tend to melt to a higher degree with the production of low-silica alkaline melts and infiltrate the surrounding peridotitic wall rock. Two mechanisms have been proposed for vein-plus-wall rock melting processes: ①mixing of vein-derived melt with low-degree partial melts of peridotite (e.g., Ma et al., 2011; Tappe et al., 2016), and ② dissolution of orthopyroxene from the wall-rock peridotite in silica-undersaturated vein derived melts (Foley, 1992; Panter et al., 2018). In this study, we favored the mechanism of dissolution of wall-rock minerals because easily fusible materials (pyroxenite and hornblendite) embedded in peridotite are more likely to melt to a higher degree than peridotite, when the latent heat is removed, the adjacent peridotite mantle will be undercooled, and the melting will be suppressed (Daniele et al., 2018; Pilet et al., 2008). In contrast, the dissolution of orthopyroxene from the wall-rock peridotite is caused by the disequilibrium between vein-derived strongly silica-undersaturated melts and wall-rock mineral orthopyroxene. Such reactions occurred below the solidus of adjacent peridotite (Foley, 1992). Experimental studies demonstrate that this reaction will evolve low-silica alkaline melts into melts with higher SiO_2 contents and lower TiO_2 and trace element contents while maintaining constant trace element patterns (Pilet et al., 2008). These features are consistent with the geochemical characteristics of the SQB diabases (42–45 wt% SiO_2). It should be noted that reactions between vein melts and wall rock will change the major element contents of a melt, but the incompatible trace element contents and isotope compositions will still be dominated by the vein composition (Tappe et al., 2006).

Amphibole and phlogopite are not stable in the convecting asthenosphere or within upwelling mantle plumes (Class and Goldstein, 1997). The presence of amphibole and phlogopite in the mantle source of the SQB diabases suggests that the melting region may have been located in the cooler lithospheric mantle. The stability field of amphibole in upper mantle rocks is restricted to pressures less than 3.0 GPa and temperatures less than 1 100 ℃ (Green et al., 2010), whereas, in fertile mantle compositions, the upper stability limit of amphibole is enhanced to ~3.8 GPa and ~1 100 ℃ (Mandler and Grove, 2016). Moreover, the transition from garnet to spinel is estimated to occur over a pressure range of 2.5–3.0 GPa (Robinson and Wood, 1998). It is likely that the diabases formed by partial melting of the spinel phlogopite (phl)-clinopyroxene (cpx)-hornblendite veins, which suggests that the melting region was located in the spinel stability field (<2.5–3.0 GPa).

The melting conditions of the SQB diabases are modeled using the algorithm of Putirka (2016), which is valid for magmas produced from a metasomatized mantle source (Natali et al., 2018). For the diabases with MgO >9 wt%, the primary magmas were reconstructed by adding olivine incrementally to the melt until it was in equilibrium with the mantle source ($Mg^\# = 87$; Natali et al., 2018) and the H_2O content of melts was assumed to be 4.5 wt% (Supplementary Table S4). The results indicate that the SQB diabases were generated within a mantle potential temperature (T_p) range of 1 258–1 300 ℃ at pressure (P) of 1.4–1.5 GPa (Table 1), within the spinel stability field. The P-T estimates are considered to be consistent for the generation of alkali magmas by the partial melting of the metasomatized mantle source (Pilet et al., 2008).

5.3.3　Mantle melting modeling

To constrain the nature of the mantle source and the role of hydrous metasomatic veins during melt generation, incongruent batch melting modeling was performed (Shaw, 1970). We approximated a metasomatic vein composed of 60% amphibole, 25% clinopyroxene and 15% phlogopite. This mineral assemblage is consistent with the mineralogy of metasomatized lithospheric mantle xenoliths hosted by alkaline volcanic rocks (Davies and Lloyd, 1989; Huang, 1993; Xu et al., 1997). The composition of the source veins was based on the AG7 clinopyroxene-hornblendite from the French Pyrenees, which are representative of metasomatic veins (Pilet et al., 2008). The effect of vein-plus-wall rock melting was simulated by dissolved orthopyroxene in alkaline melts. Model parameters and partition coefficients are given in Table 2.

The melting model results were compared with the trace element compositions of the SQB diabases (Fig. 10c). In general, partial melting of an amphibole-clinopyroxene-phlogopite vein produced normalized trace element patterns resembling the SQB diabases. At 36% melting (complete consumption of amphibole), a good fit with respect to the REE, HFSE and LILE patterns was achieved. In particular, the AG7 cpx-hornblendite yielded melts with positive Ba and Nb anomalies, as well as negative Zr-Hf and K anomalies, which were observed in the SQB diabases. However, partial melts derived from metasomatic veins displayed higher trace element concentrations than the SQB diabases. If metasomatic vein-derived silica-undersaturated melts reacted with orthopyroxene hosted in peridotitic wall rock, a parallel trace element pattern with lower trace element concentrations can be obtained. The vein-plus-wall rock melting result (30% orthopyroxene dissolved) provided a better match to the observed compositional pattern of the SQB diabases (Fig. 10c). There remained discrepancies between the model results and sample data, including that the model melts showed lower Rb, Ba, Th, U and Ti contents, which suggest that the source of the SQB diabases was probably more enriched in these elements than the AG7 cpx-hornblendite. This interpretation is supported by the fact that hydrous pyroxenites and hornblendite xenoliths found in SQB volcanism have higher Rb, Ba, Th, U and Ti contents compared with the AG7 cpx-hornblendite (Xu et al., 1997). In REE systematics (Fig. 10a), most of the SQB diabases formed a coherent group and plotted near the model curve for spinel phl-cpx-hornblendite. The overall variation in REE indicated melting degrees ranging between 20% and 36%, which is consistent with the estimated degree and experimental results (Pilet et al., 2008). Notably, melts reacting with wall rock dilute the REE concentrations but do not change the REE ratios unless the amount of orthopyroxene dissolved in those melts is unrealistically high (>80%). In summary, the partial melting model suggests that the SQB diabases can be satisfactorily explained by a high degree of partial melting of a metasomatic vein with subsequent reaction with peridotitic wall rock.

5.4　The style of mantle metasomatism

Mantle metasomatism by percolating melts or fluids with various compositions is a key mechanism that accounts for mantle enrichment and isotopic anomalies (Ackerman et al., 2013). Metasomatic agents include silicate (e.g., Zanetti et al., 1996) and carbonatitic (e. g., Lai et al., 2014; Yaxley et al., 1998) melts.

Carbonatitic melts are effective metasomatic agents in the mantle because of their low viscosity and high mobility (Hammouda and Laporte, 2000); the percolating melts react with mantle peridotite and produce specific geochemical signatures (e.g., Green and Wallace, 1988). The enrichment in LREEs, LILEs and Nb coupled with depletion in Zr, Hf and Ti, is often considered as a characteristic signature of carbonatitic metasomatism (e.g., Dai et al., 2017, 2018; Dasgupta et al., 2007; Zeng et al., 2010). Negative Zr-Hf and K anomalies observed in the SQB diabases can be interpreted as indicators of carbonatitic metasomatism. Nevertheless, carbonatitic metasomatism would increase the La/Yb values and deplete the HFSE[4+] contents (e.g., low Ti/Eu and Hf/Sm; Dai et al., 2017). As shown in Fig. 11, the SQB diabases have lower La/Yb (13-16) but higher Ti/Eu (2 700-13 000) than basalts derived from carbonated mantle sources (e.g., Dai et al., 2017, 2018; Zeng et al., 2010). Furthermore, carbonatite infiltrating the lithospheric mantle could generate metasomatized domains with elevated Nb/Ta and Zr/Hf ratios, which could then be transferred to intraplate basaltic rocks (Pfänder et al., 2012). However, the relatively low Nb/Ta (14-16) and Zr/Hf (39-42) in the SQB diabases plot within the OIB domain (Pfänder et al., 2012; Fig. 11a). These discrepancies preclude the direct participation of carbonatitic metasomatism in the genesis of the SQB diabases.

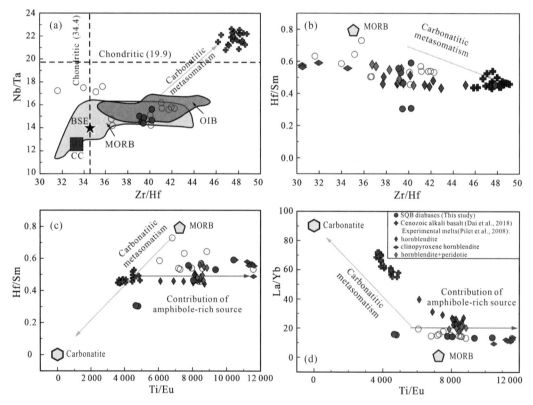

Fig. 11　Constraints on the metasomatic agent.

(a,b) Nb/Ta and Hf/Sm vs. Zr/Hf (modified after Pfänder et al., 2012). Low Hf/Sm and relatively low Nb/Ta and Zr/Hf in diabases argue against carbonate metasomatism. (c,d) Hf/Sm and La/Yb vs. Ti/Eu. Data for alkali basalt from West Qinling (Dai et al., 2018) are shown for comparison. Also shown are compositions of mid-oceanic ridge basalt (MORB; Hofmann, 1988) and oceanic carbonatite (Hoernle et al., 2002). Experimental melts of amphibole-rich sources are from Pilet et al. (2008).

Low-degree silicate melt metasomatism is a widespread phenomenon in the mantle (Niu and O'Hara, 2003; Tappe et al., 2006). Low-degree melts released from the asthenosphere are enriched in volatiles as well as incompatible elements and result in modal and cryptic metasomatism in peridotite when they migrate upward and infiltrate the lithospheric mantle (Pilet et al., 2008). As a result of the low heat content of low-degree silicate melts, their migration is likely to terminate within the lithospheric mantle (McKenzie, 1989). With decreasing pressure and temperature, fractional crystallization of metasomatic melts within the lithospheric mantle will generate a continuum of cumulate assemblages from anhydrous (clinopyroxene ± garnet ± olivine) to hydrous (amphibole + clinopyroxene ± phlogopite) veins (Foley, 1992; Pilet et al., 2011; Fig. 12a). The hydrous metasomatic veins are considered to be a potential source for alkaline magmas. There are two important aspects of the metasomatic veins model. First, the major and trace element compositions of hydrous metasomatic veins are largely controlled by the mineralogical structure of amphibole and the partition coefficients between minerals and liquid instead of the initial melt composition (Pilet et al., 2011). Second, as a result of the low solidus of amphibole (Condamine and Médard, 2014; Foley et al., 1999; Médard et al., 2006), such metasomatic veins are expected to melt to a higher degree, and possibly even to be completely consumed during mantle melting. There is no significant trace element fractionation during this process, and the composition of the generated magmas is largely inherited from their source vein (Pilet et al., 2008). Melting of hydrous metasomatic veins can explain many of features observed in the SQB diabases, such as low SiO_2 contents, high contents of incompatible elements, and fractionated REE patterns. Particularly, the pronounced negative Zr-Hf anomalies are also compatible with the metasomatic vein model. Because both Zr and Hf are moderately incompatible in amphibole, whereas the neighboring elements Sm and Nd are moderately incompatible to compatible in amphibole with $D_{Sm} > D_{Nd} > D_{Hf} > D_{Zr}$ (Tiepolo et al., 2007), the generated amphibole veins are expected to be characterized by negative Zr-Hf anomalies, consistent with the compositions of amphiboles observed in lithospheric veins or disseminated in peridotite (e.g., Ionov and Hofmann, 1995; Pilet et al., 2008; Zanetti et al., 1996). High-degree partial melting of amphiboles is responsible for the negative Zr-Hf anomalies in the generated melts. As shown in Fig. 11c, elevated Ti/Eu but low Hf/Sm values in the SQB diabases are consistent with the assumption of an amphibole-rich mantle source. Similarly, the depletion in Th and U can also be attributed to amphibole precipitated in lithospheric veins that preferentially incorporated neighboring Ba and Nb over Th and U (Tiepolo et al., 2007).

The moderately depleted Nd and Hf isotope compositions of the SQB diabases provide evidence against a long-term enriched mantle source. Metasomatic melts infiltrated the lithospheric mantle shortly prior to the onset of SQB magmatism, without enough time to produce negative $\varepsilon_{Nd}(t)$ and $\varepsilon_{Hf}(t)$ values (Tappe at al., 2006). In hydrous metasomatic veins, pyroxenitic clinopyroxene and Ti-oxides are the major host of Hf (Pearson and Nowell,

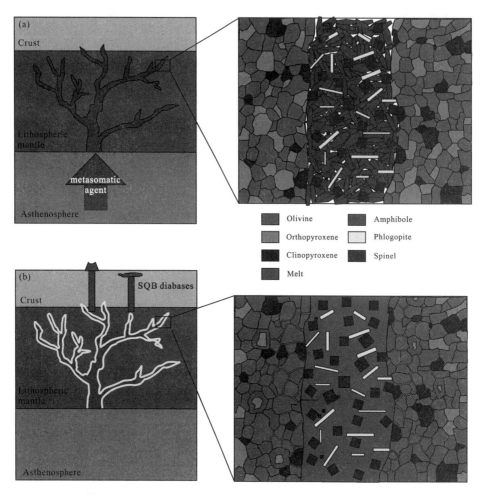

Fig. 12　Schematic illustrations for the generation of SQB diabases by the vein-plus-wall rock melting model.
（a）Silicate melts released from the asthenosphere were injected into the lithospheric mantle and produced the amphibole-rich metasomatic veins. （b）Remelting of the amphibole-rich veins produced highly silica-undersaturated melts. The wall-rock mineral orthopyroxene is in disequilibrium with the melts and will be dissolved in it. As a result, the melts will evolve into higher SiO_2 content and lower TiO_2 and trace element contents while constant trace element patterns are maintained.

2004). Melts derived from ancient metasomes should display negative $\varepsilon_{Hf}(t)$ and plot below the Nd-Hf isotope mantle array as a result of their low Lu/Hf (Nelson et al., 2019; Tappe at al., 2006), as observed in the Gerba Guracha lavas and mica-amphibole-rutile-ilmenite-diopside (MARID) xenoliths (Fitzpayne et al., 2019; Rooney et al., 2014). However, all of the SQB diabases plotted close to or above the Nd-Hf isotope mantle array (Fig. 7b), indicating the absence of a long-term enriched source and that the metasomatic activity occurred shortly prior to melting.

In summary, the SQB diabases were probably derived from a lithospheric mantle source metasomatized by low-degree silicate melts from the asthenosphere. Infiltration, cooling and fractional crystallization of these metasomatic agents within the lithospheric mantle generated a continuum of anhydrous cumulate to hydrous veins. Subsequently, a high degree of partial melting of hydrous veins accompanied by interaction with peridotitic wall-rock could explain most features observed in the SQB diabases (Fig. 12b).

6 Tectonic implications

Early Paleozoic magmatism in the SQB is dominated by alkali diabases, alkali basalts, and trachytes, but lacks tholeiites, and has been linked to variable tectonic settings, including mantle plumes (Xu et al., 2008; Zhang et al., 2002), back-arc basin (Wang et al., 2015) or rifts in passive continental margins (Dong and Santosh, 2016; Wang et al., 2017).

Our work and previous research on xenoliths (Huang et al., 1993; Xu et al., 1997) indicate that amphibole-bearing metasomes are widespread within the SQB lithospheric mantle. Such metasomes are expected to melt at lower temperatures than adjacent peridotites and produce alkaline melts (Foley, 1992). Elevated potential temperature in mantle plumes will induce the melting not only of metasomes, but also of enclosing peridotites. This process will produce large-scale tholeiitic magmas (e.g., continental flood basalts) and erase the enriched signature of melts produced from fertile mantle lithologies (Pilet, 2015). The absence of tholeiites in the SQB formed during the Early Paleozoic precludes the possible role of hot thermal plumes in the genesis of SQB alkaline magmatism. The mantle potential temperatures of 1 258−1 300 ℃ beneath the SQB at ca. 455 Ma are significantly cooler than the temperature of plume-type mantle (~1 600 ℃; Putirka, 2016). At lower mantle potential temperatures, easily fusible materials (pyroxenite and hornblendite) embedded in peridotite tend to melt to a higher degree in comparison with peridotite. Eventually, the host peridotite mantle will be undercooled and the melting of peridotite will be suppressed (Daniele et al., 2018; Pilet et al., 2008). During such processes, the melting of abundant hydrous metasomes within the lithospheric mantle will generate melts that are mainly composed of alkaline lavas without large scale tholeiites. To account for the alkaline magmatism in the SQB, a relatively low mantle potential temperature is required. The lithospheric extension or thermal perturbation are plausible explanations for the origin of the SQB diabases.

In the Ziyang-Zhuxi area of the SQB, the close temporal and spatial connection between alkali basalts and trachytes is interpreted as a bimodal rock suite that suggests continental rifting during the Silurian (Huang et al., 1992). This magmatism coincided with the onset of rift basin formation in the SQB (Gao et al., 1995; Luo and Duanmu, 2001). The parallelism between the distribution of alkaline intrusions and the strike of the northeast-dipping fault systems indicates that the emplacement of alkaline intrusions was structurally controlled. Wang et al. (2017) suggested that an asymmetric rift developed in the SQB during the Early

Paleozoic, which produced the gently dipping Chengkou-Fangxian Fault that cuts at low angles through the crust to the base of the lithosphere. Continued Early Paleozoic lithospheric extension and upwelling of the asthenosphere beneath a developing rift branch caused scavenging of the most easily fusible material in the lithosphere which gave rise to the Early Paleozoic alkaline magmatism in South Qinling.

7 Conclusions

The major and trace elements, zircon U-Pb ages and Sr-Nd-Hf isotopes revealed that the SQB alkali diabases formed at 455. 9-446. 2 Ma. The high TiO_2 contents, high Na_2O/K_2O and enriched trace element characteristics of these diabases can be explained by high-degree melting of amphibole-clinopyroxene-phlogopite veins with subsequent dissolution of orthopyroxene from the wall-rock peridotite within the spinel-stabilized lithospheric mantle. The Sr-Nd-Hf isotope signatures indicate that metasomatic melts infiltrated the lithospheric mantle shortly prior to the onset of SQB magmatism. Given the low solidus of amphibole-rich metasomes, magmatism was probably triggered by the lithospheric extension or thermal perturbation, rather than a hot thermal plume.

Acknowledgements　We thank Jianqi Wang for his assistance during XRF measurements, and Ye Liu for her help in ICP-MS trace element analyses. Chunlei Zong is thanked for her assistance with radiogenic Sr-Nd-Hf isotope analyses. We also thank Prof. Gregory Shellnutt and two anonymous reviewers for constructive comments. Additionally, we thank Kara Bogus, PhD, and Sara J. Mason, MSc, from Liwen Bianji, Edanz Editing China, for editing the English text of drafts of this manuscript. This work was jointly supported by the National Natural Science Foundation of China (Grant Nos. 41772052 and 41421002) and the Program for Changjiang Scholars and Innovative Research Team in University (IRT1281).

Appendix A. Supplementary data　Supplementary data to this article can be found online at https://doi. org/10. 1016/j.lithos.2020. 105619.

References

Ackerman, L., Spacek, P., Magna, T., Ulrych, J., Svojtka, M., Hegner, E., Balogh, K., 2013. Alkaline and carbonate-rich melt metasomatism and melting of subcontinental lithospheric mantle: Evidence from mantle xenoliths, NE Bavaria, Bohemian massif. Journal of Petrology, 54, 2597-2633.

Adam, J., Green, T., 2006. Trace element partitioning between mica- and amphibole-bearing garnet lherzolite and hydrous basanitic melt: 1. Experimental results and the investigation of controls on partitioning behaviour. Contributions to Mineralogy and Petrology, 152, 1-17.

Andersen, T., 2002. Correction of common lead in U-Pb analyses that do not report ^{204}Pb. Chemical Geology, 192(1-2), 59-79.

Chauvel, C., Lewin, E., Carpentier, M., Arndt, N.T., Marini, J.C., 2008. Role of recycled oceanic basalt and sediment in generating the Hf-Nd mantle array. Nature Geoscience, 1, 64-67.

Chen, H., Tian, M., Wu, G.L., Hu, J.M., 2014. The Early Paleozoic alkaline and mafic magmatic events in Southern Qinling Belt, Central China: Evidences for the break-up of the Paleo-Tethyan ocean. Geological Review, 60, 1437-1452 (in Chinese with English abstract).

Chen, Y.Z., Liu, S.W., Li, Q.G., Dai, J.Z., Zhang, F., Yang, P.T., Guo, L.S., 2010. Geology, Geochemistry of Langao mafic volcanic rocks in South Qinling Orogenic Belt and its tectonic implications. Acta Scientiarum Naturalium Universitatis Pekinensis, 46 (4), 607-619.

Class, C., Goldstein, S.L., 1997. Plume-lithosphere interactions in the ocean basins: Constraints from the source mineralogy. Earth & Planetary Science Letters, 150, 245-260.

Condamine, P., Médard, E., 2014. Experimental melting of phlogopite-bearing mantle at 1 GPa: Implications for potassic magmatism. Earth & Planetary Science Letters, 397, 80-92.

Dai, L.Q., Zhao, Z.F., Zheng, Y.F., An, Y.J., Zheng, F., 2017. Geochemical distinction between carbonate and silicate metasomatism in generating the mantle sources of alkali basalts. Journal of Petrology, 58, 863-884.

Dai, L.Q., Zheng, F., Zhao, Z.F., Zheng, Y.F., 2018. Geochemical insights into the lithology of mantle sources for Cenozoic alkali basalts in west Qinling, China. Lithos, 302-303, 86-98.

Dalpe, C., Baker, D.R., 1994. Partition coefficients for rare-earth elements between calcic amphibole and Ti-rich basanitic glass at 1.5 GPa, 1 100 ℃. Mineralogical Magazine, 58, 207-208.

Daniele, B., Anna, C., Enrico, B., 2018. Thermal effects of pyroxenites on mantle melting below mid-ocean ridges. Nature Geoscience, 11, 520-525.

Dasgupta, R., Hirschmann, M.M., Smith, N.D., 2007. Partial melting experiments of peridotite + CO_2 at 3 GPa and genesis of alkalic ocean island basalts. Journal of Petrology, 48, 2093-2124.

Davidson, J., Turner, S., Plank, T., 2013. Dy/Dy*: Variations arising from mantle sources and petrogenetic processes. Journal of Petrology, 54, 525-537.

Davies, G., Lloyd, F., 1989. Pb-Sr-Nd isotope and trace element data bearing on the origin of the potassic subcontinental lithosphere beneath south-west Uganda. In: Ross, J., Jaques, A. L., Ferguson, J., Green, D.H., OReilly, S.Y., Danchin, R.V., Janse, A.J.A. Kimberlites and Related Rocks. Geological Society of Australia, Sydney, 784-794.

Davis, F.A., Humayun, M., Hirschmann, M.M., Cooper, R.S., 2013. Experimentally determined mineral/melt partitioning of first-row transition elements (FRTE) during partial melting of peridotite at 3 GPa. Geochimica et Cosmochimica Acta, 104, 232-260.

Dong, Y.P., Santosh, M., 2016. Tectonic architecture and multiple orogeny of the Qinling Orogenic Belt, Central China. Gondwana Research, 29, 1-40.

Dong, Y.P., Sun, S.S., Yang, Z., Liu, X.M., Zhang, F.F., Li, W., Cheng, B., He, D.F., Zhang, G.W., 2017. Neoproterozoic subduction-accretionary tectonics of the South Qinling Belt, China. Precambrian Research, 293, 73-90.

Fitzpayne, A., Giuliani, A., Maas, R., Hergt, J., Janney, P., Phillips, D., 2019. Progressive metasomatism of the mantle by kimberlite melts: Sr-Nd-Hf-Pb isotope compositions of MARID and PIC minerals. Earth & Planetary Science Letters, 509, 15-26.

Foley, S. F., 1992. Vein-plus-wall-rock melting mechanisms in the lithosphere and the origin of potassic alkaline magmas. Lithos, 28, 435-453.

Foley, S. F., Musselwhite, D. S., van der Laan, S. R., 1999. Melt compositions from ultramafic vein assemblages in the lithospheric mantle: A comparison of cratonic and non-cratonic settings. In: Gurney, J. J., Gurney, J.L., Pascoe, M.D., Richardson, S. H. The J. B. Dawson Volume. Proceedings of the 7th International Kimberlite Conference Volume 1. Red Roof Design, Cape Twon, 238-246.

Furman, T., Graham, D., 1999. Erosion of lithospheric mantle beneath the East African Rift system: Geochemical evidence from the Kivu volcanic province. Lithos, 48, 237-262.

Gao, S., Ling, W.L., Qiu, Y.M., Lian, Z., Hartmann, G., Simon, K., 1999. Contrasting geochemical and Sm-Nd isotopic compositions of Archean metasediments from the Kongling high-grade terrain of the Yangtze craton: Evidence for cratonic evolution and redistribution of REE during crustal anatexis. Geochimica et Cosmochimica Acta, 63, 2071-2088.

Gao, S., Rudnick, R.L., Xu, W.L., Yuan, H.L., Liu, Y.S., Walker, R.J., Puchtel, I.S., Liu, X.M., Huang, H., Wang, X.R., Yang, J., 2008. Recycling deep cratonic lithosphere and generation of intraplate magmatism in the North China Craton. Earth & Planetary Science Letters, 270, 41-53.

Gao, S., Zhang, B.R., Gu, X.M., 1995. Silurian-Devonian provenance changes of South Qinling basins: Implications for accretion of the Yangtze (South China) to the North China cratons. Tectonophysics, 250, 183-197.

Green, D.H., Hibberson, W.O., Kovács, I., Rosenthal, A., 2010. Water and its influence on the lithosphere asthenosphere boundary. Nature, 467, 448-451.

Green, D.H., Wallace, M.E., 1988. Mantle metasomatism by ephemeral carbonatite melts. Nature, 336, 459-462.

Hammouda, T., Laporte, D., 2000. Ultrafast mantle impregnation by carbonatite melts. Geology, 28, 283-285.

Hauri, E.H., Wagner, T.P., Grove, T.L., 1994. Experimental and natural partitioning of Th, U, Pb and other trace elements between garnet, clinopyroxene and basaltic melts. Chemical Geology, 117, 149-166.

Hawthorne, F.C., Oberti, R., Harlow, G.E., Maresch, W.V., Martin, R.F., Schumacher, J.C., Welch, M. D., 2012. Nomenclature of the amphibole supergroup. American Mineralogist, 97, 2031-2048.

Hoernle, K., Tilton, G., Le Bas, M.J., Duggen, S., Garbe-Schonberg, D., 2002. Geochemistry of oceanic carbonatites compared with continental carbonatites: Mantle recycling of oceanic crustal carbonate. Contributions to Mineralogy and Petrology, 142, 520-542.

Hofmann, A.W., 1988. Chemical differentiation of the Earth: The relationship between mantle, continental crust, and oceanic crust. Earth & Planetary Science Letters, 90, 297-314.

Hofmann, A.W., Jochum, K.P., Seufert, M., White, W.M., 1986. Nb and Pb in oceanic basalts: New constraints on mantle evolution. Earth & Planetary Science Letters, 79, 33-45.

Hofmann, A.W., White, W.M., 1982. Mantle plumes from ancient oceanic crust. Earth & Planetary Science Letters, 57, 421-436.

Hu, J., Liu, X.C., Chen, L.Y., Qu, W., Li, H.K., Geng, J.Z., 2013. A ~2.5 Ga magmatic event at the northern margin of the Yangtze craton: Evidences from U-Pb dating and Hf isotope analysis of zircons from the Douling Complex in the South Qinling orogen. Science Bulletin, 58(28), 3564-3579.

Huang, Y. H., 1993. Mineralogical characteristics of Phlogopite-Amphibole-Pyroxenite mantle xenoliths included in the alkali mafic-ultramafic subvolcanic complex from Langao country, China. Acta Petrologica Sinica, 9, 367-378 (in Chinese with English abstract).

Huang, Y.H., Ren, Y.X., Xia, L.Q., Xia, Z.C., Zhang, C., 1992. Early Palaeozoic bimodal igneous suite on Northern Daba Mountains-Gaotan diabase and Haoping trachyte as examples. Acta Petrologica Sinica, 8, 243-256 (in Chinese with English abstract).

Ionov, D.A., Hofmann, A.W., 1995. Nb-Ta-rich mantle amphiboles and micas: Implications for subduction-related metasomatic trace element fractionations. Earth & Planetary Science Letters, 131, 341-356.

Jung, C., Jung, S., Hoffer, E., Berndt, J., 2006. Petrogenesis of Tertiary mafic alkaline magmas in the Hocheifel, Germany. Journal of Petrology, 47, 1637-1671.

Kelemen, P.B.,Yogodzinski, G.M., Scholl, D.W., 2003. Along-strike variation in the Aleutian island arc: Genesis of high $Mg^{\#}$ andesite and implications for continental crust. Geophysical Monograph Series, 138, 223-276.

Lai, S.C., Qin, J.F., Khan, J., 2014. The carbonated source region of Cenozoic mafic and ultra-mafic lavas from western Qinling: Implications for eastern mantle extrusion in the northeastern margin of the Tibetan Plateau. Gondwana Research, 25, 1501-1516.

LaTourrette, T., Hervig, R.L., Holloway, J.R., 1995. Trace element partitioning between amphibole, phlogopite, and basanite melt. Earth & Planetary Science Letters, 135, 13-30.

Ling, W.L., Ren, B.,Duan, R., Liu, X., Mao, X., Peng, L., Liu, Z., Cheng, J., Yang, H., 2008. Timing of the Wudangshan, Yaolinghe volcanic sequences and mafic sills in South Qinling: U-Pb zircon geochronology and tectonic implication. Science Bulletin, 53, 2192-2199.

Liu, Y., Liu, X.M., Hu, Z.C., Diwu, C.R., Yuan, H.L., Gao, S., 2007. Evaluation of accuracy and long-term stability of determination of 37 trace elements in geological samples by ICP-MS. Acta Petrologica Sinica, 23, 1203-1210 (in Chinese with English abstract).

Ludwig, K.R., 2003. ISOPLOT 3.0: A geochronological toolkit for Microsoft Excel. Berkeley Geochronology Center, Special Publication, 4,71.

Luo, K.L., DuanMu, H.S., 2001. Timing of Early Paleozoic basic igneous rocks in the Daba Mountain. Regional Geology of China, 20, 262-266 (in Chinese with English abstract).

Ma, G.S.K., Malpas, J., Xenophontos, C., Chan, G.H.N., 2011. Petrogenesis of latest Miocene-Quaternary continental intraplate volcanism along the northern Dead Sea fault system (Al Ghab-Homs Volcanic Field), Western Syria: Evidence for lithosphere-asthenosphere interaction. Journal of Petrology, 52, 401-430.

Mandler, B.E., Grove, T.L., 2016. Controls on the stability and composition of amphibole in the Earth's mantle. Contributions to Mineralogy and Petrology, 171, 68.

Mayer, B., Jung, S., Romer, R.L., Pfänder, J.A., Klügel, A., Pack, A., Gröner, E., 2014. Amphibole in alkaline basalts from intraplate settings: Implications for the petrogenesis of alkaline lavas from the metasomatised lithospheric mantle. Contributions to Mineralogy and Petrology, 167, 989.

Mayer, B., Jung, S., Romer, R.L., Stracke, A., Haase, K.M., Garbe, S.C.D., 2013. Petrogenesis of Tertiary hornblende-bearing lavas in the Rhön, Germany. Journal of Petrology, 54, 2095-2123.

McDonough, W.F., Sun, S.S., 1995. The composition of the Earth. Chemical Geology, 120, 223-253.

McKenzie, D. 1989. Some remarks on the movement of small melt fractions in the mantle. Earth & Planetary Science Letters, 95, 53-72.

Médard, E., Schmidt, M.W., Schiano, P., Ottolini, L., 2006. Melting of amphibole-bearing wehrlites: An experimental study on the origin of ultra-calcic nepheline-normative melts. Journal of Petrology, 47, 481-504.

Meng, Q.R., Zhang, G.W., 2000. Geologic framework and tectonic evolution of the Qinling orogen, central China. Tectonophysics, 323, 183-196.

Middlemost, E.A.K., 1994. Naming materials in the magma/igneous rock system. Earth-Science Reviews, 37, 215-224.

Morimoto, N.,Fabries, J., Ferguson, A.K., Ginzburg, I.V., Ross, M., Seifert, F.A., Zussman, J., Aoki, K., Gottardi, G., 1988. Nomenclature of pyroxenes. Mineralogical Magazine, 52, 535-550.

Natali, C., Beccaluva, L., Bianchini, G., Siena, F., 2018. Coexistence of alkaline-carbonatite complexes and high-MgO CFB in the Paranà-Etendeka province: Insights on plume-lithosphere interactions in the Gondwana realm. Lithos, 296-299, 54-66.

Nelson, W.R., Hanan, B., Graham, D.W., Shirey, S.B., Yirgu, G., Ayalew, D., Furman, T., 2019. Distinguishing plume and metasomatized lithospheric mantle contributions to post-flood basalt volcanism on the southeastern Ethiopian Plateau. Journal of Petrology, 60, 1063-1094.

Niu, Y.L., O'Hara, M.J., 2003. Origin of ocean island basalts: A new perspective from petrology, geochemistry, and mineral physics considerations. Journal of Geophysical Research, 108, 2209-2228.

Panter, K.S., Castillo, P., Krans, S., Deering, C., McIntosh, W., Valley, J.W., Kitajima, K., Kyle, P., Hart, S., Blusztajn, J., 2018. Melt origin across a rifted continental margin: A case for subduction-related metasomatic agents in the lithospheric source of alkaline basalt, northwest Ross Sea, Antarctica. Journal of Petrology, 59, 517-558.

Pearson, D.G., Nowell, G.M., 2004. Re-Os and Lu-Hf isotope constraints on the origin and age of Pyroxenites from the Beni Bousera Peridotite Massif: Implications for mixed peridotite-pyroxenite mantle sources. Journal of Petrology, 45, 439-455.

Pfänder, J.A., Jung, S., Klügel, A., Münker, C., Romer, R.L., Sperner, B., Rohrmüller, J., 2018. Recurrent local melting of metasomatised lithospheric mantle in response to continental rifting: Constraints from basanites and nephelinites/melilitites from SE Germany. Journal of Petrology, 59, 667-694.

Pfänder, J.A., Jung, S., Münker, C., Stracke, A., Mezger, K., 2012. A possible high Nb/Ta reservoir in the continental lithospheric mantle and consequences on the global Nb budget-Evidence from continental basalts from Central Germany. Geochimica et Cosmochimica Acta, 77, 232-251.

Pilet, S., 2015. Generation of low-silica alkaline lavas: Petrological constraints, models and thermal implications. In: Foulger, G. R., Lustrino, M., King, S. The Interdisciplinary Earth: In Honor of Don L. Anderson. Geological Society of America, Special Papers, 514, 514-517.

Pilet, S., Baker, M.B., Müntener, O., Stolper, E.M., 2011. Monte Carlo simulations of metasomatic enrichment in the lithosphere and implications for the source of alkaline basalts. Journal of Petrology, 52, 1415-1442.

Pilet, S., Baker, M.B., Stolper, E.M., 2008. Metasomatized lithosphere and the origin of alkaline lavas. Science, 320, 916-919.

Polat, A., Hofmann, A., Rosing, M.T., 2002. Boninite-like volcanic rocks in the 3.7 – 3.8 Ga Isua greenstone belt, West Greenland: Geochemical evidence for intra-oceanic subduction zone processes in the early earth. Chemical Geology, 184(3-4), 231-254.

Prelević, D., Foley, S.F., 2007. Accretion of arc-oceanic lithospheric mantle in the Mediterranean: Evidence from extremely high-Mg olivines and Cr-rich spinel inclusions from lamproites. Earth & Planetary Science Letters, 256, 120-135.

Prytulak, J., Elliott, T., 2007. TiO_2 enrichment in ocean island basalts. Earth & Planetary Science Letters, 263, 388-403.

Putirka, K., 2016. Rates and styles of planetary cooling on Earth, Moon, Mars and Vesta, using new models for oxygen fugacity, ferric-ferrous ratios, olivine-liquid Fe-Mg exchange, and mantle potential temperature. American Mineralogist, 101, 819-840.

Ridolfi, F., Renzulli, A., Puerini, M., 2010. Stability and chemical equilibrium of amphibole in calc-alkaline magmas: An overview, new thermobarometric formulations and application to subduction-related volcanoes. Contributions to Mineralogy and Petrology, 160(1), 45-66.

Robinson, J.C., Wood, B.J., 1998. The depth of the spinel to garnet transition at the peridotite solidus. Earth & Planetary Science Letters, 164, 277-284.

Rooney, T.O., Nelson, W.R., Dosso, L., Furman, T., Hanan, B., 2014. The role of continental lithosphere metasomes in the production of HIMU-like magmatism on the northeast African and Arabian plates. Geology, 42, 419-422.

Rooney, T.O., Nelson, W.R., Ayalew, D., Hanan, B., Yirgu, G., Kappelman, J., 2017. Melting the lithosphere: Metasomes as a source for mantle-derived magmas. Earth & Planetary Science Letters, 461, 105-118.

Rudnick, R.L., Gao, S., 2003. Composition of the continental crust. In: Rudnick, R.L. The Crust. Treatise on Geochemistry 3. Oxford: Elsevier Pergamon, 1-64.

Shaw, D.M., 1970. Trace element fractionation during anatexis. Geochimica et Cosmochimica Acta, 34, 237-243.

Stracke, A., 2012. Earth's heterogeneous mantle: A product of convection-driven interaction between crust and mantle. Chemical Geology, 330-331, 274-299.

Su, B.X., Zhang, H.F., Yang, Y.H., Sakyi, P.A., Ying, J.F., Tang, Y.J., 2012. Breakdown of orthopyroxene contributing to melt pockets in mantle peridotite xenoliths from the Western Qinling, central China: Constraints from in situ LA-ICP-MS mineral analyses. Contributions to Mineralogy and Petrology, 104, 225-247.

Sun, S.S., McDonough, W.F., 1989. Chemical and isotopic systematics of oceanic basalts: Implications for mantle composition and processes. Geological Society Special Publication, 42, 313-345.

Tappe, S., Foley, S.F., Jenner, G.A., Heaman, L.M., Kjarsgaard, B.A., Romer, R.L., Stracke, A., Joyce, N., Hoefs, J., 2006. Genesis of ultramafic lamprophyres and carbonatites at Aillik Bay, Labrador: A consequence of incipient lithospheric thinning beneath the North Atlantic Craton. Journal of Petrology,

47, 1261-1315.

Tappe, S., Smart, K. A., Stracke, A., Romer, R. L., Prelević, D., van den Bogaard, P., 2016. Melt evolution beneath a rifted craton edge: $^{40}Ar/^{39}Ar$ geochronology and Sr-Nd-Hf-Pb isotope systematics of primitive alkaline basalts and lamprophyres from the SW Baltic Shield. Geochimica et Cosmochimica Acta, 173, 1-36.

Tiepolo, M., Oberti, R., Zanetti, A., Vannucci, R., Foley, S., 2007. Trace-element partitioning between amphibole and silicate melt. In: Hawthorne, F.C., Oberti, R., Ventura, G.D., Mottana, A. Amphiboles: Crystal Chemistry, Occurrence, and Health Issues. Mineralogical Society of America and Geochemical Society, Reviews in Mineralogy and Geochemistry, 67, 417-452.

Wang, K.M., Wang, Z.Q., Zhang, Y.L., Wang, G., 2015. Geochronology and Geochemistry of Mafic Rocks in the Xuhe, Shaanxi, China: Implications for Petrogenesis and Mantle Dynamics. Acta Geologica Sinica (English Edition), 89, 187-202.

Wang, R.R., Xu, Z.Q., Santosh, M., Liang, F.H., Fu, X.H., 2017b. Petrogenesis and tectonic implications of the early Paleozoic intermediate and mafic intrusions in the South Qinling Belt, Central China: Constraints from geochemistry, zircon U-Pb geochronology and Hf isotopes. Tectonophysics, 712-713, 270-288.

Wang, X.X., Wang, T., Zhang, C.L., 2013. Neoproterozoic, Paleozoic, and Mesozoic granitoid magmatism in the Qinling Orogen, China: Constraints on orogenic process. Journal of Asian Earth Sciences, 72, 129-151.

Workman, R.K., Hart, S.R., 2005. Major and trace element composition of the depleted MORB mantle (DMM). Earth & Planetary Science Letters, 231, 0-72.

Xiang, Z.J., Yan, Q.R., Song, B., Wang, Z.Q., 2016. New evidence for the age of ultramafic to mafic dikes and alkaline volcanic complexes in the North Daba Mountains and its geological implication. Acta Geologica Sinica, 90, 896-916 (in Chinese with English abstract).

Xu, C., Campbell, I.H., Allen, C.M., Chen, Y., Huang, Z., Qi, L., Zhang, G.S., Yan, Z.F., 2008. U-Pb zircon age, geochemical and isotopic characteristics of carbonatite and syenite complexes from the Shaxiongdong, China. Lithos, 105, 118-128.

Xu, J.F., Castillo, P.R., Li, X.H., Yu, X.Y., Zhang, B.R., Han, Y.W., 2002. MORB-type rocks from the Paleo-Tethyan Mian-Lueyang northern ophiolite in the Qinling Mountains, central China: Implications for the source of the low $^{206}Pb/^{204}Pb$ and high $^{143}Nd/^{144}Nd$ mantle component in the Indian Ocean. Earth & Planetary Science Letters, 198, 323-337.

Xu, X. Y., Huang, Y. H., Xia, L. Q., Xia, Z. C., 1997. Phlogopite-Amphibole-Pyroxenite Xenoliths in Langao, Shaanxi Province: Evidences for Mantle Metasomatism. Acta Petrologica Sinica, 13, 1-13 (in Chinese with English abstract).

Yaxley, G.M., Green, D.H., Kamenetsky, V., 1998. Carbonatite metasomatism in the Southeastern Australian lithosphere. Journal of Petrology, 39, 1917-1930.

Yuan, H.L., Gao, S., Liu, X.M., Li, H.M., Gunther, D., Wu, F.Y., 2004. Accurate U-Pb age and trace element determinations of zircon by laser ablation-inductively coupled plasma mass spectrometry. Geostandards Newsletter, 28, 353-370.

Yuan, H.L., Gao, S., Luo, Y., Zong, C.L., Dai, M.N., Liu, X.M., Diwu, C.R., 2007. Study of Lu-Hf geochronology: A case study of eclogite from Dabie UHP Belt. Acta Petrologica Sinica, 23, 233-239 (in Chinese with English abstract).

Zanetti, A., Vannucci, R., Bottazzi, P., Oberti, R., Ottolini, L., 1996. Infiltration metasomatism at lherz as monitored by systematic ion-microprobe investigations close to a hornblendite vein. Chemical Geology, 134, 113-133.

Zeng, G., Chen, L.H., Xu, X.S., Jiang, S.Y., Hofmann, A.W., 2010. Carbonated mantle sources for Cenozoic intra-plate alkaline basalts in Shandong, North China. Chemical Geology, 273, 35-45.

Zhang, C.L., Gao, S., Zhang, G.W., Liu, X.M., Yu, Z.P., 2002. Geochemistry and significance of the Early Paleozoic alkaline mafic dyke in Qinling. Science in China: Series D, Geoscience, 32, 819-829 (in Chinese with English abstract).

Zhang, C.L., Shan, G., Yuan, H.L., Zhang, G.W., Yan, Y.X., Luo, J.L., Luo, J.H., 2007. Sr-Nd-Pb isotopes of the early paleozoic mafic-ultramafic dykes and basalts from south Qinling belt and their implications for mantle composition. Science in China, 50, 1293-1301(in Chinese with English abstract).

Zhang, F.Y., Lai, S.C., Qin, J.F., Zhu, R.Z., Yang, H., Zhu, Y., 2020. Geochemical characteristics and geological significance of early Paleozoic alkali diabases in North Daba Mountains. Acta Petrologica et Mineralogica, 39, 35-46 (in Chinese with English abstract).

Zhang, G.W., Zhang, B.R., Yuan, X.C., Xiao, Q.H., 2001. Qinling Orogenic Belt and Continental Dynamics. Beijing: Science Press, 1-855.

Early Paleozoic alkaline trachytes in the North Daba Mountains, South Qinling Belt: Petrogenesis and geological implications[①]

Yang Hang　Lai Shaocong[②]　Qin Jiangfeng　Zhu Renzhi　Zhao Shaowei

Zhu Yu　Zhang Fangyi　Zhang Zezhong　Wang Xingying

Abstract: Early Paleozoic alkaline magmatism is widely preserved in the North Daba Mountains, South Qinling Belt (SQB), predominately composed of basalt-diabase and trachyte-syenite. The petrogenesis of the felsic rocks and their genetic connections with the mafic members remain controversial. Here, an integrated investigation, combining geochronology and whole-rock and Sr-Nd isotopic geochemistry, is conducted to further constrain the origin and tectono-magmatic evolution of the trachytes in Pingli-Zhuxi area, North Daba Mountains. Zircon U-Pb dating for the Pingli-Zhuxi trachytes yield ages of 406.0−427.9 Ma, which are close to those of the mafic rocks (420−455 Ma) in research area. The trachytes show relatively high SiO_2(60.88−63.87 wt%) and total alkali ($Na_2O + K_2O$ = 10.19−12.24 wt%) contents, and are characterized by pronounced enrichment in LREEs and HFSEs, with insignificantly negative Eu anomalies and significantly negative Sr anomalies. All of the samples display low and variable initial $^{87}Sr/^{86}Sr$ ratios of 0.695 9−0.708 3 and narrow range initial $^{143}Nd/^{144}Nd$ ratios of 0.512 3−0.512 5 with positive $\varepsilon_{Nd}(t)$ values of +3.0 to +7.3, suggesting a depleted source. Together with the published data from coeval SQB diabases, our geochemistry evidences show regular and linear variations between mafic and felsic end-members, suggesting their closely genetic link. A plausible petrogenetic hypothesis for the genesis of the Pingli-Zhuxi trachytes implies a protracted process of fractional crystallization driven by separation of K-feldspar, plagioclase, biotite, apatite, and Ti-magnetite from a basaltic melt. Enrichment in REEs and HFSEs and similar geochemical characteristics point to that the Pingli-Zhuxi trachytes and the coeval mafic rocks originate from a cogenetic metasomatized lithosphere source in rift setting. Asthenospheric upwelling is a key factor for continental break up and lithosphere metasomatism. Collectively, melting of the metasomatized source, followed by protracted K-feldspar-dominated fractional crystallization, leading to the occurrence of the Pingli-Zhuxi trachytes in the North Daba Mountains.

---

①　Published in *International Geology Review*, 2020,63(16).

②　Corresponding author.

1　Introduction

Alkaline igneous complexes have complex mineralogical compositions and obvious petrographic variability with distinctive isotopic and geochemical features, comprising an extensive compositional range including ultramafic, mafic, felsic, and carbonatitic lithologies, which has drawn considerable attention to the worldwide academic community for decades (Armbrustmacher and Hedge, 1982; Liégeois et al., 1998; Jung et al., 2012; Saha et al., 2017; Marks et al., 2018; Andersen et al., 2018). The alkaline felsic rocks (e.g., trachyte, syenite, and rhyolite) are commonly considered to represent an evolved end-member among alkaline magmatism. Several petrogenetic models involving crustal, mantle or mixing of these two distinct end members sources have been proposed for their genesis (Vernikovsky et al., 2003; Su et al., 2007; Lucassen et al., 2013). Additionally, their tectonic settings (e.g., continental rift system, oceanic island setting, and mantle plume model) also remain a matter of debate (Zhao et al., 1995; Zhang et al., 2005). Therefore, research on these felsic rocks can not only shed light on their genetic model and magma evolution, but also contribute to understanding thermal, petrological, and tectonic structure of a region.

The Paleozoic magmatism in the North Qinling Belt (NQB) can be temporally classified into three stages (507 – 470 Ma, 460 – 422 Ma, and 415 – 400 Ma), corresponding to the bidirectional subduction to collision processes of the Shangdan (Paleo-Tethys) and Erliangping oceans (Wang et al., 2015b). However, the coeval magmatic activities in the South Qinling Belt (SQB) are complex and remain hotly debated. Several models, mainly including the opening of the Mianlue ocean (Dong et al., 2011; Chen et al., 2014), a back-arc extensional setting (Zhang, 2010; Wang et al., 2015a), and mantle plume model (Xu et al., 2008; Long, 2016), were proposed to explain the Paleozoic magmatism in SQB. The North Daba Mountains are located along the zone of collision between the SQB and the Sichuan Basin. Voluminous Early Paleozoic alkaline magmatic suites are situated in the North Daba Mountains (Table 1), which were considered as important products from the early spreading of the Mianlue ocean (Zhang et al., 2002, 2007). These alkaline rocks span a large compositional range from basalt to trachyte, with a lack of intermediate (48 – 62 wt% SiO_2) compositions, constituting a set of bimodal rocks. Previous work has mainly focused on the mafic end-members of this bimodal combination. However, the origin and tectonic setting of the felsic end-members and their genetic connections with the mafic members remain poorly understood. Detailed geochemical and Sr-Nd isotopic composition analyses in previous studies indicate that both partial melting of lower continental crust and fractional crystallization of alkaline basaltic magma are capable of generating the trachytic magmas in the North Daba Mountains (Li, 2009; Wang, 2014a; Yang et al., 2017). Models involving mantle plumes (Yan, 2005; Li, 2009),

Table 1　Summary of the Early Paleozoic alkaline rocks in the North Daba Mountains.

Location	Sample	Lithology	Method	Age/Ma	Geochemical features	Isotopic	Reference
Langao		basalt			$(La/Yb)_N$ = 15.87–32.46　Eu/Eu^* = 0.91–1.14		Xiang et al., 2010
	11LGC-8	diabase	LA-ICP-MS zircon U-Pb	441.9±3.2	MgO = 3.81%–8.07%　Eu/Eu^* = 1.07–1.56		Wang, 2014b
		gabbro-diabase, diabase, pyroxenite and basalt			NaO/K_2O = 2.37–25.00　Eu/Eu^* = 0.97–1.50		Zhang et al., 2002
						$\varepsilon_{Nd}(t)$ = +3.28 to +5.02	Zhang et al., 2007
	12GPC	trachyte	LA-ICP-MS apatite U-Pb	417±42	MgO = 0.63%–0.85%　Eu/Eu^* = 0.82–1.05	$\varepsilon_{Nd}(t)$ = +1.95 to +3.27	Wang, 2014a
Ziyang	QLhp	quartz syenite	LA-ICP-MS zircon U-Pb	435.1±1.2	SiO_2 = 63.64%–64.78%　Eu/Eu^* = 0.77–1.71	$\varepsilon_{Hf}(t)$ = 12.1±2.4 (weighted mean)	Wang et al., 2017
	QL15-3	diabase	LA-ICP-MS zircon U-Pb	440.0±0.5	$NaO+K_2O$ = 2.21%–5.00%　Eu/Eu^* = 1.10–1.18		
	DBS257/1	diabase and gabbro	SHRIMP zircon U-Pb	422.1±4.7	MgO = 3.86%–7.8%　NaO/K_2O = 1.76–5.51	$\varepsilon_{Nd}(t)$ = +0.22 to +3.91	Chen et al., 2014
	12HP	trachyte	LA-ICP-MS apatite U-Pb	377±37	MgO = 0.33%–1.85%　Eu/Eu^* = 0.79–0.95	$\varepsilon_{Nd}(t)$ = +2.83 to +3.92	Wang, 2014a
	12GQB-10	pyroxene diorite	SIMS zircon U-Pb	438.4±3.4	Na_2O = 4.71%–16.71%　$Mg^\#$ = 20–78	$\varepsilon_{Nd}(t)$ = +2.25 to +2.88	Wang, 2014b
	GQ-4-1 GT-10-1	diabases	LA-ICP-MS zircon U-Pb	455.9±1.5 446.2±1.1	MgO = 4.00%–9.40%　TiO_2 = 2.89%–5.17%　NaO/K_2O = 2.50–11.90	$\varepsilon_{Nd}(t)$ = +2.4 to +3.5　$\varepsilon_{Hf}(t)$ = +6.1 to +7.1	Zhang et al., 2020

Continued

Location	Sample	Lithology	Method	Age/Ma	Geochemical features	Isotopic	Reference
Pingli		quartz trachyte			SiO_2 = 66.06% – 77.63% Eu/Eu^* = 0.56-0.63		Chen, 2015
		selagite			SiO_2 = 57.91% – 67.60% Eu/Eu^* = 0.63-0.68		
Zhuxi	205	trachytic volcanic rocks	LA-ICP-MS zircon U-Pb	430.6±2.7	SiO_2 = 54.16%-66.29% $(La/Yb)_N$ = 5.50 – 25.82 Eu/Eu^* = 0.44-1.00		Wan et al., 2016
		trachyte			$NaO+K_2O$=9.76-11.63% Eu/Eu^* = 0.61-0.96		Yang et al., 2017
Zhenping	D0212	diabase	SHRIMP zircon U-Pb	439±6	$(La/Yb)_N$ = 11.18-15.67 Eu/Eu^* = 1.01-1.17		Zou et al., 2011
	ZP-7	diabase	SHRIMP zircon U-Pb	451±4	MgO=6.21%-6.41% Eu/Eu^* = 1.06-1.07		Xiang et al., 2016

continental rifting (Wang, 2007; Wang et al., 2009a; Zou et al., 2011), and back-arc basin extension (Wang et al., 2009b; Wang, 2014a,b) have been proposed to explain their tectonic settings. Thus, additional investigations on the Early Paleozoic felsic rocks are required to provide more reliable insights into the tectonic-magmatic history of the North Daba Mountains.

We recently identified the trachytes in the Zhuxi-Pingli area of the North Daba Mountains, South Qinling Belt. Combined with the zircon U-Pb ages, whole-rock major and trace element compositions, and Sr-Nd isotope data, we discuss the petrogenesis and tectonic settings of the trachytes, as well as their petrogenetic relationships with the coeval diabases in SQB. The paper offers important geodynamic constraints on further research of the Early Paleozoic tectonic evolution and genetic model of the felsic alkaline magmatism.

2 Geological setting and petrography

The Qinling Orogenic Belt connects the Kunlun-Qilian orogen to the west and Dabie-Sulu orogen to the east (Dong and Santosh, 2016), formed by the collision between the North China Block and the Yangtze Block during the Paleozoic and Mesozoic (Fig. 1a; Zhang et al., 1996; Yi et al., 2017). The orogen is divided into the North Qinling Belt (NQB) and South Qinling Belt (SQB) along the Shangdan suture. The NQB is identified as an active continental margin with widespread Paleozoic island-arc type magmatism and metamorphism (Sun et al., 2002; Lu et al., 2006); whereas the SQB is a passive margin and characterized by abundant Triassic granitic magmatism and metamorphism (Mattauer et al., 1985; Sun et al., 2000; Lu et al., 2006; Dong et al., 2011). The SQB has Sinian to Mesozoic sedimentary rocks overlying the Precambrian basement (Zhang et al., 2001). Paleoproterozoic to Neoproterozoic Wudang and Yunxi groups are composed of effusive alkaline basalt, alkaline trachyte, dacite, rhyolite, and related pyroclastic rocks. Neoproterozoic Yaolinghe Group is widely exposed, which consists of tholeiite, spilitic diabase, spilite, and keratophyre, as well as interbedded clastics (Huang, 1993a). Early Paleozoic magmatism in the SQB extends eastward from the North Daba Mountain to the Suizhou and Zaoyang area, mainly composed of mafic and intermediate alkaline intrusions, alkaline basalts, trachytes and a few carbonatite-syenite complexes (Xu et al., 2010a; Wang, 2014a; Song et al., 2016), most of them are intruded into surrounding Cambrian to Silurian strata and emplaced as strips along a NW trend (Wang, 2007; Li, 2009; Zhang, 2010).

As important components of SQB, the North Daba Mountains represent the Chengkou-Fangxian Fault as the southern boundary, and the Ankang Fault as the northern boundary, the east and west sides are Wudang Mountain uplift and Hannan complex, respectively (Fig. 1b). The alkaline dykes in the North Daba Mountains are NW trending (Fig. 1c). There are mainly two sets of alkaline dyke associations in the Early Paleozoic strata: ① diabase-alkali diabase

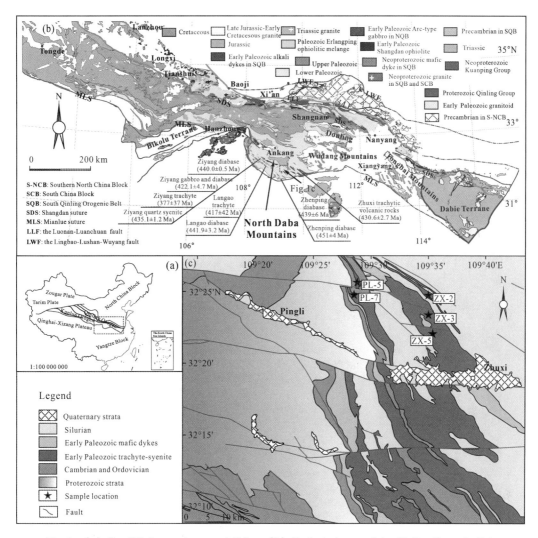

Fig. 1　(a) Simplified tectonic map of China; (b) Geological map of the Qinling Orogenic Belt
and adjacent areas (modified after Dong and Santosh, 2016); (c) Simplified geological map
showing the distribution of Early Paleozoic alkaline rocks in the Pingli-Zhuxi area
(modified after the 1:200 000 geological map of Pingli sheet).
Sample locations are marked by star symbols.

association. The rock types include olivine diabase, feldspar diabase, quartz diabase, gabbroic
diabase, alkaline diabase. They invaded into the Lower Paleozoic strata, and some ultrabasic
mantle xenoliths were found in these basic dykes (Xu et al., 1996). And ② syenite-trachyte
association. Syenite, alkaline syenite, trachyte, and minor amounts of volcanoclastic have
represented the felsic members of the regional alkaline magmatism. Besides, a few carbonatite
complexes have been reported, which are genetically related to the associated syenites (Xu et
al., 2010a; Song et al., 2016).

The Pingli-Zhuxi trachytes reported in this study occur on the western side of the North

Daba Mountains, along the southern margin of the SQB. The trachytes are light-gray in color and show a typical porphyritic texture (Fig. 2) with 60%−70% euhedral-subhedral K-feldspar, 10%−15% plagioclase, 8%−10% subhedral biotite (<5%). Accessory minerals (<5%) include Ti-magnetite, titanite, zircon, and apatite. The trachytes have immense K-feldspar phenocrysts up to 1.5−2 cm, with obvious simple twinning and trachytic texture in photomicrographs (Fig. 2d). Groundmass mainly consists of minor K-feldspar, biotite, and Ti-magnetite.

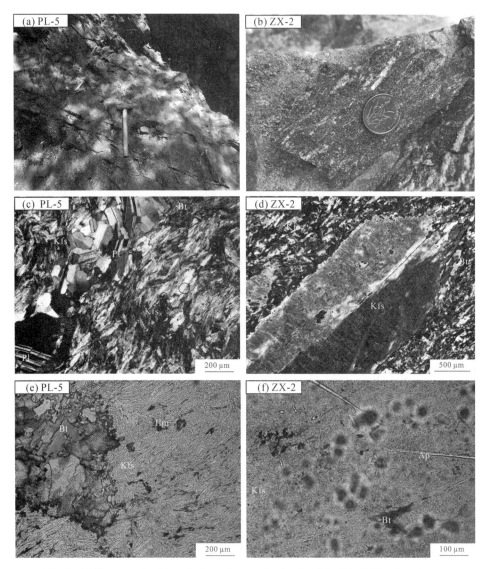

Fig. 2 Field photos (a,b) and photomicrographs (c,f) of the Pingli-Zhuxi trachytes.

Plane-polarized light: c,d; coss-polarized light: e,f. Kfs: K-feldspar;

Pl: plagioclase; Ap: apatite; Bt: biotite; Mt: magnetite.

3 Analytical methods

3.1 Zircon LA-ICP-MS U-Pb dating

Zircon grains from the Pingli-Zhuxi trachytes were separated by using conventional heavy liquid and magnetic techniques. Representative zircon grains were handpicked and mounted in epoxy resin disks, then polished and carbon coated. Internal morphology was examined by cathodoluminescence (CL) prior to U-Pb isotopic analyses. Laser ablation (LA) ICP-MS zircon U-Pb analyses were conducted on an Agilent 7500a ICP-MS equipped with a 193 nm laser, which is located at the State Key Laboratory of Continental Dynamics, Northwest University, following the methods of Yuan et al. (2004). The $^{207}Pb/^{206}Pb$ and $^{206}Pb/^{238}U$ ratios were calculated using the GLITTER program and corrected using the Harvard zircon 91500 as external calibration. These correction factors were then applied to each sample to correct for both instrumental mass bias and depth-dependent elemental and isotopic fractionation. The detailed analytical technique is described in Yuan et al. (2004). Common Pb contents were therefore evaluated by using the method described in Andersen (2002). The age calculations and plotting of concordia diagram were made using ISOPLOT (version 3.0; Ludwig, 2003). The errors quoted in tables and figures are at the 2σ level.

3.2 Major and trace elements

Thirteen samples of Pingli-Zhuxi trachytes were analyzed at the State Key Laboratory of Continental Dynamics, Northwest University, Xi'an, China. Whole-rock samples were trimmed to remove weathered surfaces, cleaned with deionized water, crushed, and then powdered through a 200-mesh screen. Major and trace elements were analyzed by X-ray fluorescence (XRF; Rikagu RIX 2100) and inductively coupled plasma-mass spectrometry (ICP-MS; Agilent 7500a), respectively. Analyses of USGS and Chinese national rock standards (BCR-2, GSR-1, and GSR-3) indicate that both analytical precision and accuracy for major elements are generally better than 5%. For the trace element analyses, sample powders were digested using an HF + HNO$_3$ mixture in high-pressure Teflon bombs at 190 ℃ for 48 h. Precision and accuracy were monitored by measurements of replicates and standard reference materials BHVO-2, AGV-2, BCR-2, and GSP-2 (Supp. Table 1). For most trace elements, the precision was greater than 10% (Liu et al., 2007).

3.3 Sr-Nd isotopic analyses

Whole-rock Sr-Nd isotopic data were obtained by using a Nu Plasma HR multi-collector mass spectrometer at the same laboratory. Sr and Nd isotopic fractionations were corrected to

$^{87}Sr/^{86}Sr = 0.119\ 4$ and $^{146}Nd/^{144}Nd = 0.721\ 9$, respectively. During the period of analysis, the NIST SRM 987 standard gave an average value of $^{87}Sr/^{86}Sr = 0.710\ 250 \pm 0.000\ 012$ (2σ, $n = 15$) and the La Jolla standard gave an average value of $^{146}Nd/^{144}Nd = 0.511\ 859 \pm 0.000\ 006$ (2σ, $n = 20$). Full details of analytical methods were present in Gao et al. (2008).

4 Results

4.1 Zircon LA-ICP-MS U-Pb dating

The zircon cathodoluminescence (CL) images and the U-Pb concordia diagrams of the Pingli-Zhuxi trachytes are shown in Fig. 3. The zircon U-Pb dating results are listed in Supp. Table 2. Zircon grains from the sample PL-7 and sample ZX-2 have crystal sizes of 40−120 μm with aspect ratios of 1:1 to 1:2 and exhibit dark and weak banded zoning in CL images. Most grains show high Th/U ratios (0.11 − 1.10), suggesting magmatic origin (Hoskin and Schaltegger, 2003). Seven analyses in sample PL-7 yielded a weighted mean $^{206}Pb/^{238}U$ age of 406.0 ± 12.0 Ma (MSWD = 5.7, $n = 7$) and seven analyses in sample ZX-2 yielded a weighted mean $^{206}Pb/^{238}U$ age of 427.9 ± 6.6 Ma (MSWD = 3.2, $n = 7$). These ages show that the trachytes formed at 406−427 Ma.

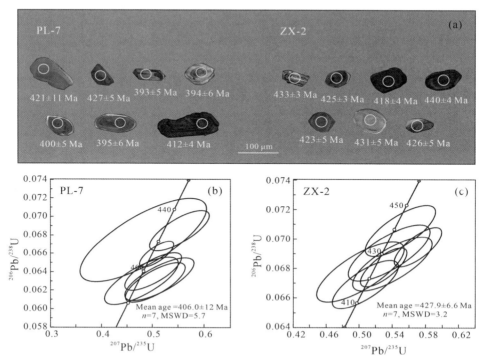

Fig. 3 (a) Zircon CL images of representative zircon grains and (b,c) LA-ICP-MS
U-Pb zircon concordia diagrams for the Pingli-Zhuxi trachytes.

4. 2 Major and trace elements geochemistry

Major and trace element data forthe Pingli-Zhuxi trachytes are listed in Supp. Table 3. All of the samples plotted in the field of trachyte on the total alkali versus SiO_2 (TAS) diagram (Fig. 4a). The Pingli-Zhuxi trachytes are characterized by relative high SiO_2 (60. 88 – 63. 87 wt%) and total alkali ($Na_2O + K_2O = 10. 19 – 12. 24$ wt%) contents, with low MgO (0. 41 – 1. 64 wt%), TiO_2 (0. 44 – 1. 27 wt%), and P_2O_5 (0. 06 – 0. 22 wt%) contents. Lower and variable Al_2O_3 contents of 11. 72–14. 25 wt% with A/CNK (molar Al_2O_3/CaO + $Na_2O + K_2O$) ratios of 0. 86 – 1. 00 (Fig. 4b) classify the trachytes as metaluminous and peralkaline. The Rittman index δ range from 6. 44 to 8. 01 indicate all the samples are chemically alkaline.

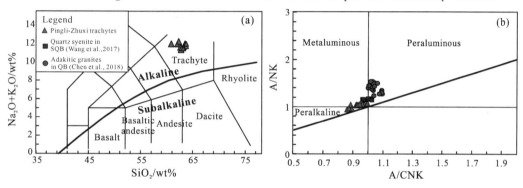

Fig. 4   (a) Total alkali vs. SiO_2 (TAS) diagram for the Pingli-Zhuxi trachytes (after Middlemost, 1994).

(b) A/CNK-A/NK diagram for the Pingli-Zhuxi trachytes.

The data of the quartz syenite (purple square) in SQB are from Wang et al. (2017);

the data of Paleozoic adakitic granites (green circle) in QOB are from Chen et al. (2018).

The trachytes yield right steep slope rare earth elements (REEs) patterns with $(La/Yb)_N$ ratios from 11. 0 to 21. 8, exhibiting clear fractionation between light rare earth elements (LREEs) and heavy rare earth elements (HREEs) on the chondrite-normalized REE diagram (Fig. 5a), as well as extremely high total REE contents of 394. 9–1 050. 2 ppm. The samples display slightly to moderately negative Eu anomalies ($\delta Eu = 0. 64 – 0. 91$). On the primitive mantle-normalized trace element spider diagram (Fig. 5b), all the samples enriched in most of large ion lithophile elements (LILEs) and high field strength elements (HFSEs), with significant negative Sr and Ba anomalies. Early Paleozoic Haoping quartz syenites identified in SQB exhibit similar major and trace elements characteristics with our samples (Figs. 5 and 6), suggesting that a similar genetic model is likely to be responsible for both units. These evolved rocks represent the felsic end-members in the Early Paleozoic alkaline magmatism. Additionally, the Pingli-Zhuxi trachytes show linear major elements variations and similar REE distributions with the SQB diabases (Figs. 5 and 6), the correlation with the felsic and mafic members will be further discussed in following sections.

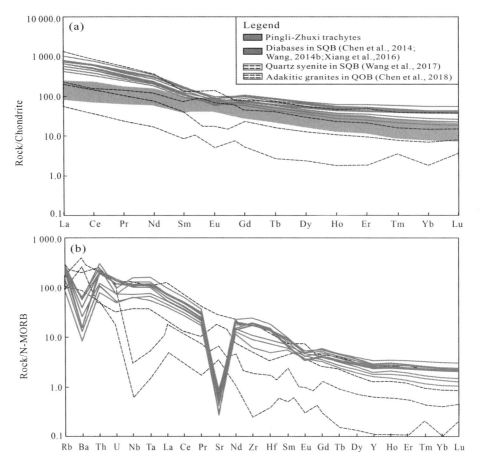

Fig. 5　(a) Chondrite-normalized rare earth element (REE) diagram and (b) N-MORB
normalized trace element spider diagram for the Pingli-Zhuxi trachytes
(normalization values after Sun and McDonough, 1989).

The data of diabases (grey shadow) in SQB are from Chen et al. (2014) and Wang (2014b) and Xiang et al.
(2016); the data of the quartz syenite (purple dashed area) in SQB are from Wang et al. (2017); the data of
Paleozoic adakitic granites (green dashed area) in the QB are from Chen et al. (2018).

4. 3　Whole-rock Sr-Nd isotopes

Thirteen samples of the Pingli-Zhuxi trachytes were chosen for whole-rock Sr-Nd isotopic analyses and the results are listed in Supp. Table 4. The initial isotopic ratios of the samples are calculated for emplacement ages of 420 Ma. All the samples have high Rb (45. 6–151. 6 ppm) and low Sr (24. 68–78. 22 ppm) contents. The samples display relatively lower and variable initial ^{87}Sr/^{86}Sr range from 0. 695 9 to 0. 708 3 and narrow range of initial ^{143}Nd/^{144}Nd ratios of 0. 512 3–0. 512 5, with moderately depleted $\varepsilon_{Nd}(t)$ values from +3. 0 to +7. 3, and one-stage Nd model ages (T_{DM1}) of 0. 55–0. 83 Ga. Notably, the varying of the initial ^{87}Sr/^{86}Sr ratios is possibly ascribed to the extremely low Sr contents and high Rb/Sr ratios in our samples, as low

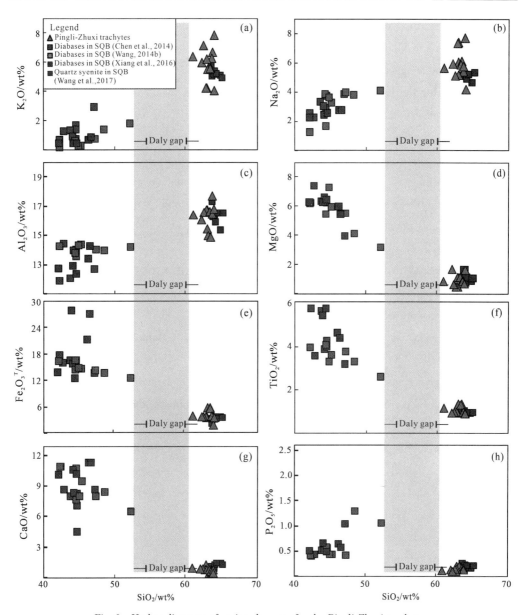

Fig. 6   Harker diagrams of major elements for the Pingli-Zhuxi trachytes.

Sr contents and high Rb/Sr ratios are considered easily to result in large calculation errors of the initial Sr isotope (Ding et al., 2011) , thus we focus on the Nd isotope results in following discussions.

5   Discussion

5. 1   Petrogenesis

Our samples were collected from several parallel outcrops in the Pingli-Zhuxi area. They were exposed as strips along a NW trend, with hundreds to thousands of meters in width and several to dozens of kilometers in length (Fig. 1c). The close temporal and spatial connections as well as similar geochemical characteristics are interpreted as the products of similar magmatism during the Early Paleozoic. Generally, three main hypotheses have been proposed to explain the genesis of the felsic rocks in association with bimodal magmatism: ① protracted fractional crystallization (FC) of basaltic magmas at low pressure (Barberi et al., 1975; Skridlaite et al., 2003; Peccerillo et al., 2007; Lucassen et al., 2013); ② partial melting of a thickened lower crust at high pressure, followed by fractional crystallization (Trua et al., 1999; Li et al., 2004; Su et al., 2007; Meng et al., 2013), and ③ mixing of mantle-derived mafic magmas and crust-derived felsic magmas (Mingram et al., 2000; Vernikovsky et al., 2003; Avanzinelli et al., 2004).

5. 1. 1　Thickened lower crust source or magma mixing model?

A central issue in the North Daba Mountains, as well as in other alkaline magmatism, is the origin of the more differentiated rocks. Partial melting of ordinary crustal rocks is deemed incapable of generating large-scale trachytic magma directly (Montel and Vielzeuf, 1997; Litvinovsky et al., 2000). However, experimental petrologic evidence has demonstrated that the trachytic magma can originate from partial melting of a thickened lower crust under high pressure (> 1.5 GPA; Meng et al., 2013). The adakitic granites in the Qinling Orogenic Belt (QOB), for example, the Huichizi Granitic Complexes, have been proposed to originate from a thickened lower crust (Li et al., 2001; Chen et al,. 2018). Likewise, the trachytes in the North Daba Mountains, with relatively high SiO_2 and low MgO contents and high $(La/Yb)_N$ values, were previously considered as the products of thickened lower crust in QOB (Wang, 2014a). However, previous tectonic studies document an extensional setting during Early Paleozoic, which is considered difficult to produce in association with large-scale trachytic magmas (Zhang et al., 2007; Chen et al., 2014). In addition, significant geochemical discrepancies are found between the trachytes and the adakitic granites. Obviously, the Pingli-Zhuxi trachytes exhibit higher $K_2O + Na_2O$ contents (> 11.5 wt%) than those of QOB adakitic granites, which belong to sub-alkaline and calc-alkaline series (Li et al., 2001; Chen et al., 2018); compared to QOB adakitic granites, apparent enrichment in Nb, Ta, and REEs contents (Fig. 5), as well as positive $\varepsilon_{Nd}(t)$ values in trachytes illustrate their disparate source region and interpreted genetic model. Therefore, the thickened lower crust model is not regarded as a feasible interpretation. This is further confirmed by some trace element ratios, such as Ce/Pb and Nb/U, which demonstrate evident differences between the crust and mantle (Hofmann et al., 1986). Oceanic basalts have high Ce/Pb (25 ± 5) and Nb/U (47 ± 10) ratios, whereas continental crust displays low Ce/Pb (< 5) and Nb/U (< 9.7) ratios (Rudnick and Gao, 2003). All of the trachytes yield extremely high Ce/Pb ($19 - 158$) and

Nb/U (36－79) ratios, in accord with the SQB diabases (Ce/Pb = 9－64, Nb/U = 36－59; Chen et al., 2014; Wang, 2014b), showing the mantle material dominates.

Mixing of mantle-derived basaltic magmas with crust-derived granitic magmas formed by melting of lower crustal material is viable to explain the occurrence of trachytic rocks. Nevertheless, as mentioned above, homogeneously radiogenic Nd isotopic compositions and slightly positive Nb-Ta anomalies in our samples suggest that involvement of a crustal component is negligible. Moreover, the absence of mafic microgranular enclaves (MMEs) or disequilibrium mineral pairs in the samples is inconsistent with the magma mixing model. In summary, neither partial melting of a thickened crust nor mixture of basaltic and granitic magma is likely to be responsible for the petrogenesis of these trachytes.

5. 1. 2   Metasomatized mantle source for the Pingli-Zhuxi trachytes

The trachytes and coexisting diabases can originate from either a homogeneous or two distinct source regions, which are commonly related to intraplate extensional tectonics (Wilson, 1989; Tschegg et al., 2011; Hagen and Cottle, 2016). The SQB diabases are set as the representative of the mafic end-members in the Early Paleozoic alkaline magmatism on account of their wide exposure and concordant geochemical features. In this study, petrological, geochronology, and geochemical results indicate a cogenetic source region of the Pingli-Zhuxi trachytes and SQB diabases. The following reasons have been proposed to explain the homology: ① The trachytes spatially coexist with the diabases (Fig. 1c). The emplacement of the diabases occurred at ca. 430－455 Ma, with a peak around 440 Ma (Wang et al., 2009a; Zou et al., 2011; Chen et al., 2014; Wang, 2014b; Xiang et al., 2016; Zhang et al., 2020), close to those of the trachytes that formed at ca. 410－435 Ma with a peak at 420 Ma (Duan, 2010; Wang, 2014a; Wan et al., 2016; Wang et al., 2017); such close temporal and spatial correlations between diabases and trachytes offer convincing evidence that these rocks are different products of a similar thermal/magmatic event. ② The trachytes and coeval SQB diabases exhibit similar distribution on rare earth element diagram (Fig. 5). Moreover, the fractional crystallization process has little effect on the ratios of some incompatible elements (e.g., Nb/Ta, Th/Ta, Th/U), which can be useful in tracing the source of these bimodally associated suites. The results show that the average Nb/Ta, Th/Ta, and Th/U ratios of the Pingli-Zhuxi trachytes are 17. 36, 1. 82, and 4. 70, similar to the coeval SQB diabases with the average ratios of 15. 32, 1. 34, and 4. 18 (Chen et al., 2014; Wang et al., 2014b; Xiang et al., 2016), respectively. ③ The trachytes show initial ^{143}Nd/^{144}Nd ratios and $\varepsilon_{Nd}(t)$ values ranging from 0. 512 3 to 0. 512 5 and +3. 0 to +7. 3, respectively, which are close to those of mafic rocks (Fig. 7a), implying that they were all from a depleted mantle source. And ④ the variation of the major and trace elements (e.g., Sc, V, Cr, Ni, Sr, and Eu) of the suite has strong covariance, indicating a genetic relationship between the trachytes and the diabases via fractional crystallization from a cogenetic basaltic magma.

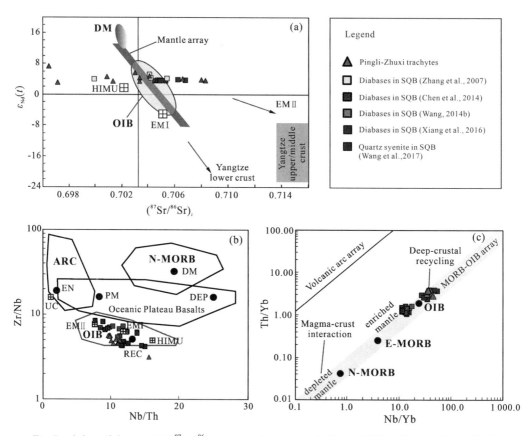

Fig. 7　(a) $\varepsilon_{Nd}(t)$ vs. initial $^{87}Sr/^{86}Sr$ diagram (after Qi and Zhou, 2008). (b) Zr/Nb vs. Nb/Th diagram (after Condie, 2005). (c) Th/Yb vs. Nb/Yb diagram (after Pearce, 2008).

DM: depleted mantle; MORB: middle oceanic ridge basalt; OIB: ocean island basalt; HIMU: high U/Pb mantle component; EMI: enriched mantle component I; EMⅡ: enriched mantle component Ⅱ; DEP: deep depleted mantle; En: enriched component; REC: recycled component. The DM and mantle array are from Zindler and Hart (1986); MORB, HIMU, EM I and EM Ⅱ are from Hart et al. (1992); OIB is from Wilson (1989); Yangtze upper/middle and lower crust are from Gao et al. (1999), Ma et al. (2000), and Chen and Jahn (1998). The isotope data of diabase in SQB (light-green square) are from Zhang et al. (2007); Other published data in SQB are same to Fig. 5.

HFSEs and REEs display similar partition coefficients for most mafic minerals, and their ratios generally remain stable during partial melting and fractional crystallization and thus can be used to reflect the nature of the mantle source (Weaver, 1991). High Nb/Ta (16.3 - 20.3) and Zr/Hf (44.1.0 - 48.3) ratios in these samples are close to those of typical OIB (Nb/Ta = 17.8, Zr/Hf = 35.9; Sun and McDonough, 1989). Based on plots in the Zr/Nb vs. Nb/Th diagram (Fig. 7b) and Th/Yb versus Nb/Yb diagram (Fig. 7c), most of the trachytes and coeval diabases plot similarly, showing an OIB affinity. One of the most remarkable features of the Pingli-Zhuxi trachytes is their highly enriched incompatible elements, especially for HFSE (e.g., Nb, Ta, Zr, and Hf), which are significantly higher than common trachytic/syenitic rocks worldwide. In this case, the enrichment of HFSE and other incompatible elements in Pingli-Zhuxi trachytes (Fig. 5) is inferred to be probably inherited from the source region. Recycled oceanic crust (ROC) is commonly invoked as a potential source component of alkaline magmas (Hofmann and White, 1982; Sobolev et al., 2007). However, taking into account the enriched features and intensely fractionated LREEs and HREEs [$(La/Yb)_N > 11$] in our samples, the incompatible-element-depleted recycled ocean crust is unlikely to be possible source material (Niu, 2009). Alternatively, a metasomatized lithosphere model opens new perspectives to explain the formation of alkaline rocks by different tectonic processes (Niu and O'Hara, 2003; Pilet, 2015). The concept of the mantle metasomatism was first proposed by Bailey (1970) to explain the enriched incompatible elements in alkaline magmatism. Hydrous mantle xenoliths rich in amphibole, phlogopite, and clinopyroxene have been observed in the North Daba Mountains (Huang, 1993b; Xu et al., 1996), which have been regarded as a potential indicator for a metasomatic process. Mantle metasomatism by percolating melts or fluids is a crucial mechanism that accounts for mantle enrichment and isotopic anomalies (Ackerman et al., 2013). Subduction-related metasomatism is characterized by significant depletion of HFSEs with very low HFSEs/LREEs ratios (Nb/La < 0.3). In contrast, the Pingli-Zhuxi trachytes are enriched in LREEs, LILEs, and HFSEs, and display high HFSEs/LREEs ratios (Nb/La = 1.25 − 1.74, average 1.50). These features are similar to those of magmatic rocks derived from mantle sources asthenosphere-derived melts (Turner et al., 1992; Li et al., 2003). Carbonatitic melts are often invoked as effective metasomatic agents in the mantle because of their low viscosity and high mobility (Hammouda and Laporte, 2000). Carbonatitic metasomatism commonly increase the $(La/Yb)_N$ ratios of mantle sources and lead to strong depletion in HFSEs (e.g., low Ti/Eu, Coltorti et al., 2007; Dai et al., 2017). However, the relatively low $(La/Yb)_N$ ratios but high Ti/Eu ratios documented in SQB diabases indicate the involvement of carbonatitic components is negligible (Zhang et al., 2020). In this regard, the asthenosphere-derived silicate melts with OIB features are considered to be a feasible metasomatic agent. Recent work has confirmed the occurrence of amphibole-clinopyroxene-phlogopite veins in the source region and proposed a relatively low mantle potential temperature of 1 258 − 1 300 ℃ beneath the SQB (Zhang et al., 2020). At lower mantle potential temperatures, easily fusible materials tend to melt to a higher degree in comparison with peridotite. Considering significant amount of HFSEs and REEs are stored within these hydrous metasomes and their fusible property in partial melting process (Médard et al., 2006), selective partial melting of metasomatized lithosphere can effectively release the HFSEs and REEs to the melts (Mayer et al., 2014). Elevated Nb/U and Ce/Pb ratios in Pingli-Zhuxi trachytes can be also explained by the presence of amphibole in the source region, for which partial melting of amphibole will strengthen the enrichment of Nb and depletion of Pb in initial melts (LaTourrette et al., 1995; Ionov and Hofmann, 1995; Li et al., 2014). In general, partial melting of an amphibole-clinopyroxene-phlogopite-bearing lithosphere source that metasomatized by silicate melts from the asthenosphere probably accounts for the enriched features in our samples.

5.1.3 "Daly gap" and fractional crystallization process

One of the notable issues in studies of bimodal magmatism, such as those in North Daba Mountains, is the explanation for lack of apparent intermediate compositions (Niu et al., 2013; Peng et al., 2015; Szymanowski et al., 2015). Silicate liquid immiscibility hypothesis was proposed by Huang et al. (1992) to explain the occurrence of coexisting mafic and felsic components accompanied by minor intermediate rocks in North Daba Mountains. Experimental data have demonstrated the feasibility that silicate liquid immiscibility during magma evolution produces contrasting Fe-rich and Si-rich liquid compositions (Roedder, 1978; Dixon and Rutherford, 1979; Philpotts, 1982). Unfortunately, only microscopic segregations have been observed; apart from the immiscible droplets of glasses trapped in the mesostasis of basalts (e.g., Philpotts, 1982; Luais, 1987; Kontak et al., 2002), silicate liquid immiscibility identified in slowly cooled plutonic environments (e.g., Skaergaard intrusion, Bushveld Complex, and Sept Iles intrusion) commonly segregated at mm- to m-scale (Charlier et al., 2011; Jakobsen et al., 2011; VanTongeren and Mathez, 2012). The general dearth of evidence for large-scale immiscibility reach the consensus that immiscibility does not play an important role on the relative rarity of intermediate compositions in extensive bimodal magmatism (Bowen, 1928).

Niu et al. (2013) considered that the "Daly Gap" is not as enigmatical as widely thought, but is a straightforward consequence of basaltic magma evolution. Previous evidence has shown that the change in composition from basalt to trachybasalt requires nearly 55% fractionation, but further differentiation to trachyte requires only a further 15% crystallization (Clague, 1978), which indicates fractional crystallization can felicitously explain the lack of intermediate SiO_2 compositions. This is confirmed by recently studies that identically suggested that such a compositional gap can be attributed to fractional crystallization process (e.g., Xu et al., 2010b; Melekhova et al., 2013; Dostal et al., 2017). At low melt fractions (0% to ~50% crystals), melt and crystals are not easily separable due to the stirring effect of magma convection and the short time that these magmas spent at low crystallinities (Huber et al., 2009; Bachmann and Huber, 2016). In contrast, the occurrence of crystal-melt separation via compaction is commonly at a high crystallinity between ~50% and 70% (Dufek and Bachmann, 2010). Most of the Pingli-Zhuxi trachytes are crystal-poor (< 10% phenocrysts), and show relatively higher SiO_2 but extremely lower MgO, Sr, and Ba contents compared to SQB mafic rocks, which indicate a high-degree crystal separation process in magma evolution. In this regard, models of interstitial melt extraction from highly crystalline magma reservoirs provide a reasonable explanation for the "Daly Gap". Importantly, further fractionation of trachytic magma has little effect on the major element composition of the residual liquids (Clague, 1978); therefore, trachytes can represent highly evolved components in a set of bimodal volcanic rocks.

The relative high SiO_2(60.88–63.87 wt%) contents with low Cr (0.98–4.11 ppm), Ni (0.28–1.64 ppm), and MgO (0.50–1.64 wt%) contents in our samples indicate they were unlikely produced by directly melting of a mantle source, but experience igneous fractionation of mafic minerals from an evolved magma (Baker et al., 1995; Frost and Frost, 1997). Apparent correlation in major element concentrations between the trachytes and coeval SQB diabases are shown on the Harker diagrams (Fig. 6a-h), indicating the trachytes are possibly generated by fractional crystallization from a basaltic magma. The samples broadly exhibit an obvious continuous trend; components that decrease in concentration across the "Daly gap" include most major elements (e.g., MgO, $Fe_2O_3^T$, CaO, TiO_2, and P_2O_5) and a few trace elements, such as Sc, V, Cr, Ni, Sr, and Ba. Slightly positive to moderate negative Eu anomalies from the diabases to the trachytes suggest the minor fractional crystallization of plagioclase (Green, 1994). Crystallization of K-feldspar will enhance these Ba, Sr, and Eu depletions and near steady-state behavior of Ga as those of which are compatible in K-feldspar (White et al., 2003; Shao et al., 2015). This is strengthened by the positive correlations of Rb and Ba with varying Sr concentrations, with an obvious trend toward K-feldspar-dominated fractionation can be observed (Fig. 8). The relatively low P_2O_5 contents and less enriched in the middle REEs compared to the diabases are considered to be related to the fractionation of apatite (Cousens et al., 2003). The low concentration of TiO_2 and $Fe_2O_3^T$ may be attributed to separation of Ti-Fe oxides. These inferred fractionating phases are in agreement with the

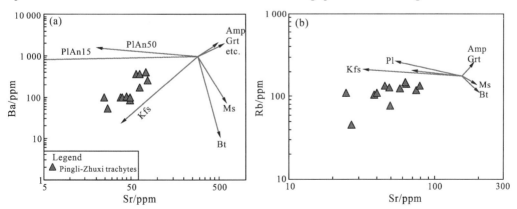

Fig. 8 (a) Ba vs. Sr diagram and (b) Rb vs. Sr diagram for the Pingli-Zhuxi trachytes
(after Jung et al., 2007; Wang et al., 2017).

occurrence of K-feldspar being the dominant phase present, with minor plagioclase, biotite, apatite, and Ti-magnetite in our samples. Additionally, we have tested the fractional crystallization hypothesis by considering plots of an incompatible element vs the ratio of the same incompatible to a less-incompatible element (e.g., Nb-Zr, Ce-Yb. Joron and Treuil, 1989; Meng et al., 2013). The results show that the Nb/Zr and Ce/Yb ratios in the Pingli-Zhuxi trachytes do not change with the variation of Nb and Ce contents, respectively (Fig. 9). In summary, fractional crystallization from basaltic magma is suggested as the dominant process for the formation of trachytic melts and occurrence of the "Daly gap".

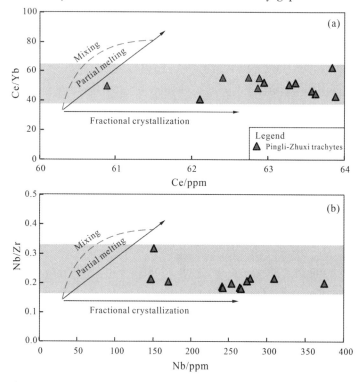

Fig. 9 (a)Ce/Yb vs. Ce and (b)Nb/Zr vs. Nb diagrams for the Pingli-Zhuxi trachytes.

5. 2 Geological implications

5. 2. 1 Tectonic setting

Alkaline volcanic combinations are enigmatic not only with regard to their petrogenesis but also in terms of their tectonic settings (Dall'Agnol et al., 2012). Alkaline rocks formed in subduction-related regimes, or those previously modified by subduction processes, normally exhibit a characteristic negative Nb anomaly and high Sr isotopic ratios and low $\varepsilon_{Nd}(t)$ values (Roger et al., 1985; Zhao et al., 1995). However, with regard to the North Daba Mountains, this tectonic model is excluded because these geochemistry characters are not observed in our samples. The tectonic discrimination diagrams show that the alkaline magmatic rocks in the

North Daba Mountains are related to a within-plate setting (Fig. 10). With the further progress of regional research, there is general agreement that the North Daba Mountains was subjected to extensional tectonics during the Late Ordovician and Early Silurian (Zhang et al., 2002). Pingli-Zhuxi trachytes and coexisting mafic rocks display slightly positive Nb anomalies and high ε_{Nd} values, similar to ocean island basalts, showing typical characters of rift- or hotspot-related alkaline rocks (Clague, 1987). Abundant fossils (e.g., graptolite) were founded in Silurian sediments of the North Daba Mountains (Luo and Duanmu, 2001; Guo, 2017), indicating a local, deep and anoxic marine environments, which provides evidence for a rift tectonic setting (Gao et al., 1995). Moreover, various alkaline volcanic activities coupled with close temporal and spatial connections, as mentioned above, further support a rift setting in the North Daba Mountains (Xu et al., 2008; Zhu et al., 2016).

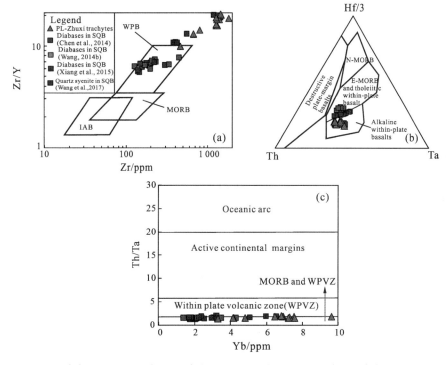

Fig. 10　(a) Zr vs. Zr/Y diagram (after Pearce and Norry, 1979) and (b) Hf/3-Th-Ta triangular diagram (after Wood, 1980) and (c) Yb vs. Th/Ta diagram (after Gorton and Schandl, 2000) for the Pingli-Zhuxi trachytes.

WPB: within plate basalt; IAB: island arc basalt; MORB: middle oceanic ridge basalt.

5. 2. 2　Ascent processes and geodynamic mechanism

Multi-depth magma chambers have been documented in many magmatic systems (Peccerillo et al., 2006; Scaillet et al., 2008). This is particularly true for intraplate alkaline magmas that usually originate from deep mantle and pass through thick sequences of lithosphere, differentiating at several depths (Marzoli et al., 2015). Previous studies show that the pyroxene phenocrysts of mafic rocks in the North Daba Mountains were mainly formed at three depths (i.e., 40−68 km, 15−20 km, and 5−9 km), suggesting that the primary magma was formed below 68 km and accumulated during episodic ascent (Xiang et al., 2010). Note that these cogenetic Early Paleozoic alkaline rocks in the North Daba Mountains range in composition from basalt to trachyte, and exhibit scattered temporal distributions from 410 Ma to 455 Ma. This is also reflected in the Pingli-Zhuxi trachytes that yielded the U-Pb ages of 406. 0 Ma and 427. 9 Ma, respectively. Calculation results for rift-related alkaline magma have shown that ascent rates decrease with magma viscosity and proportion of crystals, and a negative exponential relationship between silica and velocity of the magmas [velocity (m/s) = $19.07e^{-0.03SiO_2(wt\%)}$] is established (Kokandakar et al. 2018). In this case, we prefer that the extensive petrological compositions and the scattered temporal distributions are attributed to a protracted fractional crystallization during ascent. That is, as lithosphere thins due to active rifting, selective partial melting of the metasomatized lithosphere produced a primary basaltic magma. Magma with high ascent velocity experienced relatively low and variable evolution, more likely to generate the mafic end-members with varied MgO, Cr, and Ni contents. In contrast, magma that stalled at multi-depth reservoirs experienced protracted fractional crystallization, achieving higher degrees of differentiation and viscosity. Elevated viscosity and crystallinity hamper the ascent rate and make these more felsic magmas accumulate during ascent, resulting in delayed emplacement and possible eruption of the mafic magmatic systems.

Understanding the geodynamic mechanism in rift systems has significance for rift opening, magma emplacement, and evolution. A high heat-flux, deep-rooted mantle plume activity was considered a key factor explaining the continental break-up, lithospheric thinning, and associated large-scale intraplate magmatism in the North Daba Mountains (Zhang et al., 2002; Yan, 2005; Li and Yang, 2011). However, this hypothesis is inconsistent with the absence of radial dike swarms and tholeiitic magmatism (e.g., contemporaneous continental flood basalt) in the North Daba Mountains. Therefore, asthenospheric upwelling maybe a more plausible mechanism to explain the lithospheric and continental uplift and fracturing. Low-degree melts with high concentration incompatible elements released from the underlying asthenosphere and metasomatizes the lithosphere during ascent, giving rise to the enriched features in the source region. Driven by thermal balance, the asthenospheric upwelling commonly causes thinning of the lithosphere (Turner et al. 1992). This is confirmed by the numerical simulations which proved that the small-scale thermal convection in the asthenosphere mantle can effectively lead to a large-scale thinning of the lithosphere (Qiao et al., 2013). Heating advection from depth reduced the density of asthenosphere material and then began to ascend. The rising flow gradually cooled down and descended when interacting with the colder mantle, producing circulation driven by density balance. We simply illustrate this process in Fig. 11. With decreasing pressure, the material in lithosphere began to melt and was metasomatized by asthenosphere-derived melts, forming the primary magma with OIB characteristics. In the wake of the formation of a rift basin, the primary basaltic magma has undergone protracted fractional crystallization during ascent, eventually forming the Early Paleozoic Pingli-Zhuxi trachytes in the North Daba Mountains.

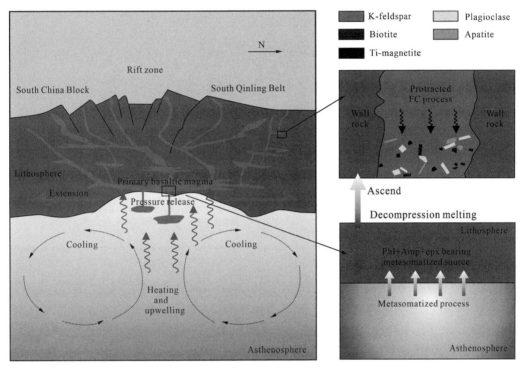

Fig. 11   Geodynamic model for the rift setting in the North Daba Mountains.
Phl: phlogopite; Amp: amphibole; Cpx: clinopyroxene.

6   Conclusions

The evidence obtained from the geochronological and geochemical investigation of the alkaline trachytes in the North Daba Mountains leads to the following conclusions:

(1) Zircon U-Pb ages reveal that the Pingli-Zhuxi trachytes formed at 406.0–427.9 Ma. The samples display narrow range initial $^{143}Nd/^{144}Nd$ ratios of 0.512 3–0.512 5 with positive $\varepsilon_{Nd}(t)$ values from +3.0 to +7.3, consistent with those of SQB diabases. They are characterized by relatively high SiO_2 (60.88–63.87 wt%) and total alkali ($Na_2O + K_2O$ = 10.19–12.24 wt%) contents, as well as elevated abundances of REEs and HFSEs.

(2) The genesis of Pingli-Zhuxi trachytes is attributed to protracted fractionation of a basaltic melt. The trachyte and SQB diabases are genetically related through fractional crystallization from a cogenetic metasomatized lithosphere source in rift setting. Variable

degrees of protracted fractional crystallization during ascent give rise to the scattered temporal distributions (410-455 Ma) in the trachytes and coexisting mafic rocks.

(3) Asthenospheric upwelling is considered to be responsible for continental break up and lithosphere metasomatism. Melting of the metasomatized source, followed by protracted fractional crystallization, gives rise to the enriched features in the Pingli-Zhuxi trachytes.

Acknowledgements We sincerely thank Dr Lianxun Wang, Dr David Lentz and an anonymous reviewer for their insightful comments, which significantly improved the shape, language, and discussion of this manuscript. We also thank Dr Robert J. Stern for his editorial working and valuable comments. This work was jointly supported by the National Natural Science Foundation of China (Grant Nos. 41772052 and 41421002) and the program for Changjiang Scholars and Innovative Research Team in University (IRT1281).

References

Ackerman, L., Spacek, P., Magna, T., Ulrych, J., Svojtka, M., Hegner, E., Balogh, K., 2013. Alkaline and carbonate-rich melt metasomatism and melting of subcontinental lithospheric mantle: Evidence from mantle xenoliths, NE Bavaria, Bohemian massif. Journal of Petrology, 54, 2597-2633.

Andersen, T., 2002. Correction of common lead in U-Pb analyses that do not report ^{204}Pb. Chemical Geology, 192(1-2), 59-79.

Andersen, T., Elburg, M., Erambert, M., 2018. Contrasting trends of agpaitic crystallization in nepheline syenite in the Pilanesberg alkaline complex, South Africa. Lithos, 312-313, 375-388.

Armbrustmacher, T.J., Hedge, C.E., 1982. Genetic implications of minor-element and Sr-isotope geochemistry of alkaline rock complexes in the wet mountains area, Fremont and Custer counties, Colorado. Contributions to Mineralogy and Petrology, 79(4), 424-435.

Avanzinelli, R., 2004. Crystallisation and genesis of peralkaline magmas from Pantelleria Volcano, Italy: An integrated petrological and crystal-chemical study. Lithos, 73(1-2), 41-69.

Bachmann, O., Huber, C., 2016. Silicic magma reservoirs in the Earth's crust. American Mineralogist, 101 (11), 2377-2404.

Bailey, D.K., 1970. Volatile flux, heat focusing and the generation of magma. Geological Journal, 2, 177-186.

Baker, M.B., Hirschmann, M.M., Ghiorso, M.S., Stolper, E.M., 1995. Compositions of near-solidus peridotite melts from experiments and thermodynamic calculations. Nature, 375(6529), 308-311.

Barberi, F., Ferrara, G., Santacroce, R., Treuil, M., Varet, J., 1975. A transitional basalt-pantellerite sequence of fractional crystallization: The Boina centre (Afar Rift, Ethiopia). Journal of Petrology, 16, 22-56.

Bowen, N.L., 1928. The Evolution of the Igneous Rocks. American Scientist, 23 (8), 1-91.

Charlier, B., Namur, O., Toplis, M.J., Schiano, P., Cluzel, N., Higgins, M.D., Vander, A.J., 2011. Large-scale silicate liquid immiscibility during differentiation of tholeiitic basalt to granite and the origin of the Daly gap. Geology, 39(10), 907-910.

Chen, H., Tian, M., Wu, G.L., Hu, J.M., 2014. The Early Paleozoic Alkaline and mafic magmatic events in

Southern Qinling Belt, Central China: Evidences for the break-up of the Paleo-Tethyan Ocean. Geological Review, 60(6), 1437-1452 (in Chinese with English abstract).

Chen, J., Jahn, B.M., 1998. Crustal evolution of southeastern China: Nd and Sr isotopic evidence. Tectonophysics, 284(1-2), 101-133.

Chen, P.P., 2015. The petrogenesis, geochemistry and genesis of volcanic rock in Zhujiayuan of Pingli county, Shaanxi Province (Master's thesis). Beijing: China University of Geosciences (in Chinese with English abstract).

Chen, Y., Hu, R., Bi, X., Dong, S., Xu, Y., Zhou, T., 2018. Zircon U-Pb ages and Sr-Nd-Hf isotopic characteristics of the Huichizi granitic complex in the north Qinling orogenic belt and their geological significance. Journal of Earth Science, 29(5), 492-507.

Clague, D.A., 1978. The oceanic basalt-trachyte association: An explanation of the Daly gap. Journal of Geology, 86(6), 739-743.

Clague, D.A., 1987. Hawaiian alkaline volcanism. Geological Society, London, Special Publications, 30(1), 227-252.

Coltorti, M., Bonadiman, C., Faccini, B., Ntaflos, T., Siena, F., 2007. Slab melt and intraplate metasomatism in Kapfenstein mantle xenoliths. Lithos, 94, 66-89.

Condie, K.C., 2005. High field strength element ratios in Archean basalts: A window to evolving sources of mantle plumes. Lithos, 79(3-4), 491-504.

Cousens, B.L., Clague, D.A., Sharp, W.D., 2003. Chronology, chemistry, and origin of trachytes from Hualalai volcano, Hawaii. Geochemistry, Geophysics, Geosystems, 4(9), 1-27.

Dai, L.Q., Zhao, Z.F., Zheng, Y.F., An, Y.J., Zheng, F., 2017. Geochemical distinction between carbonate and silicate metasomatism in generating the mantle sources of alkali basalts. Journal of Petrology, 58, 863-884.

Dall'Agnol, R., Frost, C.D., Rämö, O.T., 2012. IGCP Project "A-type granites and related rocks through time": Project vita, results, and contribution to granite research. Lithos, 151, 1-16.

Ding, S., Huang, H., Niu, Y.L., Zhao, Z.D., Yu, X.H., Mo, X.X., 2011. Geochemistry, geochronology and petrogenesis of East Kunlun high Nb-Ta rhyolites. Acta Petrologica Sinica, 27(12), 3603-3614 (in Chinese with English abstract).

Dixon, S., Rutherford, M.J., 1979. Plagiogranites as late-stage immiscible liquids in ophiolite and mid-ocean ridge suites: An experimental study. Earth & Planetary Science Letters, 45(1), 45-60.

Dong, Y.P., Santosh, M., 2016. Tectonic architecture and multiple orogeny of the Qinling Orogenic Belt, Central China. Gondwana Research, 29, 1-40.

Dong, Y.P., Zhang, G.W., Neubauer, F., Liu, X.M., Genser, J., Hauzenberger, C., 2011. Tectonic evolution of the Qinling orogen, China: Review and synthesis. Journal of Asian Earth Sciences, 41(3), 213-237.

Dostal, J., Hamilton, T.S., Shellnutt, J.G., 2017. Generation of felsic rocks of bimodal volcanic suites from thinned and rifted continental margins: Geochemical and Nd, Sr, Pb- isotopic evidence from Haida Gwaii, British Columbia, Canada. Lithos, 292-293, 146-160.

Duan, L., 2010. Detrital zircon provenance of the Silurian and Devonian in South Qinling, and the northwest margin of Yangtze terrane and its tectonic implications (unpublished Master thesis). Xi'an: Northwest University (in Chinese with English abstract).

Dufek, J., Bachmann, O., 2010. Quantum magmatism: Magmatic compositional gaps generated by melt-crystal dynamics. Geology, 38, 687-690.

Frost, C.D., Frost, B.R., 1997. Reduced rapakivi-type granites: The tholeiite connection. Geology, 25(7), 647-650.

Gao, S., Ling, W., Qiu, Y., Lian, Z., Hartmann, G., Simon, K., 1999. Contrasting geochemical and Sm-Nd isotopic compositions of Archean metasediments from the kongling high-grade terrain of the Yangtze craton: Evidence for cratonic evolution and redistribution of REE during crustal anatexis. Geochimica et Cosmochimica Acta, 63(13-14), 2071-2088.

Gao, S., Rudnick, R.L., Xu, W.L., Yuan, H.L., Liu, Y.S., Walker, R.J., Puchtel, I.S., Liu, X.M., Huang, H., Wang, X.R., Yang, J., 2008. Recycling deep cratonic lithosphere and generation of intraplate magmatism in the North China Craton. Earth & Planetary Science Letters, 270, 41-53.

Gao, S., Zhang, B.R., Gu, X.M., 1995. Silurian-Devonian provenance changes of South Qinling basins: Implications for accretion of the Yangtze (South China) to the North China cratons. Tectonophysics, 250, 183-197.

Gorton, M.P., Schandl, E.S., 2000. From continents to island arcs: A geochemical index of tectonic setting for arc-related and within-plate felsic to intermediate volcanic rocks. The Canadian Mineralogist, 38(5), 1065-1073.

Green, T.H., 1994. Experimental studies of trace-element partitioning applicable to igneous petrogenesis-Sedona 16 years later. Chemical Geology, 117(1-4), 1-36.

Guo, X.Q., Wang, Z.Q., Yan, Z., 2017. Alkali volcanism and zinc-fluorite mineralization of Pingli-Zhenping area, North Daba Mountains. Acta Geoscientica Sinica, 38(Supp.1), 21-24 (in Chinese with English abstract).

Hagen, P.G., Cottle, J.M., 2016. Synchronous alkaline and subalkaline magmatism during the late Neoproterozoic-early Paleozoic ross orogeny, Antarctica: Insights into magmatic sources and processes within a continental arc. Lithos, 262, 677-698.

Hammouda, T., Laporte, D., 2000. Ultrafast mantle impregnation by carbonatite melts. Geology, 28, 283-285.

Hart, S.R., Hauri, E.H., Oschmann, L.A., Whitehead, J.A., 1992. Mantle plumes and entrainment: Isotopic evidence. Science, 256(5056), 517-520.

Hofmann, A.W., Jochum, K.P., Seufert, M., White, W.M., 1986. Nb and Pb in oceanic basalts: New constraints on mantle evolution. Earth & Planetary Science Letters, 79, 33-45.

Hofmann, A.W., White, W.M., 1982. Mantle plumes from ancient oceanic crust. Earth & Planetary Science Letters, 57(2), 421-436.

Hoskin, P.W.O., Schaltegger, U., 2003. The composition of zircon and igneous and metamorphic petrogenesis. Rev. Miner. Geochem., 53, 27-62.

Huang, W.F., 1993a. Multiphase deformation and displacement within a basement complex on a continental margin: The Wudang Complex in the Qinling Orogen, China. Tectonophysics, 224, 305-326.

Huang, Y. H., 1993b. Mineralogical characteristics of Phlogopite-Amphibole-Pyroxenite mantle xenoliths included in the alkali mafic-ultramafic subvolcanic complex from Langao country, China. Acta Petrologica Sinica, 9, 367-378 (in Chinese with English abstract).

Huang, Y.H., Ren, Y.X., Xia, L.Q., Xia, Z.C., Zhang, C., 1992. Early Palaeozoic bimodal igneous suite on Northern Daba Mountains-Gaotan diabase and Haoping trachyte as examples. Acta Petrologica Sinica, 8, 243-256 (in Chinese with English abstract).

Huber, C., Bachmann, O., Manga, M., 2009. Homogenization processes in silicic magma chambers by stirring and mushification (latent heat buffering). Earth & Planetary Science Letters, 283, 38-47.

Ionov, D.A., Hofmann, A.W., 1995. Nb-Ta-rich mantle amphiboles and micas: Implications for subduction-related metasomatic trace element fractionations. Earth & Planetary Science Letters, 131(3-4), 341-356.

Jakobsen, J.K., Veksler, I.V., Tegner, C., Brooks, C.K., 2011. Crystallization of the Skaergaard intrusion from an emulsion of immiscible iron- and silica-rich liquids: Evidence from melt inclusions in plagioclase. Journal of Petrology, 52(2), 345-373.

Joron, J.L., Treuil, M., 1989. Hygromagmaphile element distributions in oceanic basalts as fingerprints of partial melting and mantle heterogeneities: A specific approach and proposal of an identification and modelling method. Geological Society, London, Special Publications, 42, 277-299.

Jung, S., Hoffer, E., Hoernes, S., 2007. Neo-Proterozoic rift-related syenites (Northern Damara Belt, Namibia): Geochemical and Nd-Sr-Pb-O isotope constraints for mantle sources and petrogenesis. Lithos, 96, 415-435.

Jung, S., Vieten, K., Romer, R.L., Mezger, K., Hoernes, S., Satir, M., 2012. Petrogenesis of tertiary alkaline magmas in the Siebengebirge, Germany. Journal of Petrology, 53(11), 2381-2409.

Kokandakar, G.J., Ghodke, S.S., Rathna, K., More, L.B., Nagaraju, B., Bhosle, M.V., Kumar, K.V., 2018. Density, viscosity and velocity (ascent rate) of alkaline magmas. Journal of the Geological Society of India, 91(2), 135-146.

Kontak, D.J., De, W.D.Y.M.Y., Dostal, J., 2002. Late-stage crystallization history of the Jurassic North Mountain Basalt, Nova Scotia, Canada. I. Textural and chemical evidence for pervasive development of silicate-liquid immiscibility. Canadian Mineral., 40(5), 1287-1311.

LaTourrette, T., Hervig, R.L., Holloway, J.R., 1995. Trace element partitioning between amphibole: Phlogopite, and basanite melt. Earth & Planetary Science Letters, 135(1-4), 13-30.

Li, F.J., 2009. The rock geochemistry characteristics and tectonic implications of mafic dyke swarms and syenite porphyry veins in Zhenba eastern area, the south of Shaanxi Province (unpublished Master thesis). Xi'an: Chang'an University (in Chinese with English abstract).

Li, F.J., Yang, J., 2011. Tectonic meaning of mafic dyke swarms in Zhenba eastern area in Shannan. Journal of Sichuan University of Science and Engineering (Natural Science Edition), 24(2), 238-243 (in Chinese with English abstract).

Li, W.P., Wang, T., Wang, X.X., 2001. Source of Huichizi granitoid complex pluton in northern Qinling, Central China: Constrained in element and isotopic geochemistry. Earth Science, 26(3), 269-278 (in Chinese with English abstract).

Li, X., Liu, J.Q., Sun, C.Q., Du, D.D., Wang, S., 2014. The magma source properties and evolution of

Holocene volcanoes in Tengchong, Yunnan Province, SW China. Seismology and Geology, 36(4), 991-1008 (in Chinese with English abstract).

Li, X.H., Chen, Z.G., Liu, D.Y., Li, W.X., 2003. Jurassic gabbro-granite-syenite suites from southern Jiangxi Province, SE China: Age, origin, and tectonic significance. International Geology Review, 45, 898-921.

Li, X.Y., Guo, F., Fan, W.M., Wang, Y.J., Li, C.W., 2004. Early Cretaceous trachytes of Donglingtai formation from the Xishan area in the northern north China block: Constraints on melting of lower mafic crust. Geotectonica et Metallogenia, 28(2), 155-164.

Liégeois, J.P., Navez, J., Hertogen, J., Black, R., 1998. Contrasting origin of post-collisional high-K calc-alkaline and shoshonitic versus alkaline and peralkaline granitoids: The use of sliding normalization. Lithos, 45(1-4), 1-28.

Litvinovsky, B.A., Steele, I.M., Wickham, S.M., 2000. Silicic magma formation in overthickened crust: Melting of charnockite and leucogranite at 15, 20 and 25 kbar. Journal of Petrology, 41(5), 717-737.

Liu, Y., Liu, X.M., Hu, Z.C., Diwu, C.R., Yuan, H.L., Gao, S., 2007. Evaluation of accuracy and long-term stability of determination of 37 trace elements in geological samples by ICP-MS. Acta Petrologica Sinica, 23(5), 1203-1210 (in Chinese with English abstract).

Long, J.S., 2016. The geochemical characteristics and geological significances of the basic dike swarms and syenite porphyry veins in the Ziyang, South Shaanxi (unpublished Master thesis). Xi'an: Chang'an University (in Chinese with English abstract).

Lu, F.X., Zhang, B.R., Han, Y.W., Zhong, Z.Q., Ling, W.L., Zhang, H.F., Zheng, J.P., Hou, Q.Y., 2006. The Three-dimensional Lithospheric chemical structure in Qinling-Dabie-Sulu Area. Beijing: Geological Publishing House (in Chinese).

Luais, B., 1987. Béatrice Luais. Immiscibilité entre liquids silicatés dans les mésostases et les inclusions vitreuses des andésites basiques de Santorin. Arc Egéen, Bulletin de Mineralogie, 110(1), 93-109.

Lucassen, F., Pudlo, D., Franz, G., Romer, R.L., Dulski, P., 2013. Cenozoic intra-plate magmatism in the Darfur volcanic province: Mantle source, phonolite-trachyte genesis and relation to other volcanic provinces in NE Africa. International Journal of Earth Sciences, 102(1), 183-205.

Ludwig, K.R., 2003. ISOPLOT 3.0: A geochronological toolkit for Microsoft Excel. Berkeley Geochronology Center, Special Publication, 4, 71.

Luo, K.L., Duanmu, H.S., 2001. Timing of Early Paleozoic basic igneous rocks in the Daba Mountain. Regional Geology of China, 20, 262-266 (in Chinese with English abstract).

Ma, C.Q., Ehlers, C., Xu, C.H., Li, Z.C., Yang, K.G., 2000. The roots of the Dabieshan ultrahigh-pressure metamorphic terrane: Constraints from geochemistry and Nd-Sr isotope systematics. Precambrian Research, 102(3), 279-301.

Marks, M.A.W., Schilling, J., Coulson, I.M., Wenzel, T., Markl, G., 2018. The alkaline-peralkaline Tamazeght complex, high Atlas Mountains, Morocco: Mineral chemistry and petrological constraints for derivation from a compositionally heterogeneous mantle source. Journal of Petrology, 49(6), 1097-1131.

Marzoli, A., Aka, F.T., Merle, R., Callegaro, S., N'ni, J., 2015. Deep to shallow crustal differentiation of with in-plate alkaline magmatism at Mt. Bambouto volcano, Cameroon Line. Lithos, 220, 272-288.

Mattauer, M., Matte, P., Malavieille, L., Tapponnier, P., Maluski, H., Xu, Z.Q., Lu, Y.L., Tang, Y.Q., 1985. Tectonics of the Qinling Belt: Build up and evolution of eastern Asia. Nature, 317(6037), 496-500.

Mayer, B., Jung, S., Romer, R.L., Pfänder, J.A., Klügel, A., Pack, A., Gröner, E., 2014. Amphibole in alkaline basalts from intraplate settings: Implications for the petrogenesis of alkaline lavas from the metasomatised lithospheric mantle. Contributions to Mineralogy and Petrology, 167, 989.

Médard, E., Schmidt, M.W., Schiano, P., Ottolini, L., 2006. Melting of amphibole-bearing wehrlites: An experimental study on the origin of ultra-calcic nepheline-normative melts. Journal of Petrology, 47(3), 481-504.

Melekhova, E., Annen, C., Blundy, J., 2013. Compositional gaps in igneous rock suites controlled by magma system heat and water content. Nature Geoscience, 6, 385-390.

Meng, F.C., Liu, J.Q., Cui, Y., 2013. Petrogenesis and crypto-explosive mechanism of trachyte in Yingcheng Formation of Xujiawei fault depression, Songliao Basin, NE China. Journal of Jilin University: Earth Science Edition, 43(3), 704-715 (in Chinese with English abstract).

Middlemost, E.A.K., 1994. Naming materials in the magma/ igneous rock system. Earth-Science Reviews, 37, 215-224.

Mingram, B., Trumbull, R.B., Littman, S., Gerstenberger, H., 2000. A petrogenetic study of anorogenic felsic magmatism in the cretaceous paresis ring complex, Namibia: Evidence for mixing of crust and mantle-derived components. Lithos, 54(1-2), 1-22.

Montel, J.M., Vielzeuf, D., 1997. Partial melting of metagreywackes, Part II: Compositions of minerals and melts. Contributions to Mineralogy and Petrology, 128(2-3), 176-196.

Niu, Y.L., 2009. Some basic concepts and problems on the petrogenesis of intra-plate ocean island basalts. Chinese Science Bulletin, 54, 4148-4160.

Niu, Y.L., O'Hara, M.J., 2003. Origin of ocean island basalts: A new perspective from petrology, geochemistry, and mineral physics considerations. Journal of Geophysical Research, 108(B4), 2209-2228.

Niu, Y.L., Zhao, Z.D., Zhu, D.C., Mo, X.X., 2013. Continental collision zones are primary sites for net continental crust growth: A testable hypothesis. Earth-Science Reviews, 127(2), 96-110.

Pearce, J.A., 2008. Geochemical fingerprinting of oceanic basalts with applications to ophiolite classification and the search for Archean oceanic crust. Lithos, 100(1), 14-48.

Pearce, J.A., Norry, M.J., 1979. Petrogenetic implications of Ti, Zr, Y, and Nb variations in volcanic rocks. Contributions to Mineralogy and Petrology, 69, 33-47.

Peccerillo, A., Donati, C., Santo, A.P., Orlando, A., Yirgu, G., Ayalew, D., 2007. Petrogenesis of silicic peralkaline rocks in the Ethiopian rift: Geochemical evidence and volcanological implications. Journal of African Earth Sciences, 48(2-3), 161-173.

Peccerillo, A., Frezzotti, M.L., De Astis, G., Ventura, G., 2006. Modeling the magma plumbing system of Vulcano (Aeolian Islands, Italy) by integrated fluid-inclusion geobarometry, petrology, and geophysics. Geology, 34, 17-20.

Peng, P., Wang, X.P., Lai, Y., Wang, C., Windley, B.F., 2015. Large-scale liquid immiscibility and fractional crystallization in the 1 780 Ma Taihang dyke swarm: Implications for genesis of the bimodal Xiong'er volcanic province. Lithos, 236-237, 106-122.

Philpotts, A.R., 1982. Compositions of immiscible liquids in volcanic rocks. Contributions to Mineralogy and Petrology, 80(3), 201-218.

Pilet, S., 2015. Generation of low-silica alkaline lavas: Petrological constraints, models and thermal implications. In: Foulger, G.R., Lustrino, M., King, S. The interdisciplinary earth: In Honor of Don L. Anderson, Geological Society of America Special Paper 514 and American Geophysical Union Special Publication, 71, 281-304.

Qi, L., Zhou, M.F., 2008. Platinum-group elemental and Sr-Nd-Os isotopic geochemistry of Permian Emeishan flood basalts in Guizhou Province, SW China. Chemical Geology, 248, 83-103.

Qiao, Y.C., Guo, Z.Q., Shi, Y.L., 2013. Numerical simulation of thermal convective erosion thinning mechanism in North China craton lithosphere. Beijing: Science Press, 43(4), 642-652 (in Chinese with English abstract).

Roedder, E., 1978. Silicate liquid immiscibility in magmas and in the system $K_2O\text{-}FeO\text{-}Al_2O_3\text{-}SiO_2$: An example of serendipity. Geochimica et Cosmochimica Acta, 42(11), 1597-1617.

Rogers, N.W., Hawkesworth, C.J., Parker, R.J., Marsh, J.S., 1985. The geochemistry of potassic lavas from Vulsini, central Italy and implications for mantle enrichment processes beneath the Roman region. Contributions to Mineralogy and Petrology, 90(2-3), 244-257.

Rudnick, R.L., Gao, S., 2003. Composition of the continental crust. In: Rudnick, R.L. The crust. Treatise on geochemistry, 3. Oxford: Elsevier Pergamon, 1-64.

Saha, A., Ganguly, S., Ray, J., Koeberl, C., Thöni, M., Sarbajna, C., Sawant, S.S., 2017. Petrogenetic evolution of cretaceous samchampi-samteran alkaline complex, mikir hills, northeastern india: Implications on multiple melting events of heterogeneous plume and metasomatized sub-continental lithospheric mantle. Gondwana Research, 48, 237-256.

Scaillet, B., Pichavant, M., Cioni, R., 2008. Upward migration of Vesuvius magma chamber over the past 20 000 years. Nature, 455(7210), 216-219.

Shao, F.L., Niu, Y.L., Regelous, M., Zhu, D.C., 2015. Petrogenesis of peralkaline rhyolites in an intra-plate setting: Glass House Mountains, southeast Queensland, Australia. Lithos, 216-217, 196-210.

Skridlaite, G., Wiszniewska, J., Duchesne, J.C., 2003. Ferropotassic A-type granites and related rocks in NE Poland and S Lithuania: West of the East European Craton. Precambrian Research, 124(2), 305-326.

Sobolev, A.V., Hofmann, A.W., Kuzmin, D.V., et al., 2007. The amount of recycled crust in sources of mantle derived melts. Science, 316, 412-417.

Song, W.L., Xu, C., Smith, M.P., Kynicky, J., Huang, K.J., Wei, C.W., Zhou, L., Shu, Q.H., 2016. Origin of unusual HREE-Morich carbonatites in the Qinling orogen, China. Scientific Reports, 6(1), 37377.

Su, S., Niu, Y., Deng, J., Liu, C., Zhao, G., Zhao, X., 2007. Petrology and geochronology of Xuejiashiliang igneous complex and their genetic link to the lithospheric thinning during the Yanshanian orogenesis in eastern China. Lithos, 96(1-2), 90-107.

Sun, S.S., McDonough, W.F., 1989. Chemical and isotopic systematics of oceanic basalts: Implications or mantle composition and processes. Geological Society, London, Special Publications, 42, (1), 313-345.

Sun, W.D., Li, S.G., Chen, Y.D., Li, Y.J., 2000. Zircon U-Pb dating of granitoids from South Qinling, Central China and their geological significance. Geochemica, 29, 209-216 (in Chinese with English abstract).

Sun, W.D., Li, S.G., Sun, Y., Zhang, G.W., Li, Q.L., 2002. Mid-Paleozoic collision in the North Qinling: Sm-Nd, Rb-Sr and ^{40}Ar/^{39}Ar ages and their tectonic implications. Journal of Asian Earth Sciences, 21 (1), 69-76.

Szymanowski, D., Ellis, B.S., Bachmann, O., Guillong, M., Phillips, W.M., 2015. Bridging basalts and rhyolites in the Yellowstone-snake river plain volcanic province: The elusive intermediate step. Earth & Planetary Science Letters, 415, 80-89.

Trua, T., Deniel, D., Mazzuoli, R., 1999. Crustal control in the genesis of Plio-Quaternary bimodal magmatism of the Main Ethiopian Rift (MER): Geochemical and isotopic (Sr, Nd, Pb) evidence. Chemical Geology, 155, 201-231.

Tschegg, C., Ntaflos, T., Akinin, V.V., 2011. Polybaric petrogenesis of Neogene alkaline magmas in an extensional tectonic environment: Viliga volcanic field, Northeast Russia. Lithos, 122(1-2), 13-24.

Turner, S., Sandiford, M., Foden, J., 1992. Some geodynamic and compositional constraints on "postorogenic" magmatism. Geology, 20(10), 931-934.

VanTongeren, J.A., and Mathez, E.A., 2012. Large-scale liquid immiscibility at the top of the Bushveld Complex, South Africa. Geology, 40(6), 491-494.

Vernikovsky, V.A., Pease, V.L., Vernikovskaya, A.E., Romanov, A.P., Gee, D.G., Travin, A.V., 2003. First report of early triassic A-Type granite and syenite intrusions from Taimyr: Product of the northern Eurasian superplume? Lithos, 66(1-2), 23-36.

Wan, J., Liu, C.X., Yang, C., Liu, W.L., Li, X.W., Fu, X.J., Liu, H.X., 2016. Geochemical characteristics and LA-ICP-MS zircon U-Pb age of the trachytic volcanic rocks in Zhushan area of Southern Qinling Mountains and their significance. Geological Bulletin of China, 35(7), 1134-1143 (in Chinese with English abstract).

Wang, C.Z., Yang, K.G., Xu, Y., Cheng, W.Q., 2009a. Geochemistry and LA-ICP-MS zircon U-Pb age of basic dike swarms in North Daba Mountains and its tectonic significance. Geological Science and Technology Information, 28(3), 19-26 (in Chinese with English abstract).

Wang, G., 2014a. Metallogeny of the Mesozoic and Paleozoic volcanic igneous event in Ziyang-Langao Areas, North Daba Mountain (unpublished Doctor thesis). China University of Geosciences (in Chinese with English abstract).

Wang, K.M., 2014b. Research on the petrogenesis, tectonic and metallogeny for mafic rocks in the Ziyang-Langao area, Shaanxi Province (unpublished Doctor thesis). Beijing: Chinese Academy of Geological Sciences (in Chinese with English abstract).

Wang, K.M., Wang, Z.Q., Zhang, Y.L., Wang, G., 2015a. Geochronology and geochemistry of mafic rocks in the Xu he, Shaanxi, China: Implications for petrogenesis and mantle dynamics. Acta Geologica Sinica (English Edition), 89(1), 187-202.

Wang, R.R., Xu, Z.Q., Santosh, M., Liang, F.H., Fu, X.H., 2017. Petrogenesis and tectonic implications of the Early Paleozoic intermediate and mafic intrusions in the South Qinling Belt, central China: Constraints from geochemistry, zircon U-Pb geochronology and Hf isotopes. Tectonophysics, 712, 270-288.

Wang, X.X., Wang, T., Zhang, C.L., 2015b. Granitoid magmatism in the Qinling orogen, central China and its bearing on orogenic evolution. Science China: Earth Sciences, 58(9), 1497-1512.

Wang, Y.B., 2007. Geological characteristics and significance of Early Paleozoic alkali volcanic in South-Qinling, Langao-Pingli, Shaanxi province (unpublished Master thesis). Xi'an: Chang'an University (in Chinese with English abstract).

Wang, Z.Q., Yan, Q.R., Yan, Z., Wang, T., Jiang, C.F., Gao, L.D., Li, Q.G., Chen, J.L., Zhang, Y. L., Liu, P., Xie, C.L., Xiang, Z.J., 2009b. New division of the main tectonic units of the Qinling Orogenic Belt, Central China. Acta Geologica Sinica, 83(11), 1527-1546.

Weaver, B.L., 1991. The origin of ocean island basalt end-member compositions: Trace element and isotopic constraints. Earth & Planetary Science Letters, 104, 381-397.

White, J.C., Holt, G.S., Parker, D.F., Ren, M., 2003. Traceelement partitioning between alkali feldspar and peralkalic quartz trachyte to rhyolite magma. Part I: Systematics of trace-element partitioning. American Mineralogist, 88(2-3), 316-329.

Wilson, M., 1989. Igneous petrogenesis. London: Unwin Hyman, 287-374.

Wood, D.A., 1980. The application of a Th-Hf-Ta diagram to problems of tectonomagmatic classification and to establishing the nature of crustal contamination of basaltic lavas of the British Tertiary Volcanic Province. Earth & Planetary Science Letters, 50(1), 11-30.

Xiang, Z.J., Yan, Q.R., Song, B., Wang, Z.Q., 2016. New evidence for the ages of ultramafic to mafic dikes and alkaline volcanic complexes in the North Daba Mountains and its geological implication. Acta Geologica Sinica, 90(5),896-916 (in Chinese with English abstract).

Xiang, Z.J., Yan, Q.R., Yan, Z., Wang, Z.Q., Wang, T., Zhang, Y.L., Qin, X.F., 2010. Magma source and tectonic setting of the porphyritic alkaline basalts in the Silurian Taohekou Formation, North Daba mountain: Constraints from the geochemical features of pyroxene phenocrysts and whole rocks. Acta Geologica Sinica, 26(4), 1116-1132 (in Chinese with English abstract).

Xu, C., Campbell, I.H., Allen, C.M., Chen, Y.J., Huang, Z.L., Qi, L., Zhang, G.S., Yan, Z.F., 2008. U-Pb zircon age, geochemical and isotopic characteristics of carbonatite and syenite complexes from the Shaxiongdong, China. Lithos, 105(1), 118-128.

Xu, C., Kynicky, J., Chakhmouradian, A.R., Campbell, I.H., Allen, C.M., 2010a. Trace-element modeling of the magmatic evolution of rare-earth-rich carbonatite from the Miaoya deposit, Central China. Lithos, 118(1-2), 145-155.

Xu, X.Y., Huang, Y.H., Xia, L.Q., Xia, Z.C., 1996. Characteristics of phlogopite-amphibole-pyroxenite xenoliths from Langao County, Shaanxi Province. Acta Petrologica et Mineralogica, 15, 193-202 (in Chinese with English abstract).

Xu, Y.G., Chung, S.L., Shao, H., He, B., 2010b. Silicic magmas from the Emeishan large igneous province, Southwest China: Petrogenesis and their link with the end-Guadalupian biological crisis. Lithos, 119, 47-60.

Yan, Y.X., 2005. Research on geochemistry and Sr, Nd and Pb isotope of the basic dyke swarms in Ziyang-Langao area, Shaanxi Province (unpublished Master thesis). Xi'an: Northwest University (in Chinese with English abstract).

Yang, C., Liu, C.X., Liu, W.L., Wan, J., Duan, X.F., Zhang, Z., 2017. Geochemical Characteristics of trachyte and Nb mineralization process in Tianbao township, Zhuxi County, Southern Qinling. Acta Petrologica et Mineralogical, 36(5), 605-618 (in Chinese with English abstract).

Yi, P.F., Zhang, Y.F., Zhang, G.L., Yang, T., Yao, Z., Li, Q., Gao, H.F., 2017. LA-ICP-MS zircon U-Pb ages, geochemical characteristics of Zaomulan granitic pluton in Southern Qinling Orogenic Belt and their geological implications. Geological Review, 63(6), 71-85.

Yuan, H.L., Gao, S., Liu, X.M., Li, H.M., Gunther, D., Wu, F.Y., 2004. Accurate U-Pb age and trace element determinations of zircon by laser ablation-inductively coupled plasma mass spectrometry. Geostandards Newsletter, 28, 353-370.

Zhang, C.L., Gao, S., Zhang, G.W., Liu, X.M., Yu, Z.P., 2002. Geochemistry and significance of the Early Paleozoic alkaline mafic dyke in Qinling. Science in China: Series D, Geoscience, 32, 819-829 (in Chinese with English abstract).

Zhang, C.L., Shan, G., Yuan, H.L., Zhang, G.W., Yan, Y.X., Luo, J.L., Luo, J.H., 2007. Sr-Nd-Pb isotopes of the Early Paleozoic mafic-ultramafic dykes and basalts from south Qinling belt and their implications for mantle composition. Science in China, 50, 1293-1301 (in Chinese with English abstract).

Zhang, F.Y., Lai, S.C., Qin, J.F., Zhu, R.Z., Zhao, S.W., Zhu, Y., Yang, H., 2020. Vein-plus-wall rock melting model for the origin of Early Paleozoic alkali diabases in the South Qinling Belt, Central China. Lithos, 2020, 105619.

Zhang, G.W., Meng, Q.R., Yu, Z.P., Sun, Y., Zhou, D.W., Guo, A.L., 1996. Orogenic process and dynamic characteristics of Qinling orogenic belt. Science in China: Series D, 26(3), 193-200 (in Chinese with English abstract).

Zhang, G.W., Zhang, B.R., Yuan, X.C., Xiao, Q.H., 2001. Qinling orogenic belt and continental dynamics. Beijing: Science Press, 1-855.

Zhang, H.F., Sun, M., Zhou, X.H., Ying, J.F., 2005. Geochemical constraints on the origin of Mesozoic alkaline intrusive complexes from the north China craton and tectonic implications. Lithos, 81(1-4), 297-317.

Zhang, X., 2010. The dynamic mechanism and geological significance of mafic intrusion in the Ziyang-Zhenba area, South-Qinling (unpublished Master thesis). Xi'an: Chang'an University (in Chinese with English abstract).

Zhao, J.X., Shiraishi, K., Ellis, D.J., Sheraton, J.W., 1995. Geochemical and isotopic studies of syenites from the Yamato Mountains, East Antarctica: Implications for the origin of syenitic magmas. Geochimica et Cosmochimica Acta, 59(7), 1363-1382.

Zhu, J., Wang, L.X., Peng, S.G., Peng, L.H., Wu, C.X., Qiu, X.F., 2016. U-Pb zircon age, geochemical and isotopic characteristics of the Miaoya syenite and carbonatite complex, central China. Geological Journal, 52, 938-954.

Zindler, A., Hart, S., 1986. Chemical geodynamics. Annual Review of Earth and Planetary Sciences, 14(1), 493-571.

Zou, X.W., Duan, Q.F., Tang, C.Y., Cao, L., Cui, S., Zhao, W.Q., Xia, J., Wang, L., 2011. SHRIMP zircon U-Pb dating and lithogeochemical characteristics of diabase from Zhenping area in North Daba Mountains. Geology in China, 38(2), 282-291 (in Chinese with English abstract).

Neoproterozoic metasomatized mantle beneath the western Yangtze Block, South China: Evidence from whole-rock geochemistry and zircon U-Pb-Hf isotopes of mafic rocks[①]

Zhu Yu Lai Shaocong[②] Qin Jiangfeng Zhu Renzhi Liu Min

Zhang Fangyi Zhang Zezhong Yang Hang

Abstract: Mafic rocks contain important information about the nature of the mantle source and tectonic settings. Here we present results of whole-rock geochemistry and zircon U-Pb-Hf isotopes for the Early Neoproterozoic mafic rocks (Ganyuhe hornblende gabbros and Guandaoshan gabbro-diorites) in the western Yangtze Block, South China, in order to constrain the petrogenesis and the mantle metasomatism processes under subduction setting. LA-ICP-MS zircon U-Pb dating results display concordant crystallization ages of 857.4 ± 2.0 Ma for the hornblende gabbros and 856.4 ± 2.8 Ma for the gabbro-diorites, respectively. These mafic rocks possess low SiO_2 contents (45.5 – 53.3 wt%) as well as high MgO contents (3.27 – 9.20 wt%) and $Mg^{\#}$(48.6 – 62.6) values, and show characteristics of arc-like trace element patterns with enriched Rb, Ba, Sr, K, and Pb but depleted Nb, Ta, Zr, Hf, and Ti. The prominently positive whole-rock $\varepsilon_{Nd}(t)$ (+4.9 to +5.8) and zircon $\varepsilon_{Hf}(t)$ (+8.48 to +20.6) values suggest a depleted lithospheric mantle source. We therefore infer that the mafic rocks investigated here were derived from an amphibole-bearing spinel peridotite source that was affected by metasomatism between the overlying mantle wedge and subduction-related fluids. In conjunction with the compiled data of Neoproterozoic mafic rocks from the western Yangtze Block, we propose that there exist Neoproterozoic metasomatized mantle sources beneath the western Yangtze Block during the subduction process in response to the evolution of the supercontinent Rodinia. The Neoproterozoic metasomatic agents in mantle sources include the slab fluids, sediment melts, and oceanic slab melts.

1 Introduction

Extensive research on the paleogeographic evolution of the supercontinent Rodinia can provide significant constraints for deciphering the global tectonic framework during the Precambrian era (Cawood et al., 2013, 2016, 2017; Hofmann et al., 2011; Li et al.,

---

① Published in *Journal of Asian Earth Sciences*, 2021, 206.

② Corresponding author.

2008). As one of the largest craton blocks in eastern Asia (Zhao and Cawood, 2012), the South China Block, including the Yangtze Block to the northwest and Cathaysia Block to the southeast (Wang et al., 2013; Zhao et al., 2011), is widely accepted as an indispensable medium for reconstructing the supercontinent Rodinia during the Late Mesoproterozoic to Neoproterozoic (Li et al., 1995, 1999, 2008; Wang et al., 2007, 2008, 2013, 2014a; Zhao et al., 2018; Zhou et al., 2002, 2006a,b). The widespread Neoproterozoic igneous rocks in the South China have been mainly explained to be produced by two controversial paleotectonic models, i.e., the mantle plume setting (Li et al., 1999, 2008; Wang et al., 2007, 2008; Wu et al., 2019) and subduction setting (Wang et al., 2013, 2014a; Zhao et al., 2011, 2017, 2018, 2019; Zhou et al., 2002, 2006a,b; Zhu et al., 2019a-c, 2020a). Some geologists suggested that extensive igneous event is attributed to the upwelling of a mantle plume, which supports that the South China was located at the center of the supercontinent Rodinia (Li et al., 2003, 2008, 2010; Wang et al., 2007, 2008) (Fig. 1a). In an alternative model, the other geologists proposed that the prolonged Neoproterozoic magmatism (ca. 870−740 Ma) is associated with persistent subduction processes along the super-continental margins, which further suggests that the South China was a part of Neoproterozoic circum-Rodinia subduction system (Cawood et al., 2013, 2017; Kou et al., 2018; Zhao et al., 2017, 2018, 2019; Zhu et al., 2020a) (Fig. 1b). In addition, others have suggested that widespread Late Mesoproterozoic to Neoproterozoic magmatism in South China was controlled by a plate-rift setting (Zheng et al., 2007, 2008). Considering the hot disputes on tectono-magmatic models of the assembly and dispersal of Rodinia and specific position of the South China in Rodinia, it is vital to perform further researches on the genesis and tectonic implication of Neoproterozoic magmatic rocks.

Mafic rocks can be produced in various tectonic settings, such as subduction-related arc setting and continental rift-related setting (Pirajno, 2007; Tatsumi, 1989; Wang et al., 2008; Wilson, 1989; Zhao and Asimow, 2018; Zheng et al., 2020), the geochemical characteristics of specific mafic rocks and/or rock associations, therefore, can provide effective insights into the nature of the mantle source and thermal state as well as deep geodynamic processes (Herzberg et al., 2006; Lee et al., 2009; Niu et al., 2011; Zhao and Asimow, 2018). On the western Yangtze Block, the abundant Neoproterozoic (ca. 850−740 Ma) intermediate-felsic granitoids with subordinate mafic-ultramafic rocks have been studied to decipher the nature of the mantle source and crustal evolution (Lai et al., 2015; Li et al., 2002, 2006; Meng et al., 2015; Munteanu et al., 2010; Sun et al., 2007; Zhao et al., 2019; Zhao and Zhou, 2007; Zhou et al., 2006a; Zhu et al., 2019a-c, 2020a). However, detailed studies remain limited regarding the Early Neoproterozoic (>850 Ma) mafic rocks, which bear important constraints on the nature of the mantle source and tectonic implications. In this contribution, we present analytical results of whole-rock geochemistry and zircon U-Pb-Hf isotopes for Early

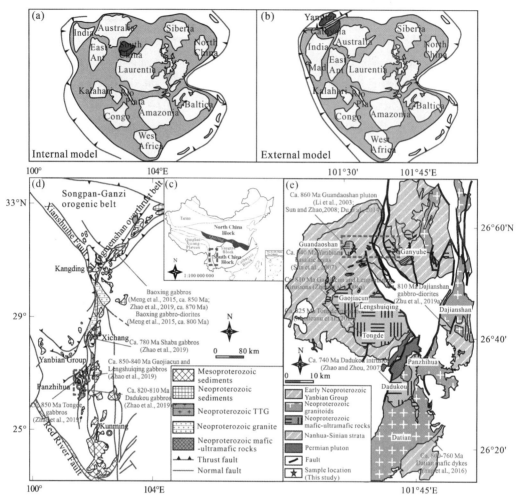

Fig. 1　The paleogeographic position of South China (a,b) (Cawood et al., 2017; Li et al., 2008) and geological map of the western Yangtze Block (c,d) (Zhao and Cawood, 2012; Zhao et al., 2019) and the Yanbian Terrane (e) (Zhu et al., 2019a).

Neoproterozoic mafic rocks, including the Ganyuhe hornblende gabbros and Guandaoshan gabbro-diorites, along the western Yangtze Block, South China. Our researches suggest that these mafic rocks originated from an amphibole-bearing spinel peridotite source that previously experienced the metasomatism between the overlying mantle and subduction-related fluids. Furthermore, we compile literature data of Neoproterozoic mafic rocks, and therefore propose that there exist the subduction components-enriched mantle sources beneath the western Yangtze Block during the Neoproterozoic.

2　Geological background and sample description

The South China Block can be divided into two Precambrian blocks, including the Yangtze Block in the northwest and Cathaysia Block in the southeast, which were welded together by the ~1 500-km-long Jiangnan Fold Belt during the Neoproterozoic (Wang et al., 2013; Zhao et al., 2011) (Fig. 1c). The Yangtze Block is tectonically separated from the North China Craton by the Triassic Qinling-Dabie orogenic belt to the north, bounded by the Songpan-Garze terrane of the Tibetan Plateau to the west, and surrounded by the Longmenshan overthrust belt to the northwest (Zhao and Cawood, 2012; Zhou et al., 2006a) (Fig. 1c,d).

The Yangtze Block comprises Archean to Early Neoproterozoic basement complexes, which are overlain by the Late Neoproterozoic to Cenozoic cover (Zhang et al., 2006; Zhao et al., 2011). Only a few localities in the northern Yangtze Block preserve the Archean Kongling basement rocks, which consist of felsic gneisses and metasedimentary rocks associated with minor amphibolite and mafic granulite (Gao et al., 1999, 2011; Guo et al., 2014; Zhang et al., 2006). In the western Yangtze Block, the oldest strata are the Late Paleoproterozoic Dahongshan, Hekou, Dongchuan, and Tong'an groups (Fan et al., 2013, 2020 and reference therein; Greentree and Li, 2008; Wang and Zhou, 2014; Wang et al., 2014b; Zhao et al., 2010; Zhu et al., 2017). The Late Paleoproterozoic to Mesoproterozoic mafic intrusions and meta-volcanic layers were dated at ca. 1.7−1.5 Ga (Fan et al., 2013, 2020 and reference therein; Lu et al., 2019; Zhao et al., 2010; Zhu et al., 2016). In addition, the Late Mesoproterozoic (ca. 1 020−1 050 Ma) A-type volcanic rocks and granites as well as S-type granites have been documented (Zhu et al., 2020b and references therein). Voluminous Neoproterozoic granitoids are widely preserved with subordinate mafic-ultramafic rocks, which show predominant zircon crystallization ages ranging from ca. 870 Ma to 750 Ma (Zhao et al., 2018 and references therein, 2019; Zhu et al., 2019a-c, 2020a). In the southwestern margin of the Yangtze Block, the Yanbian Group consists of volcanic and clastic rocks that are unconformably overlain by Sinian strata and can be subdivided into the Huangtian, Yumen, Xiaoping, and Zhagu formations from bottom to top (Du et al., 2014; Li et al., 2006; Sun and Zhou, 2008; Sun et al., 2008). Detrital zircons in the Yanbian Group display U-Pb ages ranging from ca. 1 000 Ma to 865 Ma with two peaks at ca. 920 Ma and 900 Ma (Sun et al., 2008). Previous studies have reported that the ca. 860 − 740 Ma Neoproterozoic mafic-ultramafic to intermediate igneous rocks intruded into the Yanbian Group (Du et al., 2014; Li et al., 2003; Li and Zhao, 2018; Munteanu et al., 2010; Sun and Zhou, 2008; Yang et al., 2016; Zhao and Zhou, 2007; Zhou et al., 2006a) (Fig. 1d,e). Recently, the new identifications of ca. 860−840 Ma Tongde, Gaojiacun, and Lengshuiqing gabbros were thought to originate from subduction sediment melt-enriched mantle sources (Zhao et al., 2019). Moreover, there exist ca. 810 Ma Dajianshan gabbro-diorites, ca. 800 Ma Dajianshan adakitic granites, and ca. 750 Ma Panzhihua A-type granites, which were considered as the magma response of the tectonic transition from vertically crustal growth to regionally crustal thinning under the subduction background (Zhu et al., 2019a).

In this study, two groups of Early Neoproterozoic mafic rocks, including hornblende gabbros and gabbro-diorites, were sampled from the Ganyuhe and Guandaoshan areas in the northern part of the Yanbian Terrane (Fig. 1e). In the Ganyuhe area (26°54′1″N 101°41′32″E), the hornblende gabbros display a dark and massive structure. The modal layering phenomenon was not observed in the outcrop of Ganyuhe hornblende gabbros (Fig. 2a), indicting they were not cumulate (Li et al., 2008; Xu et al., 2019). These samples are medium-grained and predominantly composed of amphibole (~ 50% − 65%), plagioclase (~ 15% − 25%), quartz (~ 5% − 10%), zircon, magnetite, and minor chlorite. The pleochroic amphiboles are subhedral to euhedral (Fig. 2b, c). The plagioclases are euhedral and display well-developed polysynthetic twins (Fig. 2b, c). In the Guandaoshan area (26° 55′ 57″ N 101°31′34″E), the medium- to coarse grained gabbro-diorites show a grey and massive structure, and mainly consist of plagioclase (~ 30% − 40%), amphibole (~ 25% − 40%), clinopyroxene (~ 5% − 10%), quartz (~ 5% − 15%), magnetite (~ 5%), zircon, and chlorite. The euhedral plagioclases are altered on the surface. The amphiboles are euhedral to subhedral and contain some plagioclase inclusions and magnetite grains (Fig. 2e, f).

3　Analytical methods

3.1　Zircon U-Pb and Lu-Hf isotopes analysis

Zircon U-Pb and Lu-Hf isotopic analyses were processed at the State Key Laboratory of Continental Dynamics (SKLCD), Northwest University, Xi'an, China. Two samples (GYH-1 and GDS-10) were selected for zircon U-Pb-Hf isotopic analyses.

Zircon grains from representative samples of hornblende gabbro (GYH-1) and gabbro-diorite (GDS-10) were separated using conventional heavy liquid and magnetic techniques. Representative zircon grains were handpicked and mounted in epoxy resin disks and then polished and carbon coated. The internal morphology of zircon grains were shown with cathodoluminescence (CL) microscopy. Laser ablation inductively-coupled plasma mass spectrometry (LA-ICP-MS) U-Pb analyses were conducted on an Agilent 7500a ICP-MS equipped with a 193 nm laser following the analytical method as described in Yuan et al. (2004). The $^{207}Pb/^{235}U$ and $^{206}Pb/^{238}U$ ratios were calculated using GLITTER program (Macquarie University), which were corrected using the Harvard zircon 91500 as an external calibration with a recommended $^{206}Pb/^{238}U$ isotopic age of 1 065.4 ± 0.6 Ma (Wiedenbeck et al., 2004). The GJ-1 is also a standard sample with a recommended $^{206}Pb/^{238}U$ age of 603.2 ± 2.4 Ma (Liu et al., 2007). Common Pb contents were subsequently evaluated using the method from Andersen (2002). The age calculations and concordia diagrams were gotten using ISOPLOT (version 3.0; Ludwig, 2003). Uncertainties are quoted at the 1σ level.

Fig. 2　Field and hand specimen photographs as well as representative photomicrographs
of the Early Neoproterozoic hornblende gabbros (a-c) and gabbro-diorites (d-f)
in the western Yangtze Block, South China.
Pl: plagioclase; Qtz: quartz; Amp: amphibole; Cpx: clinopyroxene; Mag: magnetite.

In-situ zircon Lu-Hf isotopic analyses were performed on a Neptune MC-ICP-MS. The laser repetition rate was 10 Hz at 100 mJ and the spot size was 32 μm. Instrument information can be available in Bao et al. (2017). The detailed analytical technique is depicted by Yuan et al. (2008). During the analyses, the measured values of well-characterized zircon standards (91500, GJ-1, and Monastery) were consistent with the recommended values within 2σ (Yuan et al., 2008). The obtained Hf isotopic compositions [GJ-1 standard = 0. 282 016 ± 20 (2σ, $n=84$), Monastery standard = 0. 282 735 ± 24 (2σ, $n=84$)] are also consistent with the recommended values (Yuan et al., 2008) to within 2σ. The initial $^{176}Hf/^{177}Hf$ ratios and $\varepsilon_{Hf}(t)$ values were calculated with the chondritic reservoir (CHUR) at the time of zircon growth from the magmas. The decay constant for ^{176}Lu of 1.867×10^{-11} year^{-1}, chondritic $^{176}Hf/^{177}Hf$ ratio of 0.282 785 and $^{176}Lu/^{177}Hf$ ratio of 0.033 6 were adopted (Soderlund et al., 2004). The depleted mantle model ages (T_{DM}) used for basic rocks were calculated with reference to present-day depleted mantle $^{176}Hf/^{177}Hf$ ratio of 0.283 25, similar to that of the average MORB (Nowell et al., 1998) and $^{176}Lu/^{177}Hf = 0.038$ 4 (Griffin et al., 2000). For the zircons in felsic magmatic rocks, we also calculated the "crustal" Hf model age (T_{DMC}) by assuming that its parental magma was derived from an average continental crust with $^{176}Lu/^{177}Hf = 0.015$, which was originated from the depleted mantle source (Griffin et al., 2000).

3. 2　Whole-rock geochemistry and Sr-Nd isotopes

The whole-rock major- and trace- elemental analyses were processed at the State Key Laboratory of Continental Dynamics (SKLCD), Northwest University, Xi'an, China. Fourteen samples were performed for the whole-rock major- and trace elemental analyses. The Sr-Nd isotopes were acquired from Guizhou Tuopu Resource and Environmental Analysis Center in China. Five samples were selected for whole-rock Sr-Nd isotopic analyses.

Weathered surfaces of selected samples were removed and the fresh parts were chipped and powdered to ~200 mesh using a tungsten carbide ball mill. Major and trace elements were examined by X-ray fluorescence (XRF; Rikagu RIX 2100) and inductively coupled plasma mass spectrometry (ICP-MS; Agilent 7500a), respectively. Analyses of USGS and Chinese national rock standards (BCR-2, GSR-1, and GSR-3) showed that the analytical precision and accuracy for the major elements were generally better than 5%. For the trace element analyses, sample powders were digested using an $HF + HNO_3$ mixture in high-pressure Teflon bombs at 190 ℃ for 48 h. The analytical error was < 2% and the precision was > 10% for most trace elements (Liu et al., 2007).

Sr and Nd isotopic fractionation was corrected to $^{87}Sr/^{86}Sr = 0.119$ 4 and $^{146}Nd/^{144}Nd = 0.721$ 9, respectively (Chu et al., 2009). During the analysis process, a Neptune multi-collector ICP-MS was used to measure the $^{87}Sr/^{86}Sr$ and $^{143}Nd/^{144}Nd$ isotope ratios. NIST SRM-987 and JMC-Nd were used as certified reference standard solutions for $^{87}Sr/^{86}Sr$ and $^{143}Nd/^{144}Nd$ isotopic ratios, respectively. BCR-1 and BHVO-1 were used as the reference materials.

4　Results

4. 1　Zircon U-Pb geochronology and Lu-Hf isotopic data

The zircon LA-ICP-MS U-Pb dating results and concordia diagrams for the Early

Neoproterozoic mafic rocks were shown in Table 1 and illustrated in Fig. 3. In-situ Lu-Hf isotopic analysis for the studied mafic rocks were undertaken on the same domains as the U-Pb spots. The initial $^{176}Hf/^{177}Hf$ ratios and $\varepsilon_{Hf}(t)$ values were calculated according to the zircon crystallization ages. The calculated results were displayed in Table 2, and Figs. 4 and 5.

4.1.1 Ganyuhe hornblende gabbro (GYH-1, 26°54′1″N 101°41′32″E)

The zircon grains from the Ganyuhe hornblende gabbro (GYH-1) are mostly euhedral short-prismatic crystals with aspect ratios from 1:1 to 3:2. Most grains are grey and black and display average crystal lengths of 50–150 μm. Thirty-six analyses from sample GYH-1 contain low Th (16.8–64.7 ppm) and U contents (20.4–116 ppm) and variable Th/U ratios (0.40–1.11). These 36 concordant analytical spots are characterized by a weighted mean age of 857.4±2.0 Ma (MSWD=1.0, n=36) (Fig. 3a,b).

The 36 concordant spots have relatively constant $(^{176}Hf/^{177}Hf)_i$ ratios from 0.282 492 to 0.282 707. They show depleted zircon Hf isotopic compositions with positive $\varepsilon_{Hf}(t)$ values from +8.48 to +16.0 (mean = +13.0, n=36) (Figs. 4 and 5a).

4.1.2 Guandaoshan gabbro-diorite (GDS-10, 26°55′57″N 101°31′34″E)

Zircon grains, separated from the Guandaoshan gabbro-diorite (GDS-10), are predominantly subhedral to euhedral, measuring up 50–150 μm in length with aspect ratios of approximately 1:1–2:1. The U and Th concentrations for the 23 concordant analyzed spots range from 6.12 ppm to 383 ppm and 19.6 ppm to 294 ppm, respectively, with Th/U ratios of 0.31–1.30, indicating an igneous origin (Hoskin and Schaltegger, 2003). They yield a concordant weighted mean age of 856.4±2.8 Ma (MSWD=1.4, n=23) (Fig. 3c,d), which is in agreement with the formation ages of the Guandaoshan pluton (e.g., Du et al., 2014; Li et al., 2003; Sun and Zhou, 2008).

Twenty-three analytical spots from the gabbro-diorite sample (GDS-10) display high $(^{176}Hf/^{177}Hf)_i$ ratios ranging from 0.282 539 to 0.282 863. They also contain notably positive $\varepsilon_{Hf}(t)$ values that are distributed from +8.80 to +20.6 (mean = +13.4, n=23) (Figs. 4 and 5b), which are similar to the measured values of the Guandaoshan pluton presented in Sun and Zhou (2008) [$\varepsilon_{Hf}(t)$= +11 to +17, mean = +14.2] (Fig. 4).

4.2 Whole-rock major and trace elements and Sr-Nd isotopes

The whole-rock major and trace elements of six Ganyuhe hornblende gabbro samples and eight Guandaoshan gabbro-diorite samples from the western Yangtze Block were summarized and presented in Table 3, and Figs. 6 and 7. The Sr-Nd isotopes of two hornblende gabbro samples and three gabbro-diorite samples were shown in Table 4 and Fig. 8. The initial $^{87}Sr/^{86}Sr$ isotopic (I_{Sr}) ratios and $\varepsilon_{Nd}(t)$ values were calculated according to ca. 857 Ma for the Ganyuhe hornblende gabbros and ca. 856 Ma for the Guandaoshan gabbro-diorites, respectively.

Table 1　LA-ICP-MS zircon U-Pb dating results for the Early Neoproterozoic mafic rocks in the western Yangtze Block, South China.

Analysis	Content/ppm			Ratios								Ages/Ma							
	Th	U	Th/U	$\frac{207Pb}{206Pb}$	1σ	$\frac{207Pb}{235U}$	1σ	$\frac{206Pb}{238U}$	1σ	$\frac{208Pb}{232Th}$	1σ	$\frac{207Pb}{206Pb}$	1σ	$\frac{207Pb}{235U}$	1σ	$\frac{206Pb}{238U}$	1σ	$\frac{208Pb}{232Th}$	1σ
GYH-1 (Ganyuhe hornblende gabbros, 26°54′1″N 101°41′32″E)																			
GYH-1-01	16.8	20.4	0.82	0.073 03	0.003 70	1.456 30	0.068 46	0.144 65	0.002 45	0.044 58	0.001 15	1 015	68	912	28	871	14	882	22
GYH-1-02	33.0	30.6	1.08	0.065 74	0.002 82	1.294 70	0.050 09	0.142 85	0.002 15	0.045 60	0.000 87	798	56	843	22	861	12	901	17
GYH-1-03	31.6	50.6	0.63	0.067 50	0.002 90	1.314 06	0.051 00	0.141 19	0.002 17	0.040 67	0.001 01	853	55	852	22	851	12	806	20
GYH-1-04	24.9	36.8	0.68	0.065 58	0.002 60	1.291 35	0.045 52	0.142 83	0.002 06	0.041 99	0.000 92	793	50	842	20	861	12	831	18
GYH-1-05	43.4	109.1	0.40	0.065 96	0.002 52	1.282 75	0.043 09	0.141 06	0.002 00	0.041 46	0.001 12	805	47	838	19	851	11	821	22
GYH-1-06	24.6	51.3	0.48	0.066 57	0.002 34	1.299 61	0.039 19	0.141 60	0.001 89	0.042 43	0.000 95	824	41	846	17	854	11	840	18
GYH-1-07	26.6	32.7	0.81	0.064 95	0.002 87	1.270 95	0.050 93	0.141 93	0.002 13	0.041 73	0.000 91	773	59	833	23	856	12	826	18
GYH-1-08	25.8	48.0	0.54	0.063 79	0.002 65	1.245 76	0.046 41	0.141 64	0.002 08	0.039 50	0.000 97	735	54	821	21	854	12	783	19
GYH-1-09	19.8	24.8	0.80	0.068 98	0.003 10	1.358 95	0.055 61	0.142 89	0.002 25	0.041 76	0.001 00	898	58	871	24	861	13	827	19
GYH-1-10	18.0	27.1	0.67	0.063 75	0.003 18	1.242 93	0.057 37	0.141 42	0.002 32	0.044 22	0.001 18	733	70	820	26	853	13	875	23
GYH-1-11	35.1	45.7	0.77	0.065 20	0.002 45	1.274 85	0.041 82	0.141 81	0.001 96	0.042 81	0.000 83	781	46	835	19	855	11	847	16
GYH-1-12	23.8	33.7	0.70	0.068 02	0.002 84	1.323 17	0.049 58	0.141 09	0.002 12	0.043 75	0.001 00	869	53	856	22	851	12	865	19
GYH-1-13	36.0	80.7	0.45	0.067 61	0.002 16	1.328 66	0.035 05	0.142 54	0.001 81	0.043 93	0.000 88	857	34	858	15	859	10	869	17
GYH-1-14	42.3	84.1	0.50	0.065 12	0.002 99	1.274 37	0.053 41	0.141 93	0.002 28	0.044 33	0.001 24	778	61	834	24	856	13	877	24
GYH-1-15	22.3	55.7	0.40	0.066 04	0.002 42	1.306 11	0.041 61	0.143 44	0.001 98	0.043 62	0.001 10	808	44	848	18	864	11	863	21
GYH-1-16	37.3	73.2	0.51	0.066 38	0.005 43	1.290 54	0.101 83	0.141 00	0.003 72	0.040 00	0.002 24	818	121	842	45	850	21	793	44
GYH-1-17	34.8	78.7	0.44	0.067 33	0.002 83	1.304 15	0.051 94	0.140 49	0.001 92	0.042 82	0.000 50	848	90	848	23	847	11	847	10
GYH-1-18	26.7	62.8	0.43	0.067 51	0.002 23	1.316 86	0.036 41	0.141 47	0.001 84	0.043 49	0.000 93	854	36	853	16	853	10	860	18
GYH-1-19	32.9	59.3	0.56	0.065 44	0.002 48	1.300 51	0.043 25	0.144 14	0.002 02	0.042 51	0.000 95	789	46	846	19	868	11	841	18
GYH-1-20	36.4	32.7	1.11	0.065 71	0.003 12	1.303 32	0.056 76	0.143 84	0.002 37	0.044 49	0.000 97	797	64	847	25	866	13	880	19
GYH-1-21	38.2	69.1	0.55	0.067 90	0.002 16	1.345 36	0.035 31	0.143 71	0.001 83	0.043 62	0.000 81	866	34	866	15	866	10	863	16
GYH-1-22	20.7	43.2	0.48	0.068 86	0.003 18	1.352 40	0.059 23	0.142 44	0.002 06	0.043 30	0.000 52	895	98	869	26	858	12	857	10
GYH-1-23	31.5	67.6	0.47	0.069 09	0.003 48	1.350 62	0.063 01	0.141 77	0.002 51	0.039 61	0.001 38	901	67	868	27	855	14	785	27

Continued

Analysis	Content/ppm			Ratios								Ages/Ma							
	Th	U	Th/U	$^{207}Pb/^{206}Pb$	1σ	$^{207}Pb/^{235}U$	1σ	$^{206}Pb/^{238}U$	1σ	$^{208}Pb/^{232}Th$	1σ	$^{207}Pb/^{206}Pb$	1σ	$^{207}Pb/^{235}U$	1σ	$^{206}Pb/^{238}U$	1σ	$^{208}Pb/^{232}Th$	1σ
GYH-1-24	18.5	43.2	0.43	0.067 48	0.003 04	1.312 28	0.056 04	0.141 03	0.002 00	0.042 97	0.000 51	853	96	851	25	850	11	850	10
GYH-1-25	41.4	92.5	0.45	0.071 01	0.005 63	1.398 48	0.106 64	0.142 82	0.003 76	0.039 49	0.002 14	958	113	888	45	861	21	783	42
GYH-1-26	27.1	63.8	0.43	0.064 52	0.002 98	1.254 05	0.053 01	0.140 96	0.002 27	0.043 21	0.001 37	759	62	825	24	850	13	855	27
GYH-1-27	64.7	116.2	0.56	0.066 05	0.002 00	1.286 91	0.031 56	0.141 30	0.001 75	0.041 96	0.000 69	808	31	840	14	852	10	831	13
GYH-1-28	22.9	35.0	0.65	0.071 22	0.002 91	1.403 15	0.051 26	0.142 87	0.002 13	0.044 65	0.001 03	964	50	890	22	861	12	883	20
GYH-1-29	35.5	87.4	0.41	0.067 34	0.002 18	1.321 82	0.035 68	0.142 34	0.001 83	0.044 79	0.000 95	848	35	855	16	858	10	886	18
GYH-1-30	28.4	31.3	0.91	0.062 42	0.003 93	1.235 53	0.073 88	0.143 55	0.002 88	0.043 65	0.001 37	689	93	817	34	865	16	864	27
GYH-1-31	31.7	32.6	0.97	0.069 98	0.007 09	1.360 86	0.133 90	0.141 01	0.004 55	0.042 10	0.002 38	928	150	872	58	850	26	834	46
GYH-1-32	39.8	44.5	0.89	0.066 38	0.002 49	1.309 61	0.042 96	0.143 07	0.002 01	0.042 76	0.000 79	818	45	850	19	862	11	846	15
GYH-1-33	39.1	41.2	0.95	0.064 70	0.002 59	1.264 76	0.045 08	0.141 77	0.002 06	0.044 46	0.000 84	765	51	830	20	855	12	879	16
GYH-1-34	31.9	42.6	0.75	0.064 39	0.002 64	1.279 6	0.046 96	0.144 12	0.002 13	0.043 18	0.000 93	754	52	837	21	868	12	854	18
GYH-1-35	30.9	32.1	0.96	0.064 42	0.003 35	1.260 41	0.061 01	0.141 89	0.002 44	0.043 22	0.001 03	755	73	828	27	855	14	855	20
GYH-1-36	37.5	39.9	0.94	0.066 21	0.002 74	1.299 68	0.048 29	0.142 35	0.002 12	0.041 18	0.000 83	813	53	846	21	858	12	816	16
GDS-10(Guandaoshan gabbro-diorites, 26°55′57″N101°31′34″E)																			
GDS-10-01	43.9	75.2	0.58	0.064 73	0.002 20	1.259 59	0.036 80	0.141 17	0.001 94	0.045 36	0.000 89	766	39	828	17	851	11	897	17
GDS-10-02	90.7	123.1	0.74	0.067 92	0.002 01	1.317 32	0.031 82	0.140 70	0.001 82	0.045 09	0.000 71	866	29	853	14	849	10	891	14
GDS-10-03	54.8	82.7	0.66	0.068 32	0.002 12	1.327 11	0.034 13	0.140 91	0.001 86	0.045 04	0.000 76	878	32	858	15	850	11	890	15
GDS-10-06	50.0	86.2	0.58	0.066 82	0.002 06	1.295 00	0.033 06	0.140 59	0.001 85	0.044 25	0.000 76	832	32	844	15	848	10	875	15
GDS-10-07	382.9	293.6	1.30	0.067 73	0.001 61	1.340 45	0.022 46	0.143 56	0.001 70	0.046 19	0.000 52	860	17	863	10	865	10	913	10
GDS-10-09	96.9	116.6	0.83	0.066 78	0.002 33	1.312 95	0.039 59	0.142 61	0.002 01	0.046 33	0.000 84	831	40	851	17	859	11	915	16
GDS-10-10	25.7	46.0	0.56	0.072 97	0.003 45	1.438 16	0.062 71	0.142 97	0.002 51	0.044 65	0.001 29	1 013	60	905	26	861	14	883	25
GDS-10-11	88.3	106.7	0.83	0.067 42	0.002 06	1.317 26	0.033 37	0.141 72	0.001 88	0.042 68	0.000 68	851	31	853	15	854	11	845	13
GDS-10-13	52.7	86.1	0.61	0.066 67	0.002 06	1.319 82	0.033 91	0.143 58	0.001 91	0.044 08	0.000 75	827	32	854	15	865	11	872	15
GDS-10-14	31.3	60.5	0.52	0.073 59	0.002 52	1.427 63	0.042 23	0.140 70	0.002 00	0.048 69	0.001 03	1 030	37	901	18	849	11	961	20

Continued

Analysis	Content/ppm			Ratios								Ages/Ma							
	Th	U	Th/U	$^{207}Pb/^{206}Pb$	1σ	$^{207}Pb/^{235}U$	1σ	$^{206}Pb/^{238}U$	1σ	$^{208}Pb/^{232}Th$	1σ	$^{207}Pb/^{206}Pb$	1σ	$^{207}Pb/^{235}U$	1σ	$^{206}Pb/^{238}U$	1σ	$^{208}Pb/^{232}Th$	1σ
GDS-10-15	83.6	152.5	0.55	0.068 86	0.001 77	1.346 01	0.026 02	0.141 78	0.001 74	0.046 82	0.000 66	895	21	866	11	855	10	925	13
GDS-10-16	25.1	45.8	0.55	0.068 62	0.002 68	1.341 86	0.046 80	0.141 82	0.002 15	0.043 21	0.001 05	887	47	864	20	855	12	855	20
GDS-10-17	138.9	159.0	0.87	0.068 84	0.001 95	1.339 33	0.030 43	0.141 11	0.001 82	0.043 22	0.000 63	894	26	863	13	851	10	855	12
GDS-10-18	50.6	79.2	0.64	0.068 51	0.002 11	1.351 88	0.034 72	0.143 12	0.001 91	0.043 47	0.000 74	884	31	868	15	862	11	860	14
GDS-10-21	134.7	146.8	0.92	0.066 43	0.001 79	1.316 77	0.027 80	0.143 75	0.001 81	0.044 41	0.000 60	820	24	853	12	866	10	878	12
GDS-10-25	53.0	90.7	0.58	0.067 16	0.002 74	1.327 04	0.050 95	0.143 32	0.002 00	0.043 69	0.000 52	843	87	858	22	863	11	864	10
GDS-10-27	23.1	53.9	0.43	0.066 67	0.002 56	1.298 83	0.044 40	0.141 28	0.002 12	0.044 23	0.001 13	827	46	845	20	852	12	875	22
GDS-10-28	6.1	19.6	0.31	0.068 78	0.004 27	1.364 83	0.080 68	0.143 92	0.002 90	0.047 85	0.002 49	892	89	874	35	867	16	945	48
GDS-10-29	120.3	203.3	0.59	0.069 26	0.001 80	1.362 57	0.027 15	0.142 69	0.001 79	0.046 04	0.000 66	906	22	873	12	860	10	910	13
GDS-10-32	25.5	57.8	0.44	0.065 50	0.002 36	1.273 93	0.040 28	0.141 04	0.002 05	0.043 18	0.001 01	790	42	834	18	851	12	854	20
GDS-10-34	33.6	73.4	0.46	0.067 50	0.002 18	1.319 37	0.036 35	0.141 76	0.001 95	0.045 00	0.000 91	853	35	854	16	855	11	890	18
GDS-10-35	66.3	114.3	0.58	0.067 22	0.001 92	1.331 65	0.030 76	0.143 66	0.001 87	0.045 78	0.000 68	845	27	860	13	865	11	905	13
GDS-10-36	84.1	102.7	0.82	0.072 27	0.002 21	1.404 05	0.035 77	0.140 90	0.001 91	0.042 18	0.000 67	994	30	891	15	850	11	835	13

Table 2　Zircon Lu-Hf isotopic data for the Early Neoproterozoic mafic rocks in the western Yangtze Block, South China.

Grain spot	Age /Ma	$^{176}Yb/^{177}Hf$	$2SE$	$^{176}Lu/^{177}Hf$	$2SE$	$^{176}Hf/^{177}Hf$	$2SE$	$(^{176}Hf/^{177}Hf)_i$	$f_{Lu/Hf}$	$\varepsilon_{Hf}(t)$	$2SE$	T_{DM1}/Ma
Ganyuhe hornblende gabbros (GYH-1)												
GYH1-1	857	0.025 870	0.000 127	0.001 051	0.000 004	0.282 595	0.000 022	0.282 578	-0.97	11.47	0.79	956
GYH1-2	857	0.034 043	0.000 164	0.001 340	0.000 006	0.282 674	0.000 026	0.282 652	-0.96	13.94	0.91	858
GYH1-3	857	0.026 995	0.000 258	0.001 090	0.000 012	0.282 715	0.000 020	0.282 698	-0.97	15.70	0.69	788
GYH1-4	857	0.021 953	0.000 115	0.000 877	0.000 005	0.282 634	0.000 023	0.282 620	-0.97	13.05	0.80	893
GYH1-5	857	0.031 316	0.000 261	0.001 243	0.000 008	0.282 727	0.000 019	0.282 707	-0.96	15.95	0.67	778
GYH1-6	857	0.016 359	0.000 223	0.000 683	0.000 010	0.282 608	0.000 021	0.282 597	-0.98	12.35	0.75	921
GYH1-7	857	0.019 455	0.000 151	0.000 781	0.000 007	0.282 702	0.000 017	0.282 689	-0.98	15.57	0.58	794

Continued

Grain spot	Age /Ma	$^{176}Yb/^{177}Hf$	2SE	$^{176}Lu/^{177}Hf$	2SE	$^{176}Hf/^{177}Hf$	2SE	$(^{176}Hf/^{177}Hf)_i$	$f_{Lu/Hf}$	$\varepsilon_{Hf}(t)$	2SE	T_{DM1}/Ma
GYH1-8	857	0.018 733	0.000 248	0.000 771	0.000 011	0.282 636	0.000 020	0.282 624	-0.98	13.26	0.71	885
GYH1-9	857	0.017 865	0.000 117	0.000 724	0.000 006	0.282 683	0.000 021	0.282 671	-0.98	14.96	0.73	818
GYH1-10	857	0.018 822	0.000 138	0.000 758	0.000 006	0.282 590	0.000 021	0.282 578	-0.98	11.65	0.73	949
GYH1-11	857	0.027 861	0.000 140	0.001 108	0.000 006	0.282 675	0.000 023	0.282 657	-0.97	14.23	0.81	846
GYH1-12	857	0.016 553	0.000 219	0.000 664	0.000 007	0.282 710	0.000 021	0.282 699	-0.98	15.99	0.75	777
GYH1-13	857	0.028 091	0.000 088	0.001 098	0.000 004	0.282 595	0.000 020	0.282 577	-0.97	11.41	0.72	959
GYH1-14	857	0.035 511	0.000 106	0.001 372	0.000 004	0.282 545	0.000 021	0.282 523	-0.96	9.34	0.76	1 043
GYH1-15	857	0.015 633	0.000 188	0.000 652	0.000 007	0.282 670	0.000 020	0.282 659	-0.98	14.58	0.71	833
GYH1-16	857	0.025 111	0.000 263	0.001 024	0.000 011	0.282 666	0.000 021	0.282 649	-0.97	14.02	0.75	855
GYH1-17	857	0.021 925	0.000 095	0.000 915	0.000 005	0.282 582	0.000 021	0.282 567	-0.97	11.16	0.74	968
GYH1-18	857	0.019 353	0.000 054	0.000 790	0.000 002	0.282 668	0.000 019	0.282 655	-0.98	14.37	0.68	841
GYH1-19	857	0.022 715	0.000 143	0.000 914	0.000 006	0.282 681	0.000 020	0.282 666	-0.97	14.67	0.70	829
GYH1-20	857	0.023 550	0.000 220	0.000 974	0.000 008	0.282 508	0.000 028	0.282 492	-0.97	8.48	0.97	1 075
GYH1-21	857	0.017 605	0.000 029	0.000 727	0.000 002	0.282 597	0.000 020	0.282 585	-0.98	11.92	0.71	938
GYH1-22	857	0.014 557	0.000 111	0.000 613	0.000 005	0.282 694	0.000 019	0.282 684	-0.98	15.48	0.66	797
GYH1-23	857	0.027 135	0.000 494	0.001 118	0.000 020	0.282 675	0.000 020	0.282 657	-0.97	14.23	0.71	846
GYH1-24	857	0.017 478	0.000 195	0.000 750	0.000 005	0.282 597	0.000 023	0.282 585	-0.98	11.91	0.81	938
GYH1-25	857	0.022 071	0.000 244	0.000 881	0.000 009	0.282 618	0.000 018	0.282 603	-0.97	12.48	0.63	916
GYH1-26	857	0.023 078	0.000 187	0.000 906	0.000 005	0.282 620	0.000 022	0.282 605	-0.97	12.53	0.79	914
GYH1-27	857	0.028 978	0.000 350	0.001 132	0.000 013	0.282 550	0.000 021	0.282 531	-0.97	9.78	0.73	1 024
GYH1-28	857	0.029 459	0.000 077	0.001 299	0.000 002	0.282 600	0.000 031	0.282 579	-0.96	11.39	1.10	961
GYH1-29	857	0.026 877	0.000 207	0.001 130	0.000 005	0.282 636	0.000 023	0.282 618	-0.97	12.85	0.82	902
GYH1-30	857	0.013 406	0.000 102	0.000 610	0.000 007	0.282 586	0.000 023	0.282 576	-0.98	11.65	0.82	948
GYH1-31	857	0.022 470	0.000 082	0.000 888	0.000 003	0.282 648	0.000 020	0.282 633	-0.97	13.53	0.69	874
GYH1-32	857	0.032 600	0.000 080	0.001 294	0.000 004	0.282 689	0.000 022	0.282 668	-0.96	14.54	0.79	834
GYH1-33	857	0.034 661	0.000 146	0.001 442	0.000 008	0.282 588	0.000 026	0.282 565	-0.96	10.80	0.91	985
GYH1-34	857	0.014 299	0.000 102	0.000 614	0.000 006	0.282 682	0.000 022	0.282 672	-0.98	15.07	0.77	813

Continued

Grain spot	Age /Ma	^{176}Yb/^{177}Hf	2SE	^{176}Lu/^{177}Hf	2SE	^{176}Hf/^{177}Hf	2SE	(^{176}Hf/^{177}Hf)$_i$	$f_{Lu/Hf}$	$\varepsilon_{Hf}(t)$	2SE	T_{DM1}/Ma
GYH1-35	857	0.013 037	0.000 071	0.000 565	0.000 004	0.282 582	0.000 024	0.282 573	-0.98	11.58	0.83	951
GYH1-36	857	0.015 108	0.000 108	0.000 649	0.000 007	0.282 594	0.000 020	0.282 584	-0.98	11.91	0.72	938
Guandaoshan gabbro-diorites (GDS-10)												
GDS10-1	856	0.115 207	0.000 786	0.004 541	0.000 021	0.282 774	0.000 042	0.282 700	-0.86	13.81	1.50	863
GDS10-2	856	0.020 238	0.000 343	0.001 044	0.000 011	0.282 640	0.000 027	0.282 623	-0.97	13.04	0.97	893
GDS10-3	856	0.075 065	0.000 359	0.003 107	0.000 006	0.282 700	0.000 034	0.282 650	-0.91	12.83	1.19	904
GDS10-4	856	0.129 986	0.001 010	0.005 074	0.000 035	0.282 796	0.000 039	0.282 714	-0.85	13.97	1.37	856
GDS10-5	856	0.060 891	0.000 643	0.002 757	0.000 035	0.282 729	0.000 034	0.282 685	-0.92	14.27	1.20	844
GDS10-6	856	0.065 506	0.000 260	0.002 766	0.000 006	0.282 720	0.000 035	0.282 675	-0.92	13.92	1.23	858
GDS10-7	856	0.031 219	0.000 483	0.001 450	0.000 019	0.282 807	0.000 031	0.282 784	-0.96	18.52	1.10	672
GDS10-8	856	0.083 950	0.000 341	0.003 320	0.000 032	0.282 706	0.000 032	0.282 652	-0.90	12.80	1.13	906
GDS10-9	856	0.111 551	0.000 595	0.004 465	0.000 019	0.282 721	0.000 038	0.282 649	-0.87	12.02	1.34	942
GDS10-10	856	0.058 310	0.000 502	0.002 564	0.000 011	0.282 778	0.000 036	0.282 736	-0.92	16.21	1.29	763
GDS10-11	856	0.099 093	0.000 918	0.003 981	0.000 025	0.282 710	0.000 039	0.282 646	-0.88	12.20	1.40	933
GDS10-12	856	0.034 218	0.000 728	0.001 435	0.000 028	0.282 646	0.000 028	0.282 623	-0.96	12.82	1.00	902
GDS10-13	856	0.113 578	0.000 073	0.004 437	0.000 003	0.282 743	0.000 037	0.282 671	-0.87	12.83	1.29	906
GDS10-14	856	0.088 134	0.001 050	0.003 656	0.000 032	0.282 604	0.000 041	0.282 545	-0.89	8.80	1.44	1 077
GDS10-15	856	0.032 825	0.000 503	0.001 358	0.000 021	0.282 649	0.000 030	0.282 627	-0.96	13.02	1.05	894
GDS10-16	856	0.083 066	0.000 489	0.003 515	0.000 024	0.282 749	0.000 043	0.282 693	-0.89	14.11	1.53	850
GDS10-17	856	0.059 122	0.001 590	0.002 614	0.000 062	0.282 581	0.000 047	0.282 539	-0.92	9.19	1.66	1 054
GDS10-18	856	0.050 268	0.000 468	0.002 414	0.000 011	0.282 675	0.000 041	0.282 636	-0.93	12.72	1.44	908
GDS10-19	856	0.050 391	0.000 171	0.002 200	0.000 004	0.282 657	0.000 039	0.282 621	-0.93	12.34	1.38	923
GDS10-20	856	0.042 932	0.000 354	0.001 828	0.000 011	0.282 658	0.000 037	0.282 629	-0.94	12.81	1.31	903
GDS10-21	856	0.047 636	0.000 427	0.002 463	0.000 025	0.282 641	0.000 059	0.282 601	-0.93	11.48	2.09	959
GDS10-22	856	0.058 904	0.000 324	0.002 688	0.000 014	0.282 906	0.000 050	0.282 863	-0.92	20.61	1.76	579
GDS10-23	856	0.040 206	0.000 335	0.001 760	0.000 015	0.282 665	0.000 036	0.282 637	-0.95	13.13	1.26	890

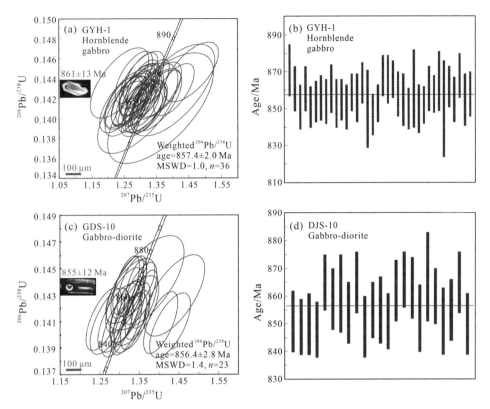

Fig. 3 Representative cathodoluminescence (CL) images and Laser ablation-inductively coupled plasma-mass spectrometry (LA-ICP-MS) U-Pb zircon concordia diagrams for the Early Neoproterozoic hornblende gabbros and gabbro-diorites in the western Yangtze Block, South China.

4.2.1 Ganyuhe hornblende gabbros

All of the hornblende gabbro samples studied here contain narrow compositional ranges of SiO_2 (45.5−46.0 wt%), Al_2O_3 (14.5−15.0 wt%), and K_2O (0.37−0.41 wt%) as well as high TiO_2 (1.34−1.39 wt%), $Fe_2O_3^T$ (13.0−13.2 wt%), Na_2O (2.30−2.46 wt%), CaO (11.4−11.7 wt%), MgO (8.85−9.20 wt%), and $Mg^\#$ (60.4−62.2) values (Table 3). A notably positive correlation between SiO_2 contents and $Mg^\#$ values is not observed (not shown), hinting that the hornblende gabbros are not cumulate (Wolf and Wyllie, 1994; Xu et al., 2019). The samples have low total rare earth element ($\sum REE$) concentrations varying from 44.2 ppm to 55.9 ppm. As shown in the chondrite-normalized REE patterns (Fig. 7a), the Ganyuhe hornblende gabbros exhibit REE patterns similar to N-MORB and insignificant Eu anomalies ($Eu/Eu^* = 0.97−1.04$). On the primitive mantle-normalized diagram (Fig. 7b), the samples display pronounced positive anomalies of large ion lithophile elements (LILEs, e.g., Rb, Ba, Sr, and K) and Pb and negative anomalies of high field strength elements (HFSEs, e.g., Nb and Ta).

Two samples (GYH-1-5 and GYH-1-6) of the Ganyuhe hornblende gabbros display

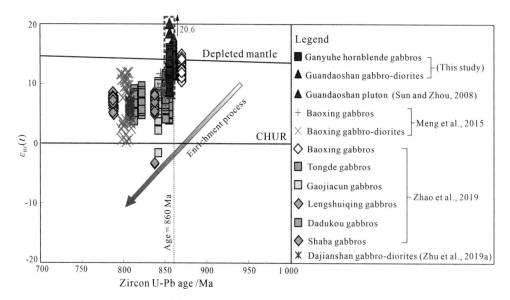

Fig. 4　Plots of zircon $\varepsilon_{Hf}(t)$ values vs. zircon U-Pb ages for the Neoproterozoic mafic rocks in the western Yangtze Block, South China.

The zircon Hf isotopes of Neoproterozoic mafic magmatism were from Meng et al. (2015), Sun and Zhou (2008), Zhao et al. (2019), and Zhu et al. (2019a).

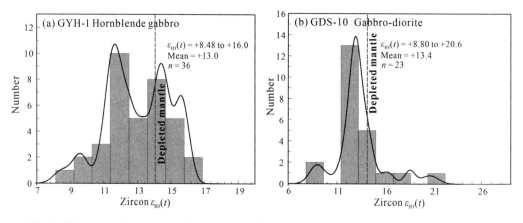

Fig. 5　Histograms of zircon $\varepsilon_{Hf}(t)$ values and Hf model age for the Early Neoproterozoic hornblende gabbros and gabbro-diorites in the western Yangtze Block, South China.

constant initial $^{87}Sr/^{86}Sr$ ratios ranging from 0. 703 461 to 0. 703 535. They yield notably depleted Nd isotopic compositions with positive $\varepsilon_{Nd}(t)$ values from +4. 9 to +5. 0 (Fig. 8).

4. 2. 2　Guandaoshan gabbro-diorites

In comparison with the Ganyuhe hornblende gabbros, the gabbro-diorites collected from the Guandaoshan area have higher SiO_2(52. 0–53. 3 wt%), Al_2O_3(18. 3–19. 2 wt%), K_2O (1. 04–1. 54 wt%), and Na_2O (3. 11–4. 70 wt%), but lower TiO_2(0. 41–0. 48 wt%), $Fe_2O_3^T$(7. 56–8. 45 wt%), CaO (8. 48–9. 62 wt%), MgO (3. 27–3. 66 wt%), and $Mg^{\#}$

Table 3 Major(wt%) and trace(ppm) element analysis results for the Early Neoproterozoic mafic rocks in the western Yangtze Block, South China.

Sample	GYH-1-1	GYH-1-2	GYH-1-3	GYH-1-5	GYH-1-6	GYH-1-7	GDS-1	GDS-2	GDS-3	GDS-4	GDS-5	GDS-6	GDS-7	GDS-8
Lithology	Ganyuhe hornblende gabbros, 26°54'1"N 101°41'32"E						Guandaoshan gabbro-diorites, 26°54'1"N 101°41'32"E							
SiO_2	45.98	46.04	45.81	45.91	45.53	45.61	52.19	53.20	52.36	53.21	52.50	53.27	52.00	52.35
TiO_2	1.34	1.35	1.37	1.37	1.36	1.39	0.41	0.41	0.46	0.43	0.43	0.42	0.48	0.48
Al_2O_3	14.49	14.56	14.91	15.02	14.82	14.85	18.65	19.11	18.52	19.06	18.29	19.14	18.87	19.17
$Fe_2O_3^{T}$	12.99	13.04	13.22	13.07	13.24	13.09	7.84	7.83	8.45	8.20	8.31	7.56	8.30	8.32
MnO	0.20	0.21	0.19	0.21	0.20	0.21	0.18	0.17	0.18	0.18	0.19	0.17	0.19	0.16
MgO	9.17	9.20	8.64	8.86	9.01	8.85	3.38	3.27	3.50	3.32	3.48	3.28	3.66	3.38
CaO	11.43	11.47	11.72	11.57	11.59	11.40	8.54	8.48	8.78	8.72	9.62	8.70	9.32	9.12
Na_2O	2.46	2.42	2.30	2.46	2.36	2.46	4.70	3.79	3.86	3.39	3.42	3.50	3.35	3.11
K_2O	0.37	0.37	0.40	0.39	0.41	0.39	1.42	1.54	1.35	1.50	1.04	1.26	1.25	1.38
P_2O_5	0.11	0.10	0.19	0.19	0.18	0.18	0.08	0.08	0.09	0.09	0.08	0.08	0.08	0.09
LOI	1.04	1.04	0.93	1.04	1.17	1.07	2.25	2.02	2.08	1.64	2.51	2.32	2.23	2.10
Total	99.58	99.80	99.68	100.09	99.87	99.50	99.64	99.90	99.63	99.74	99.87	99.70	99.73	99.66
Mg#	62.2	62.2	60.4	61.2	61.3	61.2	50.1	49.3	49.1	48.5	49.4	50.3	50.7	48.6
Na_2O/K_2O	6.65	6.54	5.75	7.20	6.31	5.76	6.31	3.31	2.46	2.86	2.26	3.29	2.78	2.68
Li	4.31	4.55	4.35	4.68	4.77	4.90	7.38	5.90	7.25	11.5	9.51	9.38	8.48	7.83
Be	0.45	0.45	0.47	0.47	0.47	0.49	0.40	0.38	0.37	0.37	0.38	0.37	0.39	0.40
Sc	45.8	45.2	46.1	42.4	44.7	43.8	23.8	22.5	25.3	24.0	23.8	23.4	26.3	24.2
V	323	325	330	323	325	326	218	209	239	223	224	223	228	226
Cr	215	217	234	181	204	186	14.8	10.7	12.6	11.2	13.6	13.4	11.1	13.4
Co	83.5	82.9	83.8	80.9	81.3	80.7	43.1	57.4	46.2	57.0	51.3	51.0	46.0	72.5
Ni	80.9	80.8	88.7	86.1	87.8	85.5	8.06	6.77	7.79	7.26	7.36	7.36	7.21	6.63

Continued

Sample	GYH-1-1	GYH-1-2	GYH-1-3	GYH-1-5	GYH-1-6	GYH-1-7	GDS-1	GDS-2	GDS-3	GDS-4	GDS-5	GDS-6	GDS-7	GDS-8
Cu	41.0	41.3	50.5	54.7	80.7	54.3	24.1	20.8	21.5	20.4	22.7	22.5	27.5	20.6
Zn	74.7	74.8	72.9	77.4	79.7	78.5	60.4	57.2	63.0	58.2	60.6	60.1	62.5	60.4
Ga	17.3	17.3	18.1	17.5	18.0	17.7	14.7	14.4	14.6	14.4	14.2	14.2	14.9	14.9
Ge	1.59	1.57	1.77	1.64	1.65	1.64	1.26	1.24	1.33	1.28	1.35	1.33	1.31	1.39
Rb	3.36	3.15	4.18	3.44	4.04	3.58	21.5	24.0	20.3	23.1	16.4	16.1	19.3	20.1
Sr	192	192	239	252	253	256	532	493	529	507	585	581	557	540
Y	23.8	23.8	24.2	25.5	25.1	26.0	12.8	12.3	13.6	12.8	13.0	12.9	15.7	13.2
Zr	64.1	62.8	60.7	61.6	58.6	65.6	25.8	42.4	28.1	20.5	35.2	35.1	29.0	22.5
Nb	1.63	1.61	1.69	1.74	1.71	1.78	0.95	1.03	1.13	1.06	0.97	0.96	1.34	1.49
Cs	0.10	0.10	0.18	0.11	0.16	0.12	0.51	0.54	0.51	0.55	0.43	0.43	0.48	0.50
Ba	84.5	84.1	81.3	118	82.6	139	644	682	661	654	484	480	584	644
La	3.65	3.63	4.77	5.10	4.95	5.18	3.10	2.99	3.21	3.11	3.04	3.03	3.52	3.81
Ce	10.5	10.4	13.2	14.3	13.9	14.8	6.98	6.98	7.52	7.22	6.79	6.78	8.75	9.04
Pr	1.73	1.74	2.08	2.28	2.23	2.33	1.00	0.99	1.09	1.04	0.98	0.99	1.26	1.28
Nd	9.46	9.40	10.7	11.9	11.7	12.2	5.00	5.00	5.50	5.21	4.98	4.99	6.21	6.28
Sm	3.05	3.06	3.19	3.56	3.57	3.65	1.49	1.45	1.59	1.52	1.50	1.50	1.76	1.66
Eu	1.13	1.12	1.18	1.23	1.26	1.26	0.62	0.61	0.65	0.63	0.60	0.60	0.72	0.68
Gd	3.71	3.71	3.79	4.15	4.14	4.29	1.71	1.66	1.87	1.76	1.74	1.73	2.06	1.89
Tb	0.65	0.65	0.66	0.71	0.71	0.73	0.30	0.29	0.33	0.31	0.31	0.31	0.37	0.33
Dy	4.12	4.15	4.18	4.51	4.43	4.58	2.07	2.02	2.23	2.11	2.14	2.13	2.49	2.20
Ho	0.87	0.87	0.88	0.94	0.93	0.96	0.45	0.43	0.48	0.45	0.46	0.46	0.54	0.49
Er	2.48	2.49	2.51	2.69	2.63	2.71	1.38	1.36	1.51	1.41	1.44	1.44	1.68	1.50
Tm	0.36	0.36	0.36	0.39	0.38	0.39	0.22	0.21	0.23	0.22	0.22	0.22	0.26	0.23

Continued

Sample	GYH-1-1	GYH-1-2	GYH-1-3	GYH-1-5	GYH-1-6	GYH-1-7	GDS-1	GDS-2	GDS-3	GDS-4	GDS-5	GDS-6	GDS-7	GDS-8
Yb	2.30	2.31	2.31	2.41	2.39	2.46	1.52	1.49	1.64	1.52	1.56	1.56	1.82	1.61
Lu	0.34	0.34	0.34	0.36	0.35	0.36	0.23	0.23	0.24	0.23	0.24	0.24	0.27	0.25
Hf	1.83	1.80	1.69	1.77	1.72	1.86	0.83	1.11	0.88	0.72	1.03	1.04	0.91	0.85
Ta	0.14	0.14	0.13	0.13	0.13	0.13	0.092	0.12	0.11	0.12	0.10	0.10	0.13	0.18
Pb	1.91	1.87	1.26	1.36	1.32	1.36	3.05	2.60	3.03	3.13	3.46	3.48	2.73	3.41
Th	0.25	0.28	0.16	0.18	0.19	0.19	0.39	0.43	0.49	0.39	0.44	0.44	0.41	0.50
U	0.064	0.067	0.042	0.049	0.048	0.051	0.12	0.14	0.15	0.13	0.14	0.14	0.14	0.17
REE	44.3	44.2	50.2	54.6	53.6	55.9	26.1	25.7	28.1	26.7	26.0	26.0	31.7	31.3
Eu/Eu*	1.03	1.02	1.04	0.98	1.00	0.97	1.19	1.20	1.15	1.17	1.14	1.15	1.15	1.17

$Mg^{\#}$ = molar $100MgO/(MgO+FeO^{T})$; $A/CNK=Al_2O_3/(CaO+K_2O+Na_2O)$ molar ratio; $Eu/Eu^* =Eu_N/(Sm_N \cdot Gd_N)^{1/2}$.

Table 4 Whole-rock Sr-Nd isotopic compositions for the Early Neoproterozoic mafic rocks in the western Yangtze Block, South China.

Sample	$^{87}Sr/^{86}Sr$	2SE	Rb/ppm	Sr/ppm	$^{143}Nd/^{144}Nd$	2SE	Nd/ppm	Sm/ppm	T_{DM2}/Ga	$\varepsilon_{Nd}(t)$	I_{Sr}
Ganyuhe hornblende gabbros											
GYH-1-5	0. 704 018	0. 000 005	3. 4	252	0. 512 799	0. 000 004	11. 9	3. 56	1. 08	+ 4. 9	0. 703 535
GYH-1-6	0. 704 026	0. 000 004	4. 0	253	0. 512 825	0. 000 003	11. 7	3. 57	1. 07	+ 5. 0	0. 703 461
Guandaoshan gabbro-diorites											
GDS-3	0. 704 562	0. 000 011	20. 3	529	0. 512 812	0. 000 010	5. 5	1. 59	1. 02	+ 5. 8	0. 703 203
GDS-4	0. 705 504	0. 000 014	23. 1	507	0. 512 794	0. 000 019	5. 2	1. 52	1. 05	+ 5. 3	0. 703 895
GDS-5	0. 704 855	0. 000 010	16. 4	585	0. 512 838	0. 000 006	5. 0	1. 50	1. 03	+ 5. 6	0. 703 864

$^{87}Rb/^{86}Sr$ and $^{147}Sm/^{144}Nd$ ratios were calculated using Rb, Sr, Sm and Nd contents analyzed by ICP-MS, referenced from DePaolo (1981).
T_{DM2} represent the two-stage model age and were calculated using present-day $(^{147}Sm/^{144}Nd)_{DM} = 0.213\ 7$, $(^{147}Sm/^{144}Nd)_{DM} = 0.513\ 15$ and $(^{147}Sm/^{144}Nd)_{crust} = 0.101\ 2$. $\varepsilon_{Nd}(t)$ values were calculated using present-day $(^{147}Sm/^{144}Nd)_{CHUR} = 0.196\ 7$ and $(^{147}Sm/^{144}Nd)_{CHUR} = 0.512\ 638$.

$$\varepsilon_{Nd}(t) = \left[(^{143}Nd/^{144}Nd)_{sample}(t)/(^{143}Nd/^{144}Nd)_{CHUR}(t) -1\right]\times 10^4.$$

$$T_{DM2} = \frac{1}{\lambda}\left\{ 1+\left[\left[(^{143}Nd/^{144}Nd)_S -\left((^{147}Sm/^{144}Nd)_S -(^{147}Sm/^{144}Nd)_{crust}\right)(e^{\lambda t}-1) -(^{143}Nd/^{144}Nd)_{DM}\right]/\left[((^{147}Sm/^{144}Nd)_{crust} -(^{147}Sm/^{144}Nd)_{DM}\right]\right\}.$$

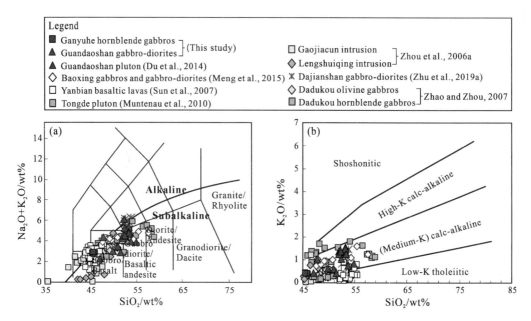

Fig. 6　Plots of Na$_2$O + K$_2$O vs. SiO$_2$(a) (Middlemost, 1994) and K$_2$O vs. SiO$_2$(b)
(Robert and Clemens, 1993) for the Neoproterozoic mafic rocks
in the western Yangtze Block, South China.
The Neoproterozoic mafic magmatism were from Du et al. (2014), Meng et al. (2015), Munteanu et al.
(2010), Sun et al. (2007), Zhao and Zhou (2007), Zhou et al. (2006a), and Zhu et al. (2019a).

(48. 6–50. 7) values (Table 3). All of the Guandaoshan gabbro-diorites possess lower total REE concentrations (25. 7–31. 7 ppm) than the Ganyuhe hornblende gabbros. The chondrite-normalized REE patterns are flat [(La/Lu)$_N$ = 1. 36–1. 61] with weakly positive Eu anomalies (Eu/Eu* = 1. 14–1. 20) (Fig. 7a). In the primitive mantle-normalized diagram (Fig. 7b), they also exhibit obvious enrichment of Ba, Sr, K, and Pb, and depletion of Nb, Ta, Zr, Hf, and Ti. In addition, the Guandaoshan gabbro-diorites are characterized by lower Cr (10. 7–14. 8 ppm) and Ni (6. 63–8. 06 ppm) and higher Sr (493–585 ppm), Ba (480–682 ppm), and Rb (16. 1–24. 0 ppm) concentrations than the Ganyuhe hornblende gabbros (e.g., Cr = 181–234 ppm, Ni = 80. 8–88. 7 ppm, Sr = 192–256 ppm, Ba = 81. 3–139 ppm, and Rb = 3. 15–4. 18 ppm) (Table 3).

Three gabbro-diorite samples (GDS-3, GDS-4, and GDS-5) show slightly variable (^{87}Sr/^{86}Sr)$_i$ values from 0. 703 203 to 0. 703 895. They also have highly positive $\varepsilon_{Nd}(t)$ values ranging from +5. 3 to +5. 8, which approach the Nd isotopic compositions of the Guandaoshan pluton [e.g., $\varepsilon_{Nd}(t)$ = +4. 8 to +5. 2; Du et al., 2014. $\varepsilon_{Nd}(t)$ = +3. 9 to +5. 1; Sun and Zhou, 2008] (Fig. 8).

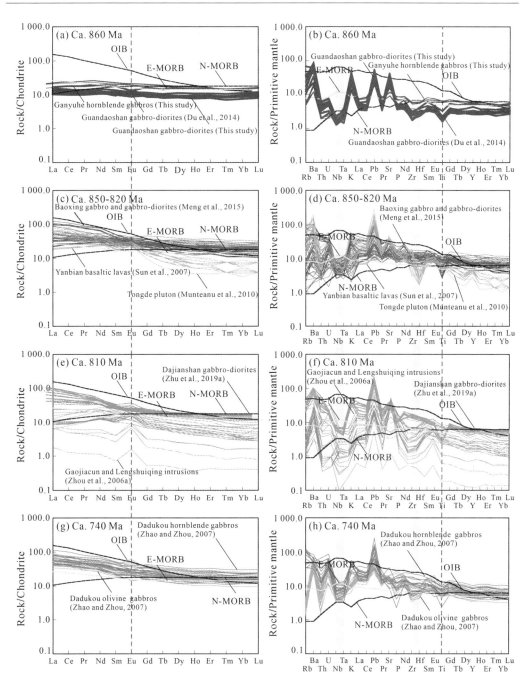

Fig. 7 Diagrams of chondrite-normalized REE patterns and primitive mantle-normalized trace element patterns for the Neoproterozoic mafic rocks in the western Yangtze Block, South China.

The primitive mantle and chondrite values are from Sun and McDonough (1989). The lower continental crust reference is from Rudnick and Gao (2003). The N-MORB and E-MORB references are from Sun and McDonough (1989). The Neoproterozoic mafic magmatism were from Du et al. (2014), Meng et al. (2015), Munteanu et al. (2010), Sun et al. (2007), Zhao and Zhou (2007), Zhou et al. (2006a), and Zhu et al. (2019a).

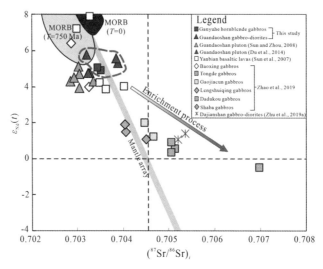

Fig. 8　Whole-rock $\varepsilon_{Nd}(t)$ vs. initial $^{87}Sr/^{86}Sr$ values diagram for the Neoproterozoic mafic rocks in the western Yangtze Block, South China.

The Sr-Nd isotopes of Neoproterozoic mafic magmatism were from Du et al. (2014), Meng et al. (2015), Sun et al. (2007), Sun and Zhou (2008), Zhao et al. (2019), and Zhu et al. (2019a).

5　Discussion

5.1　Assessment of alteration effects, crustal contamination, and fractional crystallization

Although the petrographic observations on our mafic rocks suggest some extent of alteration (Fig. 2), they possess relatively low loss-on-ignition (LOI) values (0.93–2.51 wt%, average 1.67 wt%) (Table 3), indicating minimal hydrothermal alteration of the whole-rock geochemical data (Zhao and Zhou, 2007). As shown in the trace elemental patterns diagrams (Fig. 7a,b), these samples display smooth and parallel patterns with insignificant Ce (Ce/Ce* = 0.96–1.04) and Eu (Eu/Eu* = 0.97–1.20) anomalies, which preclude extensive mobility of REEs and HFSEs (Cai et al., 2014; Polat and Hofmann, 2003; Zhang and Wang, 2016). The alteration effects in our samples thus display an insignificant role during magmatic evolution.

Mafic rocks commonly experience continental crustal contamination during magmatic emplacement (Depaolo, 1981). In general, crustal contamination may decrease Nb/La ratios, resulting into a negative relationship between SiO_2 contents and Nb/La ratios (Kou et al., 2018). Nevertheless, there is no obviously negative correlation between SiO_2 and Nb/La values (Fig. 9a), which indicates that continental crustal contamination is minimal. Further evidence against significant crustal contamination is presented by the substantially lower concentrations of

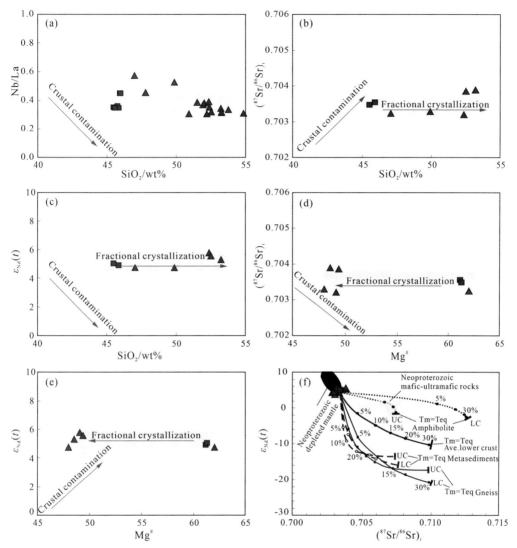

Fig. 9　Plots of Nb/La vs. SiO$_2$(a), (^{87}Sr/^{86}Sr)$_i$ vs. SiO$_2$(b), $\varepsilon_{Nd}(t)$ vs. SiO$_2$(c), (^{87}Sr/^{86}Sr)$_i$ vs. Mg$^{\#}$(d), $\varepsilon_{Nd}(t)$ vs. Mg$^{\#}$(e), and $\varepsilon_{Nd}(t)$ vs. (^{87}Sr/^{86}Sr)$_i$(f) for the Early Neoproterozoic hornblende gabbros and gabbro-diorites in the western Yangtze Block, South China. Symbols as Fig. 6.

Th (0. 16 – 0. 50 ppm), U (0. 04 – 0. 17 ppm), and Rb (3. 15 – 24. 0 ppm) than upper crust values (e.g., Th = 10. 5 ppm, U = 2. 7 ppm, and Rb = 84 ppm) (Rudnick and Gao, 2003; Zhao and Zhou, 2009). Minor crustal contamination may lead to negative Nb-Ta accompanied with positive Zr-Hf anomalies (Zhao and Zhou, 2007; Zhou et al., 2006a), but the variable degrees of Zr-Hf troughs (Fig. 7b) in our mafic rocks argue against substantial crustal contamination. If mantle-derived mafic rocks have been significantly contaminated by crustal components during the magmatic evolution, significant variations of whole-rock Sr-Nd isotopes are expected. However, the relatively constant initial $^{87}Sr/^{86}Sr$ ratios and $\varepsilon_{Nd}(t)$ values obtained here suggest that the primary magmas of our mafic rocks witnessed little crustal contamination (Zhou et al., 2006a) (Fig. 9b-e). Limited crustal contamination is further substantiated by their low initial $^{87}Sr/^{86}Sr$ (0. 703 203 – 0. 703 895) ratios and high $\varepsilon_{Nd}(t)$ (+4. 9 to +5. 8) values, as shown in the EC-AFC modelling calculations of various crustal rocks for the Neoproterozoic mafic-ultramafic rocks in the western Yangtze Block (Fig. 9f) (Zhao et al., 2018 and references therein). Therefore, these lines of geochemical evidence demonstrate that the parental magma of the mafic rocks in this study experienced negligible crustal contamination.

The mafic rocks studied here possess variable MgO (3. 27 – 9. 20 wt%) contents and $Mg^{\#}$ (48. 6 – 62. 6) values, indicating that their primary melts underwent fractional crystallization during magma ascent (Zhang and Wang, 2016). Typical mantle-derived primary melts generally possess high concentrations of Cr (>1 000 ppm) and Ni (>400 ppm) (Litvak and Poma, 2010; Wilson, 1989). The low and variable concentrations of Cr (10. 7 – 234 ppm) and Ni (6. 63 – 88. 7 ppm) in our mafic rocks also suggest variable degrees of fractional crystallization prior to emplacement. The positive correlations for Cr and Ni concentrations against MgO contents in our samples are indicative of olivine and/or clinopyroxene fractionation (Zheng et al., 2019) (Fig. 10a,b). Their CaO contents and CaO/Al_2O_3 ratios are positively correlated with MgO contents (Fig. 10c,d), indicating the fractional crystallization of clinopyroxene (Zhao and Zhou, 2007). Moreover, the fractionation of amphibole is supported by the positive relationship between SiO_2 contents and La/Sm ratios and negative relationship between SiO_2 contents and Dy/Yb ratios (Davidson et al., 2007) (Fig. 10e,f). The positive relationships between $Fe_2O_3^T$, TiO_2, and MgO contents observed in our mafic rocks suggest the fractional crystallization of Fe-Ti oxides (Fig. 10g,h). In addition, the slightly positive and negative Eu anomalies (Eu/Eu* = 0. 97 – 1. 20) (Fig. 7a) reflect insignificant accumulation or fractionation of plagioclase. In summary, the fractional crystallization process played a vital role during magmatic evolution.

5. 2 Nature of the mantle source: An amphibole-bearing mantle source metasomatized by subduction fluids

The Early Neoproterozoic hornblende gabbros and gabbro-diorites in this study have low SiO_2 (45. 5 – 53. 3 wt%) and high MgO (3. 27 – 9. 20 wt%) contents as well as $Mg^{\#}$ values (48. 6 – 62. 6) (Table 3), suggesting a mafic mantle source (Wilson, 1989). The high Zr/Nb ratios (15. 1 – 41. 4) differ from those of OIB-derived melts (Zr/Nb = 5. 83) (Sun and McDonough, 1989), ruling out an asthenosphere mantle source (Xu et al., 2016). In terms of Nd-Hf isotopes, the prominently positive whole-rock $\varepsilon_{Nd}(t)$ (+4. 9 to +5. 8) and zircon

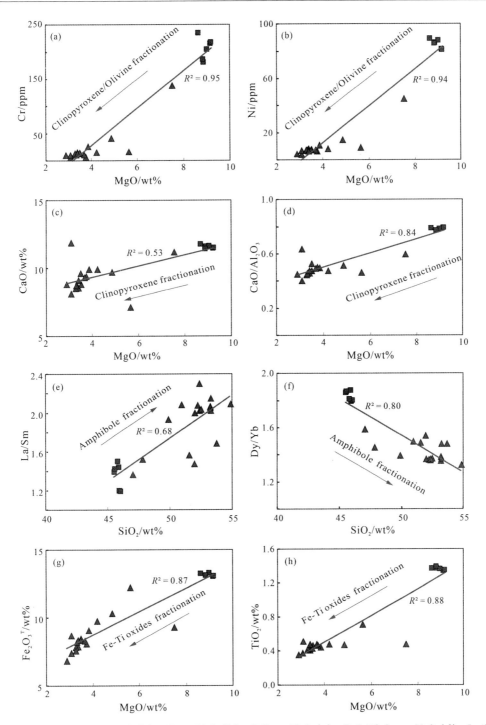

Fig. 10　Plots of Cr vs. MgO（a）, Ni vs. MgO（b）, CaO vs. MgO（c）, CaO/Al₂O₃ vs. MgO（d）, La/Sm
vs. SiO₂(e）, Dy/Yb vs. SiO₂(f）, Fe₂O₃^T vs. MgO（g）, and TiO₂ vs. MgO（h）for the Early
Neoproterozoic hornblende gabbros and gabbro-diorites in the western Yangtze Block, South China.
Symbols as Fig. 6.

$\varepsilon_{Hf}(t)$ (+ 8. 48 to + 20. 6) values are coordinated with those of Early Neoproterozoic mafic rocks that have been explained to be derived from depleted lithospheric mantle sources (e.g., Du et al., 2014; Meng et al., 2015; Sun and Zhou, 2008; Zhao et al., 2018, 2019) (Figs. 4 and 8). Thus, the parental magmas of our mafic rocks might be sourced from the depleted lithospheric mantle. Moreover, it is notable that these mafic rocks exhibit more enriched Rb, Ba, K, Pb, and Sr and depleted Nb, Ta, Zr, Hf, and Ti than the average MORB compositions (Fig. 7b). Considering the insignificant crustal contamination as demonstrated above, the enriched trace elemental signatures in our samples are likely inherited from a mantle source that had undergone metasomatism related to the infiltration of subduction components prior to its partial melting.

Numerous studies have proposed that subduction-related fluids, sediment melts, and slab melts can be incorporated into the lithospheric mantle beneath the active continental margins, subsequently leading to the generation of arc-affinity mafic-ultramafic rocks that are enriched in LILEs and depleted in HFSEs (Hanyu et al., 2006; Hawkesworth et al., 1997; Kelemen et al., 2003, 2014; Kepezhinskas et al., 1997; Pilet et al., 2008; Zheng et al., 2020). The mafic rocks in this study yield variable (Ta/La)$_N$(0. 43-0. 80) and (Hf/Sm)$_N$(0. 68-1. 10) ratios (Fig. 11a), suggesting a trend of subduction fluid-related metasomatism (LaFlèche et al., 1998). They display variable Th/Zr (0. 002-0. 022) and Rb/Y (0. 13-1. 96) ratios (Fig. 11b,c) (Kepezhinskas et al., 1997), which indicate fluid-related enrichment, as found in the Neoproterozoic mafic rocks from the western Yangtze Block (e.g., Meng et al., 2015; Sun et al., 2007; Zhao and Zhou, 2007; Zhou et al., 2006a). Previous studies have demonstrated that subduction slab-derived aqueous fluids are characterized by the enrichment of Ba, Rb, Sr, U, and Pb, whereas subducted sediment-derived melts usually display high Th and LREE contents (Hawkesworth et al., 1997; Woodhead et al., 2001). As shown in the trace element diagrams (Fig. 7b), our samples display notably positive Rb, Ba, Sr, U, and Pb anomalies, compared with those of N-MORB (Sun and McDonough, 1989). Their low and constant Th/Yb ratios (0. 07-0. 31 ppm) as well as highly variable Ba/La ratios (16. 7-228) reflect a greater contribution of hydrous fluids to mantle metasomatism, rather than sediment melts (Hanyu et al., 2006) (Fig. 11d). Hydrous characteristics of the mantle source are also supported by the abundance of amphiboles observed in our samples (Du et al., 2014; Munteanu et al., 2010) (Fig. 2). Consequently, the mafic rocks studied here are inferred to be derived from a depleted lithospheric mantle source that was variably metasomatized by subduction-related fluids.

The mineralogy of the lithospheric mantle exerts a pronounced influence on the composition of primary melts (Hunt et al., 2012). Experimental studies have proposed that fluids and/or melts released from subducting slabs can react with overlying mantle peridotite,

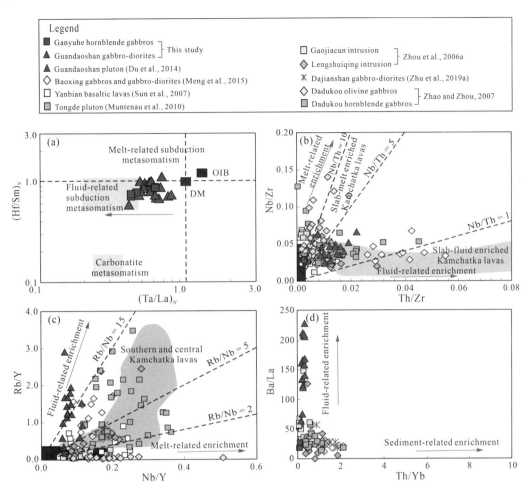

Fig. 11   Plots of (Hf/Sm)$_N$ vs. (Ta/La)$_N$ (a) (LaFlèche et al., 1998), Nb/Zr vs. Th/Zr (b)
(Kepezhinskas et al., 1997), Rb/Y vs. Nb/Y (c) (Kepezhinskas et al., 1997), and
Ba/La vs. Th/Yb (d) (Hanyu et al., 2006) for the Neoproterozoic mafic rocks
in the western Yangtze Block, South China.
Symbols as Fig. 6.

leading to the consumption of olivine and generation of new metasomatic minerals (e. g.,
garnet, clinopyroxene, amphibole, and phlogopite) (Prouteau et al., 2001; Rapp et al.,
1999). Phlogopite and amphibole are the most common hydrous mineral phases in the mantle
source, both of which can be produced by fluid-peridotite reactions (Prouteau et al., 2001).
Melts sourced from the phlogopite-bearing mantle source are usually potassic and characterized
by high Rb/Sr (>0.1) and low Ba/Rb (<20) ratios, whereas melts in equilibrium with
amphibole-bearing mantle sources are sodic and expected to have low Rb/Sr (<0.1) and high
Ba/Rb (>20) ratios (Furman and Graham, 1999; Pilet et al., 2008). The mafic rocks in
this study are Na-rich (Na$_2$O/K$_2$O=2.25-6.65) and display low Rb/Sr (0.01-0.05) and
high Ba/Rb (19.4-32.5) ratios (Fig. 12a), indicating the predominant presence of

amphibole in the mantle source. Furthermore, owing to the high distribution coefficients of HREE in garnet, a residual garnet phase in the source may lead to the notably right-sloping REE patterns and high Dy/Yb ratios (Blundy et al., 1998). However, our samples exhibit relatively flat REE patterns with low $(La/Yb)_N$ (1. 13 – 1. 70) and $(Dy/Yb)_N$ (0. 91 – 1. 25) ratios (Fig. 7a), hinting the significant presence of spinel not the garnet in the mantle source (Blundy et al., 1998; Liu and Zhao, 2019). The presence of spinel is further demonstrated by low Sm/Yb and variable La/Sm ratios in our samples (Aldanmaz et al., 2000; Liu and Zhao, 2019; Zhao and Zhou, 2007) (Fig. 12b).

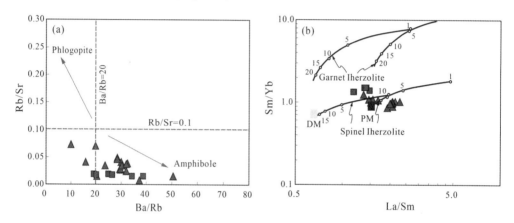

Fig. 12　Plots of Rb/Sr vs. Ba/Rb (a) (Furman and Graham, 1999) and Sm/Yb vs. La/Sm (b) (Liu and Zhao, 2019 and references therein) for the Early Neoproterozoic hornblende gabbros and gabbro-diorites in the western Yangtze Block, South China.
Symbols as Fig. 6.

In summary, we suggest that the Early Neoproterozoic mafic rocks in this study were derived from an amphibole-bearing spinel peridotite source that previously experienced the metasomatism between subduction fluids and the overlying mantle wedge.

5. 3　Neoproterozoic subduction setting and metasomatized mantle in the western Yangtze Block

Although a mantle plume setting and internal model for South China have been used to interpret the genesis of igneous rocks along the margins of the Yangtze Block (Li et al., 2002, 2003, 2006; Wang et al., 2008; Wu et al., 2019), the long-lasting magmatism (ca. 870 – 740 Ma) is distinct from the short time interval of mantle plume-related magmatism (e. g., Emeishan large igneous province, < 1 Ma; Tarim large igneous province, ~20 Ma) (Xu et al., 2017; Shellnutt, 2014). The magmatism induced by a mantle plume is characterized by widespread mafic-ultramafic rocks with OIB-like geochemical signatures (Du et al., 2014; Wang et al., 2016), such as the Gairdner dykes swarm and Amtata dykes swarm in Australia

(~210 000 km^2) (Kou et al., 2018 and references therein) and Emeishan large igneous province in South China (~300 000 km^2) (Shellnutt, 2014). However, this is inconsistent with the geological observations that the limited Neoproterozoic mafic-ultramafic rocks are subordinately distributed along the margins of the Yangtze Bock (Fig. 1d,e). The absence of large-scale OIB-like mafic-ultramafic rocks (Fig. 7) refutes a mantle-plume derivation for numerous magmatic events in the western Yangtze Block. In combination with the existence of Neoproterozoic Shimian ophiolites and oceanic slab-derived adakitic granitoids (Chen et al., 2015; Hu et al., 2017; Zhao et al., 2017; Zhou et al., 2006b), a subduction-related setting and peripheral location for South China is more reasonable for the generation of the widespread Neoproterozoic magmatic rocks.

In this study, the Early Neoproterozoic hornblende gabbros and gabbro-diorites from the western Yangtze Block were formed at ca. 860 Ma. The primitive mantle-normalized patterns of these mafic rocks display notably positive anomalies of Rb, Sr, Ba, K, and Pb, and negative anomalies of Nb, Ta, and Ti (Fig. 7b), resembling subduction-related arc magmatism (Wilson, 1989; Zhao et al., 2018). These fluid-mobile trace elements (e.g., Rb, Sr, K, and Pb) would ascend into the overlying mantle wedge, producing enriched mantle domains at sub-arc depths in subduction setting, further resulting in the formation of mafic rocks with arc-like trace elemental distribution and depleted isotopic compositions (Kelemen et al., 2014; Schmidt and Poli, 2003; Zheng et al., 2020). This is consistent with our mafic rocks that show enriched arc-like trace elemental patterns (Fig. 7b) and depleted Nd-Hf isotopic characteristics (Figs. 4 and 8). The addition of subduction components under an arc setting is also supported in the Th/Yb vs. Nb/Yb and Ti vs. Zr diagrams (Fig. 13). Furthermore, the geochemical signatures reveal an amphibole-bearing spinel peridotite source (Fig. 12), which indicates that the Early Neoproterozoic mantle sources beneath the western Yangtze Block were metasomatized by subduction-related fluids (Kelemen et al., 2014). In addition to the aforementioned mafic rocks in this study, abundant Neoproterozoic (ca. 850-740 Ma) mafic-ultramafic rocks, including gabbros, gabbro-diorites, and basaltic lavas have been documented in the western Yangtze Block (Meng et al., 2015; Munteanu et al., 2010; Sun et al., 2007; Zhao and Zhou, 2007; Zhou et al., 2006a; Zhu et al., 2019a). These mafic rocks display significant enrichment of LILEs and depletion of Nb, Ta, and Ti (Fig. 7d,f,h), suggesting a subduction-related tectonic setting. As shown in the Nb/Zr vs. Th/Zr, Rb/Y vs. Nb/Y, and Ba/La vs. Th/Yb diagrams (Fig. 11b-d), most display that the primary mantle sources were hybridized by subduction-related fluids. In addition, the subduction oceanic slab melts were introduced into the overlying mantle wedge along the western Yangtze Block (Fig. 11b,c), which have been demonstrated by the Tongde pluton and Dadukou olivine gabbros (Munteanu et al., 2010; Zhao and Zhou, 2007). Zhu et al. (2020a) recently identified ca. 850-835 Ma

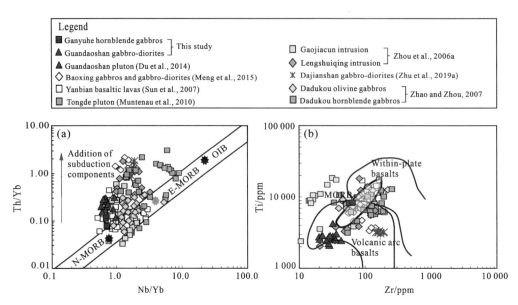

Fig. 13　Plots of Th/Yb vs. Nb/Yb (a) (Pearce, 2008) and Ti vs. Zr (b) (Pearce and Cann, 1973) for the Neoproterozoic mafic rocks in the western Yangtze Block, South China.

Symbols as Fig. 6.

Shuilu high-Mg$^{\#}$ diorites with depleted Nd-Hf isotopes as well as variable Ba contents and Rb/Y, Th/Ce, Th/Sm, Ba/La, and Th/Yb ratios, which were explained to be derived from a depleted lithospheric mantle source metasomatized by subduction fluids and sediment melts. The high zircon δ^{18}O values in the Neoproterozoic gabbros also demonstrate the presence of a subducted sediment melt-enriched mantle source (Zhao et al., 2019). In summary, the Neoproterozoic mantle sources beneath the western Yangtze Block were metasomatized by subduction compositions involving slab fluids, sediment melts, and oceanic slab melts.

6　Conclusion

In this study, the Early Neoproterozoic mafic rocks, including the Ganyuhe hornblende gabbros and Guandaoshan gabbro-diorites, were determined to have formed at 857.4 ± 2.0 Ma and 856.4 ± 2.8 Ma, respectively. These mafic rocks are geochemically characterized by enrichment in large ion lithophile elements (LILEs, e.g., Rb, Sr, and K) but depletion in high field strength elements (HFSEs, e.g., Nb, Ta, Zr, and Hf). Isotopically, they display depleted Nd-Hf isotopes compositions [e.g., $\varepsilon_{Nd}(t)$ = + 4.9 to + 5.8, $\varepsilon_{Hf}(t)$ = + 8.48 to + 20.6]. The mafic rocks studied here were derived from an amphibole-bearing spinel peridotite source that previously underwent metasomatism between the overlying mantle wedge and subduction-related fluids. In combination with a compilation of mafic rocks from the literature, we support that the Neoproterozoic mantle sources beneath the western Yangtze

Block were metasomatized by subduction-related compositions involving slab fluids, sediment melts, and oceanic slab melts.

Acknowledgements   We appreciate the help and constructive comments from two anonymous reviewers and handing editors, which significantly improve the discussion and language of this manuscript. We also thank the Dr. Esther Posner and Dr. Shuo Liu for further polish of English language. This work was jointly supported by the National Natural Science Foundation of China (Grant Nos. 41421002 and 41772052) and the program for Changjiang Scholars and Innovative Research Team in University (IRT1281).

References

Aldanmaz, E., Pearce, J.A., Thirlwall, M.F., Mitchell, J.G., 2000. Petrogenetic evolution of late Cenozoic, post-collision volcanism in western Anatolia, Turkey. Journal of Volcanology and Geothermal Research, 102, 67-95.

Andersen, T., 2002. Correction of common lead in U-Pb analyses that do not report ^{204}Pb. Chemical Geology, 192(1-2), 59-79.

Bao, Z.A., Chen, L., Zong, C.L., Yuan, H.L., Chen, K.Y., Dai, M.N., 2017. Development of pressed sulfide powder tablets for in situ, sulfur and lead isotope measurement using LA-MC-ICP-MS. International Journal of Mass Spectrometry, 421, 255-262.

Blundy, J.D., Robinson, J.A.C., Wood, B.J., 1998. Heavy REE are compatible in clinopyroxene on the spinel lherzolite solidus. Earth & Planetary Science Letters, 160 (3-4), 493-504.

Cai, Y.F., Wang, Y.J., Cawood, P.A., Fan, W.M., Liu, H.C., Xing, X.W., Zhang, Y.Z., 2014. Neoproterozoic subduction along the Ailaoshan zone, South China: Geochronological and geochemical evidence from amphibolite. Precambrian Research, 245, 13-28.

Cawood, P.A., Wang, Y.J., Xu, Y.J., Zhao, G.C., 2013. Locating South China in Rodinia and Gondwana: A fragment of greater India lithosphere? Geology, 41, 903-906.

Cawood, P.A., Strachan, R.A., Pisarevsky, S.A., Gladkochub, D.P., Murphy, J.B., 2016. Linking collisional and accretionary orogens during Rodinia assembly and breakup: Implications for models of supercontinent cycles. Earth & Planetary Science Letters, 449, 118-126.

Cawood, P.A., Zhao, G.C., Yao, J.L., Wang, W., Xu, Y.J., Wang, Y.J., 2017. Reconstructing South China in Phanerozoic and Precambrian supercontinents. Earth-Science Reviews, 186, 173-193.

Chu, Z.Y., Chen, F.K., Yang, Y.H., Guo, J.H., 2009. Precise determination of Sm, Nd concentrations and Nd isotopic compositions at the nanogram level in geological samples by thermal ionization mass spectrometry. Journal of Analytical Atomic Spectrometry, 24, 1534-1544.

Chen, Q., Sun, M., Long, X.P., Yuan, C., 2015. Petrogenesis of Neoproterozoic adakitic tonalites and high-K granites in the eastern Songpan-Ganze fold belt and implications for the tectonic evolution of the western Yangtze Block. Precambrian Research, 270, 81-203.

Davidson, J., Turner, S., Handley, H., MacPherson, C., Dosseto, A., 2007. Amphibole"sponge" in arc crust? Geology, 35 (9), 787.

DePaolo, D. J., 1981. Trace element and isotopic effects of combined wallrock assimilation and fractional crystallisation. Earth & Planetary Science Letters, 53, 189-202.

Du, L.L., Guo, J.H., Nutman, A.P., Wyman, D., Geng, Y.S., Yang, C.H., Liu, F.L., Ren, L.D., Zhou, X.W., 2014. Implications for Rodinia reconstructions for the initiation of Neoproterozoic subduction at ~860 Ma on the western margin of the Yangtze Block: Evidence from the Guandaoshan Pluton. Lithos, 196-197, 67-82.

Fan, H.P., Zhu, W.G., Li, Z.X., Zhong, H., Bai, Z.J., He, D.F., Chen, C.J., Cao, C.Y., 2013. Ca. 1. 5 Ga mafic magmatism in South China during the break-up of the supercontinent Nuna/Columbia: The Zhuqing Fe-Ti-V oxide ore-bearing mafic intrusions in western Yangtze Block. Lithos, 168-169, 85-98.

Fan, H.P., Zhu, W.G., Li, Z.X., 2020. Paleo- to Mesoproterozoic magmatic and tectonic evolution of the southwestern Yangtze Block, South China: New constraints from ca. 1.7−1.5 Ga mafic rocks in the Huili-Dongchuan area. Gondwana Research, 87, 248-262.

Furman, T., Graham, D., 1999. Erosion of lithospheric mantle beneath the East African Rift system: Geochemical evidence from the Kivu volcanic province. Lithos, 48, 237-262.

Gao, S., Ling, W.L., Qiu, Y., Zhou, L., Hartmann, G., Simon, K., 1999. Contrasting geochemical and Sm-Nd isotopic compositions of Archean metasediments from the Kongling high-grade terrain of the Yangtze craton: Evidence for cratonic evolution and redistribution of REE during crustal anatexis. Geochimica et Cosmochimica Acta, 63 (13-14), 2071-2088.

Gao, S., Jie, Y., Lian, Z., Li, M., Hu, Z.C., Guo, J.L., Yuan, H.L., Gong, H.J., Xiao, G.Q., Wei, J. Q., 2011.Age and growth of the Archean Kongling terrain, South China, with emphasis on 3. 3 Ga granitoid gneisses. Geochimica et Cosmochimica Acta Supplement, 72(12), 153-182.

Greentree, M.R., Li, Z.X., 2008. The oldest known rocks in south-western China: SHRIMP U-Pb magmatic crystallization age and detrital provenance analysis of the Paleoproterozoic Dahongshan Group. Journal of Asian Earth Sciences, 33, 289-302.

Griffin, W.L., Pearson, N.J., Belousova, E., Jackson, S.E., van Achterbergh, E., O'Reilly, S.Y., Shee, S. R., 2000. The Hf isotope composition of cratonic mantle: LAM-MC-ICP-MS analysis of zircon megacrysts in kimberlites. Geochimica et Cosmochimica Acta, 64, 133-147.

Guo, J.L., Gao, S., Wu, Y.B., Li, M., Chen, K., Hu, Z.C., Liang, Z.W., Liu, Y.S., Zhou, L., Zong, K.Q., Zhang, W., Chen, H.H., 2014. 3. 45 Ga granitic gneisses from the Yangtze Craton, South China: Implications for Early Archean crustal growth. Precambrian Research, 242, 82-95.

Hanyu, T., Tatsumi, Y., Nakai, S., Chang, Q., Miyazaki, T., Sato, K., Tani, K., Shibata, T., Yoshida, T., 2006. Contribution of slab melting and slab dehydration to magmatism in the NE Japan arc for the last 25 Myr: Constraints from geochemistry. Geochemistry, Geophysics, Geosystems, 7(8).

Hawkesworth, C.J., Turner, S.P., McDermott, F., Peate, D.W, van Calsteren, P., 1997. U-Th isotopes in arc magmas: Implications for element transfer from the subducted crust. Science, 276, 551-555.

Herzberg, C., Asimow, P.D., Arndt, N., Niu, Y., Saunders, A.D., 2006. Temperatures in ambient mantle and plumes: Constraints from basalts, picrites, and komatiites. Geochemistry, Geophysics, Geosystems, 2(2).

Hofmann, M., Linnemann, U., Rai, Becker, S., Gärtner, A., Sagawe, A., 2011. The India and South China cratons at the margin of Rodinia-synchronous Neoproterozoic magmatism revealed by LA-ICP-MS zircon analyses. Lithos, 123, 176-187.

Hoskin, P.O, Schaltegger, U., 2003. The composition of zircon and igneous and metamorphic petrogenesis. In: Hanchar, J.M., Hoskin, P.W.O. Zircon. Reviews in Mineralogy and Geochemistry, 53. Mineralogical Society of America, 27-62.

Hu, P.Y., Zhai, Q.G., Wang, J., Tang, Y., Ren, G.M., 2017. The Shimian ophiolite in the western Yangtze Block, SW China: Zircon SHRIMP U-Pb ages, geochemical and Hf-O isotopic characteristics, and tectonic implications. Precambrian Research, 298, 107-122.

Hunt, A.C., Parkinson, I.J., Harris, N.B.W., Barry, T.L., Rogers, N.W., Yondon, M., 2012. Cenozoic Volcanism on the Hangai Dome, Central Mongolia: Geochemical evidence for changing melt sources and implications for mechanisms of melting. Journal of Petrology, 53(9), 1913-1942.

Kelemen, P. B., Yogodzinski, G. M., Scholl, D.W., 2003. Along-strike variation in the Aleutian Island arc: Genesis of high Mg$^{\#}$ andesite and implications for continental crust. In: Eiler, J. Inside the subduction factory. Washington, DC: American Geophysical Union, 223-276.

Kelemen, P.B., Hanghøj, K., Greene, A.R., 2014. One view of the geochemistry of subduction-related magmatic arcs, with an emphasis on primitive andesite and lower crust. In: Turekian, H.D.H.K. Treatise on geochemistry. Oxford: Elsevier, 749-806.

Kepezhinskas, P., McDermott, F., Defant, M.J., Hochstaedter, A., Drummond, M.S., 1997. Trace element and Sr-Nd-Pb isotopic constraints on a three-component model of Kamchatka Arc petrogenesis. Geochimica et Cosmochimica Acta, 61(3), 577-600.

Kou, C.H., Liu, Y.X., Huang, H., Li, T.D., Ding, X.Z., Zhang, H., 2018. The Neoproterozoic arc-type and OIB-type mafic-ultramafic rocks in the western Jiangnan Orogen: Implications for tectonic settings. Lithos, 312-313, 38-56.

LaFlèche, M.R., Camire, G., Jenner, G.A., 1998. Geochemistry of post-Acadian, Carboniferous continental intraplate basalts from the Maritimes basin, Magdalen islands, Quebec, Canada. Chemical Geology, 148, 115-136.

Lai, S.C., Qin, J.F., Zhu, R.Z., Zhao, S.W., 2015. Neoproterozoic quartz monzodiorite-granodiorite association from the Luding-Kangding area: Implications for the interpretation of an active continental margin along the Yangtze Block (South China Block). Precambrian Research, 267(3-4), 196-208.

Lee, H.Y., Chung, S.L., Lo, C.H., Ji, J.Q., Lee, T.Y., Qian, Q., Zhang, Q., 2009. Eocene Neotethyan slab breakoff in southern Tibet inferred from the Linzizong volcanic record. Tectonophysics, 477, 20-35.

Li, Q.W., Zhao, J.H., 2018. The Neoproterozoic high-Mg dioritic dikes in south China formed by high pressures fractional crystallization of hydrous basaltic melts. Precambrian Research, 309, 198-211.

Li, X.H., Li, Z.X., Zhou, H.W., Liu, Y., Kinny, P.D., 2002. U-Pb zircon geochronology, geochemistry and Nd isotopic study of Neoproterozoic bimodal volcanic rocks in the Kangdian Rift of South China: Implications for the initial rifting of Rodinia. Precambrian Research, 113, 135-154.

Li, X.H., Li, Z.X., Zhou, H.W., Liu, Y., Liang, X.R., Li, W.X., 2003. SHRIMP U-Pb zircon age, geochemistry and Nd isotope of the Guandaoshan pluton in SW Sichuan: Petrogenesis and tectonic significance. Science in China: Series D, 46(1), 73-83.

Li, X.H., Li, Z.X., Sinclair, J.A., Li, W.X., Carter, G., 2006. Revisiting the "Yanbian Terrane":

Implications for Neoproterozoic tectonic evolution of the western Yangtze block, South China. Precambrian Research, 151(1-2), 14-30.

Li, Z.X., Zhang, L.H., Powell, C.M.A., 1995. South China in Rodinia: Part of the missing link between Aurtralia-East Antartica and Laurentia? Geology, 23, 407-410.

Li, Z.X., Li, X.H., Kinny, P.D., Wang, J., 1999. The breakup of Rodinia: Did it start with a mantle plume beneath South China? Earth & Planetary Science Letters, 173, 171-181.

Li, Z.X., Bogdanova, S.V., Collins, A.S., Davidson, A., De Waele, B., Ernst, R.E., Fitzsimons, I.C.W., Fuck, R.A., Gladkochub, D.P., Jacobs, J., Karlstrom, K.E., Lu, S., Natapov, L.M., Pease, V., Pisarevsky, S.A., Thrane, K., Vernikovsky, V., 2008. Assembly, configuration, and break-up history of Rodinia: A synthesis. Precambrian Research, 160, 179-210.

Li, X.H., Li, W.X., Li, Q.L., Wang, X.C., Liu, Y., Yang, Y.H., 2010. Petrogenesis and tectonic significance of the ~850 Ma Gangbian alkaline complex in South China: Evidence from in situ zircon U-Pb dating, Hf-O isotopes and whole-rock geochemistry. Lithos, 114,1-15.

Liu, H., Zhao, J.H., 2019. Slab breakoff beneath the northern Yangtze Block: Implications from the Neoproterozoic Dahongshan mafic intrusions. Lithos, 342-343, 263-275.

Liu, Y., Liu, X.M., Hu, Z.C., Diwu, C.R., Yuan, H.L., Gao, S., 2007. Evaluation of accuracy and long-term stability of determination of 37 trace elements in geological samples by ICP-MS. Acta Petrological Sinica, 23(5), 1203-1210 (in Chinese with English abstract).

Litvak, V.D., Poma, S., 2010. Geochemistry of mafic Paleocene volcanic rocks in the Valle del Cura region: Implications for the petrogenesis of primary mantle-derived melts over the Pampean flat-slab. Journal of South American Earth Sciences, 29, 705-716.

Lu, G.M., Wang, W., Ernst, R.E., Söderlund, U., Lan, Z.F., Huang, S.F., Xue, E.K., 2019. Petrogenesis of Paleo-Mesoproterozoic mafic rocks in the southwestern Yangtze Block of South China: Implications for tectonic evolution and paleogeographic reconstruction. Precambrian Research, 322, 66-84.

Ludwig, K.R., 2003. ISOPLOT 3.0: A geochronological toolkit for Microsoft Excel. Berkeley Geochronology Center, Special Publication, 4.

Middlemost, E.A.K., 1994. Naming materials in the magma/igneous rock system. Earth-Science Reviews, 37, 215-224.

Meng, E., Liu, F.L., Du, L.L., Liu, P.H., Liu, J.H., 2015. Petrogenesis and tectonic significance of the Baoxing granitic and mafic intrusions, southwestern china: Evidence from zircon U-Pb dating and Lu-Hf isotopes, and whole-rock geochemistry. Gondwana Research, 28(2), 800-815.

Munteanu, M., Wilson, A., Yao, Y., Harris, C., Chunnett, G., Luo, Y., 2010.The Tongde dioritic pluton (Sichuan, SW China) and its geotectonic setting: Regional implications of a local-scale study. Gondwana Research, 18(2), 455-465.

Niu, Y.L., Wilson, M., Humphreys, E.R., O'Hara, M.J., 2011. The origin of intra-plate ocean island basalts (OIB): The lid effect and its geodynamic implications. Journal of Petrology, 52, 1443-1468.

Nowell, G.M., Kempton, P.D., Noble, S.R., Fitton, J.G., Saunders, A.D., Mahoney, J.J., Taylor, R.N., 1998. High precision Hf isotope measurements of MORB and OIB by thermal ionisation mass spectrometry: Insights into the depleted mantle. Chemical Geology, 149, 211-233.

Pearce, J.A., 2008. Geochemical fingerprinting of oceanic basalts with applications to ophiolite classification and the search for Archean oceanic crust. Lithos, 100(1-4), 14-48.

Pearce, J.A., Cann, J.R., 1973. Tectonic setting of basic volcanic rocks determined using trace element analyses. Earth & Planetary Science Letters, 19, 290-300.

Pilet, S., Baker, M.B., Stolper, E.M., 2008. Metasomatized lithosphere and the origin of alkaline lavas. Science, 320, 916-919.

Pirajno, F., 2007. Mantle plumes, associated intraplate tectonomagmatic processes and ore systems. Episodes, 30, 6-19.

Polat, A., Hofmann, A.W., 2003. Alteration and geochemical patterns in the 3.7-3.8 Ga Isua greenstone belt, West Greenland. Precambrian Research, 126 (3), 197-218.

Prouteau, G., Scaillet, B., Pichavant, M., Maury, R.C., 2001. Evidence for mantle metasomatism by hydrous silicic melts derived from subducted oceanic crust. Nature, 410, 197-200.

Rapp, R.P., Shimizu, N., Norman, M.D., Applegate, G.S., 1999. Reaction between slab-derived melts and peridotite in the mantle wedge: Experimental constraints at 3.8 GPa. Chemical Geology, 160, 335-356.

Roberts, M.P., Clemens, J.D., 1993. Origin of high-potassium, calc-alkaline, I-type granitoids. Geology, 21, 825-828.

Rudnick, R.L., Gao, S., 2003. The composition of the continental crust. In: Rudnick, R.L. The Crust. Oxford: Elsevier-Pergamon, 1-64.

Schmidt, M.W., Poli, S., 2003. Generation of mobile components during subduction of oceanic crust: Treatise on. Geochemistry, 3, 567-591.

Shellnutt, G.J., 2014. The Emeishan large igneous province: A synthesis. Geoscience Frontiers, 5 (3), 369-394.

Soderlund, U., Patchett, P.J., Vervoort, J.D., Isachsen, C.E., 2004. The ^{176}Lu decay constant determined by Lu-Hf and U-Pb isotope systematics of Precambrian mafic intrusions. Earth & Planetary Science Letters, 219 (3-4), 311-324.

Sun, S.S., McDonough, W.F., 1989. Chemical and isotopic systematics of oceanic basalts: Implications for mantle composition and processes. In: Saunders, A.D., Norry, M.J. Magmatism in the Ocean Basins. Geological Society, London, Special Publications, 42, 313-345.

Sun, W.H., Zhou, M.F., Zhao, J.H., 2007. Geochemistry and Tectonic Significance of Basaltic Lavas in the Neoproterozoic Yanbian Group, Southern Sichuan Province, Southwest China. International Geology Review, 49 (6), 554-571.

Sun, W.H., Zhou, M.F., 2008. The ~860 Ma, cordilleran-type Guandaoshan dioritic pluton in the Yangtze Block, SW China: Implications for the origin of Neoproterozoic magmatism. Journal of Geology, 116(3), 238-253.

Sun, W.H., Zhou, M.F., Yan, D.P., Li, J.W., Ma, Y.X., 2008. Provenance and tectonic setting of the Neoproterozoic Yanbian group, western Yangtze Block (SW china). Precambrian Research, 167(1), 213-236.

Tatsumi, Y., 1989. Migration of fluid phases and genesis of basalt magmas in subduction zones. Journal of Geophysical Research. Solid Earth, 94, 4697-4707.

Wang, W., Zhou, M.F., 2014. Provenance and tectonic setting of the Paleo- to Mesoproterozoic Dongchuan Group in the southwestern Yangtze Block, South China: Implication for the breakup of the supercontinent Columbia. Tectonophysics, 610, 110-127.

Wang, W., Zhou, M.F., Zhao, X.F., Chen, W.T., Yan, D.P., 2014b. Late Paleoproterozoic to Mesoproterozoic rift successions in SW China: Implication for the Yangtze Block-North Australia-Northwest Laurentia connection in the Columbia supercontinent. Sedimentary Geology, 309, 33-47.

Wang, X.C., Li, X.H., Li, W.X., Li, Z.X., Tu, X.L., 2008. The Bikou basalts in the northwestern Yangtze Block, South China: Remnants of 820−810 Ma continental flood basalts? Geological Society of America Bulletin, 120(11).

Wang, X.C., Li, X.H., Li, W.X., Li, Z.X., 2007. Ca. 825 Ma komatiitic basalts in South China: First evidence for >1 500℃ mantle melts by a Rodinian mantle plume. Geology, 35, 1103-1106.

Wang, X.L., Zhou, J.C., Wan, Y.S., Kitajima, K., Wang, D., Bonamici, C., Qiu, J.S., Sun, T., 2013. Magmatic evolution and crustal recycling for Neoproterozoic strongly peraluminous granitoids from southern China: Hf and O isotopes in zircon. Earth & Planetary Science Letters, 366 (2), 71-82.

Wang, X.L., Zhou, J.C., Griffin, W.L., Zhao, G.C., Yu, J.H., Qiu, J.S., Zhang, Y.J., Xing, G.F., 2014a. Geochemical zonation across a Neoproterozoic orogenic belt: Isotopic evidence from granitoids and metasedimentary rocks of the Jiangnan orogen, China. Precambrian Research, 242 (2), 154-171.

Wang, Y.J., Zhou, Y.Z., Cai, Y.F., Liu, H.C., Zhang, Y.Z., Fan, W.M., 2016. Geochronological and geochemical constraints on the petrogenesis of the Ailaoshan granitic and migmatite rocks and its implications on Neoproterozoic subduction along the SW Yangtze Block. Precambrian Research, 283, 106-124.

Wiedenbeck, M., Hanchar, J.M., Peck, W.H., Sylvester, P., Valley, J., Whitehouse, M., Kronz, A., Morishita, Y., et al., 2004. Further characterisation of the 91500 zircon crystal. Geostandards & Geoanalytical Research, 28 (1), 9-39.

Wilson, M., 1989. Igneous petrogenesis. London: Unwin Hyman, 191-373.

Wolf, M.B., Wyllie, P.J., 1994. Dehydration-melting of amphibolite at 10kbar: The effects of temperature and time. Contributions to Mineralogy and Petrology, 115, 369-383.

Woodhead, J.D., Hergt, J.M., Davidson, J.P., Eggins, S.M., 2001. Hafnium isotope evidence for "conservative" element mobility during subduction zone processes. Earth & Planetary Science Letters, 192, 331-346.

Wu, T., Wang, X.C., Li, W.X., Wilde, S., Tian, L.Y., 2019. Petrogenesis of the ca. 820−810 Ma felsic volcanic rocks in the Bikou Group: Implications for the tectonic setting of the western margin of the Yangtze Block. Precambrian Research, 331, 105370.

Xu, Y., Yang, K.G., Polat, A., Yang, Z.N., 2016. The ~860 Ma mafic dikes and granitoids from the northern margin of the Yangtze Block, China: A record of oceanic subduction in the early Neoproterozoic. Precambrian Research, 275, 310-331.

Xu, Y.G., Zhong, Y.T., Wei, X., Chen, J., Liu, H.., Xie, W., Luo, Z.Y., Li, H.Y., He, B., Huang, X.L., Wang, Y., Chen, Y., 2017. Permian mantle plumes and Earth's surface system evolution. Bulletin of Mineralogy, Petrology and Geochemistry, 36 (3), 351-373 (in Chinese with English abstract).

Yang, Y.J., Zhu, W.G., Bai, Z., Zhong, H., Ye, X.T., Fan, H.P., 2016. Petrogenesis and tectonic implications of the Neoproterozoic Datian mafic-ultramafic dykes in the Panzhihua area, western Yangtze Block, SW China. International Journal of Earth Science, 106(1), 1-29.

Yuan, H.L., Gao, S., Liu, X.M., Li, H.M., Gunther, D., Wu, F.Y., 2004. Accurate U-Pb age and trace element determinations of zircon by laser ablation-inductively coupled plasma mass spectrometry. Geostandards & Geoanalytical Research, 28 (3), 353-370.

Yuan, H.L., Gao, S., Dai, M.N., Zong, C.L., Gunther, D., Fontaine, G.H., Liu, X.M., Diwu, C.R., 2008. Simultaneous determinations of U-Pb age, Hf isotopes and trace element compositions of zircon by excimer laser-ablation quadrupole and multiple-collector ICPMS. Chemical Geology, 247(1-2), 100-118.

Zhang, S.B., Zheng, Y.F., Wu, Y.B., Zhao, Z.F., Gao, S., Wu, F.Y., 2006. Zircon isotope evidence for >3.5 Ga continental crust in the Yangtze craton of China. Precambrian Research, 146, 16-34.

Zhang, Y.Z., Wang, Y.J., 2016. Early Neoproterozoic (~840 Ma) arc magmatism: Geochronological and geochemical constraints on the metabasites in the Central Jiangnan Orogen. Precambrian Research, 275, 1-17.

Zhao, G.C., Cawood, P.A., 2012. Precambrian geology of China. Precambrian Research, 222-223, 13-54.

Zhao, J.H., Zhou, M.F., 2007. Geochemistry of Neoproterozoic mafic intrusions in the Panzhihua district (Sichuan Province, SW China): Implications for subduction-related metasomatism in the upper mantle. Precambrian Research, 152(1), 27-47.

Zhao, J.H., Zhou, M.F., 2009. Secular evolution of the Neoproterozoic lithospheric mantle underneath the northern margin of the Yangtze Block, South China. Lithos, 107(3-4), 152-168.

Zhao, J.H., Zhou, M.F., Yan, D.P., Zheng, J.P., Li, J.W., 2011. Reappraisal of the ages of Neoproterozoic strata in South China: No connection with the Grenvillian orogeny. Geology, 39 (4), 299-302.

Zhao, J.H., Asimow, P.D., Zhou, M.F., Zhang, J., Yan, D.P., Zheng, J.P., 2017. An Andean-type arc system in Rodinia constrained by the Neoproterozoic Shimian ophiolite in South China. Precambrian Research, 296, 93-111.

Zhao, J.H., Li, Q.W., Liu, H., Wang, W., 2018. Neoproterozoic magmatism in the western and northern margins of the Yangtze Block (South China) controlled by slab subduction and subduction-transform-edge-propagator. Earth-Science Reviews, 187, 1-18.

Zhao, J.H., Asimow, P.D., 2018. Formation and evolution of a magmatic system in a rifting continental margin: The Neoproterozoic arc- and MORB-like dike swarms in South China. Journal of Petrology, 59(9), 1811-1844.

Zhao, J.H., Zhou, M.F., Wu, Y.B., Zheng, J.P., Wang, W., 2019. Coupled evolution of Neoproterozoic arc mafic magmatism and mantle wedge in the western margin of the South China Craton. Contributions to Mineralogy and Petrology, 174, 36.

Zhao, X.F., Zhou, M.F., Li, J.W., Sun, M., Gao, J.F., Sun, W.H., Yang, J.H., 2010. Late Paleoproterozoic to early Mesoproterozoic Dongchuan Group in Yunnan, SW China: Implications for tectonic evolution of the Yangtze Block. Precambrian Research, 182, 57-69.

Zheng, F., Dai, L.Q., Zhao, Z.F., Zheng, Y.F., Xu, Z., 2019. Recycling of Paleo-oceanic crust:

Geochemical evidence from Early Paleozoic mafic igneous rocks in the Tongbai orogen, Central China. Lithos, 328-329, 312-327.

Zheng, Y.F., Zhang, S.B., Zhao, Z.F., Wu, Y.B., Li, X.H., Li, Z.X., Wu, F.Y., 2007. Contrasting zircon Hf and O isotopes in the two episodes of Neoproterozoic granitoids in South China: Implications for growth and reworking of continental crust. Lithos, 96, 127-150.

Zheng, Y.F., Wu, R.X., Wu, Y.B., Zhang, S.B., Yuan, H.L., and Wu, F.Y., 2008. Rift melting of juvenile arc-derived crust: Geochemical evidence from Neoproterozoic volcanic and granitic rocks in the Jiangnan orogen, South China. Precambrian Research, 163(3-4), 351-383.

Zheng, Y.F., Xu, Z., Chen, L., Dai, L.Q., Zhao, Z.F., 2020. Chemical geodynamics of mafic magmatism above subduction zones. Journal of Asian Earth Sciences, 194, 104185.

Zhou, M.F., Yan, D., Kennedy, A.K., Li, Y.Q., Ding, J., 2002. SHRIMP U-Pb zircon geochronological and geochemical evidence for Neoproterozoic arc-magmatism along the western margin of the Yangtze Block, South China. Earth & Planetary Science Letters, 196(1-2), 51-67.

Zhou, M.F., Ma, Y.X., Yan, D.P., Xia, X.P., Zhao, J.H., Sun, M., 2006a. The Yanbian Terrane (Southern Sichuan Province, SW China): A Neoproterozoic arc assemblage in the western margin of the Yangtze Block. Precambrian Research, 144(1-2), 19-38.

Zhou, M.F., Yan, D.P., Wang, C.L., Qi, L., Kennedy, A., 2006b. Subduction-related origin of the 750 Ma Xuelongbao adakitic complex (Sichuan Province, China): Implications for the tectonic setting of the giant Neoproterozoic magmatic event in South China. Earth & Planetary Science Letters, 248(1-2), 286-300.

Zhu, W.G., Bai, Z.J., Zhong, H., Ye, X.T., Fan, H.P., 2016. The origin of the ca. 1.7 Ga gabbroic intrusion in the Hekou area, SW China: Constraints from SIMS U-Pb zircon geochronology and elemental and Nd isotopic geochemistry. Geological Magazine, 154, 286-304.

Zhu, Y., Lai, S.C., Qin, J.F., Zhu, R.Z., Zhang, F.Y., Zhang, Z.Z., Gan, B.P., 2019a. Petrogenesis and geodynamic implications of Neoproterozoic gabbro-diorites, adakitic granites, and A-type granites in the southwestern margin of the Yangtze Block, South China. Journal of Asian Earth Sciences, 183, 103977.

Zhu, Y., Lai, S.C., Qin, J.F., Zhu, R.Z., Zhang, F.Y., Zhang, Z.Z., 2019b. Geochemistry and zircon U-Pb-Hf isotopes of the 780 Ma I-type granites in the western Yangtze Block: Petrogenesis and crustal evolution. International Geology Review, 61 (10), 1222-1243.

Zhu, Y., Lai, S.C., Qin, J.F., Zhu, R.Z., Zhang, F.Y., Zhang, Z.Z., Zhao, S.W., 2019c. Neoproterozoic peraluminous granites in the western margin of the Yangtze Block, South China: Implications for the reworking of mature continental crust. Precambrian Research, 333, 105443.

Zhu, Y., Lai, S.C., Qin, J.F., Zhu, R.Z., Liu, M., Zhang, F.Y., Zhang, Z.Z., Yang, H., 2020a. Genesis of ca. 850−835 Ma high-Mg$^{\#}$ diorites in the western Yangtze Block, South China: Implications for mantle metasomatism under the subduction process. Precambrian Research, 343, 105738.

Zhu, Y., Lai, S.C., Qin, J.F., Zhu, R.Z., Zhang, F.Y., Zhang, Z.Z., 2020b. Petrogenesis and geochemical diversity of late Mesoproterozoic S-type granites in the western Yangtze Block, South China: Co-entrainment of peritectic selective phases and accessory minerals. Lithos, 352-353, 105326.

Zhu, Z.M., Tan, H.Q., Liu, Y.D., 2017. Late Paleoproterozoic Hekou Group in Sichuan, Southwest China: Geochronological framework and tectonic implications. International Geology Review, 60, 305-318.

Peritectic assemblage entrainment (PAE) model for the petrogenesis of Neoproterozoic high-maficity I-type granitoids in the western Yangtze Block, South China[①]

Zhu Yu Lai Shaocong[②] Qin Jiangfeng Zhu Renzhi Zhao Shaowei
Liu Min Zhang Fangyi Zhang Zezhong Yang Hang

Abstract: Under the condition of paucity of mantle-derived ferromagnesian components, peritectic assemblage entrainment (PAE) model can properly explain a large variation of Fe, Mg, Ti, and Ca contents as well as positive linear trends for Ti and Ca plotted against maficity (molar Fe + Mg) exhibited by more mafic I-type granitic melts. Herein we present a comprehensive study of zircon U-Pb-Hf isotopes, whole-rock major and trace elements, and Sr-Nd isotopes for newly identified Neoproterozoic high-maficity I-type granitoids collected from the Yonglang area in the western Yangtze Block, South China, in order to evaluate PAE process acting on the petrogenesis of these more mafic granitoids. Zircon U-Pb geochronological results show that the Yonglang granitoids, including the granodiorites, monzogranites, and biotite granites, yield the weighted mean $^{206}Pb/^{238}U$ ages of ca. 850−825 Ma. They are geochemically characterized by high SiO_2 (64.63−72.62 wt%) and K_2O (3.49−6.59 wt%) contents as well as variable A/CNK (0.94−1.09) and molar Fe + Mg values (0.045 2−0.121 0), and display a negative correlation of A/CNK and Fe + Mg, indicating an I-type affinity. Isotopically, they display negative whole-rock $\varepsilon_{Nd}(t)$ (−6.38 to −4.06) and highly evolved zircon $\varepsilon_{Hf}(t)$ values (−9.36 to +1.92) as well as old two-stage Nd model ages (1.69−1.85 Ga) and crustal Hf model ages (1 608−2 139 Ma), implying a crustal source. In comparison with the geochemical composition of fluid-absent experimental melts, the Yonglang I-type granitoids have Fe, Mg, Ti, and Ca contents that are too high to equilibrium with those of pure crustal melts, indicating the additional incorporation of a ferromagnesian, Ti-, and Ca-rich component into primary melts. Like the Cotoncello-Monte Capanne granitic complex in Elba Island and Verdenburg I-type granite in South Africa, the Yonglang I-type granitoids exhibit highly positive linear correlations for Ti vs. maficity and Ca vs. maficity as well as negative correlation for A/CNK vs. maficity, suggesting that these high-maficity I-type granitoids predominantly underwent the peritectic assemblage entrainment of Ca-Fe-Mg-Ti bearing minerals (e.g., clinopyroxene, plagioclase, ilmenite, etc.) during coupled melting of biotite and hornblende phases in crustal

---

① Published in *Lithos*, 2021, 402-403.

② Corresponding author.

source. The process of peritectic assemblage entrainment thus play a vital role in moulding geochemical compositions of more mafic I-type granites.

1 Introduction

As the dominant constituent of Earth's upper continental crust, granite (sensu lato) bears on significant information about the crustal evolution from its formation to destruction under the subduction and post-collisional settings (Hawkesworth and Kemp, 2006; Kemp et al., 2007; Moyen et al., 2017). Therefore, unraveling the source characteristics and compositional variability of diverse granitoids can provide indispensable insights into the crustal growth, reworking, and differentiation processes (Bailie et al., 2020; Clemens et al., 1990, 2011; Clemens and Stevens, 2012; Farina et al., 2012; Gao et al., 2016). Traditionally, the geochemical diversity exhibited by both I- and S-type granites were ascribed to a variety of geological processes, which modify the chemical composition of anatectic melt during its formation, segregation, ascent, and emplacement (Clemens and Stevens, 2012; Garcia-Arias and Stevens, 2017 and references therein). These geological processes include the fractional crystallization of mantle-derived magmas coupled with subsequent assimilation of shallow crustal wall rocks (Clemens and Stevens, 2012; Patiño Douce, 1999), magma mixing between different batches of crust- and mantle-derived melts or juvenile crust- and mature crust-derived melts (Appleby et al., 2010; Kemp et al., 2007; Perugini et al., 2008; Wang et al., 2013), disequilibrium melting process (Tang et al., 2014; Wang et al., 2018), partial melting of heterogeneous protoliths (e.g., metapelites, metagraywackes, meta-basaltic rocks, basaltic andesites, etc.) as well as different partial melting conditions [i.e., pressure-temperature-hydration (P-T-H_2O) conditions] (Clemens, 2003, 2018; Clemens and Stevens, 2012; Gao et al., 2016; Sisson et al., 2005; Zhu et al., 2019a). More recently, the peritectic assemblage entrainment (PAE) model, an unique magmatic mechanism, has been gradually devised to explain the special geochemical characteristics in more mafic granitic magmas, i.e., the high concentrations of certain elements (e.g., Fe, Mg, Ti, and Ca) that should be low solubilities in granitic melts and highly tight correlations of maficity (molar Fe + Mg) vs. Ti, Ca, Zr, etc. (Bailie et al., 2020; Clemens et al., 2011; Clemens and Stevens, 2012, 2016; Farina et al., 2012; Garcia-Arias and Stevens, 2017; Hu et al., 2019; Rong et al., 2017; Stevens et al., 2007; Villaros et al., 2009a,b, 2012, 2018; Zhu et al., 2020a). Due to apparent complexity of geochemical variation found in both S- and I-type granites, it is therefore necessary to conduct further researches on geochemical diversification of various granitic melts.

Fluid-absent partial melting of various protolith materials, such as metapelites, metagreywackes, meta-andesites, and meta-dacites, usually generate granitic melts that have low Fe, Mg, Ti, and Ca contents (Clemens and Watkins, 2001; Patiño Douce and Harris, 1998; Vielzeuf and Montel, 1994). However, various studies have revealed that some more mafic granites do not dissolve sufficient ferromagnesian contents to equilibrium with those of fluid-absent experimental melts, even at temperature around 1 000 ℃ (Clemens et al., 2011; Farina et al., 2012; Garcia-Arias and Stevens, 2017; Stevens et al., 2007). Therefore, the primary melts of these more mafic granites demand the addition of a ferromagnesian, Ti-, and Ca-rich components (Farina et al., 2012; Stevens et al., 2007). The PAE model generally explains that the well-defined quasi-linear trends of maficity with Ti and Ca are attributed to the peritectic entrainment of selective mineral assemblages (e.g., garnet, cordierite, orthopyroxene, clinopyroxene, plagioclase, ilmenite, zircon, etc.) during incongruent partial melting of heterogeneous sources (Bailie et al., 2020; Clemens et al., 2011; Farina et al., 2012; Garcia-Arias and Stevens, 2017; Stevens et al., 2007; Zhu et al., 2020a), thus can logically make up low Fe, Mg, and Ca concentrations in more mafic granitic melts (Stevens et al., 2007). Most of PAE processes have been seriously tested into the compositional variation observed in typical S-type granites around the world, such as the Mid-Ediacaran to Early Cambrian S-type granites from Cape Granite Suite (CGS) in South Africa (Clemens et al., 2011; Clemens and Stevens, 2012; Stevens et al., 2007; Villaros et al., 2009a,b, 2012), the Paleozoic fractionated S-type granites from North Queensland in Australia (Champion and Bultitude, 2013), and the Late Mesoproterozoic S-type granites in the western Yangtze Block, South China (Zhu et al., 2020a). There is, however, limited studies of PAE processes evaluating on the petrogenesis of I-type granites until recently (e.g., Bailie et al., 2020; Farina et al., 2012; Hu et al., 2019). By this, further studies on the PAE model of I-type granitic melts were necessary.

The South China Block, composed of the Yangtze and Cathaysia blocks, is one of the significant Precambrian blocks in China, which is characterized by widespread Neoproterozoic magmatism that are associated with the assembly and breakup of supercontinent Rodinia (Wang et al., 2013; Zhao et al., 2018; Zhao and Cawood, 2012 and references therein). In the western margin of the Yangtze Block, voluminous Neoproterozoic granitic intrusions and subordinate mafic-ultramafic intrusions have been studied to understand the processes of crustal growth and reworking as well as the nature of the mantle source (Zhao et al., 2018 and references therein, 2019; Zhou et al., 2002, 2006a,b; Zhu et al., 2019a-c, 2020b, 2021). Among them, various kinds of Neoproterozoic granitoids have been documented and dated at ca. 850−750 Ma, including the adakitic granitoids (Huang et al., 2009; Zhao and Zhou, 2007; Zhou et al., 2006b; Zhu et al., 2019b), peraluminous granites (Zhu et al., 2019c), A-type granites (Huang et al., 2008; Zhao et al., 2008; Zhu et al., 2019b), and I-type granitoids (Zhao et al., 2008; Zhu et al., 2019a). Therefore, these abundant granites are ideal media to test the petrogenesis and geochemical diversity of different kinds of granitic melts. In this contribution, we present a case study on newly identified Neoproterozoic high-maficity I-type granites in the Yonglang area from the western Yangtze Block to test PAE model acting on the genesis of these more mafic granites.

2 Geological background and sample description

South China Block is one of the largest craton blocks in eastern Asia, which is subdivided into the Yangtze Block to the northwest and the Cathaysia Block to the southeast by the NNE-trending, 1 500-km-long Neoproterozoic Jiangnan orogenic belt (Wang et al., 2013, 2014; Zhao et al., 2011; Zhao and Cawood, 2012) (Fig. 1a,b). The Yangtze Block is separated from the North China Craton by the Qinling-Dabie orogenic belt to the north. The Ailaoshan-Songma suture zone is located between the Indochina Block and Yangtze Block in the southwest, and the Longmenshan fault is inlaid in the northwestern margin of the Yangtze Block (Gao et al., 1999; Zhao and Cawood, 2012). The oldest rocks unit in the Yangtze Block is the Kongling Complex, which consists of the Archean to Paleoproterozoic high-grade metamorphic TTG (tonalite, trondhjemite, and granodiorite) gneisses, metasedimentary rocks, amphibolites, and mafic granulite (Gao et al., 1999, 2011). In the southwestern

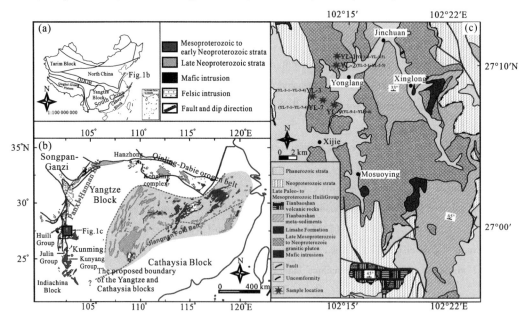

Fig. 1 Simplified geological map of the Yangtze Block, South China (a,b) (modified after Zhao and Cawood, 2012; Zhao et al., 2018) and regional geological map of the studied granites in the Yonglang area along the western Yangtze Block, South China (c) (modified after Miyi 1:200 000 geological map, SPBGMR, 1972; Zhu et al., 2016).

Yangtze Block, there are widespread Late Mesoproterozoic to Early Neoproterozoic volcanic-sedimentary sequences, including the Upper Kunyang, Upper Huili, Julin, and Yanbian groups (Chen et al., 2014; Zhao et al., 2018; Zhu et al., 2016, 2020a). Voluminous Late Mesoproterozoic A-type magmatic rocks and S-type granites were documented at ca. 1 020−1 040 Ma (Zhu et al., 2020a and references therein). In addition, numerous Neoproterozoic igneous complexes, such as the Kangding, Miyi, Tongde, Datian, and Yuanmou complexes as well as Dalu and Shuilu granitoid plutons, were well preserved and dated at ca. 860−750 Ma (Zhao et al., 2018 and references therein; Zhou et al., 2002, 2006a; Zhu et al., 2019a-c, 2020b, 2021).

In this study, our research region is located at the middle segment of the western Yangtze Block, South China (Fig. 1b,c). Voluminous Late Mesoproterozoic to Neoproterozoic granitoids, including the Jinchuan, Xinglong, and Mosuoying granitic plutons, co-exist with the widespread Neoproterozoic strata (SPBGMR, 1972) (Fig. 1c). The Neoproterozoic granitoids studied here were collected from different locations near the Yonglang town in the western Yangtze Block, South China. These granitoids are widely distributed ca. 20 km^2 and intrude into the Neoproterozoic strata and Tianbaoshan meta-sediments (SPBGMR, 1972) (Fig. 1c). The Yonglang granitic intrusions occurred as N-S trend along the Yonglang town, paralleling to the Anninghe fault in the western Yangtze Block. The Yonglang granitoids are grey to white in color, and predominantly composed of medium- to coarse-grained granodiorites, monzogranites, and biotite granites (Fig. 2a-e). The detailed boundary of different kinds of granites are difficult to identify in the field due to vegetation cover. The plagioclase phenocryst, up to 3 cm in length, can be discovered in the granodiorites (Fig. 2b). In comparison with monzogranites, biotite granites are more coarse-grained and contain more biotite and K-feldspar minerals (Fig. 2c-e). These granitoids mainly consist of K-feldspar (15%−25%), perthite (5%−10%), plagioclase (15%−25%), quartz (20%−30%), biotite (5%−10%), hornblende (5%−15%) and accessory minerals including Fe-Ti oxides and zircon (Fig. 2f-i). The plagioclases occur as euhedral granular crystals and show the polysynthetic twinning (Fig. 2h). The K-feldspars display distinctive feature of cross-hatched twinning (Fig. 2i). Some plagioclases and K-feldspars have been partially altered to sericite and kaolinite, respectively. The brown hornblendes are subhedral to euhedral crystal and the biotites display subhedral (Fig. 2f-g). Some hornblendes were altered into the biotites.

3　Analytical procedures

In this study, the in-situ zircon U-Pb-Hf isotopic analyses and whole-rock major and trace elemental analyses were carried out at the State Key Laboratory of Continental Dynamics, Northwest University, Xi'an, China. The whole-rock Sr-Nd isotopes were obtained at the Guizhou Tuopu Resource and Environment Analysis Center in China.

Fig. 2   Field and hand specimen petrographs (a-e) as well as representative microscope photographs (f-i) for the Neoproterozoic granitoids from the Yonglang area in the western Yangtze Block, South China.

Pl: plagioclase; Kfs: K-feldspar; Per: perthite; Qtz: quartz; Bt: biotite; Hbl: hornblende.

3. 1   Zircon U-Pb-Hf isotopes

We selected enough and representative zircon grains from five granitoid samples (YL-9, YL-7, YL-3, YL-2, and YL-1) taken from different locations in the vicinity of the Yonglang town along the western Yangtze Block. Zircon grains were separated using conventional heavy liquid and magnetic techniques. Representative grains were handpicked and mounted in epoxy resin disks, then polished and carbon coated. Internal morphology was studied with cathodoluminescence (CL) microscopy prior to zircon U-Pb-Hf isotopic analyses. Zircon Laser Ablation Inductively Coupled Plasma Mass Spectrometry (LA-ICP-MS) U-Pb analyses were conducted on Agilent 7500a ICP-MS equipped with a 193 nm laser, following the method of Yuan et al. (2004). The spot sizes of U-Pb analyses were 37 μm. The $^{207}Pb/^{235}U$ and $^{206}Pb/^{238}U$ ratios were computed using the GLITTER program (version 4. 0), which was corrected by Harvard zircon 91500 as an external calibration with a recommended $^{206}Pb/^{238}U$ isotopic age of 1 065. 4 ± 0. 6 Ma (Wiedenbeck et al., 2004). The GJ-1 is also a standard sample with a recommended $^{206}Pb/^{238}U$ age of 603. 2 ± 2. 4 Ma (Liu et al., 2007a). Common Pb contents were subsequently evaluated using the method described in Andersen (2002). The age calculations and plotting of concordia diagrams were made utilizing ISOPLOT (version

3. 0; Ludwig, 2003).

The in-situ zircon Lu-Hf isotopic analyses were processed using a Neptune MC-ICP-MS. The laser repetition rate was 10 Hz at 100 mJ and the spot sizes were 32 μm. The instrument information can be available from Bao et al. (2017). The detailed analytical technique is depicted by Yuan et al. (2008). The measured values of zircon standards (91500, GJ-1, and Monastery) were consistent with the recommended values within 2σ (Yuan et al., 2008). The obtained Hf isotopic compositions were 0. 282 016±20 (2σ, $n=84$) for the GJ-1 standard and 0. 282 735±24 (2σ, $n=84$) for the Monastery standard, respectively, consistent with the recommended values (Yuan et al., 2008) within 2σ. The initial ^{176}Hf/^{177}Hf ratios and $\varepsilon_{Hf}(t)$ values were calculated with the chondritic reservoir (CHUR) at the time of zircon growth from the magmas. The decay constant for ^{176}Lu of 1. 867 × 10^{-11} year^{-1}, chondritic ^{176}Hf/^{177}Hf ratio of 0. 282 785, and ^{176}Lu/^{177}Hf ratio of 0. 033 6 were adopted (Soderlund et al., 2004). The depleted mantle model ages (T_{DM}) for basic rocks were calculated with reference to present-day depleted mantle ^{176}Hf/^{177}Hf ratio of 0. 283 25, similar to that of the average MORB (Nowell et al., 1998) and ^{176}Lu/^{177}Hf = 0. 038 4 (Griffin et al., 2000). The Hf isotopic "crustal" model age (T_{DMC}) for felsic rocks were assumed that its parental magma was derived from an average continental crust with ^{176}Lu/^{177}Hf = 0. 015, which was originated from the depleted mantle source (Griffin et al., 2000). Our conclusions would be unaffected even if other decay constants were used.

3. 2   Whole-rock major and trace elements

Weathered surfaces of the Yonglang granitoid samples were removed and the fresh parts were then chipped and powdered to ~200 mesh size using a tungsten carbide ball mill. Major and trace elements were examined by X-ray fluorescence (XRF; Rikagu RIX 2100) and inductively coupled plasma mass spectrometry (ICP-MS; Agilent 7500a), respectively. Analyses of USGS and Chinese national rock standards (BCR-2, GSR-1, and GSR-3) showed that the analytical precision and accuracy for the major elements were generally better than 5%. For the trace element analyses, sample powders were digested using an HF + HNO$_3$ mixture in high-pressure Teflon bombs at 190 ℃ for 48 h. For most trace elements, the analytical error was less than 2% and the precision was greater than 10% (Liu et al., 2007b).

3. 3   Whole-rock Sr-Nd isotopes

Sr and Nd isotopic fractionation was corrected to ^{87}Sr/^{86}Sr = 0. 119 4 and ^{146}Nd/^{144}Nd = 0. 721 9, respectively (Chu et al., 2009). A Neptune multi-collector ICP-MS was used to measure the ^{87}Sr/^{86}Sr and ^{143}Nd/^{144}Nd isotope ratios. NIST SRM-987 and JMC-Nd were used as certified reference standard solutions for ^{87}Sr/^{86}Sr and ^{143}Nd/^{144}Nd isotopic ratios, respectively.

During the samples run, the La Jolla standard yielded an average value of $^{143}Nd/^{144}Nd =$ 0. 511 862 ± 5 (2σ), and the NIST SRM-987 standard yielded an average value of $^{87}Sr/^{86}Sr =$ 0. 710 236 ± 16 (2σ). BCR-1 and BHVO-1 were used as the reference materials. The average $^{87}Sr/^{86}Sr$ ratios of the BCR-1 and BHVO-1 standards are 0. 705 014 ± 3 (2σ) and 0. 703 477 ± 20 (2σ), respectively. The average $^{146}Nd/^{144}Nd$ ratios of the BCR-1 and BHVO-1 are 0. 512 615 ± 12 (2σ) and 0. 512 987 ± 23 (2σ).

4 Results

4. 1 Zircon U-Pb-Hf isotopic analyses

The typical zircon grains from five granitoid samples (YL-9, YL-7, YL-3, YL-2, and YL-1) are selected for U-Pb-Hf isotopic analyses. The zircon U-Pb isotopic data are presented in Table 1. The representative CL images and concordia diagrams are shown in Fig. 3. Most of zircon grains in the Yonglang granitoids are euhedral columnar crystals, and display well-developed oscillatory zoning, suggesting an igneous origin (Hoskin and Schaltegger, 2003). These zircon grains range in size from 50 μm to 250 μm with aspect ratios about 1:1 to 4:1, and have Th/U ratios varying from 0. 10 to 1. 09. The in-situ zircon Lu-Hf isotopes of five granitoid samples were performed on the same domains as the concordant U-Pb spots. The initial $^{176}Hf/^{177}Hf$ ratios, $\varepsilon_{Hf}(t)$ values, and crustal Hf model ages were calculated according to zircon crystallization ages. The analytic results are listed and shown in Table 2, and Figs. 4 and 5.

4. 1. 1 Sample YL-9

The magmatic zircons from sample YL-9 have Th/U ratios of 0. 11 − 1. 09. Twenty-one concordant analytic spots are characterized by $^{206}Pb/^{238}U$ ages ranging from 839 ± 9 Ma to 864 ± 11 Ma, with a weighted mean age of 852. 0 ± 3. 0 Ma (MSWD = 2. 0, $n = 21$) (Fig. 3a), which represent the crystallization age of the sample YL-9.

Fifteen Lu-Hf analytic spots were processed on the sample YL-9. The initial $^{176}Hf/^{177}Hf$ ratios span a range from 0. 282 072 to 0. 282 315 with an average of 0. 282 178. They show negative-dominantly $\varepsilon_{Hf}(t)$ values ranging from −6. 50 to +0. 93 (Figs. 4 and 5a) and ancient two-stage Hf model ages of 1 674 − 2 139 Ma ($n = 15$) (with a mean of 1 926 Ma) (Table 2).

4. 1. 2 Sample YL-7

Zircons from sample YL-7 contain medium Th (39. 9 − 233 ppm), U concentrations (104 − 617 ppm) and narrow range of Th/U ratios (0. 18 − 0. 62). All the thirty-three concordant analytic spots are characterized by $^{206}Pb/^{238}U$ ages ranging from 835 ± 9 Ma to 856 ± 9 Ma with a weighted mean age of 844. 9 ± 2. 6 Ma (MSWD = 2. 6, $n = 33$) (Fig. 3b).

Table 1 LA-ICP-MS zircon U-Pb dating results for the Neoproterozoic high-maficity I-type granitoids in the western Yangtze Block, South China.

Analysis	Content/ppm			Ratios								Age/Ma							
	Th	U	Th/U	$^{207}Pb/^{206}Pb$	1σ	$^{207}Pb/^{235}U$	1σ	$^{206}Pb/^{238}U$	1σ	$^{208}Pb/^{232}Th$	1σ	$^{207}Pb/^{206}Pb$	1σ	$^{207}Pb/^{235}U$	1σ	$^{206}Pb/^{238}U$	1σ	$^{208}Pb/^{232}Th$	1σ
YL-9																			
YL-9-01	80.38	118.16	0.68	0.069 95	0.002 11	1.383 36	0.034 40	0.143 46	0.001 87	0.042 55	0.000 81	927	61	882	15	864	11	842	16
YL-9-02	16 654.78	15 347.94	1.09	0.067 09	0.001 38	1.309 13	0.015 68	0.141 55	0.001 56	0.022 02	0.000 21	841	42	850	7	853	9	440	4
YL-9-04	76.83	185.15	0.41	0.068 51	0.002 16	1.323 32	0.034 80	0.140 11	0.001 86	0.042 87	0.000 79	884	64	856	15	845	11	848	15
YL-9-05	1 162.03	1 196.77	0.97	0.068 27	0.001 49	1.331 32	0.018 59	0.141 46	0.001 59	0.041 02	0.000 44	877	45	860	8	853	9	813	9
YL-9-06	365.84	1 162.84	0.31	0.067 64	0.001 40	1.312 31	0.015 89	0.140 72	0.001 55	0.043 57	0.000 46	858	42	851	7	849	9	862	9
YL-9-07	744.57	1 257.87	0.59	0.065 26	0.001 39	1.250 26	0.016 22	0.138 97	0.001 54	0.039 88	0.000 41	783	44	824	7	839	9	791	8
YL-9-12	213.31	642.80	0.33	0.075 30	0.001 60	1.460 32	0.018 73	0.140 65	0.001 55	0.056 37	0.000 60	1 077	42	914	8	848	9	1 108	12
YL-9-13	170.76	1 575.12	0.11	0.070 74	0.001 48	1.377 29	0.016 89	0.141 19	0.001 54	0.056 33	0.000 65	950	42	879	7	851	9	1 108	12
YL-9-14	254.7	1 919.18	0.13	0.070 03	0.001 45	1.356 39	0.016 27	0.140 47	0.001 53	0.056 18	0.000 61	929	42	870	7	847	9	1 105	12
YL-9-15	304.92	1 892.94	0.16	0.068 78	0.001 46	1.341 28	0.016 91	0.141 42	0.001 55	0.042 54	0.000 55	892	43	864	7	853	9	842	11
YL-9-16	156.43	597.23	0.26	0.068 83	0.001 72	1.330 36	0.024 18	0.140 16	0.001 64	0.047 02	0.000 72	894	51	859	11	846	9	929	14
YL-9-24	313.42	1 689.73	0.19	0.071 38	0.001 48	1.392 29	0.016 15	0.141 43	0.001 52	0.046 04	0.000 47	969	42	886	7	853	9	910	9
YL-9-25	147.49	207.72	0.71	0.075 37	0.002 03	1.490 26	0.030 52	0.143 38	0.001 73	0.043 32	0.000 71	1 078	53	926	12	864	10	857	14
YL-9-26	57.05	59.46	0.96	0.067 35	0.002 81	1.321 55	0.049 68	0.142 30	0.002 20	0.041 00	0.001 01	849	85	855	22	858	12	812	20
YL-9-27	809.51	1 236.86	0.65	0.067 52	0.001 45	1.309 38	0.016 75	0.140 63	0.001 52	0.040 56	0.000 40	854	44	850	7	848	9	804	8
YL-9-28	299.78	1 018.94	0.29	0.076 34	0.001 75	1.504 37	0.022 73	0.142 90	0.001 60	0.052 30	0.000 69	1 104	45	932	9	861	9	1 030	13
YL-9-29	109.74	305.16	0.36	0.066 38	0.001 50	1.296 96	0.018 91	0.141 69	0.001 56	0.042 55	0.000 52	818	47	844	8	854	9	842	10
YL-9-30	114.74	676.67	0.17	0.067 89	0.001 47	1.334 04	0.017 40	0.142 49	0.001 54	0.047 96	0.000 60	865	44	861	8	859	9	947	12
YL-9-31	230.01	324.79	0.71	0.070 04	0.001 57	1.358 41	0.019 32	0.140 64	0.001 54	0.044 44	0.000 50	930	45	871	8	848	9	879	10
YL-9-32	706.84	2 489.57	0.28	0.072 33	0.001 86	1.398 94	0.026 47	0.140 25	0.001 64	0.080 22	0.001 37	995	52	889	11	846	11	1 560	26
YL-9-33	42.30	226.08	0.19	0.075 58	0.001 85	1.488 64	0.025 50	0.142 82	0.001 63	0.059 34	0.001 16	1 084	48	926	10	861	10	1 165	22
YL-7																			
YL-7-01	107.1	511.08	0.21	0.071 79	0.001 65	1.369 17	0.020 17	0.138 33	0.001 52	0.055 6	0.000 77	980	46	876	9	835	9	1 094	15

Continued

Analysis	Content/ppm			Ratios								Age/Ma							
	Th	U	Th/U	$^{207}Pb/^{206}Pb$	1σ	$^{207}Pb/^{235}U$	1σ	$^{206}Pb/^{238}U$	1σ	$^{208}Pb/^{232}Th$	1σ	$^{207}Pb/^{206}Pb$	1σ	$^{207}Pb/^{235}U$	1σ	$^{206}Pb/^{238}U$	1σ	$^{208}Pb/^{232}Th$	1σ
YL-7-02	77.19	352.15	0.22	0.068 47	0.001 61	1.340 62	0.020 86	0.142 00	0.001 58	0.046 16	0.000 66	883	48	864	9	856	9	912	13
YL-7-03	95.56	354.34	0.27	0.071 25	0.001 62	1.390 33	0.020 11	0.141 52	0.001 56	0.051 74	0.000 68	965	46	885	9	853	9	1 020	13
YL-7-04	117.54	319.96	0.37	0.067 81	0.001 57	1.310 43	0.019 80	0.140 14	0.001 55	0.042 64	0.000 52	863	47	850	9	845	9	844	10
YL-7-05	63.81	207.02	0.31	0.069 99	0.001 70	1.339 32	0.022 36	0.138 77	0.001 57	0.045 09	0.000 64	928	49	863	10	838	9	891	12
YL-7-06	100.17	177.18	0.57	0.065 92	0.001 65	1.280 26	0.022 90	0.140 83	0.001 61	0.042 81	0.000 55	804	52	837	10	849	9	847	11
YL-7-07	87.81	197.86	0.44	0.065 30	0.001 56	1.253 88	0.020 47	0.139 22	0.001 57	0.039 66	0.000 52	784	49	825	9	840	9	786	10
YL-7-09	79.39	237.45	0.33	0.066 86	0.001 59	1.302 67	0.021 19	0.141 27	0.001 59	0.043 14	0.000 59	833	49	847	9	852	9	854	11
YL-7-10	45.77	111.88	0.41	0.066 64	0.001 79	1.302 33	0.026 44	0.141 70	0.001 69	0.044 04	0.000 69	826	55	847	12	854	10	871	13
YL-7-11	98.02	347.73	0.28	0.070 75	0.001 62	1.354 23	0.020 27	0.138 78	0.001 55	0.048 36	0.000 62	950	46	869	9	838	9	955	12
YL-7-12	70.78	204.26	0.35	0.067 46	0.001 65	1.292 02	0.022 22	0.138 86	0.001 59	0.042 42	0.000 61	852	50	842	10	838	9	840	12
YL-7-13	133.96	432.08	0.31	0.065 17	0.001 60	1.244 02	0.021 82	0.138 40	0.001 60	0.041 85	0.000 63	780	51	821	10	836	9	829	12
YL-7-15	232.62	616.68	0.38	0.067 33	0.001 45	1.316 51	0.017 38	0.141 76	0.001 56	0.046 51	0.000 50	848	44	853	8	855	9	919	10
YL-7-16	82.81	247.40	0.33	0.065 08	0.001 53	1.270 13	0.020 64	0.141 50	0.001 61	0.042 54	0.000 58	777	49	832	9	853	9	842	11
YL-7-17	80.29	435.99	0.18	0.065 73	0.001 50	1.259 30	0.019 07	0.138 90	0.001 57	0.049 66	0.000 78	798	47	828	9	838	9	980	15
YL-7-18	72.85	120.38	0.61	0.066 74	0.001 99	1.286 59	0.031 26	0.139 77	0.001 78	0.042 38	0.000 71	830	61	840	14	843	10	839	14
YL-7-19	138.75	442.41	0.31	0.065 20	0.001 44	1.254 15	0.017 72	0.139 46	0.001 56	0.043 73	0.000 53	781	46	825	8	842	9	865	10
YL-7-20	39.85	103.70	0.38	0.065 15	0.001 77	1.264 98	0.026 73	0.140 79	0.001 71	0.042 60	0.000 72	779	56	830	12	849	12	843	14
YL-7-21	62.45	141.97	0.44	0.063 72	0.001 61	1.236 26	0.023 24	0.140 69	0.001 66	0.042 68	0.000 62	732	53	817	11	849	11	845	12
YL-7-22	72.28	314.98	0.23	0.066 66	0.001 50	1.278 25	0.019 02	0.139 04	0.001 57	0.043 33	0.000 60	827	46	836	8	839	9	857	12
YL-7-23	94.87	303.95	0.31	0.070 51	0.001 59	1.351 08	0.020 26	0.138 94	0.001 58	0.049 34	0.000 63	943	45	868	9	839	9	974	12
YL-7-24	61.76	130.81	0.47	0.063 74	0.001 61	1.244 82	0.023 46	0.141 61	0.001 68	0.042 22	0.000 60	733	53	821	11	854	11	836	12
YL-7-25	108.76	199.83	0.54	0.066 96	0.001 61	1.282 50	0.022 14	0.138 92	0.001 63	0.043 75	0.000 59	837	49	838	10	839	10	865	11
YL-7-26	113.24	197.28	0.57	0.067 47	0.001 74	1.302 63	0.025 50	0.140 03	0.001 69	0.042 07	0.000 60	853	53	847	11	845	10	833	12

Continued

Analysis	Content/ppm			Ratios								Age/Ma							
	Th	U	Th/U	$\frac{207Pb}{206Pb}$	1σ	$\frac{207Pb}{235U}$	1σ	$\frac{206Pb}{238U}$	1σ	$\frac{208Pb}{232Th}$	1σ	$\frac{207Pb}{206Pb}$	1σ	$\frac{207Pb}{235U}$	1σ	$\frac{206Pb}{238U}$	1σ	$\frac{208Pb}{232Th}$	1σ
YL-7-28	131.88	344.21	0.38	0.064 32	0.001 52	1.229 06	0.020 58	0.138 62	0.001 62	0.066 01	0.000 83	752	49	814	9	837	9	1 292	16
YL-7-29	100.32	174.52	0.57	0.064 00	0.001 55	1.247 80	0.022 13	0.141 46	0.001 67	0.042 01	0.000 55	742	51	822	10	853	9	832	11
YL-7-30	82.98	192.65	0.43	0.063 67	0.001 53	1.238 03	0.021 57	0.141 08	0.001 66	0.044 20	0.000 60	731	50	818	10	851	9	874	12
YL-7-31	63.38	141.00	0.45	0.062 35	0.002 18	1.187 75	0.036 22	0.138 25	0.001 94	0.041 15	0.000 94	686	73	795	17	835	11	815	18
YL-7-32	54.36	126.15	0.43	0.063 63	0.001 88	1.241 60	0.030 29	0.141 62	0.001 83	0.042 37	0.000 80	729	61	820	14	854	10	839	16
YL-7-33	136.70	245.89	0.56	0.064 16	0.001 47	1.223 89	0.019 62	0.138 45	0.001 61	0.042 06	0.000 51	747	48	812	9	836	9	833	10
YL-7-34	86.85	186.43	0.47	0.063 17	0.001 51	1.224 23	0.021 28	0.140 67	0.001 66	0.042 14	0.000 58	714	50	812	10	848	9	834	11
YL-7-35	102.95	165.59	0.62	0.062 78	0.001 64	1.198 00	0.024 26	0.138 52	0.001 70	0.040 44	0.000 58	701	55	800	11	836	10	801	11
YL-7-36	117.81	364.68	0.32	0.064 58	0.001 42	1.262 95	0.018 62	0.141 97	0.001 64	0.043 46	0.000 56	761	46	829	8	856	9	860	11
YL-3																			
YL-3-01	114.79	692.64	0.17	0.063 86	0.001 42	1.241 76	0.017 86	0.141 08	0.001 58	0.044 03	0.000 83	737	46	820	8	851	9	871	16
YL-3-02	2 664.79	7 482.40	0.36	0.069 16	0.001 41	1.312 44	0.015 15	0.137 69	0.001 50	0.042 96	0.000 44	903	41	851	7	832	9	850	8
YL-3-03	3 912.22	3 617.21	1.08	0.064 69	0.001 33	1.245 66	0.014 82	0.139 69	0.001 52	0.044 27	0.000 44	764	43	821	7	843	9	876	8
YL-3-04	1 542.65	2 437.16	0.63	0.066 56	0.001 41	1.265 26	0.016 28	0.137 91	0.001 52	0.044 11	0.000 46	824	44	830	7	833	9	872	9
YL-3-06	1 409.36	1 937.15	0.73	0.064 69	0.001 37	1.239 34	0.015 80	0.138 99	0.001 53	0.041 39	0.000 43	764	44	819	7	839	9	820	8
YL-3-07	36.33	374.80	0.10	0.063 39	0.001 47	1.209 75	0.019 06	0.138 43	0.001 56	0.049 99	0.000 96	722	48	805	9	836	9	986	18
YL-3-08	122.93	262.49	0.47	0.065 45	0.001 58	1.256 56	0.021 39	0.139 26	0.001 60	0.044 02	0.000 59	789	50	826	10	841	9	871	11
YL-3-09	545.38	1 459.39	0.37	0.067 42	0.001 46	1.287 45	0.017 48	0.138 52	0.001 53	0.046 40	0.000 53	851	45	840	8	836	9	917	10
YL-3-10	144.96	428.86	0.34	0.064 36	0.001 46	1.246 68	0.018 77	0.140 51	0.001 57	0.042 61	0.000 57	753	47	822	8	848	9	843	11
YL-3-11	103.09	220.40	0.47	0.066 94	0.001 66	1.295 22	0.023 15	0.140 35	0.001 63	0.043 67	0.000 61	836	51	844	10	847	9	864	12
YL-3-12	161.09	622.31	0.26	0.070 74	0.001 63	1.359 95	0.021 01	0.139 43	0.001 57	0.051 98	0.000 63	950	46	872	9	842	9	1 024	12
YL-3-13	468.75	689.87	0.68	0.071 96	0.001 66	1.373 83	0.021 34	0.138 46	0.001 56	0.042 99	0.000 50	985	46	878	9	836	9	851	10
YL-3-14	323.36	501.17	0.65	0.065 25	0.001 72	1.244 06	0.024 72	0.138 27	0.001 63	0.041 05	0.000 61	783	54	821	11	835	9	813	12

Continued

Analysis	Content/ppm			Ratios								Age/Ma							
	Th	U	Th/U	$^{207}Pb/^{206}Pb$	1σ	$^{207}Pb/^{235}U$	1σ	$^{206}Pb/^{238}U$	1σ	$^{208}Pb/^{232}Th$	1σ	$^{207}Pb/^{206}Pb$	1σ	$^{207}Pb/^{235}U$	1σ	$^{206}Pb/^{238}U$	1σ	$^{208}Pb/^{232}Th$	1σ
YL-3-15	900.71	1 473.66	0.61	0.066 61	0.001 38	1.286 53	0.015 32	0.140 09	0.001 51	0.042 17	0.000 42	826	43	840	7	845	9	835	8
YL-3-16	1 036.35	1 497.45	0.69	0.065 14	0.001 37	1.265 22	0.015 54	0.140 87	0.001 52	0.040 75	0.000 40	779	44	830	7	850	9	807	8
YL-3-17	684.97	722.53	0.95	0.068 19	0.001 68	1.300 68	0.022 92	0.138 34	0.001 59	0.059 29	0.000 85	874	50	846	10	835	9	1 164	16
YL-3-18	176.36	392.23	0.45	0.071 65	0.001 98	1.375 34	0.029 60	0.139 21	0.001 70	0.057 29	0.001 22	976	55	879	13	840	10	1 126	23
YL-3-19	2 203.52	2 769.84	0.80	0.061 98	0.001 27	1.197 16	0.013 50	0.140 07	0.001 49	0.032 21	0.000 30	674	43	799	6	845	8	641	6
YL-3-20	241.49	599.26	0.40	0.072 41	0.001 72	1.401 18	0.023 02	0.140 34	0.001 59	0.051 86	0.000 71	997	48	889	10	847	9	1 022	14
YL-3-21	2 293.48	2 330.81	0.98	0.065 34	0.001 35	1.246 23	0.014 33	0.138 32	0.001 47	0.040 78	0.000 38	785	43	822	6	835	8	808	7
YL-3-22	1 349.38	1 884.66	0.72	0.068 00	0.001 66	1.290 32	0.022 34	0.137 61	0.001 57	0.042 34	0.000 52	869	50	841	10	831	9	838	10
YL-3-23	377.70	725.31	0.52	0.067 69	0.001 54	1.299 65	0.019 22	0.139 24	0.001 53	0.052 07	0.000 63	859	46	846	8	840	9	1 026	12
YL-3-25	674.10	2 344.32	0.29	0.066 07	0.001 38	1.257 50	0.014 96	0.138 01	0.001 46	0.045 97	0.000 48	809	43	827	7	833	8	908	9
YL-3-26	65.14	144.55	0.45	0.061 00	0.002 90	1.164 39	0.051 05	0.138 42	0.002 27	0.038 18	0.001 23	639	99	784	24	836	13	757	24
YL-3-27	825.33	1 514.07	0.55	0.066 51	0.001 40	1.270 69	0.015 37	0.138 54	0.001 47	0.044 39	0.000 44	822	43	833	7	836	8	878	8
YL-3-28	453.8	1 152.26	0.39	0.067 42	0.001 44	1.305 74	0.016 34	0.140 44	0.001 50	0.033 14	0.000 35	851	44	848	7	847	8	659	7
YL-3-29	159.25	404.34	0.39	0.066 25	0.001 58	1.259 06	0.020 58	0.137 80	0.001 53	0.040 81	0.000 54	814	49	828	9	832	9	808	11
YL-3-30	174.51	411.34	0.42	0.063 90	0.003 39	1.223 98	0.060 57	0.138 89	0.002 51	0.044 32	0.001 53	738	108	812	28	838	14	876	30
YL-3-31	145.40	344.32	0.42	0.066 28	0.001 54	1.265 67	0.019 52	0.138 47	0.001 52	0.041 79	0.000 51	815	48	831	9	836	9	828	10
YL-3-33	63.57	137.03	0.46	0.067 18	0.002 01	1.304 44	0.031 58	0.140 80	0.001 75	0.040 63	0.000 68	843	61	848	14	849	10	805	13
YL-3-34	1 572.36	1 837.93	0.86	0.066 57	0.001 60	1.262 37	0.020 83	0.137 50	0.001 53	0.041 44	0.000 46	824	49	829	9	831	9	821	9
YL-3-35	78.98	191.00	0.41	0.067 64	0.001 77	1.302 77	0.025 43	0.139 66	0.001 62	0.044 31	0.000 68	858	54	847	11	843	10	876	13
YL-3-36	86.64	193.84	0.45	0.066 92	0.001 65	1.300 83	0.022 74	0.140 96	0.001 58	0.040 82	0.000 56	835	51	846	10	850	10	809	11
YL-2																			
YL-2-01	147.53	311.40	0.47	0.069 12	0.001 64	1.334 85	0.020 39	0.140 03	0.001 51	0.040 65	0.000 47	902	48	861	9	845	9	805	9
YL-2-02	121.38	230.86	0.53	0.068 19	0.001 77	1.285 36	0.023 97	0.136 69	0.001 54	0.039 51	0.000 53	874	53	839	11	826	11	783	10

Continued

Analysis	Content/ppm			Ratios								Age/Ma							
	Th	U	Th/U	$\frac{^{207}Pb}{^{206}Pb}$	1σ	$\frac{^{207}Pb}{^{235}U}$	1σ	$\frac{^{206}Pb}{^{238}U}$	1σ	$\frac{^{208}Pb}{^{232}Th}$	1σ	$\frac{^{207}Pb}{^{206}Pb}$	1σ	$\frac{^{207}Pb}{^{235}U}$	1σ	$\frac{^{206}Pb}{^{238}U}$	1σ	$\frac{^{208}Pb}{^{232}Th}$	1σ
YL-2-04	82.88	197.54	0.42	0.076 20	0.001 86	1.459 60	0.024 00	0.138 91	0.001 53	0.046 77	0.000 60	1 100	48	914	10	839	9	924	12
YL-2-05	114.01	291.01	0.39	0.069 20	0.001 65	1.326 50	0.020 80	0.139 01	0.001 51	0.040 95	0.000 51	905	48	857	9	839	9	811	10
YL-2-06	703.04	1 346.17	0.52	0.068 23	0.001 64	1.289 09	0.020 82	0.137 02	0.001 49	0.040 14	0.000 51	876	49	841	9	828	8	796	10
YL-2-07	83.64	211.40	0.40	0.067 63	0.001 85	1.307 10	0.027 09	0.140 18	0.001 61	0.039 91	0.000 61	857	56	849	12	846	9	791	12
YL-2-08	163.06	472.33	0.35	0.066 72	0.001 55	1.254 68	0.018 70	0.136 39	0.001 46	0.037 97	0.000 46	829	48	826	8	824	8	753	9
YL-2-09	136.33	727.47	0.19	0.066 09	0.001 45	1.249 77	0.016 27	0.137 17	0.001 44	0.040 95	0.000 50	809	45	823	7	829	8	811	10
YL-2-10	286.77	984.94	0.29	0.067 04	0.001 43	1.271 70	0.015 36	0.137 61	0.001 43	0.039 94	0.000 40	839	44	833	7	831	8	792	8
YL-2-11	204.22	557.75	0.37	0.066 87	0.001 71	1.261 92	0.023 24	0.136 90	0.001 54	0.039 38	0.000 62	834	52	829	10	827	9	781	12
YL-2-12	284.61	615.63	0.46	0.066 02	0.001 51	1.275 01	0.018 80	0.140 09	0.001 50	0.035 59	0.000 42	807	47	835	8	845	8	707	8
YL-2-13	117.47	303.61	0.39	0.064 44	0.001 53	1.239 46	0.019 95	0.139 54	0.001 51	0.042 26	0.000 54	756	49	819	9	842	9	837	10
YL-2-14	436.46	2 444.54	0.18	0.065 53	0.001 47	1.233 32	0.017 45	0.136 54	0.001 45	0.036 59	0.000 47	791	46	816	8	825	8	726	9
YL-2-15	232.92	611.15	0.38	0.066 51	0.001 47	1.282 66	0.017 49	0.139 91	0.001 47	0.041 42	0.000 46	822	45	838	8	844	8	820	9
YL-2-16	110.90	339.96	0.33	0.065 00	0.001 45	1.247 93	0.017 51	0.139 30	0.001 47	0.037 28	0.000 45	774	46	823	8	841	8	740	9
YL-2-17	150.76	384.09	0.39	0.071 40	0.001 77	1.361 85	0.024 09	0.138 38	0.001 54	0.042 30	0.000 59	969	50	873	10	836	9	837	11
YL-2-18	82.88	200.62	0.41	0.065 10	0.001 60	1.236 40	0.021 48	0.137 79	0.001 52	0.040 07	0.000 54	778	51	817	10	832	9	794	11
YL-2-19	138.71	275.22	0.50	0.065 27	0.001 50	1.232 42	0.018 83	0.136 99	0.001 47	0.039 04	0.000 45	783	48	815	9	828	8	774	9
YL-2-20	128.02	401.36	0.32	0.064 70	0.001 45	1.225 00	0.017 59	0.137 35	0.001 46	0.040 14	0.000 49	765	46	812	8	830	8	796	9
YL-2-21	111.71	214.15	0.52	0.066 85	0.001 60	1.258 84	0.021 10	0.136 62	0.001 50	0.038 29	0.000 48	833	49	827	9	826	9	760	9
YL-2-22	125.59	351.55	0.36	0.064 77	0.001 52	1.223 66	0.019 66	0.137 06	0.001 48	0.038 88	0.000 52	767	49	811	9	828	8	771	10
YL-2-23	104.61	183.45	0.57	0.063 80	0.001 58	1.212 11	0.021 60	0.137 83	0.001 53	0.038 01	0.000 48	735	51	806	10	832	9	754	9
YL-2-24	162.64	476.41	0.34	0.062 54	0.001 47	1.177 21	0.019 01	0.136 55	0.001 48	0.038 63	0.000 52	693	49	790	9	825	8	766	10
YL-2-25	89.92	243.32	0.37	0.064 25	0.001 50	1.213 99	0.019 68	0.137 06	0.001 48	0.037 95	0.000 51	750	49	807	9	828	8	753	10
YL-2-26	93.15	151.86	0.61	0.064 23	0.002 32	1.218 90	0.038 64	0.137 64	0.001 90	0.036 08	0.000 72	749	75	809	18	831	11	716	14

Continued

| Analysis | Content/ppm | | | Ratios | | | | | | | | Age/Ma | | | | | | | |
	Th	U	Th/U	$\frac{^{207}Pb}{^{206}Pb}$	1σ	$\frac{^{207}Pb}{^{235}U}$	1σ	$\frac{^{206}Pb}{^{238}U}$	1σ	$\frac{^{208}Pb}{^{232}Th}$	1σ	$\frac{^{207}Pb}{^{206}Pb}$	1σ	$\frac{^{207}Pb}{^{235}U}$	1σ	$\frac{^{206}Pb}{^{238}U}$	1σ	$\frac{^{208}Pb}{^{232}Th}$	1σ
YL-2-27	268.81	747.46	0.36	0.061 64	0.001 30	1.178 87	0.015 02	0.138 71	0.001 44	0.037 95	0.000 40	662	44	791	7	837	8	753	8
YL-2-28	134.84	355.35	0.38	0.061 65	0.001 35	1.171 60	0.016 71	0.137 84	0.001 45	0.038 72	0.000 45	662	46	787	8	832	8	768	9
YL-2-29	171.64	407.95	0.42	0.063 15	0.001 37	1.213 22	0.016 87	0.139 33	0.001 46	0.038 21	0.000 43	713	45	807	8	841	8	758	8
YL-2-31	84.61	194.44	0.44	0.062 47	0.001 56	1.206 40	0.022 49	0.140 03	0.001 57	0.037 24	0.000 56	690	52	804	10	845	9	739	11
YL-2-32	153.36	415.4	0.37	0.061 63	0.001 57	1.162 91	0.022 37	0.136 82	0.001 54	0.037 84	0.000 59	661	54	783	11	827	9	751	12
YL-2-33	1 683.52	7 096.15	0.24	0.060 88	0.001 22	1.153 15	0.013 07	0.137 34	0.001 40	0.029 79	0.000 30	635	42	779	6	830	8	593	6
YL-2-34	51.04	124.2	0.41	0.060 67	0.001 59	1.159 35	0.023 38	0.138 54	0.001 58	0.039 16	0.000 62	628	55	782	11	836	9	777	12
YL-2-35	70.62	147.68	0.48	0.061 21	0.001 69	1.172 10	0.025 92	0.138 83	0.001 63	0.039 29	0.000 64	647	58	788	12	838	9	779	13
YL-2-36	111.88	283.9	0.39	0.061 12	0.001 32	1.161 90	0.016 43	0.137 81	0.001 45	0.037 64	0.000 43	644	46	783	8	832	8	747	8
YL-1																			
YL-1-01	57.38	139.19	0.41	0.065 49	0.001 78	1.233 82	0.025 56	0.136 65	0.001 59	0.040 34	0.000 65	790	56	816	12	826	9	799	13
YL-1-02	82.27	221.69	0.37	0.070 43	0.001 88	1.309 18	0.026 28	0.134 82	0.001 57	0.046 17	0.000 74	941	54	850	12	815	9	912	14
YL-1-03	97.08	258.01	0.38	0.065 53	0.001 58	1.246 82	0.020 73	0.138 00	0.001 52	0.040 37	0.000 54	791	50	822	9	833	9	800	11
YL-1-04	90.43	236.33	0.38	0.063 96	0.001 69	1.194 38	0.023 51	0.135 45	0.001 55	0.039 26	0.000 61	740	55	798	11	819	9	778	12
YL-1-05	178.77	455.77	0.39	0.064 24	0.001 48	1.225 94	0.018 46	0.138 41	0.001 50	0.039 88	0.000 48	750	48	813	8	836	8	791	9
YL-1-06	60.10	205.36	0.29	0.064 08	0.002 07	1.195 35	0.032 28	0.135 30	0.001 73	0.039 13	0.000 95	744	67	798	15	818	10	776	18
YL-1-07	131.24	496.25	0.26	0.069 48	0.001 60	1.313 86	0.019 84	0.137 16	0.001 49	0.053 99	0.000 68	913	47	852	9	829	8	1 063	13
YL-1-08	141.03	421.36	0.33	0.064 92	0.001 46	1.222 73	0.017 45	0.136 61	0.001 47	0.039 54	0.000 47	772	47	811	8	826	8	784	9
YL-1-10	94.90	217.62	0.44	0.069 20	0.001 69	1.317 80	0.022 59	0.138 12	0.001 55	0.040 58	0.000 55	905	50	854	10	834	9	804	11
YL-1-11	149.77	268.82	0.56	0.069 76	0.002 09	1.297 98	0.031 47	0.134 96	0.001 68	0.041 61	0.000 73	921	61	845	14	816	10	824	14
YL-1-12	348.61	891.13	0.39	0.066 30	0.001 42	1.255 52	0.015 84	0.137 35	0.001 45	0.038 58	0.000 40	816	44	826	7	830	8	765	8
YL-1-13	184.04	428.67	0.43	0.068 74	0.001 56	1.294 31	0.018 99	0.136 56	0.001 48	0.043 12	0.000 49	891	46	843	8	825	8	853	10
YL-1-14	66.98	155.20	0.43	0.065 83	0.001 71	1.245 36	0.023 83	0.137 20	0.001 57	0.040 68	0.000 59	801	53	821	11	829	9	806	11

Continued

Analysis	Content/ppm			Ratios								Age/Ma							
	Th	U	Th/U	$^{207}Pb/^{206}Pb$	1σ	$^{207}Pb/^{235}U$	1σ	$^{206}Pb/^{238}U$	1σ	$^{208}Pb/^{232}Th$	1σ	$^{207}Pb/^{206}Pb$	1σ	$^{207}Pb/^{235}U$	1σ	$^{206}Pb/^{238}U$	1σ	$^{208}Pb/^{232}Th$	1σ
YL-1-15	75.74	178.00	0.43	0.073 60	0.002 03	1.401 15	0.029 85	0.138 08	0.001 66	0.043 46	0.000 72	1 030	55	889	13	834	9	860	14
YL-1-16	81.11	164.42	0.49	0.066 69	0.001 71	1.254 30	0.023 67	0.136 41	0.001 56	0.040 42	0.000 56	828	53	825	11	824	9	801	11
YL-1-17	202.08	563.33	0.36	0.066 09	0.001 46	1.235 52	0.017 09	0.135 59	0.001 46	0.039 21	0.000 44	809	46	817	8	820	8	777	9
YL-1-18	104.81	359.95	0.29	0.066 51	0.001 53	1.268 09	0.019 13	0.138 28	0.001 51	0.040 70	0.000 53	823	47	832	9	835	9	806	10
YL-1-19	69.13	186.95	0.37	0.066 26	0.001 67	1.252 23	0.022 93	0.137 07	0.001 56	0.037 92	0.000 57	815	52	824	10	828	9	752	11
YL-1-20	117.07	269.11	0.44	0.064 67	0.001 54	1.226 79	0.020 03	0.137 59	0.001 53	0.037 89	0.000 49	764	49	813	9	831	9	771	10
YL-1-21	96.84	276.17	0.35	0.065 86	0.001 55	1.227 26	0.019 63	0.135 14	0.001 49	0.038 64	0.000 51	802	49	813	9	817	9	766	10
YL-1-22	97.41	226.63	0.43	0.067 18	0.002 23	1.249 11	0.035 09	0.134 84	0.001 78	0.037 22	0.000 81	843	68	823	16	815	10	739	16
YL-1-23	146.78	439.88	0.33	0.066 70	0.001 52	1.246 97	0.018 58	0.135 58	0.001 48	0.038 38	0.000 48	828	47	822	8	820	8	761	9
YL-1-24	143.06	372.95	0.38	0.068 84	0.001 58	1.280 89	0.019 18	0.134 94	0.001 48	0.039 43	0.000 47	894	47	837	9	816	8	782	9
YL-1-25	89.84	256.34	0.35	0.066 59	0.001 57	1.269 87	0.020 44	0.138 30	0.001 54	0.039 88	0.000 52	825	49	832	9	835	9	790	10
YL-1-26	161.14	327.97	0.49	0.065 97	0.001 52	1.253 84	0.019 00	0.137 84	0.001 51	0.039 82	0.000 45	805	47	825	9	833	9	789	9
YL-1-27	63.79	153.09	0.42	0.065 98	0.001 86	1.234 27	0.027 36	0.135 67	0.001 64	0.037 49	0.000 64	806	58	816	12	820	12	744	12
YL-1-28	430.00	1 994.44	0.22	0.064 96	0.001 39	1.218 53	0.015 32	0.136 03	0.001 45	0.040 71	0.000 44	773	44	809	7	822	7	807	8
YL-1-29	226.14	657.91	0.34	0.064 81	0.001 41	1.224 46	0.016 13	0.137 01	0.001 47	0.037 98	0.000 41	768	45	812	7	828	8	754	8
YL-1-30	152.03	437.53	0.35	0.066 54	0.001 55	1.263 60	0.019 62	0.137 72	0.001 52	0.037 75	0.000 48	823	48	830	9	832	9	749	9
YL-1-31	51.07	228.41	0.22	0.067 22	0.001 76	1.279 94	0.025 02	0.138 08	0.001 61	0.037 34	0.000 76	845	54	837	11	834	11	741	15
YL-1-32	111.78	217.9	0.51	0.066 48	0.001 58	1.245 55	0.020 21	0.135 88	0.001 52	0.037 57	0.000 45	821	49	821	9	821	9	745	9
YL-1-33	139.53	356.00	0.39	0.067 24	0.001 54	1.258 42	0.019 04	0.135 72	0.001 50	0.037 32	0.000 44	845	47	827	9	820	9	741	9
YL-1-34	169.58	451.10	0.38	0.064 87	0.001 46	1.238 45	0.017 89	0.138 44	0.001 51	0.038 25	0.000 44	770	47	818	8	836	8	759	9
YL-1-35	147.14	365.65	0.40	0.065 99	0.001 50	1.242 73	0.018 45	0.136 57	0.001 50	0.038 75	0.000 45	806	47	820	8	825	8	768	9
YL-1-36	128.43	209.49	0.61	0.066 85	0.001 66	1.259 89	0.022 39	0.136 66	0.001 56	0.038 85	0.000 48	833	51	828	10	826	10	770	9

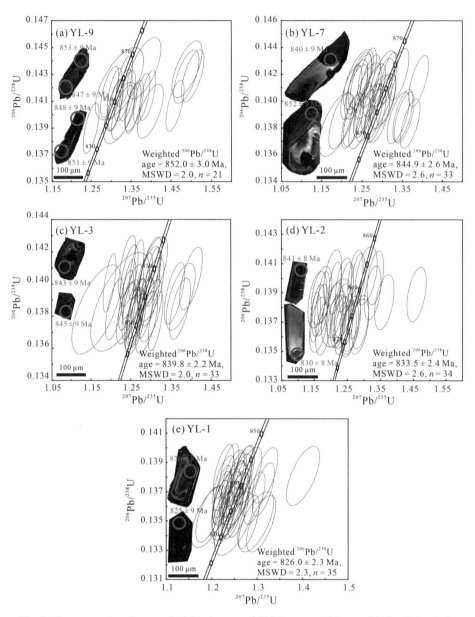

Fig. 3　Representative zircon cathodoluminescent (CL) images and Laser ablation-inductively coupled plasma-mass spectrometry (LA-ICP-MS) zircon U-Pb concordia diagrams for the Neoproterozoic Yonglang granitoids in the western Yangtze Block, South China.

Thirty-one analytic spots in the sample YL-7 were analyzed for Lu-Hf isotopic compositions. The initial ^{176}Hf/^{177}Hf ratios vary from 0.282 151 to 0.282 313, and the negative-dominantly $\varepsilon_{Hf}(t)$ values vary from -3.80 to $+1.92$ (Figs. 4 and 5b). All the thirty-one analyzed spots have old two-stage Hf model ages ($T_{DM2} = 1\ 608 - 1\ 967$ Ma, Mean = 1 852 Ma, $n = 31$) (Table 2).

Table 2　Zircon Lu-Hf isotopic data for the Neoproterozoic high-maficity I-type granitoids in the western Yangtze Block, South China.

Grain spot	Age/Ma	$^{176}Yb/^{177}Hf$	$2SE$	$^{176}Lu/^{177}Hf$	$2SE$	$^{176}Hf/^{177}Hf$	$2SE$	$(^{176}Hf/^{177}Hf)_i$	$f_{Lu/Hf}$	$\varepsilon_{Hf}(t)$	$2SE$	T_{DM1}/Ma	T_{DMC}/Ma
YL-9													
YL-9-01	850	0.010 004	0.000 251	0.000 306	0.000 008	0.282 195	0.000 015	0.282 190	−0.99	−1.99	0.52	1 471	1 857
YL-9-02	850	0.075 616	0.000 273	0.002 980	0.000 007	0.282 363	0.000 013	0.282 315	−0.91	0.93	0.47	1 396	1 674
YL-9-03	850	0.032 823	0.000 246	0.001 141	0.000 008	0.282 136	0.000 027	0.282 117	−0.97	−5.05	0.95	1 606	2 048
YL-9-04	850	0.049 784	0.000 361	0.001 715	0.000 011	0.282 113	0.000 022	0.282 085	−0.95	−6.50	0.77	1 676	2 139
YL-9-05	850	0.043 310	0.000 279	0.001 446	0.000 009	0.282 275	0.000 012	0.282 252	−0.96	−0.46	0.41	1 429	1 761
YL-9-06	850	0.048 820	0.000 275	0.001 676	0.000 012	0.282 180	0.000 022	0.282 153	−0.95	−4.10	0.77	1 579	1 989
YL-9-07	850	0.050 339	0.000 173	0.001 679	0.000 005	0.282 163	0.000 019	0.282 137	−0.95	−4.67	0.69	1 602	2 025
YL-9-08	850	0.043 922	0.000 529	0.001 389	0.000 011	0.282 260	0.000 011	0.282 238	−0.96	−0.91	0.40	1 446	1 789
YL-9-09	850	0.069 133	0.000 388	0.002 128	0.000 013	0.282 212	0.000 013	0.282 178	−0.94	−3.45	0.46	1 561	1 948
YL-9-10	850	0.038 935	0.000 080	0.001 265	0.000 024	0.282 130	0.000 004	0.282 109	−0.96	−5.40	0.84	1 622	2 070
YL-9-11	850	0.047 996	0.000 251	0.001 603	0.000 008	0.282 147	0.000 030	0.282 122	−0.95	−5.15	1.06	1 619	2 055
YL-9-12	850	0.031 847	0.000 211	0.001 029	0.000 006	0.282 296	0.000 011	0.282 279	−0.97	0.75	0.40	1 375	1 685
YL-9-13	850	0.016 125	0.000 151	0.000 505	0.000 004	0.282 080	0.000 013	0.282 072	−0.98	−6.31	0.45	1 642	2 127
YL-9-14	850	0.052 246	0.000 787	0.001 628	0.000 022	0.282 212	0.000 013	0.282 186	−0.95	−2.88	0.47	1 529	1 913
YL-9-15	850	0.057 532	0.000 516	0.002 080	0.000 017	0.282 275	0.000 029	0.282 241	−0.94	−1.18	1.03	1 469	1 806
YL-7													
YL-7-01	845	0.027 270	0.000 092	0.000 912	0.000 003	0.282 206	0.000 013	0.282 191	−0.97	−2.41	0.44	1 494	1 880
YL-7-02	845	0.021 546	0.000 281	0.000 667	0.000 008	0.282 247	0.000 013	0.282 236	−0.98	−0.66	0.45	1 421	1 770
YL-7-03	845	0.018 066	0.000 090	0.000 578	0.000 003	0.282 232	0.000 016	0.282 222	−0.98	−1.11	0.55	1 437	1 798
YL-7-04	845	0.017 501	0.000 016	0.000 553	0.000 001	0.282 234	0.000 013	0.282 225	−0.98	−1.01	0.46	1 433	1 792
YL-7-05	845	0.024 402	0.000 319	0.000 772	0.000 011	0.282 179	0.000 014	0.282 166	−0.98	−3.21	0.48	1 523	1 930
YL-7-06	845	0.031 544	0.000 098	0.000 976	0.000 002	0.282 237	0.000 017	0.282 222	−0.97	−1.36	0.60	1 453	1 814
YL-7-07	845	0.025 833	0.000 374	0.000 807	0.000 012	0.282 167	0.000 015	0.282 154	−0.98	−3.65	0.52	1 540	1 957
YL-7-08	845	0.018 199	0.000 046	0.000 583	0.000 002	0.282 211	0.000 015	0.282 202	−0.98	−1.86	0.54	1 466	1 845

Continued

Grain spot	Age/Ma	$^{176}\mathrm{Yb}/^{177}\mathrm{Hf}$	2SE	$^{176}\mathrm{Lu}/^{177}\mathrm{Hf}$	2SE	$^{176}\mathrm{Hf}/^{177}\mathrm{Hf}$	2SE	$(^{176}\mathrm{Hf}/^{177}\mathrm{Hf})_i$	$f_{\mathrm{Lu/Hf}}$	$\varepsilon_{\mathrm{Hf}}(t)$	2SE	$T_{\mathrm{DM1}}/\mathrm{Ma}$	$T_{\mathrm{DMC}}/\mathrm{Ma}$
YL-7-09	845	0.021 549	0.000 177	0.000 696	0.000 005	0.282 209	0.000 017	0.282 198	-0.98	-2.06	0.62	1 476	1 857
YL-7-10	845	0.016 821	0.000 135	0.000 541	0.000 005	0.282 196	0.000 015	0.282 188	-0.98	-2.32	0.52	1 484	1 874
YL-7-11	845	0.016 261	0.000 018	0.000 523	0.000 000	0.282 232	0.000 017	0.282 223	-0.98	-1.05	0.60	1 434	1 794
YL-7-12	845	0.041 127	0.000 424	0.001 271	0.000 014	0.282 179	0.000 015	0.282 159	-0.96	-3.74	0.53	1 553	1 963
YL-7-13	845	0.028 585	0.000 046	0.000 903	0.000 002	0.282 209	0.000 014	0.282 194	-0.97	-2.30	0.48	1 489	1 872
YL-7-14	845	0.026 289	0.000 428	0.000 885	0.000 015	0.282 327	0.000 017	0.282 313	-0.97	1.92	0.60	1 323	1 608
YL-7-15	845	0.024 075	0.000 012	0.000 752	0.000 001	0.282 170	0.000 016	0.282 158	-0.98	-3.49	0.58	1 533	1 947
YL-7-16	845	0.026 729	0.000 025	0.000 858	0.000 000	0.282 199	0.000 015	0.282 185	-0.97	-2.59	0.52	1 500	1 891
YL-7-17	845	0.033 033	0.000 052	0.001 034	0.000 004	0.282 200	0.000 014	0.282 184	-0.97	-2.73	0.51	1 508	1 900
YL-7-18	845	0.027 707	0.000 288	0.000 871	0.000 008	0.282 210	0.000 016	0.282 196	-0.97	-2.20	0.58	1 485	1 866
YL-7-19	845	0.007 126	0.000 319	0.000 210	0.000 010	0.282 160	0.000 014	0.282 157	-0.99	-3.23	0.51	1 513	1 931
YL-7-20	845	0.032 680	0.000 132	0.001 046	0.000 003	0.282 228	0.000 017	0.282 211	-0.97	-1.79	0.59	1 471	1 840
YL-7-21	845	0.028 966	0.000 051	0.000 911	0.000 002	0.282 244	0.000 015	0.282 229	-0.97	-1.06	0.55	1 440	1 795
YL-7-22	845	0.031 216	0.000 232	0.000 977	0.000 007	0.282 203	0.000 015	0.282 188	-0.97	-2.57	0.52	1 501	1 890
YL-7-23	845	0.029 474	0.000 087	0.000 913	0.000 003	0.282 194	0.000 016	0.282 179	-0.97	-2.83	0.57	1 510	1 906
YL-7-24	845	0.034 856	0.000 429	0.001 099	0.000 014	0.282 243	0.000 014	0.282 226	-0.97	-1.29	0.49	1 452	1 809
YL-7-25	845	0.033 250	0.000 030	0.001 048	0.000 001	0.282 252	0.000 016	0.282 235	-0.97	-0.93	0.56	1 437	1 787
YL-7-26	845	0.025 627	0.000 345	0.000 808	0.000 010	0.282 247	0.000 015	0.282 234	-0.98	-0.82	0.52	1 429	1 780
YL-7-27	845	0.026 705	0.000 128	0.000 837	0.000 003	0.282 229	0.000 016	0.282 215	-0.97	-1.51	0.57	1 457	1 823
YL-7-28	845	0.019 470	0.000 035	0.000 616	0.000 001	0.282 235	0.000 014	0.282 225	-0.98	-1.03	0.51	1 435	1 793
YL-7-29	845	0.024 315	0.000 100	0.000 760	0.000 003	0.282 232	0.000 017	0.282 220	-0.98	-1.31	0.60	1 448	1 810
YL-7-30	845	0.027 897	0.000 277	0.000 878	0.000 007	0.282 165	0.000 016	0.282 151	-0.97	-3.80	0.57	1 548	1 967
YL-7-31	845	0.028 975	0.000 079	0.000 933	0.000 002	0.282 186	0.000 016	0.282 171	-0.97	-3.13	0.57	1 522	1 925
YL-3													
YL-3-01	840	0.027 894	0.000 124	0.000 887	0.000 005	0.282 218	0.000 015	0.282 204	-0.97	-2.06	0.54	1 475	1 854

Continued

Grain spot	Age/Ma	$^{176}Yb/^{177}Hf$	2SE	$^{176}Lu/^{177}Hf$	2SE	$^{176}Hf/^{177}Hf$	2SE	$(^{176}Hf/^{177}Hf)_i$	$f_{Lu/Hf}$	$\varepsilon_{Hf}(t)$	2SE	T_{DM1}/Ma	T_{DMC}/Ma
YL-3-02	840	0.139 226	0.000 524	0.004 141	0.000 014	0.282 230	0.000 016	0.282 165	-0.88	-5.26	0.55	1 672	2 054
YL-3-03	840	0.139 314	0.000 967	0.004 068	0.000 029	0.282 173	0.000 024	0.282 109	-0.88	-7.20	0.85	1 753	2 175
YL-3-04	840	0.078 772	0.000 464	0.002 609	0.000 018	0.282 066	0.000 027	0.282 025	-0.92	-9.36	0.95	1 805	2 310
YL-3-05	840	0.077 266	0.001 290	0.002 348	0.000 037	0.282 233	0.000 016	0.282 196	-0.93	-3.15	0.56	1 545	1 922
YL-3-06	840	0.089 117	0.000 308	0.002 766	0.000 012	0.282 200	0.000 020	0.282 156	-0.92	-4.80	0.69	1 622	2 025
YL-3-07	840	0.033 136	0.000 027	0.001 082	0.000 002	0.282 184	0.000 016	0.282 167	-0.97	-3.47	0.56	1 534	1 942
YL-3-08	840	0.040 137	0.000 235	0.001 377	0.000 008	0.282 237	0.000 026	0.282 216	-0.96	-1.91	0.93	1 478	1 844
YL-3-09	840	0.041 126	0.000 179	0.001 332	0.000 009	0.282 223	0.000 018	0.282 202	-0.96	-2.38	0.64	1 495	1 874
YL-3-10	840	0.021 892	0.000 545	0.000 685	0.000 016	0.282 144	0.000 013	0.282 133	-0.98	-4.46	0.45	1 565	2 004
YL-3-11	840	0.031 311	0.000 074	0.000 966	0.000 002	0.282 227	0.000 015	0.282 212	-0.97	-1.81	0.54	1 467	1 838
YL-3-12	840	0.022 032	0.000 466	0.000 858	0.000 016	0.282 241	0.000 014	0.282 227	-0.97	-1.21	0.49	1 441	1 800
YL-3-13	840	0.045 683	0.000 211	0.001 588	0.000 009	0.282 246	0.000 024	0.282 221	-0.95	-1.85	0.85	1 479	1 840
YL-3-14	840	0.034 484	0.000 487	0.001 054	0.000 016	0.282 164	0.000 017	0.282 147	-0.97	-4.17	0.61	1 561	1 986
YL-3-15	840	0.057 959	0.000 846	0.001 847	0.000 029	0.282 176	0.000 017	0.282 146	-0.94	-4.62	0.58	1 595	2 014
YL-3-16	840	0.060 089	0.000 103	0.001 921	0.000 003	0.282 242	0.000 020	0.282 212	-0.94	-2.35	0.71	1 505	1 872
YL-3-17	840	0.059 133	0.000 453	0.001 923	0.000 015	0.282 199	0.000 024	0.282 169	-0.94	-3.88	0.84	1 566	1 967
YL-3-18	840	0.015 169	0.000 072	0.000 520	0.000 003	0.282 151	0.000 015	0.282 143	-0.98	-4.01	0.53	1 545	1 976
YL-3-19	840	0.066 636	0.000 927	0.002 018	0.000 024	0.282 184	0.000 017	0.282 152	-0.94	-4.51	0.58	1 594	2 007
YL-3-20	840	0.046 293	0.000 462	0.001 416	0.000 015	0.282 178	0.000 015	0.282 155	-0.96	-4.07	0.53	1 564	1 979
YL-3-21	840	0.202 996	0.000 253	0.005 901	0.000 011	0.282 207	0.000 026	0.282 114	-0.82	-8.05	0.91	1 843	2 227
YL-3-22	840	0.057 636	0.001 030	0.001 869	0.000 030	0.282 145	0.000 024	0.282 116	-0.94	-5.73	0.84	1 640	2 083
YL-3-23	840	0.073 969	0.000 989	0.002 213	0.000 028	0.282 098	0.000 021	0.282 063	-0.93	-7.77	0.73	1 730	2 211
YL-3-24	840	0.083 609	0.001 400	0.002 546	0.000 037	0.282 194	0.000 022	0.282 154	-0.92	-4.75	0.78	1 615	2 022
YL-3-25	840	0.091 495	0.001 010	0.002 980	0.000 031	0.282 214	0.000 024	0.282 167	-0.91	-4.53	0.86	1 615	2 008
YL-3-26	840	0.029 342	0.000 019	0.000 937	0.000 001	0.282 228	0.000 014	0.282 213	-0.97	-1.76	0.50	1 464	1 835

Continued

Grain spot	Age/Ma	$^{176}\text{Yb}/^{177}\text{Hf}$	2SE	$^{176}\text{Lu}/^{177}\text{Hf}$	2SE	$^{176}\text{Hf}/^{177}\text{Hf}$	2SE	$(^{176}\text{Hf}/^{177}\text{Hf})_i$	$f_{\text{Lu}/\text{Hf}}$	$\varepsilon_{\text{Hf}}(t)$	2SE	T_{DM1}/Ma	T_{DMC}/Ma
YL-3-27	840	0.046 346	0.000 289	0.001 453	0.000 005	0.282 137	0.000 018	0.282 114	-0.96	-5.56	0.62	1 624	2 072
YL-3-28	840	0.032 966	0.000 097	0.001 037	0.000 004	0.282 175	0.000 013	0.282 158	-0.97	-3.75	0.46	1 544	1 960
YL-3-29	840	0.031 442	0.000 093	0.001 057	0.000 004	0.282 236	0.000 016	0.282 219	-0.97	-1.62	0.55	1 460	1 826
YL-3-30	840	0.032 131	0.000 090	0.001 061	0.000 005	0.282 149	0.000 017	0.282 133	-0.97	-4.67	0.60	1 581	2 017
YL-3-31	840	0.033 843	0.000 062	0.001 046	0.000 002	0.282 230	0.000 018	0.282 214	-0.97	-1.79	0.62	1 467	1 837
YL-2													
YL-2-01	833	0.034 758	0.000 113	0.001 077	0.000 004	0.282 170	0.000 014	0.282 153	-0.97	-4.12	0.49	1 554	1 977
YL-2-02	833	0.041 279	0.000 163	0.001 402	0.000 006	0.282 135	0.000 025	0.282 113	-0.96	-5.70	0.89	1 623	2 076
YL-2-03	833	0.050 045	0.000 199	0.001 503	0.000 006	0.282 172	0.000 015	0.282 148	-0.95	-4.52	0.51	1 578	2 002
YL-2-04	833	0.026 343	0.000 043	0.000 820	0.000 001	0.282 221	0.000 015	0.282 208	-0.98	-2.03	0.53	1 467	1 846
YL-2-05	833	0.042 466	0.000 230	0.001 303	0.000 008	0.282 221	0.000 014	0.282 200	-0.96	-2.57	0.51	1 496	1 880
YL-2-06	833	0.027 151	0.000 112	0.000 885	0.000 004	0.282 237	0.000 011	0.282 223	-0.97	-1.52	0.40	1 448	1 815
YL-2-07	833	0.020 775	0.000 085	0.000 648	0.000 003	0.282 189	0.000 013	0.282 178	-0.98	-2.97	0.46	1 501	1 906
YL-2-08	833	0.040 808	0.000 125	0.001 262	0.000 002	0.282 224	0.000 012	0.282 204	-0.96	-2.41	0.44	1 489	1 870
YL-2-09	833	0.049 397	0.000 342	0.001 503	0.000 009	0.282 259	0.000 014	0.282 235	-0.95	-1.45	0.50	1 455	1 810
YL-2-10	833	0.024 558	0.000 250	0.000 758	0.000 006	0.282 198	0.000 013	0.282 187	-0.98	-2.75	0.45	1 494	1 891
YL-2-11	833	0.045 052	0.000 272	0.001 368	0.000 007	0.282 227	0.000 013	0.282 206	-0.96	-2.41	0.46	1 491	1 870
YL-2-12	833	0.032 074	0.000 108	0.001 018	0.000 003	0.282 238	0.000 013	0.282 222	-0.97	-1.64	0.47	1 455	1 822
YL-2-13	833	0.049 108	0.000 457	0.001 563	0.000 016	0.282 219	0.000 012	0.282 194	-0.95	-2.91	0.43	1 515	1 902
YL-2-14	833	0.043 315	0.000 174	0.001 336	0.000 005	0.282 168	0.000 012	0.282 147	-0.96	-4.48	0.41	1 573	2 000
YL-2-15	833	0.019 588	0.000 207	0.000 622	0.000 006	0.282 230	0.000 013	0.282 220	-0.98	-1.47	0.44	1 442	1 811
YL-2-16	833	0.021 405	0.000 348	0.000 653	0.000 009	0.282 217	0.000 015	0.282 207	-0.98	-1.96	0.52	1 461	1 842
YL-2-17	833	0.026 874	0.000 051	0.000 853	0.000 003	0.282 222	0.000 014	0.282 209	-0.97	-2.02	0.51	1 467	1 846
YL-2-18	833	0.033 070	0.000 087	0.001 009	0.000 002	0.282 203	0.000 014	0.282 187	-0.97	-2.87	0.50	1 503	1 899
YL-2-19	833	0.022 416	0.000 095	0.000 713	0.000 002	0.282 249	0.000 014	0.282 238	-0.98	-0.90	0.51	1 421	1 776

Continued

Grain spot	Age/Ma	^{176}Yb/^{177}Hf	2SE	^{176}Lu/^{177}Hf	2SE	^{176}Hf/^{177}Hf	2SE	$(^{176}$Hf/^{177}Hf$)_i$	$f_{Lu/Hf}$	$\varepsilon_{Hf}(t)$	2SE	T_{DM1}/Ma	T_{DMC}/Ma
YL-2-20	833	0.031 243	0.000 180	0.000 960	0.000 006	0.282 211	0.000 014	0.282 196	-0.97	-2.53	0.49	1 489	1 878
YL-2-21	833	0.026 177	0.000 106	0.000 816	0.000 004	0.282 217	0.000 012	0.282 204	-0.98	-2.17	0.42	1 472	1 855
YL-2-22	833	0.028 271	0.000 167	0.000 871	0.000 005	0.282 218	0.000 016	0.282 204	-0.97	-2.18	0.55	1 474	1 856
YL-2-23	833	0.029 514	0.000 358	0.000 939	0.000 010	0.282 170	0.000 014	0.282 155	-0.97	-3.96	0.51	1 545	1 967
YL-2-24	833	0.017 449	0.000 055	0.000 562	0.000 001	0.282 247	0.000 014	0.282 239	-0.98	-0.79	0.50	1 414	1 769
YL-2-25	833	0.022 308	0.000 161	0.000 704	0.000 005	0.282 243	0.000 013	0.282 232	-0.98	-1.11	0.47	1 429	1 789
YL-2-26	833	0.026 044	0.000 084	0.000 814	0.000 001	0.282 222	0.000 012	0.282 209	-0.98	-1.97	0.42	1 465	1 843
YL-2-27	833	0.026 142	0.000 237	0.000 885	0.000 010	0.282 167	0.000 013	0.282 153	-0.97	-4.01	0.45	1 546	1 971
YL-2-28	833	0.037 015	0.000 279	0.001 134	0.000 007	0.282 251	0.000 015	0.282 233	-0.97	-1.31	0.54	1 444	1 802
YL-2-29	833	0.064 383	0.000 796	0.002 036	0.000 021	0.282 180	0.000 015	0.282 148	-0.94	-4.82	0.52	1 601	2 021
YL-2-30	833	0.024 035	0.000 043	0.000 793	0.000 002	0.282 218	0.000 012	0.282 205	-0.98	-2.11	0.42	1 470	1 852
YL-2-31	833	0.024 077	0.000 100	0.000 750	0.000 002	0.282 208	0.000 014	0.282 196	-0.98	-2.42	0.50	1 481	1 871
YL-2-32	833	0.065 645	0.001 050	0.002 180	0.000 038	0.282 225	0.000 014	0.282 191	-0.93	-3.37	0.49	1 545	1 930
YL-2-33	833	0.022 943	0.000 116	0.000 731	0.000 004	0.282 215	0.000 013	0.282 203	-0.98	-2.13	0.46	1 469	1 853
YL-2-34	833	0.037 882	0.000 144	0.001 181	0.000 004	0.282 234	0.000 014	0.282 216	-0.96	-1.95	0.51	1 470	1 841
YL-1													
YL-1-01	825	0.020 038	0.000 075	0.000 696	0.000 004	0.282 230	0.000 016	0.282 219	-0.98	-1.72	0.55	1 446	1 821
YL-1-02	825	0.019 561	0.000 008	0.000 628	0.000 000	0.282 214	0.000 014	0.282 205	-0.98	-2.21	0.48	1 464	1 852
YL-1-03	825	0.019 150	0.000 043	0.000 614	0.000 002	0.282 202	0.000 015	0.282 192	-0.98	-2.63	0.52	1 480	1 878
YL-1-04	825	0.029 226	0.000 105	0.000 953	0.000 004	0.282 202	0.000 015	0.282 187	-0.97	-3.02	0.55	1 501	1 902
YL-1-05	825	0.019 782	0.000 071	0.000 634	0.000 002	0.282 229	0.000 016	0.282 219	-0.98	-1.69	0.57	1 444	1 819
YL-1-06	825	0.043 413	0.000 447	0.001 375	0.000 014	0.282 234	0.000 014	0.282 213	-0.96	-2.32	0.49	1 481	1 859
YL-1-07	825	0.028 035	0.000 035	0.000 932	0.000 004	0.282 177	0.000 014	0.282 162	-0.97	-3.88	0.49	1 535	1 957
YL-1-08	825	0.023 838	0.000 161	0.000 770	0.000 006	0.282 211	0.000 017	0.282 199	-0.98	-2.48	0.59	1 477	1 869
YL-1-09	825	0.027 328	0.000 037	0.000 861	0.000 002	0.282 261	0.000 019	0.282 248	-0.97	-0.81	0.66	1 413	1 764

Continued

Grain spot	Age/Ma	$^{176}Yb/^{177}Hf$	2SE	$^{176}Lu/^{177}Hf$	2SE	$^{176}Hf/^{177}Hf$	2SE	$(^{176}Hf/^{177}Hf)_i$	$f_{Lu/Hf}$	$\varepsilon_{Hf}(t)$	2SE	T_{DM1}/Ma	T_{DMC}/Ma
YL-1-10	825	0.027 117	0.000 280	0.000 926	0.000 012	0.282 184	0.000 019	0.282 170	-0.97	-3.61	0.66	1 524	1 939
YL-1-11	825	0.032 288	0.000 150	0.000 999	0.000 004	0.282 223	0.000 016	0.282 207	-0.97	-2.33	0.58	1 475	1 859
YL-1-12	825	0.028 039	0.000 039	0.000 880	0.000 002	0.282 195	0.000 018	0.282 181	-0.97	-3.18	0.64	1 506	1 912
YL-1-13	825	0.023 747	0.000 218	0.000 740	0.000 006	0.282 198	0.000 017	0.282 187	-0.98	-2.91	0.59	1 493	1 896
YL-1-14	825	0.034 041	0.000 096	0.001 084	0.000 003	0.282 256	0.000 012	0.282 239	-0.97	-1.25	0.41	1 434	1 791
YL-1-15	825	0.022 862	0.000 055	0.000 831	0.000 003	0.282 216	0.000 025	0.282 203	-0.97	-2.37	0.88	1 473	1 862
YL-1-16	825	0.022 934	0.000 230	0.000 731	0.000 007	0.282 160	0.000 013	0.282 149	-0.98	-4.23	0.47	1 545	1 978
YL-1-17	825	0.032 330	0.000 104	0.001 022	0.000 002	0.282 180	0.000 016	0.282 164	-0.97	-3.86	0.58	1 536	1 955
YL-1-18	825	0.025 222	0.000 014	0.000 811	0.000 001	0.282 210	0.000 017	0.282 197	-0.98	-2.56	0.61	1 481	1 874
YL-1-19	825	0.022 771	0.000 025	0.000 729	0.000 001	0.282 192	0.000 014	0.282 181	-0.98	-3.12	0.49	1 501	1 908
YL-1-20	825	0.036 170	0.000 124	0.001 142	0.000 003	0.282 209	0.000 014	0.282 191	-0.97	-2.97	0.50	1 503	1 899
YL-1-21	825	0.023 590	0.000 044	0.000 742	0.000 001	0.282 219	0.000 015	0.282 207	-0.98	-2.18	0.52	1 464	1 850
YL-1-22	825	0.033 112	0.000 213	0.001 034	0.000 006	0.282 233	0.000 015	0.282 217	-0.97	-1.99	0.54	1 462	1 838
YL-1-23	825	0.035 555	0.000 144	0.001 115	0.000 004	0.282 208	0.000 016	0.282 191	-0.97	-2.97	0.57	1 502	1 899
YL-1-24	825	0.014 527	0.000 068	0.000 473	0.000 002	0.282 244	0.000 016	0.282 236	-0.99	-1.00	0.58	1 414	1 776
YL-1-25	825	0.060 734	0.000 470	0.001 943	0.000 016	0.282 194	0.000 015	0.282 164	-0.94	-4.38	0.54	1 574	1 987
YL-1-26	825	0.036 629	0.000 165	0.001 136	0.000 004	0.282 224	0.000 014	0.282 207	-0.97	-2.41	0.51	1 480	1 864
YL-1-27	825	0.033 707	0.000 049	0.001 069	0.000 002	0.282 180	0.000 015	0.282 163	-0.97	-3.92	0.53	1 539	1 959
YL-1-28	825	0.027 118	0.000 071	0.000 842	0.000 002	0.282 196	0.000 013	0.282 183	-0.97	-3.11	0.45	1 503	1 908
YL-1-29	825	0.029 551	0.000 135	0.000 918	0.000 003	0.282 218	0.000 015	0.282 204	-0.97	-2.40	0.55	1 476	1 864
YL-1-30	825	0.037 293	0.000 283	0.001 143	0.000 008	0.282 223	0.000 013	0.282 205	-0.97	-2.46	0.47	1 483	1 868
YL-1-31	825	0.033 110	0.000 320	0.001 027	0.000 011	0.282 219	0.000 016	0.282 203	-0.97	-2.48	0.56	1 481	1 868
YL-1-32	825	0.024 865	0.000 059	0.000 791	0.000 002	0.282 206	0.000 015	0.282 194	-0.98	-2.67	0.52	1 485	1 881
YL-1-33	825	0.032 841	0.000 282	0.001 019	0.000 010	0.282 212	0.000 016	0.282 196	-0.97	-2.74	0.55	1 491	1 885

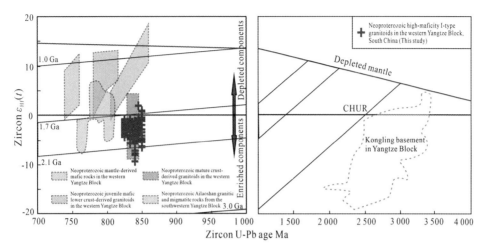

Fig. 4　The zircon $\varepsilon_{Hf}(t)$ values vs. zircon U-Pb age diagram for the Neoproterozoic Yonglang granitoids in the western Yangtze Block, South China (modified after Zhu et al., 2019c).

4.1.3　Sample YL-3

The zircons from sample YL-3 have variable Th/U ratios of 0.10−1.08. The thirty-three concordant analyzed spots show a weighted mean age of 839.8±2.2 Ma (MSWD=2.0, n = 33) (Fig. 3c).

In sample YL-3, the thirty-one analyzed spots exhibit initial $^{176}Hf/^{177}Hf$ ratios varying from 0.282 025 to 0.282 227. They yield negative $\varepsilon_{Hf}(t)$ values (−9.36 to −1.21) (Figs. 4 and 5c) and ancient two-stage Hf model ages (T_{DM2}=1 800−2 130 Ma, Mean=1 980 Ma, n = 31) (Table 2).

4.1.4　Sample YL-2

The thirty-four concordant spots in sample YL-2 were analyzed for U-Pb isotopes. They yield Th/U ratios of 0.18−0.61 and a weighted mean age of 833.5±2.4 Ma (MSWD=2.6, n =34) (Fig. 3d).

The thirty-four spots display initial $^{176}Hf/^{177}Hf$ ratios from 0.282 113 to 0.282 239. They show the enriched zircon Hf isotopes [$\varepsilon_{Hf}(t)$ = −5.70 to −0.79] (Figs. 4 and 5d) and old two-stage Hf model ages (T_{DM2}=1 769−2 076 Ma, Mean=1 881 Ma, n=34) (Table 2).

4.1.5　Sample YL-1

The magmatic zircons of sample YL-1 have Th/U ratios from 0.22 to 0.61. Among thirty-five concordant analyzed spots, they display a weighted mean age of 826.0±2.3 Ma (MSWD= 2.3, n=35) (Fig. 3e), which represent the crystallization age of the sample YL-1.

Thirty-three spots from sample YL-1 were analyzed to determine their Lu-Hf isotopic compositions. The initial $^{176}Hf/^{177}Hf$ ratios span a range of 0.282 149−0.282 248. All the thirty-three analyzed spots give negative $\varepsilon_{Hf}(t)$ values of −4.38 to −0.81 (Figs. 4 and 5e)

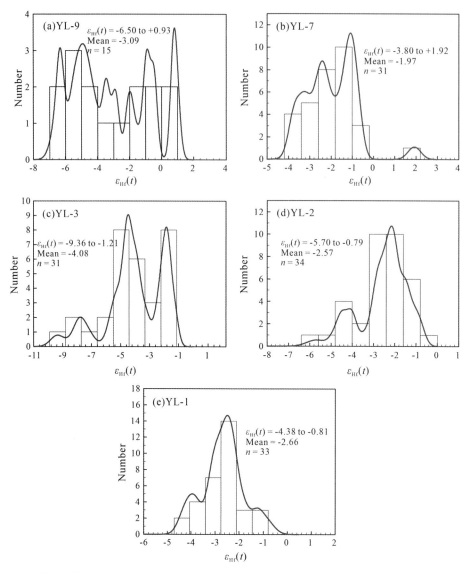

Fig. 5　Histograms of $\varepsilon_{Hf}(t)$ isotope ratios for the Neoproterozoic Yonglang granitoids
in the western Yangtze Block, South China.

and ancient T_{DM2} values of 1 764–1 987 Ma (Mean = 1 880 Ma, $n = 33$) (Table 2).

　　In general, zircon U-Pb geochronology display that the Yonglang granitoids investigated here were formed at ca. 850–825 Ma, which are corresponding to Neoproterozoic long-term magmatism along the western margin of the Yangtze Block, South China (Zhao et al., 2018 and references therein). These granitoids show the negative-dominantly zircon $\varepsilon_{Hf}(t)$ values of −9.36 to −0.27 (only three analyzed spots are positive: +0.75 to +1.92) (Fig. 4), which are distinct from the literature data of Neoproterozoic mantle-derived mafic rocks and juvenile mafic lower crust-derived granitoids in the western Yangtze Block (Zhu et al., 2019c and references therein). In addition, these granitoids exhibit ancient two-stage Hf model ages ranging from 1 608 Ma to 2 139 Ma (Table 2).

4. 2 Whole-rock major and trace elemental geochemistry

In this study, the whole-rock major and trace element results of ca. 850−825 Ma Yonglang granitoids are given in Table 3, and Figs. 6 and 7.

The granitoids collected from the Yonglang area have variable SiO_2 (64. 63−72. 62 wt%) and high K_2O (3. 49−6. 59 wt%), but low Al_2O_3 (12. 93−16. 38 wt%) and MgO (0. 32−0. 91 wt%) contents. They yield variable A/CNK [molar ratios of $Al_2O_3/(CaO + Na_2O + K_2O)$] values of 0. 94−1. 09, belonging to the metaluminous to weakly peraluminous series (Fig. 6a). On the $(Na_2O + K_2O - CaO)$ vs. SiO_2 diagram (Fig. 6b), the Yonglang granitoids fall into the alkali-calcic to calc-alkalic fields. These granitoids yield high K_2O/Na_2O ratios (1. 10−3. 43) and low $Mg^\#$ values (20−26) (Fig. 6c, d). It is notable that the Yonglang granitoids have high and variable $Fe_2O_3^T$ (2. 69−7. 30 wt%) and $FeO^T + MgO$ (2. 69−7. 30 wt%) contents as well as Fe + Mg values (0. 045 2−0. 121 0).

On the chondrite-normalized rare earth element (REE) pattern diagram (Fig. 7a), the Yonglang granitoids display relatively consistent rare earth elements (REEs) distribution patterns, showing the enrichment in light rare earth elements (LREEs) [$(La/Yb)_N = 3. 43 - 30. 0$, where N denotes normalization to chondrite] and relatively flat heavy rare earth elements (HREEs) patterns [$(Gd/Yb)_N = 1. 34 - 3. 90$]. They yield high total REE contents of 196−467 ppm and obviously negative Eu anomalies ($Eu/Eu^* = 0. 30 - 0. 80$). On the primitive mantle-normalized diagram (Fig. 7b), the Yonglang granitoids exhibit the enrichment in large ion lithophile elements (LILEs) (e.g., Rb, Th, U, Pb, and K), and the depletion of Ba and Sr as well as high field strength elements (HFSEs) (e.g., Nb, Ta, P, and Ti). In addition, they contain low Sr (107 − 176 ppm), Cr (3. 95 − 29. 6 ppm), and Ni (1. 51 − 16. 9 ppm) concentrations as well as high Y (24. 7−78. 2 ppm) and Yb (2. 15−7. 56 ppm) concentrations.

4. 3 Whole-rock Sr-Nd isotopes

Eight granitoid samples collected from Yonglang area were analyzed for whole-rock Sr-Nd isotopes. The analytic results are listed in Table 4 and Fig. 8.

All of the initial $^{87}Sr/^{86}Sr$ ratios (I_{Sr}) and $\varepsilon_{Nd}(t)$ values were calculated using the zircon U-Pb ages. The Yonglang granitoids studied here have variable initial $^{87}Sr/^{86}Sr$ ratios (0. 693 2−0. 714 5) and significantly negative $\varepsilon_{Nd}(t)$ values (−6. 38 to −4. 06), which are different from Neoproterozoic juvenile mafic lower crust-derived granitoids and mantle-derived mafic rocks in the western Yangtze Block (Fig. 8) (Zhu et al., 2019c and references therein). In addition, these granitoids display ancient two-stage Nd model ages ($T_{DM2} = 1. 69 - 1. 85$ Ga).

Table 3　Major (wt%) and trace (ppm) element analysis results for the Neoproterozoic high-maficity I-type granitoids in the western Yangtze Block, South China.

Sample	YL-9-1	YL-9-2	YL-9-3	YL-9-4	YL-7-1	YL-7-2	YL-7-3	YL-7-3R	YL-7-4	YL-3-1	YL-3-2	YL-3-3	YL-3-4	YL-3-5	YL-2-1	YL-2-2	YL-2-3	YL-2-3R	YL-2-4	YL-2-5	YL-1-2	YL-1-3	YL-1-4
Location	27°9′38″N,102°13′15″E				27°9′9″N,102°12′37″E					27°9′13″N,102°13′22″E					27°8′44″N,102°13′20″E						27°8′44″N,102°13′20″E		
Major elements / wt%																							
SiO_2	67.37	65.83	65.77	66.18	67.88	70.32	69.07	69.19	66.82	70.59	70.95	72.62	65.31	66.02	67.78	66.47	64.82	64.63	68.09	69.17	65.77	66.25	69.75
TiO_2	0.74	0.74	0.79	0.43	0.43	0.28	0.33	0.33	0.43	0.42	0.43	0.28	0.78	0.73	0.54	0.45	0.73	0.73	0.47	0.45	0.41	0.51	0.39
Al_2O_3	14.30	14.31	14.38	15.59	15.59	14.57	15.08	15.13	15.72	13.25	13.23	12.93	14.39	14.49	14.21	15.12	15.02	14.94	14.27	14.12	16.38	15.61	14.65
$Fe_2O_3^T$	6.10	6.57	7.01	4.06	4.06	2.84	3.35	3.39	4.21	3.72	3.76	2.57	7.00	6.50	5.59	4.69	6.59	6.56	5.02	4.56	4.27	4.79	3.72
MnO	0.06	0.08	0.10	0.04	0.04	0.02	0.08	0.08	0.10	0.04	0.04	0.03	0.10	0.10	0.08	0.06	0.09	0.10	0.07	0.06	0.04	0.05	0.04
MgO	0.91	0.86	0.87	0.48	0.48	0.32	0.38	0.39	0.50	0.51	0.50	0.38	0.87	0.82	0.59	0.51	0.79	0.80	0.57	0.50	0.51	0.58	0.47
CaO	2.80	3.24	3.06	3.11	2.39	1.34	1.55	1.59	1.94	1.99	2.01	1.30	3.39	3.24	2.62	2.26	3.52	3.51	2.70	2.84	2.56	2.36	2.20
Na_2O	2.76	2.81	3.23	3.23	3.23	2.39	2.53	2.56	3.12	2.11	2.11	1.81	2.80	2.83	2.74	2.67	3.18	3.15	3.17	3.13	2.97	2.85	3.05
K_2O	3.77	4.00	4.01	4.50	4.50	6.59	6.38	6.32	6.03	4.85	4.83	6.20	4.02	4.22	4.81	6.22	3.50	3.49	4.65	3.79	5.25	5.68	4.62
P_2O_5	0.18	0.22	0.24	0.11	0.11	0.07	0.08	0.08	0.10	0.09	0.09	0.05	0.22	0.21	0.14	0.13	0.21	0.20	0.14	0.12	0.11	0.12	0.11
LOI	0.95	0.96	0.30	0.82	0.82	0.78	0.77	0.77	0.74	2.15	2.16	1.61	1.01	0.78	1.03	0.92	1.43	1.43	0.97	0.83	1.34	0.77	1.12
Total	99.94	99.62	100.22	99.53	99.53	99.52	99.55	99.78	99.65	99.72	100.11	99.78	99.89	99.94	100.13	99.50	99.88	99.54	100.12	99.57	99.61	99.57	100.12
FeO^T+MgO	6.40	6.77	7.18	4.13	4.13	2.88	3.39	3.44	4.29	3.86	3.88	2.69	7.17	6.67	5.62	4.73	6.72	6.70	5.09	4.60	4.35	4.89	3.82
Major elements / molar %																							
$Mg^\#$	26	23	22	22	22	21	21	21	22	24	24	26	22	23	20	20	22	22	21	20	22	22	23
A/CNK	1.04	0.99	0.94	1.07	1.07	1.08	1.09	1.09	1.04	1.07	1.07	1.07	0.95	0.96	0.98	0.99	0.97	0.97	0.94	0.98	1.08	1.03	1.05
Fe+Mg	0.107 5	0.112 8	0.119 2	0.068 4	0.068 4	0.047 5	0.056 1	0.056 9	0.071 0	0.064 5	0.064 7	0.045 2	0.119 0	0.110 8	0.092 4	0.077 9	0.111 3	0.111 2	0.084 0	0.075 9	0.072 1	0.081 1	0.063 4
Si	1.121 3	1.095 7	1.094 7	1.101 5	1.129 8	1.170 4	1.149 6	1.151 6	1.112 2	1.174 9	1.180 9	1.208 7	1.087 1	1.098 9	1.128 2	1.106 4	1.078 9	1.075 7	1.133 3	1.151 3	1.094 7	1.102 7	1.161 0
Ti	0.009 3	0.009 3	0.009 9	0.005 4	0.005 4	0.003 5	0.004 1	0.004 1	0.005 4	0.005 3	0.005 4	0.003 5	0.009 8	0.009 1	0.006 8	0.005 6	0.009 1	0.009 1	0.005 9	0.005 6	0.005 1	0.006 4	0.004 9
Al	0.280 5	0.280 7	0.282 1	0.305 8	0.305 8	0.285 8	0.295 8	0.296 8	0.308 4	0.259 9	0.259 5	0.253 6	0.282 3	0.284 2	0.278 7	0.296 6	0.294 6	0.293 0	0.279 9	0.277 0	0.321 3	0.306 2	0.287 4
Fe	0.084 9	0.091 4	0.097 6	0.056 5	0.056 5	0.039 5	0.046 6	0.047 2	0.058 6	0.051 8	0.052 3	0.035 8	0.097 4	0.090 5	0.077 8	0.065 3	0.091 7	0.091 3	0.069 9	0.063 5	0.059 4	0.066 7	0.051 8
Mn	0.000 8	0.001 1	0.001 4	0.000 6	0.000 6	0.000 3	0.001 1	0.001 1	0.001 4	0.000 6	0.000 6	0.000 4	0.001 4	0.001 4	0.001 1	0.000 8	0.001 3	0.001 4	0.001 0	0.000 8	0.000 6	0.000 7	0.000 6
Mg	0.022 6	0.021 3	0.021 6	0.011 9	0.011 9	0.007 9	0.009 4	0.009 7	0.012 4	0.012 7	0.012 4	0.009 4	0.021 6	0.020 3	0.014 6	0.012 7	0.019 6	0.019 9	0.014 1	0.012 4	0.012 7	0.014 4	0.011 7

Continued

Sample	YL-9-1	YL-9-2	YL-9-3	YL-9-4	YL-7-1	YL-7-2	YL-7-3	YL-7-3R	YL-7-4	YL-3-1	YL-3-2	YL-3-3	YL-3-4	YL-3-5	YL-2-1	YL-2-2	YL-2-3	YL-2-3R	YL-2-4	YL-2-5	YL-1-2	YL-1-3	YL-1-4
Location	27°9′38″N,102°13′15″E				27°9′9″N,102°12′37″E					27°9′13″N,102°13′22″E					27°8′44″N,102°13′20″E						27°8′44″N,102°13′20″E		
Ca	0.0499	0.0578	0.0546	0.0555	0.0426	0.0239	0.0276	0.0284	0.0346	0.0355	0.0358	0.0232	0.0604	0.0578	0.0467	0.0403	0.0628	0.0626	0.0481	0.0506	0.0456	0.0421	0.0392
Na	0.0891	0.0907	0.0874	0.1042	0.1042	0.0771	0.0816	0.1007	0.0681	0.0584	0.0904	0.0913	0.0884	0.0862	0.1021	0.1026	0.1016	0.1023	0.0884	0.0862	0.0958	0.0920	0.0984
K	0.0800	0.0849	0.0828	0.0851	0.0955	0.1399	0.1355	0.1342	0.1280	0.1030	0.1021	0.1321	0.0743	0.0896	0.1021	0.1321	0.0743	0.0987	0.0805	0.1115	0.1206	0.0981	
P	0.0025	0.0031	0.0034	0.0034	0.0015	0.0010	0.0010	0.0014	0.0014	0.0013	0.0007	0.0031	0.0020	0.0020	0.0018	0.0030	0.0030	0.0020	0.0020	0.0017	0.0015	0.0017	0.0015

Trace elements /ppm

Sample	YL-9-1	YL-9-2	YL-9-3	YL-9-4	YL-7-1	YL-7-2	YL-7-3	YL-7-3R	YL-7-4	YL-3-1	YL-3-2	YL-3-3	YL-3-4	YL-3-5	YL-2-1	YL-2-2	YL-2-3	YL-2-3R	YL-2-4	YL-2-5	YL-1-2	YL-1-3	YL-1-4
Li	16.0	11.1	11.7	11.3	31.7	20.1	25.5	26.1	34.5	7.57	7.69	5.46	9.68	9.15	11.2	10.3	9.86	9.78	11.3	11.7	5.68	23.4	14.8
Be	3.32	3.71	3.70	3.71	3.70	2.41	2.63	2.68	2.96	2.67	2.67	1.96	3.65	3.54	3.81	3.16	3.95	3.91	3.86	4.20	3.12	2.67	2.53
Sc	10.3	19.9	16.7	16.8	10.5	7.40	9.00	9.31	11.2	7.19	7.13	4.67	22.3	19.4	21.8	17.9	20.2	20.3	18.5	14.9	14.9	10.5	5.10
V	33.6	24.3	26.9	26.9	19.3	12.2	14.5	15.2	19.9	12.9	12.7	8.27	27.8	24.3	19.2	15.1	25.4	25.5	16.2	13.9	21.0	23.0	16.6
Cr	8.22	5.59	6.30	10.2	7.68	8.15	6.04	6.27	7.39	4.35	7.37	3.95	29.6	5.52	5.64	4.36	6.42	6.38	5.62	4.64	7.37	9.38	6.57
Co	172	151	138	139	173	180	163	170	159	162	161	187	123	115	159	138	152	153	224	175	151	167	200
Ni	4.44	2.63	3.71	5.78	2.76	4.36	2.23	2.36	2.69	1.72	3.74	1.51	16.9	2.56	3.04	1.99	3.87	2.88	3.25	1.96	2.35	2.81	2.33
Cu	13.7	5.43	14.0	9.25	2.25	1.77	1.84	2.46	2.47	3.25	3.35	2.38	10.8	8.67	3.89	3.60	7.15	6.88	3.02	2.73	5.63	6.96	4.62
Zn	71.3	64.0	93.8	94.2	60.8	42.4	49.9	51.4	62.6	47.4	47.7	33.6	77.2	89.7	75.6	65.5	85.3	86.0	71.1	66.4	44.9	70.8	51.4
Ga	21.5	22.3	23.1	23.6	24.6	21.8	22.4	23.2	24.5	18.9	18.8	17.0	22.9	22.2	23.1	22.2	24.2	24.4	23.1	23.3	23.8	23.2	20.9
Ge	1.92	2.14	2.30	2.39	1.77	1.86	1.79	1.84	1.87	1.75	1.80	1.81	2.25	2.13	2.19	2.11	2.30	2.34	2.16	2.10	1.84	1.78	1.54
Rb	213	192	201	199	221	285	271	282	273	218	216	241	203	208	227	263	159	160	218	194	173	235	170
Sr	145	167	150	152	129	110	116	120	131	128	127	107	163	166	134	153	175	176	134	136	176	147	162
Y	32.1	72.8	62.9	65.4	36.8	31.4	31.1	32.2	33.1	32.2	32.2	27.7	76.2	67.3	78.2	67.0	62.2	62.2	69.4	61.1	42.2	35.9	24.7
Zr	238	249	253	279	334	324	349	347	331	215	218	165	263	274	264	224	276	272	241	248	364	378	377
Nb	17.9	18.9	19.8	19.1	19.8	12.7	15.5	16.1	19.7	15.6	15.4	10.2	20.0	18.3	19.7	16.6	19.0	19.0	16.2	16.0	15.5	18.8	14.1
Cs	6.08	5.79	8.05	7.90	5.56	4.77	5.16	5.34	5.91	8.17	8.15	6.36	6.86	6.77	8.22	7.91	5.27	5.32	7.72	7.71	1.78	5.26	2.07
Ba	551	700	587	591	1351	1484	1572	1619	1765	643	646	832	669	727	738	1369	670	678	650	560	1515	1836	1433
La	42.0	38.5	46.2	79.6	46.3	96.3	75.2	79.4	56.9	94.6	94.8	94.7	37.5	31.9	56.5	47.6	53.8	55.8	61.4	58.4	53.0	47.7	41.8
Ce	86.6	88.6	99.7	167	84.2	218	162	170	108	202	200	202	86.0	73.5	123	103	117	121	133	127	108	98.9	84.0

Continued

Sample	YL-9-1	YL-9-2	YL-9-3	YL-9-4	YL-7-1	YL-7-2	YL-7-3	YL-7-3R	YL-7-4	YL-3-1	YL-3-2	YL-3-3	YL-3-4	YL-3-5	YL-2-1	YL-2-2	YL-2-3	YL-2-3R	YL-2-4	YL-2-5	YL-1-2	YL-1-3	YL-1-4
Location	27°9'38"N,102°13'15"E				27°9'9"N,102°12'37"E					27°9'13"N,102°13'22"E					27°8'44"N,102°13'20"E						27°8'44"N,102°13'20"E		
Pr	9.72	11.9	12.4	19.7	10.2	23.8	18.5	19.4	13.2	23.2	22.8	23.0	11.6	9.59	15.0	12.7	14.5	15.0	16.3	15.7	12.6	11.5	9.34
Nd	36.9	49.6	48.1	75.6	38.4	87.3	67.4	71.0	47.3	84.0	84.0	84.1	48.4	42.1	62.4	50.3	56.0	57.5	64.9	59.6	47.6	43.6	35.2
Sm	6.96	12.4	10.9	14.1	7.38	14.3	11.2	11.8	8.60	13.3	13.3	12.9	12.2	10.6	14.1	12.1	12.1	12.3	13.6	12.8	9.75	8.84	6.47
Eu	1.18	1.48	1.42	1.50	1.47	1.37	1.44	1.49	1.59	1.13	1.13	1.19	1.55	1.53	1.37	1.63	1.58	1.61	1.37	1.35	1.81	1.79	1.58
Gd	6.26	12.5	10.8	12.8	7.16	10.8	8.99	9.42	7.70	10.2	10.2	9.62	12.2	10.8	13.9	11.9	11.5	11.6	13.1	11.9	8.92	7.99	5.61
Tb	0.95	2.18	1.84	2.03	1.15	1.37	1.23	1.28	1.14	1.29	1.28	1.17	2.16	1.90	2.35	2.00	1.89	1.91	2.19	1.95	1.37	1.19	0.79
Dy	5.54	13.4	11.3	12.1	6.78	6.81	6.40	6.66	6.36	6.62	6.58	5.83	13.6	12.0	14.7	12.4	11.4	11.5	13.2	11.6	8.09	6.88	4.53
Ho	1.12	2.72	2.32	2.41	1.32	1.16	1.14	1.19	1.18	1.18	1.17	1.01	2.80	2.46	2.92	2.47	2.30	2.32	2.64	2.30	1.57	1.31	0.87
Er	3.24	7.70	6.68	6.92	3.62	3.00	2.98	3.10	3.17	3.23	3.19	2.78	8.13	7.18	8.18	6.98	6.59	6.58	7.36	6.40	4.33	3.58	2.44
Tm	0.49	1.14	1.01	1.03	0.52	0.39	0.40	0.42	0.45	0.46	0.45	0.40	1.20	1.06	1.18	1.01	0.97	0.97	1.06	0.93	0.62	0.50	0.35
Yb	3.08	6.86	6.36	6.46	3.16	2.30	2.42	2.50	2.69	2.90	2.90	2.55	7.56	6.67	7.23	6.20	6.23	6.20	6.65	5.78	3.89	3.08	2.15
Lu	0.46	0.97	0.92	0.93	0.45	0.33	0.35	0.36	0.39	0.43	0.43	0.38	1.07	0.94	1.05	0.89	0.90	0.90	0.94	0.83	0.58	0.46	0.32
Hf	6.78	7.31	7.49	8.39	9.23	9.76	10.1	10.1	9.07	6.59	6.62	5.22	7.55	7.83	7.79	6.64	8.03	8.00	7.38	7.52	9.89	10.0	10.0
Ta	1.66	1.94	2.02	1.95	1.66	1.12	1.28	1.32	1.52	1.45	1.44	1.23	1.90	1.78	1.80	1.58	1.83	1.84	1.49	1.68	1.25	1.34	1.13
Pb	17.2	19.4	22.6	24.6	20.8	34.8	28.7	29.9	24.4	26.3	26.2	31.4	20.1	21.3	23.5	28.4	19.2	19.5	24.2	21.9	18.5	25.4	23.1
Th	21.6	18.3	18.8	32.6	17.3	78.9	50.4	53.3	23.6	49.8	51.0	51.3	14.0	12.5	23.5	23.2	21.1	22.1	27.7	28.5	19.9	17.7	15.3
U	6.15	9.11	4.98	7.83	3.55	10.8	7.52	7.86	4.84	6.14	6.78	5.44	8.80	5.06	6.93	12.4	6.29	7.17	6.10	8.93	14.0	2.75	10.6
Eu/Eu*	0.54	0.36	0.40	0.34	0.62	0.34	0.44	0.43	0.60	0.30	0.30	0.33	0.39	0.44	0.30	0.42	0.41	0.41	0.31	0.33	0.59	0.65	0.80
T_{Zr}/°C	815	808	813	814	851	853	859	858	843	816	817	794	810	815	818	802	816	815	804	814	856	855	863
REE	204	250	260	403	212	467	360	378	258	444	443	441	246	212	323	271	297	305	338	316	262	237	196

$Mg^\# = molar\ 100MgO/(MgO+FeO^T)$; $A/CNK=Al_2O_3/(CaO+K_2O+Na_2O)$ molar ratio; $Eu/Eu^* = Eu_N/(Sm_N \cdot Gd_N)^{1/2}$; $Fe+Mg$: moles $Mg+Fe$ per 100 g of rock samples; T_{Zr}/°C: the whole-rock Zr saturation temperature.

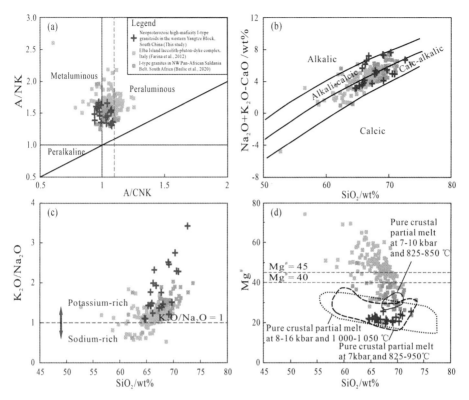

Fig. 6　The A/NK vs. A/CNK diagram（a）,（Na$_2$O + K$_2$O−CaO）vs. SiO$_2$（b）（Frost et al., 2001）,
K$_2$O/Na$_2$O vs. SiO$_2$（c）（Moyen and Martin, 2012）, Mg$^\#$ vs. SiO$_2$（d）diagrams for the Neoproterozoic
high-maficity I-type granitoids from the Yonglang area in the western Yangtze Block, South China.

The Elba Island laccolith-pluton-dyke complex in Italy are from Farina et al. (2012). The I-type granites in NW Pan-
African Saldania Belt, South Africa are from Bailie et al. (2020). In Fig. 6d, also shown are the fields of pure crustal
partial melts determined in experimental studies on dehydration melting of two-mica schist at 7−10 kbar, 825−850 ℃
(Patiño Douce and Johnston, 1991), low-K basaltic rocks at 8−16 kbar, 1 000−1 050 ℃ (Rapp and Watson, 1995), and
moderately hydrous (1.7−2.3 wt% H$_2$O) medium- to high-K basaltic rocks at 7 kbar, 825−950 ℃ (Sisson et al., 2005).

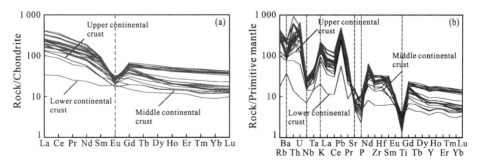

Fig. 7　Diagrams of chondrite-normalized REE patterns（a）and primitive mantle-normalized trace
element patterns（b）for the Neoproterozoic high-maficity I-type granitoids from the Yonglang area
in the western Yangtze Block, South China.

The primitive mantle and chondrite values are from Sun and McDonough (1989). The lower, middle,
and upper continental crust references are from Rudnick and Gao (2003).

Table 4　Whole-rock Sr-Nd isotopic compositions of the Neoproterozoic high-maficity I-type granitoids in the western Yangtze Block, South China.

Sample	$^{87}Sr/^{86}Sr$	2SE	Sr/ppm	Rb/ppm	$^{143}Nd/^{144}Nd$	2SE	Nd/ppm	Sm/ppm	Age/Ma	T_{DM2}/Ga	$\varepsilon_{Nd}(t)$	I_{Sr}
YL-9-1	0.763 668	0.000 012	145	213	0.511 874	0.000 004	36.9	6.96	850	1.82	−5.93	0.711 8
YL-9-2	0.753 645	0.000 009	167	192	0.512 057	0.000 005	49.6	12.4	850	1.85	−6.38	0.713 1
YL-7-1	0.772 757	0.000 012	129	221	0.511 917	0.000 004	38.4	7.38	845	1.78	−5.37	0.712 5
YL-7-2	0.784 380	0.000 009	110	285	0.511 889	0.000 005	87.3	14.3	845	1.69	−4.06	0.693 2
YL-3-3	0.787 020	0.000 005	107	241	0.511 844	0.000 004	84.1	12.9	840	1.70	−4.32	0.708 2
YL-2-1	0.773 033	0.000 003	134	227	0.512 043	0.000 003	62.4	14.1	833	1.76	−5.21	0.714 3
YL-2-2	0.774 107	0.000 004	153	263	0.512 038	0.000 003	50.3	12.10	833	1.83	−6.26	0.714 5
YL-1-3	0.769 129	0.000 005	147	235	0.511 973	0.000 003	43.6	8.84	825	1.75	−5.16	0.714 3

$^{87}Rb/^{86}Sr$ and $^{147}Sm/^{144}Nd$ ratios were calculated using Rb, Sr, Sm and Nd contents analyzed by ICP-MS, referenced from DePaolo (1981).

T_{DM2} represent the two-stage model age and were calculated using present-day $(^{147}Sm/^{144}Nd)_{DM} = 0.213\ 7$, $(^{147}Sm/^{144}Nd)_{DM} = 0.513\ 15$ and $(^{147}Sm/^{144}Nd)_{crust} = 0.101\ 2$. $\varepsilon_{Nd}(t)$ values were calculated using present-day $(^{147}Sm/^{144}Nd)_{CHUR} = 0.196\ 7$ and $(^{147}Sm/^{144}Nd)_{CHUR} = 0.512\ 638$.

$$\varepsilon_{Nd}(t) = \left[\left((^{143}Nd/^{144}Nd)_S(t) \right) / \left((^{143}Nd/^{144}Nd)_{CHUR}(t) - 1 \right) \right] \times 10^4.$$

$$T_{DM2} = \frac{1}{\lambda} \left\{ 1 + \left[\left((^{143}Nd/^{144}Nd)_S - ((^{147}Sm/^{144}Nd)_S - (^{147}Sm/^{144}Nd)_{crust})(e^{\lambda t} - 1) - (^{143}Nd/^{144}Nd)_{DM} \right) / \left[(^{147}Sm/^{144}Nd)_{crust} - (^{147}Sm/^{144}Nd)_{DM} \right] \right\}.$$

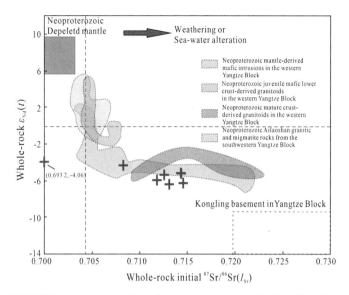

Fig. 8　Initial Sr-Nd isotopic compositions for the Neoproterozoic high-maficity I-type granitoids from the Yonglang area in the western Yangtze Block, South China (modified after Zhu et al., 2019c).

5　Discussion

5.1　Magmatic type

In this study, the Neoproterozoic Yonglang granitoids are characterized by a predominant magmatic mineralogy of quartz + plagioclase + K-feldspar + hornblende + biotite (Fig. 2), and lack Al-rich minerals such as muscovite, garnet, tourmaline, etc. This petrographic feature suggests an I-type affinity for the Yonglang granitoids (Chappell and White, 1992). The I-type affinity of Yonglang granitoids can also be evidenced by positive correlation of Rb and Th due to the fact that the Th-rich minerals (e.g., monazite) are not preferentially crystallized in metaluminous and mildly peraluminous melts (Fig. 9a) (Chappell, 1999). Furthermore, the Yonglang granitoids were plotted into I-type field in the $[Al_2O_3 - (Na_2O + K_2O)]$-CaO-$(FeO^T + MgO)$ ternary diagram (Fig. 9b), which is distinct from those of typical S-type granites from the northern Queensland in Australia and Cape Granite Suite in South Africa (Champion and Bultitude, 2013; Villaros et al., 2009a). Previous studies have suggested that I-type granitic melts gradually become more metaluminous with increasing maficity because of the predominant entrainment of clinopyroxene mineral (Clemens et al., 2011; Clemens and Stevens, 2012). Different from those of the Kosciusko and Darling S-type suites that show increasing A/CNK associated with higher values of maficity (Clemens et al., 2011), the Yonglang granitoids display the negative correlation between A/CNK and Fe + Mg values (Fig. 9c), which is similar to the evolution trend of Kosciusko and Moonbi I-type suites

(Clemens et al., 2011). In general, all of mineral and geochemical evidence support the interpretation that the Neoproterozoic Yonglang granitoids studied here belong to I-type granites. Moreover, it is worth remarking that the Yonglang granitoids are geochemically characterized by high molar Fe + Mg (0.045 2-0.121 0) and K_2O (3.49-6.59 wt%) contents.

The following parts will explore appropriatepetrogenesis of Neoproterozoic Yonglang high-K, high-maficity I-type granitoids in the western Yangtze Block, South China, including: ① magma mixing of mantle- and crust-derived magmas; ② partial melting of potassic metabasalts and basaltic andesites; and ③ peritectic assemblage entrainment.

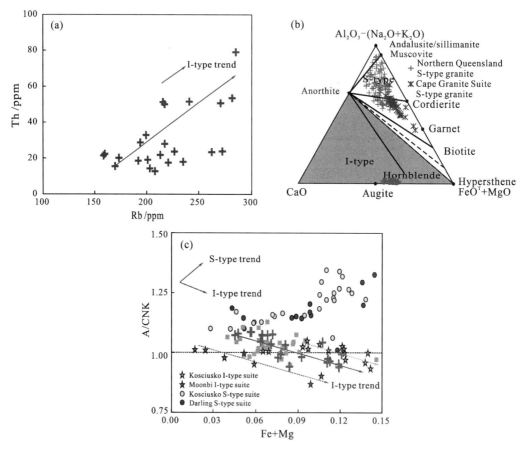

Fig. 9   The diagrams of Th vs. Rb (a), [$Al_2O_3-(Na_2O+K_2O)$]-CaO-(FeO^T+MgO) (b) (Zhu et al., 2020a), and A/CNK vs. Fe + Mg (c) (Clemens et al., 2011) for the Neoproterozoic high-maficity I-type granitoids from the Yonglang area in the western Yangtze Block, South China. In Fig. 9b, the S-type granites in Northern Queensland in Australia were from Champion and Bultitude (2013) and the S-type granites in Cape Granite Suite in South Africa were from Villaros et al. (2009a). In Fig. 9c, the Kosciusko S- and I-type suites are from southeastern Australia, the I-type Moonbi suite is from the New England batholith of New South Wales in Australia, and the S-type rocks of the Darling S-type suite is part of the Cape Granite Suite of South Africa (Clemens et al., 2011). Symbols as Fig. 6.

5. 2　Magma mixing of mantle- and crust-derived magmas

Experimental petrology has displayed that the anatectic melts formed by fluid-absent partial melting of crustal rocks (e.g., metapelites, metagreywackes, etc.) are geochemically featured by high Si and low Fe, Mg, Ca, and Ti concentrations (Fig. 10) (Stevens et al., 2007). However, the Yonglang I-type granitoids studied here acquire geochemical compositions beyond the range defined by the compositions of crustal melts (i.e., pelites, volcanoclastic greywackes, and greywackes) (Fig. 10). They display variable Fe + Mg, Ti, and Ca contents that are too high to represent pure crustal melts (Fig. 10), indicating the demand of additional ferromagnesian minerals (Farina et al., 2012). There is a general knowledge that the incorporation of mantle-derived magmas can make up high mafic components in I-type granitic melts and explain their chemical and isotopic variations (Clemens et al., 2011 and references therein). However, numerous lines of field, petrological, geochemical evidence argue against this candidate for the Yonglang high-maficity I-type granitoids.

First, the Yonglang granitoids in this study are devoid of voluminous mafic-intermediate microgranular enclaves as well as disequilibrium mineral features (e. g., crystal resorption texture and reactive overgrowth rim) (Fig. 2) (SPBGMR, 1972). There are also no significant populations of coeval mafic intrusions around the Yonglang area (Fig. 1c). The field and petrological features thus exclude their derivation from a greater proportion of admixed mantle-derived mafic melts.

Second, the Yonglang I-type granitoids have high and variable SiO_2 contents (64. 63 - 72. 62 wt%) and low $Mg^{\#}$ values (20-26) (Fig. 6d) as well as Cr (3. 95-29. 6 ppm) and Ni (1. 51-16. 9 ppm) contents, which is different from the significant involvement of mantle-derived magmas that would elevate $Mg^{\#}$, Cr, and Ni values of resultant melts (Xing et al., 2020). During the magma mixing process, elements with similar diffusivity values generally display linear patterns while elements with contrasting diffusivity will be strongly unrelated (De Campos et al., 2011; Perugini et al., 2008). Previous studies have indicated that the Na-K and Sr-Ba pairs in felsic melts have similar diffusivities while Mg and Ti have contrasting diffusivity with Ti being three orders of magnitude less mobile than Mg (Farina et al., 2012; Zhang et al., 2010). Therefore, the magma mixing process would lead to good linear correlations of Sr vs. Ba and K vs. Na and no correlation of Mg vs. Ti. In contrast, the scattered distribution in Sr vs. Ba and K vs. Na (Fig. 11a,b), coupled with the tight relationship in Mg vs. Ti ($R^2 = 0. 98$) (Fig. 11c), cast doubt on the magma mixing of mantle- and crust-derived melts for the Yonglang I-type granitoids studied here (Farina et al., 2012). The magma mixing hypothesis is also difficult to reconcile the scattered variation between K and maficity (Fig. 10a) and tight correlation between Ti and maficity ($R^2 = 0. 98$) (Fig. 10c) (Clemens et

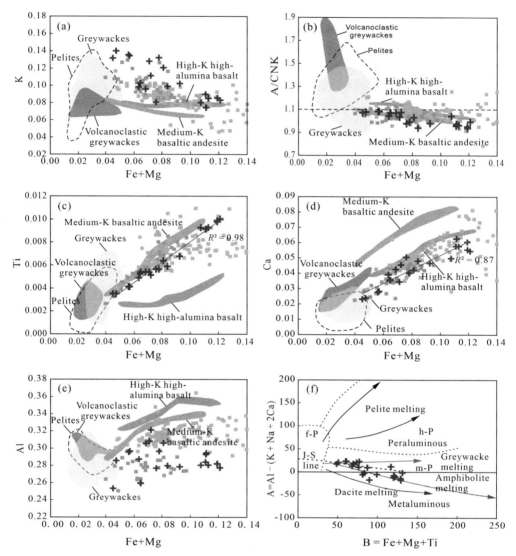

Fig. 10　Comparison between the composition of fluid-absent experimental melts and the Neoproterozoic high-maficity I-type granitoids from the Yonglang area in the western Yangtze Block, South China (a-e) (modified after Farina et al., 2012). In Fig. 10a-e, the fields are from experiments performed using different natural starting materials at $T < 950$ ℃ and pressures in the range of 0.5-1.5 GPa. The fields represent ≈ 95% of experimental melt compositions, with obvious outlier from trends having been excluded. Metagreywacke-derived melts are from Gardien et al. (1995), Montel and Vielzeuf (1997), Patiño Douce and Beard (1996), Stevens et al. (1997); metapelite-derived melts are from Patiño Douce and Harris (1998), Patiño Douce and Johnston (1991), Pickering and Johnston (1998), Stevens et al. (1997), Vielzeuf and Holloway (1988); metavolcanoclastic metagreywacke-derived melt is from Skjerlie and Johnston (1996). Melts from high-K high-alumina basalts and medium-K basaltic andesites are from Sisson et al. (2005). In Fig. 10f, the arrows represent the compositions of progressive melt fractions produced by experimental melting of different crustal rocks (Villaseca et al., 1998 and references therein). Symbols as Fig. 6.

al., 2011; Clemens and Stevens, 2012; Farina et al., 2012).

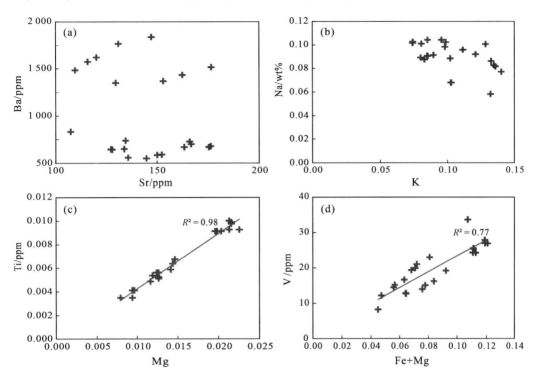

Fig. 11   The diagrams of Ba vs. Sr (a), Na vs. K (b), Ti vs. Mg (c), and V vs. Fe + Mg (d)
for the Neoproterozoic high-maficity I-type granitoids
from the Yonglang area in the western Yangtze Block, South China.

Third, the involvement of Neoproterozoic depleted mantle-derived melts in the western Yangtze Block is usually expected by significantly depleted Nd-Hf isotopic compositions. The Yonglang I-type granitoids, however, yield pronouncedly negative whole-rock $\varepsilon_{Nd}(t)$ (-6.38 to -4.06) (Fig. 8) and highly evolved zircon $\varepsilon_{Hf}(t)$ (-9.36 to $+1.92$) values (Fig. 4), which are inconsistent with the predominantly depleted Nd-Hf isotopic features of Neoproterozoic mantle-derived mafic intrusions in the western Yangtze Block (Figs. 4 and 8).

In summary, a major involvement of mantle-derived ferromagnesian components is unlikely for the genesis of more mafic I-type granitoids collected from the Yonglang area in the western Yangtze Block.

5.3   Partial melting of potassic metabasalts and basaltic andesites

Apart from high maficity index (molar Fe + Mg = 0.045 2−0.121 0), the Yonglang I-type granitoids belong to potassic melts with high K_2O (3.49−6.59 wt%) contents and K_2O/Na_2O (1.10−3.43) ratios (Fig. 6c). Nevertheless, this is contrast with the sodic experimental melts (usually $Na_2O/K_2O > 1$) derived from fluid-absent partial melting of metabasalts and eclogites (e.g., Rushmer, 1991; Sen and Dunn, 1994). Although Sisson et al. (2005) suggested that the K-rich, high-maficity granitic melts can be formed by fluid-absent partial melting of medium- to high-K basalts and basaltic andesites, the specific geochemical characteristics of Yonglang I-type granitoids can not support the interpretation of medium- to high-K basaltic protoliths.

Farina et al. (2012) pointed out that the K and Na contents in granitic magmas are related to the stoichiometry of melting reactions, thus can reflect the geochemical characteristics of melts source. The Yonglang granitoids studied here display incompatible geochemistry (e.g., K) with partial melting results from medium- to high-K basaltic rocks exemplified by Sisson et al. (2005) (e.g., Fig. 10a). Fluid-absent partial melting of medium- to high-K basaltic rocks generally generate granitic melts with high potassium contents (3.0 - 3.9 wt% K_2O, 0.07 - 0.08 molar K), while the K_2O contents of the Yonglang I-type granitoids are significantly higher (3.49 - 6.59 wt% K_2O, 0.07 - 0.14 molar K) (Fig. 10a) (Farina et al., 2012; Sisson et al., 2005). In addition, the resultant melts sourced from medium- to high-K basaltic rocks are characterized by metaluminous to strongly peraluminous features associated with increasing Fe + Mg values (Sisson et al., 2005), but the Yonglang granitoids are metaluminous to slightly peraluminous (A/CNK = 0.94 - 1.09) with variable Fe + Mg values (Fig. 10b). In terms of the contents of Ti, Ca, and Al, the Yonglang granitoids are also not exactly the same with the experimental melts produced from medium-K basaltic andesites and high-K high-alumina basalts (Fig. 10c-e) (Bailie et al., 2020; Farina et al., 2012; Sisson et al., 2005). Another argument precluding these protoliths is that there is no crop out of ancient potassic-ultrapotassic igneous rocks (e.g., Paleoproterozoic potassic to ultrapotassic basalts) in the western Yangtze Block, South China (Lu et al., 2019 and references therein).

Therefore, the partial melting of potassium-rich basaltic rocks, including the medium-K basaltic andesite and high-K high-alumina basalt, is not a reasonable mechanism for the Yonglang I-type granitoids.

5.4　Peritectic assemblage entrainment

The peritectic assemblage entrainment (PAE) model has been gradually used to explain a large spectrum of maficity (molar Fe + Mg), Ti, and Ca values in more mafic I- and S-type granitic melts (Bailie et al., 2020; Champion and Bultitude, 2003; Clemens et al., 2011; Clemens and Stevens, 2012; Farina et al., 2012; Garcia-Arias and Stevens, 2017; Stevens et al., 2007; Villaros et al., 2009a,b, 2012; Zhu et al., 2020a). The peritectic mineral phases described in PAE model are initially entrained into mafic granitic melts from magmatic source, and considered as major contributors of additional ferromagnesian components (Bailie et al.,

2020; Clemens et al., 2011; Clemens and Stevens, 2012; Farina et al., 2012; Villaros et al., 2012; Zhu et al., 2020a). The geochemical basis of PAE model is the strong correlations of low-solubility elements (e.g., Ti, Ca) plotted with maficity (molar Fe + Mg) (Clemens et al., 2011; Clemens and Stevens, 2012). Like the Cotoncello-Monte Capanne granitic complex in Elba Island and Verdenburg I-type granite in South Africa (Bailie et al., 2020; Farina et al., 2012), the Yonglang I-type granitoids in this study display high degree of positive linear correlations for Ti vs. maficity ($R^2 = 0.98$) (Fig. 10c) and Ca vs. maficity ($R^2 = 0.87$) (Fig. 10d) as well as negative correlation between A/CNK and Fe + Mg values (Fig. 10b). This imply that the Yonglang granitoids were formed by incongruent melting of hydrous phases, such as biotite and hornblende of mafic-intermediate rocks in crustal source, and have underwent the co-entrainment of peritectic selective minerals dominated by the Ca-Fe-Mg-Ti bearing phases (e.g., ferromagnesian and plagioclase minerals) (Bailie et al., 2020; Clemens et al., 2011; Clemens and Stevens, 2012; Farina et al., 2012; Gao et al., 2016; Skjerlie and Johnston, 1996; Skjerlie and Patiño Douce, 1995). The Yonglang I-type granitoids display the significantly enriched Nd-Hf isotopic features [$\varepsilon_{Nd}(t) = -6.38$ to -4.06, $\varepsilon_{Hf}(t) = -9.36$ to $+1.92$] (Figs. 4 and 8) as well as ancient two-stage Nd model ages (1.69 – 1.85 Ga) and crustal Hf model ages (1 608 – 2 139 Ma) (Table 2), indicating a predominantly crustal source. Because biotite is a main K-bearing mineral phase during crustal melting, the extremely high K:Na ratios ($K_2O/Na_2O = 1.10 – 3.43$) observed in Yonglang granitoids indicate that biotite is a major reactant phase in the melting reaction (Clemens et al., 2011; Farina et al., 2012). The obviously positive correlation defined by V (vanadium) vs. maficity ($R^2 = 0.77$) (Fig. 11d) in Yonglang granitoids also support the incongruent melting of biotite-bearing crustal source (Bailie et al., 2020). The involvement of hornblende phase in the melting reaction can be demonstrated in the Al – (K + Na + 2Ca) vs. Fe + Mg + Ti diagram (Fig. 10f) (Bailie et al., 2020; Villaseca et al., 1998). Clemens et al. (2011) have modelled that variable Bt:Hbl ratios in crustal source can lead to different slopes of Ti vs. maficity in resultant granitic melts. The Yonglang granitoids, as shown in Fig. 12a, definitely display the partial melting of crustal source containing both biotite and hornblende phases (Clemens et al., 2011; Clemens and Stevens, 2012). Therefore, all above evidence suggest that the Yonglang high-maficity I-type granitoids studied here arise from partial melting of crustal source rocks dominantly containing hornblende and biotite mineral phases.

The coupled melting reaction of biotite and hornblende in crustal source generally produces a peritectic ferromagnesian assemblage dominated by clinopyroxene and orthopyroxene, as shown by the peritectic reaction of $Bt + Hbl + Qtz + Pl^1 = melt + Pl^2 + Cpx + Opx + Ilm \pm Grt$ (Pl^1: plagioclase in reactant phase; Pl^2: plagioclase in peritectic phase) (Clemens et al., 2011; Clemens and Stevens, 2012; Skjerlie and Johnston, 1996). The peritectic garnet is usually detected in S-type granites (Stevens et al., 2007; Villaros et al., 2012; Zhu et al., 2020a), which is inconsistent with the case of Yonglang granitoids showing I-type affinity. Thus, the peritectic garnet is not compatible with Yonglang granitoids. Due to the low alkali (Na_2O + K_2O) and CaO contents in orthopyroxene mineral, the entrainment of this phase usually generates a negative correlation of Ca vs. maficity as well as positive correlation of A/CNK vs. maficity. However, the peritectic clinopyroxene lead to an opposite effect, causing a

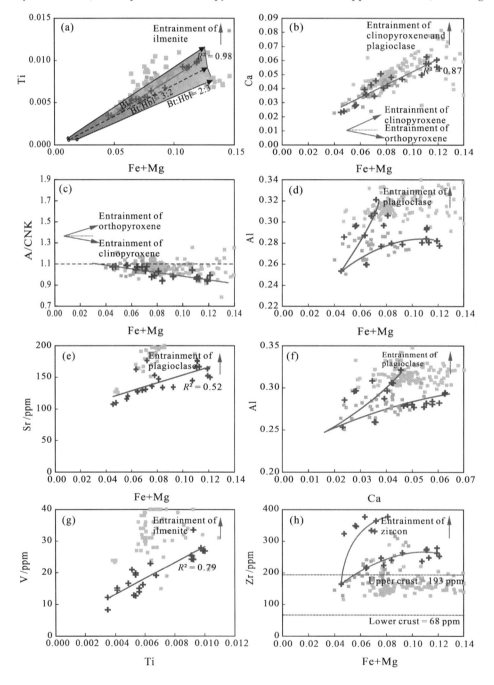

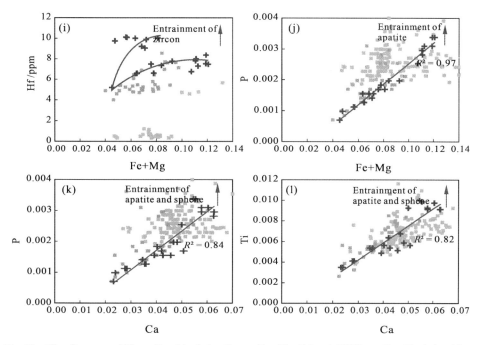

Fig. 12　The diagrams of Ti vs. Fe+Mg (a), Ca vs. Fe+Mg (b), A/CNK vs. Fe+Mg (c), Al vs. Fe+Mg (d), Sr vs. Fe+Mg (e), Al vs. Ca (f), V vs. Ti (g), Zr vs. Fe+Mg (h), Hf vs. Fe+Mg (i), P vs. Fe+Mg (j), P vs. Ca (k), and Ti vs. Ca (l) for the Neoproterozoic high-maficity I-type granitoids from the Yonglang area in the western Yangtze Block, South China, which display peritectic assemblage entrainment of clinopyroxene, plagioclase, ilmenite, zircon, apatite, and sphene.

In Fig. 12a (modified after Clemens et al., 2011), the shadow area displays the geochemical variation produced by partial melting of a rock with biotite and hornblende. The solid line represents the trend that would result from the breakdown of biotite in a metapelitic rock, producing peritectic garnet and ilmenite. The trend marked with long dashes represents mixing between the initial melt and peritectic garnet, clinopyroxene and ilmenite in the ratio of these minerals produced by partial melting of a rock with biotite and hornblende in the ratio 3:2. The final trend, marked with short dashes, represents mixing between the initial melt and peritectic orthopyroxene, clinopyroxene, garnet and ilmenite formed during partial melting of a rock containing biotite and hornblende in the ratio 2:3.

progressively increasing Ca at increasing maficity and a trend toward metaluminous with higher values of maficity (Clemens et al., 2011; Farina et al., 2012). In this study, the well-defined positive correlation of Ca vs. maficity ($R^2 = 0.87$) and negative trend of A/CNK vs. maficity indicate that the Yonglang granitoids underwent the peritectic entrainment of clinopyroxene, not the orthopyroxene phase (Fig. 12b,c). Farina et al. (2012) suggested that the coupled melting of biotite and hornblende phases in crustal source produces peritectic clinopyroxene, simultaneously generates entrained plagioclase. Considering the Ca, Al, and Sr concentrations are compatible in plagioclase mineral, the positive correlations of Ca vs. maficity, Al vs. maficity, and Sr vs. maficity observed in Yonglang granitoids reflect the entrainment of peritectic plagioclase (Fig. 12b,d,e) (Bailie et al., 2020; Farina et al., 2012; Garcia-Arias and Stevens, 2017). The entrainment of plagioclase can also be detected by positive correlation between Ca and Al (Fig. 12f) (Farina et al., 2012).

Apart from the predominant peritectic mineral phases, the incongruent melting of both biotite and hornblende phases in I-type granites also lead to the entrainment of peritectic accessory mineral phases (Bailie et al., 2020; Clemens et al., 2011; Clemens and stevens, 2012; Farina et al., 2012). The entrained ilmenite can produce the increasing Ti with increasing maficity (Clemens et al., 2011; Clemens and stevens, 2012; Farina et al., 2012). In this study, the highly positive linear correlation of Ti vs. maficity ($R^2 = 0.98$) in the Yonglang granitoids reflect the peritectic entrainment of ilmenite (Fig. 12a) (Bailie et al., 2020; Clemens and stevens, 2012; Vielzeuf and Montel, 1994). Because V has high compatibility in ilmenite ($K_D \approx 80$) (Farina et al., 2012; Latourette et al., 1991), the entrainment of peritectic ilmenite in the Yonglang granitoids is further supported by the positive correlation between Ti and V ($R^2 = 0.79$) (Fig. 12g) (Clemens and stevens, 2012; Farina et al., 2012). In addition, the Yonglang granitoids have higher whole-rock Zr (165−378 ppm) concentrations than the values of average lower crust (68 ppm) and upper crust (193 ppm) as well as typical crustal rocks (~100−200 ppm) (Miller et al., 2003 and references therein; Rudnick and Gao, 2003), implying the presence of peritectic zircon in source (Zhu et al., 2020a). The positive relationships between Zr and Hf with maficity for the Yonglang granitoids further suggest that zircons hosted in biotite have been entrained during the incongruent melting (Fig. 12h, i) (Bailie et al., 2020; Farina et al., 2012; Stevens et al., 2007; Villaros et al., 2009a, b). Moreover, the potential effect of entrainment of Ca-bearing accessary mineral phases (e.g., apatite and sphene) can be shown by the tight relationships of P vs. maficity ($R^2 = 0.97$), P vs. Ca ($R^2 = 0.84$), and Ti vs. Ca ($R^2 = 0.82$) (Fig. 12j-l) (Bailie et al., 2020; Farina et al., 2012; Villaros et al., 2009b).

In summary, the Neoproterozoic Yonglang high-maficity I-type granitoids were formed by coupled biotite and hornblende fluid-absent melting in a crustal source, simultaneously accompanied by the peritectic assemblage entrainment of clinopyroxene and plagioclase phases as well as accessory minerals (e.g., zircon, ilmenite, apatite, and sphene), as shown in Fig. 13 (modified after Villaros et al., 2018).

6　Concluding remarks

The zircon U-Pb geochronological results display that the newly identified Neoproterozoic Yonglang high-maficity I-type granitoids in the western Yangtze Block were formed at ca. 850−825 Ma. Compared with the geochemical composition of fluid-absent experimental melts, the Yonglang I-type granitoids contain higher Fe, Mg, Ti, and Ca contents than the pure crustal melts. Considering the highly positive linear correlations for Ti vs. maficity and Ca vs. maficity

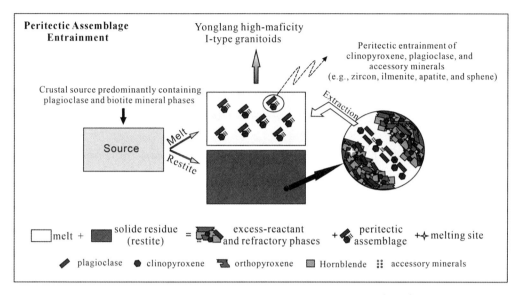

Fig. 13 Schematic representation of the peritectic assemblage entrainment (PAE) model acting on
the genesis of the Neoproterozoic high-maficity I-type granitoids from the Yonglang area
in the western Yangtze Block, South China (modified after Villaros et al., 2018).

as well as negative correlation for A/CNK vs. maficity, we support that the Yonglang high-maficity I-type granitoids were formed by the coupled melting of biotite and hornblende phases in ancient crustal source, simultaneously experienced the significantly peritectic assemblage entrainment of clinopyroxene and plagioclase as well as accessory minerals (e.g., zircon, ilmenite, apatite, and sphene), similar to the typical I-type granites controlled by PAE model elsewhere (e.g., Bailie et al., 2020; Farina et al., 2012; Hu et al., 2019). Consequently, the peritectic assemblage entrainment model is an effective mechanism for shaping the unique geochemistry of more mafic I-type granitic melts, as proposed by Clemens et al. (2011).

Acknowledgements We thank Editor Prof. Michael Roden for his editor work and constructive comments. We also appreciate two anonymous reviewers for their constructive suggestions, which significantly improve the quality of our manuscript. Prof. Chunrong Diwu is thanked for his profound suggestion and discussion. This work was jointly supported by the National Natural Science Foundation of China (Grant Nos. 41421002 and 41772052) and the program for Changjiang Scholars and Innovative Research Team in University (IRT1281).

References

Andersen, T., 2002. Correction of common lead in U-Pb analyses that do not report [204]Pb. Chemical Geology, 192(1-2), 59-79.

Appleby, S.K., Gillespie, M.R., Graham, C.M., Hinton, R.W., Oliver, G.J.H., Kelly, N.M., 2010. Do S-type granites commonly sample infracrustal sources? New results from an integrated O, U-Pb and Hf isotope study of zircon. Contributions to Mineralogy and Petrology, 160(1), 115-132.

Bailie, R., Adriaans, L., Roux, P.L., 2020. Peritectic assemblage entrainment as the main compositional driver in the I-type Vredenburg Granite, north-western Pan-African Saldania Belt, South Africa: A whole-rock chemical perspective. Lithos, 364-365, 105522.

Bao, Z.A., Chen, L., Zong, C.L., Yuan, H.L., Chen, K.Y., Dai, M.N., 2017. Development of pressed sulfide powder tablets for in situ, sulfur and lead isotope measurement using LA-MC-ICP-MS. International Journal of Mass Spectrometry, 421, 255-262.

Champion, D.C., Bultitude, R.J., 2013. The geochemical and Sr, Nd isotopic characteristics of Paleozoic fractionated S-types granites of north Queensland: Implications for S-type granite petrogenesis. Lithos, 162-163(2), 37-56.

Chappell, B.W., 1999. Aluminium saturation in I- and S-type granites and the characterization of fractionated haplogranites. Lithos, 46, 535-551.

Chappell, B.W., White, A.J.R., 1992. I- and S-type granites in the Lachlan fold belt. Transactions of the Royal Society of Edinburgh: Earth Sciences, 83.

Chen, W.T., Sun, W.H., Wang, W., Zhao, J.H., Zhou, M.F., 2014. "Grenvillian" intra-plate mafic magmatism in the southwestern Yangtze Block SW China. Precambrian Research, 242, 138-153.

Chu, Z.Y., Chen, F.K., Yang, Y.H., Guo, J.H., 2009. Precise determination of Sm, Nd concentrations and Nd isotopic compositions at the nanogram level in geological samples by thermal ionization mass spectrometry. Journal of Analytical Atomic Spectrometry, 24, 1534-1544.

Clemens, J.D., 1990. The granulite-granite connexion. In: Vielzeuf, D., Vidal, P. Granulites and Crustal Differentiation. NATO ASI Series. Dordrecht: Kluwer Academic Publishers, 25-36.

Clemens, J.D., 2003. S-type granitic magmas-petrogenetic issues, models and evidence. Earth-Science Reviews, 61, 1-18.

Clemens, J.D., 2018. Granitic magmas with I-type affinities, from mainly metasedimentary sources: The Harcourt batholith of southeastern Australia. Contributions to Mineralogy and Petrology, 173, 93.

Clemens, J.D., Stevens, G., 2012. What controls chemical variation in granitic magmas? Lithos, 134-135, 317-329.

Clemens, J.D., Stevens, G., 2016. Melt segregation and magma interactions during crustal melting: Breaking out of the matrix. Earth-Science Reviews, 160, 333-349.

Clemens, J.D., Stevens, G., Farina, F., 2011. The enigmatic sources of I-type granites: The peritectic connexion. Lithos, 126(3-4), 174-181.

Clemens, J.D., Watkins, J.M., 2001. The fluid regime of high temperature metamorphism during granitoid magma genesis. Contributions to Mineralogy and Petrology, 140, 600-606.

De Campos, C.P., Perugini, D., Ertel-Ingrisch, W., Dingwell, D.B., Poli, G., 2011. Enhancement of magma mixing efficiency by chaotic dynamics: An experimental study. Contributions to Mineralogy and Petrology, 161(6), 863-881.

DePaolo, D.J., 1981. A Neodymium and Strontium Isotopic Study of the Mesozoic Calc-Alkaline Granitic Batholiths of the Sierra Nevada and Peninsular Ranges, California. Granites and Rhyolites. Journal of Geophysical Research, 86 (B11), 10470-10488.

Farina, F., Stevens, G., Dini, A., Rocchi, S., 2012. Peritectic phase entrainment and magma mixing in the late Miocene Elba Island laccolith-pluton-dyke complex (Italy). Lithos, 153, 243-260.

Frost, B.R., Barnes, C.G., Collins, W.J., Arculus, R.J., Ellis, D.J., Frost, C.D., 2001. A geochemical classification for granitic rocks. Journal of Petrology, 42, 2033-2048.

Gao, P., Zheng, Y.F., Zhao, Z.F., 2016. Experimental melts from crustal rocks: A lithochemical constraint on granite petrogenesis. Lithos, 266-267, 133-157.

Gao, S., Ling, W.L., Qiu, Y., Zhou, L., Hartmann, G., Simon, K., 1999. Contrasting geochemical and Sm-Nd isotopic compositions of Archean metasediments from the Kongling high-grade terrain of the Yangtze craton: Evidence for cratonic evolution and redistribution of REE during crustal anatexis. Geochimica et Cosmochimica Acta, 63(13-14), 2071-2088.

Gao, S., Yang, J., Zhou, L., Li, M., Hu, Z., Guo, J., Yuan, H., Gong, H., Xiao, G., Wei, J., 2011. Age and growth of the Archean Kongling terrain, South China, with emphasis on 3.3 Ga granitoid gneisses. American Journal of Science, 311, 153-182.

Garcia-Arias, M., Stevens, G., 2017. Phase equilibrium modelling of granite magma petrogenesis: An evaluation of the magma compositions produced by crystal entrainment in the source. Lithos, 277, 131-153.

Gardien, V., Thompson, A.B., Grujic, D., Ulmer, P., 1995. Experimental melting of biotie + plagioclase + quartz ± muscovite assemblages and implications for crustal melting. Journal of Geophysical Research: Solid Earth, 100(B8), 15581-15591.

Griffin, W.L., Pearson, N.J., Belousova, E., Jackson, S.E., van Achterbergh, E., O'Reilly, S.Y., Shee, S. R., 2000. The Hf isotope composition of cratonic mantle: LAM-MC-ICP-MS analysis of zircon megacrysts in kimberlites. Geochimica et Cosmochimica Acta, 64(1), 133-147.

Hawkesworth, C.J., Kemp, A.I.S., 2006. The differentiation and rates of generation of the continental crust. Chemical Geology, 226(3-4), 134-143.

Hoskin, P.W.O., Schaltegger, U., 2003. The composition of zircon and igneous and metamorphic petrogenesis. Reviews in Mineralogy and Geochemistry, 53, 27-62.

Hu, J.Q., Li, X.W., Xu, J.F., Mo, X.X., Wang, F.Y., Yu, H.X., Shan, W., Xing, H.Q., Huang, X.F., Dong, G.C., 2019. Generation of coeval metaluminous and muscovite-bearing peraluminous granitoids in the same composite pluton in West Qinling, NE Tibetan Plateau. Lithos, 344-345, 374-392.

Huang, X.L., Xu, Y.G., Li, X.H., Li, W.X., Lan, J.B., Zhang, H.H., Liu, Y.S., Wang, Y.B., Li, H. Y., Luo. Z.Y., Yang, Q.J., 2008b. Petrogenesis and tectonic implications of Neoproterozoic, highly fractionated A-type granites from Mianning, South China. Precambrian Research, 165(3-4), 190-204.

Huang, X.L., Xu, Y.G., Lan, J.B., Yang, Q.J., Luo, Z.Y., 2009. Neoproterozoic adakitic rocks from Mopanshan in the Western Yangtze craton: Partial melts of a thickened lower crust. Lithos, 112(3), 367-381.

Kemp, A.I.S., Hawkesworth, C.J., Foster, G.L., Paterson, B.A., Woodhead, J.D., Hergt, J.M., Gray, C. M., Whitehouse, M.J., 2007. Magmatic and crustal differentiation history of granitic rocks from Hf-O isotopes in zircon. Science, 315, 980-983.

Latourette, T.Z., Burnet, D.S., Bacon, C.R., 1991. Uranium and minor-element partitioning in Fe-Ti oxides and zircons from partially melted granodiorites, Crater Lake, Oregon. Geochimica et Cosmochimica Acta,

55(2), 457-469.

Liu, X.M., Gao, S., Diwu, C.R., Yuan, H.L., Hu, Z.C., 2007a. Simultaneous in-situ determination of U-Pb age and trace elements in zircon by LA-ICP-MS in 20 μm spot size. Chinese Science Bulletin, 52(9), 1257-1264.

Liu, Y., Liu, X.M., Hu, Z.C., Diwu, C.R., Yuan, H.L., Gao, S., 2007b. Evaluation of accuracy and long-term stability of determination of 37 trace elements in geological samples by ICP-MS. Acta Petrologica Sinica, 23(5), 1203-1210 (in Chinese with English abstract).

Lu, G.M., Wang, W., Ernst, R.E., Söderlund, U., Lan, Z.F., Huang, S.F., Xue, E.K., 2019. Petrogenesis of Paleo-Mesoproterozoic mafic rocks in the southwestern Yangtze Block of South China: Implications for tectonic evolution and paleogeographic reconstruction. Precambrian Research, 322, 66-84.

Ludwig, K.R., 2003. ISOPLOT 3. 0: A geochronological toolkit for Microsoft Excel. Berkeley Geochronology Center, Special Publication, 4.

Miller, C.F., McDowell, S.M., Mapes, R.W., 2003. Hot and cold granites? Implications of zircon saturation temperatures and preservation of inheritance. Geology, 31, 529-532.

Montel, J.M., Vielzeuf, D., 1997. Partial melting of greywackes, Part II. Compositions of minerals and melts. Contributions to Mineralogy and Petrology, 128, 176-196.

Moyen, J.F., Martin, H., 2012. Forty years of TTG research. Lithos, 148, 312-336.

Moyen, J.F., Laurent, O., Chelle-Michou, C., Couzinié, S., Vanderhaeghe, O., Zeh, A., 2017. Collision vs. subduction-related magmatism: Two contrasting ways of granite formation and implications for crustal growth. Lithos, 277, 154-177.

Nowell, G.M., Kempton, P.D., Noble, S.R., Fitton, J.G., Saunders, A.D., Mahoney, J.J., Taylor, R.N., 1998. High precision Hf isotope measurements of MORB and OIB by thermal ionisation mass spectrometry: Insights into the depleted mantle. Chemical Geology, 149(3-4), 211-233.

Patiño Douce, A.E., 1999. What do experiments tell us about the relative contributions of crust and mantle to the origin of the granitic magmas. Geological Society, London, Special Publications, 168(1), 55-75.

Patiño Douce, A.E., Beard, J.S., 1995. Dehydration-melting of biotite gneiss and quartz amphibolite from 3 kbar to 15 kbar. Journal of Petrology, 36, 707-738.

Patiño Douce, A.E., Johnston, A.D., 1991. Phase equilibria and melt productivity in the pelitic system: Implications for the origin of peraluminous granitoids and aluminous granulites. Contributions to Mineralogy and Petrology, 107, 202-218.

Patiño Douce, A.E., Harris, N., 1998. Experimental constraints on Himalayan anatexis. Journal of Petrology, 39(4), 689-710.

Perugini, D., De Campos, C.P., Dingwell, D.B., Petrelli, M., Poli, G., 2008. Trace element mobility during magma mixing: Preliminary experimental results. Chemical Geology, 256, 146-157.

Pickering, J. M., Johnston, D. A., 1998. Fluid-absent melting behaviour of a two mica metapelite: Experimental constraints on the origin of Black Hill Granite. Journal of Petrology, 39(10), 1787-1804.

Rapp, R. P., Watson, E. B., 1995. Dehydration melting of metabasalt at 8 – 32 kbar: Implications for continental growth and crust-mantle recycling. Journal of Petrology, 36(4), 891-931.

Rong, W., Zhang, S.B., Zheng, Y.F., 2017. Back-reaction of peritectic garnet as an explanation for the origin of mafic enclaves in S-type granite from the Jiuling Batholith in South China. Journal of Petrology, 3, 569-598.

Rudnick, R.L., Gao, S., 2003. Composition of the continental crust. Treatise Geochemistry, 3, 1-64.

Rushmer, T., 1991. Partial melting of two amphibolites: Contrasting experimental results under fluid-absent conditions. Contributions to Mineralogy and Petrology, 107, 41-59.

Sen, C., Dunn, T., 1994. Dehydration melting of a basaltic composition amphibolite at 1.5 GPa and 2.0 GPa: Implications for the origin of adakites. Contributions to Mineralogy and Petrology, 117, 394-409.

Sichuan Provincial Bureau of Geology and Mineral Resources (SPBGMR), 1972. Regional geological survey of People's Republic of China, the Miyi Sheet map and report, scale 1:200 000 (in Chinese).

Sisson, T.W., Ratajeski, K., Hankins, W.B., Glazner, A.F., 2005. Voluminous granitic magmas from common basaltic sources. Contributions to Mineralogy and Petrology, 148, 635-661.

Skjerlie, K.P., Johnston, A.D., 1996. Vapour-absent melting from 10 to 20 kbar of crustal rocks that contain multiple hydrous phases: Implications for anatexis in the deep to very deep continental crust and active continental margins. Journal of Petrology, 37(3), 661-691.

Skjerlie, K.P., Patiño Douce, A.E., 1995. Anatexis of interlayered amphibolite and pelite at 10 kbar: Effect of diffusion of major components on phase relations and melt fraction. Contributions to Mineralogy and Petrology, 22, 62-78.

Soderlund, U., Patchett, P.J., Vervoort, J.D., Isachsen, C.E., 2004. The ^{176}Lu decay constant determined by Lu-Hf and U-Pb isotope systematics of Precambrian mafic intrusions. Earth & Planetary Science Letters, 219(3-4), 311-324.

Stevens, G., Clemens, J.D., Droop, G.T.R., 1997. Melt production during granulite-facies anatexis: Experimental data from "primitive" metasedimentary protoliths. Contributions to Mineralogy and Petrology, 128, 352-370.

Stevens, G., Villaros, A., Moyen, J.F., 2007. Selective peritectic garnet entrainment as the origin of geochemical diversity in S-type granites. Geology, 35(1), 9-12.

Sun, S.S., McDonough, W.F., 1989. Chemical and isotopic systematics of oceanic basalts: Implications for mantle composition and processes. In: Saunders, A.D., Norry, M.J. Magmatism in the Ocean Basins. Geological Society, London, Special Publications, 42, 313-345.

Tang, M., Wang, X.L., Shu, X.J., Wang, D., Yang, T., Gopon, P., 2014. Hafnium isotopic heterogeneity in zircons from granitic rocks: Geochemical evaluation and modeling of "zircon effect" in crustal anatexis. Earth & Planetary Science Letters, 389, 188-199.

Vielzeuf, D., Montel, J.M., 1994. Partial melting of metagreywackes. Part I. Fluid-absent experiments and phase relationships. Contributions to Mineralogy and Petrology, 117, 375-393.

Vielzeuf, D., Holloway, J.R., 1988. Experimental determination of the fluid-absent melting relations in the pelitic system. Consequences for crustal differentiation. Contributions to Mineralogy and Petrology, 98, 257-276.

Villaros, A., Stevens, G., Moyen, J.F., Buick, I.S., 2009a. The trace element compositions of S-type granites: Evidence for disequilibrium melting and accessory phase entrainment in the source. Contributions to Mineralogy and Petrology, 158(4), 543-561.

Villaros, A., Stevens, G., Buick, I.S., 2009b. Tracking S-type granite from source to emplacement: Clues from garnet in the Cape Granite Suite. Lithos, 112, 217-235.

Villaros, A., Buick, I.S., Stevens, G., 2012. Isotopic variations in S-type granites: An inheritance from a heterogeneous source? Contributions to Mineralogy and Petrology, 163(2), 243-257.

Villaros, A., Laurent, O., S Couzinié., Moyen, J.F., Mintrone, M., 2018. Plutons and domes: The consequences of anatectic magma extraction: Example from the southeastern French Massif Central. International Journal of Earth Sciences, 107(8), 1-24.

Villaseca, C., Barbero, L., Herreros, V., 1998. A re-examination of the typology of peraluminous granite types in intracontinental orogenic belts. Trans. R. Soc. Edinburgh, 89, 113-119.

Wang, X.L., Zhou, J.C., Wan, Y.S., Kitajima, K., Wang, D., Bonamici, C., Qiu, J.S., Sun, T., 2013. Magmatic evolution and crustal recycling for Neoproterozoic strongly peraluminous granitoids from Southern China: Hf and O isotopes in zircon. Earth & Planetary Science Letters, 366(2), 71-82.

Wang, X.L., Zhou, J.C., Griffin, W.L., Zhao, G.C., Yu, J.H., Qiu, J.S., Zhang, Y.J., Xing, G.F., 2014. Geochemical zonation across a Neoproterozoic orogenic belt: Isotopic evidence from granitoids and metasedimentary rocks of the Jiangnan orogen, China. Precambrian Research, 242(2), 154-171.

Wang, D., Wang, X.L., Cai, Y., Goldstein, S.L., Yang, T., 2018. Do Hf isotopes in magmatic zircons represent those of their host rocks? Journal of Asian Earth Sciences, 154, 202-212.

Wiedenbeck, M., Hanchar, J.M., Peck, W.H., Sylvester, P., Valley, J., Whitehouse, M., Kronz, A., Morishita, Y., et al., 2004. Further characterisation of the 91500 zircon crystal. Geostandards & Geoanalytical Research, 28(1), 9-39.

Xing, H.Q., Li, X.W., Xu, J.F., Mo, X.X., Shan, W., Yu, H.X., Hu, J.Q., Huang, X.F., Dong, G.C., 2020. The genesis of felsic magmatism during the closure of the Northeastern Paleo-Tethys ocean: Evidence from the Heri batholith in west Qinling, China. Gondwana Research, 84, 38-51.

Yuan, H.L., Gao, S., Liu, X.M., Li, H.M., Gunther, D., Wu, F.Y., 2004. Accurate U-Pb age and trace element determinations of zircon by laser ablation-inductively coupled plasma mass spectrometry. Geostandards & Geoanalytical Research, 28 (3), 353-370.

Yuan, H.L., Gao, S., Dai, M.N., Zong, C.L., Gunther, D., Fontaine, G.H., Liu, X.M., Diwu, C.R., 2008. Simultaneous determinations of U-Pb age, Hf isotopes and trace element compositions of zircon by excimerlaser-ablation quadrupole and multiple-collector ICPMS. Chemical Geology, 247(1-2), 100-118.

Zhang, Y., Ni, H., Chen, Y., 2010. Diffusion data in silicate melts. Reviews in Mineralogy and Geochemistry, 72, 311-408.

Zhao, G.C., Cawood, P.A., 2012. Precambrian geology of China. Precambrian Research, 222-223, 13-54.

Zhao, J.H., Zhou, M.F., 2007. Neoproterozoic Adakitic Plutons and Arc Magmatism along the western Margin of the Yangtze Block, South China. Journal of Geology, 115(6), 675-689.

Zhao, J.H., Zhou, M.F., Yan, D.P., Zheng, J.P., Li, J.W., 2011. Reappraisal of the ages of Neoproterozoic strata in South China: No connection with the Grenvillian orogeny. Geology, 39(4), 299-302.

Zhao, J.H., Li, Q.W., Liu, H., Wang, W., 2018. Neoproterozoic magmatism in the western and northern margins of the Yangtze Block (South China) controlled by slab subduction and subduction-transform-edge-propagator. Earth-Science Reviews, 187, 1-18.

Zhao, J.H., Zhou, M.F., Wu, Y.B., Zheng, J.P., Wang, W., 2019. Coupled evolution of Neoproterozoic arc mafic magmatism and mantle wedge in the western margin of the South China Craton. Contributions to Mineralogy and Petrology, 174, 36.

Zhao, X.F., Zhou, M.F., Li, J.W., Wu, F.Y., 2008. Association of Neoproterozoic A- and I-type granites in South China: Implications for generation of A-type granites in a subduction-related environment. Chemical Geology, 257 (s1), 1-15.

Zhou, M.F., Yan, D., Kennedy, A.K., Li, Y.Q., Ding, J., 2002. SHRIMP U-Pb zircon geochronological and geochemical evidence for Neoproterozoic arc-magmatism along the western margin of the Yangtze Block, South China. Earth & Planetary Science Letters, 196(1-2), 51-67.

Zhou, M.F., Ma, Y.X., Yan, D.P., Xia, X.P., Zhao, J.H., Sun, M., 2006a. The Yanbian Terrane (Southern Sichuan Province, SW China): A Neoproterozoic arc assemblage in the western margin of the Yangtze Block. Precambrian Research, 144(1-2), 19-38.

Zhou, M.F., Yan, D.P., Wang, C.L., Qi, L., Kennedy, A., 2006b. Subduction-related origin of the 750 Ma Xuelongbao adakitic complex (Sichuan Province, China): Implications for the tectonic setting of the giant Neoproterozoic magmatic event in South China. Earth & Planetary Science Letters, 248(1-2), 286-300.

Zhu, W.G., Zhong, H., Li, Z.X., Bai, Z.J., Yang, Y.J., 2016. SIMS zircon U-Pb ages, geochemistry and Nd-Hf isotopes of ca. 1.0 Ga mafic dykes and volcanic rocks in the Huili area, SW China: Origin and tectonic significance. Precambrian Research, 273, 67-89.

Zhu, Y., Lai, S.C., Qin, J.F., Zhu, R.Z., Zhang, F.Y., Zhang, Z.Z., 2019a. Geochemistry and zircon U-Pb-Hf isotopes of the 780 Ma I-type granites in the western Yangtze Block: Petrogenesis and crustal evolution. International Geology Review, 61(10), 1222-1243.

Zhu, Y., Lai, S.C., Qin, J.F., Zhu, R.Z., Zhang, F.Y., Zhang, Z.Z., Gan, B.P., 2019b. Petrogenesis and geodynamic implications of Neoproterozoic gabbro-diorites, adakitic granites, and A-type granites in the southwestern margin of the Yangtze Block, South China. Journal of Asian Earth Sciences, 183, 103977.

Zhu, Y., Lai, S.C., Qin, J.F., Zhu, R.Z., Zhang, F.Y., Zhang, Z.Z., Zhao, S.W., 2019c. Neoproterozoic peraluminous granites in the western margin of the Yangtze Block, South China: Implications for the reworking of mature continental crust. Precambrian Research, 333, 105443.

Zhu, Y., Lai, S.C., Qin, J.F., Zhu, R.Z., Zhang, F.Y., Zhang, Z.Z., 2020a. Petrogenesis and geochemical diversity of late Mesoproterozoic S-type granites in the western Yangtze Block, South China: Co-entrainment of peritectic selective phases and accessory minerals. Lithos, 352-353, 105326.

Zhu, Y., Lai, S.C., Qin, J.F., Zhu, R.Z., Liu, M., Zhang, F.Y., Zhang, Z.Z., Yang, H., 2020b. Genesis of ca. 850-835 Ma high-Mg$^{\#}$ diorites in the western Yangtze Block, South China: Implications for mantle metasomatism under the subduction process. Precambrian Research, 343, 105738.

Zhu, Y., Lai, S.C., Qin, J.F., Zhu, R.Z., Liu, M., Zhang, F.Y., Zhang, Z.Z., Yang, H., 2021. Neoproterozoic metasomatized mantle beneath the western Yangtze Block, South China: Evidence from whole-rock geochemistry and zircon U-Pb-Hf isotopes of mafic rocks. Journal of Asian Earth Sciences, 206, 104616.

High-K calc-alkaline to shoshonitic intrusions in SE Tibet: Implications for metasomatized lithospheric mantle beneath an active continental margin[①]

Zhu Renzhi[②]　Ewa Słaby　Lai Shaocong[②]　Chen Lihui　Qin Jiangfeng
Zhang Chao　Zhao Shaowei　Zhang Fangyi　Liu Wenhang　Mike Fowler

Abstract: High-K calc-alkaline to shoshonitic suites are widespread and generally volumetrically small but provide key information on magmatic mantle-crust interactions. Limited work has addressed the multi-stage formation of relatively high-volume high-K to shoshonitic rocks. A newly identified, Late-Cretaceous to Early-Cenozoic high-volume high-K to shoshonitic association (ca. 28 000 km^3) is described, from the southeastern Himalayan-Tibetan orogen. The mafic intrusions are shoshonitic (K_2O contents, ~3.3 wt%), have high-Mg contents (~5.1 wt%) with high-$Mg^{\#}$ (57), LREE and LILEs with low Ba/Th, Ba/La, Sm/La (<0.3), Nb/Yb and $^{208}Pb/^{206}Pb$ ratios but high Hf/Sm (>0.70), Th/Yb, Th/La (>0.2) and La/Sm, and lack an Eu anomaly. Their mafic minerals are hydrous, dominated by magnesio-hornblende ($Mg^{\#}$, ~0.69–0.73) and Mg- and K-biotite (MgO, 7.27–9.26 wt%; K_2O, 9.65–10.1 wt%). Such characteristics strongly suggest derivation from the sub-continental lithospheric mantle (SCLM), metasomatized by sediment-derived melts/fluids. The associated felsic intrusions have high SiO_2, alkali content ($Na_2O + K_2O$ contents up to 10.0 wt%) and incompatible elements (e.g., K, Rb), with ferropargasite-hastingsite and ferrobiotite-siderophyllite as the mafic phases. These characteristics point to derivation from the continental crust enriched by fluids, likely those released from the contemporaneous mafic magmas crystallising at depth. In contrast to low-volume potassic magmatism from post-collision and earliest arc-rift settings, these high-volume high-K and shoshonitic intrusions define a mantle-to-upper crust pathway at an active continental margin. Alongside geophysical data, these observations are consistent with the contemporary subduction history, arising from subduction of the Neo-Tethyan ocean slab to initial collision of India-Asia, from steepening subduction to slab rollback and breakoff.

Introduction

High-K calc-alkaline to shoshonitic rocks occur commonly in island arc and post-collision

①　Published in *Contributions to Mineralogy and Petrology*, 2021, 176(10).

②　Corresponding authors.

settings (for example: the Izu-Bonin-Mariana arc, Ishizuka et al., 2010; the Mediterranean, Variscan and Caledonian terrains of Europe, Thompson and Fowler, 1986; Conticelli and Peccerillo, 1992; Peccerillo, 1999; Duggen et al., 2005; Francalanci et al., 2000; Conticelli et al., 2009a,b, 2011, 2015; Avanzinelli et al., 2008, 2009; Fowler et al., 2008; Prelević et al., 2008, 2010; Soder et al., 2018; Janoušek et al., 2020) and the Himalayan belt from Europe to Asia (Chung et al., 1998; Pe-Piper et al., 2009; Jiang et al., 2006; Campbell et al., 2014; Table 1). Their origin is widely debated: ① those of the Izu-Bonin and Mariana arcs could result from melting of an enriched lithosphere and uppermost asthenosphere by arc-rift propagation (Ishizuka et al., 2010); ② in the post-collision example of the Tethyan orogeny their appearance has been explained in many ways: melting of veined lithosphere that was metasomatized by subduction-related fluids/melts from ancient slabs (Guo et al., 2005; Gao et al., 2007; Pe-Piper et al., 2009; Huang et al., 2010; Conticelli et al., 2009b); sediment/dunite reaction (Förster et al., 2019); subducted continental sediments or lower crust (Liu et al., 2014, 2015); subducted continental crust followed by melting and/or progressive interaction of components from both environments (Campbell et al., 2014; Soder et al., 2018; Avanzinelli et al., 2020); mantle reaction with Si-rich melts from recycled continental crustal materials (Dallai et al., 2019); melting of metasomatized lithospheric mantle, thickened lithosphere and lower crust in a complex and continuous interaction with some contribution from the asthenosphere (Jiang et al., 2002). Amongst all these suggestions, a common theme is a complicated petrogenesis involving mantle-crust interaction via subducted crustal materials.

Table 1   Summary data of high-K calc-alkaline to shoshonitic granitoids in the central Tengchong terrane, SE Tibet.

Sample	Location (N,E)	Lithology	Major mineral composition	Age /Ma	Reference
09QG-60	25°18′30″N, 98°12′45″E	Granite	Qtz(25−35 vol%) + Kf(25−35 vol%) + Pl(25−35 vol%) + Bi(5−10 vol%)	64.8±0.7	Qi et al., 2015
09QG-45	24°57′41″N, 98°19′13″E	Granite	Qtz(25−35 vol%) + Kf(25−35 vol%) + Pl(25−35 vol%) + Bi(5−10 vol%) + Ms(0−3 vol%)	76.6±2.0	Qi et al., 2015
09QG-50	25°13′50″N, 98°11′42″E	Granite porphyry	Qtz(25−35 vol%) + Kf(25−35 vol%) + Pl(25−35 vol%) + Bi(5−10 vol%)	65.3±1.0	Qi et al., 2015
09QG-55	25°13′50″N, 98°12′02″E	Granite	Qtz(25−35 vol%) + Kf(25−35 vol%) + Pl(25−35 vol%) + Bi(5−10 vol%)	64.9±0.9	Qi et al., 2015
13MB06	25°00′05″N, 98°19′56″E	Monzogranite	Qtz(20−25 vol%) + Kf(30−35 vol%) + Pl(30−35 vol%) + Bi(1−3 vol%) + Hbl(0−5 vol%)	73.8±0.5	Xie et al., 2016

Continued

Sample	Location (N,E)	Lithology	Major mineral composition	Age /Ma	Reference
13TC19	25°21′58″N, 98°18′53″E	Monzogranite	Qtz(20−25 vol%)+Kf(30−35 vol%)+ Pl(30−35 vol%)+Bi(1−3 vol%)+ Hbl(0−5 vol%)	74.3±0.5	Xie et al., 2016
13ZX06	25°05′40″N, 98°14′53″E	Monzogranite	Qtz(20−25 vol%)+Kf(30−35 vol%)+ Pl(30−35 vol%)+Bi(1−3 vol%)+ Hbl(0−5 vol%)	65.6±0.6	Xie et al., 2016
13LLS04-1	24°55′31″N, 98°15′33″E	Monzogranite	Qtz(~25 vol%)+Kf(25−35 vol%)+ Pl(30−40 vol%)+Bi(3−8 vol%)	50.5±0.4	Xie et al., 2016
13XQ02	25°03′44″N, 98°17′38″E	Granodiorite	Qtz(20−25 vol%)+Kf(30−35 vol%)+ Pl(30−35 vol%)+Bi(1−3 vol%)+ Hbl(0−5 vol%)	50.4±0.4	Xie et al., 2016
XLH-6	25°26′49″N, 98°25′46″E	Granite	Qtz(20−30 vol%)+Kf(38 vol%)+ Pl(32 vol%)+Bi(8 vol%)	73.3±0.2	Chen et al., 2015
DSP-1	25°25′36″N, 98°22′50″E	Granite	Qtz(20−30 vol%)+Kf(38 vol%)+ Pl(32 vol%)+Bi(8 vol%)	73.3±0.3	Chen et al., 2015
LLS-2	24°55′14″N, 98°15′33″E	Granite	Qtz(28 vol%)+Kf(35 vol%)+ Pl(30 vol%)+Bi(10 vol%)	53.0±0.2	Chen et al., 2015
LLS-6	24°56′49″N, 98°16′45″E	Granite	Qtz(28 vol%)+Kf(35 vol%)+ Pl(30 vol%)+Bi(10 vol%)	52.7±0.2	Chen et al., 2015
LL96	24°20.371′N, 97°50.562′	Granite	Kf(25 vol%)+Pth(20 vol%)+ Pl(28 vol%)+Qtz(20 vol%)+ Bi(7 vol%)	50.6±0.6	Zhao et al., 2016
LC2-17	24°20.404′N, 97°50.397′	Granite	Kf(15 vol%)+Pth(25 vol%)+ Pl(20 vol%)+Mc(15 vol%)+ Qtz(20 vol%)+Bi(5 vol%)	48.9±0.5	Zhao et al., 2016
LC2-09	24°20.421′N, 97°50.575′E	Granodiorite porphyry	Pth(25 vol%)+Mc(15 vol%)+ Pl(30 vol%)+Qtz(15 vol%)+ Bi(12 vol%)+Hbl(3 vol%)	50.0±0.4	Zhao et al., 2016
LC2-30	24°20.823′N, 97°50.286′E	Mafic enclave	Bi(25 vol%)+Hbl(20 vol%)+ Pl(30 vol%)+Kf(20 vol%)+ Qtz(5 vol%)	49.8±0.2	Zhao et al., 2016
GY06	25°22.695′N, 98°12.251′E	Granite	Qtz(25 vol%)+Pl(20 vol%)+ Mc(15 vol%)+Kf(25 vol%)+ Bi(12 vol%)+Hbl(3 vol%)	64.0±1.0	Zhao et al., 2017
GY17	25°21.476′N, 98°13.647E	Granite	Qtz(30 vol%)+Pl(25 vol%)+ Mc(20 vol%)+Kf(22 vol%)+ Bi(3 vol%)	65.0±1.0	Zhao et al., 2017

Continued

Sample	Location (N,E)	Lithology	Major mineral composition	Age /Ma	Reference
GY46	25°20. 715′N, 98°15. 840′E	Granite	Qtz(28 vol%) + Pl(35 vol%) + Mc(15 vol%) + Kf(20 vol%) + Bi(2 vol%)	64. 0 ± 1. 0	Zhao et al., 2017
LC28	24°27. 855′N, 97°45. 040′E	Granodiorite	Qtz(20 vol%) + Pl(30 vol%) + Mc(10 vol%) + Kf(15 vol%) + Bi(7 vol%) + Hbl(8 vol%)	64. 0 ± 1. 0	Zhao et al., 2017
JC01	24°42. 09′N, 98°04. 58′E	Granite	Qtz(25−30 vol%) + Kf(40−45 vol%) + Pl(10−15 vol%) + Bi(3−5 vol%) + Hbl(1−3 vol%)	70. 3 ± 0. 7	This study
JC31	27°42. 13′N, 98°04. 52′E	Monzodiorite	Hbl(15−28 vol%) + Bi(5−10 vol%) + Pl(40−45 vol%) + Kf(15−25 vol%) + minor Qtz	71. 0 ± 1. 0	This study
JC71	27°42. 17′N, 98°05. 20′E	Granite	Qtz(35−40 vol%) + Kf(40−45 vol%) + Pl(5 vol%) + Bi(1−3%)	65. 4 ± 0. 6	This study
JC56	27°44. 29′N, 98°01. 25E	Granodiorite	Hbl(~ 5 vol%) + Bi(15−20 vol%) + Pl(30−35 vol%) + Kf(20−25 vol%) + Qtz(10 vol%)	55. 4 ± 0. 6	This study
JC46	27°44. 09′N, 98°00. 35′E	Granite	Qtz(35−40 vol%) + Kf(40−45 vol%) + Pl(5 vol%) + Bi(1−3 vol%)	52. 5 ± 0. 6	This study
JC86	27°43. 32′N, 98°00. 31′E	Granite	Qtz(35−40 vol%) + Kf(40−45 vol%) + Pl(5 vol%) + Bi(3−5 vol%)	52. 4 ± 0. 7	This study

Hbl: Hornblende; Bi: Biotite; Qtz: Quartz; Pl: Plagioclase; Kf: K-feldspar; Ms: Muscovite; Mc: Microcline; Pth: Perthite.

In the westernmost sector of the Alpine-Himalayan collisional margin, high-K calc-alkaline to shoshonitic rocks are widespread but usually of relatively low volume. In this paper, we report a newlyidentified Late-Cretaceous to Early-Cenozoic high-volume province in the southeastern Himalayan-Tibetan orogen (Fig. 1). We use whole-rock and mineral chemistry (amphibole, biotite and apatite) as well as zircon U-Pb and Lu-Hf data and whole-rock radiogenic isotopes to investigate the source and the evolution of the magma. Together with other geochemical and geophysical data from the literature we aim to constrain the geodynamic processes operating during the transition from Neo-Tethyan oceanic subduction to initial collision of Indian- and Asian- plates, and contribute to the understanding of high-K calc-alkaline to shoshonitic magmatic systems elsewhere, regardless of the time of their formation.

Geological background and general petrography

The Tethyan tectonic realm is regarded as an important window to the various

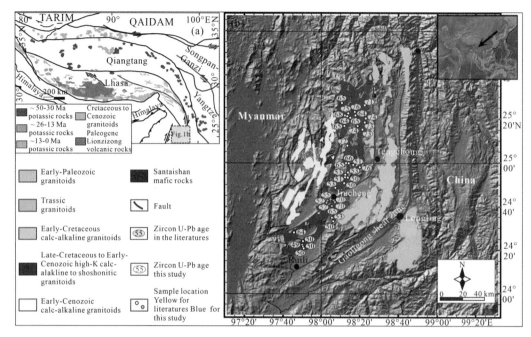

Fig. 1　(a) Continental blocks and potassic rocks of the Tibetan orogenic system
(Chung et al., 2005). (b) Magmatic distribution of the Tengchong terrane.

superimposed evolutionary processes between micro-continental and oceanic plates (Tommasini et al., 2011), which are globally widespread in space and time. This long tectonic belt extends from the Mediterranean Sea in the west to Tibet and SE Asia in the east. The Himalayan-Tibetan Tethyan realm (Fig. 1a) has a detailed record of geological processes since the early Paleozoic and is subdivided into a number of blocks, such as northern Qiangtang, southern Qiangtang, Lhasa and Himalaya, with suture zones such as Jinshajiang-Ailaoshan, Longmucuo-Shuanghu, Bangong-Nujiang and Yarlung-Zangbo. Within this domain, Cenozoic potassic to ultrapotassic igneous rocks are widespread, though usually small in volume, throughout Tibet (Chung et al., 1998, 2005; Guo et al., 2005; Jiang et al., 2006; Gao et al., 2007; Huang et al., 2010; Liu et al., 2014, 2015). They are generally considered to represent a magmatic response to the convective removal of lower lithosphere during the collision between India and Asia (Chung et al., 1998).

In southeastern Tibet, the Tengchong Block is located in the southeastern segment of the Tibetan Tethyan domain (Fig. 1a). It is bordered by the Nujiang-Lushui-Ruili fault in the east, further separated from Baoshan Block, with the Putao-Myitkyina suture in the west to separate it from the west Burma Block (Metcalfe, 2013). The presence of Permo-Carboniferous glaciomarine deposits overlying Gondwana-like fossil assemblages supports an origin from the western Australia margin, an eastern segment of the Gondwana continent. Notably, the prominent Gaoligong group, comprising amphibolites, feldspathic gneisses, quartzites, migmatites, marble and schists, has been regarded as Mesoproterozoic metamorphic basement. Upper Paleozoic to lower Mesozoic carbonates and clastic sediments overlie the basement. Three boundary faults divide the Tengchong Block: the Mingguang-Menglian, Guyong-Longchuan and Sudian-Tongbiguan magmatic belts are separated by the Dayingjiang, Gudong-Tengchong and Binlangjiang faults (Fig. 1b). Late Mesozoic to Early-Cenozoic granitoid magmatic belts are emplaced into these country rocks and are partly covered by young volcanic sequences during Tertiary-Quaternary.

This study describes high-potassium calc-alkaline to shoshonitic intrusive igneous rocks ranging in composition from monzodiorite to granodiorite to granite (~28 000 km^3, calculated using the methods of Ratchbacher et al., 2019) in the Jiucheng area of the Guyong-Longchuan magmatic belt, in the central Tengchong Block (Fig. 1b). In the field, the monzodiorites have sharp intrusive contacts with granodiorite (Fig. 2a), granodiorite usually show sharp contacts with the biotite granites and granites (Fig. 2b,d and e), and biotite granite also show sharp contacts with granite (Fig. 2c) although some are gradational (Fig. 2d). Abundant felsic veins can also be found (Fig. 2a). Monzodiorites are fine-grained, granodiorites vary from coarse- to

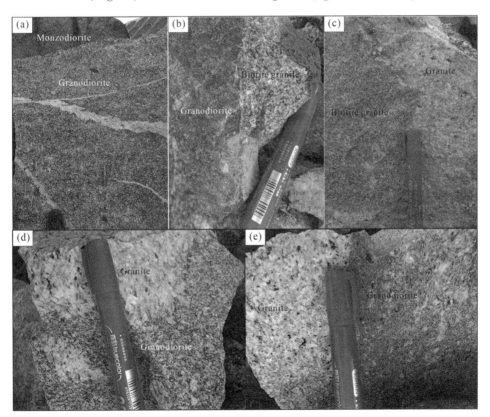

Fig. 2　Field features and contact relation of high-K calc-alkaline to shoshonitic monzodiorite, granodiorite and granite suite in the Tengchong of southeastern Tibet, SW China.

medium-grained and fine-grained (Fig. 2a-c), but both biotite granite and granite are also fine-grained (Fig. 2d-e). The monzodiorites (e.g., JC-33) mainly contain hornblende (15% - 28%), biotite (5% - 10%), plagioclase (40% - 45%), alkali feldspar (15% - 25%) and minor quartz, with accessory zircon, titanite and apatite. Most hornblende (Fig. 3c, d) is in intergrowth with biotite and titanite. Accessory minerals usually occur within the felsic minerals. Granodiorites (e.g., JC-59) mainly contain hornblende (~ 3%), biotite (15% - 20%), plagioclase (30% - 35%), alkali feldspar (20% - 25%) and quartz (10% - 15%), with abundant accessory titanite, zircon and apatite. The hornblende-bearing granite (JC-02) contains hornblende (1% - 3%), biotite (3% - 5%), plagioclase (10% - 15%), alkali feldspar (40% - 45%) and quartz (25% - 30%), with accessory zircon and apatite. A few hornblendes occur between quartz and feldspars. The leucogranite (JC-47) has biotite (3% - 5%), alkali feldspar (40% - 45%) and quartz (40% - 45%), accessory minerals including zircon and apatite. Biotites occur between quartz and feldspar, the latter are often observed as phenocrysts. Both sample JC-73 and JC-80 granites are more leucocratic and contain less biotite (< 3%), with alkali feldspar (40% - 45%) and quartz (40% - 45%), plus accessory zircon and apatite. Biotites occur between quartz and feldspar and all feldspars are perthitic. Anhedral

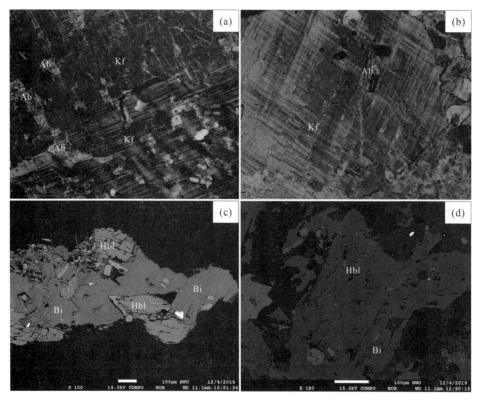

Fig. 3　Microscope and BSE image of high-K calc-alkaline to shoshonitic monzodiorite, granodiorite and granite suite in the Tengchong of southeastern Tibet, SW China.

quartz and albites often occur within feldspar phenocrysts (Fig. 3a, b).

Analytical methods

Zircon has been separated from six samples from the central Tengchong Block (Fig. 1b). Detailed analytical methods of in situ zircon U-Pb and Lu-Hf and associated bulk-rock geochemistry are given in Supplementary Methods, and Tables 1 and 2, largely after Weis et al. (2006), Yuan et al. (2008), and Bao et al. (2016). Data are presented as follows: in-situ zircon U-Pb isotopic data ($n = 6$) in Supplementary Dataset 1; in-situ zircon Lu-Hf isotopic data ($n = 84$) in the Supplementary Table 1; whole-rock geochemistry ($n = 31$) and Sr-Nd-Pb isotopes ($n = 10$) in Supplementary Table 2. Other high-potassium calc-alkaline to shoshonitic samples ($n = 71$) in the central Tengchong, from Cao et al. (2016, 2017), Chen et al. (2014), Qi et al. (2015), Xie et al. (2016), Xu et al. (2012), Zhao et al. (2016, 2017), are also summarized in Table 1.

Major element compositions of minerals, including amphibole, biotite and apatite were analyzed by using a JXA-8230 electron microprobe and in situ trace elements of amphibole were analyzed by LA-ICP-MS at the State Key Laboratory of Continental Dynamics, Northwest University, Xi'an, China. Detailed methods are presented in Supplementary Methods and data in Supplementary Tables 3 and 4.

Results

The samples in this study can be divided into two obvious groups, the mafic and felsic components. The mafic components represent shoshonitic diorites, and the felsic components represent high-K calc-alkaline to shoshonitic granodiorite-granite suites. Sample localities with age and lithology are shown in Fig. 4. Samples are selected to cover the lithological range observed from the freshest localities, all samples represent the comprehensive features of the high-K calc-alkaline to shoshonitic granitoids in central Tengchong, SE Tibet.

In-situ zircon U-Pb ages

The sample location, lithology and ages are summarized in Table 1 and zircon CL images with associated U-Pb ages are shown in Fig. 4. The zircons are generally euhedral, up to 150–350 μm long, and length/width ratios range from 1:1 to 3:1. They are colorless to light brown, prismatic, transparent to translucent and have clear oscillatory zoning. Most have high Th/U ratios (>0.4), indicating a magmatic origin.

Nineteen analyses were obtained from hornblende-bearing granite sample JC-01. Th/U values range from 0.31 to 1.05 (mean value = 0.64), with associated $^{206}Pb/^{238}U$ ages from 72.0 ± 1.9 Ma to 68.1 ± 1.5 Ma, with a weighted mean age of 70.3 ± 0.7 Ma (MSWD = 0.6,

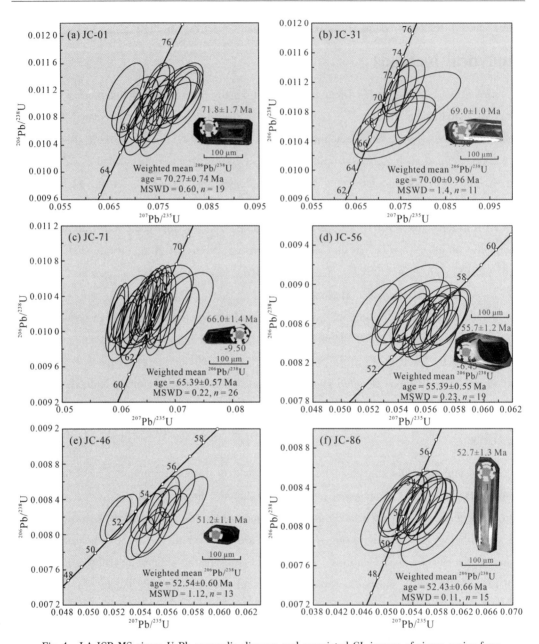

Fig. 4　LA-ICP-MS zircon U-Pb concordia diagram and associated CL images of zircon grains from the high-K calc-alkaline to shoshonitic granitoids in the Tengchong Block, SW China.

The pink circles are the U-Pb isotopic analysis spots and the yellow circles are the zircon Lu-Hf isotopic analysis spots.

$n = 19$, Fig. 4a). Eleven spot analyses of zircons were obtained from monzodiorite sample JC-31. These have moderate Th/U values = 0. 29 - 1. 13 (mean value = 0. 66), a small range of $^{206}Pb/^{238}U$ ages from 72. 0 ± 2. 0 Ma to 68. 0 ± 1. 0 Ma, with a weighted mean age of 70. 0 ± 1. 0 Ma (MSWD = 1. 4, $n = 11$, Fig. 4b). Twenty-eight zircon analyses were obtained from leucogranite sample JC-71. They have relatively low to moderate Th/U values = 0. 14 - 0. 93

(mean value = 0. 39) and their $^{206}Pb/^{238}U$ ages range from $66. 6 \pm 2. 0$ Ma to $64. 3 \pm 1. 3$ Ma, with a weighted mean age of $65. 4 \pm 0. 6$ Ma (MSWD = 0. 2, $n = 28$, Fig. 4c). Granodiorite sample JC-56 yielded 19 spot analyses. They have moderate Th/U values = 0. 21–1. 16 (mean value = 0. 70) and their $^{206}Pb/^{238}U$ ages range from $57. 0 \pm 1. 3$ Ma to $54. 2 \pm 1. 2$ Ma, with a weighted mean age of $55. 4 \pm 0. 6$ Ma (MSWD = 0. 2, $n = 19$, Fig. 4d). Sample JC-46, a hornblende-bearing granite, gave thirteen analyses and associated $^{206}Pb/^{238}U$ ages from $54. 3 \pm 1. 1$ Ma to $50. 0 \pm 1. 0$ Ma, with a weighted mean age of $52. 5 \pm 0. 6$ Ma (MSWD = 1. 1, $n = 13$, Fig. 4e). They also have moderate Th/U values = 0. 28–1. 15 (mean value = 0. 48). Finally, hornblende-bearing granite JC-86 shows a similar zircon U-Pb age range to sample JC-46, $^{206}Pb/^{238}U$ ages range from $53. 0 \pm 1. 6$ Ma to $51. 1 \pm 1. 2$ Ma, with a weighted mean age of $52. 4 \pm 0. 7$ Ma (MSWD = 0. 1, $n = 15$, Fig. 4f). Their Th/U values range from 0. 15 to 1. 35 (mean value = 0. 68).

In-situ zircon Lu-Hf isotopes

Lu-Hf isotopes analyses on the same domains are listed in Supplementary Table 1. Initial $^{176}Hf/^{177}Hf$ ratios and $\varepsilon_{Hf}(t)$ values of zircons were calculated at their crystallization ages, described above. All have evolved Hf isotopic compositions, with $\varepsilon_{Hf}(t)$ values of $-15. 4$ to $-7. 20$ for the Late-Cretaceous granite (JC-01, 19 zircons, initial $^{176}Hf/^{177}Hf$ 0. 282 231– 0. 282 525), $-11. 7$ to $-3. 11$ for the Late-Cretaceous monzodiorite JC-31 (1 zircons, initial $^{176}Hf/^{177}Hf$ 0. 282 397–0. 282 641), $-11. 1$ to $-7. 29$ for the Late-Cretaceous granite JC-71 (16 zircons, initial $^{176}Hf/^{177}Hf$ 0. 282 419–0. 282 525), $-11. 1$ to $-5. 61$ for the Early-Cenozoic granodiorite JC-56 (19 zircons, initial $^{176}Hf/^{177}Hf$ 0. 282 424 – 0. 282 579), and $-9. 19$ to $-4. 82$ for the Early-Cenozoic hornblende-bearing granite JC-86 (17 zircons, initial $^{176}Hf/^{177}Hf$ 0. 282 480–0. 282 603). The one inherited zircon grain of JC-01 has $\varepsilon_{Hf}(t)$ of $-17. 6$ (at 1 978 Ma). The two-stage zircon Lu-Hf isotopic model ages of the rocks are in the range of 1 594–1 766 Ma for Late-Cretaceous hornblende-bearing granites, 873–1 220 Ma for the Late-Cretaceous monzodiorites, 1 596–1 834 Ma for Late-Cretaceous leucogranites, 1 482– 1 829 Ma for the Early-Cenozoic granodiorite, 1 430 – 1 706 Ma for the Early-Cenozoic leucogranites.

Whole-rock major and trace element geochemistry

On K_2O vs. SiO_2(Peccerillo and Taylor, 1976) and $K_2O + Na_2O$ vs. SiO_2(Le Maitre, 2002) diagrams, the data fall in the high-K calc-alkaline to shoshonitic fields and in shoshonite, latite and rhyolite extrusive equivalents respectively, in common with similar rocks from southern and eastern Tibet and central Tengchong (Fig. 5a,b). Major and trace elements are described in more detail below.

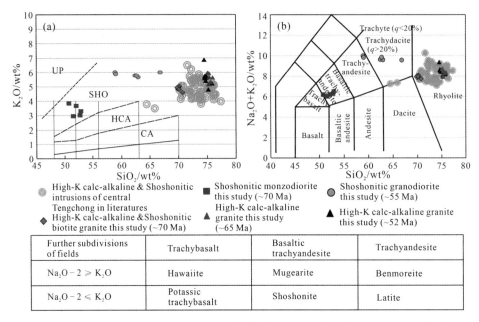

Fig. 5 K_2O vs SiO_2 and $Na_2O + K_2O$ vs. SiO_2(a,b)diagram for the high-potassium to shoshonitic rocks(Peccerillo and Taylor, 1976; Le Maltre, 2002).

Data of high-K calc-alkaline to shoshonitic granitoids in central Tengchong are summarized in Table 1.

Late-Cretaceous units

Samples of Late-Cretaceous shoshonitic monzodiorite (~ 70 Ma) from the Jiucheng area contain low concentrations of SiO_2 (50.8 – 52.9 wt%) and high concentrations of $Fe_2O_3^T$ (8.53–9.15 wt%), TiO_2 (1.17–1.25 wt%) and MgO (4.73–5.84 wt%) with high-$Mg^\#$ = 56.0–59.8. They also show high K_2O (2.91–3.78), $K_2O + Na_2O$ (6.03–6.49), and Al_2O_3 concentrations (16.2–17.1) with low A/CNK ratios of 0.71–0.75. These samples display moderate (La/Yb)$_N$ values (5.21–7.15) and low (Gd/Yb)$_N$(1.49–1.65) values and have rare-earth element (REE) patterns (Fig. 6a) with small or insignificant Eu anomalies (δEu = 0.76–0.93). Significant trace element features are relative enrichment in Rb, K and Pb and depletion in Nb, Ta and Ti (Fig. 5b).

The late-Cretaceous (~ 70 Ma) hornblende-bearing granite samples have moderate SiO_2 concentrations of 69.8–70.8 wt% and high concentrations of K_2O (4.53–4.95 wt%) with K_2O/Na_2O ratios of 1.42–1.59, but contain low concentrations of $Fe_2O_3^T$(3.14–3.34 wt%), TiO_2(0.46–0.48 wt%), CaO (2.16–2.30 wt%), and MgO (0.74–0.79 wt%), with $Mg^\#$ values of 35.3–35.7. These samples also contain 14.1–14.7 wt% Al_2O_3 and have relatively high A/CNK (molar $Al_2O_3/CaO + Na_2O + K_2O$) ratios of 0.99–1.00. In the chondrite-normalized rare earth element (REE) variation diagram (Fig. 6a), these samples have

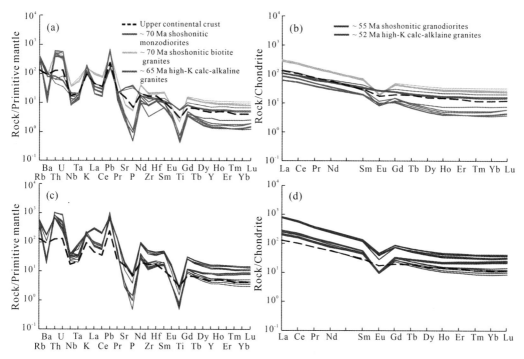

Fig. 6 Primitive-mantle-normalized trace element spider diagram and Chondrite-normalized REE
diagrams (a,b) for the high-potassic to shoshonitic granitoids in the Tengchong Block.

The primitive mantle and chondrite values are from Sun and McDonough (1989).

significant negative Eu anomalies (δEu = 0. 34−0. 40), high (La/Yb)$_N$(10. 8−11. 6), and low (Gd/Yb)$_N$(1. 47−1. 77) values, indicating possible involvement of plagioclase in magma genesis. On a primitive-mantle-normalized trace element variation diagram (Fig. 6b), these samples are notably enriched in the large ion lithophile elements (LILE; Rb, Th, U, K and Pb), and are depleted in Nb, Ta, Sr, P and Ti.

The Late-Cretaceous leucogranite (~65 Ma) samples show high SiO_2 concentrations of 75. 1−76. 0 wt% and high concentrations of K_2O (5. 09−5. 89 wt%) and K_2O/Na_2O ratios of 1. 70−2. 07, but comparatively low concentrations of $Fe_2O_3^T$(0. 90−1. 45 wt%), TiO_2(0. 09− 0. 16 wt%), CaO (1. 02−1. 28 wt%), and MgO (0. 07−0. 19 wt%), with Mg$^\#$ values of 15. 3−23. 4. They also contain Al_2O_3 values = 12. 8−13. 3 wt% with associated high A/CNK = 1. 02−1. 03. In the rare earth element (REE) variation diagram (Fig. 6a), the samples have notable negative Eu anomalies (δEu = 0. 41−0. 70), high (La/Yb)$_N$(7. 74−26. 7) and moderated (Gd/Yb)$_N$(1. 61−3. 14) values, again indicating plagioclase involvement. On Fig. 6b, these samples are enriched in LILE (K, Th, U, Rb and Pb) and are sharply depleted in P, Ti, Ba, Nb, Ta and Sr.

Early-Cenozoic units

The Early-Cenozoic granodiorite (~ 55 Ma) samples have moderate SiO_2 (58.9 − 66.8 wt%) with high K_2O (5.58 − 5.92), $K_2O + Na_2O$ (9.58 − 10.0), and high Al_2O_3 concentrations (16.7 − 19.0) with A/CNK ratios of 0.99 − 1.02. They also show low CaO (2.17 − 3.39 wt%) and MgO (0.47 − 0.83 wt%) with low-$Mg^\#$ = 23.5 − 25.0 and very low Cr (1.4 − 2.5 ppm), Ni (< 1.4 ppm). These samples display high total REE values (714 − 810 ppm) with high $(La/Yb)_N$ values (20.3 − 26.3) and $(Gd/Yb)_N$ (1.96 − 2.45) values (Fig. 6d) with significant negative Eu anomalies (δEu = 0.30 − 0.44). In the trace element patterns (Fig. 6c), the samples are significantly enriched large ion lithophile elements and depleted in P, Nb, Ta and Ti.

The Early-Cenozoic leucogranites (~ 52 Ma) samples have high SiO_2 concentrations of 74.6 − 75.9 wt% and high concentrations of K_2O (4.72 − 6.77 wt%) and K_2O/Na_2O ratios of 1.40 − 2.58, but correspondingly low concentrations of $Fe_2O_3^T$ (1.40 − 1.75 wt%), TiO_2 (0.10 − 0.17 wt%), CaO (0.63 − 1.13 wt%), and MgO (0.11 − 0.16 wt%) with $Mg^\#$ values of 15.2 − 17.7. These samples also contain moderate Al_2O_3 values = 12.5 − 13.4 wt% with associated A/CNK = 1.01 − 1.06. In the REE plots (Fig. 6d), the samples have consistent negative Eu anomalies (δEu = 0.23 − 0.29), variable high $(La/Yb)_N$ (8.35 − 26.4) and moderate $(Gd/Yb)_N$ (1.32 − 2.88) values, again suggesting plagioclase involvement in petrogenesis. On the trace element variation diagram (Fig. 6c), these samples are again enriched in the large ion lithophile elements, and are also sharply depleted in P, Ti, Ba, Nb, Ta and Sr.

Whole-rock Pb-Sr-Nd isotopes

The whole-rock Pb-Sr-Nd isotopic compositions define a restricted range of variation, described below.

The Late-Cretaceous hornblende-bearing granites (~ 70 Ma) show high initial $^{87}Sr/^{86}Sr$ ratio = 0.712 717 − 0.712 718 and low initial $^{143}Nd/^{144}Nd$ ratios = 0.511 984 − 0.511 992 with negative $\varepsilon_{Nd}(t)$ values = − 11.0 to − 10.9. Late-Cretaceous monzodiorites (~ 70 Ma) have similar high initial $^{87}Sr/^{86}Sr$ ratios = 0.710 105 − 0.710 220 and low initial $^{143}Nd/^{144}Nd$ ratios = 0.512 027 − 0.512 086 with negative $\varepsilon_{Nd}(t)$ values = − 10.2 to − 9.0. The 65 Ma (Late Cretaceous) leucogranites also have high initial $^{87}Sr/^{86}Sr$ ratio = 0.711 854 − 0.711 957 and low initial $^{143}Nd/^{144}Nd$ ratios = 0.512 055 − 0.512 144 with negative $\varepsilon_{Nd}(t)$ values (− 9.7 to − 8.0). Early-Cenozoic granodiorite (~ 55 Ma) have high initial $^{87}Sr/^{86}Sr$ ratio = 0.710 798 − 0.711 031 and low initial $^{143}Nd/^{144}Nd$ ratios = 0.512 141 − 0.512 157 with negative $\varepsilon_{Nd}(t)$

values (-8.3 to -8.0), and finally the Early-Cenozoic leucogranites (~ 52 Ma) likewise have high initial $^{87}Sr/^{86}Sr$ ratio $= 0.710\ 463 - 0.710\ 545$ and low initial $^{143}Nd/^{144}Nd$ ratios $= 0.512\ 162 - 0.512\ 166$ with negative $\varepsilon_{Nd}(t)$ values (-7.9 to 8.0). All these values lie at the enriched terminus of the mantle array and the "less-evolved" end of the field defined by other coeval potassic magmas from central Tengchong.

Late-Cretaceous to Early-Cenozoic high-K calc-alkaline to shoshonitic samples have relatively high and similar whole-rock Pb isotopic components, comparable with samples from Tengchong and elsewhere. Initial $^{206}Pb/^{204}Pb$, $^{207}Pb/^{204}Pb$, and $^{208}Pb/^{204}Pb$ ratios of Late-Cretaceous shoshonitic granite (~ 70 Ma) are $18.626 - 18.686$, 15.742, and $39.260 - 39.356$; of monzodiorites (~ 70 Ma) are $18.636 - 18.641$, $15.723 - 15.730$, and $39.183 - 39.238$; and of leucogranite (~ 65 Ma) are $18.669 - 18.672$, $15.737 - 15.744$, and $39.264 - 39.316$. The initial $^{206}Pb/^{204}Pb$, $^{207}Pb/^{204}Pb$, and $^{208}Pb/^{204}Pb$ ratios of Early-Cenozoic granodiorites (~ 55 Ma) are $18.688 - 18.707$, $15.720 - 15.737$, and $39.262 - 39.318$; and of 52 Ma hornblende-bearing granites are $18.755 - 18.765$, 15.743, and $39.325 - 39.334$. In addition, the $^{208}Pb/^{206}Pb$ ratios for ~ 70 Ma granite and monzodiorites, ~ 65 Ma leucogranite, ~ 55 Ma granodiorites and ~ 52 Ma hornblende-bearing granites are also similar, at 2.11, 2.10, $2.10 - 2.11$, 2.10, and 2.10, respectively.

Electron microprobe analyses and LA-ICP-MS in situ trace element

Hornblendes are the dominant mafic and hydrous minerals in the shoshonitic monzodiorites and are of typical magnesio-hornblende type (Fig. 7a). They (JC-33, Fig. 3c, d) are euhedral and usually intergrown with biotite. The EMPA chemical data (Supplemental Table 3, Fig. 7a) show the following ranges of composition: SiO_2($45.0\% - 53.8\%$), Al_2O_3 ($2.19\% - 8.73\%$), FeO^T($12.2\% - 16.8\%$), TiO_2($0.33\% - 0.61\%$), MgO ($11.0\% - 15.7\%$), Cr_2O_3($0.001\% - 0.57\%$), MnO($0\% - 0.31\%$), NiO($0\% - 0.05\%$), CaO ($10.4\% - 12.6\%$), Na_2O($0.13\% - 0.64\%$), K_2O($0.18\% - 0.81\%$), F($0\% - 0.21\%$) and SrO($0.10\% - 0.20\%$). LA-ICP-MS trace element (Supplemental Table 4) show that the magnesio-hornblendes share similar characters with those from normal arc crust (Smith, 2009), with slightly elevated MREE, but more enriched Rb, La, Th, U and Pb and variable Eu in the primitive mantle-normalizedand chondrite-normalized diagrams (Fig. 8a,b). Calculated equilibrium melts of amphibole have lower REE content than the whole-rock composition, suggesting crystallization of magnesio-hornblende somewhat later than REE-rich accessory minerals (Fig. 8e,f).

Hornblendes from the coeval and juxtaposed hornblende-bearing granite (JC-59, JC-88, JC-02) are typical ferro-pargasite to hastingsite type (Fig. 3c,d, Fig. 7b), with high Al_2O_3

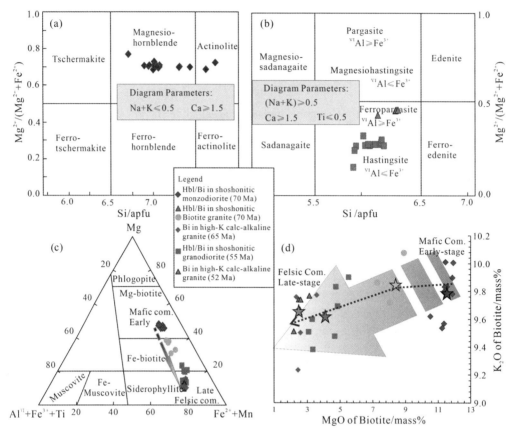

Fig. 7　Nomenclature of amphibole type diagram (a,b) (Leake et al., 1997). Nomenclature
of biotite type (c) (Froster, 1960); K_2O vs. MgO (d) of biotite content of
high-K to shoshonitic intrusions in southeastern Tibet.

(9.47% – 12.3%), FeO^T (23.4% – 29.4%), TiO_2 (0.53% – 1.34%), K_2O (1.36% – 2.03%) and Na_2O (0.87% – 1.21%), and low SiO_2, MgO (1.72% – 6.11%) and CaO (9.40% – 11.5%).

Biotites in the shoshonitic monzodiorites (JC-33) are typical magnesio-biotite (Fig. 6c), with high MgO (10.5% – 11.9%) but high K_2O (9.54% – 10.0%) content (Fig. 7c,d), low FeO^T(17.2% – 19.1%). Biotite from those granodiorite-granite suites (JC-58, JC-59, JC-47, JC-88, JC-02, JC-03 and JC-73) show decreasing K_2O with MgO, with MgO content range from 9.26% to 2.18%, FeO^T up to 29.3% and K_2O-falling to 9.23% (Fig. 7d, Supplementary Table 3).

Apatites in both shoshonitic mafic and felsic granitoids are typical low Cl fluorapatite (Fig. 8e): those in the mafic component have slightly higher Cl (0.04 wt%), P_2O_5(42.44 wt%), SrO (0.03 wt%) than in the associated felsic components.

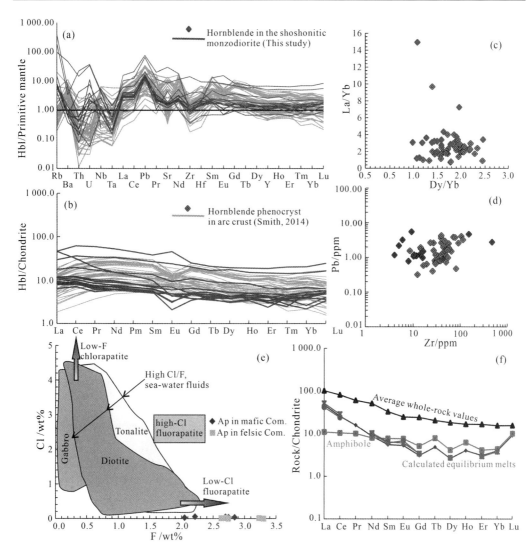

Fig. 8　Primitive-mantle-normalized trace element spiderdiagram and Chondrite-normalized REE diagrams (a,b), the La/Yb vs. Dy/Yb and Pb vs. Zr (c,d) for amphibole in the mafic component of the high-K calc-alkaline to shoshonitic intrusions in the Southeastern Tibet. Cl vs. F of apatite in high-K to shoshonitic intrusions (Zhang et al. 2017) (e); Ch Chondrite-normalized REE diagrams for whole-rock of shoshonitic mafic intrusions, associated amphibole and calculated equilibrium melts (f).

The primitive mantle and chondrite values are from Sun and McDonough (1989).

Data of amphibole from normal arc crust in the subduction zone from Smith (2014).

Discussion

Petrogenesis of the high-K calc-alkaline to shoshonitic intrusions

Late-Cretaceous to Early-Cenozoic intrusions in the Jiucheng area, central Tengchong Terrane all have notably high alkalis and K_2O/Na_2O ratios (Supplemental Table 2) and belong to the high-K calc-alkaline to shoshonitic suites of typical potassic rocks (Peccerillo and Taylor, 1976; Le Maitre et al., 2002) (Fig. 5a,b). Such features are comparable with many other potassic intrusions in the Tethyan belt from the Mediterranean to Greater Tibet (Tommasini et al., 2011; Huang et al., 2010; Avanzinelli et al., 2008, 2009, 2020). Similarly, they show "enriched" crustal isotope signatures [initial $^{87}Sr/^{86}Sr$ ca. 0.710 1 – 0.712 1, $\varepsilon_{Nd}(t)$ ca. −7.9 to 11.0], akin to coeval and local arc magmas, and similar to potassic and ultrapotassic rocks in southern Tibet and shoshonitic magmas in eastern Tibet (Fig. 9a-d). Their associated initial $^{206}Pb/^{204}Pb$, $^{207}Pb/^{204}Pb$, and $^{208}Pb/^{204}Pb$ ratios are consistent, 18.626 – 18.765, 15.720 – 15.744 and 39.183 – 39.356, respectively (Supplementary Table 2; Fig. 9e,f), and plot in the field of potassic rocks and S-type granites from the Lhasa terrane, overlapping fields of lower to upper crust and near the top field of Early-Cenozoic potassic rocks in eastern Tibet. Zircon in situ Lu-Hf isotopic compositions range from $\varepsilon_{Hf}(t) = -17.6$ to −3.11, in the continental crustal array including global lower crust and sediments (Supplementary Table 2; Fig. 10). All these features resonate with trace element characteristics such as enriched LILE and LREE with relatively depleted HFSE, similar to typical Cenozoic potassic rocks in the Tethyan realm. They will be explored further below.

Late-Cretaceous units

The high-K to shoshonite suite of the Late Cretaceous defines a separate set of trends from those of the Early-Cenozoic counterparts (Fig. 5), suggesting distinctive evolutionary trends. In the Late Cretaceous case, the samples have distinctly bimodal distributions close to the high-K liquid line of descent (LLD, Fig. 5). The shoshonitic monzodiorites have low SiO_2 (50.8–52.9 wt%) and high MgO (4.73–5.84 wt%) with high-$Mg^{\#}$ (56.0–59.8) and relatively high Cr, Ni (Cr = 88–145 ppm, Ni = 23–33 ppm) (Supplemental Table 2), and therefore are likely to be mantle-derived. Since all the rocks show "crustal" isotopic and elemental signatures, those of the mafic end-member strongly suggest a mantle influenced by crust-derived materials. Many petrogenetic models have been invoked to explain similar observations elsewhere, for example: ① mantle metasomatized by subduction-related fluids/melts, such as subducted continental crust (Liu et al., 2014; Soder et al., 2018), fluids from altered oceanic slab (Guo et al., 2005; Pe-Piper et al., 2009; Huang et al., 2010) and subducted sediments (including carbonates) (Gao et al., 2007; Conticelli et al., 2008; Liu et al., 2015); ② melting of otherwise hydrous mantle (Meen 1987); and/or ③ melting of continental crust that had been thrust into the upper mantle (Campbell et al., 2014). Fig. 10 assembles a variety of relevant diagnostic plots from the literature. Low CaO/Al_2O_3 ratios (< 0.50) (Fig. 10a), high K_2O/Na_2O and high Hf/Sm ratios (> 0.70) (Fig. 11a) argue against the possibility of metasomatism by carbonatites (Liu et al., 2015). On the other hand, high Th/Yb and La/Sm

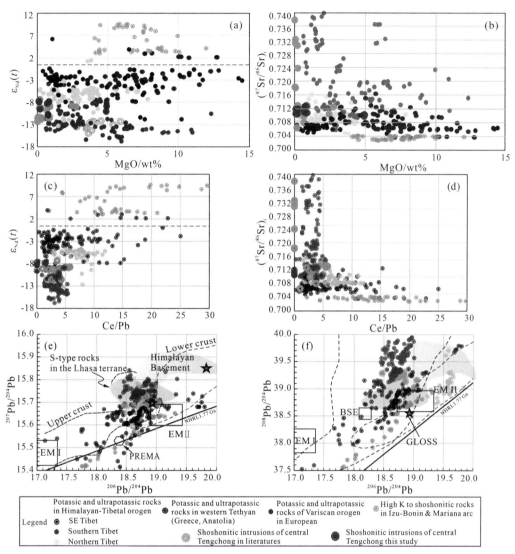

Fig. 9　Bulk-rock $\varepsilon_{Nd}(t)$ and initial $^{87}Sr/^{86}Sr$ vs. MgO (wt%) and Ce/Pb ratios (a-d),
$^{207}Pb/^{204}Pb$ vs. $^{206}Pb/^{204}Pb$ (e) and $^{208}Pb/^{204}Pb$ vs. $^{206}Pb/^{204}Pb$ (f)
from various shoshonitic provinces around the world that is summarized in Table 1.
S-type granites in the Lhasa terrane and Himalayan basement from Liu et al.(2014).

with low Ba/Th and Ba/La ratios share similar arrays with sediment-derived melts rather than
hydrous fluids (Fig. 10b-e). Compared to experimental data (Förster et al., 2019; Fig. 11e,
f), those from the shoshonitic monzodiorites have relatively high Th/Yb and Th /La (>0.2)
and their low Sm/La (<0.3) and Nb/Yb ratios fall in the field of sediment/dunite reaction
melts (with variable ratios) rather than basanite/dunite reaction melts. Also, on the Th/La vs.
$^{208}Pb/^{206}Pb$ diagram (Fig. 11e,f), all samples show a trend from GLOSS to SALATHO, which
has been regarded as the result of metasomatism by K_2O-rich sediment melts (Tommasini et

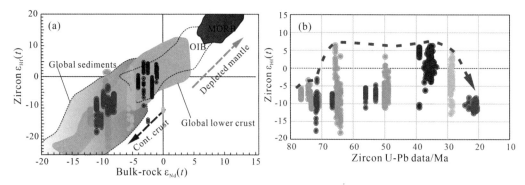

Fig. 10　Zircon $\varepsilon_{Hf}(t)$ vs. bulk-rock $\varepsilon_{Nd}(t)$ and zircon U-Pb data (a,b)

for shoshonitic magmatism in the Himalayan-Tibetan orogen.

Lower continental crust (Miller et al., 1999) and upper continental crust (Harris et al., 1990); MORB and mantle array
referenced as in Wang et al. (2014); global sediments referenced from Vervoort et al. (2011). Symbols as Fig. 9

al., 2011). A recent B isotope study based in Anatolia (Palmer et al., 2019) has suggested that such fluids result from recycling of high-pressure phengite, and that the resulting volcanism offers a time constraint on slab rollback and eventual detachment. Finally, the lack of negative Eu anomalies might be consistent with a plagioclase-free residue of melting. However Eu's behavior in this case is not unequivocal. Eu anomalies may also be caused by the differential mobility of this element in the original magma, even before plagioclase crystallises. The phenomenon of diffusive fractionation has been studied, both in experiments and numerical modelling (see Perugini et al., 2006,2015). The differential Eu-mobility in the silicate melt is decisive and we think the observed anomaly is related to such phenomenon.

Mineral chemistry offers additional insights. Both hydrous minerals, hornblende and biotite, in the monzodiorite are magnesio-type (Fig. 7a-c), which is of typical mantle-derived nature (Zhu et al., 2017). Biotite in the mafic components has higher K_2O than those from the felsic equivalent (Fig. 7d) imply that the mafic component could be metasomatized by K-rich melts/fluids in source. Typical low Cl fluorapatite in mafic component preclude the possible of sea-water fluids (Zhang et al., 2017) (Fig. 8e). It therefore seems likely that the mafic representatives of these potassic intrusions were derived from sub-continental lithospheric mantle that had interacted with sediment-derived melts.

The Late-Cretaceous high-K calc-alkaline to shoshonitic hornblende-bearing granites have high SiO_2(69. 8−76. 0 wt%) and extremely low MgO (0. 07−0. 79) with $Mg^{\#}$(15. 2−35. 7) (Supplementary Table 2), thus probably resulted from the melting of evolved continental crustal materials. The A/CNK values are weakly peraluminous (0. 99−1. 02), which together with their high CaO/Na_2O and Al_2O_3/TiO_2 and low (Al_2O_3 + MgO + $Fe_2O_3^{T}$ + TiO_2) and moderate $Al_2O_3/$(MgO + $Fe_2O_3^{T}$ + TiO_2) ratios (Supplementary Table 2), suggest derivation from a metagraywacke source (Patiño Douce, 1999). However, partial melting of

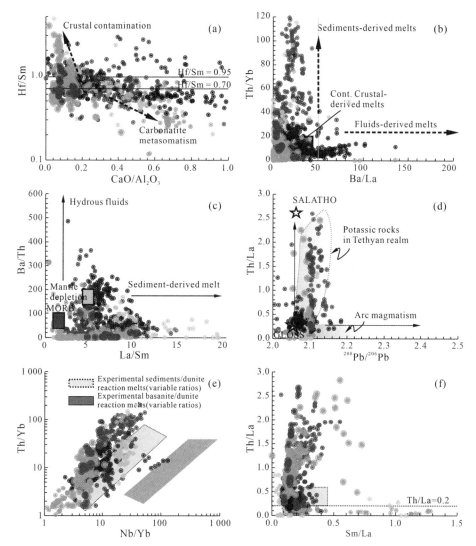

Fig. 11　Hf/Sm vs. CaO/Al$_2$O$_3$(a), Th/Yb vs. Ba/La (b), Ba/Th vs. La/Sm (c) referenced from Zhu et al. (2019). Th/La vs. ^{208}Pb/^{206}Pb (d), Th/Yb vs. Nb/Yb (e), Th/La vs. Sm/La (f) referenced from Tommasini et al. (2010).
Experimental data of sediments/dunite reaction melts and basanite/dunite reaction melts (variable ratios) refenced from Förster et al. (2019). Symbols as Fig. 9.

metagraywacke is unlikely to produce such potassium-rich (shoshonitic) melts with correspondingly high "incompatible" elements including LILE (Rb, Th) and LREE, well above those of typical upper continental crust (e. g., Zhao et al., 2016) (Fig. 6). Castro (2019) has provided experimental evidence for a link between the crystallization of sanukitoid-like (hydrous, high-K) magmas and fluid-fluxing of granite genesis in the lower crust. Such a model is attractive here, given the intimate relationships of the felsic plutons with coeval and/or earlier K- and Mg-rich mafic intrusions derived from metasomatized sub-continental lithosphere.

Early-Cenozoic units

The Early-Cenozoic high-K calc-alkaline to shoshonitic granodiorites have moderate SiO_2 (average 62. 1 wt%) but also notably high K_2O (5. 58–5. 91 wt%), total alkalis (9. 58–10. 0 wt%), REE and LILE (Fig. 6). Both LREE and HREE have twice the concentrations of normal upper continental crustal materials. A similar phenomenon is reported from the southern Bangwan area (Zhao et al., 2016), and such rocks are unlikely to have been produced by melting of normal continental crust. The extremely low MgO (average 0. 66 wt%), Cr (average 2 ppm) and Ni (average 1 ppm) exclude direct derivation from the mantle. Experimental studies have shown that intermediate SiO_2 could be derived from granulite facies rocks in the lower crust (Rapp and Watson, 1995), during which abundant amphibole and biotite likely result from water-rich melting (Milord et al., 2001). Similar to Castro's model (Castro, 2019), Lóperz et al. (2005) suggested the presence of K_2O- and LILE-rich fluids. The granites have high SiO_2(74. 5–75. 9 wt%) but relatively low Ba, La, Ce, HFSE, total REE with marked troughs at P, Ti, Ba, Nb, Ta and Eu on mantle-normalised patterns (Fig. 6c). Nevertheless, LILE and LREE are higher than normal upper continental crust, but HREE patterns are broadly similar (Fig. 6c). Overall, they are elementally very similar to the Late-Cretaceous leucogranites and are therefore likely to share a common petrogenesis.

Most of the rocks discussed above are plutonic and thus had substantial crustal residence before reaching their solidus. Thus a combination of crustal contamination with or without crystal fractionation (AFC, assimilation-fractionation crystallization) likely operated within the magmatic reservoir. Such processes often modify geochemical signatures, but rarely beyond recognition. For example, case studies of the Rogart granite (Fowler et al., 2001), Ach'Uaine appinites (Fowler and Henney, 1996) and Glen Dessary syenite (Fowler, 1992, all in NW Scotland), show that the magmas had a long history of fractionation which involved crystal-liquid separation and crustal contamination, but the data could also allow some constraints on the source composition and processes of magmagenesis. In the present study, field observation and petrography do not provide any evidence for widespread crustal contamination, such as the presence of xenoliths (e. g., Conticelli and Peccerillo, 1990; Conticelli, 1998). However, data from this study that have been least influenced by AFC processes include: ① Zircon and whole-rock isotope systematics – these have likely retained information on magmatic source, since zircon is a robust a mineral consistently used for precisely this purpose, and in the present study the zircon Hf isotopic data are coupled with the whole-rock Nd isotopes (Fig. 10). Although there is the variation in $\varepsilon_{Hf}(t)$ values for each sample, the $\varepsilon_{Hf}(t)$ value of each sample is in the range of −10. 0 to −5. 0 (not more than 5ε units) and does not correlate with $^{176}Yb/^{177}Hf$ ratios (Supplemental Table 1). Both isotope systems suggest that there are enriched and evolved ancient crust materials in the source, which is also consistent with whole-rock Sr and Pb isotopic data (Fig. 9). ② Mineral chemistry discriminators—the biotite from shoshonitic monzodiorites has higher Mg and K contents than biotite from high-K calc-alkaline to shoshonitic granodiorite and granite suites; the different hornblende compositions: while magnesio-hornblendes occurs in diorites, ferropargasites in granodiorite-granite suites (Fig. 7). These features also argue against AFC derivation of one from the other. And ③ major element discriminators — the trends between shoshonitic monzodiorite and high-K calc-alkaline granodiorite-granite are not consistent with AFC, FC or mixing, although FC-related process may occur in the generation of high-K calc-alkaline granodiorite-granite. On the basis of these observations, we suggest that the high-K calc-alkaline to shoshonitic suites are not sufficiently affected by AFC to obscure primary petrogenetic processes. The following discussion is based on this understanding.

Geodynamic implications: The transition from Neo-Tethys subduction to collision between India and Asia

The subduction of Neo-Tethyan oceanic slab under the southern margin of the Asia-Europe plate was ongoing from late Triassic and/or early Jurassic until the collision of India-Asia during Early Cenozoic. Comprehensive evidence from stratigraphic, paleobiological, geochronological, geochemical and paleomagnetic data show that the widespread Mesozoic to Early-Cenozoic magmatism in Tengchong could be the southeastern extension of Gangdese magmatic arc (Zhao et al., 2016; Zhu et al., 2017, 2018). Clockwise rotation of the Tengchong Block occurred at ~ 40 Ma along the Eastern Himalayan syntaxis and crustal detachments and strike-slip shear followed (Kornfeld et al., 2014). Late-Cretaceous to Early-Cenozoic high-K calc-alkaline to shoshonitic suites have been identified for the first time in the south margin of Asia that are related to the subduction of Neo-Tethyan ocean crust and convergence of India-Asian plates. The two different shoshonitic suites must be integrated into a regional geodynamic process that explains, in particular, the melting episodes in sub-continental lithospheric mantle (Fig. 12a).

Late-Cretaceous geodynamics

The former stage (ca. 75 – 64 Ma) was bimodal, with mafic intrusions the products of melting of sub-continental lithosphere mantle metasomatized by sediment-derived fluids/melts, and felsic intrusions produced by high-degree melting of metagraywacke, significantly influenced by fluids from the shoshonitic mafic intrusions. Bimodal magmatism is sometimes regarded as an indicator of crustal extension. A Late-Cretaceous extensional setting in the Tengchong Block may be a response to extensional collapse after significant compression during

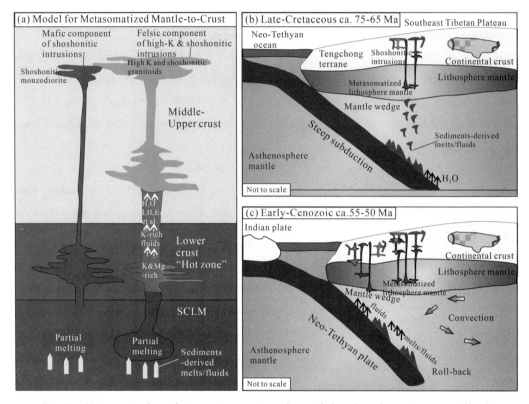

Fig. 12　Metasomatized mantle-to-crust magmatic pathways (a) and geodynamic processes(b,c) due to change from the Neo-Tethyan subduction to initial collision.

a later stage of early Cretaceous or an early stage of Late-Cretaceous convergence (Searle et al., 2007), induced by the northward subduction of Neo-Tethys. According to Sr/Y ratios of the Guyong-Husa granites in central Tengchong and their apparent relationship with geophysically-determined Moho depth (Chiaradia, 2015), Zhao et al. (2016) considered that the crust of Tengchong Block was indeed thin during this time, while that of the eastern Himalayan syntaxis and the northern part of southern Lhasa were normal and the southern part of southern Lhasa was thickened. After a period of magmatic quiescence, the rate of convergence increased dramatically and flat subduction transformed to steep subduction (Lee and Lawver, 1995). This resulted in the generation of the Gangdese magmatic arc (Chung et al., 2005) and its southeastern extension in Tengchong. During steep subduction (Fig. 12b), subducted sediment-derived melts and/or fluids play a significant role in the production of high-K calc-alkaline to shoshonitic rocks, for example the Choshi rocks (Hanyu et al., 2006), granodiorites in southeastern Lhasa (Zhu et al., 2019) and the high-K to shoshonitic suites in central Tengchong that are the subject of this study. Normal geothermal models cannot produce a slab hot enough to melt the sediment carapace-to do sotemperatures at the slab top to be 200℃ higher than normal are required (Schmidt et al., 2004; Zhu et al., 2019). However, an abrupt transition from flat to steep subduction can disturb the normal mantle wedge convective flow pattern and greatly affect slab temperatures (Behn et al., 2011). In addition, the down-going slab will take the subducted sediments deep, to form buoyant diapirs that may release fluids/melts laden with trace elements into the overlying lithospheric mantle (Behn et al., 2011; Zhu et al., 2019), to form the source of the mafic potassic magmas. Such magmas with their advected heat may ascend to the base of continental crust, where fluxed melting by exsolved aqueous fluids, forming associated felsic intrusions.

Early-Cenozoic geodynamics

Subsequently, at ca. 65 – 58 Ma, initial collision between Indian and Asian plates occurred, as supported by paleogeographic, kinematic and geodynamic data (van Hinsbergen et al., 2019). Since ca. 55 Ma, the northward drift rate of India plate had been sharply decelerated from 18 – 19 cm/y to 4 – 5 cm/y (Klootwijk et al., 1992). Together with the buoyancy of the Tengchong lithosphere, this dampened subduction and induced slab roll-back (Fig. 11c; Kohn and Parkinson, 2002). Abundant calc-alkaline arc magmatism along the continental arc margin of the Tengchong terrane had variable bulk-rock $\varepsilon_{Nd}(t)$ values and positive zircon $\varepsilon_{Hf}(t)$ values, plus low initial $^{87}Sr/^{86}Sr$ ratios tending towards the mantle array. Compositional variations are correlated with gradually increasing enrichment from west to east (Zhao et al., 2019). Thus, hydrous fluids may have been released mainly into the western Tengchong Block (Wang et al., 2014) giving way to sediment-derived potassium-rich melts, enriched with incompatible elements and REE, in the central Tengchong Block, such as the Bangwan and Jiucheng areas (Zhao et al., 2016, 2019; and this study). This model for high-K to shoshonitic intrusions is consistent with experimental result of Castro (2019) that the mantle-derived K-rich, high-$Mg^{\#}$ magmas may be coeval with, and contribute to, granite production. In addition, zircon $\varepsilon_{Hf}(t)$ data of shoshonitic rocks also show that the notably increasing $\varepsilon_{Hf}(t)$ values since ca. 70 Ma until 40 Ma (Fig. 10b), implying the more primitive mantle inputs from the later stages, closely controlled by geodynamic processes.

The general geodynamic significance of high-K calc-alkaline to shoshonitic magmatism

In general, the post-collision potassic magmatism of both European Variscan and Asian Himalayan-Tibetan orogen is widespread but low-volume (Janoušek et al., 2020; Chung et al., 1998; Guo et al., 2007). This type of K-rich (shoshonitic) magmatism often occurs at the very end of the orogeny. At this stage, because of exhumation and cooling of lower to middle crust, the widespread transfer of high-volume magmatism from mantle to crust is difficult. There are often close spatio-temporal relationship with regional deformation, such as strike-slip faults, shear zones and major thrust systems within a narrow belt over long distances. Thus, geodynamic processes at depth plus regional structure logically control pathways from mantle-derived melts to upper, more evolved magmas. Similar low-volume but widely distributed high-K to shoshonitic magmatism has occurred in oceanic arcs (Ishizuka et al., 2010), for example, Izu-Bonin-Mariana arc. These volcanoes extend more than 250 km from Io-to island to Hiyoshi, associated with earliest back-arc basin rifting, which taps a source region in the enriched lithosphere or uppermost asthenosphere. In this setting, the rocks are characterized by lower Th/Yb. Th/La, Hf/Sm, $^{87}Sr/^{86}Sr$, $^{207}Pb/^{204}Pb$, $^{208}Pb/^{204}Pb$, $^{206}Pb/^{204}Pb$, positive $\varepsilon_{Nd}(t)$ values, higher MgO contents and Ce/Pb ratios (Figs. 8 and 10) than those from orogen, post-collision setting.

Continental arc examples, such as the Late-Cretaceous rocks described here, show many similar elemental and isotopic characteristics with those in post-collision settings. Both high-volume shoshonitic rocks (in this case) and low-volume-widespread potassic rocks in southern and northern Tibet have lower MgO contents, Ce/Pb ratios, notably negative $\varepsilon_{Nd}(t)$ and $\varepsilon_{Hf}(t)$ values, higher $^{87}Sr/^{86}Sr$, Th/Yb, Th/La, Hf/Sm, $^{207}Pb/^{204}Pb$, $^{208}Pb/^{204}Pb$, $^{206}Pb/^{204}Pb$ ratios than potassic rocks in island arcs. Both have been affected by subducted crustal materials but through different mechanisms: subducted sediments vs. continental crust. In other words, crustal metasomatism, enriched mantle underplating and extensive melting of overlying continental crust could be significant and continuous along the pathway of mantle-to-crust in the generation of high-volume high-K calc-alkaline to shoshonitic magmatism on the Earth.

Conclusions

(1) Hydrous, shoshonitic, monzodioritic magmas from melting of sediment-metasomatized sub-continental lithospheric mantle provide heat, water and recycled crustal materials to induce fluxed melting of the overlying crust. This results in generation of high-K calc-alkaline to shoshonitic granodiorite-granite suites.

(2) Metasomatized mantle input to the base of a continental arc and subsequent melting at active continental margins are required to produce extensive high-K calc-alkaline to shoshonitic magmatism, in contrast to widespread but low-volume potassic magmatism else-where.

(3) The resulting mantle-to-crust magmatic pathways at active continental margins are produced by multi-stage processes, for example where dramatically increasing convergence rate results in steeper subduction of the Neo-Tethyan slab and subsequent slab roll-back induced by the initial collision of India-Asia.

(4) In both post-collisional and continental arc cases, the potassic magmatism has higher Th/Yb, Th/La, $^{206}Pb/^{204}Pb$, $^{207}Pb/^{204}Pb$, $^{208}Pb/^{204}Pb$, initial $^{87}Sr/^{86}Sr$ ratios and lower Ce/Pb, $\varepsilon_{Nd}(t)$ and $\varepsilon_{Hf}(t)$ values than in arc rifting.

Acknowledgements　We gratefully acknowledge constructive comments from Prof. Sandro Conticelli and another anonymous reviewer. We also thank the Prof. Daniela Rubatto for editorial and scientific comments to help us improve this manuscript. This work was financed by the National Natural Science Foundation of China (Grant No. 41802054) and also supported from Grant Nos. 41902046 and 41772052, Natural Science Foundation of Shaanxi Grant. 2019JQ-719 and China Postdoctoral Science Special Foundation Grant. 2019T120937 and Foundation Grant. 2018M643713. Thanks for the field work help from Senior engineer Chengmin Yan and engineer Yawei Wang in the Yunnan Bureau Geological Mineral Resource and Dr. Xiangyu Gao in the Ocean University of China.

Supplementary Information　The online version contains supplementary material available at https://doi.org/10.1007/s00410-021-01843-z.

References

Avanzinelli, R., Bianchini, G., Tiepolo, M., Jasim, A., Natali, C., Braschi, E., Dallai, L., Beccaluva, L., Conticelli, S., 2020. Subduction-related hybridization of the lithospheric mantle revealed by trace element and Sr-Nd-Pb isotopic data in composite xenoliths from Tallante (Betic Cordillera, Spain). Lithos, 352-353,105316 .

Avanzinelli, R., Elliot, T., Tommasini, S., Conticelli, S., 2008. Constraints on the genesis of the potassium-rich Italian volcanics from U/Th disequilibrium. J Petrol. 49, 195-223.

Avanzinelli, R., Lustrino, M., Mattei, M., Melluso, L., Conticelli, S., 2009. Potassic and ultrapotassic magmatism in the circum-Tyrrhenian region: Significance of carbonated pelitic vs. pelitic sediment recycling at destructive plate margins. Lithos, 113(1-2), 213-227.

Bao, Z., Yuan, H.L., Zong, C.L., et al., 2016. Simultaneous Determination of Trace Elements and Lead Isotopes in Fused Silicate Rock Powders Using a Boron Nitride Vessel and fsLA-(MC)-ICP-MS. J Anal at Spectrom, 31(4),1012-1022.

Behn, M.D., Kelemen, P.B., Hirth, G., Hacker, B.R., Massonne, H.J., 2011. Diapirs as the source of the sediment signature in arc lavas. Nat Geosci, 4(9), 641.

Campbell, I. H., Stepanov, A.S., Liang, H.Y., Allen, C.M., Norman, M.D., Zhang, Y.Q., Xie, Y.W., 2014. The origin of shoshonites: New insights from the Tertiary high-potassium intrusions of eastern Tibet. Contrib Miner Petrol, 167(3), 983.

Cao, H.W., Zou, H., Zhang, Y.H., Zhang, S.T., Zheng, L., Zhang, L.K., et al., 2016. Late Cretaceous magmatism and related metallogeny in the Tengchong area: Evidence from geochronological, isotopic and geochemical data from the Xiaolonghe Sn deposit, western Yunnan, China. Ore Geol Rev, 78, 196-212.

Cao, H.W., Pei, Q.M., Zhang, S.T., Zhang, L.K., Tang, L., Lin, J.Z., Zheng, L., 2017. Geology, geochemistry and genesis of the Eocene Lailishan Sn deposit in the Sanjiang region, SW China. J Asian Earth Sci, 137, 220-240.

Castro, A., 2019. The dual origin of I-type granites: The contribution from experiments. Post-Archean Granitic Rocks: Petrogenetic Processes and Tectonic Environments. Geol Soc London Special Publications, 491.

Chen, X.C., Hu, R.Z., Bi, X.W., Zhong, H., Lan, J.B., Zhao, C.H., et al., 2015. Petrogenesis of metaluminous A-type granitoids in the Tengchong-Lianghe tin belt of southwestern China: Evidences from zircon U-Pb ages and Hf-O isotopes, and whole-rock Sr-Nd isotopes. Lithos, 212(215), 93-110.

Chiaradia, M., 2015. Crustal thickness control on Sr/Y signatures of recent arc magmas: An Earth scale perspective. Sci Rep, 5, 8115.

Chung, S.L., Lo, C.H., Lee, T.Y., Zhang, Y., Xie, Y., Li, X., et al., 1998. Diachronous uplift of the Tibetan plateau starting 40 Myr ago. Nature, 394(6695),769.

Chung, S.L., Chu, M.F., Zhang, Y., Xie, Y., Lo, C.H., Lee, T.Y,. et al., 2005. Tibetan tectonic evolution inferred from spatial and temporal variations in post-collisional magmatism. Earth Sci Rev, 68(3-4), 173-196.

Conticelli, S., 1998. The effect of crustal contamination on ultrapotassic magmas with lamproitic affinity: Mineralogical, geochemical and isotope data from the Torre Alfina lavas and xenoliths, central Italy. Chem Geol, 149(1-2), 51-81.

Conticelli, S., Peccerillo, A., 1992. Petrology and geochemistry of potassic and ultrapotassic volcanism in central italy: Petrogenesis and inferences on the evolution of the mantle sources. Lithos, 28(3-6), 221-240.

Conticelli, S., Guarnieri, L., Farinelli, A., Mattei, M., Avanzinelli, R., Bianchini, G., et al., 2009a. Trace elements and Sr-Nd-Pb isotopes of K-rich, shoshonitic, and calc-alkaline magmatism of the western Mediterranean region, genesis of ultrapotassic to calc-alkaline magmatic associations in a post-collisional geodynamic setting. Lithos, 107(1), 68-92.

Conticelli, S., Marchionni, S., Rosa, D., Giordano, G., Boari, E., Avanzinelli, R., 2009b. Shoshonite and sub-alkaline magmas from an ultrapotassic volcano: Sr-nd-pb isotope data on the Roccamonfina volcanic rocks, Roman magmatic province, southern Italy. Contrib Miner Petrol, 157(1), 41-63.

Conticelli, S., Avanzinelli, R., Marchionni, S., Tommasini, S., Melluso, L., 2011. Sr-Nd-Pb isotopes from the Radicofani volcano, central Italy: Constraints on heterogeneities in a veined mantle responsible for the shift from ultrapotassic shoshonite to basaltic andesite magmas in a post-collisional setting. Mineral Petrol, 103(1-4), 123-148.

Conticelli, S., Avanzinelli, R., Ammannati, E., Casalini, M., 2015. The role of carbon from recycled sediments in the origin of ultrapotassic igneous rocks in the central mediterranean. Lithos, 232, 174-196.

Dallai, L., Bianchini, G., Avanzinelli, R., Natali, C., Conticelli, S., 2019. Heavy oxygen recycled into the lithospheric mantle. Sci Rep, 9(1), 8793.

Duggen, S., Hoernle, K., Van D.B.P., et al., 2005. Post-collisional transition from subduction- to intraplate-type magmatism in the westernmost Mediterranean: Evidence for continental-edge delamination of subcontinental lithosphere. J Petrol, 6, 1155-1201.

Förster, M.W., Prelević, D., Buhre, S., Mertz-Kraus, R., Foley, S.F., 2019. An experimental study of the role of partial melts of sediments versus mantle melts in the sources of potassic magmatism. J Asian Earth Sci, 177, 76-88.

Forster, M.D., 1960. Interpretation of the composition of trioctahedral micas. U S Geol Survey Prof Paper, 254-B, 11-49.

Fowler, M.B., 1992. Elemental and O-Sr-Nd isotope geochemistry of the Glen Dessarry syenite, NW Scotland. J Geol Soc, 149(2), 209-220.

Fowler, M.B., Henney, P.J., 1996. Mixed Caledonian appinite magmas: Implications for lamprophyre fractionation and high Ba-Sr granite genesis. Contrib Miner Petrol, 126(1-2), 199-215.

Fowler, M.B., Henney, P.J., Darbyshire, D.P.F., Greenwood, P.B., 2001. Petrogenesis of high Ba-Sr granites: The Rogart pluton, Sutherland. J Geol Soc London, 158, 521-534.

Fowler, M.B., Kocks, H., Da Rbyshire, D., Greenwood, P.B., 2008. Petrogenesis of high Ba-Sr plutons from the northern Highlands terrane of the British Caledonian province. Lithos, 105(1-2), 129-148.

Francalanci, L., Innocenti, F., Manetti, P., Savas, M.Y., 2000. Neogene alkaline volcanism of the afyon-isparta area, Turkey: Petrogenesis and geodynamic implications. Miner Petrol, 70(3-4), 285-312.

Gao, Y., Hou, Z., Kamber, B.S., Wei, R., Meng, X., Zhao, R., 2007. Lamproitic rocks from a continental collision zone: Evidence for recycling of subducted Tethyan oceanic sediments in the mantle beneath southern Tibet. J Petrol, 48(4), 729-752.

Guo, Z., Hertogen, J.A.N., Liu, J., Pasteels, P., Boven, A., Punzalan, L.E.A., et al., 2005. Potassic magmatism in western Sichuan and Yunnan provinces, SE Tibet, China: Petrological and geochemical constraints on petrogenesis. J Petrol, 46(1), 33-78.

Hanyu, T., Tatsumi, Y., Nakai, S.I., Chang, Q., Miyazaki, T., Sato, K., et al,. 2006. Contribution of slab melting and slab dehydration to magmatism in the NE Japan arc for the last 25 Myr: Constraints from geochemistry. Geochem Geophys Geosyst, 7(8).

Harris, N.B.W., Inger, S., Xu, R., 1990. Cretaceous plutonism in Central Tibet: An example of post-collision magmatism? J Volcanol Geoth Res, 44, 21-32.

Huang, X.L., Niu, Y., Xu, Y.G., Chen, L.L., Yang, Q.J., 2010. Mineralogical and geochemical constraints on the petrogenesis of post-collisional potassic and ultrapotassic rocks from western Yunnan, SW China. J Petrol, 51(8), 1617-1654.

Ishizuka, O., Yuasa, M., Tamura, Y., Shukuno, H., Stern, R.J., Naka, J., et al., 2010. Migrating shoshonitic magmatism tracks Izu-Bonin-Mariana intra-oceanic arc rift propagation. Earth Planet Sci Lett, 294(1-2), 111-122.

Janoušek, V., Hanl, P., Svojtka, M., Hora, J.M., David, B., 2020. Ultrapotassic magmatism in the heyday of the Variscan orogeny: The story of the Tebí pluton, the largest durbachitic body in the bohemian massif. Int J Earth Sci, 109, 1767-1810.

Jiang, Y.H., Jiang, S.Y., Ling, H.F., Zhou, X.R., Rui, X.J., Yang, W.Z., 2002. Petrology and geochemistry of shoshonitic plutons from the western Kunlun orogenic belt, Xinjiang, northwestern China: Implications for granitoid geneses. Lithos, 63(3-4), 165-187.

Jiang, Y.H., Jiang, S.Y., Ling, H.F., Dai, B.Z., 2006. Low-degree melting of a metasomatized lithospheric mantle for the origin of Cenozoic Yulong monzogranite-porphyry, east Tibet: Geochemical and Sr-Nd-Pb-Hf isotopic constraints. Earth Planet Sci Lett, 241(3-4), 617-633.

Klootwijk, C.T., Gee, J.S., Peirce, J.W., Smith, G.M., 1992. Neogene evolution of the Himalayan-Tibetan region: Constraints from ODP Site 758, northern Ninetyeast Ridge; bearing on climatic change. Palaeogeogr Palaeoclimatol Palaeoecol, 95(1-2), 95-110.

Kohn, M.J., Parkinson, C.D., 2002. Petrologic case for Eocene slab breakoff during the Indo-Asian collision. Geology, 30(7), 591-594.

Kornfeld, D., Eckert, S., Appel, E., Ratschbacher, L., Sonntag, B.L., Pfänder, J.A., et al., 2014. Cenozoic clockwise rotation of the Tengchong block, southeastern Tibetan Plateau: A paleomagnetic and geochronologic study. Tectonophysics, 628, 105-122.

Le Maitre, R.W., 2002. Igneous rocks: A classification and glossary of terms. Cambridge: Cambridge University Press, 236.

Leake, B.E., Woolley, A.R., Arps, C.E.S., Birch, W.D., Gilbert, M.C., Grice, J.D., Hawthorne, F.C., et al., 1997. Nomenclature of Amphiboles: Report of the subcommittee on amphiboles of the international mineralogical association, commission on new minerals and mineral names. Can Mineral, 35, 219-246.

Lee, T.Y., Lawver, L.A., 1995. Cenozoic plate reconstruction of Southeast Asia. Tectonophysics, 251(1-4), 85-138.

Liu, D, Zhao, Z., Zhu, D.C., Niu, Y., DePaolo, D.J., Harrison, T.M., et al., 2014. Postcollisional potassic and ultrapotassic rocks in southern Tibet: Mantle and crustal origins in response to India-Asia collision and convergence. Geochim Cosmochim Acta, 143, 207-231.

Liu, D., Zhao, Z., Zhu, D.C., Niu, Y., Widom, E., Teng, F.Z., et al., 2015. Identifying mantle carbonatite metasomatism through Os-Sr-Mg isotopes in Tibetan ultrapotassic rocks. Earth Planet Sci Lett, 430, 458-469.

Lóperz, S., Castro, A., García-Casco, A., 2005. Production of granodiorite melt by interaction between hydrous magma and tonalitic crust. Experimental constraints and implication for the generation of Archaean TTG complexes. Lithos, 79, 229-250.

Meen, J.K., 1987. Formation of shoshonites from calcalkaline basalt magmas: Geochemical and experimental constraints from the type locality. Contrib Mineral Petrol, 97(3), 333-351.

Metcalfe, I., 2013. Gondwana dispersion and Asian accretion: Tectonic and palaeogeographic evolution of eastern Tethys. J Asian Earth Sci, 66, 1-33

Miller, C., Schuster, R., Kotzli, U., et al., 1999. Post-collisional potassic and ultrapotassic magmatism in SW Tibet: Geochemical and Sr-Nd-Pb-O isotopic constraints for mantle source characteristics and petrogenesis. J Petrol, 40, 1399-1424.

Milord, I, Sawyer, E.W., Brown, M., 2001. Formation of diatexite migmatite and granite magma during anatexis of semipelitic metasedimentary rocks: An example from St. Malo, France. J Petrol, 42, 487-505.

Palmer, M.R., Ersoy, E.Y., Akal, C., Uysal, I., Genç, Ç.C., Banks, L.A., Cooper, M.J., et al., 2019. A short, sharp pulse of potassium-rich volcanism during continental collision and subduction. Geology, 47, 1079-1082.

Patiño Douce, A.E., 1999. What do experiments tell us about the relative contributions of crust and mantle to the origin of the granitic magmas. Geol Soc London, 168, 55-75.

Peccerillo, A., 1999. Multiple mantle metasomatism in central-southern Italy: Geochemical effects, timing and geodynamic implications. Geology, 47, 1079-1082.

Peccerillo, A., Taylor, S.R., 1976. Geochemistry of Eocene calc-alkaline volcanic rocks from Kastamonu area, Northern Turkey. Contrib Mineral Petrol, 58, 63-81.

Pe-Piper, G., Piper, D.J., Koukouvelas, I., Dolansky, L.M., Kokkalas, S., 2009. Postorogenic shoshonitic rocks and their origin by melting underplated basalts: The Miocene of Limnos, Greece. Geol Soc Am Bull, 121(1-2), 39-54.

Perugini, D., Campos, C.D., Petrelli, M., Dingwell, D.B., 2015. Concentration variance decay during magma mixing: A volcanic chronometer. Sci Rep, 5(1), 14225.

Perugini, D., Petrelli, M., Poli, G., 2006. Diffusive fractionation of trace elements by chaotic mixing of magmas. Earth Planet Sci Lett, 243(3), 669-680.

Prelević, D., Stracke, A., et al., 2010. Hf isotope compositions of Mediterranean lamproites: Mixing of melts from asthenosphere and crustally contaminated mantle lithosphere. Lithos, 119(3-4), 297-312.

Prelević, D., Foley, S.F., Romer, R., Conticelli, S., 2008. Mediterranean Tertiary lamproites derived from multiple source components in postcollisional geodynamics. Elsevier, 72, 2125-2156

Qi, X., Zhu, L., Grimmer, J.C., Hu, Z., 2015. Tracing the Transhimalayan magmatic belt and the Lhasa Block southward using zircon U-Pb, Lu-Hf isotopic and geochemical data: Cretaceous-Cenozoic granitoids in the Tengchong block, Yunnan, China. J Asian Earth Sci, 110, 170-188.

Rapp, R.P., Watson, E.B., 1995. Dehydration melting of metabasalt at 8 − 32 kbar: Implications for continental growth and crust-mantle recycling. J Petrol, 36(4), 891-931.

Ratschbacher, B.C., Paterson, S.R., Fischer, T.P., 2019. Spatial and depth-dependent variations in magma volume addition and addition rates to continental arcs: Application to global CO_2 fluxes since 750 Ma. Geochem Geophys Geosyst, 20, 2997-3018.

Schmidt, M.W., Vielzeuf, D., Auzanneau, E., 2004. Melting and dissolution of subducting crust at high pressures: The key role of white mica. Earth Planet Science Lett, 228(1-2), 65-84.

Searle, M.P., Noble, S.R., Cottle, J.M., Waters, D.J., Mitchell, A.H.G., et al., 2007. Tectonic evolution of the Mogok metamorphic belt, Burma (Myanmar) constrained by U-Th-Pb dating of metamorphic and magmatic rocks. Tectonics, 26(3).

Smith, D.J., 2014. Clinopyroxene precursors to amphibole sponge in arc crust. Nat Commun, 5, 4329

Smith, D.J., Petterson, M.G., Saunders, A.D., Millar, I. L., Jenkin, G.R.T., Toba, T., et al., 2009. The petrogenesis of sodic island arc magmas at savo volcano, solomon islands. Contrib Mineral Petrol, 158(6), 785-801.

Soder, C.G., Romer, R.L., 2018. Post-collisional potassic-ultrapotassic magmatism of the Variscan Orogen: Implications for mantle metasomatism during continental subduction. J Petrol, 59(6), 1007-1034.

Sun, S.S., McDonough, W.F., 1989. Chemical and isotopic systematics of oceanic basalts: Implications for mantle composition and processes. In: Saunders, A.D., Norry, M.J. Magmatism in the Ocean Basins. Geological Society of London, Special Publication, 42, 313-345.

Thompson, R.N., Fowler, M.B., 1986. Subduction-related shoshonitic and ultrapotassic magmatism: A study of Siluro-Ordovician syenites from the Scottish Caledonides. Contrib Mineral Petrol, 94(4), 507-522.

Tommasini, S., Avanzinelli, R., Conticelli, S., 2011. The Th/La and Sm/La conundrum of the Tethyan realm lamproites. Earth Planet Sci Lett, 301(3-4), 469-478.

van Hinsbergen, D.J., Lippert, P.C., Li, S., Huang, W., Advokaat, E.L., Spakman, W., 2019. Reconstructing Greater India: Paleogeographic, kinematic, and geodynamic perspectives. Tectonophysics,

760, 69-94.

Vervoort, J.D., Plank, T., Prytulak, J., 2011. The Hf-Nd isotopic composition of marine sediments. Geochim et Cosmochim Acta, 75(20), 5903-5926

Wang, Y., Zhang, L., Cawood, P.A., Ma, L., Fan, W., Zhang, A., et al., 2014. Eocene supra-subduction zone mafic magmatism in the Sibumasu Block of SW Yunnan: Implications for Neotethyan subduction and India-Asia collision. Lithos, 206, 384-399.

Weis, D., Kieffer, B., Maerschalk, C., et al., 2006. High-precision isotopic characterization of USGS reference materials by TIMS and MC-ICP-MS. Geochem Geophys Geosyst, 7(8), Q08006

Xie, J.C., Zhu, D.C., Dong, G., Zhao, Z.D., Wang, Q., Mo, X., 2016. Linking the Tengchong Terrane in SW Yunnan with the Lhasa Terrane in southern Tibet through magmatic correlation. Gondwana Res, 39, 217-229.

Xu, Y.G., Yang, Q.J., Lan, J.B., Luo, Z. Y., Huang, X.L., Shi, Y.R., Xie, L.W., 2012. Temporal-spatial distribution and tectonic implications of the batholiths in the Gaoligong-Tengliang-Yingjiang area, western Yunnan: Constraints from zircon U-Pb ages and Hf isotopes. J Asian Earth Sci, 53, 151-175.

Yuan, H.L., Gao, S., Dai, M.N., Zong, C.L., Gunther, D., Fontaine, G.H., Liu, X.M., Diwu, C.R., 2008. Simultaneous determinations of U-Pb age, Hf isotopes and trace element compositions of zircon by excimer laser-ablation quadrupole and multiple-collector ICPMS. Chem Geol, 247, 100-118.

Zhang, C., Koepke, Juergen Horn, Ingo, Holtz, Francois, et al., 2017. Apatite in the dike-gabbro transition zone of mid-ocean ridge: Evidence for brine assimilation by axial melt lens. Am Mineral, 102(3/4), 558-570.

Zhao, S.W., Lai, S.C., Qin, J.F., Zhu, R.Z., 2016. Petrogenesis of eocene granitoids and microgranular enclaves in the western Tengchong Block: Constraints on eastward subduction of the Neo-tethys. Lithos, 264, 96-107.

Zhao, S.W., Lai, S.C., Qin, J.F., Zhu, R.Z., Wang, J.B., 2017. Geochemical and geochronological characteristics of late Cretaceous to early Paleocene granitoids in the Tengchong Block, southwestern China: Implications for crustal anatexis and thickness variations along the eastern Neo-Tethys subduction zone. Tectonophysics, 694, 87-100.

Zhao, S.W., Lai, S.C., Pei, X.Z., Qin, J.F., Zhu, R.Z., Tao, N., Gao, L., 2019. Compositional variations of granitic rocks in continental margin arc: Constraints from the petrogenesis of Eocene granitic rocks in the Tengchong Block, SW China. Lithos, 326, 125-143.

Zhu, R.Z., Lai, S.C., Santosh, M., Qin, J.F., Zhao, S.W., 2017. Early Cretaceous Na-rich granitoids and their enclaves in the Tengchong Block, SW China: Magmatism in relation to subduction of the Bangong-Nujiang Tethys ocean. Lithos, 286(287), 175-190.

Zhu, R.Z., Lai, S.C., Qin, J.F., Zhao, S.W., Santosh, M., 2018. Strongly peraluminous fractionated S-type granites in the Baoshan Block, SW China: Implications for two-stage melting of fertile continental materials following the closure of Bangong-Nujiang Tethys. Lithos, 316, 178-198.

Zhu, R.Z., Lai, S.C., Qin, J.F., Zhao, S.W., Santosh, M., 2019. Petrogenesis of high-K calc-alkaline granodiorite and its enclaves from the SE Lhasa block, Tibet (SW China): Implications for recycled subducted sediments. Geol Soc Am Bull, 131(7-8), 1224-1238.

Petrogenetic evolution of Early Paleozoic trachytic rocks in the South Qinling Belt, central China: Insights from mineralogy, geochemistry, and thermodynamic modeling[①]

Yang Hang Lai Shaocong[②] Qin Jiangfeng Zhang Fangyi Zhu Renzhi
Zhu Yu Liu Min Zhao Shaowei Zhang Zezhong

Abstract: Early Paleozoic volcanic-intrusive rocks in the South Qinling Belt, central China, consist of bimodal alkaline suites ranging in composition from basalt to trachyte, with an apparent scarcity of intermediate silicic components (the so-called "Daly gap"). A series of trachytic rocks have been identified in the Quanxi area of the South Qinling Belt. In this work, we combine mineralogical evidence with whole-rock major- and trace-element data, Sr-Nd-Pb isotope data, and the results of thermodynamic modeling to gain a better understanding of the petrogenesis and evolutionary process of the Quanxi trachytic rocks, as well as their compositional discontinuity with coeval mafic rocks. Our samples can be clearly classified into two lithological groups based on their phenocryst content: Group 1, phenocryst-poor trachytes (< 5 vol% phenocrysts; SiO_2 = 63.91 – 66.13 wt%) and Group 2, trachytic tufflavas (30 – 45 vol% pyroclasts; SiO_2 = 53.34 – 59.03 wt%). All of the samples exhibit typical oceanic island basalt-like enrichment in light rare earth elements and high field strength elements, as well as marked depletion of Ba, Sr, and Ti. A cogenetic metasomatized lithosphere source for the Quanxi trachytic rocks and contemporaneous mafic rocks is indicated by their homogeneous isotopic features [$\varepsilon_{Nd}(t)$ = + 2.3 to + 3.4; initial $^{206}Pb/^{204}Pb$ = 17.92 – 18.69; initial $^{207}Pb/^{204}Pb$ = 15.53 – 15.57; initial $^{208}Pb/^{204}Pb$ = 37.96 – 38.99] and linear geochemical variations, with protracted fractional crystallization regarded as the key mechanism for the compositional variations among these cogenetic rocks. The results of Rhyolite-MELTS modeling predict 72% – 80% crystallization involving plagioclase, pyroxene, biotite, apatite, and Fe-Ti oxide, to yield the magmatic compositions from an evolved mafic progenitor to the Quanxi trachytes under conditions of low pressure (1.5 – 2 kbar), high H_2O content (2 – 3 wt%), and high f_{O_2} (FMQ – 0.5). Some incompatible elements in trachytic tufflavas deviate markedly from the liquid lines of descent from coeval mafic rocks to the Quanxi trachytes: This deviation is attributed to a combination of the involvement of large proportions of alkali-feldspar pyroclasts and post-magmatic hydrothermal alteration. The presence of the "Daly gap" in early

---

① Published in *Lithos*, 2022, 418-419.

② Corresponding author.

Paleozoic alkaline suites can be explained by an accelerated differentiation rate (i.e., $dSiO_2/dt$) in the SiO_2-intermediate field, which is driven by the simultaneous separation of a considerable scale of SiO_2-poor phases.

1　Introduction

Silicic alkaline igneous rocks contain a broad range of lithological and mineralogical variation with phonolitic, trachytic, and rhyolitic compositions. They are genetically linked to extensional events and represent the evolved end-members of continental rift systems, oceanic island settings, and continental intraplate suites (Barberi et al., 1975; Peccerillo et al., 2003; Choi, 2020). Unusual mineralogy and elevated abundances of alkalis and rare metals [e.g., Nb, Ta, and rare earth elements (REEs)] are prevalent in these rocks, meaning that they possess the great potential for hosting deposits with considerable industrial and economical applications (Dostal, 2017; Hussain et al., 2020). Their formation, storage condition, and chemical differentiation have long been a topic of debate in igneous geology and volcanology (Andújar and Scaillet, 2012; Chandler and Spandler, 2020). Several alternative models, including partial melting of the lower crust (Su et al., 2007; Karsli et al., 2018), low-degree melting of the metasomatized mantle (Bailey, 1987; Lewis et al., 2016), and liquid immiscibility or protracted crystal fractionation from a mafic progenitor (Barberi et al., 1975; Charlier et al., 2011, 2013; Namur et al., 2011), have been proposed to explain the generation of such silica- and alkali-enriched magmas. An unanswered question is whether their highly evolved compositions were generated by extensive magma differentiation, or merely reflect the partial melting of their source rocks. Furthermore, silicic alkaline rocks often occur in association with large-scale mafic rocks and exhibit apparent discontinuities in chemical composition (i.e., the "Daly gap"), with typical bimodal characteristics (Peccerillo et al., 2003; Szymanowski et al., 2015). There are two main gaps in our understanding of the origin of bimodal series: ① the genetic affinity between the felsic and mafic members, and ② the nature of the compositional gap. The main hypotheses for the origin of such gaps are partial melting of the crust; the immiscibility of different magma types; crystallization of specific mineral phases; and inability of intermediate melts to rise up because of viscosity, density, and volatile relationships (Peccerillo et al., 2003; Suneson and Lucchitta, 1983 and references therein). To summarize, these bimodal rocks could be the product of separate evolution of cogenetic magma, or could arise from various sources with no direct consanguinity. Research on the genetic relationship between silicate rocks and associated mafic rocks will help to improve understanding of their petrogenesis and the formation of the "Daly gap".

Early Paleozoic alkaline rocks with obvious bimodal affinity are widely exposed in the

South Qinling Belt (SQB), and are dominated by basalts, diabases, gabbros, syenites, and trachytes (Table 1; Huang, 1993; Yang et al., 2020a; Zhang et al., 2020a,b; Wang et al., 2021). The majority of previous studies have focused on the alkaline ultramafic-mafic complexes, including the aspects of their source nature, storage condition, and tectonic environment (Xiang et al., 2016; Zhang et al., 2020a,b; Zhang, et al., 2002, 2007, 2020). The silicic alkaline igneous complexes have been less well studied. The special geochemical characteristics, e.g., alkali and SiO_2 enrichment, MgO depletion, high field strength elements (HFSEs) enrichment, and Sr-Nd depletion $[\varepsilon_{Nd}(t) = +2.0$ to $+7.4]$ suggest that these highly differentiated rocks originated from an enriched mantle source and formed in an extensional environment (Wang et al., 2017; Wang et al., 2021; Yang et al., 2020a). The formation of these felsic rocks is hypothesized to have resulted from immiscibility of silicate liquids (Huang, 1992) or extensive fractional crystallization from an initial melt with a basaltic composition (Wang et al., 2017; Yang et al., 2020a). However, at present there are insufficient constraints on their evolutionary process, particularly identification of their parental magma, quantitative data on the crystalline minerals, and estimation of storage conditions. In addition, although apparent spatial and temporal correlations have been recognized in SQB alkaline volcanic-intrusive suites (Huang, 1992; Wang et al., 2017; Yang et al., 2020a; Wang et al., 2021), some aspects of the relationship between the mafic and felsic members are not well understood, e.g., their genetic connection and the lack of rocks with 48−62 wt% SiO_2. Thus, further research is required to constrain the origin and evolution of the felsic rocks, their genetic relationship with coeval mafic magmas, and possible explanations for the scarcity of intermediate compositions in the SQB volcanic-intrusive sequence.

The trachytes and associated volcaniclastic rocks constitute the majority of outcrops in the Quanxi area and were formed predominantly as lava flows. In this study, we extend previous data sets of the dominant coeval mafic rocks in SQB and concentrate on the newly identified Quanxi trachytic rocks in the Quanxi area. By combining whole-rock geochemical, Sr-Nd-Pb isotopic, mineralogical, and thermodynamic data, we gain a more comprehensive understanding of the petrogenesis and evolutionary process of these trachytic rocks, as well as the presence of the "Daly gap" in the SQB alkaline suites.

2 Geological background and field geology

The Qinling orogenic belt is situated in the central part of the Asian continent (Dong et al., 2015; Ratschbacher et al., 2003), and was formed by multistage collisions between the Yangtze Block and the North China Craton during the Paleozoic and Mesozoic (Fig. 1a; Zhang et al., 1996). The belt can be further subdivided into the North Qinling belt (NQB) and the SQB along the Shangdan suture (Fig. 1b). The SQB has been identified as a portion of the

Table 1　Summaries of the early Paleozoic alkaline rocks in the South Qingling Belt.

Location	Sample	Lithology	Method	Age/Ma	Geochemical features	Isotopic	Reference
Langao	11LGC-8	Diabase	LA-ICP-MS zircon U-Pb	441.9±3.2	MgO=3.81-8.07 wt%, Eu/Eu*=1.07-1.56		Wang, 2014b
		Diabase, pyroxenite and basalt			Na₂O/K₂O=2.37-25.00, Eu/Eu*=0.97-1.50	$\varepsilon_{Nd}(t)$ = +3.28 to +5.02	Zhang et al., 2002, 2007
	12GPC	Trachyte	LA-ICP-MS apatite U-Pb	417±42	MgO=0.63-0.85 wt%, Eu/Eu*=0.82-1.05	$\varepsilon_{Nd}(t)$ = +1.95 to +3.27	Wang, 2014a
Ziyang	QLhp	Quartz syenite	LA-ICP-MS zircon U-Pb	435.1±1.2	SiO₂=63.64-64.78 wt%, Eu/Eu*=0.77-1.71	$\varepsilon_{Hf}(t)$=12.1±2.4 (weighted mean)	Wang et al., 2017
	QL15-3	Diabase	LA-ICP-MS zircon U-Pb	440.0±0.5	Na₂O+K₂O=2.21-5.00 wt%, Eu/Eu*=1.10-1.18		
	DBS257/1	Diabase and gabbro	SHRIMP zircon U-Pb	422.1±4.7	MgO=3.86-7.8 wt%, Na₂O/K₂O=1.76-5.51	$\varepsilon_{Nd}(t)$ = +0.22 to +3.91	Chen et al., 2014
	12HP	Trachyte	LA-ICP-MS apatite U-Pb	377±37	MgO=0.33-1.85 wt%, Eu/Eu*=0.79-0.95	$\varepsilon_{Nd}(t)$ = +2.83 to +3.92	Wang, 2014a
	12GQB-10	Pyroxene diorite	SIMS zircon U-Pb	438.4±3.4	Na₂O=4.71-16.7 wt%, Mg#=20-78	$\varepsilon_{Nd}(t)$ = +2.25 to +2.88	Wang, 2014b
	GQ-4-1 GT-10-1	Diabases	LA-ICP-MS zircon U-Pb	455.9±1.5 446.2±1.1	MgO=4.00-9.40 wt%, TiO₂=2.89-5.17 wt%, Na₂O/K₂O=2.50-11.90	$\varepsilon_{Nd}(t)$ = +2.4 to +3.5	Zhang et al., 2020a
Pingli	PL-7	Trachyte	LA-ICP-MS zircon U-Pb	406.0±12	SiO₂=60.88-62.88 wt%, Eu/Eu*=0.64-0.91	$\varepsilon_{Hf}(t)$ = +6.1 to +7.1	Yang et al., 2020a
	205	Trachytic volcanic rocks	LA-ICP-MS zircon U-Pb	430.6±2.7	SiO₂=54.16-66.29 wt%, (La/Yb)$_N$=5.50-25.82, Eu/Eu*=0.44-1.00	$\varepsilon_{Nd}(t)$ = +3.0 to +7.3	Wan et al., 2016
Zhuxi	ZX-2	Trachyte	LA-ICP-MS zircon U-Pb	427.9±6.6	SiO₂=62.10-63.87 wt%, Eu/Eu*=0.64-0.87	$\varepsilon_{Nd}(t)$ = +3.0 to +4.5	Yang et al., 2020a
	16PL28-1	Trachytic volcanic rocks	LA-ICP-MS apatite U-Pb	434±10	SiO₂=55.21-66.26 wt%, Eu/Eu*=0.43-1.05		Wang et al., 2021
Zhenping	D0212	Diabase	SHRIMP zircon U-Pb	439±6	(La/Yb)$_N$=11.18-15.67, Eu/Eu*=1.01-1.17		Zou et al., 2011
	ZP-7	Diabase	SHRIMP zircon U-Pb	451±4	MgO=6.21-6.41 wt%, Eu/Eu*=1.06-1.07		Xiang et al., 2016

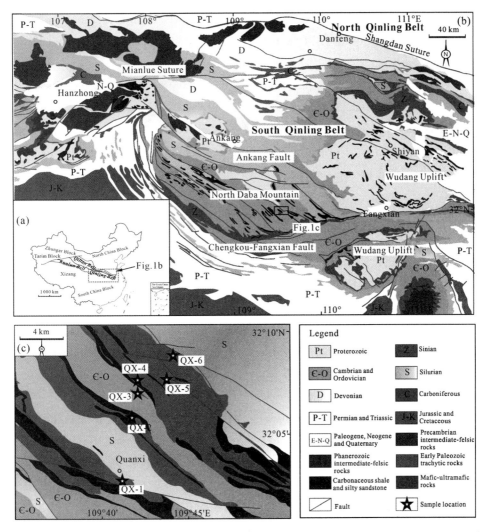

Fig. 1 (a) Simplified tectonic map of China; (b) Geological map of the Qinling Orogenic Belt and adjacent areas (modified after Wang et al., 2017); (c) Simplified geological map showing the distribution of Early Paleozoic alkaline rocks in the Quanxi area (modified after the 1 : 200 000 geological map of Pingli sheet).

Sample locations are marked by star symbols.

northern margin of the Yangtze Block (NYB) prior to the opening of the Mianlue Ocean (Zhang et al., 1996) and is characterized by abundant Triassic granitic magmatism and metamorphism (Dong et al., 2011). The basement of the SQB consists of Precambrian low-grade metamorphic rocks (the Douling, Wudang, and Yaolinghe groups), and is overlain by Sinian to Mesozoic sedimentary rocks (Wang et al., 2017). The Douling complex forms the Paleoproterozoic crystalline basement, and is primarily composed of biotite gneiss, amphibolite, and schist (Hu et al., 2013). Neoproterozoic basement complexes in the SQB mainly consist of the Wudang and Yaolinghe groups (Dong et al., 2017). The Wudang group is a volcanic-sedimentary sequence with a greenschist metamorphic overprint, whereas the Yaolinghe group mainly consists of tholeiitic basalt, spilitic diabase, alkali trachyte, and spilite (Huang, 1993).

The early Paleozoic extension and rifting of the SQB were related to the opening of the Mianlue Ocean (Dong et al., 2011; Ratschbacher et al., 2003), which was accompanied by extensive bimodal magmatism (Wang et al., 2017; Yang et al., 2020a). Mafic members make up the majority (~80 vol%) of the bimodal suite and mainly consist of intrusive rocks (i.e., diabases, gabbros and pyroxenites) and minor basalts (Chen et al., 2014; Zhang et al., 2002, 2007, 2020a, b). Most of the intrusive rocks were concordantly intruded into surrounding Cambrian to Silurian strata and emplaced as NW-SE trending dykes several to hundreds of meters wide and hundreds to thousands of meters long (Fig. 1b). The felsic members are adjacent to the mafic rocks and formed in their earthen side (Pingli-Zhuxi-Quanxi area) along the North Daba Mountains. The main lithologies of these felsic rocks include trachytes, syenites, and pyroclastic rocks (Fig. 1c). They are interbedded with each other and exhibit extrusive contact with surrounding Silurian strata and locally contact with coeval pyroxene rocks (Chen et al., 2014). Additionally, a few Nb-REEs-containing carbonatite complexes, which are genetically related to the associated alkaline rocks, have been discovered in the Miaoya and Shaxiongdong areas (Su et al., 2021).

The Quanxi trachytic rocks are important components of the SQB alkaline volcanic-intrusive suites and cover an area of ~80 km^2 with a NW-SE trend (Fig. 1c). These volcanic rocks range in chemical composition from trachyandesite to trachyte and their eruption was accompanied by the formation of a series of faults (Fig. 1c). The phenocryst-poor trachytes exhibit relatively low vales of Porphyritic Index (the total phenocrysts content) of 3-5 vol% with typical trachytic textures (Fig. 2c, d). The trachytes mainly consist of euhedral alkali feldspars and accessory subhedral Fe-Ti oxides, with the matrix dominated by microcrystalline alkali feldspars and quartz. Alkali feldspars are most commonly found as megacrysts (up to 1.5 cm in size) with typical Carlsbad twinning (Fig. 2c-e). Trachytic tufflavas show a transition from lava to pyroclastic rock with no obvious pseudofluidal structure and are composed primarily of volcanic pyroclasts (30-45 vol%) and aphanitic matrix (55-70 vol%). Crystal clasts make up the majority of pyroclasts (70-80 vol%) and are primarily composed of alkali feldspars with minor mica-group minerals and titanites (Fig. 2f, g). Alkali feldspars in the tufflavas are mostly subhedral to anhedral and display obvious plastic deformation (Fig. 2g), in contrast to those in trachytes. The lithic fragments are mainly composed of trachytic rocks and cut through by randomly distributed calcite-dominated veins (Fig. 2h). The matrix consists of microcrystalline alkali feldspars and minor tuffaceous components (Fig. 2f, g).

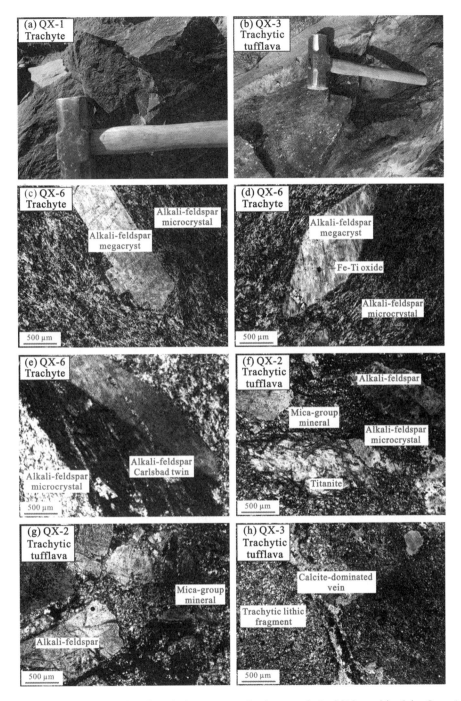

Fig. 2　Field photographs (a,b) and photomicrographs (cross-polarized light, c-h) of the Quanxi trachytic rocks. (c,d) Alkali feldspar megacrystal and alkali feldspar microcrystal forming typical trachytic texture in the trachyte; (e) Alkali feldspar crystals displaying Carlsbad twin in the trachyte; (f,g) Alkali feldspar dominated clasts and matrix in trachytic tufflava; (h) Trachytic lithic fragment in the trachytic tufflava.

3 Analytical methods

3.1 TESCAN Integrated Mineral Analyzer (TIMA) and electron microprobe analysis (EMPA)

A TESCAN Integrated Mineral Analyzer (TIMA) housed in the Northwest University, Xi'an, was used for automated mineralogical, modal and textural analyses (Hrstka et al., 2018). This includes the collection of backscattered electron (BSE) and EDS data on a regular grid of 10 mm point spacing. The individual points are grouped based on a similarity search algorithm, and areas of coherent BSE and EDS data are merged to produce segments (i.e., mineral grains). Individual spectra from points within each segment are summed. Data from each segment are then compared against a classification scheme to identify the mineral and assign its chemistry and density. The results are plotted as a map showing the distribution of minerals within the sample.

The compositions of rock-forming minerals were analyzed using a JXA-8230 electron microprobe at the State Key Laboratory of Continental Dynamics, Northwest University, Xi'an, China. The microprobe analyses were carried out with an accelerating voltage of 15 kV, a sample current of 10 nA, and a beam diameter of 1 μm. Microprobe standards of natural and synthetic phases were supplied by SPI Company: jadeite for Si, Al, and Na, diopside for Ca, olivine for Mg, sanidine for K, hematite for Fe, rhodonite for Mn, and rutile for Ti.

3.2 Major and trace elements

Twenty-two samples were selected for whole-rock major and trace element analyses at the State Key Laboratory of Continental Dynamics, Northwest University in Xi'an, China. Samples were powdered to 200 mesh size using a tungsten carbide ball mill. Major elements were determined on lithium borate fusion disks using X-ray fluorescence spectrometry (XRF). Analyses of USGS and Chinese national rock standards (BCR-2, GSR-1 and GSR-3) indicate that both the analytical precision and accuracy for major elements were generally better than 5%. Trace element contents were determined by inductively coupled plasma mass spectrometry (ICP-MS); sample powders were digested using an $HF + HNO_3$ mixture in high-pressure Teflon bombs at 190 ℃ for 48 h. The detailed sample-digesting procedure for ICP-MS analysis was described by Liu et al. (2007). Precision and accuracy were monitored by measurements of replicates and the standard reference materials BHVO-2, AGV-2, BCR-2 and GSP-2. Analytical precision for most trace elements was better than 10%.

3.3 Sr-Nd-Pb isotopic analyses

High precision isotopic (Sr, Nd, Pb) measurements were carried out at Nanjing FocuMS

Technology Co. Ltd (www.focums.com). Geological rock powders were mixed with 0.5 ml 60 wt% HNO_3 and 1.0 ml 40 wt% HF in high-pressure PTFE bombs. These bombs were steel-jacketed and placed in the oven at 195 ℃ for 3 days. Digested samples were dried down on a hotplate, reconstituted in 1.5 ml of 0.2 N HBr+0.5 N HNO_3 before ion exchange purification. Raw data of isotopic ratios were internally corrected for mass fractionation by normalizing to $^{86}Sr/^{88}Sr = 0.1194$ for Sr, $^{146}Nd/^{144}Nd = 0.7219$ for Nd, $^{205}Tl/^{203}Tl = 2.3885$ for Pb with exponential law. International isotopic standards (NIST SRM 987 for Sr, JNdi-1 for Nd, and NIST SRM 981 for Pb) were periodically analyzed to correct instrumental drift. Geochemical reference materials of USGS BCR-2, BHVO-2, AVG-2, RGM-2 were treated as quality control (isotopic results of the reference materials were listed in Supplementary Table S3). These isotopic results agreed with previous publications within analytical uncertainty (Weis et al., 2006).

4 Results

4.1 Mineral features

The results of qualitative and quantitative TIMA analysis of the minerals in the Quanxi trachytic rocks by TIMA (Figs. 3 and 4) are basically consistent with the microscopic observations. Albite is the most abundant mineral in the Quanxi trachytes (~ 54 vol%), occurring as a matrix mineral and occasionally as phenocrysts. In contrast, orthoclase is mostly present as microcrystal and is found only in the matrix. The matrix also contains a notable proportion of quartz (~ 16 vol%), as well as minor amounts of biotite and magnetite (Fig. 3). The Quanxi trachytic tufflavas have a mineral assemblage that is generally similar to that of trachyte, but with secondary minerals (calcite, biotite, and muscovite) in veins (Fig. 4). The crystal pyroclasts include albite, orthoclase, biotite, and titanite. The mineral assemblage of the tufflava matrix is dominantly abundant microcrystalline albite with minor orthoclase (Fig. 4).

Representative mineral compositions of the Quanxi trachytic rocks, obtained by EPMA, are listed in Supplementary Table S1. Despite apparent discrepancies in morphology, the feldspar compositions in both the Quanxi trachytes and the trachytic tufflavas overlap. All of the alkali feldspars analyzed are composed of either near endmember albite ($Ab_{92.5} Or_{7.5}$-$Ab_{99.6} Or_{0.3}$) or orthoclase ($Or_{92.5} Ab_{7.5}$-$Or_{98.8} Ab_{1.1}$), and there is no discernible compositional difference between cores and rims among these feldspars. The albites contain a high concentration of Na_2O (9.5 - 12.3 wt%), but low concentrations of K_2O (0.1 - 1.3 wt%), CaO (< 0.2 wt%), and SrO (0.2 - 0.7 wt%). The orthoclases (both crystal pyroclasts and microcrystals) contain abundant K_2O (16.0 - 17.2 wt%) and low Na_2O (0.1 - 0.9 wt%), CaO (< 0.6 wt%), and SrO (< 0.3 wt%) contents. The biotites of the Quanxi trachytic

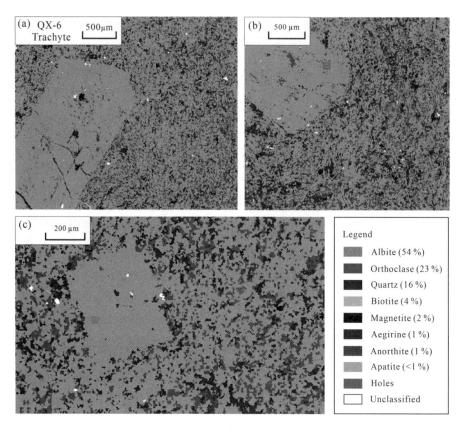

Fig. 3   TESCAN Integrated Mineral Analyzer (TIMA) mineral maps of the Quanxi trachytes.

tufflavas generally exhibit high FeO (21. 6–22. 5 wt%) and K_2O (9. 8–10. 4 wt%) abundance as well as low SiO_2 (34. 1–36. 1 wt%) levels. Muscovites in tufflavas are mostly fine-scale aggregates of sericite and occur in association with calcite veins with high amounts of Al_2O_3 (27. 1–28. 2 wt%) and SiO_2 (47. 0–50. 3 wt%). Titanites are also ubiquitous in the trachytic tufflavas, with variable amounts of TiO_2 (31. 5–35. 7 wt%), CaO (26. 0–28. 2 wt%), SiO_2 (29. 6–31. 6 wt%), and Al_2O_3 (1. 3–4. 5 wt%).

4. 2   Major and trace elements

Major- and trace-element data for 22 whole-rock samples from the Quanxi area are listed in Supplementary Table S2 and illustrated in Figs. 5, 6, and 7. All samples contain high alkali contents ($Na_2O + K_2O$ = 10. 07–12. 64 wt%) and are classified as trachyte and trachyte andesite in the total alkali-silica diagram (Fig. 5a). The trachytes range from weakly metaluminous to peralkaline with A/CNK [molar ratio of Al_2O_3/($CaO + Na_2O + K_2O$)] ratios of 0. 87–0. 95 (Fig. 5b). They are characterized by relatively high silica levels (SiO_2 = 63. 91–66. 13 wt%) but low contents of MgO (0. 62–0. 77 wt%), TiO_2 (0. 83–1. 07 wt%), and CaO (0. 40–0. 76 wt%), with Na_2O/K_2O ratios of 0. 5–1. 4 (Fig. 6 and Supplementary Table S2). In contrast, the

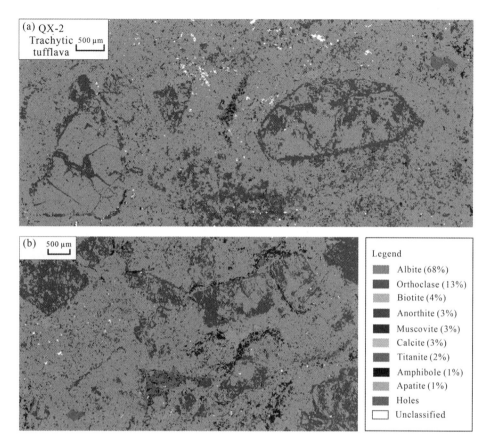

Fig. 4   TESCAN Integrated Mineral Analyzer (TIMA) mineral maps of
the Quanxi trachytic tufflavas.

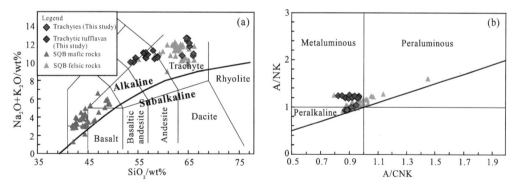

Fig. 5   (a) Total alkali vs. SiO$_2$(TAS) diagram (after Middlemost, 1994) and

(b) A/CNK-A/NK diagram of the Quanxi trachytic rocks.

The published data of SQB felsic rocks are from Wang et al. (2017), Yang et al. (2020a, b), and Wang et al.
(2021). The published data of SQB mafic rocks are from Chen et al. (2014), Wang (2014b), Xiang et al.
(2016), and Zhang et al. (2020a).

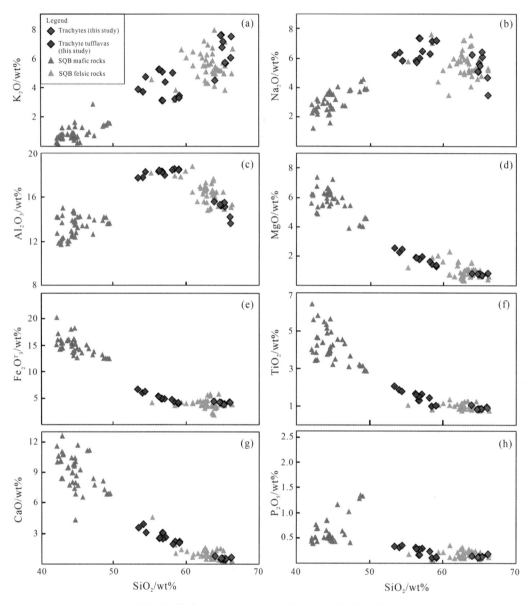

Fig. 6 Harker diagrams of the Quanxi trachytic rocks.
The published data in SQB are the same as Fig. 5.

trachytic tufflavas are predominantly metaluminous (Fig. 5b) and contain relatively lower amounts of SiO_2 (53. 34 - 59. 03 wt%) and higher contents of Al_2O_3 (17. 78 - 18. 60 wt%) , CaO (1. 94 - 3. 94 wt%) , and MgO (1. 23 - 2. 50 wt%) , with variable Na_2O/K_2O ratios of 1. 1 - 2. 4.

The distribution patterns of primitive mantle-normalized trace elements in the Quanxi trachytic rocks are characterized by negative Ba, Sr, and Ti anomalies with strong enrichment of HFSEs (e.g., Nb, Ta, Zr, and Hf). However, the abundances of Ba (906. 1 - 1 819. 8 ppm) and Sr (602. 1 - 1 336. 5 ppm) in tufflavas are markedly higher than those in trachytes

（ Ba = 19. 6 – 50. 2 ppm; Sr = 18. 8 – 74. 2 ppm）, consistent with the lithological differences （ Fig. 7a）. In the chondrite-normalized REEs diagram （ Fig. 7b）, all samples exhibit high total REE contents of 572. 5 – 1 469. 0 ppm and notable enrichment of light REEs （ LREEs） with steep slopes （ La_N/Yb_N = 16. 54 – 34. 17） and slightly negative Eu anomalies （0. 62 – 0. 96）. In contrast to the SQB mafic rocks （ 127. 2 – 334. 1 ppm; Eu/Eu^* = 1. 01 – 1. 38）, the Quanxi trachytic rocks have a higher but approximately parallel distribution of REEs, as well as variable Eu anomalies of 0. 62 – 0. 92 （ Fig. 7b）.

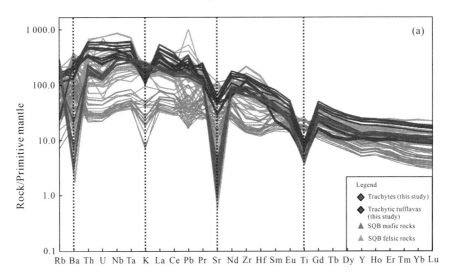

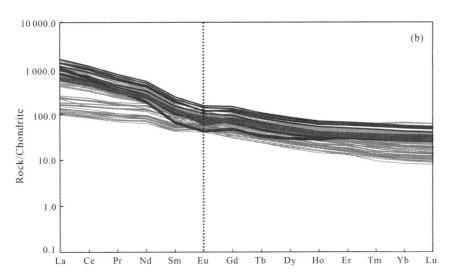

Fig. 7 （ a） Primitive-mantle normalized trace element distribution pattern of the Quanxi trachytic rocks and （ b） chondrite-normalized rare earth element （ REE） pattern of the Quanxi trachytic rocks.

The chondrite and primitive mantle data are from Sun and McDonough （ 1989） and McDonough and Sun （ 1995）, respectively. The published data in SQB are the same as Fig. 5.

4.3　Whole-rock Sr-Nd-Pb isotope

Whole-rock Sr-Nd-Pb isotopic data for the Quanxi trachytic rocks are listed in Supplementary Table S3 and presented as Sr-Nd and Pb-Pb isotope correlation diagrams in Figs. 8 and 9, respectively. Initial isotopic ratios were recalculated using an emplacement age of 420 Ma. The trachytes display variable initial $^{87}Sr/^{86}Sr$ ratios of 0.698 03 - 0.719 47; In contrast, the trachytic tufflavas yield a narrow range of initial $^{87}Sr/^{86}Sr$ ratios of 0.703 88 - 0.704 43. All samples exhibit uniform $^{143}Nd/^{144}Nd$ ratios of 0.512 21 - 0.512 27, with moderately depleted $\varepsilon_{Nd}(t)$ values of +2.3 to +3.4, which are comparable to those of the SQB mafic rocks (Fig. 8).

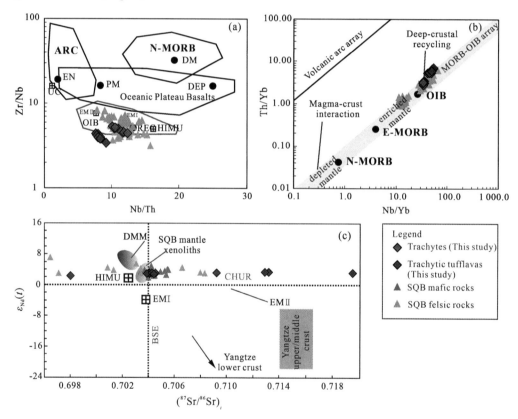

Fig. 8　(a) Zr/Nb vs. Nb/Th diagram and (b) Th/Yb vs. Nb/Yb diagram and (c) $\varepsilon_{Nd}(t)$ vs. $(^{87}Sr/^{86}Sr)_i$ diagram of the Quanxi trachytic rocks.

The isotopic data were calculated for the emplacement age of 420 Ma. DM: depleted mantle; MORB: middle oceanic ridge basalt; OIB: ocean island basalt; HIMU: high U/Pb mantle component; EMI: enriched mantle component I; EM Ⅱ: enriched mantle component Ⅱ; DEP: deep depleted mantle; En: enriched component; REC: recycled component. Data source: The DM and mantle array are from Zindler and Hart (1986); MORB, HIMU, EMI and EMⅡ are from Hart et al. (1992); OIB is from Wilson (1989); The Sr-Nd isotopic data of SQB mafic rocks are from Zhang et al. (2007) and Wang (2014a); the Sr-Nd isotopic data of SQB mafic rocks are from Yang et al. (2020a); hydrous mantle xenoliths of SQB are from Xu et al. (1997). Other published data in SQB are the same as Fig. 5.

The initial $^{206}Pb/^{204}Pb$, $^{207}Pb/^{204}Pb$, and $^{208}Pb/^{204}Pb$ ratios of the Quanxi trachytes are 18.05－18.27, 15.55, and 37.96－38.32, respectively, and of the trachytic tufflavas are 17.92 － 18.69, 15.53 － 15.57, and 38.28 － 38.99, respectively. The Pb isotopic characteristics are similar to those of the NYB and SQB basement rocks, but distinct from those of the Taihua Group (Fig. 9).

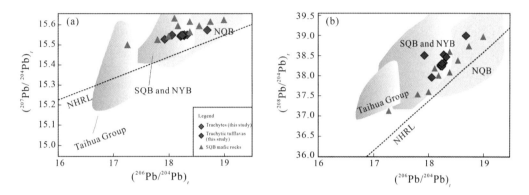

Fig. 9   Initial Pb isotopic ratios of the Quanxi trachytic rocks.

The isotopic data were calculated for the emplacement age of 420 Ma. The fields of the Taihua Group and the basement materials of NQB, SQB, and NYB are modified by Yang et al.(2020b); the Pb isotopic data of SQB mafic rocks are from Zhang et al. (2007).

5   Discussion

5.1   Genesis of the Quanxi trachytes

The Quanxi trachytes are phenocryst-poor lavas with relatively high SiO_2 (63.9－66.1 wt%) and total alkali ($Na_2O + K_2O = 10.1－12.6$ wt%) concentrations, as well as low MgO, CaO, TiO_2, Cr, and Ni contents (Fig. 6; Supplementary Table S2). These geochemical characteristics mean that these trachytes represent highly evolved compositions within the SQB alkaline suites. Experimental and geochemical studies have demonstrated that evolved silicate magmas can be generated by partial melting of lower crustal materials (e.g., Dai et al., 2017; Kaszuba and Wendlandt, 2000). Such melts commonly have high Al_2O_3 (> 15 wt%) and Sr (>300 ppm) levels with low Y (< 20 ppm) and Yb (< 2 ppm) contents, resulting in distinctively elevated Sr/Y ratios (> 40, Ding, 2011; Qin et al., 2015). However, these features have not been observed in Quanxi trachytes, which exhibit moderate Al_2O_3 contents (13.64－15.62 wt%), low Sr levels (18.8－74.2 ppm), and extremely low Sr/Y ratios (<1.3). Furthermore, the positive Nb-Ta anomalies and high average Nb/U (60.3) and Ce/Pb (25.1) ratios measured in the present study are markedly different from those of ordinary crustal rocks, which have low Nb/U (<9.7) and Ce/Pb (<5) ratios (Rudnick and

Gao, 2003), implying a negligible contribution of crustal components to the Quanxi trachytes. The depleted Nd and Pb isotopic compositions of the Quanxi trachytes [initial $^{206}Pb/^{204}Pb =$ 18.05-18.28; $^{207}Pb/^{204}Pb = 15.55$; $^{207}Pb/^{204}Pb = 37.96-38.32$; $\varepsilon_{Nd}(t) = +2.3$ to $+3.1$] strongly indicate a mantle source.

The possibility of derivation of felsic melts by direct partial melting of upper mantle ultramafic rocks has been supported by observational and experimental evidence (Yoder, 1973; Bailey, 1987). Indeed, a growing body of experimental data provide support that low degree melting (< 5%) of an alkali- and volatile-enriched mantle source at relatively low pressure (< 1.5 GPa) can produce silica- and alkali-enriched melts, such as those of trachytic or phonolitic compositions (e.g., Falloon et al., 1997; Draper and Green, 1999; Laporte et al., 2014). Such felsic melts are expected to exhibit high $Mg^#$ values (> 50; Laporte et al., 2014), and typically contain mantle xenoliths similar to those found in Heldburg phonolites (Grant et al., 2013). Additionally, given the insufficiency of feldspar fractionation at mantle pressures, the Ba and Sr concentrations of these magmas are thought to be elevated, with values of several hundred to thousands of parts per million (Irving and Price, 1981). These characteristics, however, differ appreciably from those of the Quanxi trachytes, which exhibit markedly low $Mg^#$ values (26.6-29.4) and Sr (18.8-74.2 ppm) and Ba (19.6-50.2 ppm) contents: These values are all considered too low for primary mantle melts. In addition, mantle xenoliths containing amphibole, clinopyroxene, and phlogopite have been reported in SQB mafic rocks (Huang, 1993; Xu et al., 1997), but not in our samples or other silicic rocks in the SQB. These observations rule out the possibility of their formation as a result of direct mantle melting. Alternatively, the Quanxi trachytes may have been derived from a mafic parental magma via magmatic differentiation mechanism, through either liquid immiscibility or protracted fractional crystallization processes.

5.2　Genetic relationship between Quanxi trachytic rocks and SQB mafic rocks

Prior to determining the origin and evolution of the Quanxi trachytes, identification of their mafic progenitor is crucial. The early Paleozoic mafic rocks in SQB are the obvious candidate for the progenitor. The trachytic rocks and mafic rocks in SQB were emplaced as NW-SE trending dykes and sills, consistent with the regional structure (Fig. 1c). These rocks mostly have concordant contacts with the surrounding lower Paleozoic sedimentary strata, corresponding to an extensional setting. The formation age of the felsic rocks has been constrained to 430-440 Ma (Wan et al., 2016; Wang et al., 2017; Wang et al., 2021), contemporaneous with the SQB mafic rocks (435-455 Ma, Zou et al., 2011; Chen et al., 2014; Wang 2014b; Xiang et al., 2016; Zhang et al., 2020a). Trace-element ratios such as

Zr/Nb, Nb/Th, Nb/Yb, and Th/Yb are least likely to be affected by alteration and can thus be applied to assess petrologic relationships among igneous rocks (Hofmann et al., 1986). Our samples display Zr/Nb, Nb/Th, Nb/Yb, and Th/Yb ratios similar to those of the SQB mafic rocks, and the majority of them plot in the oceanic island basalt field (Fig. 8a,b). The Nd and Pb isotopic compositions of the Quanxi trachytic rocks overlap with those of the SQB mafic rocks (Figs. 8c and 9), but the trachytes exhibit a wide range of Sr isotopic compositions (^{87}Sr/^{86}Sr = 0. 698 03 − 0. 719 47; Fig. 8c). Crustal contamination, which commonly has notable effects on the Sr isotopic composition of mantle rocks because of elevated ^{87}Sr/^{86}Sr ratios and high Sr abundances in crustal rocks (DePaolo, 1981), can be excluded in this case. Indeed, variations of the ^{87}Sr/^{86}Sr ratios in residual liquids of magmas subjected to fractional crystallization and/or liquid immiscibility can be disregarded in the majority of cases. However, the Sr isotopic system can become sensitive during magma differentiation in a high-Rb/Sr rhyolitic system (Christensen and DePaolo, 1993; Mahood and Halliday, 1988). Cavazzini (1994) proposed that elevated Rb/Sr ratios in magma can result in a marked increase in the ^{87}Sr/^{86}Sr ratio even for a short evolution time (< 1 Myr). The Quanxi trachytes exhibit extremely low Sr (18. 8 − 74. 2 ppm) contents with high Rb/Sr (1. 8 − 8. 6) ratios. Among samples with unusually high initial ^{87}Sr/^{86}Sr ratios, the trachytes with lower Sr contents generally exhibit higher ^{87}Sr/^{86}Sr ratios, consistent with Sr isotopic variations caused by magma differentiation in a high-Rb/Sr system (Cavazzini, 1994). Thus, the wide Sr variations in the Quanxi trachytes can most likely be attributed to high-Rb/Sr magmatic evolution and so cannot be used to track the nature of their source region. Even so, the close spatial and temporal association and the Nd and Pb isotopic similarities of the Quanxi trachytic rocks and the SQB mafic rocks strongly suggest a common mantle source region. Additionally, the marked LREE and HFSE enrichment in the SQB alkaline suites (Fig. 7), and the results of petrological studies of mantle xenoliths hosted in alkali basalt (e.g., amphibole and phlogopite, Huang, 1993; Xu et al., 1997), have been interpreted to possibly indicate a cogenetic metasomatized mantle source (Yang et al., 2020a; Zhang et al., 2020a).

The silicate liquid immiscibility hypothesis was proposed by Huang et al. (1992) to explain the coexistence of mafic and felsic components accompanied by minor intermediate rocks in the SQB. Charlier et al. (2011, 2013) demonstrated the viability of producing coherent Si-rich and Fe-rich melts from initial basaltic magmas via liquid immiscibility. However, liquid immiscibility has been examined only at relatively small scales (e.g., within meters), and even for particularly large examples such as the Bushveld Complex, the phenomenon of immiscibility can only just be identified at the hundred-meter scale (VanTongeren and Mathez, 2012). Therefore, whether large-scale segregation of Fe-rich and Si-rich components can occur in nature is not yet proven. Additionally, the Quanxi trachytes and coeval mafic rocks do not plot in the appropriate fields of liquid immiscibility (Fig. 10) , but are more consistent with the field of natural rhyolitic melts. Thus, we conclude that the formation of these trachytes cannot be explained by silicate liquid immiscibility. As an alternative hypothesis, the coherent major-element trends (Fig. 6) and the similarity of trace-element distributions (Fig. 7) strongly suggest that the SQB alkaline suites are cogenetic via fractional crystallization. This genetic relationship is clearly illustrated by the "fractional crystallization" trend in the Nb/La vs. SiO_2 and $\varepsilon_{Nd}(t)$ vs. SiO_2 diagrams (Fig. 11). To summarize, protracted fractional crystallization is the most likely mechanism for the petrogenesis of the Quanxi trachytic rocks and the compositional variations within the SQB alkaline suites.

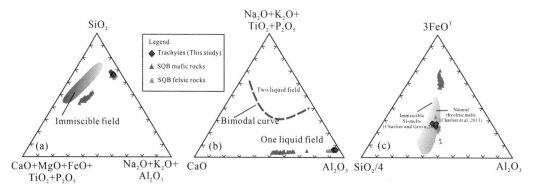

Fig. 10　(a) SiO_2 vs. ($CaO + MgO + FeO + TiO_2 + P_2O_5$) vs. ($Na_2O + K_2O + Al_2O_3$) and (b) CaO vs. Al_2O_3 vs. ($TiO_2 + P_2O_5 + Na_2O + K_2O$) diagram, and (c) $SiO_2/4$ vs. $3FeO^T$ vs. Al_2O_3 diagram of Quanxi trachytes, SQB felsic rocks and SQB mafic rocks. Fig. 10a is adapted from Charlier et al. (2011). Fig. 10b is after Charlier and Grove (2012). Fig. 10b is after Charlier et al. (2013). The published data in SQB are the same as Fig. 5.

5. 3　Geochemistry and thermodynamic modeling of magma differentiation

The Quanxi trachytes are generally phenocryst-poor and exhibit homogeneous whole-rock chemistry and phenocryst compositions, with a small proportion of mafic minerals. We consider that the whole-rock compositions of the Quanxi trachytes are representative of the evolved melt (liquid) compositions and can be applied to assess the magma differentiation from SQB mafic end-members. Apparent correlation of major- and trace-element concentrations between the Quanxi trachytes and coeval SQB mafic rocks is demonstrated in the Harker diagrams and trace element distribution patterns (Figs. 6 and 7). A process governed by separation of pyroxene, feldspar and apatite could account for the lower levels of MgO, CaO, P_2O_5, Ba, and Sr as well as the higher REE levels and parallel patterns (Fig. 7b) of the trachytes relative to the SQB mafic rocks. Positive to moderately negative Eu anomalies (Supplementary Table S2) from the mafic rocks to the trachytes indicate separation of plagioclase. Additionally, the reduced Ti/Ti^*

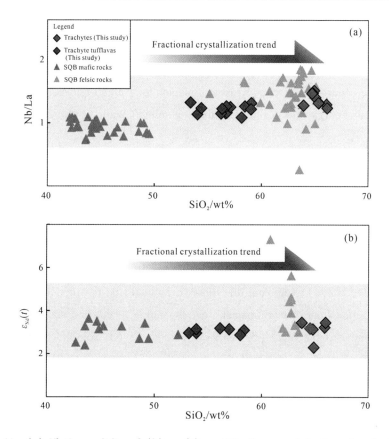

Fig. 11   (a) Nb/La vs. SiO₂ and (b) $\varepsilon_{Nd}(t)$ vs. SiO₂ diagrams of the Quanxi trachytic rocks.
The published data in SQB are the same as Fig. 5.

ratio in trachytes compared with SQB mafic rocks might be a result of Fe-Ti oxide fractionation. For trace elements, alkaline melts commonly have high solubilities of Zr and other HFSEs, and suppression of zircon crystallization allows build-up of extreme Zr concentrations in the evolved alkaline melts (Chandler and Spandler, 2020). Notable enrichments in Zr and other incompatible trace elements were detected throughout the whole differentiation process from the SQB mafic rocks to the Quanxi trachytes. The Zr concentrations of trachytes increase continually until ~65 wt% SiO_2, at which point they progressively stabilize, suggesting that slight zircon saturation occurred during the last stage of magmatic evolution. Zircon saturation temperatures calculated from samples with $SiO_2 > 65$ wt% are primarily between 900 ℃ and 930 ℃. Based on the degree of enrichment of incompatible trace elements, these compositional variations require ~80% fractional crystallization of SQB mafic rocks containing ~200 ppm Zr to form trachytic melts with ~1 000 ppm Zr. The levels of K_2O are positively correlated with those of SiO_2(Fig. 6a), whereas the amounts of Na_2O and Al_2O_3 decrease with increasing SiO_2 through nearly the entire trachyte interval (Fig. 6b, c), reflecting fractionation of Na-rich feldspar. Mineralogical data show that the feldspar megacrysts in the Quanxi trachytes have uniform chemical compositions with high concentrations of SiO_2 (68. 0 – 73. 8 wt%) and Na_2O (9. 4 – 12. 3 wt%) , implying that these alkali feldspars were formed in a homogeneous, highly evolved environment during magma evolution. These inferred fractionating phases are in agreement with the occurrence of clinopyroxene, plagioclase, alkali-feldspar, apatite, biotite, apatite, zircon, and Fe-Ti oxide in the SQB alkaline suites (Wang et al. , 2017; Yang et al. , 2020a; Fig. 2).

We further constrained the conditions of magma differentiation by analysis with the program Rhyolite-MELTS (Gualda et al. , 2012). Magmatic evolution process from primary basaltic magma to trachytic magma is considered to have taken place in multiple stages: An early stage of high-pressure fractional crystallization and later continuous magmatic differentiation at shallow depths. The pyroxene phenocrysts of SQB mafic rocks were crystallized at multiple depths (e. g. , 40 – 68 km, 25 – 40 km, 15 – 20 km, and 5 – 9 km; Xiang et al. , 2010; Zhang et al. , 2020b), implying that those rocks could be the product of high-pressure fractional crystallization from primary magma, corresponding to the first stage. This process is also reflected in the variable MgO (3. 11 – 9. 38 wt%), Cr (1. 39 – 303. 32 ppm), and Ni (0. 58 – 173. 88 ppm) contents of the SQB mafic rocks (Chen et al. , 2014; Wang, 2014b; Xiang et al. , 2016; Zhang et al. , 2020a). The absence of high-pressure minerals (e. g. , pyroxene and amphibole) and the prevalence of low-temperature minerals (e. g. , feldspar) in the Quanxi trachytes imply that their formation was governed by protracted magma differentiation at shallow depths. Thus, the whole-rock composition of an evolved mafic sample (XHK-10-2, $SiO_2 = 49. 52$ wt%, $MgO = 4. 52$ wt%; Zhang et al. , 2020a) is inferred to yield information on the closest parental composition from which the trachytic magma originated. Simulations were performed under fO_2 buffers at fayalite-magnetite-quartz (FMQ) – 0. 5 and isobaric conditions at different pressures (1. 5 – 2 kbar). Regional work described a relatively H_2O-rich condition in the mantle source of the SQB alkaline suites, based on the presence of hydrous metasomes such as amphibole and phlogopite (Huang, 1993; Xu, 1997; Zhang et al. , 2020a). Moreover, a relatively low mantle potential temperature in the SQB was estimated by Zhang et al. (2020a). Based on these considerations, in our modeling the temperature was set at decreasing intervals from 1 100 ℃ to 820 ℃ and the range of initial water content was 2. 0 – 3. 0 wt%. The results of the MELTS modeling are provided in Supplementary Table S4, and Figs. 12 and 13. The results suggest that, under these conditions, the Quanxi trachytes might be the result of 72% – 80% crystallization of a separation assemblage of 50% – 53% plagioclase, 15% – 17% clinopyroxene, 10% – 11% orthopyroxene, 10% – 11% magnetite, 5% ilmenite, 3% – 5% biotite and 3% apatite, from an evolved mafic melt. The estimated compositions of feldspar rapidly change from calcium-rich to sodium-rich as the magma evolves, whereas potassium feldspar is almost absent (Fig. 13b). Based on our mineralogical

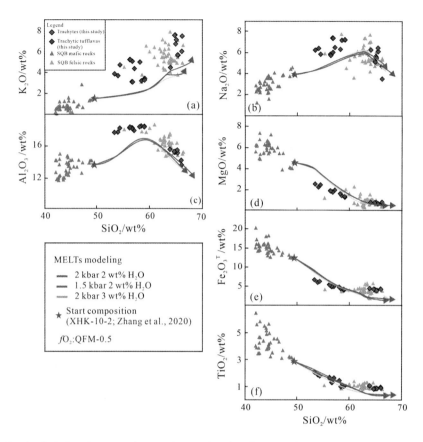

Fig. 12 (a-f) Harker diagrams showing the simulated evolution of the residual melt compositions by Rhyolite-MELTS from an evolved mafic composition (XHK-10-2; Zhang et al., 2020a) to the Quanxi trachytes, at different conditions of oxygen fugacity, pressure and water content.

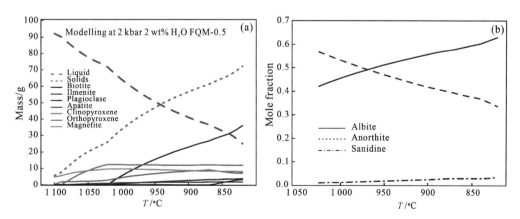

Fig. 13 (a) Mass variation as a function of temperature; and (b) Variation in feldspar composition as a function of temperature, estimated from the best fit Rhyolite-MELTS output (2 kbar; 2 wt% H_2O; FQM-0.5).

data and those from other regional investigations (Wang et al., 2021; Yang et al., 2020a,b), we conclude that feldspar mainly involves sodium-rich plagioclases in earlier stages and alkali feldspars in later stages of magma evolution. The phenocryst-poor nature of the Quanxi trachytes indicates that the melt and crystalline mineral phases were sufficiently separated, with residual alkali feldspar megacrysts surviving in the separated melts because of their low density and unique morphology (Young, 1981). The petrogenesis and evolutionary model of the Quanxi trachytes are briefly described in Fig. 14.

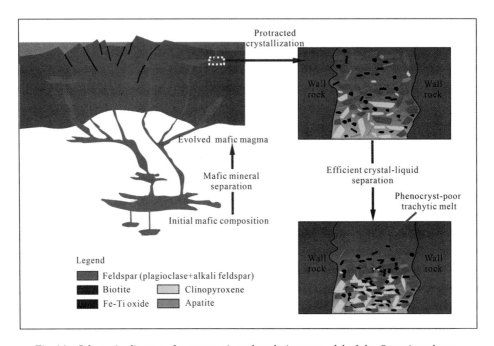

Fig. 14　Schematic diagram of petrogenesis and evolutionary model of the Quanxi trachytes.

5. 4　Formation of the Quanxi trachytic tuffaceous lavas

Field observations, mineral compositions, and isotopic evidence strongly suggest that the crystal-rich trachytic tufflavas and phenocryst-poor trachytes are petrologically related and share a common mantle source, as described above. The variations in the MgO, $Fe_2O_3^T$, and TiO_2 contents of the tufflavas are consistent with the liquid lines of descent from the SQB mafic rocks to the Quanxi trachytes estimated by MELTS modeling (Fig. 12d-f), whereas the variations in Na_2O, K_2O, and Al_2O_3 contents clearly deviate from this tendency (Fig. 12a-c). Additionally, the markedly elevated Sr, Ba, Nb, and REE contents in tufflavas relative to those in trachytes (Fig. 6) do not fit the trend of fractional crystallization beginning with the SQB mafic progenitor. These observations indicate that the formation of Quanxi crystal-rich tuffaceous lava with intermediate SiO_2 composition was not solely controlled by fractional crystallization.

Numerous silicic eruptions zoned from crystal-poor to crystal-rich (up to 50 vol% crystals) have been considered to be the products of the intermediate "mushes" co-erupted with the voluminous crystal-poor evolved rocks, reflecting efficient melt-crystal separation coupled with synchronous mineral accumulation (Bachmann and Huber, 2016; Deering et al., 2011; Ellis et al., 2014). In this work, the homogeneous isotopic characteristics and mineral compositions, together with the wide range of crystal proportions in the Quanxi trachytes and trachytic tufflavas, provide preliminary support for this petrogenetic model. Given that highly compatible and incompatible elements can fractionate sufficiently between crystals and liquids, the characteristics of the cumulate process could be distinguished by the variations of some distinctive elements, such as Sr, Ba, and Eu, which are preferentially incorporated into crystallizing phases but depleted in liquids in feldspar-dominated mush systems at shallow depths. However, despite evident discrepancies in the elemental abundances, the expected relationships of such cumulate-related elements have not been observed between the Quanxi trachytes and the tufflavas (Fig. 7). In addition, the minerals in our samples do not exhibit typical characteristics of crystal accumulation in mush chambers, such as prevalence of glomerocrysts and disequilibrium texture, as described by Ellis et al. (2014) and Feng et al. (2021). These features argue against a cumulate origin of the tufflavas. Alternatively, although quartz (~16 vol%) is found exclusively in the matrix of the Quanxi trachytes, qualitative and quantitative analysis of minerals revealed no substantial variation in matrix composition between the trachytes and the tufflavas, which are dominated by albites and orthoclases. Thus, the whole-rock compositional divergence between trachytes and tufflavas should have resulted primarily from the phenocryst/crystal pyroclast diversity. Because alkali feldspar is the main repository of alkalis and also contains Ba, Rb, and Sr (White et al., 2003), addition of alkali feldspar to a melt can elevate the Na_2O, K_2O, Al_2O_3, Ba, and Sr concentrations without affecting the other trace elements. Therefore, the abundant alkali feldspar pyroclasts in the Quanxi trachytic tufflavas, as revealed in the TIMA mineral maps (Fig. 4), were responsible for the elevated abundances of related elements. In addition to alkali feldspar, the relatively high levels of titanites and mica-group minerals in the trachytic tufflavas (Fig. 4) can explain the variations in MgO, $Fe_2O_3^T$, and TiO_2 from the tufflavas to the trachytes. However, the tufflavas have markedly higher Nb (217. 5 – 415. 5 ppm) and REE (665. 5 – 1 469. 0 ppm) contents than the more evolved trachytes (Nb = 160. 8 – 202. 2 ppm; REEs = 572. 5 – 670. 7 ppm). This result is difficult to explain solely by the involvement of crystal pyroclasts. The tufflavas exhibit some unusual petrographic details, such as randomly distributed calcite-biotite-muscovite veinlets and replacement of K-feldspars by albites with metasomatic relict texture (Fig. 4). Chemically, the tufflavas have obviously higher Na_2O/K_2O ratios (1. 1 – 2. 4) compared to those in the trachytes (Na_2O/K_2O = 0. 5 – 1. 4). These features suggest that post-magmatic hydrothermal alteration might have induced further enrichment of Na_2O and other incompatible elements in the Quanxi trachytic tufflavas. This explanation is supported by the reports of Nb-REE bearing hydrothermal minerals (e. g., titanites, aeschynites, monazites, and columbites) in the Zhuxi and Zhujiayuan trachytic rocks (Nie et al., 2019; Wang et al., 2021), which have close geographically and geochemical connections with the Quanxi trachytic rocks.

5. 5　Formation of the "Daly gap" in the SQB alkaline suites

A notable paucity of intermediate SiO_2 compositions, termed the "Daly Gap", has been documented in a variety of volcanic settings worldwide, such as the Gedemsa volcano in Ethiopia and Masset Formation in Canada (Dostal et al., 2017; Peccerillo et al., 2003). This bimodal distribution has also been observed in the SQB alkaline suites (Huang, 1992; Yang et al., 2020a,b). Notably, although the chemical compositions of the trachytic pyroclastics in the SQB can fill the "Daly Gap" to a certain extent (Wan et al., 2016; Wang et al., 2021), this does not imply that there was ever a continuous liquid line of descent from mafic to felsic compositions because the geochemical features of these pyroclastics are not solely controlled by fractional crystallization.

This compositional discontinuity has been suggested to be produced in a cogenetic suite of rocks via magma differentiation (e. g., Melekhova et al., 2013; Peccerillo et al., 2003). Based on the discussion above, mantle-derived magma differentiation is deemed to have been the dominant factor determining the compositional variations in the SQB alkaline suites, and a high degree of fractional crystallization (estimated by MELTS to be 72%−80%) is believed to have occurred during formation of the Quanxi trachytes. The shortage of rocks of intermediate compositions might reflect the onset of pronounced SiO_2-poor phase fractionation, such as spinel, Fe-Ti oxide, and apatite, which can trigger a rapid increase in the SiO_2 content (Garland et al., 1995; White et al., 2009). Experimental evidence also indicates that large compositional variations over small temperature intervals markedly suppress the preservation of intermediate compositions (Reubi and Blundy, 2009). To test this hypothesis, we applied thermodynamic parameters to discuss the effect of the differentiation process on the generation of such component discontinuances. The thermodynamic modeling reported in this study indicates that SiO_2 does not change linearly with falling temperature (Fig. 15a) or melt fraction (Fig. 15b), suggesting a heterogeneous crystallization process at different stages of magmatic evolution. To visualize the differentiation process, we introduce an equation presented by Mushkin et al. (2002) to estimate the slewing rate of magma differentiation, from the outputs created by Rhyolite-MELTS:

$$[\,dSiO_2/dt\,] \approx [\,dSiO_2/dH\,] \cdot (T_m-T_0) \cdot V^{2/3} \qquad (1)$$

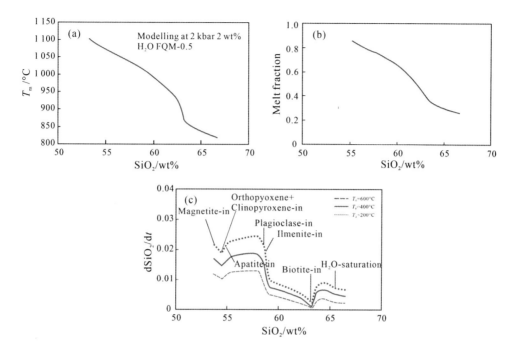

Fig. 15　The predicted melt temperature (a), and the predicted melt fraction (b), and the rate of
SiO$_2$ change in the silicate melt during magma differentiation (c) vs. SiO$_2$, estimated
from the best fit Rhyolite-MELTS output (2 kbar; 2 H$_2$O; FQM-0.5) at wall rock
temperatures of 600 ℃, 400 ℃, and 200 ℃.

where dSiO$_2$/dt denotes the rate of change of SiO$_2$(i.e., SiO$_2$ variation as a function of time),
H is the released heat at each melting temperature (T_m), and V is the melt volume. The
temperature (T_0) of wall rock was set at several levels (200 ℃, 400 ℃, and 600 ℃). The
models predict that faster rates of differentiation occur between ~55 wt% and ~59 wt% SiO$_2$
(Fig. 15c), meaning that there is an abrupt change as magma differentiation crosses the
intermediate interval. This change has been suggested to be the result of contemporaneous
crystallization of various minerals such as magnetite, apatite, and pyroxene over a narrow
temperature interval, predicted by Rhyolite-MELTS modeling (Fig. 13a). Especially, the
fractionation of magnetite is considered to exert the most critical role in this process because:
① separation of such Fe-Ti oxide commonly has significant effect on the variation of SiO$_2$ in
residual melt and ② the magnetite occupies large proportion of the crystalline phases in the
early stage of magma evolution which corresponds to the rapid rise in the SiO$_2$ (Fig. 13a;
Supplementary Table S4). The rapid decrease in differentiation rate at ~59 wt% SiO$_2$ is
assumed to be the result of Na-rich plagioclase crystallization, a process that has no obvious
effect on SiO$_2$ variation. Biotite in the crystallization assemblage triggers a rapid increase in
dSiO$_2$/dt at ~63 wt% SiO$_2$, although the increase is inconspicuous and impersistent because of

the limited crystalline proportion of biotite (Fig. 13a). To summarize, the occurrence of the "Daly Gap" in the SQB was most likely the result of an increased differentiation rate at intermediate compositions during magma differentiation, which could have markedly reduced the volume of intermediate products and rapidly changed the melt composition to an evolving field.

6 Conclusions

The Quanxi trachytic complex inthe SQB is composed of phenocryst-poor trachytes and crystal-rich trachytic tufflavas. Both of these rock types contain high concentrations of REEs and HFSEs, indicating an affinity with oceanic island basalt-like rock. On the basis of the major- and trace-element data, isotope chemistry, and stratigraphic relationships of the SQB mafic rocks, we propose that protracted fractional crystallization starting from a basaltic parental magma can explain the formation of the Quanxi trachytes. Using Rhyolite-MELTS modeling, we expect 72%−80% crystallization of plagioclase, pyroxene, biotite, apatite, and Fe-Ti oxide at conditions of $1.5-2$ kbar, FMQ-0.5, and $2-3$ wt% water. The trachytic tufflavas deviate markedly from the liquid lines of descent from the SQB mafic rocks to the Quanxi trachytes: This deviation is attributed to a combination of the large proportions of alkali feldspar pyroclasts and post-magmatic hydrothermal alteration. The "Daly Gap" in the SQB alkaline suites was most likely caused by an elevated differentiation rate in the SiO_2-intermediate field, resulting from the separation of several SiO_2-poor phases such as apatite and Fe-Ti oxide during magma differentiation.

Acknowledgements We thank Prof. Gregory Shellnutt and two anonymous reviewers for constructive comments. We also thank Lucy Muir, PhD, from Liwen Bianji (Edanz) (www. liwenbianji.cn), for editing the English text of a draft of this manuscript. This work was jointly supported by the National Natural Science Foundation of China (Grant Nos. 41772052 and 41421002).

Appendix A. Supplementary data Supplementary data to this article can be found online at https://doi.org/10.1016/j.lithos.2022.106683.

References

Andújar, J., Scaillet, B., 2012. Relationships between pre-eruptive conditions and eruptive styles of phonolite-trachyte magmas. Lithos, 152, 122-131.

Bachmann, O., Huber, C., 2016. Silicic magma reservoirs in the Earth's crust. Am. Mineral., 101, 2377-2404.

Bailey, D.K., 1987. Mantle metasomatism: Perspective and prospect. Geol. Soc. Spec. Publ., 30, 1-13.

Barberi, F., Ferrara, G., Santacroce, R., Treuil, M., Varet, J., 1975. A transitional basalt-pantellerite sequence of fractional crystallization, the Boina centre (Afar Rift, Ethiopia). J. Petrol., 16, 22-56.

Cavazzini, G., 1994. Increase of $^{87}Sr/^{86}Sr$ in residual liquids of high-Rb/Sr magmas that evolve by fractional crystallization. Chem. Geol., 118, 321-326.

Chandler, R., Spandler, C., 2020. The igneous petrogenesis and rare metal potential of the peralkaline volcanic complex of the southern peak range, central Queensland, Australia. Lithos, 358-359, 105386.

Charlier, B, Grove, T.L., 2012. Experiments on liquid immiscibility along tholeiitic liquid lines of descent, Contrib. Mineral. Petrol., 164, 27-44.

Charlier, B., Namur, O., Grove, T. L., 2013. Compositional and kinetic controls on liquid immiscibility in ferrobasalt-rhyolite volcanic and plutonic series. Geochim. Cosmochim. Acta, 113, 79-93.

Charlier, B., Namur, O., Toplis, M.J., Schiano, P., Cluzel, N., Higgins, M.D., Vander, A.J., 2011. Large-scale silicate liquid immiscibility during differentiation of tholeiitic basalt to granite and the origin of the Daly gap. Geology, 39, 907-910.

Chen, H., Tian, M., Wu, G.L., Hu, J.M., 2014. The Early Paleozoic alkaline and mafic magmatic events in Southern Qinling Belt, Central China: Evidences for the break-up of the Paleo-Tethyan Ocean. Geol. Rev., 60, 1437-1452 (in Chinese with English abstract).

Choi, S.H., 2020. Geochemistry and petrogenesis of quaternary volcanic rocks from Ulleung Island, South Korea. Lithos, 380-381, 105874.

Christensen, J.N., DePaolo, D.J., 1993. Time scales of large volume silicic magma systems: Sr isotope systematics of phenocrysts and glass from the Bishop Tuff, Long Valley, California. Contrib. Mineral. Petrol., 113,100-114.

Dai, F.Q., Zhao, Z.F., Zheng, Y.F., 2017. Partial melting of the orogenic lower crust: Geochemical insights from post-collisional alkaline volcanics in the Dabie orogen. Chem. Geol., 454, 25-43.

Deering, C.D., Bachmann, O., Vogel, T.A., 2011. The Ammonia Tanks Tuff: Erupting a melt-rich rhyolite cap and its remobilized crystal cumulate. Earth Planet. Sci. Lett., 310, 518-525.

DePaolo, D.J., 1981. Trace element and isotopic effects of combined wall rock assimilation and fractional crystallization. Earth Planet. Sci. Lett., 53, 189-202.

Ding, L.X., Ma, C.Q., Li, J.W., Robinson, P.T., Deng, X.D., Zhang, C., 2011. Timing and genesis of the adakitic and shoshonitic intrusions in the Laoniushan complex, southern margin of the North China Craton: Implications for post-collisional magmatism associated with the Qinling Orogen. Lithos. 126, 212-232.

Dong, Y.P., Sun, S.S., Yang, Z., Liu, X.M., Zhang, F.F., Li, W., Cheng, B., He, D.F., Zhang, G.W., 2017. Neoproterozoic subduction-accretionary tectonics of the South Qinling Belt, China. Precambrian Res., 293, 73-90.

Dong, Y.P., Zhang, X.N., Liu, X.M., Li, W., Chen, Q., Zhang, G.W., Zhang, H.F., Yang, Z., Sun, S. S., Zhang, F.F., 2015. Propagation tectonics and multiple accretionary processes of the Qinling orogen.J. Asian Earth Sci., 104, 84-98.

Dong, Y.P., Zhang, G.W., Neubauer, F., Liu, X.M., Genser, J., Hauzenberger, C., 2011. Tectonic evolution of the Qinling orogen, China: Review and synthesis. J. Asian Earth Sci., 41, 213-237.

Dostal, J., 2017. Rare Earth Element Deposits of Alkaline Igneous Rocks. Resources, 6, 34-45.

Dostal, J., Hamilton, T.S., Shellnutt, J.G., 2017. Generation of felsic rocks of bimodal volcanic suites from thinned and rifted continental margins: Geochemical and Nd, Sr, Pb-isotopic evidence from Haida Gwaii,

British Columbia, Canada. Lithos, 292-293, 146-160.

Draper, D.S., Green, T.H., 1999. *P-T* phase relations of silicic, alkaline, aluminous liquids: New results and applications to mantle melting and metasomatism. Earth Planet. Sci. Lett., 170, 255-268.

Ellis, B.S., Bachmann, O., Wolff, J.A., 2014. Cumulate fragments in silicic ignimbrites: The case of the Snake River Plain. Geology, 42, 431-434.

Falloon, T.J., Green, D.H., O'Neill, H.St.C., Hibberson, W.O., 1997. Experimental tests of low degree peridotite partial melt compositions: Implications for the nature of anhydrous near-solidus peridotite melts at 1 GPa. Earth Planet. Sci. Lett., 152, 149-162.

Feng, Z., Sun, D.Y., Gou, J., 2021. Differentiation of magma composition: Reactivation of mush and melt reaction in a magma chamber. Lithos., 388-389. 106066.

Garland, F.E., Hawkesworth, C.J., Mantovani, M.S.M., 1995. Description and petrogenesis of Parana rhyolites. J. Petrol., 36, 1193-1227.

Grant, T.B., Milke, R., Pandey, S., Jahnke, H., 2013. The Heldburg Phonolite, Central Germany: Reactions between phonolite and xenocrysts from the upper mantle and lower crust. Lithos, 182-183, 86-101.

Gualda, G.A., Ghiorso, M.S., Lemons, R.V., Carley, T.L., 2012. Rhyolite-MELTS: A modified calibration of MELTs optimized for silica-rich, fluid-bearing magmatic systems. J. Petrol., 53, 875-890.

Hart, S.R., Hauri, E.H., Oschmann, L.A., Whitehead, J.A., 1992. Mantle plumes and entrainment: Isotopic evidence. Science, 256, 517-520.

Hofmann, A.W., Jochum, K.P., Seufer, M., White, W.M., 1986. Nb and Pb in oceanic basalts: New constraints on mantle evolution. Earth Planet. Sci. Lett., 79, 33-45.

Hrstka, T., Gottlieb, P., Skala, R., Breiter, K., Motl, D., 2018. Automated mineralogy and petrology-applications of TESCAN Integrated Mineral Analyzer (TIMA). J. Geosci., 63, 47-63.

Hu, J., Liu, X.C., Chen, L.Y., Qu, W., Li, H.K., Geng, J.Z., 2013. A ~2.5 Ga magmatic event at the northern margin of the Yangtze craton: Evidences from U-Pb dating and Hf isotope analysis of zircons from the Douling Complex in the south Qinling orogen. Sci. Bull., 58, 3564-3579.

Huang, Y.H., 1993. Mineralogical characteristics of Phlogopite-Amphibole-Pyroxenite mantle xenoliths included in the alkali mafic-ultramafic subvolcanic complex from Langao County, China. Acta Petrol. Sin., 9, 367-378 (in Chinese with English abstract).

Huang, Y.H., Ren, Y.X., Xia, L.Q., Xia, Z.C., Zhang, C., 1992. Early Palaeozoic bimodal igneous suite on Northern Daba Mountains-Gaotan diabase and Haoping trachyte as examples. Acta Petrol. Sin., 8, 243-256 (in Chinese with English abstract).

Irving, A.J., Price, R.C., 1981. Geochemistry and evolution of lherzolite-bearing phonolitic lavas from Nigeria, Australia, East Germany and New Zealand. Geochim. Cosmochim. Acta, 45, 1309-1320.

Jung, S., Hoffer, E., Hoernes, S., 2007. Neo-Proterozoic rift-related syenites (Northern Damara Belt, Namibia): Geochemical and Nd-Sr-Pb-O isotope constraints for mantle sources and petrogenesis. Lithos, 96, 415-435.

Karsli, O., Aydin, F., Uysal, I., Dokuz, A., Kumral, M., Kandemir, R., Budakoglu, M., Ketenci, M., 2018. Latest cretaceous "A_2-type" granites in the Sakarya Zone, NE Turkey: Partial melting of mafic lower crust in response to roll-back of Neo-Tethyan oceanic lithosphere. Lithos, 302-303, 312-328.

Kaszuba, J.P., Wendlandt, R.F., 2000. Effect of carbon dioxide on dehydration melting reactions and melt compositions in the lower crust and the origin of alkaline rocks. J. Petrol., 41, 363-386.

Laporte, D., Lambart, S., Schiano, P. Ottolini, L., 2014. Experimental derivation of nepheline syenite and phonolite liquids by partial melting of upper mantle peridotites. Earth Planet. Sci. Lett., 404, 319-331.

Lewis, A., Trond, T., H Péter, Chris, H., Susan, W., Stephanie, W., Fernando C., 2016. A mantle-derived origin for Mauritian trachytes. J. Petrol., 9, 1-31.

Liu, Y., Liu, X.M., Hu, Z.C., Diwu, C.R., Yuan, H.L., Gao, S., 2007. Evaluation of accuracy and long-term stability of determination of 37 trace elements in geological samples by ICP-MS. Acta Petrol. Sin., 23, 1203-1210 (in Chinese with English abstract).

Mahood, G.A., Halliday, A.N., 1988. Generation of high-silica rhyolite: A Nd, Sr and O isotopic study of Sierra La Primavera, Mexican Neovolcanic Belt. Contrib. Mineral. Petrol., 100, 183-191.

McDonough, W.F., Sun, S.S., 1995. The composition of the Earth. Chem. Geol., 120, 223-253.

Melekhova, E., Annen, C., Blundy, J., 2013. Compositional gaps in igneous rock suites controlled by magma system heat and water content. Nat. Geosci., 6, 385-390.

Middlemost, E.A.K., 1994. Naming materials in the magma/igneous rock system. Earth Sci. Rev., 37, 215-224.

Mushkin, M., Stein, M., Halicz, L., Navon, O., 2002. The Daly gap: Low-pressure fractionation and heat-loss from cooling magma chamber. Geochim. Cosmochim. Acta, 66, A539.

Namur, O., Charlier, B., Toplis, M.J., Higgins, M.D., Hounsell, V., Liégeois, J.P., Auwera, J.V., 2011. Differentiation of tholeiitic basalt to A-type granite in the Sept Iles layered intrusion, Canada. J. Petrol., 52, 487-539.

Nie, X., Wang, Z., Chen, L., Yin, J., Xu, H., Fan, L., Wang, G., 2019. Mineralogical constraints on Nb-REE mineralization of the Zhujiayuan Nb (-REE) deposit in the North Daba Mountain, South Qinling, China. Geol. J., 1-19.

Peccerillo, A., Barberio, M.R., Yirgu, G., Ayalew, D., Barbieri, M., Wu, T.W., 2003. Relationships between mafic and peralkaline silicic magmatism in continental rift settings: A petrological, geochemical and isotopic study of the Gedemsa volcano, central Ethiopian rift. J. Petrol., 11, 2003-2032.

Qin, Z., Wu, Y., Siaebel, W., Gao, S., Wang, H., Abdallsamed, M.I.M., Zhang, W., Yang, S., 2015. Genesis of adakitic granitoids by partial melting of thickened lower crust and its implications for early crustal growth: A case study from the Huichizi pluton, Qinling orogen, central China. Lithos, 238, 1-12.

Ratschbacher, L., Hacker, B.R., Calvert, A., Webb, L.E., Grimmer, J.C., McWilliams, M.O., Ireland, T., Dong, S.W., Hu, J.M., 2003. Tectonic of the Qinling (Central China): Tectonostratigraphy, geochronology, and deformation history. Tectonophysics, 366, 1-53.

Reubi, O., Blundy, J., 2009. A dearth of intermediate melts at subduction zone volcanoes and the petrogenesis of arc andesites. Nature, 461, 1269-1273.

Rudnick, R.L., Gao, S., 2003. Composition of the continental crust. In: Rudnick, R.L. The Crust. Treatise on Geochemistry, 3. Oxford: Elsevier Pergamon, 1-64.

Su, J.H., Zhao, X.F., Li, X.C., Su, Z.K., Liu, R., Qin, Z.J., Chen, M., 2021. Fingerprinting REE mineralization and hydrothermal remobilization history of the carbonatite-alkaline complexes, Central China: Constraints from in situ elemental and isotopic analyses of phosphate minerals. Am. Mineral., 106, 1545-1558.

Sun, S.S., McDonough, W.F., 1989. Chemical and isotopic systematics of oceanic basalts: Implications for mantle composition and processes. Geol. Soc. Spec. Publ., 42, 313-345.

Suneson, N.H., Lucchitta, I., 1983. Origin of bimodal volcanism, southern Basin and Range province, west-central Arizona. Geol. Soc. Am. Bull., 94, 1005-1019.

Szymanowski, D., Ellis, B.S., Bachmann, O., Guillong, M., Phillips, W.M., 2015. Bridging basalts and rhyolites in the Yellowstone-snake river plain volcanic province: The elusive intermediate step. Earth Planet. Sci. Lett. 415, 80-89.

VanTongeren, J.A., Mathez, E.A., 2012. Large-scale liquid immiscibility at the top of the Bushveld Complex, South Africa. Geology, 40, 491-494.

Wan, J., Liu, C.X., Yang, C., Liu, W.L., Li, X.W., Fu, X.J., Liu, H.X., 2016. Geochemical characteristics and LA-ICP-MS zircon U-Pb age of the trachytic volcanic rocks in Zhushan area of Southern Qinling Mountains and their significance. Geol. Bull. China, 35, 1134-1143 (in Chinese with English abstract).

Wang, G., 2014a. Metallogeny of the Mesozoic and Paleozoic volcanic igneous event in Ziyang-Langao Areas, North Daba Mountain (Ph.D. thesis). Beijing: China University of Geosciences (in Chinese with English abstract).

Wang, K.M., 2014b, Research on the petrogenesis, tectonic and metallogeny for mafic rocks in the Ziyang-Langao area, Shaanxi Province (Ph.D. thesis). Beijing: Chinese Academy of Geological Sciences (in Chinese with English abstract).

Wang, K., Wang, L.X., Ma, C.Q., Zhu, Y.X., She, Z.B., Deng, X., Chen, Q., 2021. Mineralogy and geochemistry of the Zhuxi Nb-rich trachytic rocks, South Qinling (China): Insights into the niobium mineralization during magmatic-hydrothermal processes. Ore Geol. Rev., 138, 104346.

Wang, R.R., Xu, Z.Q., Santosh, M., Liang, F.H., Fu, X.H., 2017. Petrogenesis and tectonic implications of the Early Paleozoic intermediate and mafic intrusions in the South Qinling Belt, central China: Constraints from geochemistry, zircon U-Pb geochronology and Hf isotopes. Tectonophysics, 712, 270-288.

Weis, D., Kieffer, B., Maerschalk, C., Barling, J., Jong, J.D., Williams, G.A., Hanano, D., Pretorius, W., Mattielli, N., Scoates, J., Goolaerts, A., Friedman, R., Mahoney, J., 2006. High-precision isotopic characterization of USGS reference materials by TIMS and MC-ICP-MS. Geochem. Geophys. Geosyst, 7, 139-149.

White, J.C., Holt, G.S., Parker, D.F., Ren, M., 2003. Trace- element partitioning between alkali feldspar and peralkalic quartz trachyte to rhyolite magma. Part I : Systematics of trace-element partitioning. Am. Mineral., 88, 316-329.

White, J. C., Parker, D. F., Minghua, R., 2009. The origin of trachyte and pantellerite from Pantelleria, Italy: Insights from major element, trace element, and thermodynamic modelling. J. Volcanol. Geotherm. Res., 179, 33-55.

Wilson, M., Hyman, U., 1989. Igneous petrogenesis. London: Unwin Hyman, 287-374.

Xiang, Z.J., Yan, Q.R., Yan, Z., Wang, Z.Q., Wang, T., Zhang, Y.L., Qin, X.F., 2010. Magma source and tectonic setting of the porphyritic alkaline basalts in the Silurian Taohekou Formation, North Daba mountain: Constraints from the geo- chemical features of pyroxene phenocrysts and whole rocks. Acta Geol. Sin., 26, 1116-1132 (in Chinese with English abstract).

Xiang, Z.J., Yan, Q.R., Song, B., Wang, Z.Q., 2016. New evidence for the age of ultra- mafic to mafic dikes and alkaline volcanic complexes in the North Daba Mountains and its geological implication. Acta Geol. Sin., 90, 896-916 (in Chinese with English abstract).

Xu, X.Y., Huang, Y.H., Xia, L.Q., Xia, Z.C., 1997. Phlogopite-Amphibole-Pyroxenite Xenoliths in Langao, Shaanxi Province: Evidences for Mantle Metasomatism. Acta Petrol. Sin., 13, 1-13 (in Chinese with English abstract).

Yang, H., Lai, S.C., Qin, J.F., Zhu, R.Z., Zhao, S.W., Zhu, Y., Zhang, F.Y., Zhang, Z.Z., Wang, X. Y., 2020a. Early Palaeozoic alkaline trachytes in the North Daba Mountains, South Qinling Belt: Petrogenesis and geological implications. Int. Geol. Rev., 63, 2037-2056.

Yang, Y.Z., Wang, Y., Siebel, W., Zhang, Y.S., Chen, F., 2020b. Zircon U-Pb-Hf, geochemical and Sr-Nd-Pb isotope systematics of late Mesozoic granitoids in the Lantian-Xiaoqinling region: Implications for tectonic setting and petrogenesis. Lithos, 374-375, 105709.

Yoder, H.S., 1973. Contemporaneous basaltic and rhyolitic magmas. Am. Mineral., 58, 153-171.

Young, D.A., 1981. Alkali feldspars. In: Mineralogy. Boston: Springer, 2-7.

Zhang, C.L., Gao, S., Zhang, G.W., Liu, X.M., Yu, Z.P., 2002. Geochemistry and significance of the Early Paleozoic alkaline mafic dyke in Qinling. Sci. China: Series D, Geoscience, 32, 819-829 (in Chinese with English abstract).

Zhang, C.L., Shan, G., Yuan, H.L., Zhang, G.W., Yan, Y.X., Luo, J.L., Luo, J.H., 2007. Sr-Nd-Pb isotopes of the early Paleozoic mafic-ultramafic dykes and basalts from south Qinling belt and their implications for mantle composition. Sci. China, 50, 1293-1301(in Chinese with English abstract).

Zhang, F.Y., Lai, S.C., Qin, J.F., Zhu, R.Z., Zhao, S.W., Zhu, Y., Yang, H., 2020a. Vein-plus-wall rock melting model for the origin of Early Paleozoic alkali diabases in the South Qinling Belt, Central China. Lithos, 370-371, 105619.

Zhang, F.Y., Lai, S.C., Qin, J.F., Zhu, R.Z., Zhao, S.W., Yang, H., Zhu, Y., Zhang, Z.Z., 2020b. Magma source and evolution process of Early Paleozoic basalts in the South Qinling Belt. Acta Petrol. Sin., 36, 2149-2162 (in Chinese with English abstract).

Zhang, G.W., Meng, Q.R., Yu, Z.P., Sun, Y., Zhou, D.W., Guo, A.L., 1996. Orogenic process and dynamic characteristics of Qinling orogenic belt. Science in China: Series D, 26, 193-200 (in Chinese with English abstract).

Zindler, A., Hart, S., 1986. Chemical geodynamics. Annu. Rev. Earth Planet. Sci., 14, 493-571.

Zou, X.W., Duan, Q.F., Tang, C.Y., Cao, L., Cui, S., Zhao, W.Q., Xia, J., Wang, L., 2011. SHRIMP zircon U-Pb dating and lithogeochemical characteristics of diabase from Zhenping area in North Daba Mountain. Geol. China, 38, 282-291 (in Chinese with English abstract).

Westward migration of high-magma addition rate events in SE Tibet[①]

Zhu Renzhi[②]   Lai Shaocong[②]   Scott R Paterson

Peter Luffi   Zhang Bo   Lance R Pompe

Abstract: Arc magmatism is an important process in the formation and evolution of the continental crust. Various arcs developed in the southern margin of Eurasian continent that recorded the final formation and growth of these micro-continent before the final India-Eurasia amalgamation. We have examined the often neglected Tengchong arc segment in southeastern Tibet using compiled zircon U-Pb-Hf and whole-rock geochemistry with GPS location and lithology control. The aim is to better understand the driving mechanism of spatial migration and temporal evolution of multi-stage high-MAR (magma addition rate) events and its role in formation and growth of Tengchong arc segment.

Results indicate three magmatic "flare-ups" during east-to-west arc migration, from ~131 - 111 Ma (eastern), ~76 - 64 Ma (central), to ~55 - 49 Ma (western), during amalgamation of Lhasa-Tengchong and Qiangtang-Baoshan blocks and final collision of India-Asia plate following subduction of Neo-Tethys. Zircon Hf isotopes and geochemical analyses shows the significant increasing juvenile and/or mantle-derived materials and the range of isotopes also broaden during flare-ups, indicating the melting of diverse lithospheric and upper plate domains. Through comparison with geophysical parameters, these pulses are closely coupled with the arc migration along with changes of crustal thickness, but not correlated with angle and rates of convergence. The spatial arc migration of the multi-stage high-MAR events in Tengchong arc segment was potentially driven by slab steepening and break-off following the initial collision, and abrupt changes of subduction slab dynamics. These processes are well coupled with multi-stage interactions of crust-mantle and transitional Moho-depths.

1   Introduction

Episodic high-volume magmatism in arcs lasting ca. 5-40 m.y., also known as "magmatic flare-ups", have played a key role in building oceanic and continental crust as well as driving long-term surface processes such as weathering, erosion and greenhouse events (Paterson and Ducea, 2015; Ducea et al., 2015; Lee et al., 2012; Cao et al., 2016; Chapman et al.,

①   Published in *Tectonophysics*, 2022, 830.

②   Corresponding authors.

2021; Li et al., 2022). Continental arcs are some of the most important locations for mountain building, ore deposit formation, water resources, climate change, and various geological hazards (Lee et al., 2012; Li et al., 2019). These processes are closely associated with the spatial and temporal evolution and distribution of episodic high-MAR events (Paterson and Ducea, 2015; Ducea et al., 2015; Li et al., 2022).

The mechanism(s) driving flare-ups are either related to external forcing and/or internal, upper plate processes (Paterson and Ducea, 2015). One upper plate process driving flare-ups may be episodic mantle processes (Anderson, 1982; Gurnis, 1988). Alternatively, upper-plate processes, including episodic addition of melt-fertil materials by underthrusting of retroarc lower crust (DeCelles et al., 2009), crustal/lithospheric thickening during subduction (Ducea and Barton, 2007), and the role of arc migration into subduction-related metasomatized material (Chapman and Ducea, 2019) have been suggested as drivers of arc flare-ups.

Flare-ups may also be driven by episodic increases in mantle magma addition into the arc (Martinez-Ardila et al., 2019). Attia et al. (2020) suggests increased mantle magma input not only drives arc flare-ups but also represent significant continental crust formation in continental arcs. Flare-ups of the Mesozoic Median Batholith could be from the underlying mantle and triggered by an external dynamic mantle process (Schwartz et al., 2017). Both fertile upper-lithospheric and juvenile magma from the mantle could provide significant contributions to flare-ups.

As an interesting comparison to Cordilleran systems, a series of magmatic arcs developed from the Mediterranean, along the Himalayan system to Sumatra during the long Tethyan closure before final India-Asian welding. During the Cretaceous to Early-Cenozoic, an Andean-style continental arc associated with the northward subduction of Neo-Tethyan oceanic lithosphere developed on the southern margin of the Eurasian continent (Fig. 1a) (Searle et al., 1987). It has been proposed for the Neo-Tethyan arc system that the changes in slab dynamic parameters in the subduction zone and subsequent amalgamation could induce episodic magmatism with both juvenile mantle-like and fertile continental crustal characteristics (Zhu et al., 2011; Zhang et al., 2019; Shafaii Moghadam et al., 2020; Guo et al., 2022).

In this contribution, we focus on the Tengchong arc, an often overlooked arc segment at the southeastern end of the Tibetan Plateau. Similar to the Kohistan and Gangdese arcs in the Southern Tibetan Plateau, the Tengchong arc segment records the formation and evolution of continental crust in the southern margin of the Eurasian plate before the final welding of India-Asia. The westward spatial distribution of episodic magmatic arcs (predominately preserved as granitoid batholiths) through temporal evolution from Early-Cretaceous to early-Cenozoic in southeastern Himalayan-Tibetan orogenic system is shown in Fig. 1b. Using zircon U-Pb geochronology, Lu-Hf isotopes, bulk-rock geochemistry, crustal thickness and other

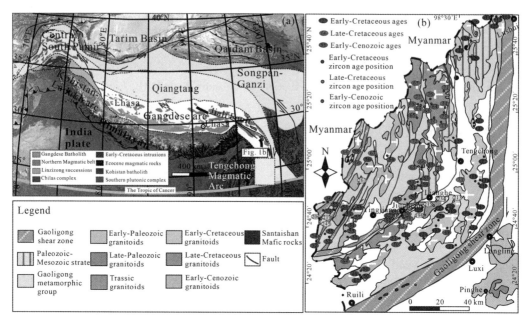

Fig. 1　(a) Distribution of main continental blocks in the Tibet-Pamir orogenic system
(Chung et al., 2005; Chapman et al., 2018). (b) The geological and mainly magmatic
map of Tengchong terrane in the SE Tibet orogen (Xu et al., 2012).

geophysical parameters, we not only trace the role of the crust-mantle source contribution of
Tengchong arc magmatism but explore insights into the role of deep crust-mantle dynamic
processes during subduction and subsequent collision of two continental plates. We also
calculate the evolution of Moho depth through space and time and reconstruct the geochronology
and evolving isotopes. We then synthesize these data with the aim of better understanding the
mechanisms of migration and multi-stage flare-ups, the interaction of the crust and mantle
through space and time and its role in formation of this continental boundary.

2　Geological background

The Greater Tibetan Plateau, including Pamir and Tibet plateau (Chapman et al.,
2018), is regarded as part of a single contiguous orogenic plateau (Himalaya) consisting of a
series of allochthonous Gondwanan continental fragments that were accreted to Asia during the
early-Mesozoic (Allegre et al., 1984; Burtman and Molnar, 1993; Robinson et al., 2012;
Chapman et al., 2018), culminating in the India collision with Asia during the Late-
Cretaceous to Early-Cenozoic (Hu et al., 2016; Bouihol et al., 2013; Chung et al., 2005).
In this orogenic system, the fragments include the Songpan-Ganzi, Qiangtang, Lhasa and
Himalayan terranes, separated by the Jinsha, Bangong-Nujiang, and Indus-Yarlung sutures
(Yin and Harrison, 2000). The Qiangtang terrane, from north to south, is laterally equivalent
to the central Pamir terrane, the south Pamir terrane, and the Karakoram terrane (Chapman et

al., 2018), but has no direct equivalence with the Lhasa terrane in Pamir (Robinson et al., 2012). The Lhasa terrane is laterally equivalent to the Tengchong terrane of southeastern Tibet (Xie et al., 2016), although rotated ca. 87° at ca. 40 Ma (Kornfeld et al., 2014; Xu et al., 2015) (Fig. 1a). These accreted processes are typical of the Cordillera and/or Andes, where long belts of deformation and magmatism are associated with the subduction of oceanic plates beneath continental plates (DeCelles et al., 2009). Multi-stage continent-continental collision events resulted in the final convergence and uplift of the Tibetan Plateau (Yin and Harrison, 2000), to form the highest mountains on Earth.

The Tengchong arc segment forms the southeastern extension of the Lhasa terrane (Xu et al., 2015; Xie et al., 2016; Fig. 1b) and is separated from the eastern Baoshan terrane by the Nujiang-Longling-Ruili fault (NLRF) and from the western Burma terrane by the Putao-Myitkyina suture zone (Xu et al., 2015; Metcalfe, 2013). The terrane contains Mesoproterozoic metamorphic basement belonging to the Gaoligong Mountain Group and upper Paleozoic clastic sedimentary rocks and carbonates. Mesozoic to Tertiary granitoids were emplaced into these strata (Zhu et al., 2015; Zhu et al.,2018a) and were then covered by Tertiary-Quaternary volcano-sedimentary sequences (YNBGMR, 1991). The Tengchong arc segment is typical of old arcs that have been eroded down to plutonic levels, where little coeval volcanic rocks remain: older units are covered by recent late-Cenozoic to Quaternary volcanic sequences. Today there is little arc-related volcanic rocks preserved in this region compared to the extensive granitoid plutons.

A series of N-S and NE-SW trending faults developed in the Tengchong Terrane (Fig. 1b). The Nujiang-Lushui-Longling-Ruili fault and Gaoligong shear zone separates the tectonic mélangeto the west and the Cambrian-Ordovician granitoids to the east (Zhao et al., 2016; Zhu et al., 2018b). Three faults divide the Tengchong Terrane into three magmatic belts: Mingguang-Menglian, Guyong-Longchuan and Sudian-Tongbiguan magmatic belts, separated by the Dayingjiang, Gudong-Tengchong and Binlangjiang faults (Xie et al., 2016; YNBGMR, 1991) (Fig. 1b). The three magmatic belts are ~120 km in width and ~240 km in length, each with a ca. 30−40 km wide trench parallel band.

The three 30−40 km wide ribbons of arc magmatism reflect a westward migration of arc magmatism through time (Fig. 1b). The eastern magmatic belt is located in the eastern Tengchong terrane and is composed mostly of three batholiths: ①The Pianma-Mingguang-Gudong batholith intruded into Paleozoic to Mesozoic strata and Gaoligong metamorphic group, containing the Pianma diorite-granodiorite-monzogranite association (Zhu et al., 2017b), Mingguang-Gudong highly fractionated I-type granites (Zhu et al., 2015), and Dajujie alkaline granites (Zhu et al., 2018b). ②The Xinhua-Menglian batholith intruding the Gaoligong metamorphic group consisting mostly of granodiorite-monzogranites and associated abundant mafic magmatic enclaves (Xie et al., 2016; Zhu et al., 2017a). And ③ the Xiaotang-Mangdong batholith containing mainly mozodiorite-tonalite-granodiorite and associated mafic magmatic enclaves (Cong et al., 2011; Zhu et al., 2017a). The central magmatic belt is located in the central Tengchong terrane and is separated by the eastern Tengchong-Dayingjiang and western Binliangjiang faults. From north to south, the Guyong, Lailishan, Jiucheng, Bangwan and Longchuan plutons intrude the Paleozoic to Mesozoic strata and are composed mainly of granodiorite and monzogranites with minor mafic intrusions and mafic enclaves in some granodiorites (Chen et al., 2015; Zhao et al., 2016; Zhu et al., 2019). The western belt is located in the western Tengchong terrane and is composed mainly of meta-gabbroic rocks, amphiboles, diorites, granodiorites and monzogranites with associated mafic enclaves, and small amounts of mafic-ultramafic rocks as lenses within the gneisses (Wang et al., 2014, 2015; Ma et al., 2021; Zhao et al., 2019).

3　Data and results: Westward spatial distribution of three-stage intensive magmatism through temporal evolution in the Tengchong arc segment

To explore the spatial and temporal distribution of continental arc magmatism and its causes, comprehensive geochronologic ($n = 82$), in situ zircon Hf isotopic ($n = 820$) and geochemical ($n = 328$) data of the three Tengchong "flare-ups" are compiled and examined (Figs. 2 and 3). The data were mapped with detailed spatial control using MATLAB (Fig. 4) and the Mohodepth and elevation (Fig. 6) were calculated using methods by Chapman et al. (2015) and Hu et al. (2016)'s method and most recently by Luffi (2019), with findings summarized in the figures and supplementary dataset.

We use volumetric magmatic flux in this paper defined using Darcy's law, as a volume passing through a designated area over a period of time (e.g., $km^3/km^2/yr = m/s$). Paterson et al. (2011) also defined the terms total added volume (km^3) as the volumetric amount of material added and volumetric addition rate (km^3/yr) as the total added volume per time. Volumetric addition rates are referred to as a magmatic flux by some, but little information has been preserved in paleoarcs about the actual volumetric magmatic flux (Ratschbacher et al., 2019): Authors typically use more readily determined areal measurements to determine areal additions (km^2/yr). In this case, we use km^2/yr to compare the magmatic areal addition rates of the multi-stage magmatism in the Tengchong arc segment.

Fig. 1 shows a spatiotemporal trend as episodic magmatism migrates from east to west in the Tengchong arc segment, with an early-Cretaceous flare-up in the east, a late-Cretaceousto early-Cenozoic flare-up in the west. Before final collision in the early Cenozoic, this east -to-

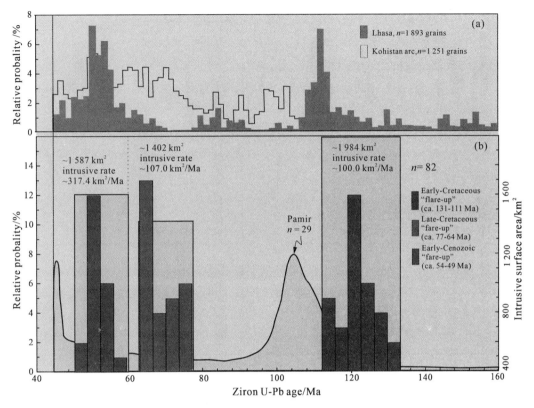

Fig. 2　The mainly geochronological distribution of magmatic events in the Tibet-Pamir orogenic system
(a) modified from Pamir (Chapman et al., 2018) and Kohistan-Lhasa (Bouilhol et al., 2013)
and based on background values in above associated literatures, and
(b) magmatic flare-up and lull of Tengchong arc segment.

west represented a north to south migration (see Kornfeld et al., 2014; Xu et al., 2015). For ease of presentation, we use the present orientation. The temporal evolution of the Tengchong arc segment is shown in Fig. 2. In the Tengchong arc segment, the three high-MAR events occurred from east (ca. 131−111 Ma) to ~76−64 Ma (central), to ~55−49 Ma (western). The associated lulls range from ca. 110 Ma to 80 Ma. The eastern flare-up resulted in an areal addition of ca. 1 984 km^2 and addition rate of ca. 100 km^2 per million years, the central flare-up resulted in an areal addition of ca. 1 402 km^2 and addition rate of ca. 107 km^2 per million years, and the western flare-up resulted in an areal addition of ca. 1 587 km^2 and addition rate of ca. 317. 4 km^2 per million years.

Major, trace element and isotopic compositions of the three high MAR events display very distinctive features from eastern, central to western parts (Figs. 3, 4 and 5), as follows: The early Cretaceous flare-up shows variable SiO_2(47. 7−77. 9 wt%), MgO (0. 03−7. 07 wt%), Mg#(7. 9−65. 4), K/Na ratios (0. 11−2. 51), total Fe+Mg (0. 73%−15. 8%), A/CNK and Na/Ta ratios (3. 53 − 22. 8), but low La/Yb and Sr/Y ratios. Average SiO_2 values

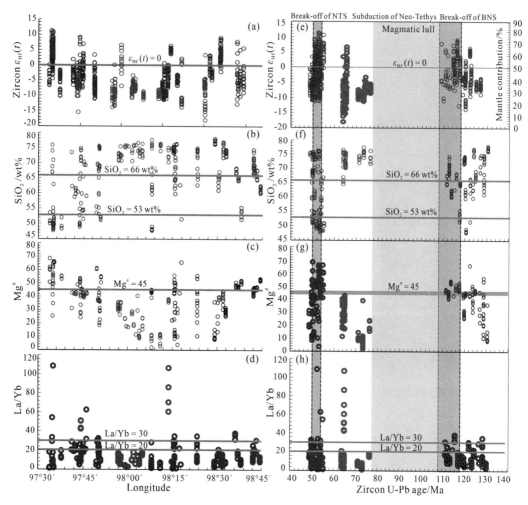

Fig. 3　Major high-MAR magmatic composition trend with spatial-temporal evolution.

(67. 5 wt%) are similar to typical continental arc compositions, such as the Coast Mountains and Sierra Nevada batholiths (Ducea et al. , 2015) , and the continental crust (Taylor and McLennan, 1985; Rudnick and Gao, 2003) . The variable zircon Hf isotopic compositions range between − 15 and + 10; the Late-Cretaceous flare-up displays high SiO_2 (average 72. 5 wt%) , K/Na ratios (most >1, up to 2. 76 wt%) and A/CNK values (most >1) , low MgO (0. 01−2. 76 wt%) and $Mg^{\#}$ (<45) , total Fe + Mg (<8%) , Nb/Ta ratios (most <15) and La/Yb ratios with a few extremely high samples. Recently identified mafic intrusions have high MgO (~5. 0 wt%) and $Mg^{\#}$ (~56) , typical of high-potassium to shoshonitic series (Zhu et al. , 2021) . Their zircon Hf isotopes show predominantly negative components that evolved from ancient continental crust; The early-Cenozoic magmatic flare-up of the western Tengchong arc segment has strongly variable major element compositions: SiO_2 (45 − 76 wt%) , K/Na (0. 1−2. 5) , $Mg^{\#}$ (3−70) , total Fe + Mg (1%−22%) but Nb/Ta >8. 3 and A/CNK <1. 1.

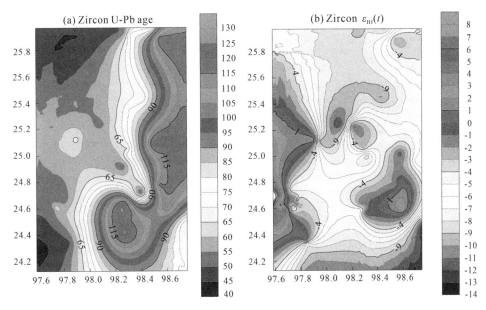

Fig. 4　Mapping of in situ zircon U-Pb age and Hf isotopic composition trend
with spatial and temporal evolution of Tengchong arc.

Average values are lower than typical continental crust and arcs, being closer to that of island arcs (Jagoutz and Kelemen, 2015; Ducea et al., 2015). It has higher La/Yb and Sr/Y ratios than the previous two-stage magmatic episode. Zircon Hf composition also displays large variation from -12 to $+12$, indicating the strong interaction of crust and mantle.

The spatiotemporal pattern of the three-stage period of intensive magmatism in the Tengchong arc segment progressed from east to west, with both the earliest and latest episodes showing strongly variable compositions from mafic to felsic materials with the central segment in contrast having a notably more evolved composition. Experimental results suggest the Lu-Hf isotopic system is relatively stable when zircon is exposed to metamorphic and/or metasomatized events (Daniel et al., 2013). Both spatial and temporal in situ zircon Lu-Hf isotopic data also display variable patterns from east to west and from early and late Cretaceous to early-Cenozoic (Fig. 6). Both early-Cretaceous and early-Cenozoic magmatic events show extremely variable zircon Hf isotopic compositions from negative to positive indicating the strong interaction of depleted mantle-derived and evolved continental components (Kemp et al., 2007) and a significant mantle contribution (ca. >70%). In contrast, the late-Cretaceous flare-up in the central zone shows a significant crustal and evolved zircon Hf signature (Fig. 3). Recently Zhu et al. (2021) identified the high Mg and K dioritic intrusions that were contacted and coeval with these granitic rocks, the former was produced from the metasomatized lithosphere mantle and played the key role in the extension melting of the overlying crust to produce these crustal signature materials.

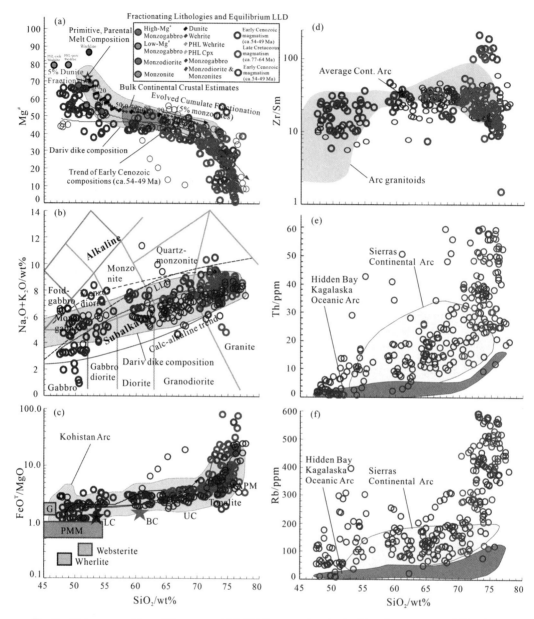

Fig. 5　Major composition of three-stage high-MAR magmatism in the Tengchong terrane of SE Tibet.

Fig. 5a-d referenced Jagoutz (2009, 2010) and Jagoutz and Kelemen (2015); Fig. 5e-f referenced Kay et al. (2019). The data of Nevada, Aleutian, and Gangdese arc from https://georoc.mpch-mainz.gwdg.de/georoc/.

The whole-rock geochemical compositions and stable in situ zircon Hf isotopic components can provide robust evidence to understand the nature and mechanisms of the observed episodic magmatism (Jagoutz and Behn, 2013; Jagoutz and Kelemen, 2015; Chapman et al., 2018; Chapman and Ducea, 2019). We compile representative major and trace element variation with SiO$_2$ and compare them with compositions from typical arc magmatism around the world

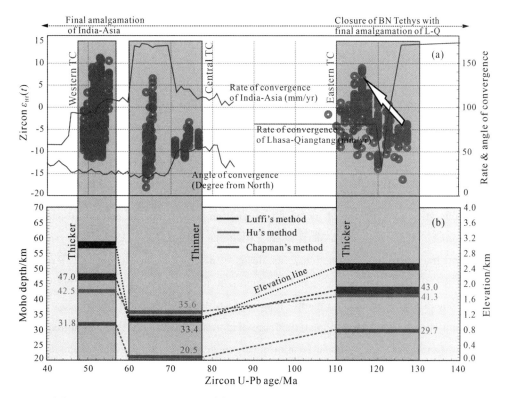

Fig. 6 (a) Zircon U-Pb age vs. zircon $\varepsilon_{Hf}(t)$ value of Tengchong arc magmatism, convergence rate of Neo-Tethys and angle, the convergence data of India-Asia from Lee and Lawver (1995) and Lhasa-Qiangtang Block referenced from Young et al. (2019). (b) Zircon U-Pb age of Tengchong arc magmatism vs. Moho depth and elevation of Tengchong arc segment (Chapman et al., 2015; Hu et al., 2016; Luffi et al., 2019).

(Fig. 5), such as the Kohistan arc, Coast Mountains and Sierra Nevada continental arcs, Hidden Bay Kagalaska oceanic arc and the Gangdese continental arc. The data indicates that both the early-Cretaceous and early-Cenozoic flare-ups have close affinity with continental arcs rather than oceanic arcs. Late-Cretaceous magmatism has more enriched compositions than normal continental arcs and displays higher incompatible element compositions than normal continental crust.

4 Discussion

4.1 Driven mechanism of the multi-stage high-MAR events in Tengchong arc segment: External, internal or interaction?

The driving forces of flare-ups are caused by possibly being either upper plate (lithospheric mantle or crust) (Ducea and Barton, 2007; DeCelles et al., 2009; Chapman and Ducea, 2019; Martinez-Ardila et al., 2019; Attia et al., 2020) or lower plate (slab or tectonic) (Schwartz et al., 2017; Shafaii et al., 2020). Various mechanisms have been proposed to explain episodic magma flare-ups through space and time. In Cordilleran orogenic systems, Ducea and Barton (2007) suggest crustal/lithospheric thickening during subduction is the key to trigger the high MAR magmatism primarily derived from an upper lithospheric source in Cordilleran arcs. Duceaet al. (2015) and DeCelles et al. (2009) proposed a model that argues that ca. 25－50 Ma episodic high-MAR magmatism events are fueled by cycles of upper-plate processes, where continental crust shortens by thrusting behind the arc triggering intensive magmatism resulting in dense melt residues. These residues eventually sink into the mantle beneath the arc and induce the renewal of the cycle.

Arc migration into regions of increased subduction-related mantle metasomatism may also be important for triggering a high-MAR event (Chapman and Ducea, 2019). The zircon Lu-Hf isotopic and trace element information from Mesozoic magmatism in the central Sierra Nevada (Attia et al., 2020), however, records a dominant juvenile source for magmas throughout MAR episodes showing increasing mantle input in Cordilleran arcs. Most likely, both the upper plate crustal and mantle-derived materials are important to support the generation of MAR episodes in Cordilleran arc systems. In Fiordland, meanwhile, zircon geochronologic, isotopic and associated bulk-rock geochemical data from the Mesozoic Median Batholith shows the flare-up has high Sr/Y magmatism from the underlying mantle and the Zealandia HMA event was triggered by an external dynamic mantle process (Schwartz et al., 2017).

Major flare-ups along the southern margin of Asia plate during the subduction of Neo-Tethys, from west to east, major arcs including Iran, Kohistan, Lhasa (Gangdese), Tengchong, and Sumatra arcs. ① Iran arc, major episodes occurred at 110－80 Ma, 75－50 Ma, 50－35 Ma, 35－20 Ma and 15－10 Ma, but the 35－20 Ma and 15－10 Ma episodes show notably less volume than others, so the major pulses focused on first to third pulses. The data indicated that the first to second episodic magmas dominated by the underlying mantle, and third episodic magmas show increasing contributions from the crust with highest contribution from continental crust (Shafaii et al., 2020). These episodes mainly accompanied changes from extension, collision to extension in subduction zone dynamics (Shafaii et al., 2020). ② Kohistan arc, an ongoing 120 Ma of magmatic evolution, with 150－80 Ma and 80－50 Ma main episodes of distinct geochemical signatures involving the slab and the sub-arc mantle components that are controlled by overall geodynamic of the Neo-Tethyan slab from subduction to continental arc collision (Jagoutz et al., 2019). At ca. 50 Ma, there is notably continental crust contributions due to collision with India (Bouihol et al., 2013; Jagoutz et al., 2019). ③ The Lhasa magmatic arc, it can be divided into two parts, including northern-central and southern Lhasa (Gangdese arc). The Gangdese arc has three high flux events that peak at ca. 93 Ma, 50 Ma and 15 Ma (Chapman and Kapp, 2017), each of them is characterized by northward migration of magmatism and more evolved isotopic compositions of magmatism located farther north from the Indus-Yarlung suture zone and mainly related to the subduction of Neo-Tethys and the collision of India-Asia. The northern-central Lhasa shows three major episodes at ca. 150−110 Ma, 90−85 Ma, and 52 Ma, with major continental crust contributions and related mantle melting inputs in the late-stage, which resulted from the dynamic process of slab break-off following the Qiangtang-Lhasa and India-Asia collision (Zhu et al., 2011). And ④ the Sumatra arc that exhibits major magmatic pulses during ca. 102−85 Ma and 52 Ma, similar to both tempos and isotopic compositions of magmatism in Gangdese arc, these magmatic events most likely correlated with repeated steepening and shallowing of the slab dip, rather than India-Asian convergence rates (Zhang et al., 2019).

The Tengchong arc segment displays a well coupled spatio-temporal evolution (Fig. 3), where younger high-MAR events are to the west (Figs. 1 and 2). The oldest (east) and youngest (west) high MAR events could be driven by a similar mechanism due to that they have similar petrochemical and isotopic features. Similar to most arcs in the southern margin of Asia plate, such as Iran, Kohistan, Gangdese and Sumatra arcs, both show similar high MAR events during the early Cenozoic at ca. 55−50 Ma, these early Cenozoic pulses display very closely correlation with the final initial collision of India-Eurasia plate (Zhu et al., 2011; Chapman et al., 2017; Jagoutz et al., 2019). Both are closely involved with the Neo-Tethyan subduction and final welding of the India-Asia continent. The coeval pulses in the Iran and Sumatra arcs may relate to the steepening and extension of Neo-Tethyan slab dip (Zhang et al., 2019; Shafaii et al., 2020). But, as for the trend of geochemical and isotopic compositions, the ca. 55−49 Ma magmatic pulse in the Tengchong arc segment show similar character to those of Kohistan and Gangdese arcs: Both have highly geochemical and isotopic variations such as highly variable Si content and $Mg^{\#}$ and zircon $\varepsilon_{Hf}(t)$ values (most samples from ca. −10 to +10, even more variable) (Zhu et al., 2011; Bouilhol et al., 2013; Jagoutz et al., 2019, and this study) (Fig. 3). Chapman and Kapp (2017) suggested that this high flux events in the Gangdese arc is characterized by northward migration of magmatism and more evolved isotopic compositions of magmatism located farther north from the suture zone. Undoubtedly, both Kohistan, Gangdese and Tengchong arc rocks at ca. 55−50 Ma recorded a broader range of geochemical and radiogenic isotopic composition, which were reflective of the wider arc and the age of the lithospheric provinces encountered (Chapman et al., 2021 reference therein). During the initial collision of India-Asia, crustal thickening was achieved through both magmatic and tectonic activities, where more evolved continental crust merged into the ca. 55−50 Ma magmatic pulse, which resulted into an enhanced evolved isotopic shift in both Kohistan and Gangdese arcs. The phenomenon that the flare ups shifts to more evolved isotopic ratios has been proposed elsewhere, including some crustal materials, but also plenty of mantle-derived materials, such as Cretaceous Sierra Nevada batholith, Cretaceous Peninsular Ranges batholith, Cretaceous Median batholith, Jurassic and Triassic arcs in the Korean Peninsula, Paleogene Coast Mountains batholith (Chapman et al., 2021 reference therein). The presence of significant high positive zircon $\varepsilon_{Hf}(t)$ values (up to $+12.0$), low SiO_2 content, and high-$Mg^{\#}$ (up to 70) imply significant contributions from juvenile mantle-derived materials. The upwelling mantle materials could be a key control process to induce the melting of overlying mafic crust to generate the extensive granitic magmatism (Zhu et al., 2017a). Isotopic shifts can well imply the variable contributions from mantle and/or crustal materials (Chapman and Ducea, 2019; Chapman et al., 2021). Evolved isotopic shift may result from retroarc understrusting and introduction of isotopically evolved, melt-fertile continental crust or lithosphere into the arc source (DeCelles et al., 2009, 2015; DePaolo et al., 2019; Ducea, 2002; Ducea and Barton, 2007; Chapman et al., 2021 reference). Arcs constructed on juvenile lithosphere would not show the same shift (e.g., Coast Mountain batholith, Cecil et al., 2019). Similarly, we argue that the ca. 55 – 49 Ma pulse in the central to western Tengchong arc segment may correlate with the initial collision of India-Asia, and source lithospheric mantle, continental arc materials and slab-derived fluids/melts (Wang et al., 2014; Zhao et al., 2016). These melt-fertile crustal materials increased the range of isotopic ratios in both Kohistan, Gangdese and Tengchong arcs (Zhu et al., 2011; Bouilhol et al., 2013; Jagoutz et al., 2019, and this study), where the slab break-off following the initial collision played a key role in this flare-up (Zhu et al., 2011).

The pulses at ca. 130 – 110 Ma (Figs. 2 and 3), are different from any arcs in the southern margin of Asia plate, but similar to those magmatic pulses in the central and northern Lhasa that could be associated with the southward subduction of Bangong-Nujiang Tethys (Zhu et al., 2011; Zhu et al., 2017a). Both show similar temporal and evolutionary trend of geochemical and isotopic composition. In the initial stage of the early Cretaceous flare-up events, evolved isotopic compositions with negative zircon $\varepsilon_{Hf}(t)$ values predominated, which indicate major contributions from evolved crustal materials. Arc migration may be induced by steepening of slab dip, and in this setting, subducted recycled sediments and metasomatized mantle wedge may result in the production of high Mg-K dioritic and granodioritic magmas (Zhu et al., 2019; Ma et al., 2021). Subsequently, with slab rollback and break-off, more mantle-derived magmas may upwell ca. 122 – 120 Ma (Zhu et al., 2017a) and result in the magmatism with higher MgO and $Mg^{\#}$, enhanced positive zircon $\varepsilon_{Hf}(t)$ values (up to $+10$) (Figs. 3 and 6), and a broader range of composition and isotopes than before. The early-Cretaceous high MAR events has increasing zircon $\varepsilon_{Hf}(t)$ values up to $+10.0$ implying increasing mantle contributions during this "flare-up", suggesting possible upwelling mantle materials induced this flare-up, and evolving from dominated fertile crustal isotopic signatures to significant mantle isotopic controls. This could be interpreted as related to the tectonic processes upon closure of Qiangtang-Baoshan and Lhasa-Tengchong, from soft-collision to slab break-off of Bangong-Nujiang Tethyan ocean (Zhu et al., 2015). During slab break-off, a slab window may result in the partial melting of various sources region including upwelling asthenospheric mantle, enriched lithospheric mantle, and overlying crust (Ferrari, 2004). The sub-arc mantle wedge could possibly contain melts from both sub-slab asthenosphere enriched in HFSE and normal sub-arc mantle depleted in HFSE as supported by more primitive isotopic compositions (Ferrari, 2004). In which case arc magmatism shifts to greater contributions of mantle and/or juvenile source.These signatures also indicated the varied sources involved in the magmatism following the collision and thicken crust (Fig. 6).

The pulse at ca. 76−64 Ma can be compared with the coeval pulse of back arc magmatism in the Iran arc (Shafaii et al., 2020), the difference is the distinct isotopic composition. In the Iran case, the magmatic rocks show mantle-like signatures not affected by AFC processes (Shafaii et al., 2020). But in the Tengchong case, the late-Cretaceous magmatism shows notably crustal signatures (Zhu et al., 2021; Fig. 3). Two possible model could be considered, the melting of evolved continental crust and/or a metasomatized and fertile lithospheric mantle source combining with overlying crust. Zhu et al. (2021) identified the coeval high Mg and K dioritic magmas with have notable crustal signatures including negative $\varepsilon_{Hf}(t)$ values and high K and Th from nearby, coeval intrusions. The high Mg and K diorites can be compared with sanukitoids produced from the mantle wedge metasomatized by water-rich subducted sediments during steepening subduction (Zhu et al., 2021). Therefore, we infer that the late-Cretaceous pulse could be formed through the change from shallow to steep subduction (Zhu et al., 2021), resulting in the fertile lithospheric mantle and the fertile mantle-derived magmas underplating the base of the continental crust and inducing melting of overlying evolved continental crust. Therefore, sediments subduction/accretion may play a role in enhancing the flare-up (Clift and Vannucchi, 2004; Ducea et al., 2015).

This case is similar to the situation in Cordilleran arcs where subduction-related metasomatism is an important element in enhancing a high MAR event (Chapman and Ducea, 2019). Both the Gangdese and Tengchong have late-Cretaceous intensive magmatism with crustal signatures (Zhu et al., 2011; Zhu et al., 2021), prior to the early-Cenozoic flare-up. Increasing subduction-related metasomatism was identified (Zhu et al., 2021), therefore, these cumulatively metasomatized-related magmatism in the southern margin of the Asian continent could be important fuel for refertilization and hydration in the deep lithosphere to enhance the largest flare-up after 55 Ma following the final closure of Neo-Tethys.

Various lines of evidence have helped to establish correlations between intensive magmatic episodes and various processes, such as cycle of closely linked upper-plate shortening by thrusting (DeCelles et al., 2009), dynamic mantle process (Schwartz et al., 2017), dynamic transformation in the orogen (Guo et al., 2022), and arc migration (Chapman and Ducea, 2019). In Iran and Sumatra arcs, there is no clear relationship between the subduction velocity, convergence rates and magmatic pulses. Similarly, the multi-stage intensive magmatism in the Tengchong case also lacks good correlation with these geophysical parameters. The evolution of early Cretaceous magmatism shows no relationship with the rate of convergence of Lhasa-Qiangtang (Fig. 6a), which changes at the flare-up peak, not flare-up initiation. For the late Cretaceous flare-up, major shifts in rate and angle of convergence of India-Asia plate (Fig. 6a), occur well after flare-up initiation. The early Cenozoic magmatic pulse occurred ca 5 m.y. after the decreasing rate of convergence and shows no correlation to angle of convergence of the India-Asia plate (Fig. 6a).

Both the latest results by Luffi et al. (2019) and others (e.g., Chapman et al., 2015; Hu et al., 2016) confirm thickening during the early-Cretaceous, thinning in the late-Cretaceous followed by thickening again in the early-Cenozoic high MAR events. The two crustal thickening events correlate with the amalgamation of Lhasa-Tengchong and Qiangtang-Baoshan and the final welding of the India-Asia plate, but also likely reflect magma additions to the crust in the high MAR events during the early-Cretaceous and early Cenozoic (Cao and Paterson, 2016).

4. 2 The role of multi-stage "high MAR events" in continental formation and growth

Mature continental arcs share similar major element compositions with those of the continental crust (Rudnick and Gao, 2003), producing on average upper crustal materials similar to the bulk continental crust (Ducea et al., 2015). Most scholars consider continental arcs to be the primary factory in making new continental crust, with significant input of juvenile contributions from the mantle via basaltic melts (Ducea et al., 2015; Jagoutz and Schnidt, 2012) after removal of mafic to ultramafic roots, for example the Kohistan arc.

These high MAR events, emplaced as plutons and volcanism (most has been removed during uplift and erosion due to the tectonic activity), forming a major component of the Tengchong arc segment making up 50%−70% of the area exposed in this region. From a view of both area and volume of these high MAR events, it is a significant component in the formation of the Tengchong arc segment. The residue of these arcs is granulitic at pressures below ca. 15 kbar, and could become a dense (3. 6 g/cm^{-3}), bi-mineralic garnet pyroxenite (eclogite) at higher pressures (deeper levels) (Ducea, 2002). The high La/Yb and Sr/Y ratios (Fig. 3) precluded the formation of cyclic garnet-bearing "arclogites", instead resulting in amphibolite to granulite facies and garnet-free residues in the thin arcs (Ducea et al.,

2020). Therefore, it is less likely that foundering-prone, sizable roots (Ducea, 2002) were generated by high-MAR magmatism in the Tengchong, as readily observed in the modern Andes. The multi-stage high MAR events record the multi-stage crustal thickness through the formation of the Tengchong arc segment, and this Tengchong case resembles continental arcs that have not experienced delamination and instead preserve thicker sections of mafic lower crust (Chapman et al., 2021).

The three-stage high-MAR events in the Tengchong arc have average andesite to dacite compositions. Al-in-Hbl calculated results from diorite-tonalite-granodiorite-monzogranite-granites of early-Cretaceous and gabbro-diorite-granodiorite-monzogranite of early-Cenozoic high-MAR events have a depth range from ~22 km to 5 km for early Cretaceous pluton, and 24 km to 9 km for early Cenozoic pluton, respectively (Zhu et al., 2017a; Zhao et al., 2016, 2019), indicating these high-MAR rocks were mainly emplaced in the middle crust. Ducea et al. (2015) has suggested that high MAR events in the subduction systems can involve ~50% recycled upper plate crust and mantle lithosphere, and the remaining ~50% comes from the mantle wedge. Similarly, in the Tengchong case, although the tonalite-granodiorite-monzogranites dominate, early Cretaceous high-Mg diorites to quartz diorites (Zhu et al, 2017a,b; Ma et al., 2021), late Cretaceous high-Mg monzodiorites (Zhu et al., 2021), and early Cenozoic gabbro-diorites (Wang et al., 2014, 2015; Ma et al., 2021; Zhao et al., 2019), are closely associated with these intermediate to felsic rocks and represent the significant mantle contributions. According to Hf isotopic calculation (Zhu et al., 2018a), at least, the mantle contributions can reach up to 50% for early Cretaceous and early Cenozoic flare-up. Therefore, these high MAR events produced plutons that accumulated within the continental crust, representing growth of the continental arc segment from various contributions, including various continental crust materials, slab-derived melts/fluids, and significant materials from both juvenile and metasomatized mantle materials, resulting in a significant contribution to the growth of the continental crust.

5 Conclusions

Three high-MAR events in the east (ca. 131−111 Ma), middle (ca. 76−64 Ma), to the west (ca. 55−49 Ma) took place in the westward migrating Tengchong arc segment of southeastern Tibet. Both the early-Cretaceous and Cenozoic magmatic episodes have variable compositions and significant mantle contributions, similar to typical continental arcs. These pulses are closely correlated with the changes of crustal thickness, but no correlation with angle and rates of convergence of the plates.

The multistage high MAR events have typical low isotopic characteristics that closely correlate to the interaction of crust-mantle driven by collision-related crustal thickening, but the subsequent increasing isotope trend indicates the increasing contribution from juvenile mantle-derived materials, potentially driven by slab roll-back and break-off. The central magmatic flare-up was produced by an increase in subduction-related metasomatism, which provided enough fuel to trigger the following early-Cenozoic flare-up in the Himalayan-Tibetan orogenic system.

The flare-up produced plutons and batholiths located within the middle continental crust and constitute an area of ~5 000 km^2, greater than 50% of the entire segment, with a granulite facies residue in the deep lithosphere. This has played a major role in the formation and growth of the Tengchong arc segment in the SE Tibetan Plateau.

Acknowledgements　Dr. Ren-Zhi Zhu was inspired by Professors Paterson and Ducea, 2015 where young scientists were motivated to pursue research in "Spatiotemporal magmatism in continental arcs" and develop and synthesize associated datasets, benefiting our society and educational system. Thanks for Dr Bo Zhang help to construct the zircon U-Pb and Lu-Hf isotopic mapping. It was supported by the National Natural Science Foundation of China (Grant No. 41802054), China Postdoctoral Science Special Foundation Grant. 2019T120937 and Foundation Grant. 2018M643713, Natural Science Foundation of Shaanxi Grant. 2019JQ-719 and Shaanxi Postdoctoral Science Foundation.

Appendix A. Supplementary data　Dataset of whole geochronology, geochemistry and in situ zircon Hf isotopes for Cretaceous-Cenozoic magmatism in the Tengchong arc, SE Tibet. Supplementary data to this article can be found online at https:// doi. org/10.1016/j. tecto. 2022.229308.

References

Allegre, C.O., Courtillot, V., Tapponnier, P., Hirn, A., Mattauer, M., Coulon, C., Burg, J.P., 1984. Structure and evolution of the Himalaya-Tibet orogenic belt. Nature, 307(5946), 17.

Anderson, D.L., 1982. Isotopic evolution of the mantle: The role of magma mixing. Earth Planet. Sci. Lett., 57(1), 1-12.

Ardila, A.M.M., Paterson, S.R., Memeti, V., Parada, M.A., Molina, P.G., 2019. Mantle driven cretaceous flare-ups in Cordilleran arcs. Lithos, 326, 19-27.

Attia, S., Cottle, J.M., Paterson, S.R., 2020. Erupted zircon record of continental crust formation during mantle driven arc flare-ups. Geology, 48(5), 446-451.

Bouilhol, P., Jagoutz, O., Hanchar, J.M., Dudas, F.O., 2013. Dating the India-Eurasia collision through arc magmatic records. Earth Planet. Sci. Lett., 366, 163-175.

Burtman, V.S., Molnar, P.H., 1993. Geological and Geophysical Evidence for Deep Subduction of Continental Crust beneath the Pamir, 281. Geological Society of America.

Cao, W., Paterson, S., 2016. A mass balance and isostasy model: Exploring the interplay between magmatism, tectonism and surface erosion in continental arcs. Geochem. Geophys. Geosyst.

Cecil, M.R., Ferrer, M.A., Riggs, N. R., Marsaglia, K., Kylander-Clark, A., Ducea, M.N., Stone, P., 2019. Early arc development recorded in Permian-Triassic plutons of the northern Mojave Desert region, California, USA. Bulletin, 131(5-6), 749-765.

Chapman, J.B., Ducea, M.N., 2019. The role of arc migration in Cordilleran orogenic cyclicity. Geology, 47 (7), 627-631.

Chapman, J. B., Kapp, P., 2017. Tibetan magmatism database. Geochem. Geophys. Geosyst., 18(11), 4229-4234.

Chapman, J.B., Ducea, M.N., DeCelles, P.G., Profeta, L., 2015. Tracking changes in crustal thickness during orogenic evolution with Sr/Y: An example from the north American Cordillera. Geology, 43, 919-922.

Chapman, J.B., Ducea, M. N., Kapp, P., Gehrels, G. E., DeCelles, P.G., 2017. Spatial and temporal radiogenic isotopic trends of magmatism in Cordilleran orogens. Gondwana Res., 48, 189-204.

Chapman, J.B., Scoggin, S.H., Kapp, P., Carrapa, B., Ducea, M.N., Worthington, J., Gadoev, M., 2018. Mesozoic to Cenozoic magmatic history of the Pamir. Earth Planet. Sci. Lett., 482, 181-192.

Chapman, J.B., Shields, J., Ducea, M. N., Paterson, S., Attia, S., Ardill, K., 2021. The causes of continental arc flare ups and drivers of episodic magmatic activity in Cordilleran orogenic systems. Lithos, 398-399, 106307.

Chen, X.C., Hu, R.Z., Bi, X.W., Zhong, H., Lan, J.B., Zhao, C.H., Zhu, J.J., 2015. Petrogenesis of metaluminous A-type granitoids in the Tengchong-Lianghe tin belt of southwestern China: Evidences from zircon U-Pb ages and Hf-O isotopes, and whole-rock Sr-Nd isotopes. Lithos, 212, 93-110.

Chung, S.L., Chu, M.F., Zhang, Y., Xie, Y., Lo, C.H., Lee, T.Y., Wang, Y., 2005. Tibetan tectonic evolution inferred from spatial and temporal variations in post-collisional magmatism. Earth Sci. Rev., 68 (3-4),173-196.

Clift, P. D., Vannucchi, P., 2004. Controls on tectonic accretion versus erosion in subduction zones: Implications for the origin and recycling of the continental crust. Rev. Geophys., 42(2).

Cong, F., Lin, S., Zou, G., Li, Z., Xie, T., Peng, Z., Liang, T., 2011. Magma mixing of granites at Lianghe: In-situ zircon analysis for trace elements, U-Pb ages and Hf isotopes. Sci. China Earth Sci., 54 (9), 1346-1359.

Daniel, C.G., Pfeifer, L.S., Jones III, J.V., McFarlane, C.M., 2013. Detrital zircon evidence for non-Laurentian provenance, Mesoproterozoic (ca.1 490−1 450 Ma) deposition and orogenesis in a reconstructed orogenic belt, northern New Mexico, USA: Defining the Picuris orogeny. GSA Bull., 125 (9-10), 1423-1441.

DeCelles, P.G., Ducea, M.N., Kapp, P., Zandt, G., 2009. Cyclicity in cordilleran orogenic systems. Nature Geosci. 2(4), 251-257.

DePaolo, D. J., Harrison, T. M., Wielicki, M., Zhao, Z., Zhu, D. C., Zhang, H., Mo, X., 2019. Geochemical evidence for thin syn-collision crust and major crustal thickening between 45 Ma and 32 Ma at the southern margin of Tibet. Gondwana Res., 73, 123-135.

Ducea, M.N., 2002. Constraints on the bulk composition and root foundering rates of continental arcs: A California arc perspective. J. Geophys. Res. Solid Earth, 107(B11), ECV-15.

Ducea, M.N., Barton, M.D., 2007. Igniting flare-up events in Cordilleran arcs. Geology, 35 (11), 1047-1050.

Ducea, M.N., Paterson, S.R., DeCelles, P.G., 2015. High-volume magmatic events in subduction systems. Elements, 11(2), 99-104.

Ducea, M.N., Chapman, A.D., Bowman, E., Balica, C., 2020. Arclogites and their role in continental evolution. Part 2: Relationship to batholiths and volcanoes, density and foundering, remelting and long-term storage in the mantle. Earth Sci. Rev., 214, 103476.

Ferrari, L., 2004. Slab detachment control on mafic volcanic pulse and mantle heterogeneity in Central Mexico. Geology, 32(1), 77-80.

Guo, X., Li, C., Gao, R., Li, S., Xu, X., Lu, Z., Xiang, B., 2022. The India-Eurasia convergence system: Late Oligocene to early Miocene passive roof thrusting driven by deep-rooted duplex stacking. Geosyst. Geoenviron., 1(1).

Gurnis, M., 1988. Large-scale mantle convection and the aggregation and dispersal of supercontinents. Nature, 332(6166), 695-699.

Hu, X., Garzanti, E., Wang, J., Huang, W., An, W., Webb, A., 2016. The timing of India-Asia collision onset-Facts, theories, controversies. Earth Sci. Rev., 160, 264-299.

Jagoutz, O.E., 2010. Construction of the granitoid crust of an island arc. Part II: A quantitative petrogenetic model. Contrib. to Mineral. Petrol., 160(3), 359-381.

Jagoutz, O., Behn, M.D., 2013. Foundering of lower island-arc crust as an explanation for the origin of the continental Moho. Nature, 504(7478), 131.

Jagoutz, O., Bouilhol, P., Schaltegger, U., Müntener, O., 2019. The isotopic evolution of the Kohistan Ladakh arc from subduction initiation to continent arc collision. Geological Society, London, Special Publications, 483(1), 165-182.

Jagoutz, O.E., Burg, J.P., Hussain, S., Dawood, H., Pettke, T., Iizuka, T., Maruyama, S., 2009. Construction of the granitoid crust of an island arc. Part I: Geochronological and geochemical constraints from the plutonic Kohistan (NW Pakistan). Contrib. to Mineral. Petrol., 158(6), 739-755.

Jagoutz, O., Kelemen, P.B., 2015. Role of arc processes in the formation of continental crust. Annu. Rev. Earth Planet. Sci., 43, 363-404.

Jagoutz, O., Schmidt, M.W., 2012. The formation and bulk composition of modern juvenile continental crust: The Kohistan arc. Chem. Geol., 298, 79-96.

Kay, S.M., Jicha, B.R., Citron, G.L., Kay, R.W., Tibbetts, A.K., Rivera, T.A., 2019. The calc-alkaline Hidden Bay and Kagalaska plutons and the construction of the central Aleutian oceanic arc crust. J. Petrol., 60(2), 393-439.

Kemp, A.I.S., Hawkesworth, C.J., Foster, G.L., Paterson, B.A., Woodhead, J.D., Hergt, J.M., Whitehouse, M.J., 2007. Magmatic and crustal differentiation history of granitic rocks from Hf-O isotopes in zircon. Science, 315(5814), 980-983.

Kornfeld, D., Eckert, S., Appel, E., Ratschbacher, L., Sonntag, B.L., Pfänder, J.A., Liu, D., 2014. Cenozoic clockwise rotation of the Tengchong Block, southeastern Tibetan Plateau: A paleomagnetic and geochronologic study. Tectonophysics, 628, 105-122.

Lee, T.Y., Lawver, L.A., 1995. Cenozoic plate reconstruction of Southeast Asia. Tectonophysics, 251(1-4), 85-138.

Lee, H.Y., Chung, S.L., Ji, J., Qian, Q., Gallet, S., Lo, C.H., Zhang, Q., 2012. Geochemical and Sr-Nd isotopic constraints on the genesis of the Cenozoic Linzizong volcanic successions, southern Tibet. J. Asian Earth Sci., 53, 96-114.

Li, S., Suo, Y., Li, X., Zhou, J., Santosh, M., Wang, P., Zhang, G., 2019. Mesozoic tectono-magmatic response in the East Asian ocean-continent connection zone to subduction of the Paleo-Pacific Plate. Earth Sci. Rev., 192, 91-137.

Li, S., Li, X., Zhou, J., Cao, H., Liu, L., Liu, Y., Jiang, Z., 2022. Passive magmatism on Earth and Earth-like planets. Geosyst. Geoenviron., 1(1).

Luffi, P., 2019. Paleo-Mohometry: Assessing the Crust Thickness of Ancient Arcs Using Integrated Geochemical Data. In: Goldschmidt Conference, Paper no. 2075, Barcelona, Spain.

Ma, P.F., Xia, X.P., Lai, C.K., Cai, K.D., Yang, Q., 2021. Evolution of the Tethyan Bangong-Nujiang Ocean and its SE Asian connection: Perspective from the early cretaceous high-Mg granitoids in SW China. Lithos, 388-389(5), 106074.

Metcalfe, I., 2013. Gondwana dispersion and Asian accretion: Tectonic and palaeogeographic evolution of eastern Tethys. J. Asian Earth Sci., 66, 1-33.

Paterson, S.R., Ducea, M.N., 2015. Arc magmatic tempos: Gathering the evidence. Elements, 11(2), 91-98.

Paterson, S.R., Okaya, D., Memeti, V., Economos, R., Miller, R.B., 2011. Magma addition and flux calculations of incrementally constructed magma chambers in continental margin arcs: Combined field, geochronologic, and thermal modeling studies. Geosphere, 7(6), 1439-1468.

Ratschbacher, B.C., Paterson, S.R., Fischer, T., 2019. Spatial and depth-dependent variations in magma volume addition and addition rates to continental arcs: Application to global CO_2 fluxes since 750 Ma. Geochem. Geophys. Geosyst., 20, 2997-3018.

Robinson, A.C., Ducea, M., Lapen, T.J., 2012. Detrital zircon and isotopic constraints on the crustal architecture and tectonic evolution of the northeastern Pamir. Tectonics, 31(2).

Rudnick, R.L., Gao, S., 2003. Composition of the continental crust. Treat. Geochem., 3, 659.

Schwartz, J.J., Klepeis, K.A., Sadorski, J.F., Stowell, H.H., Tulloch, A.J., Coble, M.A., 2017. The tempo of continental arc construction in the Mesozoic Median Batholith, Fiordland, New Zealand. Lithosphere, 9(3), 343-365.

Searle, M.P., Windley, B.F., Coward, M.P., Cooper, D.J.W., Rex, A.J., Rex, D., Kumar, S., 1987. The closing of Tethys and the tectonics of the Himalaya. Geol. Soc. Am. Bull., 98(6), 678-701.

Shafaii Moghadam, H., Li, Q.L., Li, X.H., Stern, R.J., Levresse, G., Santos, J.F., Hassannezhad, A., 2020. Neotethyan subduction ignited the Iran arc and backarc differently. J. Geophys. Res. Solid Earth, 125(5), e2019JB018460.

Taylor, S.R., McLennan, S.M., 1985. The continental crust: Its composition and evolution.

Wang, Y., Li, S., Ma, L., Fan, W., Cai, Y., Zhang, Y., Zhang, F., 2015. Geochronological and geochemical constraints on the petrogenesis of Early Eocene metagabbroic rocks in Nabang (SW Yunnan) and its implications on the Neotethyan slab subduction. Gondwana Research, 27(4), 1474-1486.

Wang, Y., Zhang, L., Cawood, P.A., Ma, L., Fan, W., Zhang, A., Bi, X., 2014. Eocene supra-subduction zone mafic magmatism in the Sibumasu Block of SW Yunnan: Implications for Neotethyan subduction and India-Asia collision. Lithos, 206, 384-399.

Xie, J.C., Zhu, D.C., Dong, G., Zhao, Z.D., Wang, Q., Mo, X., 2016. Linking the Tengchong Terrane in SW Yunnan with the Lhasa Terrane in southern Tibet through magmatic correlation. Gondwana Res., 39, 217-229.

Xu, Z., Wang, Q., Cai, Z., Dong, H., Li, H., Chen, X., Burg, J.P., 2015. Kinematics of the Tengchong Terrane in SE Tibet from the late Eocene to early Miocene: Insights from coeval mid-crustal detachments and strike-slip shear zones. Tectonophysics, 665, 127-148.

Yin, A., Harrison, T.M., 2000. Geologic evolution of the Himalayan-Tibetan orogen. Annu. Rev. Earth Planet. Sci., 28(1), 211-280.

Young, A., Flament, N., Maloney, K., Williams, S., Matthews, K., Zahirovic, S., et al., 2019. Global kinematics of tectonic plates and subduction zones since the late Paleozoic era. Geosci. Front., 10(3), 989-1013.

Yunnan Bureau Geological Mineral Resource (YBGMR), 1991. Regional geology of Yunnan Province. Beijing: Geol. Publ. House, 1-729 (in Chinese with English abstract).

Zhao, S.W., Lai, S.C., Qin, J.F., Zhu, R.Z., 2016. Petrogenesis of eocene granitoids and microgranular enclaves in the western Tengchong Block: Constraints on eastward subduction of the Neo-tethys. Lithos, 264, 96-107.

Zhang, X., Chung, S.L., Lai, Y.M., Ghani, A.A., Murtadha, S., Lee, H.Y., Hsu, C.C., 2019. A 6 000-km-long Neo-Tethyan arc system with coherent magmatic flare-ups and lulls in South Asia. Geology, 47(6), 573-576.

Zhao, S.W., Lai, S.C., Pei, X.Z., Qin, J.F., Zhu, R.Z., Tao, N., Gao, L., 2019. Compositional variations of granitic rocks in continental margin arc: Constraints from the petrogenesis of Eocene granitic rocks in the Tengchong Block, SW China. Lithos, 326, 125-143.

Zhu, D.C., Zhao, Z.D., Niu, Y., Mo, X.X., Chung, S.L., Hou, Z.Q., Wu, F.Y., 2011. The Lhasa Terrane: Record of a microcontinent and its histories of drift and growth. Earth Planet. Sci. Lett., 301(1-2), 241-255.

Zhu, R.Z., Lai, S.C., Qin, J.F., Zhao, S.W., 2015. Early-cretaceous highly fractionated I-type granites from the northern Tengchong Block, western Yunnan, SW China: Petrogenesis and tectonic implications. J.Asian Earth Sci., 100, 145-163.

Zhu, R.Z., Lai, S.C., Santosh, M., Qin, J.F., Zhao, S.W., 2017a. Early cretaceous Na-rich granitoids and their enclaves in the Tengchong Block, SW China: Magmatismin relation to subduction of the Bangong-Nujiang Tethys Ocean. Lithos, 286-287, 175-190.

Zhu, R.Z., Lai, S.C., Qin, J.F., Zhao, S.W., Wang, J.B., 2017b. Late Early-cretaceous quartz diorite-granodiorite-monzogranite association from the Gaoligong belt, southeastern Tibet Plateau: Chemical variations and geodynamic implications. Lithos, 288-289, 311-325.

Zhu, R.Z., Lai, S.C., Qin, J.F., Zhao, S.W., 2018a. Petrogenesis of late Paleozoic-to-early Mesozoic granitoids and metagabbroic rocks of the Tengchong Block, SW China: Implications for the evolution of the eastern Paleo-Tethys. Int. J. Earth Sci., 107(2), 431-457.

Zhu, R.Z., Lai, S.C., Qin, J.F., Zhao, S.W., Santosh, M., 2018b. Strongly peraluminous fractionated S-type granites in the Baoshan Block, SW China: Implications for two-stage melting of fertile continental materials following the closure of Bangong- Nujiang Tethys. Lithos, 316, 178-198.

Zhu, R.Z., Lai, S.C., Qin, J.F., Zhao, S.W., Santosh, M., 2019. Petrogenesis of high-K calc-alkaline granodiorite and its enclaves from the SE Lhasa Block, Tibet (SW China): Implications for recycled subducted sediments. GSA Bull., 131(7-8), 1224-1238.

Zhu, R.Z., Słaby, E., Lai, S.C., Chen, L.H., Qin, J.F., Zhang, C., ..., Fowler, M., 2021. High-K calc-alkaline to shoshonitic intrusions in SE Tibet: Implications for metasomatized lithospheric mantle beneath an active continental margin. Contrib. Mineral. Petrol., 176(10), 1-19.

Magma mixing for the genesis of Neoproterozoic Mopanshan granitoids in the western Yangtze Block, South China[①]

Zhu Yu　Lai Shaocong[②]　Qin Jiangfeng　Zhu Renzhi　Zhao Shaowei

Liu Min　Zhang Fangyi　Zhang Zezhong　Yang Hang

Abstract: Crust- and/or mantle-derived igneous rocks in the Neoproterozoic continental magmatic arc along the western Yangtze Block (South China) have been extensively studied, but the crust-mantle interaction that might have been involved in the formation of these rocks was poorly understood. In this paper, we present an integrated study on petrology, whole-rock compositions, in-situ plagioclase Sr and zircon U-Pb-Hf isotopes for granodiorites and monzogranites from the Neoproterozoic Mopanshan pluton in the western Yangtze Block. Our objective is to provide vital constraint on the magma mixing process in the formation of these rocks. The new LA-ICP-MS zircon U-Pb dating results reveal that both granodiorites and monzogranites of the Mopanshan pluton were coeval and formed at ca. 820 Ma. They are calc-alkaline to high-K calc-alkaline ($SiO_2 = 66.06 - 70.30$ wt%, $K_2O = 2.24 - 3.34$ wt%) and metaluminous to slightly peraluminous (A/CNK = 0.99 - 1.10) rocks, and characterized by middle-upper crustal trace element patterns (e.g., enrichment in Rb, K, Th, U, and Pb, and depletion in Nb, Ta, Sr, and Ti). The magma mixing process is supported by following ample evidence, including ① the disequilibrium mineral textures and an abruptly increased An values (up to 60) as well as significant decrease of $(^{87}Sr/^{86}Sr)_i$ ratios from core to rim in plagioclase crystals, ② higher $Mg^{\#}$ values (45-50 > 40) than those of experimental melts from basalts at the same silica contents, and ③ positive and negative whole-rock $\varepsilon_{Nd}(t)$ (−0.72 to +1.01) values as well as variable whole-rock $(^{87}Sr/^{86}Sr)_i$ (0.702 778−0.705 404) and zircon $\varepsilon_{Hf}(t)$ (+1.76 to +7.72) values. The special geochemical characteristics and element modeling support that the Mopanshan pluton was generated by magma mixing of ancient crust-derived melts and relatively mafic melts from metasomatized mantle source. The adakitic signatures observed in the Mopanshan pluton were attributed to subsequent hornblende-dominant intra-crustal fractional crystallization, rather than derived from a thickened lower crust source. In combination with regional geology and our compilation for Nd-Hf isotopes of Neoproterozoic igneous rocks, the Mopanshan pluton is magmatic response of intensive crust-mantle interaction induced by underplating of voluminous mantle-derived magma in a back-arc extension setting. This research

---

①　Published in *Journal of Asian Earth Sciences*, 2022, 231.

②　Corresponding author.

highlights that the crust-mantle interaction is significant for the petrological and geochemical diversity of Neoproterozoic granitoids from the western Yangtze Block.

1 Introduction

Granites are one of dominant components of continental crust, thus can preserve important information about the origin and evolution of continental crust as well as crust-mantle interaction and differentiation (Jagoutz and Klein, 2018; Moyen et al., 2021). Knowledge of their petrogenetic mechanism is, therefore, significant for deciphering essential information about crustal growth and reworking as well as deep crust-mantle interaction (Jagoutz and Klein, 2018; Zheng, 2021). Beneath an active continental margin, various granitoids (e.g., dioritic, tonalitic, granodioritic, granitic rocks) could be originated from distinct sources, from subduction slab, heterogeneous mantle wedge, to juvenile or ancient continental crust (Winter, 2014). Furthermore, diverse magmatic processes, including the recharging, evacuating, and fractional crystallization (REFC) as well as melting, assimilation, storage, and homogenization (MASH) (Annen et al., 2006; Lee et al., 2014), can shape the geochemical compositions of arc igneous rocks. By this token, the continental arc magmatism is characterized by multisource and multistage process (Winter, 2014).

Situated at the western margin of the Yangtze Block, South China, a Neoproterozoic continental magmatic arc is characterized by intensive granitic, mafic, and migmatitic rocks, which represent the magmatic products of the assembly and breakup of supercontinent Rodinia (Li et al., 2021a; Zhao et al., 2018). These widespread rocks, predominantly dating at 870−750 Ma (Fig. 1), have been interpreted to be derived from different magmatic sources, including the subduction oceanic slab (Zhao et al., 2018; Zhao and Zhou, 2007a; Zhou et al., 2006a), metasomatized mantle wedge (Zhao and Zhou, 2007b; Zhao et al., 2018, 2019; Zhou et al., 2006b; Zhu et al., 2020a, 2021a), juvenile mafic lower crust (Lai et al., 2015; Zhao et al., 2018), and ancient continental crust (Lai and Zhu, 2020; Li et al., 2021a; Zhu et al., 2019a). Nevertheless, the crust-mantle interaction that might have been involved in the formation of Neoproterozoic igneous rocks remains unclear. Here we conduct a comprehensive study of petrology, whole-rock compositions, in-situ plagioclase Sr and zircon U-Pb-Hf isotopes on the granodiorites and monzogranites from the Neoproterozoic Mopanshan pluton in the western Yangtze Block, which were previously explained to be adakitic rocks and originated from a thickened lower crust source (Huang et al., 2009). Our objective is to provide particular constraint on crust-mantle interaction during the Neoproterozoic. The systematic petrological observation, plagioclase mineralogical chemistry and in-situ Sr isotopes, zircon U-Pb-Hf isotopes, whole-rock geochemistry, and element modeling support that the Neoproterozoic

Mopanshan pluton was predominantly formed by the magma mixing of ancient crust-derived felsic melts and metasomatized mantle-derived mafic melts. Both granodiorites and monzogranites of the Mopanshan pluton stand for the magmatic products of intensive crust-mantle interaction induced by underplating of voluminous mantle-derived magma in a back-arc extension setting.

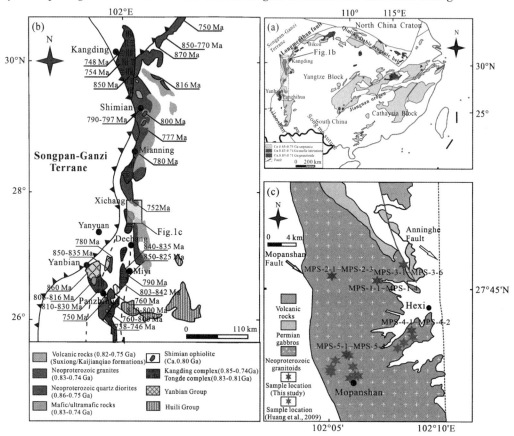

Fig. 1　Geological map for the South China (a), the western Yangtze Block (b), and Neoproterozoic Mopanshan pluton (c).

Fig. 1a is after Zhou et al. (2006b). Fig. 1b and published ages of Neoproterozoic igneous rocks are after Lai and Zhu (2020) and references therein. Fig. 1c is after Huang et al. (2009).

2　Geological setting and samples

2.1　Geological background

The Yangtze Block is tectonically separated from the North China Craton by the Triassic Qinling-Dabie orogenic belt to the north, from the Cathysia Block by the Neoproterozoic Jiangnan Fold Belt to the southeast, and is bounded by the Songpan-Ganzi terrane to the west (Fig. 1a) (Wang et al., 2013; Zhao et al., 2011; Zhao and Cawood, 2012). The Yangtze Block comprises Archean to Early Neoproterozoic basement complexes, which are overlain by the Neoproterozoic to Mesozoic sequences (Zhao et al., 2018). The oldest Archean Kongling basement rocks cover an area of 360 km^2 in the northern Yangtze Block, and are composed of felsic gneisses and metasedimentary rocks as well as minor amphibolite and mafic granulite (Gao et al., 1999, 2011; Guo et al., 2014). In the southwestern Yangtze Block, the newly identified Mesoarchean igneous rocks, including 3 092 – 3 071 Ma trondhjemites and ca. 2 920 Ma potassic granites, are oldest magmatic record so far (Zhao et al., 2020). The oldest strata in the southwestern Yangtze Block include the Late Paleoproterozoic Dahongshan, Hekou, Dongchuan, and Tong'an groups (Greentree and Li, 2008; Wang and Zhou, 2014; Wang et al., 2014; Zhao et al., 2010). The Late Paleoproterozoic to Mesoproterozoic mafic intrusions and meta-volcanic layers were formed at ca. 1.7 – 1.5 Ga (Zhu et al., 2016, 2021a). There also exist widespread Late Mesoproterozoic to Early Neoproterozoic volcanic-sedimentary sequences in the southwestern Yangtze Block, such as the Upper Kunyang, Upper Huili, Julin, and Yanbian groups (Zhao et al., 2018; Zhu et al., 2016). In addition, extensive Late Mesoproterozoic A-type magmatic rocks and S-type granites have been documented at 1 040 – 1 020 Ma (Zhu et al., 2016; Zhu et al., 2020b). Voluminous Neoproterozoic granitic and mafic-ultramafic intrusive complexes, such as the Pengguan, Kangding, Shimian, Miyi, Tongde, Datian, and Yuanmou complexes as well as Moutuo, Jiaoziding, Yonglang, Dalu, Shuilu granitoid plutons, were well preserved and show predominant zircon U-Pb ages of ca. 870 – 750 Ma (Hu et al., 2020; Li et al., 2018; Qi and Zhao, 2020; Zhao et al., 2018 and references therein, 2019; Zhu et al., 2019a-c, 2020a, 2021a,b).

2.2 Samples and petrological characteristics

In this study, the Mopanshan pluton, covering an area of ~ 280 km^2, crops out at the middle segment of the western Yangtze Block (Fig. 1b), and is cut by the NS-trending Mopanshan Fault to the west and the Anninghe Fault to the east (Huang et al., 2009) (Fig. 1c). It is situated at the west part of the Hexi town and consists of granodiorites and monzogranites. Field observations display no obvious contact among different types of rocks. We collected the granodiorites (MPS-4 ~ MPS-5) in the south of the Mopanshan pluton and monzogranites (MPS-1 ~ MPS-3) in the north part (Fig. 1c). Both granodiorites and monzogranites are medium- to fine-grained granitic texture. The granodiorites are grey and composed of 20% – 25% quartz, 10% – 20% K-feldspar, 35% – 50% plagioclase, 10% – 15% hornblende, 5% – 10% biotite, 0% – 5% magnetite, titanite, apatite, and zircon (Fig. 2a-d). Most plagioclase crystals are subhedral to euhedral tabular with obviously concentric zonation and some have resorption planes pointing to dissolution and regrowth (Fig. 2a-d), indicating magma mixing process. The euhedral plagioclases display core-rim or core-mantle-rim textures (Fig. 2a-d). Some plagioclases have irregular resorption faces and two-stage growth margin (Fig. 2b,c).

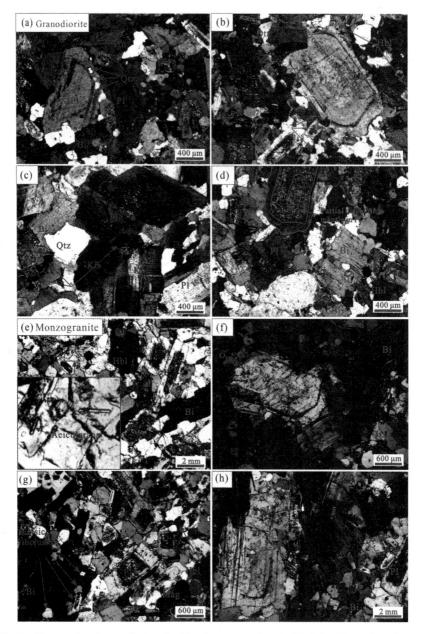

Fig. 2　Photographs of granodiorites (a-d) and monzogranites (e-h) of the Mopanshan pluton from the western Yangtze Block, South China.

Tiny quartz crystals occur at the rim of plagioclase (a). Plagioclase displays an overgrowth rim (b). Plagioclase shows complex oscillatory zoning and resorption texture (c). Plagioclase shows core-mantle-rim texture, and the mantle of plagioclase is partial dissolved (d). Tiny acicular apatite is detected (e). The overgrowth rim is observed in the plagioclase, and a tiny plagioclase crystal is inlaid at the margin of this plagioclase crystal (f). Quartz crystals as inclusions occur in biotite (g). Plagioclases display complex textures with resorption surface and oscillatory zoning (g-h). Hbl: hornblende; Kfs: K-feldspar; Pl: plagioclase; Qtz: quartz; Apa: apatite; Ttn: titanite; Mag: magnetite; Bi: biotite.

Both hornblendes and biotites are subhedral. Quartz crystals are subhedral, and filled in the interstitial of plagioclase and biotite (Fig. 2a,d). The mantle of plagioclase is partial dissolved (Fig. 2d), reflecting the reaction between mafic and felsic melts. The monzogranites are greyish white and medium- to fine-grained texture, which consist of 25%−35% quartz, 25%−30% K-feldspar, 30%−35% plagioclase, 5%−10% hornblende, and 5%−10% biotite (Fig. 2e-h). Accessory minerals are magnetite, zircon, and acicular apatite. Like the granodiorites, the overgrowth rim of plagioclase in the monzogranites are also observed (Fig. 2f). Some plagioclases have oscillatory zoning and sieve texture in their cores (Fig. e,g and h), implying mixing process. Some felsic inclusions (i. e., tiny quartz crystals) are detected in the interior of hornblende, biotite, and plagioclase (Fig. 2e,g and h).

3　Analytical methods

In this study, zircon U-Pb-Hf isotopic analyses, whole-rock major- and trace-elemental analyses, and electron microprobe analysis (EMPA) of plagioclases were processed at the State Key Laboratory of Continental Dynamics, Northwest University, Xi'an, China. Whole-rock Sr-Nd isotopes were performed in the Guizhou Tuopu Resource and Environmental Analysis Center in China. In-situ plagioclase Sr isotopes were performed at Nanjing Hongchuang Geological Exploration Technology Service Co., Ltd. (NHEXTS), Nanjing, China. The detailed descriptions have been documented in Supplementary Analytical Method.

4　Analytical results

4. 1　LA-ICP-MS zircon U-Pb dating

Zircon CL images and LA-ICP-MS U-Pb dating results for three representative samples of the Mopanshan pluton, including one granodiorite (MPS-5) and two monzogranite samples (MPS-3 and MPS-2), were summarized in Supplementary Table 1, and illustrated in Figs. 3 and 4. Zircon grains are predominantly euhedral prismatic crystals with length/width ratios of 1 :1−3 :1. These grains are $100-200$ μm long and display grey and black. Most of them exhibit clear oscillatory zoning (Fig. 3), indicating a magmatic origin (Hoskin and Schaltegger, 2003).

4.1.1　Mopanshan granodiorite (MPS-5)

For granodiorite sample MPS-5 (27°42′2″N 102°6′3″E) (Fig. 1c), zircon grains display variable Th contents (45. 2−716 ppm), U contents (43. 3−1 026 ppm), and Th/U ratios (0. 41−1. 90). A total of twenty-three concordant spots were conducted and plotted in a group showing a weighted mean age of 819. 1 ± 2. 5 Ma (MSWD = 1. 6, $n = 23$) (Fig. 4a,b), which represent the emplacement time of granodiorite.

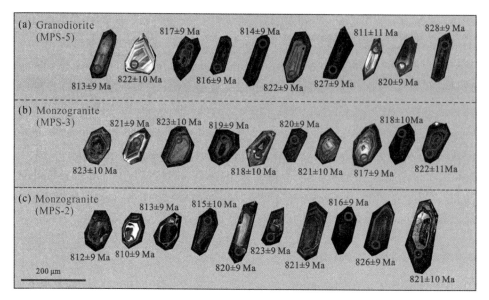

Fig. 3   Zircon cathodoluminescence (CL) images of representative grains for Neoproterozoic Mopanshan granodiorites (a) and monzogranites (b, c) in the western Yangtze Block, South China.

4.1.2   Mopanshan monzogranite (MPS-3)

The sample MPS-3 (27°45′46″N 102°10′41″E) is a monzogranite from the Mopanshan pluton (Fig. 1c). Zircon grains of sample MPS-3 have relatively constant Th contents of 76.4–267 ppm, U contents of 91.1–372 ppm, and Th/U ratios of 0.70–1.13. Among twenty concordant analyses spots, they yield a concordant weighted mean age of 820.3 ± 2.2 Ma (MSWD = 0.1, $n = 20$) (Fig. 4c, d).

4.1.3   Mopanshan monzogranite (MPS-2)

Zircon grains, separated from the monzogranite sample MPS-2 (27°45′33″N 102°6′55″E) (Fig. 1c), have constant Th (52.9–350 ppm) and U (92.8–551 ppm) concentrations as well as high Th/U ratios (0.37–1.83). Twenty-eight reliable analyses yield a concordant $^{206}Pb/^{238}U$ weighted age of 819.1 ± 2.1 Ma (MSWD = 1.4, $n = 28$) (Fig. 4e, f).

4.2   Whole-rock major and trace element geochemistry

The whole-rock major and trace elements of granodiorites and monzogranites from different locations of the Mopanshan pluton were presented in Table 1, and Figs. 5 and 6.

4.2.1   Mopanshan granodiorite

All of the granodiorites display moderate SiO_2(66.06–66.74 wt%) and low K_2O (2.25–2.44 wt%) contents as well as high Na_2O/K_2O (1.62–1.75) ratios, belonging to the calc-alkaline series (Fig. 5a). They contain high contents of Al_2O_3(16.52–17.09 wt%), CaO (4.08–4.15 wt%), Na_2O (3.85–3.98 wt%), and A/CNK [molar ratio of $Al_2O_3/(CaO + Na_2O + K_2O)$] (1.01–1.03) values, pointing to the weakly peraluminous series (Fig. 5b).

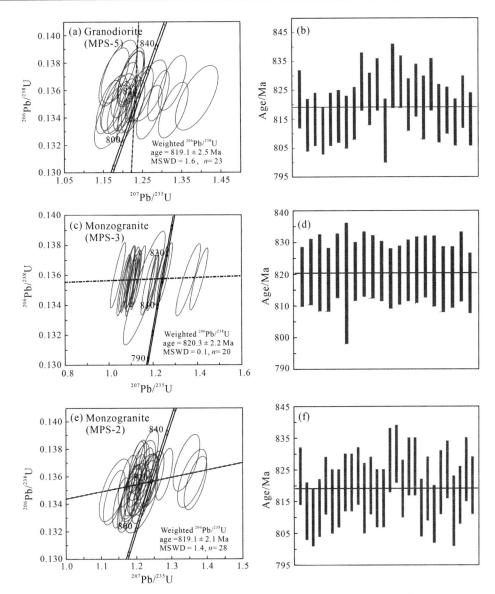

Fig. 4　Laser ablation-inductively coupled plasma-mass spectrometry (LA-ICP-MS) zircon U-Pb
Concordia diagrams for Neoproterozoic Mopanshan granodiorites (a,b) and
monzogranites (c-f) in the western Yangtze Block, South China.

They also have slightly high $Fe_2O_3^T$ (3. 39−3. 65 wt%) and MgO (1. 19−1. 57 wt%) contents as well as $Mg^{\#}$ (45−50) values (Fig. 5c). On the $FeO^T/(FeO^T + MgO)$ vs. SiO_2 diagram (Fig. 5d), the granodiorites are plotted into the field of magnesian series. These granodiorites have constant total rare earth element (ΣREE) concentrations from 101 ppm to 149 ppm. In the chondrite-normalized REE pattern diagram (Fig. 6a), they show fractionated REE patterns with the enrichment of light rare earth elements (LREEs) [$(La/Yb)_N = 12. 4−21. 3$] and relative depletion of high rare earth elements (HREEs) [$(Dy/Yb)_N = 1. 29−1. 46$] as well as

Table 1　Major (wt%) and trace (ppm) element analysis results for the Neoproterozoic Mopanshan granodiorite-monzogranite association in the western Yangtze Block, South China.

Sample	MPS-5-1	MPS-5-2	MPS-5-3	MPS-5-4	MPS-4-1	MPS-4-1R	MPS-4-2	MPS-3-1	MPS-3-2	MPS-3-3	MPS-3-4	MPS-3-5	MPS-3-6	MPS-3-6R	MPS-2-1	MPS-2-2	MPS-2-3	MPS-1-1	MPS-1-2	MPS-1-3	MPS-1-4
Lithology	Granodiorite							Monzogranite													
Location	27°42'2"N 102°6'3"E				27°42'44"N 102°7'55"E			27°45'46"N 102°10'41"E							27°45'20"N 102°5'45"E			27°45'17"N 102°7'22"E			
SiO_2	66.16	66.06	66.27	66.74	66.67	66.62	66.23	68.64	69.06	69.80	68.01	69.07	69.35	69.37	69.62	70.30	70.09	68.34	69.81	69.30	69.00
TiO_2	0.40	0.39	0.38	0.38	0.46	0.46	0.46	0.39	0.40	0.36	0.40	0.38	0.37	0.37	0.30	0.30	0.30	0.37	0.33	0.32	0.33
Al_2O_3	17.09	17.04	16.67	16.78	16.56	16.52	16.54	15.39	15.48	15.09	15.56	15.24	15.34	15.34	15.25	15.37	15.38	15.11	15.07	15.50	15.58
$Fe_2O_3^T$	3.55	3.52	3.40	3.39	3.62	3.65	3.61	2.88	2.88	2.65	2.92	2.94	2.63	2.62	2.53	2.53	2.55	2.80	2.63	2.60	2.66
MnO	0.06	0.06	0.06	0.06	0.05	0.05	0.05	0.04	0.05	0.05	0.05	0.05	0.05	0.05	0.04	0.04	0.04	0.05	0.04	0.04	0.04
MgO	1.27	1.25	1.21	1.19	1.55	1.57	1.56	1.10	1.09	1.04	1.14	1.15	1.07	1.07	0.89	0.90	0.93	1.05	1.02	0.99	1.01
CaO	4.15	4.12	4.08	4.11	4.08	4.09	4.11	3.18	3.21	2.88	3.35	3.09	3.13	3.13	2.10	2.15	2.14	2.48	2.63	2.55	2.60
Na_2O	3.95	3.95	3.98	3.94	3.85	3.91	3.89	3.81	3.81	3.80	3.91	3.87	3.99	4.03	4.32	4.06	3.98	5.21	3.93	4.22	4.13
K_2O	2.44	2.41	2.29	2.25	2.28	2.28	2.27	3.06	3.06	3.04	2.91	2.85	2.66	2.67	3.17	3.30	3.34	2.31	2.24	2.41	2.40
P_2O_5	0.13	0.13	0.13	0.12	0.13	0.12	0.13	0.11	0.11	0.10	0.12	0.11	0.10	0.10	0.09	0.09	0.09	0.11	0.09	0.11	0.10
LOI	0.74	0.97	1.17	0.83	1.00	1.01	0.98	1.16	1.16	1.07	1.22	1.16	0.97	0.98	1.36	1.37	1.30	1.88	1.99	1.67	1.83
Total	99.94	99.90	99.64	99.79	100.25	100.28	99.83	99.76	100.31	99.88	99.59	99.91	99.66	99.73	99.67	100.41	100.14	99.71	99.78	99.71	99.68
$Mg^\#$	46	45	45	45	50	50	50	47	47	48	48	48	49	49	45	45	46	47	48	47	47
A/CNK	1.02	1.03	1.01	1.02	1.02	1.01	1.01	1.00	1.00	1.02	0.99	1.01	1.01	1.01	1.06	1.09	1.09	0.97	1.10	1.09	1.10
Li	21.1	21.7	24.8	20.0	16.6	16.4	16.3	19.6	16.8	18.3	18.6	19.5	16.9	16.8	13.5	13.6	12.5	16.2	15.1	14.2	14.8
Be	1.33	1.35	1.36	1.35	1.30	1.30	1.30	1.44	1.58	1.49	1.68	1.55	1.72	1.72	1.29	1.30	1.22	1.31	1.23	1.24	1.28
Sc	6.44	6.39	6.15	6.06	7.09	7.02	7.01	5.08	4.99	5.07	5.87	4.80	4.77	4.72	4.33	4.29	4.33	5.03	4.52	4.42	4.61
V	38.1	38.0	36.8	36.2	47.9	48.8	48.9	35.7	32.0	35.4	35.9	30.9	31.7	31.4	26.2	25.3	26.0	28.5	26.8	27.2	28.3

Continued

Sample	MPS-5-1	MPS-5-2	MPS-5-3	MPS-5-4	MPS-4-1	MPS-4-1R	MPS-4-2	MPS-3-1	MPS-3-2	MPS-3-3	MPS-3-4	MPS-3-5	MPS-3-6	MPS-3-6R	MPS-2-1	MPS-2-2	MPS-2-3	MPS-1-1	MPS-1-2	MPS-1-3	MPS-1-4
Lithology	Granodiorite							Monzogranite													
Location	27°42'2"N 102°6'3"E				27°42'44"N 102°7'55"E			27°45'46"N 102°10'41"E							27°45'20"N 102°5'45"E			27°45'17"N 102°7'22"E			
Cr	11.0	11.1	11.0	11.7	14.0	13.9	14.0	10.3	9.5	10.6	11.6	9.5	9.6	9.6	9.70	9.27	9.50	6.07	5.76	5.73	5.93
Co	127	128	140	183	161	161	161	40	33	36	35	40	40	40	167	166	159	146	162	122	137
Ni	4.89	4.81	4.71	4.83	5.77	5.70	5.67	5.22	4.58	8.14	5.11	4.87	4.89	4.88	3.97	3.86	3.73	3.06	2.85	2.85	2.91
Cu	12.2	12.3	12.1	10.7	4.15	4.25	4.19	3.3	3.1	3.2	3.1	2.96	2.92	2.80	4.09	3.88	5.33	2.96	2.46	2.85	2.64
Zn	55.9	55.6	54.2	53.9	56.1	55.8	55.4	46.5	42.3	47.0	45.4	42.6	43.1	42.9	36.1	36.1	35.6	44.4	41.2	41.2	42.2
Ga	20.0	20.0	19.4	19.7	20.0	20.0	19.9	18.7	18.1	19.1	18.7	18.3	18.8	18.7	17.4	17.5	17.1	17.5	17.4	18.0	18.4
Ge	1.06	1.08	1.04	1.06	1.08	1.08	1.09	1.12	1.17	1.11	1.19	1.15	1.17	1.15	1.18	1.16	1.31	0.96	0.89	0.93	0.94
Rb	65.9	65.0	69.6	61.4	61.6	61.5	61.2	85.3	85.1	82.6	82.2	86.5	79.9	78.9	87.3	88.9	74.0	62.6	65.2	68.4	68.0
Sr	357	359	360	351	373	374	373	331	314	343	332	327	328	328	218	221	189	298	278	285	286
Y	14.5	14.5	13.3	12.2	12.2	12.0	12.1	12.1	12.2	12.1	13.9	12.2	12.5	12.3	13.0	13.4	13.0	15.1	12.3	14.1	14.8
Zr	196	215	186	179	197	184	185	172	147	170	175	161	161	154	162	165	173	166	160	156	161
Nb	5.22	5.23	4.89	4.66	5.10	5.09	5.12	5.89	6.05	5.97	6.95	6.36	6.65	6.50	5.31	5.21	5.47	5.60	4.90	4.84	5.01
Cs	1.25	1.24	1.20	1.23	0.94	0.95	0.95	1.31	1.17	1.33	1.19	1.09	1.17	1.15	1.08	1.09	1.07	1.89	1.77	1.76	1.80
Ba	763	754	695	670	745	747	746	812	696	777	661	768	636	633	810	822	745	627	591	669	661
La	26.4	25.1	21.5	34.8	31.1	30.5	29.5	27.9	28.3	28.0	27.4	32.2	29.5	28.8	24.8	25.7	26.2	30.9	17.2	31.7	40.4
Ce	51.8	49.1	42.4	66.2	61.1	59.9	57.8	54.7	55.7	54.5	54.2	61.9	57.7	56.0	47.6	49.8	50.2	61.3	34.2	61.6	77.7
Pr	5.90	5.63	4.88	7.32	6.79	6.65	6.49	5.99	6.00	5.98	5.95	6.59	6.23	6.06	5.34	5.60	5.59	6.89	3.94	6.89	8.61
Nd	22.3	21.5	18.9	26.3	25.1	24.7	24.2	22.3	22.3	22.5	22.5	24.0	23.0	22.4	19.6	20.5	20.3	25.2	15.2	25.1	30.8
Sm	4.03	3.93	3.53	4.09	4.19	4.15	4.12	3.85	3.79	3.88	3.99	3.92	3.91	3.84	3.43	3.59	3.49	4.35	2.92	4.18	4.91

Continued

Sample	MPS-5-1	MPS-5-2	MPS-5-3	MPS-5-4	MPS-4-1	MPS-4-1R	MPS-4-2	MPS-3-1	MPS-3-2	MPS-3-3	MPS-3-4	MPS-3-5	MPS-3-6	MPS-3-6R	MPS-2-1	MPS-2-2	MPS-2-3	MPS-1-1	MPS-1-2	MPS-1-3	MPS-1-4
Lithology	Granodiorite							Monzogranite													
Location	27°42′2″N 102°6′3″E				27°42′44″N 102°7′55″E			27°45′46″N 102°10′41″E							27°45′20″N 102°5′45″E			27°45′17″N 102°7′22″E			
Eu	1.07	1.06	1.00	1.02	1.04	1.04	1.03	0.97	0.91	0.98	0.93	0.92	0.92	0.92	0.87	0.88	0.83	0.94	0.87	0.95	0.98
Gd	3.43	3.42	3.03	3.30	3.45	3.40	3.42	3.18	3.14	3.20	3.34	3.16	3.17	3.11	2.91	3.00	2.94	3.63	2.56	3.43	3.93
Tb	0.47	0.46	0.42	0.42	0.44	0.44	0.44	0.42	0.42	0.42	0.45	0.42	0.42	0.42	0.40	0.41	0.40	0.49	0.37	0.45	0.50
Dy	2.66	2.65	2.40	2.32	2.43	2.41	2.41	2.33	2.31	2.31	2.51	2.28	2.29	2.31	2.25	2.31	2.26	2.67	2.14	2.50	2.73
Ho	0.51	0.51	0.46	0.44	0.45	0.44	0.45	0.43	0.43	0.43	0.47	0.43	0.44	0.43	0.43	0.44	0.43	0.51	0.42	0.48	0.50
Er	1.44	1.45	1.30	1.26	1.25	1.24	1.23	1.20	1.22	1.22	1.36	1.22	1.26	1.22	1.24	1.26	1.25	1.43	1.22	1.34	1.40
Tm	0.21	0.21	0.19	0.18	0.17	0.17	0.17	0.17	0.18	0.17	0.20	0.18	0.18	0.18	0.18	0.18	0.18	0.21	0.18	0.19	0.19
Yb	1.36	1.37	1.25	1.17	1.12	1.11	1.11	1.14	1.18	1.13	1.34	1.20	1.23	1.20	1.17	1.20	1.18	1.31	1.15	1.22	1.22
Lu	0.20	0.20	0.19	0.18	0.17	0.17	0.17	0.17	0.18	0.17	0.21	0.19	0.19	0.18	0.18	0.18	0.18	0.20	0.17	0.18	0.18
Hf	4.88	5.30	4.65	4.44	4.93	4.61	4.64	4.52	3.94	4.48	4.64	4.35	4.36	4.17	4.23	4.39	4.47	4.36	4.21	4.11	4.18
Ta	0.50	0.51	0.50	0.51	0.52	0.51	0.51	0.54	0.55	0.54	0.62	0.61	0.60	0.59	0.59	0.59	0.58	0.58	0.55	0.49	0.51
Pb	11.8	11.9	11.2	11.4	12.1	12.3	12.0	12.1	13.2	11.4	12.5	13.5	12.5	12.4	10.8	11.4	6.81	11.6	11.0	11.7	12.1
Th	6.30	6.75	5.93	7.17	6.68	6.51	6.55	8.45	9.36	7.73	9.63	9.86	10.24	10.05	7.77	7.79	8.16	8.47	6.08	7.47	8.95
U	0.94	1.17	1.11	0.90	0.86	0.86	0.87	1.66	1.92	2.07	1.80	2.44	2.25	2.24	1.09	1.08	1.35	1.10	0.87	0.90	0.94
REE	122	117	101	149	139	136	133	125	126	125	125	139	130	127	110	115	115	140	83	140	174
Eu/Eu*	0.88	0.89	0.94	0.85	0.83	0.84	0.84	0.85	0.81	0.85	0.78	0.80	0.80	0.82	0.84	0.82	0.79	0.73	0.97	0.77	0.68
Sr/Y	24.5	24.8	27.1	28.8	30.6	31.1	30.9	27.3	25.8	28.4	24.0	26.7	26.3	26.6	16.7	16.6	14.5	19.8	22.7	20.1	19.3

$Mg^{\#} = $ molar $100MgO/(MgO+FeO^{T})$; $A/CNK=Al_2O_3/(CaO+K_2O+Na_2O)$ molar ratio; $Eu/Eu^{*} = Eu_N/(Sm_N \cdot Gd_N)^{1/2}$.

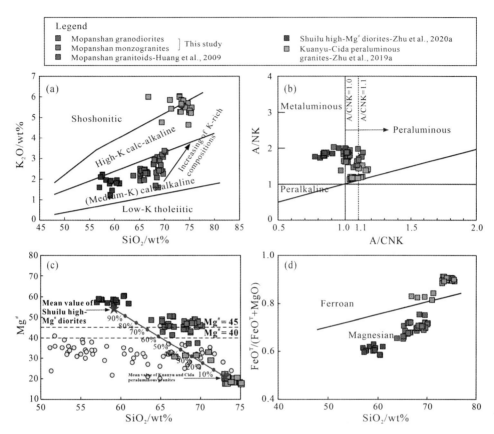

Fig. 5   The K_2O vs. SiO_2(a) (Roberts and Clemens, 1993), A/NK vs. A/CNK (b) (Frost et al., 2001), $Mg^{\#}$ vs. SiO_2(c) (Chen et al., 2013), and $FeO^T/(FeO^T + MgO)$ vs. SiO_2(d) (Frost et al., 2001) diagrams for the Neoproterozoic Mopanshan granodiorite-monzogranite association in the western Yangtze Block, South China.

The published Mopanshan pluton is from Huang et al. (2009). The Shuilu high-$Mg^{\#}$ diorites are from Zhu et al. (2020a). The Kuanyu-Cida peraluminous granites are from Zhu et al. (2019a). In Fig. 5c, the yellow circles represent synthetic melts from basalt melting experiments (Sen and Dunn, 1994; Rapp and Watson, 1995). The simple binary mixing line is after Ersoy and Helvaci (2010). For interpretation of the references to colour in this figure legend, the reader is referred to the web version of this article.

weakly negative Eu anomalies ($Eu/Eu^* = 0.83 - 0.94$). In the primitive mantle-normalized diagram (Fig. 6b), they are enriched in large ion lithophile elements (LILEs, e.g., Rb, Sr, K, and Ba), and Pb, and depleted in high field strength elements (HFSEs, e.g., Nb, Ta, Ti, and P), resembling with the trace element patterns of middle-upper continental crust (Rudnick and Gao, 2003). In addition, they have moderate Sr (351–374 ppm) and Y (12.0–14.5 ppm) contents as well as Sr/Y (24.5–31.1) ratios.

4.2.2　Mopanshan monzogranite

The monzogranites have high contents of $SiO_2 = 68.01 - 70.30$ wt% and $K_2O = 2.24 - 3.34$ wt% as well as $Na_2O/K_2O = 1.19 - 2.26$ ratios, plotting into the field of calc-alkaline to

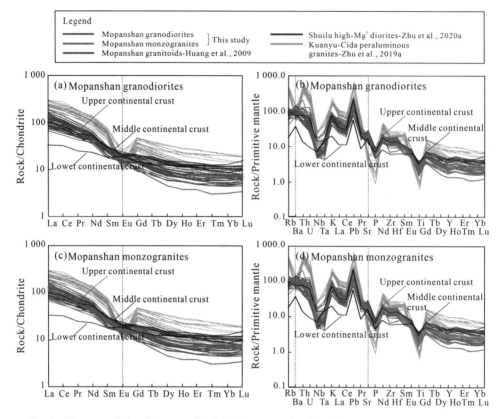

Fig. 6   Diagrams of chondrite-normalized REE patterns (a,c) and primitive mantle-normalized trace element patterns(b,d) for the Neoproterozoic Mopanshan granodiorites and monzogranites in the western Yangtze Block, South China.

The primitive mantle and chondrite values are from Sun and McDonough (1989). The lower, middle, and upper continental crust references are from Rudnick and Gao (2003). The published Mopanshan pluton is from Huang et al. (2009). The Shuilu high-Mg$^{\#}$ diorites are from Zhu et al. (2020a). The Kuanyu-Cida peraluminous granites are from Zhu et al. (2019a).

high-K calc-alkaline series (Fig. 5a). They possess relatively low $Al_2O_3 = 15.07-15.58$ wt% and $CaO = 2.10-3.35$ wt% as well as variable $Na_2O = 3.80-5.21$ wt% and A/CNK = 0.97−1.10 values, indicating the metaluminous to weakly peraluminous feature (Fig. 5b). These monzogranites contain moderate $Fe_2O_3^T = 2.53-2.94$ wt% and MgO = 0.89−1.15 wt% contents as well as similar $Mg^{\#}(45-49)$ values to the granodiorites ($Mg^{\#} = 45-50$) (Fig. 5c). All of the monzogranites also display magnesian feature with $FeO^T/(FeO^T + MgO)$ ratios of 0.69−0.72 (Fig. 5d). The monzogranites contain ΣREE concentrations ranging from 83 ppm to 174 ppm and share similar REE patterns to the granodiorites (Fig. 6c), showing the enriched LREEs $[(La/Yb)_N = 10.7-23.7]$ and slightly depleted HREEs $[(Dy/Yb)_N = 1.24-1.50]$ as well as negative Eu anomalies ($Eu/Eu^* = 0.68-0.97$). Like the granodiorites, the monzogranites also display similar trace element patterns to the middle-upper crustal compositions, which show the positive anomalies of Rb, Th, U, K, and Pb, and negative anomalies of Nb, Ta, P, and Ti (Fig. 6d). These monzogranites have low Sr contents from 183 ppm to 343 ppm and moderate Y contents from 12. 1 ppm to 15. 1 ppm, yielding moderate Sr/Y ratios of 14. 5–28. 4.

4. 3　Whole-rock Sr-Nd isotopes

Whole-rock Sr-Nd isotopic data of the granodiorites and monzogranites from the Mopanshan pluton were displayed in Fig. 7 and Table 2. All the initial $^{87}Sr/^{86}Sr$ isotopic ratios $[(^{87}Sr/^{86}Sr)_i]$ and $\varepsilon_{Nd}(t)$ values were calculated according to the LA-ICP-MS zircon U-Pb ages (ca. 820 Ma).

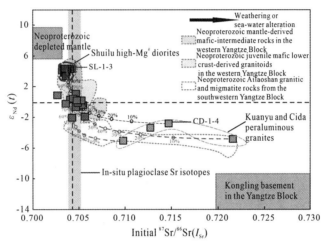

Fig. 7　Initial Sr-Nd isotopic compositions (after Zhu et al., 2019a) for the Neoproterozoic Mopanshan granodiorite-monzogranite association in the western Yangtze Block, South China.
Symbols as Fig. 5.

Both granodiorites and monzogranites of the Mopanshan pluton display relatively uniform Sr-Nd isotopic compositions (Fig. 7). The four granodiorite samples have low initial $^{87}Sr/^{86}Sr$ ratios ranging from 0. 704 736 to 0. 705 004 as well as positive and negative $\varepsilon_{Nd}(t)$ values ranging from −0. 72 to +1. 01. Their two-stage Nd model ages were calculated ranging from 1. 32 Ga to 1. 44 Ga. The six monzogranite samples yield variable initial $^{87}Sr/^{86}Sr$ ratios from 0. 702 778 to 0. 705 404 and heterogeneous $\varepsilon_{Nd}(t)$ values from −0. 29 to +0. 37 as well as old two-stage Nd model ages from 1. 34 Ga to 1. 41　Ga.

4. 4　Zircon Lu-Hf isotopic compositions

Zircon Lu-Hf isotopes for granodiorite sample (MPS-5) and monzogranite sample (MPS-2) were processed on same domains as the U-Pb spots. The initial $^{176}Hf/^{177}Hf$ ratios, $\varepsilon_{Hf}(t)$ values and two-stage depleted mantle model ages (T_{DM2}) were calculated according to zircon crystallization ages (ca. 820 Ma). The results were shown in Supplementary Table 2 and Fig. 8.

Table 2　Whole-rock Sr-Nd isotopic compositions of the Neoproterozoic Mopanshan granodiorite-monzogranite association in the western Yangtze Block, South China.

Sample	$^{87}Sr/^{86}Sr$	2SE	Sr/ppm	Rb/ppm	$^{143}Nd/^{144}Nd$	2SE	Nd/ppm	Sm/ppm	T_{DM2}/Ga	$\varepsilon_{Nd}(t)$	$(^{87}Sr/^{86}Sr)_i\,(I_{Sr})$	$(^{143}Nd/^{144}Nd)_i$
Granodiorite												
MPS-5-1	0.711 007	0.000 003	357	65.9	0.512 131	0.000 003	22.3	4.03	1.44	-0.72	0.704 749	0.511 543
MPS-5-2	0.711 141	0.000 006	359	65.0	0.512 166	0.000 004	21.5	3.93	1.40	-0.17	0.705 004	0.511 572
MPS-4-1	0.710 348	0.000 005	373	61.6	0.512 174	0.000 004	25.1	4.19	1.32	1.01	0.704 751	0.511 632
MPS-4-2	0.710 297	0.000 006	373	61.2	0.512 148	0.000 004	24.2	4.12	1.37	0.28	0.704 736	0.511 595
Monzogranite												
MPS-3-1	0.712 841	0.000 005	331	85.3	0.512 150	0.000 003	22.3	3.85	1.38	0.18	0.704 111	0.511 589
MPS-3-5	0.713 420	0.000 006	327	86.5	0.512 131	0.000 002	24.0	3.92	1.36	0.37	0.704 435	0.511 599
MPS-2-2	0.717 212	0.000 003	218	87.3	0.512 134	0.000 002	19.6	3.43	1.41	-0.29	0.703 629	0.511 565
MPS-2-3	0.716 054	0.000 005	189	74.0	0.512 177	0.000 003	20.3	3.5	1.34	0.75	0.702 778	0.511 619
MPS-1-1	0.712 526	0.000 002	298	62.6	0.512 150	0.000 004	25.2	4.35	1.38	0.17	0.705 404	0.511 589
MPS-1-2	0.712 835	0.000 005	278	65.2	0.512 213	0.000 004	15.2	2.92	1.38	0.17	0.704 883	0.511 589

$^{87}Rb/^{86}Sr$ and $^{147}Sm/^{144}Nd$ ratios were calculated using Rb, Sr, Sm and Nd contents analyzed by ICP-MS, referenced from DePaolo(1981).

T_{DM2} represent the two-stage model age and were calculated using present-day $(^{147}Sm/^{144}Nd)_{DM} = 0.213\ 7$, $(^{147}Sm/^{144}Nd)_{DM} = 0.513\ 15$ and $(^{147}Sm/^{144}Nd)_{crust} = 0.101\ 2$. $\varepsilon_{Nd}(t)$ values were calculated using present-day $(^{147}Sm/^{144}Nd)_{CHUR} = 0.196\ 7$ and $(^{147}Sm/^{144}Nd)_{CHUR} = 0.512\ 638$.

$$\varepsilon_{Nd}(t) = [(^{143}Nd/^{144}Nd)_S(t)/(^{143}Nd/^{144}Nd)_{CHUR}(t)-1]\times10^4.$$

$$T_{DM2} = \frac{1}{\lambda}\{1+[[(^{143}Nd/^{144}Nd)_S-((^{147}Sm/^{144}Nd)_S-((^{147}Sm/^{144}Nd)_{crust})(e^{\lambda t}-1)-(^{143}Nd/^{144}Nd)_{DM}]/[(^{147}Sm/^{144}Nd)_{crust}-(^{147}Sm/^{144}Nd)_{DM}]\}.$$

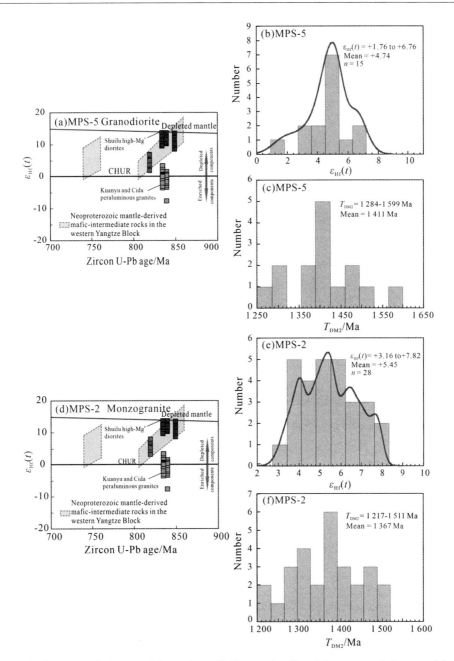

Fig. 8　Diagrams of zircon $\varepsilon_{Hf}(t)$ vs. zircon U-Pb ages (a,d) and histograms of zircon $\varepsilon_{Hf}(t)$ values (b,e) and zircon Hf model ages (c,f) for the Neoproterozoic Mopanshan granodiorite-monzogranite association in the western Yangtze Block, South China. Symbols as Fig. 5.

　　The fifteen analyses spots in zircons of granodiorite sample MPS-5 have variable $(^{176}Hf/^{177}Hf)_i$ ratios from 0.282 331 to 0.282 472. They yield variably positive $\varepsilon_{Hf}(t)$ values ranging from +1.76 to +6.76 (mean = +4.74, n = 15) and two-stage Hf model ages ranging from 1 284 Ma to 1 599 Ma (mean = 1 411 Ma)(Fig. 8a-c). Zircon grains from monzogranite sample

MPS-2 contain slightly high $(^{176}Hf/^{177}Hf)_i$ ratios (0.282 377 − 0.282 494). The twenty-eight analyses spots also have depleted Hf isotopes $[\varepsilon_{Hf}(t) = +3.16$ to $+7.82$, mean $= +5.45$, $n = 28]$ and slightly old two-stage Hf model ages ($T_{DM2} = 1\ 217 − 1\ 511$ Ma, mean $= 1\ 367$ Ma) (Fig. 8d-f).

4.5　Plagioclase EMPA data

The plagioclases of granodiorites and monzogranites from the Mopanshan pluton were selected to analyze mineral chemistry. Representative plagioclase EMPA data were listed in Supplementary Table 3 and illustrated in Figs. 9 and 10.

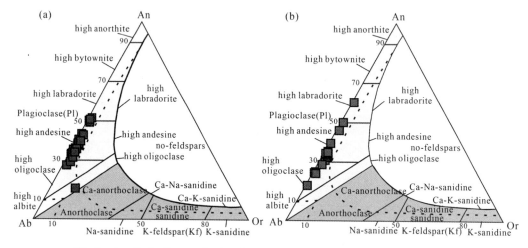

Fig. 9　An-Ab-Or diagram for the plagioclase from the Neoproterozoic Mopanshan granodiorite-monzogranite association in the western Yangtze Block, South China. Symbols as Fig. 5.

In the An-Ab-Or diagram (Fig. 9), the plagioclase crystals in the monzogranites are the oligoclase, andesite, and labradorite, which display more wider range compositions ($An_{18~60}$) than the plagioclases in granodiorites (An_{16-51}). In granodiorite sample MPS-4, the plagioclase crystal displays prominent core-rim zoning, showing obvious high calcium rim (An_{50}) (Fig. 10a,b). In monzogranite sample MPS-3, the plagioclase crystal shows sieve texture, and has "M" shape compositional variation, with a low-Ca core (An_{35}) enclosed by a dramatic high-Ca mantle (An up to 60), which is, in turn, surrounded by low-Ca rim (An_{18-25}) (Fig. 10d,e).

4.6　In-situ plagioclase Sr isotopes

The in-situ plagioclase Sr isotopes of granodiorites and monzogranites in the Mopanshan pluton were listed in Supplementary Table 4 and illustrated in Figs. 10 and 11.

All the in-situ plagioclase $(^{87}Sr/^{86}Sr)_i$ ratios are from 0.703 863 to 0.705 172, which are within the range of whole-rock $(^{87}Sr/^{86}Sr)_i$ ratios (0.702 778 − 0.705 404) (Fig. 7). The plagioclase crystal MPS-3-1 from the monzogranites displays core-mantle-rim texture and

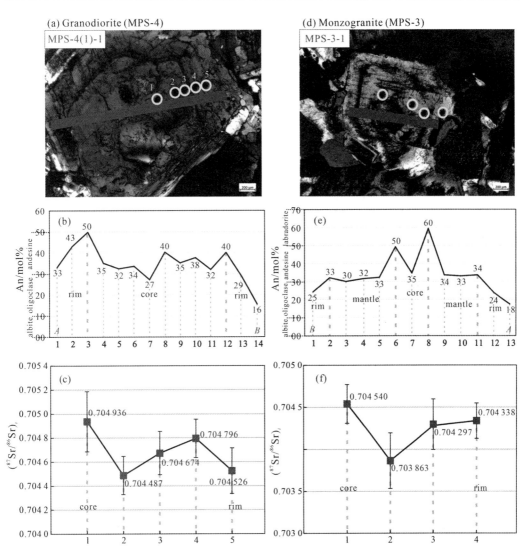

Fig. 10 An values (red lines) and in-situ Sr isotopes (yellow circles) of representative plagioclase crystals from the Neoproterozoic Mopanshan granodiorites (a-c) and monzogranites (d-e) in the western Yangtze Block, South China.

For interpretation of the references to colour in this figure legend, the reader is referred to the web version of this article.

experienced significant decrease of $(^{87}Sr/^{86}Sr)_i$ ratios from core to rim (Fig. 10f), indicating the addition of relative mafic melts. In the granodiorites, the plagioclase crystals MPS-4(1)-1, MPS-4(2)-1, and MPS-4(2)-2 also show abrupt decrease of $(^{87}Sr/^{86}Sr)_i$ ratios from core to rim (Figs. 10c and 11d, f), suggesting the periodically recharge of more mafic and depleted magma (Yan et al., 2020). The plagioclases MPS-4(1)-3 and MPS-4(2)-3 display obvious overgrowth rim and oscillatory zoning, which yield decrease $(^{87}Sr/^{86}Sr)_i$ ratios across the whole crystals (Fig. 11b, g). The rim of plagioclase crystal MPS-4(2)-4 have evident lower $(^{87}Sr/^{86}Sr)_i$ ratios than the core part (Fig. 11j).

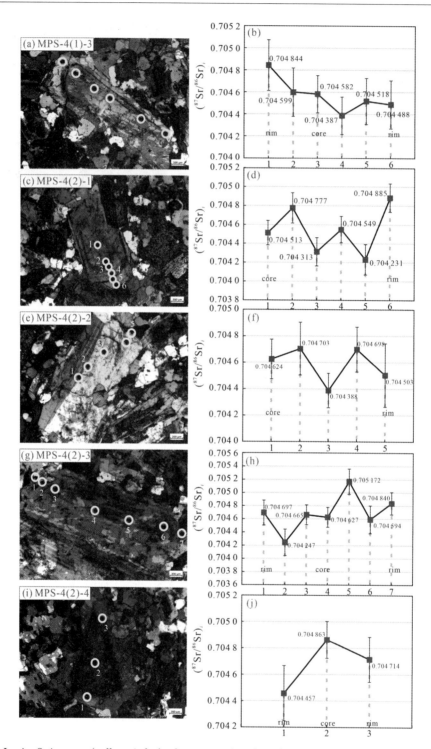

Fig. 11　In-situ Sr isotopes (yellow circles) of representative plagioclase crystals from the Neoproterozoic Mopanshan granodiorites in the western Yangtze Block, South China.

For interpretation of the references to colour in this figure legend, the reader is referred to the web version of this article.

5　Discussion

5. 1　Petrogenesis of Mopanshan granitoids

5. 1. 1　Magma mixing between mafic and felsic melts

In this study, both granodiorites and monzogranites of the Mopanshan pluton possess variable Sr (183−374 ppm) and Y (12. 0−15. 1 ppm) concentrations as well as Sr/Y ratios (14. 5−31. 1), which were partly plotted into the field of adakitic rocks (Defant and Drummond, 1990) (Fig. 12a). Based on the adakitic signatures, Huang et al. (2009) proposed that the Mopanshan pluton was formed by the partial melting of thickened continental lower crust. However, several lines of evidence cast doubt on a thickened lower crust source for the Mopanshan pluton, and lead us to support that magma mixing between relatively mafic magmas and felsic melts is more practicable.

(1) Petrological evidence. A thickened lower crust source ignores the special mineral phenomenon observed in the Mopanshan pluton that is usually ascribed to magma mixing between mafic and felsic melts (e.g., Baxter and Feely, 2002; Chen et al., 2013, 2016a,b; Jiang et al., 2018; Vernon, 2014), such as the disequilibrium texture in plagioclases, existence of acicular apatite, and presence of felsic inclusions in biotite and plagioclase crystals. Crystallization within a closed system can not lead to structural disequilibrium observed in plagioclases of the Mopanshan pluton. The injection of relatively mafic melts into felsic magmas in an open system is manifested by the existence of resorption surface and sieve-texture in plagioclase crystals (Chen et al., 2016a, b; Jiang et al., 2018; Wang et al., 2019). The core-mantle or core-mantle-rim structure, overgrowth rim, resorption surface, and sieve-texture can be detected in the Mopanshan pluton (Fig. 2a-d, e-g). The partial dissolution mantle of plagioclase reflects the reaction between silicic and relatively mafic magma (Perugini and Poli, 2012; Yan et al., 2020). In addition, the fine-grained mafic mineral phases occur at the mantle-rim connection of the plagioclase (Fig. 2a, c), implying the chemical or thermal change owing to the fresh injection of mafic compositions (Baxter and Feely, 2002). The acicular apatites within the feldspar phases in the Mopanshan pluton also indicate that the relatively mafic melts were injected into the earlier cool felsic melts at the borders of magma conduits (Vernon, 2014) (Fig. 2e).

(2) Plagioclase mineral chemistry evidence. In general, the recharge of a more mafic and depleted magma will cause higher An contents and lower $(^{87}Sr/^{86}Sr)_i$ ratios in plagioclase crystals from core to rim (Jiang et al., 2018; Yan et al., 2020). The sharp increasing An values (up to 50−60) across single plagioclase crystal of the Mopanshan pluton suggest an input of Ca-rich mafic magma into pre-exist magma system (Fig. 10b, e). All the plagioclase

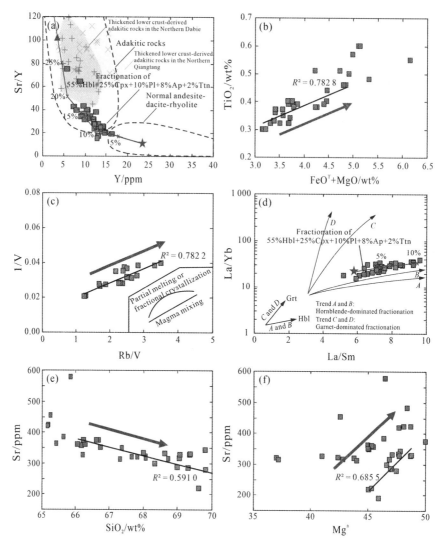

Fig. 12　Diagrams of Sr/Y vs. Y (a), TiO$_2$ vs. FeO^T+MgO (b), 1/V vs. Rb/V (c) (Schiano et al., 2010), La/Yb vs. La/Sm (d) (Chen et al., 2018), Sr vs. SiO$_2$(e), Sr vs. Mg$^\#$(f) (Chen et al., 2016a) for the Neoproterozoic Mopanshan granodiorite-monzogranite association in the western Yangtze Block, South China.

In Fig. 12a, the trend line represents residual liquids after Raleigh fractionation of hornblende (55%) + clinopyroxene (25%) + plagioclase (10%) + apatite (8%) + titanite (2%) (Ersoy and Helvaci, 2010). The partition coefficients are from Rollinson (1993). The thickened lower crust-derived adakitic rocks in the Northern Dabie and Northern Qiangtang are from Wang et al. (2007) and Lai et al. (2007), respectively. In Fig. 12d, the lines A and B represent the hornblende-dominant fractionation (Ersoy and Helvaci, 2010), the lines C and D represent the dominant fractionation crystallization of garnet and clinopyroxene (Chen et al., 2018). In Fig. 12a, d, the parental magma (purple pentagram) is assumed to have Y = 23.5 ppm, Sr/Y = 11.5, La/Yb = 21.5, and La/Sm = 5.85, which represent the magma mixing results of ~40% ancient crust-derived felsic melts (i.e., Kuanyu-Cida peraluminous granites) and ~60% metasomatized mantle-derived mafic melts (i.e., Shuilu high-Mg$^\#$ diorites). Symbols as Fig. 5. For interpretation of the references to colour in this figure legend, the reader is referred to the web version of this article.

crystals in the Mopanshan pluton underwent sudden decrease of $(^{87}Sr/^{86}Sr)_i$ ratios from core to rim (Figs. 10 and 11), which demonstrate the subsequent injection of relatively mafic magma in an open system.

(3) Major element evidence. $Mg^#$ value is a useful parameter for discriminating purely crust-derived magma from those that have involved the mantle-derived component. Experimental studies proposed that the dehydration melting of basalts in the lower crust are generally characterized by low $Mg^#$ values, regardless of the degrees of partial melting (Sen and Dunn, 1994; Rapp and Watson, 1995). Therefore, the high $Mg^#$ (45−50 > 40) values of the Mopanshan pluton are in stark contrast to those of experimental melts derived from purely crustal rocks at the same silica contents (Fig. 5c), which indicate that a sole thickened mafic lower crust source is unreasonable, and the addition of high-Mg lithospheric mantle-derived mafic magma is required (Chen et al., 2013, 2016a; Ma et al., 2019). Previous studies have suggested that the increasing Fe + Mg contents would be expected due to the basification of relatively felsic magma (Farner et al., 2014). In this study, the Mopanshan pluton displays positive correlation of $FeO^T + MgO$ and TiO_2 contents (Fig. 12b), which also in favor of magma mixing between mantle-derived mafic magma and crust-derived felsic melts (Clemens and Stevens, 2012). In addition, the K_2O increases with increasing SiO_2 contents (Fig. 5a), suggesting the involving of potassium-rich crustal compositions in magmatic system.

(4) Trace element evidence. It is notable that the Mopanshan pluton overlaps the adakitic rocks and non-adakitic rocks fields (Fig. 12a), which is different from those of typical thickened lower crust-derived adakitic rocks (e.g., Lai et al., 2007; Wang et al., 2007). A thickened lower crust source is inconsistent with the fact that the trace element patterns of the Mopanshan pluton are similar to middle-upper crustal trends (Fig. 6). The Nb/Ta ratios of the Mopanshan pluton (8.84−11.3, average value = 10.1) is obviously higher than the average value of lower crust (8.3) (Sun and McDonough, 1989), ruling out the possibility of thickened mafic lower crust source. Furthermore, Schiano et al. (2010) suggested that magma mixing usually leads to a positive linear relationship in diagrams of $1/C^{compatible\ element}$ vs. $C^{incompatible\ element}/C^{compatible\ element}$, whereas partial melting or fractional crystallization produce a hyperbolic curve on these diagrams. As shown in the 1/V vs. Rb/V diagram (Fig. 12c), the Mopanshan pluton investigated here displays a positively linear correlation ($R^2 = 0.7822$), further pointing out the magma mixing process.

(5) Isotopic evidence. A simple derivation from thickened lower crust is contradictory with the fact that the Mopanshan pluton displays positive and negative whole-rock $\varepsilon_{Nd}(t)$ values (−0.72 to +1.01), which is commonly agree with mixing of heterogeneous source or different batches magmas in an open system (Fig. 7). The maximum zircon $\varepsilon_{Hf}(t)$ (+1.76 to +7.82, up to +7.82) values of the Mopanshan pluton is comparable with those of the Neoproterozoic

metasomatized mantle-derived mafic-intermediate rocks (Fig. 8a,d) (Zhao et al., 2018), which support the addition of mantle-derived magma. Furthermore, zircons from more felsic monzogranites show slightly higher $\varepsilon_{Hf}(t)$ values than less evolved granodiorites (Fig. 8), indicating the later replenishment of more mafic compositions from depleted source during magma mixing process. On the whole, the whole-rock Sr-Nd isotopes and zircon Hf isotopes of the Mopanshan pluton are intermediate between those of Neoproterozoic metasomatized mantle-derived mafic-intermediate rocks and ancient crust-derived granitoids (Figs. 7 and 8), further supporting a hybrid origin (Jiang et al., 2018).

A sole thickened lower crust source fails to explain the mineral-scale structural and compositional disequilibrium as well as special geochemical signatures as listed above, we therefore suggest that the magma mixing involving mantle-derived mafic and crust-derived felsic melts should be supported for the formation of the Mopanshan pluton.

As mentioned above, both granodiorites and monzogranites (our samples and published data) of the Mopanshan pluton are characterized by variable concentrations of Sr (183 − 577 ppm) and Y (6. 06 − 16. 6 ppm) as well as Sr/Y ratios (14. 5 − 75. 0) (Huang et al., 2009; This study), which were partly plotted into the field of adakitic series (Fig. 12a). Because the subducting oceanic slab and thickened lower crust sources for the Mopanshan pluton have been discarded (Huang et al., 2009; This study), we suggest that the adakitic signatures observed in the Mopanshan pluton were attributed to a process involving the magma mixing and subsequent fractional crystallization of ferromagnesian mineral phases, resembling with the adakite-like rocks elsewhere under both arc and intra-continental setting (e. g., Chiaradia, 2009; Chen et al., 2013; 2016a; Ma et al., 2019; Meng et al., 2019; Sun et al., 2018). Because some ferromagnesian minerals (e.g., hornblende) and accessory minerals (e. g., apatite) have low D_{Sr} values and high D_Y values (Rollinson, 1993), the fractional crystallization of these minerals could result in high Sr and low Y concentrations, magnifying the adakitic geochemical signatures for residual melts (Chen et al., 2013; Ma et al., 2019). As shown in the modeling of Sr/Y vs. Y (Fig. 12a), the Mopanshan pluton follows a continuous variation trend from low Sr/Y and high Y concentrations (non-adakitic field) to high Sr/Y ratios and low Y concentrations (adakitic field), reflecting the fractional crystallization of 55% hornblende + 25% clinopyroxene + 10% plagioclase + 8% apatite + 2% titanite. The hornblende preferentially incorporates the middle rare earth elements (MREEs, e.g., Sm, Eu, Gd, Tb, Dy, and Ho) (Rollinson, 1993), both granodiorites and monzogranites of the Mopanshan pluton display moderately fractionated LREE patterns relative to MREE $[(La/Sm)_N = 3. 80-5. 48]$ (Fig. 6a,c), thus indicating the fractionation of MREE-enriched minerals (e.g., hornblende, clinopyroxene, and apatite) (Ma et al., 2019). The hornblende-predominant crystallization fractionation can also be evidenced by slightly positive correlation between La/Sm and La/Yb ratios (Chen et al., 2018) (Fig. 12d). Moreover, the variable Eu anomalies (Eu/Eu* = 0. 68-0. 97) (Fig. 6a, c), coupled with linear relationship between SiO$_2$ and Sr (Fig. 12e), suggest the involvement of plagioclase fractionation.

To sum up, we propose that the Neoproterozoic Mopanshan pluton was formed by the magma mixing of mafic and felsic melts. The subsequent hornblende-dominated fractional crystallization is responsible for the adakitic signatures (i.e., highly variable Sr and Y contents as well as Sr/Y ratios) observed in the Mopanshan pluton.

5. 1. 2　Source characteristics of mafic and felsic end-members

Macroscale evidence lead us to support that the Mopanshan pluton was predominantly generated by magma mixing between relatively mafic magma and felsic melts. Therefore, we present the following discussions about the characteristics of mafic and felsic end-members.

The Mopanshan pluton has positive $\varepsilon_{Nd}(t)$ values (up to +1. 01) as well as high $\varepsilon_{Hf}(t)$ (up to +7. 82) and Mg$^{\#}$(up to 50) values, pointing out the incorporation of mantle-derived mafic melts. The positive relationship between Sr concentrations and Mg$^{\#}$ values in intermediate-felsic rocks imply that the enriched LILEs (e.g., Sr) were inherited from parental mafic melts sourced from a lithospheric mantle that had been metasomatized by subducted slab-derived components (Chen et al., 2013, 2016a). This is consistent with the fact that the Mopanshan pluton displays increasing Mg$^{\#}$ values accompanied with increasing Sr concentrations (Fig. 12f). We therefore suggest that the Shuilu high-Mg$^{\#}$ diorites in the west Yangtze Block can be considered as a suitable mafic end-member (Zhu et al., 2020a). These high-Mg$^{\#}$ diorites contain moderate SiO$_2$(57. 08-61. 12 wt%), high Mg$^{\#}$(56-60), $\varepsilon_{Nd}(t)$ (+3. 26 to +4. 26), and $\varepsilon_{Hf}(t)$ (+8. 43 to +13. 6) values, and have been interpreted to be derived from the depleted mantle source metasomatized by subducting fluids and sediment melts without significant crustal contamination (Zhu et al., 2020a).

The Mopanshan pluton displays a slight evolution trend from calc-alkaline to high-K calc-alkaline series (Fig. 5a), indicating the increasing incorporation of potassium-rich crustal compositions. Most of the Mopanshan granodiorites and monzogranites are high-SiO$_2$ peraluminous rocks (A/CNK = 0. 99-1. 10) (Fig. 5a, b), also suggesting the involvement of evolved, high-K crust-derived melts. The more felsic monzogranites in the Mopanshan pluton contain variable contents of molar Al$_2$O$_3$/(MgO + FeO^T) and molar CaO/(MgO + FeO^T), Al$_2$O$_3$/(FeO^T + MgO + TiO$_2$), Al$_2$O$_3$ + FeO^T + MgO + TiO$_2$, Rb/Ba, Rb/Sr, which display a trend from basaltic melts to metagreywacke melts, thus indicating the addition of crust-derived components (Fig. 13). The involvement of evolved crustal compositions can also be supported by the similar trace element patterns to middle-upper crustal compositions (Rudnick and Gao, 2003) (Fig. 6) as well as negative whole-rock $\varepsilon_{Nd}(t)$ values (Fig. 7) and old two-stage Hf model ages （up to 1 599 Ma） （Fig. 8c,f）. Therefore, we propose that the Kuanyu-Cida peraluminous granites from the western Yangtze Block, which have high SiO_2 （66.9 – 75.6 wt%） and A/CNK （1.04–1.18） values, low $Mg^\#$（17–33）, $\varepsilon_{Nd}(t)$ （−5.1 to −2.9）, and $\varepsilon_{Hf}(t)$ （−7.75 to +3.31） values, can be taken as a presumed felsic end-member candidate （Zhu et al., 2019a）. These peraluminous granites were originated from heterogeneous metasedimentary in ancient crustal source without addition of mantle-derived components （Zhu et al., 2019a）.

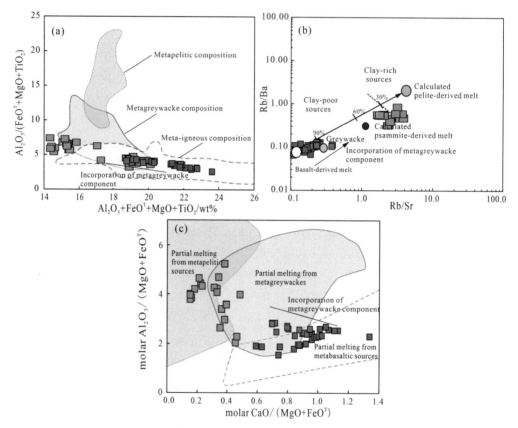

Fig. 13　Diagrams of $Al_2O_3/(FeO^T+MgO+TiO_2)$ vs. $(Al_2O_3+FeO^T+MgO+TiO_2)$ （a）, Rb/Ba vs. Rb/Sr （b） （Patiño Douce, 1999）, and molar $Al_2O_3/(MgO+FeO^T)$ vs. molar $CaO/(MgO+FeO^T)$ （c）

（Altherr et al., 2000） for the Neoproterozoic Mopanshan granodiorite-monzogranite association

in the western Yangtze Block, South China.

Symbols as Fig. 5.

Based on the hypothetical mafic and felsic end members as mentioned above, we process a classic major elements two end-members mixing modeling follow the equation of $C_m = C_a(1 - x) + C_b x$ （Słaby and Martin, 2008）. The C_m represents the contents of hybrid magma, and C_a and C_b represent the compositions of felsic and mafic end-members, respectively, x is the mass fraction of felsic end-member. In this study, the average compositions of the metasomatized mantle-derived Shuilu High-Mg$^{\#}$ diorites were used as mantle-derived mafic end-member and the average contents of the ancient crust-derived Kuanyu-Cida peraluminous granites as felsic end-member. The average major element contents of the Mopanshan granodiorites, monzogranites, granodiorites + monzogranites, plot on a straight line that pass through the origin (Fig. 14), suggesting that the Mopanshan pluton can be produced by the magma mixing of felsic melts and relatively mafic melts. The mass-balance calculations suggest that the proportions of felsic end-member are ~38% for the Mopanshan pluton (Fig. 14c). In addition, the magma mixing of metasomatized mantle- and ancient crust-derived melts can be modeled by whole-rock Sr-Nd isotopes, indicating the ~30%-45% ancient crust-derived contributions for the Mopanshan pluton (Fig. 7). Therefore, all the model calculations support that the magma mixing between mafic mantle-derived mafic and ancient crust-derived felsic melts is reasonable for the formation of the Mopanshan pluton.

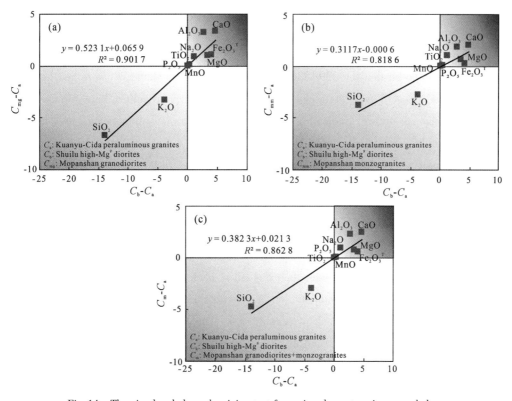

Fig. 14　The simple whole-rock mixing test for major elements using mass-balance calculation of equation of $C_m = C_a(1-x) + C_b x$ (Słaby and Martin, 2008).

The major elements of the Mopanshan pluton plot on a straight line that pass through the origin, indicating that they were formed by magma mixing process of modeled end-member melts (c). Some of whole-rock major elements slightly scatter may be caused by subsequent crystal-liquid fractionation or more complex mixing behavior (Jiang et al., 2018). The detailed descriptions are shown in the text.

5. 2　Petrological and geodynamic implications

As demonstrated from petrological, mineralogical, and geochemical evidence, the Mopanshan pluton may not represent the typical adakitic rocks originated from a thickened lower crust source. In contrast, both granodiorites and monzogranites of the Mopanshan pluton are products of magma mixing of mafic and felsic melts, and subsequently underwent fractional crystallization of ferromagnesian mineral phases. We thus propose that a combined study including petrological observation and mineralogical chemistry as well as whole-rock geochemistry is significant before concluding on magmatic source and geodynamic implication during the study of Neoproterozoic igneous rocks in the western Yangtze Block.

Voluminous igneous, metamorphic, sedimentary, geophysical evidence have supported that the Neoproterozoic widespread magmatism along the margins of the Yangtze Block were formed under long-lived subduction setting (e.g., Gao et al., 2016; Li et al., 2018a, 2019, 2021a,b; Sun et al., 2008, 2009; Wang et al., 2016; Zhao et al., 2018). Neoproterozoic tectonic transition process from subduction to retreat of oceanic slab has also been proposed. Hu et al. (2020) proposed that the 839−836 Ma rhyolites in the Shimian area were geochemically standing for the oldest A-type granitoids so far. These A_2-type rhyolites were formed in the back-arc extensional setting, indicating that the Neoproterozoic subduction-related back-arc extension probably initiated at ca. 840 Ma. Cui et al. (2015) suggested that the 809−796 Ma Mianning diabases were formed by magma mixing between lithospheric mantle and deep asthenosphere mantle sources under the slab tear or break-off background. In addition, the subsequent 810−750 Ma A_2-type granites further demonstrate the occurrence of subduction-induced back-arc extension as well as slab tearing and breakoff (Zhao et al., 2008, 2018; Zhu et al., 2019b). In this study, the Mopanshan pluton was formed at ca. 820 Ma, in a geochronological view, which was later than the initial back-arc extension (ca. 840 Ma) (Hu et al., 2020). As the continuous subduction of Neoproterozoic oceanic slab, the gravitational sinking would have resulted rollback of oceanic slab. The resultant back-arc extension could lead to asthenosphere upwelling, and back-arc extension-induced mantle melting act as a precursor for later magmatism in the western Yangtze Block (Hu et al., 2020; Zhao et al., 2018). We compiled the evolution of whole-rock Nd and zircon Hf isotopes of Neoproterozoic distinct source igneous rocks as shown in Fig. 15. The partial melting of Neoproterozoic metasomatized lithospheric mantle source in the western Yangtze Block began at ca. 870 Ma during the early stage of subduction process, which can be demonstrated by positive zircon $\varepsilon_{Hf}(t)$ values (up to +20.6) and whole-rock $\varepsilon_{Nd}(t)$ values (up to +6.0) of ca. 870−850 Ma hornblende gabbros, gabbros, gabbro-diorites, and diorites (Fig. 15a,d) (e.g., Zhao et al., 2019; Zhu et al., 2021a). At the same time, the significant melting of juvenile and ancient crustal sources have not happened. Therefore, there is a stage of hot mantle and cold crust at ca. 870−850 Ma (Fig. 15). After ca. 850−840 Ma, the western Yangtze Block entered into the stage of hot mantle and hot crust (Fig. 15). Apart from the melting of metasomatized mantle source, the widespread ancient and juvenile crustal sources were heated and melted by underplated mantle-derived basaltic magma, which can be evidenced by high and variable SiO_2 contents as well as positive and negative whole-rock $\varepsilon_{Nd}(t)$ (up to −6.38) and zircon $\varepsilon_{Hf}(t)$ values (up to −9.36) observed in intermediate-felsic rocks (Fig. 15b-c, e-f) (e.g., Zhu et al., 2019a, 2021b). In this regard, the underplating of voluminous high-temperature mantle-derived mafic magmas under back-arc extension background can provide heat, and trigger

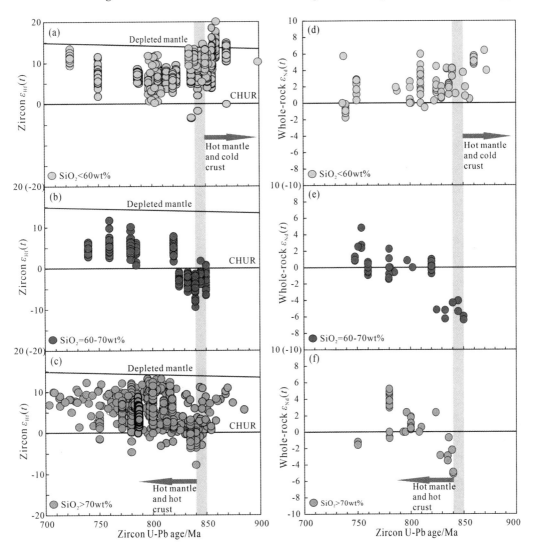

Fig. 15 The comparison of zircon Hf isotopes and whole-rocks Nd isotopes for the Neoproterozoic igneous rocks in the western Yangtze Block, South China (after Lai et al., 2022).

large-scale ancient and juvenile crustal melting and extensive crust-mantle interaction, resulting the formation of hybrid magmas parental to the Mopanshan pluton studied here and many other Neoproterozoic mafic dykes- mafic enclaves-diorites-tonalites-granodiorites- monzogranites association in the Kangding-Luding-Shimian areas along the western margin of the Yangtze Block (Our data in preparation). We therefore support that the subduction-related back-arc extension plays an important role into intensive intra-crustal melting and crust-mantle interaction in the western Yangtze Block during the Neoproterozoic. The crust-mantle interaction is significant for the petrological and geochemical diversity of Neoproterozoic granitoids.

6　Conclusion

Comprehensive mineralogical and geochemical studies suggest that a model of partial melting of thickened continental lower crust is not pertinent for the Neoproterozoic Mopanshan pluton. Both granodiorites and monzogranites of the Mopanshan pluton were generated by magma mixing between metasomatized mantle-derived mafic melts and ancient crust-derived felsic melts. Similar to those of adakite-like granitoids elsewhere in both arc and intra- continental setting, the fractionation of hornblende-dominated phases after magma mixing process contributed to the adakitic signatures observed in the Mopanshan pluton. In conjunction with regional geodynamic setting and published Nd-Hf isotopes of Neoproterozoic igneous rocks, we propose that Neoproterozoic subduction-related back-arc extension induced the intensive intra-crustal melting and crust-mantle interaction, resulting into the parental magmas of the Mopanshan pluton and other Neoproterozoic igneous rocks in the western Yangtze Block. The petrological and geochemical diversities of Neoproterozoic magmatism were in relationship with the intensive crust-mantle interaction in a back-arc extension setting.

Acknowledgements　We would like to thank the editorial working and language polish of Prof. Meifu Zhou, which significantly improve the English expression and struct of this manuscript. We also acknowledge the constructive comments from Prof. Huan Li and another anonymous reviewer. This work was jointly supported by the National Natural Science Foundation of China (Grant Nos. 41421002 and 41772052), the Project funded by China Postdoctoral Science Foundation (Grant Nos. 2021M702647), and the program for Changjiang Scholars and Innovative Research Team in University (Grant IRT1281).

Appendix A. Supplementary data　Supplementary data to this article can be found online at https://doi.org/10.1016/j.jseaes.2022.105227.

References

Altherr, R., Holl, A., Hegner, E., Langer, C., Kreuzer, H., 2000. High-potassium, calc-alkaline I-type plutonism in the European Variscides: Northern Vosges (France) and northern Schwarzwald (Germany). Lithos, 50(1-3), 51-73.

Annen, C., Blundy, J., Sparks, R.S.J., 2006. The Genesis of Intermediate and Silicic Magmas in Deep Crustal Hot Zones. Journal of Petrology, 47, 505-539.

Baxter, S., Feely, M., 2002. Magma mixing and mingling textures in granitoids: Examples from the Galway Granite, Connemara, Ireland. Mineralogy and Petrology, 76, 63-74.

Chen, B., Jahn, B.M., Suzuki, K., 2013. Petrological and Nd-Sr-Os isotopic constraints on the origin of high-Mg adakitic rocks from the North China Craton: Tectonic implications. Geology, 41(1), 91-94.

Chen, C.J., Chen, B., Li, Z., Wang, Z.Q., 2016a. Important role of magma mixing in generating the Mesozoic monzodioritic-granodioritic intrusions related to Cu mineralization, Tongling, East China: Evidence from petrological and in situ Sr-Hf isotopic data. Lithos, 248-251, 80-93.

Chen, M., Sun, M., Buslov, M.M., Cai, K.D., Zhao, G.C., Kulikova, A.V., Rubanova, E.S., 2016b. Crustal melting and magma mixing in a continental arc setting: Evidence from the Yaloman intrusive complex in the Gorny Altai terrane, Central Asian Orogenic Belt. Lithos, 252-253, 76-91.

Chen, S., Niu, Y.L., Xue, Q.Q., 2018. Syn-collisional felsic magmatism and continental crust growth: A case study from the North Qilian Orogenic Belt at the northern margin of the Tibetan Plateau. Lithos, 308-309, 53-64.

Chiaradia, M., 2009. Adakite-like magmas from fractional crystallization and melting-assimilation of mafic lower crust (Eocene Macuchi arc, Western Cordillera, Ecuador). Chemical Geology, 265, 468-487.

Chu, Z.Y., Chen, F.K., Yang, Y.H., Guo, J.H., 2009. Precise determination of Sm, Nd concentrations and Nd isotopic compositions at the nanogram level in geological samples by thermal ionization mass spectrometry. Journal of Analytical Atomic Spectrometry, 24, 1534-1544.

Clemens, J.D., Stevens, G., 2012. What controls chemical variation in granitic magmas? Lithos, 134-135, 317-329.

Cui, X.Z., Jiang, X.S., Wang, J., Wang, X.C., Zhuo, J.W., Deng, Q., Liao, S.Y., Wu, H., Jiang, Z.F., Wei, Y.A., 2015. Mid-Neoproterozoic diabase dykes from Xide in the western Yangtze Block, South China: New evidence for continental rifting related to the breakup of Rodinia supercontinent. Precambrian Research, 268, 339-356.

Defant, M.J., Drummond, M.S., 1990. Derivation of some modern arc magmas by melting of young subducted lithosphere. Nature, 347, 662-665.

Ersoy, Y., Helvaci, C., 2010. FC-AFC-FCA and mixing modeler: A Microsoft Excel spreadsheet program for modeling geochemical differentiation of magma by crystal fractionation, crustal assimilation and mixing. Computers & Geosciences, 36(3), 383-390.

Farner, M.J., Lee, C.T.A., Putirka, K.D., 2014. Mafic-felsic magma mixing limited by reactive process: A case study of biotite-rich rinds on mafic enclaves. Earth & Planetary Science Letters, 393, 49-59.

Frost, B.R., Barnes, C.G., Collins, W.J., Arculus, R.J., Ellis, D.J., Frost, C.D., 2001. A Geochemical Classification for Granitic Rocks. Journal of Petrology, 42, 2033-2048

Gao, R., Chen, C., Wang, H.Y., Lu, Z.W., Brown, L., Dong, S.W., Feng, S.Y., Li, Q.S., Li, W.H., Wen, Z.P., Li, F., 2016. SINOPROBE deep reflection profile reveals a Neoproterozoic subduction zone beneath Sichuan basin. Earth & Planetary Science Letters, 454, 86-91.

Gao, S., Ling, W.L., Qiu, Y., Zhou, L., Hartmann, G., Simon, K., 1999. Contrasting geochemical and Sm-Nd isotopic compositions of Archean metasediments from the Kongling high-grade terrain of the Yangtze craton: Evidence for cratonic evolution and redistribution of REE during crustal anatexis. Geochimica et Cosmochimica Acta, 63(13-14), 2071-2088.

Gao, S., Yang, J., Zhou, L., Li, M., Hu, Z., Guo, J., Yuan, H., Gong, H., Xiao, G., Wei, J., 2011. Age and growth of the Archean Kongling terrain, South China, with emphasis on 3.3 Ga granitoid gneisses. American Journal of Science, 311, 153-182.

Guo, J.L., Gao, S., Wu, Y.B., Li, M., Chen, K., Hu, Z.C., Liang, Z.W., Liu, Y.S., Zhou, L., Zong, K.Q., Zhang, W., Chen, H.H., 2014. 3.45 Ga granitic gneisses from the Yangtze Craton, South China: Implications for Early Archean crustal growth. Precambrian Research, 242, 82-95.

Greentree, M.R., Li, Z.X., 2008. The oldest known rocks in south-western China: SHRIMP U-Pb magmatic crystallization age and detrital provenance analysis of the Paleoproterozoic Dahongshan Group. Journal of Asian Earth Sciences, 33, 289-302.

Hoskin, P.W.O., Schaltegger, U., 2003. The composition of zircon and igneous and metamorphic petrogenesis. Reviews in Mineralogy and Geochemistry, 53, 27-62.

Hu, P.Y., Zhai, Q.G., Wang, J., Tang, Y., Ren, G.M., Zhu, Z.C., Wang, W., Wu, H., 2020. U-Pb zircon geochronology, geochemistry, and Sr-Nd-Hf-O isotopic study of Middle Neoproterozoic magmatic rocks in the Kangdian Rift, South China: Slab rollback and backarc extension at the northwestern edge of the Rodinia. Precambrian Research, 347, 105863.

Huang, X.L., Xu, Y.G., Lan, J.B., Yang, Q.J., Luo, Z.Y., 2009. Neoproterozoic adakitic rocks from Mopanshan in the Western Yangtze craton: Partial melts of a thickened lower crust. Lithos, 112(3), 367-381.

Jagoutz, O., Klein, B., 2018. On the importance of crystallization-differentiation for the generation of SiO_2-rich melts and the compositional build-up of arc (and continental) crust. American Journal of Science, 318 (1), 29-63.

Jiang, D.S., Xu, X.S., Xia, Y., Erdmann, S., 2018. Magma mixing in a granite and related rock association: Insight from its mineralogical, petrochemical, and "reversed isotope" features. Journal of Geophysical Research: Solid Earth, 123, 2262-2285.

Lai, S.C., Qin, J.F., Li, Y.F., 2007. Partial melting of thickened Tibetan crust: Geochemical evidence from Cenozoic adakitic volcanic rocks. International Geology Review, 49, 357-373.

Lai, S.C., Qin, J.F., Zhu, R.Z., Zhao, S.W., 2015. Neoproterozoic quartz monzodiorite-granodiorite association from the Luding-Kangding area: Implications for the interpretation of an active continental margin along the Yangtze Block (South China Block). Precambrian Research, 267(3-4), 196-208.

Lai, S.C, Zhu, Y, 2020. Petrogenesis and geodynamic implications of Neoproterozoic typical intermediate-felsic magmatism in the western margin of the Yangtze Block, South China. Journal of Geomechanics, 26 (5), 759-790 (in Chinese with English abstract).

Lai, S.C., Zhu, R.Z., Qin, J.F., Zhao, S.W., Zhu, Y., 2022. Granitic magmatism in eastern Tethys domain (western China) and their geodynamic implications. Acta Geologica Sinica: English Edition, 96(2),

401-415.

Lee, C., Lee, T.C., Wu, C.T., 2014. Modeling the compositional evolution of recharging, evacuating, and fractionating (REFC) magma chambers: Implications for differentiation of arc magmas. Geochimica et Cosmochimica Acta, 143, 8-22.

Li, C., Zhou, L., Zhao, Z., Zhang, Z., Zhao, H., Li, X., 2018. In-situ Sr isotopic measurement of scheelite using fs-LA-MC-ICPMS. Journal of Asian Earth Sciences, 160 (7), 38-47.

Li, H., Zhou, Z. K., Algeo, T. J., Wu, J. H., Jiang, W. C., 2019. Geochronology and geochemistry of tuffaceous rocks from the Banxi Group: Implications for Neoproterozoic tectonic evolution of the southeastern Yangtze Block, South China. Journal of Asian Earth Sciences, 177, 152-176.

Li, J.Y., Tang, M., Lee, C., Wang, X.L., Gu, Z.D., Xia, X.P., Wang, D., Du, D.H., Li, L.S., 2021a. Rapid endogenic rock recycling in magmatic arcs. Nature Communications, 12(1), 3533.

Li, Z.M.G., Chen, Y.C., Zhang, Q.W.L., Liu, J.H., Wu, C.M., 2021. U-Pb dating of metamorphic monazite of the Neoproterozoic Kang-Dian Orogenic Belt, southwestern China. Precambrian Research, 361, 106262.

Ma, X.H., Cheng, C.J., Zhao, J.X., Qiao, S.L., Zhou, Z.H., 2019a. Late Permian intermediate and felsic intrusions in the eastern Central Asian Orogenic Belt: Final-stage magmatic record of Paleo-Asian Oceanic subduction? Lithos, 326-327, 265-278.

Meng, Y.K., Xiong, F.H., Xu, Z.Q., Ma, X.X., 2019. Petrogenesis of Late Cretaceous mafic enclaves and their host granites in the Nyemo region of southern Tibet: Implications for the tectonic-magmatic evolution of the Central Gangdese Belt. Journal of Asian Earth Sciences, 176, 27-41.

Moyen, J.F., Janoušek, V., Laurent, O., Bachmann, O., Jacob, J.B., Farina, F., Fiannacca, P., Villaros, A., 2021. Crustal melting vs. fractionation of basaltic magmas: part 1, the bipolar disorder of granite petrogenetic models. Lithos, 402-403, 106291.

Patiño Douce, A.E., 1999. What do experiments tell us about the relative contributions of crust and mantle to the origin of the granitic magmas. Geol. Soc. London Spec. Publ., 168 (1), 55-75.

Perugini, D., Poli, G., 2012. The mixing of magmas in plutonic and volcanic environments: Analogies and differences. Lithos, 153, 261-277.

Qi, H., Zhao, J.H., 2020. Petrogenesis of the Neoproterozoic low-δ^{18}O granitoids at the western margin of the Yangtze block in South China. Precambrian Research, 351, 105953.

Rapp, R.P., Watson, E.B., 1995. Dehydration melting of metabasalt at 8-32 kbar: Implications for continental growth and crust-mantle recycling. Journal of Petrology, 36(4), 891-931.

Rollinson, H., 1993. Using Geochemical Data: Evaluation, Presentation, Interpretation. London: Longman Scientific and Technical, 108-111.

Roberts, M.P., Clemens, J.D., 1993, Origin of high-potassium, calc-alkaline, I-type granitoids. Geology, 21, 825-828.

Rudnick, R.L., Gao, S., 2003. Composition of the continental crust. Treatise Geochem, 3, 1-64.

Schiano, P., Monzier, M., Eissen, J.P., Martin, H., Koga, K.T., 2010. Simple mixing as the major control of the evolution of volcanic suites in the Ecuadorian Andes. Contributions to Mineralogy and Petrology, 160, 297-312.

Sen, C., Dunn, T., 1994. Dehydration melting of a basaltic composition amphibolite at 1. 5 GPa and 2. 0 GPa: Implications for the origin of adakites. Contributions to Mineralogy and Petrology, 117, 394-409.

Słaby, E., Martin, H., 2008. Mafic and felsic magma interaction in granites: The Hercynian Karkonosze Pluton (Sudetes, Bohemian Massif). Journal of Petrology, 49, 353-391.

Sun, X., Lu, Y.J., McCuaig, T.C., Zheng, Y.Y., Chang, H.F., Guo, F., Xu, L.J., 2018. Miocene Ultrapotassic, High-Mg Dioritic, and Adakite-like Rocks from Zhunuo in Southern Tibet: Implications for Mantle Metasomatism and Porphyry Copper Mineralization in Collisional Orogens. Journal of Petrology, 29 (3), 341-386.

Sun, S.S., McDonough, W.F., 1989. Chemical and isotopic systematics of oceanic basalts: implications for mantle composition and processes. In: Saunders, A.D., Norry, M.J. Magmatism in the Ocean Basins. Geological Society, London, Special Publications, 42, 313-345.

Sun, W.H., Zhou, M.F., Yan, D.P., Li, J.W., Ma, Y.X., 2008. Provenance and tectonic setting of the Neoproterozoic Yanbian Group, western Yangtze Block (SW China). Precambrian Research, 167(s1-2), 213-236.

Sun, W.H., Zhou, M.F., Gao, J.F., Yang, Y.H., Zhao, X.F., Zhao, J.H., 2009. Detrital zircon U-Pb geochronological and Lu-Hf isotopic constraints on the Precambrian magmatic and crustal evolution of the western Yangtze Block, SW China. Precambrian Research, 172(1-2), 0-126.

Vernon, R.H., 2014. Microstructures of microgranitoid enclaves and the origin of S-type granitoids. Australian Journal of Earth Sciences, 61(2), 227-239.

Wang, K.X., Yu, C.D., Yan, J., Liu, X.D., Liu, W.H., Pan, J.Y., 2019. Petrogenesis of Early Silurian granitoids in the Longshoushan area and their implications for the extensional environment of the North Qilian Orogenic Belt, China. Lithos, 342-343, 152-174.

Wang, Q., Wyman, D.A., Xu, J., Jian, P., Zhao, Z., Li, C., Xu, W., Ma, J., He, B., 2007. Early Cretaceous adakitic granites in the Northern Dabie Complex, central China: Implications for partial melting and delamination of thickened lower crust. Geochim. Cosmochim. Acta, 71 (10), 2609-2636.

Wang, W., Zhou, M.F., 2014. Provenance and tectonic setting of the Paleo- to Mesoproterozoic Dongchuan Group in the southwestern Yangtze Block, South China: Implication for the breakup of the supercontinent Columbia. Tectonophysics, 610, 110-127.

Wang, W., Zhou, M.F., Zhao, X.F., Chen, W.T., Yan, D.P., 2014. Late Paleoproterozoic to Mesoproterozoic rift successions in SW China: Implication for the Yangtze Block-North Australia-Northwest Laurentia connection in the Columbia supercontinent. Sedimentary Geology, 309, 33-47.

Wang, X.L., Zhou, J.C., Wan, Y.S., Kitajima, K., Wang, D., Bonamici, C., Qiu, J.S., Sun, T., 2013. Magmatic evolution and crustal recycling for Neoproterozoic strongly peraluminous granitoids from Southern China: Hf and O isotopes in zircon. Earth & Planetary Science Letters, 366(2), 71-82.

Wang, Y.J., Zhou, Y.Z., Cai, Y.F., Liu, H.C., Zhang, Y.Z., Fan, W.M., 2016. Geochronological and geochemical constraints on the petrogenesis of the Ailaoshan granitic and migmatite rocks and its implications on Neoproterozoic subduction along the SW Yangtze Block. Precambrian Research, 283, 106-124.

Winter, J.D., 2014. Principles of Igneous and Metamorphic Petrology. Second Edition.

Yan, L.L., He, Z.Y., Xu, X.S., 2020. Magma recharge processes of the Yandangshan volcanic-plutonic caldera complex in the coastal SE China: Constraint from inter-grain variation of Sr isotope of plagioclase. Journal of Asian Earth Sciences, 201, 104511.

Zhao, G.C., Cawood, P.A., 2012. Precambrian geology of China. Precambrian Research, s 222-223, 13-54.

Zhao, J.H., Zhou, M.F., 2007a. Neoproterozoic Adakitic Plutons and Arc Magmatism along the western Margin of the Yangtze Block, South China. Journal of Geology, 115(6), 675-689.

Zhao, J.H., Zhou, M.F., 2007b. Geochemistry of Neoproterozoic mafic intrusions in the Panzhihua district (Sichuan Province, SW China): Implications for subduction-related metasomatism in the upper mantle. Precambrian Research, 152(1), 27-47.

Zhao, J. H., Zhou, M. F., Yan, D. P., Zheng, J. P., Li, J. W., 2011. Reappraisal of the ages of Neoproterozoic strata in South China: No connection with the Grenvillian orogeny. Geology, 39 (4), 299-302.

Zhao, J.H., Li, Q.W., Liu, H., Wang, W., 2018. Neoproterozoic magmatism in the western and northern margins of the Yangtze Block (South China) controlled by slab subduction and subduction-transform-edge-propagator. Earth-Science Reviews, 187, 1-18.

Zhao, J.H., Zhou, M.F., Wu, Y.B., Zheng, J.P., Wang, W., 2019. Coupled evolution of Neoproterozoic arc mafic magmatism and mantle wedge in the western margin of the South China Craton. Contributions to Mineralogy and Petrology, 174(4).

Zhao, T.Y., Li, J., Liu, G.C., Cawood, P.A., Zi, J.W., Wang, K., Feng, Q.L., Hu, S.B., Zeng, W.T., Zhang, H., 2020. Petrogenesis of Archean TTGs and potassic granites in the southern Yangtze Block: Constraints on the early formation of the Yangtze Block. Precambrian Research, 347, 105848.

Zhao, X.F., Zhou, M.F., Li, J.W., Wu, F.Y., 2008. Association of Neoproterozoic A- and I-type granites in South China: Implications for generation of A-type granites in a subduction-related environment. Chemical Geology, 257 (s1), 1-15.

Zhao, X. F., Zhou, M. F., Li, J. W., Sun, M., Gao, J. F., Sun, W. H., Yang, J. H., 2010. Late Paleoproterozoic to early Mesoproterozoic Dongchuan Group in Yunnan, SW China: Implications for tectonic evolution of the Yangtze Block. Precambrian Research, 182, 57-69.

Zheng, Y.F., 2021. Preface to the origin of granites and related rocks. Lithos, 402-403, 106380.

Zhou, M.F., Yan, D.P., Wang, C.L., Qi, L., Kennedy, A., 2006a. Subduction-related origin of the 750 Ma Xuelongbao adakitic complex (Sichuan Province, China): Implications for the tectonic setting of the giant Neoproterozoic magmatic event in South China. Earth & Planetary Science Letters, 248(1-2), 286-300.

Zhou, M.F., Ma, Y.X., Yan, D.P., Xia, X.P., Zhao, J.H., Sun, M., 2006b. The Yanbian Terrane (Southern Sichuan Province, SW China): A Neoproterozoic arc assemblage in the western margin of the Yangtze Block. Precambrian Research, 144, 19-38.

Zhu, W.G., Zhong, H., Li, Z.X., Bai, Z.J., Yang, Y.J., 2016. SIMS zircon U-Pb ages, geochemistry and Nd-Hf isotopes of ca. 1.0 Ga mafic dykes and volcanic rocks in the Huili area, SW China: Origin and tectonic significance. Precambrian Research, 273, 67-89.

Zhu, Y., Lai, S.C., Qin, J.F., Zhu, R.Z., Zhang, F.Y., Zhang, Z.Z., Zhao, S.W., 2019a. Neoproterozoic peraluminous granites in the western margin of the Yangtze Block, South China: Implications for the reworking of mature continental crust. Precambrian Research, 333, 105443.

Zhu, Y., Lai, S.C., Qin, J.F., Zhu, R.Z., Zhang, F.Y., Zhang, Z.Z., Gan, B.P., 2019b. Petrogenesis and geodynamic implications of Neoproterozoic gabbro-diorites, adakitic granites, and A-type granites in the southwestern margin of the Yangtze Block, South China. Journal of Asian Earth Sciences, 183, 103977.

Zhu, Y., Lai, S.C., Qin, J.F., Zhu, R.Z., Zhang, F.Y., Zhang, Z.Z., 2019c. Geochemistry and zircon U-Pb-Hf isotopes of the 780 Ma I-type granites in the western Yangtze Block: Petrogenesis and crustal evolution. International Geology Review, 61 (10), 1222-1243.

Zhu, Y., Lai, S.C., Qin, J.F., Zhu, R.Z., Liu, M., Zhang, F.Y., Zhang, Z.Z., Yang, H., 2020a. Genesis of ca. 850−835 Ma high-Mg$^{\#}$ diorites in the western Yangtze Block, South China: Implications for mantle metasomatism under the subduction process. Precambrian Research, 343, 105738.

Zhu, Y., Lai, S.C., Qin, J.F., Zhu, R.Z., Zhang, F.Y., Zhang, Z.Z., 2020b. Petrogenesis and geochemical diversity of late Mesoproterozoic S-type granites in the western Yangtze Block, South China: Co-entrainment of peritectic selective phases and accessory minerals. Lithos, 352-353, 105326.

Zhu, Y., Lai, S.C., Qin, J.F., Zhu, R.Z., Liu, M., Zhang, F.Y., Zhang, Z.Z., Yang, H., 2021a. Neoproterozoic metasomatized mantle beneath the western Yangtze Block, South China: Evidence from whole-rock geochemistry and zircon U-Pb-Hf isotopes of mafic rocks. Journal of Asian Earth Sciences, 206, 104616.

Zhu, Y., Lai, S.C., Qin, J.F., Zhu, R.Z., Zhao, S.W., Liu, M., Zhang, F.Y., Zhang, Z.Z., Yang, H., 2021b. Peritectic assemblage entrainment (PAE) model for the petrogenesis of Neoproterozoic high-maficity I-type granitoids in the western Yangtze Block, South China. Lithos, 402-403, 106247.

Episodic provenance changes in Middle Permian to Middle Jurassic foreland sediments in southeastern Central Asian Orogenic Belt: Implications for collisional orogenesis in accretionary orogens[①]

Liu Min   Lai Shaocong[②]   Zhang Da   Di Yongjun   Zhou Zhiguang

Qin Jiangfeng   Zhu Renzhi   Zhu Yu   Zhang Fangyi

Abstract: The final collisional orogenesis in accretionary orogens is crucial for reconstructing the whole orogenic circle and understanding the material and energy recycling along convergent margins. However, it is always not straightforward to be addressed, particularly when the final suturing following the ocean closure occurred softly, for instance, the Central Asian Orogenic Belt (CAOB). In this study, a comprehensive dataset of detrital zircon U-Pb ages and Hf isotopes for the Middle Permian to Middle Jurassic foreland sediments from the Linxi region, southeastern CAOB, is synthesized. We observed multiple abrupt changes in sedimentary provenance over time from this dataset: ① the ca. 275 Ma sediments (Upper Zhesi Fm.) were entirely sourced from the local NAO, whereas the ca. 270−258 Ma sediments (Middle-Upper Linxi Fm.) received an additional detritus contribution from the northern NCC, ② the ca. 257−236 Ma sediments (Lower-Middle Xingfuzhilu Fm.) returned to a pure NAO source, but the slightly postdated ca. 227 Ma sediments (Upper Xingfuzhilu Fm.) show provenance affinities of both NAO and northern NCC, and ③ the ca. 171 Ma sediments (Middle Xinmin Fm.) were exclusively derived from the Late Paleozoic to Early Mesozoic magmas in the NAO and northern NCC with no Early Paleozoic and Precambrian detritus. The above episodic shifts in sedimentary provenance observed in the Linxi region, in conjunction with other regional geological evidence, provide key constraints on the pre- to post-collisional evolution in the southeastern CAOB, including the transition from waning subduction to initial collision, subsequent slab break-off and intracontinental contraction, and the post-collisional extension before the superposition of the Paleo-Pacific tectonic regime. Our results show that understanding the spatio-temporal changes of sedimentary provenance of the foreland basins systems during the plate convergence and orogenic evolution can aid in exploring the final collisional dynamics and evolutionary history in ancient accretionary orogens.

---

① Published in *Tectonophysics*, 2022, 840.

② Corresponding author.

1 Introduction

Reconstructing the terminal collision after the complete consumption of the oceanic lithosphere in accretionary orogens is the prerequisite to understanding material recycling and energy exchange processes along the convergent margins (Condie, 2007; Cawood et al., 2009). However, it could be rather challenging if the final suturing process takes place in a "soft" manner without typical continent-continent collisional signatures (such as large-scale crustal shortening/thickening and high to ultra-high grade metamorphism; Song et al., 2014), as exemplified by the Appalachian orogen in North America (Draut and Clift, 2013). The same is true of the giant Central Asian Orogenic Belt (CAOB) that developed between the North China, Tarim, and Siberian cratons in central-east Asia (Windley et al., 2007; Cawood et al., 2009; Xiao et al., 2015; Song et al., 2015, 2021; Eizenhöfer and Zhao, 2018).

A general view is that the CAOB was built by the amalgamation of numerous orogenic components, including arcs, microcontinents, seamounts, oceanic crust fragments, and accretionary complexes during the prolonged subduction of the Paleo-Asian Ocean (PAO) from Neoproterozoic to Early Mesozoic (e.g., Windley et al., 2007; Xiao et al., 2015; Şengör et al., 2018). In the southeastern section of CAOB, the PAO ultimately closed through a soft suturing of the opposing accretionary prisms along the Solonker suture zone (SSZ) (Fig. 1a) (Sengör and Natal'in, 1996; Xiao et al., 2015; Fu et al., 2018; Song et al., 2018, 2021; Eizenhöfer and Zhao, 2018). Although the life span of this final collision has been roughly constrained to the Middle Permian to Middle-Late Triassic period (Xiao et al., 2009; Li et al., 2014a, 2017a; Eizenhöfer and Zhao, 2018; Song et al., 2021), how it evolved remains enigmatic, especially considering that different deep dynamics, including slab break-off, moderate crustal thickening, and post-orogenic delamination, might be involved (Jian et al., 2010; Li et al., 2014a, 2017a). Thus, reconstructing the last episode of the orogenic evolution in the southeastern CAOB requires new solid evidence and a more comprehensive understanding of the available information.

Generally, foreland basin systems associated with the retroarc, peripheral (or collisional), and retreating collisional settings dominate the sedimentary accumulation on convergent margins during the plate convergence and orogenic evolution (Supplementary Fig. S1) (e.g., Garzanti et al., 2007; DeCelles, 2012). Also, major geodynamic processes on convergent margins are usually invoked where abrupt changes in sedimentary provenance emerge (Cawood et al., 2012; Gehrels, 2014). Thus, evaluating the spatio-temporal provenance changes of the foreland sediments on convergent margins can aid in reconstructing the final orogenic evolution in ancient accretionary orogens. For the southeastern CAOB, in association with the pre- to post-collisional evolution, different foreland basin systems had

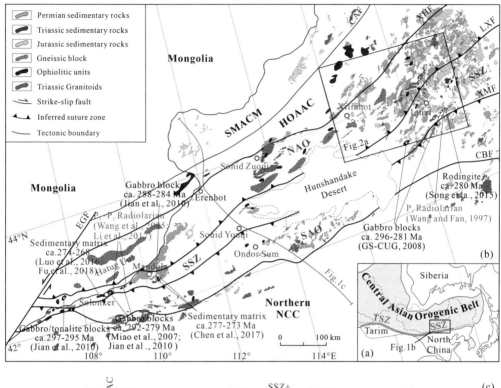

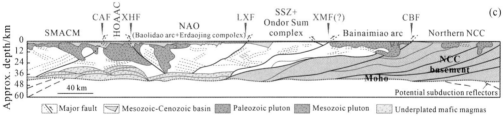

Fig. 1 (a) Sketch map of the Central Asian Orogenic Belt (CAOB) modified from Jahn (2004),
showing the locations of the TSZ (Tianshan suture zone) and SSZ (Solonker suture zone) in southern
CAOB. (b) Tectonic subdivision of the southeastern CAOB (Jian et al., 2010; Xiao et al., 2015).
(c) Structural cross section interpreted from the SinoProbe deep seismic reflection profile
across the SSZ (Zhang et al., 2014b).

SAO: southern accretionary orogen; NAO: northern accretionary orogen; HOAAC: Hegenshan ophiolite-arc-accretion
complex; SMACM: Southern Mongolian active continental margin; CBF: Chifeng-Bayan Obo fault; XMF: Xar Moron
fault; LXF: Linxi fault; XHF: Xilinhot fault; CAF: Chagan'aobao-Arongqi fault; EGF: East Gobi fault.

contributed to the clastic deposits in the frontal region of the northern accretionary orogen
(NAO), north side of the SSZ. In this paper, we focused on a comprehensive dataset of
detrital zircon U-Pb ages and Lu-Hf isotopes of the Middle Permian to Middle Jurassic
sediments from the Linxi region of the frontal NAO (Fig. 1b), and identified episodic changes
in sedimentary provenance from ca. 275 Ma to ca. 171 Ma. These results, together with other
regional geological evidence, provide significant insights into the last episode of orogenic
evolution in the southeastern CAOB.

2　Geological background

From south to north, the southeastern segment of CAOB is composed of several nearly ENE-trending orogenic components, namely the northern margin of North China craton (NCC), southern accretionary orogen (SAO), SSZ, NAO, Hegenshan ophiolite-arc-accretion complex, and Southern Mongolian active continental margin (Fig. 1b, c) (Jian et al., 2008, 2010; Eizenhöfer et al., 2015; Xiao et al., 2015). Anatomically, the NAO and SAO constitute a nearly N-S symmetrical geometry with the SSZ serving as the axis (Fig. 1b). The NAO primarily comprises the Paleozoic Baolidao arc belt with entrainment of some Precambrian basement rocks (Xilinhot Block) (Chen et al., 2000; Jian et al., 2008; Y. Li et al., 2017a). The Baolidao arc includes the ca. 498−420 Ma and ca. 382−274 Ma middle-K to high-K plutonic rocks (Chen et al., 2000; Jian et al., 2008; Eizenhöfer and Zhao, 2018), and is interpreted to be generated by the prolonged northward subduction of the PAO beneath the Southern Mongolian active continental margin (e.g., Xiao et al., 2015). The Hutag Uul terrane in Mongolia (Fig. 1b), which is considered to be the western continuation of the NAO (Badarch et al., 2002; Jian et al., 2010), also contains lots of Late Paleozoic subduction-related felsic plutons and volcanic-clastic rocks, with a rich assemblage of shallow-marine fossils (Lamb and Badarch, 1997; Badarch et al., 2002). Similarly, a large-scale Early Paleozoic arc belt (called the Bainaimiao arc, ca. 475−417 Ma) has also been documented in the SAO, it was initially built on a Precambrian microcontinent with Tarim affinity and was ultimately accreted onto the northern NCC by the southward subduction of the PAO by the latest Silurian (Zhang et al., 2014a; Eizenhöfer and Zhao, 2018; Ma et al., 2019; Liu et al., 2020). Besides, Extensive Late Paleozoic arc magmatic intrusions exist along the northern margin of the NCC, demonstrating intense southward subduction of the PAO during the Late Paleozoic (e.g., Zhang et al., 2009, 2010).

The SSZ, which strikes eastward from Solon Obo in the northern Inner Mongolia to Changchun and Yanji areas in NE China, is widely interpreted as the location where the PAO ultimately closed in the southeastern CAOB (e.g., Sengör and Natal'in, 1996; Jian et al., 2010; Xiao et al., 2003, 2015; Li et al., 2017a; Fu et al., 2018; Eizenhöfer and Zhao, 2018; Song et al., 2021). The SSZ defines a "soft collision" between the opposing Erdaojing and Ondor Sum subduction-accretion complexes without deep continental subduction, large-scale crustal shortening/thickening, and high-grade metamorphism typical of continent-continent collision (Song et al., 2015; Eizenhöfer et al., 2015; Fu et al., 2018; Eizenhöfer and Zhao, 2018). The Erdaojing subduction-accretion complex stacked by the northward subduction of the PAO is composed of diverse blocks of the ophiolitic fragment, tectonic mélange, and metamorphic rock that encapsulated (as faulted lenses mostly) in the foliated

Late Paleozoic tuffaceous sandstones and schists (Xiao et al., 2003; Liu et al., 2013; Fu et al., 2018). The Ondor Sum subduction-accretion complex, which recorded the multi-staged southward subduction of the PAO beneath the northern NCC (e.g., Zhang et al., 2009, 2014a; Xiao et al., 2015; Eizenhöfer and Zhao, 2018), dominantly consists of the ophiolitic fragments, mélanges, and metamorphic rocks that juxtaposed in a south-directed thrusting stack system (de Jong et al., 2006; Jian et al., 2008; Y. Li et al., 2014a,b; Xiao et al., 2015). Ophiolitic mélanges in the Solonker region, western part of the SSZ, typically show a block-in-matrix fabric with variable-sized basalts, cherts, sandstones, and ultramafic-mafic blocks that are embedded in a fine-grained clastic to foliated siliceous-argillaceous matrix (e.g., Fu et al., 2018). Geochronological investigations indicate that the blocks are mostly Early Permian in age (Miao et al., 2007; Jian et al., 2010; Song et al., 2015; Fu et al., 2018), consistent with the formation ages of the radiolarian fossils along the SSZ (Wang and Fan, 1997; Wang et al., 2005; Li et al., 2017a-d), as well as the maximum depositional ages of the sedimentary matrix of the ophiolitic mélanges in the Solonker-Mandula region (Fig. 1b) (Luo et al., 2016; Chen et al., 2017; Fu et al., 2018).

In recent years, some collision-related magmatic rocks have been identified along the SSZ and its adjacent areas. For instance, Jian et al. (2010) reported that the ca. 255–248 Ma adakite-sanukitoid association in the Mandula region was formed by decompression melting of the metasomatized mantle peridotite during the asthenosphere upwelling caused by the slab break-off following the final collision, similar to the genesis of the Plio-Quaternary post-collisional basaltic to dacitic magmas in the southern Philippines (Sajona et al., 2000). Also, S. Li et al. (2016a, 2017) suggested that the sublinear-distributed ca. 255–251 calc-alkaline granitoids along the southern Xilinhot area were intrinsically magmatic responses to the above-mentioned slab break-off phase, but the slightly posted high Sr/Y granitoids (ca. 251–245) in Linxi region were resulted from the following intracontinental contraction process. The Middle Triassic high-Mg andesites in Linxi region may have recorded the melting of the fossil PAO slab in upper mantle after the collision (Liu et al., 2012). In addition, extensive intermediate to low P-T metamorphism correlated with the final closure of the PAO basin are also documented along the SSZ (Zhang et al., 2016a).

3 Pre- to post-collisional sediments in the Linxi region and samples

The intense bidirectional subduction of the PAO since Carboniferous finally led to the "soft" collision of the northern and southern accretionary complexes (wedges) along the SSZ in the southeastern CAOB in the latest Permian to Early-Middle Triassic (e.g., S. Li et al., 2016a; Eizenhöfer and Zhao, 2018). In this regard, different foreland basin systems, including retroarc, collisional, and retreating collisional ones, should have successively developed on the frontal regions of both NAO and SAO during the convergent and following final suturing along the SSZ during this period. For the frontal NAO, sediments from the above different foreland basin systems are expected to show distinctive provenance signatures. For example, the forearc basin sediments from the Early Permian retroarc basin system in the frontal NAO should lack the detritus typical of the northern NCC because of the PAO barrier. However, the northern NCC detritus can be transported to the latest Permian to Early-Middle Triassic collisional (peripheral) foreland basin system of the frontal NAO when the final collision occurred. Furthermore, major changes in deep dynamics, such as slab break-off and crustal thickening, may also result in significant provenance changes in the peripheral foreland basin in the frontal NAO. The Linxi region in the frontal NAO preserves a relatively complete outcrop of the Middle Permian to Early Mesozoic strata (Fig. 2a), including the Middle Permian Zhesi formation (Fm.), Upper Permian Linxi Fm., Triassic Xingfuzhilu Fm., and Middle Jurassic Xinmin Fm. from bottom to top (Fig. 2a, b). These units show various rock assemblages and structures and are interpreted to be progressively deposited on the northern SSZ in association with the pre- to post-collisional evolution in southeastern CAOB (Han et al., 2012, 2015; Li et al., 2014a; Eizenhöfer et al., 2014, 2015; Zhu and He, 2017; Eizenhöfer and Zhao, 2018). Thus, the provenances of the above Middle Permian to Middle Jurassic sediments in the Linxi region provides important information for reconstructing final collisional orogenesis in the southeastern CAOB.

The Middle Permian Zhesi Fm. sediments in the Linxi region is dominated by interbeds of bioclastic limestones, clastic and siliceous rocks in variable thickness, and massive sandstones, conglomerates, and tuffaceous siltstone interbedded with mudstones, overall consistent with a neritic to littoral sedimentation (BGMRIM, 1996; Han et al., 2012). In most cases, the Zhesi Fm. comfortably overlies the Lower Permian Dashizhai Fm. marine strata (Supplementary Fig. S2), yet a thin-bedded volcanic gravel layer can be occasionally observed between them, which may be suggestive of a short sedimentary hiatus (BGMRIM, 1996; Bao et al., 2006). Together with the fact that most outcrops of the Middle Permian Zhesi Fm. in this region are tectonically located between the Baolidao arc belt and the SSZ (Fig. 2a), in accord with a forearc position, it is likely that the Middle Permian Zhesi Fm. sediments in the Linxi region were deposited in the forearc basin of the retroarc setting associated with the northward subduction of the PAO.

The Upper Permian Linxi Fm. uncomfortably overlies the Zhesi Fm. in the Linxi region (Fig. 2a). It is overall characterized by dark-gray slates interbedded with terrestrial sandstones, conglomerates, and siltstones (BGMRIM, 1996; Shen et al., 2006; Han et al., 2012). Eizenhöfer et al. (2014) further proposed a turbiditic origin for it based on the field

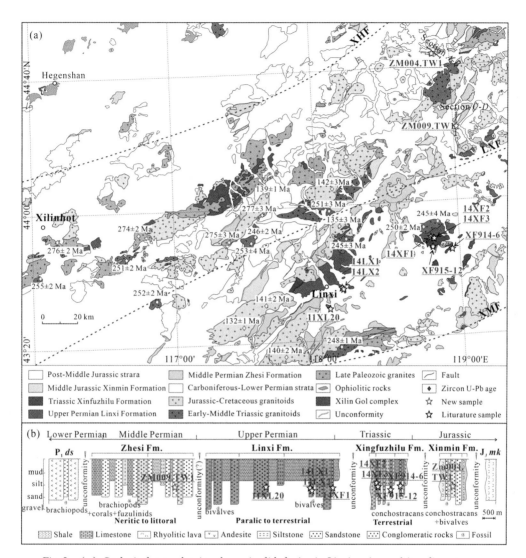

Fig. 2 (a) Geological map showing the major lithologies in Linxi region and its adjacent areas (modified after the 1∶50 000 Geological Map of the Solonker-Xilin Gol region complied by the Tianjin Institute of Geology and Mineral Resource). (b) Integrated stratigraphic column for the Middle Permian to Middle Jurassic strata in the frontal NAO, modified from BGMRIM (1996), Zhu (2015), and Zhu and He (2017).

Cross sections *A-B* and *C-D* respectively showing the field relationships of Upper Zhesi Fm. and Middle Xinmin Fm. sandstones with other lithostructural units in the Linxi region are given in Supplementary Fig. S2. Zircon U-Pb ages for the Early Permian to Early-Middle Triassic magmatic rocks are from S. Li et al. (2016a, b, 2017), Shi et al. (2004), Y. Li et al. (2014), and Wu et al. (2011).

occurrence of conglomeratic arkose, signs of cross-stratification in greywackes, and vertical bedding of the fine-grained succession. The Triassic Xingfuzhilu Fm. in the Linxi region, which uncomfortably overlies the Linxi Fm., was misleadingly referred to as a typical red clastic sequence on the top of the "original" Linxi Fm. in the early days (Zhu, 2015). In fact, it is an independent stratigraphic unit which can be further subdivided into three members: the lower part comprises conglomeratic rocks with andesitic gravel and coarse-grained sandstones; the middle part is dominated by sandstones and siltstones with parallel and cross beddings; the upper part is composed of mudstones, sandstones, and shales with limestone concretions (Li et al., 2014a,b). Overall, there is no marine sequence in the Upper Permian Linxi Fm. and the Triassic Xingfuzhilu Fm., consistent with the regional cessation of the marine deposition after the Middle Permian (e.g., Shen et al., 2006; Li et al., 2014a,b). Combined with the Middle Permian termination of arc magmatism in the Baolidao belt to the north (Li et al., 2016a,b), we proposed that both Upper Permian Linxi Fm. and Triassic Xingfuzhilu Fm. in the Linxi region were possibly deposited in the foreland basin of the collisional setting associated with the final orogenic process, although the provenance signatures might be different.

The Middle Jurassic Xinmin Fm. sediments in the Linxi region uncomfortably cover the elder Xingfuzhilu Fm., Linxi Fm., Zhesi Fm., and Dashizhai Fm. units with angular unconformities in most cases (Fig. 2a). Also, it is uncomfortably covered by the Upper Jurassic Manketouebo Fm. volcanic rocks and intruded by the Late Jurassic syenogranites in the field (Supplementary Fig. S2). Overall, the Middle Jurassic Xinmin Fm. in this region comprises a lower part of intermediate-acid volcanic rocks thin sandstone layers, a middle part of sandstones with coal seams, and an upper part of acid volcanic rocks (BGMRIM, 1996; Zhou et al., 2018a). Together with ① the post-collisional collapse in the southern CAOB in the Late Triassic (e.g., Davis et al., 2009; Xu et al., 2013) and ② the onset of westward subduction of the Paleo-Pacific Ocean (PPO) (e.g., Wu et al., 2011; Xu et al., 2013; Liu et al., 2021), the Middle Jurassic Xinmin Fm. sediments in the Linxi region were mostly likely deposited in the intraplate extensional basin corresponding to the PPO subduction beneath the NE China since the Late Mesozoic.

To track the provenance shifts during the accumulation of the Middle Permian Zhesi Fm., Upper Permian Linxi Fm., Triassic Xingfuzhilu Fm., and Middle Jurassic Xinmin Fm. sediments during the pre- to post-collisional orogenic evolution in the Linxi region, new and previously reported detrital zircon U-Pb ages and Hf isotopic data of a total of 10 sandstone samples from these four formations are synthesized and analyzed in the study (Li et al., 2014a; Eizenhöfer et al., 2014; Zhu and He, 2017). Detailed petrographic characteristics, maximum depositional ages and data sources are summarized in Supplementary Table S1. Sampling locations and approximate stratigraphic positions of each sample are shown in Fig. 2a,b.

4　Analytical methods

Detrital zircon U-Pb dating of the new samples (ZM009.TW1 and ZM004.TW1) was conducted using a Neptune MC-ICP-MS (Thermo Fisher Ltd., USA) coupled with a

193 nm New Wave UP193FX ArF laser ablation system (ESI Ltd., USA) at the Institute of Geology and Mineral Resources, Tianjin, China. In-situ zircon Lu-Hf isotopic analysis was carried out by a Nu Plasma II MC-ICP-MS (Nu Instrument Ltd., UK) equipped with a 193 nm RESOLution M-50 ArF laser ablation system (ASI Ltd., Australia) at the State Key Laboratory of Continental Dynamics, Northwest University, Xi'an, China. More details for zircon U-Pb dating and in-situ Lu-Hf analytical procedures can be found in the Yuan et al. (2008) and Bao et al. (2017). Our newly acquired data and the previously reported data are synthetically presented in Supplementary Table S2-S3. Notably, we used the $^{207}Pb/^{206}Pb$ ages for zircons older than 1.0 Ga and the $^{206}Pb/^{238}U$ ages for the zircons younger than 1.0 Ga. Only zircon ages with <10% discordance are included in the further data interpretation.

The cumulative distribution curve (CDC) is widely used as an estimate of the distribution function of the detrital zircon U-Pb ages as it can minimize the effect of the sampling uncertainty (the random error or statistical uncertainty during the sampling process) (e.g., Andersen et al., 2018). The kernel density estimate (KDE) plot is also a common visualization approach for the frequency distribution of detrital zircon age data. In this study, the CDC, coupled with corresponding KDE plot, were applied to characterize the detrital zircon U-Pb age distribution pattern of the Middle Permian to Middle Jurassic sandstone samples from the Linxi region. Furthermore, we performed an inter-sample statistical comparison of these samples using the "*detzrcr*" software of Andersen et al. (2018), including the $I-O$ overlap matrix of detrital zircon U-Pb and Hf model ages [illustrated as $(I-O)_{U-Pb}$ and $(I-O)_{Hf}$ values, respectively] and the upper versus lower quartile plots of zircon U-Pb and Hf model ages, to quantify the degrees of difference in terms of their detrital zircon populations. The $I-O$ parameter is a measurement of the "inter-sample" overlapping degree of the CDC confidence intervals (both U-Pb and Hf model ages). It is not subject to the "minimum number of analyses" and allows comparisons of different-sized U-Pb and Hf datasets. Also, the $I-O$ parameter takes values from 1.00 to 0.00, with 1.00 indicating not overlapped and 0.00 indicating that the CDCs are indistinguishable. Here, we cautiously assume that the $I-O$ parameter with values >0.05 signifies a non-trivial difference. In addition, the upper vs. lower quartile plots of U-Pb ages and Hf model ages were integrated for further evaluation and sample grouping (Andersen et al., 2018). Prior to the inter-sample comparison, all samples were put into a chronological order according to their maximum deposition ages (MDAs) calculated by the YC 2σ (weighted mean age of the youngest cluster with 3 or more zircons overlapping in 2σ error) (e.g., Dickinson and Gehrels, 2009).

5 Results

5.1 Results of inter-sample comparison

The CDCs and corresponding KDE plots of all samples from the Linxi region show different age peaks with varying relative abundance of the Precambrian ages (Fig. 3a,b), indicating variable degrees of difference in their detrital zircon populations. Hence, we use the $I-O$ overlap matrix of detrital zircon U-Pb and Hf model ages to further quantify the degrees of difference (Fig. 4a). Overall, the Upper Linxi Fm. samples 14LX1, 14LX2, and 14XF1 (ca. 271−258 Ma) mutually yield low $(I-O)_{U-Pb}$ values of 0.00 and low $(I-O)_{Hf}$ values of 0.00, suggesting a high degree of similarity of their detritus sources (Fig. 4a). The same is true for the Lower-Middle Xingfuzhilu Fm. samples 14XF2, 14XF3, and XF915-12 (ca. 257−236 Ma), they mutually yield low $(I-O)_{U-Pb}$ values of 0.00 and low $(I-O)_{Hf}$ values of 0.00−0.01 (Fig. 4a). However, high values of $(I-O)_{U-Pb}$ and $(I-O)_{Hf}$ can be observed between samples from different units (Fig. 4a): ① the Upper Zhesi Fm. sample ZM009.TW1 (ca. 275 Ma) yields a $(I-O)_{U-Pb}$ value of 0.32 and a $(I-O)_{Hf}$ value of 0.03 with the Middle Linxi Fm. sample 11XL20 (ca. 270 Ma); ②the Middle Linxi Fm. sample 11XL20 yields $(I-O)_{U-Pb}$ value of 0.00−0.09 and $(I-O)_{Hf}$ values of 0.26−0.31 with the ca. 271−258 Ma Upper Linxi Fm. samples; ③ the Upper Linxi Fm. samples yield $(I-O)_{U-Pb}$ values of 0.04−0.06 and $(I-O)_{Hf}$ values of 0.13−0.29 with the ca. 257−236 Ma Lower-Middle Xingfuzhilu Fm. samples; ④ the Lower-Middle Xingfuzhilu Fm. samples yield $(I-O)_{U-Pb}$ values of 0.04−0.06 and $(I-O)_{Hf}$ values of 0.20−0.30 with the Upper Xingfuzhilu Fm. sample XF914-6 (ca. 227 Ma); and ⑤ the Upper Xingfuzhilu Fm. sample XF914-6 yields a $(I-O)_{U-Pb}$ value of 0.29 and a $(I-O)_{Hf}$ value of 0.24. These results suggest that multiple abrupt changes in the detritus sources of the Middle Permian to Middle Jurassic sediments occurred over time in the Linxi region. These variations are also well illustrated in the plots of upper versus lower quartile zircon U-Pb ages and Hf model ages (Fig. 4b,c). Thus, we grouped the samples with indistinguishable detrital zircon distributions and summarized their detrital zircon U-Pb ages and Hf isotopic compositions in Section 5.2 and Fig. 5 based on the above comparison results.

5.2 Detrital zircon U-Pb ages and Hf isotopes

5.2.1 Upper Zhesi Fm. sediments (ca. 275 Ma)

Detrital zircon U-Pb age spectra of the Upper Zhesi Fm. sample ZM009.TW1 ($n = 96$) range from 270 Ma to 1 597 Ma and forms a Late Paleozoic peak at ca. 286 Ma as well as an Early Paleozoic peak at ca. 443 Ma (Fig. 5a). These two age peaks have variable zircon $\varepsilon_{Hf}(t)$ values of −7.3 to +8.6 and −14.3 to +8.5, respectively (Fig. 5h). The Precambrian zircons

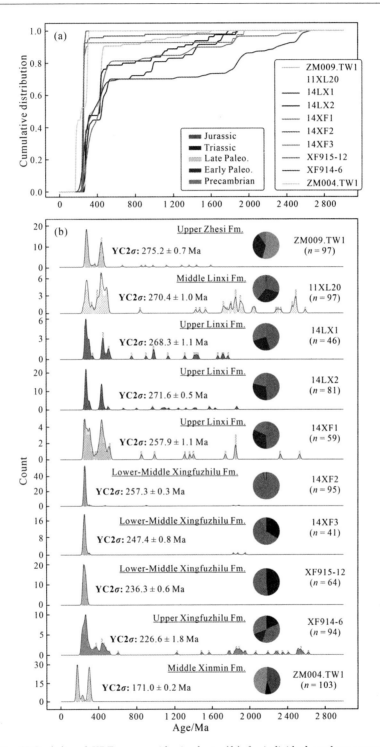

Fig. 3   CDCs (a) and KDE curves with pie charts (b) for individual sandstone samples,
showing the variations in detrital zircon populations.

The YC 2σ (youngest age cluster with 2 or more grains in 1σ error) refers to maximum depositional age (MDA).

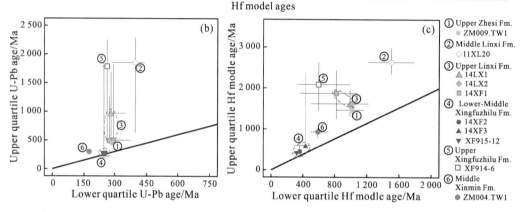

Fig. 4 (a) I–O overlap matrixes showing the overlap degrees of confidence intervals of zircon U-Pb age and Hf model age distributions; (b) upper quartile vs. lower quartile plots of zircon U-Pb age; (c) upper quartile vs. lower quartile plots of zircon Hf model age.

See text for detailed explanations.

($n=9$) are dominantly Neo- and Meso-proterozoic in age (664–1 597 Ma) and yield variable zircon $\varepsilon_{Hf}(t)$ values of -10.0 to $+9.1$ (Fig. 5a, g and h). Unlike the detrital zircon age spectra of the Lower Zhesi Fm. sediments that characterized by a unimodal Early Paleozoic age peak (ca. 445 Ma) (Chen et al., 2017), the Upper Zhesi Fm. sediments typically include a much higher relative abundance ($\sim 53\%$) of the Late Paleozoic grains (Fig. 5a), indicating a significant change in the sedimentary sources.

5.2.2 Middle Linxi Fm. sediments (ca. 270 Ma)

Detrital zircons from the Middle Linxi Fm. sample 11XL20 ($n=97$) yield U-Pb ages of 247–2 594 Ma, with a Late Paleozoic age peak of ca. 284 Ma and an Early Paleozoic age peak of ca. 443 Ma similar to that of the Upper Zhesi Fm. sample ZM009.TW1 (Fig. 5b). However, the Early Paleozoic age peak of the Middle Linxi Fm. sample shows more enriched zircon $\varepsilon_{Hf}(t)$ values of -23.6 to $+0.1$ (Fig. 5h). Moreover, a broad Precambrian age plateau of 846–2 594 Ma with two peak centered at 1 851 Ma and 2 488 Ma also have been documented in the Middle Linxi Fm. sample 11XL20 (Fig. 5b, g). These two Precambrian age peaks have

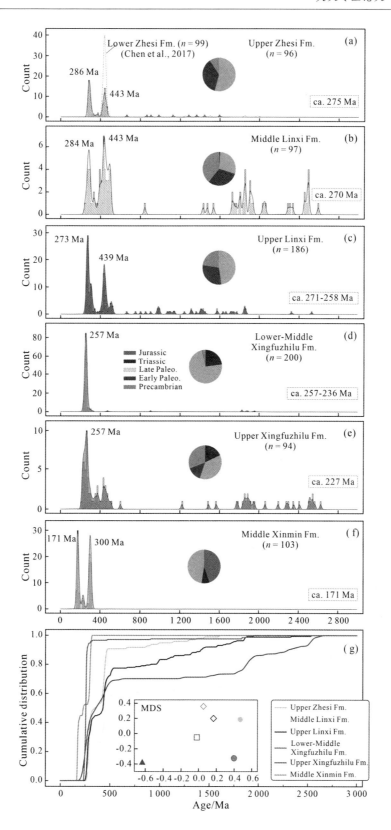

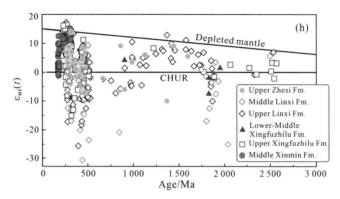

Fig. 5   Detrital zircon U-Pb age and Hf isotope distributions for the grouped Upper Zhesi Fm., Middle
Linxi Fm., Upper Linxi Fm., Lower-Middle Xingfuzhilu Fm., Upper Xingfuzhilu Fm., and Middle
Xinmin Fm. samples from the Linxi region: (a-f) KDE curves with pie charts CDCs showing
the major age peaks; (g) CDCs and multidimensional scaling (MDS) map illustrating the
overall dissimilarities; (e) Distributions of zircon $\varepsilon_{Hf}(t)$ values and U-Pb ages.

enriched zircon $\varepsilon_{Hf}(t)$ values of -30.6 to $+0.4$ and -17.0 to $+0.2$, respectively (Fig. 5h).

5.2.3   Upper Linxi Fm. sediments (ca. 271-258 Ma)

A total of 186 detrital zircons from the Upper Linxi Fm. samples 14LX1, 14LX2, and 14XF1 also form two major age peaks: ① a Late Paleozoic peak (ca. 273 Ma) with zircon $\varepsilon_{Hf}(t)$ of -17.2 to $+14.0$, and ②an Early Paleozoic peak (ca. 439 Ma) with zircon $\varepsilon_{Hf}(t)$ values of -17.1 to $+8.0$ (Fig. 5c). Overall, the Upper Linxi Fm. has a lower relative abundance of the Precambrian zircons ($\sim 22\%$) than that of the Middle Linxi Fm. ($\sim 39\%$), though their age plateau and zircon Hf isotopic compositions are to some content similar (Fig. 5c,g and h).

5.2.4   Lower-Middle Xingfuzhilu Fm. sediments (ca. 257-236 Ma)

Detrital zircon U-Pb ages of the Lower-Middle Xingfuzhilu Fm. samples 14XF2, 14XF3, and XF915-12 ($n=200$) vary from 233 Ma to 1 950 Ma and show a dominant Late Permian peak at ca. 257 Ma (Fig. 5d). However, unlike those Late Paleozoic age peaks of the underlying Zhesi Fm. and Linxi Fm. samples, this age peak is characterized by much more depleted zircon $\varepsilon_{Hf}(t)$ values of $+9.4$ to $+16.5$ (Fig. 5h). In addition, the low relative abundance and restricted zircon $\varepsilon_{Hf}(t)$ values of the Precambrian zircons (-7.2 to $+4.5$) of Lower-Middle Xingfuzhilu Fm. samples the are also different from that of the Zhesi Fm. and Linxi Fm. ones (Fig. 5d,g and h).

5.2.5   Upper Xingfuzhilu Fm. sediments (ca. 227 Ma)

The slightly younger Upper Xingfuzhilu Fm. sample XF914-6 ($n = 94$) has a Late Paleozoic peak (ca. 257 Ma) similar to the Lower-Middle Xingfuzhilu Fm. samples, however, its zircon isotopic compositions are slightly more enriched [$\varepsilon_{Hf}(t) = -4.4$ to $+14.9$] by comparison (Fig. 5e). Moreover, the Upper Xingfuzhilu Fm. sample also contains a much higher relative abundance of both Precambrian ages (~31%) and Early Paleozoic ages (~20%) than that of the Lower-Middle Xingfuzhilu Fm. samples (Fig. 5e,g and h).

5. 2. 6   Middle Xinmin Fm. sediments (ca. 171 Ma)

Detrital zircon U-Pb age spectra of the Middle Xinmin Fm. sample ZM004.TW1 ($n = 103$) is characterized by a much narrower age range of 169-317 Ma (Fig. 5f). Even so, two major age peaks can be observed: a Late Paleozoic one (ca. 300 Ma) with $\varepsilon_{Hf}(t)$ values of -1.3 to +12.5 and an Early Mesozoic one (ca. 171 Ma) with $\varepsilon_{Hf}(t)$ values of +2.9 to +15.2 (Fig. 5h). The Early Paleozoic and Precambrian zircons that widely documented in the underlying Xingfuzhilu Fm., Linxi Fm., and Zhesi Fm. samples no longer occur in the Middle Xinmin Fm. (Fig. 5a-g).

6   Discussion

6. 1   Provenance of the Precambrian zircons: Xilinhot Block or northern NCC?

The varying relative abundances of the Neoproterozoic to Archean zircons (604-2 625 Ma) in all samples excepting the youngest Middle Xinmin Fm. one are suggestive of different degrees of influxes of the Precambrian basement detritus (Fig. 5). In general, the Paleoproterozoic to Archean ages in the southeastern CAOB are typical of the NCC basement in the south side of the SSZ, whereas the Neoproterozoic to Mesoproterozoic ages (Xilinhot Block) are more common across in the north side of the SSZ, including the NAO and the HOAAC (e.g., Yang et al., 2006; Darby and Gehrels, 2006; Y. Li et al., 2011; Sun et al., 2013). However, it should be noted that the Baolidao arc belt and Erdaojing subduction-accretionary complex in the NAO also contain discrete components of the Paleoproterozoic to Archean ages that possibly recycled from other Precambrian continental fragments or terranes in the CAOB (Y. Li et al., 2017a,b; Li et al., 2021), which indicates that the NCC basement south of the SSZ is not the only potential source for the Paleoproterozoic and Archean detrital ages. Similarly, the Neoproterozoic to Mesoproterozoic detritus could also be derived from the south side of the SSZ, since recent studies show that the Bainaimiao arc in the SAO was essentially built on an exotic terrane with Neoproterozoic to Mesoproterozoic basement (Zhang et al., 2014a; Ma et al., 2019; Zhou et al., 2020). Nevertheless, the Precambrian zircons derived from the south side of the SSZ are characterized by much more variable zircon $\varepsilon_{Hf}(t)$ values than that of the Precambrian zircons from the north side of the SSZ (Fig. 6a), which can be used to distinguish the source provenance of the Precambrian detrital zircon in the studied samples from the Linxi region.

As illustrated in Fig. 6a, the Precambrian zircons of the Upper Zhesi Fm. sediments show a limited age spectrum of 663-1 597 Ma with restricted zircon $\varepsilon_{Hf}(t)$ values of -10.0 to +9.1,

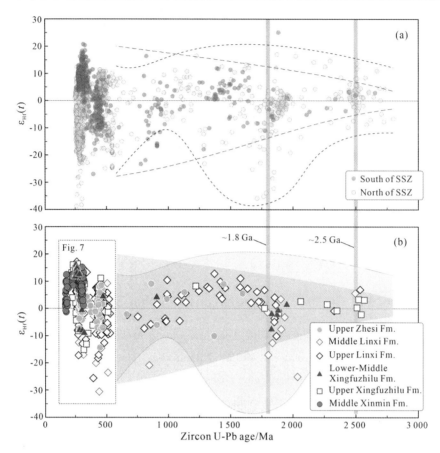

Fig. 6　(a) Distributions of zircon $\varepsilon_{Hf}(t)$ values and U-Pb ages for the Paleozoic sedimentary and igneous rocks and Precambrian basement complexes from the north and south sides of the SSZ; (b) Plot of zircon $\varepsilon_{Hf}(t)$ values vs. U-Pb ages.

Data sources: South of SSZ (Eizenhöfer et al., 2015; Zhang et al., 2007a, b, 2009, 2014a, 2016b; Zhou et al., 2020; Liu et al., 2020), North of SSZ (Chen et al., 2009; Eizenhöfer et al., 2014; Hu et al., 2015; Y. Li et al., 2017a, b; Liu et al., 2013; Yu et al., 2017; Sun et al., 2013; Li et al., 2021).

which is more consistent with a derivation from the local Xilinhot Block on north side of the SSZ (Fig. 6b). In contrast, the unconformably overlaid Middle Linxi Fm. sediments contain a high proportion of the Paleoproterozoic to Archean zircons, which typically show highly variable $\varepsilon_{Hf}(t)$ values of − 25.1 to + 5.6 (Fig. 6b). These characteristics coincide with the Precambrian detritus derived from the NCC basement on the south side of the SSZ, although one Neoproterozoic zircon with a highly negative $\varepsilon_{Hf}(t)$ value of −20.9 was more likely sourced from the local Xilinhot basement (Fig. 6b). As for the Precambrian detrital zircons in the Upper Linxi Fm., Lower-Middle Xingfuzhilu Fm., and Upper Xingfuzhilu Fm., all of them exhibit limited variations in the zircon $\varepsilon_{Hf}(t)$ values and thus are more consistent with a derivation from the local Xilinhot basement, rather than the NCC basement on the south side of the SSZ (Fig. 6b).

The above provenance analysis of the Precambrian detrital zircons reveal a significant input of the NCC basement detritus into the Middle Linxi Fm. sediments, although the local Xilinhot basement remained a dominant Precambrian detritus source for the Middle Permian to Triassic sediments in the Linxi region. Taking into consideration the MDA of the studied Middle Linxi Fm. sediments (ca. 270 Ma), this additional provenance input of the NCC basement detritus from the south side of the SSZ suggests that the wide PAO basin completely separating the NAO and SAO may no longer exist (or partially closed) by the end of the Early Permian.

6. 2   Primary "age components" and provenance changes

The main structure of detrital zircon populations of the Middle Permian to Middle Jurassic sediments in the Linxi region are constituted by the Early Paleozoic, Late Paleozoic, and Early Mesozoic components in different proportions and combinations (Fig. 5a-g). Thereinto, the Early Paleozoic and Late Paleozoic "age components" show wide age ranges and are chronologically coincident with arc magmatic events in response to the bidirectional subduction of the PAO in the southeastern CAOB (e.g., Zhang et al., 2009, 2014a; Chen et al., 2009; Shi et al., 2016; Li et al., 2016b). It is noteworthy that Late Paleozoic arc magmas from northern NCC arc belt (including the Bainaimiao arc) on the south side of the SSZ have much more enriched zircon Hf isotopic compositions than the coeval arc magmas in the local Baolidao arc belt (Fig. 7), though there is no substantial difference between the Early Paleozoic arc magmas. Likewise, the Early Mesozoic "age components" are chronologically consistent with the collision-related and post-collisional magmatic rocks on both sides of the SSZ (e.g., Zhang et al., 2014b; Li et al., 2016a, 2017a-d; Tang et al., 2020). However, the Early Mesozoic collision-related and post-collisional magmatic rocks in the NAO, north side of the SSZ, have more depleted zircon Hf isotopic compositions than the coeval magmatic rocks from the south side of the SSZ (Fig. 7). These differences in zircon Hf isotopic compositions can aid in tracing the source provenance of the Paleozoic to Early Mesozoic detrital zircons in the studied Middle Permian to Middle Jurassic sediments from the Linxi region.

As shown in Fig. 7a, the Late Paleozoic detrital zircons from the Upper Zhesi Fm. sediments, which have slightly enriched $\varepsilon_{Hf}(t)$ values of -14.3 to $+8.5$, overlaps the compositional range of the Late Paleozoic Baolidao arc magmas, indicating a derivation from the local Baolidao arc belt on the north side of the SSZ. Conversely, the Late Paleozoic detrital zircons from the Middle Linxi Fm. sediments show highly enriched $\varepsilon_{Hf}(t)$ values of -17.0 to $+0.4$, more consistent with the compositional range of the Late Paleozoic arc magmas from the south side of the SSZ (Fig. 7b). This observation suggests that the Middle Linxi Fm. sediments likely received additional detritus input from the south side of the SSZ. Furthermore, the majority of the Late Paleozoic detrital zircons from the studied Upper Linxi Fm. sediments show

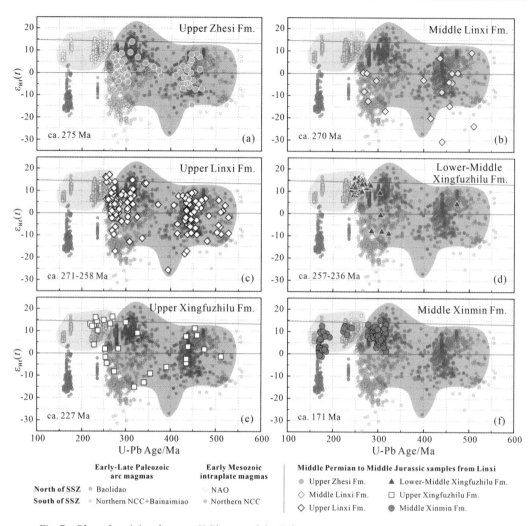

Fig. 7　Plots of $\varepsilon_{Hf}(t)$ values vs. U-Pb ages of the Paleozoic to Mesozoic detrital zircons from the Middle Permian to Middle Jurassic sandstone samples from the Linxi region.

The Paleozoic to Mesozoic co-magmatic zircons from the north and south sides of the SSZ are shown for comparison. Data sources: South of SSZ (Liu et al., 2020; Zhang et al., 2007a,b,2009, 2014a,b, 2016b; Zhou et al., 2018b, 2020; Qian et al., 2017), North of SSZ (Chai et al., 2020; Chen et al., 2009,2016; Eizenhöfer et al., 2015; Hu et al., 2015; Liu et al., 2009,2013; Li et al., 2017a,b,2021; S. Li et al., 2016b; Shi et al., 2016; Yang et al., 2020; Yu et al., 2017; Sun et al., 2013; Tang et al., 2020).

slightly enriched Hf isotopes similar to that of the Late Paleozoic arc magmas from the local Baolidao arc belt (Fig. 7c). Despite that, there still exist some Early Permian grains with highly enriched zircon $\varepsilon_{Hf}(t)$ values (as low as -17.2), denoting an additional input of detritus from the arc magmas on the south side of the SSZ (Fig. 7c). The Late Paleozoic detrital zircons from the studied Lower-Middle Xingfuzhilu Fm. sediments comprise abundant end-Permian to Middle Triassic detrital zircons with highly depleted $\varepsilon_{Hf}(t)$ values of $+8.6$ to $+16.5$, and minor Early Permian to Ordovician zircons with slightly enriched $\varepsilon_{Hf}(t)$ values of

-8.6 to +11.6 (Fig. 7d). These characteristics are overall consistent with a derivation from the local Baolidao arc belt in the NAO. However, the Upper Xingfuzhilu Fm. contains some Middle-Late Permian zircons whose $\varepsilon_{Hf}(t)$ values are similar to that of the coeval magmatic rocks on the south side of the SSZ (Fig. 7e), indicating an extra detritus contribution from the south. As for the Middle Xinmin Fm. sediments, they are composed of three distinct age clusters, in which the Late Carboniferous to Early Permian one and Middle-Late Triassic one were possibly derived from the NAO given their relatively depleted zircon $\varepsilon_{Hf}(t)$ values (Fig. 7f). However, the youngest Early-Middle Jurassic cluster particularly shows a more enriched zircon $\varepsilon_{Hf}(t)$ values of -1.3 to +12.5 (Fig. 7f), suggesting that they may have received an additional detritus input from the northern NCC.

The above observations, together with the provenances of the Precambrian zircons (Fig. 6), indicate that multiple changes in the sedimentary provenance progressively occurred during the deposition of the Middle Permian to Middle Jurassic sediments in the Linxi region of the frontal NAO: ① the Upper Zhesi Fm. sediments were exclusively derived from the NAO; ② the Middle and Upper Linxi Fm. sediments received additional input of detritus from the south side of the SSZ; ③ the Lower-Middle Xingfuzhilu Fm. sediments were sourced from the NAO, but the slightly postdated Upper Xingfuzhilu Fm. sediments received an extra detritus contribution from the south side of the SSZ; and ④ the youngest Middle Xinmin Fm. sediments were dominantly originated from the Late Paleozoic arc magmas and Early Mesozoic intraplate magmas in the NAO, though minor detritus from the northern NCC may also have participated.

6.3 Linking provenance changes with pre- to post-collisional orogenic evolution in southern CAOB

The terminal collision along the SSZ is proposed to be a "soft" suturing process between the Erdaojing subduction-accretion complex to the north and the Ondor Sum subduction-accretion complex to the south (Fig. 1), marking the onset of the last episode of orogenic evolution in southeastern CAOB (Jian et al., 2010; Xiao et al., 2015; Li et al., 2017a; Eizenhöfer and Zhao, 2018). Given that the prolonged arc magmatism on both sides of the SSZ terminated after ca. 275 Ma (e.g., Jian et al., 2010; Xiao et al., 2015; S. Li et al., 2016b), it is likely that this "soft" collision had already been initiated by the Middle Permian. Furthermore, the earliest geological records of the tectonic superimposition of the Paleo-Pacific regime in the southeastern CAOB are exemplified by the ca. 186-174 Ma deformation of the Heilongjiang Complex and the coeval low-middle K calc-alkaline volcanic rocks in the Jilin-Heilongjiang region (Zhou et al., 2009; Xu et al., 2013). Thus, it can be speculated that the pre- to post-collisional evolution of the southeastern CAOB may have lasted for as long as 90 Ma.

As noted earlier, the provenance characteristics of the Middle Permian to Middle Jurassic sediments in the Linxi region of the frontal NAO can aid in reconstructing the pre- to post-collisional orogenic evolution in the southeastern CAOB. Similar to the older Lower Zhesi Fm. sediments (Chen et al., 2017), the Upper Zhesi Fm. sediments (ca. 275 Ma) in the Linxi region were deposited in the forearc basin associated with the northward subduction of the PAO, and acquired detritus exclusively from the local NAO, including the Paleozoic Baolidao arc magmas and the Precambrian basement of the Xilinhot Block (Figs. 6b and 7a). This feature suggests that the NAO was still separated from the SAO and northern NCC by the wide PAO basin before ca. 275 Ma (Fig. 8a), which prevented the transportation of detritus from the south side of the SSZ. The above inference is reinforced by the formation age of the last arc magmatic flare-up in the NAO (ca. 281–275), which is characterized by slightly enriched whole-rock Sr-Nd and zircon Hf isotopes as well as high zircon $\delta^{18}O$ values, similar to that of the earlier arc magmas in the Baolidao arc belt (Fig. 8b) (Liu et al., 2009; Li et al., 2016b).

Significantly, the first arrival of detritus from south side of the SSZ to the Linxi region possibly occurred at ca. 270 Ma, as evidenced by the first occurrence of detrital zircons with highly enriched Hf isotopic compositions in the Middle Linxi Fm. sediments (Figs. 6b and 7b), denoting the input of detritus with northern NCC affinity. Such a conspicuous provenance change from Upper Zhesi Fm. to Middle Linxi Fm., together with the regional unconformity between the Zhesi Fm. and the Linxi Fm. in the Linxi region, are consistent with the mixing of the Late Permian Cathayian Flora with the Angaran Flora in southeastern CAOB (e.g., Wang and Liu, 1986; Huang, 1993; Deng et al., 2009), as well as the beginning of the Middle Permian magmatic lull in the NAO (Fig. 8b) (S. Li et al., 2016a). The above lines of evidence corroborate the initiation of the final collisional phase along the SSZ by at least ca. 270 Ma, which switched the depositional region of the frontal NAO from the former retroarc setting to a collisional-related one (Fig. 8a). Provenance results show that the ca. 271–257 Ma Upper Linxi Fm. sediments received detritus from both sides of the SSZ, with the NAO being the dominant source (Figs. 6b and 7c). This deposition stage was chronologically coincident with the above mentioned Middle Permian magmatic lull in the NAO (Fig. 8b) (S. Li et al., 2016a). According to S. Li et al. (2014a, 2016a, 2017a), this magmatic hiatus was caused by the initial "soft" collision between the SAO and NAO, which resulted in a remnant sea with distal marine sedimentation along the SSZ during this period. Thus, we propose that the ca. 271–257 Ma sediments (Upper Linxi Fm.) in the Linxi region were in association with this collisional process (Fig. 8a), but the specific cause for the provenance difference between the Middle and Upper Linxi Fm. sediments still needs more evidence.

Slab break-off is considered to be a natural consequence of collisional orogenesis caused

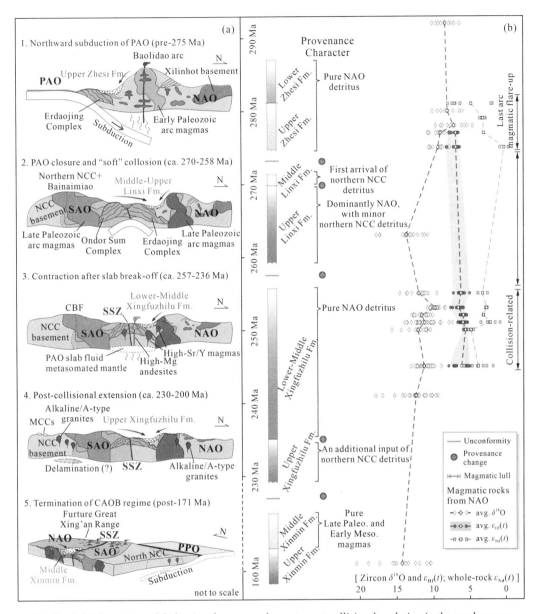

Fig. 8 (a) Schematic model showing the proposed pre- to post-collisional evolution in the southeastern
CAOB, and (b) corresponding spatio-temporal shifts in sedimentary provenances and variations
in magma composition observed in the Linxi region and its adjacent areas.

by the intense dragging force of continuously subducted slab (e.g., Andrews and Billen, 2009;
van Hunen and Allen, 2011). In the southeastern CAOB, recent studies propose that break-off
of the subducted PAO slab occurred soon after the initial collision, which leads to the formation
of a liner-distributed magmatic flare-up near the SSZ, including the ca. 253−250 Ma E-MORB-
like diabase, sanukitoid, and anorthosite in Mandula area and the ca. 255−251 Ma calc-
alkaline I-type granitoids in southern Xilinhot (Jian et al., 2010; Li et al., 2016a).

Furthermore, the continuous convergence force ultimately closed the remnant sea in the Early-Middle Triassic period, triggering the ca. 251–245 high Sr/Y and high-Mg adakitic magmas in the frontal NAO (Liu et al., 2012; Wang et al., 2015; Li et al., 2016a, 2017a), as well as extensive intermediate-low P-T metamorphism and moderate crustal thickening along the SSZ (Zhang et al., 2016a). The ca. 251–245 high Sr/Y magmas, which show geochemical and Nd-Hf-O isotopic systematics highly distinct to the earlier arc magmas in the Baolidao arc belt (Fig. 8b), were generated in response to the crustal thickening process after the PAO closure (Li et al., 2017a). As aforementioned, the ca. 257–236 Ma sediments in the Linxi region (Lower-Middle Xingfuzhilu Fm.) were exclusively sourced from the NAO and contain a much lower proportion of the Paleozoic arc detritus than the uncomfortably underlaid Upper Linxi Fm. sediments (Fig. 7c, d). We suggest that this provenance change can be attributed to the transition from the syn-collision to slab break-off and following contraction tectonics in the southeastern CAOB (Fig. 8a), though the onset of contraction is not well embodied in the provenance characteristics of the Lower-Middle Xingfuzhilu Fm. sediments.

The post-collisional extension evolution caused by the lithospheric delamination in the southeastern CAOB was likely occurred by the Middle-Late Triassic, as suggested by the transition of regional deformation pattern from N-S or NE-SW contraction to NE-SW extension, for example, the occurrence of the Late Triassic metamorphic core complex in Sonid Zuoqi region (Davis et al., 2004). Also, the widespread Middle-Late Triassic A-type granitoids and alkaline complexes in both NAO and northern NCC stand in sharp contrast to the aforementioned Early-Middle Triassic high Sr/Y and high-Mg adakitic magmas in this region (Wu et al., 2011; Zhang et al., 2012, 2014b). In this regard, we speculate that the ca. 227 Ma sediments (Upper Xingfuzhilu Fm.) in the Linxi region, which typically contain much more Paleozoic arc detritus than the Lower-Middle Xingfuzhilu Fm. sediments (Fig. 7d, e), were deposited in the foreland basin that associated with the post-collisional extension in the southeastern CAOB (Fig. 8a). The continuum of the NE-SW extensional deformation in the Early Jurassic suggests that the post-collisional extension process may have lasted to the Early Jurassic (e.g., Davis et al., 2009; Hu et al., 2010). Afterwards, extensive Middle-Late Jurassic arc magmatism with highly variable zircon $\delta^{18}O$ values (Fig. 8b), coupled with widespread large-scale folds and thrusts emerged in the whole NE China, marking the beginning of the tectonic superposition caused by the westward subduction of the PPO (Davis et al., 1998; Ma et al., 2002; Wu et al., 2011; Xu et al., 2013; Zhang et al., 2014b; Liu et al., 2021). Thus, we propose that the provenance change and the sedimentary interval between the Middle Xinmin Fm. (ca. 171 Ma) and Upper Xingfuzhilu Fm. (ca. 227 Ma) in the Linxi region were deposited in the extensional basin that associated with the transition from Paleo-Asian tectonic regime to Paleo-Pacific tectonic regime, denoting the terminus of the last episode of orogenic evolution of the southeastern CAOB (Fig. 8a).

6.4 Implications for tracking the final orogenic evolution in accretionary orogens

As the orogenic cycle draws to a close, a collisional phase is almost inevitable for all accretionary orogens when the oceanic lithosphere is completely consumed (Cawood et al., 2009). But in cases like the archipelago-type CAOB, the final collisional orogenesis usually takes place in a manner of a "soft" collision/ amalgamation between two opposing accretionary prisms that stacked during the protracted bidirectional oceanic plate subduction (Draut and Clift, 2013; Xiao et al., 2015; Eizenhöfer and Zhao, 2018). Resultantly, typical collisional signatures, such as the high to ultra-high metamorphic rocks and large-scale crustal uplift and thickening, are generally absent. Moreover, the varying geothermal gradient and overprinting structure associated with the post-collisional evolution may also obscure the earlier collision-related geological information (Marrone et al., 2021). Thus, the spatial-temporal assessment of the pre- to post-collisional orogenic evolution of an ancient accretionary orogen is always challenging.

The Middle Permian to Middle Jurassic sediments from the Linxi region of the frontal NAO show apparent provenance changes at the varying geodynamic stages of the pre- to post-collisional orogenic evolution of the southeastern CAOB (Figs. 7 and 8). In particular, the timing of the initial collision along the SSZ (ca. 270 Ma) predicted by the first arrival of northern NCC detritus from the south side of the SSZ to the north that recorded by the Middle Linxi Fm. in the Linxi region is in accordance with not only the cessation of the prolonged arc magmatism but also the occurrence of the mixing of Cathayian Flora with Angaran Flora in the southeastern CAOB (e.g., Deng et al., 2009; S. Li et al., 2016a). Moreover, the abrupt decrease of the Paleozoic arc magmatic detritus in the ca. 257−236 Ma deposition (Lower-Middle Xingfuzhilu Fm.) in the Linxi region coincides with the slab break-off and following intracontinental contraction tectonics that documented by the linear distributed collision-related magmatism and high Sr/Y granitoids in the southeastern CAOB (Jian et al., 2010; Li et al., 2016a, 2017). The life span of the post-collisional extension (ca. 227−171 Ma) that inferred from the provenances of Upper Xingfuzhilu Fm. (beginning) and Middle Xinmin Fm. (ending) sediments in the Linxi region is also in agreement with the currently known Mesozoic tectonic scenario of the NE China (e.g., Xu et al., 2013; Zhang et al., 2014b; Liu et al., 2021). The above observations on the pre- to post-collisional evolution of the southeastern CAOB suggest that the final collisional orogenesis in accretionary orogens may be comparable to the typical collisional orogens like the Himalayas in terms of geodynamic processes (Zhu et al., 2015 and references therein), though the deep conditions are much different. In other words, the spatio-temporal variations of sedimentary provenance of the foreland basins systems during the plate convergence and orogenic evolution provides important time-stamps for the geodynamic processes that involved in the pre- to post-collisional evolution in accretionary orogens, which, together with other geological evidence, can aid in reconstructing the final collision orogenesis in accretionary orogens, including the ones whose final suturing process occurred in a rather "soft" manner.

7 Conclusions

A comprehensive dataset of detrital zircon U-Pb ages and Hf isotopic data for the Middle Permian to Middle Jurassic foreland sediments in the Linxi region of the frontal NAO, southeastern CAOB, reveals episodic provenance changes, which, in combination with other regional geological records, provide powerful constraints on the pre- to post-collisional orogenic evolution in southeastern CAOB.

(1) The ca. 275 Ma Upper Zhesi Fm. was exclusively sourced from the local NAO, whereas the ca. 270 – 258 Ma Middle-Upper Linxi Fm. received an additional input of northern NCC detritus from the south side of the SSZ, corroborating the initial collision by the Middle Permian.

(2) The ca. 257 – 236 Ma Lower-Middle Xingfuzhilu Fm. returned to a pure NAO provenance with a low proportion of the Paleozoic arc detritus, coincident with the transition from syn-collision to slab break-off and following intraplate contraction tectonics.

(3) The ca. 227 Ma Upper Xingfuzhilu Fm., which contains more Paleozoic arc detritus and show provenance affinities of both NAO and northern NCC, possibly recorded the tectonic transition from intracontinental contraction to post-collisional extension.

(4) The ca. 171 Ma Middle Xinmin Fm. was exclusively sourced from the Late Paleozoic to Early Mesozoic magmatic rocks in both NAO and northern NCC, marking the terminus of the post-collisional extension in southeastern CAOB and the onset of the Paleo-Pacific tectonic regime.

Acknowledgements This work was co-supported by the China Geological Survey (Grant Nos.1212011085490, 1212010881204, and 1212010881207), the National Natural Science Foundation of China (Grant Nos. 41421002 and 41902049), and the Geological Survey Achievement Transformation Fund of China University of Geosciences (Beijing). The authors thank G.Q. Xiong, H.T. Zhao, and Z.A. Bao for assisting with the sample collection and zircon analyses. The authors also appreciate the Editor-in-Chief Ling Chen and the Guest Editor Qin Wang for the editorial handling. Constructive and insightful comments from Boris Natal'in and another anonymous reviewer significantly improved the quality of this manuscript.

Appendix A. Supplementary data Supplementary data to this article can be found online at https://doi.org/10.1016/j.tecto.2022.229550.

References

Andersen, T., Kristoffersen, M., Elburg, M.A., 2018. Visualizing, interpreting and comparing detrital zircon age and Hf isotope data in basin analysis: A graphical approach. Basin Research, 30, 132-147.

Andrews, E.R., Billen, M.I., 2009. Rheologic controls on the dynamics of slab detachment. Tectonophysics, 464, 60-69.

Badarch, G., Cunningham, W. D., Windley, B. F., 2002. A new terrane subdivision for Mongolia: Implications for the Phanerozoic crustal growth of Central Asia. Journal of Asian Earth Sciences, 21, 87-104.

Bao, Q.Z., Zhang, C.J., Wu, Z.L., Wang, H., Li, W., Su, Y.Z., Sang, J.H., Liu, Y.S., 2006. Carboniferous-Permian marine lithostratigraphy and sequence stratigraphy in Xi Ujimqin Qi, southeastern Inner Mongolia, China. Geological Bulletin of China, 25, 572-579 (in Chinese with English abstract).

Bao, Z.A., Chen, L., Zong, C.L., Yuan, H.L., Chen, K.Y., Dai, M.N., 2017. Development of pressed sulfide powder tablets for in situ sulfur and lead isotope measurement using LA-MC-ICP-MS. International Journal of Mass Spectrometry, 421, 255-262.

Bureau of Geology Mineral Resources of Inner Mongolia (BGMRIM), 1996. Rock and Stratum of Inner Mongolia Autonomous Region. Wuhan: China University of Geosciences Press (in Chinese).

Cawood, P.A., Hawkesworth, C.J., Dhuime, B., 2012. Detrital zircon record and tectonic setting. Geology, 40, 875-878.

Cawood, P. A., Kröner, A., Collins, W. J., Kusky, T. M., Mooney, W. D., Windley, B. F., 2009. Accretionary orogens through Earth history. Geological Society, London, Special Publications, 318, 1-36.

Chai, H., Ma, Y.F., Santosh, M., Hao, S.L., Luo, T.W., Fan, D.Q., Gao, B., Zong, L.B., Mao, H., Wang, Q.F., 2020. Late Carboniferous to early Permian oceanic subduction in central Inner Mongolia and its correlation with the tectonic evolution of the southeastern Central Asian Orogenic Belt. Gondwana Research, 84, 245-259.

Chen, B., Jahn, B.M., Tian, W., 2009. Evolution of the Solonker suture zone: Constraints from zircon U-Pb ages, Hf isotopic ratios and whole-rock Nd-Sr isotope compositions of subduction- and collision-related magmas and forearc sediments. Journal of Asian Earth Sciences, 34, 245-257.

Chen, B., Jahn, B.M., Wilde, S., Xu, B., 2000. Two contrasting Paleozoic magmatic belts in northern Inner Mongolia, China: Petrogenesis and tectonic implications. Tectonophysics, 328, 157-182.

Chen, Y., Zhang, Z.C., Li, K., Li, Q.G., Luo, Z.W., 2017. Provenance of the Middle Permian Zhesi Formation in central Inner Mongolia, northern China: Constraints from petrography, geochemistry and detrital zircon U-Pb geochronology: Zhesi Formation Provenance and Tectonic Setting. Geological Journal, 52, 92-109.

Chen, Y., Zhang, Z.C., Li, K., Yu, H.F., Wu, T.R., 2016. Geochemistry and zircon U-Pb-Hf isotopes of Early Paleozoic arc-related volcanic rocks in Sonid Zuoqi, Inner Mongolia: Implications for the tectonic evolution of the southeastern Central Asian Orogenic Belt. Lithos, 264, 392-404.

Condie, K.C., 2007. Accretionary orogens in space and time. In: Geological Society of America Memoirs. Geological Society of America, 145-158.

Darby, B.J., Gehrels, G., 2006. Detrital zircon reference for the North China block. Journal of Asian Earth Sciences, 26, 637-648.

Davis, G.A., Meng, J., Cao, W., Du, X., 2009. Triassic and Jurassic tectonics in the eastern Yanshan Belt, North China: Insights from the controversialDengzhangzi Formation and its neighboring units. Earth Science Frontiers, 16, 69-86.

Davis, G.A., Wang, C., Zheng, Y., Zhang, J., Zhang, C., Gehrels, G., 1998. The enigmatic Yinshan fold-and-thrust belt of northern China: New views on its intraplate continental styles. Geology, 26, 43-46.

Davis, G.A., Xu, B., Zheng, Y.D., Zhang, W.J., 2004. Indosinian extension in the Solonker suture zone: The Sonid Zuoqi metamorphic core complex, Inner Mongolia, China. Earth Science Frontiers, 11, 135-144.

de Jong, K., Xiao, W.J., Windley, B.F., Masago, H., Lo, C.H., 2006. Ordovician $^{40}Ar/^{39}Ar$ phengite ages from the blueschist-facies Ondor Sum subduction-accretion complex (Inner Mongolia) and implications for the early Paleozoic history of continental blocks in China and adjacent areas. American Journal of Science, 306, 799-845.

DeCelles, P.G., 2012. Foreland basin systems revisited: Variations in response to tectonic settings. In: Busby, C., Pérez, A. Tectonics of Sedimentary Basins: Recent Advances. Chichester: John Wiley, 405-426.

Deng, S., Wan, C., Yang, J., 2009. Discovery of a Late Permian Angara-Cathaysia mixed flora from Acheng of Heilongjiang, China, with discussions on the closure of the Paleoasian Ocean. Science in China: Series D, Earth Sciences, 52, 1746-1755.

Dickinson, W.R., Gehrels, G.E., 2009. Use of U-Pb ages of detrital zircons to infer maximum depositional ages of strata: A test against a Colorado Plateau Mesozoic database. Earth & Planetary Science Letters, 288, 115-125.

Draut, A.E., Clift, P.D., 2013. Differential preservation in the geologic record of intra-oceanic arc sedimentary and tectonic processes. Earth-Science Reviews, 116, 57-84.

Eizenhöfer, P.R., Zhao, G.C., 2018. Solonker Suture in East Asia and its bearing on the final closure of the eastern segment of the Palaeo-Asian Ocean. Earth-Science Reviews, 186, 153-172.

Eizenhöfer, P.R., Zhao, G.C., Sun, M., Zhang, J., Han, Y.G., Hou, W.Z., 2015. Geochronological and Hf isotopic variability of detrital zircons in Paleozoic strata across the accretionary collision zone between the North China craton and Mongolian arcs and tectonic implications. Geological Society of America Bulletin, 127, 1422-1436.

Eizenhöfer, P.R., Zhao, G.C., Zhang, J., Sun, M., 2014. Final closure of the Paleo-Asian Ocean along the Solonker Suture Zone: Constraints from geochronological and geochemical data of Permian volcanic and sedimentary rocks: Xar Moron Manuscript. Tectonics, 33, 441-463.

Fu, D., Huang, B., Kusky, T.M., Li, G.Z., Wilde, S.A., Zhou, W.X., Yu, Y., 2018. A Middle Permian Ophiolitic Mélange Belt in the Solonker Suture Zone, Western Inner Mongolia, China: Implications for the Evolution of the Paleo-Asian Ocean. Tectonics, 37, 1292-1320.

Garzanti, E., Doglioni, C., Vezzoli, G., Ando, S., 2007. Orogenic belts and orogenic sediment provenance. Journal of Geology, 115, 315-334.

Gehrels, G., 2014. Detrital zircon U-Pb geochronology applied to tectonics. Annual Review of Earth and Planetary Sciences, 42, 127-149.

Han, G.Q., Liu, Y.J., Neubauer, F., Genser, J., Zhao, Y.L., Wen, Q.B., Li, W., Wu, L.N., Jiang, X. Y., Zhao, L.M., 2012. Provenance analysis of Permian sandstones in the central and southern Da Xing'an Mountains, China: Constraints on the evolution of the eastern segment of the Central Asian Orogenic Belt. Tectonophysics, 580, 100-113.

Han, J., Zhou, J.B., Wang, B., Cao, J.L., 2015. The final collision of the CAOB: Constraint from the zircon U-Pb dating of the Linxi Formation, Inner Mongolia. Geoscience Frontiers, 6, 211-225.

Hu, C.S., Li, W.B., Xu, C., Zhong, R.C., Zhu, F., 2015. Geochemistry and zircon U-Pb-Hf isotopes of the granitoids of Baolidao and Halatu plutons in Sonidzuoqi area, Inner Mongolia: Implications for petrogenesis and geodynamic setting. Journal of Asian Earth Sciences, 97, 294-306.

Hu, J.M., Zhao, Y., Liu, X.W., Xu, G., 2010. Early Mesozoic deformations of the eastern Yanshan thrust belt, northern China. International Journal of Earth Sciences, 99, 785-800.

Huang, B.H., 1993. Carboniferous and Permian Systems and Floras in the Da Hinggan Range. Beijing: Geological Publishing House (in Chinese).

Jahn, B.M., 2004. The Central Asian Orogenic Belt and growth of the continental crust in the Phanerozoic. Geological Society, London, Special Publications, 226, 73-100.

Jian, P., Liu, D.Y., Kröner, A., Windley, B.F., Shi, Y.R., Zhang, W., Zhang, F.Q., Miao, L.C., Zhang, L.Q., Tomurhuu, D., 2010. Evolution of a Permian intraoceanic arc-trench system in the Solonker suture zone, Central Asian Orogenic Belt, China and Mongolia. Lithos, 118, 169-190.

Jian, P., Liu, D.Y., Kröner, A., Windley, B.F., Shi, Y.R., Zhang, F.Q., Shi, G.H., Miao, L.C., Zhang, W., Zhang, Q., Zhang, L.Q., Ren, J.S., 2008. Time scale of an early to mid-Paleozoic orogenic cycle of the long-lived Central Asian Orogenic Belt, Inner Mongolia of China: Implications for continental growth. Lithos, 101, 233-259.

Lamb, M.A., Badarch, G., 1997. Paleozoic Sedimentary Basins and Volcanic-Arc Systems of Southern Mongolia: New Stratigraphic and Sedimentologic Constraints. International Geology Review, 39, 542-576.

Li, G.Z., Wang, Y.J., Li, C.Y., Bai, Y.M., Xue, J.P., Zhao, G.M., Bo, H.J., Liang, Y.S., Liu, W., 2017. Discovery of Early Permian radiolarian fauna in the Solon Obo ophiolite belt, Inner Mongolia and its geological significance. Chinese Science Bulletin, 62, 400-406 (in Chinese with English abstract).

Li, J.L., Liu, J.G., Wang, Y.J., Zhu, D.C., Wu, C., 2021. Late Carboniferous to Early Permian ridge subduction identified in the southeastern Central Asian Orogenic Belt: Implications for the architecture and growth of continental crust in accretionary orogens. Lithos, 384-385, 105969.

Li, S., Chung, S.L., Wilde, S.A., Jahn, B.M., Xiao, W.J., Wang, T., Guo, Q.Q., 2017. Early-Middle Triassic high Sr/Y granitoids in the southern Central Asian Orogenic Belt: Implications for ocean closure in accretionary orogens: Tracing the Fate of Accretionary Orogens. Journal of Geophysical Research: Solid Earth, 122, 2291-2309.

Li, S., Chung, S.L., Wilde, S.A., Wang, T., Xiao, W.J., 2016a. Linking magmatism with collision in an accretionary orogen. Scientific Reports, 6, 25751.

Li, S., Wilde, S.A., He, Z.J., Jiang, X.J., Liu, R.Y., Zhao, L., 2014. Triassic sedimentation and postaccretionary crustal evolution along the Solonker suture zone in Inner Mongolia, China: Post-accretionary crustal evolution. Tectonics, 33, 960-981.

Li, S., Wilde, S.A., Wang, T., Xiao, W.J., Guo, Q.Q., 2016b. Latest Early Permian granitic magmatism in southern Inner Mongolia, China: Implications for the tectonic evolution of the southeastern Central Asian Orogenic Belt. Gondwana Research, 29, 168-180.

Li, Y.L., Brouwer, F.M., Xiao, W.J., Wang, K.L., Lee, Y.H., Luo, B.J., Su, Y.P., Zheng, J.P., 2017a. Subduction-related metasomatic mantle source in the eastern Central Asian Orogenic Belt: Evidence from amphibolites in the Xilingol Complex, Inner Mongolia, China. Gondwana Research, 43, 193-212.

Li, Y.L., Brouwer, F.M., Xiao, W.J., Zheng, J.P., 2017b. A Paleozoic fore-arc complex in the eastern Central Asian Orogenic Belt: Petrology, geochemistry and zircon U-Pb-Hf isotopic composition of paragneisses from the Xilingol Complex in Inner Mongolia, China. Gondwana Research, 47, 323-341.

Li, Y.L., Zhou, H.W., Brouwer, F.M., Xiao, W.J., Wijbrans, J.R., Zhong, Z.Q., 2014. Early Paleozoic to Middle Triassic bivergent accretion in the Central Asian Orogenic Belt: Insights from zircon U-Pb dating of ductile shear zones in central Inner Mongolia, China. Lithos, 205, 84-111.

Li, Y. L., Zhou, H. W., Brouwer, F. M., Xiao, W. J., Zhong, Z. Q., Wijbrans, J. R., 2011. Late Carboniferous-Middle Permian arc/forearc-related basin in Central Asian Orogenic Belt: Insights from the petrology and geochemistry of the Shuangjing Schist in Inner Mongolia, China. Island Arc, 20, 535-549.

Liu, J.F., Li, J.Y., Chi, X.G., Qu, J.F., Hu, Z.C., Fang, S., Zhang, Z., 2013. A late-Carboniferous to early early-Permian subduction-accretion complex in Daqing pasture, southeastern Inner Mongolia: Evidence of northward subduction beneath the Siberian paleoplate southern margin. Lithos, 177, 285-296.

Liu, J.L., Ni, J.L., Chen, X.Y., Craddock, J.P., Zheng, Y.Y., Ji, L., Hou, C.R., 2021. Early Cretaceous tectonics across the North Pacific: New insights from multiphase tectonic extension in Eastern Eurasia. Earth-Science Reviews, 217, 103552.

Liu, M., Lai, S.C., Zhang, D., Zhu, R.Z., Qin, J.F., Xiong, G.Q., Wang, H.R., 2020. Constructing the latest Neoproterozoic to Early Paleozoic multiple crust-mantle interactions in western Bainaimiao arc terrane, southeastern Central Asian Orogenic Belt. Geoscience Frontiers, 11, 1727-1742.

Liu, W., Pan, X.F., Liu, D.Y., Chen, Z.Y., 2009. Three-step continental-crust growth from subduction accretion and underplating, through intermediary differentiation, to granitoid production. International Journal of Earth Sciences, 98, 1413-1439.

Liu, Y.S., Wang, X.H., Wang, D.B., He, D.T., Zong, K.Q., Gao, C.G., Hu, Z.C., Gong, H.J., 2012. Triassic high-Mg adakitic andesites from Linxi, Inner Mongolia: Insights into the fate of the Paleo-Asian ocean crust and fossil slab-derived melt-peridotite interaction. Chemical Geology, 328, 89-108.

Luo, Z. W., Xu, B., Shi, G. Z., Zhao, P., Faure, M., Chen, Y., 2016. Solonker ophiolite in Inner Mongolia, China: A late Permian continental margin-type ophiolite. Lithos, 261, 72-91.

Ma, S.X., Wang, Z.Q., Zhang, Y.L., Sun, J.X., 2019. Bainaimiao arc as an exotic terrane along the Northern margin of the North China Craton: evidences from petrography, zircon U-Pb dating, and geochemistry of the early Devonian deposits. Tectonics, 38.

Ma, Y.S., Wu, M.L., Zeng, Q.L., 2002. The Mesozoic-Cenozoic compression and extension transformation process and ore-forming process in Yanshan and adjacent area. Acta Geosci. Sin., 23, 115-122 (in

Chinese with English abstract).

Marrone, S., Monié, P., Rossetti, F., Aldega, L., Bouybaouene, M., Charpentier, D., Lucci, F., Phillips, D., Theye, T., Zaghloul, M.N., 2021. Timing of Alpine Orogeny and Postorogenic Extension in the Alboran Domain, Inner Rif Chain, Morocco. Tectonics, 40, e2021TC006707.

Miao, L.C., Zhang, F., Fan, W.M., Liu, D.T., 2007. Phanerozoic evolution of the Inner Mongolia-Daxinganling orogenic belt in North China: Constraints from geochronology of ophiolites and associated formations. Geological Society, London, Special Publications, 280, 223-237.

Qian, X.Y., Zhang, Z.C., Chen, Y., Yu, H.F., Luo, Z.W., Yang, J.F., 2017. Geochronology and geochemistry of Early Paleozoic igneous rocks in Zhurihe area, Inner Mongolia and their tectonic significance. Earth Science, 42, 1472-1494 (in Chinese with English abstract).

Sajona, F.G., Maury, R., Pubellier, M., Leterrier, J., Bellon, H., Cotton, J., 2000. Magmatic source enrichment by slab-derived melts in a young post-collision setting, central Mindanao (Philippines). Lithos, 54, 173-206.

Sengör, A.M.C., Natal'in, B.A., 1996. Paleotectonics of Asia: Fragments of a synthesis. In: The Tectonic Evolution of Asia. New York: Cambridge University Press.

Şengör, A.M.C., Natal'in, B.A., Sunal, G., van der Voo, R., 2018. The Tectonics of the Altaids: Crustal Growth during the Construction of the Continental Lithosphere of Central Asia Between ~ 750 Ma and ~ 130 Ma Ago. Annu. Rev. Earth Planet. Sci., 46, 439-494.

Shen, S.Z., Zhang, H., Shang, Q.H., Li, W.Z., 2006. Permian stratigraphy and correlation of Northeast China: A review. Journal of Asian Earth Science, 26, 304-326.

Shi, Y.R., Jian, P., Kröner, A., Li, L.L., Liu, C., Zhang, W., 2016. Zircon ages and Hf isotopic compositions of Ordovician and Carboniferous granitoids from central Inner Mongolia and their significance for early and late Paleozoic evolution of the Central Asian Orogenic Belt. Journal of Asian Earth Sciences, 117, 153-169.

Shi, Y.R., Liu, D.Y., Zhang, Q., Jian, P., Zhang, F.Q., Miao, L.C., Shi, G.H., Zhang, L.Q., Tao, H.L., 2004. SHRIMP Dating of Diorites and Granites in Southern Suzuoqi, Inner Mongolia. Acta Geol. Sin., 78, 789-799 (in Chinese with English abstract).

Song, D.F., Xiao, W.J., Collins, A.S., Glorie, S., Han, C.M., Li, Y.C., 2018. Final Subduction Processes of the Paleo-Asian Ocean in the Alxa Tectonic Belt (NW China): Constraints From Field and Chronological Data of Permian Arc-Related Volcano-Sedimentary Rocks. Tectonics, 37, 1658-1687.

Song, D.F., Xiao, W.J., Windley, B.F., Mao, Q.G., Ao, S.J., Wang, H.Y.C., Li, R., 2021. Closure of the Paleo-Asian Ocean in the Middle-Late Triassic (Ladinian-Carnian): Evidence From Provenance Analysis of Retroarc Sediments. Geophys Res Lett, 48, e2021GL094276.

Song, S.G., Niu, Y.L., Su, L., Zhang, C., Zhang, L.F., 2014. Continental orogenesis from ocean subduction, continent collision/subduction, to orogen collapse, and orogen recycling: The example of the North Qaidam UHPM belt, NW China. Earth-Science Reviews, 129, 59-84.

Song, S.G., Wang, M.M., Xu, X., Wang, C., Niu, Y.L., Allen, M.B., Su, L., 2015. Ophiolites in the Xing'an-Inner Mongolia accretionary belt of the CAOB: Implications for two cycles of seafloor spreading and accretionary orogenic events. Tectonics, 34, 2221-2248.

Sun, L.X., Ren, B.F., Zhao, F.Q., Gu, Y.C., Li, Y.F., Liu, H., 2013. Zircon U-Pb dating and Hf isotopic compositions of the Mesoproterozoic granitic gneiss in Xilinhot Block, Inner Mongolia. Geological Bulletin of China, 32, 327-340 (in Chinese with English abstract).

Tang, Z.Y., Sun, D.Y., Mao, A.Q., 2020. Geochemistry of Late Mesozoic volcanic rocks in the central Great Xing'an Range, NE China: Petrogenesis and crustal growth in comparison with adjacent areas. International Geology Review, 62, 1-28.

van Hunen, J., Allen, M.B., 2011. Continental collision and slab break-off: A comparison of 3-D numerical models with observations. Earth & Planetary Science Letters, 302, 27-37.

Wang, H., Wang, Y.J., Chen, Z.Y., Li, Y.X., Su, M.R., Bai, L.B., 2005. Discovery of Permian radiolarians from the Bayanaobao area, Inner Mongolia. Journal of Stratigraphy, 29, 368-371 (in Chinese with English abstract).

Wang, Y.J., Fan, Z.Y., 1997. Discovery of Permian radiolarians in ophiolite belt on northern side of Xar Moron river, Nei Monggol and its geological significance. Acta Palaeontologica Sinica, 36, 58-69 (in Chinese with English abstract).

Wang, Q., Liu, X., 1986. Paleoplate tectonics between Cathaysia and Angaraland in Inner Mongolia of China. Tectonics, 5, 1073-1088.

Wang, Z.J., Xu, W.L., Pei, F.P., Wang, Z.W., Li, Y., Cao, H.H., 2015. Geochronology and geochemistry of middle Permian-Middle Triassic intrusive rocks from central-eastern Jilin Province, NE China: Constraints on the tectonic evolution of the eastern segment of the Paleo-Asian Ocean. Lithos, 238, 13-25.

Windley, B.F., Alexeiev, D., Xiao, W.J., Kröner, A., Badarch, G., 2007. Tectonic models for accretion of the Central Asian Orogenic Belt. Journal of the Geological Society, 164, 31-47.

Wu, F.Y., Sun, D.Y., Ge, W.C., Zhang, Y.B., Grant, M.L., Wilde, S.A., Jahn, B.M., 2011. Geochronology of the Phanerozoic granitoids in northeastern China. Journal of Asian Earth Sciences, 41, 1-30.

Xiao, W.J., Windley, B.F., Hao, J., Zhai, M.G., 2003. Accretion leading to collision and the Permian Solonker suture, Inner Mongolia, China: Termination of the central Asian orogenic belt. Tectonics, 22, 1-8.

Xiao, W.J., Windley, B.F., Huang, B.C., Han, C.M., Yuan, C., Chen, H.L., Sun, M., Sun, S., Li, J.L., 2009. End-Permian to mid-Triassic termination of the accretionary processes of the southern Altaids: implications for the geodynamic evolution, Phanerozoic continental growth, and metallogeny of Central Asia. International Journal of Earth Sciences, 98, 1189-1217.

Xiao, W.J., Windley, B.F., Sun, S., Li, J.L., Huang, B.C., Han, C.M., Yuan, C., Sun, M., Chen, H.L., 2015. A Tale of Amalgamation of Three Permo-Triassic Collage Systems in Central Asia: Oroclines, Sutures, and Terminal Accretion. Annual Review of Earth and Planetary Sciences, 43, 477-507.

Xu, B., Charvet, J., Chen, Y., Zhao, P., Shi, G.Z., 2013. Middle Paleozoic convergent orogenic belts in western Inner Mongolia (China): Framework, kinematics, geochronology and implications for tectonic evolution of the Central Asian Orogenic Belt. Gondwana Research, 23, 1342-1364.

Yang, J.H., Wu, F.Y., Shao, J.A., Wilde, S.A., Xie, L.W., Liu, X.M., 2006. Constraints on the timing of uplift of the Yanshan Fold and Thrust Belt, North China. Earth & Planetary Science Letters, 246, 336-352.

Yang, Z.L., Zhang, X.H., Yuan, L.L., 2020. Construction of an island arc and back-arc basin system in eastern Central Asian Orogenic belt: Insights from contrasting Late Carboniferous intermediate intrusions in Central Inner Mongolia, North China. Lithos, 372-373, 105672.

Yu, Q., Ge, W.C., Zhang, J., Zhao, G.C., Zhang, Y.L., Yang, H., 2017. Geochronology, petrogenesis and tectonic implication of Late Paleozoic volcanic rocks from the Dashizhai Formation in Inner Mongolia, NE China. Gondwana Research, 43, 164-177.

Yuan, H.L., Gao, S., Dai, M.N., Zong, C.L., Gunther, D., Fontaine, G.H., Liu, X.M., Diwu, C.R., 2008. Simultaneous determinations of U-Pb age, Hf isotopes and trace element compositions of zircon by excimer laser-ablation quadrupole and multiple-collector ICP-MS. Chemical Geology, 247, 100-118.

Zhang, J.R., Wei, C.J., Chu, H., Chen, Y.P., 2016. Mesozoic metamorphism and its tectonic implication along the Solonker Suture Zone in central Inner Mongolia, China. Lithos, 261, 262-277.

Zhang, S.H., Zhao, Y., Davis, G.A., Ye, H., Wu, F., 2014b. Temporal and spatial variations of Mesozoic magmatism and deformation in the North China Craton: Implications for lithospheric thinning and decratonization. Earth-Science Reviews, 131, 49-87.

Zhang, S.H., Zhao, Y., Liu, J.M., Hu, J.M., Song, B., Liu, J., Wu, Hai., 2010. Geochronology, geochemistry and tectonic setting of the Late Paleozoic-Early Mesozoic magmatism in the northern margin of the North China Block: A preliminary review. Acta Petrologica et Mineralogica, 29, 824-842 (in Chinese with English abstract).

Zhang, S.H., Zhao, Y., Liu, J.M., Hu, Z.C., 2016. Different sources involved in generation of continental arc volcanism: The Carboniferous-Permian volcanic rocks in the northern margin of the North China Block. Lithos, 240-243, 382-401.

Zhang, S.H., Zhao, Y., Song, B., Hu, J.M., Liu, S.W., Yang, Y.H., Chen, F.K., Liu, X.M., Liu, J., 2009. Contrasting Late Carboniferous and Late Permian-Middle Triassic intrusive suites from the northern margin of the North China craton: Geochronology, petrogenesis, and tectonic implications. Geological Society of America Bulletin, 121, 181-200.

Zhang, S.H., Zhao, Y., Song, B., Liu, D.Y., 2007a. Petrogenesis of the Middle Devonian Gushan diorite pluton on the northern margin of the North China Block and its tectonic implications. Geological Magazine, 144, 553-568.

Zhang, S.H., Zhao, Y., Song, B., Yang, Y.H., 2007b. Zircon SHRIMP U-Pb and in-situ Lu-Hf isotope analyses of a tuff from Western Beijing: Evidence for missing Late Paleozoic arc volcano eruptions at the northern margin of the North China Block. Gondwana Research, 12, 157-165.

Zhang, S.H., Zhao, Y., Ye, H., Hou, K.J., Li, C.F., 2012. Early Mesozoic alkaline complexes in the northern North China Craton: Implications for cratonic lithospheric destruction. Lithos, 155, 1-18.

Zhang, S.H., Zhao, Y., Ye, H., Liu, J.M., Hu, Z.C., 2014a. Origin and evolution of the Bainaimiao arc belt: Implications for crustal growth in the southern Central Asian orogenic belt. Geological Society of America Bulletin, 26, 1275-1300.

Zhou, G.W., Lin, M., Qiu, G.C., Wang, F., 2018. The redefinition of Xinmin Formation in Zhamuqin area of Horqin Right Wing Middle Banner of Inner Mongolia and its geological significance. Geological Bulletin of China, 37, 1579-1587 (in Chinese with English abstract).

Zhou, H., Zhao, G.C, Han, Y.G, Wang, B., 2018. Geochemistry and zircon U-Pb-Hf isotopes of Paleozoic intrusive rocks in the Damao area in Inner Mongolia, northern China: Implications for the tectonic evolution of the Bainaimiao arc. Lithos, 314-315, 119-139.

Zhou, H., Zhao, G.C., Han, Y.G., Wang, B., Pei, X.Z., 2020. Tectonic origin of the Bainaimiao arc terrane in the southern Central Asian orogenic belt: Evidence from sedimentary and magmatic rocks in the Damao region. Geological Society of America Bulletin, 133, 802-818.

Zhou, J.B., Wilde, S.A., Zhang, X.Z., Zhao, G.C., Zheng, C.Q., Wang, Y.J., Zhang, X.H., 2009. The onset of Pacific margin accretion in NE China: Evidence from the Heilongjiang high-pressure metamorphic belt. Tectonophysics, 478, 230-246.

Zhu, D.C., Wang, Q., Zhao, Z.D., Chung, S.L., Cawood, P.A., Niu, Y.L., Liu, S.A., Wu, F.Y., Mo, X.X., 2015. Magmatic record of India-Asia collision. Scientific Reports, 5, 14289.

Zhu, J.B., 2015. The Upper Carboniferous-Lower Triassic Sedimentary Environment and Tectonic Setting of Southeast Inner Mongolia. Beijing: China University of Geosciences (in Chinese with English abstract).

Zhu, J.B., He, Z.J., 2017. Detrital zircon records of Upper Permian-Middle Triassic sedimentary sequence in the Linxi area, Inner Mongolia and constraints on timing of final closure of the Paleo-Asian Ocean (eastern segment). Acta Geologica Sinica, 91, 232-248 (in Chinese with English abstract).

Petrogenesis of Early Cretaceous alkaline basalts in the West Qinling: Constraints from olivine chemistry[①]

Yang Zhen   Lai Shaocong[②]   Qin Jiangfeng   Zhu Renzhi

Liu Min   Zhang Fangyi   Yang Hang   Zhu Yu

Abstract: Olivine crystals in alkaline basalts preserves important information about the Oxygen fugacity and thermal condition for the mantle-derived magma. This paper presents new olivine mineral chemistry and whole-rock geochemistry for Early-Cretaceous alkali basalts in the Ganjia area, northeastern margin of the Tibetant Plateau. The olivine phenocrysts of Ganjia basalts have low Ni (74–2 261 ppm) and high Ca contents (1 217–4 443 ppm), similar to the olivines in MORB, suggesting that they were crystallized from peridotite-derived basaltic magma. Al-in-olivine/ spinel thermometer results indicate crystallization temperatures of olivine around 1 202 ± 46 ℃, corresponding to a mantle potential temperature of 1 279 ± 47 ℃, with low oxygen fugacity of FMQ −1 to FMQ 0, suggesting a reduced mantle source. Whole-rock geochemical compositions of the Ganjia basalts were characterized by low SiO_2 contents (45. 39–48. 70 wt%), high alkali contents (3. 88–5. 11 wt%), positive Nb and Ta anomalies, and enrichment in LREEs, HFSEs and LILEs, similar to those of OIB rocks, with high ratios of $(Dy/Yb)_N$ (1. 38–1. 54) and $(La/Yb)_N$ (5. 37– 7. 42) ratios, indicating garnet residue. The above results suggest that the Early Cretaceous intraplate alkaline basalts in West Qinling were formed by low degree of partial melting of peridotite, and that no high-temperature mantle plume existed. The Qinling Orogenic Belt may have been affected by the subduction of the Paleo-Pacific. We propose that the Ganjia basalts originated by the interaction between the upwelling asthenosphere and a lithospheric source.

1   Introduction

Discrimination of the source lithology of basalts plays a critical role in our understanding of the origin of mantle-derived magmas and magmatic processes and can provide valuable information about the crustal material or mantle metasomatism that contributes to mantle heterogeneity (Herzberg, 2011; Yang et al., 2016). Nevertheless, the mechanisms of mantle melting and the lithology of mantle sources are still under debate because their signals are

①   Published in *Geological Journal*, 2023, 58(2).

②   Corresponding author.

strongly influenced by the melting and crystallization conditions of the basaltic magmas and the compositional variety of the source rocks (Niu et al., 2011; Yang and Zhou, 2013). Alkaline basalt is the representative intraplate magma, representing the product of partial melting of mantle peridotite in the deep mantle (Herzberg and O'Hara, 2002; Walter, 1998). Whereas, recent studies have suggested that basaltic magma produced by partial melting of mantle peridotite is inconsistent with the natural alkaline basalts with respect to several specific major elements (Dai et al., 2018). There is increasing evidence that alkali basalts can form via partial melting of several different source lithologies, such as olivine-free pyroxenite, carbonated peridotite, peridotite + amphibole veins, hornblendite or commixture between these sources (Dai et al., 2014; Dasgupta et al., 2007; Lambart et al., 2016; Liu and Ren., 2013; Pilet et al., 2008). However, trace elements and traditionally used radiogenic isotopes (Sr-Nd-Pb-Hf isotopes) could not reveal whether the recycled material was still present as a lithological unit in the mantle source (pyroxenite), or whether it was only imprinted onto the peridotite in the form of a geochemical signal (Ammannati et al., 2016; Herzberg et al., 2011). Olivine is the most common mineral phase of peridotite, making up 50% to >90% of the rock. Simultaneously, it is usually the first silicate mineral to crystallize from almost all mantle-derived melt compositions, and preserves nearly primary information about the mantle melts (Foley et al., 2013). The concentration of some compatible trace element (e.g., Ni, Mn, Cr, Co) depends critically upon the modal abundance of olivine in the residual mantle source (De Hoog et al., 2010; Straub et al., 2008). On the other hand, the distribution of other trace elements (e.g., Ca, Li, Ti) in olivine is mainly determined by their original abundance in the mantle source (Foley et al., 2011).Therefore, the features of transition metal elements in olivine phenocrysts emerged as an important approach for investigating the lithology of the mantle source; incompatible trace element are more sensitive to melting processes, whereas compatible trace element are more sensitive to source compositions (Geng at al.,2019; Herzberg, 2006; Le Roux et al., 2011; Zhang et al., 2016).

Intraplate basaltic magmatism was emplaced in the Ganjia area, West Qinling Orogenic Belt. These volcanic rocks contain fresh olivine and clinopyroxene phenocrysts. These basalts and olivine phenocrysts provide a valuable opportunity to investigate the petrogenesis of alkaline basalts and explore the potential controlling processes. In this paper, we report a combined study of mineral chemistry and whole-rock major and trace element data for Early Cretaceous intraplate alkaline basalts of the Ganjia area, West Qinling. These data were used to investigate the magmatic processes and the characteristics of deep mantle source, as well as to identify the mantle lithology of these alkaline volcanic rocks. The results confirmed that the source lithology of the Ganjia basalts was normal mantle peridotite.

2　Geological background and petrography

2.1　Geological background

The West Qinling Orogenic Belts are important tectonic units in the northeastern margin of the Tibetan Plateau (Fig. 1a). It was built through the collision among the Paleo-Asian tectonic regime, the Paleo-Tethyan tectonic regime and the Pacific tectonic regime during the process of the major orogens of China are converged at the West Qinling Orogenic Belts (Lai et al., 2022).

Traditionally, West Qinling Orogenic Belts and East Qinling Orogenic Belts are typically separated by the Foping Dome and Huicheng Basin (Fig. 1b). The West Qinling Orogenic Belt has been viewed as a westward extension of the Qinling Orogenic Belt. It is tectonically bounded by A'nimaque-Mianlue Suture to the south, the Qilian Orogenic Belt to the north, the Qaidam Basin and East Kunlun Orogenic Belt along the Wahongshan-Wenquan Fault to the west, and the East Qinling Orogenic Belt to the east (Fig. 1c; Dong, Zhang, Neubauer et al., 2011; Dong et al., 2016). The A'nimaque-Mianlue Suture is widely interpreted as a symbol for northward subduction of the remnant of the Paleo-Tethys Ocean (Bian et al., 2004). Although the timing of the initial collision remains contentious at present, it is widely accepted that the post-orogenic magmatism and intracontinental convergence occurred as early as the Late Triassic (Dong et al., 2013, 2016). Under the influence of Late Triassic tectonic events, the West Qinling Orogen Belt has underwent complex structural deformation after the closure of the Qinling Ocean (Zhang et al., 2004). The occurrences of small volumes of volcanic rocks and strike-slip faults are significant geological signatures formed during late Mesozoic tectonic events.

Substantial granitites and intermediate-felsic rocks of the Triassic age are spatially and temporally widespread in the West Qinling Orogenic Belt (Dong, Zhang, Neubauer et al., 2011; Feng et al., 2003). There are a series of volcanic rocks in Xiahe and the adjacent area as well, such as Cretaceous alkali basalt in Madang, Cretaceous basalt and andesite in Duofutun, and Cenozoic melilite and carbonatite in the Lixian area (Table 1).

2.2　Field geology and petrography

Volcanic rocks outcrop in Ganjia town (E102°34′11″, N35°23′51″), about 20 km to the northwest of Xiahe County (Fig. 1d), distributed within the margin of the E-W-directed Cenozoic basin, and cover an area of about 3 km^2 (Lu et al., 2021). These volcanic rocks mainly consist of basalt, basaltic breccia, and a small amount of basalt andesite. The basalts have a typical pillow structure. Delamination of the volcanic rocks can be clearly observed in field outcrops (Fig. 2a). The volcanic rocks consist of basalts and basalt-andesite. In addition, these volcanic rocks unconformably overlay the Middle and Lower Permian Daguan Formation,

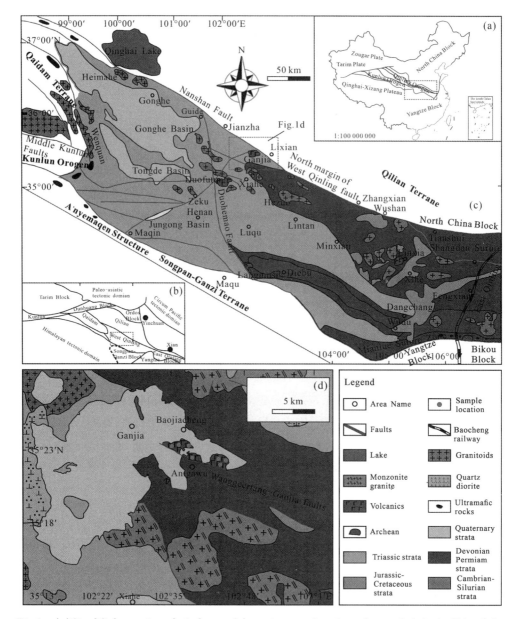

Fig. 1　(a) Simplified tectonic geological map of the major tectonic units and orogenic belts in China (after Zhang et al., 2020). (b) Geological map of the main tectonic units neighboring the West Qinling Orogenic Belt. (c) The distributions of Cretaceous alkali basalts in West Qinling is modified after Li et al. (2013). (d) Geological sketch map of Ganjia area based on the 1 : 200 000 geological map of the Linxia sheet. Sample locations are marked by grey circles.

and their overlying layers are red Quaternary sediments (Ding et al., 2013). The Daguan Formation predominantly consists of conglomerates, phyllitic tuffaceous slates, and massive breccia interbedded with yellowish-gray silty slate. Its lower layer is the Permian Shiguan Formation, which mainly consists of gray-green siliceous marlstone, siliceous limestone, silty

Table 1　Summary of Early Cretaceous alkaline basalts in the West Qinling.

Location	Lithology	Method	Age/Ma	Petrogenesis	Reference
Ganjia	Alkaline basalt	U-Pb	101.1±1	Partial melting of metasomatic hornblende veins in mantle peridotite	Dai, 2014
Ganjia	Alkaline Basalt	U-Pb	220.5±4.2	A low degree of partial melting of asthenosphere mantle	Yan et al. 2012
Ganjia	Alkaline Basalt			Lithosphere metasomatized by asthenosphere-derived silicate melts	Huang et al., 2013
Madang	Alkaline Basalt			Partial melting of mixed asthenosphere mantle	Lai et al., 2007
Hongqiang	Alkaline Basalt	Ar-Ar	112±0.56	Partial melting of asthenosphere mantle containing ancient recycling oceanic crust	Fan et al., 2007
Hongqiang	Alkaline Basalt	U-Pb	104.8±0.99	A low degree of partial melting of asthenosphere mantle	Ding et al., 2013
Rizhou	Alkaline Basalt	U-Pb	106.27±1.3	A low degree of partial melting of asthenosphere mantle	Pu et al., 2019
Duofutun	Alkaline Basalt	$^{40}Ar/^{39}Ar$	112 Ma	Asthenospheric mantle with input of delaminated lithospheric component	Zhang et al., 2018
Duofutun	Alkaline Basalt	$^{40}Ar/^{39}Ar$	103±2	Small degrees of partial melting of asthenosphere mantle	Li et al., 2013
Duofutun	Alkaline Basalt	U-Pb	105.8±0.9	Small degrees of partial melting of asthenosphere mantle	Li et al., 2013
Duofutun	Alkaline Basalt	$^{40}Ar/^{39}Ar$	86.9±2.8	Eastward flow of asthenospheric mantle beneath the Tibetan Plateau	Zhao, 2009
Duofutun	Alkaline Basalt	$^{40}Ar/^{39}Ar$	96.21±2.10	Partial melting of an asthenospheric mantle	Hu et al., 2012
Duofutun	Alkaline Basalt	Fossil		Enriched mantle mixed with HIMU	Qi et al., 2012

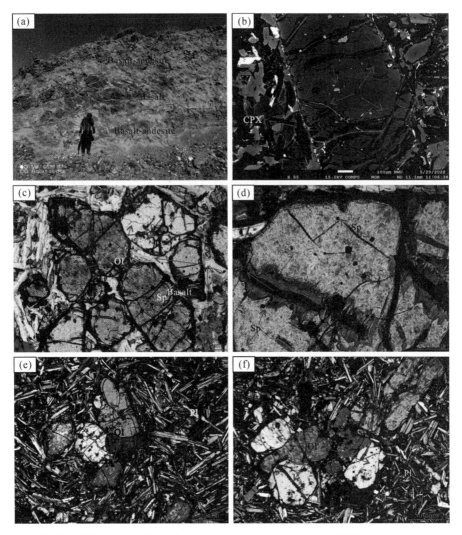

Fig. 2　Field photographs and microtextures of representative alkali basalt samples
from the West Qinling Orogen Belt.

(a) Field occurrence of volcanic rocks. (b) Backscattered electron image of olivine and clinopyroxene. (c) Euhedral olivine with spinel surrounded by plagioclase. (d) Serpentinized olivine and spinel inclusions. (e,f) Olivine phenocrysts and Plagioclase in the matrix. Cpx: clinopyroxene; Ol: olivine; Pl: plagioclase; Srp: serpentine; Sp: spinel.

slates, and dark purple medium to coarse-grained quartz arkose. The distribution patterns and formation of these volcanic rocks in this area were closely related with the local active faults (Fan et al., 2007). A series of NW-NWW directed faults developed in the Ganjia Basin, called the Wangeertang-Gangjia Fault. This unique geological pattern resulted in the occurrence of fissure eruption of these volcanic rocks along the faults (Fig. 1d).

The Wangeertang-Gangjia Fault is the westward extension of the northern margin of western Qinling fault (NWQF), and is a left-lateral fault with a length of about 66 km. This fault formed during the Caledonian period and was intensely active until the Late Yanshanian period. Previous geophysical data have shown that the NWQF is a translithospheric fault, which cuts through the lithosphere and reaches the upper mantle (Kang et al., 1990; Zhang, 2012).

The samples for this study are predominantly composed of basalt. Clinopyroxene is the most abundant mineral in the Ganjia basalts (~15 vol%), occurring as phenocryst. Olivine is also a major phenocryst (3−5 vol%). In contrast, plagioclases are presented as microcrystal and found only in the matrix. Additionally, magnetite and titanite can be observed in some samples (Fig. 2e). The plagioclases are euhedral acicular or tabular grains (0.3−1 mm). Some of the basalts do not contain olivine phenocrysts and only have clinopyroxene phenocrysts. The olivines are euhedral to subhedral, and some are fractured (Fig. 2c). The grain sizes of olivine phenocrysts range from 0.3 mm to 0.5 mm (Fig. 2b). Additionally, the rims of olivines are slightly altered, and some olivines are replaced by serpentine (Fig. 2d). Plastic deformation features of olivines, such as kink bands and wave-like extinction, were not observed (Fig. 2f). The basalts exhibit a porphyritic texture with massive structure. Vesicular structures and amygdaloidal structures are observed in minor basalts and the almond structures are mainly filled with calcite and chlorite. In addition, a small proportion of samples are slightly altered. These samples showed minor alteration with chlorite replacement of clinopyroxene and clay minerals replacing plagioclase.

3 Analytical method

Fifteen samples were selected for whole-rock major and trace element analyses at the State Key Laboratory of Continental Dynamics, Northwest University in Xi'an, China. Fresh chips of whole samples were powered to a grain size of 200 mesh size using a tungsten carbide ball mill. Major elements were determined on lithium borate fusion disks using X-ray fluorescence spectrometry (XRF). Analyses of USGS and Chinese national rock standards (BCR-2, GSR-1 and GSR-3) indicate that both the analytical precision and accuracy for major elements were generally better than 5%. Trace element contents were measured by inductively coupled plasma mass spectrometry (ICP-MS); sample powders were digested using an $HF + HNO_3$ mixture in high-pressure Teflon bombs at 190 ℃ for 48 h. The procedures for ICP-MS analysis was described in detail by Liu et al. (2007). Precision and accuracy were monitored by measurements of replicates and the standard reference materials BHVO-2, AGV-2, BCR-2 and GSP-2. Analytical precision for most trace elements was better than 10%.

Microprobe analyses were carried out on a pilished thin section using a JXA-8230 electron microprobe at the State Key Laboratory of Continental Dynamics, Northwest University. The operating conditions included an acceleration voltage of 15 kV, a beam current of 10 nA, and a beam diameter of 1 μm. Microprobe standards of natural and synthetic phases were supplied by

SPI company: jadeite for Si, Al, and Na, diopside for Ca, olivine for Mg, sanidine for K, hematite for Fe, rhodonite for Mn, and rutile for Ti.

High precision isotopic (Sr, Nd) measurements were carried out at Nanjing FocuMS Technology Co. Ltd. (www.focums.com). Geological rock powders were mixed with 0.5 ml 60 wt% HNO_3 and 1.0 ml 40 wt% HF in high-pressure PTFE bombs. These bombs were steel-jacketed and placed in the oven at 195 ℃ for 3 days. Digested samples were dried down on a hotplate and reconstituted in 1.5 ml of 0.2 N HBr + 0.5 N HNO_3 before ion exchange purification. Raw data of isotopic ratios were internally corrected for mass fractionation by normalizing to $^{86}Sr/^{88}Sr = 0.119\,4$ for Sr, $^{146}Nd/^{144}Nd = 0.721\,9$ for Nd, $^{205}Tl/^{203}Tl = 2.388\,5$ for Pb with exponential law. International isotopic standards (NIST SRM 987 for Sr, JNdi-1 for Nd) were periodically analyzed to correct instrumental drift. Geochemical reference materials of USGS BCR-2, BHVO-2, AVG-2, RGM-2 were treated as quality control). These isotopic results agreed with previous publications within analytical uncertainty (Weis et al., 2006).

4　Results

4.1　Whole-rock major and trace elements

Whole-rock major and trace element data for the Ganjia basalt are presented in Table S1. The basalts were relatively fresh, and had moderate loss on ignition (LOI), varying from 2.35 wt% to 4.62 wt%. The low LOI and fresh plagioclase in the matrix show that the basalt samples did not undergo post-eruption alteration. The Ganjia basalt has the characteristics of low SiO_2 contents (45.39−48.70 wt%), high alkali contents ($Na_2O + K_2O = 3.88-5.11$ wt%), and high CaO contents (5.89−8.98 wt%). These basalt samples belonged to the sodic series with $Na_2O = 2.65-3.96$ wt%, $K_2O = 0.93-1.78$ wt%, and $Na_2O/K_2O = 1.73-4.26$. MgO contents ranged from 4.50 wt% to 9.29 wt%, and the $Mg^{\#}$ [$Mg^{\#} = 100Mg^{2+}/(Mg^{2+} + Fe^{2+})$] ranged from 49.15 to 63.66. CaO contents varied from 5.89 wt% to 8.98 wt%, and Al_2O_3 contents varied from 15.23 wt% to 17.30 wt%. CaO/Al_2O_3 ratios were between 0.34 and 0.86, and were positively correlated with the MgO contents (Table S1). The TiO_2 contents ranged from 1.95% to 2.55 wt%, higher than those of island arc basalts (0.58−0.85 wt%) and typical continental tholeiitic basalts (Pearce, 2008; Pearce and Cann, 1973). These results show a close affinity with intraplate alkaline basalts and continental rift alkaline basalts (Wilson et al., 1989). The majority of samples plotted in the basalt area in the total alkalis versus SiO_2(TAS) diagram (Fig. 3), based on the nomenclature of Le Bas et al. (1986). Moreover, all of them were alkaline basalts according to the sub-alkaline alkaline classification standard ($Na_2O + K_2O$ = 3.88−5.11 wt%). These results suggest that the Ganjia basalt is a set of sodic alkaline basalts.

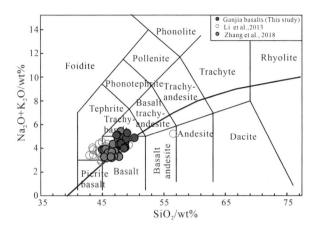

Fig. 3　Total alkali vs. SiO$_2$(TAS) diagram for the alkaline basalt
in the West Qinling Orogenic Belt (after Le Bas et al., 1986).

The data in this study have removed the LOI and normalized the oxide contents to 100%. The published data
on alkaline basalt in the West Qinling Orogenic Belt are from Li et al. (2013), Zhang et al. (2018).

The compatible trace element Ni showed clear correlations with MgO contents, as well as the trend between Cr and MgO. The concentrations of Cr and Ni were 130−353. 9 ppm and 50−160 ppm, respectively. The chondrite-normalized rare earth element (REE) patterns are given in Fig. 4a. Nearly all of the samples exhibit highly enrichment in light REE (LREEs) [(La/Yb)$_N$ = 5. 37−7. 42]. The Ganjia basalts showed consistent chondrite-normalized sub-parallel REE patterns (∑REE = 105. 2−133. 8 ppm), and the average was 116. 9 ppm (Table S1). The (Dy/Yb)$_N$ values ranged from 1. 38 to 1. 54, with a mean value of 1. 43. The samples had slight Eu anomalies (Eu/Eu* = 0. 95−1. 01) (Fig. 4b). In the primitive mantle-normalized multi-element spider diagram (Fig. 4b), all the samples showed consistent characteristics of positive Nb-Ta-Ti anomalies but negative Rb-Th-P anomalies. In general, the geochemical characteristics of the Ganjia basalts were similar to the trace element distribution patterns for intracontinental alkali basalts and ocean island basalts (OIB).

4. 2　Mineral chemistry of olivine phenocrysts and spinel

The electron microprobe data of olivine and spinel in the basalts are presented in Table S2 and Table S3, respectively. Fresh olivine phenocrysts were a common phase observed in most of the Ganjia basalt samples. Compositionally, the olivine phenocrysts were rich in SiO$_2$ (37. 88−41. 57 wt%), MgO (37. 43−47. 18 wt%), CaO (0. 17−0. 92 wt%), and FeO (13. 30−3. 19 wt%), with low NiO contents (0. 01−0. 29 wt%). The CaO and Al$_2$O$_3$ contents were higher than those of typical olivine in mantle xenoliths (CaO < 0. 1 wt%, Al$_2$O$_3$ < 0. 03 wt%; Foley et al., 2013). The Fo of the olivines varied from 74. 21 to 85. 65, with an average of 82. 95. According to the classification standard of olivine, these olivines can be classified

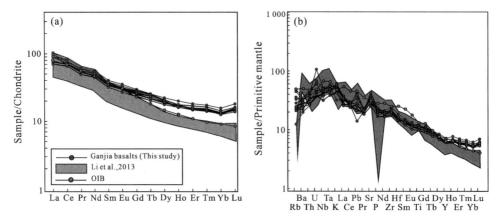

Fig. 4　(a) Chondrite-normalized rare earth element (REE) patterns. (b) Primitive-mantle normalized spidergrams of the Early Cretaceous alkali basalts from the West Qinling Orogenic Belt.

The published data is from Li et al. (2013). The compositions of OIB are from Sun and McDonough (1989).

chrysolite. In addition, the olivines containing spinel had the highest Fo value (85.9). These results suggest that the crystallization of spinel occurred very early and may have been simultaneous with the crystallization of olivine. The spinels were predominantly composed of Al_2O_3(25.1–47.1 wt%), Cr_2O_3(13.8–22.6 wt%), and FeO^T(15.5–35.2 wt%). The $Cr^{\#}$ of the spinels varied from 0.16 to 0.37, with a mean value of 0.22.

4.3　Olivine crystallization temperature

Previous studies for the olivine crystallization temperatures of primitive magma have concentrated on about olivine-melt thermometry (Putirka, 2005,2007). Coogan et al. (2014) calibrated a new thermometer based on the Al content of olivine in equilibrium with Cr-spinel, which is applicable to nearly all primitive basalts and could provide more precise estimates of olivine crystallization temperatures in different geodynamic background. It is necessary evaluate whether the olivine and melt are in equilibrium before olivine to evaluate the thermal state and lithological nature of mantle sources. According to Fig.5, all samples were plotted within the equilibrium area, indicating that the olivine were generally in equilibrium with the melt. The olivine crystallization temperatures was calculated using the formula proposed by Coogan et al. (2014).The formulas were given as follows：

$$T_1(\mathrm{K}) = 1\,000/0.575 + 0.884Cr^{\#} - 0.897\ln K_d$$

$$K_d = Al_2O_3^{\,olivine}/Al_2O_3^{\,spinel}$$

$$Cr^{\#} = Cr / (Cr + Al)$$

where all the temperature terms are in ℃, and Kdol-liq Fe-Mg are given from 0.30 to 0.34 (Putirka, 2016).

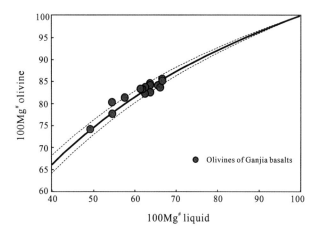

Fig. 5 $100Mg^\#$ liquid vs. $100Mg^\#$ olivine for the Ganjia basalt, West Qinling Orogenic Belt.

The thermometry results for west Qinling olivines showed that the crystallization temperatures of the olivine phenocrysts were 1 106−1 270 ℃, with a mean value of 1 202 ± 46.4 ℃. The estimated result and the EMPA data for calculation of olivine crystallization temperature were listed in Table S3.

4.4   Whole-rock Sr-Nd isotopes

Sr-Nd isotope data are given in Table S4 and Fig. 6. The initial isotopic ratios of all samples were calculated for emplacement age of 106 Ma. The Ganjia basalts displayed variable initial $(^{87}Sr/^{86}Sr)_i$ ratios from 0.705 327 to 0.703 480 and $\varepsilon_{Nd}(t)$ values of +4.8 to +9.1. Additionally, all the samples have relatively homogeneous $(^{143}Nd/^{144}Nd)_i$ ratios, varying from 0.512 745 to 0.512 970, similar to those of OIB rocks. However, the Ganjia basalts have lower $^{87}Sr/^{86}Sr$ but higher $^{143}Nd/^{144}Nd$ compared to those of Cenozoic basalts in East China (Fig. 6).

5   Discussion

5.1   Crustal contamination and fractional crystallization

The positive correlations of Ni and Cr with MgO contents imply the fractionation of olivine or Cr-rich spinel. This inference is also in accordance with the occurrence of minerals in the rocks. The P_2O_5 contents increased linearly with decreasing MgO content, which suggests the parental magmas underwent the fractionation crystallization of apatite (Fig. 7h). Fractional crystallization of plagioclase is most likely insignificant, as suggested by the insignificant Eu anomalies. The coupling of the concentrations of TiO_2, and FeO^T with MgO can be readily ascribed to the fractionation of Ti-Fe oxides (Fig. 7c).

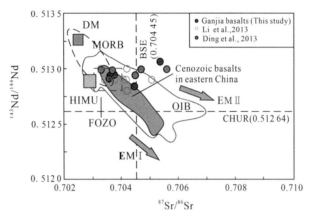

Fig. 6　Plot of Sr-Nd isotopic compositions for the Ganjia basalts.
MORB, DM, OIB, HIMU, EM I and EM II fields are from Zindler and Hart (1986); FOZO fields
is from Hauri et al. (1994). The fields of Cenozioc basalts in eastern China are from Zou et al. (2000).

Mantle-derived magma must transit through the thick continental crust during its ascent, which may result in the interchange of materials and heat conduction with the surrounding crust. Therefore, it is necessary to consider whether the rock was contaminated by crustal materials before the nature of the source region for these basalts can be determined. Nb/U and Ce/Pb have been viewed as effective index to evaluate the influence of crustal assimilation and contamination (Pfänder et al., 2018), because these trace elements are not fractionated during the process of mantle melting and therefore represent the compositional variation of the mantle (Hofmann et al., 1986). Generally, OIB and intracontinental basalts have higher Ce/Pb (25 ± 5) and Nb/U (47±10), whereas the crust has lower Ce/Pb (5), Nb/U (<10) and Nb/Ta (12-13, Rudnick and Gao, 2003; Taylor and McLennan, 1995). However, these ratios of incompatible elements in mantle-derived magmas change significantly if the magmas undergo assimilation and contamination of continental crust. The ratios of Ce/Pb, Nb/U, and Nb/Ta for these samples were 4.42-20.27, 15.33-39.94, and 14.57-15.44, respectively (Table S1). According to Fig. 8a, they plot within the PM area, whereas the basalts tended to display low Ce/Pb and especially low Nb/U. In Fig. 8b, the majority of these samples were plot in PM area, as well. Although lower Ce/Pb and Nb/U values may indicate a crustal assimilation, this process simultaneously results in a correlated variation in the Ce/Pb and Nb/U ratios, as well as evidently lower Nb/Ta ratios and $\varepsilon_{Nd}(t)$ value of the basalt (Pfänder et al., 2014, 2018). However, this trend was not observed. Qi et al. (2012) and Li et al. (2013) reported $\varepsilon_{Nd}(t)$ values of the basalts are +4.2 to +7.2 and +6.1 to +9.1, respectively. Moreover, the $\varepsilon_{Nd}(t)$ valuess of Ganjia basalts are +4.8 to +9.1. These values indicate depletion. Furthermore, $\varepsilon_{Nd}(t)$ was not clearly correlated with the Ce/Pb, Nb/U, and Nb/Ta ratios. The positive correlation between Nb/U and Ce/Pb, Nb/Ta may be induced by the recycled crustal

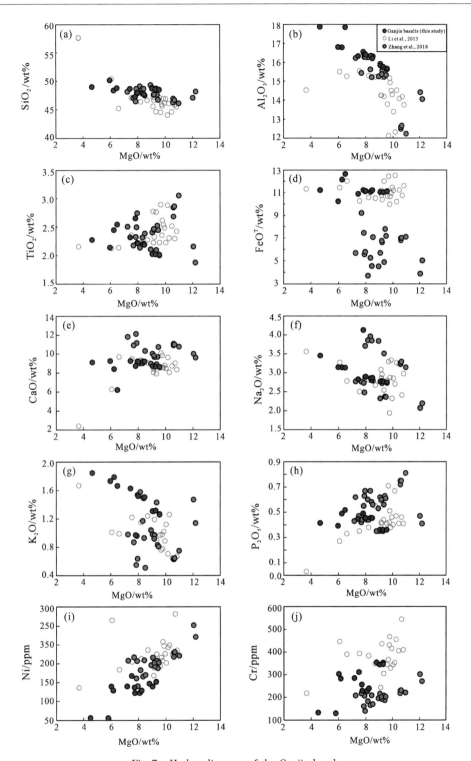

Fig. 7　Harker diagrams of the Ganjia basalt.

Also shown for comparison are published data for alkali basalts adjacent Ganjia area from Li et al. (2013) and Zhang et al. (2018). The data from this study have been removed the LOI and normalized the oxide contents to 100%.

materials in the mantle source. However, the lithology of the mantle source is peridotite but not pyroxene. In addition, the incorporation of recycled crustal materials usually result in a significant decrease in Th/La of basalts (Chen et al., 2012). The Th/La ratio of Ganjia basalts is 0.138, which is similar to the depleted mantle (0.14). Therefore, the recycled crustal materials are not the reason of the signature of decreasing in Nb/U and Ce/Pb ratios. Accordingly, we propose that the variations of these ratios can be caused by mantle source heterogeneity rather than crustal assimilation.

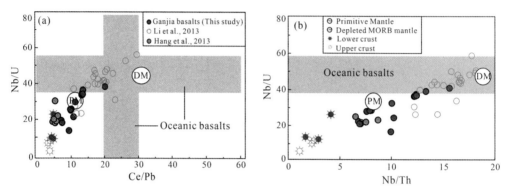

Fig. 8 (a) Nb/U vs. Ce/Pb for the Ganjia basalt, West Qinling Orogenic Belt.
(b) Nb/U vs. Nb/Th for the Ganjia basalt.

Blue circles represent the range of our data. The ranges of Ce/Pb and Nb/U in oceanic basalts are marked by the grey bars (Hofmann et al., 1986) and the OIB compositions with HIMU and EM affinities. Average continental crust (CC) and lower continental crust (LC) values are from Rudnick and Gao (2003). The published data for alkali basalts in West Qinling are from Li et al. (2013) and Huang et al. (2013).

5.2 Oxygen fugacity of magma

Oxygen fugacity (f_{O_2}) is an important parameter influencing the P-T position of the mantle solidus, the compositions of mantle-derived melts and fluids, and a series of geophysical properties of the mantle (Mallmann and O'Neill, 2013). Ballhaus et al. (1991) reported a olivine-spinel-orthopyroxene oxybarometer based on the redox reaction of $6Fe_2SiO_4(Ol) + O_2 = 3Fe_2Si_2O_6(Opx) + 2Fe_2O_4(Sp)$. The formula for the oxybarometer is as follows:

$$\Delta lg(f_{O_2})^{FMQ} = 0.27 + 2\,505/T - 400P/T - 6lg(XOl\ Fe) - 3\,200(1 - XOl\ Fe)^2/T + 2lg(XSp\ Fe^{2+}) + 4lg(XSp\ Fe^{3+}) + 2\,630(XSp\ Al)^2/T$$

where T is in K, and XOl Fe, XSp Al and XSp Fe^{2+} are the mole fractions of cations in olivine and spinel, respectively.

The crystallization temperature of olivine can be obtained using the Al-in-olivine thermometer proposed by Coogan et al. (2014). The error of this method is ±0.4 log units. The pressure of this experiment varied from 0.3 GPa to 2.7 GPa and the temperature was between 1 040 ℃ and 1 300 ℃. The oxybarometer results indicate that the oxygen fugacity of

Ganjia basalts varied from FMQ −1 to FMQ 0, which suggests formation in a relatively reducing environment, consistent with the normal deep mantle (Frost and McCammon, 2008). The estimated results can be found in Table S3.

5.3 Source composition

5.3.1 Presence of garnets in mantle sources

The variables $(La/Yb)_N$ and $(Dy/Yb)_N$ can be effective indicators to discriminate whether garnet was present in the mantle source. Garnet has a higher phase/melt partition coefficient for Yb than for Dy, whereas spinel has similar phase/melt partition coefficients for both of these elements (McKenzie and O'Nions, 1991). Therefore, $(La/Yb)_N$ is mainly controlled by the degree of partial melting during the process of mantle melting. However, the ratios of medium REE vs. heavy REE and $(Dy/Yb)_N$ are mainly controlled by whether garnet residues existed in the mantle source. A melt formed by a garnet-free source generally has $(Dy/Yb)_N < 1.06$ (Blundy et al., 1998; Chang et al., 2009). In addition, garnet has a higher phase/melt partition coefficient for Yb than Gd. Thus, $(Gd/Yb)_N$ ratios are also an effective indicator for discriminating the fingerprint of garnet from spinel in the source. $(Gd/Yb)_N$ ratios of magmas derived from garnet-bearing source will be readily higher than 1 (Allen et al., 2013). $(Gd/Yb)_N$ ratios of the Ganjia basalts ranging from 1.80 to 2.13 probably resulted from the presence of garnet in the partial melting residue. The $(Dy/Yb)_N$ values of the Ganjia basalt range from 1.38 to 1.54 (Table S1), which also suggests that garnet act as a fundamental part in the mantle source.

5.3.2 Peridotite rather than pyroxenite in the mantle source

For a long time, it was commonly accepted that mantle peridotite was the predominant source lithology that melted to produce basaltic magmas (Green and Ringwood, 1963; O'Hara and Yoder, 1967). Nevertheless, experimental petrology indicates that pyroxenite may be a potential source lithology for basaltic magmas (Sobolev et al., 2005). Olivine is the most abundant mineral phase in the upper mantle, and it is typically the earliest silicate phase to crystallize from almost all primary magmas. Thus, the concentration of compatible trace elements in olivine such as Ni, Mn, Cr and Co act as a good proxy to explore the lithological nature of magmatic process and mantle sources (Sobolev et al., 2005, 2007). This new progress provided a new perspective for us to study the mantle source of the Ganjia basalts base on olivine phenocryst composition.

The pyroxenites were interpreted as formed by recycled ocean crust, based on experimental petrology showing that Si-rich melts of subducted basaltic ocean crust react with mantle peridotite to produce olivine-free pyroxenite (Yaxley et al., 1998). This reaction consumes olivine in favor of orthopyroxene, eventually producing an olivine-free pyroxenite

(Gómez-Ulla et al., 2017; Howarth and Harris., 2017; Neave et al., 2018). Such a pyroxenite source releases more Ni contents than does "normal" peridotite during the process of partial melting because of the lower bulk partition coefficient for Ni. Ni content is hosted mostly in pyroxene (lower D_{Ni}) during the process of partial melting, which results in magmas with higher Ni contents; therefore, high-Ni olivine crystallized later from such high-Ni melts. For peridotite, Ni content mainly enters into olivine during the process of partial melting, which leads to the decrease of Ni content in the melt. Therefore, the olivine crystallized from this low-Ni melt also has the characteristic of low Ni content. In contrast, the D_{Ca} in pyroxene is much higher than that in olivine. Greater Ca content enters into the pyroxene to form a low-Ca melt during the process of partial melting; olivine crystallized from such a melt has low Ca and high Ni contents. The lithology of the mantle source for Hawaii is olivine-free pyroxenite. However, the mantle source for MORB is peridotite. High Ni content in olivines is attributed to a much higher D_{Ni} in olivine relative to pyroxene, as a result of the greater compatibility of Ni in pyroxene than in olivine. Therefore, Hawaiian olivines are characterized by high Ni contents, low Ca contents, and high Fe/Mn ratios. However, MORB olivines have the opposite trend compared to Hawaiian olivines.

Olivines from the Ganjia basalt have systematically lower Ni contents than those reported for Hawaiian olivines, and the majority plot within the MORB (Fig. 9a). The olivines show a narrow range of Ni contents (74 – 2 261 ppm) with decreasing Fo. Overall, the Ca contents (1 217 – 4 443 ppm), generally exceeded the field for Hawaiian olivines (Fig. 9b). The major elements data of olivines are provided in Table S2. The olivines from the Ganjia basalt were generally characterized by low Ni and high Ca contents. Therefore, the olivines of the Ganjia basalts are classified as MORB olivines, and the lithology of mantle sources is peridotite.

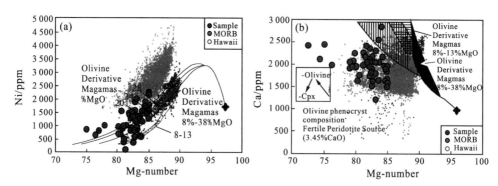

Fig. 9　(a) The diagrams of Mg-number vs. Ni for olivine phenocrysts of the Ganjia basalts
(modified after Herzberg, 2011). (b) The diagrams of Mg-number vs. Ca
for olivine phenocrysts of the Ganjia basalts (modified after Herzberg, 2011).
Mg-number = 100MgO∕ (MgO + FeO). The published data of olivines for MORB and
Hawaii are from Sobolev et al. (2005, 2007).

Carbonated peridotite can also form alkaline basalts (Zeng et al., 2010). However, previous studies have shown that although these rocks have variable alkali contents, all of them have high CaO/Al_2O_3 ratios (≥ 0.6), as well as evidence of co-crystallization of clinopyroxene and olivine (Dasgupta et al., 2006, 2007). The CaO/Al_2O_3 ratios of the Ganjia basalt is 0.53 ± 0.05, however, which is much lower than the range for a carbonated mantle source. It is instead interpreted to have originated from a low degree of partial melting of deep mantle peridotite.

5.3.3 Mantle potential temperature

The mantle potential temperature (T_p) refers to the hypothetical temperature a parcel of mantle would have if it rises adiabatically to the planetary surface without melting (Putirka, 2016). It can provide unique insights to explore information about mantle convection and thermal state in the deep mantle.

In this study, the mantle potential temperature was calculated using the formula proposed by Liu et al. (2017). The olivine liquidus temperature was determined at a lower crust depth (T_1) based on an independent method (the Al-in-olivine thermometer) to recover the mantle potential temperature. The correlation between the olivine liquidus temperature at 1 atm (T_1) and T_p was used and calibrated by

$$T_1^{Ol/L} = T_p^{Ol/L} - 54P + 2P^2$$
$$T_p = 1.049T_1^{Ol/L} - 0.000\,19(T_1^{Ol/L})^2 + 1.487 \times 10^{-7}(T_1^{Ol/L})^3$$

where both temperature terms are in ℃. The temperature is calculated by olivines which have a high Fo (≥ 83).

The calculated temperatures for the Ganjia basalt varied from 1 229 ℃ to 1 330 ℃, with an average value of 1 279 ± 47 ℃, slightly lower than the mantle potential temperature of intraplate magmatism beneath continents (1 350−1 450 ℃; Lee et al., 2009). The calculated data and more details are given in Table S3. Furthermore, this temperature range is consistent with the thermal state of the mantle beneath the West Qinling Orogenic Belt (Yu et al., 2001; Hoskin and Black, 2010), which was deduced from deep geophysical data and mantle xenoliths. The results also show that the Ganjia basalts were derived from the asthenosphere source.

5.3.4 Sr-Nd isotopes

The Ganjia basalts display lower $^{87}Sr/^{86}Sr$ and higher $^{143}Nd/^{144}Nd$ ratios relative to depleted mantle, plotting into the Hawaii OIB fields in Fig. 6. Additionally, the Sr-Nd isotopic systematics is distinct from that of Cenozoic basalts in East China. (Zou et al., 2000). Except for the GJ-1, the deleted Sr-Nd isotopic compositions of Ganjia basalts plot in the field near to the FOZO (Hart et al., 1992) are indicative of involvement of FOZO-like component in magma source.

6　Tectonic implication

Li et al. (2013) and Zhang et al. (2018) reported the ^{40}Ar/^{39}Ar plateau age of 112 Ma and 103 ± 2 Ma defined by hornblende grains and whole rock, respectively. Pu et al. (2019) reported a zircon U-Pb age of 106.27 ± 1.3 Ma. These data indicate that the Ganjia basalts erupted during the Early Cretaceous. The Qinling orogenic belt was formed by the collision of the North China Block and South China Block in the Early Mesozoic. During the Late Jurassic to Cretaceous, eastern Asian was affected by the subduction of the Paleo-Pacific beneath the North China Block. Recent studies indicate the potential influence of the subduction of Paleo-Pacific plate in the Qinling area (Qiu, Kong et al., 2022; Qin, Yan et al., 2022; Zhang et al., 2022). The abundances and ratios of certain immobile elements (such as Nb, Ta, Zr, and Hf) are important indicators to distinguish different geotectonic backgrounds (Wood, 1980). According to (Fig. 10a), nearly all the target basalts plotted in the WPA (intraplate alkaline basalts) area. The Hf/3-Th-Ta diagram (Fig. 10b) shows that the majority of these samples plotted in the WPA area, although three sample plotted in the WPB (intraplate tholeiitic basalt) area. The Ganjia basalts showed no significant negative Nb-Ta-Ti anomalies, which differed from the characteristics of basalt at the boundaries of convergent plates. Thus, the Ganjia basalts formed in an intraplate environment during the Early Cretaceous. Early Cretaceous magmatism in the West Qinling Orogenic Belt is dominated by alkali basalts, which are spatially associated with syn-depositional faulting (Qinghai BGMR, 1991). Three models have been proposed to explain the formation of alkaline basalt: ① splash mantle plumes

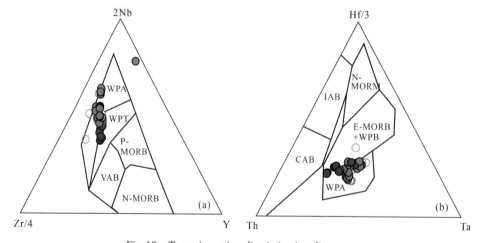

Fig. 10　Tectonic setting discrimination diagrams.

(a) 2Nb-Zr/4-Y diagram of the alkali basalts. (b) Hf/3-Th-Ta diagram of the alkaline basalts. CAB: calc-alkali basalts; N-MORB: normal mid-ocean ridge basalt; E-MORB: enriched mantle; IAB: island arc basalt; P-MORB: mantle plume; VAB: volcanic arc basalt; WPA: intraplate alkaline basalt; WPB: intraplate basalt; WPT: intraplate tholeiite. The published data are the same as Fig. 7.

（Kuritani et al., 2019）, ② passive upwelling-induced decompression melting of the asthenosphere caused by the thickness lithospheric rifting （Hammond et al., 2013）, and ③ edge-driven convection caused by steps in lithosphere structure combined with plate movement （Dai et al., 2021; Ma et al., 2011; Missenard and Cadoux et al., 2011; Moufti et al., 2012）. Nevertheless, most intraplate volcanic rocks in the West Qinling Orogenic Belt occur with small volumes, and the mantle potential temperatures beneath the Ganjia area vary from 1 215 ℃ to 1 330 ℃, lower than the temperature of plume-type mantle（~1 600 ℃; Putirka, 2016）. Therefore, no high-temperature mantle plume existed in the Ganjia area. Instead, the model of passive mantle upwelling of asthenosphere with a small degree of melting accompanied by thinning of the continental lithosphere is more consistent with the geological framework of West Qinling Orogenic Belt at that time. The composition of the volcanic rocks varied from intermediate-acidic calc-alkaline andesite and basaltic andesite at Duofutun to alkaline basalts at Ganjia. The trend of these compositions likely indicates the involvement of detach lithosphere resulting from gravitational instability following crustal shortening （Fig. 11）. In response to the crustal shortening, the thickened lithospheric root was removed. The removed root fall into the asthenosphere causing the thermal interface of asthenospheric mantle was uplifted （Yan et al., 2012）. The interaction between the upwelling asthenosphere and lithosphere around the asthenosphere-lithosphere boundary layer could readily contribute enough materials and heat flux to the primary magmas （Gao et al., 2004; Pang et al., 2012）. Moreover, the Wanggeertang-Ganjia Fault in the Ganjia basin is the westward extension of the NWQF, which a translithospheric fault that cuts through the lithosphere and reaches the upper mantle （Shi, 2011; Zhang, 2012）. The faults caused the lithosphere to break up and form a

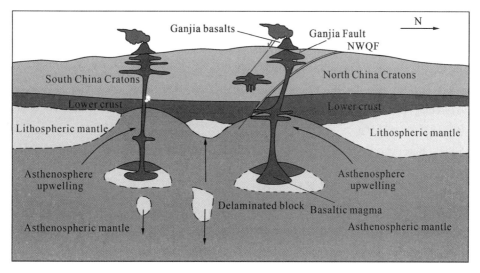

Fig. 11　Schematic diagram of petrogenesis and evolutionary model of Ganjia basalts.

variety of small blocks in the West Qinling Orogenic Belt (Xin, 2013). The magma migrated rapidly upward along the local fault, without significant interchange of material or heat conduction with surrounding crust. The upwelling asthenosphere and translithospheric fault jointly contributed to the Cretaceous magmatism in the Ganjia area.

7 Conclusion

(1) The mantle potential temperature beneath the Ganjia area was 1 215–1 330 ℃, and the crystallization temperature of the olivine was 1 106–1 270 ℃.

(2) The oxybarometer results showed that the oxygen fugacity of Ganjia basalts varied from FMQ −1 to FMQ 0, which indicate a relatively reduced environment.

(3) The olivine was characteried by low Ni and high Ca contents, which indicate that the lithology of the mantle source for the Ganjia basalts was peridotite.

(4) The generation of the Ganjia basalt was related to the interaction between upwelling asthenosphere and a lithospheric source.

Acknowledgement　This work was supported by the National Natural Science Foundation of China under (Grant No. 42172056).

References

Allen, M.B., Kheirkhah ,M., Neill ,I., Emami, M.H., Mcleod, C.L., 2013. Generation of arc and within-plate chemical signatures in collision zone magmatism: Quaternary lavas from Kurdistan Province, Iran. Journal of Petrology, 54, 887-911.

Ammannati, E., Jacob, D.E., Avanzinelli, R., Foley, S.F., Conticelli, S., 2016. Low Ni olivine in silica-undersaturated ultrapotassic igneous rocks as evidence for carbonate metasomatism in the mantle. Earth & Planetary Science Letters, 444, 64-74.

Ballhaus, C., Berry, R.F., Green, D.H., 1991. High pressure experimental calibration of the olivine-orthopyroxene-spinel oxygen geobarometer: Implications for the oxidation state of the upper mantle. Contributions to Mineralogy and Petrology, 107(1), 27-40.

Bian, Q.T., Li, D.H., Pospelov, I., Yin, L.M., Li, H.S., Zhao, D.S., et al., 2004. Age, geochemistry and tectonic setting of Buqingshan ophiolites, North Qinghai-Tibet Plateau, China. Journal of Asian Earth Sciences, 23(4), 577-596.

Blundy, J.D., Robinson, J.A.C., Wood, B.J., 1998. Heavy REE are compatible in clinopyroxene on the spinel lherzolite solidus. Earth & Planetary Science Letters, 160, 493-504.

Chang, J.M., Feeley, T.C., Deraps, M.R., 2009. Petrogenesis of basaltic volcanic rocks from the Pribilof Islands, Alaska, by melting of metasomatically enriched depleted lithosphere, crystallization differentiation, and magma mixing. Journal of Petrology, 50, 2249-2286.

Chen, L.H., Zeng, G., Hu, S.L., Yu, X., Chen, X.Y., 2012. Crustal recycling and genesis of continental alkaline basalts: Case study of the Cenozoic alkaline basalts from Shandong Province, East China. Geological Journal of China Universities, 18(1),16-27.

Coogan, L.A., Saunders, A.D., Wilson, R.N., 2014. Aluminum-in-olivine thermometry of primitive basalts: Evidence of an anomalously hot mantle source for large igneous provinces. Chemical Geology, 368, 1-10.

Dai, H.K., Beñat, O., Zheng, J.P., Griffin, W.L., Afonso, J.C., Xiong, Q., O'Reilly, S.Y., 2021. Melting dynamics of Late Cretaceous lamprophyres in Central Asia suggest a mechanism to explain many continental intraplate basaltic suite magmatic provinces. Journal of Geophysical Research: Solid Earth, 126, e2021JB021663.

Dai, L.Q., 2014a. A geochemical study of Early Cretaceous postsollisional mafic igneous rocks from the Qinling-Hongan-Dabie orogens [Ph. D. thesis]. Hefei: University of Science and Technology of China.

Dai, L.Q., Zhao, Z.F., Zheng, Y.F., 2014b. Geochemical insights into the role of metasomatic hornblendite in generating alkali basalts. Geochemistry, Geophysics, Geosystems, 15, 3762-3779.

Dai, L.Q., Zheng, F., Zhao, Z.F., Zheng, Y.F., 2018. Geochemical insights into the lithology of mantle sources for Cenozoic alkali basalts in West Qinling, China. Lithos, 302-303, 86-98.

Dasgupta, R., Hirschmann, M.M., Smith, N. D., 2007. Partial melting experiments of peridotite + CO_2 at 3 GPa and genesis of alkalic Ocean Island Basalts. Journal of Petrology, 47(4), 647-671.

Dasgupta, R., Hirschmann, M.M., Stalker K., 2006. Immiscible transition from carbonate-rich to silicate-rich melts in the 3GPa melting interval of eclogite + CO_2 and genesis of silica-undersaturated Ocean Island Lavas. Journal of Petrology, 47(4), 647-671.

De Hoog, J.C.M., Gall, L., Cornell, D., 2010. Trace element geochemistry of mantle olivine and applications to mantle petrogenesis and geo-thermo-barometry. Chemical Geology, 270, 196-215.

Ding, Y., Yu, X.H., Mo, X. X., Li, X.W., Huang, X.F., Wei, P., 2013. Geochronology, geochemistry and petrogenesis of the Hongqiang basalts from Northeast Qinghai-Tibetan Plateau. Earth Science Frontiers, 20, 180-191 (in Chinese with English Abstract).

Dong, Y., Zhang, G.W., Neubauer, F., Liu, X., Genser, J., Hauzenberger, C., 2011. Tectonic evolution of the Qinling orogen, China: Review and synthesis. Journal of Asian Earth Sciences, 41(3), 213-237.

Dong, Y.P., Liu, X.M., Neubauer, F., Zhang, G.W., Li, W., 2013. Timing of paleozoic amalgamation between the North China and South China Blocks: Evidence from detrital zircon U-Pb ages. Tectonophysics, 586.

Dong, Y.P., Yang, Z., Liu, X.M., Sun, S.S., Zhang, G.W., 2016. Mesozoic intracontinental orogeny in the Qinling Mountains, central China. Gondwana Research, 30, 144-158.

Dong, Y.P., Zhang, G.W., Hauzenberger, C., Neubauer, F., Yang, Z., Liu, X.M., 2011a. Paleozoic tectonics and evolutionary history of the Qinling orogen: Evidence from geochemistry and geochronology of ophiolite and related volcanic rocks. Lithos, 122, 39-56.

Fan, L.Y., Wang, Y.J., Li, X.Y., 2007. Geochemical characteristics of late mesozoic mafic volcanic rocks from Westem Qinling and its tectonic implications. Journal of Mineralogy and Petrology, 27, 63-72 (in Chinese with English Abstract).

Feng, Y.M., Cao, X.D., Zhang, E.P., Hu, Y.X., Pan, X.P., Yang, J.L., Jia, Q.Z., Li, W.M., 2003. Tectonic evolution framework and nature of the West Qinling orogenic belt. Northwestern Geology, 36, 1-10 (in Chinese with English abstract).

Foley, S.F., Jacob, D.E., O'Neill, H.S.C., 2011. Trace element variations in olivine phenocrysts from

Ugandan potassic rocks as clues to the chemical characteristics of parental magmas. Contributions to Mineralogy and Petrology, 162, 1-20.

Foley, S.F., Prelevic, D., Rehfeldt, T., Jacob, D. E., 2013. Minor and trace elements in olivines as probes into early igneous and mantle melting processes. Earth & Planetary Sciences Letters, 363(2), 181-191.

Frost, D.J., Mccammon, C.A., 2008. The redox state of earth's mantle. Annual Review of Earth and Planetary Sciences, 36(1), 389-420.

Gao, S., Rudnick, R.L., Yuan, H.L., Liu, X.M., Liu, Y.S., Xu, W.L., Ling, W.L., Ayers, J., Wang, X.C., Wang, Q.H., 2004. Recycling lower continental crust in the North China craton. Nature, 432, 892-897.

Geng, X.L., Foley, S.F., Liu, Y.S., Wang, Z.C., Hu, Z.C., Zhou, L., 2019. Thermal-chemical conditions of the north China mesozoic lithospheric mantle and implication for the lithospheric thinning of cratons. Earth & Planetary Science Letters, 516, 1-11.

Gómez-Ulla, A., Sigmarsson, O., Gudfinnsson, G. H., 2017. Trace element systematics of olivine from historical eruptions of Lanzarote, Canary Islands: Constraints on mantle source and melting mode. Chemical Geology, 449, 99-111.

Green, D. H., Ringwood, A. E., 1963. Mineral assemblages in a model mantle composition. Journal of Geophysical Research, 68(3), 937-945.

Hammond, J.O.S., Kendall, J.M., Stuart, G.W., Ebinger, C.J., Wright, T. J., 2013. Mantle upwelling and initiation of rift segmentation beneath the afar depression. Geology, 41(6), 635-638.

Hart, S.R., Hauri, E.H., Oschmann, L.A., Whitehead, J.A., 1992. Mantle plumes and entrainment: Isotopic evidence. Science, 256, 517-520.

Hauri, E.H., Whitehead, J.A., Hart, S.R., 1994. Fluid dynamic and geochemical aspects of entrainment in mantle plumes. Journal of Geophysical Research, 99, 24275-24300.

Herzberg, C., 2006. Petrology and thermal structure of the Hawaiian plume from Mauna kea volcano. Nature, 444(7119), 605-609.

Herzberg, C., 2011. Identification of source lithology in the Hawaiian and Canary islands: Implications for origins. Journal of Petrology, 1, 113-146.

Herzberg, C., O'Hara, M. J., 2002. Plume-associated ultramafic magmas of phanerozoic age. Journal of Petrology, 43 (10), 1857-883.

Hofmann, A.W., Jochum, K.P., Seufer, M., White, W.M., 1986. Nb and Pb in oceanic basalts: New constraints on mantle evolution. Earth & Planetary Science Letters, 79, 33-45.

Hoskin, P.W.O., Black, L.P., 2010. Metamorphic zircon formation by solid-state recrystallization of protolith igneous zircon. Journal of Metamorphic Geology, 18(4), 423-439.

Howarth, G.H., Harris, C., 2017. Discriminating between pyroxenite and peridotite sources for continental flood basalts (CFB) in southern Africa using olivine chemistry. Earth & Planetary Science Letters, 475, 143-151.

Huang, X.F., Yu, X.H., Mo, X.X., Li, X.W., Ding, Y., Wei, P., He, W.Y., Yu, J.C., 2013. The discovery of OIB-type potassic tholeiitic from the Ganjia Area in West Qinling: Implications for the Late Mesozoic continental rift of West Qinling. Earth Science Froniters, 20(3), 204-216.

Hu, X.J., Guo, A.L., Zong, C.L.,Zhang, C.F., Guo, Y.Y., Zhang, L., 2012. $^{40}Ar/^{39}Ar$ isotopic dating, geochemistry and their tectonic implications of Duofutun Na-rich mafic volcanic rocks, the Northeastern margin of the Qinghai-Tibet Plateau. Journal of Northwest University: Natural Science Edition, 42(3), 102 (in Chinese with English abstract).

Kang, L.X., 1990. Main characteristics of seismic topography on fault zone along north fringe of Qinling Mountain.Crustal Deformation and Earthquake, 10(1), 9 (in Chinese with English abstract).

Kuritani, T., Xia, Q.K., Kimura, J.I., Liu, J., Shimizu, K., Ushikubo, T., et al., 2019. Buoyant hydrous mantle plume from the mantle transition zone. Scientific Reports, 9, 6549

Lai, S. C., Zhang, G. W., Li, Y. F., Qin, J. F., 2007. Petrogenesis and significance of deep dynamics of sodic alkaline basalts in Madang, eastern margin of Qinghai-Tibet Plateau. Science China, 37(A01), 8 (in Chinese with English abstract).

Lai, S.C., Zhu, R.Z., Qin, J.F., Zhao,S.W., Zhu, Y., 2022. Granitic magmatism in eastern Tethys domain (Western China) and their geodynamic implications. Acta Geologica Sinica: English Edition, 96(2), 15.

Lambart, S., Stolper, E., Baker, M., et al., 2016. The role of pyroxenite in basalt genesis: Melt-px, a melting parameterization for mantle pyroxenites between 0. 9 and 5 gpa. Journal of Geophysical Research: Solid Earth, 121(8), 5708-5735.

LeBas, M.J., Maitre, R.W., Streckeisen, A., Zanettin, B., 1986. A chemical classification of volcanic rocks based on the total alkali-silica diagram. Journal of Petrology, 27(3), 745-750.

Le Roux, V., Dasgupta, R., Lee, C. T. A, 2011. Mineralogical heterogeneities in the earth's mantle: Constraints from Mn, Co, Ni and Zn partitioning during partial melting. Earth & Planetary Science Letters, 307(3-4), 395-408.

Lee, C.T.A., Luffi, P., Plank, T., et al., 2009. Constraints on the depths and temperatures of basaltic magma generation on Earth and other terrestrial planets using new thermobarometers for mafic magma. Earth & Planetary Science Letters, 279(1-2), 20-33.

Li, X.W., Mo, X.X., Yu, X.H., Ding, Y., Huang, X.F., Wei, P., Wen, Y.H., 2013. Geochronological, geochemical and Sr-Nd-Hf isotopic constraints on the origin of the cretaceous intraplate volcanism in west Qinling, Central China: Implications for asthenosphere-lithosphere interaction. Lithos, 177, 381-401.

Liu. J., Xia, Q.K., Kuritani, T., Hanski. E., Yu. H.R., 2017. Mantle hydration and the role of water in the generation of large igneous provinces. Nature Communications, 8(1), 1824.

Liu, J.Q., Ren, Z.Y., 2013. Diversity of Source Lithology and its Identification for Basalts: A Case Study of the Hainan Basalts. Geotectonica et Metallogenia, 37(3), 18 (in Chinese with English abstract).

Liu, Y., Liu, X.M., Hu, Z.C.,Diwu, C.R., Yuan, H.L., Gao, S., 2007. Evaluation of accuracy and long-term stability of determination of 37 trace elements in geological samples by ICP-MS. Acta Petrologica Sinica. 23, 1203-1210 (in Chinese with English abstract).

Lu, S.M, Wang, A.G, Wen, A.G., Zhang, B., 2021. Evidence of new activities along the western margin fault zone of Ganjia Basin. China Earthquake Engineering Journal, 43(5), 1045-1053 (in Chinese with English abstract).

Ma, G.S.K., Malpas, J., Xenophontos, C., Chan, G.H.N., 2011. Petrogenesis of latest Miocene-quaternary continental intraplate volcanism along the Northern Dead Sea Fault System (Al Ghab-Homs Volcanic Field),

Western Syria: Evidence for lithosphere-asthenosphere interaction. Journal of Petrology, 52, 401-430.

Mallmann, G., O'Neill, H.S.C., 2013. Calibration of an empirical thermometer and oxybarometer based on the partitioning of Sc, Y and V between olivine and silicate melt. Journal of Petrology, (6), 933-949.

Mckenzie, D., O'Nions, R.K., 1991. Partial melt distributions from inversion of rare earth element concentrations. Journal of Petrology, 32, 1021-1091.

Missenard, Y., Cadoux, A., 2012. Can Moroccan Atlas lithospheric thinning and volcanism be induced by edge-driven convection? Terra Nova, 24, 27-33.

Moufti, M.R., Moghazi, A.M., Ali, K.A., 2012. Geochemistry and Sr-Nd-Pb isotopic composition of the Harrat Al-Madinah Volcanic Field, Saudi Arabia. Gondwana Research, 21, 670-689.

Neave, D.A., Shorttle, O., Oeser, M., Weyer, S., Kobayashi, K., 2018. Mantle-derived trace element variability in olivines and their melt inclusions. Earth & Planetary Science Letters: A Letter Journal Devoted to the Development in Time of the Earth and Planetary System, 483, 90-104.

Niu, Y.L., Marjorie, W., Humphreys, E.R., O'Hara, M.J., 2011. The origin of intra-plate ocean island basalts (OIB): The lid effect and its geodynamic implications. Journal of Petrology, 7-8, 1443-1468.

O'Hara, M.J., Yoder, H.S., 1967. Formation and fractionation of basic magmas at high pressures. Scottish Journal of Geology, 3(1), 67-117.

Pang, K.N., Chung, S.L., Zarrinkoub, M.H., Mohammadi, S.S., Yang, H.M., Chu, C.H., Lee, H.Y., Lo, C.H., 2012. Age, geochemical characteristics and petrogenesis of late Cenozoic intraplate alkali basalts in the Lut-Sistan region, eastern Iran. Chemical Geology, 306-307, 40-53.

Pearce, J.A., 2008. Geochemical fingerprinting of oceanic basalts with applications to ophiolite classification and the search for Archean oceanic crust. Lithos, 00, 18-14.

Pearce, J.A., Cann, J.R., 1973. Tectonic setting of basic volcanic rocks determined using trace element analyses. Earth & Planetary Science Letters, 19, 290-300.

Pfänder, J.A., Jung, S., Klügel, A., et al., 2018. Recurrent local melting of metasomatised lithospheric mantle in response to continental rifting: Constraints from basanites and nephelinites/melilitites from SE Germany. Journal of Petrology, 59, 667-694.

Pfänder, J.A., Sperner, B., Ratschbacher, L., Fischer, A., Meyer, M., Leistner, M., Schaeben, H., 2014. High-resolution $^{40}Ar/^{39}Ar$ dating using a mechanical sample transfer system combined with a high-temperature cell for step heating experiments and a multicollector ARGUS noble gas mass spectrometer. Geochemistry, Geophysics, Geosystems, 15, 2713.

Pilet, S., Baker, M.B., Stolper, E.M., 2008. Metasomatized lithosphere and the origin of alkaline lavas. Science, 320, 916-91.

Pu, W.F., Zhong, X., Xu, Y.B., Wang, S.H., Wang, H.T., 2019. Geochronology, geochemistry and tectonic significance of Tengbu-Rizhou Early Cretaceous basalts in Xiahe, West Qinling, Gansu Province. Earth Science Frontiers, 2019, 26(5), 304-316 (in Chinese with English abstract).

Putirka, K.D., 2005. Mantle potential temperatures at Hawaii, Iceland, and the Mid-Ocean Ridge system, as inferred from olivine phenocrysts: Evidence for thermally driven mantle plumes. Geochemistry, Geophysics, Geosystems, 6.

Putirka, K.D., 2016. Rates and styles of planetary cooling on Earth, Moon, Mars, and Vesta, using new models for oxygen fugacity, ferric-ferrous ratios, olivine-liquid Fe-Mg exchange, and mantle potential temperature. The American Mineralogist, 101,819-840.

Putirka, K.D., Perfit, M., Ryerson, F.J., Jackson, M.G., 2007. Ambient and excess mantle temperatures, olivine thermometry, and active vs. passive upwelling. Chemical Geology, 241(3-4), 177-206.

QBGMR (Qinghai Bureau of Geology and Mineral Resources), 1991. Regional Geology of Qinghai Province. Beijing: Geological Publishing House, 662 (in Chinese).

Qi, S.X., Deng, J.F., Chen. J., Fu, J., Shi, L.C., 2012. The determination and significance of lower Cretaceous continental rift environment volcanic rock in Tongren area, Qinghai Province. Northwestern Geology, 45(1), 13 (in Chinese with English abstract).

Qiu, L., Kong, R.Y., Yan, D.P., Mu, H.X., Sun, W.H., Sun, S.H., Han, Y.G., Li, C.M., Zhang, L. L., Cao, F.D., Ariser, S., 2022. Paleo-Pacific plate subduction on the eastern Asian margin: Insights from the Jurassic foreland system of the overriding plate. GSA Bulletin, 134 (9-10), 2305-2320.

Qiu, L., Yan, D.P., Ma, H.B., Sun, S,H., Deng, H.L., Zhou, Z.C., Zhu, J.C., Zhang, Q.H., Shi, H. T., Wang, X., 2022. Late Cretaceous geodynamics of the Palaeo-Pacific plate inferred from basin inversion structures in the Songliao Basin (NE China). Terra Nova, 34, 465- 473.

Rudnick, R.L., Gao, S., 2003. Composition of the Continental Crust. In: Heinrich, D.H., Karl, K.T. Treatise on Geochemistry, 3. Oxford: Pergamon, 1-64.

Shi, H.T., 2011.The Structral Framework of Duohemao, Qinghai Province [MSc. thesis]. Chinese University of Geoscience.

Sobolev, A. V., Hofmann, A. W., Kuzmin, D. V., Yaxley, G. M., Arndt, N. T., Chung, S. L., et al., 2007.The amount of recycled crust in sources of mantle-derived melts. Science, 316, 412-417.

Sobolev, A.V., Hofmann, A.W., Sobolev, A.V., Nikogosian, I.K., 2005. An olivine-free mantle source of Hawaiian shield basalts. Nature, 434, 590-597.

Straub, S.M., La Gatta, A.B., Martin-Del Pozzo, A.L., Langmuir, C.H., 2008. Evidence from high-Ni olivines for a hybridized peridotite/pyroxenite source for orogenic andesites from the central Mexican volcanic belt. Geochemistry, Geophysics, Geosystems, 9, Q03007.

Sun, S. S., McDonough, W. F., 1989. Chemical and isotopic systematics of oceanic basalts: Implications for mantle composition and processes. Geological Society Special Publication, 42, 313-345.

Taylor, S. R., McLennan, S. M., 1995. The geochemical evolution of the continental crust. Reviews of Geophysics, 33, 241-265.

Walter, M. J., 1998. Melting of garnet peridotite and the origin of komatiite and depleted lithosphere. Journal of Petrology, 39, 29-60.

Weis, D., Kieffer, B., Maerschalk, C., Barling, J., Jong, J.D., Williams, G.A., Hanano, D., Pretorius, W., Mattielli, N., Scoates, J., Goolaerts, A., Friedman, R., Mahoney, J.,2006. High-precision isotopic characterization of USGS reference materials by TIMS and MC-ICP-MS. Geochemistry, Geophysics, Geosystems, 7, 139-149.

Wilson, M., 1989. Igneous Petrogenesis: A Global Tectonic Approach. London: Unwin Hyman, 1-466.

Wood, D. A., 1980. The application of a Th-Hf-Ta diagram to problems of tectonomagmatic classification. Earth & Planetary Science Letters, 50(1), 1-30.

Xin, D., 2013. Cretaceous volcanic geology and tectonic significance of Dogarmo, Qinghai [MSc. thesis]. Chinese University of Geoscience.

Yan, Z., Guo, X.Q., Fu, C.L., Wang, T., Wang, Z.Q., Li, J.L., 2012. Petrology, geochemistry and SHRIMP U-Pb dating of zircons from Late Triassic OIB-basalts in the conjunction of the Qinling-Kunlun-orogens. Earth Science Frontiers, 2012, 19(5), 164-176 (in Chinese with English abstract).

Yang, Z. F., Li, J., Liang, W. F., Luo, Z. H., 2016. On the chemical markers of pyroxenite contributions in continental basalts in eastern China: Implications for source lithology and the origin of basalts. Earth-Science Reviews, 157, 18-31.

Yang, Z. F., Zhou, J. H., 2013. Can we identify source lithology of basalt? Scientific Reports, 3, 1856.

Yaxley, G. M., Green, D. H., Kamenetsky, V., 1998. Carbonatite metasomatism in the south-eastern Australian lithosphere. Journal of Petrology, 39, 1917-1930.

Yu, X. H., Mo, X. X., Liao, Z. L., Zhao, X., Su, Q., 2001. Temperature and pressure conditions of Garnet dipyroxene and Garnet Iherzolite inclusion in the Western Qinling. Science China, 31, 128-133 (in Chinese with English abstract).

Zeng, G., Chen, L.H., Xu, X.S., Jiang, S.Y., Hofmann, A.W., 2010. Carbonated mantle sources for Cenozoic intra-plate alkaline basalts in Shandong, North China. Chemical Geology, 273(1-2), 35-45.

Zhang, B., 2012. The study of new activities on western segment of northern margin of western Qingling fault and Laji Shan fault [MSc. Thesis]. Lanzhou: Lanzhou Institute of Seismology, China Earthquake Administration.

Zhang, F.F., Wang, Y.J, Cawood, P.A., Dong, Y.P., 2018. Geochemistry, $^{40}Ar/^{39}Ar$ geochronology, and geodynamic implications of early cretaceous basalts from the western qinling orogenic belt, China. Journal of Asian Earth Sciences, 151, 62-72.

Zhang, F.Y., Lai, S.C., Qin, J.F., Zhu, R.Z., Yang, H., 2020. Vein-plus-wall rock melting model for the origin of Early Paleozoic alkali diabases in the South Qinling belt, Central China. Lithos, 307/301, 105619.

Zhang, G.W., Dong, Y.P., Lai, S.C., 2004. Mianlue tectonic zone and Mianlue suture zone on southern margin of Qinling-Dabie orogenic belt. Science in China: Series D, 47, 300-316.

Zhang, L.Y., Li, N., Prelevic, D., 2016. The research status of olivine trace elements in-situ analysis and perspectives of its application. Acta Petrologica Sinica, 32(6), 1877-1890 (in Chinese with English abstract).

Zhang, W., Wang, F., Wu, L., Yang, L., Tan, X., Shi, W., Xu, X.W., 2022. Reactivated margin of the western north china craton in the late cretaceous: Constraints from zircon (U-Th)/He thermochronology of Taibai Mountain. Tectonics, 41, e2021TC007058.

Zhao, W.T., 2009. Petrological, Geochemical compositions and the origin of the Mesozoic basalts at the northeastern Tibetan Plateau [MSc. thesis]. Chinese University of Geoscience.

Zindler, A., Hart, S., 1986. Chemical geodynamics. Annu. Rev. Earth & Planetary Science Letters, 14, 493-571.

Zou, H., Zindler, A., Xu, X.S., Qi, Q., 2000. Major, trace element, and Nd, Sr and Pb isotope studies of Cenozoic basalts in SE China: Mantle sources, regional variations, and tectonic significance. Chemical Geology, 171, 33-47.

U-Pb zircon geochronology, geochemistry, and Sr-Nd-Pb-Hf isotopic composition of the Late Cretaceous monzogranite from the north of the Yidun Arc, Tibetan Plateau Eastern, SW China: Petrogenesis and tectonic implication[①]

Gan Baoping　Lai Shaocong[②]　Qin Jiangfeng　Zhu Renzhi　Zhu Yu

Abstract: The Yidun Arc belt, a Triassic volcanic arc, was the result of Late Triassic large-scale subduction orogenic process of Paleo-Tethys and exposed the voluminous intermediate-acid intrusions during Mesozoic. In this paper, we mainly present petrography and petrology, major and trace element, Sr-Nd-Pb-Hf isotopic compositions, and zircon U-Pb ages for the monzogranites from the Queer Mountains area at the northern of Yidun Arc. The Queer Mountains monzogranites were emplaced at 93.6 ± 0.9 Ma (MSWD $= 6.8$, $n = 34$, 2σ), revealing that the monzogranites are the products of magmatism during the period of Late Cretaceous ("Yanshanian"). The monzogranites have high SiO_2, Al_2O_3, and K_2O contents and enrichment in light rare earth elements (LREEs) and large-ion lithophile elements (LILEs) but depleted in high field strength elements contents (HFSEs), with relatively high variable $(La/Yb)_N$ values ($1.38 - 11.79$) and evidently negative Eu anomalies ($Eu/Eu^* = 0.07 - 0.35$). The samples have high initial $^{87}Sr/^{86}Sr$ ($0.708\,421 - 0.715\,698$), with negative $\varepsilon_{Nd}(t)$ (-3.57 to -0.53) and $\varepsilon_{Hf}(t)$ (-3.2 to -0.4) values, and the Nd and Hf of two-stage model ages are $1.03 - 1.24$ and $1.18 - 1.35$ Ga, respectively. The $(^{206}Pb/^{204}Pb)_t$, $(^{207}Pb/^{204}Pb)_t$, and $(^{208}Pb/^{204}Pb)_t$ values of the monzogranites vary from $18.723\,4$ to $18.843\,6$, from $15.703\,1$ to $15.726\,1$, and from $39.050\,7$ to $39.278\,0$, respectively. The geochemistry signatures indicate that the granite samples may be mainly derived from the melts of upper crust. According to above geochemical signatures, the mature of monzogranites source was mainly a metapelite source, and they should be generated from the Mesoproterozoic crust material affinity to Yangtze Craton, later underwent extensive crustal contamination and fractional crystallization of K-feldspar during emplacement. These monzogranites maybe undergo crystallization temperatures from 730 ℃ to 804 ℃. According to the previous studies and coeval magmatic-tectonic thermal event of the Yidun Arc area and this study, the Queer Mountains monzogranite was most likely formed in a post-collisional extensional tectonic setting after the syn-collision in Yidun Arc region.

---

① Published in *Arabian Journal of Geosciences*, 2018, 11.
② Corresponding author.

Introduction

As widely accepted, the Sanjiang Orogenic Belt (SOB), which mainly includes Nujiang, Lancangjiang, and Jinshajiang, is lying at the strong collision region between Eurasia and Gondwanaland in SW China (Lü et al., 1993; Mo et al., 1993; Zhong, 2000; Qu et al., 2002; Hou et al., 2003, 2004). Besides, the SOB is also closely associated with the Tethyan giant metallogenic belt (Mo et al., 1993; Hou et al., 2004, 2007; Xu et al., 2006; Zaw et al., 2007; Wang et al., 2014a; Li et al., 2017). It consists mainly of two ophiolitic suture zones, volcanic arcs, and several terranes (Fig. 1a,b; Wang et al., 2000; Zhong, 2000; Jian et al., 2009). The Yidun Arc is a major tectonic unit part of the Sanjiang area (Liu et al., 2006b) and a Mo-Cu-Ag polymetallic ore-forming belts, too (Hou, 1993; Hou et al., 2007; Qu et al., 2002; Zaw et al., 2007; Leng et al., 2012, 2014; Wang et al., 2014b; Peng et al., 2014; Li et al., 2017), and generated by the subduction of the Paleo-Tethys Ocean crustal slabs (Mo et al., 1993; Hou et al., 2007; Lai et al., 2010; Wang et al., 2014a). In this complex area, the complex evolution of the Paleo-Tethys has caused numerous important tectonic-magmatic thermal events, and the framework of temporal and spatial evolvement has been established (Hou and Mo, 1993; Hou et al., 2003; Lü et al., 1993; Wang et al., 2000; Pan et al., 2003; Reid et al., 2007; Peng et al., 2014; Hu et al., 2005; Zi et al., 2012).

The Yidun Arc's formation and evolution not only had a significant constraint effect for the Tethys tectonic evolution but also play the most vital role in clarifying ore-forming of a collision-orogenic belt (Qu et al., 2002). There is a consensus that the Yidun continental arc of Triassic was formed by the westward subduction of Garzê-Litang Ocean around ca. 235−210 Ma (Hou and Mo, 1993; Hou et al., 2003; Mo et al., 1993; Li et al., 2007; Fang et al., 2017; Li et al., 2017). After the Yidun Arc underwent a complex subduction to collision orogeny to extension tectonic evolution with some magmatic events (Yang et al., 2015), felsic to basic intrusive rocks formed in Late Triassic-Cretaceous (Yang et al., 2016). The numerous Mesozoic calc-alkaline rocks were exposed in Yidun Arc. Meanwhile, these rocks were intruded into the metamorphosed Paleozoic rocks and volcano-sedimentary sequences (Fig. 1c; Reid et al., 2007). Data from previous research indicate that the magmatic events in this area mainly focused on the period of Indosinian (from Late Triassic to Cretaceous) (Lü et al., 1993; Hou et al., 2001, 2003, 2004; Qu et al., 2002; Reid et al., 2005a, 2007; Liu et al., 2006b; Wang et al., 2011, 2013a,b; He et al., 2013; Peng et al., 2014; Wu et al., 2014a; Lai and Zhao, 2015). Hou et al. (2001) established a temporal and spatial coordinate for the intermediate-acid volcanics exposed in the Yidun Arc area from the Mesozoic to Cenozoic, and approximately four genetic types of the granites were divided: 238−206 Ma for island arc granites, 138−206 Ma for syn-collision granites, 76−135 Ma for post-collision granites (Pan et

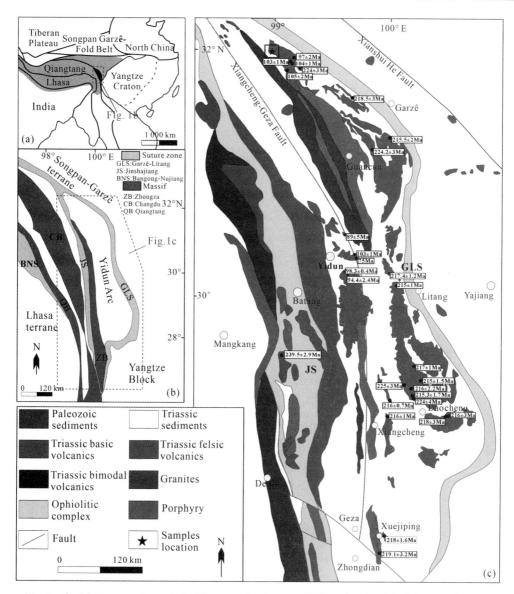

Fig. 1 (a,b)The tectonic geological location sketch map of Yidun Arc (modified from Reid et al., 2005a,b;Hou et al., 2007; Leng et al., 2012; Peng et al., 2014). (c)Simplified stratigraphic map of the Yidun Arc belt (modified after He et al., 2013; Peng et al., 2014; Wu et al., 2017). The location of samples and age data of the granitoids from Qu et al. (2002), Reid et al. (2005a,b, 2007), Liu et al.(2006b, 2016), Yan et al. (2006), Weislogel (2008), Leng et al. (2012, 2014), He et al. (2013), Li et al. (2014), Peng et al. (2014), and Wu et al. (2014a).

al., 2003), and 65－15 Ma for Himalayan granites. However, Reid et al. (2007) only divided the granitoids of the Yidun Arc into three stages from Early-Middle Triassic to Late Triassic to Cretaceous and lacked of record for Jurassic granitic.

The Yidun Arc experienced massive subduction-related orogeny in Late Triassic and subsequently underwent the syn-collision orogeny and post-orogenic extension during the

Jurassic-Cretaceous (Qu et al., 2002; Hou et al., 2003, 2004; Li, 2007; Reid et al., 2007; Yang et al., 2015; Wang et al., 2016). Finally, it underwent the superimposition and reformation of the tectonic deformation of the inner-continental in Cenozoic (Deng et al., 2010a,b, 2014; Liu et al., 2013). In recent years, the researchers have focused on the ore deposit and related granitic intrusions of the Indosinian orogeny in this area, and abundant data have been accumulated. These studies mainly are related to the timing of igneous tectonic thermal events and Cu-Mo-Ag polymetallic mineralization deposits of the Yidun area (Hou and Luo, 1992; Hou et al., 1995, 2003; Qu et al., 2002; Pan et al., 2003; Reid et al., 2005a, 2007; Hou et al., 2007; Li, 2007; Long, 2007; Wang et al., 2013a,b; Leng et al., 2012; Peng et al., 2014; Wang et al., 2014a; Cao et al., 2015; Yu and Li, 2016). Previous researches have paid little attention to the geochemical features, petrogenesis, and geodynamic implication of the Cretaceous volcanics in the Yidun Arc region (Wang et al., 2011, 2013a, b). Because of the lack of detailed geochronology and geochemical data for the Queer Mountains granites, discussing and comprehensively dissecting their petrogenesis and tectonic environment are difficult.

Here, we put forward a new zircon U-Pb data, detailed petrography and petrology descriptions, and geochemical and Sr-Nd-Pb-Hf isotopic composition for the Queer Mountains monzogranites of Yidun Arc. This study's main aims are intended to confirm the emplacement age for the monzogranites, then to constrain the petrogenesis and the nature of primary magma, and finally to discuss their geodynamic implication of the Yidun Arc in the Late Mesozoic.

Regional geology setting

In terms of tectonic, the Yidun Arc tectonic belt, a NNW-trending, is located on the southwestern side of the Songpan-Garzê Fold Belt and the east side of the Qiangtang Block, SW China (Fig. 1a; Hou et al., 2003; Reid et al., 2005a,b, 2007; Liu et al., 2006a,b; Wang et al., 2014a; Wang et al., 2016), and the left and right sides are the Jinshajiang suture zone and the Garzê-Litang suture zone, respectively (Fig. 1b; Roger et al., 2008, 2010; Wang et al., 2011; He et al., 2013). Both of them are important constraint on the evolution of Tethys Ocean (Wang et al., 2000; Zhong, 2000). Meanwhile, this belt develops a lots fracture of the NW-SE trending, which is basically parallel to the Jinshajiang and Garzê-Litang mélange (Cai, 2012; Wu et al., 2014a). The Garzê-Litang suture zone exposed a large number of the granitic plutons during the Middle-Late Triassic (Zhong, 2000; He et al., 2013). And the Jinshajiang suture zone is noticeable for a series of Indosinian granitoids from north to south and shows a line spread. The belts would provide a significance geological record for the evolution of the Paleo-Tethys Ocean subduction collision (Lai et al., 2010; He et al., 2013), particularly, to study the Mesozoic magma evolution in eastern Tibet area, SW China (Reid et al., 2005a; Leng et al., 2012; Peng et al., 2014).

The oldest rocks in Yidun Arc are exposed on the Zhongza micro-continent and are dominated by the weak metamorphic Paleozoic carbonate rocks, clastic rocks, minor mafic volcanic rocks, and the metamorphic basement of Neoproterozoic (Chang, 1997; Roger et al., 2008, 2010; Wu, 2015; Wu et al., 2017) (Fig. 1c). The Triassic strata with intercalated flysch are the most widely distributed in Yidun Arc area (Hou and Luo, 1992; Hou et al., 2003, 2007; Reid et al., 2007), and the exposed area accounts for more 80% of the total area (Wu, 2015). Thereinto, the Middle and Upper Triassic sequences of clastic rocks were the most prevalent strata and mingled with some felsic plutonic, which invaded the Paleozoic meta-sedimentary strata in Yidun Arc area (Hou and Mo, 1991, 1993; Hou et al., 2003; Pan et al., 2003; Reid et al., 2005a; Leng et al., 2012; Yang et al., 2015). And these rocks have a relatively low metamorphic grade (Reid et al., 2005b). Here, exposed voluminous granite and granodiorite plutonic mainly formed in Triassic (ca. 230−206 Ma) (Hou et al., 2001; Reid et al., 2007; Leng et al., 2012; Wu et al., 2017; Yang et al., 2017). Additionally, the sedimentary strata of the Jurassic and Cretaceous are obviously deficient in this area. However, some Cretaceous granites (ca. 107−77 Ma) and Late Jurassic basic igneous rocks are exposed (Qu et al., 2002; Liu et al., 2006a; Reid et al., 2007; Wu et al., 2014b, 2017). The Yidun Arc underwent regionally extensive deformation from Late Triassic to Jurassic (Hou and Mo, 1991; Xu et al., 1992; Roger et al., 2004, 2010; Reid et al., 2005a, 2007). Later, the Tertiary experienced the collision of the Indian plate and the Asian plate and formed numerous strike-slip faults of the NW-SE trending in the Yidun Arc (Wang and Burchfiel, 2000; Reid et al., 2005a).

The Queer Mountains intrusions are located to ~35 km northeastern of the Dege County, Sichuan Province, SW China. The complex body mainly contains monzogranite, syenite granite, and granodiorite exposed approximately 825 km^2 with spread along N-S trending (Fig. 1c), and it is the largest garanitoid intrusive in the northeastern section of Yidun Arc. The wall rocks mainly are comprised of the inter-bed sandstone and slate of Lanashan Formation (T$_3$l) (Yang et al., 2015), and the intrusive contact is relatively obvious between the wall rocks and granitoids.

Samples analytical methods

Ten samples are selected from Queer Mountains area in Dege County, Sichuan Province in this paper. The locations of the ten samples were collected from 31°57.069′N, 98°54.132′E (Fig. 1b). The samples are gray, medium to coarse grain porphyaceous texture, massive structure (Fig. 2a,b) and are composed of anhedral-subhedral of K-feldspar (45%), subhedral and columnar of plagioclase (25%), graininess anhedral-subhedral of quartz

(25%), and black subhedral of biotite (5%). In addition, accessory minerals in the sample are mainly zircon, titanite, and apatite (Fig. 2c,d). Minority biotite minerals also show evidence of alteration, and the margin may have experienced chloritization.

Fig. 2 (a,b) Field, hand-specimen, and microscopic photographs of monzogranites from the Queer Mountains area in the north of the Yidun Arc. (c) Plane-polarized light. (d) Cross-polarized light.
Kfs: K-feldspar; Pl: plagioclase; Qtz: quartz; Bi: biotite.

In this study, the sample data analysis (mainly includes major and trace elements, zircon U-Pb, and Sr-Nd-Pb-Hf isotopic) was carried out at the State Key Laboratory of Continental Dynamics, Northwest University in Xi'an, China.

Zircon U-Pb dating

The zircon grains were separated by the Hebei Regional Geological Survey, using conventional heavy liquid and magnetic techniques. Typical zircon grains were handpicked and mounted in epoxy resin disks and then carefully polished and finally coated with carbon. Internal morphology was examined using cathodoluminescent (CL) prior to U-Pb analyses. Laser, ablation inductively coupled plasma mass spectrometry (LA-ICP-MS) zircon U-Pb analyses were conducted on an Agilent 7500a ICP-MS equipped with a 193-nm laser, following the method of Yuan et al. (2004). The $^{207}Pb/^{235}U$ and $^{206}Pb/^{238}U$ ratios were calculated by using the GLITTER program and corrected by using the Harvard zircon 91500 as an external standard calibration.

Common Pb contents were, therefore, evaluated by using the method described in Andersen (2002). Age calculations and plotting concordia diagrams were completed using Isoplote (version 3.0; Ludwig, 2003). The errors quoted in tables and figures are at the 2σ level.

Major and trace elements

The samples were removed and some weathered surfaces and the fresh parts were cleaned with deionized water and then powdered through a 200-mesh screen using a tungsten carbide ball mill. Major elements analyzed by X-ray fluorescence (XRF; Rikagu RIX 2100) and inductively coupled plasma mass spectrometer (ICP-MS; Agilent 7500a), using USGS and Chinese national rock standards (BCR-2, GSR-1 and GSR-3), showed that the accuracy and precision of experimental analysis for the major elements were generally better than 5%. For the trace element analyses, sample powders were digested using an HF + HNO$_3$ mixture in high-pressure Teflon bombs at 190 ℃ for 48 h. For the trace elements, the analytical error was less than 2% and the precision better than 10% (Liu et al., 2007a).

Sr-Nd-Pb-Hf isotopic

For the Sr-Nd-Pb-Hf isotopic data, analyses were obtained using a Nu Plasma HR multi-collector mass spectrometer. Sr and Nd isotopic fractionations were corrected to $^{87}Sr/^{86}Sr = 0.1194$ and $^{146}Nd/^{144}Nd = 0.7219$. During the period of analysis, the NIST SRM 987 standard yielded an average value of $^{87}Sr/^{86}Sr = 0.710\ 250 \pm 12$ (2σ, $n = 15$) and the La Jolla standard gave a mean value of $^{146}Nd/^{144}Nd = 0.511\ 859 \pm 6$ (2σ, $n = 20$). Whole-rock Pb was separated by anion exchange in HCl-Br columns, and Pb isotopic fractionation was corrected to $^{205}Tl/^{203}Tl = 2.3875$. Over the period of analysis, 30 measurements of NBS 981 gave average values of $^{206}Pb/^{204}Pb = 16.937 \pm 1$ (2σ), $^{207}Pb/^{204}Pb = 15.491 \pm 1$ (2σ), and $^{208}Pb/^{204}Pb = 36.696 \pm 1$ (2σ). BCR-2 standard yielded average values of $^{206}Pb/^{204}Pb = 18.742 \pm 1$ (2σ), $^{207}Pb/^{204}Pb = 15.620 \pm 1$ (2σ), and $^{208}Pb/^{204}Pb = 38.705 \pm 1$ (2σ). Whole-rock Hf was also separated by a single anion exchange column, and 22 measurements of JCM 475 yielded an average value of $^{176}Hf/^{177}Hf = 0.282\ 161\ 3 \pm 0.000\ 001\ 3$ (2σ) (Yuan et al., 2007).

Results

Zircon U-Pb results

Zircon data from the monzogranites (QES-20) of the Queer Mountains area in Dege County, Sichuan Province (Fig. 1c) are listed in Table 1 and shown in Fig. 3.

The grains are mostly colorless or light gray, transparent, euhedral-subhedral in shape and elongate prisms of 50−200 μm, and the length-to-width ratios of 1 : 1 to 4 : 1. These zircons

Table 1　Results of zircon LA-ICP-MS U-Pb results for the monzogranite from the Queer Mountains, northern of the Yidun Arc.

Analysis	Th	U	Th/U	Isotopic ratios								Isotopic age /Ma							
				$^{207}Pb/^{206}Pb$	2σ	$^{207}Pb/^{235}U$	2σ	$^{206}Pb/^{238}U$	2σ	$^{208}Pb/^{232}Th$	2σ	$^{207}Pb/^{206}Pb$	2σ	$^{207}Pb/^{235}U$	2σ	$^{206}Pb/^{238}U$	2σ	$^{208}Pb/^{232}Th$	2σ
QES-20-01	759	6 985	0.1	0.048 57	0.001 01	0.094 02	0.001 73	0.014 04	0.000 14	0.004 45	0.000 04	127	50	91	2	90	1	90	1
QES-20-02	724	6 060	0.1	0.049 06	0.001 05	0.095 31	0.001 83	0.014 09	0.000 14	0.004 46	0.000 04	151	52	92	2	90	1	90	1
QES-20-03	2 374	14 475	0.2	0.050 13	0.000 96	0.100 58	0.001 68	0.014 55	0.000 14	0.004 59	0.000 04	201	46	97	2	93	1	93	1
QES-20-04	645	3 531	0.2	0.048 51	0.001 09	0.094 03	0.001 90	0.014 06	0.000 14	0.004 45	0.000 04	124	54	91	2	90	1	90	1
QES-20-05	602	2 833	0.2	0.049 45	0.001 26	0.101 97	0.001 86	0.014 95	0.000 15	0.004 59	0.000 07	169	24	99	2	96	1	93	1
QES-20-06	609	2 646	0.2	0.051 37	0.001 26	0.105 16	0.002 36	0.014 85	0.000 15	0.004 67	0.000 04	257	58	102	2	95	1	94	1
QES-20-07	919	7 835	0.1	0.051 08	0.001 14	0.104 67	0.001 43	0.014 86	0.000 15	0.006 09	0.000 08	244	15	101	1	95	1	123	2
QES-20-08	1 262	9 118	0.1	0.048 46	0.000 91	0.100 87	0.001 62	0.015 10	0.000 15	0.004 78	0.000 05	122	45	98	1	97	1	96	1
QES-20-09	1 080	2 265	0.5	0.050 11	0.001 20	0.098 13	0.001 59	0.014 20	0.000 14	0.004 31	0.000 05	200	20	95	1	91	1	87	1
QES-20-10	2 218	11 169	0.2	0.048 88	0.001 12	0.097 32	0.002 02	0.014 44	0.000 14	0.004 57	0.000 04	142	55	94	2	92	1	92	1
QES-20-11	1 080	5 099	0.2	0.048 42	0.001 04	0.093 78	0.001 79	0.014 05	0.000 14	0.004 45	0.000 04	120	52	91	2	90	1	90	1
QES-20-12	1 998	5 570	0.4	0.050 64	0.001 68	0.103 39	0.003 24	0.014 81	0.000 16	0.004 67	0.000 04	224	79	100	3	95	1	94	1
QES-20-13	880	4 608	0.2	0.048 55	0.001 09	0.096 02	0.001 95	0.014 34	0.000 14	0.004 54	0.000 04	126	54	93	2	92	1	92	1
QES-20-14	802	2 782	0.3	0.050 27	0.001 46	0.104 48	0.002 83	0.015 07	0.000 16	0.004 75	0.000 04	208	69	101	3	96	1	96	1
QES-20-15	802	6 003	0.1	0.048 35	0.001 06	0.099 02	0.001 31	0.014 85	0.000 15	0.004 83	0.000 06	116	15	96	1	95	1	97	1
QES-20-16	1 453	10 142	0.1	0.048 29	0.000 91	0.094 22	0.001 52	0.014 15	0.000 14	0.004 49	0.000 04	113	45	91	1	91	1	91	1
QES-20-17	549	1 777	0.3	0.051 20	0.002 03	0.106 27	0.003 70	0.015 05	0.000 19	0.004 65	0.000 12	250	57	103	3	96	1	94	2
QES-20-18	892	13 522	0.1	0.052 34	0.001 21	0.108 31	0.001 64	0.015 01	0.000 15	0.008 22	0.000 13	300	18	104	2	96	1	165	3
QES-20-19	427	3 626	0.1	0.047 82	0.001 68	0.100 34	0.003 35	0.015 22	0.000 17	0.004 83	0.000 11	91	80	97	3	97	1	97	2

Continued

Analysis	Th	U	Th/U	Isotopic ratios								Isotopic age /Ma							
				$\frac{^{207}Pb}{^{206}Pb}$	2σ	$\frac{^{207}Pb}{^{235}U}$	2σ	$\frac{^{206}Pb}{^{238}U}$	2σ	$\frac{^{208}Pb}{^{232}Th}$	2σ	$\frac{^{207}Pb}{^{206}Pb}$	2σ	$\frac{^{207}Pb}{^{235}U}$	2σ	$\frac{^{206}Pb}{^{238}U}$	2σ	$\frac{^{208}Pb}{^{232}Th}$	2σ
QES-20-20	675	4 126	0.2	0.047 84	0.001 18	0.097 51	0.001 71	0.014 78	0.000 15	0.004 89	0.000 08	91	23	94	2	95	1	99	2
QES-20-21	1 036	8 143	0.1	0.048 29	0.000 89	0.097 46	0.001 53	0.014 64	0.000 14	0.004 64	0.000 04	113	44	94	1	94	1	94	1
QES-20-22	385	3 183	0.1	0.047 95	0.001 68	0.092 63	0.003 10	0.014 01	0.000 15	0.004 45	0.000 10	97	80	90	3	90	1	90	2
QES-20-23	825	2 875	0.3	0.048 71	0.001 12	0.100 14	0.002 07	0.014 91	0.000 15	0.004 72	0.000 04	134	55	97	2	95	1	95	1
QES-20-24	608	2 114	0.3	0.050 54	0.001 47	0.097 92	0.002 65	0.014 05	0.000 15	0.004 43	0.000 04	220	69	95	2	90	1	89	1
QES-20-25	1 089	4 823	0.2	0.049 24	0.001 12	0.103 26	0.001 54	0.015 21	0.000 15	0.004 45	0.000 06	159	18	100	1	97	1	90	1
QES-20-26	1 005	3 209	0.3	0.047 86	0.001 77	0.097 62	0.003 46	0.014 79	0.000 16	0.004 70	0.000 05	92	84	95	3	95	1	95	1
QES-20-27	565	6 276	0.1	0.050 36	0.001 17	0.105 24	0.001 64	0.015 16	0.000 15	0.006 30	0.000 10	212	19	102	2	97	1	127	2
QES-20-28	954	3 364	0.3	0.051 20	0.001 15	0.103 43	0.001 51	0.014 65	0.000 15	0.004 73	0.000 05	250	16	100	1	94	1	95	1
QES-20-29	424	1 314	0.3	0.050 74	0.002 11	0.099 45	0.003 98	0.014 21	0.000 16	0.004 48	0.000 05	229	98	96	4	91	1	90	1
QES-20-30	867	8 034	0.1	0.051 27	0.001 18	0.104 39	0.001 60	0.014 77	0.000 15	0.006 29	0.000 09	253	18	101	1	95	1	127	2
QES-20-31	467	1 098	0.4	0.051 08	0.001 89	0.105 54	0.003 71	0.014 98	0.000 17	0.004 72	0.000 04	245	87	102	3	96	1	95	1
QES-20-32	915	7 276	0.1	0.047 62	0.000 85	0.096 45	0.001 47	0.014 69	0.000 14	0.004 67	0.000 04	80	43	93	1	94	1	94	1
QES-20-33	1 846	4 694	0.4	0.049 78	0.001 13	0.102 47	0.001 52	0.014 93	0.000 15	0.004 05	0.000 04	185	17	99	1	96	1	82	1
QES-20-34	808	4 476	0.2	0.048 57	0.001 18	0.098 67	0.002 18	0.014 73	0.000 15	0.004 67	0.000 04	127	59	96	2	94	1	94	1

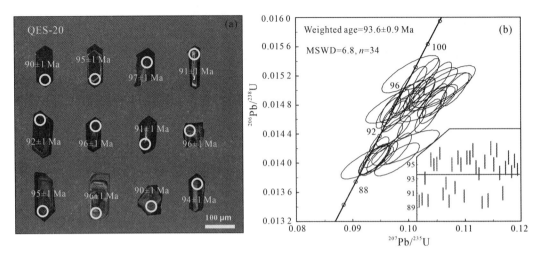

Fig. 3 (a,b)Cathodoluminescence (CL) images and LA-ICP-MS U-Pb zircon concordia diagrams of representative zircon grains of the monzogranites from Queer Mountains region in the north Yidun Arc.

have an obviously oscillatory zoning (Fig. 3a), indicating magma crystallization. In the CL images, some grains display discontinuous oscillatory zoning, which may represent different stages of crystallization. The zircons almost display Th/U ratios of 0. 1 – 0. 5, revealing they have a magmatic origin.

Thirty-six reliable analytical data were issued by sample QES-20. Two spots yield older $^{206}Pb/^{238}Pb$ ages from 100 ± 1 Ma to 103 ± 1 Ma, which suggest the ages of inherent or xenocryst zircon. The remaining 34 concordant spots have highly variable U content of 1 098 – 14 475 ppm, Th content of 385 – 2 218 ppm, and Th/U ratios of 0. 1 – 0. 5. In addition, they have concordant $^{206}Pb/^{238}Pb$ ages of 90 – 97 Ma and define a weighted mean age of 93. 6 ± 0. 9 Ma (MSWD = 6. 8, n = 34, 2σ) (Fig. 3b), which is interpreted as the age of the monzogranite in Queer Mountains region.

Major and trace element

The results of major and trace element data of ten granite samples are listed in Table 2. The rocks from the Queer Mountains area show SiO_2 = 70. 89 – 77. 23 wt%, TiO_2 = 0. 06 – 0. 23 wt%, Al_2O_3 = 12. 67 – 15. 83 wt%, $Fe_2O_3^T$ = 0. 78 – 1. 67 wt%, MgO = 0. 09 – 0. 30 wt%, Na_2O = 3. 77 – 2. 72 wt%, K_2O = 3. 96 – 6. 45 wt%, and $Mg^\#$ = 20. 0 – 29. 5. In the plots of $Na_2O + K_2O$ vs. SiO_2 classification diagram of Middlemost (1994), the Queer Mountains granites and granodiorite samples include the monzogranite in this study and granodiorite from Liu et al., (2016). These samples belong to granite and granodiorites, respectively (Fig. 4a). The monzogranites belong to the high-K calc-alkaline and shoshonite series (Fig. 4b). The A/CNK ratios of 1. 01 to 1. 10 indicate that these rock sample monzogranites from Queer Mountains belong to the peraluminous composition (Fig. 4c).

Table 2 Major(wt%) and trace(ppm) element analysis results
for the monzogranite of the Queer Mountains area.

Sample	QES04	QES05	QES09	QES11	QES12	QES13	QES18	QES23	QES26	QES27
Lithology	Granite									
SiO_2	71. 32	73. 39	73. 60	74. 00	73. 83	77. 23	73. 89	70. 89	73. 53	74. 84
TiO_2	0. 11	0. 14	0. 13	0. 14	0. 12	0. 06	0. 09	0. 19	0. 19	0. 23
Al_2O_3	15. 83	13. 98	14. 40	13. 79	14. 21	12. 67	13. 80	15. 50	14. 25	12. 72
$Fe_2O_3^T$	1. 25	1. 47	1. 41	1. 42	1. 34	0. 78	1. 12	1. 58	1. 29	1. 67
MnO	0. 03	0. 03	0. 03	0. 04	0. 05	0. 02	0. 03	0. 02	0. 02	0. 02
MgO	0. 21	0. 22	0. 23	0. 17	0. 18	0. 09	0. 12	0. 27	0. 23	0. 30
CaO	0. 99	1. 10	0. 97	0. 92	0. 98	0. 78	0. 65	1. 55	1. 38	0. 99
Na_2O	3. 44	3. 26	3. 26	3. 21	3. 45	3. 77	3. 11	3. 37	3. 25	2. 72
K_2O	6. 45	5. 20	5. 69	5. 47	5. 24	3. 96	5. 99	6. 07	5. 74	5. 71
P_2O_5	0. 06	0. 06	0. 05	0. 06	0. 05	0. 03	0. 04	0. 06	0. 05	0. 06
LOI	0. 55	0. 68	0. 56	0. 93	0. 92	0. 56	0. 69	0. 35	0. 41	0. 35
Total	100. 24	99. 53	100. 33	100. 15	100. 37	99. 95	99. 53	99. 85	100. 34	99. 61
$Mg^\#$	28. 1	25. 9	27. 5	21. 8	23. 8	21. 2	20. 0	28. 5	29. 4	29. 5
A/CNK	1. 10	1. 08	1. 08	1. 07	1. 08	1. 06	1. 08	1. 04	1. 01	1. 02
Li	94. 5	94. 4	98. 9	38. 4	108	40. 4	56. 3	58. 2	47. 8	61. 3
Be	8. 02	7. 62	7. 34	6. 63	7. 39	7. 00	5. 50	5. 57	4. 82	4. 02
Sc	3. 56	3. 82	3. 57	3. 77	4. 06	2. 62	2. 97	4. 17	3. 28	4. 14
V	5. 46	6. 62	6. 32	5. 68	4. 97	2. 40	3. 87	8. 59	8. 32	10. 3
Cr	2. 44	2. 98	2. 46	3. 11	3. 14	1. 30	1. 82	4. 86	2. 86	5. 82
Co	80. 7	167	116	105	98. 6	130	133	59. 6	92. 5	121
Ni	2. 20	2. 54	2. 68	2. 26	2. 16	1. 71	2. 00	3. 48	2. 80	3. 94
Cu	0. 63	1. 41	1. 91	1. 16	0. 53	0. 93	0. 62	1. 07	0. 94	1. 17
Zn	26. 5	42. 3	33. 8	41. 6	33. 7	22. 0	31. 9	28. 6	23. 2	28. 9
Ga	22. 7	21. 4	21. 1	21. 0	23. 0	20. 5	20. 3	20. 1	20. 0	18. 2
Ge	1. 74	1. 66	1. 70	1. 59	1. 93	1. 86	1. 58	1. 36	1. 35	1. 37
Rb	457	380	407	380	469	343	429	299	292	311
Sr	77. 7	72. 9	72. 9	65. 4	56. 0	24. 2	54. 5	114	107	84. 3
Y	45. 1	45. 3	40. 4	49. 2	66. 9	75. 5	35. 7	34. 3	34. 7	36. 8
Zr	116	137	134	136	120	74. 1	127	179	146	193
Nb	22. 3	22. 4	22. 7	24. 3	29. 2	29. 3	19. 9	12. 4	12. 6	14. 7
Cs	25. 9	22. 3	24. 6	11. 3	24. 0	10. 9	13. 7	15. 9	8. 48	9. 33
Ba	341	294	305	264	208	65. 6	257	441	410	330
La	36. 8	40. 5	39. 4	40. 0	36. 7	16. 0	27. 7	45. 4	42. 5	54. 1
Ce	73. 3	81. 2	77. 9	80. 2	75. 6	33. 0	56. 6	86. 5	80. 8	102
Pr	8. 37	9. 23	8. 91	9. 27	8. 77	3. 87	6. 56	9. 60	8. 85	11. 0
Nd	29. 6	32. 5	31. 7	32. 7	31. 0	14. 5	23. 1	32. 8	30. 2	37. 3
Sm	6. 98	7. 37	7. 12	7. 76	7. 97	5. 08	5. 48	6. 46	6. 24	7. 23

Continued

Sample	QES04	QES05	QES09	QES11	QES12	QES13	QES18	QES23	QES26	QES27
Eu	0.44	0.42	0.42	0.40	0.32	0.14	0.36	0.70	0.62	0.50
Gd	6.59	6.93	6.57	7.34	8.08	6.96	5.14	5.89	5.76	6.54
Tb	1.14	1.18	1.08	1.27	1.54	1.60	0.89	0.93	0.95	1.03
Dy	7.11	7.22	6.53	7.87	10.1	11.6	5.60	5.64	5.86	6.15
Ho	1.44	1.45	1.30	1.58	2.10	2.58	1.14	1.13	1.16	1.22
Er	4.26	4.26	3.80	4.64	6.41	7.92	3.45	3.31	3.35	3.53
Tm	0.66	0.66	0.58	0.71	1.02	1.26	0.55	0.50	0.50	0.52
Yb	4.33	4.29	3.86	4.70	6.79	8.28	3.78	3.30	3.13	3.29
Lu	0.62	0.61	0.56	0.68	0.97	1.17	0.54	0.47	0.43	0.47
Hf	3.89	4.40	4.48	4.55	4.33	3.41	4.15	4.79	4.10	5.21
Ta	4.03	3.43	3.99	3.82	7.04	2.30	3.62	1.81	1.87	1.82
Pb	52.6	42.6	46.1	46.5	44.4	46.2	47.0	33.7	33.0	32.2
Th	40.8	45.5	45.2	43.0	37.0	26.1	37.4	31.4	31.7	36.0
U	5.41	11.4	19.2	8.68	18.0	13.4	9.50	5.70	4.22	5.03
δEu	0.20	0.18	0.19	0.16	0.12	0.07	0.20	0.35	0.31	0.22
ΣREE	181.7	197.9	189.8	199.2	197.3	114.0	140.9	202.6	190.4	234.9
Rb/Sr	5.88	5.21	5.59	5.81	8.38	14.17	7.88	2.62	2.73	3.69
$(La/Yb)_N$	6.10	6.78	7.32	6.10	3.87	1.38	5.25	9.87	9.73	11.79
$T_{Zr}/℃$	761	778	776	777	767	730	772	792	776	804

$Mg^{\#} = 100Mg/(Mg + Fe)$; T_{Zr} is calculated from zircon saturation thermometry (Watson and Harrison, 1983).

In the chondrite-normalized rare earth element pattern diagrams (Fig. 5a), the samples exhibit low rare earth elements (REE) contents of 114 - 234.9 ppm, and enriched in light REEs and depleted in HREEs, with significantly fractionated REE [$(La/Yb)_N = 1.38 - 11.79$] and negative Eu anomalies ($δEu = 0.07 - 0.35$). In primitive mantle normalized trace element spider diagrams (Fig. 5b), the monzogranites display a peak in U, Th, and Pb and a trough in Ba, Nb, Sr, and Eu. These traits are similar to the evolution model of the lower continental crust. The monzogranites have relatively high Rb content of 292 - 469 ppm and low Sr content of 24.2 - 114 ppm, and Rb/Sr ratios range from 2.62 to 14.17 (Table 2).

Sr-Nb-Pb-Hf isotope data

The results of whole-rock Sr-Nb-Pb-Hf isotopic composition of the Queer Mountains monzogranites are listed in Tables 3, 4, and 5. The isotopic data for all samples were calculated at $t = 93$ Ma (U-Pb).

The three monzogranite samples display high initial $^{87}Sr/^{86}Sr(t)$ ratios of 0.708 421 - 0.715 698, with negative $\varepsilon_{Nd}(t)$ value of -3.57 to -0.53 and Nd two-stage model stage of

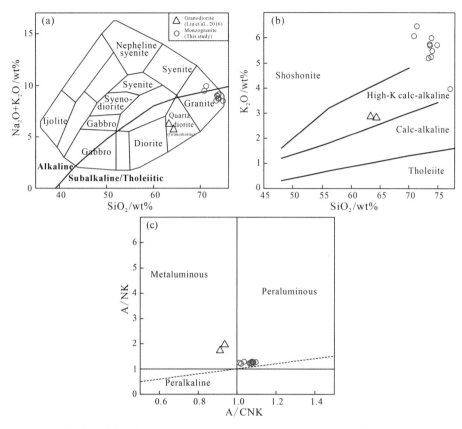

Fig. 4　Major element compositions of samples from the Queer Mountains.
(a) Total alkali vs. silica (TAS) diagram (after Middlemost, 1994), (b) K_2O vs. SiO_2 diagram,
and (c) A/NK vs. A/CNK classification diagram after Frost et al. (2001).

1.03-1.24 Ga (Table 3) and negative $\varepsilon_{Hf}(t)$ values of -3.2 to -0.4, with Hf two-stage model ages of 1.18 - 1.35 Ga (Table 5). The calculated initial Pb isotope ratios were $(^{206}Pb/^{204}Pb)_t = 18.7234-18.8436$, $(^{207}Pb/^{204}Pb)_t = 15.7031-15.7261$, and $(^{208}Pb/^{204}Pb)_t = 39.0507-39.2780$ (Table 4).

Discussion

Formation age of granite

The Queer Mountains granitic complex mainly contains monzogranite, syenite granite, and a few granodiorite. Although the predecessor carried out the study of geochronology for the Queer Mountains complex rock mass, it is still unclear. The age dating indicates 224 ± 3 Ma (Late Triassic) for the Queer Mountains granodiorites (Liu et al., 2006a), 167.5-67.9 Ma for the syenogranites by Chengdu institute of geology and mineral resource, Sichuan, in 1 978 Ma to 1 983 Ma, 88 Ma for the Queer Mountains syenogranite (Lü et al., 1993), 102.6±

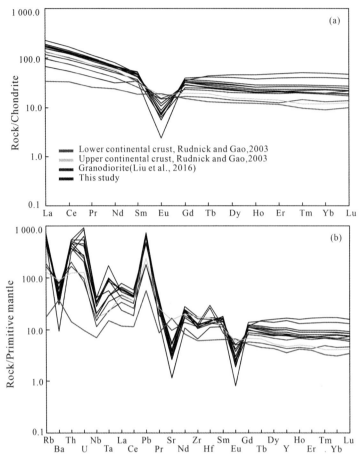

Fig. 5　Chondrite-normalized REE pattern (a) and primitive mantle (PM) normalized trace element
spider (b) diagrams of the monzogranite from the Queer Mountains region.
The primitive mantle and chondrite values are from Sun and McDonough (1989)

1. 1 Ma and 105. 9±1. 3 Ma for the Queer Mountains' coarse grain porphyaceous biotite moyite
and the fine-grain biotite monzogranite, respectively (Wu et al., 2014a), 102 Ma and 97 ±
2 Ma for the Queer Mountains pluton (Yan et al., 2006; Gao, 2015), 213 ± 10 Ma for the
Garzê plutons, 104. 7±2 Ma for the Queershan pluton, and 94. 4±2. 4 Ma for the Haizi granite
plutons (Reid et al., 2007). By collecting previous research data, we can conclude that the
Mesozoic magmatic events' peak value of the Yidun Arc mainly focused on the Triassic
(～220 Ma) and Cretaceous (～105 Ma) (Table 6). Although Jurassic magmatic activities
have been recorded in the region (Qu et al., 2003; Liu et al., 2007b; Zhao et al., 2007),
the research degree is relatively deficient. This study showing that the formation ages of the
Queer Mountains monzogranite are 93. 6±0. 9 Ma indicated that these rocks are the products of
early magmatic activity in Late Cretaceous.

By collecting previous research and this study, we found that the complex intrusions of
the Queer Mountains mainly were the diorite-granite in the Indosinian and Yanshanian period,

Table 3 Whole-rock Rb-Sr and Sm-Nd isotopic data for the monzogranite of the Queer Mountains, northern of the Yidun Arc.

Sample	$^{87}\mathrm{Sr}/^{86}\mathrm{Sr}$	2σ	Rb /ppm	Sr /ppm	$^{87}\mathrm{Rb}/^{86}\mathrm{Sr}$	$^{147}\mathrm{Sm}/^{144}\mathrm{Nd}$	$^{143}\mathrm{Nd}/^{144}\mathrm{Nd}$	2σ	Nd /ppm	Sm /ppm	T_{DM2} /Ga	$\varepsilon_{\mathrm{Nd}}(t)$ ($t=93$ Ma)	$^{87}\mathrm{Sr}/^{86}\mathrm{Sr}(t)$ ($t=93$ Ma)	$^{143}\mathrm{Nd}/^{144}\mathrm{Nd}(t)$ ($t=93$ Ma)
QES09	0.736 612	0.00 0006	72.9	407	16.199	0.135 776	0.512 290	0.000 005	31.7	7.12	1.19	-2.88	0.715 206	0.512 207
QES18	0.745 906	0.000 005	54.5	429	22.859	0.143 406	0.512 259	0.000 009	23.1	5.48	1.24	-3.57	0.715 698	0.512 172
QES23	0.718 459	0.000 005	114	299	7.596	0.119 062	0.512 400	0.000 005	32.8	6.46	1.03	-0.53	0.708 421	0.512 328

$^{87}\mathrm{Rb}/^{86}\mathrm{Sr}$ and $^{147}\mathrm{Sm}/^{144}\mathrm{Nd}$ ratios were calculated using Rb, Sr, Sm, and Nd contents analysed by ICP-MS.

T_{DM2} represents the two-stage model age and was calculated using present-day $(^{147}\mathrm{Sm}/^{144}\mathrm{Nd})_{\mathrm{DM}}=0.213\ 7$, $(^{147}\mathrm{Sm}/^{144}\mathrm{Nd})_{\mathrm{DM}}=0.513\ 15$, and $(^{147}\mathrm{Sm}/^{144}\mathrm{Nd})_{\mathrm{crust}}$ $=0.101\ 2$. $\varepsilon_{\mathrm{Nd}}(t)$ values were calculated using present-day $(^{147}\mathrm{Sm}/^{144}\mathrm{Nd})_{\mathrm{CHUR}}=0.196\ 7$ and $(^{147}\mathrm{Sm}/^{144}\mathrm{Nd})_{\mathrm{CHUR}}=0.512\ 638$.

$$\varepsilon_{\mathrm{Nd}}(t)=\left[\left(\left(^{143}\mathrm{Nd}/^{144}\mathrm{Nd}\right)_{\mathrm{s}}(t)/\left(^{143}\mathrm{Nd}/^{144}\mathrm{Nd}\right)_{\mathrm{CHUR}}(t)\right)-1\right]\times10^{4}.$$

$$T_{\mathrm{DM2}}=\frac{1}{\lambda}\left\{1+\left[\left(\left(^{143}\mathrm{Nd}/^{144}\mathrm{Nd}\right)_{\mathrm{s}}-\left(\left(^{147}\mathrm{Sm}/^{144}\mathrm{Nd}\right)_{\mathrm{s}}-\left(^{147}\mathrm{Sm}/^{144}\mathrm{Nd}\right)_{\mathrm{crust}}\right)\left(e^{\lambda t}-1\right)-\left(^{143}\mathrm{Nd}/^{144}\mathrm{Nd}\right)_{\mathrm{DM}}\right]/\left[\left(^{147}\mathrm{Sm}/^{144}\mathrm{Nd}\right)_{\mathrm{crust}}-\left(^{147}\mathrm{Sm}/^{144}\mathrm{Nd}\right)_{\mathrm{DM}}\right]\right\},\ \lambda=6.54\times10^{-12}/\mathrm{a}.$$

Table 4 Whole-rock Pb isotopic data for the monzogranite of the Queer Mountains, northern of the Yidun Arc.

Sample	U /ppm	Th /ppm	Pb /ppm	$^{206}\mathrm{Pb}/^{204}\mathrm{Pb}$	2σ	$^{207}\mathrm{Pb}/^{204}\mathrm{Pb}$	2σ	$^{208}\mathrm{Pb}/^{204}\mathrm{Pb}$	2σ	$\dfrac{^{238}\mathrm{U}}{^{204}\mathrm{Pb}}$	$\dfrac{^{232}\mathrm{Th}}{^{204}\mathrm{Pb}}$	$\left(^{206}\mathrm{Pb}/^{204}\mathrm{Pb}\right)(t)$ ($t=93$ Ma)	$\left(^{207}\mathrm{Pb}/^{204}\mathrm{Pb}\right)(t)$ ($t=93$ Ma)	$\left(^{208}\mathrm{Pb}/^{204}\mathrm{Pb}\right)(t)$ ($t=93$ Ma)
QES09	19.2	45.2	46.1	19.207	0.000 2	15.743	0.000 2	39.580	0.000 2	26.96	65.12	18.815 4	15.723 8	39.278 0
QES18	9.5	37.4	47.0	19.033	0.000 4	15.735	0.000 3	39.365	0.000 9	13.02	52.57	18.843 6	15.726 1	39.121 0
QES23	5.7	31.4	33.7	18.881	0.000 2	15.711	0.000 2	39.335	0.000 6	10.86	61.39	18.723 4	15.703 1	39.050 7

U, Th, and Pb concentrations were analysed by ICP-MS. Initial Pb isotopic ratios were calculated for 93 Ma using single-stage model.

**Table 5 Whole-rock Lu-Hf isotopic data for the monzogranite of
the Queer Mountains, northern of the Yidun Arc.**

Sample	Lu /ppm	Hf /ppm	$^{176}Lu/^{177}Hf$	$^{176}Hf/^{177}Hf$	2σ	$\varepsilon_{Hf}(t)$ ($t=93$ Ma)	T_{DM2} /Ga	$f_{Lu/Hf}$
QES09	0.56	4.48	0.017 748	0.282 656	0.000 008	−3.2	1.35	−0.47
QES18	0.54	4.15	0.018 475	0.282 705	0.000 005	−1.5	1.24	−0.44
QES23	0.47	4.79	0.013 932	0.282 728	0.000 004	−0.4	1.18	−0.58

Lu and Hf concentrations were analysed by ICP-MS.

T_{DM2} was calculated using present-day $(^{176}Lu/^{177}Hf)_{DM}=0.038\ 4$ and $(^{176}Hf/^{177}Hf)_{DM}=0.283\ 25$.

$f_{CC}=-0.55$, and $f_{DM}=0.16$. f_S is the sample's $f_{Lu/Hf}$.

$\varepsilon_{Hf}(t)$ values were calculated using present-day $(^{176}Lu/^{177}Hf)_{CHUR}=0.033\ 2$ and $(^{176}Hf/^{177}Hf)_{CHUR}=0.282\ 772$.

$$T_{DM1}=\frac{1}{\lambda}\ln\left\{1+\left[(^{176}Hf/^{177}Hf)_S-(^{176}Hf/^{177}Hf)_{DM}\right]/\left[(^{176}Lu/^{177}Hf)_S-(^{176}Lu/^{177}Hf)_{DM}\right]\right\}.$$

$$T_{DM2}=T_{DM1}-(T_{DM1}-t)\left[(f_{CC}-f_S)/(f_{CC}-f_{DM})\right], \lambda=1.876\times10^{-11}/a.$$

respectively. The ages span range is relatively large for the Queer Mountains granites, but the monzogranite mainly belongs to the product of the magmatism of Late Cretaceous (Yanshanian). This conclusion is consistent with the previous studies (Liu et al., 2006a; Yan et al., 2006).

Petrogenesis of the granites

Late Cretaceous monzogranites of Queer Mountains area have high SiO_2 and Al_2O_3, low TiO_2, $Fe_2O_3^T$, MgO, and Na_2O. The samples appertain high-K, calc-alkaline and slightly peraluminous series (Fig. 4) and corundum (C) content of $0.31-1.56$ wt% in Normative-CIPW calculation. The monzogranites are strongly depleted in Ba, Nb, Sr, and Eu (Fig. 5b). These major and trace elements anomaly feature may be due to the fractionation of early crystallized mineral during the evolution of granitic magma (Zhao et al., 2016). As shown in Fig. 6, the Queer Mountains monzogranites display significant negative correlations of major elements with SiO_2 contents. These variations in feature suggest that the fractional crystallization has a significant effect in the evolution of magma. We still need further confirmation which minerals have dominated in the proceeding crystallization.

As shown in Fig. 7a, the obvious negative correlation of Sr and Ba contents (omitted) vs. SiO_2 contents, then the strongly depleted Eu (Fig. 5a), indicated the significant fractionation of the plagioclase and K-feldspar during magma evolution (Rollinson, 1993; Peng et al., 2015). Furthermore, SiO_2 contents have a positive correlation with Y concentrations in some typical arc lavas (Rohrlach and Loucks, 2005). These samples obviously indicate that hornblende is an insignificant mineral during magma evolution (Fig. 7b). Besides, in log-log diagrams (Fig. 7e, f) of Sr vs. Rb and Ba, especially negative Ba anomalies indicate that

Table 6　The Mesozoic (Triassic and Jurassic) granitoid distribution in the Yidun Arc area, SW China.

Location	Sample	Lithology	Method	Age/Ma	Isotope	Model age/Ga	Reference
Queer Mountains		Granitoids	Zircon U-Pb	97±2		$\varepsilon_{Nd}(t)=-5.3$ to -6.3, $T_{DM}=1.23-1.61$	Liu et al., 2006a; Yan et al., 2006
	QES-7N	Monzogranite		102.6±1.1		$\varepsilon_{Hf}(t)=-0.2$ to 0.63, $T_{DM2}=1.13-1.19$	Wu et al., 2014
	QES-11N	Moyite		105.9±1.3		$\varepsilon_{Hf}(t)=-0.56$ to 1.43, $T_{DM2}=1.07-1.2$	
	MD5007	Granodiorite		224±3			Liu et al., 2016
	PM07215	Monzogranite		103±1			
	QES-20	Monzogranite		93.6±0.9	$I_{Sr}=0.709\ 421-$ $0.715\ 698$	$\varepsilon_{Nd}(t)=-3.57$ to -0.53, $T_{DM2}=1.03-1.24$; $\varepsilon_{Hf}(t)=-3.2$ to -0.4, $T_{DM2}=1.18-1.35$	This study
Daocheng	1007-6	Granite		217.4±1.1		$\varepsilon_{Hf}(t)=-7.1$ to -0.1, $T_{DM2}=1.26-1.7$	Wang et al., 2016
Haizishan	1008-3, 1008-6, 1009-3, 1009-6	Biotite monzogranite		98.3±0.4		$\varepsilon_{Hf}(t)=-12.1$ to 2.5, $T_{DM2}=1.0-1.93$	
Daocheng	DC0910XC0903 DC0922	Granites (I-type)		215±1.5; 216±2.2; 216±0.7	$I_{Sr}=0.705\ 88-0.710\ 16$	$\varepsilon_{Nd}(t)=-5.7$ to -7.8, $T_{DM2}=1.46-1.62$; $\varepsilon_{Hf}(t)=-9.8$ to -2.9, $T_{DM2}=1.44-1.88$	He et al., 2013
Xuejiping	XC07-23	Porphyry		218±1.6	$I_{Sr}=0.705\ 10-0.705\ 90$	$\varepsilon_{Nd}(t)=-3.8$ to -2.1, $T_{DM}=1.01-1.15$	Leng et al., 2012
Lianlong		Granite	Rb-Sr	89±5	$I_{Sr}=0.708\ 3$	$\varepsilon_{Nd}(t)=-5.95$ to -5.15, $T_{DM}=1.2$	Qu et al., 2002
Rongyicuo	D030B1	Granite	$^{40}Ar/^{39}Ar$	75	$I_{Sr}=0.710\ 10-0.725\ 00$	$\varepsilon_{Nd}(t)=-5.58$ to -5.44	
Xiasai	XSG-5	Granite (A-type)	Zircon U-Pb	103±1		$\varepsilon_{Hf}(t)=-2.7$ to 0.6, $T_{DM}=0.9-1.1$	Li et al., 2014
Dongco	CX3035-2	Granitoid	SHRIMP	224±3		$\varepsilon_{Nd}(t)=-3.27$, $T_{DM}=1.21$	Liu et al., 2006b
Xiangcheng	CX3033-1	Dioritic porphyries		222±3		$\varepsilon_{Nd}(t)=-8.10$, $T_{DM}=1.87$	

Continued

Location	Sample	Lithology	Method	Age/Ma	Isotope	Model age/Ga	Reference
Suwalong	457	Biotite granite	Zircon U-Pb	240±6.2		$\varepsilon_{Hf}(t) = -1.1$ to -7.5, $T_{DM} = 1.30-1.69$	Reid et al., 2005a, 2007
	460	Granodiorite		239±5.8			
Baimaxue-shan	YA13	Granite		245.2±3.4			
Garzê	406b	Granodiorite		218.5±3			
Daocheng	YA05	Granodiorite		215.2±3.4		$\varepsilon_{Hf}(t) = -0.5$ to -6.7, $T_{DM} = 1.28-1.63$	
Queershan	426	Biotite granite		104.7±2			
Haizi	YA32	Biotite granite		94.4±2.4		$\varepsilon_{Hf}(t) = 0$ to -3.7, $T_{DM} = 1.24-2.42$	
Disuga	DSG07-7	Basaltic andesite		219.8±1.9	$I_{Sr} = 0.705\,70\sim0.706\,00$	$\varepsilon_{Nd}(t) = -2.7$ to $+0.7$, $T_{DM} = 0.9-1.1$;	Leng et al., 2014
	DSG07-3	Andesite		218.1±4.8		$\varepsilon_{Hf}(t) = 0$ to 4, $T_{DM} = 0.7-0.9$	
	DSG07-5	Trachyandesite		220.8±5.3			
Xuejiping	HN11-4	Trachyandesite		219.1±3.2			Peng et al., 2014
Dongcuo	10YD-65	Monzogranite		217±1		$\varepsilon_{Hf}(t) = -22.7$ to -1.25, $T_{DM2} = 1.35-2.53$	
	SJ-209	Monzogranite		224±4			
Maxionggou	SJ-199	Granodiorite		216±1		$\varepsilon_{Hf}(t) = -1.08$ to $+3.31$, $T_{DM2} = 1.06-1.34$	
shengmu	SJ-235	Monzogranite		218±3			
	SJ-241	Granodiorite		216±3		$\varepsilon_{Hf}(t) = -7.32$ to -1.32, $T_{DM2} = 1.34-1.76$	
Cuojiaoma	E017	Granite		215.1±2			Weislogel, 2008
	E018	Granite		224.5±2			

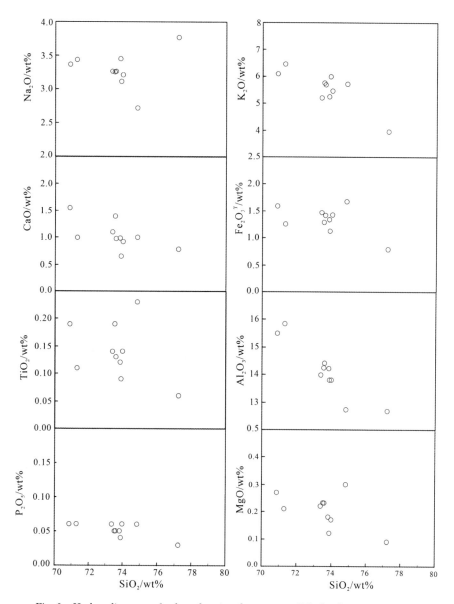

Fig. 6   Harker diagrams of selected major elements vs. SiO_2 for the monzogranites
from the Queer Mountains in the north Yidun Arc.

K-feldspar is likely to play an important role than plagioclase during the upwelling of magma. In a diagram of SiO_2 vs. $\varepsilon_{Nd}(t)$ and I_{Sr} (Fig. 7c, d), the monzogranite sample data present an obvious linear correlation, suggesting that the crustal contamination could have an impact during magma ascending.

The isotope data from the Queer Mountains granites have a negative $\varepsilon_{Nd}(t) = -3.57$ to -0.53, with the $T_{DM2} = 1.03 - 1.24$ Ga, and high initial $^{87}Sr/^{86}Sr$ (Table 3) and negative $\varepsilon_{Hf}(t) = -3.2$ to -0.4, with the $T_{DM2} = 1.18 - 1.35$ Ga (Table 5), revealing that the primary

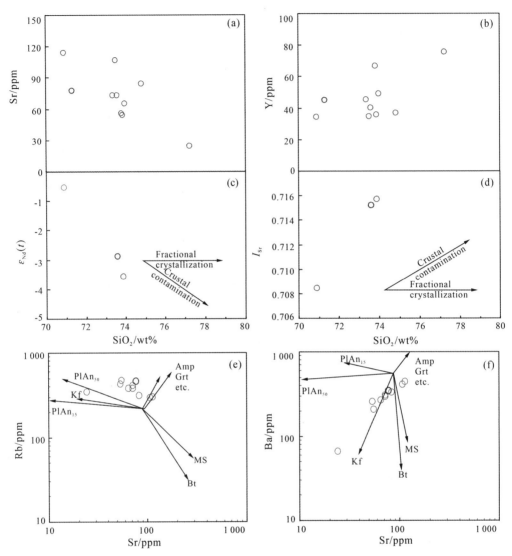

Fig. 7 (a) SiO₂ vs. Sr, (b) SiO₂ vs. Y, (c) SiO₂ vs. $\varepsilon_{Nd}(t)$, (d) SiO₂ vs. I_{Sr}, (e, f) Sr vs. Rb and Ba diagram, respectively (after Janoušek et al., 2004).

Pl: plagioclase; Kf: K-feldspar; Bt: biotite; Ms: muscovite; Grt: granite; Amp: amphibole.

magma impossibility originated from the depleted mantle. In Fig. 8, the samples are plotted in the upper continental crust end-members, obviously different Songpan-Garzê granitoids, which indicates that these monzogranites were mainly derived from the upper crust. Furthermore, the monzogranites have average high values and small span Pb isotopic ratios ($^{206}Pb/^{204}Pb)_t = 18.723\,4-18.843\,6$, ($^{207}Pb/^{204}Pb)_t = 15.703\,1-15.726\,1$, and ($^{208}Pb/^{204}Pb)_t = 39.050\,7-39.278\,0$ (Table 4). Pb isotopic feature indicates that the Yidun terrane should have an affinity with the basement of Yangtze Block (Fig. 9). What is more, the Pb isotopic compositions of Xuejiping porphyry, Southern Yidun volcanic rocks, and Queer Mountains monzogranites also have resemble property of the Yangtze Block basement (Leng et al., 2012, 2014; this study), but they are obviously different to the Mesozoic granitoids of North China (Zhang, 1995; this study), and all samples are plotted in the upper crust line (Fig. 9). The monzogranites also show the characteristics of the trace element evolution pattern of the upper crust (Fig. 5). Besides, Xuejiping porphyries were derived from a metasomatized mantle and underwent limited crustal contamination (Leng et al., 2012). Moreover, the Daocheng granites originated from the lower crust and minor mantle basaltic magma mixed (He et al., 2013). These traits are markedly different from the Queer Mountains granites (Fig. 8).

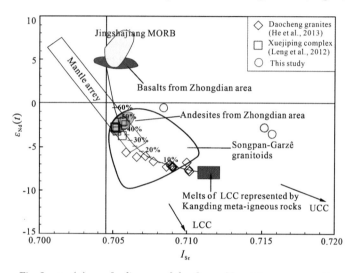

Fig. 8　$\varepsilon_{Nd}(t)$ vs. I_{Sr} diagram of the Queer Mountains monzogranite

from the north Yidun Arc (after Leng et al., 2012; He et al., 2013).

The trends of lower continental crust (LCC) and upper continental crust (UCC) are from Jahn et al. (1999).

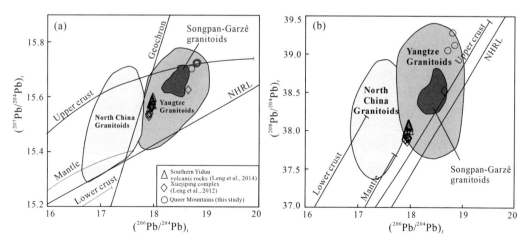

Fig. 9　(a) $(^{207}Pb/^{206}Pb)_t$ vs. $(^{206}Pb/^{204}Pb)_t$ diagram; (b) $(^{208}Pb/^{204}Pb)_t$ vs. $(^{206}Pb/^{204}Pb)_t$ diagram

(modified from Leng et al., 2012).

Based on the above isotopic characteristic and Table 6, it can be seen that the Daocheng granites and Xuejiping granite porphyry have similar source area features and mainly focuses on igneous lower crust in the Triassic. While the characteristics of the granite source region of the Queer Mountains are mainly the meta-sedimentary upper crust during the Cretaceous. The Cretaceous granites in the Yidun Arc were distributed from northern to southern; meanwhile, the affinity of their source displays the trend from sedimentary to igneous crust (Yang et al., 2015). Then, the tectonic framework and magmatism mechanisms of the north to south segments of Yidun Arc were obviously different (Leng et al., 2012, 2014). Combined the above isotopic characteristics, it is believed that the magma of Queer Mountains granite in the northern Yidun Arc was likely to derive from the Mesoproterozoic crust material, and the basement may be strongly associated with the base properties of the Yangtze Block.

Predecessor research shows that the peraluminous granites mainly caused by the meta-sedimentary rocks (Le Fort et al., 1987; White and Chappell, 1988) or metaigneous rocks (Miller, 1985; Patiño Douce, 1995) in the lower crust. The granites have low CaO/Na_2O ratios (0.21−0.46) and high ratios of Al_2O_3/TiO_2(55−211) additionally high ratios of Rb/Sr (2.6−14.2) and Rb/Ba (0.7−5.2). Sylvester (1998) research shows that the peraluminous granites have a relatively lower CaO/Na_2O ratio (< 0.3) and originated from the melts of pelites. As shown by geochemical diagrams (Fig. 10a,b), the monzogranites could be derived from a source of the clay-rich metapelite and formed by the partial melting.

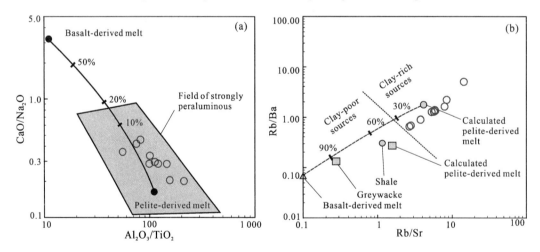

Fig. 10　(a) Al_2O_3/TiO_2 vs. CaO/Na_2O diagrams and (b) Rb/Sr vs. Rb/Ba diagrams of the Queer Mountains monzogranite, after Sylvester (1998).

The Zr contents decrease with increasing SiO_2 contents for the monzogranite shows that zircon separated during the evolution of fractionation melt (Li et al., 2007; Zhong et al., 2007; Liu et al., 2009; omitted). Zircon saturation thermometry can simply calculate the temperatures of magma source region (Watson and Harrison 1983). In this paper, simple zircon saturation temperatures (T_{Zr}) calculation indicate that the crystallization temperature of granite is about 730−804 ℃ (Table 2).

Tectonic setting and significance

Previous research has suggested that the Yidun Arc is related to the Late-Paleozoic Paleo-Tethys oceanic subduction in Middle Triassic (Wang et al., 2000; Zhong, 2000; Reid et al., 2005a; Lai et al., 2009). And the Yidun Arc experienced the large scale subduction orogeny in the late Indosinian and resulted in numerous magmatic activity and tectonic thermal events and formed a magmatic arc (Lü et al., 1993; Hou et al., 2001; Reid et al., 2005a, 2007). At present, the Yidun Arc had experienced a complex and changeable evolution process from the subduction orogeny (ca. 238−210 Ma) to the arc-extension collision orogeny (ca. 207−138 Ma) and to the post-collision extension (ca. 138−73 Ma) (Hou et al., 2001, 2003, 2004; Reid et al., 2007; Li et al., 2014). Numerous granitic bodies and volcanic rocks are exposed and the magmatism of related-subduction is recorded in the Yidun Arc area in Mesozoic, particularly in Late Triassic (Reid et al., 2007; Leng et al., 2012; He et al., 2013). Although the Yidun tectonic belt has a special tectonic history in the SOB area (Hou et al., 2003), the precise significant geodynamics and tectonic evolution setting remain unclear in the period of Mesozoic.

Using igneous rock associations and petrogeochemistry to probe tectonic movement, orogenic belt evolution has gradually become a weighty content for study on continental tectonics (Qu et al., 2002). In Rb vs. Y+Nb diagrams (Fig. 11a), the monzogranites fall into WPG to syn-COLG fields. However, the granodiorite samples fall within the VAG field. In the R_1 vs. R_2 diagrams (Fig. 11b), the monzogranite samples mainly concentrated in the syn-COLG field. And the granodiorite samples mainly fall within the fields of pre-plate collision. Combined with the above discussion and referred to relevant coeval regional geological studies, the monzogranite formed in a post-collision, namely, an extension tectonic setting. The Yidun Arc area may undergo a complex and systematic tectonic transition in Mesozoic.

Many scholars have different views with respect to the geodynamic setting of the Cretaceous volcanic rocks in the Yidun area. Some scholars deem these rock masses formed in a post-collisional tectonic environment after the collision between the Yidun Arc and Yangtze Block around ca. 200 Ma (Qu et al., 2002; Li et al., 2017). Others argued that their formations are related to the collision of the Qiangtang Terrane and Lhasa Tarrane (Kapp et al., 2005, 2007; Wang et al., 2014a, b; Yang et al., 2016, 2017). Nevertheless, Garzê-Litang ocean has basically reached agreement on the westward subduction at ca. 216 Ma, and the Yidun Arc Cretaceous granites may be formed in an extensional tectonic setting.

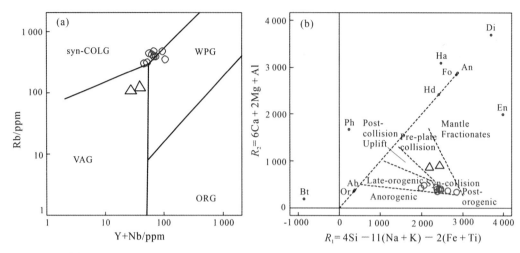

Fig. 11 (a)Rb vs. Y + Nb (after Pearce et al., 1984; Pearce, 1996) and (b)R_2($= 6Ca + 2Mg + Al$) vs. $R_1[= 4Si - 11(Na + K) - 2(Fe + Ti)]$ (after Batchelor and Bowden, 1985) diagrams for the Queer Mountains in the north Yidun Arc, SW China. Symbols are as in Fig. 4.

Together the collected geological data and this study, we have believed that the complex intrusions of Queer Mountains in the Yidun Arc, including the diorites and granites, were the product of the magmatism of the Indosinian and Yanshanian period, respectively. It is believed that the Mesozoic magmatic event peak value of the Yidun Arc mainly focused on the ~220 Ma and ~105 Ma (Table 6), and the early diorites were the product of the related-subduction during the Triassic period. In this period, the Late Triassic granite was distributed along the suture zone of Garzê-Litang might be a response to this geological process resulted in a relatively high temperature (Hou et al., 2001; He et al., 2013; Zhu et al., 2011; Peng et al., 2014). The study of massive tectonic-magmatic activity of the Jurassic to Early Cretaceous period is comparatively deficient. This paper further supports that the monzogranites from the Queer Mountains formed in a post-collision extensional tectonic setting until the Cretaceous period in the northern Yidun Arc (Hou et al., 2001; Liu et al., 2006a; Yan et al., 2006; Wang et al., 2014b; Wu, 2015; Liu et al., 2016).

Conclusions

(1)Zircon U-Pb dating reveals that the Queer Mountains monzogranites were erupted at ca. 93 Ma. They were the product of the magmatism in Late Cretaceous (Yanshanian).

(2)The Queer Mountains monzogranite exhibits the geochemistry characteristics of relatively high SiO_2, Al_2O_3, Al_2O_3/TiO_2, and enriched LREEs and LILEs but depleted in HFSEs and negative Eu anomalies. The monzogranites have high initial $^{87}Sr/^{86}Sr$ ratios (0. 708 421–0. 715 698) and negative $\varepsilon_{Nd}(t)$ ($-3. 57$ to $-0. 53$) and $\varepsilon_{Hf}(t)$ ($-3. 2$ to

-0.4) values and reveal that the monzogranite was predominantly derived from the Mesoproterozoic crust material. Pb isotopic compositions imply the contribution of the upper crustal composition and these signatures have affinity with Yangtze Craton. The fractional crystallization of K-feldspar was experienced during magma evolution and at crystallization temperatures between 730 ℃ and 804 ℃.

(3) Based on previous geological data and this study, we have proposed that the Queer Mountains monzogranite might be generated in a post-collisional extensional setting in Late Cretaceous.

Acknowledgements　　We are grateful to the anonymous reviewers for providing constructive comments and suggestions. Financial support for this study was jointly provided by Natural Science Foundation Innovation Group (Grant No. 41421002) and State Key Laboratory of Continental Dynamics, Northwest University.

References

Andersen, T., 2002. Correction of common lead in U-Pb analyses that do not report ^{204}Pb. Chem Geol, 192(1-2), 59-79.

Batchelor, R.A., Bowden, P., 1985. Petrogenetic interpretation of granitoid rock series using multicationic parameters. Chem Geol ,48(1), 43-55.

Cai, X.Q., 2012. Lithogeochemistry of granites and associated uranium mineralization in Queer Mountains. Chengdu: Master degree paper of Chengdu University of Technology (in Chinese with English abstract).

Cao, K., Xu, J.F., Chen, J.L., Huang, X.X., Ren, J.B., Zhao, X.D., Liu, Z.X., 2015. Double layer structure of the crust beneath the Zhongdian arc, SW China: U-Pb geochronology and Hf isotope evidence. J Asian Earth Sci ,115, 455-467.

Chang, C., 1997. Geology and tectonics of Qinghai-Xizang Plateau solid earth sciences research in China. Beijing: Science Press, 153.

Deng, J., Yang, L.Q., Ge, L.S., Yuan, S.S., Wang, Q.F., Zhang, J., Gong, Q.J., Wang, C.M., 2010a. Character and post-ore changes, modifications and preservation of Cenozoic alkali-rich porphyry gold metallogenic system in western Yunnan, China. Acta Petrol Sin, 26(6), 1633-1645 (in Chinese with English abstract).

Deng, J., Hou, Z.Q., Mo, X.X., Yang, L.Q., Wang, Q.F., Wang, C.M., 2010b. Superimposed orogenesis and metallogenesis in Sanjiang Tethys. Mineral Deposits , 29(1), 37-42 (in Chinese with English abstract).

Deng, J., Wang, Q.F., Li, G.J., Santosh, M., 2014. Cenozoic tectonomagmatic and metallogenic processes in the Sanjiang region, southwestern China. Earth Sci Rev, 138, 268-299.

Fang, X.Y., Peng, T.P., Fan, W.M., Gao, J.F., Liu, B.B., Zhang, J.Y., 2017. Origin of the Middle-Late Triassic intermediate-acid intrusive rocks in the Yidun terrane and its geological significance. Geochimica, 46(5), 413-434 (in Chinese with English abstract).

Frost, B.R., Barnes, C.G., Collins, W.J., Arculus, R.J., Ellis, D.J., Frost, C.D., 2001. A geochemical classification for granitic rocks. J Petrol ,42(11), 2033-2048.

Gao, L., 2015. Brief study on the petrology and litho-geochemistry characters and its origin of Queershan granitoid batholith of Dege County in Northwestern Sichuan. Chengdu: Master degree paper of Chengdu University of Technology (in Chinese with English abstract).

He, D.F., Zhu, W.G., Zhong, H., Ren, T., Bai, Z.J., Fan, H.P., 2013. Zircon U-Pb geochronology and elemental and Sr-Nb-Hf isotopic geochemistry of the Daocheng granitic pluton from the Yidun Arc, SW China. J Asian Earth Sci, 68(1), 1-17.

Hou, Z.Q., 1993. The tectono-magmatic evolution of Yidun island-arc and geodynamic setting of the formation of Kuroko-type massive sulphide deposits in Sanjiang region, southwestern China. Resour Geol, 17, 336-350.

Hou, Z.Q., Luo, Z.W., 1992. Origin of the andesite in Yidun island arc, Sanjiang region. Acta Petrol Mineral, 11, 1-14 (in Chinese with English abstract).

Hou, Z.Q., Mo, X., 1991. A tectono-magmatic evolution of Yidun island arc in Sanjiang region, China. Contri Geol Qinghai-Xizang (Tibet) Plateau, 21, 153-165 (in Chinese).

Hou, Z.Q., Mo, X.X., 1993. Geology, geochemistry and genetic aspects of kuroko-type volcanogenic massive sulphide deposits in Sanjiang region, southwestern China. Explor Min Geol, 2, 17-29.

Hou, Z.Q., Hou, L.W., Ye, Q.T., Liu, F.L., Tang, G.Q., 1995. Tectono-magmatic evolution of Yidun Island Arc and volcanogenic massive sulfide deposits in Sanjiang region, S. W. China. Beijing: Seismological Press, 1-218 (in Chinese).

Hou, Z.Q., Qu, X.M., Yang, Y.Q., 2001. Collision orogeny in the Yidun Arc: Evidence from granites in the Sanjiang region, China. Acta Geol Sin, 75, 484-497 (in Chinese with English abstract).

Hou, Z.Q., Yang, Y.Q., Wang, H.P., Qu, X.M., Lü, Q.T., Huang, D.H., Wu, X.Z., Tang, S.H., Zhao, J.H., 2003. Collision-orogenic progress and mineralization system of Yidun Arc. Beijing: Geological Publishing House, 345 (in Chinese).

Hou, Z.Q., Yang, Y.Q., Qu, X.M., Huang, D.H., Lü, Q.T., Wang, H.P., Yu, J.J., Tang, S.H., 2004. Tectonic evolution and mineralization systems of the Yidun Arc orogen in Sanjiang region, China. Acta Geol Sin, 78(1), 109-119 (in Chinese with English abstract).

Hou, Z.Q., Zaw, K., Pan, G.T., Mo, X.X., Xu, Q., Hu, Y.Z., Li, X.Z., 2007. Sanjiang Tethyan metallogenesis in SW China: Tectonic setting, metallogenic epochs and deposit types. Ore Geol Rev, 31, 48-87.

Hu, J.M., Meng, Q.R., Shi, Y.R., Qu, H.J., 2005. SHRIMP U-Pb dating of zircons from granitoid bodies in the Songpan-Ganzi terrane and its implications. Acta Petrol Sin, 21, 867-880 (in Chinese with English abstract).

Jahn, B.M., Wu, F.Y., Lo, C.H., Tsai, C.H., 1999. Crust-mantle interaction induced by deep subduction of the continental crust: Geochemical and Sr-Nd isotopic evidence from post-collisional mafic-ultramafic intrusions of the northern Dabie complex, Central China. Chem Geol, 157, 119-146.

Janoušek, V., Finger, F., Roberts, M., Fryda, J., Pin, C., Dolejš, D., 2004. Deciphering the petrogenesis of deeply buried granites: Whole-rock geochemical constraints on the origin of largely underplated felsic granulites from the Moldanubian zone of the Bohemian Massif. Trans R Soc Edinb Earth Sci, 95, 141-159.

Jian, P., Liu, D., Kroener, A., Zhang, Q., Wang, Y., Sun, X., Zhang, W., 2009. Devonian to Permian plate tectonic cycle of the Paleo-Tethys Orogen in southwest China (I): Geochemistry of ophiolites, arc/back-arc assemblages and within-plate igneous rocks. Lithos, 113, 748-766.

Kapp, P., Yin, A., Harrison, T.M., Ding, L., 2005. Cretaceous-Tertiary shortening, basin development, and volcanism in central Tibet. Geol Soc Am Bull, 117, 865-878.

Kapp, P., DeCelles, P.G., Gehrels, G.E., Heizier, M., Ding, L., 2007. Geological records of the Lhasa-Qiangtang and Indo-Asian collisions in the Nima area of Central Tibet. Geol Soc Am Bull, 119, 917-932.

Lai, S.C., Zhao, S.W., 2015. Geochemistry and petrogenesis of quart diorite in Tagong area of northwest Sichuan. J Earth Sci Environ, 37(3), 1-13 (in Chinese with English abstract).

Lai, S.C., Qin, J.F., Zang, W.J., Li, X.F., 2009. Geochemistry of the Permian basalt and its relationship with east Palaeo-Tethys evolution in Xiangyun area, Yunnan Province. J Northwest Univ: Natural Science Edition, 39(3), 444-452 (in Chinese with English abstract).

Lai, S.C., Qin, J.F., Li, X.F., Zang, W.J., 2010 Geochemistry and Sr-Nd-Pb isotopic features of the Ganlongtang-Nongba ophiolite from the Changning-Menglian suture zone. Acta Petrol Sin, 26(11), 3195-3205 (in Chinese with English abstract).

Le Fort, P., Cuney, M., Deniel, C., France-Lanord, C., Sheppard, S.M.F., Upreti, B.N., Vidal, P., 1987. Crustal generation of the Himalayan leucogranites. Tectonophysics, 134, 3957.

Leng, C.B., Zhang, X.C., Hu, R.Z., Wang, S.X., Zhong, H., Wang, W.Q., Bi, X.W., 2012. Zircon U-Pb and molybdenite Re-Os geochronology and SrNd-Pb-Hf isotopic constraints on the genesis of the Xuejiping porphyry copper deposit in Zhongdian, Northwest Yunnan, China. J Asian Earth Sci, 60(22), 31-48.

Leng, C.B., Huang, Q.Y., Zhang, X.C., Wang, S.X., Zhong, H., Hu, R.Z., Bi, X.W., Zhu, J.J., Wang, X.S., 2014. Petrogenesis of the late Triassic volcanic rocks in the southern Yidun Arc, SW China: Constraints from the geochronology, geochemistry, and Sr-Nd-Pb-Hf isotopes. Lithos, 190-191 (2), 363-382.

Li, W.C., 2007. The tectonic evolution of the Yidun island arc and the metallogenic model of the Pulang porphyry copper deposit Yunnan, SW China. Beijing: China University of Geoscience for Doctoral Degree Paper, 1-123 (in Chinese with English abstract).

Li, X.H., Li, Z.X., Li, W.X., Liu, Y., Yuan, C., Wei, G.J., Qi, C.S., 2007. U-Pb zircon, geochemical and Sr-Nd-Hf isotopic constraints on age and origin of Jurassic I- and A-type granites from central Guangdong, SE China: A major igneous event in response to foundering of a subducted flatslab? Lithos, 96, 186-204.

Li, Y.J., Wei, J.H., Chen, H.Y., Li, H., Chen, C., Hou, B.J., 2014. Petrogenesis of the Xiasai Early Cretaceous A-type granite from the Yidun Island Arc Belt, SW China: Constraints from zircon U-Pb age, geochemistry and Hf isotope. Geotecton Metallog, 38(4),939-953 (in Chinese with English abstract).

Li, W.C., Yu, H.J., Gao, X., Liu, X.L., Wang, J.H., 2017. Review of Mesozoic multiple magmatism and porphyry Cu-Mo (W) mineralization in the Yidun Arc, eastern Tibet Plateau. Ore Geol Rev, 90, 795-812.

Liu, S.W., Wang, Z.Q., Yan, Q.R., Li, Q.G., Zhang, D.H., Wang, J.G., 2006a. Geochemistry and petrogenesis of Queershan granitoids, western Sichuan Province. Acta Geol Sin, 80(9), 1355-1363 (in

Chinese with English abstract).

Liu, S.W., Wang, Z.Q., Yan, Q.R., Li, Q.G., Zhang, D.H., Wang, J.G., Yang, B., Gu, L.B., Zhang, F.S., 2006b. Indosinian tectonic setting of the southern Yidun Arc: Constraints from shrimp zircon chronology and geochemistry of dioritic porphyries and granites. Acta Geologica Sin, 80(3), 387-399.

Liu, Y., Liu, X.M., Hu, Z.C., Diwu, C.R., Yuan, H.L., Gao, S., 2007a. Evaluation of accuracy and long-term stability of determination of 37 trace elements in geological samples by ICP-MS. Acta Petrol Sin, 23 (5), 1203-1210 (in Chinese with English abstract).

Liu, Y., Deng, J., Li, C.F., Shi, G.H., Zheng, A.L., 2007b. REE composition in scheelite and scheelite Sm-Nd dating for the Xuebaoding W-Sn-Be deposit in Sichuan. Chin Sci Bull, 52, 2543-2550.

Liu, S., Hu, R.Z., Gao, S., Feng, C.X., Huang, Z.L., Lai, S.C., Yuan, H.L., Liu, X.M., Coulson, I. M., Feng, G.Y., Wang, T., Qi, Y.Q., 2009. U-Pb zircon, geochemical and Sr-Nd- Hf isotopic constraints on the age and origin of Early Palaeozoic I-type granite from the Tengchong-Baoshan Block, Western Yunnan Province, SW China. J Asian Earth Sci, 36, 168-182.

Liu, J.T., Yang, L.Q., Lü, L., 2013. Pulang reduced porphyry copper deposit in the Zhongdian area, Southwest China: Constrains by the mineral assemblages and the ore-forming fluid compositions. Acta Petrol Sin, 29(11), 3914-3924 (in Chinese with English abstract).

Liu, H.Y., Wang, M., Yan, W.B., Zeng, Q., Zhao, L., Liao, B.Y., Yang, J.H., Zhang, Z.H., Zhang, T. Y., Luo, W., 2016. Zircon U-Pb and geochemistry of Que'ershan granitic complex in western Sichuan and metallogenic specialization. Global Geol, 35(4), 968-981 (in Chinese with English abstract).

Long, S.N., 2007. Ore-forming evolution and geochemistry character of typical deposit of the Yidun Arc in Sanjiang region, western Sichuan province. Beijing: China University of Geoscience for Master Degree Paper, 1-39 (in Chinese with English abstract).

Lü, B.X., Wang, Z., Zhang, N. D., Duan, J.Z., Gao, Z.Y., Shen, G.F., Pan, C.Y., Yao, P., 1993. Granitoid in the Sanjiang Region (Nujiang-LancangjiangJinsha Jiang Region) and their metallogenic specialization. Beijing: Geological Publishing House, 1-328 (in Chinese).

Ludwig, K.R., 2003. ISOPLOT 3.0: A geochronological toolkit for Microsoft Excel. Berkeley Geochronology Center, Special Publication, 4.

Miller, C.F., 1985. Are strongly peraluminous magmas derived from pelitic sedimentary sources? J Geol, 93 (6), 673-689.

Middlemost, E.A.K., 1994. Naming materials in the magma/igneous rock system. Earth Sci Rev, 37(3-4), 215-224.

Mo, X.X., Lu, F.X., Shen, S.Y., Zhu, Q.W., Hou, Z.Q., 1993. Volcanism and metallogeny in the Sanjiang Tethys. Beijing: Geological Publishing House, 1-250 (in Chinese with English abstract).

Pan, G.T., Xu, Q., Hou, Z.Q., Wang, L.Q., Du, D.X., Mo, X.X., Li, D.M., Wang, M.J., Li, X.Z., Jiang, X.S., Hu, Y.S., 2003. Archipelagic orogenesis, metallogenic systems and assessment of the mineral resources along the Nujiang-Lancangjiang-Jinshajiang area in southwestern China. Beijing: Geological Publishing House, 1-420 (in Chinese).

Patiño Douce, A.E., 1995. Experimental generation of hybrid silicic melts by reaction of high-Al basalt with metamorphic rocks. J Geol, 100, 623-639.

Pearce, J.A., 1996. Sources and settings of granitic rocks. Episodes, 19, 120-125.

Pearce, J.A., Harris, N. B.W., Tindle, A.G., 1984. Trace element discrimination diagrams for the tectonic interpretation of granitic rocks. J Petrol, 25, 956-983.

Peng, T.P., Zhao, G.C., Fan, W.M., Peng, B.X., Mao, Y.S., 2014. Zircon geochronology and Hf isotopes of Mesozoic intrusive rocks from the Yidun terrane, Eastern Tibetan Plateau: Petrogenesis and their bearings with Cu mineralization. J Asian Earth Sci, 80(2), 18-33.

Peng, T.P., Fan, W.M., Zhao, G.C., Peng, B.X., Xia, X.P., Mao, Y.S., 2015. Petrogenesis of the early Paleozoic strongly peraluminous granites in the western south China Block and its tectonic implications. J Asian Earth Sci, 98(98), 399-420.

Qu, X.M., Hou, Z.Q., Zhou, S.G., 2002. Geochemical and Nd, Sr isotopic study of the post-orogenic granites in the Yidun Arc belt of northern Sanjiang region, southwestern China. Resour Geol, 52(2), 163-172.

Qu, X.M., Hou, Z.Q., Tang, S.H., 2003. Age of intraplate volcanism in the back-arc area of Yidun island arc and its significance. Petrol Min, 22, 131-137 (in Chinese with English abstract).

Reid, A.J., Fowler, A.P., Phillips, D., Wilson, C.J.L., 2005a. Thermochronology of the Yidun Arc, central eastern Tibetan Plateau: Constraints from ^{40}Ar/^{39}Ar K-feldspar and apatite fission track data. J Asian Earth Sci, 25(6), 915-935.

Reid, A.J., Wilson, C.J.L., Liu, S., 2005b. Structural evidence for the Permo-Triassic tectonic evolution of the Yidun Arc, eastern Tibetan Plateau. J Struct Geol, 27, 119-137.

Reid, A., Wilson, C.J.L., Shun, L., Pearson, N., Belousova, E., 2007. Mesozoic plutons of the Yidun Arc, SW china, U/Pb geochronology and Hf isotopic signature. Ore Geol Rev, 31(1-4), 88-106.

Roger, F., Malavieille, J., Leloup, P.H., Calassou, S., Xu, Z., 2004. Timing of granite emplacement and cooling in the Songpan-Garzê Fold Belt (eastern Tibetan Plateau) with tectonic implications. J Asian Earth Sci, 22(5), 465-481.

Roger, F., Jolivet, M., Malavieille, J., 2008. Tectonic evolution of the Triassic fold belts of Tibet. Compt Rendus Geosci, 340, 180-189.

Roger, F., Jolivet, M., Malavieille, J., 2010. The tectonic evolution of the Songpan-Ganzi (North Tibet) and adjacent areas from Proterozoic to present: A synthesis. J Asian Earth Sci, 39, 254-269.

Rohrlach, B.D., Loucks, R.R., 2005. Multi-million-year cyclic ramp-up of volatiles in a lower crustal magma reservoir trapped below the Tampakan copper-gold deposit by Mio-Pliocene crustal compression in the southern Philippines. In: Porter, T.M. Super porphyry copper & gold deposits-a global perspective, 2. Adelaide: PCG Publishing, 269-407.

Rollinson, H.R., 1993. Using geochemical data: Evaluation, presentation, interpretation. London: Longman, 1-352.

Sun, S.S., McDonough, W.F., 1989. Chemical and isotopic systematics of oceanic basalts: Implications for mantle composition and processes. Geol Soc Lond Spec Publ, 42(1), 313-345.

Sylvester, P.J., 1998. Post collisional strongly peraluminous granites. Lithos, 45, 29-44.

Wang, E., Burchfiel, B.C., 2000. Late Cenozoic to Holocene deformation in southwestern Sichuan and adjacent Yunnan, China, and its role in formation of the southeastern part of the Tibetan Plateau. In:

Geissman John, W., Glazner Allen, F. Special focus on the Himalaya. Geological Society of America Bulletin, Boulder, CO, United States, 413-423.

Wang, X.F., Metcalfe, I., Jian, P., He, L.Q., Wang, C.S., 2000. The Jinshajiang-Ailaoshan suture zone, China: Tectonostratigraphy, age and evolution. J Asian Earth Sci, 18, 675-690.

Wang, B.Q., Zhou, M.F., Li, J.W., Yan, D.P., 2011. Late Triassic porphyritic intrusions and associated volcanic rocks from the Shangri-La region, Yidun terrane, Eastern Tibetan Plateau: Adakitic magmatism and porphyry copper mineralization. Lithos, 127, 24-38.

Wang, B. Q., Zhou, M. F., Chen, W. T., Gao, J. F., Yan, D. P., 2013a. Petrogenesis and tectonic implications of the Triassic volcanic rocks in the northern Yidun Terrane, Eastern Tibet. Lithos, 175-176, 285-301.

Wang, B.Q., Wang, W., Chen, W.T., Gao, J.F., Zhao, X.F., Yan, D.P., Zhou, M.F., 2013b. Constraints of detrital zircon U-Pb ages and Hf isotopes on the provenance of the Triassic Yidun Group and tectonic evolution of the Yidun Terrane, Eastern Tibet. Sediment Geol, 289, 74-98.

Wang, X.S., Bi, X.W., Leng, C.B., Zhong, H., Tang, H.F., Chen, Y.W., Yin, G.H., Huang, D.Z., Zhou, M.F., 2014a. Geochronology and geochemistry of Late Cretaceous igneous intrusions and Mo-Cu-(W) mineralization in the southern Yidun Arc, SW China: Implications for metallogenisis and geodynamic setting. Ore Geol Rev, 61, 73-95.

Wang, X.S., Hu, R.Z., Bi, X.W., Leng, C.B., Pan, L.C., Zhu, J.J., Chen, Y.W., 2014b. Petrogenesis of Late Cretaceous I-type granites in the southern Yidun terrane: New constraints on the Late Mesozoic tectonic evolution of the eastern Tibetan Plateau. Lithos, 208-209, 202-219.

Wang, N., Wu, C.L., Qin, H.P., Lei, M., Guo, W.F., Zhang, X., Chen, H.J., 2016. Zircon U-Pb geochronology and Hf isotopic characteristics of the Daocheng granite and Haizishan granite in the Yiduan Arc, Western Sichuan, and their geological significance. Acta Geol Sin, 90(11), 3227-3245.

Watson, E.B., Harrison, T.M., 1983. Zircon saturation revisited: Temperature and composition effects in a variety of crustal magma types. Earth Planet Sci Lett, 64, 295-304.

Weislogel, A.L., 2008. Tectonostratigraphic and geochronologic constraints on evolution of the northeast Paleotethys from the Songpan-Ganzi complex, central China. Tectonophysics, 451, 331-345.

White, A.J.R., Chappell, B.W., 1988. Some supracrustal (S-type) granites of the Lachlan Fold Belt. Trans R Soc Edinb Earth Sci, 79, 169-181.

Wu, T., 2015. Early Mesozoic magmatism and tectonic evolution of Yidun Arc belt, eastern Tibet Plateau. Wuhan: China University of Geoscience for Doctoral Degree Paper, 1-168 (in Chinese with English abstract).

Wu, T., Xiao, L., Gao, R., Yang, H.J., Yang, G., 2014a. Petrogenesis and tectonic setting of the Queershan composite granitic pluton, Eastern Tibetan Plateau: Constraints from geochronology, geochemistry and Hf isotope data. Sci China Earth Sci, 44(8), 1791-1806 (in Chinese without English abstract).

Wu, T., Xiao, L., Ma, C.Q., Pirajno, F., Sun, Y., Zhan, Q.Y., 2014b. A mafic intrusion of "arcaffinity" in a post-orogenic extensional setting: A case study from Ganluogou gabbroin the northern Yidun Arc Belt, eastern Tibetan Plateau. J Asian Earth Sci, 94, 139-156.

Wu, T., Xiao, L., Wilde, S.A., Ma, C.Q., Zhou, J.X., 2017. A mixed source for the Late Triassic Garzê-Daocheng granitic belt and its implications for the tectonic evolution of the Yidun Arc belt, eastern Tibetan Plateau. Lithos, 288, 214-230.

Xu, Z., Hou, L., Wang, Z., 1992. Orogenic processes of the Songpan-Garzê Orogenic Belt of China. Beijing: Geological Publishing House, 1-190 (in Chinese).

Xu, X.W., Cai, X.P., Qu, W.J., Song, B.C., Qin, K.Z., Zhang, B.L., 2006. Later Cretaceous granitic porphyritic Cu-Mo mineralization system in the Hongshan area, northwestern Yunnan and its significances for tectonics. Acta Geol Sin, 80, 1422-1433 (in Chinese with English abstract).

Yan, Q.R., Wang, Z.Q., Liu, S.W., Shi, Y.R., Li, Q.G., Yan, Z., Wang, T., Wang, J.G., Zhang, D. H., Zhang, H.Y., 2006. Eastern margin of the Tibetan Plateau a window to the complex geological history from the Proterozoic to the Cenozoic revealed by SHRIMP analyses. Acta Geol Sin, 80(9), 1287-1294 (in Chinese with English abstract).

Yang, L.Q., Gao, X., He, W.Y., 2015. Late Cretaceous porphyry metallogenic system of the Yidun Arc, SW China. Acta Petrol Sin, 31(11), 3155-3170 (in Chinese with English abstract).

Yang, L.Q., Deng, J., Wang, Z.L., Guo, L.N., Li, R.H., Groves, D.I., Danyushevsky, L., Zhang, C., Zheng, X.L., Zhao, H., 2016. Relationships between gold and pyrite at the Xincheng gold deposit, Jiaodong Peninsula, China: Implications for gold source and deposition in a brittle epizonal environment. Econ Geol, 111, 105-126.

Yang, L.Q., Deng, J., Gao, X., He, W.Y., Meng, J.Y., Santosh, M., Yu, H.J., Yang, Z., Wang, D., 2017. Timing of formation and origin of the Tongchanggou porphyry-skarn deposit: Implications for Late Cretaceous Mo-Cu metallogenes in the southern Yidun Terrane. SE Tibetan Plateau Ore Geol Rev, 81, 1015-1032.

Yu, H.J., Li, W.C., 2016. Geochronology and geochemistry of Xiuwacu intrusions, NW Yunnan: Evidences for two-period magmatic activity and mineralization. Acta Petrol Sin, 32(8), 2265-2280 (in Chinese with English abstract).

Yuan, H.L., Gao, S., Liu, X.M., Li, H.M., Gunther, D., Wu, F.Y., 2004. Accurate U-Pb age and trace element determinations of zircon by laser ablation-inductively coupled plasma mass spectrometry. Geostand Geoanal Res, 28(3), 353-370.

Yuan, H.L., Gao, S., Luo, Y., Zong, C.L., Dai, M.N., Liu, X.M., Diwu, C.R., 2007. Study of Lu-Hf geochronology: A case study of eclogite from Dabie UHP Belt. Acta Petrol Sin, 23(2), 233-239 (in Chinese with English abstract).

Zaw, K., Peters, S.G., Cromie, P., Burrett, C., Hou, Z.Q., 2007. Nature, diversity of deposit types and metallogenic relations of South China. Ore Geol Rev, 31(1-4), 3-47.

Zhang, L.G., 1995. Block-geology of eastern Asia lithosphere-isotope geochemistry and dynamics of upper mantle, basement and granite. Beijing: Science Press, 252 (in Chinese with English abstract).

Zhao, Y.J., Yuan, C., Zhou, M.F., Yan, D.P., Long, X.P., Cai, K.D., 2007. Postorogenic extension of Songpan-Garzê orogen in Early Jurassic: Constraints from Niuxingou monzodiorite and Siguniangshan A-type granite of western Sichuan, China. Geochemica, 36, 139-142 (in Chinese with English abstract).

Zhao, S.W., Lai, S.C., Gao, L., Qin, J.F., Zhu, R.Z., 2016. Evolution of the Proto-Tethys in the Baoshan

Block along the east Gondwana margin: Constraints from early Palaeozoic magmatism. Int Geol Rev, 59 (1), 1-15.

Zhong, D., 2000. Palaeotethysides in West Yunnan and Sichuan, China. Beijing: Science Press, 1-248 (in Chinese).

Zhong, H., Zhu, W.G., Chu, Z.Y., He, D.F., Song, X.Y., 2007. Shrimp U-Pb zircon geochronology, geochemistry, and Nd-Sr isotopic study of contrasting granites in the Emeishan large igneous province, SW China. Chem Geol, 236, 112-133.

Zhu, J.J., Hu, R.Z., Bi, X.W., Zhong, H., Chen, H., 2011. Zircon U-Pb ages, Hf-O isotopes and whole-rock Sr-Nd-Pb isotopic geochemistry of granitoids in the Jinshajiang suture zone, SW China: Constraints on petrogenesis and tectonic evolution of the Paleo-Tethys Ocean. Lithos, 126(3), 248-264.

Zi, J.W., Cawood, P.A., Fan, W.M., Wang, Y.J., Tohver, E., McCuaig, T.C., Peng, T.P., 2012. Triassic collision in the Paleo-Tethys Ocean constrained by volcanic activity in SW China. Lithos, 144-145, 145-160.